Perspectives in Electronic Structure Theory

Roman F. Nalewajski

Perspectives in Electronic Structure Theory

Springer

Prof. Dr. Roman F. Nalewajski
Jagiellonian University
Department of Theoretical Chemistry
ul. R. Ingardena 3
30-060 Krakow
Poland
nalewajs@chemia.uj.edu.pl

ISBN 978-3-642-20179-0 ISBN 978-3-642-20180-6 (eBook)
DOI 10.1007/978-3-642-20180-6
Springer Heidelberg Dordrecht London New York

Library of Congress Control Number: 2011934249

Printed on acid-free paper

Springer is part of Springer Science+Business Media (www.springer.com)

Preface

This book is primarily intended as a textbook introducing to the reader the basic elements of the quantum theory of the electronic structure of molecular systems, including in its first two parts the basic axioms of the nonrelativistic quantum mechanics and rudiments of the wave function and density based theories. Its remaining two parts, of a more monographic character, contain the *Information Theory* (IT) description and some elements of the modern theory of chemical reactivity, respectively. The basic aim of this book is to present in a single text alternative outlooks on the molecular electronic structure, including the basic principles and techniques of the contemporary conceptual and computational quantum chemistry, covering also the insights provided by IT. Together these complementary perspectives enhance the depth of our understanding of the electronic/geometric structure of molecules and provide a full "vocabulary" to tackle diverse conditions, which influence their reactivity behavior. Indeed, only the insights from several different point of view amount to a real understanding of the problem. The emphasis is on the concepts involved and the key ideas encountered in these alternative approaches in the molecular quantum mechanics, and on the interpretation of calculated results in chemical terms: the bonded atoms and molecular fragments, the chemical bonds that connect these building blocks of molecules, and on their responses in a changing environment, which shape the reactivity preferences of reactants.

Explanation and understanding of chemical phenomena ultimately call for the quantum mechanical description provided by the modern quantum chemistry. The latter uses ideas and concepts that differ substantially from their classical analogs. A precise formulation of these generalized physical concepts, which requires some new mathematical tools, is the subject of Part I of this book. The depth and rigor of this physical/mathematical supplement have been dictated by the main didactic purpose of this text: to introduce all tools necessary for understanding the abstract ideas of the modern theory of the molecular structure and chemical reactivity. The foundations of quantum mechanics are covered using the familiar axiomatic approach, with only an introductory summary of the key experiments that led to their formulation.

The reader can familiarize himself with these novel ideas in the simplest problem of the stationary (bonded) states of the hydrogen-like atom presented in the part closing chapter.

The main theories of the molecular electronic structure are presented in Part II. In its opening chapter it examines available techniques of reducing the complexity of solving the molecular Schrödinger equation. In particular, the rudiments of the adiabatic separation of the electronic and nuclear motions are given and the elements of the approximate perturbational and variational approaches for determining the electronic quantum states are outlined. This brief overview also covers the basics of the orbital approximation and the idea of a pseudopotential, which effectively removes the chemically inactive electrons of the atomic inner shells from an explicit treatment in molecular calculations. The subsequent exposition of the principal *Wave Function Theories* (WFT), in which the system wave-function (probability amplitude) defines the quantum state of the molecule, covers the *Self-Consistent-Field Molecular Orbital* (SCF MO) theory, major *Configuration-Interaction* (CI) techniques for dealing with the Coulomb correlation problem, and rudiments of the *Valence Bond* (VB) treatment, which gives a more chemical understanding of molecules compared to its chief rival, the *Molecular Orbital* (MO) description and currently experiences a notable revival.

The following presentation of theoretical basis of the modern techniques of the *Density-Functional Theory* (DFT), in which the electron density or the density matrix constitute the system basic *state*-variables, covers the famous *Hohenberg–Kohn* (HK) theorems and some of their refinements/extensions, the basic elements of the ground-state *Kohn–Sham* (KS) theory and the associated ensemble approach to excited states. The theory of the density functional for the exchange-correlation energy is summarized, including the rudiments of the adiabatic connection and some more recent developments in the field of the density-matrix and orbital-dependent functionals, time-dependent DFT and alternative approaches to the molecular *van der Waals* (vdW) interactions. This short exposition also introduces the main concepts of the density-based reactivity theory: the hardness and softness responses of the electron distribution in molecules in the complementary *electron-following* (EF) and *electron-preceding* (EP) perspectives.

The additional insights from IT are presented in the monographic Part III of this textbook. Its dominating theme is the electron distribution as a source and carrier of information in molecules. First, the basic elements are summarized in the part opening chapter, to be followed by a brief exposition of the information principles in molecular quantum mechanics. The local IT probes of the presence of the direct chemical bonds are formulated and the importance of the nonadditive (interference) information tools is emphasized. In particular, the *Electron Localization Function* (ELF) and the *Contra-Gradience* (CG) bond criterion are used to explore the molecular electronic structure and the IT variational principles are used to derive the so called stockholder scheme for dividing the molecular electron density into the associated atomic pieces. Various *Charge Transfer* (CT) and *Polarization* (P) displacements accompanying the formation of chemical bonds in molecular systems are examined, including the equilibrium redistribution of electrons among the bonded *Atoms-in-Molecules* (AIM) and the molecular promotion of the latter.

Alternative bond-multiplicity descriptors and the electron localization criteria are introduced and molecules are interpreted as communication systems. This concept, developed within the standard SCF MO description, gives rise to the *Orbital Communication Theory* (OCT) of the chemical bond (Nalewajski 2010) an extension of the bond Communication Theory in atomic resolution (Nalewajski 2006). They both use the standard entropic descriptors of information channels in exploring patterns of the chemical bonds in molecules and their constituent parts, as well as the bond covalent/ionic composition.

The molecularly promoted AIM are only slightly modified, compared to their free (separated) analogs, mainly in the outer (valence) shell of electrons. These "external" electrons are responsible for the AIM chemical behavior and the equilibrium bonding pattern they exhibit in the field exerted by the framework of the practically unchanged atomic-cores. This bonding shell of the (delocalized) electrons is also crucial for the propagation of information in the molecule among the system constituent AIM and the *Atomic Orbitals* (AO) the latter contribute to the bonding subspace of the occupied MO, which ultimately determine the system network of chemical bonds. Using the standard tools of IT (summarized in the opening chapter of Part III) in treating these information scattering phenomena due to "communications" via the system chemical bonds provides a novel perspective on the origins and multiplicity of the system chemical bonds, as well as on the entropic nature of their covalent and ionic composition. In particular, the IT multiplicities of the localized chemical bonds are generated, the bond-coupling phenomena in molecular subsystems are discussed and the interference effects due to the multiple information scattering in molecules are examined. The new indirect (*through-bridge*) bonding mechanism is identified, which complements the familiar direct (*through-space*) chemical interactions in molecular systems, and its origins due to the implicit dependencies between AO in the molecular bonding subspace are explored.

The chemical concepts are discussed in a more depth in Part IV. It first provides a survey of alternative perspectives on diverse phenomena conditioning the chemical reactivity, stressing the importance of the conceptual approaches for a more chemical understanding of these bond-forming/bond-breaking processes. The distinction between the "horizontal" (involving displacements of the system electron density) and "vertical" (for the fixed electron distribution) changes in the molecular electronic structure is made and the responses of molecular fragments in the fragment-constrained equilibria are described in terms of the subsystem charge sensitivities. These perturbation–response relations are summarized for all admissible representations of the molecular/subsystem states, covering both the EF perspective of the Born–Oppenheimer approximation and the complementary EP picture, in the spirit of modern DFT. The illustrative case of the bimolecular reactive system is discussed in a more detail and alternative measures of the adiabatic coupling between the electronic and geometrical degrees-of-freedom of the molecular and reactive systems, including the novel compliant theoretical framework, are identified and modeled.

Finally, several qualitative approaches to reactivity phenomena are summarized. They cover recent IT probes of the elementary reaction mechanisms, chemical reactivity indices provided by the alternative hardness/softness (Fukui function) descriptors of molecules and their fragments, e.g., reactants in the *Donor–Acceptor* (DA) systems, as well as the associated equilibrium and stability criteria of molecules and the maximum hardness and the *Hard/Soft Acids and Bases* (HSAB) principles of chemistry. The importance of the complementary internal and external eigenvalue problem of quantum-mechanical observables for a compact description of the electronic processes in molecules and reactants is stressed and alternative hardness-decoupling schemes are examined.

This joint exposition of a variety of perspectives on the electronic structure of molecular systems, which are usually presented in separate texts, aims at comparing these diverse philosophies of treating the subject in the unifying language of the (nonrelativistic) molecular quantum mechanics and IT. Such presentation should help in uncovering the mutual relations between the specific concepts and techniques of these complementary approaches by extracting their common roots in the molecular quantum mechanics, in the frameworks of both the molecular states involved and the associated probability/density distributions.

The book may serve as both the classroom and reference text of the classical and modern ideas in the field of the chemical bond and reactivity theories. This text has evolved from teaching both the graduate and undergraduate courses in quantum chemistry, density-functional and reactivity theories, as well as the IT of molecular systems. It is intended for graduate and advanced undergraduate students and chemical researchers interested in the new ways of looking at the subject. It is hoped that a significant diversity of the student backgrounds have been accommodated in this textbook/monograph of the contemporary ways of thinking about classical issues in the theory of the electronic structure and reactivity behavior of molecules.

Cracow
June 2011

Roman F. Nalewajski

References

Nalewajski RF (2010) Information origins of the chemical bond. Nova Science Publishers, New York

Nalewajski RF (2006) Information theory of molecular systems. Elsevier, Amsterdam

Contents

Acronyms

A	Acidic, acceptor reactant, fragment
a.u.	Atomic units
AB	Acid–base, interaction, complex
AC	Adiabatic connection, of KS DFT
ACFDT	Adiabatic-connection fluctuation-dissipation theorem, DFT method
AIM	Atoms-in-molecules
AO	Atomic orbital(s)
APSG	Antisymmetrized product of strongly-orthogonal geminals, method
Av	Average value
B	Basic, reactant, fragment
BEBO	Bond-energy–bond-order, energy surface, reaction coordinate
BEF	Binary entropy function
BO	Born–Oppenheimer approximation
BOVB	Breathing orbital VB, theory
BSSE	Basis set superposition error
c	Coulomb correlation, hole, energy, potential
CAS	Complete active space, method
CBO	Charge-and-bond-order (1-density) matrix
CC	Coupled-cluster, approximation, method
CCCI	Correlation consistent CI, method, CI extension of GVB
CCSD(T)	CC theory including singly-, doubly-, and triply-excited configurations
CDFT	Conceptual DFT, approach
CEPA	Coupled electron pair approximation, method
CF	Coulson–Fischer, wave function
CG	Contra-gradience, criterion of bond localization
CI	Configuration interaction, theory, method
CID	Limited SR CI method involving *double*-excitations

CIDQ	Limited SR CI method involving *double-* and *quadruple-* excitations
CIDQH	Limited SR CI method involving *double-*, *quadruple-* and *hextuple-* excitations
CISD	Limited SR CI method involving *single-* and *double-*excitations
CISTQ	Limited SR CI method involving *single-*, *double-*, *triple-*, and *quadruple-*excitations
CM	Center of mass
CPKS	Coupled-perturbed KS, method
CPMET	Coupled-pair many electron theory
CS	Charge sensitivities
CSA	Charge sensitivity analysis
CT	Charge transfer, stage of bond formation or reaction, configurations
CTCB	Communication theory of the chemical bond
D	Donor, reactant, fragment
DA	Donor–acceptor, interaction, complex, reactive system
DCACP	Dispersion-corrected atom-centered potentials, DFT method
DFPT	Density functional perturbation theory, see GL theory
DFT	Density functional theory, method
DFT-SAPT	Combined DFT/SAPT, method
DHF	Dirac–Hartree–Fock, method
DIM	Diatomics-in-molecules, method, PES
DMFT	Density-matrix functional theory
E	Electrophilic, site in molecules, attacking agents
EE	Electronegativity-equalization, principle of Sanderson, equations
EF	Electron following, perspective
EHK	Ensemble HK, theory, functional
EKT	Extended Koopmans theorem
ELF	Electron-localization function
EP	Electron preceding, perspective
EPI	Extreme physical information, principle
ESP	Electrostatic potential
EXX	Exact-exchange, approach in DFT
F	Fisher information, intrinsic accuracy
FCI	*Full*-CI, scheme
FDT	Fluctuation-dissipation theorem
FE	Frontier electron, densities
FF	Fukui function, electronic descriptors
FO	Frontier orbitals, theory, electrons, densities
g	*Gerade*, even parity
GEA	Gradient expansion approximation, of *xc*-energy in DFT
GFF	Geometric Fukui Function

GGA	Generalized gradient approximation, of correlation functionals in DFT
GL	Görling–Levy theory, see DFPT
GTO	Gaussian-type orbitals
GVB	Generalized VB, method
GVB-PPA	GVB method using PPA
H	Hartree, theory, method
H	Heisenberg picture of time evolution
HF	Hartree–Fock, theory, method
HFl	HF limit, energy reference
HGGA	*Hyper*-GGA, hybrid gradient functionals in DFT
HK	Hohenberg–Kohn, theorems, method, principle
Hl	Hartree limit, energy reference
HL	Heitler–London, theory, wave-function
Hn	Hellmann, pseudopotential
HOMO	Highest occupied molecular orbital, frontier orbital
HSAB	Hard-soft-acids-and-bases, principle
I	Interaction picture of time evolution
IEPA	Independent electron pair approximation, method
INM	Internal normal modes
IRC	Intrinsic reaction coordinate
IRM	Interreactant modes, see also ISM
ISM	Intersubsystem Modes
IT	Information theory, Information theoretic
K	Kullback symmetrized cross-entropy, divergence
KL	Kullback–Leibler cross-entropy, entropy deficiency, missing information, information distance, directed divergence
KLI	Krieger–Li–Yafrate, method
KS	Kohn–Sham, method, orbitals
KW	Kołos–Wolniewicz, results for H_2
LCAO MO	Linear combinations of AO representing MO
LCGTO	Linear combinations of GTO representing MO
LCSTO	Linear combinations of STO representing MO
LDA	Local density approximation, of functionals
LEPS	London–Eyring–Polanyi–Sato, PES
LO	Localized orbitals
LR	Linear-response, approximation, function
LSDA	Local spin-density approximation of DFT
LUMO	Lowest unoccupied molecular orbital
MBPT	*Many*-body PT
ME	Maximum entropy, rule
MEC	Minimum energy coordinates, in the compliant approach
MED	Minimum entropy deficiency, principle
MEP	Minimum energy path

MGGA	*Meta*-GGA, gradient functionals in DFT
MH	Maximum hardness, principle
MIM	Molecules-in-molecules, method
MO	Molecular orbitals, theory
MP	Møller–Plesset, method
MR (MC) SCF	Multireference (configuration) SCF, method
MRCI	Multireference CI, method
MRVB	Multireference VB, theories
N	Nucleophilic, sites in molecules, attacking agents
NFF	Nuclear FF, geometric descriptors
NG	Natural geminals
NLDA	Nonlocal density approximation, of functionals
NO	Natural orbitals
NOCV	Natural orbitals for chemical valence
NREL	Nonrelativistic limit, energy reference
NSO	Natural spin-orbitals
OAO	Symmetrically (Löwdin) orthogonalized AO
OCT	Orbital communication theory, of the chemical bond
OEP	Optimized effective potential, method in DFT
OEP_x	x-only OEP approach
OPM	Optimized potential method, in orbitally-dependent DFT
OPW	Orthogonalized plane wave, method
P	Polarization (promotion), stage of bond formation or reaction
PES	Potential energy surface
PK	Phillips–Kleinman, local pseudo-potential
PNM	Populational normal modes
PNO	Pseudonatural orbitals
PPA	Perfect pairing approximation, in VB theory
PPP	Pariser–Parr–Pople, semiempirical SCF MO theory
PT	Perturbation theory
QMC	Quantum Monte Carlo, method, results
QSPR	Quantitative structure–property relationships
R	Radical, sites in molecules, attacking agents
RGVB	Resonating GVB, theory
RHF	Spin-restricted HF, theory
RPA	Random phase approximation
RPA DFT	RPA-ACFDT method in DFT
RS	Range separation, in electron interactions
S	Schrödinger picture of time evolution
S	Shannon entropy
SAL	Separated atoms limit, dissociation energy reference, wave-function
SAPT	Symmetry adapted perturbation theory
SBC	Symmetric binary channel

SCF MO	Self-consistent-field MO method
SCVB	Spin-coupled extension of VB, method
SE	Schrödinger equation
SI	Self-interaction, error, hole
SO	Spin orbital(s)
SPA	Single-pole approximation, of TDDFT
SR CI	*Single*-reference CI, method
SR SCF	*Single*-reference SCF, HF method
SRL	Separated reactant limit, reference
STO	Slater-type orbitals
TDDFT	Time dependent DFT, theory, method
TF	Thomas–Fermi, theory
TFD	Thomas–Fermi–Dirac, theory
TS	Transition state, concept of Slater, complex in absolute rate theory
TST	Transition state theory, of reaction rates
u	*ungerade*, odd parity
UHF	Spin-unrestricted HF theory
VB	Valence-bond, theory, structures
VB CI	Valence bond CI, method
VB CIS	Limited CI involving *single*-excitations from VBSCF reference, method
VB CISD	Limited CI involving *single*- and *double*-excitations from VBSCF reference, method
VBSCF	Valence bond SCF, method
vdW	van der Waals, interactions
vW	von Weizsäcker, nonhomegeneity correction
WFT	Wave-function theory
x	Exchange (Fermi) correlation, hole, energy, potential
xc	Exchange-correlation, hole, energy, potential

Part I
Foundations of Quantum Mechanics

Chapter 1
Sources

Abstract A brief presentation of the experimental origins of quantum mechanics is given. The key experiments leading to contradictions with accepted physical theories of matter and radiation, signaling a need for a thorough revision of classical mechanics and electrodynamics, are surveyed. The early attempts to resolve these controversies, formulated at the beginning of twentieth century and often named as the Old Quantum Theory, which mark the genesis of the modern quantum mechanics, are summarized. The specificity of the classical description of physical processes is briefly outlined and main suggestions addressed to a more general mechanics describing the elementary particles, atoms, and molecules are enumerated. The particle diffraction experiment is examined in some detail to pinpoint the essence of the wave–particle duality and to identify the key elements of the quantum description: the initial and final experiments, as well as the free evolution of the system dynamic state which separates them, without any interference from the measuring apparatus. The internal angular momentum of an elementary particle, called spin, is introduced. The emphasis in this historical background is on the development of the classical concepts into their more general quantum counterparts, rather than on their discontinuity in the two theories. On one hand, the classical (approximate) mechanics, in which some very small quantities such as the quantum of the physical action – measured by the Planck constant – are approximated by zero, provides the *geometric* optics limit of the quantum (exact) mechanics. On the other hand, the quantum description has to use the classical concepts due to a macroscopic character of the measuring devices, which adds to the intimate relationship between the two formulations.

1.1 Experimental Origins and Old Quantum Theory

At the current state of our understanding of matter the modern quantum mechanics plays a fundamental role in describing phenomena and processes in the surrounding world, particularly at the *microscopic* level of photons, elementary particles, atoms,

R.F. Nalewajski, *Perspectives in Electronic Structure Theory*,
DOI 10.1007/978-3-642-20180-6_1, © Springer-Verlag Berlin Heidelberg 2012

and molecules. It should be emphasized, however, that the complete theory of *macroscopic* objects, of dimensions perceived by our senses, also requires the quantum mechanical description of interactions between their constituent atoms and molecules since the quantum nature of these microscopic particles can be manifested also at the macroscopic level. Clearly, in the limit of very large masses and energies of macroscopic objects the predictions of quantum mechanics must be identical with those resulting from its classical analog. Thus, when supplemented by the laws of statistical thermodynamics the quantum mechanics gives rise to the complete description of the natural world.

It was born in the atmosphere of severe confusion at the beginning of twentieth century, when the accepted physical theories were challenged by numerous dilemmas resulting from a series of remarkable new experimental observations, which could not be explained by the classical mechanics and electrodynamics. The physics at the end of nineteenth century distinguished the categories of matter and radiation, and used separate laws to describe them: Newton's mechanics, to predict motions of material bodies, and the Maxwell equations of the electromagnetic theory of radiation, which unites the electric, magnetic, and optical phenomena. We recall at this point that the so-called *wave* optics becomes the *geometric* optics in the limit of infinitely small wavelength, $\lambda \to 0$, i.e., for infinitely large frequency, $\nu \to \infty$, of the monochromatic radiation.

Let us now briefly summarize the key stages of the development of quantum ideas in physics (see, e.g., van der Waerden 1968) with the experiment and intuitive insight ultimately leading to a new philosophy of science (Heisenberg 1949, 1958; Yourgrau and van der Merve 1979; Bohm 1980) with the *exact* determinism of classical predictions being replaced by the *statistical* determinism of quantum laws. This "revolution" has also led to a dramatically different way of thinking about the process of measurement, to a discovery of the universal character of the *particle–wave dualism* of both the radiation and matter, and a new definition of the mechanical state of microscopic systems. The crisis of classical physics was indeed observed first on the subatomic and atomic/molecular scales, in processes involving interactions of such objects with electromagnetic radiation, a diffraction of radiation and elementary particles, etc.

We begin this short survey with the problem of the black-body radiation, at equilibrium in the given temperature T, which could not be explained by the classical electrodynamics and eventually led to formulation in 1901 of the famous Planck's hypothesis of the *energy quantization*. The question was this: how much energy is present as radiation in the given volume of an empty space of a cavity in an object held at the definite temperature T, and how it is distributed as a function of the radiation frequency? The quantity describing such a distribution is called the radiation energy density $u(\nu, T)$, which measures the energy of the monochromatic radiation of frequency ν per unit volume of the cavity, in thermal equilibrium at absolute temperature T. The Rayleigh–Jeans law of 1900, $u(\nu, T) \propto \nu^2 T$, derived using the classical electrodynamics and statistical thermodynamics, is correct only for low frequencies (in the infrared region of the electromagnetic radiation spectrum) and it dramatically fails for high frequencies (in the ultraviolet region), where the experimental data show a sharp drop in the energy distribution, with $u \rightarrow 0$ in the *geometric* optics limit of $v \rightarrow \infty$. This classical distribution has been obtained by first calculating the number of elementary oscillators (cavity standing waves) of the electromagnetic field, each corresponding to a particular frequency of radiation, and then ascribing them an average energy $k_B T$, where the Boltzmann constant $k_B = 1.381 \times 10^{-23}$ [J K^{-1}], in accordance with the classical energy *equi*-partition principle.

In order to overcome this discrepancy, also known as the *ultraviolet catastrophe*, which could not be explained by classical means, Planck has proposed that the energy of the elementary radiation oscillator of frequency v, is restricted to integral multiples of the *energy quantum*, $hv \equiv \hbar\omega$, where the new universal constant h has a dimension of the mechanical action [energy $\times$ time]; here, the radiation angular frequency $\omega = 2\pi v$ [radians/s] and the symbol $\hbar = h/2\pi$. In other words, this finite "grain" of the oscillator energy constitutes the smallest amount by which the oscillator energy can be increased or lowered. Hence, the energy absorbed by the elementary oscillators of the surrounding cavity can also be absorbed or emitted in integral multiples of such energy quanta, for all frequencies allowed by the cavity standing-wave boundary conditions, as implied by the condition of a thermal equilibrium in the black-body radiation problem: $\Delta E = hv$. This quantum (non-classical) assumption gives rise to the celebrated *Planck's distribution law*:

$$u(v, T) \propto v^3 [\exp(hv/k_B T) - 1]^{-1}, \tag{1.1}$$

which is in perfect agreement with experimental observations for the Planck constant (quantum portion of the physical action) $h = 6.626 \times 10^{-34}$ [Js] or $\hbar = h/2\pi = 1.055 \times 10^{-34}$ [Js].

It should be emphasized that this assumption was incompatible with the principles of classical physics. Thus, the agreement with experiment has been achieved only by introducing into the framework of the contemporary physics, in which the oscillator energy and mechanical action constitute the continuous dynamical quantities, the artificial "discrete" quantum condition, incompatible with the basic principles of the classical theory.

This energy quantization has been generalized in 1905 by Einstein into hypothesis of the elementary, localized (indivisible) portions of the electromagnetic energy, defining the radiation particles called *photons*, each containing Planck's portion of the energy: $E = hv$. This assumption provides the complete explanation of the *photoelectric effect* discovered by Hertz in 1886 and 1887. Photoelectrons are produced instantaneously, when the light of a frequency higher than some threshold value v_0 strikes any substance. This phenomenon is governed by the two laws formulated by Lenard in 1899–1902: (1) the number of photoelectrons is proportional to the intensity of the incident radiation; (2) their maximum velocity v and hence also the kinetic energy are affected only by the radiation frequency, and not by its intensity as predicted by the classical, wave theory of radiation. In Einstein's hypothesis the photoelectron energy of motion originates entirely from a single photon, representing a localized corpuscle of the energy, and satisfies the energy conservation

$$\frac{1}{2}m_e v^2 = h\nu - h\nu_0, \tag{1.2}$$

where m_e denotes the mass of an electron and the threshold energy $\Phi = h\nu_0$ measures the so-called *work function* of the irradiated substance.

The electromagnetic radiation thus exhibits a dual character. On one hand, in the diffraction (interference) experiments, it behaves as a wave characterized by the frequency ν [s^{-1}] or wave length $\lambda = c/\nu$, where c stands for the velocity of light in vacuum. On the other hand, as the localized particle of energy, it should be characterized by the linear momentum p. Using the relativistic expression for the energy, $E = m_f c^2 = p_f c = h\nu$, where m_f stands for the photon mass of motion (its rest mass vanishes), one obtains the relativistic expression for the photon momentum:

$$p_f = h\nu/c = h/\lambda \quad \text{or} \quad p_f = \hbar(2\pi/\lambda) \equiv \hbar k, \tag{1.3}$$

where k [m^{-1}] stands for the photon wave number.

In 1922 this corpuscular nature of radiation has been confirmed experimentally by Compton in the X-ray photon scattering by electrons. The collisions between photon (particle of radiation) and electron (particle of matter) have been shown to be governed by the conservation of the system energy and linear momentum, the two laws that govern any perfectly elastic collisions, e.g., of the billiard balls in the macroscopic world. It also follows from this experiment that any measurement of the particle position, effected by a scattering of light, influences the particle linear momentum; the more precise is this experiment, i.e., the shorter the wave of the incident radiation, the more perturbed is the particle motion after collision with the photon. This implies that in the microscopic world the measuring device and the object of measurement are not absolutely separable as it is implicitly assumed in the classical theory.

A second challenge to the established theory came from the atomic physics. In 1911 Rutherford had demonstrated, by scattering the α-radiation particles (nuclei of the helium atoms) on thin layers of heavy metals, that each atom contains the positively charged, heavy nucleus, with the estimated diameter of the order 10^{-15} [m], surrounded by light, negatively charged electrons, with the estimated diameter of the atom as a whole of the order 10^{-10} [m]. He also guessed that electrons are moving along the circular or elliptic trajectories around the nuclear attractor. This "planetary" model of an atom was in an obvious conflict with the accepted classical electrodynamics, which predicted that electrons moving on a circular orbit, thus being accelerated, should radiate electromagnetic energy and ultimately collapse onto the nucleus. Therefore, the very stability of such a "classical" atomic model has been put in doubt.

To remove this troubling inconsistency, in 1913 Bohr has followed the Planck approach of incorporating in the classical theory subsidiary quantum conditions which contradicted it. He has achieved an excellent agreement with the available experimental data for the hydrogen atom by assuming that in the circular motion of an electron allowed are only specific, *stationary* orbits, on which the particle energy remains fixed. These *stationary* energy levels $\{E_n\}$ and corresponding radii $\{r_n\}$ are identified by the orbit *quantum number* $n = 1, 2, \ldots$. The energy is emitted/absorbed in the discrete manner, not continuously as predicted by the classical electrodynamics, only when electron makes a transition between the two stationary orbits. Emission takes place when electron "jumps" from an outer orbit, exhibiting larger radius, to an inner orbit of smaller radius, identified by the higher and lower values of n, respectively. Accordingly, the inner $\rightarrow$ outer transitions are possible only after absorbing the energy from an incident radiation. Bohr has used Planck's relation between the transition energy and frequency of the emitted/absorbed radiation:

$$\Delta E_{n\rightarrow n'} = E_{n'} - E_n = h\nu_{n\rightarrow n'}. \tag{1.4}$$

Bohr's quantum conditions, which determine the stationary orbits, can be formulated as those for the allowed, discrete values of the length of the electron angular momentum $\boldsymbol{l}_n = \boldsymbol{r}_n \times \boldsymbol{p}_n$,

$$l_n = |\boldsymbol{l}_n| = m_e v_n r_n = n\hbar, \tag{1.5}$$

where $\boldsymbol{r}_n$ denotes the electron position vector on nth orbit, and $\boldsymbol{p}_n = m_e \boldsymbol{v}_n$ stands for its linear momentum.

This model has been subsequently developed in 1915 and 1916 by Sommerfeld and Planck, who introduced the elliptic orbits and the spatial quantization of the angular momentum. This generalized planetary model still gave wrong predictions already for helium atom (*two*-electron system), which signaled that this *Old Quantum Theory* was far from the final formulation of the new, generalized mechanics of microscopic objects. It should be realized, however, that new physical ideas are always arrived at by understanding the novel in terms of the familiar. Clearly, Bohr's quantization rules, successful as they were, entail assumptions which are in conflict with the classical physics. For example, the latter predicts that an electron on the circular orbit should emit radiation and this contradicts the assumed stationary character of such a trajectory. Although it was clear already at the time of its invention that this *ad hoc* synthesis of the quantum elements with the classical theory has hardly any future as the consistent physical theory, Bohr's planetary model has turned out to be quite successful in explaining the observed series of spectral lines emitted by hydrogen. The predictive power of the model was quite limited, however, since – despite later improvements – it dramatically failed to explain the spectral data of *many* electron atoms.

Since the *micro*-objects escape perception by human sense organs, their observation always requires the measurement devices, the *macro*-objects which translate their interactions with the *micro*-objects in terms of macroscopic quantities. This points out to a subtle relationship between the quantum mechanics and classical physics. In his celebrated *Correspondence Principle* Bohr has recognized that quantum mechanics must be consistent with classical mechanics. The classical limit corresponds to very large energies (quantum numbers), when such minute quantities as the Planck constant can be formally treated as zeros, in the $h \to 0$ limit.

In 1924 the quantum condition (1.5) of Bohr's model has gained a convincing interpretation in de Broglie's hypothesis of the *universal* character of the *particle–wave dualism*, which was first observed in the electromagnetic radiation. He suggested that the relations between corpuscular (E, p) and wave (ν, λ) attributes of material particles, which exhibit a nonzero rest mass, are the same as for photons, for which the rest mass vanishes (1.3). Therefore, there should also be a new, wave facet of electrons, linked to their more familiar corpuscular aspect by the associated relations:

$$E_e = h\nu_e, \quad p_e = h\nu_e/c = h/\lambda_e. \tag{1.6}$$

The existence of such *matter waves* has been confirmed experimentally in 1927 by Davisson and Germer, who diffracted the electron beam on a crystal. This development has quantitatively verified the preceding relations thus demonstrating that the particle–wave duality constitutes a universal characteristic of nature, i.e., of all objects in the microworld, or the *micro-objects* for short, rather than being a monopoly of light. Apparently, in this scale of the linear dimensions 10^{-8}–10^{-15} [m], the differences between the material and radiation particles are significantly blurred. The hope was that in the final version of the quantum theory this important discovery will find a consistent synthesis and a more explicit dynamical expression. At this time it has not been understood yet as to how de Broglie's waves propagate and how they influence the motion of individual particles. They do offer, however, a solid basis for explaining Bohr's quantum condition of (1.5). More specifically, rewriting it in terms of the electron de Broglie's wavelength of an electron moving on nth stationary orbit, $\lambda_n = h/p_n$ (1.6), gives: $2\pi r_n = n\lambda_n$. This condition thus represents the classical criterion for the standing wave along the whole perimeter of the electron circular orbit. In other words, only on the stationary orbits of Bohr the constructive interference of de Broglie's (traveling) waves explains the stability of the electron distribution. Accordingly, the destructive interference of the de Broglie waves in an atom disallows any orbit which fails to satisfy this quantum condition.

Since science is concerned only with observable things one has to let the micro-particle to respond to some outside influence, in order to observe it. As we have already argued above, when examining the implications of the Compton experiment, the measurement process inadvertently modifies the state of the micro-object. A careful examination of the limitations imposed by this influence on the accuracies Δx and Δp_x of the simultaneous determination of the particle position (Cartesian) coordinate x and its conjugate linear momentum p_x, respectively, has led Heisenberg to formulate in 1926 and 1927 his famous *Uncertainty Principle*, also known as the *Principle of Indeterminacy*, which states that the limiting value of the product of these two indeterminacies has a very small but finite value of the order of Planck's constant:

$$\Delta x \, \Delta p_x \geq \hbar. \tag{1.7}$$

The specific multiple of $\hbar$ in r.h.s. of the preceding inequality depends on the adopted measure of the measurement precision. For example, the *standard deviation* σ_A of physical quantity A, $\Delta A \cong \sigma_A = \left\langle (A - \langle A \rangle)^2 \right\rangle^{\frac{1}{2}} = \left(\langle A^2 \rangle - \langle A \rangle^2 \right)^{\frac{1}{2}}$, where $\langle A \rangle$ is the average, statistical expectation value of A and $\langle A^2 \rangle$ denotes the average value of its square, can be used to quantify the accuracy of such measurements. We shall use this familiar descriptor of a random variable later in this book, when formulating the Uncertainty Principle in terms of concepts of the molecular wave mechanics.

This limit to the fineness of our power to observe the atomic objects and the smallness of their accompanying disturbance in an act of measurement introduces the *absoluteness* to the distinction between the micro- and macro-objects. This limit can never be surpassed by an improved technique or increased skill of an observer, since a fraction of a photon is never observed. It is inherent in natural world and the dual particle–wave behavior, "anomalous" from the classical perspective, is not peculiar to light, but it is universally present in all material particles as well.

1.2 Classical–Mechanical Description and a Need for Its Revision in Generalized Mechanics

A necessity for a departure from the classical mechanics and its causality is thus clearly demonstrated by the experimental observations. The classical concepts have been proved to be inadequate to describe the molecular, atomic, and subatomic events. The uncertainty principle denies an observer the ability to simultaneously measure the conjugate components of the position and momentum vectors of micro-objects with arbitrary high precision. This contradicts the basic assumption of the classical mechanics, in the canonical formulation of the Hamilton equations of motion, where the exact knowledge of such quantities is required for the very definition of the particle dynamic state. According to the Heisenberg principle of indeterminacy such simultaneously (sharply) unobserved quantities are *unknowable*. Therefore, one is forced to resign from the classical concept of the particle trajectory, e.g., Bohr's orbit, which is unobservable thus belonging in the micro-world to a "metaphysical" rather than physical category.

Hence, the precise description of the time evolution of a micro-object, which requires an exact knowledge of its position and momentum at the given time, is unavailable in the quantum theory. This restriction does not reflect our technical inability of a precise measurement, but rather it signifies the incompatibility of the two observations involved. Such physical quantities, which cannot be sharply defined simultaneously, are called the *complementary observables*. As we shall see later in the book, besides the complementary pair of the particle position and momentum, (x, p_x), there is a number of such relations in quantum physics: energy and time, (E, t), any two Cartesian components of the angular momentum, e.g., (l_x, l_y), etc.

The uncertainty relations give rise to *statistical* predictions of the quantum theory, in contrast to the *deterministic* predictions of the classical physics. In the macroscale of objects perceived by our senses, the statistical distribution of the alternative outcomes of a measurement, represented by the normal (Gaussian) distribution, can be made infinitely sharp in the limit of the Dirac *delta* function (Dirac 1967), which can be thought of as representing the ordinary Gauss curve of the probability theory in the limit of its vanishing variance. Therefore, the *statistical* (*multiple*-valued) determinism of quantum mechanics constitutes a natural extension of its limiting form in the *strict* (*single*-valued) determinism of the classical theory.

According to Bohr's *Complementarity Principle* both coexisting wave and particle aspects of all objects in the microworld are essential for their full description. However, the precise specification of one complementary observable rules out any specification of the other. Should the particle momentum be known exactly, $\Delta p_x \to 0$, one would then have no knowledge of its position whatsoever, $\Delta x \to \infty$; accordingly, when the object position is sharply defined, $\Delta x \to 0$, one looses all the knowledge about its momentum: $\Delta p_x \to \infty$. The principle operates not only in these limiting cases, but it also covers all intermediate, finite precisions of specifying the pairs of complementary observables. The more the precise localization of an electron (or photon) in space, when its momentum is not well specified, the more the particle-like behavior. Accordingly, the wave-like character is uncovered, when the particle localization is not well specified, i.e., when its momentum is determined more precisely.

As further articulated by Bohr and his Copenhagen School, all physical quantities such as position, momentum, angular momentum, energy, etc., have to be specified by measurement, which conveys information to our senses. It has to contain amplification mechanisms by which microscopic effects are translated into macroscopic effects accessible to our understanding. Indeed, all experiments in the atomic, nuclear, and subnuclear scales in the final analysis are described in classical terms, related to attributes of the macroscopic measuring apparatus. This emphasizes a unique, intimate relationship between the quantum mechanics and its classical limit, with the former being destined to use the concepts of the latter to describe the behavior of the micro-objects.

The indeterminacy principle also implies a relativity of the quantum description with respect to the adopted method of measurement, since the specific experimental device uncovers its own "projection" of the observed "reality." This also constitutes a natural extension of the classical relativity of the description of physical phenomena with respect to the adopted reference frame. This feature signifies a deeper, fully objective approach, which resigns from the subjective classical idealization of the exact separability of the observed object and the measuring device. It is implicitly assumed in the classical theory that the progress of a physical process is independent of the experimental observations, which monitor its current stage. In other words, classical theory claims a lack of interference of the measuring device into the state of the probed mechanical system, i.e., the absolute separability of these two subsystems of an experimental arrangement.

Clearly, the physical objects evolve freely when undisturbed by an act of measurement, but finally we have to bring them into contact with the experimental apparatus to monitor their current (final) state. The progress of classical process is assumed to be independent whether they are observed experimentally or not, but in the realm of quantum mechanics the experimental monitoring is not without an influence, sometimes decisive, on the behavior of the observed micro-object. In the macroworld this influence can be practically neglected. For example, the perturbation of the airplane trajectory created by the photons of the illuminating radar radiation is nonexistent for all practical reasons. To summarize, the impression of the unequivocal determinism in the Newtonian mechanics is created by the very high masses and energies of the classical objects. It hardly implies the universality of this limiting macroconjecture of the absolute separability of the object and measuring device, to also cover the microworld where such small perturbations do matter.

The classical description also assumes the possibility of limitless gathering of simultaneous measurement information, i.e., the availability of the precise values of all mechanical properties of all constituent particles at the given time. In other words, this approach assumes that in principle at a given time all objects can be absolutely localized in space and their momenta can be determined with arbitrary precision, as can be any physical property of the dynamical system under consideration. Clearly, for practical reasons only, we are unable to reach this level of the precise specification of the mechanical microstate of all atoms/molecules in a macroscopic amount of matter. However, as claimed in the classical statistical thermodynamics, such detailed data are in principle knowable with arbitrary precision. Only due to the obvious "technical" difficulties of reaching this goal, and in view of the implications of the *Law of Large Numbers*, which renders such information irrelevant, we resort to familiar methods of the statistical mechanics in predicting the *average* descriptors of the system macrostate.

Let us briefly summarize the main suggestions addressed to the generalized mechanics capable of describing the behavior of micro-objects. As we have already argued in the preceding section, the relation between this, yet unknown, new mechanics and its classical analog should be similar to the relation between the *wave-* and *geometrical* optics; the former becomes the latter in the formal short-wave limit of $\lambda \to 0$ ($\nu \to \infty$), which is a characteristic of de Broglie's wave of a macro-object, when the free particle would not be diffracted but going along a straight rectilinear path, just as we expect classically. The new mechanics should thus include the classical mechanics as its limiting case for very large energies and hence also large values of its quantum numbers – or equivalently – in the formal limit of the vanishing quantum of the physical action: $h \to 0$. This can be argued more precisely by observing that the wave aspect of matter will be hidden from our sight, if de Broglie's wavelength λ is much lower than a characteristic length d involved in describing the motion of a body of momentum p: $\lambda/d = h/(dp) \ll 1$.

Thus, the $\lambda \to 0$ and $h \to 0$ limits are equivalent in identifying the range of applications of the classical mechanics. This postulate is known as Bohr's Correspondence Principle.

In contrast to old quantum theories, the general quantum theory must be internally consistent, i.e., all its experimental consequences must follow from the same axiomatic basis. It has to be capable of explaining all known experimental facts, rather than a narrow selection of such data. In the new mechanics we have to refrain from the classical definition of the system dynamic state, which uses the complementary observables. The new definition must instead be based only on the strictly knowable state parameters, which can be simultaneously determined with utmost precision. Clearly, such a positivistic attitude is a prerequisite of any sound physical theory.

The new definition of the mechanical state must be complete so that the results of all possible experiments performed on the microsystem can be extracted from it. In particular, it must offer means to predict the possible outcomes (spectrum) $\{a_i\}$ of any single measurement of quantity A, as well as the frequencies m_i (or probabilities) $\{p_i = m_i/m\}$ of these experimentally allowed values of the measured physical quantity in many repetitions $m = \sum_i m_i$ of the given experiment, performed on systems in the same dynamical state. This information on a multitude of measurements performed on replicas of the system then suffices to determine the statistical expectation value of the measured physical quantity:

$$\langle A \rangle = \sum\nolimits_i p_i \, a_i. \tag{1.8}$$

1.3 Implications from the Particle Diffraction Experiment

Let us consider the double-slit interference of photons or electrons, in analogy with Young's optical experiment. In this experimental arrangement the monochromatic stream of quantum particles falls on the opaque diaphragm with two slits O_1 and O_2. This experiment is crucial for distinguishing whether a perturbation traveling in space is of the particle or wave character.

The intensities $I_1(x)$ and $I_2(x)$ of two streams of the noninteracting particles passing through the openings O_1 and O_2, respectively, when the other slit is closed, upon reaching the screen $\mathcal{E}$ would produce the sum of such individual intensities (probabilities), $I_1(x) + I_2(x)$. The superposition of the corresponding waves $\psi_1(x) = |\psi_1(x)| \exp[\mathrm{i}\phi_1(x)]$ and $\psi_2(x) = |\psi_2(x)| \exp[\mathrm{i}\phi_2(x)]$,

$$\psi(x) = \psi_1(x) + \psi_2(x), \tag{1.9}$$

gives rise to the screen intensity distribution exhibiting the interference effects,

$$\begin{aligned} I(x) = |\psi(x)|^2 = \psi(x)\psi^*(x) = |\psi_1(x)|^2 + |\psi_2(x)|^2 + 2|\psi_1(x)||\psi_2(x)|\cos[\phi_1(x) - \phi_2(x)] \\ \equiv [I_1(x) + I_2(x)] + 2[I_1(x)I_2(x)]^{\frac{1}{2}}\cos[\phi_1(x) - \phi_2(x)] \equiv I^{add}(x) + I^{nadd}(x), \end{aligned} \tag{1.10}$$

because of the last, nonadditive (oscillatory) term $I^{nadd}(x)$. Above, we have identified the intensity of wave by the squared modulus of the scalar wave field $\psi(x)$, by analogy to the intensities of the electric, $E(x)$, or magnetic, $H(x)$, fields.

It has been established experimentally that the interference fringes are the statistical result of a very large number of independent particles hitting the screen, when each particle retains its individuality being finally deposited on a single grain of the photographic plate of the screen, at apparently random positions, hitting also the regions no classical particle could reach. The same interference pattern appears when a beam of particles goes through the slits simultaneously, and when single particles are scattered, one at a time, with the impact locations being observed in seemingly random fashion, now here, now there, over a length of time. The statistical determinism in this scattering of micro-objects, which give the impression of being truly indeterminable and chaotic, is only revealed after very many repetitions of such elementary, single-particle experiments, when the interference pattern finally emerges.

The appearance of interference depends critically on *both* slits being open, and it vanishes when one of them is closed, i.e., when a single particle goes definitely through one slit or the other, giving after many repetitions the separate distributions $I_1(x)$ or $I_2(x)$ on the screen. One thus concludes that the observance of interference denies us the determination of the slit through which the particle has actually passed. The interference pattern cannot be explained in the corpuscular representation, as a result of some collective effect of interactions between the beam particles. More specifically, by diminishing the density of the incident stream of particles, and hence also the number of particles passing through the slits in unit time, one changes such interactions, and this should affect the interference pattern on the screen. However, the experiment does not exhibit any influence of this kind; the diffraction pattern remains the same even in the limit of a single particle passing the slits at a time. The attempts to explain this phenomenon in the wave representation alone also fail, as the interference intensities, i.e., the wave determinism of the particle distribution is uncovered only after many repetitions of the single-particle scatterings performed at the specified dynamical conditions of the incident beam.

These apparent contradictions illustrate the wave–particle dualism of the micro-objects. Indeed, in accordance with the Heisenberg indeterminacy principle, it is impossible to simultaneously, sharply specify the particle momentum $p = h/\lambda$, which implies the knowledge of the interference pattern, and its position, which presupposes the knowledge of the slit, through which the particle has passed, when the other slit remains closed.

Therefore, there is a distinct *wave* causality in this at first glance "random" scattering of independent particles so that de Broglie's wave $\psi(x, t)$, or the *wave (state) function* for short, indeed describes in a statistical sense a movement of a single particle, with the wave intensity $I(x, t) = |\psi(x, t)|^2$ (1.10) measuring the chance of finding it hitting the screen at location x at time t. This probabilistic interpretation of the waves of matter is due to Born, who proposed in 1927 to call the intensity $I(x, t)$ the *probability density* of observing the particle at specified localization at the given time. As we shall see later in the book, in the modern quantum mechanics this identification forms a basis for interpreting the system wave function, which carries the complete information about the dynamic state of the micro-object. It should also be emphasized that this function itself, the solution of the Schrödinger wave equation formulated in 1926, which governs the dynamics of microsystems, cannot be treated as a measure of the likelihood of finding a particle at the given position, since for that it should be positive everywhere, being then incapable of the destructive interference, which is the observed fact.

The *double*-slit diffraction of microparticles identifies two types of experiments involved in establishing the classical attributes of quantum systems. Let us examine the consecutive stages of a general setup in a thought experiment shown in Fig. 1.1. We denote the initial and final states (wave functions) of the quantum system, at time $t_0 \equiv 0$ and $t > 0$, respectively, by $\psi(x, t_0)$ and $\psi(x, t)$. The classical attributes of the initial state are determined by performing the so-called *initial experiment*, which in fact creates $\psi(x, t_0)$, e.g., the monochromatic beam of particles of the specified momentum. Thus, this first category of experiment in quantum mechanics always refers to the *future*, by preparing the quantum state the time evolution of which we intend to study.

In the period $t_0 \to t$ the system evolves freely, $\psi(x, t_0) \to \psi(x, t)$, without any perturbing influence from measuring devices. This wave deterministic process will be described by the Schrödinger equation of motion, which in the modern quantum mechanics replaces the Newton (Hamilton) equations of motion of the classical theory. As we shall see later in the book, this evolution of the state function in the

$$\begin{bmatrix} \text{initial} \\ \text{experiment} \end{bmatrix} \Rightarrow \begin{bmatrix} \psi(t_0) \\ \text{initial} \\ \text{state} \end{bmatrix} \Rightarrow \begin{bmatrix} \psi(t) \\ \text{final} \\ \text{state} \end{bmatrix} \xrightarrow{\text{free evolution}} \begin{bmatrix} \text{final experiments:} \\ \text{measurements of } A \quad \text{spectrum, probabilities} \\ \{\hat{A}\psi(t)\} \quad \Rightarrow \quad \{a_i\}, \{p_i\} \end{bmatrix}$$

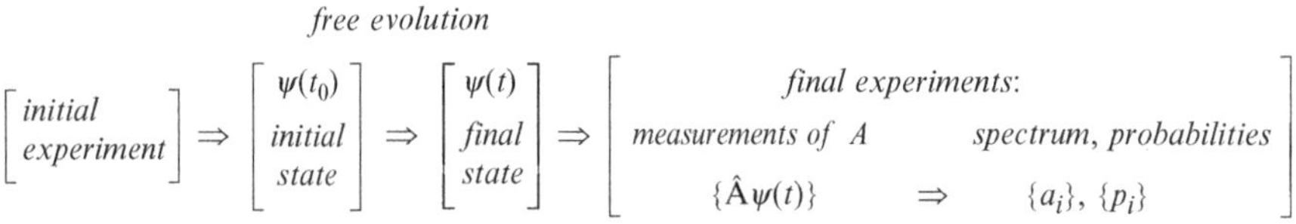

Fig. 1.1 Qualitative diagram of the *initial* and *final* experiments involved in preparing the initial state $\psi(t_0)$ and extracting the classical attributes of the final state $\psi(t)$ reached after free (undisturbed by measurement) evolution in the time interval $t_0 \to t$. The initial experiment arrangement, including the particle collimating slits and an appropriate velocity selector, transforms the polychromatic electron beam into its monochromatic component, thus preparing the initial state $\psi(t_0)$. In the time interval $t_0 \to t$ the system evolves freely, without any intervention from the measuring devices, in the specified dynamical conditions, e.g., when the particle motion is influenced by the force field generated by the external potential $v(x)$, in accordance with the *strictly* deterministic laws of quantum dynamics: $\psi(t_0) \to \psi(t)$. The statistically distributed classical attributes of the final state $\psi(t)$ are then extracted by performing the final experiment, using, e.g., the *double*-slit arrangement or a crystal as the measuring apparatus, which diffracts electrons to the movable detector or a photographic plate. This position-extraction experiment is an illustrative example of a general measurement-event of any physical observable A. The process of extracting the observed values $\{a_i\}$ (spectrum) of A in the *single*-particle experiments performed on the final state $\psi(t)$ has been symbolically depicted in the diagram as performance of the relevant mathematical operation $\hat{A}$ on $\psi(t)$, $\hat{A}\psi(t)$, with the operator $\hat{A}$ being specific for the measured quantity A. The observed spectrum $\{a_i\}$ of A and the associated probabilities $\{p_i = m_i/m\}$ can be determined only after many $m = (\sum_i m_i) \to \infty$ repetitions of the *single*-electron scatterings, with m_i denoting the frequency of observing a_i

specified dynamical conditions is strictly deterministic, with the given initial state $\psi(x, t_0)$ giving rise to a single final state $\psi(x, t)$.

The aim of the *final experiment* is to determine the classical descriptors of the quantum system in state $\psi(x, t)$. It should be stressed that after the particle has been localized on the screen, by using the photographic plate or some clever monitoring device, its dynamical state has been inadvertently and irreversibly destroyed as a result of the interaction with such an apparatus. Indeed the particle's precise localization denies us of any knowledge about the particle momentum. Thus, the final experiment can have implications only to the very *past* event, when the micro-object reaches the screen.

Due to the particle–wave duality, the link between $\psi(x, t)$ and possible outcomes of the final experiment is generally of the "*one-to-many*" type, thus giving rise to statistical predictions of specific values of classical descriptors of the system final state. Indeed, we cannot a priori predict, where the scattered electron hits the screen, but the final interference pattern, obtained after numerous repetitions of the single-electron diffractions, uniquely identifies the probability distribution $|\psi(x, t)|^2$ of the final state. It should be emphasized that only very numerous repetitions of the single-particle "experiment" together constitute the complete final experiment in quantum mechanics.

The preceding discussion prompts us to revise our ideas of causality (Heisenberg 1949, 1958; Born 1964; Bohm 1980; see also: Penrose 1989). Causality applies only to the micro-objects which are left undisturbed. Therefore, only the free-evolution in the chain of events depicted in Fig. 1.1 represents the causal stage, while the final measurement produces a disturbance in the state of the object serious enough to destroy any causal connection between the *separate* results of observations monitoring the object final state.

The statistical predictions and the indeterminism of quantum laws are a property inherent in nature, and should not be regarded as resulting from our temporary ignorance, which could be removed by some future theory, better and more complete. Although the modern quantum theory provides a thoroughly rational, coherent, and extremely successful description of micro-objects of the subatomic and atomic/molecular levels, one should not dogmatically rule out its future improvements and extensions, e.g., on the subnuclear level. However, as much as the quantum mechanics was forced upon the modern science by the physical rather than metaphysical necessity, these developments have to address future experimental findings, which could not be explained by the quantum theory. Indeed, as history teaches us, no matter how complete the description of the dynamical state may seem today, sooner or later new experimental facts will require us to improve the theoretical model and arrive at an even more general description, more detailed and usually more complex.

For example, all empirical evidence, including the *Stern–Gerlach experiment* and atomic spectra, points to the need for attributing to many elementary particles, notably electrons, protons, and neutrons, the intrinsic angular momentum, or *spin*, and the associated magnetic moment. Therefore, such particles can hardly be treated as mass points without any internal structure. Hence, for the complete specification of their dynamic states one has to provide the relevant spin quantum numbers, which fix these internal degrees-of-freedom of such micro-objects. These new dynamical variables of entirely nonclassical origin have to be specified besides the remaining simultaneously measurable observables.

1.4 Particle Spin

In 1925 Uhlenbeck and Goudsmit hypothesized the existence of yet another internal attribute of atoms and elementary particles, called *spin* angular momentum and the associated intrinsic magnetic dipole moment, which complement such properties of these micro-objects as mass, electric dipole moment, moment of inertia, electric charge, etc. This internal state variable has been originally introduced to simplify the classification of atomic spectra. This goal has been achieved, when one envisaged the existence of the internal angular momentum $\mathbf{s}$ of an electron, called the spin, the length of which is quantized by the *half* integral quantum number $s = \frac{1}{2}: s = |\mathbf{s}| = [s(s+1)]^{\frac{1}{2}}\hbar$ (Fig. 1.2).

Confirmation of this experimental conjecture came in 1928 from the relativistic quantum theory of Dirac. The existence of the electronic spin also transpires from

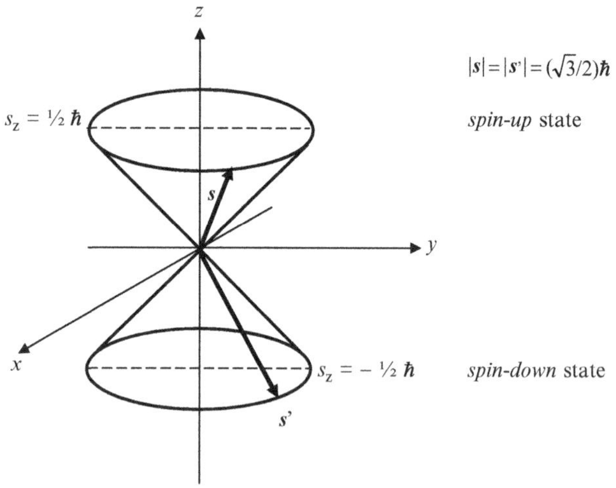

Fig. 1.2 The electron spin $\mathbf{s}$ can be characterized in quantum mechanics by two simultaneously observable attributes: its length $s = |\mathbf{s}| = [s(s+1)]^{\frac{1}{2}}\hbar = (\sqrt{3}/2)\hbar$, for the half-integral spin quantum number $s = \frac{1}{2}$, and its projection on the specified axis, say axis "z" of the Cartesian coordinate system: $s_z = \sigma\hbar$, where $\sigma = \pm s$. These two observables do not strictly specify the spin vector, but rather they define the whole family of admissible vector directions determining the cone surfaces shown in the diagram. The length and a single projection exhaust the complete list of simultaneously observed properties of any angular momentum in quantum mechanics. In other words, the direction of the angular momentum of the microparticle is not an observable

the earlier Stern–Gerlach experiment of 1921 in which a beam of silver atoms, containing a single, outermost spin-unpaired electron, produce two traces corresponding to the spin-up ($s_z = {}^1\!/_2\hbar$) and spin-down ($s_z = -{}^1\!/_2\hbar$) states (Fig. 1.2) of their valence electron, after being deflected in a nonuniform magnetic field.

These two spin states of a single electron can be uniquely specified by the quantum numbers determining the two simultaneously measurable attributes of the spin vector: $\mathscr{s}$, for its length, and $\sigma = \pm\mathscr{s}$, for its projection along the specified direction, say the "z" axis in Fig. 1.2: $s_z = \sigma\hbar$. They can be symbolically represented as the following "state vectors," in which one provides an explicit or symbolic specification of the state spin quantum numbers within the arrow-like symbol of Dirac:

$$\textit{spin-up}\ \text{state:} \quad |\alpha\rangle = |\mathscr{s}, \sigma = +\mathscr{s}\rangle = |{}^1\!/_2, +{}^1\!/_2\rangle = |+\rangle,$$

$$\textit{spin-down}\ \text{state:} \quad |\beta\rangle = |\mathscr{s}, \sigma = -\mathscr{s}\rangle = |{}^1\!/_2, -{}^1\!/_2\rangle = |-\rangle.$$

1.5 Birth of Modern Quantum Mechanics

The consistent quantum mechanics (see, e.g., Messiah 1961; Davydov 1965; Dirac 1967; Merzbacher 1967; Cohen-Tannoudji et al. 1977; Fock 1986), which explains the origins of the quantization of the physical observables and introduces the generalized dynamics of quantum states, has emerged in 1926–1927 in two equivalent forms: the *Matrix Mechanics* of Heisenberg and the *Wave Mechanics* of Schrödinger. Although using quite different mathematical apparatuses, the matrix algebra and differential equations, respectively, these two rival theories gave rise to identical physical predictions, in complete agreement with all experimental data. It was clear, therefore, that these two approaches represent the same physical theory, as indeed demonstrated later by Schrödinger and Dirac (see, e.g., Buckley and Peat 1979).

Heisenberg discovered the need for a generally *noncommutative* multiplication of physical quantities in quantum mechanics, which gives rise to the position–momentum indeterminacy. The analogies with systems in classical mechanics, which are governed by the linear equations of motion, a consequence of the superposition relationships between states of vibrating strings or membranes, have led Schrödinger to establish the basic equations of the Wave Mechanics. The resulting equation of state is also linear in the unknowns, because of the assumption of the quantum superposition principle. In Heisenberg's approach the quantum states and physical observables are represented by the matrix vectors and square matrices, respectively, while in Schrödinger's treatment they are accordingly associated with functions and differential operators. The important contributions to the final form of the modern quantum theory have also been made by other members of the Göttingen School, Born and Jordan, and by Dirac and Pauli, who invented the relativistic version of the quantum theory.

These revolutionary departures from principles of the classical theory, and particularly in the form of the quantum superposition of states demanding indeterminacy in the results of observations, are necessary to provide a sensible physical interpretation and to explain all known experimental facts. These new ideas find their expression through the introduction of a new mathematical formalisms as well as novel axioms and rules of manipulation. The two original formulations of the modern quantum mechanics can be united in a more general and abstract form of the quantum theory, which includes both the wave mechanics and matrix theory as its special cases. This "geometric" formulation requires the complex linear vector space, called the *Hilbert space*, in which vectors represent state functions. Both n-dimensional and $n \to \infty$ spaces are invoked, including the indenumerably infinite case of vectors corresponding to continuous variables. The matrix and wave function theories then appear as corresponding to different choices of the basic vectors in the Hilbert-space, which define the chosen reference frame for concepts and equations of quantum mechanics. This is similar to the relationship between the form of equations in classical physics and the adopted coordinate system in which they are formulated. With the increased elegance and mathematical abstractness of this unifying geometric formulation one also gains a great deal of understanding.

The geometric approach using Dirac's vector notation is the method chosen in the present short presentation of the principles of quantum mechanics. Its relation to the two original formulations will be briefly explored, emphasizing their equivalence in predicting the possible outcomes of experiments and the dynamical equation of motion. Since the wave mechanics appears to be conceptually simpler in chemical applications and directly connecting to the particle–wave dualism, a stronger emphasis will be made on this (nonrelativistic) version of the quantum theory. However, for reasons of convenience, in specific problems covered by the book the matrix theory will also be applied. In this study an emphasis is put on the conceptual developments rather than specific applications. For the solvable problems in quantum mechanics and quantum chemistry the reader is referred to specific textbooks and monographs (e.g., Flügge 1974; Szabo and Ostlund 1982; Atkins 1983; Levine 1983; McQuarrie 1983; Johnson and Pedersen 1986).

References

Atkins PW (1983) Molecular quantum mechanics. Oxford University Press, Oxford

Bohm D (1980) Causality and chance in modern physics. University of Pennsylvania Press, Philadelphia

Born M (1964) Natural philosophy of cause and chance. Dover, New York

Buckley P, Peat FD (eds) (1979) A question of physics: conversations in physics and biology. University of Toronto Press, Toronto

Cohen-Tannoudji C, Diu B, Laloë F (1977) Quantum mechanics. Hermann, Wiley, Paris

Davydov AS (1965) Quantum mechanics. Pergamon, Oxford

Dirac PAM (1967) The principles of quantum mechanics. Clarendon, Oxford
Flügge S (1974) Practical quantum mechanics. Springer, New York
Fock VA (1986) Fundamentals of quantum mechanics. Mir, Moscow
Heisenberg W (1949) The physical principles of the quantum theory. Dover, New York
Heisenberg W (1958) Physics and philosophy: the revolution in modern science. Dover, New York
Johnson CS, Pedersen LG (1986) Problems and solutions in quantum chemistry and physics. Dover, New York
Levine IN (1983) Quantum chemistry. Allyn and Bacon, Boston
McQuarrie DA (1983) Quantum chemistry. University Science Books and Oxford University Press, London
Merzbacher E (1967) Quantum mechanics. Wiley, New York
Messiah A (1961) Quantum mechanics. North Holland, Amsterdam
Penrose R (1989) The emperor's new mind: concerning computers, minds, and the laws of physics. Oxford University Press, New York, p 225
Szabo A, Ostlund NS (1982) Modern quantum chemistry: introduction to advanced electronic structure theory. Macmillan, New York
van der Waerden BL (ed) (1968) Sources of quantum mechanics. Dover, New York
Yourgrau W, van der Merve A (eds) (1979) Perspectives in quantum theory. Dover, New York

Chapter 2
Mathematical Apparatus

Abstract The mathematical tools of quantum mechanics are summarized. This overview, which makes no attempt to be mathematically complete and rigorous, is intended as an introduction for readers unfamiliar with the subject. We begin with some geometrical analogies of the basic concepts and techniques of the mathematical formalism used to treat the extended Hilbert space of the quantum-mechanical states, the abstract vector space spanned by the *state* vectors or the associated wave functions of the physical system of interest. Dirac's vector notation, which greatly simplifies manipulations on these mathematical objects, and the alternative representations of the singular delta "function" are given. The linear operators acting on the state vectors as well as their adjoints are defined and the basis set representations of vectors and operators are introduced. The eigenvalue problem of the linear *self*-adjoint (Hermitian) operators is examined in some detail and the complete set of the commuting observables is defined. The two most important (continuous) bases of vectors for representing quantum states of a single particle, defined by the eigenvectors of the particle *position* and *momentum* operators, respectively, are explored. In particular, the position representation of the momentum operator, as well as the momentum representation of the position operator, are examined in some detail. Next, the discrete energy representation is briefly examined and the unitary transformation of states and operators is discussed. Finally, the functional derivatives are introduced and the associated Taylor expansion of functionals is formulated. The localized displacements of the functional argument function are defined using Dirac's delta function and the rules of functional differentiation are outlined stressing analogies to familiar operations performed on functions of many variables. The *chain* rule transformations of functional derivatives are summarized.

R.F. Nalewajski, *Perspectives in Electronic Structure Theory*,
DOI 10.1007/978-3-642-20180-6_2,

2.1 Geometrical Analogies

The ordinary *three*-dimensional *physical* space R^3 is spanned by the orthonormal *basis* $\{\boldsymbol{i}, \boldsymbol{j}, \boldsymbol{k}\} \equiv \boldsymbol{e}^{(3)}$ (a *row* vector of vector elements), consisting of three unit vectors $\{\boldsymbol{e}_i, i = 1 \equiv x, 2 \equiv y, 3 \equiv z\}$ along the mutually perpendicular axes $\{x, y, z\}$, respectively, in the Cartesian coordinate system. The orthogonality of different basis vectors, $i \neq j$, expressed by the vanishing scalar product $\boldsymbol{e}_i \cdot \boldsymbol{e}_j = 0$, and their unit length (normalization), $\boldsymbol{e}_i \cdot \boldsymbol{e}_i = |\boldsymbol{e}_i|^2 \equiv e_i^2 = 1$, can be combined into the *orthonormality* relations expressed in terms of Kronecker's delta,

$$\boldsymbol{e}_i \cdot \boldsymbol{e}_j = \delta_{i,j} = \{1, \text{ for } i = j; 0, \text{ for } i \neq j\}, \tag{2.1a}$$

defining the *three*-dimensional, *unit*-metric tensor represented by the identity matrix $\mathbf{I}^{(3)} = \{\delta_{i,j}\}$:

$$\boldsymbol{e}^{(3)} \cdot \boldsymbol{e}^{(3)} \equiv \boldsymbol{e}^{(3)\mathrm{T}} \boldsymbol{e}^{(3)} = \mathbf{I}^{(3)}, \tag{2.1b}$$

where $\boldsymbol{e}^{(3)\mathrm{T}}$ denotes the *transposed* (T), *column* vector of transposed vector elements.

Any vector in R^3 can be expanded in this reference system,

$$\boldsymbol{A} = \boldsymbol{A}_x + \boldsymbol{A}_y + \boldsymbol{A}_z \equiv \sum_{i=1}^{3} \boldsymbol{A}_i = \boldsymbol{i}a_x + \boldsymbol{j}a_y + \boldsymbol{k}a_z \equiv \sum_{i=1}^{3} \boldsymbol{e}_i a_i = \boldsymbol{e}^{(3)} \boldsymbol{a}^{(3)\mathrm{T}}, \tag{2.2}$$

with the *row* vector of coordinates $\boldsymbol{a}^{(3)} = \{a_i = \boldsymbol{e}_i \cdot \boldsymbol{A}\} = [a_x, a_y, a_z]$, measuring the lengths $\{a_i = |\boldsymbol{A}_i|\}$ of projections $\{\boldsymbol{A}_i\}$ of $\boldsymbol{A}$ onto the corresponding axes, providing the *matrix representation* of $\boldsymbol{A}$ in the adopted basis set: $\boldsymbol{A} \leftrightarrow \boldsymbol{a}^{(3)}$.

It should be also observed that in the preceding equation the resolution of $\boldsymbol{A}$ into its projections $\{\boldsymbol{A}_i\}$ along the directions of basic vectors $\boldsymbol{e}^{(3)}$ in this coordinate system can be also interpreted as a result of acting on $\boldsymbol{A}$ with the *projection operator* $\hat{\mathrm{P}}(R^3)$ onto the whole R^3 space,

$$\hat{\mathrm{P}}(R^3) = \sum_{i=1}^{3} (\boldsymbol{e}_i \boldsymbol{e}_i \cdot) \equiv \sum_{i=1}^{3} \hat{\mathrm{P}}(\boldsymbol{e}_i), \tag{2.3}$$

defined by the sum of individual projectors $\{\hat{\mathrm{P}}(\boldsymbol{e}_i)\}$ onto the specified axes. Indeed, the following identity directly follows from (2.2):

$$\boldsymbol{A} = \sum_{i=1}^{3} \boldsymbol{e}_i a_i = \left(\sum_{i=1}^{3} \boldsymbol{e}_i \boldsymbol{e}_i \cdot \right) \boldsymbol{A} = \hat{\mathrm{P}}(R^3) \boldsymbol{A} = \sum_{i=1}^{3} \hat{\mathrm{P}}(\boldsymbol{e}_i) \boldsymbol{A} \equiv \sum_{i=1}^{3} \boldsymbol{A}_i. \tag{2.4}$$

The preceding relation also implies that the projection of any vector $\boldsymbol{A}$ in R^3, or $\boldsymbol{A}(R^3)$ for short, amounts to multiplying it by the unity (identity) operation

$\hat{\mathrm{P}}(R^3) = 1$: $\hat{\mathrm{P}}(R^3)\boldsymbol{A}(R^3) = \boldsymbol{A}(R^3)$. Clearly, the sum of projections onto any two basis vectors $\hat{\mathrm{P}}(\boldsymbol{e}_i, \boldsymbol{e}_j) = \hat{\mathrm{P}}(\boldsymbol{e}_i) + \hat{\mathrm{P}}(\boldsymbol{e}_j)$ defines the projection onto the plane defined by these two axes:

$$\hat{\mathrm{P}}(\boldsymbol{e}_i, \boldsymbol{e}_j)\boldsymbol{A} = \hat{\mathrm{P}}(\boldsymbol{e}_i)\boldsymbol{A} + \hat{\mathrm{P}}(\boldsymbol{e}_j)\boldsymbol{A} = \boldsymbol{A}_i + \boldsymbol{A}_j \equiv \boldsymbol{A}_{(i,j)}. \tag{2.5}$$

This overall projection onto the whole physical space allows one to interpret the scalar product of two vectors $\boldsymbol{A}$ and $\boldsymbol{B}$ in R^3 in terms of their coordinates $\boldsymbol{a}^{(3)}$ and $\boldsymbol{b}^{(3)}$, respectively:

$$\boldsymbol{A} \cdot \boldsymbol{B} = \boldsymbol{A} \cdot \hat{\mathrm{P}}(R^3)\boldsymbol{B} = \sum_{i=1}^{3} (\boldsymbol{A} \cdot \boldsymbol{e}_i)(\boldsymbol{e}_i \cdot \boldsymbol{B}) = \sum_{i=1}^{3} a_i b_i = \boldsymbol{a}^{(3)}\boldsymbol{b}^{(3)\mathrm{T}}. \tag{2.6}$$

As seen from this example, the coordinate-resolved expression results directly from placing the identity operator $\hat{\mathrm{P}}(R^3) = 1$ between the two vectors in the scalar product. Obviously, this formal manipulation has no effect on the product value.

The characteristic property of projections is that the effect of a singular projection is identical to that of the subsequent repetition of the same projection. This immediately implies the *idempotency* property of the projection operators,

$$\hat{\mathrm{P}}(R^3)\hat{\mathrm{P}}(R^3) \equiv [\hat{\mathrm{P}}(R^3)]^2 = \hat{\mathrm{P}}(R^3), \quad [\hat{\mathrm{P}}(\boldsymbol{e}_i, \boldsymbol{e}_j)]^2 = \hat{\mathrm{P}}(\boldsymbol{e}_i, \boldsymbol{e}_j), \quad [\hat{\mathrm{P}}(\boldsymbol{e}_i)]^2 = \hat{\mathrm{P}}(\boldsymbol{e}_i), \tag{2.7}$$

where we have identified the square of an operator as a double execution of the operation it symbolizes. One can straightforwardly verify these identities using the orthonormality relations of (2.1a, 2.1b), which also imply that the product of projections into the mutually orthogonal subspaces identically vanishes, e.g.,

$$\hat{\mathrm{P}}(\boldsymbol{i})\hat{\mathrm{P}}(\boldsymbol{k}) = \hat{\mathrm{P}}(\boldsymbol{i})\hat{\mathrm{P}}(\boldsymbol{j}) = \hat{\mathrm{P}}(\boldsymbol{j})\hat{\mathrm{P}}(\boldsymbol{k}) = \hat{\mathrm{P}}(\boldsymbol{i},\boldsymbol{j})\hat{\mathrm{P}}(\boldsymbol{k}) = 0. \tag{2.8}$$

These observations can be naturally generalized into the n-dimensional *Euclidean* space R^n, spanned by n orthonormal basic vectors $\boldsymbol{e}^{(n)} = \{\boldsymbol{e}_i, i = 1, 2, \ldots, n\}$, $\boldsymbol{e}^{(n)\mathrm{T}} \cdot \boldsymbol{e}^{(n)} = \mathbf{I}^{(n)}$, also including the $n \to \infty$ limit. In particular, the matrix representations of vectors and the coordinate-resolved expression for the scalar product of vectors $\boldsymbol{A}(R^n)$ and $\boldsymbol{B}(R^n)$ directly follow from applying the projector onto the whole space R^n,

$$\hat{\mathrm{P}}(R^n) = \sum_{i=1}^{n} (\boldsymbol{e}_i \boldsymbol{e}_i \cdot) \equiv \sum_{i=1}^{n} \hat{\mathrm{P}}(\boldsymbol{e}_i), \tag{2.9}$$

$$\begin{aligned} \boldsymbol{A}(R^n) &= \hat{\mathrm{P}}(R^n)\boldsymbol{A}(R^n) = \sum_{i=1}^{n} \hat{\mathrm{P}}(\boldsymbol{e}_i)\boldsymbol{A}(R^n) = \sum_{i=1}^{n} \boldsymbol{e}_i[\boldsymbol{e}_i \cdot \boldsymbol{A}(R^n)] \\ &= \sum_{i=1}^{n} \boldsymbol{e}_i a_i = \sum_{i=1}^{n} \boldsymbol{A}_i = \boldsymbol{e}^{(n)}\boldsymbol{a}^{(n)\mathrm{T}}, \end{aligned} \tag{2.10}$$

$$A(R^n) \cdot B(R^n) = A(R^n) \cdot \hat{\mathbf{P}}(R^n) B(R^n) = \sum_{i=1}^{n} (A \cdot e_i)(e_i \cdot B)$$
$$= \sum_{i=1}^{n} a_i b_i = a^{(n)} b^{(n)\mathrm{T}}. \tag{2.11}$$

In particular, for two identical vectors $A(R^n) = B(R^n)$ one obtains the following expression for the vector length (norm):

$$A = |A| = \sqrt{A^2} = \left(\sum_{i=1}^{n} a_i^2\right)^{1/2} \geq 0. \tag{2.12}$$

One similarly defines the projection operators into various subspaces in R^n, e.g., its complementary, mutually orthogonal parts $P^m = \{e_i, i = 1, 2, \ldots, m\} \equiv \mathcal{P}$ and $Q^{n-m} = \{e_i, j = m+1, m+2, \ldots, n\} \equiv \mathcal{Q}$:

$$\hat{\mathbf{P}}(P^m) \equiv \hat{\mathbf{P}}_{\mathcal{P}} = \sum_{i \in \mathcal{P}} \hat{\mathbf{P}}(e_i), \quad \hat{\mathbf{P}}(Q^{n-m}) \equiv \hat{\mathbf{P}}_{\mathcal{Q}} = \sum_{j \in \mathcal{Q}} \hat{\mathbf{P}}(e_j), \quad \hat{\mathbf{P}}_{\mathcal{P}} \hat{\mathbf{P}}_{\mathcal{Q}} = 0,$$
$$A(R^n) = (\hat{\mathbf{P}}_{\mathcal{P}} + \hat{\mathbf{P}}_{\mathcal{Q}}) A(R^n) = A_{\mathcal{P}} + A_{\mathcal{Q}}, \tag{2.13}$$

where $A_{\mathcal{P}}$ and $A_{\mathcal{Q}}$ stand for the projections of $A(R^n)$ into the P^m and Q^{n-m} subspaces, respectively.

The scalar product of (2.11) can be also given the (linear) *functional* interpretation. In mathematics the linear functional $F[\varphi]$ of the argument φ, e.g. a function or vector, is a linear operation performed on the argument, which gives the scalar quantity F, $F[\varphi] = F$, e.g., the definite integral $I[f] = \int_{x_1}^{x_2} f(x)\,dx = I$. The same property can be associated with the (discrete) scalar product, say a projection of the argument vector $A \equiv \vec{A}$ onto another vector $B \equiv \vec{B}$:

$$B \cdot A = \vec{B} \cdot \vec{A} \equiv \overleftarrow{B}[\vec{A}], \tag{2.14}$$

where $\overleftarrow{B}[\vec{V}]$ denotes the functional of the vector argument $\vec{V}$ giving the value of its scalar product with the vector $\vec{B}$. The latter thus defines the functional $\overleftarrow{B}[X]$ itself, denoted as the "reversed" vector, by specifying the *direction* onto which the argument vector X is to be projected.

It can be then demonstrated that these scalar product functionals also span the vector space, called the *dual space*, since any combination of such quantities represents another linear functional of the same type. Let us examine these reversed "vector" quantities (functionals) associated with the independent basis vectors $\{e_i = \vec{e}_i\}$. They represent the *dual* basis "vectors" $\{\overleftarrow{e}_i[\vec{V}] \equiv \overleftarrow{e}_i\}$ of the scalar product functionals. Indeed, any combination of them also belongs to this dual space, e.g.,

$$C_i\overleftarrow{e}_i[\vec{V}] + C_j\overleftarrow{e}_j[\vec{V}] = (C_i\vec{e}_i + C_j\vec{e}_j) \cdot \vec{V} \equiv \vec{W} \cdot \vec{V} = \overleftarrow{W}[\vec{V}], \tag{2.15}$$

and to every vector $\vec{A}$ corresponds its functional analog $\overleftarrow{A}$ in the dual space, since the vector is uniquely specified by the complete set of its scalar products (components) with all independent vectors $\boldsymbol{e}^{(n)}$:

$$\vec{A} = \sum_{i=1}^{n} a_i\vec{e}_i = \sum_{i=1}^{n} \vec{A}_i \Rightarrow \overleftarrow{A}[\vec{V}] = \sum_{i=1}^{n} a_i\overleftarrow{e}_i[\vec{V}] = \sum_{i=1}^{n} \overleftarrow{A}_i[\vec{V}]. \tag{2.16}$$

It also follows from these relations that in Euclidean space this correspondence is linear: the linear combination of vectors in R^n is represented in the associated dual space by the associated combination, with the same expansion coefficients, of the corresponding dual-space functionals.

It should be emphasized that the dual-space elements, the "reversed" vectors, represent mathematical quantities (functionals of vectors) quite different from the original (argument) vectors on which they act.

2.2 Dirac's Vector Notation and Delta Function

In accordance with the *Superposition Principle* of quantum mechanics (Dirac 1967), any combination of states represents an admissible quantum state of the given molecular or atomic system. This property is also typical of ordinary vectors, $C_A\, \boldsymbol{A} + C_B\, \boldsymbol{B} = \boldsymbol{C}$, where the numerical coefficients C_A and C_B determine the relative participation of both vectors in the combination. We shall use this analogy in the vector notation of Dirac, in which the quantum states Ψ and Φ are denoted as arrowed "*ket*" symbols $|\Psi\rangle$, $|\Phi\rangle, \ldots$, called *state vectors*. Their linear combination $C_\Psi|\Psi\rangle + C_\Phi|\Phi\rangle = |\Theta\rangle$ determines another state $|\Theta\rangle$. When these states are functions of the continuous parameter $x \in [\xi, \zeta\,]$, $|\Psi\rangle = |\Psi(x)\rangle \equiv |\,x\rangle$, this summation of vector states is generalized into its continuous (integral) analog:

$$|\Theta\rangle = \int_{\xi}^{\zeta} c(x)\ |x\rangle\ dx. \tag{2.17}$$

Here, the combination coefficients $\{c(x), c(x'), \ldots\}$ are in general complex since the quantum states are complex entities. The resultant state $|\Theta\rangle$ of the given combination is said to be *dependent* upon the component states $\{|x\rangle,\ |x'\rangle,\ \ldots\}$. These *independent* state vectors cannot be expressed as combinations, with nonvanishing coefficients, of the remaining states in this basis set.

In the quantum kinematics it is the *direction* of the state vector $|\Psi\rangle$ that matters and uniquely identifies the quantum state Ψ. Therefore, the opposite state vectors along the same direction, e.g., $|\Psi\rangle$ and $-|\Psi\rangle$, in fact represent the same state Ψ, and any combination of the state with itself, $C_1|\Psi\rangle + C_2|\Psi\rangle = (C_1 + C_2)\,|\Psi\rangle = C|\Psi\rangle \equiv Me^{i\phi}\,|\Psi\rangle$, where M and ϕ stand for the modulus and phase of the complex coefficient C, also denote the same state Ψ. As we shall see later in this chapter, the length (norm) of the state vectors in quantum mechanics will be fixed by the appropriate *normalization* requirement resulting from the probabilistic interpretation of quantum states. In case of the square integrable wave functions it calls for $M = 1$, but the phase ϕ will be left undetermined as immaterial and having no physical meaning.

This property of the quantum superposition rule distinguishes it from the corresponding classical principle, e.g., that for combining vibrations of a string or a membrane, in which the combination of a state with itself gives another state exhibiting different amplitude. There is also another important distinction between the quantum and classical kinematics: in quantum mechanics the state vector of the vanishing norm (length), which thus has no specified direction in the vector space of quantum states, does not exist and thus has no physical meaning, while the classical vibration of the vanishing amplitude everywhere does in fact represent the real physical state of rest of a string or a membrane.

It was shown in the preceding section that to any vector space the dual space of the "reversed" vectors, the entities of quite different mathematical variety (functionals), can be ascribed through the concept of the scalar product (projection) of the vectors themselves. The dual space to the *ket*-space of state vectors $\{|\Psi_i\rangle\}$ is called the *bra*-space of the reversed "vectors" (functionals) $\{\langle\Psi_i|\}$, with the one-to-one (antilinear) correspondence: $\langle\Psi_i| \leftrightarrow |\Psi_i\rangle$, $(\langle\Psi_i| + \langle\Psi_j|) \leftrightarrow (|\Psi_i\rangle + |\Psi_j\rangle)$, $C^*\langle\Psi| \leftrightarrow C|\Psi\rangle$, etc., where C^* denotes the *complex* conjugate of C. In the original terminology of Dirac the *bra*- "vector" $\langle\Psi|$ represents the *conjugate-imaginary* of the associated *ket*-vector $|\Psi\rangle$. Again, the basic difference between the elements of the two vector spaces, with the "*bras*" in fact representing the functionals acting on "kets," it is improper to regard the *bra*-"vectors" as the *complex conjugates* of the corresponding *ket*-vectors.

In Dirac's notation the bra $\langle\Phi|$ and ket $|\Psi\rangle$ symbols are examples of an *incomplete* "bracket," while the result of $\langle\Phi|$ acting on $|\Psi\rangle$ gives the *complete* bracket of the *scalar product* of $|\Psi\rangle$ and $|\Phi\rangle$, $\langle\Phi|\Psi\rangle \equiv \Phi[|\Psi\rangle]$, which measures the projection of $|\Psi\rangle$ on $|\Phi\rangle$. The complete bracket generates the complex number. This association also explains the English nomenclature of the "bra" and "ket" symbols. This definition also implies that in contrast to the Euclidean space the complex numbers of the projections of $|\Psi\rangle$ on $|\Phi\rangle$ and of $|\Phi\rangle$ on $|\Psi\rangle$, respectively, are not equal in general, one representing the complex conjugate of the other:

$$\langle\Phi|\Psi\rangle = \Phi[|\Psi\rangle] \equiv \langle\Psi|\Phi\rangle^* = \Psi[|\Phi\rangle]^*. \tag{2.18}$$

One also observes that this linear functional of the ket vector:

$$\Phi[|C_1\Psi_1 + C_2\Psi_2\rangle] = C_1\Phi[|\Psi_1\rangle] + C_2\Phi[|\Psi_2\rangle], \tag{2.19}$$

is antilinear with respect to the bra vector, which determines the direction on which the projection is made:

$$\langle C_1\Phi_1 + C_2\Phi_2|\Psi\rangle = C_1{}^*\Phi_1[|\Psi\rangle] + C_2{}^*\Phi_2[|\Psi\rangle]. \tag{2.20}$$

Any vector in the ket space has its unique analog in the dual space of the bra "vectors" (functionals). There is a close analogy with the Euclidean space, in which the scalar product functional has also been used to define the dual "vector". Indeed the vector is uniquely defined by its projections on all (independent, orthonormal) vectors $\{|X_i\rangle = |i\rangle\}$, possibly including indenumerable vectors $\{|X(x)\rangle \equiv |x\rangle\}$ labeled by the continuous parameter(s) x. The set of projections $\{\langle\Phi|X_i\rangle = \langle X_i|\Phi\rangle^*\}$ thus uniquely determines the original ket $|\Phi\rangle$ associated with the functional $\Phi[] = \langle\Phi|$.

The "orthonormality" relations for the continuous basis vectors $\{|x\rangle\}$ are expressed in terms of the continuous analog of the Koronecker delta $\delta_{i,j} = \langle i|j\rangle$, called the *Dirac delta* "function" $\delta(x' - x) = \langle x|x'\rangle$. For any function $f(x)$ of the continuous argument(s) x this kernel satisfies the following "projection" identity:

$$f(x) = \int \delta(x' - x) f(x')\, dx'. \tag{2.21}$$

This equation indicates that this singular function represents the kernel of the integral operator $\int dx'\delta(x' - x)$, which acting on function $f(x')$ generates $f(x)$. Moreover, since the integral of the preceding equation formally expresses the functional $f(x) = f[f(x')]$, Dirac's delta can also be interpreted as the functional derivative (see Sect. 2.7):

$$\delta(x' - x) = \frac{\delta f(x)}{\delta f(x')}. \tag{2.22}$$

We shall discuss other properties of this mathematical entity later in this section.

The Dirac delta function $\delta(x' - x)$ of (2.21) represents the *unity*-normalized, $\int\delta(x' - x)dx' = 1$, infinitely sharp distribution centered at $x' = x$, exhibiting vanishing values at any finite distance from this point. It can be thus envisaged as the limiting form of the ordinary Gaussian (normal) distribution of the probability theory in the limit of the vanishing variance:

$$\delta(x' - x) = \lim_{\sigma\to 0} \frac{1}{\sqrt{2\pi\sigma^2}} \exp\left(-\frac{(x' - x)^2}{2\sigma^2}\right). \tag{2.23}$$

Alternatively, one can use any complete, say discrete, set of orthonormal basis functions $\{\chi_i(x)\}$, $\int\chi_i^*(x)\chi_j(x)dx = \delta_{i,j}$, to generate the analytical representation of this singular function. Indeed, expanding $f(x)$ in terms of the complete (orthonormal) basis set $\{\chi_i(x)\}$ gives:

$$f(x) = \sum_i \chi_i(x)c_i = \sum_i \chi_i(x)[\int \chi_i^*(x')f(x')\,dx']$$
$$= \int \Big\{\sum_i \chi_i^*(x')\chi_i(x)\Big\}f(x')\,dx'. \tag{2.24}$$

Hence, comparing the last equation with (2.21) gives the *closure* relation:

$$\delta(x' - x) = \sum_i \chi_i(x)\chi_i^*(x'). \tag{2.25a}$$

For the continuous orthonormal basis set $\{u_\alpha(x)\}$ labeled by the continuous index α, $\int u_\alpha^*(x)\, u_{\alpha'}(x)\, dx = \delta(\alpha' - \alpha)$, one similarly finds

$$\delta(x' - x) = \int u_\alpha(x)\; u_\alpha^*(x')\; d\alpha. \tag{2.25b}$$

When the complete basis set is "mixed," containing the discrete and continuous parts, $\{\chi_i(x), u_\alpha(x)\}$, with $\int u_\alpha^*(x)\, \chi_i(x)\, dx = 0$, this closure relation reads

$$\delta(x' - x) = \sum_i \chi_i(x)\chi_i^*(x') + \int u_\alpha(x)\; u_\alpha^*(x')\; d\alpha. \tag{2.25c}$$

Another important example of the continuous analytical representation of Dirac's delta originates from the Fourier-transform relations, e.g., between the wave function in the position and momentum representations of quantum mechanics (see Sect. 2.6),

$$\Phi(k) = \frac{1}{\sqrt{2\pi}}\int \exp(-ikx)f(x)dx \quad \text{and} \quad f(x) = \frac{1}{\sqrt{2\pi}}\int \exp(ik'x)\Phi(k')dk',$$
$$i = \sqrt{-1}. \tag{2.26}$$

Substituting the second, inverse transformation into the first one then gives

$$\Phi(k) = \frac{1}{2\pi}\int \Phi(k')\;\{\int \exp[ix(k' - k)]\,dx\}\,dk' \tag{2.27}$$

and hence

$$\delta(k' - k) = \frac{1}{2\pi}\int \exp[ix(k' - k)]\,dx. \tag{2.28}$$

The singular Dirac delta function $\delta(x' - x) \equiv \delta(z)$ satisfies the following identities:

$$\delta(z) = \delta(-z), \quad z\delta(z) = 0,$$

$$\delta(az) = |a|^{-1}\delta(z), \quad f(x')\delta(x' - x) = f(x)\delta(x' - x),$$

$$\int \delta(x' - x)\,\delta(x - x'')dx = \delta(x' - x''),$$

$$\delta(x^2 - a^2) = (2|a|)^{-1}[\delta(x - a) + \delta(x + a)]. \tag{2.29}$$

Of interest also are the related properties of the derivative of Dirac's delta "function," $\delta\,'(z) \equiv d\delta(z)/dz$,

$$\int f(z)\delta'(z)dz = -f'(0) \quad \text{or} \quad \int f(z)\delta'(-z)dz = f'(0), \; z\delta'(z) = -\delta(z). \tag{2.30}$$

2.3 Linear Operators and Their Adjoints

The complex number resulting from the scalar product between two state vectors is the result of applying the functional represented by its *bra* factor to its *ket* argument. When the linear action of a mathematical object on ket results in another ket, i.e., when it attributes in the linear fashion the uniquely specified *result*-vector $|\Psi'\rangle$ to the given *argument*-vector $|\Psi\rangle$, it is said to define the linear *operator* $\hat{A}$:

$$\hat{A}|\Psi\rangle = |\hat{A}\Psi\rangle \equiv |\Psi'\rangle, \quad \hat{A}|C_1\Psi_1 + C_2\Psi_2\rangle = C_1\hat{A}|\Psi_1\rangle + C_2\hat{A}|\Psi_2\rangle. \tag{2.31}$$

The operator is defined when its action on every ket is determined; it becomes zero, $\hat{A} = 0$, when its action on every ket $|\Psi\rangle$ gives zero. Thus, two operators are equal when they produce equal results when applied to every ket.

The linear operators can be added and multiplied:

$$(\hat{A} + \hat{B})|\Psi\rangle = \hat{A}|\Psi\rangle + \hat{B}|\Psi\rangle, \quad (\hat{A}\hat{B})|\Psi\rangle = \hat{A}(\hat{B}|\Psi\rangle) \equiv \hat{A}\hat{B}|\Psi\rangle. \tag{2.32}$$

In general, they do not commute, giving rise to nonvanishing *commutator*

$$[\hat{A}, \hat{B}] \equiv \hat{A}\hat{B} - \hat{B}\hat{A} \neq 0. \tag{2.33}$$

A multiplication by a number is a trivial case of a linear operation, which commutes with all linear operators. It can be easily verified that commutators satisfy the following identities:

$$[\hat{A}, \hat{B}] = -[\hat{B}, \hat{A}], \quad [\hat{A}, \hat{B} + \hat{C}] = [\hat{A}, \hat{B}] + [\hat{A}, \hat{C}],$$

$$[\hat{A}, \hat{B}\hat{C}] = [\hat{A}, \hat{B}]\hat{C} + \hat{B}[\hat{A}, \hat{C}], \qquad [\hat{A}, [\hat{B}, \hat{C}]] + [\hat{B}, [\hat{C}, \hat{A}]] + [\hat{C}, [\hat{A}, \hat{B}]] = 0. \tag{2.34}$$

Linear operators can also act on the bra vectors, with the latter always put to the left of the operator, giving other bras. Indeed, the symbol $\hat{A}\langle\Phi|$ has no meaning of the bra vector (functional), since its action on the ket vector $|\Psi\rangle$ gives another operator, $(\hat{A}\langle\Phi|)|\Psi\rangle = \hat{A}\langle\Phi|\Psi\rangle = \langle\Phi|\Psi\rangle\hat{A}$, thus representing an alien object in the present mathematical formalism. However, it can be straightforwardly demonstrated, again using the scalar product functional as the link to the definition of (2.31), that $\langle\Phi|\hat{A} = \langle\Phi'|$. Indeed, since $\hat{A}$ is linear and the scalar product depends linearly on the ket, the scalar products $\Phi[\hat{A}|\Psi\rangle] = \langle\Phi|(\hat{A}|\Psi\rangle)$ for the specified $\langle\Phi|$ and $\hat{A}$, associate with every ket $|\Psi\rangle$ in the vector space a number which depends linearly on $|\Psi\rangle$. This new linear functional thus defines a new bra vector $\langle\Phi'|$, which can be regarded as a result of $\hat{A}$ acting on $\langle\Phi|$:

$$\langle\Phi|(\hat{A}|\Psi\rangle) = (\langle\Phi|\hat{A})|\Psi\rangle = \langle\Phi'|\Psi\rangle. \tag{2.35}$$

Therefore, the linear operators act either on bras to their left or on kets to their right. In other words, the position of parentheses in the above matrix element of $\hat{A}$ is of no importance:

$$\langle\Phi|(\hat{A}|\Psi\rangle) = (\langle\Phi|\hat{A})|\Psi\rangle = \langle\Phi|\hat{A}|\Psi\rangle. \tag{2.36}$$

The operation $\langle\Phi|\hat{A} = \langle\Phi'|$ is linear, because for arbitrary $|\Psi\rangle$ and $\langle\Omega| = C_1\langle\Phi_1| + C_2\langle\Phi_2|$ one obtains:

$$\begin{aligned}(\langle\Omega|\hat{A})|\Psi\rangle &= \langle\Omega|(\hat{A}|\Psi\rangle) = C_1\langle\Phi_1|(\hat{A}|\Psi\rangle) + C_2\langle\Phi_2|(\hat{A}|\Psi\rangle) \\ &= C_1(\langle\Phi_1|\hat{A})|\Psi\rangle + C_2(\langle\Phi_2|\hat{A})|\Psi\rangle,\end{aligned} \tag{2.37}$$

and hence $\langle\Omega|\hat{A} = C_1\langle\Phi_1|\hat{A} + C_2\langle\Phi_2|\hat{A}$.

It can be directly verified that the product of the ket and bra vectors, $|\Psi\rangle\ \langle\Phi|$, represents an *operator*. When acting on ket $|\Xi\rangle$ it generates another ket vector along $|\Psi\rangle$, $|\Psi\rangle\ \langle\Phi|\Xi\rangle = \langle\Phi|\Xi\rangle\ |\Psi\rangle$, while the result of its action on bra $\langle\Omega|$ produces another bra vector (functional), proportional to $\langle\Phi|$: $\langle\Omega|\Psi\rangle\ \langle\Phi|$. It thus defines the linear operator:

$$|\Psi\rangle\ \langle\Phi|C_1\Xi_1 + C_2\Xi_2\rangle = C_1\langle\Phi|\Xi_1\rangle|\Psi\rangle + C_2\langle\Phi|\Xi_2\rangle|\Psi\rangle,$$

$$(C_1\langle\Omega_1| + C_2\langle\Omega_2|)|\Psi\rangle\ \langle\Phi| = C_1\langle\Omega_1|\Psi\rangle\ \langle\Phi| + C_2\langle\Omega_2|\Psi\rangle\ \langle\Phi|. \tag{2.38}$$

In particular, the operator $|X_i\rangle\ \langle X_i|$ defined by the normalized vector $|X_i\rangle \equiv |i\rangle$ and its *bra* conjugate amounts to the projection onto the $|i\rangle$ direction:

$$|i\rangle\ \langle i|\Psi\rangle \equiv \hat{P}_i|\Psi\rangle = \langle i|\Psi\rangle\ |i\rangle \equiv \Psi_i|i\rangle, \tag{2.39}$$

where Ψ_i stands for ith component of $|\Psi\rangle$ in the $|\boldsymbol{i}\rangle = \{|i\rangle\}$ representation (row vector). The projector idempotency then directly follows:

$$\hat{\mathrm{P}}_i^{\,2} = |i\rangle\langle i|i\rangle\langle i| = |i\rangle\langle i| = \hat{\mathrm{P}}_i. \tag{2.40}$$

When this discrete (countable) basis set spans the complete space, the sum of all such projectors, i.e., the projection on the whole space, amounts to the identity operation,

$$\hat{\mathrm{P}} = |\boldsymbol{i}\rangle\langle\boldsymbol{i}| = \sum_i \hat{\mathrm{P}}_i = 1, \tag{2.41a}$$

where $\langle\boldsymbol{i}|$ stands for the column vector of bras associated with the row vector of the basis kets $|\boldsymbol{i}\rangle$, because then $\hat{\mathrm{P}}|\Psi\rangle = |\Psi\rangle$. Similarly, when the complete basis set $|\boldsymbol{x}\rangle = \{|x\rangle\}$ is noncountable in character, with the orthonormality relations expressed by Dirac's delta "function" of (2.21), the summation is replaced by the integral over the continuous parameter(s),

$$\hat{\mathrm{P}} \equiv |\boldsymbol{x}\rangle\langle\boldsymbol{x}| = \int |x\rangle\langle x|\,dx = \int \hat{\mathrm{P}}(x)\,dx = 1, \tag{2.41b}$$

where we have again interpreted $|\boldsymbol{x}\rangle$ and $\langle\boldsymbol{x}|$ as the (*continuous*) row and column vectors, respectively. Finally, when the complete (mixed) basis contains both the discrete part $|\boldsymbol{\alpha}\rangle = \{|\alpha\rangle\}$ and the indenumerable subspace $|\boldsymbol{y}\rangle = \{|y\rangle\}$, $|\boldsymbol{m}\rangle = [|\boldsymbol{\alpha}\rangle, |\boldsymbol{y}\rangle]$ the identity operator of the complete overall projection operator includes both the discrete and continuous projections:

$$\hat{\mathrm{P}} \equiv |\boldsymbol{m}\rangle\langle\boldsymbol{m}| = |\boldsymbol{\alpha}\rangle\langle\boldsymbol{\alpha}| + |\boldsymbol{y}\rangle\langle\boldsymbol{y}| = \sum_\alpha \hat{\mathrm{P}}_\alpha + \int \hat{\mathrm{P}}(y)dy = 1. \tag{2.41c}$$

The (antilinear) one-to-one correspondence between kets and bras associates with every linear operator $\hat{\mathrm{A}}$ its *adjoint* (linear) operator $\hat{\mathrm{A}}^\dagger$ by the requirement that the bra associated with the ket $\hat{\mathrm{A}}|\Psi\rangle = |\hat{\mathrm{A}}\Psi\rangle \equiv |\Psi'\rangle$ is given by the result of action of $\hat{\mathrm{A}}^\dagger$ on the bra associated with $|\Psi\rangle$:

$$\langle\Psi'| = \langle\hat{\mathrm{A}}\Psi| \equiv \langle\Psi|\hat{\mathrm{A}}^\dagger. \tag{2.42}$$

Hence, since $\langle\Phi|\hat{\mathrm{A}}\Psi\rangle = \langle\hat{\mathrm{A}}\Psi|\Phi\rangle^*$ one obtains:

$$\langle\Phi|\hat{\mathrm{A}}\Psi\rangle \equiv \langle\Phi|\hat{\mathrm{A}}|\Psi\rangle = \langle\hat{\mathrm{A}}\Psi|\Phi\rangle^* \equiv \langle\Psi|\hat{\mathrm{A}}^\dagger|\Phi\rangle^*. \tag{2.43}$$

Moreover, because $(\hat{\mathrm{A}}^\dagger)^\dagger = \hat{\mathrm{A}}$ and hence $\langle\hat{\mathrm{A}}^\dagger\Phi| = \langle\Phi|\hat{\mathrm{A}}$, the adjoint operators can be alternatively defined by the identity:

$$\langle\hat{\mathrm{A}}^\dagger\Phi|\Psi\rangle = \langle\Phi|\hat{\mathrm{A}}|\Psi\rangle = \langle\Phi|\hat{\mathrm{A}}\Psi\rangle. \tag{2.44}$$

Next, it is easy to show that $(\lambda\hat{\mathrm{A}})^\dagger = \lambda^*\hat{\mathrm{A}}^\dagger$ and $(\hat{\mathrm{A}}+\hat{\mathrm{B}})^\dagger = \hat{\mathrm{A}}^\dagger + \hat{\mathrm{B}}^\dagger$. To determine the adjoint of the product of two operators one observes that the ket $|\Omega\rangle = \hat{\mathrm{A}}\hat{\mathrm{B}}|\Psi\rangle \equiv \hat{\mathrm{A}}|\Theta\rangle$ is associated with the bra

$$\langle\Omega| = \langle\Psi|(\hat{\mathrm{A}}\,\hat{\mathrm{B}})^\dagger = \langle\Theta|\hat{\mathrm{A}}^\dagger = \langle\Psi|\hat{\mathrm{B}}^\dagger\,\hat{\mathrm{A}}^\dagger, \tag{2.45}$$

where we have realized that the bra associated with $|\Theta\rangle, \langle\Theta| = \langle\Psi|\hat{\mathrm{B}}^\dagger$. Hence, $(\hat{\mathrm{A}}\hat{\mathrm{B}})^\dagger = \hat{\mathrm{B}}^\dagger\hat{\mathrm{A}}^\dagger$. This change of order, when one takes the adjoint of a product of operators, can be generalized to an arbitrary number of them: $((\hat{\mathrm{A}}\hat{\mathrm{B}}\ldots\hat{\mathrm{C}})^\dagger = \hat{\mathrm{C}}^\dagger\ldots\hat{\mathrm{B}}^\dagger\hat{\mathrm{A}}^\dagger$. One also observes that the following identity is satisfied for commutators: $[\hat{\mathrm{A}},\hat{\mathrm{B}}]^\dagger = [\hat{\mathrm{B}}^\dagger,\hat{\mathrm{A}}^\dagger]$.

We can now summarize the mutual relations between the mathematical entities hitherto introduced in terms of the general *Hermitian* conjugation denoted by the adjoint symbol "$\dagger$". In the Dirac notation the ket $|\Psi\rangle$ and its associated bra $\langle\Psi|$ are said to be Hermitian conjugates of each other: $\langle\Psi| = |\Psi\rangle^\dagger$ and $\langle\Psi|^\dagger = |\Psi\rangle$. Moreover, the operators $\hat{\mathrm{A}}$ and $\hat{\mathrm{A}}^\dagger$ are also related by the Hermitian conjugation. As we have observed in the preceding equation the hermitian conjugation of the product of operator factors changes the order in the product of the adjoint operators. This rule holds for other entities as well. For example, the Hermitian conjugate of $\hat{\mathrm{A}}|\Psi\rangle$ gives:

$$(\hat{\mathrm{A}}|\Psi\rangle)^\dagger = |\hat{\mathrm{A}}\Psi\rangle^\dagger = |\Psi\rangle^\dagger\hat{\mathrm{A}}^\dagger = \langle\Psi|\hat{\mathrm{A}}^\dagger. \tag{2.46}$$

Similarly,

$$\begin{aligned}&(|\Psi\rangle\,\langle\Phi|)^\dagger = (\langle\Phi|^\dagger)(|\Psi\rangle^\dagger) = |\Phi\rangle\,\langle\Psi|,\quad (\langle\Phi|\Psi\rangle)^\dagger = (|\Psi\rangle^\dagger)(\langle\Phi|^\dagger) = \langle\Psi|\Phi\rangle,\\ &(\lambda\langle\Phi|\Psi\rangle|\Psi\rangle\,\langle\Phi|)^\dagger = |\Phi\rangle\,\langle\Psi|\langle\Psi|\Phi\rangle\lambda^* = \lambda^*\langle\Psi|\Phi\rangle|\Phi\rangle\,\langle\Psi|,\ \text{etc.}\end{aligned} \tag{2.47}$$

Thus, to obtain the adjoint (Hermitian conjugate) of any expression composed of constants, kets, bras and linear operators, one replaces the constants by their complex conjugates, kets by the associated bras, bras by the associated kets, operators by their adjoints and reverses the order of factors in the products. However, as we have observed in the last line of (2.47), the position of constants, λ^*, $\langle\Psi|\Phi\rangle$, etc., is of no importance.

2.4 Basis Set Representations of Vectors and Operators

Selection of the complete (orthonormal) basis of the reference ket vectors in the vector space of the system quantum-mechanical states, either discrete $|\boldsymbol{i}\rangle = \{|i\rangle\}$, $\langle i|j\rangle = \delta_{i,j}$, or the continuous infinity of vectors $|\boldsymbol{x}\rangle = \{|x\rangle\}$, $\langle x|x'\rangle = \delta(x'-x)$, defines the specific *representation* in which both the vectors and operators can be expressed. By convention the basis vectors $|\boldsymbol{i}\rangle$ and $|\boldsymbol{x}\rangle$ are arranged as the *row* vectors. Accordingly, their Hermitian conjugates define the respective column vectors of the bra basis: $|\boldsymbol{i}\rangle^{\dagger} = \langle \boldsymbol{i}|$ and $|\boldsymbol{x}\rangle^{\dagger} = \langle \boldsymbol{x}|$.

Using the closure relations of (2.41a), (2.41b) and the above orthonormality relations for these basis vectors gives the associated expansions of any ket $|\Psi\rangle$:

$$\begin{aligned} |\Psi\rangle &= \sum_i |i\rangle \langle i|\Psi\rangle = \sum_i |i\rangle \Psi_i = |\boldsymbol{i}\rangle \langle \boldsymbol{i}|\Psi\rangle \equiv |\boldsymbol{i}\rangle \boldsymbol{\Psi}^{(i)}, \\ |\Psi\rangle &= \int |x\rangle \langle x|\Psi\rangle\, dx = \int |x\rangle \Psi(x)\, dx \equiv |\boldsymbol{x}\rangle \langle \boldsymbol{x}|\Psi\rangle \equiv |\boldsymbol{x}\rangle \boldsymbol{\Psi}^{(x)}. \end{aligned} \tag{2.48a}$$

The components $\{\Psi_i\}$ or $\{\Psi(x), \Psi(x'), \ldots\}$, by convention arranged vertically as the *column* vectors, $\boldsymbol{\Psi}^{(i)} = \langle \boldsymbol{i}|\Psi\rangle$ and $\boldsymbol{\Psi}^{(x)} = \langle \boldsymbol{x}|\Psi\rangle$, provide the representations of the ket $|\Psi\rangle$ in the basis sets $|\boldsymbol{i}\rangle$ and $|\boldsymbol{x}\rangle$, respectively. In the mixed basis set case of (2.41c) the expansion of ket $|\Psi\rangle$ in $|\boldsymbol{m}\rangle$ will contain both the discrete and continuous components:

$$\begin{aligned} |\Psi\rangle = |\boldsymbol{m}\rangle \langle \boldsymbol{m}|\Psi\rangle \equiv |\boldsymbol{m}\rangle \boldsymbol{\Psi}^{(m)} &= |\boldsymbol{\alpha}\rangle \langle \boldsymbol{\alpha}|\Psi\rangle + |\boldsymbol{y}\rangle \langle \boldsymbol{y}|\Psi\rangle = |\boldsymbol{\alpha}\rangle\, \boldsymbol{\Psi}^{(\alpha)} + |\boldsymbol{y}\rangle\, \boldsymbol{\Psi}^{(y)} \\ &= \sum_\alpha |\alpha\rangle \langle \alpha|\Psi\rangle + \int |y\rangle \langle y|\Psi\rangle\, dy. \end{aligned} \tag{2.48b}$$

The associated expansions of the bra vector $\langle \Phi|$ in terms of the reference bra vectors $\langle \boldsymbol{i}|$, $\langle \boldsymbol{x}|$, and $\langle \boldsymbol{m}|$, respectively, directly follow from applying the corresponding *unity*-projections of (2.41a)–(2.41c) to $\langle \Phi|$ (from the right):

$$\begin{aligned} \langle \Phi| &= \sum_i \langle \Phi|i\rangle \langle i| = \sum_i \Phi_i^* \langle i| = \langle \Phi|\boldsymbol{i}\rangle \langle \boldsymbol{i}| \equiv \boldsymbol{\Phi}^{(i)\dagger} \langle \boldsymbol{i}|, \\ \langle \Phi| &= \int \langle \Phi|x\rangle \langle x|\, dx = \int \Phi^*(x)\, \langle x|\, dx = \langle \Phi|\boldsymbol{x}\rangle \langle \boldsymbol{x}| \equiv \boldsymbol{\Phi}^{(x)\dagger} \langle \boldsymbol{x}|, \\ \langle \Phi| &= \langle \Phi|\boldsymbol{m}\rangle \langle \boldsymbol{m}| \equiv \boldsymbol{\Phi}^{(m)\dagger} \langle \boldsymbol{m}| = \langle \Phi|\boldsymbol{\alpha}\rangle \langle \boldsymbol{\alpha}| + \langle \Phi|\boldsymbol{y}\rangle \langle \boldsymbol{y}| \equiv \boldsymbol{\Phi}^{(\alpha)\dagger} \langle \boldsymbol{\alpha}| + \boldsymbol{\Phi}^{(y)\dagger} \langle \boldsymbol{y}|. \end{aligned} \tag{2.49}$$

Therefore, the vector components $\boldsymbol{\Phi}^{(i)\dagger} = \langle \Phi|\boldsymbol{i}\rangle$, $\boldsymbol{\Phi}^{(x)\dagger} = \langle \Phi|\boldsymbol{x}\rangle$ and $[\{\Phi_\alpha^*\}, \Phi^*(y)] = [\langle \Phi|\boldsymbol{\alpha}\rangle, \langle \Phi|\boldsymbol{y}\rangle]$, when arranged horizontally as the associated *row* vectors, constitute the corresponding representations of $\langle \Phi|$ in these three types of the basis set. Again, the continuous representation of the bra vector, e.g., the complex conjugate wave function $\Phi^*(x) = \langle \Phi|x\rangle$, can also be regarded as the continuous *row* vector with components $[\Phi^*(x), \Phi^*(x'), \ldots]$.

In these three types of the basis sets, the linear operator $\hat{\mathrm{A}}$ is accordingly represented by the square matrix and/or the continuous kernel, respectively,

$$\mathbf{A}(\boldsymbol{i},\boldsymbol{i}') \equiv \langle \boldsymbol{i}|\hat{\mathrm{A}}|\boldsymbol{i}'\rangle = \{A_{i,i'} = \langle i|\hat{\mathrm{A}}|i'\rangle\} \equiv \mathbf{A}^{(\boldsymbol{i})},$$
$$\mathbf{A}(\boldsymbol{x},\boldsymbol{x}') \equiv \langle \boldsymbol{x}|\hat{\mathrm{A}}|\boldsymbol{x}'\rangle = \{A(x,x') = \langle x|\hat{\mathrm{A}}|x'\rangle\} \equiv \mathbf{A}^{(\boldsymbol{x})},$$
$$\mathbf{A}(\boldsymbol{m},\boldsymbol{m}') \equiv \langle \boldsymbol{m}|\hat{\mathrm{A}}|\boldsymbol{m}'\rangle = \begin{bmatrix} \mathbf{A}(\boldsymbol{\alpha},\boldsymbol{\alpha}') = \langle \boldsymbol{\alpha}|\hat{\mathrm{A}}|\boldsymbol{\alpha}'\rangle & \mathbf{A}(\boldsymbol{\alpha},\boldsymbol{y}') = \langle \boldsymbol{\alpha}|\hat{\mathrm{A}}|\boldsymbol{y}'\rangle \\ \mathbf{A}(\boldsymbol{y},\boldsymbol{\alpha}') = \langle \boldsymbol{y}|\hat{\mathrm{A}}|\boldsymbol{\alpha}'\rangle & \mathbf{A}(\boldsymbol{y},\boldsymbol{y}') = \langle \boldsymbol{y}|\hat{\mathrm{A}}|\boldsymbol{y}'\rangle \end{bmatrix} \equiv \mathbf{A}^{(\boldsymbol{m})}. \tag{2.50}$$

The adjoint operator $\hat{\mathrm{A}}^{\dagger}$ is similarly represented by the corresponding Hermitian conjugates of these "matrices,"

$$\langle \boldsymbol{i}|\hat{\mathrm{A}}^{\dagger}|\boldsymbol{i}'\rangle = \langle \hat{\mathrm{A}}\boldsymbol{i}|\boldsymbol{i}'\rangle = \langle \boldsymbol{i}'|\hat{\mathrm{A}}|\boldsymbol{i}\rangle^* = \mathbf{A}(\boldsymbol{i}',\boldsymbol{i})^* = [\mathbf{A}(\boldsymbol{i},\boldsymbol{i}')^*]^{\mathrm{T}} = \mathbf{A}(\boldsymbol{i},\boldsymbol{i}')^{\dagger} \equiv \mathbf{A}^{\dagger(\boldsymbol{i})},$$
$$\langle \boldsymbol{x}|\hat{\mathrm{A}}^{\dagger}|\boldsymbol{x}'\rangle = \langle \boldsymbol{x}'|\hat{\mathrm{A}}|\boldsymbol{x}\rangle^* = \langle \boldsymbol{x}|\hat{\mathrm{A}}|\boldsymbol{x}'\rangle^{\dagger} = \mathbf{A}(\boldsymbol{x},\boldsymbol{x}')^{\dagger} = [\mathbf{A}(\boldsymbol{x},\boldsymbol{x}')^*]^{\mathrm{T}} \equiv \mathbf{A}^{\dagger(\boldsymbol{x})},$$
$$\langle \boldsymbol{m}|\hat{\mathrm{A}}^{\dagger}|\boldsymbol{m}'\rangle = \langle \boldsymbol{m}'|\hat{\mathrm{A}}|\boldsymbol{m}\rangle^* = \langle \boldsymbol{m}|\hat{\mathrm{A}}|\boldsymbol{m}'\rangle^{\dagger} = \mathbf{A}(\boldsymbol{m},\boldsymbol{m}')^{\dagger} = [\mathbf{A}(\boldsymbol{m},\boldsymbol{m}')^*]^{\mathrm{T}}$$
$$= \begin{bmatrix} \mathbf{A}(\boldsymbol{\alpha},\boldsymbol{\alpha}')^{\dagger} = \langle \boldsymbol{\alpha}'|\hat{\mathrm{A}}|\boldsymbol{\alpha}\rangle^* & \mathbf{A}(\boldsymbol{\alpha},\boldsymbol{y}')^{\dagger} = \langle \boldsymbol{y}'|\hat{\mathrm{A}}|\boldsymbol{\alpha}\rangle^* \\ \mathbf{A}(\boldsymbol{y},\boldsymbol{\alpha}')^{\dagger} = \langle \boldsymbol{\alpha}'|\hat{\mathrm{A}}|\boldsymbol{y}\rangle^* & \mathbf{A}(\boldsymbol{y},\boldsymbol{y}')^{\dagger} = \langle \boldsymbol{y}'|\hat{\mathrm{A}}|\boldsymbol{y}\rangle^* \end{bmatrix} \equiv \mathbf{A}^{\dagger(\boldsymbol{m})}. \tag{2.51}$$

Hence, the Hermitian (*self*-adjoint) operator $\hat{\mathrm{A}}$ of the physical observable A, for which $\hat{\mathrm{A}}^{\dagger} = \hat{\mathrm{A}}$, is represented by the Hermitian matrix/kernel: $\mathbf{A}^{\dagger(\boldsymbol{b})} = \mathbf{A}^{(\boldsymbol{b})}$, $\boldsymbol{b} = \boldsymbol{i}, \boldsymbol{x}, \boldsymbol{m}$.

The relations between vectors of (2.31) and (2.42) are thus transformed into the corresponding equations in terms of the basis set components. For example, (2.31) then reads:

$$\hat{\mathrm{A}}|\Psi\rangle = |\Psi'\rangle \leftrightarrow \mathbf{A}^{(\boldsymbol{b})}\,\boldsymbol{\Psi}^{(\boldsymbol{b})} = \boldsymbol{\Psi}'^{(\boldsymbol{b})}, \quad \boldsymbol{b} = \boldsymbol{i},\boldsymbol{x},\boldsymbol{m}, \quad \text{i.e.,}$$
$$\sum_{i'} A_{i,i'}\,\Psi_{i'} = \Psi_i', \quad \int A(x,x')\,\Psi(x')\,dx' = \Psi'(x),$$
$$\sum_{\alpha'} A_{\alpha,\alpha'}\Psi_{\alpha'} + \int A(\alpha,y')\,\Psi(y')\,dy' = \Psi_\alpha' \quad \text{and}$$
$$\sum_{\alpha'} A(y,\alpha')\,\Psi_{\alpha'} + \int A(y,y')\,\Psi(y')dy' = \Psi'(y). \tag{2.52}$$

The corresponding basis set transcriptions of (2.42) similarly give:

$$\langle \Psi'| = \langle \Psi|\hat{\mathrm{A}}^{\dagger} \Leftrightarrow (\hat{\mathrm{A}}|\Psi\rangle = |\Psi'\rangle)^{\dagger} \quad \leftrightarrow \quad \boldsymbol{\Psi}^{(\boldsymbol{b})\dagger}\mathbf{A}^{\dagger(\boldsymbol{b})} = \boldsymbol{\Psi}'^{\dagger(\boldsymbol{b})}, \quad \boldsymbol{b} = \boldsymbol{i},\boldsymbol{x},\boldsymbol{m}, \quad \text{i.e.,}$$
$$\sum_{i'} \Psi_{i'}{}^* A_{i',i}{}^* = \Psi'_i{}^*, \quad \int \Psi(x')^* A(x',x)^* dx' = \Psi'(x)^*,$$
$$\sum_{\alpha'} \Psi_{\alpha'}{}^*(A_{\alpha',\alpha})^* + \int \Psi(y')^*\,A(y',\alpha)^*\,dy' = (\Psi_\alpha{}')^* \quad \text{and}$$
$$\sum_{\alpha'} \Psi_{\alpha'}{}^* A(\alpha',y)^* + \int \Psi(y')^* A(y',y)^*\,dy' = \Psi'(y)^*. \tag{2.53}$$

It should be emphasized that the basis set representations of the state vector are fully equivalent to the state specification by the vector itself. For example (see Sect. 2.6), when the continuous basis set is labeled by the position of a particle in space, $x = \boldsymbol{r}$, or its momentum, $x = \boldsymbol{p}$, the associated representations $\boldsymbol{\Psi}^{(\boldsymbol{r})} \equiv \Psi(\boldsymbol{r}) = \langle \boldsymbol{r}|\Psi\rangle$ and $\boldsymbol{\Psi}^{(\boldsymbol{p})} \equiv \Psi(\boldsymbol{p}) = \langle \boldsymbol{p}|\Psi\rangle$, called the wave functions in the position ($\boldsymbol{r}$) and momentum ($\boldsymbol{p}$) representations, respectively, provide alternative specifications of the quantum state of the particle, which uniquely establish the direction of the ket $|\Psi\rangle$ in the system Hilbert space.

2.5 Eigenvalue Problem of Linear Hermitian Operators

For the linear operator to represent the physically observable quantity in quantum mechanics it has to be *self-adjoint*, i.e., its hermitian conjugate (adjoint) must be identical with the operator itself: $\hat{A}^\dagger = \hat{A}$. Only such *Hermitian* operators can be associated with the measurable quantities of physics. They satisfy the following scalar product identity [see (2.43)]:

$$\langle\Phi|\hat{A}|\Psi\rangle = \langle\Psi|\hat{A}|\Phi\rangle^* = \langle\Phi|\hat{A}|\Psi\rangle^\dagger. \tag{2.54}$$

The projector $\hat{P}_\Psi = |\Psi\rangle\langle\Psi|$ provides an example of the Hermitian operator: $\hat{P}_\Psi^\dagger = \hat{P}_\Psi$. One also observes that the change of order of the adjoint factors in the Hermitian conjugate of the product of two operators implies that the product of the *commuting* Hermitian operators also represents the Hermitian operator. Indeed, when $[\hat{A}, \hat{B}] = 0$, $(\hat{A}\hat{B})^\dagger = \hat{B}^\dagger\hat{A}^\dagger = \hat{B}\hat{A} = \hat{A}\hat{B}$.

In quantum mechanics the *eigenvalue problem* of the linear Hermitian operator $\hat{A}$ corresponding to the physical quantity A is of paramount importance in determining the outcomes of its measurement. It is defined by the following equation:

$$\hat{A}|\Psi_i\rangle = a_i|\Psi_i\rangle \quad \text{or} \quad \langle\Psi_i|\hat{A}^\dagger = \langle\Psi_i|a_i^* = \langle\Psi_i|\hat{A} = \langle\Psi_i|a_i, \tag{2.55a}$$

where a_i denotes ith *eigenvalue* (a number) and $|\Psi_i\rangle \equiv |a_i\rangle$ and $\langle\Psi_i| \equiv \langle a_i|$ stand for the associated *eigen-ket(bra)*, i.e., the eigenvector belonging to a_i. Therefore, the action of $\hat{A}$ on its eigenvector does not affect the direction of the latter, with only its length being multiplied by the corresponding eigenvalue.

A trivial example is the multiplication by a number a. This operator has just one eigenvalue, this number itself: any ket is an eigenket and any bra is an eigenbra corresponding to this eigenvalue. One observes that this number has to be real for such a *number operator* to be self-adjoint [see (2.55a)].

In quantum theory the Hermitian operator $\hat{A}$, the eigenvectors of which form a basis in the state space, is called an *observable*. The projections onto all such eigenstates amounts to the identity operations of (2.41a)–(2.41c). The projector $\hat{P}_\Psi$ is an example of the quantum-mechanical observable, which exhibits only two eigenvectors. Indeed, for an arbitrary ket $|\Phi\rangle$ the two functions $|1\rangle \equiv \hat{P}_\Psi|\Phi\rangle$ and $|0\rangle \equiv (1 - \hat{P}_\Psi)|\Phi\rangle$ can be shown to satisfy the eigenvalue problem of $\hat{P}_\Psi$:

$$\hat{P}_\Psi|1\rangle = \hat{P}_\Psi^2|\Phi\rangle = \hat{P}_\Psi|\Phi\rangle = |1\rangle, \quad \hat{P}_\Psi|0\rangle = (\hat{P}_\Psi - \hat{P}_\Psi^2)|\Phi\rangle = 0|0\rangle, \tag{2.56}$$

where we have used the idempotency property of projection operators [(2.40)]. Therefore, the two state vectors $\{|1\rangle, |0\rangle\}$ are the eigenvectors of $\hat{P}_\Psi$ corresponding to eigenvalues $\{1, 0\}$. Since every ket in the state space can be expanded in these two eigenstates, $|\Phi\rangle = |1\rangle + |0\rangle$, they form the basis in the state space, $|1\rangle\langle 1| + |0\rangle\langle 0| = 1$, thus confirming that $\hat{P}_\Psi$ is an observable.

The eigenbra problem is similarly defined by the Hermitian conjugate of (2.55a):

$$\langle\Psi_i|\hat{A} = \langle\Psi_i|a_i^*. \tag{2.55b}$$

It then follows from the Hermitian character of $\hat{A}$ that all its eigenvalues are real numbers. It suffices to multiply (2.55a) by $\langle\Psi_i|$ (from the left) and (2.55b) by $|\Psi_i\rangle$ (from the right), subtract the resulting equations and use (2.54) (for $\Phi = \Psi = \Psi_i$) to obtain the identity:

$$0 = (a_i - a_i^*)\,\langle\Psi_i|\Psi_i\rangle \Rightarrow a_i = a_i^*. \tag{2.57}$$

The eigenvalues can be degenerate, when several independent eigenvectors $\{|\Psi_{i,j}\rangle\} = \{|\Psi_{i,1}\rangle, |\Psi_{i,2}\rangle, \ldots, |\Psi_{i,g}\rangle\} \equiv \{|i_j\rangle, j = 1, 2, \ldots, g\}$ belong to the same eigenvalue a_i:

$$\hat{A}|i_1\rangle = a_i|i_1\rangle,\ \hat{A}|i_2\rangle = a_i|i_2\rangle,\ \ldots,\ \hat{A}|i_g\rangle = a_i|i_g\rangle; \tag{2.58}$$

here the number g of such linearly independent (mutually orthogonal) components determines the *multiplicity* of such degenerate eigenvalue. It then directly follows from the linear character of $\hat{A}$ that any combination of such states, say $|\Psi\rangle = C_1|i_1\rangle + C_2|i_2\rangle + \ldots C_g|i_g\rangle$, also represents the eigenvector of $\hat{A}$ belonging to this eigenvalue:

$$\hat{A}|\Psi\rangle = C_1\hat{A}|i_1\rangle + C_2\hat{A}|i_2\rangle + \cdots + C_g\hat{A}|i_g\rangle = a_i|\Psi\rangle. \tag{2.59}$$

The Hermitian character of the linear operator also implies that eigenvectors $|\Psi_i\rangle \equiv |i\rangle$ and $|\Psi_j\rangle \equiv |\,j\rangle$, which belong to different eigenvalues $a_i \neq a_j$, respectively, are automatically orthogonal. Indeed, the associated eigenvalue equations, $\hat{A}|i\rangle = a_i|i\rangle$ and $\langle j|\hat{A} = \langle j|a_j$, give by an analogous manipulation involving a multiplication of the former by $\langle j|$, of the latter by $|i\rangle$, and a subtraction of the resulting equations,

$$0 = (a_i - a_i)\,\langle j|i\rangle \Rightarrow \langle j|i\rangle = 0. \tag{2.60}$$

In the degenerate case, each vector belonging to the subspace $\{|i_k\rangle\}$ of eigenvalue a_i is thus orthogonal to every vector belonging to the subspace $\{|j_l\rangle\}$ of eigenvalue a_j: $\langle i_k|j_l\rangle = 0$. Inside each degenerate subspace the vectors can always be constructed as othonormal, $\langle i_k|i_l\rangle = \delta_{k,l}$, by choosing appropriate combinations of the initial independent (normalized but nonorthogonal) state vectors.

For the given representation in the Hilbert space, say, specified by the discrete orthonormal basis $|\boldsymbol{i}\rangle$, the eigenvalue equation (2.55a) assumes the form of the separate systems of algebraic equations for each eigenvalue, which can be summarized in the joint matrix form [see (2.52)]:

$$\mathbf{A}^{(i)}\,\boldsymbol{\Psi}^{(i)} = \mathbf{a}\,\boldsymbol{\Psi}^{(i)}, \tag{2.61}$$

with the operator represented by the square matrix $\mathbf{A}^{(i)} = \{\langle i|\hat{\mathrm{A}}|i'\rangle\}$, the diagonal matrix $\mathbf{a} = \{a_s\delta_{s,s'} = \langle\Psi_s|\hat{\mathrm{A}}|\Psi_s\rangle\delta_{s,s'}\}$ grouping the eigenvalues $\{a_s\}$ corresponding to eigenvectors $|\boldsymbol{\Psi}\rangle = \{|\Psi_s\rangle \equiv |s\rangle\}$ determined by the corresponding columns $\boldsymbol{\Psi}_s^{(i)} = \langle\boldsymbol{i}|s\rangle$ of the rectangular matrix $\boldsymbol{\Psi}^{(i)} = \{\boldsymbol{\Psi}_s^{(i)}\} = \langle\boldsymbol{i}|\boldsymbol{\Psi}\rangle = \{\langle i|s\rangle\}$ grouping the relevant expansion coefficients (projections).

Moreover, since both $|\boldsymbol{\Psi}\rangle$ and $|\boldsymbol{i}\rangle$ form bases in the Hilbert space, the overall projection $|\boldsymbol{\Psi}\rangle\langle\boldsymbol{\Psi}| = |\boldsymbol{i}\rangle\langle\boldsymbol{i}| = 1$ and hence

$$\begin{aligned}\boldsymbol{\Psi}^{(i)}\,\boldsymbol{\Psi}^{(i)\dagger} &= \langle\boldsymbol{i}|\boldsymbol{\Psi}\rangle\,\langle\boldsymbol{\Psi}|\boldsymbol{i}\rangle = \langle\boldsymbol{i}|\boldsymbol{i}\rangle = \{\delta_{i,i'}\} = \mathbf{I}^{(i)} \quad \text{and}\\ \boldsymbol{\Psi}^{(i)\dagger}\,\boldsymbol{\Psi}^{(i)} &= \langle\boldsymbol{\Psi}|\boldsymbol{i}\rangle\,\langle\boldsymbol{i}|\boldsymbol{\Psi}\rangle = \langle\boldsymbol{\Psi}|\boldsymbol{\Psi}\rangle = \{\delta_{s,s'}\} = \mathbf{I}^{(\boldsymbol{\Psi})}.\end{aligned} \tag{2.62}$$

Thus, the basis set components of eigenvectors, $\boldsymbol{\Psi}^{(i)}$, define the unitary matrix: $(\boldsymbol{\Psi}^{(i)})^\dagger = (\boldsymbol{\Psi}^{(i)})^{-1}$. Hence, the multiplication, from the right, of both sides of (2.61) by $\boldsymbol{\Psi}^{(i)\dagger}$ allows one to rewrite this matrix equation as the unitary (similarity) transformation ("rotation"), which diagonalizes the Hermitian matrix $\mathbf{A}^{(i)}$, the basis set representation of the Hermitian operator $\hat{\mathrm{A}}$, to its eigenvector representation $\mathbf{a} = \langle\boldsymbol{\Psi}|\hat{\mathrm{A}}|\boldsymbol{\Psi}\rangle \equiv \mathbf{A}^{(\boldsymbol{\Psi})}$:

$$\boldsymbol{\Psi}^{(i)\dagger}\mathbf{A}^{(i)}\boldsymbol{\Psi}^{(i)} = (\boldsymbol{\Psi}^{(i)})^{-1}\mathbf{A}^{(i)}\,\boldsymbol{\Psi}^{(i)} = \mathbf{a}. \tag{2.63}$$

This is the standard numerical procedure, which is routinely applied in the computer programs for the finite basis set determination of eigenvalues of Hermitian matrices.

When dealing with problems of the simultaneous measurements of physical quantities in quantum mechanics, one encounters the *common* eigenvalue problem of several mutually commuting observables. It can be straightforwardly demonstrated that the commutation of operators $\hat{\mathrm{A}}$ and $\hat{\mathrm{B}}$, $[\hat{\mathrm{A}}, \hat{\mathrm{B}}] = 0$, implies the existence of their common eigenvectors, which form the basis in the space of state vectors. In other words, for the case of the discrete spectrum of eigenvalues $\{a_i\}$ and $\{b_j\}$ of these two operators, there exist the common eigenvectors $\{|a_i, b_j\rangle\}$ of $\hat{\mathrm{A}}$ and $\hat{\mathrm{B}}$, which satisfy the simultaneous eigenvalue problems of these two operators:

$$\hat{\mathrm{A}}|a_i, b_j\rangle = a_i|a_i, b_j\rangle \quad \text{and} \quad \hat{\mathrm{B}}|a_i, b_j\rangle = b_j|a_i, b_j\rangle. \tag{2.64}$$

Indeed, when $|a_i\rangle$ denotes the eigenvector of $\hat{A}, \hat{A}|a_i\rangle = a_i|a_i\rangle$, and $[\hat{A}, \hat{B}] = 0$, applying $\hat{B}$ to both sides of this eigenvalue equation gives: $\hat{B}\hat{A}|a_i\rangle = \hat{A}(\hat{B}|a_i\rangle) = a_i(\hat{B}|a_i\rangle)$. Therefore, $\hat{B}|a_i\rangle$ is also the eigenvector of $\hat{A}$ belonging to the same eigenvalue a_i. Hence, for the nondegenerate eigenvalue $a_i, \hat{B}|a_i\rangle$ must be collinear with $|a_i\rangle$, since there is only one independent eigenstate corresponding to a_i, identified by the direction of $|a_i\rangle$. Hence, $\hat{B}|a_i\rangle$ is then proportional to $|a_i\rangle$, thus also satisfying the eigenvalue equation of $\hat{B}$,

$$\hat{B}|a_i\rangle = b_j|a_i\rangle \Rightarrow |a_i\rangle = |a_i, b_j\rangle. \tag{2.65}$$

For the degenerate eigenvalue $a_i, \hat{B}|a_i\rangle$ gives a vector belonging to the subspace $\{|a_i\rangle_k\}$ of a_i, so that such eigenvalue subspace of $\hat{A}$ remains globally invariant under the action of $\hat{B}$. One also observes that for such a pair of commuting operators, the two eigenvectors for different eigenvalues of one operator, say $|a_i\rangle$ and $|a_j\rangle$ of $\hat{A}, a_i \neq a_j$, give the vanishing matrix element of the other operator: $\langle a_i|\hat{B}|a_j\rangle = 0$. This directly follows from their vanishing commutator which implies

$$\langle a_i|[\hat{A}, \hat{B}]|a_j\rangle = (a_i - a_j)\langle a_i|\hat{B}|a_j\rangle = 0 \quad \Rightarrow \quad \langle a_i|\hat{B}|a_j\rangle = 0, \tag{2.66}$$

where we have recognized the Hermitian character of $\hat{A}$.

In fact the commutation of two operators constitutes both the necessary and sufficient condition for the two operators to have the common eigenvectors. The above demonstration of the sufficient criterion can be supplemented by the inverse theorem of the necessary condition that the existence of the common eigenvalue problem of the two operators implies that they commute. Since the common eigenvectors $\{|a_i, b_j\rangle\}$ constitute the basis (complete) set one can expand any ket

$$|\Psi\rangle = \sum_{i,j} |a_i, b_j\rangle\langle a_i, b_j|\Psi\rangle \equiv \sum_{i,j} |a_i, b_j\rangle C_{i,j}. \tag{2.67}$$

Therefore:

$$\begin{aligned}[\hat{A}, \hat{B}]|\Psi\rangle &= \sum_{i,j} C_{i,j}[\hat{A}\hat{B} - \hat{B}\hat{A}]|a_i, b_j\rangle = \sum_{i,j} C_{i,j}(a_i b_j - b_j a_i)|a_i, b_j\rangle = 0 \\ &\Rightarrow \quad [\hat{A}, \hat{B}] = 0.\end{aligned} \tag{2.68}$$

The minimum set of the mutually commuting observables $\{\hat{A}, \hat{B}, \ldots, \hat{C}\}$, which uniquely specify the direction of the state vector $|\Psi\rangle$, is called the complete set of commuting observables. Hence, there exists a unique orthonormal basis of their common eigenvectors and the corresponding eigenvalues (a_i, b_j, …, c_k) provide the complete specification of the state under consideration: $|\Psi\rangle = |a_i, b_j, \ldots, c_k\rangle$. One should realize, however, that for a given molecular system there exist several such sets of observables. We shall encounter their examples in the next section.

2.6 Position and Momentum Representations

Two important cases of the continuous basis sets in the vector space of quantum states of a single (spinless) particle combine all state vectors corresponding to its sharply specified position $\boldsymbol{r} = (x, y, z)$ or momentum $\boldsymbol{p} = (p_x, p_y, p_z)$. These states, $\{|\boldsymbol{r}\rangle\}$ and $\{|\boldsymbol{p}\rangle\}$, respectively, labeled by the respective three continuous coordinates are the eigenvectors of the particle position and momentum operators, $\hat{\mathbf{r}} = (\hat{\mathrm{x}}, \hat{\mathrm{y}}, \hat{\mathrm{z}})$ and $\hat{\mathbf{p}} = (\hat{\mathrm{p}}_x, \hat{\mathrm{p}}_y, \hat{\mathrm{p}}_z)$,

$$\hat{\mathbf{r}}|\boldsymbol{r}'\rangle = \boldsymbol{r}'|\boldsymbol{r}'\rangle, \quad \langle\boldsymbol{r}|\boldsymbol{r}'\rangle = \delta(\boldsymbol{r}' - \boldsymbol{r}) = u_{\boldsymbol{r}'}(\boldsymbol{r}), \quad \int |\boldsymbol{r}\rangle\langle\boldsymbol{r}|\, d\boldsymbol{r} = 1;$$

$$\hat{\mathbf{p}}|\boldsymbol{p}'\rangle = \boldsymbol{p}'|\boldsymbol{p}'\rangle, \quad \langle\boldsymbol{p}|\boldsymbol{p}'\rangle = \delta(\boldsymbol{p}' - \boldsymbol{p}) = u_{\boldsymbol{p}'}(\boldsymbol{p}), \quad \int |\boldsymbol{p}\rangle\langle\boldsymbol{p}|\, d\boldsymbol{p} = 1. \tag{2.69}$$

The Dirac deltas $\{\delta(\boldsymbol{r}' - \boldsymbol{r})\}$ and $\{\delta(\boldsymbol{p}' - \boldsymbol{p})\}$ in these equations define the continuous basis functions $\{u_{\boldsymbol{r}'}(\boldsymbol{r})\}$ and $\{u_{\boldsymbol{p}'}(\boldsymbol{p})\}$ for expanding the particle wave functions $\Psi(\boldsymbol{r}') = \langle\boldsymbol{r}'|\Psi\rangle$ and $\Psi(\boldsymbol{p}') = \langle\boldsymbol{p}'|\Psi\rangle$ in these two bases:

$$\Psi(\boldsymbol{r}') = \int \langle\boldsymbol{r}'|\boldsymbol{r}\rangle\langle\boldsymbol{r}|\Psi\rangle\, d\boldsymbol{r} = \int u_{\boldsymbol{r}'}{}^*(\boldsymbol{r})\ \Psi(\boldsymbol{r})\ d\boldsymbol{r},$$

$$\Psi(\boldsymbol{p}') = \int \langle\boldsymbol{p}'|\boldsymbol{p}\rangle\langle\boldsymbol{p}|\Psi\rangle\, d\boldsymbol{p} = \int u_{\boldsymbol{p}'}{}^*(\boldsymbol{p})\ \Psi(\boldsymbol{p})\ d\boldsymbol{p}. \tag{2.70}$$

Indeed, these two equations express the basic integral property of Dirac's delta function [(2.21)]:

$$\Psi(\boldsymbol{r}') = \int \delta(\boldsymbol{r} - \boldsymbol{r}')\,\Psi(\boldsymbol{r})\,d\boldsymbol{r} \quad \text{and} \quad \Psi(\boldsymbol{p}') = \int \delta(\boldsymbol{p} - \boldsymbol{p}')\,\Psi(\boldsymbol{p})\,d\boldsymbol{p}.$$

They also identify the function "coordinates" as the corresponding projections in the function space spanned by the bases $\{u_{\boldsymbol{r}'}(\boldsymbol{r})\}$ and $\{u_{\boldsymbol{p}'}(\boldsymbol{p})\}$, respectively.

The orthogonality relation between quantum states $|\Psi\rangle$ and $|\Phi\rangle$ can thus be expressed as the isomorphic relations between the corresponding wave functions:

$$\begin{aligned}\langle\Psi|\Phi\rangle &= \int \langle\Psi|\boldsymbol{r}\rangle\langle\boldsymbol{r}|\Phi\rangle\, d\boldsymbol{r} = \int \Psi^*(\boldsymbol{r})\,\Phi(\boldsymbol{r})\,d\boldsymbol{r}\\ &= \int \langle\Psi|\boldsymbol{p}\rangle\langle\boldsymbol{p}|\Phi\rangle\, d\boldsymbol{p} = \int \Psi^*(\boldsymbol{p})\,\Phi(\boldsymbol{p})\,d\boldsymbol{p} = 0.\end{aligned} \tag{2.71}$$

It also follows from (2.69) that the basis functions $u_{\boldsymbol{r}'}(\boldsymbol{r})$ and $u_{\boldsymbol{p}'}(\boldsymbol{p})$ are themselves wave functions of quantum states with the sharply defined position $\boldsymbol{r} = \boldsymbol{r}'$ and momentum $\boldsymbol{p} = \boldsymbol{p}'$, respectively. There is one-to-one correspondence between wave functions and the associated state vectors they represent, e.g.,

$$u_{\boldsymbol{r}'}(\boldsymbol{r}) \Leftrightarrow |\boldsymbol{r}'\rangle, \quad u_{\boldsymbol{p}'}(\boldsymbol{p}) \Leftrightarrow |\boldsymbol{p}'\rangle, \quad \Psi(\boldsymbol{r}) \Leftrightarrow |\Psi\rangle, \quad \Psi(\boldsymbol{p}) \Leftrightarrow |\Psi\rangle. \tag{2.72}$$

Of interest also are the relations between the wave functions in the momentum and position representations. They are summarized by the Fourier transformations of (2.26), which in three dimensions read in terms of the wave vector $\boldsymbol{k} = \boldsymbol{p}/\hbar$,

$$\Psi(\boldsymbol{k}) = (2\pi)^{-3/2}\int \exp(-i\boldsymbol{k}\cdot\boldsymbol{r})\Psi(\boldsymbol{r})\,d\boldsymbol{r} \text{ or } \Psi(\boldsymbol{p}) = (2\pi\hbar)^{-3/2}\int \exp(-\frac{i}{\hbar}\boldsymbol{p}\cdot\boldsymbol{r})\Psi(\boldsymbol{r})\,d\boldsymbol{r},$$
$$\Psi(\boldsymbol{r}) = (2\pi)^{-3/2}\int \exp(i\boldsymbol{k}\cdot\boldsymbol{r})\Psi(\boldsymbol{k})\,d\boldsymbol{k} = (2\pi\hbar)^{-3/2}\int \exp(\frac{i}{\hbar}\boldsymbol{p}\cdot\boldsymbol{r})\Psi(\boldsymbol{p})\,d\boldsymbol{p}. \tag{2.73}$$

Substituting one transform into the other then generates the following analytical representations of the Dirac deltas [see (2.28)]:

$$\delta(\boldsymbol{r}' - \boldsymbol{r}) = (2\pi\hbar)^{-3}\int \exp\left[\frac{i}{\hbar}\boldsymbol{p}\cdot(\boldsymbol{r}' - \boldsymbol{r})\right]\,d\boldsymbol{p},$$
$$\delta(\boldsymbol{p}' - \boldsymbol{p}) = (2\pi\hbar)^{-3}\int \exp\left[\frac{i}{\hbar}(\boldsymbol{p}' - \boldsymbol{p})\cdot\boldsymbol{r}\right]\,d\boldsymbol{r}. \tag{2.74}$$

Hence, by transcribing (2.73) in terms of corresponding state vectors,

$$\Psi(\boldsymbol{p}) = \langle\boldsymbol{p}|\Psi\rangle = \int \langle\boldsymbol{p} \mid \boldsymbol{r}\rangle\langle\boldsymbol{r}|\Psi\rangle\,d\boldsymbol{r} = \int u_{\boldsymbol{p}}^*(\boldsymbol{r})\,\Psi(\boldsymbol{r})\,d\boldsymbol{r},$$
$$\Psi(\boldsymbol{r}) = \langle\boldsymbol{r}|\Psi\rangle = \int \langle\boldsymbol{r} \mid \boldsymbol{p}\rangle\langle\boldsymbol{p}|\Psi\rangle\,d\boldsymbol{p} = \int u_{\boldsymbol{r}}^*(\boldsymbol{p})\,\Psi(\boldsymbol{p})\,d\boldsymbol{p}, \tag{2.75}$$

one identifies the following representation of basis vectors of one representation in terms of vectors comprised in the other basis set:

$$u_{\boldsymbol{p}}(\boldsymbol{r}) = \langle\boldsymbol{r}|\boldsymbol{p}\rangle = (2\pi\hbar)^{-3/2}\exp(\frac{i}{\hbar}\boldsymbol{p}\cdot\boldsymbol{r}) \quad \text{and}$$
$$u_{\boldsymbol{r}}(\boldsymbol{p}) = \langle\boldsymbol{p}|\boldsymbol{r}\rangle = u_{\boldsymbol{p}}(\boldsymbol{r})^* = (2\pi\hbar)^{-3/2}\exp(-\frac{i}{\hbar}\boldsymbol{p}\cdot\boldsymbol{r}). \tag{2.76}$$

Let us now examine the associated representations of the position and momentum operators in these two continuous basis sets. We first observe that these operators are the continuous diagonal when represented in the basis set of their own eigenvectors [see (2.69)]:

$$\langle\boldsymbol{r}''|\hat{\mathbf{r}}|\boldsymbol{r}'\rangle = \boldsymbol{r}'\langle\boldsymbol{r}''|\boldsymbol{r}'\rangle = \boldsymbol{r}'\delta(\boldsymbol{r}' - \boldsymbol{r}''), \quad \langle\boldsymbol{p}''|\hat{\mathbf{p}}|\boldsymbol{p}'\rangle = \boldsymbol{p}'\langle\boldsymbol{p}''|\boldsymbol{p}'\rangle = p'\delta(\boldsymbol{p}' - \boldsymbol{p}''). \tag{2.77}$$

Therefore, the action of the position operator on the wave function in the position representation amounts to a straightforward multiplication by the position vector:

$$\int \langle \boldsymbol{r}''|\hat{\mathbf{r}}|\boldsymbol{r}'\rangle \langle \boldsymbol{r}'|\Psi\rangle \, d\boldsymbol{r}' = \int \boldsymbol{r}'\delta(\boldsymbol{r}' - \boldsymbol{r}'')\Psi(\boldsymbol{r}')d\boldsymbol{r}' = \boldsymbol{r}''\Psi(\boldsymbol{r}''). \tag{2.78}$$

The action of the momentum operator on the wave function in the momentum representation similarly represents the multiplication by the momentum vector:

$$\int \langle \boldsymbol{p}''|\hat{\mathbf{p}}|\boldsymbol{p}'\rangle \langle \boldsymbol{p}'|\Psi\rangle \, d\boldsymbol{p}' = \int \boldsymbol{p}'\delta(\boldsymbol{p}' - \boldsymbol{p}'')\Psi(\boldsymbol{p}')d\boldsymbol{p}' = \boldsymbol{p}''\,\Psi(\boldsymbol{p}''). \tag{2.79}$$

Next, let us establish the form of the *momentum* operator in the *position* representation. It can be recognized by examining the position representation of the ket $\hat{\mathbf{p}}|\Psi\rangle$,

$$\langle \boldsymbol{r}|\hat{\mathbf{p}}|\Psi\rangle = \int \langle \boldsymbol{r}|\boldsymbol{p}\rangle\langle \boldsymbol{p}|\hat{\mathbf{p}}|\Psi\rangle \, d\boldsymbol{p} = (2\pi\hbar)^{-3/2} \int \exp\left(\frac{i}{\hbar}\boldsymbol{p}\cdot\boldsymbol{r}\right)\boldsymbol{p}\,\Psi(\boldsymbol{p}) \, d\boldsymbol{p}. \tag{2.80}$$

Hence, by comparing the previous equation with the last (2.73) gives:

$$\langle \boldsymbol{r}|\hat{\mathbf{p}}|\Psi\rangle = -i\hbar\nabla_{\boldsymbol{r}}\langle \boldsymbol{r}|\Psi\rangle \equiv \hat{\mathbf{p}}(\boldsymbol{r})\Psi(\boldsymbol{r}), \tag{2.81}$$

where the differential vector operator $\nabla_{\boldsymbol{r}} = \boldsymbol{i}\partial/\partial x + \boldsymbol{j}\partial/\partial y + \boldsymbol{k}\partial/\partial z \equiv \partial/\partial\boldsymbol{r}$ stands for the *position* gradient. Therefore, the action of the momentum operator in the position representation amounts to performing the differential operation $\hat{\mathbf{p}}(\boldsymbol{r}) = -i\hbar\nabla_{\boldsymbol{r}}$ on the wave function $\Psi(\boldsymbol{r})$. Hence, the matrix element $\langle\Phi|\hat{\mathbf{p}}|\Psi\rangle$ in this representation is determined by the associated integral in terms of the position wave functions:

$$\langle\Phi|\hat{\mathbf{p}}|\Psi\rangle = \int \langle\Phi|\boldsymbol{r}\rangle\langle \boldsymbol{r}|\hat{\mathbf{p}}|\Psi\rangle \, d\boldsymbol{r} = -i\hbar \int \Phi^*(\boldsymbol{r}) \, \nabla_{\boldsymbol{r}}\Psi(\boldsymbol{r}) \, d\boldsymbol{r}. \tag{2.82}$$

One could alternatively calculate the kernel $\hat{\mathbf{p}}(\boldsymbol{r},\boldsymbol{r}') = \langle \boldsymbol{r}|\hat{\mathbf{p}}|\boldsymbol{r}'\rangle$ (the continuous matrix element) of the momentum operator, in terms of which the operation of (2.81) reads:

$$\langle \boldsymbol{r}|\hat{\mathbf{p}}|\Psi\rangle = \int \langle \boldsymbol{r}|\hat{\mathbf{p}}|\boldsymbol{r}'\rangle\langle \boldsymbol{r}'|\Psi\rangle \, d\boldsymbol{r}' = \int \hat{\mathbf{p}}(\boldsymbol{r},\, \boldsymbol{r}')\Psi(\boldsymbol{r}') \, d\boldsymbol{r}'. \tag{2.83}$$

By twice inserting the closure relation into this matrix element, and using the analytical expression for the Dirac delta (2.74) one then finds:

$$\begin{aligned}
\langle \boldsymbol{r}|\hat{\mathbf{p}}|\boldsymbol{r}'\rangle &= \iint \langle \boldsymbol{r}|\boldsymbol{p}\rangle\langle \boldsymbol{p}|\hat{\mathbf{p}}|\boldsymbol{p}'\rangle\langle \boldsymbol{p}'|\boldsymbol{r}'\rangle \, d\boldsymbol{p}\, d\boldsymbol{p}' \\
&= \iint u_{\boldsymbol{r}}^*(\boldsymbol{p})\,\boldsymbol{p}\delta(\boldsymbol{p}' - \boldsymbol{p})\, u_{\boldsymbol{r}'}(\boldsymbol{p}') \; d\boldsymbol{p}' d\boldsymbol{p} \\
&= (2\pi\hbar)^{-3} \int \exp\left[\frac{i}{\hbar}\boldsymbol{p}\cdot(\boldsymbol{r}' - \boldsymbol{r})\right]\boldsymbol{p}\, d\boldsymbol{p} = i\hbar\nabla_{\boldsymbol{r}'}\delta(\boldsymbol{r}' - \boldsymbol{r}).
\end{aligned} \tag{2.84}$$

Substituting this result to (2.83), after integration by parts [see (2.30)], gives the same result as in (2.82):

$$\int \hat{\mathrm{p}}(\boldsymbol{r}, \boldsymbol{r}')\Psi(\boldsymbol{r}')\, d\boldsymbol{r}' = i\hbar \int \nabla_{r'}\delta(\boldsymbol{r}' - \boldsymbol{r})\Psi(\boldsymbol{r}')\, d\boldsymbol{r}'.$$
$$= -i\hbar \int \delta(\boldsymbol{r}' - \boldsymbol{r})\nabla_{r'}\Psi(\boldsymbol{r}')\, d\boldsymbol{r}' = -i\hbar\nabla_{r}\Psi(\boldsymbol{r}). \tag{2.85}$$

One similarly derives the remaining kernel providing the momentum representation of the position operator,

$$\hat{\mathrm{r}}(\boldsymbol{p}, \boldsymbol{p}') = \langle \boldsymbol{p}|\hat{\mathrm{r}}|\boldsymbol{p}'\rangle = \iint \langle \boldsymbol{p}|\boldsymbol{r}\rangle\langle \boldsymbol{r}|\hat{\mathrm{r}}|\boldsymbol{r}'\rangle\langle \boldsymbol{r}'|\boldsymbol{p}'\rangle\, d\boldsymbol{r}\, d\boldsymbol{r}' = \iint u_p^*(\boldsymbol{r})\, \boldsymbol{r}\delta(\boldsymbol{r}' - \boldsymbol{r})\, u_{p'}(\boldsymbol{r}')\, d\boldsymbol{r}\, d\boldsymbol{r}'$$
$$= (2\pi\hbar)^{-3} \int \exp\left[\frac{i}{\hbar}(\boldsymbol{p}' - \boldsymbol{p}) \cdot \boldsymbol{r}\right] \boldsymbol{r}\, d\boldsymbol{r} = -i\hbar\nabla_{p'}\delta(\boldsymbol{p}' - \boldsymbol{p}), \tag{2.86}$$

where the *momentum* gradient $\nabla_p = \boldsymbol{i}\partial/\partial p_x + \boldsymbol{j}\partial/\partial p_y + \boldsymbol{k}\partial/\partial p_z \equiv \partial/\partial \boldsymbol{p}$. It gives rise to the following momentum representation of the ket $\hat{\mathrm{r}}|\Psi\rangle$:

$$\langle \boldsymbol{p}|\hat{\mathrm{r}}|\Psi\rangle = \int \langle \boldsymbol{p}|\hat{\mathrm{r}}|\boldsymbol{p}'\rangle\langle \boldsymbol{p}'|\Psi\rangle\, d\boldsymbol{p}' = \int \hat{\mathrm{r}}(\boldsymbol{p}, \boldsymbol{p}')\Psi(\boldsymbol{p}')\, d\boldsymbol{p}'$$
$$= -i\hbar \int \nabla_{p'}\delta(\boldsymbol{p}' - \boldsymbol{p})\Psi(\boldsymbol{p}')\, d\boldsymbol{p}' = i\hbar \int \delta(\boldsymbol{p}' - \boldsymbol{p})\nabla_{p'}\Psi(\boldsymbol{p}')\, d\boldsymbol{r}'$$
$$= i\hbar\nabla_p\Psi(\boldsymbol{p}) \equiv \hat{\mathrm{r}}(\boldsymbol{p})\Psi(\boldsymbol{p}). \tag{2.87}$$

Therefore, the action of the position operator in the momentum space coincides with the differential operation $\hat{\mathrm{r}}(\boldsymbol{p}) = i\hbar\nabla_p$ performed on the wave function $\Psi(\boldsymbol{p})$.

The same result directly follows from inserting the closure identity into the initial scalar product of the preceding equation:

$$\langle \boldsymbol{p}|\hat{\mathrm{r}}|\Psi\rangle = \int \langle \boldsymbol{p}|\boldsymbol{r}\rangle\langle \boldsymbol{r}|\hat{\mathrm{r}}|\Psi\rangle\, d\boldsymbol{r} = (2\pi\hbar)^{-3/2} \int \exp(-\frac{i}{\hbar}\boldsymbol{p} \cdot \boldsymbol{r})\boldsymbol{r}\, \Psi(\boldsymbol{r})\, d\boldsymbol{r}. \tag{2.88}$$

Hence, by comparing this expression with the second (2.73) again gives:

$$\langle \boldsymbol{p}|\hat{\mathrm{r}}|\Psi\rangle = i\hbar\nabla_p\langle \boldsymbol{p}|\Psi\rangle = \hat{\mathrm{r}}(\boldsymbol{p})\Psi(\boldsymbol{p}). \tag{2.89}$$

2.7 Energy Representation and Unitary Transformations

The energy representation of quantum states and operators is defined by the basis set of the (orthonormal) eigenvectors $\{|E_n\rangle\}$ of the system energy operator, the Hamiltonian $\hat{\mathrm{E}} \equiv \hat{\mathrm{H}}$,

$$\hat{\mathrm{H}}|E_n\rangle = E_n|E_n\rangle. \tag{2.90}$$

They represent the stationary states, with the sharply specified energy. Here, for the sake of simplicity we have assumed the discrete spectrum of the allowed energy levels $\{E_n\}$.

In the position representation $\boldsymbol{\xi} = \{|\xi\rangle, |\xi'\rangle, \ldots\} = \{|\xi\rangle\}$, $\langle\xi'|\xi\rangle = \delta(\xi - \xi')$, where ξ groups the system coordinates, the eigenkets $\{|E_n\rangle\}$ of the energy basis set are represented by the associated wave functions $\{\varphi_{E_n}(\xi) = \langle\xi|E_n\rangle\} = \langle\boldsymbol{\xi}|E_n\rangle$ of the continuous *column* vector, while the corresponding eigenbras define the associated continuous *row* vector: $\{\varphi^*_{E_n}(\xi) = \langle\xi|E_n\rangle^* = \langle E_n|\xi\rangle\} = \langle E_n|\boldsymbol{\xi}\rangle$. In this position basis the Hamiltonian $\hat{\mathrm{H}}$ is similarly represented by the continuous (diagonal) matrix: $\hat{\mathrm{H}} \Rightarrow \{\hat{\mathrm{H}}(\xi,\xi') = \langle\xi|\hat{\mathrm{H}}|\xi'\rangle = \hat{\mathrm{H}}(\xi)\delta(\xi' - \xi)\}$. In the position representation the energy eigenvalue problem of (2.90) reads:

$$\int \langle\xi|\hat{\mathrm{H}}|\xi'\rangle\langle\xi'|E_n\rangle\, d\xi' = E_n\langle\xi|E_n\rangle \tag{2.91}$$

or

$$\int \hat{\mathrm{H}}(\xi,\xi')\varphi_{E_n}(\xi')\, d\xi' = \int \hat{\mathrm{H}}(\xi')\delta(\xi' - \xi)\varphi_{E_n}(\xi')\, d\xi' = \hat{\mathrm{H}}(\xi)\varphi_{E_n}(\xi) = E_n\varphi_{E_n}(\xi). \tag{2.92}$$

The orthonormality of the energy eigenvectors (discrete spectrum), $\langle E_n|E_n\rangle = \delta_{E_m,E_n}$, can be also expressed in terms of the associated wave functions:

$$\langle E_m|E_n\rangle = \int \langle E_m|\xi\rangle\langle\xi|E_n\rangle\, d\xi = \int \varphi^*_{E_m}(\xi)\ \varphi_{E_n}(\xi)\, d\xi = \delta_{E_m,E_n}. \tag{2.93}$$

Any state vector $|\Psi\rangle$ is thus equivalently represented either by the components $\{\Psi_{E_n} = \langle E_n|\Psi\rangle = \int \varphi^*_{E_n}(\xi)\Psi(\xi)\, d\xi\} \equiv \boldsymbol{\Psi}^{(E)}$ in the energy representation or by the wave function $\Psi(\xi) = \langle\xi|\Psi\rangle \equiv \boldsymbol{\Psi}^{(\xi)}$ in the position representation. They are related *via* the following transformations:

$$\begin{aligned}\Psi(\xi) = \langle\xi|\Psi\rangle &= \sum_m \langle\xi|E_m\rangle\langle E_m|\Psi\rangle = \sum_m \varphi_{E_m}(\xi)\ \Psi_{E_m} \text{ or } \boldsymbol{\Psi}^{(\xi)} = \mathbf{T}(\boldsymbol{\xi},\boldsymbol{E})\boldsymbol{\Psi}^{(E)},\\ \Psi_{E_n} = \langle E_n|\Psi\rangle &= \int \langle E_n|\xi\rangle\langle\xi|\Psi\rangle\, d\xi = \int \varphi^*_{E_n}(\xi)\Psi(\xi)\, d\xi \text{ or } \boldsymbol{\Psi}^{(E)} = \mathbf{T}(\boldsymbol{E},\boldsymbol{\xi})\boldsymbol{\Psi}^{(\xi)}.\end{aligned} \tag{2.94}$$

Thus, the energy eigenfunctions $\{\varphi_{E_m}(\xi)\}$, with the continuous (discrete) position (energy) labels, transform the energy representation of the state vector to its associated position representation. Accordingly, the complex conjugate functions $\{\varphi^*_{E_m}(\xi)\}$ are seen to define the reverse transformation of the state vector, from its position representation to the energy representation.

Therefore, should one regard the coefficients of these mutually reverse transformations as elements of the corresponding transformation matrices identified by the discrete $\{E_n\}$ and continuous $\{\xi\}$ indices,

$$\{\varphi_{E_m}(\xi)\} \equiv \mathbf{T}(\boldsymbol{\xi}, \boldsymbol{E}) = \langle \boldsymbol{\xi} | \boldsymbol{E} \rangle,$$
$$\{\varphi^*_{E_m}(\xi)\} \equiv \mathbf{T}(\boldsymbol{E}, \boldsymbol{\xi}) = \langle \boldsymbol{E} | \boldsymbol{\xi} \rangle = \mathbf{T}(\boldsymbol{\xi}, \boldsymbol{E})^\dagger, \tag{2.95}$$

their mutual reciprocity relations imply:

$$\mathbf{T}(\boldsymbol{\xi}', \boldsymbol{E})\,\mathbf{T}(\boldsymbol{E}, \boldsymbol{\xi}) = \langle \boldsymbol{\xi}' | \boldsymbol{E} \rangle \langle \boldsymbol{E} | \boldsymbol{\xi} \rangle = \langle \boldsymbol{\xi}' | \boldsymbol{\xi} \rangle = \delta(\boldsymbol{\xi} - \boldsymbol{\xi}')$$
$$\Rightarrow \mathbf{T}(\boldsymbol{E}, \boldsymbol{\xi}) = \mathbf{T}(\boldsymbol{\xi}, \boldsymbol{E})^\dagger = \mathbf{T}(\boldsymbol{\xi}, \boldsymbol{E})^{-1};$$
$$\mathbf{T}(\boldsymbol{E}, \boldsymbol{\xi})\,\mathbf{T}(\boldsymbol{\xi}, \boldsymbol{E}') = \langle \boldsymbol{E} | \boldsymbol{\xi} \rangle \langle \boldsymbol{\xi} | \boldsymbol{E}' \rangle = \langle \boldsymbol{E} | \boldsymbol{E}' \rangle = \delta_{\boldsymbol{E},\boldsymbol{E}'} = \mathbf{I}.$$
$$\Rightarrow \mathbf{T}(\boldsymbol{\xi}, \boldsymbol{E}) = \mathbf{T}(\boldsymbol{E}, \boldsymbol{\xi})^\dagger = \mathbf{T}(\boldsymbol{E}, \boldsymbol{\xi})^{-1}. \tag{2.96}$$

One thus concludes that each of these mutually inverse transformation matrices is the Hermitian conjugate of the other thus defining the *unitary* transformations ("rotations") of one orthonormal basis set into another.

To summarize, the system energy, with discrete (or continuous/mixed) set of eigenvalues, constitutes the independent variable of the energy representation. The square of the modulus of the component $\Psi_{E_n} = \langle E_n | \Psi \rangle$ measures the (conditional) probability $W(E_n|\Psi)$ of observing the system in state $|\Psi\rangle$ at the specified energy:

$$W(E_n|\Psi) = |\langle E_n|\Psi\rangle|^2 = \langle \Psi|E_n\rangle \langle E_n|\Psi\rangle,$$
$$\sum_n W(E_n|\Psi) = \sum_n \langle \Psi|E_n\rangle\langle E_n|\Psi\rangle = \langle \Psi|\Psi\rangle = 1. \tag{2.97}$$

As we have already observed in (2.75) of the preceding section, the wave functions (2.76) define another pair of such mutually reverse transformations:

$$u_{\boldsymbol{r}}(\boldsymbol{p}) = \langle \boldsymbol{p}|\boldsymbol{r}\rangle \equiv \mathbf{t}(\boldsymbol{p}, \boldsymbol{r}) \quad \text{and} \quad u_{\boldsymbol{p}}(\boldsymbol{r}) = \langle \boldsymbol{r}|\boldsymbol{p}\rangle \equiv \mathbf{t}(\boldsymbol{r}, \boldsymbol{p}),$$
$$\int \mathbf{t}(\boldsymbol{p}, \boldsymbol{r})\,\mathbf{t}(\boldsymbol{r}, \boldsymbol{p}')d\boldsymbol{r} = \delta(\boldsymbol{p} - \boldsymbol{p}'), \quad \int \mathbf{t}(\boldsymbol{r}, \boldsymbol{p})\,\mathbf{t}(\boldsymbol{p}, \boldsymbol{r}')d\boldsymbol{p} = \delta(\boldsymbol{r} - \boldsymbol{r}'). \tag{2.98}$$

They also define the unitary kernels,

$$\mathbf{t}(\boldsymbol{p}, \boldsymbol{r}) = \mathbf{t}(\boldsymbol{r}, \boldsymbol{p})^\dagger = \mathbf{t}(\boldsymbol{r}, \boldsymbol{p})^{-1} \quad \text{and} \quad \mathbf{t}(\boldsymbol{r}, \boldsymbol{p}) = \mathbf{t}(\boldsymbol{p}, \boldsymbol{r})^\dagger = \mathbf{t}(\boldsymbol{p}, \boldsymbol{r})^{-1}, \tag{2.99}$$

of the integral transformations between the position and momentum representations:

$$\int \mathbf{t}(\boldsymbol{p}, \boldsymbol{r})\,\boldsymbol{\Psi}(\boldsymbol{r})\,d\boldsymbol{r} \equiv \mathbf{T}(\boldsymbol{p}, \boldsymbol{r})\,\boldsymbol{\Psi}(\boldsymbol{r}) = \boldsymbol{\Psi}(\boldsymbol{p}) \quad \text{or} \quad \hat{\mathrm{T}}(\boldsymbol{p}, \boldsymbol{r})\Psi(\boldsymbol{r}) = \Psi(\boldsymbol{p}),$$
$$\int \mathbf{t}(\boldsymbol{r}, \boldsymbol{p})\,\boldsymbol{\Psi}(\boldsymbol{p})\,d\boldsymbol{p} \equiv \mathbf{T}(\boldsymbol{r}, \boldsymbol{p})\,\boldsymbol{\Psi}(\boldsymbol{p}) = \boldsymbol{\Psi}(\boldsymbol{r}) \quad \text{or} \quad \hat{\mathrm{T}}(\boldsymbol{r}, \boldsymbol{p})\Psi(\boldsymbol{p}) = \Psi(\boldsymbol{r}). \tag{2.100}$$

Above, $\mathbf{T}(\boldsymbol{p}, \boldsymbol{r})$ represents the integral operator $\widehat{\mathrm{T}}(\boldsymbol{p},\boldsymbol{r})$ defined by the kernel $\mathbf{t}(\boldsymbol{p}, \boldsymbol{r})$, which replaces the arguments of the wave function: $\boldsymbol{r} \to \boldsymbol{p}$, etc.

It follows from the preceding equations that these transformations are unitary:

$$\mathbf{T}(\boldsymbol{r},\boldsymbol{p}) = \mathbf{T}(\boldsymbol{p},\boldsymbol{r})^{-1} = \mathbf{T}(\boldsymbol{p},\boldsymbol{r})^{\dagger} \quad \text{or} \quad \widehat{\mathrm{T}}^{-1}(\boldsymbol{p},\boldsymbol{r}) = \widehat{\mathrm{T}}^{\dagger}(\boldsymbol{p},\boldsymbol{r}) = \widehat{\mathrm{T}}(\boldsymbol{r},\boldsymbol{p}), \qquad (2.101)$$

where the inverse operator $\widehat{\mathrm{T}}^{-1}(\boldsymbol{p},\boldsymbol{r})$ replaces the variables in the wave function it acts upon in the inverse order: $\boldsymbol{p} \to \boldsymbol{r}$. Therefore, the reciprocity relations of (2.98) in fact express the unitary character of the above (integral) transformation operators,

$$\begin{aligned} \widehat{\mathrm{T}}(\boldsymbol{r},\boldsymbol{p})\widehat{\mathrm{T}}^{\dagger}(\boldsymbol{r},\boldsymbol{p}) &= \widehat{\mathrm{T}}(\boldsymbol{r},\boldsymbol{p})\widehat{\mathrm{T}}(\boldsymbol{p},\boldsymbol{r}) = 1 \quad \text{and} \\ \widehat{\mathrm{T}}(\boldsymbol{p},\boldsymbol{r})\widehat{\mathrm{T}}^{\dagger}(\boldsymbol{p},\boldsymbol{r}) &= \widehat{\mathrm{T}}(\boldsymbol{p},\boldsymbol{r})\widehat{\mathrm{T}}(\boldsymbol{r},\boldsymbol{p}) = 1, \end{aligned} \qquad (2.102)$$

because the double exchange of variables $\boldsymbol{p} \to \boldsymbol{r} \to \boldsymbol{p}$ amounts to identity operation on the wave function $\Psi(\boldsymbol{p})$ and the *double* exchange $\boldsymbol{r} \to \boldsymbol{p} \to \boldsymbol{r}$ operation performed of $\Psi(\boldsymbol{r})$ leaves it unchanged.

Transitions from one set of independent variables to another are called the *canonical* transformations. They have been shown to correspond to unitary operators, which also transform the matrix representations of the quantum-mechanical operators to a new set of variables. Indeed, by unitary transforming both sides of the momentum representation of (2.31),

$$\int \hat{\mathrm{A}}(\boldsymbol{p},\boldsymbol{p}')\Psi(\boldsymbol{p}')\, d\boldsymbol{p}' = \Psi'(\boldsymbol{p}),$$

and using the identity (2.102) one obtains

$$\begin{aligned} [\widehat{\mathrm{T}}(\boldsymbol{r},\boldsymbol{p})\hat{\mathrm{A}}(\boldsymbol{p},\boldsymbol{p}')\widehat{\mathrm{T}}^{\dagger}(\boldsymbol{r}',\boldsymbol{p}')][\widehat{\mathrm{T}}(\boldsymbol{r}',\boldsymbol{p}')\Psi(\boldsymbol{p}')] &\equiv \hat{\mathrm{A}}(\boldsymbol{r},\boldsymbol{r}')\Psi(\boldsymbol{r}') \\ &= \widehat{\mathrm{T}}(\boldsymbol{r},\boldsymbol{p})\Psi'(\boldsymbol{p}) \equiv \Psi'(\boldsymbol{r}). \end{aligned} \qquad (2.103)$$

Hence, the canonically transformed resultant vector $\Psi'(\boldsymbol{p})$ in the new variables becomes: $\widehat{\mathrm{T}}(\boldsymbol{r},\boldsymbol{p})\ \Psi'(\boldsymbol{p}) = \Psi'(\boldsymbol{r})$. It results from the transformed vector $\widehat{\mathrm{T}}(\boldsymbol{r}',\boldsymbol{p}')\ \Psi(\boldsymbol{p}') = \Psi(\boldsymbol{r}')$ by the action of the transformed operator

$$\widehat{\mathrm{T}}(\boldsymbol{r},\boldsymbol{p})\hat{\mathrm{A}}(\boldsymbol{p},\boldsymbol{p}')\widehat{\mathrm{T}}^{\dagger}(\boldsymbol{r}',\boldsymbol{p}') = \widehat{\mathrm{T}}(\boldsymbol{r},\boldsymbol{p})\hat{\mathrm{A}}(\boldsymbol{p},\boldsymbol{p}')\widehat{\mathrm{T}}^{-1}(\boldsymbol{p}',\boldsymbol{r}') = \hat{\mathrm{A}}(\boldsymbol{r},\boldsymbol{r}') \qquad (2.104)$$

with the preceding equation thus expressing a general transformation law for changing representations of linear operators.

Another important type of the unitary operators is represented by the *phase* transformation $\hat{\mathrm{S}}(\boldsymbol{x}) = \exp[i\hat{\alpha}(\boldsymbol{x})]$. It involves the linear Hermitian operator $\hat{\alpha}(\boldsymbol{x})$, the function of the same list of variables as those of the wave function itself.

This transformation of $\Psi(\boldsymbol{x})$ modifies the wave function without affecting its set of the independent state variables.

All physical predictions of quantum mechanics can be shown to remain unaffected by the unitary transformations of states and operators, since they are related to specific *invariants* of such operations. These invariant properties include the linear and Hermitian character of quantum-mechanical observables, all algebraic relations between them, e.g., the commutation rules, spectrum of eigenvalues and the matrix elements of operators.

The diversity of unitary transformations is not limited to those changing a description of the system quantum-mechanical states at the given time (quantum *kinematics*): $\Psi(\boldsymbol{x}) = \Psi(\boldsymbol{x},\ t = 0)$. In the next chapter, we shall examine other examples of unitary transformations of wave functions and operators, which generate different *pictures* of the quantum-mechanical *dynamics*, e.g., the evolution of quantum states with time in the *Schrödinger picture*:

$$\Psi(\boldsymbol{x}, t) = \hat{\mathrm{U}}(t)\Psi(\boldsymbol{x}), \quad \hat{\mathrm{U}}^{\dagger}(t)\hat{\mathrm{U}}(t) = 1. \tag{2.105}$$

2.8 Functional Derivatives

The functional of the state vector argument or of its continuous basis representation – the wave function – gives the scalar. The representative example of such a mathematical entity is the definite integral, e.g., the scalar product of two wave functions. It may additionally involve various derivatives of the function argument. For simplicity, let us assume the functional F of a single function $f(x)$ of the continuous variable x,

$$F[f] = \int \mathcal{L}[x, f(x), f'(x), \ldots]dx. \tag{2.106}$$

This functional attributes to the argument function f the scalar $F = F[f]$. It is defined by the functional density $\mathcal{L}[x, f(x), f'(x), \ldots]$, which in a general case depends on the current value of x, the argument function itself, $f(x)$, and its derivatives: $f'(x) = df(x)/dx$, etc.

An important problem, which we shall often encounter in this book, is to find the functional variation $\delta F = F[f + \delta f] - F[f]$ due to a small modification of the argument function, $\delta f = \varepsilon h$, where ε is a small parameter and h stands for the displacement function (perturbation). The first differential of the functional is the component of δF that depends on δf linearly:

$$\delta^{(1)}F = \int \frac{\delta F}{\delta f(x)}\, \delta f(x)\, dx, \tag{2.107}$$

with the (local) coefficient before $\delta f(x)$ in the integrand defining the *first functional derivative* of F with respect to f at point x. It is seen to transform the local displacements of the argument function into the first differential of the functional. This expression can be viewed as the continuous generalization of the familiar differential of the function of several variables: $d^{(1)}f(x_1, x_2, \ldots) = \sum_i (\partial f/\partial x_i)\, dx_i$.

The global shift δf in the functional argument can be viewed as composed of local manipulations on f which are conveniently expressed in terms of the Dirac delta function: $\delta f(x) = \int \delta f(x'-x)\, dx'$, where $\delta f(x' - x) = \delta f(x')\delta(x' - x) = \varepsilon h(x')\delta(x' - x) \equiv \varepsilon h(x' - x)\}$. Here, $\delta f(x' - x)$ stands for the localized displacement of the argument function, centered around x, in terms of which the first functional derivative, itself the functional of f, reads:

$$\frac{\delta F}{\delta f(x)} = \lim_{\varepsilon \to 0} \frac{F[f(x') + \varepsilon h(x' - x)]_{x'} - F[f(x')]_{x'}}{\varepsilon} \equiv g[f; x], \tag{2.108}$$

where subscript x' in the functional symbol symbolizes integration over this argument [see (2.106)].

One similarly introduces *higher functional derivatives*, which define the consecutive terms in the functional Taylor–Volterra expansion (Volterra 1959; Gelfand and Fomin 1963):

$$\begin{aligned} \delta F[f] &= \int \frac{\delta F}{\delta f(x)}\, \delta f(x)\, dx + \frac{1}{2} \iint \delta f(x)\, \frac{\delta^2 F}{\delta f(x)\, \delta f(x')}\, \delta f(x')\, dx\, dx' + \ldots \\ &\equiv \delta^{(1)} F[f] + \delta^{(2)} F[f] + \ldots \end{aligned} \tag{2.109}$$

For example, in the localized perspective on modifying the argument function of the functional, one interprets its second functional derivative as the limiting ratio:

$$\frac{\delta^2 F}{\delta f(x)\, \delta f(x')} = \frac{\delta g[f; x]}{\delta f(x')} = \lim_{\varepsilon \to 0} \frac{g[f(x''') + \varepsilon h(x''' x'); x]_{x'''} - g[f; x]}{\varepsilon}. \tag{2.110}$$

In (2.109) it determines the continuous transformation of the *two*-point displacements of the argument function, $\delta f(x'' - x)\delta f(x''' - x')$, centered around x and x', respectively, into the second differential $\delta^{(2)}F[f]$. The latter again parallels the familiar expression for the second differential of a function of several variables: $d^{(2)}f(x_1, x_2, \ldots) = ½\sum_i\sum_j (\partial^2 f/\partial x_i \partial x_j)\, dx_i\, dx_j$.

The rules of the functional differentiation thus represent the local, function generalization of those characterizing the differentiation of functions. The functional derivatives of the sum and product of two functionals, respectively, read:

$$\begin{aligned} \frac{\delta}{\delta f(x)} \{aF[f] + bG[f]\} &= a \frac{\delta F}{\delta f(x)} + b \frac{\delta G}{\delta f(x)}, \\ \frac{\delta}{\delta f(x)} \{F[f]\, G[f]\} &= G \frac{\delta F}{\delta f(x)} + F \frac{\delta G}{\delta f(x)}. \end{aligned} \tag{2.111}$$

The *chain rule* transformation of functional derivatives also holds. Consider the *composite* functional $F[f] = F[f[g]] \equiv \bar{F}[g]$. Substituting the first differential of $f(x) = f[g; x]$,

$$\delta^{(1)} f[g; x] = \int \frac{\delta f(x)}{\delta g(x')} \delta g(x')\, dx', \tag{2.112}$$

into $\delta^{(1)} F[f]$ of (2.108) gives:

$$\delta^{(1)} \bar{F}[g] = \int \frac{\delta \bar{F}}{\delta g(x')} \delta g(x')\, dx' = \int \frac{\delta F}{\delta f(x)} \left[\int \frac{\delta f(x)}{\delta g(x')} \delta g(x')\, dx' \right] dx. \tag{2.113}$$

Hence, the functional derivative of the composite functional follows from the chain rule

$$\frac{\delta \bar{F}}{\delta g(x')} = \int \frac{\delta F}{\delta f(x)} \frac{\delta f(x)}{\delta g(x')} dx. \tag{2.114}$$

One similarly derives the chain rules for *implicit* functionals. When functional $F[f, g]$ is held constant, the variations of the two argument functions are not independent, since the relation $F[f, g] = const.$ implies the associated functional relation between them, e.g., $g = g[f]_F$. The vanishing first differential,

$$\delta^{(1)} F[f, g] = \int \left[\left(\frac{\partial F}{\partial f(x)} \right)_g [\delta f(x)]_F + \left(\frac{\partial F}{\partial g(x)} \right)_f [\delta g(x)]_F \right] dx = 0, \quad \text{or}$$

$$\int \left(\frac{\partial F}{\partial f(x)} \right)_g [\delta f(x)]_F\, dx = -\int \left(\frac{\partial F}{\partial g(x')} \right)_f [\delta g(x')]_F\, dx', \tag{2.115}$$

is determined by the *partial* functional derivatives, a natural local extension of the ordinary partial derivatives of a function of several variables, e.g.,

$$\left(\frac{\partial F}{\partial f(x)} \right)_g = \lim_{\varepsilon \to 0} \frac{F[f(x') + \varepsilon h(x' - x), g]_{x'} - F[f, g]}{\varepsilon}. \tag{2.116}$$

Finally, differentiating both sides of Eq. (2.115) with respect to one of the argument functions for constant F gives the following implicit chain rules:

$$\left(\frac{\partial F}{\partial f(x)} \right)_g = -\int \left(\frac{\partial F}{\partial g(x')} \right)_f \left(\frac{\partial g(x')}{\partial f(x)} \right)_F dx',$$

$$\left(\frac{\partial F}{\partial g(x')} \right)_f = -\int \left(\frac{\partial F}{\partial f(x)} \right)_g \left(\frac{\partial f(x)}{\partial g(x')} \right)_F dx. \tag{2.117}$$

These relations parallel familiar manipulations of derivatives in the classical thermodynamics.

For the fixed value of the composite functional $F[f[u], g[u]] = \tilde{F}[u] = const.$ one similarly finds:

$$\left(\frac{\partial g(x')}{\partial f(x)}\right)_{\tilde{F}} = \int \left(\frac{\partial g(x')}{\partial u(x'')}\right)_{\tilde{F}} \left(\frac{\partial u(x'')}{\partial f(x)}\right)_{\tilde{F}} dx'',$$
$$\left(\frac{\partial f(x)}{\partial g(x')}\right)_{\tilde{F}} = \int \left(\frac{\partial f(x)}{\partial u(x'')}\right)_{\tilde{F}} \left(\frac{\partial u(x'')}{\partial g(x')}\right)_{\tilde{F}} dx''. \quad (2.118)$$

Let us further assume that functions $f(x)$ and $g(x)$ are unique functionals of each other, $f(x) = f[g; x]$ and $g(x') = g[f; x']$. Substitution of (2.112) into

$$\delta^{(1)} g[f; x''] = \int \frac{\delta g(x'')}{\delta f(x)} \delta f(x)\, dx, \quad (2.119)$$

then gives:

$$\delta^{(1)} g[f; x''] = \int \frac{\delta g(x'')}{\delta f(x)} \delta f(x) dx = \iint \frac{\delta g(x'')}{\delta f(x)} \frac{\delta f(x)}{\delta g(x')} \delta g(x') dx' dx. \quad (2.120)$$

This equation identifies the Dirac delta function as the functional derivative of the function at one point with respect to its value at another point, as also implied by (2.107):

$$\int \frac{\delta g(x'')}{\delta f(x)} \frac{\delta f(x)}{\delta g(x')} dx = \frac{\delta g(x'')}{\delta g(x')} = \delta(x'' - x'), \quad (2.121)$$

where we have applied the functional chain rule. The preceding equation also defines the mutually inverse functional derivatives:

$$\frac{\delta g(x')}{\delta f(x)} = \left(\frac{\delta f(x)}{\delta g(x')}\right)^{-1}. \quad (2.122)$$

Let us assume the functional (2.106) in the typical form including the dependence of its density on the argument function itself and its first n derivatives: $f^{(i)}(x) = d^i f(x)/dx^i$, $i = 1, 2, \ldots, n$:

$$\mathcal{L}(x) = \mathcal{L}(x, f(x), f^{(1)}(x), f^{(2)}(x), \ldots, f^{(n)}(x)). \quad (2.123)$$

The functional derivative of $F[f]$ is then given by the following general expression:

$$\frac{\delta F}{\delta f(x)} = \frac{\partial \mathcal{L}(x)}{\partial f(x)} - \frac{d}{dx}\left(\frac{\partial \mathcal{L}(x)}{\partial f^{(1)}(x)}\right) + \frac{d^2}{dx^2}\left(\frac{\partial \mathcal{L}(x)}{\partial f^{(2)}(x)}\right) - \cdots + (-1)^n \frac{d^n}{dx^n}\left(\frac{\partial \mathcal{L}(x)}{\partial f^{(n)}(x)}\right). \tag{2.124}$$

The first term in the r.h.s. of the preceding equation defines the so-called *variational derivative*. It determines the functional derivative of the *local* functionals, the densities of which depend solely upon the argument function itself.

This development can be straightforwardly generalized to cover functionals of functions in three dimensions. Consider, e.g., the functional of $f(\boldsymbol{r})$ depending on the position vector in the physical space: $\boldsymbol{r} = (x, y, z)$. Equation (2.124) can be then extended to cover the $f = f(\boldsymbol{r})$ case by replacing the operator d/dx by its three-dimensional analog – the gradient $\nabla \equiv \partial/\partial \boldsymbol{r}$. For example, for $\mathcal{L}(\boldsymbol{r}, f(\boldsymbol{r}), |\nabla f(\boldsymbol{r})|)$ the functional derivative of $F[f]$ is given by the expression:

$$\frac{\delta F}{\delta f(\boldsymbol{r})} = \frac{\partial \mathcal{L}(\boldsymbol{r})}{\partial f(\boldsymbol{r})} - \nabla\left(\frac{\partial \mathcal{L}(\boldsymbol{r})}{\partial |\nabla f(\boldsymbol{r})|}\right). \tag{2.125}$$

Similarly, for

$$\tilde{F}[f] = F[f] + \int \ell(\Delta f(\boldsymbol{r}))\, d\boldsymbol{r} \equiv \int \tilde{\mathcal{L}}(\boldsymbol{r}, f(\boldsymbol{r}), |\nabla f(\boldsymbol{r})|, \Delta f(\boldsymbol{r}))\, d\boldsymbol{r}, \quad \Delta = \nabla^2,$$

$$\frac{\delta \tilde{F}}{\delta f(\boldsymbol{r})} = \frac{\delta F}{\delta f(\boldsymbol{r})} + \Delta\left(\frac{\partial \tilde{\mathcal{L}}(\boldsymbol{r})}{\partial \Delta f(\boldsymbol{r})}\right). \tag{2.126}$$

References

Dirac PAM (1967) The principles of quantum mechanics. Clarendon, Oxford
Gelfand IM, Fomin SV (1963) Calculus of variations. Englewood Cliffs, Prentice-Hall
Volterra V (1959) Theory of functionals. Dover, New York

Chapter 3
Basic Concepts and Axioms

Abstract The postulates of quantum mechanics are formulated using the mathematical tools of the preceding chapter. First, the axioms related to the quantum kinematics are summarized, dealing with a variety and physical meaning of quantum states at the specified time. They include alternative definitions and interpretations of the wave functions of microobjects as amplitudes of the particle probability distributions in the configuration or momentum spaces. As an illustrative example the electron densities are then discussed. The superposition principle is formulated, and the symmetry implications of indistinguishability of identical particles in quantum mechanics are examined. The links between the quantum states and outcomes of the physical measurements are then surveyed and the physical observables are attributed to quantum mechanical operators, linear and Hermitian, and their specific forms in the position and momentum representations are introduced. The eigenvalues of the quantum mechanical operator are postulated to determine a variety of all possible results of a single experiment measuring the physical property the operator represents, while the operator expectation value represents the average value of this quantity in a very large number of repeated measurements performed on the system in the same quantum state. The eigenstates of the quantum mechanical operator are shown to correspond to the sharply specified value of the physical property under consideration, while other quantum states exhibit distributions of its allowed eigenvalues. The statistical mixtures of quantum states are defined in terms of the density operator and the ensemble averages of physical observables in such mixed states are examined. The simultaneous sharp measurement of several physical observables is linked to the mutual commutation of their operators and the quantum mechanical formulation of the general Principle of Indeterminacy is given. Properties of the electron angular momentum and spin operators are examined.

In the dynamical development, the pictures of time evolution in quantum mechanics are introduced through the alternative time-dependent unitary transformations of the state vectors/operators. The Schrödinger equation is explored in some detail, with the emphasis placed upon the stationary states, time dependence of expectation values, conservation laws, the probability current, and continuity equation. The

R.F. Nalewajski, *Perspectives in Electronic Structure Theory*,
DOI 10.1007/978-3-642-20180-6_3, © Springer-Verlag Berlin Heidelberg 2012

correspondence between the quantum and classical dynamics is established through the Ehrenfest principle. Finally, the rudiments of the Heisenberg and interaction pictures of quantum dynamics are briefly summarized.

3.1 *N*-Electron Wave Functions and Their Probabilistic Interpretation

In the canonical formulation of classical mechanics, the system dynamics is formulated in terms of the Hamilton function $E = H(\boldsymbol{Q}, \boldsymbol{P})$ expressing the system energy E in terms of its generalized coordinates $\boldsymbol{Q} = \{Q_\alpha\}$ and their conjugated momenta $\boldsymbol{P} = \{P_\alpha\}$, $\alpha = 1, 2, \ldots, f$, with f denoting the system number of dynamical *degrees of freedom*. Together these conjugate dynamical variables uniquely specify the system mechanical state. Indeed, the knowledge of $\boldsymbol{Q}(t)$ and $\boldsymbol{P}(t)$ at the specified time $t = t_0$ allows one to determine the exact time evolution of these state parameters, via the Hamilton equations of motion:

$$\dot{Q}_\alpha = \frac{dQ_\alpha}{dt} = \frac{\partial H}{\partial P_\alpha}, \quad \dot{P}_\alpha = \frac{dP_\alpha}{dt} = -\frac{\partial H}{\partial Q_\alpha}, \quad \alpha = 1, 2, \ldots, f. \tag{3.1}$$

Since these are the *first*-order differential equations, their solutions $\{\boldsymbol{Q}(t), \boldsymbol{P}(t)\}$ are uniquely specified when the values of these classical state variables are fixed at $t = t_0$. Thus, knowing the state $\{\boldsymbol{Q}(t_0), \boldsymbol{P}(t_0)\}$ of the classical system at this time, one can in principle predict with certainty the system mechanical state at $t \neq t_0$, i.e., precisely determine the outcome of any measurement at an earlier or later stage of the system time evolution.

As we have argued in Chap. 1, this classical specification of the mechanical state is inapplicable in the quantum theory, due to the simultaneous indeterminacy of coordinates and momenta of microobjects (the Heisenberg principle). Indeed, since the state variables must be precisely specified, either the position coordinates or the components of the canonically conjugated momenta of the system particles should be used to uniquely characterize its quantum state. Therefore, at the given time t, which in the simplest (nonrelativistic) formulation of the quantum theory plays the role of a parameter, the quantum state corresponding to the state vector $|\Psi(t)\rangle$ is represented by the wave functions in either the position or momentum representations,

$$\Psi(\boldsymbol{Q}; t) = \langle \boldsymbol{Q} | \Psi(t) \rangle \quad \text{or} \quad \Psi(\boldsymbol{P}; t) = \langle \boldsymbol{P} | \Psi(t) \rangle; \tag{3.2a}$$

here, the representation basis sets $\{|\boldsymbol{Q}\rangle\}$ and $\{|\boldsymbol{P}\rangle\}$ correspond to the position and momentum eigenstates, respectively, in which these molecular variables are known precisely. For quantum particles these classical state "coordinates" should be also supplemented with all nonclassical, internal (*spin*) degrees of freedom for each particle, $\boldsymbol{\Sigma} = \{\Sigma_k\}$. Therefore, the full specification of the mechanical state of the given quantum system, in either the *position-spin* or *momentum-spin* representations, is embodied in the corresponding wave functions:

$$\Psi(\boldsymbol{Q},\boldsymbol{\Sigma};\, t) = \langle \boldsymbol{Q},\boldsymbol{\Sigma}|\Psi(t)\rangle \quad \text{or} \quad \Psi(\boldsymbol{P},\boldsymbol{\Sigma};\, t) = \langle \boldsymbol{P},\boldsymbol{\Sigma}|\Psi(t)\rangle. \tag{3.2b}$$

Since the theoretical description of the electronic structure of molecules is the main objective of this book, in what follows we shall focus on a general (atomic or molecular) N-electron system, with the list of the (coordinate/momenta)-spin variables in the Cartesian coordinates:

$$\begin{aligned} |\boldsymbol{Q},\boldsymbol{\Sigma}\rangle &\equiv |\boldsymbol{q}^N\rangle = |\{\boldsymbol{q}_k\}\rangle \equiv |\mathcal{Q}^N\rangle, \quad \boldsymbol{q}_k = (\boldsymbol{r}_k,\, \sigma_k), \\ |\boldsymbol{P},\boldsymbol{\Sigma}\rangle &\equiv |\boldsymbol{u}^N\rangle = |\{\boldsymbol{u}_k\}\rangle \equiv |\mathcal{P}^N\rangle, \quad \boldsymbol{u}_k = (\boldsymbol{p}_k,\, \sigma_k), \quad k = 1,2,\ldots,N; \end{aligned} \tag{3.3}$$

here $\boldsymbol{r}_k = (x_k, y_k, z_k), \boldsymbol{p}_k = (p_{x_k}, p_{y_k}, p_{z_k})$ and σ_k, respectively, denote the continuous position, momentum vectors of kth electron and its discrete spin orientation variable $\sigma_k \in (+½, -½)$ (see Fig. 1.2).

Therefore, the vector space of the N-electron system is spanned by all basis vectors in either the position $\{|\mathcal{Q}^N\rangle\}$ or momentum $\{|\mathcal{P}^N\rangle\}$ representations. In what follows we shall call this vector space the *molecular Hilbert space*. The specific state of such an N-electron system in time t will be denoted by the ket $|\Psi^N(t)\rangle$. Since each basis vector is specified by the three position/momentum coordinates and one spin variable for each electron, the overall dimensionality of either the *position*-spin or *momentum*-spin spaces is $4N$. The basis vectors $|\mathcal{Q}^N\rangle$ and $|\mathcal{P}^N\rangle$ are then identified by corresponding points in these configurational spaces. It should be observed that in the classical mechanics the system state was uniquely specified at the given time by selecting the point in the $6N$-dimensional position–momentum *phase space* of N particles.

Moreover, the corresponding position-spin or momentum-spin data for the atomic nuclei are also required for the complete specification of the molecular state. However, as we shall argue in Part II of this book, due to a huge difference in masses between the (light) electrons and (heavy) nuclei, the dynamics of the former can be to a good approximation described by examining their fast movements in the effective potential generated by the "frozen" nuclear framework, with the fixed positions of nuclei playing the role of parameters in the electronic structure theory. In this *adiabatic approximation* of Born and Oppenheimer the nuclei, sources of the external potential in which electrons move, thus determine the assumed molecular geometry.

After these short preliminaries, we are now in a position to formulate the important postulate of quantum mechanics, due to Born, which provides the physical interpretation of the wave functions of (3.2a) and (3.2b):

Postulate I: The (normalized) quantum mechanical state of the molecular system containing N-electrons in time t, $\langle\Psi^N(t)|\Psi^N(t)\rangle \equiv \|\Psi^N(t)\|^2 = 1$, where $\|\Psi^N(t)\|$

stands for the norm ("length") of the state vector, is uniquely specified by the orientation of the state-vector $|\Psi^N(t)\rangle$ in the molecular Hilbert space or equivalently by its equivalent representations (wave functions) in the position or momentum basis sets, respectively,

$$\Psi(\mathcal{Q}^N;t) = \langle \mathcal{Q}^N | \Psi^N(t) \rangle \quad \text{or} \quad \Psi(\mathcal{P}^N;t) = \langle \mathcal{P}^N | \Psi^N(t) \rangle. \tag{3.2c}$$

These in general complex-valued functions determine the probability amplitudes of simultaneously observing at this time the specified positions/momenta and spin orientations of all N electrons, with the corresponding probability densities being determined by the squares of the wave function moduli:

$$\begin{aligned} p(\mathcal{Q}^N;t) &= |\langle \mathcal{Q}^N|\Psi^N(t)\rangle|^2 = |\Psi(\mathcal{Q}^N;t)|^2 \equiv P[\mathcal{Q}^N|\Psi^N(t)], \\ \int p(\mathcal{Q}^N;t)\, d\mathcal{Q}^N &= \int \langle \Psi^N(t)|\mathcal{Q}^N\rangle\langle \mathcal{Q}^N|\Psi^N(t)\rangle\, d\mathcal{Q}^N = \langle \Psi^N(t)\,|\Psi^N(t)\rangle = 1; \\ \pi(\mathcal{P}^N;t) &= |\langle \mathcal{P}^N|\Psi^N(t)\rangle|^2 = |\Psi(\mathcal{P}^N;t)|^2 \equiv P[\mathcal{P}^N|\Psi^N(t)], \\ \int \pi(\mathcal{P}^N;t)\, d\mathcal{P}^N &= \int \langle \Psi^N(t)|\mathcal{P}^N\rangle\langle \mathcal{P}^N|\Psi^N(t)\rangle\, d\mathcal{P}^N = \langle \Psi^N(t)\,|\Psi^N(t)\rangle = 1. \end{aligned} \tag{3.4}$$

Here, the generalized "integration" symbol $\int d\mathcal{Q}^N$ actually denotes the definite integrations over the position coordinates and summations over the spin variables of all electrons:

$$\int d\mathcal{Q}^N \equiv \int d\boldsymbol{q}_1 \ldots d\boldsymbol{q}_N \equiv \int d\boldsymbol{r}_1 \ldots d\boldsymbol{r}_N \sum_{\sigma_1} \cdots \sum_{\sigma_N}, \tag{3.5a}$$

The related operation in the momentum-spin space similarly reads:

$$\int d\mathcal{P}^N \equiv \int d\boldsymbol{u}_1 \ldots d\boldsymbol{u}_N \equiv \int d\boldsymbol{p}_1 \ldots d\boldsymbol{p}_N \sum_{\sigma_1} \cdots \sum_{\sigma_N}. \tag{3.5b}$$

In fact, the normalization conditions of this postulate, for the position-spin and momentum-spin probability densities $p(\mathcal{Q}^N;t)$ and $\pi(\mathcal{P}^N;t)$, respectively, express the unit probability of the sure event that at the specified time t all electrons are located somewhere in the physical or momentum spaces, and assume one of its allowed spin orientations. We have also indicated in (3.4) that the probability densities $P[\mathcal{Q}^N|\Psi^N(t)]$ and $P[\mathcal{P}^N|\Psi^N(t)]$ of the particle positions and momenta, respectively, are *conditional* upon the specified quantum state. Indeed, these densities represent the conditional probabilities of observing the basis set events corresponding to the wave function arguments $\mathcal{Q}^N$ or $\mathcal{P}^N$ (*variables*), given the molecular state $|\Psi^N(t)\rangle$ (the *parameter*): $p(\mathcal{Q}^N;t) = P[\mathcal{Q}^N|\Psi^N(t)]$ and $\pi(\mathcal{P}^N;t) = P[\mathcal{P}^N|\Psi^N(t)]$. The normalization relations thus involve the integrations/summations of these conditional probabilities over the *variable* states $\{|\mathcal{Q}^N\rangle\}$ and $\{|\mathcal{P}^N\rangle\}$, respectively, for the fixed *parameter* state $|\Psi^N(t)\rangle$. The integrands of these sum rules thus provide the associated probabilities of the particles being simultaneously found in their specified, infinitesimal ranges of coordinates $d\mathcal{Q}^N = \{d\boldsymbol{r}_k, \sigma_k\}$ or momenta $d\mathcal{P}^N = \{d\boldsymbol{p}_k, \sigma_k\}$, i.e., of the system particles occupying the corresponding volumes of the position or momentum spaces for their specified spin orientations.

This physical interpretation of the quantum mechanical wave functions has far reaching implications for their admissible analytical form. First, the normalization condition excludes the functions which become infinite over a finite region of space, since then Born's interpretation would be untenable. Clearly, the Dirac-delta wave functions of (2.69), which correspond to precise localizations or momenta of electrons, are not excluded since their infinite values extend only over the infinitesimal volumes of space, thus giving rise to the finite normalization integral. However, for the finite, constant probability densities, e.g., $p(\mathcal{Q}^N; t) = const. > 0$, this integral may become infinite, when the movements of electrons are not confined to finite regions of space. In such cases, this density provides only a *relative* measure of probability.

Another implication of the Born probability interpretation is that the wave functions must be single valued. Indeed, $\Psi(\mathcal{Q}^N; t)[\text{or } \Psi(\mathcal{P}^N; t)]$ must generate the *unique* representation of the quantum state $|\Psi^N(t)\rangle$. Additional constraints on their admissible forms are imposed by the form of the quantum mechanical operators. As we have established in Sect. 2.6, the position operator in the momentum representation and the momentum operator in the position representation correspond to differential operators (gradients), e.g., $\hat{\mathrm{p}}(\boldsymbol{r}) = -i\hbar\nabla$. For these operations to be mathematically meaningful, the wave functions on which these observables act must be continuous. Sometimes, the additional condition of the continuous first derivative is also invoked, since the action of the kinetic energy operator of a single particle in the position representation, $\hat{\mathrm{T}}(\boldsymbol{r}) = \hat{\mathrm{p}}^2(\boldsymbol{r})/2m = -(\hbar^2/2m)\Delta$, involves a double differentiation of the wave function embodied in the Laplacian operator $\Delta = \nabla^2$. However, this condition is too severe, since the expectation value of the kinetic energy, when transformed by parts,

$$T = \langle\Psi|\hat{\mathrm{T}}|\Psi\rangle = \int \Psi^*(\boldsymbol{r})\hat{\mathrm{T}}(\boldsymbol{r})\Psi(\boldsymbol{r})d\boldsymbol{r} = (\hbar^2/2m)\int |\nabla\Psi(\boldsymbol{r})|^2 d\boldsymbol{r}, \tag{3.6}$$

remains well defined even for the discontinuous derivatives of the wave function. For example, such discontinuity is encountered for some excessively ill-behaved potentials $V(\boldsymbol{r})$ of forces acting on the particle, e.g., in the *particle-in-the-box* problem, when it jumps from zero to infinity in an infinitesimal distance.

To summarize, in quantum mechanics only such well-behaved wave functions have the physical meaning of probability amplitudes implied by Postulate I. The Born interpretation thus imposes a restriction on the "acceptable" solutions of the differential equations of quantum mechanics, e.g., the crucial Schrödinger equations for determining the system stationary states and their quantum dynamics. Only such well-behaved wave functions may represent the dynamical states of physical systems.

The constraints of the wave function finiteness, single valuedness, and continuity, supplemented by the boundary conditions appropriate for the physical problem in question, give rise to the *quantization* of physical properties, e.g., the system energy (see Sect. 2.7). Indeed, only for some energy levels, the eigenvalues $\{E_n\}$ of the system Hamiltonian, it is possible to construct the well-behaved eigenfunction. For example, in a system with boundaries, when the movement of particles is confined to some finite region of space, the energy is quantized and the less confining is the potential, the less separation is predicted between the neighboring energy levels.

As a result of the Heisenberg uncertainty principle the physically admissible wave functions may penetrate, i.e., exhibit finite values, in the classically forbidden regions, where the total energy is below the potential energy level, $E < V$, thus generating the nonzero probability of finding a particle in such locations. For example, the motion of the quantum mechanical harmonic oscillator is not confined to the classical region between the turning points of the parabolic Hooke potential, and the quantum particles may tunnel through the finite potential barriers. In these classically-forbidden positions the microparticle formally exhibits the negative kinetic energy. This does not imply, however, that the *average* kinetic energy, represented by the expectation value of (3.6), becomes negative in such states. Indeed, the average value over both the (dominating) region of space, where the kinetic energy is positive, and the classically inaccessible (marginal) regions, where it is negative, is always positive. It should be observed, however, that it would be meaningless to speak of the precise kinetic energy of the localized particle anyway, since its momentum is completely unknown!

The electron density $\rho(\boldsymbol{r})$ of locating any of the system N electrons at point $\boldsymbol{r}$ can be obtained from the N-electron probability density $p(\mathbb{Q}^N;t)$ of Eq. (3.4) by the appropriate integration/summation over the remaining arguments of the wave function, i.e., over all admissible events satisfying the condition $\boldsymbol{r}_k = \boldsymbol{r}$, $k = 1, 2, \ldots, N$, enforced by the relevant Dirac deltas in the integrand,

$$\begin{aligned}\rho(\boldsymbol{r};t) &= \sum_{k=1}^{N} \int \delta(\boldsymbol{r}_k - \boldsymbol{r})\, p(\mathbb{Q}^N;t)\, d\mathbb{Q}^N \\ &\equiv \int \Psi^*(\mathbb{Q}^N;t)\, \hat{\rho}(\boldsymbol{r}) \Psi(\mathbb{Q}^N;t)\, d\mathbb{Q}^N = N \int \delta(\boldsymbol{r}_1 - \boldsymbol{r})\, p(\mathbb{Q}^N;t)\, d\mathbb{Q}^N.\end{aligned} \tag{3.7}$$

In the preceding equation we have introduced the electron density *operator* $\hat{\rho}(\boldsymbol{r}) = \sum_k \delta(\boldsymbol{r}_k - \boldsymbol{r})$ and recognized that due to the indistinguishability of electrons, i.e., impossibility to recognize which electron is which, all contributions in the sum of the first line of the equation must be identical. Indeed, we cannot follow the precise trajectories of the separate electrons, due to the incompatibility of its position and momentum, so that their specific identities (hypothetical labels) remain unknown. Clearly, the integral of the electron density over all locations in space must satisfy the sum rule

$$\int \rho(\boldsymbol{r};t) d\boldsymbol{r} = \int \sum_{k=1}^{N} \left[\int \delta(\boldsymbol{r}_k - \boldsymbol{r}) d\boldsymbol{r} \right] p(\mathbb{Q}^N;t)\, d\mathbb{Q}^N = N \int p(\mathbb{Q}^N;t)\, d\mathbb{Q}^N = N. \tag{3.8}$$

One similarly obtains the corresponding *spin* densities of electrons, of detecting at the specified location $\boldsymbol{r}$ electrons with the specified spin $\sigma = (-½, +½)$, the condition enforced by the corresponding Dirac and Kronecker deltas, which together identify the point $\boldsymbol{q} = (\boldsymbol{r}, \sigma)$ in the four-dimensional position-spin space:

$$\begin{aligned}\rho(\boldsymbol{q};t) \equiv \rho^{\sigma}(\boldsymbol{r};t) &= \sum_{k=1}^{N}\int \delta(\boldsymbol{r}_k-\boldsymbol{r})\,\delta_{\sigma_k,\sigma}p(\mathcal{Q}^N;t)\,d\mathcal{Q}^N \\ &\equiv \int \Psi^*(\mathcal{Q}^N;t)\,\hat{\rho}(\boldsymbol{q})\Psi(\mathcal{Q}^N;t)\,d\mathcal{Q}^N \\ &= N_\sigma\int \delta(\boldsymbol{r}_\sigma-\boldsymbol{r})\,p(\mathcal{Q}^N;t)\,d\mathcal{Q}^N, \\ \rho(\boldsymbol{r};t) &= \sum\nolimits_\sigma \rho(\boldsymbol{q};t); \quad \int \rho(\boldsymbol{q};t)\,d\boldsymbol{r} = N_\sigma; \\ \int \rho(\boldsymbol{q};t)\,d\boldsymbol{q} &= \sum\nolimits_\sigma \int \rho(\boldsymbol{q};t)\,d\boldsymbol{r} = N,\end{aligned} \tag{3.9}$$

where N_σ stands for the number of electrons exhibiting the spin orientation σ.

In a similar manner, one determines the *many*-electron densities or their respective spin components and the associated operators in the position representation. For example, the spinless *two*-electron density, $\rho_2(\boldsymbol{r}, \boldsymbol{r}'; t)$, of observing one electron (of all N electrons) at $\boldsymbol{r}$ and another electron (of all the remaining $N-1$ electrons) at $\boldsymbol{r}'$ is given by the following expression:

$$\begin{aligned}\rho_2(\boldsymbol{r},\boldsymbol{r}';t) &= \sum_{k=1}^{N}\sum_{l\neq k}\int \delta(\boldsymbol{r}_k-\boldsymbol{r})\delta(\boldsymbol{r}_l-\boldsymbol{r}')p(\mathcal{Q}^N;t)\,d\mathcal{Q}^N \\ &\equiv \int \Psi^*(\mathcal{Q}^N;t)\,\hat{\rho}_2(\boldsymbol{r},\boldsymbol{r}')\Psi(\mathcal{Q}^N;t)\,d\mathcal{Q}^N \\ &= N(N-1)\int \delta(\boldsymbol{r}_1-\boldsymbol{r})\delta(\boldsymbol{r}_2-\boldsymbol{r}')p(\mathcal{Q}^N;t)\,d\mathcal{Q}^N, \\ \iint \rho_2(\boldsymbol{r},\boldsymbol{r}';t)\,d\boldsymbol{r}\,d\boldsymbol{r}' &= N(N-1).\end{aligned} \tag{3.10}$$

Again, this *two*-electron distribution can be decomposed into the spin-resolved components:

$$\begin{aligned}\rho_2(\boldsymbol{q},\boldsymbol{q}';t) \equiv \rho^{\sigma,\sigma'}(\boldsymbol{r},\boldsymbol{r}';t) &= \sum_{k=1}^{N}\sum_{l\neq k}\int \delta(\boldsymbol{r}_k-\boldsymbol{r})\delta(\boldsymbol{r}_l-\boldsymbol{r}')\,\delta_{\sigma_k,\sigma}\delta_{\sigma_l,\sigma'}p(\mathcal{Q}^N;t)\,d\mathcal{Q}^N \\ &\equiv \int \Psi^*(\mathcal{Q}^N;t)\,\hat{\rho}_2(\boldsymbol{q},\boldsymbol{q}')\Psi(\mathcal{Q}^N;t)\,d\mathcal{Q}^N, \\ \rho_2(\boldsymbol{r},\boldsymbol{r}';t) &= \sum_{\sigma}\sum_{\sigma'}\rho^{\sigma,\sigma'}(\boldsymbol{r},\boldsymbol{r}';t), \\ \iint \rho_2(\boldsymbol{q},\boldsymbol{q}';t)\,d\boldsymbol{r}\,d\boldsymbol{r}' &= \begin{cases} N_\sigma(N_\sigma-1), & \sigma'=\sigma \\ N_\sigma N_{\sigma'}, & \sigma'\neq\sigma \end{cases}, \\ \iint \rho_2(\boldsymbol{q},\boldsymbol{q}';t)\,d\boldsymbol{q}\,d\boldsymbol{q}' &= \sum_{\sigma}\sum_{\sigma'}\iint \rho^{\sigma,\sigma'}(\boldsymbol{r},\boldsymbol{r}';t)\,d\boldsymbol{r}\,d\boldsymbol{r}' = N(N-1).\end{aligned} \tag{3.11}$$

Of interest also is the *pair* density in which the permuted *two*-electron localization events $(\boldsymbol{r}_k = \boldsymbol{r}) \wedge (\boldsymbol{r}_l = \boldsymbol{r}')$ and $(\boldsymbol{r}_k = \boldsymbol{r}') \wedge (\boldsymbol{r}_l = \boldsymbol{r})$ are regarded as physically identical and thus counted only once:

$$\Gamma(\boldsymbol{r},\boldsymbol{r}';t) = \int \Psi^*(\mathcal{Q}^N;t)\,\hat{\Gamma}(\boldsymbol{r},\boldsymbol{r}')\Psi(\mathcal{Q}^N;t)\,d\mathcal{Q}^N,$$
$$\hat{\Gamma}(\boldsymbol{r},\boldsymbol{r}') = \sum_{k=1}^{N-1}\sum_{l=k+1}^{N} \delta(\boldsymbol{r}_k - \boldsymbol{r})\delta(\boldsymbol{r}_l - \boldsymbol{r}'), \tag{3.12}$$
$$\iint \Gamma(\boldsymbol{r},\boldsymbol{r}';t)\,d\boldsymbol{r}\,d\boldsymbol{r}' = N(N-1)/2 = \binom{N}{2}.$$

This distribution of the physically *indistinguishable* electronic pairs satisfies the pair normalization of the preceding equation (Löwdin 1955a, b), which differs from that adopted for the *two*-electron distribution of (3.10) (McWeeny 1989). This change in the normalization simplifies the expression for the average electron repulsion energy,

$$V_{e,e}(N;t) = \int \Psi^*(\mathcal{Q}^N;t)\hat{\mathrm{V}}_{e,e}(N)\Psi(\mathcal{Q}^N;t)d\mathcal{Q}^N, \tag{3.13}$$

the expectation value of the associated (multiplicative) operator in position representation, $\hat{\mathrm{V}}_{e,e}(N)$, which measures the interelectron Coulomb interaction for the sharply specified locations of all N electrons:

$$\hat{\mathrm{V}}_{e,e}(N) = \sum_{k=1}^{N-1}\sum_{l=k+1}^{N} \frac{1}{|\boldsymbol{r}_k - \boldsymbol{r}_l|} \equiv \sum_{k=1}^{N-1}\sum_{l=k+1}^{N} g(k,l). \tag{3.14}$$

In terms of the above *two*-electron densities, the expectation value of the electron repulsion energy of (3.13) thus reads:

$$\begin{aligned} V_{e,e}(N;t) &= \frac{1}{2}\iint |\boldsymbol{r}-\boldsymbol{r}'|^{-1}\rho_2(\boldsymbol{r},\boldsymbol{r}';t)\,d\boldsymbol{r}\,d\boldsymbol{r}' \\ &= \iint |\boldsymbol{r}-\boldsymbol{r}'|^{-1}\Gamma(\boldsymbol{r},\boldsymbol{r}';t)\,d\boldsymbol{r}\,d\boldsymbol{r}'. \end{aligned} \tag{3.15}$$

Clearly, by using the corresponding Kronecker deltas of the spin variables of electrons [see (3.11)], one could similarly define the spin components of the pair density as well.

The extension of these concepts into the corresponding momentum-spin densities is straightforward. For example, the spinless *one*- and *two*-electron densities in the momentum space of N electrons become:

$$\pi(\boldsymbol{p};t) = \sum_{k=1}^{N}\int \delta(\boldsymbol{p}_k-\boldsymbol{p})\,\pi(\mathcal{P}^N;t)\,d\mathcal{P}^N \equiv \int \Psi^*(\mathcal{P}^N;t)\,\hat{\pi}(\boldsymbol{p})\Psi(\mathcal{P}^N;t)\,d\mathcal{P}^N$$
$$= N\int \delta(\boldsymbol{p}_1-\boldsymbol{p})\,\pi(\mathcal{P}^N;t)\,d\mathcal{P}^N \equiv NP(\boldsymbol{p};t)$$
$$\int \pi(\boldsymbol{p};t)\,d\boldsymbol{p} = N; \tag{3.16}$$

$$\begin{aligned}\pi_2(\boldsymbol{p},\boldsymbol{p}';t) &= \sum_{k=1}^{N}\sum_{l\neq k}\int \delta(\boldsymbol{p}_k-\boldsymbol{p})\delta(\boldsymbol{p}_l-\boldsymbol{p}')\pi(\mathcal{P}^N;t)\,d\mathcal{P}^N \\ &\equiv \int \Psi^*(\mathcal{P}^N;t)\,\hat{\pi}_2(\boldsymbol{p},\boldsymbol{p}')\Psi(\mathcal{P}^N;t)\,d\mathcal{P}^N \\ &= N(N-1)\int \delta(\boldsymbol{p}_1-\boldsymbol{p})\delta(\boldsymbol{p}_2-\boldsymbol{p}')\pi(\mathcal{P}^N;t)\,d\mathcal{P}^N, \\ \iint \pi_2(\boldsymbol{p},\boldsymbol{p}';t)\,d\boldsymbol{p}\,d\boldsymbol{p}' &= N(N-1). \end{aligned} \tag{3.17}$$

Consider now the expectation value of the kinetic energy of N-electrons in the momentum representation,

$$\begin{aligned}T_e(N;t) &= \langle \Psi^N(t)|\hat{\mathrm{T}}(N)|\Psi^N(t)\rangle \equiv \int \Psi^*(\mathcal{P}^N;t)\,\hat{\mathrm{T}}(\mathcal{P}^N)\Psi(\mathcal{P}^N;t)d\mathcal{P}^N \\ &= \int T(\mathcal{P}^N)\pi(\mathcal{P}^N;t)d\mathcal{P}^N, \end{aligned} \tag{3.18}$$

where the (multiplicative) kinetic energy operator $\hat{\mathrm{T}}(\mathcal{P}^N) = T(\mathcal{P}^N)$ measures the system kinetic energy when the momenta of all N electrons are sharply specified:

$$\hat{\mathrm{T}}(\mathcal{P}^N) = \frac{1}{2m}\sum_{k=1}^{N}\boldsymbol{p}_k^2 \equiv \sum_{k=1}^{N} T(\boldsymbol{p}_k) = T(\mathcal{P}^N) = NT(\boldsymbol{p}). \tag{3.19}$$

Therefore, the expectation value of (3.18) is given by the following mean value expression involving the *one*-electron density in momentum space:

$$T_e(N;t) = N\int T(\boldsymbol{p})\,P(\boldsymbol{p};t)d\boldsymbol{p} = \int T(\boldsymbol{p})\,\pi(\boldsymbol{p};t)d\boldsymbol{p}. \tag{3.20}$$

3.2 Superposition Principle, Expectation Values, and Indistinguishability of Identical Particles

The superposition principle of Sect. 2.2 is formally summarized by another basic axiom of quantum mechanics:

Postulate II: Any combination $|\Psi\rangle = \sum_i C_i|\Psi_i\rangle$ of the admissible quantum states $\{|\Psi_i\rangle\}$, where $\{C_i\}$ denotes generally complex factors, also represents a possible quantum state of the system under consideration. The squares of moduli of these expansion coefficients determine the normalized conditional probabilities $\{P(\Psi_i|\Psi) = |C_i|^2\}$ of observing state Ψ_i given the state Ψ: $\sum_i P(\Psi_i|\Psi) = 1$.

As an illustration let us consider the basis eigenvectors $|\boldsymbol{i}\rangle = \{|\Psi_i\rangle \equiv |a_i\rangle\}$ of the quantum observable $\hat{\mathrm{A}}$ (2.55a), which for reasons of simplicity we assume to correspond to the discrete spectrum of eigenvalues $\{a_i\}$. Expanding a general state vector $|\Psi\rangle$ in this basis set (2.48a) then gives the following components of its (column) vector representation: $\boldsymbol{\Psi}(\boldsymbol{i}) = \langle \boldsymbol{i}|\Psi\rangle = \{C_i = \langle\Psi_i|\Psi\rangle\} \equiv \boldsymbol{C}$. Hence [see (2.39)], the corresponding conditional probabilities read:

$$\begin{aligned} P(\Psi_i|\Psi) = |C_i|^2 = C_i C_i^* &= \langle\Psi_i|\Psi\rangle\langle\Psi|\Psi_i\rangle = \langle\Psi_i|\hat{\mathrm{P}}_\Psi|\Psi_i\rangle \\ = C_i^* C_i &= \langle\Psi|\Psi_i\rangle\langle\Psi_i|\Psi\rangle = \langle\Psi|\hat{\mathrm{P}}_i|\Psi\rangle = P(\Psi|\Psi_i). \end{aligned} \tag{3.21}$$

It follows from this equation that the conditional probabilities between two quantum states can be considered as the expectation values in the *variable* state of the projection operator onto the *reference* state, which plays the role of a parameter. Their normalization then directly follows from the basis set closure of (2.41a):

$$\Sigma_i P(\Psi_i|\Psi) = \langle\Psi|\sum_i \hat{\mathrm{P}}_i|\Psi\rangle = \langle\Psi|\Psi\rangle = 1. \tag{3.22}$$

As we shall see in Sect. 3.3, the conditional probabilities of (3.21) also reflect relative frequencies of possible outcomes $\{a_i = \langle\Psi_i|\hat{\mathrm{A}}|\Psi_i\rangle\}$ of the experiments measuring the physical quantity A. Indeed, the eigenvector representation of $\hat{\mathrm{A}}$ is given by the diagonal matrix $\mathbf{A} = \langle\boldsymbol{i}|\hat{\mathrm{A}}|\boldsymbol{i}\rangle = \{A_{m,n} = \langle\Psi_m|\hat{\mathrm{A}}|\Psi_n\rangle = a_m\delta_{m,n}\}$. Therefore, the statistical average (expectation) value $\langle A\rangle$ in state $|\Psi\rangle$ is given by the relevant mean value expression:

$$\langle A\rangle = \Sigma_m P(\Psi_m|\Psi)\, a_m = \Sigma_m\Sigma_n C_m^* A_{m,n}\, C_n = \langle\Psi|\boldsymbol{i}\rangle\langle\boldsymbol{i}|\hat{\mathrm{A}}|\boldsymbol{i}\rangle\langle\boldsymbol{i}|\Psi\rangle = \langle\Psi|\hat{\mathrm{A}}|\Psi\rangle. \tag{3.23}$$

We have already encountered such a statistical (ensemble) interpretation in (2.97), when defining the probability $W(E_n|\Psi)$ of observing the specified energy level E_n in the given quantum state $|\Psi\rangle$.

In the case of a degenerate eigenvalue a_i the probability of observing it in state $|\Psi\rangle$ is given by the sum of contributions $P(\Psi_{i,j}|\Psi)$ originating from all independent component states for this eigenvalue, $\{|\Psi_{i,j}\rangle = |i_j\rangle,\ j = 1,\ 2,\ g\}$ [see (2.58)]: $P(a_i) = \sum_{j=1}^{g} |\langle i_j|\Psi\rangle|^2$.

The superposition principle can be straightforwardly extended into the continuous basis sets $|\boldsymbol{x}\rangle = \{|x\rangle\}$, e.g., the position and momentum representations of Sect. 2.6: any continuous combination [see (2.17)] also represents a possible quantum state of the system with $\{P(x|\Psi) = |c(x)|^2\}$ now providing the conditional probability density of observing $|x\rangle$ given $|\Psi\rangle$, and hence also of all its physical observables in the reference state $|\Psi\rangle$, with the relevant normalization condition $\int P(x|\Psi)\,dx = 1$. Indeed, since $\{c(x)\} = \boldsymbol{\Psi}(\boldsymbol{x}) = \langle \boldsymbol{x}|\Psi\rangle$ (continuous column-vector),

$$\int P(x|\Psi)dx = \int \langle\Psi|x\rangle\langle x|\Psi\rangle dx = \langle\Psi|\boldsymbol{x}\rangle\langle\boldsymbol{x}|\Psi\rangle = \langle\Psi|\Psi\rangle = 1, \tag{3.24}$$

where we have used the closure relation of (2.41b). Best illustration of this continuous version of the superposition principle is Postulate I itself. Indeed, as implied by (2.70) and (3.2c), the wave functions in the position and momentum representations constitute the expansion coefficients in the basis sets consisting of the eigenstates of the position and momentum operators, respectively, and hence the squares of their moduli are in fact the conditional probabilities of observing in $|\Psi^N(t)\rangle$ the sharply specified locations and momenta of the system constituent particles:

$$P[\mathcal{Q}^N|\Psi(t)] = |\Psi(\mathcal{Q}^N;t)|^2 = p(\mathcal{Q}^N;t), \quad P[\mathcal{P}^N|\Psi(t)] = |\Psi(\mathcal{P}^N;t)|^2 = \pi(\mathcal{P}^N;t). \tag{3.25}$$

Consider next the expression for the average kinetic energy (3.6) of a spinless particle, corresponding to the quantum observable $\hat{\mathrm{T}} = \hat{\mathbf{p}}^2(\boldsymbol{r})/2m$. The relevant expansion is again that in terms of the eigenstates $\{|\boldsymbol{p}\rangle\}$ of the particle momentum (2.75), $c(\boldsymbol{p}) = \Psi(\boldsymbol{p}) = \langle\boldsymbol{p}|\Psi\rangle$, which also mark the eigenstates of $\hat{\mathrm{T}}$ corresponding to the eigenvalues $\{T(\boldsymbol{p}) = \boldsymbol{p}^2/2m\}$. The associated conditional probability density is therefore the momentum density of (3.16), $P(\boldsymbol{p}|\Psi) = |\Psi(\boldsymbol{p})|^2 = \pi(\boldsymbol{p})$, which gives rise to the following mean value expression for the expectation value of the kinetic energy in state $|\Psi\rangle$ [see also (3.20)]:

$$\langle T\rangle_\Psi = \int T(\boldsymbol{p})P(\boldsymbol{p}|\Psi)\,d\boldsymbol{p} = \int T(\boldsymbol{p})\,\pi(\boldsymbol{p})\,d\boldsymbol{p}. \tag{3.26}$$

In the mixed basis set case, $|\boldsymbol{m}\rangle = (\{|\alpha\rangle\}, \{|y\rangle\})$, the expansion is generated by the identity projector of (2.41c). The squares of expansion coefficients, $\{C_\alpha = \langle\alpha|\Psi\rangle\}$ and $\{c(x) = \langle x|\Psi\rangle\}$, thus determine the corresponding conditional probabilities of observing the representation discrete and continuous eigenvalues, respectively,

$$P(\alpha|\Psi) = |\langle\alpha|\Psi\rangle|^2 \quad \text{and} \quad P(y|\Psi) = |\langle y|\Psi\rangle|^2, \tag{3.27}$$

with the normalization condition [see the closure relation of (2.41c)]:

$$\sum_\alpha P(\alpha|\Psi) + \int P(y|\Psi)dy = \langle\Psi|\boldsymbol{m}\rangle\langle\boldsymbol{m}|\Psi\rangle = \langle\Psi|\Psi\rangle = 1. \tag{3.28}$$

An important property of the wave functions of identical particles is embodied in their symmetry properties with respect to the operation exchanging the spin-position (or momentum-position) variables of two particles. The physical meaning of the quantum state is not affected by such an operation since the identical particles, e.g., electrons in a molecule, are indistinguishable due to the basic inability to follow their classical trajectories in quantum mechanics (the Heisenberg Principle of Indeterminacy). Therefore, should we mentally associate some labels distinguishing electrons at the specified time, their identity afterwards would be still completely unknown. Clearly, the objective laws of quantum mechanics cannot depend upon such a subjective act of attributing these identity labels to electrons.

This physical invariance with respect to exchanging two identical particles, say electrons k and l, symbolized by the associated permutation operator $\hat{\mathrm{X}}(k,l)$, is also reflected by the symmetry of the system Hamiltonian $\hat{\mathrm{H}}(\mathcal{Q}^N)$ with respect to such an operation [see (2.104)],

$$\begin{aligned} \hat{\mathrm{X}}(k,l)\,\hat{\mathrm{H}}(\mathcal{Q}^N)\hat{\mathrm{X}}(k,l)^{-1} &= \hat{\mathrm{H}}(\mathcal{Q}^N) \quad \text{or} \\ \hat{\mathrm{X}}(k,l)\,\hat{\mathrm{H}}(\mathcal{Q}^N) &= \hat{\mathrm{H}}(\mathcal{Q}^N)\,\hat{\mathrm{X}}(k,l). \end{aligned} \tag{3.29}$$

The conservation in such an operation of the probability densities of Postulate I,

$$\begin{aligned} p(\boldsymbol{q}_1,\ldots\boldsymbol{q}_k,\ldots,\boldsymbol{q}_l,\ldots,\boldsymbol{q}_N;t) &= p(\boldsymbol{q}_1,\ldots\boldsymbol{q}_l,\ldots,\boldsymbol{q}_k,\ldots,\boldsymbol{q}_N;t) \quad \text{or} \\ \pi(\boldsymbol{u}_1,\ldots\boldsymbol{u}_k,\ldots,\boldsymbol{u}_l,\ldots,\boldsymbol{u}_N;t) &= \pi(\boldsymbol{u}_1,\ldots\boldsymbol{u}_l,\ldots,\boldsymbol{u}_k,\ldots,\boldsymbol{u}_N;t), \end{aligned} \tag{3.30}$$

thus requires preservation of the squares of the moduli of the associated wave functions. It is assured, when the wave functions themselves are either symmetrical or antisymmetrical with respect to such a permutation of two identical particles:

$$\begin{aligned} \hat{\mathrm{X}}(k,l)\Psi(\mathcal{Q}^N;t) &= \pm\Psi(\mathcal{Q}^N;t) \equiv X_\xi\Psi(\mathcal{Q}^N;t) \quad \text{or} \\ \hat{\mathrm{X}}(k,l)\Psi(\mathcal{P}^N;t) &= \pm\Psi(\mathcal{P}^N;t) \equiv X_\xi\Psi(\mathcal{P}^N;t). \end{aligned} \tag{3.31}$$

Thus, in view of the commutation relation (3.29), $[\hat{\mathrm{H}}(\mathcal{Q}^N),\hat{\mathrm{X}}(k,l)] = 0$, the eigenfunctions of the Hamiltonian of a system of identical particles also satisfy the simultaneous eigenvalue problem (3.31) (see Sect. 2.5) of the particle exchange operator $\hat{\mathrm{X}}(k,l)$, which exhibits only two eigenvalues: $X_\xi = \pm 1$. This symmetry or antisymmetry feature of the wave function reflects the identity of the particles involved. This permutational symmetry of quantum states is conserved in time.

These symmetry properties of the admissible wave functions of identical particles can be summarized in the following postulate of Pauli:

Postulate III: The physical wave functions of the system identical particles must be either symmetric or antisymmetric with respect to the permutation of their position-spin $\{\boldsymbol{q}_k\}$ or momentum-spin $\{\boldsymbol{u}_k\}$ variables. Those particles for which the wave functions are symmetric are called bosons and those for which they are antisymmetric are called fermions.

Thus, the elementary particles existing in nature are divided into two categories: the particles corresponding to $X = +1$, i.e., the symmetric wave functions, called *bosons*, and those associated with $X = -1$, i.e., the antisymmetric wave functions, called *fermions*. All currently known particles obey the following *empirical* rule related to their spin quantum number $\boldsymbol{S}$ (see Sect. 1.4): particles of *half*-integral spin (e.g., electrons, positrons, protons, neutrons, muons) are fermions, while those of the integral spin (e.g., photons, mesons) are bosons. It also holds for the composite particles such as the atomic nuclei, which are known to be composed of nucleons (neutrons and protons), which are fermions. Thus, the spin of the nucleus as a whole, is reflected by the parity of the number of nucleons: the nuclei with an even number of fermions, e.g., ^{4}He isotope, are bosons, while those containing an odd number of nucleons, e.g., ^{3}He isotope, are fermions, since the resultant spin of such composite particles is integral in the first case and *half*-integral in the other case.

There are also macroscopic consequences of the particle spin identity in the statistical mechanics, which predicts the physical properties of systems composed of a very large number of particles as averages over the ensembles corresponding to alternative thermodynamic equilibria. The statistical weight of a macroscopic state is then proportional to a number of the microscopic states, through which it can be realized, a variety of which strongly depends on the particle identity.

In the classical, *Maxwell–Boltzmann statistics*, the identical particles were in fact treated as if they are different. Indeed, the microscopic states with identical list of states of individual (identical) particles were considered distinct, when the permutation of particles among these states was different. In the quantum statistical mechanics the above symmetrization postulate intervenes, so that an admissible microscopic state is now solely identified by the enumeration of individual particle states which form it, their actual ordering being insignificant. This gives rise to different predictions compared with those resulting from the classical statistical mechanics.

The consequence of the antisymmetrization rule for the wave function of fermions implies that two identical fermions cannot "occupy" the same quantum state, a restriction known as the *Pauli Exclusion Principle*. There are no such occupation restrictions implied by the symmetrization rule for bosons, so that an individual state is accessible to any number of such integral spin particles. Different statistical averages result: the bosons obey the *Bose–Einstein statistics*, while fermions – the *Fermi–Dirac statistics*, which explains the nomenclature adopted to distinguish these two categories of quantum particles. Thermodynamic differences between them are amplified at low temperatures: the *Bose condensation* is observed for systems composed of identical bosons, with particles accumulating in the lowest energy individual states; by the Pauli exclusion rule this effect is prohibited in systems of identical fermions.

All physical predictions for quantum objects are expressed in terms of the probability amplitudes (see Postulate I), which represent the scalar products of two state vectors, or matrix elements of an operator. The symmetrization requirement of Postulate III causes special *interference effects* between the so-called "direct" and "exchange" processes, to appear in the conditional probabilities (see Postulate II) of specific outcomes of experiments performed on systems of identical particles. The formal postulates related to measurement processes, single or repeated, performed on quantum systems are the subject of the next section.

3.3 Results of Physical Measurements

In this section, we shall further elaborate on the physical implications of the mathematical concepts of the quantum mechanical description, which has been introduced in the preceding chapter, by specifically addressing the link between this abstract formalism and the results of measurements. As in previous sections we shall focus on the position and momentum wave functions and the associated operators representing the physical properties of the microsystems. In what follows both the results of a single experiment and the average values of a large number of repetitions of the same experiment performed on systems in the same initial quantum state will be tackled by the corresponding postulates of quantum mechanics.

3.3.1 Classical Observables in Position and Momentum Representations

As we have already remarked in Sect. 2.5, each physical quantity A is represented in quantum mechanics by its linear and Hermitian operator $\hat{\mathrm{A}}$, the eigenvalue problem of which plays the fundamental role in predicting the outcomes of physical measurements. This correspondence is formalized in terms of the following axiom:

Postulate IV.1: To every mechanical quantity A there corresponds in quantum mechanics the associated operator $\hat{\mathrm{A}}$ called an observable. It has to be linear, to satisfy the requirements of the Superposition Principle (Postulate II), and Hermitian (self-adjoint), for its eigenvalues to be real. Their eigenvectors form the bases in the vector space of all quantum states of the physical system.

The prescription for constructing the position/momentum representations of the quantum mechanical observables are known as the *Jordan rules*. Consider the classical quantities, which can be expressed as functions of the particle positions and momenta, $A = A(\{\boldsymbol{r}_k\},\{\boldsymbol{p}_k\})$, or equivalently in terms of the conjugated Cartesian coordinates, $A = A(\{x_\alpha\}, \{p_\beta\})$. The Jordan rules summarize the results of Sect. 2.6 by attributing to such functions the corresponding functions of the position and momentum operators:

$$\hat{\mathrm{A}} = A(\{\hat{\mathbf{r}}_k\}, \{\hat{\mathbf{p}}_k\}) \quad \text{or} \quad \hat{\mathrm{A}} = A(\{\hat{\mathrm{x}}_\alpha\}, \{\hat{\mathrm{p}}_\beta\}). \tag{3.32}$$

In the *position representation* $\{\boldsymbol{r}_k\} = \{x_\alpha\}$, the coordinate operator $\hat{x}_\alpha$ denotes the multiplication by x_α, $\hat{x}_\alpha = x_\alpha$. Similarly for any function of the particle coordinates, e.g., the position vector $\boldsymbol{r} = x\boldsymbol{i} + y\boldsymbol{j} + z\boldsymbol{k}$ or the potential energy $V(\{x_\alpha\})$, there corresponds the associated multiplicative operators:

$$\hat{\mathbf{r}}(\boldsymbol{r}) = \hat{x}\,\boldsymbol{i} + \hat{y}\,\boldsymbol{j} + \hat{z}\,\boldsymbol{k} = x\boldsymbol{i} + y\boldsymbol{j} + z\boldsymbol{k} = \boldsymbol{r},$$
$$\hat{V}(\{x_\alpha\}) = V(\{\hat{x}_\alpha\}) = V(\{x_\alpha\}), \text{etc.} \tag{3.33}$$

The elementary momentum operators in this representation,

$$\hat{\mathbf{p}}_k(\boldsymbol{r}_k) = -i\hbar\nabla_{\boldsymbol{r}_k} \equiv -i\hbar\nabla_k, \quad \hat{p}_\alpha(x_\alpha) = -i\hbar\partial/\partial x_\alpha, \tag{3.34}$$

similarly determine the quantum mechanical operator of any function of the particle momenta, e.g., the kinetic energy $T = T(\{\boldsymbol{p}_k\}) = \sum_k \boldsymbol{p}_k^2/2m_k$:

$$\hat{T}(\{\boldsymbol{r}_k\}) = T(\{\hat{\mathbf{p}}_k(\boldsymbol{r}_k)\}) = \Sigma_k \hat{\mathbf{p}}_k^2(\boldsymbol{r}_k)/2m_k = -\sum_k \frac{\hbar^2}{2m_k}\Delta_k. \tag{3.35}$$

These rules are sufficient to generate the quantum operator in the position representation for any physical quantity encountered in the classical mechanics, e.g., that of the orbital angular momentum of a single particle:

$$\boldsymbol{l} = \boldsymbol{r} \times \boldsymbol{p} = \begin{vmatrix} \boldsymbol{i} & \boldsymbol{j} & \boldsymbol{k} \\ x & y & z \\ p_x & p_y & p_z \end{vmatrix} \rightarrow \hat{\mathbf{l}}(\{x_\alpha\}) = \boldsymbol{i}\hat{l}_x + \boldsymbol{j}\hat{l}_y + \boldsymbol{k}\hat{l}_z$$
$$= -i\hbar \begin{vmatrix} \boldsymbol{i} & \boldsymbol{j} & \boldsymbol{k} \\ x & y & z \\ \partial/\partial x & \partial/\partial y & \partial/\partial z \end{vmatrix}, \tag{3.36}$$

or the operator attributed to the system Hamilton function $H(\{\hat{\mathbf{r}}_k\}, \{\hat{\mathbf{p}}_k(\boldsymbol{r}_k)\}) = E(\{\hat{\mathbf{r}}_k\}, \{\hat{\mathbf{p}}_k(\boldsymbol{r}_k)\}) = \hat{H}(\{x_\alpha\})$, the system Hamiltonian in the position representation:

$$\hat{H}(\{x_\alpha\}) = \hat{T}(\{x_\alpha\}) + \hat{V}(\{x_\alpha\}) = -\sum_k \frac{\hbar^2}{2m_k}\Delta_k + V(\{x_\alpha\}), \tag{3.37}$$

where the Laplacian $\Delta_k = \nabla_k^2 = \partial^2/\partial x_k^2 + \partial^2/\partial y_k^2 + \partial^2/\partial z_k^2$.

These rules can be straightforwardly transcribed into the associated prescriptions for the *momentum representation* $\{\boldsymbol{p}_k\} = \{p_\alpha\}$, in which (see again Sect. 2.6) $\hat{p}_\alpha = p_\alpha$, or $\hat{\mathbf{p}}(\boldsymbol{p}) = \hat{p}_x\boldsymbol{i} + \hat{p}_y\boldsymbol{j} + \hat{p}_z\boldsymbol{k} = p_x\boldsymbol{i} + p_y\boldsymbol{j} + p_z\boldsymbol{k} = \boldsymbol{p}$, and

$$\hat{x}_\alpha(p_\alpha) = i\hbar\partial/\partial p_\alpha \quad \text{or} \quad \hat{\mathbf{r}}_k(\boldsymbol{p}_k) = i\hbar\nabla_{\boldsymbol{p}_k}. \tag{3.38}$$

Therefore, in this representation the kinetic energy corresponds to the multiplicative operator $\hat{T}(\{\boldsymbol{p}_k\}) = T(\{\boldsymbol{p}_k\}) = \sum_k \boldsymbol{p}_k^{\,2}/2m_k$, while the potential energy function generates the associated differential operator:

$$\hat{V}(\{\hat{x}_\alpha(p_\alpha)\}) = V(\{i\hbar\partial/\partial p_\alpha\}). \tag{3.39}$$

3.3.2 *Possible Outcomes of a Single Measurement*

In accordance with the discussion in Sect. 2.5 the possible outcomes of individual measurements of the physical quantity A are related to its quantum mechanical operator $\hat{A}$ via the

Postulate IV.2: The result of a single measurement of the physical quantity A is one of the eigenvalues $\{a_i\}$ of its observable $\hat{A}$ in the eigenvalue problem (2.55a). In position/momentum representations, it reads:

$$\hat{A}(\{x_\alpha\})\Psi_i(\{x_\alpha\}) = a_i\Psi_i(\{x_\alpha\}), \quad \hat{A}(\{p_\alpha\})\Psi_i(\{p_\alpha\}) = a_i\Psi_i(\{p_\alpha\}), \tag{3.40}$$

where $\Psi_i(\{x_\alpha\})$ and $\Psi_i(\{p_\alpha\})$ denote the corresponding eigenfunctions associated with the eigenvalue a_i.

Since the set of eigenvectors $\{|\Psi_i\rangle\}$ of the quantum mechanical observable $\hat{A}$ forms the complete basis in the system vector space (see Sect. 2.5), any state $|\Psi\rangle$ can be expressed as their combination, with the squares of the moduli of the expansion coefficients determining the conditional probabilities of observing $|\Psi_i\rangle$ in state $|\Psi\rangle$ (Postulate II):

$$|\Psi\rangle = \Sigma_i|\Psi_i\rangle\langle\Psi_i|\Psi\rangle = \Sigma_i|\Psi_i\rangle C_i, \quad P(\Psi_i|\Psi) = |C_i|^2 \le 1. \tag{3.41}$$

The $P(\Psi_i|\Psi) = 1$, and hence $\{P(\Psi_{j\neq i}|\Psi) = 0\}$, marks the eigenvector itself, $|\Psi\rangle = |\Psi_i\rangle$, when we know with certainty that the eigenvalue a_i (nondegenerate) has been observed. Therefore, a general combination of the preceding equation is reduced after the measurement of A to a single eigenvector of $\hat{A}$, the one corresponding to the observed eigenvalue. This "contraction" of $|\Psi\rangle$ into $|\Psi_i\rangle$ marks the irreversible intervention of the measuring device. Indeed, as we have emphasized in Chap. 1, any experiment performed on the microobject inadvertently modifies its state.

This contraction of a combination of eigenstates into its single member has to be modified in the case of the degenerate eigenvalue a_i of the physical quantity A. Such a result of the experiment implies that the state immediately after the measurement is now the normalized projection $\left(\langle\Psi|\hat{P}(a_i)|\Psi\rangle\right)^{-1/2}\hat{P}(a_i)|\Psi\rangle$ of the initial state $|\Psi\rangle$ into the eigensubspace associated with a_i, $\{|\Psi_{i,j}\rangle = |i_j\rangle, j = 1, 2, g\}$, which is effected by the subspace projector $\hat{P}(a_i) = \sum_{j=1}^{g}|i_j\rangle\langle i_j|$.

In quantum mechanics the set of eigenvalues $\{a_i\}$ thus determines the spectrum of all possible outcomes of the single measurement of A. Since the eigenvalues of the square of the observable $\hat{A}^2$ are given by the squares of eigenvalues of $\hat{A}$, for the same set of eigenstates,

$$\hat{A}^2|\Psi_i\rangle = a_i\hat{A}|\Psi_i\rangle = a_i^2|\Psi_i\rangle, \tag{3.42}$$

the square of the dispersion in A, $\sigma_A^2 = \langle A^2\rangle - \langle A\rangle^2$ [see (3.23)], observed in the repeated measurements of A in the eigenstate $|\Psi_i\rangle$, represented by the eigenfunctions $\Psi_i(\{x_\alpha\})$ or $\Psi_i(\{p_\alpha\})$, identically vanishes:

$$\sigma_A^2 = \langle A^2\rangle_i - \langle A\rangle_i^2 = \langle\Psi_i|\hat{A}^2|\Psi_i\rangle - \langle\Psi_i|\hat{A}|\Psi_i\rangle^2 = a_i^2 - a_i^2 = 0. \tag{3.43}$$

Therefore, in the eigenstate of $\hat{A}$ the physical quantity A is sharply specified, and each single measurement of this physical property in this state always gives the same result a_i, as reflected by the conditional probabilities: $P(\Psi_i|\Psi_i) = 1$ and $P(\Psi_{j\neq i}|\Psi_i) = 0$.

As we have already demonstrated in (2.60), the eigenstates corresponding to different eigenvalues are automatically orthogonal. However, for the degenerate eigenvalues, several eigenstates correspond to the same eigenvalue (2.58), so they have to be orthogonalized to safeguard their linear independence. This orthogonalization is performed by taking appropriate linear combinations of generally nonorthogonal state vectors, which satisfy the conditions of their mutual orthogonality.

As schematically shown in Fig. 3.1, the prescription to make any pair of degenerate state vectors to be mutually "perpendicular" is not unique. Thus, the specific orthogonalization scheme can be selected for reasons of convenience. For the sake of simplicity consider two normalized state vectors $|\boldsymbol{a}_i\rangle = \{|i_1\rangle, |i_2\rangle\}$ of the *doubly* degenerate eigenvalue a_i, $g = 2$, which define the *overlap* matrix of their scalar products:

$$\mathbf{S} = \langle\boldsymbol{a}_i|\boldsymbol{a}_i\rangle = \{\langle i_j|i_{j'}\rangle; j, j' = 1, 2\} \equiv \begin{bmatrix} 1 & S \\ S & 1 \end{bmatrix}, \tag{3.44}$$

where for definiteness we assume $S > 0$. In the (nonsymmetric) *Schmidt orthogonalization* scheme (see Fig. 3.1) of transforming the original vectors $\{|i_1\rangle, |i_2\rangle\}$ to the mutually orthogonal set $|\bar{\boldsymbol{a}}_i\rangle = \{|\bar{i}_1\rangle, |\bar{i}_2\rangle\}$, one leaves one of these vectors unchanged, say $|\bar{i}_1\rangle = |i_1\rangle$, and "rotates" the other, $|\bar{i}_2\rangle = \mathcal{N}_2(|i_2\rangle + C|i_1\rangle)$, where $\mathcal{N}_2$ is the normalization constant and C denotes the mixing coefficient, until the two vectors become mutually orthogonal: $\langle i_1|\bar{i}_2\rangle = 0$. This condition then gives $C = -S$, while the normalization $\langle\bar{i}_2|\bar{i}_2\rangle = 1$ implies $\mathcal{N}_2 = (1 - S^2)^{-1/2}$, and hence

$$|\bar{i}_2\rangle = (1 - S^2)^{-1/2}(|i_2\rangle - |i_1\rangle S) = \mathcal{N}_2(|i_2\rangle - |i_1\rangle\langle i_1|i_2\rangle) = \mathcal{N}_2(|i_2\rangle - \hat{P}_i|i_2\rangle). \tag{3.45a}$$

This expression can be straightforwardly extended to a general case of the normalized state vector $|\psi\rangle$ Schmidt orthogonalized with respect to the given subspace $|\boldsymbol{\varphi}\rangle = (|\varphi_1\rangle, |\varphi_2\rangle, \ldots, |\varphi_r\rangle)$ of the orthonormal states:

$$|\bar{\psi}\rangle = \mathcal{N}[|\psi\rangle - \sum_{i=1}^{r} |\varphi_i\rangle\langle\varphi_i \mid \psi\rangle] = \mathcal{N}[|\psi\rangle - \hat{\mathrm{P}}_{\varphi}|\psi\rangle]. \tag{3.45b}$$

Alternatively, as also shown in the figure, one could manipulate the two nonorthogonal vectors simultaneously in a symmetrical way, so that both orthogonalized vectors $|\tilde{\boldsymbol{a}}_i\rangle = \{|\tilde{i}_1\rangle, |\tilde{i}_2\rangle\}$, strongly resemble their initial, nonorthogonal analogs. In the *Löwdin orthogonalization* scheme, this transformation is effected through the symmetric matrix $\mathbf{S}^{-1/2}$, $|\tilde{\boldsymbol{a}}_i\rangle = |\boldsymbol{a}_i\rangle\mathbf{S}^{-1/2}$, defined by the eigenvalue problem of the overlap matrix $\mathbf{S}$, i.e., its diagonalization in the orthogonal transformation:

$$\mathbf{O}^{\mathrm{T}}\mathbf{S}\mathbf{O} = \mathbf{s} = \{s_m\delta_{m,n}\},\ \mathbf{S}^{-1/2} = \mathbf{O}\mathbf{s}^{-1/2}\mathbf{O}^{\mathrm{T}},\ \mathbf{s}^k = \{(s_m)^k\delta_{m,n}\},\ \mathbf{O}\mathbf{O}^{\mathrm{T}} = \mathbf{I}. \tag{3.46}$$

Indeed, the orthogonality of the symmetrically rotated vectors $|\tilde{\boldsymbol{a}}_i\rangle$ then directly follows from the orthogonal transformation $\mathbf{O}$ which diagonalizes the overlap matrix:

$$\langle\tilde{\boldsymbol{a}}_i|\tilde{\boldsymbol{a}}_i\rangle = \mathbf{S}^{-1/2}\langle\boldsymbol{a}_i|\boldsymbol{a}_i\rangle\,\mathbf{S}^{-1/2} = \mathbf{S}^{-1/2}\mathbf{S}\,\mathbf{S}^{-1/2} = \mathbf{S}^0 = \mathbf{I}. \tag{3.47}$$

These matrix equations apply to any number of the orthogonalized vectors or wave functions. In the latter case, the overlap matrix is defined to be the corresponding integrals between nonorthogonal functions, e.g., $\boldsymbol{\chi}(\boldsymbol{r}) = \{\chi_t(\boldsymbol{r})\}$ (row vector), when $\mathbf{S}_{\chi} = \langle\boldsymbol{\chi}|\boldsymbol{\chi}\rangle = \{S_{r,t} = \int\chi_r^*(\boldsymbol{r})\chi_t(\boldsymbol{r})\,d\boldsymbol{r}\}$: $\tilde{\boldsymbol{\chi}}(\boldsymbol{r}) = \boldsymbol{\chi}(\boldsymbol{r})\mathbf{S}_{\chi}^{-1/2}$.

The specific forms of these matrices for the metric of (3.44) read:

$$\begin{aligned}
\mathbf{s} &= \begin{bmatrix} 1+S & 0 \\ 0 & 1-S \end{bmatrix} = \begin{bmatrix} s_1 & 0 \\ 0 & s_2 \end{bmatrix}, \quad \mathbf{O} = \frac{1}{\sqrt{2}}\begin{bmatrix} 1 & 1 \\ 1 & -1 \end{bmatrix}, \\
\mathbf{S}^{-1/2} &= \begin{bmatrix} a & b \\ b & a \end{bmatrix}, \quad a = \frac{1}{2}\left(\frac{1}{\sqrt{s_1}} + \frac{1}{\sqrt{s_2}}\right), \quad b = \frac{1}{2}\left(\frac{1}{\sqrt{s_1}} - \frac{1}{\sqrt{s_2}}\right).
\end{aligned} \tag{3.48}$$

3.3.3 Expectation Value of Repeated Measurements and Heisenberg Uncertainty Principle

The average result of the repeated measurements of A in quantum mechanics, performed on the system in the same initial quantum state $|\Psi\rangle$, has already

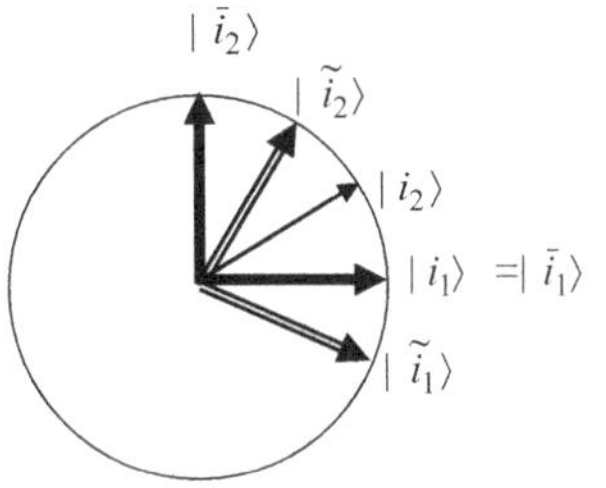

Fig. 3.1 The diagrammatic representation of the mutually nonorthogonal state vectors $\{|i_1\rangle, |i_2\rangle\}$, and the two sets of their orthogonalized (mutually "perpendicular") analogs: the Schmidt (nonsymmetrically) orthogonalized vectors $|\bar{\boldsymbol{a}}_i\rangle = \{|\bar{i}_1\rangle, |\bar{i}_2\rangle\}$, and the Löwdin (symmetrically) orthogonalized vectors $|\tilde{\boldsymbol{a}}_i\rangle = \{|\tilde{i}_1\rangle, |\tilde{i}_2\rangle\}$. The two sets are related by the unitary (rotation) transformation $\mathbf{U} = \langle \bar{\boldsymbol{a}}_i | \tilde{\boldsymbol{a}}_i \rangle$: $|\tilde{\boldsymbol{a}}_i\rangle = |\bar{\boldsymbol{a}}_i\rangle \mathbf{U}$

been established in (3.23). It can be formally stated in the form of a separate postulate:

Postulate IV.3: The statistically average result of a very large number $m \to \infty$ of repeated measurements of the physical quantity A performed on the microsystem in the same initial state $|\Psi\rangle$ is given by the expectation value of its quantum mechanical operator $\hat{\mathrm{A}}$:

$$\begin{aligned}\langle A \rangle_\Psi &= \sum_i P(\Psi_i|\Psi)\, a_i = \langle \Psi|\hat{\mathrm{A}}|\Psi\rangle = \int \Psi^*(\mathcal{Q}^N;t)\, \hat{\mathrm{A}}(\mathcal{Q}^N)\Psi(\mathcal{Q}^N;t)\; d\mathcal{Q}^N \\ &= \int \Psi^*(\mathcal{P}^N;t)\, \hat{\mathrm{A}}(\mathcal{P}^N)\Psi(\mathcal{P}^N;t) d\mathcal{P}^N.\end{aligned} \tag{3.49}$$

It has been demonstrated in (3.43) that in the eigenstate $|\Psi_i\rangle$ this quantity is sharply specified with $\langle A \rangle_{\Psi_i} = a_i$, $\langle A^2 \rangle_{\Psi_i} = a_i^2$, etc. The same conclusion applies to all physical observables which commute with $\hat{\mathrm{A}}$, since all these operators have a common set of eigenvectors (see Sect. 2.5). However, in a general quantum state of (3.41), one will detect a dispersion in the measured values of A, with a statistically distributed results $\{a_i\}$ appearing with frequencies $\{m_i = mP(\Psi_i|\Psi)\}$ proportional to the conditional probabilities $\{P(\Psi_i|\Psi)\}$ of observing the specified eigenstates (see caption of Fig. 1.1).

We are now in a position to provide a general formulation of the Heisenberg Principle of Indeterminacy in quantum mechanics (see Chap. 1). As specific measures of the simultaneous accuracies of the physical quantities A and B we adopt their dispersions (standard deviations), $\sigma_X = \langle (X - \langle X \rangle)^2 \rangle^{1/2} = (\langle X^2 \rangle - \langle X \rangle^2)^{1/2}$, $X = A, B$, with the corresponding expressions in terms of the quantum mechanical expectation values:

$$\sigma_X{}^2 = \langle X^2 \rangle_\Psi - \langle X \rangle_\Psi^2 = \langle \Psi|(\hat{\mathrm{X}} - \langle X \rangle_\Psi)^2|\Psi\rangle \equiv \langle \Psi|\hat{\Delta}_X^2|\Psi\rangle, \quad X = A, B. \tag{3.50}$$

We further observe that the displacement operators $\hat{\Delta}_A$ and $\hat{\Delta}_B$ are both Hermitian, as are the observables $\hat{A}$ and $\hat{B}$ themselves, and the following commutator identity is satisfied:

$$[\hat{A}, \hat{B}] = [\hat{\Delta}_A, \hat{\Delta}_B], \tag{3.51}$$

since the average values $\langle X \rangle_\Psi$ (numbers) commute with every operator [see (2.34)].

We shall now demonstrate that the following inequality is satisfied by the simultaneous indeterminacies of the physical quantities A and B:

$$\sigma_A{}^2 \sigma_B{}^2 \geq -\frac{1}{4} \langle \Psi | [\hat{A}, \hat{B}] | \Psi \rangle^2. \tag{3.52}$$

It constitutes the quantum mechanical formulation of the Heisenberg Uncertainty Principle, which indeed predicts the simultaneous sharp specification of the commuting observables.

In order to prove this inequality let us introduce the physically meaningful, i.e., exhibiting a finite norm, auxiliary state vector $|\Phi(\lambda)\rangle$ depending on real parameter λ:

$$|\Phi(\lambda)\rangle = (\lambda \hat{\Delta}_A - i \hat{\Delta}_B) |\Psi\rangle. \tag{3.53}$$

The square of its norm (positive) then determines the quadratic function $f(\lambda)$:

$$\begin{aligned} \langle \Phi(\lambda) | \Phi(\lambda) \rangle = \| \Phi(\lambda) \|^2 &= \langle \Psi | (\lambda \hat{\Delta}_A - i \hat{\Delta}_B)^\dagger (\lambda \hat{\Delta}_A - i \hat{\Delta}_B) | \Psi \rangle \\ &= \langle \Psi | (\lambda \hat{\Delta}_A + i \hat{\Delta}_B)(\lambda \hat{\Delta}_A - i \hat{\Delta}_B) | \Psi \rangle \\ &= \langle \Psi | (\lambda^2 \hat{\Delta}_A^2 - i\lambda [\hat{\Delta}_A, \hat{\Delta}_B] + \hat{\Delta}_B^2) | \Psi \rangle \\ &= \sigma_A{}^2 \lambda^2 - i \langle \Psi | [\hat{A}, \hat{B}] | \Psi \rangle \lambda + \sigma_B{}^2 \\ &\equiv a\lambda^2 + b\lambda + c > 0. \end{aligned} \tag{3.54}$$

For $a = \sigma_A{}^2 > 0$ this inequality can be satisfied only when there are no solutions of the associated quadratic equation $a\lambda^2 + b\lambda + c = 0$, i.e., when $\Delta = b^2 - 4\,ac < 0$ or

$$-\langle \Psi | [\hat{A}, \hat{B}] | \Psi \rangle^2 < 4\, \sigma_A^2 \sigma_B^2, \tag{3.55}$$

which completes the proof.

Consider the illustrative example of the position–momentum relation (1.7). In position representation (3.34), $\hat{A} = x, \hat{B} = -i\hbar \partial / \partial x$, so that their commutator acting on the continuous function $f(x)$ gives:

$$[\hat{A}, \hat{B}] f = -i\hbar x (\partial f / \partial x) + i\hbar x (\partial f / \partial x) + i\hbar f = i\hbar f \quad \text{or} \quad [\hat{x}, \hat{p}_x] = i\hbar. \tag{3.56}$$

Thus, these two physical quantities are incompatible, with the limit of the product of their lowest (simultaneous) inaccuracies being determined by (3.52):

$$\sigma_x \sigma_{p_x} > \frac{1}{2}\hbar. \tag{3.57}$$

These predictions agree with the constant (position-independent) probability of finding a particle at the specified location in space in the state described by the basis function $u_p(\boldsymbol{r})$ (2.76), corresponding to the sharply specified momentum: $\sigma_p \to 0$. It indeed implies that all localization events are then equally probable, i.e., we are then completely ignorant about the particle position: $\sigma_r \to \infty$. In accordance with the Heisenberg principle of (3.57) only the infinite position indeterminacy gives the finite product when multiplied by the infinitesimal momentum uncertainty $\sigma_p \to 0$.

3.3.4 Ensemble Averages in Mixed States

Only certain idealized systems, isolated from their environment, are completely described by a single state vector $|\Psi\rangle$ or a single wave function $\Psi(x)$. The wave function of an isolated system depends only on its internal coordinates x and carries the maximum information about the state of the microsystem available in quantum physics. The full specification of quantum state of the microobject is through the state vector belonging to the basis set of the simultaneous eigenvectors of the system complete set of the mutually commuting observables $\{\hat{\mathrm{A}}, \hat{\mathrm{B}}, \ldots\}$, which diagonalize the matrix representations of these operators, $|\Psi\rangle \in \{|\Psi_n\rangle = |a_k, b_l, \ldots\rangle\}$. Their eigenvalues $(a_k, b_l, \ldots)$ then provide the complete identification of the direction of the state vector $|a_k, b_l, \ldots\rangle$ in the molecular Hilbert space.

However, microobjects can be coupled to their surroundings. For example, the particles at constant temperature are in contact with the thermostat (heat "bath") and the open systems, exhibiting fluctuating (fractional, continuously changing) number of particles, are coupled to the external particle "reservoir(s)." The state of the closed system *interacting* with its environment will also depend on the external degrees of freedom describing the latter. Therefore, the formalism of quantum mechanics must also admit all intermediate stages of an imprecise definition of the system state, which cannot be linked to a single state vector (wave function). Such generalized states are called the *mixed states*, while the systems with the specified wave function are said to be in the *pure state*.

As in statistical mechanics, the incomplete information about the system calls for the concept of an *ensemble* of quantum states, in which the admissible pure states appear with some probability. The ensemble consists of a very large number of replicas of the same system. For example, a system in the thermodynamic equilibrium at temperature T has a probability of being in its energy eigenstate $|E_n\rangle$ proportional to $\exp(-E_n/k_BT)$, where k_B is the Boltzmann constant. This probability describes the frequency of such a state among members of the *canonical* ensemble. Similarly, the systems in the *grand-canonical* ensemble describing the system in thermodynamic equilibrium with the heat bath at temperature T and the particle reservoir characterized by the chemical potential μ will exhibit the probability proportional to $\exp(\mu N_i - E_{n,i})/k_B T$ of observing the eigenvalue $E_{n,i}$ of the Hamitonian $\hat{\mathrm{H}}(N_i)$, for the specified (integral) number of electrons N_i.

Therefore, such an imprecise definition of the quantum mechanical state can be interpreted as the statistical mixture of the admissible states $\{|\psi_1\rangle, |\psi_2\rangle, \ldots\}$ of the system replicas in the ensemble, which appear with the associated (*external*) probabilities $\{p_1, p_2, \ldots\}$, $\sum_\alpha p_\alpha = 1$. The individual states in the mixture do not have to be orthogonal, e.g., in the grand ensemble, when we mix eigenstates of different Hamiltonians, but they are always assumed to be normalized.

The statistical mixture should not be confused with the expansion of a single wave function $|\Phi\rangle$ in the (orthonormal) basis set, say $\{|\Psi_n\rangle\}$,

$$|\Phi\rangle = \sum_n |\Psi_n\rangle\langle\Psi_n|\Phi\rangle = \sum_n |\Psi_n\rangle c_n, \tag{3.58}$$

where $|c_n|^2$ generates the conditional probability $P(\Psi_n|\Phi)$ of observing in state $|\Phi\rangle$ the physical attributes of $|\Psi_n\rangle$. Indeed, this does not imply that $|\Phi\rangle$ is the mixture of $|\Psi_1\rangle$ with the probability $P(\Psi_1|\Phi)$, and $|\Psi_2\rangle$ with the probability $P(\Psi_2|\Phi)$, etc. The square of the modulus of $\Phi(x)$, which generates the probability distribution $\rho(x) = \Phi^*(x)\Phi(x)$, then includes the crucial *interference* terms between different basis functions, $c_n^* c_m \Psi_n^*(x)\Psi_m(x)$, which are not present in the statistical mixture of the same basis functions. Thus, the probability weighted sum of distributions $\{\rho_n(x) = \Psi_n^*(x)\Psi_n(x)\}$, generated by each state in the basis set, $\rho_{ens.}(x) = \sum_n p_n \rho_n(x)$, cannot reproduce the true probability density $\rho(x)$. In other words, it is not possible to describe a statistical mixture by an "average" state vector in the form of the combination of states of (3.58): $\rho_{ens.}(x) \neq \rho(x)$.

The two levels of probabilities are thus involved in determining the results of measurements performed on systems in their mixed quantum states. On one hand, there is the *intrinsic* quantum mechanical probability of finding in each (pure) state $|\psi_\alpha\rangle$ a specific eigenvalue a_k of the observable $\hat{\mathrm{A}}$, $\hat{\mathrm{A}}|\Psi_k\rangle = a_k|\Psi_k\rangle$, given by the square of the modulus of the expansion coefficient $C_{k,\alpha} = \langle\Psi_k|\psi_\alpha\rangle$, $P_{k,\alpha} = |C_{k\alpha}|^2$ (Postulate II), which determines the quantum mechanical *expectation value*

$$\langle A\rangle_\alpha = \langle\psi_\alpha|\hat{\mathrm{A}}|\psi_\alpha\rangle = \sum_k a_k P_{k,\alpha}. \tag{3.59}$$

Notice that these eigenstates generate the diagonal representation of $\hat{\mathrm{A}}$, $\mathbf{A}^{(\Psi)} = \{A_{m,n} = \langle\Psi_m|\hat{\mathrm{A}}|\Psi_n\rangle = a_n\delta_{m,n}\}$.

On the other hand, the additional level of the *external* probabilities $\{p_\alpha\}$ of observing the individual states $\{|\psi_\alpha\rangle\}$ in the ensemble intervenes in the mixed quantum mechanical states. They define the associated *density operator* given by the sum of the externally weighted projections onto the quantum states being mixed,

$$\hat{D} = \sum_{\alpha} |\psi_{\alpha}\rangle p_{\alpha} \langle \psi_{\alpha}| = \sum_{\alpha} p_{\alpha} \hat{P}_{\alpha}, \tag{3.60a}$$

Its matrix representation in the basis set of eigenstates of $\hat{A}$,

$$\mathbf{D}^{(\boldsymbol{\Psi})} = \{D_{m,n} = \langle \Psi_m|\hat{D}|\Psi_n\rangle = \sum_{\alpha} \langle \Psi_n|\psi_{\alpha}\rangle p_{\alpha} \langle \psi_{\alpha}|\Psi_m\rangle\}, \tag{3.60b}$$

determines the *ensemble* average value of A:

$$\begin{aligned}
A_{ens.} &= \sum_{\alpha} p_{\alpha} \langle A \rangle_{\alpha} = \sum_{\alpha} p_{\alpha} \{\sum_{k} a_k P_{k,\alpha}\} \\
&= \sum_{n}\sum_{m} \left\{ \sum_{\alpha} \langle \Psi_n|\psi_{\alpha}\rangle p_{\alpha} \langle \psi_{\alpha}|\Psi_m\rangle \right\} \langle \Psi_m|\hat{A}|\Psi_n\rangle \\
&= \sum_{n}\sum_{m} \langle \Psi_n|\hat{D}|\Psi_m\rangle \langle \Psi_m|\hat{A}|\Psi_n\rangle \equiv \sum_{n}\sum_{m} D_{n,m} A_{m,n} \\
&\equiv \mathrm{tr}[\mathbf{D}^{(\boldsymbol{\Psi})}\mathbf{A}^{(\boldsymbol{\Psi})}] = \sum_{n} \langle \Psi_n|\hat{D}\,\hat{A}|\Psi_n\rangle \equiv \mathrm{tr}(\hat{D}\,\hat{A}) \\
&= \mathrm{tr}[\mathbf{A}^{(\boldsymbol{\Psi})}\mathbf{D}^{(\boldsymbol{\Psi})}] = \sum_{m} \langle \Psi_m|\hat{A}\,\hat{D}|\Psi_m\rangle = \mathrm{tr}(\hat{A}\,\hat{D}).
\end{aligned} \tag{3.60c}$$

The Hermitian (nonidempotent!) density operator $\hat{D}$ involves the probability weighted projections $\{\{\hat{P}_{\alpha}\}\}$ onto the individual states being mixed, while the *trace* operation (tr) denotes the summation of all diagonal elements of the matrix representations of operators in the adopted basis. It also follows from the definition of $\hat{D}$ that its expectation value in state $|\Phi\rangle$

$$\langle \Phi|\hat{D}|\Phi\rangle = \sum_{\alpha} p_{\alpha} \langle \psi_{\alpha}|\Phi\rangle \langle \Phi|\psi_{\alpha}\rangle = \sum_{\alpha} p_{\alpha}\, P(\psi_{\alpha}|\Phi) \geq 0, \tag{3.61}$$

and hence $\hat{D}$ is a positive operator.

It can be also verified that the trace of the product of operators is invariant with respect to the cyclic permutations of factors in the product [see (3.60c)],

$$\mathrm{tr}\,(\mathbf{AB}\ldots\mathbf{CD}) = \mathrm{tr}(\mathbf{DAB}\ldots\mathbf{C}),\ \text{etc.}, \tag{3.62}$$

and to a change $\boldsymbol{\Psi} \to \boldsymbol{\Phi}$ in the (orthonormal) basis set:

$$\begin{aligned}
\mathrm{tr}\,\hat{A} &= \sum_{n} \langle \Psi_n|\hat{A}|\Psi_n\rangle \equiv \mathrm{tr}\mathbf{A}_{\boldsymbol{\Psi}} \\
&= \sum_{n}\sum_{m}\sum_{m'} \langle \Psi_n|\Phi_m\rangle \langle \Phi_m|\hat{A}|\Phi_{m'}\rangle \langle \Phi_{m'}|\Psi_n\rangle \\
&= \sum_{n}\sum_{m}\sum_{m'} \langle \Phi_{m'}|\Psi_n\rangle \langle \Psi_n|\Phi_m\rangle \langle \Phi_m|\hat{A}|\Phi_{m'}\rangle \\
&= \sum_{m}\sum_{m'} \langle \Phi_{m'}|\Phi_m\rangle \langle \Phi_m|\hat{A}|\Phi_m\rangle \\
&= \sum_{m} \langle \Phi_m|\hat{A}|\Phi_m\rangle \equiv \mathrm{tr}\,\mathbf{A}_{\boldsymbol{\Phi}},
\end{aligned} \tag{3.63}$$

where we have used the closure relations $\sum_m |\Phi_m\rangle\langle\Phi_m| = \sum_n |\Psi_n\rangle\langle\Psi_n| = 1$ and the orthonormality of basis functions $\langle\Phi_{m'}|\Phi_m\rangle = \delta_{m,m'}$. One also observes that

$$\begin{aligned}\mathrm{tr}\,\hat{\mathrm{D}} &= \sum_n \langle\Psi_n|\hat{\mathrm{D}}|\Psi_n\rangle = \sum_\alpha\sum_n p_\alpha\langle\psi_\alpha|\Psi_n\rangle\langle\Psi_n|\psi_\alpha\rangle \\ &= \sum_\alpha p_\alpha\langle\psi_\alpha|\psi_\alpha\rangle = \sum_\alpha p_\alpha = 1.\end{aligned} \tag{3.64}$$

Obviously, the pure state, e.g., $|\psi_\alpha\rangle$, can be viewed as the limiting case of the ensemble, when $p_\alpha = 1$ and $\{p_{\beta\neq\alpha} = 0\}$, so that $\hat{\mathrm{D}} = \hat{\mathrm{P}}_\alpha$. Only in the pure quantum state the density operator is idempotent, $\hat{\mathrm{D}}^2 = \hat{\mathrm{D}}$ (idempotency of $\hat{\mathrm{P}}_\alpha$), so that $\mathrm{tr}\hat{\mathrm{D}}^2 = \mathrm{tr}\hat{\mathrm{D}} = 1$. The corresponding inequality for the mixed state reads: $\mathrm{tr}\hat{\mathrm{D}}^2 < 1$.

When describing parts of a physical system the concept of the *partial trace* emerges. Assume that the global system, (1) + (2), consists of distinct subsystems (1) and (2), described by their associated Hilbert spaces $\mathcal{H}(1) = \{|\Psi_i(1)\rangle\}$ and $\mathcal{H}(2) = \{|\Phi_m(2)\rangle\}$, the tensor product of which spans the Hilbert space of the system as a whole:

$$\mathcal{H}(1,2) = \{|\Psi_i(1)\rangle|\Phi_m(2)\rangle \equiv |\Psi_i(1)\Phi_m(2)\rangle\} = \mathcal{H}(1)\otimes\mathcal{H}(2). \tag{3.65}$$

We now introduce the partial traces of the system density operator $\hat{\mathrm{D}}$, which define the effective density operators for each subsystem: $\hat{\mathrm{D}}(1)$ and $\hat{\mathrm{D}}(2)$. This is effected by contractions of the matrix representation of $\hat{\mathrm{D}}$ in $\mathcal{H}(1, 2)$,

$$\mathbf{D}(1,2) = \{\langle\Psi_i(1)\Phi_m(2)|\hat{\mathrm{D}}|\Psi_{i'}(1)\Phi_{m'}(2)\rangle \equiv D_{i,m;i',m'}(1,2)\}, \tag{3.66}$$

by partial trace summations over $m = m'$ in one subsystem or $i = i'$ of the other subsystem:

$$\begin{aligned}\mathbf{D}(1) &= \sum_m \langle\Psi_i(1)\Phi_m(2)|\,\hat{\mathrm{D}}\,|\Psi_{i'}(1)\Phi_m(2)\rangle \equiv \mathrm{tr}_2\mathbf{D}(1,2) \equiv \{\,D_{i,i'}(1)\},\\ \mathbf{D}(2) &= \sum_i \langle\Psi_i(1)\Phi_m(2)|\,\hat{\mathrm{D}}\,|\Psi_i(1)\Phi_{m'}(2)\rangle \equiv \mathrm{tr}_1\mathbf{D}(1,2) \equiv \{\,D_{m,m'}(2)\}\,.\end{aligned} \tag{3.67}$$

Let $A(1)$ be a physical quantity of subsystem (1) with the corresponding observable $\hat{\mathrm{A}}(1)$ acting in $\mathcal{H}(1)$, which is represented in $\mathcal{H}(1, 2)$ by the matrix:

$$\begin{aligned}\mathbf{A}^{1,2}(1) &= \{\langle\Psi_i(1)\Phi_m(2)|\hat{\mathrm{A}}(1)|\Psi_{i'}(1)\Phi_{m'}(2)\rangle \equiv A_{i,m;i',m'}(1)\\ &= \langle\Psi_i(1)|\hat{\mathrm{A}}(1)|\Psi_{i'}(1)\rangle\langle\Phi_m(2)|\Phi_{m'}(2)\rangle = A_{i,i'}(1)\delta_{m,m'}(2)\}\\ &\equiv \mathbf{A}(1)\otimes\mathbf{I}(2).\end{aligned} \tag{3.68}$$

The ensemble average value of $A(1)$ [see (3.60a)–(3.60c)] now reads:

$$\begin{aligned} A_{ens}(1) &= \mathrm{tr}[\mathbf{D}(1,2)\mathbf{A}^{1,2}(1)] \\ &= \sum_i \sum_m [\sum_{i'} \sum_{m'} D_{i,m;i',m'}(1,2) A_{i',m';i,m}(1)] \\ &= \sum_i \sum_m [\sum_{i'} \sum_{m'} D_{i,m;i',m'}(1,2) A_{i',i}(1)\delta_{m',m}(2)] \\ &= \sum_i \sum_{i'} [\sum_m D_{i,m;i',m}(1,2)] A_{i',i}(1)] \\ &= \sum_i \sum_{i'} [\sum_m D_{i,i'}(1) A_{i',i}(1)] \\ &= \mathrm{tr}[\mathbf{D}(1)\mathbf{A}(1)]. \end{aligned} \tag{3.69}$$

Therefore, the partial trace concept enables one to calculate the ensemble average of the subsystem quantity $A(1)$ as if this part of the whole physical system were isolated in the effective mixed state of (1) in the system as a whole, defined by the density operator $\mathbf{D}(1)$, which already involves the partial trace over the states of the other subsystem.

3.4 Angular Momentum and Spin Operators

In (3.36) we have used the Jordan rules to generate the quantum mechanical observable $\hat{\mathbf{l}}(\{x_\alpha\}) = -i\hbar \boldsymbol{r} \times \nabla$ corresponding in the position representation to the particle angular momentum $\boldsymbol{l} = \boldsymbol{r} \times \boldsymbol{p}$, e.g., that of the electron moving around nucleus in an atom. This equation also defines the associated component operators, obtained by expanding the determinant of the vector product:

$$\begin{aligned} \hat{\mathrm{l}}_x &= \hat{\mathrm{y}}\hat{\mathrm{p}}_z - \hat{\mathrm{z}}\hat{\mathrm{p}}_y = -i\hbar(y\partial/\partial z - z\partial/\partial y), \\ \hat{\mathrm{l}}_y &= \hat{\mathrm{z}}\hat{\mathrm{p}}_x - \hat{\mathrm{x}}\hat{\mathrm{p}}_z = -i\hbar(z\partial/\partial x - x\partial/\partial z), \\ \hat{\mathrm{l}}_z &= \hat{\mathrm{x}}\hat{\mathrm{p}}_y - \hat{\mathrm{y}}\hat{\mathrm{p}}_x = -i\hbar(x\partial/\partial y - y\partial/\partial x). \end{aligned} \tag{3.70}$$

They give rise to the following commutation relations:

$$[\hat{\mathrm{l}}_x, \hat{\mathrm{l}}_y] = i\hbar\hat{\mathrm{l}}_z, \quad [\hat{\mathrm{l}}_y, \hat{\mathrm{l}}_z] = i\hbar\hat{\mathrm{l}}_x, \quad [\hat{\mathrm{l}}_z, \hat{\mathrm{l}}_x] = i\hbar\hat{\mathrm{l}}_y, \ [\hat{\mathbf{l}}^2, \hat{\mathrm{l}}_x] = [\hat{\mathbf{l}}^2, \hat{\mathrm{l}}_y] = [\hat{\mathbf{l}}^2, \hat{\mathrm{l}}_z] = 0. \tag{3.71}$$

It thus follows from the first three relations of this equation that for the finite angular momentum $|\boldsymbol{l}| > 0$ its three components cannot be simultaneously determined precisely; clearly, for $|\boldsymbol{l}| = 0$ they are all vanishing: $l_x = l_y = l_z = 0$. The remaining relations indicate that only the length $|\boldsymbol{l}| = (\boldsymbol{l}^2)^{1/2}$ of the angular momentum and one of its components, say l_z, can be simultaneously sharply defined. Indeed, the analysis of the quantized eigenvalue problems of these operators, which can be found in any textbook of quantum mechanics, gives:

$$l^2 = l(l+1)\hbar^2, \quad l = 0, 1, 2, \ldots; \quad l_z = m\hbar, \quad m = -l, -l+1, \ldots, 0, \ldots, l-1, l. \tag{3.72}$$

The commutation relations can be straightforwardly derived using the commutator identities of (2.34) and the known commutators involving the position $\{\hat{x}_i\}$ and momentum $\{\hat{p}_i\}$ observables [see (3.56)]:

$$[\hat{x}_i, \hat{x}_j] = [\hat{p}_i, \hat{p}_j] = 0, \quad [\hat{x}_i, \hat{p}_j] = i\hbar\delta_{i,j}. \tag{3.73}$$

For example,

$$\begin{aligned}[\hat{l}_x, \hat{l}_y] &= \hat{l}_x\hat{l}_y - \hat{l}_y\hat{l}_x = [\hat{y}\hat{p}_z - \hat{z}\hat{p}_y, \hat{z}\hat{p}_x - \hat{x}\hat{p}_z] \\ &= [\hat{y}\hat{p}_z, \hat{z}\hat{p}_x] + [\hat{z}\hat{p}_y, \hat{x}\hat{p}_z] \\ &= (\hat{x}\hat{p}_y - \hat{y}\hat{p}_x)[\hat{z}, \hat{p}_z] = i\hbar\hat{l}_z.\end{aligned} \tag{3.74}$$

However, the origin of the spin angular momenta (see Sect. 1.4) is not classical, so that the Jordan rules do not apply in constructing their operators. Consider a single electron as an example. We shall now derive the matrix representations of the spin operator $\hat{\mathbf{s}} = \boldsymbol{i}\hat{s}_x + \boldsymbol{j}\hat{s}_y + \boldsymbol{j}\hat{s}_z$ in the basis set of the two allowed spin states $|\boldsymbol{\xi}\rangle \equiv (|\alpha\rangle, |\beta\rangle)$ (see Fig. 1.2) by postulating that these nonclassical angular momentum operators satisfy the same commutator relations as their classical analogs:

$$[\hat{s}_x, \hat{s}_y] = i\hbar\hat{s}_z, \quad [\hat{s}_y, \hat{s}_z] = i\hbar\hat{s}_x, \quad [\hat{s}_z, \hat{s}_x] = i\hbar\hat{s}_y, \ [\hat{\mathbf{s}}^2, \hat{s}_x] = [\hat{\mathbf{s}}^2, \hat{s}_y] = [\hat{\mathbf{s}}^2, \hat{s}_z] = 0. \tag{3.75}$$

In other words, we again recognize that, as in the classical case, only the length and one of the components of the spin angular momentum can be simultaneously specified. This is exactly what is observed in the experiment (see Fig. 1.2).

We first observe that the two spin states of an electron are then represented by the associated spin wave functions (column vectors):

$$\begin{aligned}&\boldsymbol{\alpha}(\boldsymbol{\xi}) = \langle\boldsymbol{\xi}|\alpha\rangle = \{\langle\sigma|\alpha\rangle\} = \begin{bmatrix}1\\0\end{bmatrix}, \quad \boldsymbol{\beta}(\boldsymbol{\xi}) = \langle\boldsymbol{\xi}|\beta\rangle = \{\langle\sigma|\beta\rangle\} = \begin{bmatrix}0\\1\end{bmatrix}, \\ &\langle\beta|\alpha\rangle = \sum_\sigma \langle\beta|\sigma\rangle\langle\sigma|\alpha\rangle = \boldsymbol{\beta}^\dagger(\boldsymbol{\xi})\,\boldsymbol{\alpha}(\boldsymbol{\xi}) = 0, \\ &\langle\alpha|\alpha\rangle = \sum_\sigma \langle\alpha|\sigma\rangle\langle\sigma|\alpha\rangle = \boldsymbol{\alpha}^\dagger(\boldsymbol{\xi})\,\boldsymbol{\alpha}(\boldsymbol{\xi}) = \langle\beta|\beta\rangle = \sum_\sigma \langle\beta|\sigma\rangle\langle\sigma|\beta\rangle = \boldsymbol{\beta}^\dagger(\boldsymbol{\xi})\boldsymbol{\beta}(\boldsymbol{\xi}) = 1.\end{aligned} \tag{3.76}$$

To simplify notation, we introduce the dimensionless Pauli operator,

$$\hat{\boldsymbol{\sigma}} = 2\hat{\mathbf{s}}/\hbar = \boldsymbol{i}\hat{\sigma}_x + \boldsymbol{j}\hat{\sigma}_y + \boldsymbol{j}\hat{\sigma}_z, \tag{3.77}$$

in terms of which the first three commutation relations of (3.75) read:

$$[\hat{\sigma}_x, \hat{\sigma}_y] = 2i\hat{\sigma}_z, \qquad [\hat{\sigma}_y, \hat{\sigma}_z] = 2i\hat{\sigma}_x, \qquad [\hat{\sigma}_z, \hat{\sigma}_x] = 2i\hat{\sigma}_y. \tag{3.78}$$

The same relations must be satisfied by the matrix representations of the spin components $\{\hat{\sigma}_i\}$ in the basis $|\boldsymbol{\xi}\rangle$, called the Pauli matrices.

Since $[\hat{\boldsymbol{\sigma}}^2, \hat{\sigma}_z] = 0$, these two operators are represented by the diagonal matrices in this basis set $|\boldsymbol{\xi}\rangle$ of their common eigenvectors:

$$\hat{\boldsymbol{\sigma}}^2|\alpha\rangle = 3|\alpha\rangle, \quad \hat{\boldsymbol{\sigma}}^2|\beta\rangle = 3|\beta\rangle; \quad \hat{\sigma}_z|\alpha\rangle = |\alpha\rangle, \quad \hat{\sigma}_z|\beta\rangle = -|\beta\rangle.$$

These matrices include the corresponding eigenvalues as diagonal elements:

$$\boldsymbol{\sigma}^2 = \langle\boldsymbol{\xi}|\hat{\boldsymbol{\sigma}}^2|\boldsymbol{\xi}\rangle = \begin{bmatrix} 3 & 0 \\ 0 & 3 \end{bmatrix} \quad \text{and} \quad \boldsymbol{\sigma}_z = \langle\boldsymbol{\xi}|\hat{\sigma}_z|\boldsymbol{\xi}\rangle = \begin{bmatrix} 1 & 0 \\ 0 & -1 \end{bmatrix}. \tag{3.79}$$

In order to determine the Pauli matrices representing the remaining spin components,

$$\boldsymbol{\sigma}_x = \langle\boldsymbol{\xi}|\hat{\sigma}_x|\boldsymbol{\xi}\rangle = \begin{bmatrix} a_{1,1} & a_{1,2} \\ a_{2,1} & a_{2,2} \end{bmatrix} \quad \text{and} \quad \boldsymbol{\sigma}_y = \langle\boldsymbol{\xi}|\hat{\sigma}_y|\boldsymbol{\xi}\rangle = \begin{bmatrix} b_{1,1} & b_{1,2} \\ b_{2,1} & b_{2,2} \end{bmatrix}, \tag{3.80}$$

we first use two commutation relations of (3.78):

$$[\boldsymbol{\sigma}_x, \boldsymbol{\sigma}_z] = -2i\boldsymbol{\sigma}_y \Rightarrow \begin{bmatrix} 0 & -2a_{1,2} \\ 2a_{2,1} & 0 \end{bmatrix} = -2i\begin{bmatrix} b_{1,1} & b_{1,2} \\ b_{2,1} & b_{2,2} \end{bmatrix}, \tag{3.81}$$

$$[\boldsymbol{\sigma}_y, \boldsymbol{\sigma}_z] = 2i\boldsymbol{\sigma}_x \Rightarrow \begin{bmatrix} 0 & -2b_{1,2} \\ 2b_{2,1} & 0 \end{bmatrix} = 2i\begin{bmatrix} a_{1,1} & a_{1,2} \\ a_{2,1} & a_{2,2} \end{bmatrix}. \tag{3.82}$$

Hence, $a_{1,1} = a_{2,2} = b_{1,1} = b_{2,2} = 0$, $b_{1,2} = -ia_{1,2}$, $b_{2,1} = ia_{2,1}$. The remaining two matrix elements then result from the third commutation rule,

$$[\boldsymbol{\sigma}_x, \boldsymbol{\sigma}_y] = 2i\boldsymbol{\sigma}_z \Rightarrow 2i\begin{bmatrix} a_{1,2}a_{2,1} & 0 \\ 0 & -a_{1,2}a_{2,1} \end{bmatrix} = 2i\begin{bmatrix} 1 & 0 \\ 0 & -1 \end{bmatrix}, \tag{3.83}$$

which implies $a_{1,2}a_{2,1} = 1$. Therefore, by setting $a_{1,2} = a_{2,1} = 1$, one arrives at the following explicit forms of the Pauli matrices in (3.80):

$$\boldsymbol{\sigma}_x = \begin{bmatrix} 0 & 1 \\ 1 & 0 \end{bmatrix} \quad \text{and} \quad \boldsymbol{\sigma}_y = \begin{bmatrix} 0 & -i \\ i & 0 \end{bmatrix}. \tag{3.84}$$

Their nondiagonal character reflects the fact that these observables are not sharply defined simultaneously with the two spin parameters defining the basis set $|\boldsymbol{\xi}\rangle$.

It thus directly follows from these explicit representations of the Pauli operators that their actions on the spin functions of (3.76) give:

$$\boldsymbol{\sigma}_x\boldsymbol{\alpha} = \boldsymbol{\beta},\ \ \boldsymbol{\sigma}_x\boldsymbol{\beta} = \boldsymbol{\alpha};\ \ \boldsymbol{\sigma}_y\boldsymbol{\alpha} = i\boldsymbol{\beta},\ \ \boldsymbol{\sigma}_y\boldsymbol{\beta} = -i\boldsymbol{\alpha};\ \ \boldsymbol{\sigma}_z\boldsymbol{\alpha} = \boldsymbol{\alpha},\ \ \boldsymbol{\sigma}_z\boldsymbol{\beta} = -\boldsymbol{\beta}. \tag{3.85}$$

3.5 Pictures of Time Evolution

After establishing the basic concepts of the quantum *kinematics*, dealing with the quantum objects at the given time $t = t_0$, we now turn to alternative formulations of the quantum *dynamics*, which determines the evolution of the microsystems in time. The possibility of such different formulations arises because the basic mathematical entities of the theory, such as state vectors and operators, are not directly accessible to physical measurement. As we have seen in the preceding sections of this chapter, only the eigenvalues of the quantum observables and the scalar products of state vectors have direct experimental implications. They respectively determine the spectrum of all possible outcomes of single measurements of the physical quantity to which the operator corresponds and their associated probabilities in a very large number of repetitions of experiments carried on the same quantum state of the physical system in question. Therefore, as long as these experimental predictions remain the same, the alternative formulations of the quantum dynamics, called state *pictures*, remain acceptable and fully equivalent physical theories.

As we have seen in Sect. 2.7, the unitary operators $\hat{\mathrm{U}}$, for which $\hat{\mathrm{U}}^{\dagger} = \hat{\mathrm{U}}^{-1}$, have the desired property of not affecting the eigenvalues of the transformed operators $\hat{\mathrm{A}}' = \hat{\mathrm{U}}\hat{\mathrm{A}}\hat{\mathrm{U}}^{\dagger}$ and the scalar products between the transformed vectors $|\Psi'\rangle = \hat{\mathrm{U}}|\Psi\rangle$ and $|\Phi'\rangle = \hat{\mathrm{U}}|\Phi\rangle : \langle\Phi'|\Psi'\rangle = \langle\Phi|\hat{\mathrm{U}}^{\dagger}\hat{\mathrm{U}}|\Psi\rangle = \langle\Phi|\Psi\rangle$. The range of unitary operators is not limited to their time-independent form, which we have examined in Sect. 2.7, giving rise to different descriptions of the quantum object at the specified time $t = t_0$. The unitary transformations can be also used to express a change of quantum states with time, i.e., the alternative dynamical pictures of quantum mechanics.

For example, in the *Schrödinger* (*S*) picture, when the spectrum of the operator eigenvalues does not depend on time, one uses the time-independent operators $\hat{\mathrm{A}} \equiv \hat{\mathrm{A}}_S$ so that the evolution of quantum objects in time is embodied in the time-dependent state vector $|\Psi_S(t)\rangle$, generated from the initial state $|\Psi(t_0)\rangle$ by the action of the unitary operator $\hat{U}(t-t_0)$ of the time evolution $t_0 \rightarrow t$ of $|\Psi(t_0)\rangle$:

$$|\Psi_S(t)\rangle = \hat{U}(t-t_0)|\Psi(t_0)\rangle, \quad \hat{U}(t-t_0)^\dagger = \hat{U}(t-t_0)^{-1} \equiv \hat{U}(t_0-t),$$
$$\hat{U}(t-t_0)\hat{U}(t_0-t) = 1 \quad \text{and} \quad \hat{U}(0) = 1, \tag{3.86}$$

where the inverse evolution $t \rightarrow t_0$ of $|\Psi_S(t)\rangle$ recovers the state vector at $t = t_0$:

$$\hat{U}(t_0-t)|\Psi_S(t)\rangle = |\Psi(t_0)\rangle. \tag{3.87}$$

It also directly follows from the unitary character of the time evolution operator that the normalization of state vectors is conserved in time:

$$\langle\Psi_S(t)|\Psi_S(t)\rangle = \langle\Psi(t_0)|\hat{U}(t-t_0)^\dagger\hat{U}(t-t_0)|\Psi(t_0)\rangle = \langle\Psi(t_0)|\Psi(t_0)\rangle. \tag{3.88}$$

In the *Heisenberg (H) picture,* the state vectors do not change in time, but the operators become time dependent. Therefore, the operator of the inverse time evolution in (3.86) marks the unitary transformation of $|\Psi_S(t)\rangle$ into the time-independent vector of the Heisenberg picture: $|\Psi(t_0)\rangle \equiv |\Psi_H\rangle$. The time-dependent operators are then given by the transformation:

$$\hat{A}_H(t) = \hat{U}(t_0-t)\hat{A}_S\hat{U}(t_0-t)^{-1} = \hat{U}(t_0-t)\hat{A}_S\hat{U}(t-t_0). \tag{3.89}$$

When the quantum object is composed of interacting subsystems, its time-independent energy operator of the Schrödinger picture, the Hamiltonian $\hat{H}$, can be partitioned into the contribution representing the energy of the noninteracting subsystems, $\hat{H}_0$, and their mutual interaction, $\hat{V}$,

$$\hat{H} = \hat{H}_0 + \hat{V}. \tag{3.90}$$

The quantum dynamics of such composite systems can be best expressed in the *Interaction (I)* picture, in which both the state vectors and operators are time dependent. The relevant time-dependent unitary operator, which transforms these mathematical entities from the above Schrödinger picture, depends solely on $\hat{H}_0$:

$$\hat{S}(t) = \exp\left(\frac{i}{\hbar}\hat{H}_0 t\right). \tag{3.91}$$

Here, the exponential operator is defined by its power series expansion:

$$\hat{B}(t) \equiv \exp(\hat{A}t) = \sum_{n=0}^{\infty}\frac{(\hat{A}t)^n}{n!}, \tag{3.92a}$$

giving rise to the time derivative:

$$\frac{d\hat{\mathrm{B}}(t)}{dt} = \sum_{n=0}^{\infty} \frac{nt^{n-1}\hat{\mathrm{A}}^n}{n!} = \hat{\mathrm{A}} \sum_{n=1}^{\infty} \frac{(\hat{\mathrm{A}}t)^{n-1}}{(n-1)!} = \hat{\mathrm{A}} \sum_{m=0}^{\infty} \frac{(\hat{\mathrm{A}}t)^m}{m!} = \hat{\mathrm{A}}\exp(\hat{\mathrm{A}}t). \tag{3.92b}$$

The state vectors and operators in the I-picture of Quantum Mechanics are defined by the following transformations of their corresponding S-picture analogs:

$$|\Psi_I(t)\rangle = \hat{\mathrm{S}}(t)|\Psi_S(t)\rangle, \quad \hat{\mathrm{A}}_I(t) = \hat{\mathrm{S}}(t)\hat{\mathrm{A}}_S\hat{\mathrm{S}}(t)^{-1}. \tag{3.93}$$

In the remaining part of this chapter we shall explore in some detail the time evolution of quantum states in the Schrödinger picture and examine some of its physical implications. In the final Sect. 3.7 we summarize the related dynamical equations in the alternative pictures of quantum dynamics.

3.6 Schrödinger Picture: Dynamics of Wave Functions and Density Operators

Let us determine the explicit form of the unitary operator $\hat{\mathrm{U}}(t - t_0)$ of (3.86). The relevant *equation of motion* for quantum states in this dynamical picture is the subject of

Postulate V: The time evolution of the state vector $|\Psi_S(t)\rangle \equiv |\Psi(t)\rangle$ is governed by the Schrödinger equation:

$$i\hbar \frac{d|\Psi(t)\rangle}{dt} = \hat{\mathrm{H}}|\Psi(t)\rangle, \tag{3.94}$$

where the Hamiltonian $\hat{\mathrm{H}}$ is the observable associated with the system total energy.

The corresponding wave equations, either in the position-spin or the momentum-spin representations, determine the dynamics of the associated wave functions:

$$i\hbar \frac{d\Psi(\mathcal{Q}^N, t)}{dt} = \hat{\mathrm{H}}(\mathcal{Q}^N)\Psi(\mathcal{Q}^N, t) \quad \text{or} \quad i\hbar \frac{d\Psi(\mathcal{P}^N, t)}{dt} = \hat{\mathrm{H}}(\mathcal{P}^N)\Psi(\mathcal{P}^N, t). \tag{3.95}$$

Substituting (3.86) into (3.94) gives:

$$\left[i\hbar \frac{d\hat{\mathrm{U}}(t-t_0)}{dt} - \hat{\mathrm{H}}\,\hat{\mathrm{U}}(t-t_0)\right]|\Psi(t_0)\rangle = 0 \text{ or } i\hbar \frac{d\hat{\mathrm{U}}(t-t_0)}{dt} = \hat{\mathrm{H}}\,\hat{\mathrm{U}}(t-t_0). \tag{3.96}$$

The formal solution of this differential equation is thus given by the following evolution operator [see also (3.92a) and (3.92b)]:

$$\hat{\mathrm{U}}(t-t_0)=\exp\left(-\frac{i}{\hbar}(t-t_0)\hat{\mathrm{H}}\right)\equiv\exp\left(-\frac{i}{\hbar}\tau\,\hat{\mathrm{H}}\right)\equiv\hat{\mathrm{U}}(\tau). \tag{3.97}$$

Hence, the operator of the reverse evolution, from t to t_0,

$$\hat{\mathrm{U}}(\tau)^{\dagger}=\hat{\mathrm{U}}(\tau)^{-1}=\exp\left(\frac{i}{\hbar}\tau\,\hat{\mathrm{H}}\right)=\hat{\mathrm{U}}(-\tau). \tag{3.98}$$

It can be easily verified by the differentiation with respect to time, using the derivative (3.92b) of the exponential operator (3.92a), that the action of this unitary operator is equivalent to the dynamical Schrödinger equation (3.94).

We now briefly examine the implications of Schrödinger's time evolution for the mixed states. The unitary character of the time evolution operator then directly implies that if the system at the initial time $t=t_0$ has probability p_k of being in the state $|\psi_\alpha\rangle=|\psi_\alpha(t_0)\rangle$, then, at a subsequent time t, it has the same probability of being in the evolved state $|\psi_\alpha(t)\rangle$. Indeed, the density operator at time t [see (3.60a)–(3.60c)],

$$\hat{\mathrm{D}}(t)=\sum_\alpha p_\alpha(t)|\psi_\alpha(t)\rangle\langle\psi_\alpha(t)|=\sum_\alpha p_\alpha(t)\hat{\mathrm{P}}_\alpha(t), \tag{3.99}$$

gives

$$p_\alpha(t)=\langle\psi_\alpha(t)|\hat{\mathrm{D}}(t)|\psi_\alpha(t)\rangle=\langle\psi_\alpha|\hat{\mathrm{D}}|\psi_\alpha\rangle=p_\alpha, \tag{3.100}$$

since the matrix elements of operators are invariants of the unitary transformations.

Before we examine the *equation of motion* for $\hat{\mathrm{D}}(t)=\sum_\alpha p_\alpha\hat{\mathrm{P}}_\alpha(t)$ let us first derive it for the projection operator $\hat{\mathrm{P}}_\alpha(t)$ onto the pure state $|\psi_\alpha(t)\rangle$. Using the Schrödinger equation (3.94) for $|\psi_\alpha(t)\rangle$ and its Hermitian conjugate gives:

$$\begin{aligned}\frac{d}{dt}\hat{\mathrm{P}}_\alpha(t)&=\left(\frac{d|\psi_\alpha(t)\rangle}{dt}\right)\langle\psi_\alpha(t)|+|\psi_\alpha(t)\rangle\left(\frac{d\langle\psi_\alpha(t)|}{dt}\right)\\&=\frac{1}{i\hbar}\left(\hat{\mathrm{H}}|\psi_\alpha(t)\rangle\langle\psi_\alpha(t)|-|\psi_\alpha(t)\rangle\langle\psi_\alpha(t)|\hat{\mathrm{H}}\right)=\frac{1}{i\hbar}[\hat{\mathrm{H}},\hat{\mathrm{P}}_\alpha(t)].\end{aligned} \tag{3.101}$$

Multiplying the preceding equation by $p_\alpha(t)=p_\alpha$ and summing over all states in the statistical mixture of $\hat{\mathrm{D}}(t)$ gives the related dynamics of the density operator itself:

$$i\hbar\frac{d}{dt}\hat{\mathrm{D}}(t)=[\hat{\mathrm{H}},\hat{\mathrm{D}}(t)]. \tag{3.102}$$

3.6.1 Energy Representation and Stationary States

The explicit form of the time-dependent wave function

$$\Psi(Q^N; t - t_0) \equiv \Psi(Q^N; \tau) = \exp\left(-\frac{i}{\hbar}\hat{\mathrm{H}}(Q^N)\tau\right)\Psi(Q^N; t_0), \tag{3.103}$$

can be obtained in the energy representation of Sect. 2.7, i.e., for the orthonormal basis set of the eigenfunctions $\{\psi_n(Q^N) = \langle Q^N \mid \psi_n\rangle\}$ of the system Hamiltonian $\hat{\mathrm{H}}(Q^N) = \langle Q^N|\hat{\mathrm{H}}|Q^N\rangle$:

$$\hat{\mathrm{H}}(Q^N)\,\psi_n(Q^N) = E_n\,\psi_n(Q^N). \tag{3.104}$$

Indeed, by expanding the wave function in this energy basis set,

$$\begin{aligned} \Psi(Q^N; t_0) &= \sum_n C_n\,\psi_n(Q^N), \\ C_n = \langle\psi_n \mid \Psi(t_0)\rangle &= \int \psi_n^*(Q^N)\Psi(Q^N; t_0)\,dQ^N, \end{aligned} \tag{3.105}$$

and using the power series for the exponential evolution operator (3.92a) and its derivative (3.92b), one finds the wave function after the time interval $\tau = t{-}t_0$:

$$\begin{aligned} \Psi(Q^N; \tau) &= \sum_{k=0}^{\infty}\frac{1}{k!}\left(-\frac{i}{\hbar}\hat{\mathrm{H}}(Q^N)\,\tau\right)^k \sum_n C_n\,\psi_n(Q^N) \\ &= \sum_n C_n\,\psi_n(Q^N)\sum_{k=0}^{\infty}\frac{1}{k!}\left(-\frac{i}{\hbar}E_n\tau\right)^k \\ &= \sum_n C_n \exp\left(-\frac{i}{\hbar}E_n\tau\right)\psi_n(Q^N) \\ &\equiv \sum_n u_n(\tau)\,\psi_n(Q^N) \equiv \sum_n C_n\,\Psi_n(Q^N; \tau). \end{aligned} \tag{3.106}$$

In the preceding expansion, the time-dependent wave function is expressed in terms of the time-dependent eigenfunctions of the Hamiltonian,

$$\begin{aligned} \Psi_n(Q^N; \tau) &= \psi_n(Q^N)\exp\left(-\frac{i}{\hbar}E_n\tau\right) \equiv \psi_n(Q^N)\exp(-i\omega_n\tau) \\ &= \langle Q^N \mid \Psi_{E_n}(\tau)\rangle, \end{aligned} \tag{3.107}$$

which represent the *stationary states* of the system, for its sharply specified energies $\{E_n\}$. Such states are given by the product of the time-independent *amplitude* $\psi_n(Q^N)$, determined by the eigenvalue problem of (3.104), and the time-dependent *phase* factor $\exp(-i\omega_n\tau)$, which does not contribute to the associated (time independent) probability distribution,

$$p_n(\mathcal{Q}^N;\tau) = \left|\psi_n(\mathcal{Q}^n)\exp\left(-\frac{i}{\hbar}E_n\tau\right)\right|^2 = |\psi_n(\mathcal{Q}^n)|^2 = \psi_n^*(\mathcal{Q}^n)\psi_n(\mathcal{Q}^n), \quad (3.108)$$

which is seen to be determined solely by the state amplitude.

The time-dependent coefficients $\{u_n(\tau) = \langle\psi_n \mid \Psi(\tau)\rangle = C_n\exp(-i\omega_n\tau\}$ in (3.106) provide the energy representation of state $|\Psi(\tau)\rangle$. Since the conditional probability $P(\psi_n|\Psi(\tau)) = |u_n(\tau)|^2 = |C_n|^2$, we thus conclude that the time evolution of the state vector in the S-picture represents its "rotation" in the Hilbert space, which conserves in time the probabilities of observing the system stationary states. We also observe that for the combination of (3.106) to retain the stationary character it must be limited only to the subspace corresponding to a single degenerate eigenvalue E_n, with all its components thus exhibiting the same phase factor.

To summarize, the stationary states, in which the system energy is sharply defined, are distinguished by several special features. The energy determines uniquely the time-dependent factor of the wave function, so that the probability distribution and its current (see Section 3.6.3) are time independent. Moreover, the expectation values of any physical observable $\hat{A}(\mathcal{Q}^N)$, which does not depend on time explicitly, are conserved:

$$\langle A\rangle = \int \Psi_n^*(\mathcal{Q}^N;\tau)\hat{A}(\mathcal{Q}^N)\Psi_n(\mathcal{Q}^N;\tau)d\mathcal{Q}^N = \int \psi_n^*(\mathcal{Q}^N)\hat{A}(\mathcal{Q}^N)\psi_n(\mathcal{Q}^N)d\mathcal{Q}^N = const. \quad (3.109)$$

These average values thus become sharply defined, equal to a single eigenvalue of $\hat{A}(\mathcal{Q}^N)$, $\langle A\rangle = a_i$, when the latter commutes with the system Hamiltonian. Also, when these two observables do not commute, the conditional probability $P(\varphi_j|\Psi_n)$ of finding a given eigenvalue a_j, where $\varphi_j(\mathcal{Q}^N)$ represents the eigenstate of $\hat{A}(\mathcal{Q}^N)$,

$$\hat{A}(\mathcal{Q}^N)\,\varphi_k(\mathcal{Q}^N) = a_k\,\varphi_k(\mathcal{Q}^N), \quad (3.110)$$

given by the square of the modulus of the relevant expansion coefficient, the projection of Ψ_n into φ_k, also remains constant in time:

$$P(\varphi_j|\Psi_n) = \left|\int \varphi_j^*(\mathcal{Q}^N)\Psi_n(\mathcal{Q}^N;\tau)\, d\mathcal{Q}^N\right|^2 = const. \quad (3.111)$$

The Schrödinger equation emphasizes the crucial role of the system energy operator in determining the system dynamics, similar to that played by the Hamilton function in classical mechanics [see (3.1)]. In general, the precise specification of the system energy does not identify the stationary quantum state uniquely. Indeed, for this to be the case one also requires the eigenvalues $\boldsymbol{a} = \{a_i\}$ of the complete set of the commuting observables $\{\hat{\mathrm{A}}_i\}$, which also commute with the system Hamiltonian (see Sect. 2.5):

$$[\hat{\mathrm{H}}, \hat{\mathrm{A}}_i] = [\hat{\mathrm{A}}_i, \hat{\mathrm{A}}_j] = 0, \quad i,j = 1,2,\ldots,s. \tag{3.112}$$

Together with the sharply defined energy E_n they provide the complete description of their common eigenvectors:

$$\hat{\mathrm{H}}|E_n, \boldsymbol{a}\rangle = E_n|E_n, \boldsymbol{a}\rangle, \quad \{\hat{\mathrm{A}}_i|E_n, \boldsymbol{a}\rangle = a_i|E_n, \boldsymbol{a}\rangle\}. \tag{3.113}$$

It follows from (3.102) that in the energy representation the dynamics of the diagonal elements of the density operator $\hat{\mathrm{D}}(\tau), D_{n,n}(\tau) = \langle\psi_n|\hat{\mathrm{D}}(\tau)|\psi_n\rangle$, representing the *population* of state $|\psi_n\rangle$ in the ensemble, predicts:

$$i\hbar\frac{d}{d\tau}D_{n,n}(\tau) = \langle\psi_n|\hat{\mathrm{H}}\hat{\mathrm{D}}(\tau) - \hat{\mathrm{D}}(\tau)\hat{\mathrm{H}}|\psi_n\rangle = E_n\langle\psi_n|\hat{\mathrm{D}}(\tau) - \hat{\mathrm{D}}(\tau)|\psi_n\rangle = 0. \tag{3.114}$$

For its *off*-diagonal matrix element $D_{m,n}(\tau) = \langle\psi_m|\hat{\mathrm{D}}(\tau)|\psi_n\rangle$, representing *coherences* between states $|\psi_m\rangle$ and $|\psi_n\rangle$ in the ensemble, one similarly finds:

$$\begin{aligned} i\hbar\frac{d}{dt}D_{m,n}(\tau) &= \langle\psi_m|\hat{\mathrm{H}}\hat{\mathrm{D}}(\tau) - \hat{\mathrm{D}}(\tau)\hat{\mathrm{H}}|\psi_n\rangle \\ &= (E_m - E_n)\langle\psi_n|\hat{\mathrm{D}}(\tau)|\psi_n\rangle \\ &= (E_m - E_n)D_{m,n}(\tau) \end{aligned} \tag{3.115a}$$

or

$$\frac{d}{d\tau}[\ln D_{m,n}(\tau)] = -\frac{i}{\hbar}(E_m - E_n). \tag{3.115b}$$

Therefore, in the stationary-state representation $D_{n,n}(\tau) = const.$ and

$$D_{m,n}(\tau) = \exp\left(-\frac{i}{\hbar}(E_m - E_n)\tau\right)D_{m,n}(0). \tag{3.116}$$

In the remaining part of this section we shall explore some physical implications of the dynamical Schrödinger equation.

3.6.2 Time Dependence of Expectation Values and Ehrenfest Principle

Since the Schrödinger equation (3.94) is of the first order in t, the state $|\Psi(t)\rangle$ at any subsequent time $t > t_0$ is uniquely determined given the initial state $|\Psi(t_0)\rangle$. Therefore, there is no indeterminacy in the free evolution of quantum systems. The irreversibility arises only in an act of measurement, which unpredictably modifies the system state. Thus, between the two measurements the evolution of quantum states is perfectly deterministic.

It also follows from the linear and homogeneous character of this equation that its solutions are linearly superposable. More specifically, the linear combination at the initial time $|\Psi(t_0)\rangle = C_1|\Psi_1(t_0)\rangle + C_2|\Psi_2(t_0)\rangle$ becomes $|\Psi(t)\rangle = C_1|\Psi_1(t)\rangle + C_2|\Psi_2(t)\rangle$ at $t > t_0$, so that the correspondence between $|\Psi(t_0)\rangle$ and $|\Psi(t)\rangle$ is marked by preservation of the coefficients before their components during time evolution. Another manifestation of this property is the conservation in time of the ensemble probabilities (3.100).

Next, let us examine the time evolution of the mean (expectation) values of the physical observables. As we have already observed in (3.88), the preservation in time of the state normalization is assured by the unitary character of the time evolution operator of (3.86). Thus, in the mean value of the physical quantity A, which in general case may explicitly depend on time, $\hat{\mathrm{A}} = \hat{\mathrm{A}}(t)$, only the explicit time dependency of the wave function and that of the observable do matter, since the implicit dependence through the coordinates (or momenta) has already been eliminated by integration in the expectation value of (3.109). Using the relevant Hilbert space expression and the Schrödinger equation (3.94) then gives:

$$\begin{aligned}\frac{d\langle\Psi(t)|\hat{\mathrm{A}}(t)|\Psi(t)\rangle}{dt} &= \left(\frac{d}{dt}\langle\Psi(t)|\right)\hat{\mathrm{A}}|\Psi(t)\rangle + \langle\Psi(t)|\hat{\mathrm{A}}\left(\frac{d}{dt}|\Psi(t)\rangle\right) + \langle\Psi(t)|\frac{\partial\hat{\mathrm{A}}}{\partial t}|\Psi(t)\rangle \\ &= \frac{1}{i\hbar}\langle\Psi(t)|[\hat{\mathrm{A}},\hat{\mathrm{H}}]|\Psi(t)\rangle + \langle\Psi(t)|\frac{\partial\hat{\mathrm{A}}}{\partial t}|\Psi(t)\rangle \equiv \frac{1}{i\hbar}\langle[\hat{\mathrm{A}},\hat{\mathrm{H}}]\rangle + \left\langle\frac{\partial\hat{\mathrm{A}}}{\partial t}\right\rangle.\end{aligned} \tag{3.117}$$

Therefore, for the physical observables, which do not depend explicitly on time,

$$\frac{d\langle A\rangle}{dt} = \frac{i}{\hbar}\langle[\hat{\mathrm{H}},\hat{\mathrm{A}}]\rangle, \tag{3.118}$$

and hence the observable commuting with the Hamiltonian represents the system *constant of motion*.

Consider the illustrative example of a motion in one dimension, in the potential $V(x)$, of the spinless particle described by the Hamiltonian $\hat{\mathrm{H}}(x) = V(x) + \hat{\mathrm{p}}_x^2/2m$.

We first examine the time dependence of the particle average position $\langle x\rangle$. Using (2.34) and (3.56) in the preceding equation gives:

$$\frac{d\langle x\rangle}{dt}=\frac{i}{\hbar}\langle[\hat{\mathrm{H}},\hat{\mathrm{x}}]\rangle=\frac{i}{2m\hbar}\langle[\hat{\mathrm{p}}_x^2,\hat{\mathrm{x}}]\rangle=\frac{i}{2m\hbar}\langle[\hat{\mathrm{p}}_x[\hat{\mathrm{p}}_x,\hat{\mathrm{x}}]+[\hat{\mathrm{p}}_x,\hat{\mathrm{x}}]\hat{\mathrm{p}}_x\rangle=\frac{\langle p_x\rangle}{m}. \quad (3.119)$$

Therefore, the relation between the expectation values of the position and momentum is the same as that between their classical analogs: $v_x = dx/dt = p_x/m$.

One similarly arrives at the second Newton's law of classical dynamics, $dp_x/dt = F_x = -dV(x)/dx$, where F_x stands for the force acting on the particle, by examining the time evolution of $\langle p_x\rangle$:

$$\frac{d\langle p_x\rangle}{dt}=\frac{i}{\hbar}\langle[\hat{\mathrm{H}},\hat{\mathrm{p}}_x]\rangle=\frac{i}{\hbar}\langle[V(x),\hat{\mathrm{p}}_x]\rangle=\frac{i}{\hbar}\left\langle i\hbar\frac{\partial V}{\partial x}\right\rangle=-\left\langle\frac{\partial V}{\partial x}\right\rangle=\langle F_x\rangle. \quad (3.120)$$

Accordingly, for the movement of a quantum particle in three dimensions, in the potential $V(\boldsymbol{r})$ generating the classical force field $\boldsymbol{F}(\boldsymbol{r}) = -\nabla V(\boldsymbol{r})$, one finds

$$\frac{d\langle\boldsymbol{p}\rangle}{dt}=-\langle\nabla V\rangle=\langle\boldsymbol{F}\rangle. \quad (3.121)$$

This correspondence between the quantum relations in terms of the expectation (mean) values of physical quantities and the associated equations of classical mechanics expresses the *Ehrenfest principle* of quantum mechanics. In any quantum state $|\Psi\rangle$ the time dependencies of the expectation values of the position and momentum operators are seen to follow the corresponding relations between the associated classical quantities. This rule complements the related Correspondence Principle of Bohr (see Chap. 1) that the quantum description becomes classical in the limit of high energies and very large quantum numbers, when one can safely neglect the finite value of the quantum of action: $\hbar \to 0$.

3.6.3 Probability Current and Continuity Equation

Let us again assume the system composed of a single (spinless) particle. In the position representation, the state $|\psi(t)\rangle$ is represented by the normalized wave function $\psi(\boldsymbol{r}; t) = \langle\boldsymbol{r}|\psi(t)\rangle$ which generates the probability density

$$p(\boldsymbol{r};t)=|\psi(\boldsymbol{r},t)|^2=\langle\psi(t)|\boldsymbol{r}\rangle\langle\boldsymbol{r}|\psi(t)\rangle\equiv\langle\psi(t)|\hat{\rho}(\boldsymbol{r})|\psi(t)\rangle=\rho(\boldsymbol{r};t). \quad (3.122)$$

It directly follows from the Schrödinger equation (3.94) that the square of the norm of the wave function, i.e., the integral of $p(\boldsymbol{r}, t)$ over the whole physical space, remains constant in time and equal to 1 for the normalized quantum state. This does not imply, however, that $p(\boldsymbol{r}, t)$ is also locally conserved over time. Indeed, the stream of probability may transport the particles from one region of space to another. It is our goal in this section to establish the appropriate expression for the local probability current.

It should be recalled that in the electromagnetism the charge (volume) density $\rho_{el}(\boldsymbol{r}; t)$ is linked to the flux of the vector current density $\boldsymbol{J}_{el}(\boldsymbol{r}; t)$ through the local continuity equation,

$$\frac{\partial}{\partial t}\rho_{el}(\boldsymbol{r};t) = -\nabla \cdot \boldsymbol{J}_{el}(\boldsymbol{r};t), \tag{3.123}$$

where the left-hand part of the equation expresses the net change of the density in the fixed, infinitesimal volume around $\boldsymbol{r}$, and the right-hand part represents the flux across the surface, which defines this volume element. We are now searching for an analogous equation expressing the local probability balance in the quantum mechanics, i.e., the appropriate definition of the probability current $\boldsymbol{j}(\boldsymbol{r}; t)$. The negative divergence of this yet unknown vector will then measure the flux of particles leaving the local volume element.

The system Hamiltonian in the position representation,

$$\hat{\mathrm{H}}(\boldsymbol{r}) = V(\boldsymbol{r}) + \frac{\hat{\mathbf{p}}^2(\boldsymbol{r})}{2m} = V(\boldsymbol{r}) - \frac{\hbar^2}{2m}\Delta, \tag{3.124}$$

with the real potential $V(\boldsymbol{r})$ for $\hat{\mathrm{H}}(\boldsymbol{r})$ to be Hermitian, gives the dynamical Schrödinger equation in the form:

$$i\hbar\frac{\partial\psi(\boldsymbol{r};t)}{\partial t} = V(\boldsymbol{r})\,\psi(\boldsymbol{r};t) - \frac{\hbar^2}{2m}\Delta\psi(\boldsymbol{r};t). \tag{3.125}$$

Multiplying, from the left, both sides of this equation by $\psi^*(\boldsymbol{r}; t)$, and of the complex conjugate Schrödinger equation by $\psi(\boldsymbol{r}; t)$, subtracting the resulting equations and dividing by $i\hbar$ then give:

$$\frac{\partial[\psi^*\psi]}{\partial t} = -\frac{\hbar}{2mi}[\psi^*\Delta\psi - \psi\Delta\psi^*]. \tag{3.126}$$

This equation can be then transformed into the continuity-type equation (3.123),

$$\frac{\partial}{\partial t}p(\boldsymbol{r};t) = -\nabla \cdot \boldsymbol{j}(\boldsymbol{r};t), \tag{3.127}$$

with the probability current

$$
\begin{aligned}
j(\boldsymbol{r};t) &= \frac{\hbar}{2mi}[\psi^*(\boldsymbol{r};t)\nabla\psi(\boldsymbol{r};t) - \psi(\boldsymbol{r};t)\nabla\psi^*(\boldsymbol{r};t)] \\
&= \frac{1}{m}\mathrm{Re}[\psi^*(\boldsymbol{r};t)\frac{\hbar}{i}\nabla\psi(\boldsymbol{r};t)],
\end{aligned}
\tag{3.128}
$$

$$
\begin{aligned}
\nabla\cdot\boldsymbol{j} &= \frac{\hbar}{2mi}[(\nabla\psi^*)\cdot(\nabla\psi) + \psi^*(\nabla^2\psi) - (\nabla\psi)\cdot(\nabla\psi^*) - \psi(\nabla^2\psi^*)] \\
&= \frac{\hbar}{2mi}[\psi^*\Delta\psi - \psi\,\Delta\psi^*].
\end{aligned}
$$

The form of the probability current (3.128) indicates that it is determined by the expectation (mean) value in state $|\psi(t)\rangle$ of the Hermitian operator

$$
\hat{\mathbf{j}}(\boldsymbol{r}) = \frac{1}{2m}[|\boldsymbol{r}\rangle\langle\boldsymbol{r}|\hat{\mathbf{p}} + \hat{\mathbf{p}}|\boldsymbol{r}\rangle\langle\boldsymbol{r}|], \tag{3.129}
$$

which represents the symmetrized product of operators for the probability density, $\hat{\rho}(\boldsymbol{r}) = |\boldsymbol{r}\rangle\langle\boldsymbol{r}|$, and particle velocity, $\hat{\mathbf{v}} = \hat{\mathbf{p}}/m$. Indeed, such a product is also associated with the physical meaning of the current density vector of a classical fluid.

To conclude this section, let us express the complex wave function $\psi(\boldsymbol{r}, t)$ in terms of its (real) modulus $R(\boldsymbol{r}; t)$ and phase $\Phi(\boldsymbol{r}; t)$:

$$
\psi(\boldsymbol{r},t) = R(\boldsymbol{r};t)\exp[i\Phi(\boldsymbol{r};t)]. \tag{3.130}
$$

It then directly follows from (3.122) and (3.128) that

$$
\begin{aligned}
p(\boldsymbol{r};t) &= R^2(\boldsymbol{r};t) \quad \text{and} \\
j(\boldsymbol{r};t) &= \frac{\hbar}{m}R^2(\boldsymbol{r};t)\nabla\Phi(\boldsymbol{r};t) = p(\boldsymbol{r};t)\nabla[\frac{\hbar}{m}\Phi(\boldsymbol{r};t)].
\end{aligned}
\tag{3.131}
$$

3.7 Heisenberg and Interaction Pictures of Quantum Dynamics

We conclude this short outline of the formal framework of quantum dynamics with a summary of the relevant *equations of motion* in the H- and I-pictures of Sect. 3.5. As we have already indicated in (3.89) the operators $\{\hat{\mathrm{A}}_H\}$ in the Heisenberg picture generally depend on time, even if their analogs in the Schrödinger picture $\{\hat{\mathrm{A}}_S\}$ do not. However, for the conservative system, the Hamiltonian $\hat{\mathrm{H}}_S$ of which does not depend on time, and an observable $\hat{\mathrm{A}}_S$ representing a constant of motion (commuting with $\hat{H}_S$), the evolution operator $\hat{U}(t-t_0) \equiv \hat{U}(\tau)$ of (3.97) commutes with $\hat{A}_S$ so that

$$\hat{A}_H(\tau) = \hat{U}^{-1}(\tau)\hat{A}_S\hat{U}(\tau) = \hat{U}^{-1}(\tau)\hat{U}(\tau)\hat{A}_S = \hat{A}_S. \tag{3.132}$$

The operators for such physical properties are thus equal in both dynamical pictures, and in particular $\hat{H}_H = \hat{H}_S$.

For an arbitrary observable $\hat{A}_S(\tau)$ one finds using (3.96), its adjoint, and (3.89):

$$\begin{aligned}\frac{d}{dt}\hat{A}_H(\tau) = \frac{1}{i\hbar}\Big[\hat{U}^{-1}(\tau)\hat{A}_S(\tau)\hat{H}_S(\tau)\hat{U}(\tau) - \hat{U}^{-1}(\tau)\hat{H}_S(\tau)\hat{A}_S(\tau)\hat{U}(\tau)\Big]\\ +\hat{U}^{-1}(\tau)\frac{d\hat{A}_S(\tau)}{dt}\hat{U}(\tau).\end{aligned} \tag{3.133}$$

Inserting next the unity factor $\hat{U}(\tau)\hat{U}^{-1}(\tau) = 1$ between $\hat{H}_S$ and $\hat{A}_S$ in the first two terms of the right hand side in the preceding equation finally gives

$$\begin{aligned}\frac{d}{dt}\hat{A}_H(\tau) = \frac{1}{i\hbar}\Big\{ [\hat{U}^{-1}(\tau)\hat{A}_S(\tau)\hat{U}(\tau)]\,[\hat{U}^{-1}\hat{H}_S(\tau)\hat{U}(\tau)]\\ -[\hat{U}^{-1}(\tau)\hat{H}_S(\tau)\hat{U}(\tau)]\,[\hat{U}^{-1}(\tau)\hat{A}_S(\tau)\hat{U}(\tau)]\Big\} + \hat{U}^{-1}(\tau)\frac{d\hat{A}_S(\tau)}{dt}\hat{U}(\tau)\end{aligned} \tag{3.134}$$

and hence

$$i\hbar\frac{d}{d\tau}\hat{A}_H(\tau) = [\hat{A}_H(\tau), \hat{H}_H(\tau)] + i\hbar\left(\frac{d}{d\tau}\hat{A}_S(\tau)\right)_H. \tag{3.135}$$

It was Schrödinger who first discovered the dynamical equation bearing his name. The subsequent Heisenberg picture has established the evolution of matrices representing operators $\{\hat{A}_H(t)\}$, hence the name *Matrix Mechanics* (see Chap. 1), to be later shown to be fully equivalent to the Schrödinger *Wave Mechanics*.

For the physical observables $\hat{A}_S$, which do not depend explicitly on time, the last term in (3.135) vanishes. Moreover, since the expectation value is invariant to the unitary transformation linking the two pictures,

$$\langle A(t)\rangle = \langle\Psi_S(t)|\hat{A}_S|\Psi_S(t)\rangle = \langle\Psi_H|\hat{A}_H(t)|\Psi_H\rangle. \tag{3.136}$$

Since in the last term only the operator depends on time

$$\begin{aligned}\frac{d}{dt}\langle A(t)\rangle &= \langle\Psi_H|\frac{d\hat{A}_H(t)}{dt}|\Psi_H\rangle = \frac{1}{i\hbar}\langle\Psi_H|[\hat{A}_H(\tau),\hat{H}_H(\tau)]|\Psi_H\rangle \\ &= \frac{1}{i\hbar}\langle\Psi_S(t)|[\hat{A}_S,\hat{H}_S]|\Psi_S(t)\rangle = \frac{i}{\hbar}\left\langle[\hat{H},\hat{A}]\right\rangle_S,\end{aligned} \tag{3.137}$$

where we have again recognized that commutators and expectation values are invariants of the unitary transformation between the two pictures.

We have thus recovered (3.118) for the time evolution of expectation values in the Schrödinger dynamics. Notice, however, that (3.135) is more general than (3.118), providing the relation between *operators*, instead of their *expectation values*. Indeed, an advantage of the Heisenberg picture is that it gives rise to equations which are formally similar to those in classical mechanics. For example, the Heisenberg picture generalization of the Ehrenfest principle relations of (3.119) and (3.120) reads:

$$\frac{d\hat{x}_H(t)}{dt} = \frac{\hat{p}_{x,H}(t)}{m} \quad \text{and} \quad \frac{d\hat{p}_{x,H}(t)}{dt} = -\frac{\partial V(\hat{x}_H,t)}{\partial \hat{x}_H}. \tag{3.138}$$

Finally, let us examine the equation of motion in the interaction picture introduced in Sect. 3.5, with the unitary operator of (3.91), determined by the noninteracting Hamiltonian $\hat{H}_0$, now transforming the vectors and operators of the Schrödinger picture into their interaction picture analogs. Substituting the reverse transformation to that of (3.93),

$$|\Psi_S(t)\rangle = \hat{S}^{-1}(t)|\Psi_I(t)\rangle = \exp\left(-\frac{i}{\hbar}\hat{H}_0 t\right)|\Psi_I(t)\rangle, \tag{3.139}$$

into the Schrödinger equation (3.94) gives the corresponding dynamical equation in the I-picture:

$$i\hbar\frac{d|\Psi_I(t)\rangle}{dt} = \hat{V}_I|\Psi_I(t)\rangle, \tag{3.140a}$$

with the time evolution now governed by the transformed interaction part $\hat{V}$ of the Hamiltonian (3.90):

$$\hat{V}_I = \hat{S}(t)\hat{V}\hat{S}^{-1}(t) = \exp\left(\frac{i}{\hbar}\hat{H}_0 t\right)\hat{V}\exp\left(-\frac{i}{\hbar}\hat{H}_0 t\right). \tag{3.140b}$$

Therefore, in the interaction picture, the time dependence of operators (3.93) reads:

$$\hat{A}_I(t) = \hat{S}(t)\hat{A}_S\hat{S}^{-1}(t) = \exp\left(\frac{i}{\hbar}\hat{H}_0 t\right)\hat{A}\exp\left(-\frac{i}{\hbar}\hat{H}_0 t\right), \tag{3.141}$$

where the observable $\hat{A} = \hat{A}_S$ is time independent. It can be also expressed by the equivalent expression obtained by the differentiation with respect to time of the preceding equation [see also (3.92a) and (3.92b)]:

$$i\hbar \frac{d}{dt}\hat{A}_I(t) = [\hat{A}_I(t), \hat{H}_0]. \tag{3.142}$$

Therefore, in the interaction picture both state vectors and operators are changing with time: the time evolution of the former is described by the Schrödinger-like (3.140a) and (3.140b), while the latter evolve in time in accordance with the Heisenberg-like (3.142). This form of quantum dynamics thus represents an intermediate level between the Schrödinger and Heisenberg pictures in treating dynamics of quantum objects. Operators depend on time as do operators in the Heisenberg picture for the noninteracting physical system described by the noninteracting Hamiltonian $\hat{H}_0$, while the Schrödinger-like time dependence of the state vectors (or wave functions) is determined solely by the interaction operator $\hat{V}_I$.

References

Löwdin P-O (1955a) Phys Rev 97:1474
Löwdin P-O (1955b) Phys Rev 97:1490
McWeeny R (1989) Methods of molecular quantum mechanics. Academic, London

Chapter 4
Hydrogen-Like Atom

Abstract As an illustration of the basic principles of the Schrödinger wave mechanics presented in the preceding chapter the bonded (stationary) states and the corresponding energy levels of the *one*-electron (hydrogenic) atom are determined analytically. First, the Hamiltonian of this *two*-particle, central-potential system is separated into parts describing the free movement of the *Center-of-Mass* (CM) and the internal motion of electron relative to nucleus, respectively. In the Cartesian CM coordinates $\boldsymbol{R} = (X, Y, Z)$ the eigenstates of the CM problem are the *plane waves* representing the common eigenvectors with the operator of the system overall momentum $\boldsymbol{P}$. The separation of the spherical coordinates $\boldsymbol{R} = (R, \theta, \phi)$ allows one to uniquely specify the *spherical-waves* of the CM motion as simultaneous eigenvectors of the compatible attributes of the CM angular momentum $\boldsymbol{L}$, viz., the square of its length (L^2) and the selected coordinate (L_z), thus expressing them as products of the associated spherical harmonic (angular part) and the spherical Bessel function (radial part). The analogous separation of the *internal* spherical coordinates $\boldsymbol{r} = (r, \vartheta, \varphi)$ expresses the eigenvectors (orbitals) of the relative-motion Hamiltonian as products of the angular functions representing the simultaneous eigenfunctions of the compatible (commuting) observables l^2 and l_z associated with the electron orbital angular momentum $\boldsymbol{l}$, called the orbital spherical harmonics (the associated Legendre polynomials), and the corresponding radial functions (the Laguerre polynomials). Selected properties of these stationary states and the atomic shell structure they determine are discussed, the relation to Bohr's model of the Old Quantum Theory is examined and the system of atomic units (a.u.), convenient in molecular applications, is introduced.

4.1 Separation of Hamiltonian and Center-or-Mass Motion

The hydrogen-like atom consists of an electron of mass m_e in position $\boldsymbol{r}_e$, which exhibits the elementary negative charge $-e$, moving around the positively charged nucleus $+Ze$ of mass M_n in position $\boldsymbol{R}_n$. It represents one of the very few prototype

R.F. Nalewajski, *Perspectives in Electronic Structure Theory*,
DOI 10.1007/978-3-642-20180-6_4, © Springer-Verlag Berlin Heidelberg 2012

systems, the stationary states of which can be determined analytically by solving the associated eigenvalue problem of the system Coulombic Hamiltonian.

It is convenient to separate the movement of the *Center-of-Mass* (CM) of this *two*-particle system, $M = m_e + M_n$, with the coordinates $\boldsymbol{R} = (m_e\boldsymbol{r}_e + M_n\boldsymbol{R}_n)/M$ and momentum $\boldsymbol{P} = \boldsymbol{p}_e + \boldsymbol{P}_n$, where $\boldsymbol{p}_e$ and $\boldsymbol{P}_n$ denote the momenta of its two constituent parts, from the internal motion described by the electron position relative to nucleus, $\boldsymbol{r} = \boldsymbol{r}_e - \boldsymbol{R}_n$, and the associated relative momentum $\boldsymbol{p} = (M_n\boldsymbol{p}_e - m_e\boldsymbol{P}_n)/M \equiv \mu\dot{\boldsymbol{r}} \cong \boldsymbol{p}_e$, where the system reduced mass $\mu = m_eM_n/M \cong m_e$, due to the dominant mass of the heavy nucleus. This allows one to separate the contributions due to these two sets of coordinates/momenta in the classical Hamiltonian function combining the kinetic energies of individual particles and the potential energy due to their Coulomb interaction, $V(r) = -e^2/r$, which depends only on the interparticle distance $r = |\boldsymbol{r}|$:

$$\begin{aligned} H(\boldsymbol{r}_e, \boldsymbol{p}_e, \boldsymbol{R}_n, \boldsymbol{P}_n) &= \frac{1}{2m_e}p_e^2 + \frac{1}{2M_n}P_n^2 - \frac{e^2}{r} \\ &= \frac{1}{2M}P^2 + \left[\frac{1}{2\mu}p^2 - V(r)\right] = H_{\mathrm{CM}}(\boldsymbol{P}) + h(\boldsymbol{p}, \boldsymbol{r}), \end{aligned} \tag{4.1}$$

It should be observed that the CM movement is free (there is no potential of forces acting on CM in H_{CM}) so that $\boldsymbol{P}$ is conserved in time.

These additive contributions to the classical Hamiltonian function give rise to the corresponding energy operators in the position representation [see (3.35) and (3.37)]:

$$\begin{aligned} \hat{\mathrm{H}}(\boldsymbol{R}, \boldsymbol{r}) &= \frac{\hat{\mathbf{P}}^2(\boldsymbol{R})}{2M} + \left[\frac{\hat{\mathbf{p}}^2(\boldsymbol{r})}{2\mu} + V(r)\right] = \frac{-\hbar^2}{2M}\Delta_{\boldsymbol{R}} + \left[\frac{-\hbar^2}{2\mu}\Delta + V(r)\right] \\ &\equiv \hat{\mathrm{H}}_{\mathrm{CM}}(\boldsymbol{R}) + \hat{\mathrm{h}}(\boldsymbol{r}). \end{aligned} \tag{4.2}$$

Here, the separate Hamiltonians $\hat{\mathrm{H}}_{\mathrm{CM}}(\boldsymbol{R})$ and $\hat{\mathrm{h}}(\boldsymbol{r})$, respectively, denote the energy operators of the free movement of CM and of the relative motion of the electron in the field of its nuclear attractor.

Therefore, the stationary Schrödinger equation (3.104)

$$\hat{\mathrm{H}}(\boldsymbol{R}, \boldsymbol{r})\Phi(\boldsymbol{R}, \boldsymbol{r}) = E\Phi(\boldsymbol{R}, \boldsymbol{r}), \tag{4.3a}$$

where the amplitude wave function is given by the product

$$\Phi(\boldsymbol{R}, \boldsymbol{r}) = \Psi_{\mathrm{CM}}(\boldsymbol{R})\psi(\boldsymbol{r}) \tag{4.3b}$$

which separates the two sets of coordinates, reduces into two simpler eigenvalue problems for the two additive energy components:

$$\hat{\mathrm{H}}_{\mathrm{CM}}(\boldsymbol{R})\Psi_{\mathrm{CM}}(\boldsymbol{R}) = E_{\mathrm{CM}}\Psi_{\mathrm{CM}}(\boldsymbol{R}) \quad \text{and} \quad \hat{\mathrm{h}}(\boldsymbol{r})\psi(\boldsymbol{r}) = \varepsilon\psi(\boldsymbol{r}). \tag{4.4}$$

The system total energy is then given by the sum of their eigenvalues,

$$E = E_{\mathrm{CM}} + \varepsilon, \tag{4.3c}$$

measuring the sharply defined kinetic energy E_{CM} of the free motion of the system as a whole, and the internal energy ε of the relative motion of electron around nucleus.

Obviously, in the Cartesian coordinate system $\boldsymbol{R} = (X, Y, Z)$ the kinetic energy operator $\hat{\mathrm{H}}_{\mathrm{CM}}(\boldsymbol{R})$ commutes with the system overall momentum operator $\hat{\mathbf{P}}(\boldsymbol{R})$, since the square of an operator commutes with the operator itself. Therefore, the solutions of the first of these two Schrödinger equations can be sought as eigenfunctions of $\hat{\mathbf{P}}(\boldsymbol{R})$, i.e., as the states corresponding to the sharply specified momentum, represented by the plane waves of (2.76):

$$\Psi_{\mathrm{CM}}(\boldsymbol{R}) = (2\pi\hbar)^{-3/2} \exp(i\boldsymbol{K}\cdot\boldsymbol{R}), \quad \boldsymbol{K} = \boldsymbol{P}/\hbar, \quad E_{\mathrm{CM}} = \hbar^2 K^2/(2M). \tag{4.5}$$

Indeed, operators $\hat{\mathrm{H}}_{\mathrm{CM}}$ and $\hat{\mathbf{P}}$ constitute one of the complete sets of observables for the free motion of this CM "particle," with their common eigensolutions thus providing the full description of this global movement state in quantum mechanics.

4.2 Free Motion in Spherical Coordinates

The alternative set of the complete set of observables commuting with $\hat{\mathrm{H}}_{\mathrm{CM}}(\boldsymbol{R})$ involves the compatible pair of operators associated with the system overall angular momentum $\boldsymbol{L} = \boldsymbol{R} \times \boldsymbol{P}$, say $\hat{\mathbf{L}}^2$ and $\hat{\mathrm{L}}_Z$, $[\hat{\mathbf{L}}^2, \hat{\mathrm{L}}_Z] = 0$ [see (3.71)], which can also be shown to commute with the CM Hamiltonian:

$$[\hat{\mathbf{L}}^2, \hat{\mathrm{H}}_{\mathrm{CM}}] = [\hat{\mathrm{L}}_Z, \hat{\mathrm{H}}_{\mathrm{CM}}] = 0. \tag{4.6}$$

Expressing $L = |\boldsymbol{L}|$ in terms of the lengths of the two defining vectors and the angle α between them gives:

$$L^2 = (RP\sin\alpha)^2 = R^2P^2[1-(\cos\alpha)^2] = R^2P^2 - (\boldsymbol{R}\cdot\boldsymbol{P})^2 \equiv R^2(P^2 - P_R{}^2), \tag{4.7}$$

where $P_R = (\boldsymbol{R}/R)\cdot\boldsymbol{P} \equiv \boldsymbol{e_R}\cdot\boldsymbol{P}$ measures the *radial* component of the total momentum $\boldsymbol{P}$, i.e., its projection onto the unit vector $\boldsymbol{e_R} = \boldsymbol{R}/R$. It can then be verified that the kinetic energy of the CM

$$H_{\mathrm{CM}}(\boldsymbol{P}) = \frac{1}{2M}P^2 = \frac{1}{2M}P_R^2 + \frac{1}{2MR^2}L^2. \tag{4.8}$$

Let us further recall that all quantum mechanical observables must be Hermitian. Therefore, as the momentum operator does not commute with $\boldsymbol{e_R}$ [see (3.56)], in forming the quantum operator corresponding to P_R, one has to symmetrize the defining product, $P_R = ½(\boldsymbol{e_R}\cdot\boldsymbol{P} + \boldsymbol{P}\cdot\boldsymbol{e_R})$, which assures the Hermitian character of the associated operator for the radial component of the overall momentum in the position representation:

$$\hat{\mathrm{P}}_R(\boldsymbol{R}) = \tfrac{1}{2}[\boldsymbol{e_R}\cdot\hat{\mathbf{P}}(\boldsymbol{R}) + \hat{\mathbf{P}}(\boldsymbol{R})\cdot\boldsymbol{e_R}] = -\frac{i\hbar}{2}(\boldsymbol{e_R}\cdot\nabla_{\boldsymbol{R}} + \nabla_{\boldsymbol{R}}\cdot\boldsymbol{e_R}). \tag{4.9}$$

The first part in parentheses measures the component of $\hat{\mathbf{P}}(\boldsymbol{R})$ in direction $\boldsymbol{R}$, which in the spherical coordinates $\boldsymbol{R} = (R, \theta, \phi)$ amounts to the radial differentiation operator

$$\boldsymbol{e_R}\cdot\hat{\mathbf{P}}(\boldsymbol{R}) = -i\hbar\frac{\boldsymbol{R}}{R}\cdot\frac{\partial}{\partial\boldsymbol{R}} = -i\hbar\frac{\partial\boldsymbol{R}}{\partial R}\cdot\frac{\partial}{\partial\boldsymbol{R}} = -i\hbar\frac{\partial}{\partial R}. \tag{4.10}$$

Therefore, the action of $\hat{\mathrm{P}}_R(\boldsymbol{R})$ on the continuous function $f(\boldsymbol{R})$ gives:

$$\begin{aligned}\hat{\mathrm{P}}_R f &= -\frac{i\hbar}{2}\left(\frac{\partial f}{\partial R} + \boldsymbol{e_R}\cdot\nabla_{\boldsymbol{R}} f + f\,\nabla_{\boldsymbol{R}}\cdot\boldsymbol{e_R}\right) = -i\hbar\left(\frac{\partial f}{\partial R} + \frac{f}{R}\right)\\ &= \left(-i\hbar\frac{1}{R}\frac{\partial}{\partial R}R\right)f,\end{aligned} \tag{4.11}$$

thus identifying the radial momentum operator

$$\hat{\mathrm{P}}_R(\boldsymbol{R}) = -i\hbar\frac{1}{R}\frac{\partial}{\partial R}R. \tag{4.12}$$

To summarize, in the adopted spherical coordinates the CM Hamiltonian, which represents in quantum mechanics the physical quantity of (4.8) reads:

$$\hat{\mathrm{H}}_{\mathrm{CM}}(R,\theta,\phi) = \frac{-\hbar^2}{2M}\Delta_{\boldsymbol{R}} = \frac{1}{2M}\hat{\mathrm{P}}_R^2 + \frac{1}{2MR^2}\hat{\mathrm{L}}^2. \tag{4.13}$$

Using next the explicit form of the Laplacian in spherical coordinates,

$$\Delta_{\boldsymbol{R}} = \nabla_{\boldsymbol{R}}^2 = \left(\frac{1}{R}\frac{\partial}{\partial R}R\right)^2 + \frac{1}{R^2}\left[\frac{1}{\sin\theta}\frac{\partial}{\partial\theta}\left(\sin\theta\frac{\partial}{\partial\theta}\right) + \frac{1}{\sin^2\theta}\frac{\partial^2}{\partial\phi^2}\right], \tag{4.14}$$

one identifies

$$\hat{\mathrm{L}}^2(\theta,\phi) = -\hbar^2\left[\frac{1}{\sin\theta}\frac{\partial}{\partial\theta}\left(\sin\theta\frac{\partial}{\partial\theta}\right) + \frac{1}{\sin^2\theta}\frac{\partial^2}{\partial\phi^2}\right]. \tag{4.15}$$

In addition, by a straightforward chain-rule transformation of derivatives in (3.70), one finds:

$$\hat{\mathrm{L}}_Z(\phi) = -i\hbar\frac{\partial}{\partial\phi}. \tag{4.16}$$

Therefore, it directly follows from (4.13) and (4.15) that the first commutation relation of (4.6) indeed holds, $\left[\hat{\mathrm{L}}^2,\ \hat{\mathrm{H}}_{\mathrm{CM}}\right] = 0$, since $\hat{\mathrm{L}}^2$ commutes with itself and does not act on the radial coordinate R, thus also commuting with the first, radial part of $\hat{\mathrm{H}}_{\mathrm{CM}}(R,\theta,\phi)$. The second commutation relation of (4.6) directly follows from the commutation relations between observables representing the Cartesian components of $\boldsymbol{L}$ [see (2.34)]:

$$\begin{aligned}
[\hat{\mathrm{L}}^2,\hat{\mathrm{L}}_Z] &= [\hat{\mathrm{L}}_X^2 + \hat{\mathrm{L}}_Y^2 + \hat{\mathrm{L}}_Z^2,\hat{\mathrm{L}}_Z] = [\hat{\mathrm{L}}_X^2 + \hat{\mathrm{L}}_Y^2,\hat{\mathrm{L}}_Z] = [\hat{\mathrm{L}}_X^2,\hat{\mathrm{L}}_Z] + [\hat{\mathrm{L}}_Y^2,\hat{\mathrm{L}}_Z] \\
&= \hat{\mathrm{L}}_X[\hat{\mathrm{L}}_X,\hat{\mathrm{L}}_Z] + [\hat{\mathrm{L}}_X,\hat{\mathrm{L}}_Z]\hat{\mathrm{L}}_X + \hat{\mathrm{L}}_Y[\hat{\mathrm{L}}_Y,\hat{\mathrm{L}}_Z] + [\hat{\mathrm{L}}_Y,\hat{\mathrm{L}}_Z]\hat{\mathrm{L}}_Y \\
&= -i\hbar(\hat{\mathrm{L}}_X\hat{\mathrm{L}}_Y + \hat{\mathrm{L}}_Y\hat{\mathrm{L}}_X) + i\hbar(\hat{\mathrm{L}}_Y\hat{\mathrm{L}}_X + \hat{\mathrm{L}}_X\hat{\mathrm{L}}_Y) = 0,
\end{aligned} \tag{4.17}$$

where we have used the elementary commutators of (3.71):

$$[\hat{\mathrm{L}}_X,\hat{\mathrm{L}}_Z] = -i\hbar\hat{\mathrm{L}}_Y \quad \text{and} \quad [\hat{\mathrm{L}}_Y,\hat{\mathrm{L}}_Z] = i\hbar\hat{\mathrm{L}}_X.$$

It thus follows from (4.6) that the eigenfunctions $\Psi(R,\theta,\phi)$ of $\hat{\mathrm{H}}_{\mathrm{CM}}(R,\theta,\phi)$ should also satisfy the following simultaneous eigenvalue problems:

$$\begin{aligned}
&\hat{\mathrm{H}}_{\mathrm{CM}}(R,\theta,\phi)\Psi(R,\theta,\phi) = E_{\mathrm{CM}}\Psi(R,\theta,\phi), \quad E_{\mathrm{CM}} = \hbar^2K^2/(2M); \\
&\hat{\mathrm{L}}^2(\theta,\phi)\Psi(R,\theta,\phi) = L^2\Psi(R,\theta,\phi) \quad \text{and} \\
&\hat{\mathrm{L}}_Z(\phi)\Psi(R,\theta,\phi) = L_Z\Psi(R,\theta,\phi).
\end{aligned} \tag{4.18}$$

Hence, these common eigenfunctions can be written as products of the radial factor $f_{K,l}(R)$ and one of the angular momentum eigenfunctions $\{Y_l^m(\theta,\phi)\}$, called the *spherical harmonics*,

$$\Psi_{\mathrm{CM}}(R,\theta,\phi) = f_{K,l}(R)Y_l^m(\theta,\phi). \tag{4.19a}$$

The latter represent the common eigenfunctions of the two compatible angular momentum observables:

$$\begin{aligned}
&\hat{\mathrm{L}}^2(\theta,\phi)Y_l^m(\theta,\phi) = l(l+1)\hbar^2Y_l^m(\theta,\phi), \quad l = 0,1,2,\ldots, \\
&\hat{\mathrm{L}}_Z(\phi)Y_l^m(\theta,\phi) = m\hbar\,Y_l^m(\theta,\phi), \quad m = -l,-l+1,\ldots,l-1,l,
\end{aligned} \tag{4.20}$$

where the integral quantum numbers l and m determine the allowed spectrum of these physical quantities: $L^2 = l(l+1)\hbar^2$ and $L_Z = m\hbar$.

Inserting the product function of (4.19a) into the Schrödinger equation (4.18) for the CM motion gives the following radial equation for $f_{K,l}(R)$:

$$\left[-\left(\frac{1}{R}\frac{d^2}{dR^2}R\right)+\frac{l(l+1)}{R^2}\right]f_{K,l}(R)=K^2 f_{K,l}(R), \tag{4.21}$$

where we have used the identity

$$\left(\frac{1}{R}\frac{d}{dR}R\right)^2=\frac{1}{R}\frac{d^2}{dR^2}R. \tag{4.22}$$

Upon substituting $z = KR$, this differential equation is transformed into the *spherical Bessel equation*,

$$\frac{d^2 f_{K,l}(z)}{dz^2}+\frac{2}{z}\frac{df_{K,l}(z)}{dz}+\left[1-\frac{l(l+1)}{z^2}\right]f_{K,l}(z)=0, \tag{4.23}$$

the regular solutions of which define the *spherical Bessel functions*:

$$j_l(KR)=\left(-\frac{R}{K}\right)^l\left(\frac{1}{R}\frac{d}{dR}\right)^l j_0(KR), \quad j_0(KR)=\frac{\sin(KR)}{KR}, \tag{4.24}$$

satisfying the following orthogonality relation for the continuous spectrum of K:

$$\int_0^\infty j_l(KR)j_{l'}(K'R)\,R^2 dR=\frac{\pi}{2K^2}\delta(K-K')\delta_{l,l'}. \tag{4.25}$$

To summarize, it is natural in the spherical coordinate system to specify the stationary states of the free motion of the CM in the hydrogen-like atom as product of the spherical Bessel function and the spherical harmonic:

$$\Psi_{\mathrm{CM}}(R,\theta,\phi)=j_l(KR)Y_l^m(\theta,\phi). \tag{4.19b}$$

4.3 Eigenfunctions of Angular Momentum Operators

The spherical harmonics can be similarly factorized into eigenfunction of $\hat{\mathrm{L}}_Z(\phi)$,

$$\Phi_m(\phi)=(2\pi)^{-1/2}\exp(im\,\phi), \tag{4.26}$$

and the remaining factor $\Theta_l^m(\theta)$:

$$Y_l^m(\theta,\phi) = \Theta_l^m(\theta)\Phi_m(\phi). \tag{4.27}$$

Substituting this expression into (4.20) and using the explicit form (4.15) of the operator $\hat{\mathrm{L}}^2(\theta,\phi)$ then give the following differential equation for $\Theta_l^m(\theta)$:

$$\frac{1}{\sin\theta}\frac{\partial}{\partial\theta}\left(\sin\theta\frac{\partial\Theta_l^m(\theta)}{\partial\theta}\right) + \left[l(l+1) - \frac{m^2}{\sin^2\theta}\right]\Theta_l^m(\theta) = 0. \tag{4.28}$$

The subsequent substitution $-1 \le x = \cos\theta \le 1$ then transforms the previous equation into a more familiar form of the differential equation defining the *associated Legendre polynomials*,

$$\frac{d}{dx}\left[(1-x^2)\frac{d\Theta_l^m(x)}{dx}\right] + \left[l(l+1) - \frac{m^2}{1-x^2}\right]\Theta_l^m(x) = 0. \tag{4.29}$$

For $m = 0$ it reduces to the differential equation defining the *Legendre polynomials* of order l, $\Theta_l^0(x) \equiv P_l(x)$,

$$\frac{d}{dx}\left[(1-x^2)\frac{dP_l(x)}{dx}\right] + l(l+1)\,P_l(x) = 0. \tag{4.30}$$

Its solutions can be written in the compact (*Rodrigues*) form:

$$P_l(x) = \frac{1}{2^l l!}\frac{d^l}{dx^l}(x^2-1)^l. \tag{4.31}$$

The remaining *associated* Legendre polynomials of degree l and order $|m| \le l$, $\Theta_l^m(x) \equiv N_{l,m}P_l^{|m|}(x)$, which satisfy (4.28) for $m \neq 0$, can then be obtained from these polynomials by repeated differentiations with respect to x:

$$P_l^{|m|}(x) = (-1)^{|m|}(1-x^2)^{|m|/2}\frac{d^{|m|}P_l(x)}{dx^{|m|}}. \tag{4.32}$$

The normalization constant $N_{l,m}$ reflecting the proportionality between $\Theta_l^m(x)$ and $P_l^{|m|}(x)$ is to be determined from the following condition:

$$\int_{-1}^{1}\left(\Theta_l^m(x)\right)^2 dx = 1 \Rightarrow N_{l,m} = \left[\frac{(2l+1)(l-|m|)!}{2(l+|m|)!}\right]^{1/2}. \tag{4.33}$$

The spherical harmonics $\{Y_l^m(\theta,\phi) \equiv Y_l^m(\Omega)\}$ satisfy the usual orthonormality conditions,

$$\int_0^{\pi} \sin\theta\, d\theta \int_0^{2\pi} d\phi\, Y_l^{m^*}(\theta,\phi)\, Y_{l'}^{m'}(\theta,\phi) = \int Y_l^{m^*}(\Omega)\, Y_{l'}^{m'}(\Omega)\, d\Omega = \delta_{l,l'}\delta_{m,m'}, \quad (4.34)$$

involving the integration over the whole range of 4π steradians of the solid angle Ω, i.e., over all possible directions of the unit vector in the physical space:

$$\int_0^{\pi} \sin\theta\, d\theta \int_0^{2\pi} d\phi = \int_{-1}^{1} d\cos\theta \int_0^{2\pi} d\phi = \int d\Omega = 4\pi. \quad (4.35)$$

They are automatically satisfied when the two factors in (4.27) are chosen to obey the associated partial relations:

$$\int_0^{\pi} \sin\theta\, d\theta\, \Theta_l^m(\theta)\Theta_{l'}^m(\theta) = \delta_{l,l'} \quad \text{and} \quad \int_0^{2\pi} d\phi\, \Phi_m^*(\phi)\Phi_{m'}(\phi) = \delta_{m,m'}. \quad (4.36)$$

Clearly, the same type of spherical functions describes the eigenstates of the internal (orbital) angular momentum $\boldsymbol{l} = \boldsymbol{r} \times \boldsymbol{p}$, associated with the relative motion of electron around the atomic nucleus. The corresponding internal spherical harmonics now depend on the angular coordinates specifying the direction of the relative position vector $\boldsymbol{r} = (r, \vartheta, \varphi)$ of the system electron,

$$Y_l^m(\vartheta,\varphi) = \Theta_l^m(\vartheta)\Phi_m(\varphi), \quad (4.37)$$

and satisfy the associated eigenvalue problems of the compatible operators of the orbital angular momentum of (3.70) and (3.71):

$$\begin{aligned} \hat{\mathrm{l}}^2(\vartheta,\varphi)Y_l^m(\vartheta,\varphi) &= l(l+1)\hbar^2 Y_l^m(\vartheta,\varphi), \quad && l = 0,1,2,\ldots, \\ \hat{\mathrm{l}}_z(\varphi)Y_l^m(\vartheta,\varphi) &= m\hbar\, Y_l^m(\vartheta,\varphi), && m = -l,-l+1,\ldots,l-1,l, \end{aligned} \quad (4.38)$$

The associated commutation relations are given by (3.71) and those involving the internal Hamiltonian $\hat{\mathrm{h}}(\boldsymbol{r}) = \hat{\mathrm{h}}(r,\vartheta,\varphi)$:

$$[\hat{\mathrm{l}}^2, \hat{\mathrm{h}}] = [\hat{\mathrm{l}}_z, \hat{\mathrm{h}}] = 0, \quad (4.39)$$

where in full analogy to (4.13)

$$\hat{\mathrm{h}}(r,\vartheta,\varphi) = \frac{-\hbar^2}{2\mu}\Delta(r,\vartheta,\varphi) + V(r) = \frac{1}{2\mu}\hat{\mathrm{p}}_r^2 + \frac{1}{2\mu r^2}\hat{\mathrm{l}}^2(\vartheta,\varphi) + V(r); \quad (4.40)$$

here, the operator of the radial component of the orbital momentum [see (4.12)],

$$\hat{\mathrm{p}}_r(r) = -i\hbar\frac{1}{r}\frac{\partial}{\partial r}r, \tag{4.41}$$

and the operators of the orbital angular momentum in the spherical coordinates of the system electron read [see (4.15) and (4.16)]:

$$\hat{\mathrm{l}}^2(\vartheta,\varphi) = -\hbar^2\left[\frac{1}{\sin\vartheta}\frac{\partial}{\partial\vartheta}\left(\sin\vartheta\frac{\partial}{\partial\vartheta}\right)+\frac{1}{\sin^2\vartheta}\frac{\partial^2}{\partial\varphi^2}\right],\quad \hat{\mathrm{l}}_z(\varphi) = -i\hbar\frac{\partial}{\partial\varphi}. \tag{4.42}$$

It thus follows from (4.39) that the eigenfunctions of the internal Schrödinger equation (4.4) can be factorized in the form analogous to that in (4.19a):

$$\psi_{n,l,m}(r,\vartheta,\varphi) = R_{n,l}(r)Y_l^m(\vartheta,\varphi). \tag{4.43}$$

These functions represent the simultaneous eigenstates of the associated three (internal) commuting observables:

$$\begin{aligned}
&\hat{\mathrm{h}}(r,\vartheta,\varphi)\psi_{n,l,m}(r,\vartheta,\varphi) = \varepsilon_n\psi_{n,l,m}(r,\vartheta,\varphi)\\
&\hat{\mathrm{l}}^2(\vartheta,\varphi)\psi_{n,l,m}(r,\vartheta,\varphi) = l(l+1)\hbar^2\psi_{n,l,m}(r,\vartheta,\varphi), \quad l = 0,1,2,\ldots,\\
&\hat{\mathrm{l}}_z(\varphi)\psi_{n,l,m}(r,\vartheta,\varphi) = m\hbar\psi_{n,l,m}(r,\vartheta,\varphi), \quad m = -l,-l+1,\ldots,l-1,l.
\end{aligned} \tag{4.44}$$

4.4 Radial Eigenfunctions and Energy Levels

To obtain the radial functions $\{R_{n,l}(r)\}$ and the admissible energy levels $\{\varepsilon_n\}$ of the *bonded*, stationary states of the internal motions of the electron around the nucleus in the hydrogen-like atom, when $\varepsilon_n < 0$, one substitutes the product function of (4.43) into the first eigenvalue problem of the preceding equation. This gives the radial Schrödinger equation in the form [compare (4.21)]:

$$\left[-\frac{\hbar^2}{2\mu}\left(\frac{1}{r}\frac{d^2}{dr^2}r\right)+\frac{\hbar^2 l(l+1)}{2\mu r^2}-\frac{Ze^2}{r}+|\varepsilon_n|\right]R_{n,l}(r) = 0. \tag{4.45}$$

It can be subsequently simplified by the substitution $U_{n,l}(r) = rR_{n,l}(r)$ and by the introduction of the redefined coefficients in this differential equation: the energy parameter $\kappa_n{}^2 = 2\mu|\varepsilon_n|/\hbar^2$, the energy-scaled (dimensionless) radial distance $\rho_n = 2\kappa_n r$, and a reduced measure of the nuclear charge $\zeta_n = Z/(\kappa_n a_0)$, where $a_0 = \hbar^2/(\mu e^2) = 0.5292 \times 10^{-10}$ m denotes the radius of the first Bohr's orbit in the hydrogen atom,

$$\frac{d^2U_{n,l}(\rho_n)}{d\rho_n^2} - \frac{l(l+1)}{\rho_n^2}U_{n,l}(\rho_n) + \left(\frac{\zeta_n}{\rho_n} - \frac{1}{4}\right)U_{n,l}(\rho_n) = 0. \tag{4.46}$$

In the asymptotic region of very large distances $\rho_n \to \infty$ it thus reduces to a simple differential equation

$$\frac{d^2U_{n,l}(\rho_n)}{d\rho_n^2} = \frac{U_{n,l}(\rho_n)}{4}, \quad \rho_n \to \infty, \tag{4.47}$$

the general solution of which reads: $U_{n,l}(\rho_n) \sim A\exp(-\rho_n/2) + B\exp(\rho_n/2)$, where A and B are integration constants. For this radial function to be finite in this limit $B = 0$, so that $U_{n,l}(\rho_n) \sim A\exp(-\rho_n/2)$ ($\rho_n \to \infty$).

In the other extreme region of $\rho_n \to 0$ the radial equation (4.46) becomes

$$\frac{d^2U_{n,l}(\rho_n)}{d\rho_n^2} = \frac{l(l+1)}{\rho_n^2}U_{n,l}(\rho_n), \quad \rho_n \to 0. \tag{4.48}$$

Inserting into the preceding equation the trial function $U_{n,l}(\rho_n) = {\rho_n}^{\xi}$ then gives the following quadratic equation for the critical exponent ξ:

$$\xi(\xi - 1) = l(l+1) \quad \Rightarrow \quad \{\xi_1 = -l, \xi_2 = l+1\}, \tag{4.49}$$

thus predicting the general solution near the nucleus in the form $U_{n,l}(\rho_n) \sim A'{\rho_n}^{-l} + B'{\rho_n}^{l+1}$. The well-behaving (finite) solution thus results only for $A' = 0$: $U_{n,l}(\rho_n) \sim B'{\rho_n}^{l+1}$ ($\rho_n \to 0$).

The above analysis suggests the following general form of the radial wave function,

$$U_{n,l}(\rho_n) = \exp(-\rho_n/2){\rho_n}^{l+1}V(\rho_n), \tag{4.50a}$$

which automatically guarantees the correct behavior in both these asymptotic regions, including the additional (finite) factor $V(\rho_n)$ defined by the power series:

$$V(\rho_n) = \sum_{i=0}^{\infty} a_i \rho_n^i. \tag{4.50b}$$

Its substitution into (4.46) gives the following differential equation for determining this unknown radial factor:

$$\left[\rho_n \frac{d^2}{d\rho_n^2} + (2l + 2 - \rho_n)\frac{d}{d\rho_n} - (l + 1 - \zeta_n)\right]V(\rho_n) = 0. \tag{4.51}$$

As $V(\rho_n)$ represents the power series this differential equation effectively determines the (energy dependent) recursive relation between the coefficients $\{a_i\}$. Indeed, the left-hand side of this equation also constitutes the power series which vanishes only when coefficients at all powers of ρ_n are simultaneously equal to zero. This requirement generates the following recursion relation between the neighboring coefficients in (4.50b) for the representative term ${\rho_n}^k$:

$$\frac{a_{k+1}}{a_k} = \frac{(k+l+1)-\zeta_n}{(k+1)(k+2l+2)}. \tag{4.52}$$

This power series thus begins with the constant term $a_0 \neq 0$ and it must terminate at some *finite* maximum power. Indeed, if it failed to do so, in the limit of very large values of k, i.e., $k \to \infty$, $a_{k+1}/a_k \to 1/k$, which is characteristic of the power series expansion of the function $\exp(\rho_n) = \sum_{i=0}^{\infty} \frac{1}{i!}\rho_n^i$. Thus, should the power series in (4.50b) fail to terminate, the radial wave function $U_{n,l}(\rho_n)$ would become infinite (ill-behaved) at $\rho_n \to \infty$, diverging as $\exp(\rho_n/2)$.

Therefore, the truncation of this series into the polynomial is the crucial requirement for the radial wave function to well behave at large distances. A reference to (4.52) shows that the series will indeed become the polynomial of degree $k = j$ when $a_{j+1}/a_j = 0$, which takes place only for

$$(j+l+1) \equiv n = \zeta_n \quad \text{or} \quad n^2 = {\zeta_n}^2 = Z^2/({\kappa_n}^2 {a_0}^2) = Z^2\hbar^2/(2\mu|\varepsilon_n|{a_0}^2). \tag{4.53}$$

As, by definition, j is a non-negative integer and $l = 0, 1, \ldots$ [see (4.44)], the *principal* quantum number n, which identifies the electronic "shells," must also be a positive integer $n = 1, 2, \ldots$. It is subject to the restriction $n > l$, since the degree of the polynomial after which the series expansion terminates $j = n - (l+1) \geq 0$, $V(\rho_n) = V_{n,l}(\rho_n)$, so that there are n values of the angular momentum quantum number l consistent with the given n: $l = 0, 1, \ldots, n-1$.

To summarize, the radial wave function of the internal states of the *one*-electron atom reads:

$$U_{n,l}(\rho_n) = A_{n,l}\exp(-\rho_n/2){\rho_n}^{l+1} V_{n,l}(\rho_n) = A_{n,l}\exp(-\rho_n/2){\rho_n}^{l+1} \sum_{i=0}^{n-l-1} a_i \rho_n^i, \tag{4.54}$$

where $A_{n,l}$ stands for the appropriate normalization constant. The polynomials $V_{n,l}(\rho_n)$, the solutions of the differential equation (4.51), are known as the *associated Laguerre polynomials*:

$$V_{n,l}(\rho_n) = L_{n-l-1}^{2l+1}(\rho_n) = \sum_{i=0}^{n-l-1} \frac{(-1)^i[(n+l)!]^2}{i!(n-l-1-i)!(2l+1+i)!}\rho_n^i. \tag{4.55}$$

Again, the associated Laguerre polynomials $L_p^q(\rho_n)$ of degree p and order q, are compactly represented by the following formulas of Rodrigues in terms of the *Laguerre polynomials* $L_p(\rho_n)$ of degree p:

$$L_p^{q=0}(\rho_n) = L_p(\rho_n) = \exp(\rho_n)\frac{d^p}{d\rho_n^p}[\rho_n^p \exp(-\rho_n)], \tag{4.56}$$

$$L_p^q(\rho_n) = (-1)^q \frac{d^q}{d\rho_n^q} L_{q+p}(\rho_n). \tag{4.57}$$

Thus, the condition of the well-behaved wave function at infinity gives rise to a quantization of the internal energy $\varepsilon_n = -|\varepsilon_n|$ of electron in the hydrogen-like atom:

$$|\varepsilon_n| = \frac{Z^2}{2n^2}\left(\frac{\mu e^4}{\hbar^2}\right) \cong \frac{Z^2}{2n^2}\left(\frac{m_e e^4}{\hbar^2}\right) \equiv \frac{Z^2}{2n^2}\text{hartrees} \equiv \frac{Z^2}{n^2}\text{rydbergs}, \tag{4.58}$$

$$1 \text{ hartree} = 2 \text{ rydbers} = 2|\varepsilon_0| = 27.21161 \text{ eV} = 4.359814 \times 10^{-18} \text{ J}, \tag{4.59}$$

where we have introduced two popular units of energy used in atomic and molecular physics. For $Z = 1$ this energy spectrum reproduces that following from the historically first quantum model of the hydrogen atom proposed by Bohr in the Old Quantum Theory. One also observes that the scaling factor $\kappa_n = Z/(na_0)$ of the radial distance ρ_n is shell-specific.

4.5 Orbital Degeneracy and Electron Distribution

The energy spectrum of (4.58) becomes very dense for large values of the principal quantum number, with $\varepsilon_\infty = 0$, and becomes continuous for the nonbonded (scattering) states, for $\varepsilon > 0$, when the electron can exhibit the infinite separation from the nucleus. Therefore, such energy-continuum states of the hydrogen-like atom describe the ionization processes, involving a removal of the system electron.

The wave functions of (4.43) define the admissible (linearly independent) bonded states of electron in the hydrogen-like atom. Since the value of the allowed internal energy of (4.58) depends solely on the principal quantum number n the number of combinations of the remaining quantum numbers, the *secondary* (orbital) quantum number l and *magnetic* (azimuthal) quantum number m, which are consistent with the given value of n, determines the system overall orbital degeneracy. For each value of the quantum number associated with the length of the orbital angular momentum, $l = 0, 1, \ldots, n - 1$, which identifies specific atomic "subshells," there are $2l + 1$ admissible values of the azimuthal quantum number m (4.44) determining the spatial orientation of the angular momentum vector (Fig. 1.2). Hence, the total orbital degeneracy g_n of the given eigenvalue ε_n in hydrogen-like atom, i.e., the number of independent stationary (bonded) electronic states belonging to this energy level:

$$g_n = \sum_{l=0}^{n-1} (2l + 1) = n^2. \quad (4.60)$$

This *orbital*-degeneracy is doubled if the two spin states of an electron, $\alpha(\sigma)$ or $\beta(\sigma)$, depending on the discrete spin variable $\sigma = (-½, ½)$, are taken into account, as each *Atomic Orbital* (AO) $\psi_{n,l,m}(r, \vartheta, \varphi)$ can be combined with any of these spin functions into the corresponding *Spin Orbitals* (SOs)

$$\psi^{\sigma}_{n,l,m}(r, \vartheta, \varphi, \sigma) = \begin{cases} \psi^{+}_{n,l,m}(r, \vartheta, \varphi, \sigma) = \psi_{n,l,m}(r, \vartheta, \varphi)\alpha(\sigma) \\ \psi^{-}_{n,l,m}(r, \vartheta, \varphi, \sigma) = \psi_{n,l,m}(r, \vartheta, \varphi)\beta(\sigma) \end{cases}. \quad (4.61)$$

Hence, in hydrogen-like atom all energy levels with $n > 1$ exhibit some *orbital* degeneracy, while the ground $1s$ state,

$$\psi_{1,0,0}(r, \vartheta, \varphi) = \left(\frac{Z^3}{\pi a_0^3}\right)^{1/2} \exp(-Zr/a_0), \quad (4.62)$$

exhibits only the double *spin* degeneracy.

The appearance of the degenerate quantum states can be often ascribed to some apparent symmetry in the physical system. For example, the degeneracy with respect to the magnetic quantum number m reflects the *central* potential feature of the *one*-electron atom. It originates from the absence of the preferred spatial direction and hence from the invariance with regard to rigid rotations about the origin. The degeneracy of states corresponding to different values of l consistent with the given n is peculiar to the Coulomb potential. Any departure from the strict $1/r$ dependence, e.g., in *many*-electron atoms, will remove this ("accidental") degeneracy.

The atomic orbitals of (4.43) are complex for $m \neq 0$, because of the $\Phi_m(\varphi)$ factor in $Y_l^m(\vartheta, \varphi)$ (4.37), with only $m = 0$ functions,

$$\psi_{n,0,0}(r) \equiv ns, \quad \psi_{n,1,0}(r, \vartheta) \equiv np_z, \quad \psi_{n,2,0}(r, \vartheta) \equiv nd_{z^2}, \text{ etc.}, \quad (4.63)$$

which do not depend on the spherical angle φ, are automatically real. However, one can always transform the pair of the complex–conjugate orbital factors $\Phi_m(\varphi) = \Phi^*_{-m}(\varphi)$ for $m > 0$ into two *real* combinations by extracting their real and imaginary parts:

$$\begin{aligned} \mathrm{Re}[\Phi_m(\varphi)] &= \frac{1}{2}[\Phi_m(\varphi) + \Phi_{-m}(\varphi)] = \cos(m\varphi), \\ \mathrm{Im}[\Phi_m(\varphi)] &= \frac{1}{2i}[\Phi_m(\varphi) - \Phi_{-m}(\varphi)] = \sin(m\varphi). \end{aligned} \quad (4.64)$$

Since such real combinations involve functions with the same length of the angular momentum, this physical quantity still remains sharply specified in these combined states. However, the real orbitals are no longer eigenfunctions of the z-component of the angular momentum, as they combine functions with different eigenvalues of this observable. Therefore, in a single measurement of l_z, one has probability ½ of observing either $l_z = m\hbar$ or $l_z = -m\hbar$ and hence $\langle l_z \rangle = 0$ in such linear combinations $\psi_{n,l,\pm m}$ of the complex orbitals $\psi_{n,l,m}$.

The AO parity, i.e., the symmetry (g) or antisymmetry (u) property of $\psi_{n,l,m}$ with respect to the *inversion* operator $\hat{\mathrm{i}}$, which reverses the internal Cartesian coordinates, $\hat{\mathrm{i}}(x, y, z) = (-x, -y, -z)$, is determined solely by the associated property of the spherical harmonic factor $Y_l^m(\vartheta, \varphi)$, since such operation of reversing directions of the coordinate system does not affect the radial distance r. Indeed, in the spherical coordinates $\hat{\mathrm{i}}(r, \vartheta, \varphi) = (r, \pi - \vartheta, \varphi + \pi)$ and hence the action of $\hat{\mathrm{i}}$ on $\Phi_m(\varphi)$ gives:

$$\hat{\mathrm{i}} \exp(im\varphi) = [\exp(i\pi)]^m \exp(im\varphi) = (-1)^{|m|} \exp(im\varphi). \tag{4.65}$$

Thus, the magnetic quantum number m itself determines the parity of $\Phi_m(\varphi)$, which is symmetric (g) [antisymmetric (u)] with respect to inversion for the even (odd) values of m.

Next, let us examine the parity of the other, ϑ-dependent part of the angular function, $\Theta_l^m(x) \equiv N_{l,m} P_l^{|m|}(x)$, $x = \cos\vartheta$. Since $\hat{\mathrm{i}} \cos\vartheta = \cos(\pi - \vartheta) = -\cos\vartheta$ and the associated Legendre polynomial of degree l and order m, $P_l^{|m|}(x)$, is obtained by differentiating $(l + |m|)$-times the *even* function $(x^2 - 1)^l$ of the argument x in (4.31) and (4.32), the action of the inversion operation on this angular factor of the wave function gives:

$$\hat{\mathrm{i}} P_l^{|m|}(\cos\vartheta) = (-1)^{l+|m|} \exp(im\varphi) P_l^{|m|}(\cos\vartheta). \tag{4.66}$$

It thus follows from the preceding two equations that the overall parity of the angular function is determined by the parity of the orbital quantum number l:

$$\hat{\mathrm{i}} Y_l^m(\vartheta, \varphi) = (-1)^{l+2|m|} Y_l^m(\vartheta, \varphi) = (-1)^l Y_l^m(\vartheta, \varphi). \tag{4.67}$$

Atomic orbitals posses a number of nodal surfaces on which $\psi_{n,l,m} = 0$, as indeed required to satisfy the orthogonality relations, which guarantee the linear independence of AO. For this purpose it is customary to examine the spatial properties of the *real* AO (4.64),

$$\begin{aligned}\psi_{n,l,\pm m}(r, \vartheta, \varphi) &\propto r^l \exp(\kappa_n r) L_{n-l-1}^{2l+1}(2\kappa_n r) P_l^m(\cos\vartheta) \begin{cases} \cos(m\varphi) \\ \sin(m\varphi) \end{cases} \\ &= R_{n,l}(r) Y_l^{x^a, y^b, z^c},\end{aligned} \tag{4.68}$$

e.g., $\psi_{n,1,\pm1} = \{np_x, np_y\}$, $\psi_{n,2,\pm1} = \{nd_{xz}, nd_{yz}\}$, $\psi_{n,2,\pm2} = \{nd_{xy}, nd_{x^2-y^2}\}$, etc. The angular functions $Y_l^{x^a,y^b,z^c}$ of the real AO are simple functions of the respective integer powers $\{a, b, c\}$ of the electron Cartesian coordinates, which are indicated in their symbolic notation.

By examining the individual factors in the preceding expression, one first realizes that there are $l - m$ values of ϑ for which $P_l^m(\cos\vartheta)$ vanishes and the real/imaginary parts of (4.64) vanish at m values of the azimuth. Moreover, the associated Laguerre polynomial vanishes at $n - l - 1$ values of r; for $l \neq 0$ the radial factor r^l has also the "node" at $r = 0$. Hence, disregarding the latter, the total number of nodal surfaces in AO at finite distances is $n - 1$, including $n - l - 1$ radial and l angular surfaces.

It thus follows from these considerations that only the $l = n - 1$ AO, e.g., 1s, 2p, 3d, 4f, etc., have zero radial nodal-surfaces, thus exhibiting only one maximum in their *radial* probability density, which is customarily used to represent the distribution of electrons in atoms. More specifically, using the probability density of finding the electron at point $\boldsymbol{r} = (r, \vartheta, \varphi) = (r, \Omega)$,

$$\rho_{n,l,m}(r,\vartheta,\varphi) = |\psi_{n,l,m}(r,\vartheta,\varphi)|^2 = R_{n,l}^2(r)\left|Y_l^m(\vartheta,\varphi)\right|^2, \tag{4.69}$$

one finds from (4.34) the associated radial probability of locating the electron in the infinitesimal radial range, between the concentric spheres of radii r and $r + dr$,

$$P_{n,l}(r,dr) = r^2 R_{n,l}^2(r)\ dr \int \left|Y_l^m(\Omega)\right|^2 d\Omega = r^2 R_{n,l}^2(r)\ dr, \tag{4.70}$$

where we have recognized the angular normalization of (4.34). Hence, the *radial* probability density reads:

$$\rho_{n,l}^{rad.}(r) \equiv \frac{dP(r,dr)}{dr} = r^2 R_{n,l}^2(r). \tag{4.71}$$

For example, for the ground state of the hydrogen-like atom (4.62), for which $R_{1,0}(r) \propto \exp(-Zr/a_0)$ and hence $\rho_{1,0}^{rad.}(r) \propto r^2 \exp(-2Zr/a_0)$, the maximum of the radial distribution is observed at $r_{max.}(Z) = a_0/Z$. This radial probability density also predicts the following average values of r and r^2:

$$\langle r(Z)\rangle = \int_0^\infty r\rho_{1,0}^{rad.}(r)dr = 4\left(\frac{Z}{a_0}\right)^3 \int_0^\infty \exp\left(-\frac{2Zr}{a_0}\right) r^3 dr = \frac{3}{2}\frac{a_0}{Z}, \tag{4.72}$$

$$\langle r^2(Z)\rangle = \int_0^\infty r^2\rho_{1,0}^{rad.}(r)dr = 4\left(\frac{Z}{a_0}\right)^3 \int_0^\infty \exp\left(-\frac{2Zr}{a_0}\right) r^4 dr = 3\left(\frac{a_0}{Z}\right)^2, \tag{4.73}$$

where we have used the typical integral $\int_0^\infty y^n \exp(-by)\, dy = n!/b^{n+1}$. Hence, the square of the dispersion σ_r in the radial distance of this *one*-electron atom reads:

$$\sigma_r^2 = \left\langle r^2(Z) \right\rangle - \langle r(Z) \rangle^2 = \frac{3}{4}\left(\frac{a_0}{Z}\right)^2 \quad \text{or} \quad \sigma_r = \left(\frac{\sqrt{3}}{2}\right)\left(\frac{a_0}{Z}\right). \tag{4.74}$$

Therefore, in the hydrogen atom the maximum radial probability is found at $r_{max.}(Z = 1) = a_0$ as already predicted by Bohr. It should be emphasized, however, that the latter model has invoked the classical ("flat") planetary picture of the electron movements around the nucleus, while the quantum-mechanical perspective predicts the correct spherical distribution of the electron probability density around the nucleus.

The radial densities for the remaining AO in this prototype atomic system are well known and available in practically every textbook of quantum chemistry or elementary quantum mechanics. Let us only recall here that with the increasing principal quantum number $n = 1, 2, 3$, which determines the successive electronic shells, the average distance from the nucleus increases. The atomic subshells, identified by the alternative values of the orbital quantum number l consistent with the given principal quantum number n, exhibit the decreasing trend with increasing l in their most probable and average distances from the nucleus, e.g., $\langle r_{3d} \rangle < \langle r_{3p} \rangle < \langle r_{3s} \rangle$. This observation reflects the intervention of the orthogonality constraints with respect to the stationary states exhibiting the same symmetry and lower energy, for which the electron is on average distributed closer to the nucleus. These requirements effectively shift the probability of the outer subshells away from the nucleus. Indeed, the $3s$ orbital must be orthogonal to both $1s$ and $2s$ states, the $3p$ state is only constrained by its orthogonality to the $2p$ subshell, while $3d$ orbital has no lower-lying analog. Therefore, in the given electronic shell n, the $l = n - 1$ and $l = 0$ subshells always exhibit the minimum and maximum average distance from the nucleus, respectively.

These prototype analytical solutions for the *one*-electron atom can be also regarded as determining a general pattern of the shell structures in N-electron atoms ($N > 1$), in which electrons, occupying N lowest SO, are moving in the effective potential due to the nucleus and the remaining electrons. As this effective attraction by the "screened" nucleus is then no longer of the $1/r$ type, the accidental degeneracy of the hydrogen-like atom is lifted and the subshell energies in *many*-electron atoms depend on both n and l, $\varepsilon = \varepsilon_{n,l}$. In such atomic systems the configuration of the *outer*-most (most polarizable) *valence* shell electrons is decisive for determining the atom propensity to form chemical bonds with other atoms. In such bond-forming processes the distributions of the *inner*-shell electrons remain practically unaffected ("frozen").

It should be finally observed that these "exact" solutions of the Schrödinger equation for the *one*-electron atom, also determining the gross features of the electronic structure of *many*-electron systems, still require several corrections which must be taken into account to relate theoretical predictions to the experimental data. For example, corrections are due to the coupling between the spin and orbital angular momenta and the high speed of the electron, which call for the relativistic approach, the hyperfine structural effects reflect the magnetic properties of the nucleus, and the Lamb shift accounts for the interaction between the electron and electromagnetic field.

4.6 Atomic Units

When describing objects and processes in the atomic scale, it is convenient to use the system of *atomic units* (a.u.), which greatly simplify equations and expressions in molecular quantum mechanics [see (4.58), (4.72), (4.73)]. For example, the proportionality constant in the Coulomb Law determining the potential $V(r)$ of (4.1), $k_C = (4\pi\varepsilon_0)^{-1}$, where ε_0 stands for the electric permittivity of the free space, becomes unity in a.u., $k_C = 1$ a.u. so that $V(r) = -k_C Ze^2/r = -Ze^2/r = -Z/r$ (a.u.), where we have recognized that the magnitude of the electronic charge (proton charge) determines in a.u. the unit of electric charge: $e = 1$. Thus the a.u. of electric permittivity equals $4\pi\varepsilon_0$, or the vacuum permittivity $\varepsilon_0 = (4\pi)^{-1}$ a.u.

This system will be used in the remaining part of the book, unless specified otherwise. It is based upon the underlying units of length, mass, time, and electric charge, which subsequently determine the associated units of the remaining physical quantities, e.g., energy, physical action, angular momentum, etc. Some of these units are summarized in Table 4.1, where the expressions in terms of the universal constants and corresponding values in the *Systeme International d'Unités* (SI) are also given.

Table 4.1 Atomic units

Property	Unit	Symbol	SI value
Action and angular momentum	Planck's constant	$\hbar$	1.0546×10^{-34} J s
Electric charge	Charge of proton	e	1.6022×10^{-19} C
Electric permittivity	$4\pi\varepsilon_0$	$e^2/(E_h\, a_0)$	1.1127×10^{-10} F m^{-1}
Energy	Hartree, double magnitude of the ground-state energy of hydrogen atom for $\mu = m_e$, i.e., $M_n \to \infty$	$E_h = k_C\, e^2/a_0 = k_C^2 m_e\, e^4/\hbar^2$	4.3598×10^{-18} J
k_C	Constant in Coulomb Law	$k_C = E_h\, a_0/e^2$	8.9875×10^9 J m C^{-2}
Length	The first Bohr's radius	$a_0 = \hbar^2/(k_C\, m_e e^2)$	5.2918×10^{-10} m
Mass	Rest mass of electron	m_e	9.1095×10^{-31} kg
Probability density		a_0^{-3}	6.7483×10^{30} m^{-3}
Time	Time in which one electron on the first Bohr's orbit travels the angle distance of 1 radian	$\tau_0 = a_0/v_0 = \hbar^3/(k_C^2 m_e e^4)$	2.4189×10^{-17} s
Velocity	Speed of electron on the first Bohr's orbit	$v_0 = a_0/\tau_0 = \hbar/(m_e a_0) = k_C\, e^2/\hbar$	2.1877×10^6 m s^{-1}

Part II
Theories of Electronic Structure

Chapter 5
Approximating Molecular Schrödinger Equation

Abstract Theoretical basis of the approximate *perturbational* and *variational* approaches in quantum chemistry is outlined and the *adiabatic* separation of the fast (electronic) motions from slow (nuclear) movements in molecular systems is established. The rudiments of the *Ritz* method, a linear variant of the variational treatment, are summarized and the criteria for an effective mixing of quantum states are formulated. The illustrative applications of the perturbative and variational methods to helium atom are discussed and compared. The elements of the *orbital approximation* of the *many*-electron wave functions are introduced and selected properties of the *Slater determinant*, defined by the antisymmetrized product of the occupied *spin orbitals*, are examined in the context of the *Pauli exclusion principle*. The relevant expression for the expectation value of the electronic energy in orbital theories is derived and the *Slater–Condon* rules for matrix elements of the electronic Hamiltonian between determinantal wave functions are given. The additional possibilities of reducing the complexity of the molecular electronic Schrödinger equation by using the *pseudopotentials* are briefly outlined. These *core* potentials reflect a resultant influence of the "frozen" (chemically inactive) *inner*-shell electrons and the system nuclei in the effective Schrödinger equation for the (chemically active) *valence* shell electrons of constituent atoms, coordinates of which are treated explicitly in the approximate wave functions.

5.1 Rudiments of Perturbational and Variational Approaches

The stationary (time-independent) Schrödinger equation, i.e., the eigenvalue problem of the system Hamiltonian, can be solved analytically only for simple model systems. The quantum mechanical determination of the electronic structure of molecules, and particularly the complicated systems of interest in contemporary chemistry, requires adequate approximate methods of sufficient accuracy. In recent decades a remarkable progress of applying quantum mechanics to diverse problems in physics, chemistry, and molecular biology was possible due to spectacular

R.F. Nalewajski, *Perspectives in Electronic Structure Theory*,
DOI 10.1007/978-3-642-20180-6_5, © Springer-Verlag Berlin Heidelberg 2012

developments in the approximate theories of molecular electronic structure, covering original and sometimes ingenious new concepts and efficient algorithms, as well as a steadily increasing capability of modern computers and new software techniques of the advanced computational tools of modern quantum chemistry and solid state physics.

It is the main purpose of this chapter to summarize the main strategies used in reducing the complexity of the molecular Schrödinger equation and approximating its electronic wave function. It is intended to provide an overview of the successive levels of reducing the immense computational complexity of treating the coupled N-electron and m-nuclei problem of the molecular quantum mechanics. These perturbational and variational methods use the adiabatic, *Born–Oppenheimer* (BO) separation of the electronic and nuclear motions in molecules, as well as the orbital (Slater determinant) approximation of the trial N-electron wave functions, which automatically satisfy the requirements of the Pauli exclusion principle.

5.1.1 Perturbation Theory

It is the often encountered scenario in quantum theory that we have to estimate solutions of the Schrödinger equation for a more complicated (perturbed) *real* system from the known solutions of a simpler (unperturbed) *model* system, e.g., the stationary states and the associated energy levels of an anharmonic oscillator from the known (analytical) results for the harmonic oscillator. This goal summarizes the basic purpose of the *perturbation theory* (PT), which has also been used in classical mechanics. Its simplest variant within the Rayleigh–Schrödinger theory, for the nondegenerate energy levels and time-independent perturbations, will be summarized below.

Let us assume that the Hamiltonian $\hat{H}$ of the real (perturbed) system can be expressed as the sum of the simpler, model Hamiltonian $\hat{H}^0$, representing the associated unperturbed system the eigensolutions of which are assumed to be available, and the *perturbation* $\hat{h} \equiv \lambda \hat{h}'$ including weak interactions compared with those already comprised in $\hat{H}^0$. The perturbation approach can be then used to generate corrections to eigensolutions of $\hat{H}^0$, due to a presence of the perturbation, to approximate the exact eigensolutions of $\hat{H}$. Formally, this assumption of a relative "smallness" of $\hat{h}$ can be expressed by the condition involving the perturbation parameter λ, $|\lambda| \ll 1$,

$$\hat{H} = \hat{H}^0 + \lambda \hat{h}' = \hat{H}^0 + \hat{h}(\lambda) \equiv \hat{H}(\lambda). \tag{5.1}$$

It controls the *order* of corrections to the known unperturbed solutions,

$$\hat{H}^0 \left| n^{(0)} \right\rangle = E_n^{(0)} \left| n^{(0)} \right\rangle, \quad n = 0, 1, 2, \ldots, \quad \left\langle n^{(0)} \mid m^{(0)} \right\rangle = \delta_{n,m}, \tag{5.2}$$

for the nondegenerate energy levels $E_0^{(0)} < E_1^{(0)} < E_2^{(0)} < ...$, with $n = 0$ corresponding to the ground state of the model system, introduced to approximate the unknown stationary states of the perturbed system:

$$\hat{\mathrm{H}}(\lambda)|n(\lambda)\rangle = E_n(\lambda)|n(\lambda)\rangle, \quad n = 0, 1, 2, \ldots, \quad \langle n(\lambda) \mid m(\lambda)\rangle = \delta_{n,m}. \tag{5.3}$$

These corrections appear as coefficients in the corresponding power series expansions of the perturbed eigenstates and the associated eigenvalues,

$$\begin{aligned} |n(\lambda)\rangle &= \sum_{i=0}^{\infty} \left|n^{(i)}\right\rangle \lambda^i \equiv \left|n^{(0)}\right\rangle + \sum_{i=1}^{\infty} \left|\Delta n^{(i)}\right\rangle, \\ E_n(\lambda) &= \sum_{j=0}^{\infty} E_n^{(j)} \lambda^j \equiv E_n^{(0)} + \sum_{j=1}^{\infty} \Delta E_n^{(j)}, \end{aligned} \tag{5.4}$$

which define the kth-order corrections to the nth unperturbed state,

$$\left|\Delta n^{(k)}\right\rangle = \lambda^k \left|n^{(k)}\right\rangle \quad \text{and} \quad \Delta E_n^{(k)} = \lambda^k E_n^{(k)}, \quad k = 1, 2, \ldots$$

They can be determined by substituting these expansions into (5.3):

$$\sum_{i=0}^{\infty} \lambda^i (\hat{\mathrm{H}}^0 + \lambda \hat{\mathrm{h}}') \left|n^{(i)}\right\rangle = \sum_{i=0}^{\infty} \sum_{j=0}^{\infty} \lambda^{i+j} E_n^{(j)} \left|n^{(i)}\right\rangle. \tag{5.5}$$

Indeed, by comparing the coefficients at the given power k of the enhancement parameter λ in both sides of the preceding equation, one arrives at the following system of equations determining the corrections to the nth unperturbed state and its energy:

$$\begin{aligned}
\lambda^0 :& \quad \hat{\mathrm{H}}^0 \left|n^{(0)}\right\rangle = E_n^{(0)} \left|n^{(0)}\right\rangle; \\
\lambda^1 :& \quad \hat{\mathrm{H}}^0 \left|n^{(1)}\right\rangle + \hat{\mathrm{h}}' \left|n^{(0)}\right\rangle = E_n^{(0)} \left|n^{(1)}\right\rangle + E_n^{(1)} \left|n^{(0)}\right\rangle; \\
\lambda^2 :& \quad \hat{\mathrm{H}}^0 \left|n^{(2)}\right\rangle + \hat{\mathrm{h}}' \left|n^{(1)}\right\rangle = E_n^{(0)} \left|n^{(2)}\right\rangle + E_n^{(1)} \left|n^{(1)}\right\rangle + E_n^{(2)} \left|n^{(0)}\right\rangle; \\
\lambda^p :& \quad \hat{\mathrm{H}}^0 \left|n^{(p)}\right\rangle + \hat{\mathrm{h}}' \left|n^{(p-1)}\right\rangle = \sum_{j=0}^{p} E_n^{(j)} \left|n^{(p-j)}\right\rangle.
\end{aligned} \tag{5.6}$$

As expected, the λ^0-equation repeats the eigenvalue problem (5.2) of the unperturbed Hamiltonian. The subsequent elimination of corrections from these equations recognizes the completeness of the unperturbed solutions $\{|n^{(0)}\rangle\}$, which allows one to expand any state of the system, including all unknown corrections $\{|n^{(p)}\rangle\}$ or $\{|\Delta n^{(p)}\rangle\}$, in this basis set.

For example, one can expand the *first*-order correction $|n^{(1)}\rangle$,

$$\left|n^{(1)}\right\rangle = \sum_{j=0}^{\infty} \left|j^{(0)}\right\rangle c_{j,n}^{(1)} = \sum_{j\neq n} \left|j^{(0)}\right\rangle c_{j,n}^{(1)}, \quad c_{j,n}^{(1)} = \left\langle j^{(0)}\middle|n^{(1)}\right\rangle, \tag{5.7}$$

or the resultant state $\hat{\mathrm{h}}'\left|n^{(0)}\right\rangle$:

$$\hat{\mathrm{h}}'\left|n^{(0)}\right\rangle = \sum_{j=0}^{\infty} \left|j^{(0)}\right\rangle h_{j,n}, \quad h_{j,n} = \left\langle j^{(0)}\middle|\hat{\mathrm{h}}'\middle|n^{(0)}\right\rangle. \tag{5.8}$$

In (5.7) we have recognized that $c_{n,n}^{(1)} = \left\langle n^{(0)}\middle|n^{(1)}\right\rangle = 0$, since the *direction* of the unperturbed state vector $|n^{(0)}\rangle$ can be modified only by combining this state with the remaining states $\{|j^{(0)}\rangle, j \neq n\}$, which are orthogonal to $|n^{(0)}\rangle$.

Projecting λ^1-equation (5.6) onto $|n^{(0)}\rangle$ and $|k^{(0)}\rangle$, $k \neq n$, respectively, gives the associated equations for determining the *first*-order corrections we seek:

$$\begin{aligned}\left\langle n^{(0)}\middle|\hat{\mathrm{H}}^0\middle|n^{(1)}\right\rangle + \left\langle n^{(0)}\middle|\hat{\mathrm{h}}'\middle|n^{(0)}\right\rangle &= E_n^{(0)}\left\langle n^{(0)}\middle|n^{(1)}\right\rangle + h_{n,n} = h_{n,n}\\ &= E_n^{(0)}\left\langle n^{(0)}\middle|n^{(1)}\right\rangle + E_n^{(1)}\left\langle n^{(0)}\middle|n^{(0)}\right\rangle = E_n^{(1)},\end{aligned} \tag{5.9}$$

$$\begin{aligned}\left\langle k^{(0)}\middle|\hat{\mathrm{H}}^0\middle|n^{(1)}\right\rangle + \left\langle k^{(0)}\middle|\hat{\mathrm{h}}'\middle|n^{(0)}\right\rangle &= E_k^{(0)}\left\langle k^{(0)}\middle|n^{(1)}\right\rangle + h_{k.n} = E_k^{(0)}c_{k,n}^{(1)} + h_{k,n}\\ &= E_n^{(0)}\left\langle k^{(0)}\middle|n^{(1)}\right\rangle + E_n^{(1)}\left\langle k^{(0)}\middle|n^{(0)}\right\rangle = E_n^{(0)}c_{k,n}^{(1)}.\end{aligned} \tag{5.10}$$

A straightforward rearrangements of these equations then give the following explicit expressions for the *first*-order corrections to $E_n^{(0)}$,

$$E_n^{(1)} = \left\langle n^{(0)}\middle|\hat{\mathrm{h}}'\middle|n^{(0)}\right\rangle = h_{n,n} \quad \text{or} \quad \Delta E_n^{(1)} = \left\langle n^{(0)}\middle|\hat{\mathrm{h}}\middle|n^{(0)}\right\rangle, \tag{5.11}$$

and to $|n^{(0)}\rangle$:

$$c_{k,n}^{(1)} = h_{k,n}/[E_n^{(0)} - E_k^{(0)}] \quad \text{or} \quad \left|\Delta n^{(1)}\right\rangle = \sum_{k\neq n} \frac{\left\langle k^{(0)}\middle|\hat{\mathrm{h}}\middle|k^{(0)}\right\rangle}{E_n^{(0)} - E_k^{(0)}}\left|k^{(0)}\right\rangle. \tag{5.12}$$

When determining the *second*-order corrections one similarly expands

$$\left|n^{(2)}\right\rangle = \sum_{j=0}^{\infty} \left|j^{(0)}\right\rangle c_{j,n}^{(2)} = \sum_{j\neq n} \left|j^{(0)}\right\rangle c_{j,n}^{(2)}, \quad c_{j,n}^{(2)} = \left\langle j^{(0)}\middle|n^{(2)}\right\rangle, \tag{5.13}$$

again realizing that $c_{n,n}^{(2)} = \left\langle n^{(0)}\middle|n^{(2)}\right\rangle = 0$. The corresponding projections of the λ^2-equation (5.6) onto $\left|n^{(0)}\right\rangle$ and $|k^{(0)}\rangle$, $k \neq n$, respectively, gives the relevant equations determining the expansion coefficients $\{c_{j,n}^{(2)}\}$ and the *second*-order energy:

$$\begin{aligned}
&\langle n^{(0)}|\hat{\mathrm{H}}^{0}|n^{(2)}\rangle + \langle n^{(0)}|\hat{\mathrm{h}}'|n^{(1)}\rangle = E_n^{(0)}\langle n^{(0)}|n^{(2)}\rangle + \sum_{l\neq n} c_{l,n}^{(1)}\langle n^{(0)}|\hat{\mathrm{h}}'|l^{(0)}\rangle \\
&= \sum_{l\neq n} c_{l,n}^{(1)} h_{n,l} = E_n^{(0)}\langle n^{(0)}|n^{(2)}\rangle + E_n^{(1)}\langle n^{(0)}|n^{(1)}\rangle + E_n^{(2)}\langle n^{(0)}|n^{(0)}\rangle = E_n^{(2)},
\end{aligned} \tag{5.14}$$

$$\begin{aligned}
\langle k^{(0)}|\hat{\mathrm{H}}^{0}|n^{(2)}\rangle + \langle k^{(0)}|\hat{\mathrm{h}}'|n^{(1)}\rangle &= E_k^{(0)}\langle k^{(0)}|n^{(2)}\rangle + \sum_{l\neq n} c_{l,n}^{(1)}\langle k^{(0)}|\hat{\mathrm{h}}'|l^{(0)}\rangle \\
&= E_k^{(0)} c_{k,n}^{(2)} + \sum_{l\neq n} c_{l,n}^{(1)} h_{k,l} \\
&= E_n^{(0)}\langle k^{(0)}|n^{(2)}\rangle + E_n^{(1)}\langle k^{(0)}|n^{(1)}\rangle + E_n^{(2)}\langle k^{(0)}|n^{(0)}\rangle \\
&= E_n^{(0)} c_{k,n}^{(2)} + E_n^{(1)} c_{k,n}^{(1)}.
\end{aligned} \tag{5.15}$$

Subsequent substitution to (5.14) of the known *first*-order solutions gives the following expression for the *second*-order correction to $E_n^{(0)}$:

$$\begin{aligned}
E_n^{(2)} &= \langle n^{(0)}|\hat{\mathrm{h}}'|n^{(1)}\rangle = \sum_{l\neq n} h_{n,l} c_{l,n}^{(1)} = \sum_{l\neq n} h_{n,l} h_{l,n}/[E_n^{(0)} - E_l^{(0)}] \quad \text{or} \\
\Delta E_n^{(2)} &= \langle n^{(0)}|\hat{\mathrm{h}}|\Delta n^{(1)}\rangle = \sum_{l\neq n} |\langle n^{(0)}|\hat{\mathrm{h}}|l^{(0)}\rangle|^2/[E_n^{(0)} - E_l^{(0)}].
\end{aligned} \tag{5.16}$$

A similar rearrangement of (5.15) gives the expansion coefficients

$$c_{k,n}^{(2)} = \frac{1}{E_n^{(0)} - E_k^{(0)}}\left[\sum_{l\neq n} \frac{h_{k,l} h_{l,n}}{E_n^{(0)} - E_l^{(0)}} - \frac{h_{k,n} h_{n,n}}{E_n^{(0)} - E_k^{(0)}}\right]$$

determining the associated correction to $|n^{(0)}\rangle$:

$$|\Delta n^{(2)}\rangle = \sum_{k\neq n}\left(\sum_{l\neq n} \frac{\langle k^{(0)}|\hat{\mathrm{h}}|l^{(0)}\rangle\langle l^{(0)}|\hat{\mathrm{h}}|n^{(0)}\rangle}{(E_n^{(0)} - E_l^{(0)})(E_n^{(0)} - E_k^{(0)})} - \frac{\langle k^{(0)}|\hat{\mathrm{h}}|n^{(0)}\rangle\langle n^{(0)}|\hat{\mathrm{h}}|n^{(0)}\rangle}{(E_n^{(0)} - E_k^{(0)})^2}\right)|k^{(0)}\rangle. \tag{5.17}$$

Obviously, one could similarly extract the *higher* order corrections, but the above explicit expressions for the *first*- and *second*-order corrections are sufficient for most applications included in this book.

5.1.2 Variational Method

The alternative *variational* method of determining the approximate solutions of the time-independent Schrödinger equation guarantees that the successive approximations of increasing accuracy approach from above the exact energy level E_0 of the molecular ground state $|\psi_0\rangle$. In other words, this exact eigenvalue represents the lower bound of all approximate estimates of the system average energy: $\langle E\rangle \geq E_0$.

Indeed, the eigenstates $\{|\psi_n\rangle\}$ of the system Hamiltonian (the quantum mechanical observable),

$$\hat{\mathrm{H}}|\psi_n\rangle = E_n|\psi_n\rangle, \quad n = 0, 1, 2, \ldots, \quad E_0 \leq E_1 \leq E_2 \leq \ldots, \tag{5.18}$$

form the basis of the energy representation (see Sect. 2.7) in the molecular Hilbert space, so that any approximate state $|\phi\rangle$ can be expanded in this set:

$$|\phi\rangle = \sum_n |\psi_n\rangle\langle\psi_n|\phi\rangle = \sum_n |\psi_n\rangle C_n, \tag{5.19}$$

with $|C_n|^2 = P(\psi_n|\phi)$ measuring the conditional probability of observing E_n in state $|\phi\rangle$ (see Postulate II of Sect. 3.2). Hence, any approximate estimate of the system average energy can be expressed as the mean value of the exact energy levels (see also Postulate IV.3 of Sect. 3.3.3):

$$\langle E\rangle_\phi = \langle\phi|\hat{\mathrm{H}}|\phi\rangle = \sum_n P(\psi_n|\phi)E_n, \quad \sum_n P(\psi_n|\phi) = 1.$$

We thus conclude that $\langle E\rangle_\phi = E_0$ can be reached only for $P(\psi_0|\phi) = 1$ and $\{P(\psi_{n>0}|\phi) = 0\}$, and hence $|\phi\rangle = |\psi_0\rangle$. Any deviation from this exact solution implies a finite probability of observing one of the higher (excited) energy levels, and hence $\langle E\rangle_\phi > E_0$. These deductions constitute the essence of the Rayleigh–Ritz variational principle of quantum mechanics: for any approximate state $|\phi\rangle$ the average energy

$$\langle E\rangle_\phi \geq E_0. \tag{5.20}$$

Thus, the more accurately $|\phi\rangle$ approximates $|\psi_0\rangle$, the lower $\langle E\rangle_\phi$ level, and hence the smaller $\langle E\rangle_\phi - E_0$ error gap.

This general statement gives rise to the efficient computational technique, the *variational method*, which dominates the modern quantum mechanical calculations of molecular electronic structure. The main idea behind this computational tool is to use the parametrically defined trial state including several *variational parameters* $\boldsymbol{\lambda} = \{\lambda_t, t = 1, 2, \ldots, s\}$, $|\phi\rangle = |\phi(\lambda_1, \lambda_2, \ldots, \lambda_s)\rangle$. The domain of their admissible values then determines the whole range of the approximate (variational) states. In accordance with the variational principle, the best approximation of the molecular ground state in this family of trial states is then obtained for the optimum values of variational parameters $\boldsymbol{\lambda}^{opt.}$ which correspond to the minimum of the system average energy $\langle E(\boldsymbol{\lambda})\rangle_\phi = \langle\phi|(\boldsymbol{\lambda})|\hat{\mathrm{H}}|\phi(\boldsymbol{\lambda})\rangle$:

$$\min_{\boldsymbol{\lambda}}\langle E(\boldsymbol{\lambda})\rangle_\phi = \langle E(\boldsymbol{\lambda}^{opt})\rangle_\phi \geq E_0 \quad \text{or}$$
$$\left.\frac{\partial\langle E(\boldsymbol{\lambda})\rangle_\phi}{\partial\lambda_t}\right|_{\boldsymbol{\lambda}^{opt.}} = 0, \quad t = 1, 2, \ldots, s. \tag{5.21}$$

Both *linear* and *nonlinear* parameters $\boldsymbol{\lambda}$ are used to provide the trial state vectors or the associated wave functions exhibiting a sufficient variational flexibility, so that they are capable to adjust to the interactions embodied in the system Hamiltonian, in order to lower the energy, and thus to resemble the most the true ground state of the molecular system in question. The former, e.g., the coefficients multiplying the adopted set of the ("frozen") *basis functions*, are more easily handled, giving rise to a system of linear *secular equations* for determining the optimum values of the expansion coefficients. The latter, e.g., the exponents of the *Slater*-type orbitals (STO) or *Gaussian*-type orbitals (GTO), the popular analytical functions used to approximate the atomic or molecular orbitals, although relatively more efficient in modifying the trial wave functions, are more difficult to handle, requiring more advanced, nonlinear optimization techniques.

Consider the illustrative application of this procedure to the hydrogen-like atom of Chap. 4, for simplicity adopting a.u. of Sect. 4.6. Suppose that we take the trial wave function in the general form of a parametric family of the spherically symmetric, exponentially decaying functions defined by a single nonlinear variational parameter λ, $\phi(r, \vartheta, \varphi; \lambda) = N(\lambda)\exp(-\lambda r)$, with $N(\lambda)$ standing for the appropriate normalization factor [see (4.62)]: $N(\lambda) = (\lambda^3/\pi)^{1/2}$. It gives rise to the average electronic energy, the expectation value of the Hamiltonian (4.40),

$$E(\lambda) = \frac{1}{2}\lambda^2 - \lambda Z. \tag{5.22}$$

The optimum value of λ, which identifies the best approximation to the ground state, is then obtained for the minimum of $E(\lambda)$, $dE(\lambda)/d\lambda|_{opt.} = \lambda^{opt.} - Z = 0$, or $\lambda^{opt.} = Z$, thus correctly predicting the true ground state of (4.62).

Not knowing the true asymptotic behavior of the ground state at large distances from the nucleus, one could alternatively try the spherical Gaussian function $\varphi(r, \vartheta, \varphi; \xi) = N(\xi)\exp(-\xi r^2)$ as an approximate representation of the ground state wave function in this *one*-electron atom, which gives:

$$E(\xi) = \frac{3}{2}\xi - \sqrt{\frac{8\xi}{\pi}}Z, \quad \xi^{opt.} = \frac{8}{9\pi}Z^2, \quad E(\xi^{opt.}) = \frac{4}{3\pi}Z^2 \cong -0.424Z^2. \tag{5.23}$$

Therefore, on the basis of the variational criterion one concludes that the exponential form of the variational wave function provides a better representation of the electronic wave function in the *one*-electron atom, since it generates lower energy compared with that resulting from the optimum Gaussian state.

The *linear* variant of the variational approach is known as the *Ritz method*. The trial state $|\varphi\rangle$ is then given as the linear combination of the adopted basis states $|\boldsymbol{\chi}\rangle = \{|\chi_p\rangle, p = 1, 2, \ldots, w\}$ (*row* vector) defined by the expansion coefficients $\boldsymbol{C} = \langle\boldsymbol{\chi}|\varphi\rangle = \{C_p\}$ (*column* vector):

$$|\varphi\rangle = \sum_{p=1}^{w} |\chi_p\rangle C_p \equiv |\boldsymbol{\chi}\rangle \boldsymbol{C}. \tag{5.24}$$

However, since w basis functions define w linearly independent combinations $|\boldsymbol{\varphi}\rangle = \{|\varphi^{(s)}\rangle, s = 1, 2, \ldots, w\}$ (*row* vector), we can generalize the above expression:

$$|\varphi^{(s)}\rangle = \sum_{p=1}^{w} |\chi_p\rangle C_{p,s} = |\boldsymbol{\chi}\rangle \boldsymbol{C}^{(s)} \equiv |\varphi_{s-1}\rangle, \quad s = 1, 2, \ldots, w, \tag{5.25}$$

or in the joint, matrix notation:

$$|\boldsymbol{\varphi}\rangle = |\boldsymbol{\chi}\rangle \mathbf{C}, \;\; \mathbf{C} = (\boldsymbol{C}^{(1)}|\boldsymbol{C}^{(2)}|\ldots|\boldsymbol{C}^{(s)}|\ldots|\boldsymbol{C}^{(w)}) \equiv (\boldsymbol{C}_0|\boldsymbol{C}_1|\ldots|\boldsymbol{C}_{s-1}|\ldots|\boldsymbol{C}_{w-1}). \tag{5.26}$$

In general, the basis vectors give rise to a nonunit metric tensor defined by the *overlap matrix* $\mathbf{S} = \langle\boldsymbol{\chi}|\boldsymbol{\chi}\rangle = \{S_{p,q} = \langle\chi_p|\chi_q\rangle\}$, while the Hamiltonian is representated by the *energy matrix* $\mathbf{H} = \langle\boldsymbol{\chi}|\hat{\mathrm{H}}|\boldsymbol{\chi}\rangle = \{H_{p,q} = \langle\chi_p|\hat{\mathrm{H}}|\chi_q\rangle\}$.

In what follows we shall assume that the optimum combinations are ordered in accordance with their increasing energies $\{\langle E^{(s)}\rangle = \langle\varphi^{(s)}|\hat{\mathrm{H}}|\varphi^{(s)}\rangle \equiv \langle E_{s-1}\rangle\}$:

$$[\langle E^{(1)}\rangle \equiv \langle E_0\rangle] \le [\langle E^{(2)}\rangle \equiv \langle E_1\rangle] \le \ldots \le [\langle E^{(w)}\rangle \equiv \langle E_{w-1}\rangle]. \tag{5.27}$$

The optimum combination $|\varphi^{(1)}\rangle = |\varphi_0\rangle$ corresponding to the lowest energy $\langle E^{(1)}\rangle \equiv \langle E_0\rangle$ will then approximate the system ground state $|\psi_0\rangle$, while the remaining orthonormal combinations will approach the corresponding excited states.

In the last three equations, we have relabeled the *upper* indices of the eigenvectors, the associated columns in the $(w \times w)$ square matrix $\mathbf{C}$ grouping the combination coefficients, the linear variational parameters of the Ritz method, and the associated energy estimates into the corresponding subscripts conforming to the customary labeling of the molecular energy levels of (5.18), with $|\varphi_0\rangle$ and $\langle E_0\rangle$ denoting the *ground* state approximations and the remaining states corresponding to successive *excited* states:

$$|\varphi^{(s)}\rangle \equiv |\varphi_{s-1}\rangle, \quad \mathbf{C} = \{\, \boldsymbol{C}^{(s)} \equiv \boldsymbol{C}_{s-1}\}, \quad \{\langle E^{(s)}\rangle \equiv \langle E_{s-1}\rangle\}.$$

Let us first consider a single combination of (5.24). The expectation value of the system energy in state $|\varphi\rangle$ reads: $\langle E\rangle = \left\langle\tilde{\varphi}\middle|\hat{\mathrm{H}}\middle|\tilde{\varphi}\right\rangle = \langle\varphi|\hat{\mathrm{H}}|\varphi\rangle/\langle\varphi|\varphi\rangle$, where the denominator $\langle\varphi \mid \varphi\rangle = \sum_{p=1}^{w}\sum_{q=1}^{w} C_p^* S_{p,q} C_q = \boldsymbol{C}^{\dagger}\mathbf{S}\boldsymbol{C}$ is due to the normalization constant of $|\varphi\rangle$ in the *normalized* trial state $|\tilde{\varphi}\rangle = \langle\varphi|\varphi\rangle^{-1/2}|\varphi\rangle$. Hence,

$$\langle E\rangle\langle\varphi|\varphi\rangle = \langle E\rangle\sum_{p=1}^{w}\sum_{q=1}^{w} C_p^* S_{p,q} C_q = \langle\varphi|\hat{\mathrm{H}}|\varphi\rangle. \tag{5.28}$$

One further observes that the expansion coefficients $\boldsymbol{C}$ are in general complex numbers. Therefore, the unknowns in this linear variational problem consist of their real and imaginary parts, $\boldsymbol{C} = \mathrm{Re}(\boldsymbol{C}) + i\mathrm{Im}(\boldsymbol{C})$, where: $\mathrm{Re}(\boldsymbol{C}) = (\boldsymbol{C} + \boldsymbol{C}^*)/2$ and $\mathrm{Im}(\boldsymbol{C}) = (\boldsymbol{C} - \boldsymbol{C}^*)/2i$. Thus, one can alternatively designate the coefficients $\boldsymbol{C}$ and their complex conjugates $\boldsymbol{C}^*$ as independent variational parameters, since they uniquely identify both parts of $\boldsymbol{C}$. In fact, due to the Hermitian character of $\mathbf{H}$ and the symmetrical character of the metric $\mathbf{S}$, the secular equations for the optimum values of the linear variational parameters derived from the independent variations of $\boldsymbol{C}^*$ and $\boldsymbol{C}$, respectively, are identical.

The optimum solutions must minimize the system energy function $\langle E(\boldsymbol{C}^*, \boldsymbol{C})\rangle$ [see (5.21)]:

$$\left.\frac{\partial\langle E(C^*, C)\rangle}{\partial C_p^*}\right|_{\min} = 0 \quad \text{and} \quad \left.\frac{\partial\langle E(C^*, C)\rangle}{\partial C_p}\right|_{\min} = 0, \quad p = 1, 2, \ldots, w. \tag{5.29}$$

Differentiating (5.28) with respect to C_p^* and taking into account the condition of the energy minimum of (5.29) then gives:

$$\frac{\partial\langle E(C^*, C)\rangle}{\partial C_p^*}\sum_{p=1}^{w}\sum_{q=1}^{w} C_p^* S_{p,q} C_q + \langle E\rangle\sum_{q=1}^{w} S_{p,q} C_q = \langle E\rangle\sum_{q=1}^{w} S_{p,q} C_q = \sum_{q=1}^{w} H_{p,q} C_q \text{ or}$$

$$\sum_{q=1}^{w}\left(H_{p,q} - \langle E\rangle S_{p,q}\right) C_q = 0, \quad p = 1, 2, \ldots, w. \tag{5.30}$$

This system of the *secular* (linear, homogeneous) *equations* has in fact only $w - 1$ independent unknowns. The additional, nonhomogeneous equation required to specify $\boldsymbol{C}$ uniquely is provided by the normalization condition for the combination in question:

$$\langle\varphi|\varphi\rangle = \sum_{p=1}^{w}\sum_{q=1}^{w} C_p^* S_{p,q} C_q = \boldsymbol{C}^{\dagger}\boldsymbol{S}\boldsymbol{C} = 1. \tag{5.31}$$

It then directly follows from the Cramer rules of Algebra that the necessary condition for the physically meaningful, nontrivial solutions $\boldsymbol{C} \neq \boldsymbol{0}$ of these secular equations is the vanishing determinant of coefficients before the unknowns in these homogeneous equations, called the *secular determinant*:

$$\begin{vmatrix} H_{1,1}-\langle E\rangle S_{1,1} & H_{1,2}-\langle E\rangle S_{1,2} & \dots & H_{1,w}-\langle E\rangle S_{1,w} \\ H_{2,1}-\langle E\rangle S_{2,1} & H_{2,2}-\langle E\rangle S_{2,2} & \dots & H_{2,w}-\langle E\rangle S_{2,w} \\ \dots\dots & \dots\dots & \dots & \dots\dots \\ H_{w,1}-\langle E\rangle S_{w,1} & H_{w,2}-\langle E\rangle S_{w,2} & \dots & H_{w,w}-\langle E\rangle S_{w,w} \end{vmatrix} \equiv \left|H_{p,q}-\langle E\rangle S_{p,q}\right| = 0. \tag{5.32}$$

Hence, by expanding the determinant one arrives at the equation of degree w for the unknown $\langle E\rangle$. Its ordered solutions $\left\{\langle E^{(s)}\rangle \equiv \langle E_{s-1}\rangle\right\}$ (5.27) approximate the exact energy levels of the system ground and the first $(w - 1)$ excited states (5.18).

To summarize, one first solves (5.32) for the approximate energy levels $\{\langle E^{(s)}\rangle\}$, the knowledge of which is required to uniquely specify the coefficients of the secular equations (5.30) supplemented by (5.31). Selecting $\langle E\rangle = \langle E^{(s)}\rangle$ in these equations gives the coefficients $\boldsymbol{C}^{(s)}$ determining $|\varphi^{(s)}\rangle$, etc.

Fortunately, this rather cumbersome procedure in terms of determinants can be recast in the form of the standard *matrix diagonalization* problem, which is easily handled in computer calculations. For this purpose, we arrange the energy estimates $\{\langle E^{(s)}\rangle\}$ as diagonal elements of the *eigenvalue matrix* $\mathbf{E} = \{E_{s,s'} = \langle E^{(s)}\rangle\delta_{s,s'}\}$ and rewrite the secular equations (5.30) for sth combination of (5.25):

$$\sum_{q=1}^{w}\left(H_{p,q}-\left\langle E^{(s)}\right\rangle S_{p,q}\right)C_{q,s} = \sum_{q=1}^{w} H_{p,q}C_{q,s} - \sum_{q=1}^{w}\sum_{s'=1}^{w} S_{p,q}C_{q,s'}E_{s',s} = 0 \text{ or}$$
$$\mathbf{HC{=}SCE}. \tag{5.33}$$

This equation must be supplemented by the matrix equation combining the relevant orthonormality requirements for the optimum combinations, which are summarized by the requirement of the unit metric tensor defined by $|\boldsymbol{\varphi}\rangle = |\boldsymbol{\chi}\rangle\mathbf{C}$,

$$\langle\boldsymbol{\varphi}|\boldsymbol{\varphi}\rangle = \mathbf{C}^{\dagger}\langle\boldsymbol{\chi}\mid\boldsymbol{\chi}\rangle\mathbf{C} = \mathbf{C}^{\dagger}\mathbf{SC} = \mathbf{I}. \tag{5.34}$$

As already shown in Sect. 3.3.2, the nonorthogonal basis vectors $|\boldsymbol{\chi}\rangle$ can be transformed into the symmetrically orthogonalized analogs $|\tilde{\boldsymbol{\chi}}\rangle = |\boldsymbol{\chi}\rangle\,\mathbf{S}^{-1/2}$ of Löwdin, strongly resembling the original basis vectors $|\boldsymbol{\chi}\rangle$, which can be subsequently "rotated" in the unitary transformation $\mathbf{U}$ to the final optimum combinations we seek:

$$|\boldsymbol{\varphi}\rangle = |\boldsymbol{\chi}\rangle\mathbf{C} \equiv \left(|\boldsymbol{\chi}\rangle\mathbf{S}^{-1/2}\right)\mathbf{U} = |\tilde{\boldsymbol{\chi}}\rangle\mathbf{U}, \quad \mathbf{UU}^{\dagger}{=}\mathbf{U}^{\dagger}\mathbf{U} = \mathbf{I}. \tag{5.35}$$

This way of arriving at orthonormal combinations thus automatically satisfies (5.34). In this Löwdin orthogonalized representation the only unknown part of

$\mathbf{C} = \mathbf{S}^{-1/2}\mathbf{U}$ is its $\mathbf{U} = \mathbf{S}^{1/2}\mathbf{C}$ factor, where we have used the relation $\mathbf{S}^{-1/2}\mathbf{S}^{1/2} = \mathbf{S}^0 = \mathbf{I}$ (3.47). A straightforward transcription of (5.33) multiplied from the left by $\mathbf{S}^{-1/2}$ then gives:

$$(\mathbf{S}^{-1/2}\mathbf{H}\mathbf{S}^{-1/2})(\mathbf{S}^{1/2}\mathbf{C}) \equiv \tilde{\mathbf{H}}\mathbf{U} = (\mathbf{S}^{-1/2}\mathbf{S})\mathbf{C}\mathbf{E} = (\mathbf{S}^{1/2}\mathbf{C})\mathbf{E} = \mathbf{U}\mathbf{E}. \qquad (5.36)$$

Hence, by multiplying the preceding equation from the left by $\mathbf{U}^\dagger$ finally gives:

$$\mathbf{U}^\dagger\tilde{\mathbf{H}}\mathbf{U} = \mathbf{E}. \qquad (5.37)$$

The determination of the optimum coefficients **C**, the linear variational parameters of the Ritz method, and of the associated average energy estimates **E** is thus simultaneously accomplished by the diagonalization in the unitary transformation **U** of the Hermitian matrix $\tilde{\mathbf{H}} = \mathbf{S}^{-1/2}\mathbf{H}\mathbf{S}^{-1/2}$. The latter constitutes the matrix representation of the Hamiltonian in the symmetrically orthogonalized basis set $|\tilde{\boldsymbol{\chi}}\rangle$,

$$\tilde{\mathbf{H}} = \langle\tilde{\boldsymbol{\chi}}|\hat{\mathrm{H}}|\tilde{\boldsymbol{\chi}}\rangle = \mathbf{S}^{-1/2}\langle\boldsymbol{\chi}|\hat{\mathrm{H}}|\boldsymbol{\chi}\rangle\,\mathbf{S}^{-1/2} = \mathbf{S}^{-1/2}\mathbf{H}\mathbf{S}^{-1/2}, \qquad (5.38)$$

where we have observed that $\langle\tilde{\boldsymbol{\chi}}| = \left(|\boldsymbol{\chi}\rangle\mathbf{S}^{-1/2}\right)^\dagger = \mathbf{S}^{-1/2}\langle\boldsymbol{\chi}|$, since $\mathbf{S}^{-1/2}$ is the real, symmetric matrix. This linear variational procedure thus amounts to the standard algorithmic problem in the matrix algebra.

We conclude this section by examining general criteria for an effective mixing of quantum states in the linear combination of (5.24). In textbooks on quantum chemistry such an analysis is carried out in the context of mixing AO into *Molecular Orbitals* (MO), when the prototype chemical bond is being formed, say between atoms A and B. To simplify these qualitative considerations, we reduce the problem to two AO states $|\boldsymbol{\chi}\rangle = (|A\rangle, |B\rangle)$, originating from atoms A and B, respectively, which are assumed to be normalized but nonorthogonal (overlapping):

$$\mathbf{S} = \begin{bmatrix} 1 & S \\ S & 1 \end{bmatrix}, \quad \mathbf{H} = \begin{bmatrix} \alpha_A & \beta \\ \beta & \alpha_B \end{bmatrix}, \qquad (5.39)$$

where for definiteness we put $S = \langle A \mid B\rangle > 0$ and $\alpha_A = \langle A|\hat{\mathrm{H}}|A\rangle \le \alpha_B = \langle B|\hat{\mathrm{H}}|B\rangle < 0$ (Fig. 5.1). The (negative) *Coulomb* integrals $\{\alpha_p\}$ reflect the energy levels associated with the individual AO and hence the corresponding negative ionization potentials (see the Koopmans theorem of Sect. 6.1.2 and the Janak theorem of Sect. 7.3.6), $\alpha_p \cong -I_p$, $p = A, B$, while the *resonance* integral $\beta = \langle A|\hat{\mathrm{H}}|B\rangle = \langle B|\hat{\mathrm{H}}|A\rangle$ measures their mutual interaction (coupling) in the bond formation process. In the semiempirical theories of the molecular electronic structure, it was adequately approximated as being proportional to the AO overlap integral S and an average value (Av), arithmetic, geometric, or harmonic, of the corresponding diagonal elements of the Hamiltonian: $\beta \propto S\mathrm{Av}(\alpha_A, \alpha_B) \equiv S\langle\alpha\rangle \approx -S\mathrm{Av}(I_A, I_B) < 0$.

It then directly follows from the eigenvalue equation (5.32)

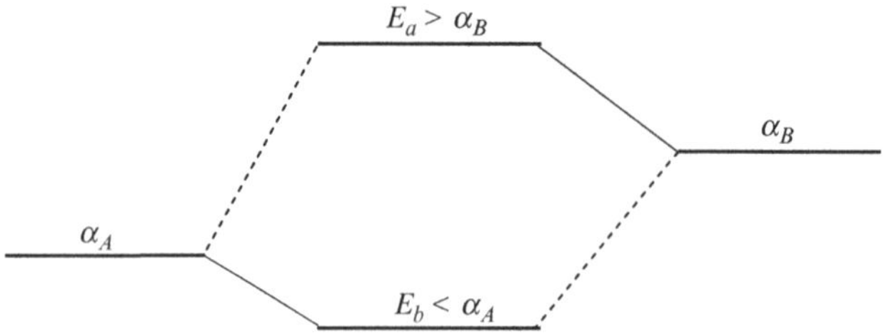

Fig. 5.1 A qualitative diagram of the chemical interaction between two AO

$$\begin{vmatrix} \alpha_A - \langle E\rangle & \beta - \langle E\rangle S \\ \beta - \langle E\rangle S & \alpha_B - \langle E\rangle \end{vmatrix} = 0 \text{ or } (\alpha_A - \langle E\rangle)(\alpha_B - \langle E\rangle) = (\beta - \langle E\rangle S)^2 > 0, \quad (5.40)$$

that the two AO energy levels "repel" each other as a result of their quantum mechanical coupling into MO. More specifically, the preceding equation allows the two optimum MO energies, which are simultaneously either above or below both AO energies:

$$[\alpha_A - \langle E\rangle > 0 \text{ and } \alpha_B - \langle E\rangle > 0] \Rightarrow \langle E\rangle = E_b < \alpha_A \quad \text{or}$$
$$[\alpha_A - \langle E\rangle < 0 \text{ and } \alpha_B - \langle E\rangle < 0] \Rightarrow \langle E\rangle = E_a > \alpha_B.$$

As a result the two MO energy estimates are obtained: the bonding level $\langle E^{(b)}\rangle = E_b < \alpha_A$ and the antibonding level $\langle E^{(a)}\rangle = E_a > \alpha_B$, which are also shown in the schematic diagram of Fig. 5.1.

For a general case of nonequal AO energy levels, the secular equation (5.40) gives the following expression for the *bonding energy*:

$$\alpha_A - E_b = (\beta - E_b S)^2/(\alpha_B - E_b), \quad (5.41)$$

which satisfies the following inequalities:

$$0 < \alpha_A - E_b < (\beta - E_b S)^2/(\alpha_B - \alpha_A), \quad (5.42)$$

since $\alpha_B - E_b > \alpha_B - \alpha_A > 0$ (see Fig. 5.1).

We thus conclude from the preceding equation that the larger the difference between the energy levels of the mixed states the smaller the bonding effect of their interaction. Indeed the strongest bonding results for $\alpha_B = \alpha_A = \alpha$ when

$$E_b = (\alpha + \beta)/(1 + S) < \alpha, \ |\varphi_b\rangle = (|A\rangle + |B\rangle)/(2 + 2S)^{-1/2}$$
$$E_a = (\alpha - \beta)/(1 - S) > \alpha, \ |\varphi_a\rangle = (|A\rangle - |B\rangle)/(2 - 2S)^{-1/2}. \quad (5.43)$$

It follows from these equations that for the overlapping AO the antibonding effect $E_a - \alpha$ always exceeds its bonding companion $\alpha - E_b$. This explains why no net chemical bonding results in this simple orbital description from the interaction between the two fully occupied AO, e.g., in He_2 and Be_2. Moreover, due to an approximate proportionality relation $\beta \propto S\langle\alpha\rangle$, the r.h.s. of (5.42), which marks the upper limit of the bonding effect,

$$\alpha_A - E_b < (\beta - E_b S)^2/(\alpha_B - \alpha_A) \approx S^2(\langle\alpha\rangle - E_b)^2/(\alpha_B - \alpha_A),$$

identically vanishes, when there is no overlap between AO. For example, at finite separations between atoms, this can be due to the symmetry restrictions in the valence shell or the "narrowness" of electron distributions in the *inner* shells of both atoms. Together with the nucleus these chemically inactive electrons of the *inner* shells define the atomic "*core*," which remains largely unaffected by the chemical bonds formed in the valence shell. At very large internuclear distances, in the *separated atoms* limit, the AO overlap also vanishes, so that no chemical interaction is predicted. We thus conclude that a large AO overlap is conducive for a strong chemical bonding originating from the orbital interaction in a molecule.

Let us next consider the *squared* secular equation for the unknown coefficients of the combination:

$$(\alpha_A - \langle E\rangle)C_A + (\beta - \langle E\rangle SC_B = 0 \Rightarrow (\alpha_A - \langle E\rangle)^2 C_A^2 = (\beta - \langle E\rangle S)^2 C_B^2. \quad (5.44)$$

Using the expression for $(\beta - \langle E\rangle S)^2$ from (5.40) gives the following ratio of the squares of coefficients, reflecting a relative participation (conditional probability) of AO in the combination,

$$C_A^2/C_B^2 = (\alpha_B - \langle E\rangle)/(\alpha_A - \langle E\rangle) = |\alpha_B - \langle E\rangle|/|\alpha_A - \langle E\rangle|. \quad (5.45)$$

Indeed, for $\alpha_B = \alpha_A$ both AO participate equally in MO and $C_A = \pm C_B$, in accordance with (5.43). In a general case of Fig. 5.1, one predicts for $\langle E\rangle = E_b$: $\alpha_B - E_b > \alpha_A - E_b$, so that orbital $|A\rangle$ dominates the bonding combination $|\varphi_b\rangle$: $C_A^2 > C_B^2$. One similarly predicts a stronger similarity of $|\varphi_a\rangle$ to $|B\rangle$ for $\langle E\rangle = E_a$, since then $|\alpha_B - E_a| < |\alpha_A - E_a|$. Therefore, with increasing gap $\alpha_B - \alpha_A$ of the AO energies the bonding combination $|\varphi_b\rangle$ more strongly resembles $|A\rangle$ and the antibonding combination $|\varphi_a\rangle$ becomes more like $|B\rangle$.

5.2 Adiabatic Separation of Electronic and Nuclear Motions

To a good approximation, when describing the state of (light) electrons in a molecule, one can treat the system (heavy) nuclei as being at rest, in view of the drastic difference in masses of these two *micro*-objects. Indeed, the motions of the former are very fast compared with the slow movements of the latter. This physical intuition suggests that for the nuclear dynamics the instantaneous positions of electrons are unimportant, with only the average effect of their fast motions influencing the forces acting on nuclei in the molecular system under consideration.

The formal basis of this separation of the electronic and nuclear degrees of freedom in the molecular (stationary) quantum mechanics is the familiar *adiabatic* approximation of Born and Oppenheimer (1927).

Consider the molecular wave function $\Psi(\mathbf{q}, \mathbf{Q})$ of N electrons at positions $\mathbf{r} = \{\boldsymbol{r}_i\}$ exhibiting the spin orientations $\boldsymbol{\sigma} = \{\sigma_i\}$, or in the combined notation $\mathbf{q} \equiv (\mathbf{r}, \boldsymbol{\sigma}) = \{\boldsymbol{r}_i, \sigma_i\} \equiv \{\boldsymbol{q}_i\}$, and m nuclei of masses $\{M_\alpha\}$ and charges $\{Z_\alpha\}$ in positions $\mathbf{R} = \{\boldsymbol{R}_\alpha\}$ with spins $\boldsymbol{\Sigma} = \{\Sigma_\alpha\}$, which determine the corresponding position-spin variables $\mathbf{Q} \equiv (\mathbf{R}, \boldsymbol{\Sigma}) = \{\boldsymbol{R}_\alpha, \Sigma_\alpha\} \equiv \{\boldsymbol{Q}_\alpha\}$. It generates the associated probability distribution of the *joint*, electronic-nuclear events: $P(\mathbf{q}, \mathbf{Q}) = |\Psi(\mathbf{q}, \mathbf{Q})|^2$, which satisfies the relevant overall and partial normalizations:

$$\iint P(\mathbf{q}, \mathbf{Q})\, d\mathbf{q}\, d\mathbf{Q} = \int \pi(\mathbf{Q})\, d\mathbf{Q} = \int \rho(\mathbf{q})\, d\mathbf{q} = 1, \tag{5.46}$$

where $\pi(\mathbf{Q})$ and $\rho(\mathbf{q})$ denote the partially integrated nuclear and electronic probability distributions, respectively.

The essence of the adiabatic approximation lies in extracting from this joint distribution the probability density of the heavy (slow) nuclei as the reference (parameter) distribution:

$$P(\mathbf{q}, \mathbf{Q}) = \pi(\mathbf{Q}) \frac{P(\mathbf{q}, \mathbf{Q})}{\pi(\mathbf{Q})} \equiv \pi(\mathbf{Q}) p(\mathbf{q}|\mathbf{Q}), \quad \int p(\mathbf{q}|\mathbf{Q})\, d\mathbf{q} = 1. \tag{5.47}$$

In the *conditional* probability density of electrons, $p(\mathbf{q}|\mathbf{Q})$, the nuclear variables thus play the role of parameters, as indeed reflected by the above normalization condition. This further implies the associated factorization of the system wave function in terms of the nuclear, $\chi(\mathbf{Q})$, and electronic, $\varphi(\mathbf{q}|\mathbf{Q})$, functions,

$$\Psi(\mathbf{q}, \mathbf{Q}) \cong \chi(\mathbf{Q}) \varphi(\mathbf{q}|\mathbf{Q}). \tag{5.48}$$

They accordingly represent the nuclear and (*conditional*) electronic amplitudes of the associated probability distributions:

$$\pi(\mathbf{Q}) = |\chi(\mathbf{Q})|^2 \quad \text{and} \quad p(\mathbf{q}|\mathbf{Q}) = |\varphi(\mathbf{q}|\mathbf{Q})|^2. \tag{5.49}$$

Therefore, in the *Born–Oppenheimer* (BO) approximation, the nuclear wave function is not *explicitly* dependent upon the electronic positions, while the electronic state $\varphi(\mathbf{q}|\mathbf{Q})$ is defined for the fixed geometry of the molecular system, defined by specified, parametric positions of the nuclei. The relevant orthonormality relations satisfied by different adiabatic states $\{\chi_i\}$ and $\{\varphi_r\}$ thus read:

$$\int \chi_i^*(\mathbf{Q}) \chi_j(\mathbf{Q})\, d\mathbf{Q} \equiv \langle \chi_i \mid \chi_j \rangle_{\mathbf{Q}} = \delta_{i,j} \quad \text{and}$$

$$\int \varphi_r^*(\mathbf{q}|\mathbf{Q}) \varphi_t(\mathbf{q}|\mathbf{Q})\, d\mathbf{q} \equiv \langle \varphi_r(\mathbf{Q}) \mid \varphi_t(\mathbf{Q}) \rangle_{\mathbf{q}} = \delta_{r,t}. \tag{5.50}$$

The molecular (Coulombic) Hamiltonian in the position representation,

$$\begin{aligned}\hat{\mathrm{H}}(\mathbf{q},\mathbf{Q}) &= \hat{\mathrm{T}}_n(\mathbf{Q}) + [\hat{\mathrm{T}}_e(\mathbf{q}) + V_{ne}(\mathbf{q},\mathbf{Q}) + V_{ee}(\mathbf{q}) + V_{nn}(\mathbf{Q})] \\ &\equiv \hat{\mathrm{T}}_n(\mathbf{Q}) + \hat{\mathrm{H}}_e(\mathbf{q},\mathbf{Q}) \equiv \hat{\mathrm{T}}_n(\mathbf{Q}) + \hat{\mathrm{H}}^e(\mathbf{q},\mathbf{Q}) + V_{nn}(\mathbf{Q}),\end{aligned} \tag{5.51}$$

groups operators of the following (dominating) contributions to the molecular energy (a.u.):

kinetic energy of nuclei: $\hat{\mathrm{T}}_n(\mathbf{Q}) = -\sum_{\alpha=1}^{m}\frac{1}{2M_\alpha}\Delta_\alpha, \quad \Delta_\alpha = \nabla^2_{\boldsymbol{R}_\alpha};$

kinetic energy of electrons: $\hat{\mathrm{T}}_e(\mathbf{q}) = -\frac{1}{2}\sum_{j=1}^{N}\Delta_j, \quad \Delta_j = \nabla^2_{\boldsymbol{r}_j};$

nuclear-electron attraction energy: $V_{ne}(\mathbf{q},\mathbf{Q}) = -\sum_{j=1}^{N}\left(\sum_{\alpha=1}^{m}\frac{Z_\alpha}{|\boldsymbol{R}_\alpha - \boldsymbol{r}_j|}\right) \equiv \sum_{j=1}^{N} v(\boldsymbol{r}_j,\mathbf{Q});$

electron repulsion energy: $V_{ee}(\mathbf{q}) = \sum_{i=1}^{N-1}\sum_{j=i+1}^{N}\frac{1}{|\boldsymbol{r}_i - \boldsymbol{r}_j|} \equiv \sum_{i=1}^{N-1}\sum_{j=i+1}^{N} g(i,j), \; g(i,j) = 1/r_{i,j};$

nuclear repulsion energy: $V_{nn}(\mathbf{Q}) = \sum_{\alpha=1}^{m-1}\sum_{\beta=\alpha+1}^{m}\frac{Z_\alpha Z_\beta}{|\boldsymbol{R}_\alpha - \boldsymbol{R}_\beta|} = \sum_{\alpha=1}^{m-1}\sum_{\beta=\alpha+1}^{m}\frac{Z_\alpha Z_\beta}{R_{\alpha,\beta}}.$

Above, $v(\boldsymbol{r}, \mathbf{Q})$ denotes the external potential for an electron in position $\boldsymbol{r}$ due to the nuclei in their "frozen" positions $\{\boldsymbol{R}_\alpha\}$.

The *electronic Hamiltonian* $\hat{\mathrm{H}}_e(\mathbf{q},\mathbf{Q})$ defined in (5.51) groups all these terms except the nuclear kinetic energy operator. Since the nuclear repulsion energy does not affect the electronic states, representing just the irrelevant additive constant in $\hat{\mathrm{H}}_e(\mathbf{q},\mathbf{Q})$, it is sometimes neglected in the Schrödinger equation for electrons, defined by the eigenvalue problem of the redefined electronic Hamiltonian $\hat{\mathrm{H}}^e(\mathbf{q},\mathbf{Q}) \equiv \hat{\mathrm{H}}_e(\mathbf{q},\mathbf{Q}) - V_{nn}(\mathbf{Q})$.

Therefore, in the BO approximation of (5.48), the molecular states $\Psi_{r,k}(\mathbf{q},\mathbf{Q}) = \varphi_r(\mathbf{q}|\mathbf{Q})\chi_k(\mathbf{Q})$ must satisfy the stationary Schrödinger equation:

$$[\hat{\mathrm{T}}_n(\mathbf{Q}) + \hat{\mathrm{H}}_e(\mathbf{q},\mathbf{Q})]\varphi_r(\mathbf{q}|\mathbf{Q})\chi_k(\mathbf{Q}) = E_{r,k}\varphi_r(\mathbf{q}|\mathbf{Q})\chi_k(\mathbf{Q}), \tag{5.52}$$

where $E_{r,k}$ stands for the molecular energy in the adiabatic state combining rth electronic and kth nuclear states. Since both factors depend, at least parametrically, on the nuclear positions the action of the nuclear kinetic energy operator on adiabatic wave function gives:

$$\begin{aligned}\hat{\mathrm{T}}_n(\varphi_r\chi_k) &= -\sum_{\alpha=1}^{m}\frac{1}{2M_\alpha}[\chi_k(\Delta_\alpha\varphi_r) + 2(\nabla_\alpha\varphi_r)\cdot(\nabla_\alpha\chi_k) + \varphi_r(\Delta_\alpha\chi_k)] \\ &= (\hat{\mathrm{T}}_n\varphi_r)\chi_k - \sum_{\alpha=1}^{m}\frac{1}{M_\alpha}(\nabla_\alpha\varphi_r)\cdot(\nabla_\alpha\chi_k) + \varphi_r(\hat{\mathrm{T}}_n\chi_k).\end{aligned} \tag{5.53}$$

The adiabatic approximation assumes that the kinetic energy operator $\hat{\mathrm{T}}_n(\mathbf{Q})$ constitutes a minor perturbation compared with the electronic Hamiltonian $\hat{\mathrm{H}}_e(\mathbf{q},\mathbf{Q})$. One can therefore envisage the perturbation approach constructed on the basis of the unperturbed Hamiltonian $\hat{\mathrm{H}}^0 = \hat{\mathrm{H}}_e(\mathbf{q},\mathbf{Q})$, in which there are no gradient operations with respect to nuclear positions. Therefore, in *zeroth* order approximation the nuclear positions are treated as parameters and one neglects the second term in r.h.s. of (5.53) as negligible, eventually to be taken into account in *higher* orders of PT. In other words, one assumes that nuclear gradient of the electronic wave function is generally small compared with the associated action of the electronic Hamiltonian. One could also neglect the first (small) Laplacian term, as in the original BO approach, but this contribution can be easily accounted for without any serious complication of the emerging formalism.

Therefore, neglecting only the second term in (5.53), which involves the scalar product of the nuclear gradients of both factors in the adiabatic form of the molecular wave function, multiplying from the left (5.52) by φ_r^*, and "integrating" the result over the electronic position-spin variables $\mathbf{q}$, denoted by $\langle\rangle_{\mathbf{q}}$, then give the following effective Schrödinger equation for the nuclear function $\chi_k(\mathbf{Q})$:

$$\begin{aligned}&[\hat{\mathrm{T}}_n(\mathbf{Q}) + \langle\varphi_r(\mathbf{q}|\mathbf{Q})|\hat{\mathrm{H}}_e(\mathbf{q},\mathbf{Q})|\varphi_r(\mathbf{q}|\mathbf{Q})\rangle_{\mathbf{q}} + \langle\varphi_r(\mathbf{q}|\mathbf{Q})|\,\hat{\mathrm{T}}_n(\mathbf{Q})\,|\varphi_r(\mathbf{q}|\mathbf{Q})\rangle_{\mathbf{q}}]\chi_k(\mathbf{Q})\\ &\quad \equiv \{\hat{\mathrm{T}}_n(\mathbf{Q}) + [E_r^e(\mathbf{Q}) + T_r^n(\mathbf{Q})]\}\chi_k(\mathbf{Q}) \equiv [\hat{\mathrm{T}}_n(\mathbf{Q}) + U_r^{adiab.}(\mathbf{Q})]\chi_k(\mathbf{Q})\\ &\quad = E_{r,k}\chi_k(\mathbf{Q}).\end{aligned} \tag{5.54}$$

This equation contains the effective adiabatic potential in the electronic state φ_r, $U_r^{adiab.}(\mathbf{Q})$, the dominant component of which is the average electronic energy, the associated expectation value of the electronic Hamiltonian:

$$E_r^e(\mathbf{Q}) = \langle\varphi_r(\mathbf{q}|\mathbf{Q})|\hat{\mathrm{H}}_e(\mathbf{q},\mathbf{Q})|\varphi_r(\mathbf{q}|\mathbf{Q})\rangle_{\mathbf{q}}, \tag{5.55}$$

called the *Potential Energy Surface* (PES). It parametrically depends on nuclear positions (molecular geometry) and carries the influence of the average electronic distribution on the system nuclei. It follows from (5.54) that adiabatic potential also includes a (small) diagonal correction due to $\hat{\mathrm{T}}_n(\mathbf{Q})$ in state φ_r,

$$T_r^n(\mathbf{Q}) = \langle\varphi_r(\mathbf{q}|\mathbf{Q})|\,\hat{\mathrm{T}}_n(\mathbf{Q})\,|\varphi_r(\mathbf{q}|\mathbf{Q})\rangle_{\mathbf{q}}, \tag{5.56}$$

which has been neglected in the original, *crude*-adiabatic BO approximation.

It thus follows from the nuclear Schrödinger equation (5.54) that it requires the knowledge of the whole electronic PES $E_r^e(\mathbf{Q})$ of electrons in the specified adiabatic state $\varphi_r(\mathbf{q}|\mathbf{Q})$, the eigenfunction of the electronic Schrödinger equation:

$$\hat{\mathrm{H}}_e(\mathbf{q},\mathbf{Q})\varphi_r(\mathbf{q}|\mathbf{Q}) = E_r^e(\mathbf{Q})\varphi_r(\mathbf{q}|\mathbf{Q}). \tag{5.57}$$

Clearly, the parametric dependence of the electronic energy on nuclear coordinates can be extracted only from a very large number of solutions of the preceding equation, for a sufficient multitude of the fixed nuclear configurations $\{\mathbf{Q}^{(i)}\}$, by an analytical interpolation of the known energies $\{E_r^e(\mathbf{Q}^{(i)})\}$, points on the resulting PES.

To summarize, solving the molecular Schrödinger equation in the adiabatic (BO) approximation first involves solving the *fixed*-nuclei, electronic eigenvalue problem for a large number of molecular geometries, in order to extract the effective potential of forces acting on the system nuclei, averaged over the instantaneous positions of the *fast*-moving electrons. In the second, nuclear stage one uses this effective adiabatic potential to solve the nuclear Schrödinger equation (5.54), which generates the amplitude functions of the nuclear probability distributions and the molecular energy levels containing the kinetic energy of the *slowly* moving nuclei.

As we have already mentioned earlier in this section, the nonadiabatic effects can be accounted for in the *higher* order of the perturbation theory in which the kinetic energy of nuclei represents the perturbation to the unperturbed, electronic Hamiltonian. Therefore, the electronic states $\{\varphi_r(\mathbf{q}|\mathbf{Q})\}$ span the complete basis of the *zeroth* order solutions [see (5.57)], in terms of which the nonadiabatic states can be expanded. Consider the dominating, *first*-order corrections to the adiabatic electronic state $\varphi_r(\mathbf{q}|\mathbf{Q})$ (see Sect. 5.1):

$$\Delta\varphi_r^{(1)}(\mathbf{q}\,|\mathbf{Q}) = \sum_{t \neq r} c_{t,r}^{(1)}(\mathbf{Q})\ \varphi_t(\mathbf{q}\,|\mathbf{Q}). \tag{5.58}$$

It follows from (5.12) that this expansion coefficient is given by the following ratio:

$$c_{t,r}^{(1)}(\mathbf{Q}) = \frac{\langle \varphi_t(\mathbf{q}|\mathbf{Q})|\hat{\mathrm{T}}_n(\mathbf{Q})|\varphi_r(\mathbf{q}|\mathbf{Q})\rangle_{\mathbf{q}}}{E_r^e(\mathbf{Q}) - E_t^e(\mathbf{Q})}. \tag{5.59}$$

The adiabatic approximation is thus adequate only, when the numerator in this expression is small compared with the denominator. Indeed, the degeneracy or *near*-degeneracy of electronic states (small value of the denominator) would generate a large nonadiabatic correction thus contradicting the basic assumption of the adiabatic approximation. The same would be true for a large value of the numerator, signifying a strong nuclear-motion coupling between electronic states.

Therefore, the adiabatic approximation breaks down when for some molecular geometries several electronic states exhibit very close values of the electronic energy. This is the case in the familiar Jahn–Teller effect (removal of the electronic degeneracy by spontaneous distortion of the molecule) and the related Renner effect, due to the vibronic coupling between electronic and nuclear motions, which have profound structural and spectroscopic implications. Let us recall that the Jahn–Teller theorem states that in any nonlinear system there exists some vibrational mode that removes the degeneracy of an electronically (orbitally) degenerate state by lowering the system symmetry. The vibronic coupling between the degenerate electronic states of linear molecules and the deformation (bending) vibrations of the nuclei are responsible for splitting these energy levels in the Renner effect. A proper quantum mechanical description of these processes calls for an explicit dependence of electronic states on nuclear coordinates in the nonadiabatic molecular wave functions, which are customarily represented as linear combinations of several adiabatic states [see (5.58)]. The nuclear motions, of the paramount importance for molecular dynamics (e.g., Murrell et al. 1984; Murrell and Bosanac 1989) and spectroscopy (e.g., Longuet-Higgins 1961), are not covered by this book.

5.3 Orbital Approximation of Electronic Wave Functions

The quantum theory of electronic structure of molecules is based upon the *one*-electron approach to electronic functions of *many*-electron systems, known as the *orbital approximation.* It has greatly influenced the existing terminology of quantum chemistry and the chemical concepts used in interpretations of diverse chemical processes. It ascribes to each electron in the system the *one*-electron function called the *spin orbital* (SO, see Sect. 4.5).

Let us recall that the internal stationary state of the hydrogen-like atom discussed in Chap. 4 has been described by a single SO, $\psi(\boldsymbol{q}) = \varphi(\boldsymbol{r})\zeta(\sigma)$, given by the product of the spatial function, the *orbital* $\varphi(\boldsymbol{r})$, and one of the two admissible *spin functions* $\zeta(\sigma) = \{\alpha(\sigma) = \langle\sigma|\alpha\rangle,\ \beta(\sigma) = \langle\sigma|\beta\rangle\}$ of an electron [see (3.76)]. When the same orbital is used to generate two SO, thus describing a pair of electrons with the opposite spin orientations, as in (3.76), one adopts the so-called *spin-restricted* version of the orbital approximation. Accordingly, in the *spin-unrestricted* description of such two spin-paired electrons, one uses different orbitals for different spins:

$$\{\psi^{+}(\boldsymbol{q}) \equiv \varphi_{\alpha}(\boldsymbol{r})\alpha(\sigma), \quad \psi^{-}(\boldsymbol{q}) \equiv \varphi_{\beta}(\boldsymbol{r})\beta(\sigma)\}. \tag{5.60}$$

Let us now examine the Slater (1929, 1931, 1960) method of constructing in the orbital approximation the N-electron wave functions $\Psi(\mathbf{q}|\mathbf{Q}) \equiv \Psi(N)$, which automatically satisfy the basic requirement of the Fermi–Dirac statistics, the Pauli postulate of their antisymmetry with respect to an exchange of any two indistinguishable fermions. Should the electronic states be exactly independent, the N-electron wave function would then be exactly given by the product of N orthonormal SO attributed to each particle,

$$\Psi(N) = \prod_{i=1}^{N} \psi_i(\boldsymbol{q}_i) \equiv \prod_{i=1}^{N} \psi_i(i) = \psi_1(1)\psi_2(2)\ldots\psi_i(i)\ldots\psi_j(j)\ldots\psi_N(N). \tag{5.61}$$

Indeed, the N-electron probability distribution would then be given by the product of distributions of independent *one*-electron events:

$$p(N) = |\Psi(N)|^2 = \prod_{i=1}^{N} |\psi_i(i)|^2 = \prod_{i=1}^{N} p_i(i). \tag{5.62}$$

Obviously, due to a finite electric charge, electrons repel each other, so that this independent particle approximation can at best be considered only as a first step in a more adequate treatment, which recognizes the dependence (correlation) between their instantaneous positions. Besides this *Coulomb* correlation the electron probabilities must also reflect the constraints imposed by the antisymmetry principle of Pauli, thus additionally exhibiting the *Fermi* (exchange) correlation, which severely conditions the simultaneous probability distributions of the spin-like electrons in the physical space.

The product trial function of (5.61), which has been used as the variational wave function in the Hartree (1928) method, clearly violates this antisymmetry requirement, since each electron is distinguished by the identity of the SO to which it has been individually ascribed. Thus, the permutation $\hat{\mathrm{P}}(i,j)$ of electrons i and j, of exchanging the wave function arguments $\boldsymbol{q}_i$ and $\boldsymbol{q}_j$, instead of changing only the sign of $\Psi(N)$ transforms it into an entirely different function, in which electrons are attributed to different SO:

$$\hat{\mathrm{P}}(i,j)\Psi(N) = \psi_1(1)\psi_2(2)\ldots\psi_j(i)\ldots\psi_i(j)\ldots\psi_N(N) \neq -\Psi(N). \tag{5.63}$$

This shortcoming can be remedied by the appropriate antisymmetrization operation $\hat{\mathrm{A}}$ performed on the product function of (5.61). It is effected by combining all product functions obtained by permuting all N electrons between all N occupied SO. Each permutation P is now identified by the *list of electrons*

$$\boldsymbol{l}(P) = \{l_i(P)\} = [l_1(P), l_2(P), \ldots, l_N(P)], \quad l_i(P) \in (1, 2, \ldots, N),$$

attributed to orbitals ψ_i in the ordered list $\{\psi_i\} = (\psi_1, \psi_2, \ldots, \psi_N)$. Thus, the permutation $\boldsymbol{l}(P) = (4, 2, \ldots, 1)$ symbolizes the product function $\psi_1(4)\,\psi_2(2)\ldots\psi_N(1)$, etc. One could alternatively identify the current permutation P by the *list of orbitals* $\{\psi_j(P)\}$, identified by their labels

$$\boldsymbol{k}(P) = \{k_j(P)\} = [k_1(P), k_2(P), \ldots, k_N(P)], \quad k_j(P) \in (1, 2, \ldots, N),$$

which are attributed to the ordered list of electrons $\{j\} = (1, 2, \ldots, N)$. Thus, the permutation $\boldsymbol{k}(P) = (4, 2, \ldots, 1)$ stands for the product function $\psi_4(1)\,\psi_2(2)\ldots\psi_1(N)$.

For the chosen type of permuting the products of SO, one then introduces the appropriate sign convention for each of $N!$ permutations in the antisymmetrized combination. In order to enforce the change of sign of the wave function, when the current permutation is obtained by an *odd* number of elementary replacements of pairs of electrons in the ordered permutation $P_0 = (1, 2, \ldots, N)$ of (5.61), one introduces the permutation parity $p(P)$, which counts the number of such pair exchanges required to bring the current permutation P to the initial permutation P_0, with $p(P_0) = 0$, and puts the sign $(-1)^{p(P)}$ before the associated product function.

The antisymmetric combination of such $N!$ product functions corresponding to either all permutations of electrons among the ordered list of SO, or all permutations of SO among the ordered list of electrons, thus determines the Slater determinant:

$$\begin{aligned}\Psi_A(N) = \hat{\mathrm{A}}\Psi(N) &= \frac{1}{\sqrt{N!}}\sum_P (-1)^{p(P)}\psi_1(l_1(P))\,\psi_2(l_2(P))\ldots\psi_N(l_N(P))\\ &= \frac{1}{\sqrt{N!}}\sum_P (-1)^{p(P)}\psi_{k_1(P)}(1)\,\psi_{k_2(P)}(2)\ldots\psi_{k_N(P)}(N)\\ &= \frac{1}{\sqrt{N!}}\begin{vmatrix}\psi_1(1) & \psi_1(2) & \ldots\ldots & \psi_1(N)\\ \psi_2(1) & \psi_2(2) & \ldots\ldots & \psi_2(N)\\ \ldots\ldots & \ldots\ldots & \ldots\ldots & \ldots\ldots\\ \psi_N(1) & \psi_N(2) & \ldots\ldots & \psi_N(N)\end{vmatrix}\\ &\equiv |\psi_1\psi_2\ldots\psi_N| \equiv \det(\psi_1\psi_2\ldots\psi_N).\end{aligned} \tag{5.64}$$

Here, the constant before the determinant assures the normalization for the orthonormal set of SO: $\int \psi_i^*(\boldsymbol{q})\,\psi_j(\boldsymbol{q})\,d\boldsymbol{q} = \langle i \mid j\rangle = \delta_{i,j}$. Since exchanging two electrons amounts to the permutation of two columns in this determinantal wave function, the correct result of a change of sign of $\Psi_A(N)$ is obtained, $\hat{\mathrm{P}}(i,j)\Psi_A(N) = -\Psi_A(N)$, as indeed required by the Pauli antisymmetry postulate for fermions. One also observes that this form of wave function automatically satisfies the Pauli exclusion principle that two electrons cannot be described by identical SO. More specifically, should this be the case, the two *rows* in the Slater determinant would then be identical, thus automatically implying $\Psi_A(N) = 0$.

A more subtle implication also follows, when two spin-like electrons near-coalesce in the same position, when $\boldsymbol{q}_i \cong \boldsymbol{q}_j$. This limiting proximity of two electrons exhibiting the same spin orientation gives rise to two identical *columns* in $\Psi_A(N)$, thus again predicting $\Psi_A(N) \cong 0$. In other words, the probability of such an event becomes very small indeed. This implies that spin-like electrons are statistically correlated, avoiding nearby positions in space. This effect is called the *Fermi* or *exchange* correlation between electrons. It should be emphasized that no such restrictions on the instantaneous positions of electrons intervene for the electrons with opposite spins, since then spatial coalescence of two electrons does not imply equality of their position-spin variables: $\boldsymbol{q}_i = (\boldsymbol{r}, \uparrow) \neq \boldsymbol{q}_j = (\boldsymbol{r}, \downarrow)$. Therefore, electrons with different orientations of their spins exhibit only the Coulomb correlation, resulting from their electric charge, while the movements of the spin-like electrons are influenced by both the Fermi and Coulomb correlations.

Thus, in the orbital approximation of the Slater determinant (5.64), the spin-like electrons are not independent, being already Fermi correlated by the exchange symmetry of the electronic wave function. It should be emphasized, however, that this variational wave function ignores completely the Coulomb correlation of all electrons. Therefore, the latter effect should be relatively more important in interactions between electrons exhibiting different spin states, since the spin-like electrons have already been Fermi correlated. It could be also expected that accounting for this missing effect within the spin-restricted approach should most influence the simultaneous probabilities of two electrons occupying the same orbital, the movements of which are confined to the same part of space, probed by the square of their common spatial function (orbital).

It should be observed that the correct symmetry of the analogous orbital wave function for the set of N identical *bosons* would call for the related symmetrization operation $\hat{\mathrm{S}}$ performed on the product wave function:

$$\begin{aligned}\Psi_S(N) = \hat{\mathrm{S}}\Psi(N) &= \frac{1}{\sqrt{N!}} \sum_P \psi_1[l_1(P)]\psi_2[l_2(P)]\ldots\psi_N[l_N(P)]\\ &= \frac{1}{\sqrt{N!}} \sum_P \psi_{k_1(P)}(1)\psi_{k_2(P)}(2)\ldots\psi_{k_N(P)}(N).\end{aligned} \tag{5.65}$$

Indeed, this symmetrical combination of the permuted product functions satisfies the symmetry postulate for bosons, $\hat{\mathrm{P}}(i,j)\Psi_S(N) = \Psi_S(N)$, since such an operation only exchanges two product functions in the sum of all $N!$ terms of $\Psi_S(N)$.

It should be realized that the set $\boldsymbol{\psi}$ of singly occupied SO defining the Slater determinant is not unique. Indeed, any unitary transformation $\boldsymbol{\psi}' = \boldsymbol{\psi}\mathbf{T}$, $\mathbf{TT}^\dagger = \mathbf{I}$, which preserves the orbital orthonormality, replaces the rows $\{\psi_i\}$ of the original determinant (5.64), $\Psi_A = \det(\boldsymbol{\psi})$, with their combinations $\{\psi_k{}' = \sum_j \psi_j T_{j,k}\}$ in the transformed determinant $\Psi_A{}' = \det(\boldsymbol{\psi}')$. It thus follows from the elementary properties of determinants that these two functions are identical: $\Psi_A{}' = \Psi_A$. The two sets of SO which define them are called the *equivalent orbitals*.

One thus encounters various types of *molecular orbitals* (MO) in the theory of electronic structure, selected for their numerical or interpretative convenience. For example, in the two most popular computational methods, formulated within the *Hartree–Fock* (HF) (Fock 1930) and *Kohn–Sham* (KS) (Kohn and Sham 1965) theories, the two canonical sets of orbitals are introduced, which are delocalized throughout the whole molecule and reflect the system spatial symmetry. They provide a useful orbital picture of the spectroscopic and electron ionization phenomena, satisfying important theorems linking their energies and decay behavior with the molecular ionization potentials. The *Natural Orbitals* (NO) of the *Configuration Interaction* (CI) theory similarly generate a compact representation of the Coulomb correlation effects. Finally, the *Localized Orbitals* (LO), describing the diatomic chemical bonds and lone electronic pairs, are useful in providing the orbital interpretations of the near-additivity of several molecular properties and in explaining the remarkable invariance of the given type of σ bonds in different molecular environments. It should be also noticed that the very criteria for the orbital localization are not unique either, so that a variety of alternative sets of the physically equivalent LO have been reported in scientific literature (e.g., Boys and Foster 1960; Edmiston and Ruedenberg 1963), which generate the same determinantal wave function of the molecular system as a whole.

The orbital approximation provides a firm basis for the classification and understanding of electronic states and configurations in atomic and molecular systems. Since to a good approximation the length of the resultant spin **S** of all electrons and its z-component S_z are sharply defined simultaneously with the system electronic energy E_e, the electronic wave functions are required to be eigenfunctions of the associated quantum mechanical operators $\hat{S}^2$ and $\hat{S}_z$, which commute with the electronic Hamiltonian. In the spin-unrestricted form the Slater determinant does not generally satisfy this requirement, while the spin-restricted functions

$$\Psi_A(N = 2p) = \left|\varphi_1^+\varphi_1^-\varphi_2^+\varphi_2^- \cdots \varphi_p^+\varphi_p^-\right|, \tag{5.66}$$

$$\Psi_A(N = 2p + q) = \left|\varphi_1^+\varphi_1^-\varphi_2^+\varphi_2^- \cdots \varphi_p^+\varphi_p^-\varphi_{p+1}^+\varphi_{p+2}^+ \cdots \varphi_{p+q}^+\right|, \tag{5.67}$$

are eigenfunctions of these two resultant-spin operators, corresponding to the quantum numbers S and M_S determining the associated eigenvalues: $|\boldsymbol{S}|^2 = S(S+1)\hbar^2$ and $S_z = M_S\hbar$, $M_S = -S, -S+1, \ldots, S-1, S$. Hence, the state spin-multiplicity $2S + 1$ determines the overall degeneracy of the electronic state with respect to alternative orientations of the resultant spin. For example, the multiplicities of the representative wave functions of (5.66) and (5.67), which correspond to $S = 0$ and $S = q/2$, respectively, are 1 and $q + 1$.

The occupation numbers of shells and orbitals define the system *electron configuration*. When the (doubly occupied) spin-restricted orbitals of (5.66) involve all symmetry-related (degenerate) orbitals of each electronic subshell, this wave function is said to describe the *closed-shell* state of the molecule. Accordingly the *open-shell* state is either characterized by the singly occupied MO, as in (5.67), or it involves doubly occupied subset of the symmetry-related (degenerate) orbitals of the occupied electronic subshell(s).

5.4 Matrix Elements of Electronic Hamiltonian in Orbital Approximation

In order to apply the Slater determinants in the *variational* determination of the approximate electronic states, we have to derive the associated expression for the expectation value of the system electronic energy in the orbital approximation. Moreover, when mixing different determinantal wave functions in a more accurate CI variant, capable of accounting for the Coulomb correlation between electrons, one needs the related expressions for the matrix elements of the electronic Hamiltonian in such a basis set of N-electron functions. A short derivation of these missing elements in the independent electron approximation is the main goal of this section.

Let us first verify the normalization constant of the Slater determinant. Expanding the Ψ_A and Ψ_A^* determinants in the normalization integral gives:

$$\langle \Psi_A | \Psi_A \rangle = \int \ldots \int \Psi_A^*(\mathbf{q}) \Psi_A(\mathbf{q})\, d\mathbf{q}$$
$$= \frac{1}{N!} \sum_P \sum_{P'} (-1)^{p(P)+p(P')} \prod_{j=1}^{N} \langle k_j(P) \mid k_j(P') \rangle. \quad (5.68)$$

Therefore, for the orthonormal orbitals $\{\psi_k(\boldsymbol{q}) = \varphi_k(\mathbf{r})\zeta_k(\sigma) \equiv \langle \boldsymbol{q}|k\rangle\}$, when $\langle k|l\rangle = \delta_{k,l}$, one obtains a nonvanishing contribution in this sum only when for all electrons $k_j(P) = k_j(P')$, i.e., when the lists of orbitals $\{k_j(P)\}$ and $\{k_j(P')\}$ in permutations P and P' are identical, i.e., when $P = P'$, and hence

$$\langle \Psi_A | \Psi_A \rangle = \frac{1}{N!} \sum_P (-1)^{2p(P)} = \frac{N!}{N!} = 1,$$

where we have recognized that there are $N!$ distinct permutations involved in the Slater determinant of (5.64).

Let us now separately combine all *one*- and *two*-electron contributions in the electronic Hamiltonian of (5.51):

$$\hat{\mathrm{H}}^e(\mathbf{q}, \mathbf{Q}) = \sum_{i=1}^{N} \hat{\mathrm{h}}(i) + \sum_{i=1}^{N-1} \sum_{j=i+1}^{N} g(i,j) \equiv \hat{\mathfrak{F}}(\mathbf{q}, \mathbf{Q}) + \hat{\mathfrak{G}}(\mathbf{q})$$
$$\equiv \hat{\mathfrak{F}}(N) + \hat{\mathfrak{G}}(N); \quad (5.69)$$

here, the *one*-electron Hamiltonian $\hat{\mathrm{h}}(i)$ groups the operators of the kinetic energy of ith electron and its attraction energy to all nuclei in their specified, fixed positions, which generate the external potential $v(i)$ (5.51),

$$\hat{\mathrm{h}}(i) = -\frac{1}{2}\Delta_i + v(i), \quad (5.70)$$

while the multiplicative operator $g(i,j)$ corresponds to the Coulomb repulsion between the indicated pair of electrons. Thus, the expectation value of the electronic energy in the state (5.64) is determined by the trivial nuclear-repulsion contribution and the sums of *one*- and *two*-electron contributions in the N-electron system:

$$\langle E_e \rangle_{\Psi_A} = \langle \Psi_A \mid \hat{\mathrm{H}}_e | \Psi_A \rangle = \left\langle \Psi_A \mid \hat{\mathrm{H}}^e | \Psi_A \right\rangle + V_{nn} \equiv \langle E^e \rangle_{\Psi_A} + V_{nn},$$
$$\langle E^e \rangle_{\Psi_A} = \left\langle \Psi_A \middle| \hat{\mathcal{F}} \middle| \Psi_A \right\rangle + \left\langle \Psi_A \mid \hat{\mathcal{G}} | \Psi_A \right\rangle \equiv \left\langle \hat{\mathcal{F}} \right\rangle_{\Psi_A} + \left\langle \hat{\mathcal{G}} \right\rangle_{\Psi_A}. \quad (5.71)$$

Consider first the *one*-electron energy $\langle \mathcal{F} \rangle_{\Psi_A}$. Expanding the two determinants as in (5.68) gives:

$$\langle \mathcal{F} \rangle_{\Psi_A} = \frac{1}{N!} \sum_P \sum_{P'} (-1)^{p(P)+p(P')} \sum_{i=1}^{N} \langle k_i(P) | \hat{\mathrm{h}}(i) | k_i(P') \rangle \prod_{j \neq i} \langle k_j(P) | k_j(P') \rangle. \quad (5.72)$$

Again, a nonvanishing product of the overlap integrals in this expression can appear only when the two permutations are identical: $P = P'$. One also realizes that due to indistinguishability of N electrons in the Slater determinant, each of them gives the same contribution as the representative electron "1" so that the above expression can be further simplified:

$$\langle \mathcal{F} \rangle_{\Psi_A} = \frac{N}{N!} \sum_P (-1)^{2p(P)} \langle k_1(P) \mid \hat{\mathrm{h}}(1) | k_1(P) \rangle = \frac{1}{(N-1)!} \sum_P \bar{h}_{k_1(P),k_1(P)}. \quad (5.73)$$

The above summation over permutations can be replaced by the equivalent summation over N different choices of spin orbital ψ_{k_1} describing electron 1, which defines the matrix elements $\{\bar{h}_{k_1,k_1}\}$ in the SO basis. These *one*-electron integrals should be then multiplied by their multiplicity in all permutations P, equal to the number $(N - 1)!$ of all permutations of the remaining $(N - 1)$ occupied SO $\{k_j \neq k_1\}$ among $(N - 1)$ electrons $(2, 3, \ldots, N)$. Hence,

$$\langle \mathcal{F} \rangle_{\Psi_A} = \frac{(N-1)!}{(N-1)!} \sum_{k=1}^{N} \bar{h}_{k,k} = \sum_{k=1}^{N} \bar{h}_{k,k}. \quad (5.74)$$

One similarly arrives at the corresponding expression for the *two*-electron energy $\langle \mathcal{G} \rangle_{\Psi_A}$. Expanding the determinantal wave functions and taking into account the indistinguishability of electrons give:

$$\begin{aligned}\langle \mathcal{G} \rangle_{\Psi_A} &= \frac{1}{N!} \sum_P \sum_{P'} (-1)^{p(P)+p(P')} \sum_{i=1}^{N-1} \sum_{j=i+1}^{N} \langle k_i(P)k_j(P) \mid g(i,j)|k_i(P')k_j(P')\rangle \\ &\quad \times \prod_{l \neq (i,j)} \langle k_l(P) \mid k_l(P')\rangle \\ &= \frac{1}{2(N-2)!} \sum_P \sum_{P'} (-1)^{p(P)+p(P')} \langle k_1(P)k_2(P) \mid g(1,\, 2)|k_1(P')k_2(P')\rangle \\ &\quad \times \prod_{l \neq (1,2)} \langle k_l(P) \mid k_l(P')\rangle,\end{aligned}$$

where we have recognized that each of the $N(N - 1)/2$ electronic pairs gives the same contribution as the representative pair (1, 2). A subsequent examination of the overlap integrals in the product indicates that all SO for remaining electrons $l \neq (1, 2)$ in permutations P and P' must be identical for electrons (3, 4, ..., N). Therefore, the nonvanishing contributions arise only when the two permutations are identical, $P = P'$, or when they differ only in orbitals describing electrons 1 and 2: $P = \hat{\mathrm{P}}(1,\, 2)P'$. In the former case, the parities of both permutations are equal, giving rise to factor $(-1)^{p(P)+p(P')} = (-1)^{2p(P)} = 1$, while in the latter case they differ by one exchange of two electrons, so that $(-1)^{p(P)+p(P')} = -1$. Moreover, for each choice of the two SO describing electrons 1 and 2, we thus have $(N - 2)!$ permutations of the remaining $(N - 2)$ orbitals $\{k_l \neq (k_1,\, k_2)\}$ among $(N - 2)$ electrons (3, 4, ..., N). Therefore, the preceding expression can be expressed in terms of contributions from *two*-electron integrals:

$$\begin{aligned}\langle \mathcal{G} \rangle_{\Psi_A} &= \frac{(N-2)!}{2(N-2)!} \sum_{k=1}^{N} \sum_{l=1}^{N} [\langle k(1)l(2)\,|\,g(1,2)|k(1)l(2)\rangle - \langle k(1)l(2)\,|\,g(1,2)|l(1)k(2)\rangle] \\ &\equiv \frac{1}{2} \sum_{k=1}^{N} \sum_{l=1}^{N} [\langle kl|g|kl\rangle - \langle kl|g|lk\rangle] \equiv \frac{1}{2} \sum_{k=1}^{N} \sum_{l=1}^{N} [(kk|ll) - (kl|lk)] \\ &\equiv \frac{1}{2} \sum_{k=1}^{N} \sum_{l=1}^{N} [\bar{J}_{k,l} - \bar{K}_{k,l}] = \sum_{k=1}^{N-1} \sum_{l=k+1}^{N} [\bar{J}_{k,l} - \bar{K}_{k,l}].\end{aligned} \tag{5.75}$$

The *two*-electron integrals for the specified pair of SO describing the states of the representative electrons "1" and "2,"

$$\begin{aligned}\bar{J}_{k,l}[\psi_k, \psi_l] &= \langle k(1)l(2) \mid g(1,\, 2)|k(1)l(2)\rangle = \iint |\psi_k(1)|^2 g(1,2) \left|\psi_l^2(1)\right|^2 d\boldsymbol{q}_1 d\boldsymbol{q}_2 \\ &\equiv \bar{J}_{k,l}, \\ \bar{K}_{k,l}[\psi_k, \psi_l] &= \langle k(1)l(2) \mid g(1,\, 2)|l(1)k(2)\rangle \\ &= \iint \psi_k^*(1)\psi_l(1) g(1,\, 2)\, \psi_l^*(2)\psi_k(2) d\boldsymbol{q}_1 d\boldsymbol{q}_2 \equiv \bar{K}_{k,l},\end{aligned} \tag{5.76}$$

are called the *Coulomb* and *exchange* integrals, respectively. The former is indeed seen to measure the Coulomb interaction between the charge distributions of electrons occupying SO ψ_k and ψ_l, respectively.

Since *two*-electron "integrations" involve summation over spin variables, the exchange integrals identically vanish for the two electrons with opposite spins, due to orthogonality of their spin functions (3.76),

$$\bar{K}_{k,l}[\psi_k, \psi_l] = \left(\iint \varphi_k^*(\boldsymbol{r}_1)\varphi_l(\boldsymbol{r}_1)g(\boldsymbol{r}_1, \boldsymbol{r}_2)\, \varphi_l^*(\boldsymbol{r}_2)\varphi_k(\boldsymbol{r}_2)d\boldsymbol{r}_1 d\boldsymbol{r}_2 \right) \left[\sum_{\sigma_1} \zeta_k^*(\sigma_1)\zeta_l(\sigma_1) \right] \\ \times \left[\sum_{\sigma_2} \zeta_l^*(\sigma_2)\zeta_k(\sigma_2) \right] = K_{k,l}[\varphi_k, \varphi_l]\delta_{\zeta_k,\zeta_l} \equiv K_{k,l}\delta_{\zeta_k,\zeta_l}, \quad (5.77)$$

where $K_{k,l}$ stands for the exchange integral defined by the specified pair of the spatial functions (*orbitals*).

It should be observed that no such restriction intervenes in calculating the Coulomb integrals:

$$\bar{J}_{k,l}[\psi_k, \psi_l] = \left(\iint |\varphi_k(\boldsymbol{r}_1)|^2 g(\boldsymbol{r}_1, \boldsymbol{r}_2) |\varphi_l^2(\boldsymbol{r}_2)|^2 d\boldsymbol{r}_1 d\boldsymbol{r}_2 \right) \left[\sum_{\sigma_1} |\zeta_k(\sigma_1)|^2 \right] \left[\sum_{\sigma_2} |\zeta_l(\sigma_2)|^2 \right] \\ = J_{k,l}[\varphi_k, \varphi_l] \equiv J_{k,l}, \quad (5.78)$$

where the sums in the square brackets are both equal to 1 by the normalization condition of the spin states [see (3.76)]:

$$\langle \alpha|\alpha \rangle = \sum_\sigma \langle \alpha|\sigma \rangle \langle \sigma|\alpha \rangle = \sum_\sigma |\alpha(\sigma)|^2 = \langle \beta|\beta \rangle = \sum_\sigma \langle \beta|\sigma \rangle \langle \sigma|\beta \rangle \\ = \sum_\sigma |\beta(\sigma)|^2 = 1. \quad (5.79)$$

It also follows from (5.77) and (5.78) that $J_{k,k} = K_{k,k}$, $J_{k,l} = J_{l,k}$, and $K_{k,l} = K_{l,k}$, since the value of the electron repulsion energy must be independent of the subjectively assigned labels of electrons. This justifies the final expression in (5.75), involving only the *off*-diagonal terms in the double summation $\sum_{k<l}$.

The same is true for the *one*-electron integrals $\bar{h}_{k,k} \equiv \bar{h}_{k,k}[\psi_k]$:

$$\bar{h}_{k,k}[\psi_k] = \int \psi_k^*(\boldsymbol{q})\, \hat{\mathrm{h}}(\boldsymbol{r})\, \psi_k(\boldsymbol{q})\, d\boldsymbol{q} = \left(\int \varphi_k^*(\boldsymbol{r})\, \hat{\mathrm{h}}(\boldsymbol{r})\, \varphi_k(\boldsymbol{r})\, d\boldsymbol{r} \right) \left[\sum_\sigma |\zeta_k(\sigma)|^2 \right] \\ = h_{k,k}[\varphi_k] \equiv h_{k,k}. \quad (5.80)$$

For example, for the wave function (5.66), one obtains the following functional for the expectation value of the electronic energy:

$$\langle E^e \rangle_{\Psi_A} = 2\sum_{i=1}^{p} h_{i,i} + \sum_{i=1}^{p}\sum_{j=1}^{p} (2J_{i,j} - K_{i,j}). \tag{5.81}$$

Clearly, its numerical value depends on the shapes of the p doubly occupied orbitals defining the associated *one-* and *two*-electron integrals. In the closed-shell ground state configuration $[1s^2]$ of the helium atom, one thus finds

$$\langle E^e(\mathrm{He}) \rangle_{\Psi_0} = 2h_{1s,1s} + J_{1s,1s}, \tag{5.82}$$

while for the ground state of beryllium, defined by configuration $[1s^2 2s^2]$, one predicts

$$\langle E^e(\mathrm{Be}) \rangle_{\Psi_0} = 2h_{1s,1s} + 2h_{2s,2s} + J_{1s,1s} + J_{2s,2s} + 4J_{1s,2s} - 2K_{1s,2s}. \tag{5.83}$$

The same result can be heuristically derived by summing the elementary *one*-electron energies of all N electrons, the expectation values of the Hamiltonian (5.70), and the repulsion energies in each of the $N(N-1)/2$ different electronic pairs. Indeed, the average interaction energy between two (indistinguishable) electrons (1, 2) occupying spin orbitals ψ_i and ψ_j is given by the expectation value of the $g(1, 2)$ in the Slater determinant $\Psi_A(2) = |\psi_i\psi_j| = 2^{-1/2}[\psi_i(1)\psi_j(2) - \psi_i(2)\psi_j(1)]$:

$$\langle \Psi_A(2)|g(1,2)|\Psi_A(2)\rangle = \bar{J}_{i,j} - \bar{K}_{i,j} = \begin{cases} J_{i,j}, & \text{for different spin states} \\ J_{i,j} - K_{i,j}, & \text{for identical spin states} \end{cases}.$$

Thus, for the two spin-paired electrons in He atom, when $\bar{K}_{i,j} = 0$, one reproduces the result of (5.82). It can be also easily verified that for beryllium atom in the ground state $\Psi_A(4) = |1s^+1s^-2s^+2s^-|$, one recovers (5.83).

The expectation value $\left\langle \Psi_A \mid \hat{\mathrm{H}}^e | \Psi_A \right\rangle = \left\langle \Psi_0 \mid \hat{\mathrm{H}}^e | \Psi_0 \right\rangle$ (5.71) represents a particular, ground state case of a general *diagonal* matrix element of the electronic Hamiltonian, $\left\langle \Psi_n \mid \hat{\mathrm{H}}^e | \Psi_n \right\rangle$, for any antisymmetric electronic state (Slater determinant) Ψ_n specified by alternative choices of N occupied, orthonormal SO. The same energy formulas also apply to excited electron configurations $\Psi_{n>0}$ obtained by replacing some of the SO occupied in Ψ_0, by the *virtual* orbitals, unoccupied in Ψ_0. Thus, given the modified list of SO occupied in Ψ_n, *occd.*$[n] \equiv [n]$, a general formula for the expectation value of the electronic energy remains unaffected:

$$\begin{aligned} \langle E^e \rangle_{\Psi_n} &= \left\langle \Psi_n \mid \hat{\mathrm{H}}^e | \Psi_n \right\rangle = \sum_k^{[n]} \bar{h}_{k,k} + \frac{1}{2} \sum_k^{[n]} \sum_l^{[n]} [\bar{J}_{k,l} - \bar{K}_{k,l}] \\ &\equiv \sum_k^{[n]} \left\{ \langle k|\hat{\mathrm{h}}|k\rangle + \sum_l^{[n]} \langle kl|\, g\, |kl - lk\rangle \right\}, \quad \langle \Psi_n|\Psi_n\rangle = \prod_i^{[n]} \langle i \mid i\rangle = 1. \end{aligned} \tag{5.84}$$

Finally, let us examine the *off-diagonal* matrix elements $\left\langle \Psi_n \mid \hat{\mathrm{H}}^e | \Psi_{n'} \right\rangle$ between two electron configurations, differing in the list of the occupied SO. Due to the orthogonality of SO such configurations can be shown to be also automatically orthogonal: $\langle \Psi_n \mid \Psi_{n'} \rangle = 0$ [see (5.68)]. The sets of the occupied SO in these two states may still exhibit some common SO, be it in different positions (rows) of two determinantal functions. Therefore, for definiteness, we assume that by appropriate exchange of rows in one of these two Slater determinants, the two configurations have been brought to the *maximum coincidence* form, in which the rows of the common SO of both configurations appear in the same positions in both determinants. We already know from the properties of the Slater determinant that such exchanges can at best change the sign of the wave function, which can be diagnosed from the known number of the *row* permutations in the original Slater determinant required for reaching this maximum coincidence. This sign can then be used to multiply the matrix element obtained from the maximum coincidence rules.

There are three general types of differences between such prearranged lists of SO in both Slater determinants, giving rise to the associated expressions for the matrix element of the electronic Hamiltonian. They can be derived in a way analogous to that used to derive the diagonal element, by expanding both determinants in terms of the permuted products of SO, applying the SO orthonormality relations, and recognizing the indistinguishability of electrons. The relevant cases are summarized by the following *Slater–Condon rules* (Slater 1929; Condon 1930):

1. Configurations Ψ_n and $\Psi_{n'}$ differ only in a single SO, with ψ_p of the former being replaced by ψ_r in the latter, as a result of the electron excitation $\psi_p \rightarrow \psi_r$,

$$\left\langle \Psi_n \middle| \hat{\mathrm{H}}^e \middle| \Psi_{n'} \right\rangle = \langle p|\hat{\mathrm{h}}|r\rangle + \sum_j^{[n]} \langle pj|g|rj - jr\rangle,$$

$$\langle \Psi_n | \Psi_{n'} \rangle = \langle p \mid r \rangle \prod_{i \neq p}^{[n]} \langle i \mid i \rangle = 0; \tag{5.85}$$

2. Configurations Ψ_n and $\Psi_{n'}$ differ only in two SO, as a result of the double excitation $(\psi_p \rightarrow \psi_r, \psi_q \rightarrow \psi_s)$ or $(\psi_p, \psi_q) \rightarrow (\psi_r, \psi_s)$,

$$\left\langle \Psi_n \middle| \hat{\mathrm{H}}^e \middle| \Psi_{n'} \right\rangle = \langle pq|g|rs - sr\rangle,$$

$$\langle \Psi_n | \Psi_{n'} \rangle = \langle p | r \rangle \langle q | s \rangle \prod_{i \neq (p,q)}^{[n]} \langle i \mid i \rangle = 0. \tag{5.86}$$

3. Configurations Ψ_n and $\Psi_{n'}$ differ in more than two SO, thus reflecting the triple or higher excitations,

$$\left\langle \Psi_n \middle| \hat{\mathrm{H}}^e \middle| \Psi_{n'} \right\rangle = 0, \quad \langle \Psi_n | \Psi_{n'} \rangle = 0. \tag{5.87}$$

To summarize, the determinantal functions corresponding to the system ground and excited configurations, in which electrons have been excited from some Ψ_0-occupied to the corresponding Ψ_0-virtual SO of the molecule, form the orthonormal basis of N-electron functions. The Slater–Condon rules allow one to express their average energies and the coupling matrix elements between them in terms of the elementary *one-* and *two*-electron integrals involving SO, the elementary *one*-electron functions of the orbital approximation.

5.5 Example: Helium Atom

As an illustration we shall now apply the approximate methods to estimate the ground state energy of the helium atom, when its two electrons occupy the lowest orbital 1*s*, from the known solutions for the *one*-electron atom [(4.58) and (4.62)]. The a.u. are used throughout.

5.5.1 Perturbation Approximation

The internal (Coulomb) Hamiltonian of two electrons moving in the field of the helium nucleus ($Z = 2$) determines the *perturbed* Hamiltonian (a.u.),

$$\begin{aligned} &\hat{\mathrm{H}}(\boldsymbol{r}_1,\boldsymbol{r}_2) = \hat{\mathrm{H}}^0(\boldsymbol{r}_1,\boldsymbol{r}_2) + \hat{\mathrm{h}}(\boldsymbol{r}_1,\boldsymbol{r}_2), \quad &&\hat{\mathrm{h}}(\boldsymbol{r}_1,\boldsymbol{r}_2) = g(1,2) = |\boldsymbol{r}_1 - \boldsymbol{r}_2|^{-1} = r_{1,2}{}^{-1}, \\ &\hat{\mathrm{H}}^0(\boldsymbol{r}_1,\boldsymbol{r}_2) = \hat{\mathrm{H}}_1(\boldsymbol{r}_1) + \hat{\mathrm{H}}_2(\boldsymbol{r}_2), &&\hat{\mathrm{H}}_i(\boldsymbol{r}_i) = -\frac{1}{2}\Delta_i - \frac{Z}{r_i}, \quad i = 1,2, \end{aligned} \tag{5.88}$$

with the electron repulsion operator representing the *perturbation* and the separable Hamiltonian $\hat{\mathrm{H}}^0(\boldsymbol{r}_1,\boldsymbol{r}_2)$ given by the sum of the hydrogen-like operators $\{\hat{\mathrm{H}}_i(\boldsymbol{r}_i)\}$ of two electrons determining the *unperturbed* Hamiltonian. This assumption thus defines the unperturbed solutions:

$$E_0^{(0)} = -Z^2 = -4, \quad \Psi_0^{(0)}(\boldsymbol{r}_1,\boldsymbol{r}_2) = \psi_{1s}(\boldsymbol{r}_1)\psi_{1s}(\boldsymbol{r}_2) = \frac{Z^3}{\pi}\exp[-Z(r_1 + r_2)], \tag{5.89}$$

since the hydrogen-like solutions for each electron read:

$$\psi_{1s}(\boldsymbol{r}_i) = \sqrt{\frac{Z^3}{\pi}}\exp(-Zr_i), \quad \hat{\mathrm{H}}_i(\boldsymbol{r}_i)\,\psi_{1s}(\boldsymbol{r}_i) = -\frac{Z^2}{2}\psi_{1s}(\boldsymbol{r}_i), \quad i = 1,2. \tag{5.90}$$

This energy estimate should be compared with the experimental value $E_0 = -2.9037$.

The first correction to this crude estimate [see (5.11)] determines the expectation value in state $\Psi_0^{(0)}$ of the repulsion energy operator (perturbation):

$$\Delta E_0^{(1)} = \left\langle \Psi_0^{(0)} \middle| \hat{\mathrm{h}} \middle| \Psi_0^{(0)} \right\rangle = \iint \psi_{1s}^*(\boldsymbol{r}_1)\psi_{1s}^*(\boldsymbol{r}_2)\frac{1}{r_{1,2}}\psi_{1s}(\boldsymbol{r}_1)\psi_{1s}(\boldsymbol{r}_2)\,d\boldsymbol{r}_1\,d\boldsymbol{r}_2. \quad (5.91)$$

To calculate this integral we first assume the spherical coordinates of electron "1" relative to the nucleus, $\boldsymbol{r}_1 = (r_1, \vartheta_1, \varphi_1)$, adopt the relative spherical coordinates of electron "2" with respect to electron "1," $\boldsymbol{r}_{1,2} = \boldsymbol{r}_2 - \boldsymbol{r}_1 = (r_{1,2}, \theta, \phi)$, and use Carnot's cosine formula to express the *inter*-electron separation:

$$r_{1,2} = (r_1^2 + r_2^2 - 2r_1r_2\cos\theta)^{1/2}. \quad (5.92)$$

Hence, the *first*-order correction of (5.91) reads:

$$\Delta E_0^{(1)} = \frac{Z^6}{\pi^2}\left(\int_0^{\pi}\sin\vartheta_1\,d\vartheta_1\int_0^{2\pi}d\varphi_1\right)\left(\int_0^{2\pi}d\phi\right)$$
$$\times\left(\int_0^{\infty}r_1^2e^{-2Zr_1}\int_0^{\infty}r_2^2e^{-2Zr_2}\int_0^{\pi}(r_1^2+r_2^2-2r_1r_2\cos\theta)^{-1/2}\sin\theta\,d\theta\,dr_1dr_2\right). \quad (5.93)$$

We then substitute $x = \cos\theta$:

$$\int_0^{\pi}(r_1^2+r_2^2-2r_1r_2\cos\theta)^{-1/2}\sin\theta\,d\theta = \int_{-1}^{1}(r_1^2+r_2^2-2r_1r_2x)^{-1/2}dx$$
$$= -\frac{1}{r_1r_2}\int_{r_{1,2}(x=-1)}^{r_{1,2}(x=1)}dr_{1,2} = \begin{cases}\frac{2}{r_1}, & r_1 > r_2\\ \frac{2}{r_2}, & r_1 < r_2\end{cases}. \quad (5.94)$$

Using this result and typical integral $\int_0^{\infty} y^n\exp(-ay)dy = n!/a^{n+1}$in (5.93) gives:

$$\Delta E_0^{(1)} = 16Z^6\int_0^{\infty}r_1^2e^{-2Zr_1}\left(\frac{1}{r_1}\int_0^{r_1}r_2^2e^{-2Zr_2}dr_2 + \int_{r_1}^{\infty}r_2e^{-2Zr_2}dr_2\right)dr_1$$
$$= 4Z^3\int_0^{\infty}r_1^2e^{-2Zr_1}\left[\frac{1}{r_1} - \left(Z+\frac{1}{r_1}\right)e^{-2Zr_1}\right]dr_1 = \frac{5Z}{8}. \quad (5.95)$$

Thus, in the *first* order of the perturbation theory, one estimates the ground state energy of the helium atom as being much closer to the experimental value,

$$E_0^{(1)} = -Z^2 + 5Z/8 = -2.75. \quad (5.96)$$

5.5.2 Variational Estimates

The simplest trial wave function of this *two*-electron system in the spin-restricted (*R*) approximation is given by the product function of (5.89), when one replaces *Z* with an effective nuclear charge ζ, a nonlinear variational parameter,

$$\Phi(\boldsymbol{r}_1,\boldsymbol{r}_2;\zeta) = \frac{\zeta^3}{\pi}\exp[-\zeta(r_1+r_2)] = \psi_{1s}(\boldsymbol{r}_1;\zeta)\psi_{1s}(\boldsymbol{r}_2;\zeta) \equiv \Phi^R(\boldsymbol{r}_1,\boldsymbol{r}_2;\zeta), \quad (5.97)$$

where the normalized trial orbital $\psi_{1s}(\boldsymbol{r}_i;\zeta)$ satisfies the energy eigenvalue equation for the *one*-electron atom described by Hamiltonian $\hat{\mathrm{H}}(\boldsymbol{r};\zeta) = -\frac{1}{2}\Delta - \frac{\zeta}{r}$,

$$\psi_{1s}(\boldsymbol{r}_i;\zeta) = \sqrt{\frac{\zeta^3}{\pi}}\exp(-\zeta r_i), \quad \hat{\mathrm{H}}(\boldsymbol{r}_i;\zeta)\,\psi_{1s}(\boldsymbol{r}_i;\zeta) = -\frac{\zeta^2}{2}\psi_{1s}(\boldsymbol{r}_i;\zeta), \quad i = 1,2. \quad (5.98)$$

Since each electron in the helium atom experiences a diminished attraction compared with that in He^+, due to a partial screening of the nucleus by the other electron, one expects the optimum value of this variational parameter to be in the range $1 < \zeta < 2$.

In order to express the average electronic energy, the expectation value of the Hamiltonian $\hat{\mathrm{H}}(\boldsymbol{r}_1,\boldsymbol{r}_2)$ (5.88), we first express the latter in terms of the effective *one*-electron Hamiltonians $\{\hat{\mathrm{H}}(\boldsymbol{r}_i;\zeta)\}$:

$$\hat{\mathrm{H}}(\boldsymbol{r}_1,\boldsymbol{r}_2) = \hat{\mathrm{H}}(\boldsymbol{r}_1;\zeta) + \hat{\mathrm{H}}(\boldsymbol{r}_2;\zeta) + (\zeta - \mathrm{Z})(r_1^{-1} + r_2^{-1}) + r_{12}^{-1}. \quad (5.99)$$

Hence, using the eigenvalues of (5.98) and the electron repulsion energy of (5.95) for $Z = \zeta$, gives

$$\begin{aligned} E(\zeta) &= \iint \Phi^*(\boldsymbol{r}_1,\boldsymbol{r}_2;\zeta)\hat{\mathrm{H}}(\boldsymbol{r}_1,\boldsymbol{r}_2)\,\Phi(\boldsymbol{r}_1,\boldsymbol{r}_2;\zeta)\,d\boldsymbol{r}_1 d\boldsymbol{r}_2 \\ &= \frac{\zeta^6}{\pi^2}\iint e^{-2\zeta(r_1+r_2)}[-\zeta^2 + (\zeta - \mathrm{Z})(r_1^{-1} + r_2^{-1}) + r_{12}^{-1}]\,d\boldsymbol{r}_1 d\boldsymbol{r}_2 \\ &= -\zeta^2 + 2(\zeta - \mathrm{Z})\zeta + 5\zeta/8 = \zeta^2 - 27\zeta/8. \end{aligned} \quad (5.100)$$

The optimum value of the effective nuclear charge is then determined by the variational condition:

$$\left.\frac{dE(\zeta)}{d\zeta}\right|_{\zeta^{opt.}} = 2\zeta^{opt.} - 27/8 = 0 \quad \text{or} \quad \zeta^{opt.} = 27/16 \cong 1.69 < \mathrm{Z}. \quad (5.101)$$

It gives the following estimate of the helium ground state energy (Hylleraas 1828):

$$E(\zeta^{opt.}) = -(27/16)^2 \cong -2.848 > E_0 = -2.9037. \tag{5.102}$$

It should be realized that the full electronic wave function for the helium atom in the spin-*restricted* variant also includes the spin singlet function of two electrons,

$$\begin{aligned}\Psi[(\boldsymbol{r}_1, \sigma_1), (\boldsymbol{r}_2, \sigma_2)] &= \Psi^R(\boldsymbol{q}_1, \boldsymbol{q}_2)\\ &= \Phi^R(\boldsymbol{r}_1, \boldsymbol{r}_2)\frac{1}{\sqrt{2}}[\alpha(\sigma_1)\beta(\sigma_2) - \beta(\sigma_1)\alpha(\sigma_2)].\end{aligned} \tag{5.103}$$

In the spin-*unrestricted* (*U*) approximation, using different orbitals for different spins, the spatial function depends on two nonlinear variational parameters:

$$\Phi^U(\boldsymbol{r}_1, \boldsymbol{r}_2; \zeta_1, \zeta_2) = \frac{1}{\sqrt{2}}[\psi_{1s}(\boldsymbol{r}_1; \zeta_1)\psi_{1s'}(\boldsymbol{r}_2; \zeta_2) + \psi_{1s'}(\boldsymbol{r}_1; \zeta_2)\psi_{1s}(\boldsymbol{r}_2; \zeta_1)]. \tag{5.104}$$

Its symmetrization with respect to the exchange of the position vectors of two electrons is required by the Pauli exclusion principle. The product of the spatial (symmetric) wave function and the singlet (antisymmetric) spin function is then antisymmetric with respect to the permutation of two electrons. The optimum values of these exponents of the two 1*s* orbitals, $\zeta_1^{opt.} = 1.19$ and $\zeta_2{}^{opt.} = 2.18$, give $E(\zeta_1{}^{opt.}, \zeta_2{}^{opt.}) = -2.876 < E(\zeta^{opt.})$ thus offering a better approximation (variational flexibility) of the ground state wave function compared with the spin-restricted analog.

Indeed, it follows from (5.104) that two electrons in the spin-unrestricted state correlate (radially) their movements around the nucleus: when one electron occupies a more compact orbital, thus being on average closer to the nucleus, the other electron occupies the more diffuse orbital, thus exhibiting larger average separation from the nucleus. Therefore, the average *inter*-electron distance in the spin unrestricted state is expected to be larger compared with that in the spin-restricted case, in which both electrons are kept within confines of the same orbital.

5.6 Idea of a Pseudopotential

As we have already observed at the end of Sect. 5.1.2, the atomic cores are predicted to remain largely inactive (invariant) in the chemical processes of the bond forming and/or bond breaking. Indeed, it directly follows from the AO-mixing criteria that the system chemical bonds must be shaped by the *valence* shells of constituent atoms. This observation is particularly important for heavy *many*-electron atoms, in which the number of the *inner*-shell electrons (n) is much larger than the complementary number of the chemically active, *valence* shell electrons: $N - n \ll n$. It is thus tempting to formulate the chemical bond theory focused solely on the quantum states of the valence electrons, since such a reduction of the problem gives rise to a significant conceptual and mathematical simplification while still retaining all significant sources of the chemical binding in molecular systems.

It is evident from the orthogonality relations between orbitals, which generate the familiar nodal structure of valence orbitals in the regions of atomic cores, that any change of the orbitals in one set affects the shapes of orbitals in the other set. Therefore, such constraints appear to prevent any formulation of the "valence-only" theory, in which the optimized valence orbitals satisfy the orthogonality constraints to the "frozen" core orbitals. However, the Pauli antisymmetrization postulate for fermions (Sect. 5.3) is automatically satisfied by the determinantal wave function, no matter whether the orbitals defining the Slater determinant are mutually orthogonal or not. Thus, the requirement that the valence orbitals be orthogonal to the core orbitals is not actually needed to satisfy the exclusion principle. The pseudopotential theory makes use of this very property in designing the "valence-only" theory of molecular electronic structure.

Achieving this goal calls for a nontrivial replacement of the valence core orthogonality requirement by an equivalent theoretical concept, which turns out to represent an additional (nonclassical) operator or the associated local potential term in the effective Hamiltonian of the valence electron, called the *pseudopotential* (Hellmann 1935; Gombas 1967; Slater 1960, 1974; Szasz 1985). It explicitly depends on the shapes and energies of the core orbitals. The *pseudoorbitals*, the eigenfunctions of this effective Hamiltonian, are determined as if the core did not exist at all, its presence being felt exclusively through the pseudopotential. The lowest pseudoorbital exhibits no nodes in the atomic core regions and generates the maximum probability density in the system valence shell.

The first attempt in this direction was made by Hellmann (Hn) (Hellmann 1935), who introduced the very idea of replacing the orthogonality requirement by the pseudopotential within the statistical Thomas–Fermi (TF) model (Thomas 1927; Fermi 1928; Gombas 1949; March 1975). He also proposed the use in atomic and molecular calculations of the model atomic potential for the valence electron, say in Na atom, including the Coulomb attraction $V_q(r)$ due to the core net charge $q = Z - n$ and the model pseudopotential $V_p(r)$:

$$V_{\mathrm{Hn}}(r) = V_q(r) + V_p(r) = -\frac{q}{r} + A\frac{\exp(-\kappa r)}{r}. \tag{5.105}$$

Here, the adjustable parameters A and κ are determined by fitting the predicted eigenvalues of the associated effective Hamiltonian for the valence electron,

$$\hat{\mathrm{H}}_{\mathrm{Hn}}(r) = -(1/2)\Delta + V_{\mathrm{Hn}}(r) \equiv \hat{\mathrm{H}}_{\mathrm{Na}}^{val.}(1), \tag{5.106}$$

to the atom experimental energy spectrum in the valence shell regime.

Notice that the pseudopotential part of this effective potential generates at small distances from the nucleus the repulsive wall, which prevents the valence orbital (nodeless in the core range) to assume large values in the region occupied by the inner-shell electrons. In other words, this repulsive barrier prevents the valence electron from falling into the core, despite the lack of the core–valence orthogonality. Thus, in the pseudopotential approach, the usual *geometrical* constraints of the orbital orthogonality are replaced by equivalent *physical* barriers preventing the valence orbitals to collapse into the inner-shell region.

Hellmann has applied the same idea to atoms with many valence electrons. For example, by determining the model potential of (5.105) to fit as accurately as possible the valence energy spectrum of Mg^+, containing a single valence electron, one writes the valence-only Hamiltonian for the neutral atom, containing two valence electrons, in the following form:

$$\hat{H}^{val.}_{Mg}(1,\,2) = \hat{H}_{Hn}(1) + \hat{H}_{Hn}(2) + 1/r_{12}. \tag{5.107}$$

The molecular applications, e.g., to $Na_2 = Na_A - Na_B$, is also valid. Since Na^+ cores do not participate in this σ bond, the effective Hamiltonian for the two valence electrons in the molecular sodium can be simplified as follows:

$$\hat{H}^{val.}_{Na_2}(1,\,2) = \sum_{i=1}^{2}\left[-\frac{1}{2}\Delta_i + V_{Hn}(r_{iA}) + V_{Hn}(r_{iB})\right] + \frac{1}{r_{12}}, \quad r_{iX} = |\boldsymbol{r}_i - \boldsymbol{R}_X|. \tag{5.108}$$

The pseudopotential method generates the exact "valence only" formalism for atoms and molecules. It can be formulated either as a model (semiempirical) procedure or as the ab initio theory in the spirit of the Phillips and Kleinman (PK) (Phillips and Kleinman 1959, 1960) treatment, which defines the *local* pseudopotential corresponding to the *pseudopotential operator* representing the Pauli exclusion principle (core–valence orthogonality). The latter aspect will be addressed in Sect. 6.1.5.

Having outlined the basic idea of the local pseudopotential, without attempting at this stage to present the ab initio theory in a comprehensive way, let us complete this section with just a short comment on some of the method's most attractive aspects. The effects of the exclusion principle and orbital orthogonality have been shown to be *exactly* representable in the form of the associated effective local pseudopotential. This localization of operators in the PK theory is very much in spirit of that later used in the semiempirical methods of quantum chemistry and in the modern DFT, to establish the effective Hamiltonian determining the KS orbitals. The local pseudopotentials are more suitable for an analytical representation, in the form of model potentials, both semiempirical and those having ab initio justifications as their background. It brings a deeper, physical understanding of the core–valence separation problem. Indeed, the local pseudopotential has permitted a plausible physical interpretation and opened the way to wide range of applications and modeling. It has provided the causal picture of the quantum states of valence electrons in atomic or molecular systems, as moving in the resultant potential generated by atomic cores, including both their electrostatic potentials and the "Pauli" term mimicking the valence–core orthogonality constraint.

The pseudopotential theory was first applied in 1930 and 1940 to problems of the solid state physics, to build up the quantum theory of metals, and then – starting from 1960 – it has been developed as an alternative theory of atoms and molecules. This conceptually appealing approach is mathematically coherent and elegant, particularly in its ab initio formulation. Its conceptual simplicity, still combined with remarkable accuracy, facilitates a subsequent modeling of atomic and molecular phenomena. These models also include those derived from the ab initio theory. A good exposition of the method origins and capabilities is given in the monograph by Szasz (1985), the chief proponent of the molecular applications of the pseudopotential theory.

References

Born M, Oppenheimer JR (1927) Ann Phys 64:457
Boys SF, Foster J (1960) Rev Mod Phys 32:305
Condon EU (1930) Phys Rev 36:1121
Edmiston C, Ruedenberg K (1963) Rev Mod Phys 32:457
Fermi E (1928) Z Phys 48:73
Fock VA (1930) Z Phys 61:126
Gombas P (1949) The statistical theory of atoms and its applications. Springer, Vienna
Gombas P (1967) Pseudopotentials. Springer, Vienna
Hartree DR (1928) Proc Camb Philos Soc 24:89
Hellmann H (1935) J Chem Phys 3:61
Hylleraas EA (1828) Z Physik 48:469
Kohn W, Sham LJ (1965) Phys Rev 140A:1133
Longuet-Higgins HC (1961) Adv Spectrosc 2:429
March NH (1975) Self-consistent fields in atoms. Pergamon, Oxford
Murrell JN, Bosanac SD (1989) Introduction to the theory of atomic and molecular collisions. Wiley, New York
Murrell JN, Carter S, Farantos SC, Huxley P, Varandas AJC (1984) Molecular potential energy functions. Wiley, New York
Phillips JC, Kleinman L (1959) Phys Rev 116:287
Phillips JC, Kleinman L (1960) Phys Rev 118:1153
Slater JC (1929) Phys Rev 34:1293
Slater JC (1931) Phys Rev 38:1109
Slater JC (1960) Quantum theory of atomic structure, vols 1 and 2. McGraw-Hill, New York
Slater JC (1974) Quantum theory of molecules and solids, vol 4: The self-consitent field for molecules and solids. McGraw-Hill, New York
Szasz L (1985) Pseudopotential theory of atoms and molecules. Wiley, New York
Thomas LH (1927) Proc Camb Philos Soc 23:542

Chapter 6
Wave Function Methods

Abstract A hierarchy of the *Self-Consistent Field* (SCF) theories of the molecular electronic structure is surveyed. First, the rudiments of the *Hartree* approach using the trial wave function in the form of the product of the occupied Molecular Orbitals (MO) describing independent one-electron states and providing the reference in defining the electron *exchange-correlation* effects are given. The *Hartree–Fock* (HF) method adopting the Slater determinant (antisymmetrized product) as the variational wave function, which constitutes a natural reference for determining the electron *Coulomb correlation* effects, and its analytical implementation in the finite basis set of AO, called SCF LCAO MO theory, are summarized. The Koopmans theorem is discussed and the concepts of Slater's *transition state* (TS) in electronic excitations and of the *local pseudopotential* of Phillips and Kleinman (PK) are introduced. Typical errors in SCF calculations are identified and the electron correlation problem is formulated in terms of the conditional two-electron densities and the associated correlation holes, the sum-rules of which are examined. Alternative *Configuration Interaction* (CI) strategies for determining the *static* and/or *dynamic* Coulomb correlation effects, formally based upon the MO expansion theorem for molecular electronic states, are reviewed. Both the *Single-Reference* (SR) SCF (HF) and *Multireference* (MR) SCF (MR SCF) or Multiconfigurational (MC) SCF (MC SCF) and the *Complete-Active-Space* (CAS) SCF (CAS SCF) wave functions can be used to generate the excited configurations to be included in the subsequent CI expansion, giving rise to SR CI and MR CI approaches, respectively. Several *single-* and *multi* reference CI methods are identified, including the alternative variants using either the *full* CI (FCI) or a limited expansion in terms of the *single* (S), *double* (D), *triple* (T), *quadruple* (Q), and in general *n-tuple* electron excitations from the HF/SCF or MR SCF wave functions, e.g., the *variational* SR techniques: CID, CISD, CISTQ, etc. The *size*-consistency and *size*-extensivity requirements of such approximate variational treatments are commented upon and the problem of choosing an effective orbital set for subsequent CI calculations is addressed. The reduced density matrices are introduced and the associated concepts of the correlated one- and two-electron functions, called the *Natural Orbitals* (NO) and *Natural Geminals* (NG),

DOI 10.1007/978-3-642-20180-6_6,

respectively, are defined together with their pseudoapproximations in the limited CI approaches.

The simplest variant of the *Many-Body Perturbation Theory* (MBPT), the *Møller–Plesset* (MP) theory, is examined and the Brillouin and McDonald theorems are formulated. A hierarchy of expressions for the electron correlation energy and CI coefficients in the intermediate-normalization representation is derived. The CI theories of the correlated electronic pairs are summarized, including several *separated*-pair approximations, e.g. the *Independent Electron Pair Approximation* (IEPA), and that using the *Antisymmetrized Product of Strongly Orthogonal Gemminals* (APSG), as well as selected *coupled*-pair approaches, e.g., the *Coupled Electron Pair Approximation* (CEPA), the related *Coupled-Pair Many Electron Theory* (CPMET), and the *Coupled Cluster* (CC) approximation. The second-quantization formalism of the electron correlation theory, in terms of the electron *creation* and *annihilation operators* or their local field analogs, acting in the molecular *Fock space* is introduced and the associated representations of the one- and two-electron terms in the molecular electronic Hamiltonian are examined. The cluster expansion of *many*-electron wave functions is introduced and the simplest case of the CC method is examined.

This overview of the standard ab initio MO theories is then followed by the elements of the *chemical* quantum theory of molecular systems provided by the modern *Valence Bond* (VB) methods originating from the classical *Heitler–London* (HL) treatment of the hydrogen molecule. The origins of the theory and a variety of the covalent and ionic VB structures constructed directly from the valence-shell AO's of constituent atoms, are traditionally presented using the illustrative case of H_2. The importance of the electron pairing and of the VB exchange integral in terms of AO for the interpretation of the origins of the chemical bonding is emphasized, various physical factors shaping the optimum orbitals are examined, and the associated estimates of the bonding energy of H_2 are summarized. The equivalence of the VB and CID theories in the *minimum basis* description of the hydrogen molecule is demonstrated and the AO-expansion theorem is formulated, which provides a formal basis for the VB treatments of general molecular systems. The semilocalized AO's of Coulson and Fischer are introduced. They are shown to absorb in the covalent HL-type function the effects due to ionic resonant structures. The *Perfect Pairing Approximation* (PPA) of the *Generalized VB* (GVB) approach of Goddard et al., corresponding to a single (dominant) Lewis structure, is introduced and the use of Rumer diagrams in selection of the linearly independent (canonical) set of VB structures from a multitude of admissible spin couplings in a molecule is illustrated for the π-electron systems in butadiene and benzene. Finally, a brief summary of the modern ab initio VB methods is given. Both *single*- and *many*-reference techniques of determining the optimized orbitals are surveyed. The former include the GVB method using both the *Perfect-Pairing* (PP) wave function (GVB-PP) and the *Spin-Coupled* (SC) algorithm, which makes no prior assumptions about the dominant spin-coupling pattern. Both schemes can be subsequently improved by adding the (nonorthogonal) CI stage, e.g., within the *Correlation Consistent* CI (CCCI) extension of GVB approach and the SCVB generalization of the SC method. The MR VB theories, using different orbitals for different structures, e.g., the *Resonating* GVB (RGVB) or *Breathing Orbital* VB (BOVB) variants, are introduced; they are essential to adequately describe some molecular states in terms of the *broken* symmetry VB functions.

6.1 Self-Consistent Field Theories

First variational theories of electronic structure of the N-electron atomic or molecular system described by the electronic Hamiltonian (5.69) have used as trial wave functions either the product of SO (5.61), describing the independent (distinguishable) spinless particles, or the Slater determinant (5.64), describing the exchange-correlated (indistinguishable) fermions. The former approach marks the *Hartree* theory (Hartree 1928), historically first quantum mechanical approach to *many*-electron systems, which still serves as the reference for defining the overall (Coulomb + Fermi) electron correlation effects, while the latter approach gives rise to the *Hartree–Fock* (HF) theory (Fock 1930; Froese-Fischer 1977), which constitutes the reference in extracting the Coulomb correlation energy. The analytical (Ritz) realization of the HF method, originally proposed by Roothaan and Hall, is known as SCF MO theory. In this section, we provide a short overview of the basic elements of these theories, including the relevant Euler equations for the optimum orbitals and, in the final part, the rudiments of the PK pseudopotential theory.

6.1.1 Hartree Method

The expectation value of the electronic Hamiltonian in the product state (5.61) defined by the N-lowest (singly occupied) SO $\{\psi_i(\boldsymbol{q}_i) = \psi_i(i) = \varphi_i(\boldsymbol{r}_i)\zeta_i(\sigma_i)\}$, with the spatial parts $\{\varphi_i(\boldsymbol{r})\} \equiv \boldsymbol{\varphi}(\boldsymbol{r})$ defining the associated (orthonormal) MO,

$$\langle\varphi_i|\varphi_j\rangle \equiv \langle i|j\rangle = \int \varphi_i^*(\boldsymbol{r})\,\varphi_j(\boldsymbol{r})\,d\boldsymbol{r} = \delta_{i,j}, \tag{6.1}$$

and electronic spin states $\{\zeta_i(\sigma_i) \in [\alpha_i(\sigma_i), \beta_i(\sigma_i)]\}$, can be expressed in terms of the corresponding one- and two-electron integrals (5.73)–(5.75)]:

$$\begin{aligned}\langle E^e(N)\rangle_\Psi &= \langle\Psi(N)|\hat{\mathrm{H}}^e(N)|\Psi(N)\rangle = \left\langle \prod_{k=1}^{N}\psi_k(k)\left|\sum_{i=1}^{N}\hat{\mathrm{h}}(i)+\sum_{i=1}^{N-1}\sum_{j=i+1}^{N}g(i,j)\right|\prod_{l=1}^{N}\psi_l(l)\right\rangle\\
&= \sum_{i=1}^{N}\langle\psi_i(i)|\hat{\mathrm{h}}(i)|\psi_i(i)\rangle\prod_{k\neq i}\langle\psi_k(k)|\psi_k(k)\rangle\\
&+\sum_{i=1}^{N-1}\sum_{j=i+1}^{N}\langle\psi_i(i)\psi_j(j)|g(i,j)|\psi_i(i)\psi_j(j)\rangle\prod_{k\neq(i,j)}\langle\psi_k(k)|\psi_k(k)\rangle\\
&= \sum_{i=1}^{N}\bar{h}_{i,i}[\psi_i]+\sum_{i=1}^{N-1}\sum_{j=i+1}^{N}\bar{J}_{i,j}[\psi_i,\psi_j]\equiv E_{\mathrm{H}}[\boldsymbol{\psi}]\\
&= \sum_{i=1}^{N}h_{i,i}[\varphi_i]+\frac{1}{2}\sum_{i=1}^{N}\sum_{j=1}^{N}(1-\delta_{i,j})J_{i,j}[\varphi_i,\varphi_j]\equiv E_{\mathrm{H}}[\boldsymbol{\varphi}]. \qquad (6.2)\end{aligned}$$

Again, due to the normalization of spin functions (5.79), the integrals in terms of SO $\boldsymbol{\psi}(\boldsymbol{q})$ are equal to those in terms of their spatial functions (MO) [see (5.78) and (5.80)]: $\bar{h}_{i,i}[\psi_i] = h_{i,i}[\varphi_i]$ and $\bar{J}_{i,j}[\psi_i,\psi_j] = J_{i,j}[\varphi_i,\varphi_j]$. For example, for $N = 2p$, i.e., p doubly occupied orbitals $\{\varphi_i(\boldsymbol{r}_i),\ i = 1, 2, \ldots, p\}$ of the spin-restricted approximation [compare (5.66)],

$$\Psi(N) = \varphi_1^+(\boldsymbol{r}_1)\varphi_1^-(\boldsymbol{r}_2)\varphi_2^+(\boldsymbol{r}_3)\varphi_2^-(\boldsymbol{r}_4)\ \ldots\ \varphi_p^+(\boldsymbol{r}_{N-1})\varphi_p^-(\boldsymbol{r}_N), \qquad (6.3)$$

one finds [compare (5.81)]:

$$E_{\mathrm{H}}[\boldsymbol{\varphi}] = 2\sum_{i=1}^{p}h_{i,i}[\varphi_i]+\sum_{i=1}^{p}\sum_{j=1}^{p}(2-\delta_{i,j})J_{i,j}[\varphi_i,\varphi_j]. \qquad (6.4)$$

In accordance with the variational principle of quantum mechanics, the optimum Hartree (H) orbitals $\boldsymbol{\varphi}_{\mathrm{H}}$ have to minimize the auxiliary energy functional including the Lagrange terms associated with conditions of their orthonormality,

$$\delta\left\{E_{\mathrm{H}}[\boldsymbol{\varphi}']-\sum_{i=1}^{p}\sum_{j=1}^{p}\lambda_{k,l}\left(\langle\varphi_i'\mid\varphi_j'\rangle-\delta_{i,j}\right)\right\}\Big|_{\boldsymbol{\varphi}_{\mathrm{H}}}\equiv\delta\mathcal{E}_{\mathrm{H}}[\boldsymbol{\varphi}_{\mathrm{H}};\boldsymbol{\lambda}]=0, \qquad (6.5)$$

where $\boldsymbol{\lambda} = \{\lambda_{i,j}\}$ groups the Lagrangian multipliers enforcing these constraints. Examining the complex conjugate of the preceding equation then reveals that $\lambda_{i,j}^*$ enforces the subsidiary condition $\langle\varphi_j'|\varphi_i'\rangle = \delta_{j,i}$, and hence $\lambda_{i,j}^* = \lambda_{j,i}$, or $\boldsymbol{\lambda} = \boldsymbol{\lambda}^\dagger$. Thus, the Lagrangian multipliers define the Hermitian matrix which can be diagonalized in the unitary transformation: $\mathbf{U}^\dagger\boldsymbol{\lambda}\mathbf{U} = \boldsymbol{\varepsilon}_{\mathrm{H}} = \{\varepsilon_i\delta_{i,j}\}$, $\mathbf{U}^\dagger\mathbf{U} = \mathbf{U}\mathbf{U}^\dagger = \mathbf{I}$. In this representation of the *canonical* orbitals $\boldsymbol{\varphi}_{\mathrm{H}} = \boldsymbol{\varphi}'\mathbf{U} = \{\varphi_k\}$, the variation principle of (6.5) reads:

$$\delta\{E_{\mathrm{H}}[\boldsymbol{\varphi}] - \sum_{k=1}^{p} \varepsilon_k \langle \varphi_k \mid \varphi_k \rangle\} \equiv \delta \mathcal{E}_{\mathrm{H}}[\boldsymbol{\varphi}_{\mathrm{H}}; \boldsymbol{\varepsilon}_{\mathrm{H}}] = 0. \tag{6.6}$$

We also recall at this point that for generally complex orbitals the variations $\delta\boldsymbol{\varphi}^*$ and $\delta\boldsymbol{\varphi}$ ultimately represent the independent displacements of the real and imaginary parts of MO (see also Sect. 5.1.2). Moreover, due to the Hermitian character of the electronic Hamiltonian the corresponding Euler equations for the optimum shapes of MO resulting from these two variations must be identical. Therefore, in what follows we assume that in taking the variation of $\mathcal{E}_{\mathrm{H}}[\boldsymbol{\varphi}; \boldsymbol{\varepsilon}_{\mathrm{H}}]$, to derive equations to be satisfied by the optimum orbitals, the complex-conjugate orbitals are being infinitesimally modified: $\boldsymbol{\varphi}^* \to \boldsymbol{\varphi}^* + \delta\boldsymbol{\varphi}^*$.

The integral $J_{k,l}[\varphi_k, \varphi_l]$ [see (5.78)] in the average electronic energy of (6.4) in the canonical representation $\boldsymbol{\varphi}_{\mathrm{H}}$ stands for the average Coulomb repulsion between one electron in state φ_k and another electron in state φ_l,

$$\begin{aligned}
J_{k,l}[\varphi_k, \varphi_l] &= \iint |\varphi_k(1)|^2 \frac{1}{r_{1,2}} |\varphi_l(2)|^2 d\boldsymbol{r}_1 d\boldsymbol{r}_2 = \iint \rho_k(1) \frac{1}{r_{1,2}} \rho_l(2) d\boldsymbol{r}_1 d\boldsymbol{r}_2 \\
&\equiv \langle k(1) l(2) | g | k(1) l(2) \rangle_{1,2} \\
&= \int \varphi_k^*(1) \left[\int \varphi_l^*(2) \frac{1}{r_{1,2}} \varphi_l(2)\, d\boldsymbol{r}_2 \right] \varphi_k(1)\, d\boldsymbol{r}_1 = \int \rho_k(1) \left[\int \frac{\rho_l(2)}{r_{1,2}} d\boldsymbol{r}_2 \right] d\boldsymbol{r}_1 \\
&\equiv \langle k(1) | \hat{\mathrm{J}}_l(1) | k(1) \rangle_1 .
\end{aligned} \tag{6.7}$$

In the preceding equation we have introduced the (multiplicative) *Coulomb operator* $\hat{\mathrm{J}}_l(1)$, which measures the average electrostatic potential in the position of electron "1" due to the probability distribution $\rho_l = |\varphi_l|^2$ of electron "2." Obviously, interchanging the orbital indices or electronic labels, which only name the integration variables of the definite integrals, has no effect on the value of the integral itself:

$$J_{k,l}[\varphi_k, \varphi_l] = \langle k(1) | \hat{\mathrm{J}}_l(1) | k(1) \rangle_1 = \langle l(1) | \hat{\mathrm{J}}_k(1) | l(1) \rangle_1 = J_{l,k}[\varphi_l, \varphi_k]. \tag{6.8}$$

The stationary condition of (6.6), that the variation $\delta\mathcal{E}_{\mathrm{H}}[\boldsymbol{\varphi}; \boldsymbol{\varepsilon}_{\mathrm{H}}]$ linear in $\delta\boldsymbol{\varphi}^*$ vanishes for the optimum canonical orbitals $\boldsymbol{\varphi} = \boldsymbol{\varphi}_{\mathrm{H}}$, marks the local extremum of this MO functional:

$$\sum_{k=1}^{N} \langle \delta k(1) | \hat{\mathrm{h}}(1) + \sum_{l=1}^{N} (1 - \delta_{k,l}) \hat{\mathrm{J}}_l(1) - \varepsilon_k | k(1) \rangle_1 = 0. \tag{6.9}$$

Since variations $\{\langle \delta\varphi_k |\}$ are arbitrary, this equation can be satisfied only when

$$\{[\hat{\mathrm{h}}(1) + \sum_{l=1}^{N} (1 - \delta_{k,l}) \hat{\mathrm{J}}_l(1)] - \varepsilon_k\} | k(1) \rangle \equiv \{\hat{\mathrm{F}}_{\mathrm{H}}(1) - \varepsilon_k\} | k(1) \rangle = 0, \tag{6.10}$$

where we have defined the effective one-electron Hamiltonian $\hat{\mathrm{F}}_{\mathrm{H}}(1)$ of the Hartree method. Its eigenvalue problem,

$$\hat{\mathrm{F}}_{\mathrm{H}}(1)\,\varphi_k(1) = \varepsilon_k\,\varphi_k(1), \qquad k = 1, 2, \ldots, N, \tag{6.11}$$

thus determines the optimum canonical orbitals $\boldsymbol{\varphi}_{\mathrm{H}} = \{\varphi_k\}$, which define the best approximation of the system electronic wave function in the family of N-electron functions delineated by the variational product of (5.61).

It also follows from this effective one-electron Schrödinger equation that the Lagrangian multiplier

$$\begin{aligned}\varepsilon_k &= \langle\varphi_k|\hat{\mathrm{F}}_{\mathrm{H}}|\varphi_k\rangle = \langle k(1)|[\hat{\mathrm{h}}(1) + \sum_{l=1}^{N}(1-\delta_{k,l})\hat{\mathrm{J}}_l(1)|k(1)\rangle_1 \\ &= h_{k,k}[\varphi_k] + \sum_{l=1}^{N}(1-\delta_{k,l})J_{k,l}[\varphi_k,\varphi_l]\end{aligned} \tag{6.12}$$

measures the corresponding *orbital energy* of an electron occupying kth MO, moving in the effective external potential

$$V_{\mathrm{H}}(\boldsymbol{r}) = v(\boldsymbol{r}) + \sum_{l=1}^{N}(1-\delta_{k,l})\hat{\mathrm{J}}_l(1) \equiv v(\boldsymbol{r}) + v_{\mathrm{H}}(\boldsymbol{r}), \tag{6.13}$$

which combines the external potential $v(\boldsymbol{r})$ due to the system nuclei and the resultant *electrostatic potential* (ESP) $v_{\mathrm{H}}(\boldsymbol{r}) = v_{\mathrm{H}}[\boldsymbol{r};\,\boldsymbol{\varphi}_{\mathrm{H}}]$ generated by the remaining electrons, averaged over their instantaneous positions.

Since the Hartree effective Hamiltonian depends upon the MO themselves, $\hat{\mathrm{F}}_{\mathrm{H}} = \hat{\mathrm{F}}_{\mathrm{H}}[\boldsymbol{\varphi}_{\mathrm{H}}]$, the Hartree equations (6.11) have to be solved iteratively by using the optimum orbitals $\boldsymbol{\varphi}_{\mathrm{H}}^{(n)}$ from the previous iteration to define the next approximation of the Hartree operator, $\hat{\mathrm{F}}_{\mathrm{H}}\left[\boldsymbol{\varphi}_{\mathrm{H}}^{(n)}\right]$, which generates better orbitals $\boldsymbol{\varphi}_{\mathrm{H}}^{(n+1)}$, etc.,

$$\ldots \to \boldsymbol{\varphi}_{\mathrm{H}}^{(n)}\Big\} \to \Big\{\hat{\mathrm{F}}_{\mathrm{H}}\left[\boldsymbol{\varphi}_{\mathrm{H}}^{(n)}\right] \to \boldsymbol{\varphi}_{\mathrm{H}}^{(n+1)}\Big\} \to \Big\{\hat{\mathrm{F}}_{\mathrm{H}}\left[\boldsymbol{\varphi}_{\mathrm{H}}^{(n+1)}\right] \to \ldots,$$

until the field of electronic interactions is *self-consistent*, $v_{\mathrm{H}}\left[\boldsymbol{r};\,\boldsymbol{\varphi}_{\mathrm{H}}^{(m+1)}\right] \cong v_{\mathrm{H}}\left[\boldsymbol{r};\boldsymbol{\varphi}_{\mathrm{H}}^{(m)}\right]$, or $\boldsymbol{\varphi}_{\mathrm{H}}^{(m+1)} \cong \boldsymbol{\varphi}_{\mathrm{H}}^{(m)}$, to within the assumed tolerance threshold.

The average electronic energy in the *Hartree limit* (Hl), for an infinite variational flexibility of orbitals $\boldsymbol{\varphi}_{\mathrm{Hl}}$ in the product function describing the independent, spinless particles, $E_{\mathrm{Hl}} = E_{\mathrm{Hl}}[\boldsymbol{\varphi}_{\mathrm{Hl}}]$, then determines the reference for extracting the overall electron correlation energy. In the next section we shall examine the related *Hartree–Fock* (HF) approximation, using the Slater determinant as variational wave function, thus describing the *exchange*-correlated fermions. Clearly, the optimum canonical Hartree–Fock orbitals $\boldsymbol{\varphi}_{\mathrm{HF}}$, already reflecting the exchange correlation between electrons, will slightly differ from their corresponding Hartree analogs $\boldsymbol{\varphi}_{\mathrm{H}}$, but these two sets strongly resemble one another: $\boldsymbol{\varphi}_{\mathrm{H}} \cong \boldsymbol{\varphi}_{\mathrm{HF}}$.

6.1.2 Hartree–Fock Theory

For the determinantal variational wave function, the average electronic energy is given by the known functional of the (orthonormal) occupied SO $\boldsymbol{\psi}$ (see Sect. 5.4):

$$\langle E^e[\boldsymbol{\psi}]\rangle_{\Psi_A} = \sum_{i=1}^{N} \bar{h}_{i,i}[\psi_i] + \frac{1}{2}\sum_{i=1}^{N}\sum_{j=1}^{N} (\bar{J}_{i,j}[\psi_i, \psi_j] - \bar{K}_{i,j}[\psi_i, \psi_j]) \equiv E_{\mathrm{HF}}[\boldsymbol{\psi}], \quad (6.14)$$

where the exchange integral $\bar{K}_{i,j}[\psi_i, \psi_j] = K_{i,j}[\varphi_i, \varphi_j]\delta_{\zeta_i,\zeta_j}$ of (5.77) identically vanishes when the two SO involve different spin states of the two electrons. The best variational approximation of the system ground state is thus obtained by the optimum SO which minimize this average electronic energy subject to the orthonormality constraints of MO [see (6.5)]:

$$\delta\left\{E_{\mathrm{HF}}[\boldsymbol{\psi}] - \sum_{i=1}^{p}\sum_{j=1}^{p} \nu_{k,l}\Big(\langle\psi_i \mid \psi_j\rangle - \delta_{i,j}\Big)\right\} \equiv \delta\mathcal{E}_{\mathrm{HF}}[\boldsymbol{\psi}; \boldsymbol{\nu}] = 0. \quad (6.15)$$

For simplicity, in what follows the closed-shell configuration of (5.66) is assumed, for which [see (5.81)]

$$E_{\mathrm{HF}}[\boldsymbol{\psi}] = E_{\mathrm{HF}}[\boldsymbol{\varphi}] = 2\sum_{i=1}^{p} h_{i,i}[\varphi_i] + \sum_{i=1}^{p}\sum_{j=1}^{p} (2J_{i,j}[\varphi_i, \varphi_j] - K_{i,j}[\varphi_i, \varphi_j]). \quad (6.16)$$

Turning now to the canonical representation of HF orbitals, $\boldsymbol{\varphi}_{\mathrm{HF}} = \boldsymbol{\varphi}\mathbf{U} = \{\phi_k\}$, in which the matrix of Lagrangian multipliers enforcing the orthonormality constraints becomes diagonal, $\mathbf{U}^{\dagger}\boldsymbol{\nu}\mathbf{U} = \mathbf{e} = \{e_i\delta_{i,j}\}$, $\mathbf{U}^{\dagger}\mathbf{U} = \mathbf{U}\mathbf{U}^{\dagger} = \mathbf{I}$, the relevant variational principle becomes:

$$\delta\left\{E_{\mathrm{HF}}[\boldsymbol{\varphi}_{\mathrm{HF}}] - \sum_{k=1}^{p} e_k(\langle\phi_k \mid \phi_k\rangle\right\} \equiv \delta\mathcal{E}_{\mathrm{HF}}[\boldsymbol{\varphi}_{\mathrm{HF}}; \mathbf{e}] = 0. \quad (6.17)$$

Again, to facilitate a compact expression for the linear variation of the system electronic energy corresponding to the variation $\delta\boldsymbol{\varphi}_{\mathrm{HF}}^{*}$ of the complex-conjugate orbitals one formally expresses the exchange integral as the expectation value of the effective *exchange operator* [compare (6.7)] defined by the following result of its action on the one-electron function $f(\boldsymbol{r}_1) \equiv f(1)$:

$$\hat{\mathrm{K}}_l(1)f(1) \equiv \left(\int \phi_l^*(2) \frac{1}{r_{1,2}} f(2) d\boldsymbol{r}_2 \right) \phi_l(1). \tag{6.18}$$

It follows from this definition that this integral operator exchanges two electrons in the product of the nonconjugate orbitals, $\phi_l(2)f(1) \rightarrow \phi_l(1)f(2)$, which explains its name. The exchange integral can be then expressed as the expectation value:

$$\begin{aligned} K_{k,l}[\phi_k, \phi_l] &= \iint \phi_k^*(1)\phi_l^*(2) \frac{1}{r_{1,2}} \phi_l(1)\phi_k(2) d\boldsymbol{r}_1 d\boldsymbol{r}_2 = \langle k(1)l(2)|g(1,2)|l(1)k(2)\rangle_{1,2} \\ &= \langle k(1)|\hat{\mathrm{K}}_l(1)|k(1)\rangle_1 = \langle l(1)|\hat{\mathrm{K}}_k(1)|l(1)\rangle_1. \end{aligned} \tag{6.19}$$

Finally, calculating the variation $\delta\mathcal{E}_{\mathrm{HF}}[\boldsymbol{\varphi}_{\mathrm{HF}}; \mathbf{e}]$ corresponding to $\delta\boldsymbol{\varphi}_{\mathrm{HF}}^*$ gives the following condition for the local extremum of the auxiliary functional of (6.17),

$$\sum_{k=1}^{p} \langle \delta k(1)|\hat{\mathrm{h}}(1) + \sum_{l=1}^{p} [2\hat{\mathrm{J}}_l(1) - \hat{\mathrm{K}}_l(1)] - e_k|k(1)\rangle = 0. \tag{6.20}$$

For arbitrary variations of MO it can be satisfied only when

$$\begin{aligned} &\left\{ [\hat{\mathrm{h}}(1) + \sum_{l=1}^{p} [2\hat{\mathrm{J}}_l(1) - \hat{\mathrm{K}}_l(1)] \right\} \phi_k(1) \equiv \hat{\mathrm{F}}(1)\, \phi_k(1) = e_k \phi_k(1), \\ &k = 1, 2, \ldots, p. \end{aligned} \tag{6.21}$$

Again, since the effective *Fock operator*

$$\begin{aligned} \hat{\mathrm{F}}(\boldsymbol{r}) &= -\frac{1}{2}\Delta + v(\boldsymbol{r}) + \sum_{l=1}^{p} [2\hat{\mathrm{J}}_l(\boldsymbol{r}) - \hat{\mathrm{K}}_l(\boldsymbol{r})]\} \equiv \hat{\mathrm{h}}(\boldsymbol{r}) + \hat{\mathrm{J}}(\boldsymbol{r}) - \hat{\mathrm{K}}(\boldsymbol{r}) \\ &\equiv -\frac{1}{2}\Delta + [v(\boldsymbol{r}) + v_{\mathrm{HF}}(\boldsymbol{r})] \equiv -\frac{1}{2}\Delta + V_{\mathrm{HF}}(\boldsymbol{r}), \end{aligned} \tag{6.22}$$

depends on the orbitals it is supposed to determine, $\hat{\mathrm{F}} = \hat{\mathrm{F}}[\boldsymbol{\varphi}_{\mathrm{HF}}]$, one has to solve the above HF equations iteratively:

$$\ldots \rightarrow \boldsymbol{\varphi}_{\mathrm{HF}}^{(n)}\} \rightarrow \left\{ \hat{\mathrm{F}}\left[\boldsymbol{\varphi}_{\mathrm{HF}}^{(n)}\right] \rightarrow \boldsymbol{\varphi}_{\mathrm{HF}}^{(n+1)} \right\} \rightarrow \left\{ \hat{\mathrm{F}}\left[\boldsymbol{\varphi}_{\mathrm{HF}}^{(n+1)}\right] \rightarrow \ldots, \right.$$

until one reaches the *self-consistent field* (SCF) of the two-electron contribution $v_{\mathrm{HF}}(\boldsymbol{r})$ to the effective external potential $V_{\mathrm{HF}}(\boldsymbol{r})$, called the Coulomb-exchange potential:

$$v_{\mathrm{HF}}\left[\boldsymbol{r}; \boldsymbol{\varphi}_{\mathrm{HF}}^{(m+1)}\right] \cong v_{\mathrm{HF}}\left[\boldsymbol{r}; \boldsymbol{\varphi}_{\mathrm{HF}}^{(m)}\right] \quad \text{or} \quad \boldsymbol{\varphi}_{\mathrm{HF}}^{(m+1)} \cong \boldsymbol{\varphi}_{\mathrm{HF}}^{(m)}.$$

The physical meaning of the diagonal Lagrangian multiplier e_k is again revealed by multiplying (6.21) from the left by $\phi_k^*(1)$ and integrating over positions of electron "1":

$$e_k = \langle\phi_k|\hat{\mathrm{F}}|\phi_k\rangle = \langle k(1)|\,[\hat{\mathrm{h}}(1) + \sum_{l=1}^{p}\,[2\hat{\mathrm{J}}_l(1) - \hat{\mathrm{K}}_l(1)]\,|k(1)\rangle_1$$
$$= h_{k,k}[\phi_k] + \sum_{l=1}^{p}(2J_{k,l}[\phi_k,\phi_l] - K_{k,l}[\phi_k,\phi_l]). \qquad (6.23)$$

This orbital energy of *k*th MO in HF theory thus combines the corresponding kinetic and nuclear attraction energies, given by the expectation value of the one-electron operator $\hat{\mathrm{h}}(1)$, and the effective Coulomb-exchange interactions with all remaining electrons. It should be observed that in (6.23) the *self*-interaction of the electron with itself is exactly eliminated by the identity $J_{k,k} = K_{k,k}$, whereas in (6.12) it is removed by the Kronecker-delta factor.

The *Koopmans theorem* links the approximate estimate of the system ionization potentials I_k, the energy required to remove the electron occupying ϕ_k,

$$I_k^{\mathrm{HF}} = E_k^+[\boldsymbol{\varphi}^+] - E^0[\boldsymbol{\varphi}^0], \qquad (6.24\mathrm{a})$$

with the orbital energy e_k; here $\boldsymbol{\varphi}^+$ denotes the optimum MO of the HF method occupied in the *cation* X_k^+, the $(N-1)$-electron system in state $|\phi_1^+\phi_1^- \ldots \phi_k^+ \ldots \phi_p^+\phi_p^-|$, and $\boldsymbol{\varphi}^0$ groups the optimum MO of the *neutral N*-electron system X^0 in state $\left|\phi_1^+\phi_1^- \ldots \phi_k^+\phi_k^- \ldots \phi_p^+\phi_p^-\right|$. By assuming that the HF MO $\boldsymbol{\varphi}^0$ of the neutral system X^0 are to a good approximation preserved after ionization, $\boldsymbol{\varphi}^+ \approx \boldsymbol{\varphi}^0$, i.e., neglecting the *orbital relaxation* accompanying the ionization process, one can express the electronic energy $E_k^+[\boldsymbol{\varphi}^0]$ of the resulting cation in terms of the same one- and two-electron integrals $\{h_{k,k}[\phi_k^0]\}$, $\{J_{k,l}[\phi_k^0,\phi_l^0]\}$, and $\{K_{k,l}[\phi_k^0,\phi_l^0]\}$ as those used for expressing the energy $E^0[\boldsymbol{\varphi}^0]$ of the neutral system. Their difference then approximates the HF ionization potential of (6.24a):

$$I_k^{\mathrm{HF}} \approx E_k^+[\boldsymbol{\varphi}^0] - E^0[\boldsymbol{\varphi}^0] \equiv \Delta E_k(\Delta\boldsymbol{n}_k;\boldsymbol{\varphi}^0) = -e_k[\boldsymbol{\varphi}^0], \qquad (6.24\mathrm{b})$$

where $\boldsymbol{n} = \{n_k\}$ groups the MO occupation numbers and $\Delta\boldsymbol{n}_k = (0, \ldots, \Delta n_k = -1, 0, \ldots, 0)$ reflects a removal of the electron occupying ϕ_k.

It should be observed that since in this ionization process a change in the occupation number of ϕ_k, $\Delta n_k = -1$, the above formulation of the Koopmans theorem can be also interpreted as the *finite*-difference approximation of the energy derivative with respect to the system occupation number of ϕ_k, or the overall number of electrons, $dN = dn_k$,

$$\partial\Delta E_k(\Delta\boldsymbol{n}_k;\boldsymbol{\varphi}^0)/\partial n_k \approx \Delta E_k(\Delta\boldsymbol{n}_k;\boldsymbol{\varphi}^0)/\Delta n_k = e_k. \qquad (6.25)$$

Thus, the canonical orbital energy approximately measures the slope of the electronic energy with respect to the MO occupation number.

In addition to the Coulomb correlation neglected in the orbital approximation of the HF theory, the Koopmans theorem does not account for the orbital relaxation. Fortunately, there is a substantial cancelation of these errors in electron removal process, since the magnitude of correlation energy increases monotonically with a growing number of electrons. Thus, the sum of a diminished correlation and the (neglected) orbital relaxation energy in the cation roughly reproduces a larger correlation error in the neutral system. The orbital relaxation error gradually disappears in large systems, for large N, when the removal of a single electron causes a relatively minor perturbation of the whole system. In such cases, the change in the Coulomb correlation of electrons is also relatively small. Accordingly, these errors become relatively large in the two-electron systems, e.g., the helium atom and hydrogen molecule. This cancelation of the orbital relaxation and electron Coulomb correlation errors in the electron removal processes explains a surprisingly good performance of Koopmans' theorem in the closed-shell systems. It also provides a physical justification for calling the canonical MO of HF method the *spectroscopic orbitals*.

The molecular electron density [see (3.7)] is the sum of the corresponding MO contributions:

$$\begin{aligned}\rho(\boldsymbol{r}) &= \sum_{k=1}^{N} |\phi_k(\boldsymbol{r})|^2 = \sum_{k=1}^{N} \rho_k(\boldsymbol{r}) \\ &= \sum_{\sigma=\uparrow,\downarrow} \left[\sum_{l=1}^{N_\sigma} |\phi_{l\sigma}(\boldsymbol{r})|^2\right] = \sum_{\sigma=\uparrow,\downarrow} \left[\sum_{l=1}^{N_\sigma} \rho_{l\sigma}(\boldsymbol{r})\right] \equiv \sum_{\sigma=\uparrow,\downarrow} \rho_\sigma(\boldsymbol{r}),\end{aligned} \tag{6.26}$$

where $\phi_{l\sigma}(\boldsymbol{r})$ denotes lth orbital describing the electron with the spin orientation $\sigma = \uparrow, \downarrow$, $\rho_\sigma(\boldsymbol{r})$ stands for the corresponding spin density (3.9), and $N_\sigma = \int \rho_\sigma(\boldsymbol{r}) d\boldsymbol{r}$ is the overall number of electrons with spin σ,

$$\sum_{\sigma=\uparrow,\downarrow} N_\sigma = N.$$

Compared with the system *promolecule* M^0, the hypothetical combination of all constituent (free) atoms in their molecular positions, it exhibits an accumulation of the electronic probability density in the bonding regions of the molecule M, between the nuclei of the bonded AIM. The promolecular reference can be used to extract the effects due to the bond formation, e.g., in form of the *density difference* function

$$\Delta\rho(\boldsymbol{r}) = \rho(\boldsymbol{r}) - \rho^0(\boldsymbol{r}), \tag{6.27}$$

where ρ^0 denotes the promolecular electron density.

The delocalized (canonical) MO, which reflect the molecular geometry, have been widely used in physical interpretations of the bonding patterns of molecular systems, their structural preferences, and in rationalizing diverse phenomena of electronic spectroscopy. They are also successfully applied in diagnosing trends in chemical reactivity, particularly in organic chemistry. The most important in these applications are the *Frontier Orbitals* (FO), including the *Highest Occupied* MO (HOMO) and the *Lowest Unoccupied* MO (LUMO). The former roughly determines how the energetically most accessible (first) *ionization* process, the electron *removal* to the system environment when the system acts as the chemical *base* (electron donor), affects its electron distribution, whereas the latter gives an approximate representation of the charge redistribution accompanying the electron *addition* to the system in molecular complexes or in the *electron affinity* phenomena when the system acts as the chemical *acid* (electron acceptor).

The normalized responses of the molecular electron density to an addition/removal of a single electron are known as the *Fukui function* (FF) descriptors (Parr and Yang 1984). It follows from these intuitive considerations that they are dominated by the topology of the corresponding FO densities. In the *donor–acceptor* interactions between molecules the FO of both reactants also play a crucial role in shaping their reactivity preferences.

To conclude this section, we observe that the HF equations for the optimum shapes of orbitals can be also derived through the functional derivatives of Sect. 2.8. As we have observed in Sect. 6.1.1, there are two groups of independent functions to be optimized when determining the extremum of the auxiliary energy functionals of (6.14) and (6.16), $\boldsymbol{\varphi}_{\mathrm{HF}} = \{\phi_l\}$ and $\boldsymbol{\varphi}^*_{\mathrm{HF}} = \left\{\phi^*_k\right\}$, since they are linearly related to independent components $\mathrm{Re}(\boldsymbol{\varphi}_{\mathrm{HF}})$ and $\mathrm{Im}(\boldsymbol{\varphi}_{\mathrm{HF}})$ of the (complex) canonical orbitals. This calls for the separate optimizations of $\boldsymbol{\varphi}^*_{\mathrm{HF}}$ and $\boldsymbol{\varphi}_{\mathrm{HF}}$, each giving the same set of equations for the optimum solutions. It can be easily verified that the vanishing functional derivative of the auxiliary functional of (6.17), $\mathcal{E}_{\mathrm{HF}}[\boldsymbol{\varphi}_{\mathrm{HF}};\,\mathbf{e}] = \mathcal{E}_{\mathrm{HF}}\left[\boldsymbol{\varphi}^*_{\mathrm{HF}},\,\boldsymbol{\varphi}_{\mathrm{HF}};\,\mathbf{e}\right]$, with respect to, say, ϕ^*_k, directly gives the HF equation for this MO:

$$\begin{aligned}\delta\mathcal{E}_{\mathrm{HF}}\left[\boldsymbol{\varphi}^*_{\mathrm{HF}},\boldsymbol{\varphi}_{\mathrm{HF}};\mathbf{e}\right]/\delta\phi^*_k(\boldsymbol{r}) &= \delta E_{\mathrm{HF}}\left[\boldsymbol{\varphi}^*_{\mathrm{HF}},\boldsymbol{\varphi}_{\mathrm{HF}};\mathbf{e}\right]/\delta\phi^*_k(\boldsymbol{r}) - e_k\delta\langle\phi_k|\phi_k\rangle/\delta\phi^*_k(\boldsymbol{r})\\ &= \hat{\mathrm{h}}(\boldsymbol{r})\phi_k(\boldsymbol{r}) + \sum_{l=1}^{N}\left(\hat{\mathrm{J}}_l(\boldsymbol{r}) - \hat{\mathrm{K}}_l(\boldsymbol{r})\right)\phi_k(\boldsymbol{r}) - e_k\phi_k(\boldsymbol{r})\\ &\equiv [\hat{\mathrm{h}}(\boldsymbol{r}) + \hat{\mathrm{J}}(\boldsymbol{r}) - \hat{\mathrm{K}}(\boldsymbol{r})]\phi_k(\boldsymbol{r}) - e_k\phi_k(\boldsymbol{r})\\ &= \hat{\mathrm{F}}(\boldsymbol{r})\phi_k(\boldsymbol{r}) - e_k\phi_k(\boldsymbol{r}) = 0. \end{aligned} \tag{6.28}$$

In the ground state of the N-electron system, only N lowest SO are occupied, with the remaining higher (empty) SO determining the system *virtual* orbitals. The latter have no physical meaning, since they do not contribute to the self-consistent field of electronic interactions and to expectation values of physical properties of the molecule. Thus, the ground state virtual orbitals become "physical", when they are fully or partly occupied, e.g., in the electronically excited states or in the CI description of the ground state. Notice that when one uses the LUMO as a probe in the electron-absorbing phenomena, one mentally populates this orbital in the intermolecular interactions with at least a fraction of an electron, thus making it physically meaningful in reactivity descriptions.

6.1.3 Transition-State Concept

The HF equations (6.21) can be compactly written in terms of the system electron density $\rho(\boldsymbol{r})$ (6.26) and the spin components $\{\gamma_\sigma(\boldsymbol{r},\boldsymbol{r}')\}$ of the overall (spinless) one-electron *density matrix*:

$$\gamma(\boldsymbol{r},\boldsymbol{r}') = \sum_{k=1}^{N} \phi_k(\boldsymbol{r})\phi_k^*(\boldsymbol{r}') = \sum_{\sigma=\uparrow,\downarrow} \left[\sum_{l=1}^{N_\sigma} \phi_{l\sigma}(\boldsymbol{r})\phi_{l\sigma}^*(\boldsymbol{r}')\right] = \sum_{\sigma=\uparrow,\downarrow} \gamma_\sigma(\boldsymbol{r},\boldsymbol{r}'). \quad (6.29)$$

Indeed, a straightforward transformation of HF equations for the optimum orbitals of spin σ gives:

$$\begin{aligned}
&\left[\hat{\mathrm{h}}(\boldsymbol{r}) + \sum_{\sigma=\uparrow,\downarrow}\sum_{l=1}^{N\sigma} \hat{\mathrm{J}}_{l\sigma}(\boldsymbol{r})\right]\phi_{l\sigma}(\boldsymbol{r}) - \sum_{l=1}^{N\sigma}\hat{\mathrm{K}}_{l\sigma}(\boldsymbol{r})\,\phi_{l\sigma}(\boldsymbol{r}) \\
&\quad \equiv [\hat{\mathrm{h}}(\boldsymbol{r}) + \hat{\mathrm{J}}(\boldsymbol{r}) - \hat{\mathrm{K}}_\sigma(\boldsymbol{r})]\phi_{l\sigma}(\boldsymbol{r}) \\
&\quad = \left(-\frac{\nabla^2}{2} + v(\boldsymbol{r}) + \int \frac{\rho(\boldsymbol{r}')}{|\boldsymbol{r}-\boldsymbol{r}'|}d\boldsymbol{r}'\right)\phi_{l\sigma}(\boldsymbol{r}) - \int \frac{\gamma_\sigma(\boldsymbol{r},\boldsymbol{r}')}{|\boldsymbol{r}-\boldsymbol{r}'|}\phi_{l\sigma}(\boldsymbol{r}')\,d\boldsymbol{r}' = e_{l\sigma}\phi_{l\sigma}(\boldsymbol{r}), \\
&\qquad \sigma = \uparrow,\downarrow\,.
\end{aligned} \quad (6.30)$$

Any set of N occupied orthonormal orbitals $\{\phi_{l\sigma}\}$ that satisfy these equations makes the total energy stationary. The energy expression (6.14) refers to the special case of the ground state (full) occupation pattern of N lowest orbitals, $\boldsymbol{n}_{g.s.} = [(1,1, \ldots, 1,)\ 0, 0, \ldots] = \{n_{l\sigma}\}$, of the whole (infinite) set of orbitals $\boldsymbol{\varphi}_{\mathrm{HF}} = [(\phi_1, \phi_2, \ldots, \phi_N), \phi_{N+1}, \phi_{N+2}, \ldots] = \{\phi_{l\sigma}\}$, arranged in such a way that $e_1 \le e_2 \le e_3 \le \ldots$, including the infinite set of *virtual* SO $\{\phi_{N+1}, \phi_{N+2}, \ldots\}$, which are not occupied in the ground-state configuration. Hence, for any fixed vector $\boldsymbol{n}$ of electron occupations obtained via the associated electron excitations from the ground state occupied to virtual subspaces, with

$$\begin{aligned}
&\sum_{k=1}^{\infty} n_k = \sum_{\sigma=\uparrow,\downarrow}\sum_{l=1}^{\infty} n_{l,\sigma} = \sum_{\sigma=\uparrow,\downarrow} N_\sigma = N, \\
&\rho^{(\boldsymbol{n})}(\boldsymbol{r}) = \sum_{k=1}^{\infty} n_k\left|\phi_k^{(\boldsymbol{n})}(\boldsymbol{r})\right|^2 \equiv \sum_{k=1}^{\infty} n_k\rho_k^{(\boldsymbol{n})}(\boldsymbol{r}) = \sum_{\sigma=\uparrow,\downarrow}\left[\sum_{l=1}^{\infty} n_{l\sigma}\left|\phi_{i\sigma}^{(\boldsymbol{n})}(\boldsymbol{r})\right|^2\right] = \sum_{\sigma=\uparrow,\downarrow}\rho_\sigma^{(\boldsymbol{n})}(\boldsymbol{r}), \\
&\gamma^{(\boldsymbol{n})}(\boldsymbol{r},\boldsymbol{r}') = \sum_{k=1}^{\infty} n_k\phi_k^{(\boldsymbol{n})}(\boldsymbol{r})\phi_k^{(\boldsymbol{n})*}(\boldsymbol{r}') \equiv \sum_{k=1}^{\infty} n_k\gamma_k^{(\boldsymbol{n})}(\boldsymbol{r},\boldsymbol{r}') \\
&\quad = \sum_{\sigma=\uparrow,\downarrow}\left[\sum_{l=1}^{\infty} n_{l\sigma}\phi_{l\sigma}^{(\boldsymbol{n})}(\boldsymbol{r})\phi_{l\sigma}^{(\boldsymbol{n})*}(\boldsymbol{r}')\right] = \sum_{\sigma=\uparrow,\downarrow}\gamma_\sigma^{(\boldsymbol{n})}(\boldsymbol{r},\boldsymbol{r}'),
\end{aligned} \quad (6.31)$$

the HF electronic energy is given by the following expression in terms of the self-consistent canonical orbitals for the specified electron occupation vector $\boldsymbol{n}$, $\boldsymbol{\varphi}^{(\boldsymbol{n})} = \left\{\phi_k^{(\boldsymbol{n})} = \phi_{l\sigma}^{(\boldsymbol{n})}\right\}$, the associated electron density $\rho^{(\boldsymbol{n})}(\boldsymbol{r})$, spin densities $\{\rho_\sigma^{(\boldsymbol{n})}(\boldsymbol{r})\}$, and spin density matrices $\{\gamma_\sigma^{(\boldsymbol{n})}(\boldsymbol{r},\boldsymbol{r}')\}$:

$$E_{\mathrm{HF}}(\boldsymbol{n}) = -\frac{1}{2}\sum_{k=1}^{\infty} n_k \int \phi_k^{(\boldsymbol{n})*}(\boldsymbol{r})\,\Delta\,\phi_k^{(\boldsymbol{n})}(\boldsymbol{r})\,d\boldsymbol{r} + \int \rho^{(\boldsymbol{n})}(\boldsymbol{r})\,v(\boldsymbol{r})\,d\boldsymbol{r}$$
$$+\frac{1}{2}\iint \frac{\rho^{(\boldsymbol{n})}(\boldsymbol{r})\rho^{(\boldsymbol{n})}(\boldsymbol{r}')}{|\boldsymbol{r}-\boldsymbol{r}'|}\,d\boldsymbol{r}\,d\boldsymbol{r}' - \frac{1}{2}\sum_{\sigma=\uparrow,\downarrow}\iint \frac{\gamma_\sigma^{(\boldsymbol{n})}(\boldsymbol{r},\boldsymbol{r}')\gamma_\sigma^{(\boldsymbol{n})}(\boldsymbol{r}',\boldsymbol{r})}{|\boldsymbol{r}-\boldsymbol{r}'|}\,d\boldsymbol{r}\,d\boldsymbol{r}'. \tag{6.32}$$

In the ordinary *ground state* (g.s.) HF calculations in each iterative step only the orbitals with N-lowest energies are used to form the Slater determinant, but one is by no means restricted to this choice: any occupation vector for which the procedure converges will determine the self-consistent solutions defining the determinantal approximation of an *excited state* (e.s.) of the N-electron system under consideration. It should be stressed, however, that the Slater determinants corresponding to different occupation vectors are not orthogonal, since they represent eigenfunctions of different Fock operators, with their effective electron interactions being defined by different sets of canonical MO. For the same reason the optimum, self-consistent orbitals in different electron configurations are not identical: $\phi_k^{(\boldsymbol{n})} \neq \phi_k^{(\boldsymbol{n}')}$.

The generalized self-consistent HF energy of (6.32), with the fully relaxed orbitals for the assumed occupation vector $\boldsymbol{n} = \{n_k \equiv n_{k\sigma}\}$, thus becomes a function of the continuous occupation variables $\{0 \le n_k \le 1\}$. As first shown by Slater the partial derivative of this function with respect to the orbital occupation gives exactly its orbital energy for the assumed occupation vector [compare the approximate Koopmans' relation of (6.25)]:

$$\left.\frac{\partial E_{\mathrm{HF}}(\boldsymbol{n})}{\partial n_k}\right|_{\boldsymbol{n}} = e_k(\boldsymbol{n}) = \left.\frac{\partial E_{\mathrm{HF}}(\boldsymbol{n})}{\partial n_{k\sigma}}\right|_{\boldsymbol{n}} = e_{k\sigma}(\boldsymbol{n})$$
$$= -\frac{1}{2}\int \phi_k^{(\boldsymbol{n})*}(\boldsymbol{r})\Delta\phi_k^{(\boldsymbol{n})}(\boldsymbol{r})\,d\boldsymbol{r} + \int \rho_k^{(\boldsymbol{n})}(\boldsymbol{r})\,v(\boldsymbol{r})\,d\boldsymbol{r}$$
$$+\iint \frac{\rho_k^{(\boldsymbol{n})}(\boldsymbol{r})\rho^{(\boldsymbol{n})}(\boldsymbol{r}')}{|\boldsymbol{r}-\boldsymbol{r}'|}\,d\boldsymbol{r}\,d\boldsymbol{r}' - \iint \frac{\gamma_{k\sigma}^{(\boldsymbol{n})}(\boldsymbol{r},\boldsymbol{r}')\gamma_\sigma^{(\boldsymbol{n})}(\boldsymbol{r}',\boldsymbol{r})}{|\boldsymbol{r}-\boldsymbol{r}'|}\,d\boldsymbol{r}\,d\boldsymbol{r}'. \tag{6.33}$$

This relation has been subsequently rediscovered and somewhat extended by Janak (1978) in the framework of KS DFT.

It is tempting to use two independent SCF calculations, for the ground state occupations $\boldsymbol{n}_{g.s.} = \boldsymbol{n}_0$ and the *singly* excited electron configuration $\boldsymbol{n}_{e.s.} = \boldsymbol{n}_{p\to q}$ corresponding to an electron transfer from the ground state occupied orbital ϕ_p to ground state virtual orbital ϕ_q, to calculate the approximate excitation energy as the difference of the corresponding total electronic energies,

$$\Delta E_{p\to q} \approx E_{\mathrm{HF}}(\boldsymbol{n}_{p\to q}) - E_{\mathrm{HF}}(\boldsymbol{n}_0), \tag{6.34}$$

where:

$$\begin{aligned}\Psi_{g.s.}(N) &= \det(\phi_1, \ldots, \phi_{p-1}, \phi_p, \phi_{p+1}, \ldots, \phi_N) \equiv \Psi_0(N),\\ \Psi_{e.s.}(N) &= \det(\phi_1, \ldots, \phi_{p-1}, \phi_{p+1}, \ldots, \phi_N, \phi_q) = \Psi_{p\to q}(N) \equiv \Psi_p^q(N).\end{aligned} \tag{6.35}$$

However, besides requiring two separate SCF calculations, this recipe is not sound numerically, since the two states are not orthogonal anyway and it determines the small quantity of the excitation energy as difference of two approximate large numbers.

Slater's (1974) concept of the *transition state* (TS) allows one to determine the approximate $p \to q$ excitation energy in a *single* SCF calculation as the difference of orbital energies (small numbers) for the hypothetical system exhibiting the half occupations of the two MO involved in the $p \to q$ transition and the full occupations of the remaining orbitals defining the Slater determinant $\Psi_{p\to q}(N)$ in (6.35):

$$\boldsymbol{n}_{\mathrm{TS}} = \begin{bmatrix} \phi: & 1 & \ldots & p-1 & p & p+1 & \ldots & N & q \\ n_\phi: & 1 & \ldots & 1 & 1/2 & 1 & \ldots & 1 & 1/2 \end{bmatrix}. \tag{6.36}$$

The electron density of TS is thus given by the following expression,

$$\rho_{\mathrm{TS}}(\boldsymbol{r}) = \sum_{l\neq p}^{N} |\phi_l(\boldsymbol{r})|^2 + \frac{1}{2}[|\phi_p(\boldsymbol{r})|^2 + |\phi_q(\boldsymbol{r})|^2] = \frac{1}{2}[\rho_0(\boldsymbol{r}) + \rho_{p\to q}(\boldsymbol{r})], \tag{6.37}$$

where $\rho_0(\boldsymbol{r})$ and $\rho_{p\to q}(\boldsymbol{r})$ denote the electron densities in the ground and excited states, respectively.

Therefore, $\rho_{\mathrm{TS}}(\boldsymbol{r})$ does not correspond to a single Slater determinant. Indeed, it represents the ensemble average (see Sect. 3.3.4) of the electron densities in the initial and final electronic states of this electron excitation [see (6.35)], corresponding to the density operator (3.60a)

$$\hat{\mathrm{D}}_{\mathrm{TS}} = |\Psi_0\rangle\frac{1}{2}\langle\Psi_0| + |\Psi_{p\to q}\rangle\frac{1}{2}\langle\Psi_{p\to q}|. \tag{6.38}$$

Slater's TS thus represents an ensemble, i.e., the statistical mixture of two electronic states, and not the pure (single) quantum mechanical state.

Since this concept invokes the continuous orbital occupations of orbitals p and q, which are involved in this electron transfer, linked by the closure relation $n_p + n_q = 1$ or $-\delta n_p = \delta n_q = \delta n = 1/2$, one can expand the HF energies in the initial and final electronic states involved in this electron excitation as the corresponding power series in the associated TS vector of displacements in the orbital occupations,

$$\delta \boldsymbol{n}_{\mathrm{TS}} = \boldsymbol{n}_{\mathrm{TS}} - \boldsymbol{n}_0$$
$$= \begin{bmatrix} \phi: & 1 & \dots & p-1 & p & p+1 & \dots & N & q \\ \delta n_\phi: & 0 & \dots & 0 & -1/2 & 0 & \dots & 0 & +1/2 \end{bmatrix}, \tag{6.39}$$

with $\boldsymbol{n}_0 = \boldsymbol{n}_{\mathrm{TS}} - \delta \boldsymbol{n}_{\mathrm{TS}}$ and $\boldsymbol{n}_{p\to q} = \boldsymbol{n}_{\mathrm{TS}} + \delta \boldsymbol{n}_{\mathrm{TS}}$. Thus, expanding the electronic energies of these two states around the TS configuration $\boldsymbol{n}_{\mathrm{TS}}$ generates the following Taylor series expressions for these two self-consistent energies:

$$E_{\mathrm{HF}}(\boldsymbol{n}_0) = E_{\mathrm{HF}}(\boldsymbol{n}_{\mathrm{TS}} - \delta \boldsymbol{n}_{\mathrm{TS}}) = E_{\mathrm{HF}}(\boldsymbol{n}_{\mathrm{TS}}) - \delta n \left(\left.\frac{\partial E_{\mathrm{HF}}}{\partial n_q}\right|_{\boldsymbol{n}_{\mathrm{TS}}} - \left.\frac{\partial E_{\mathrm{HF}}}{\partial n_p}\right|_{\boldsymbol{n}_{\mathrm{TS}}} \right)$$
$$+ \frac{(\delta n)^2}{2} \left\{ \left.\frac{\partial^2 E_{\mathrm{HF}}}{\partial n_p^2}\right|_{\boldsymbol{n}_{\mathrm{TS}}} - 2\left.\frac{\partial^2 E_{\mathrm{HF}}}{\partial n_p \partial n_q}\right|_{\boldsymbol{n}_{\mathrm{TS}}} + \left.\frac{\partial^2 E_{\mathrm{HF}}}{\partial n_q^2}\right|_{\boldsymbol{n}_{\mathrm{TS}}} \right\} + \dots,$$
$$E_{\mathrm{HF}}(\boldsymbol{n}_{p\to q}) = E_{\mathrm{HF}}(\boldsymbol{n}_{\mathrm{TS}} + \delta \boldsymbol{n}_{\mathrm{TS}}) = E_{\mathrm{HF}}(\boldsymbol{n}_{\mathrm{TS}}) + \delta n \left(\left.\frac{\partial E_{\mathrm{HF}}}{\partial n_q}\right|_{\boldsymbol{n}_{\mathrm{TS}}} - \left.\frac{\partial E_{\mathrm{HF}}}{\partial n_p}\right|_{\boldsymbol{n}_{\mathrm{TS}}} \right)$$
$$+ \frac{(\delta n)^2}{2} \left\{ \left.\frac{\partial^2 E_{\mathrm{HF}}}{\partial n_p^2}\right|_{\boldsymbol{n}_{\mathrm{TS}}} - 2\left.\frac{\partial^2 E_{\mathrm{HF}}}{\partial n_p \partial n_q}\right|_{\boldsymbol{n}_{\mathrm{TS}}} + \left.\frac{\partial^2 E_{\mathrm{HF}}}{\partial n_q^2}\right|_{\boldsymbol{n}_{\mathrm{TS}}} \right\} + \dots, \tag{6.40}$$

A subsequent subtraction of these expansions and use of (6.33) finally give:

$$\Delta E_{p\to q} = E_{\mathrm{HF}}(\boldsymbol{n}_{p\to q}) - E_{\mathrm{HF}}(\boldsymbol{n}_0) = \left.\frac{\partial E_{\mathrm{HF}}}{\partial n_q}\right|_{\boldsymbol{n}_{\mathrm{TS}}} - \left.\frac{\partial E_{\mathrm{HF}}}{\partial n_p}\right|_{\boldsymbol{n}_{\mathrm{TS}}} + O(\delta \boldsymbol{n}_{\mathrm{TS}}^3)$$
$$= e_q(\boldsymbol{n}_{\mathrm{TS}}) - e_p(\boldsymbol{n}_{\mathrm{TS}}) + O(\delta \boldsymbol{n}_{\mathrm{TS}}^3). \tag{6.41}$$

To summarize, the $p \to q$ excitation energy can be estimated, neglecting very small terms of order $\delta \boldsymbol{n}_{\mathrm{TS}}^3$, as the difference of energies of two orbitals involved in the transition, obtained from a single SCF calculation for the TS occupations $\boldsymbol{n}_{\mathrm{TS}}$ of (6.36).

6.1.4 Analytical Realization of Hall and Roothaan

In atoms, for which the angular parts of AO are given by the spherical harmonics, the radial self-consistent field can be determined numerically. However, in all molecular applications the analytical realization of the HF method, which adopts the Ritz variant of the variational method (Sect. 5.1.2) to determine the optimum MO approximated as linear combinations (LC) of the fixed *basis set* of the AO functions. For example, the canonical AO of the system constituent AIM are selected as basis functions in the LCAO MO approach, or some arbitrary functions selected for reasons of numerical convenience, e.g., the exponential *Slater-type orbitals* (STO) or *Gaussian-type orbitals* (GTO), are used to expand MO in the associated LCSTO (e.g., Harris 1967; Clementi and Roetti 1974) and LCGTO (e.g., Shavitt 1963; Boys 1968; Pople 1976; Dunning and Hay 1977; Huzinaga et al. 1984; Poirier et al. 1985) calculations, respectively. The former,

$$s_{s,\alpha}^{l,m}(\boldsymbol{r}_\alpha) = N \exp(-\zeta_s r_\alpha)\, Y_l^m(\theta, \varphi), \tag{6.42a}$$

where N stands for the normalization constant, $\boldsymbol{r}_a = \boldsymbol{r} - \boldsymbol{R}_a = x_\alpha \boldsymbol{i} + y_\alpha \boldsymbol{j} + z_\alpha \boldsymbol{k} = (r_a, \theta, \varphi)$ is the position vector of an electron relative to the atomic nucleus α in the fixed position $\boldsymbol{R}_\alpha = X_\alpha \boldsymbol{i} + Y_\alpha \boldsymbol{j} + Z_\alpha \boldsymbol{k}$ (BO approximation), provide more compact expansions, since they exhibit generally correct analytical properties at both $r_\alpha \to 0$ (cusp at nucleus) and $r_\alpha \to \infty$ (exponential decay). The latter,

$$g_{s,\alpha}^{k,l,m}(\boldsymbol{r}_\alpha) \equiv N x_\alpha^k y_\alpha^l z_\alpha^m \exp(-\mu_s r_\alpha^2), \tag{6.42b}$$

exhibit incorrect behavior in both these limits, so that several GTO are required to adequately represent a single STO. However, GTO give rise to analytical expressions for the crucial (muticenter) electron repulsion integrals, for which Slater orbitals require a time-consuming numerical integration. For this reason the SCF LCGTO MO calculations dominate all molecular applications of the HF theory. This Hall–Roothaan analytical realization is customarily denoted as the SCF method.

Thus, in the SCF LCAO MO approach, the fixed set of basis functions of the Ritz method, $\boldsymbol{\chi} = (\chi_1, \chi_2, \ldots, \chi_w)$, represents the AO of constituent AIM, themselves represented as combinations of either the primitive GTO centered on atomic nuclei or their formal *contractions* combining subsets of primitive GTO, defined to limit the computational effort and maximize the variation flexibility of MO. The basis functions available in standard programs for molecular calculations range from the *minimum set*, of AO occupied in the ground states of all constituent (free) atoms, to *extended basis sets*, including the *split valence* contractions of the valence shell orbitals and some *polarization functions* corresponding to higher values of the orbital quantum number l, compared with those characterizing the AO functions of the minimum set. Such extended bases generate the required variety of both the radial (μ) and angular exponents (k, l, m) of the primitive GTO (6.42b), thus allowing for the radial expansion/contraction of bonded atoms in the molecular environment and their angular deformations (polarizations) in presence of the remaining AIM. It should be emphasized that in each of these variants several GTO-expansions of each AO/SO can be selected, which adds to the range of the basis set options available in modern software systems of the ab initio calculations (e.g., GAMESS, GAUSSIAN).

The HF orbitals are thus expanded in the adopted basis set $\boldsymbol{\chi}$, $\boldsymbol{\varphi}_{\mathrm{HF}} = \boldsymbol{\chi}\mathbf{C}$, which also determines the associated *overlap matrix* $\mathbf{S} = \langle \boldsymbol{\chi} | \boldsymbol{\chi} \rangle = \{S_{s,t} = \langle \chi_s | \chi_t \rangle\}$, defining the metric tensor of this function space. In accordance with the development of

Sect. 5.1.2, in order to formulate the relevant *secular equations* for the optimum LCAO MO coefficients $\mathbf{C} = \{C_{s,k}\}$ one also requires the *energy* (*Fock*) *matrix* $\mathbf{F} = \langle \boldsymbol{\chi}|\hat{\mathrm{F}}|\boldsymbol{\chi}\rangle = \{F_{s,t} = \langle \chi_s|\hat{\mathrm{F}}|\chi_t\rangle\}$, the basis set representation of the Fock operator [(6.22) and (6.28)], which defines the effective one-electron Hamiltonian of the HF method. For the closed-shell configuration of (5.66) expressing the matrix element $F_{s,t}$ in terms of the elementary one- and two-electron integrals in the chosen AO basis set,

$$\begin{aligned}&\{h_{s,t} = \langle \chi_s(1)|\hat{\mathrm{h}}(1)|\chi_t(1)\rangle_1\} \quad \text{and}\\ &\left\{(st|uv) \equiv \langle \chi_s(1)\chi_u(2)|g(1,2)|\chi_t(1)\chi_v(2)\rangle_{1,2}\right\},\end{aligned} \tag{6.43}$$

then gives:

$$\begin{aligned}F_{s,t} &= \langle \chi_s(1)|\hat{\mathrm{h}}(1)|\chi_t(1)\rangle_1 + \sum_{k=1}^{p} \langle \chi_s(1)|2\hat{\mathrm{J}}_k(1) - \hat{\mathrm{K}}_k(1)|\chi_t(1)\rangle_1\\ &= h_{s,t} + \sum_{u=1}^{w}\sum_{v=1}^{w}\left(2\sum_{k=1}^{p} C_{v,k}C_{u,k}^{*}\right)\left[(st|uv) - \frac{1}{2}(sv|ut)\right]\\ &\equiv h_{s,t} + \sum_{u=1}^{w}\sum_{v=1}^{w} P_{v,u}(st\,\|\,uv).\end{aligned} \tag{6.44}$$

In the preceding equation we have introduced the elements of the *charge and-bond-order* (CBO, density) matrix,

$$\mathbf{P}(\mathbf{C}_o) = \{P_{v,u}\} = 2\mathbf{C}_o\mathbf{C}_o^{\dagger} = \mathbf{C}\mathbf{n}\mathbf{C}^{\dagger}, \tag{6.45}$$

where the rectangular ($w \times p$) matrix $\mathbf{C}_o$ groups the expansion coefficients of the p (doubly) *occupied* MO, i.e., the first p columns of $\mathbf{C} = (\mathbf{C}_o, \mathbf{C}_v)$ with the remaining columns $\mathbf{C}_v$ corresponding to the virtual (empty) MO, and the diagonal matrix of MO occupations $\mathbf{n} = \{n_k\delta_{k,l}\}$ with $n_k = 2$ for occupied MO and $n_k = 0$ for virtual MO in the *spin-restricted* HF (RHF) theory. They are the only component of the Fock matrix $\mathbf{F}(\mathbf{C}_o) = \mathbf{F}(\mathbf{P})$, which changes from one iteration to another, while the overlap matrix and the elementary integrals of (6.43) are calculated once at the beginning of the SCF procedure and used in all iterations to construct the current Fock matrix $\mathbf{F}(\mathbf{C}^{(k)}) = \mathbf{F}(\mathbf{P}^{(k)})$.

We also recall that in the basis set of the Löwdin orthogonalized functions (see Sect. 5.1.2),

$$\tilde{\boldsymbol{\chi}} = \boldsymbol{\chi}\mathbf{S}^{-1/2},\quad \boldsymbol{\varphi}_{\mathrm{HF}} = (\boldsymbol{\chi}\mathbf{S}^{-1/2})\,\mathbf{U} = \tilde{\boldsymbol{\chi}}\mathbf{U},\quad \mathbf{U} = \langle\tilde{\boldsymbol{\chi}}|\boldsymbol{\varphi}_{\mathrm{HF}}\rangle,\quad \mathbf{U}\mathbf{U}^{\dagger} = \mathbf{U}^{\dagger}\mathbf{U} = \mathbf{I}, \tag{6.46}$$

the transformed Fock matrix $\tilde{\mathrm{F}} = \langle\tilde{\boldsymbol{\chi}}|\hat{\mathrm{F}}|\tilde{\boldsymbol{\chi}}\rangle = \mathbf{S}^{-1/2}\mathbf{F}\mathbf{S}^{-1/2}$ defines the eigenvalue problem for determining the optimum (canonical) HF solutions in the form of the diagonalization of $\tilde{\mathbf{F}}$ in the unitary transformation $\mathbf{U}$ [see (5.37)]:

$$\mathbf{U}^\dagger\tilde{\mathbf{F}}\mathbf{U} = \mathbf{e} = (e_k\delta_{k,l}), \qquad \mathbf{C} = \mathbf{S}^{-1/2}\mathbf{U}. \tag{6.47}$$

Therefore, in each iteration the new Fock matrix is constructed using the CBO matrix obtained in the previous iteration; it is then transformed to the orthogonal representation and diagonalized, to determine the next approximation of the LCAO MO coefficients and hence also of the CBO matrix, the orbital energies, etc. The time-determining step in the SCF calculations is the calculation of all electron-repulsion integrals $\{(st|uv)\}$, the number of which dramatically increases with the dimension of the basis set. It scales like $O(N^4)$ with the number of electrons.

It also follows from the preceding equation that

$$\begin{aligned}\mathbf{P} &= \mathbf{CnC}^\dagger = (\mathbf{S}^{-1/2}\mathbf{U})\ \mathbf{n}\ (\mathbf{U}^\dagger\mathbf{S}^{-1/2}) \\ &= \mathbf{S}^{-1/2}\langle\tilde{\boldsymbol{\chi}}|\boldsymbol{\varphi}_{\mathrm{HF}}\rangle\ \mathbf{n}\ \langle\boldsymbol{\varphi}_{\mathrm{HF}}|\tilde{\boldsymbol{\chi}}\rangle\mathbf{S}^{-1/2} = \mathbf{S}^{-1}\langle\boldsymbol{\chi}|\boldsymbol{\varphi}_{\mathrm{HF}}\rangle\ \mathbf{n}\ \langle\boldsymbol{\varphi}_{\mathrm{HF}}|\boldsymbol{\chi}\rangle\mathbf{S}^{-1} \\ &= 2\mathbf{S}^{-1}\langle\boldsymbol{\chi}|\boldsymbol{\varphi}_o\rangle\langle\boldsymbol{\varphi}_o|\boldsymbol{\chi}\rangle\mathbf{S}^{-1} = 2\mathbf{S}^{-1}\left\langle\boldsymbol{\chi}|\hat{\mathrm{P}}^o_{\varphi}|\boldsymbol{\chi}\right\rangle\mathbf{S}^{-1}.\end{aligned} \tag{6.48}$$

Thus, for the orthogonalized AO basis set, when $\mathbf{S} = \mathbf{S}^{-1/2} = \mathbf{S}^{-1} = \mathbf{I}$, the CBO matrix constitutes the AO representation of the projector into the *bonding subspace* consisting of all (doubly) occupied MO $\boldsymbol{\varphi}_o$ or the (singly) occupied SO $\{\boldsymbol{\varphi}_{o,\sigma}\}$ for both spin orientations of an electron:

$$\hat{\mathrm{P}}^o_{\varphi} = |\boldsymbol{\varphi}_o\rangle\langle\boldsymbol{\varphi}_o| = \sum\nolimits_\sigma |\boldsymbol{\varphi}_{o,\sigma}\rangle\langle\boldsymbol{\varphi}_{o,\sigma}| = \hat{\mathrm{P}}^\alpha_{\varphi} + \hat{\mathrm{P}}^\beta_{\varphi}.$$

6.1.5 Local Pseudopotential

The pseudopotentials, representing in the valence-only calculations the presence of atomic cores which practically do not participate in the bond-forming/breaking processes, have been introduced in Sect. 5.6. Here, we present the PK idea of the *local* pseudopotential (Phillips and Kleinman 1960), for the simplest case of the $N = n + 1$ electron system with a single valence electron occupying the normalized *pseudoorbital* ψ, nonorthogonal to the core orbitals, moving in the effective field due to the nuclei and n *inner*-shell electrons occupying the normalized and mutually orthogonal orbitals $\boldsymbol{\varphi}_c = (\varphi_1, \varphi_2, \ldots, \varphi_n)$, $\langle\varphi_i|\varphi_j\rangle = \delta_{i,j}$.

We first observe that the "frozen" core scenario calls for the Schmidt orthogonalization (Sect. 3.3.2) of the *outer*-shell pseudoorbital describing the valence electron with respect to all *inner*-shell orbitals $\boldsymbol{\varphi}_c$, which does not affect these "frozen" reference states. This way of making the core (closed) shells being "felt" by the valence electrons has also been used by Herring (1940) within the *Orthogonalized Plane Wave* (OPW) method for determining the wave function of valence electrons in crystals. Following Phillips and Kleinman we Schmidt-orthogonalize the pseudoorbital [see (3.45b)] into the new valence electron state:

$$\varphi = \mathcal{N}\left[\psi - \sum_{i=1}^{n} \varphi_i \langle \varphi_i | \psi \rangle\right] \equiv \mathcal{N}\left[\psi - \sum_{i=1}^{n} \varphi_i \alpha_i\right], \ \mathcal{N} = \left(1 - \sum_{i=1}^{n} |\alpha_i|^2\right)^{-1/2}. \quad (6.49)$$

In the framework of HF theory, the (orthogonal) canonical MO $\boldsymbol{\varphi} = (\boldsymbol{\varphi}_c, \varphi \equiv \varphi_{n+1})$ satisfy the Fock equations (6.21), $\hat{\mathrm{F}}\varphi_i = e_i\varphi_i, \quad i = 1, 2, \ldots, n+1$, with $e \equiv e_{n+1}$ denoting the orbital energy of the valence electron. It should be observed that in action of the Fock operator on the valence orbital φ the Coulomb and exchange operators due to φ exactly cancel each other, since $\hat{\mathrm{J}}_\varphi \varphi = \hat{\mathrm{K}}_\varphi \varphi$, and hence

$$\hat{\mathrm{F}}[\boldsymbol{\varphi}]\varphi = \hat{\mathrm{h}}\varphi + \sum_{i=1}^{n} (\hat{\mathrm{J}}_i - \hat{\mathrm{K}}_i)\varphi = \hat{\mathrm{F}}[\boldsymbol{\varphi}_c]\varphi. \quad (6.50)$$

Therefore, inserting (6.49) into the eigenvalue problem for the canonical valence MO, $\hat{\mathrm{F}}\varphi = e\varphi$, gives the equivalent (effective) Schrödinger equation for the pseudoorbital ψ,

$$\{\hat{\mathrm{F}}[\boldsymbol{\varphi}_c] + V_p\}\, \psi = e\psi, \quad (6.51)$$

including the local pseudopotential

$$V_p(\boldsymbol{r}) = \sum_{i=1}^{n} \frac{\alpha_i (e - e_i)\varphi_i(\boldsymbol{r})}{\psi(\boldsymbol{r})}. \quad (6.52)$$

Thus, the valence shell electron can be rigorously described by the (non-orthogonal to the core) pseudoorbital, provided that the effective potential V_{HF} of (6.22) will be supplemented with the pseudopotential V_p, which replaces the valence-core orthogonality requirement, thus making the presence of the inner shell electrons felt by the valence electron. This ab initio formulation can be extended into a general case of many valence electrons, including the correlated treatment of both the inner and outer electrons (Szasz 1985).

The equivalence of both descriptions is also directly seen when one compares the associated Slater determinants: $\det(\varphi_1, \varphi_2, \ldots, \varphi_n, \varphi)$ and $\det(\varphi_1, \varphi_2, \ldots, \varphi_n, \psi)$. Indeed, since ψ in the second determinant constitutes the linear combination of functions defining the first determinant (6.49) it then directly follows from the known invariance properties of determinants that both these functions in fact determine the same state of all $N = n + 1$ electrons.

6.2 Beyond HF Theory: Electron Coulomb Correlation

The HF theory, which can be also formulated in the relativistic version, exactly accounts for the exchange (x) correlation between the spin-like electrons, i.e., that part of the interdependence between the particle instantaneous positions, which originates from the Fermi statistics. However, since the product function of the truly independent particle approximation constitutes the source of the Slater determinant of (5.64), by acting on it with the antisymmetrizer $\hat{\mathrm{A}}$, the HF method totally neglects the Coulomb (*c*) correlation between all electrons, due to a finite electronic charge. Since the spin-like, statistically correlated electrons already "avoid" each other in the determinantal wave function, the Pauli principle of the wave-function antisymmetry effectively accounts also for a large portion of their Coulomb correlation. Hence, the largest error of this missing correlation effect can be expected to originate from electrons exhibiting the opposite spin orientations, which remain statistically uncorrelated at the HF level, e.g., from electron pairs occupying the same spatial orbitals in the spin-*Restricted* HF (RHF) approach.

In atoms one distinguishes both the *radial* and *angular* correlation effects. The former has already been observed in the spin-*Unrestricted* HF (UHF), variational treatment of helium atom (Sect. 5.5.2). Namely, when one electron occupied the more compact spherical (1*s*) orbital φ_2 ($\zeta_2 = 2.18$), distributed closer to the nucleus, the other electron preferred to occupy a more diffused orbital φ_1 ($\zeta_1 = 1.19$), exhibiting a larger average distance from the nucleus. Therefore these two spin-paired electrons, which are not correlated by the Fermi statistics, indeed exhibit a distinct effect of avoiding each other radially, due to their mutual Coulomb repulsion. This effect explains the observed lowering of the total UHF electronic energy, $E_0^{\mathrm{UHF}} = -2.876$ a.u., compared with the corresponding RHF energy level, $E_0^{\mathrm{RHF}} = -2.848$ a.u., where both electrons are constrained to occupy the same AO: $\varphi_1 = \varphi_2 = \varphi$ ($\zeta = 1.69$).

However, the exact value of the ground state energy in He, $E_0 = -2.904$ a.u., indicates that if we neglect a very small relativistic correction, $|\Delta E_{rel.}| = 0.0001$ a.u., there still remains a substantial angular correlation, which has not been accounted in the above *spherical* UHF approach. Indeed, due to the mutual repulsion the two electrons at their fixed radial distances should prefer the opposite positions relative to the nucleus. This effect can be accounted for when one lifts the spherical, 1*s*-type constraints [(5.90) and (5.98)] of the two AO φ_1 and φ_2, which define the spatial part (5.104),

$$\Phi^{\mathrm{UHF}}(\boldsymbol{r}_1, \boldsymbol{r}_2) = \frac{1}{\sqrt{2}}[\varphi_1(\boldsymbol{r}_1)\varphi_2(\boldsymbol{r}_2) + \varphi_1(\boldsymbol{r}_2)\varphi_2(\boldsymbol{r}_1)], \tag{6.53}$$

of the singlet UHF wave function (5.103). For example, one can approximate these orbitals by the mutually orthogonal *sp*-type hybrids along the *z* axis:

$$h_z = \mathcal{N}_1(\varphi_1 + Cp_z), \quad h_{-z} = \mathcal{N}_2(\varphi_2 - Cp_z), \quad p_z = \mathcal{N}_z z \exp(-\zeta r), \tag{6.54}$$

where $\mathcal{N}_1$, $\mathcal{N}_2$, and $\mathcal{N}_z$ stand for the corresponding normalization constants. One observes that the symmetric combination of the associated product functions, $h_z(\boldsymbol{r}_1)h_{-z}(\boldsymbol{r}_2) + h_z(\boldsymbol{r}_2)h_{-z}(\boldsymbol{r}_1)$, then contains the previous (spherical) $\Phi^{\mathrm{UHF}}(\boldsymbol{r}_1, \boldsymbol{r}_2)$ function of (6.53) and the $[p_z^2]$ configuration, $p_z(1)p_z(2)$.

However, in order to keep the helium atom spherically symmetrical, we have to treat the three axes of the coordinate system on equal footing, which calls for the following trial function of the UHF approximation, extended by the *Configuration Interaction* (CI) between the ground state electron configuration $[\varphi_1\varphi_2]$ and the doubly excited configurations $(\varphi_1\varphi_2) \to (p^2) = \left\{ [p_x^2] + [p_y^2] + [p_z^2] \right\}$, in which the two electrons occupy the ground-state–virtual $2p$ orbitals,

$$\begin{aligned}\Phi^{\mathrm{UHF/CI}}(\boldsymbol{r}_1,\boldsymbol{r}_2) &= \mathcal{N}\left\{\Phi^{\mathrm{UHF}}(\boldsymbol{r}_1,\boldsymbol{r}_2) - \frac{2C^2}{3}[p_x(\boldsymbol{r}_1)p_x(\boldsymbol{r}_2) + p_y(\boldsymbol{r}_1)p_y(\boldsymbol{r}_2) + p_z(\boldsymbol{r}_1)p_z(\boldsymbol{r}_2)]\right\} \\ &= \mathcal{N}\left\{\Phi^{\mathrm{UHF}}[1s,1s'] - \frac{2C^2}{3}\left(\Phi[p_x^2] + \Phi[p_y^2] + \Phi[p_z^2]\right)\right\},\end{aligned} \tag{6.55}$$

where $\mathcal{N}$ denotes the overall normalization factor, containing only two variational parameters: the linear coefficient C and the nonlinear exponent ζ of $2p$ orbitals.

This angularly correlated wave function gives the best variational estimate of the ground state energy $E_0^{\mathrm{UHF/CI}} = -2.895$ a.u. Therefore, this relatively simple trial wave function is already capable of accounting for almost 80% of the Coulomb correlation energy: (2.862 – 2.895)/(2.862 – 2.904) = 0.79, where we have used the known energy level in the RHF *limit*, $E_0^{\mathrm{RHFL}} = -2.862$ a.u., representing the energy estimate for the "saturated," very large basis set, which practically generates the full variational flexibility of the atomic orbital.

This example suggests a systematic way for including the Coulomb correlation through CI. One first creates the orthonormal excited configurations by replacing in the HF Slater determinant the selected ground state occupied SO by the equinumerous list of the ground state virtual SO (see the Slater–Condon rules of Sect. 5.4), and then mixes them with the ground state determinant in the variational wave function of the CI theory, which combines the N-electron wave functions for the ground and excited configurations, all of them derived from the fixed set of the optimum orbitals determined in a single (ground state) SCF calculation. For example, in the RHF approach to helium atom, for which $\varphi_1 = \varphi_2 = \varphi = 1s$, the radial correlation effect in He can be introduced to the ground-state HF wave function $\Phi^{\mathrm{RHF}}(\boldsymbol{r}_1,\boldsymbol{r}_2) = \varphi(\boldsymbol{r}_1)\varphi(\boldsymbol{r}_2) \equiv \Phi[1s^2]$ by mixing it with the doubly excited function $\Phi[2s^2] = \phi(\boldsymbol{r}_1)\phi(\boldsymbol{r}_2)$, $\phi = 2s$, resulting from the excitation of the spin-paired electrons occupying $1s$ orbital into the virtual $2s$ state: $(\varphi^+,\varphi^-) \to (\phi^+,\phi^-)$. Furthermore, to account for the angular correlation, one should also include the doubly excited configurations $\Phi[p_i^2]$, $i = x, y, z$, $(\varphi^+,\varphi^-) \to (p_i^+, p_i^-)$:

$$\Phi^{\mathrm{RHF/CI}} = \mathcal{N}'\left\{\Phi[1s^2] + C_1\Phi[2s^2] + C_2\left(\Phi[p_x^2] + \Phi[p_y^2] + \Phi[p_z^2]\right)\right\}. \tag{6.56}$$

The RHF approach often fails to properly describe the dissociation of chemical bonds, e.g., in H_2 or F_2. This so-called *static* (near-degeneracy) *correlation* error can be corrected by an inclusion in the $\Phi^{\mathrm{RHF/CI}}$ wave function of the excited configuration $\Phi\left[\sigma_u^2\right]$, obtained by exciting two valence electrons occupying the bonding σ_g MO in $\Phi_0^{\mathrm{RHF}} = \Phi\left[\sigma_g^2\right]$ to the antibonding MO σ_u, nearly degenerated with σ_g at very large internuclear separations: $\left(\sigma_g^+, \sigma_g^-\right) \to \left(\sigma_u^+, \sigma_u^-\right)$.

The formal basis for expanding the exact wave function of N electrons in terms of Slater determinants involving all SO, both the ground state occupied and virtual, comes from the so-called *expansion theorem* of quantum theory, which forms the basis for several methods in computational quantum chemistry. When the basis set of one-electron functions $\boldsymbol{\chi}(\boldsymbol{r})$ of the SCF MO method is nearly complete, the associated (orthonormal) MO, $\boldsymbol{\varphi}(\boldsymbol{r}) = \boldsymbol{\chi}(\boldsymbol{r})\mathbf{C}$, determine a practically complete set of SO in the RHF theory, $\{\psi_k(\boldsymbol{r}, \sigma)\} = \{\boldsymbol{\varphi}(\boldsymbol{r})\alpha(\sigma), \boldsymbol{\varphi}(\boldsymbol{r})\beta(\sigma)\} = \{\psi_k(\boldsymbol{q})\}$, in terms of which any one-electron wave function can be expanded. For example, for the exact quantum state of a single electron, one obtains the familiar expansion:

$$\Psi(1) = \sum_k c_k\, \psi_k(1), \quad c_k = \int \psi_k^*(\boldsymbol{q}_1)\Psi(\boldsymbol{q}_1)\, d\boldsymbol{q}_1 \equiv \langle \psi_k(1)|\Psi(1)\rangle_1 \qquad (6.57)$$

The same procedure can be repeated for expanding the exact state of two electrons,

$$\begin{aligned} \Psi(1,2) &= \sum_k c_k(2)\, \psi_k(1), \\ c_k(2) &= \int \psi_k^*(\boldsymbol{q}_1)\Psi(\boldsymbol{q}_1, \boldsymbol{q}_2)\, d\boldsymbol{q}_1 \equiv \langle \psi_k(1)|\Psi(1,2)\rangle_1, \end{aligned} \qquad (6.58)$$

followed by the expansion of the coefficient function

$$c_k(2) = \sum_l d_{k,l}\, \psi_l(2), \quad d_{k,l} = \int \psi_l^*(\boldsymbol{q}_2)c_k(\boldsymbol{q}_2)\, d\boldsymbol{q}_2 \equiv \langle \psi_l(2)|c_k(2)\rangle_2. \qquad (6.59)$$

Substituting (6.59) into (6.58) finally gives:

$$\Psi(1,2) = \sum_k \sum_l d_{k,l}\psi_k(1)\psi_l(2), \qquad (6.60)$$

with the proper exchange symmetry for fermions subsequently enforced by the antisymmetrizer $\hat{\mathrm{A}}$ of (5.64):

$$\Psi_A(1,2) = \sum_k \sum_l d_{k,l}\hat{\mathrm{A}}\{\psi_k(1)\,\psi_l(2)\} = \sum_k \sum_l d_{k,l}\,\det(\psi_k\,\psi_l). \qquad (6.61)$$

The two-electron wave function can be thus exactly expanded in terms of two-electron Slater determinants constructed from the complete set of SO, thus giving credence to the CI expansion of the exact quantum state in He or H_2.

This expansion procedure can be straightforwardly extended to a general N-electron case:

$$\begin{aligned}\Psi_A(1,2,\ldots,N) &= \sum_{k_1}\sum_{k_2}\cdots\sum_{k_N} d_{k_1,k_2,\ldots,k_N}\hat{\mathrm{A}}\{\psi_{k_1}(1)\,\psi_{k_2}(2)\,\ldots\,\psi_{k_N}(N)\}\\ &= \sum_{k_1}\sum_{k_2}\cdots\sum_{k_N} d_{k_1,k_2,\ldots,k_N}\det\{\psi_{k_1}\psi_{k_2}\,\ldots\,\psi_{k_N}\}.\end{aligned} \tag{6.62}$$

Therefore, the exact state of N-electron systems can be always expanded as linear combination of Slater determinants of N electrons built from the complete set of SO. There is no guarantee, though, that such an expansion is fast convergent. Indeed, much of the effort in the formal CI theory (Löwdin 1959; Shavitt 1977) has been devoted to improve a generally slow convergence of this determinantal expansion.

6.2.1 *Errors in SCF MO Calculations*

Although the effects neglected in the HF approximation amount to a very small fraction (of the order of 1%) of the system total electronic energy, they may be crucial for even qualitatively correct prediction of the energy differences of chemical interest, e.g., energies of the ionization, dissociation, activation, and isomerization processes or conformational barriers. Indeed, this relatively minute correlation error often results in incorrect conclusions regarding dissociation energies, electronic spectra, and energy differences on PES. Such applications of the quantum theory call for the *chemical accuracy*, of the order of 1 kcal/mole $\approx$ 1.6 m Hartree, which escapes the approximations present in the SCF method. Fortunately, the theory systematic errors cancel in some energy differences, so that even at this low level of the theory one obtains quite satisfactory predictions of many molecular properties, e.g., selected conformational barriers, structural parameters (bond lengths and angles), and molecular bonding patterns. Indeed, one encounters quite frequent examples, where small basis sets and modest studies can provide important chemical insights.

In this section we identify the main sources of errors in SCF method, which effectively limit a range of its adequate applications. As shown in Fig. 6.1, the analytical SCF method using a finite number of basis functions exhibits the *basis set* error, relative to the *HF limit* (HFl) corresponding to the complete (infinite) basis set, i.e., the full variational flexibility of the function space in the Ritz method: $\Delta E_{basis} = E_{\mathrm{HFl}} - E_{\mathrm{SCF}} < 0$. In general, the *minimum basis set*, consisting of AO occupied in the ground states of constituent (free) atoms, has several interpretative advantages and generates the most physical net charges of bonded atoms predicted from Mulliken's populational analysis (Mulliken 1935, 1955, 1962). However, this small basis set often favors some conformations of molecular systems, thus introducing a nonsystematic errors in predicted energy barriers. Splitting the valence AO into the independent *short*- and *long*-range contractions in the *extended*

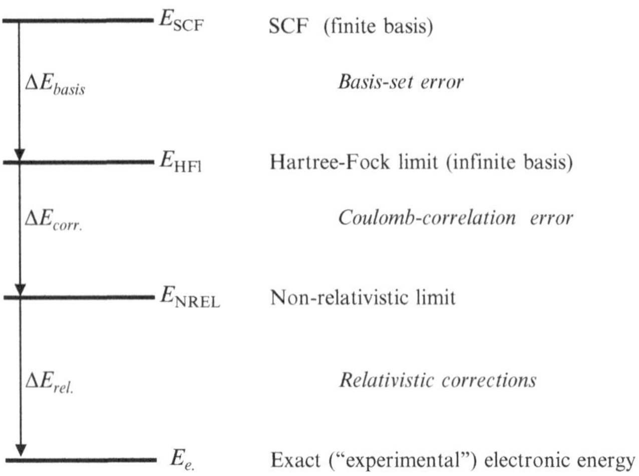

Fig. 6.1 Main sources of errors in SCF MO calculations

basis set properly describes the radial polarization of bonded atoms, while an additional inclusion of polarization functions allows for their angular adjustments in the molecular framework.

The *basis set superposition* error can also falsify the interaction energies between molecules at finite distances, since the basis functions of one reactant improves the quality of the basis function of the other reactant. To minimize this effect in the predicted interaction energy, given by the difference between the energy of the molecular complex ("supermolecule") and the sum of energies of both molecular reactants, the *counterpoise correction* of Boys and Bernardi (1970) is applied in estimating the reference energy of the two separated molecules, by using the full molecular basis set of both reactants when determining the energy of both the interacting and nearly separate reactants.

As we have already remarked in the preceding section, the other important source of error in the SCF calculations is the neglected correlation between electrons, due to their finite electric charges. This *Coulomb correlation* error, $\Delta E_{corr.} = E_{\mathrm{NREL}} - E_{\mathrm{HFl}} < 0$, is to a good approximation constant, when one compares the energies of conformations preserving the number of electronic pairs, and even more so, when the number of neighboring pairs of electrons is conserved. The main Coulomb correlation error is due to the spin-paired electrons confined to the same space occupied by the localized MO. Therefore, by preserving the number of such strongly correlated electronic pairs, one roughly preserves the overall correlation error, which then cancels out in the energy difference as the method systematic deviation. Good examples are provided by the inversion of ammonia and the "free" rotation around a single (σ) C–C bond in ethane, for which the number of the electronic pairs occupying the localized MO is the same at each stage of the conformation change.

This correlation error manifests itself strongly, when the pair-number criterion is not satisfied, e.g., in the electron ionization/attachment processes and the bond-breaking phenomena, e.g., in the bond dissociation or the "hindered" internal rotation in ethylene. It should be also observed that the RHF theory cannot explain the stability of F_2 molecule, giving the negative sign of the predicted bond dissociation energy. This is because it neglects the *dynamic* (avoidance) Coulomb correlation between the congested electronic pairs in the valence shell and fails to properly describe the *Separated Atoms Limit* (SAL), $F^0 + F^0$, giving instead a mixture of the SAL and the *ion* pair $F^+ + F^-$ state. The same *static* correlation deficiency of the RHF variant is observed in its description of the H_2 dissociation.

Finally, the *relativistic corrections*, not covered by this book, are defined as the difference between the exact (experimental) value of the electronic energy and that in the nonrelativistic limit (NREL): $\Delta E_{rel.} = E_e - E_{\mathrm{NREL}} < 0$. It combines several small energy contributions originating from the week interactions missing in the Coulombic molecular Hamiltonian of (5.51), e.g. *spin–orbit* or *spin–spin* interactions, of great importance for both the fine structure of atomic spectra and NMR and ESR experiments, the dependence of the particle mass on its speed, etc. They all require the relativistic formulation of quantum theory, in the form of the Dirac equation of state or the Pauli approximation to the Breit equation, which constitutes an approximate generalization of the Dirac equation to *many*-electron systems. The relativistic generalization of the SCF theory is known as the Dirac–Hartree–Fock (DHF) approach. Clearly, various terms in the Breit–Pauli Hamiltonian, which naturally follow from the relativistic extension of the quantum theory, can be also applied as perturbations to the nonrelativistic energy operator, and the standard perturbation theory can be then used to determine the associated corrections to the system energy.

In atoms the average correlation error per electron, $\Delta e_{corr.} = \Delta E_{corr.}/N$, stays roughly preserved with N: $\Delta e_{corr.} \approx -0.05$ a.u. The magnitude of the overall relativistic correction increases with the atomic number. For example in helium atom $|\Delta E_{rel.}(Z = 2)| = 0.0001$ a.u. $<< |\Delta E_{corr.}(Z = 2)| = 0.0420$ a.u., but starting from silicon atom it exceeds the correlation energy: $|\Delta E_{rel.}(Z = 14)| = 0.584$ a.u. $>$ $|\Delta E_{corr.}| = 0.494$ a.u. In most energy differences in chemistry, e.g., in conformational energy barriers, the relativistic energy is less important, since it is dominated by the *inner*-shell electrons of atomic cores, which remain practically unaffected by the bond-forming/breaking processes taking place in the valence shell. Therefore, neglecting the relativistic terms has only a minor effect on the energetical descriptors of bond dissociations or conformational changes. For example, the relativistic correction to the dissociation energy of F_2, $D_e = 0.062$ a.u., has been estimated as $\Delta D_{e,rel.} = 0.001$ a.u. Similarly, for the dissociation of NaCl, $D_e = 0.16$ a.u., the relativistic correction $\Delta D_{e,rel.} = 0.002$ a.u. is much below the correlation error $\Delta D_{e,corr.} = 0.04$ a.u. However, the relativistic effects may strongly influence other physical properties of atoms and molecules, e.g., the bonding energies of the core electrons and atomic radii and bond lengths, particularly in systems containing heavy atoms.

6.2.2 Static and Dynamic Correlation

The spatial RHF wave function for the ground state of $H_2 = H_A\text{–}H_B$, $\Phi_0^{\mathrm{RHF}}(A-B) = \phi_b(\boldsymbol{r}_1)\phi_b(\boldsymbol{r}_2)$, in which the spin-paired electrons occupy the bonding MO, $\phi_b = \sigma_g = N_b(1s_A + 1s_B)$, $N_b = (2 + 2\langle 1s_A|1s_B\rangle)^{-1/2}$, is not capable of a proper description of the dissociation into separated atoms, $H_A + H_B$, described by the spatial wave function $\Phi_0^{atom.}(\infty) = \left(1/\sqrt{2}\right)\,[1s_A(\boldsymbol{r}_1)1s_B(\boldsymbol{r}_2) + 1s_A(\boldsymbol{r}_2)1s_B(\boldsymbol{r}_1)]$. Indeed, expressing the occupied MO in terms of both AO gives in the SAL:

$$\Phi_0^{\mathrm{RHF}}(A-B) = \frac{1}{2}\{[1s_A(\boldsymbol{r}_1)1s_B(\boldsymbol{r}_2) + 1s_A(\boldsymbol{r}_2)1s_B(\boldsymbol{r}_1)] + [1s_A(\boldsymbol{r}_1)1s_A(\boldsymbol{r}_2) + 1s_B(\boldsymbol{r}_1)1s_B(\boldsymbol{r}_2)]\} = \frac{1}{\sqrt{2}}[\Phi_0^{atom.}(\infty) + \Phi_0^{ion.}(\infty)]. \tag{6.63}$$

It follows from this limiting RHF wave function that it describes the equal mixture of the dissociation into atoms, represented by $\Phi_0^{atom.}(\infty)$, and into ionic pair: $(H_A^- + H_B^+)$ or $(H_B^- + H_A^+)$, represented by $\Phi_0^{ion.}(\infty)$, thus giving $E_0^{\mathrm{RHF}}(\infty)$ distinctly above $E_0(\infty) = -1$ a.u. The UHF approximation, which already includes a fraction of the Coulomb correlation, correctly predicts the dissociation into atoms in this limit.

The crucial configuration, which remedies this shortcoming of the RHF treatment in the improved CI wave function is the *double* excitation to the antibonding MO $\phi_a = \sigma_u = N_a(1s_A - 1s_B), N_b = [2 - 2\langle 1s_A|1s_B\rangle]^{-1/2}$:

$$\Phi_0^{\mathrm{CI}}(A-B) = C_1^{\mathrm{CI}}\left|\overset{+}{\phi_b}\,\overset{-}{\phi_b}\right| + C_2^{\mathrm{CI}}\left|\overset{+}{\phi_a}\,\overset{-}{\phi_a}\right|. \tag{6.64}$$

In fact, at large internuclear distances, the HF configuration $|\phi_b^+\phi_b^-|$ and the *doubly* excited state $|\phi_a^+\phi_a^-|$ become degenerate, thus exhibiting comparable participation in this CI combination. Therefore, inclusion of both these functions in the CI expansion appears to be crucial for the correct description of the dissociation of this covalent bond. At shorter bond lengths $R \approx R_e$, where the equilibrium internuclear distance $R_e \cong 1.4$ a.u., this degeneracy is removed and hence $|C_1^{\mathrm{CI}}| \approx 1$ and $|C_2^{\mathrm{CI}}| \approx 0$, which explains a good description of the hydrogen molecule already in the RHF approximation.

This long-range, near-degeneracy Coulomb correlation is called *static*, since it is more related to the symmetry requirement of the wave function, rather than the instantaneous interaction between electrons. It calls for the inclusion of the missing $|\phi_a^+\phi_a^-|$ configuration in addition to the RHF determinant $|\phi_b^+\phi_b^-|$. This static correlation error introduces a substantial deviation of the bonding energy,

$$\Delta E_{bond}(R_e) = E_e(R_e) - E_e(\infty) \equiv -D_e < 0, \tag{6.65}$$

overestimated relative to $E_e^{\mathrm{RHF}}(\infty)$ dissociation limit by about 6.4 eV, compared with the experimental value reproduced exactly by the theoretical calculations of Kołos and Wolniewicz (KW) (1964, 1965, 1968): $\Delta E_{bond}^{exp.}(R_e^{exp.} = 1.4006) = \Delta E_{bond}^{\mathrm{KW}}(R_e^{\mathrm{KW}} = 1.4006) = -0.1745\,\mathrm{a.u.}$ This error of the RHF wave function is already drastically reduced at the minimum CI level of (6.64), in which this energy difference is underestimated by only 1 eV. The CI expansion involving 28 most important excited configurations gives still better prediction: $\Delta E_{bond}^{\mathrm{CI}(28)}(1.40) = -0.1672\,\mathrm{a.u.}$ The bonding energy predicted in the HFL, $\Delta E_{bond}^{\mathrm{HFL}}(1.40) = -0.1336\,\mathrm{a.u.}$, allows one to estimate the overall Coulomb correlation error in H_2 for $\Delta E_{rel.}^{bond} \cong 0$ at about $\Delta E_{corr.}^{bond} \cong -0.041\,\mathrm{a.u.} = -\,1.116\,\mathrm{eV}$.

The concept of the *dynamical* correlation refers to the instantaneous avoidance of electrons due to a strong Coulomb repulsion between them in atoms and molecules, thus resulting in their tendency to assume positions, which maximize the instantaneous distances between them. This dynamical effect gives rise to both the radial and angular correlation effects in atoms, which we have discussed before. To account for this phenomenon rather extensive CI has to be included in the ab initio (*first* principle) calculations. This dynamic correlation error is crucial for the adequate determination of the atomic electron affinities measured by the difference between the energy of the neutral atom X(0) and its anion X(−1): $A_{\mathrm{X}} = E_{\mathrm{X(0)}} - E_{\mathrm{X}\,(-1)}$. For example, the HF prediction for oxygen, $A_{\mathrm{O}}^{\mathrm{HF}} = -0.54\,\mathrm{eV}$ (!), is qualitatively incorrect, predicting the neutral atom to be more stable than the anion; the experimental value $A_{\mathrm{O}}^{\mathrm{exp.}} = 1.46\,\mathrm{eV}$ can be reproduced only by the trial wave functions of both species using large CI expansions.

The dynamical correlation also explains the failure of the SCF method to predict the stability of the F_2 molecule, $\Delta E_{bond}^{\mathrm{SCF}}(F_2) = 1.37\,\mathrm{eV}$(!), the result qualitatively incorrect compared with $\Delta E_{bond}^{\mathrm{exp.}}(F_2) = -1.68\,\mathrm{eV}$. The reason for this failure is the dynamical correlation between the electrons of the *lone*-pairs on both atoms and the *bonding* pair of electrons. Only using the multideterminant trial function and the extended basis sets including all orbitals up to $g(l = 4)$ AO, one can reproduce theoretically this experimental value with the chemical accuracy.

Large dynamical correlation error should be also present in all multiple bonded molecules, e.g., in N_2: $D_e^{\mathrm{RHF}}(N_2) = 5.3\,\mathrm{eV}$, compared with $D_e^{exp.}(N_2) = 9.9\,\mathrm{eV}$. Again, using the extended basis set (to g-functions) and the adequate level of CI required for the correct description of the dissociation into atoms, and reoptimization of orbitals in the CI formalism within the so-called *multiconfiguration SCF* (MC SCF or MR SCF) procedure gives the satisfactory value of the dissociation energy:$D_e^{\mathrm{MCSCF}}(N_2) = 9.8\,\mathrm{eV}$. The minimum level of CI, to account for the dynamical correlation between the outer electrons involved in the metal–ligand chemical bonds, is generally required also in complexes of the transitional metal ions, since the coordination bonds in these compounds involve different shells of the central atom: nd, $(n + 1)s$, and $(n + 1)p$.

The variational determinantal function(s), for which the orbitals determining the subsequent CI expansion are optimized in the relevant SCF procedure, determine the so-called *reference functions*. This set includes either a single Slater determinant, e.g., in the standard HF method for the *closed*-shell states, called the *single-reference* (SR) SCF, or several determinants, e.g., in the HF approach to the *open*-shell states or in MC SCF method, in which the MO are optimized for the CID-type combination of Slater determinants, including the ground state configuration and the most important *double* excitations from it. The latter approach can be thus called the *multi-reference* (MR) *SCF* (MR SCF) or *multiconfigurational SCF* (MC SCF) technique. Accordingly, the CI expansions of the Coulomb-correlated wave functions can be also classified as either SR CI, when they originate from the SR SCF orbitals, and MR CI, when the MR SCF orbitals are used to generate the excited configurations (Shavitt 1977).

The MR approaches are crucial for the correct description of the open-shell states to generate the correct *orbital* and *spin* symmetry of the trial wave function, in view of the degeneracy of orbitals and the alternative spin orientations of electrons in the open shells. In the MR variant, in which the ground state orbitals are optimized for a combination of several Slater determinants $\{\Phi_i^0\}$, $\Phi_0^{\mathrm{MCSCF}} = \sum_i C_{i,0}^{\mathrm{MCSCF}} \Phi_i^0$, the excited configurations derived from one determinant may repeat the configurations generated from another determinant, so that a careful selection of the *independent* configurations is required. One of the popular variants of the MC SCF method is the *Complete Active Space* SCF (CAS SCF) technique (Roos and Siegbahn 1977). In this CI method the reference set $\{\Phi_i^0\}$ contains all configurations involving the complete set of orbitals active (populated) at each stage of the chemical reaction under consideration, thus determining the subset of the *active* MO in this process, i.e., exhibiting *fractional* electron occupations. The remaining part of the fully occupied MO determines the subset of the reaction *inactive orbitals* of the CAS SCF wave function. One also delineates within some appropriately chosen "energy window" the *external orbitals*, the excitation to which should be vital to represent the most important changes in the Coulomb correlation between electrons. The independent excited configurations originating from all occupied MO, active and inactive, to these external MO are then used in the associated (limited) MR CI expansion.

The number n_c of all configurations in the given basis set grows dramatically with the basis set size $w > N$, which also marks the number of all MO determined at the SCF stage giving rise to $2w$ SO. The former is thus equal to the number of alternative choices of N occupied SO from the overall set of $2w$ functions, given by the familiar combinatorial formula

$$n_c = \binom{2w}{N} = \frac{(2w)!}{N!(2w-N)!}. \tag{6.66}$$

For example, already for rather small system containing only $N = 10$ electrons, say water molecule, and 14 SO derived from its 7 (*doubly* occupied) MO obtained in the minimum basis set, $w = 7$, one obtains $n_c = 1{,}001$, while the moderately extended, *split* valence basis, for $w = 13$ MO, i.e., 26 SO, generates $n_c = 5\,311\,735$ configurations. Therefore, the *Full* CI (FCI) calculations, involving all admissible excitations in the adopted basis set, are possible only for small molecules. The limited CI realizations of the method assume the excited configurations in the given energy window around the HOMO and LUMO (Fig. 6.2) and require rather stringent selection of the most important configurations, probed using standard perturbational techniques.

One also realizes that a direct diagonalization of the CI Hamiltonian matrix,

$$\mathbf{H}^{\mathrm{CI}} = \left\{ \langle \Phi_s(N) | \hat{\mathrm{H}}^e(N) | \Phi_t(N) \rangle = H^{\mathrm{CI}}_{s,t} \right\},$$

is impossible, due to a limited size of the operational memory of contemporary digital computers. The special, sometimes ingenious algorithms for extracting CI eigenvectors and eigenvalues for a small number of the lowest electronic states have been designed to achieve this goal when millions of configurations are included in the limited CI expansion (Shavitt 1977; Roos and Siegbahn 1977).

However, cutting the value of the excitation energy and a variety/multiplicity of excitations included in the variational CI theory using the chosen subset of configurations introduces the so-called *size consistency* error into the predicted interaction energies, which has to be properly compensated for. The alternative Møller–Plesset (MP) theory (Møller and Plessett 1934), in which the limited CI coefficients are determined from PT, is free from this shortcoming. We shall discuss such typical CI approaches in detail in other sections of this chapter, limiting the present discussion to a general survey of problems encountered in practical realizations of such advanced numerical procedures.

In general, a single Slater determinant does not exhibit the proper spatial and spin symmetry required of an adequate description of spectroscopic states. The CI wave functions $\Psi^{\mathrm{CI}}_{S,M_S}$ should correspond to the sharply specified length and the projection of the resultant spin $\boldsymbol{S}$, satisfying the associated eigenvalue problems for the overall spin of N electrons:

$$\hat{\mathrm{S}}^2 \Psi^{\mathrm{CI}}_{S,M_S} = S(S+1)\hbar^2 \Psi^{\mathrm{CI}}_{S,M_S}, \quad \hat{\mathrm{S}}_z \Psi^{\mathrm{CI}}_{S,M_S} = M_S \hbar \Psi^{\mathrm{CI}}_{S,M_S},$$
$$M_S = -S, -S+1, \ldots, S-1, S. \tag{6.67}$$

Therefore, the CI expansion of the given spin state should be limited to configurations exhibiting the same spin characteristics, given by the appropriate combinations of determinantal functions. This severely limits the "length" of the expansion itself but requires advanced algorithms for generating such *spin-adapted* configurations. Examples of such spin-adapted configurations generated by the FO are shown in Fig. 6.2.

For the Coulomb (spin-independent) Hamiltonian, the $(2S+1)$ states corresponding to different values of the (orientation) quantum number M_S for to the specified value of the (length) spin quantum number S, are degenerate so it suffices to consider a single state, say $\Psi^{\mathrm{CI}}_{S,S}$, to determine the corresponding energy.

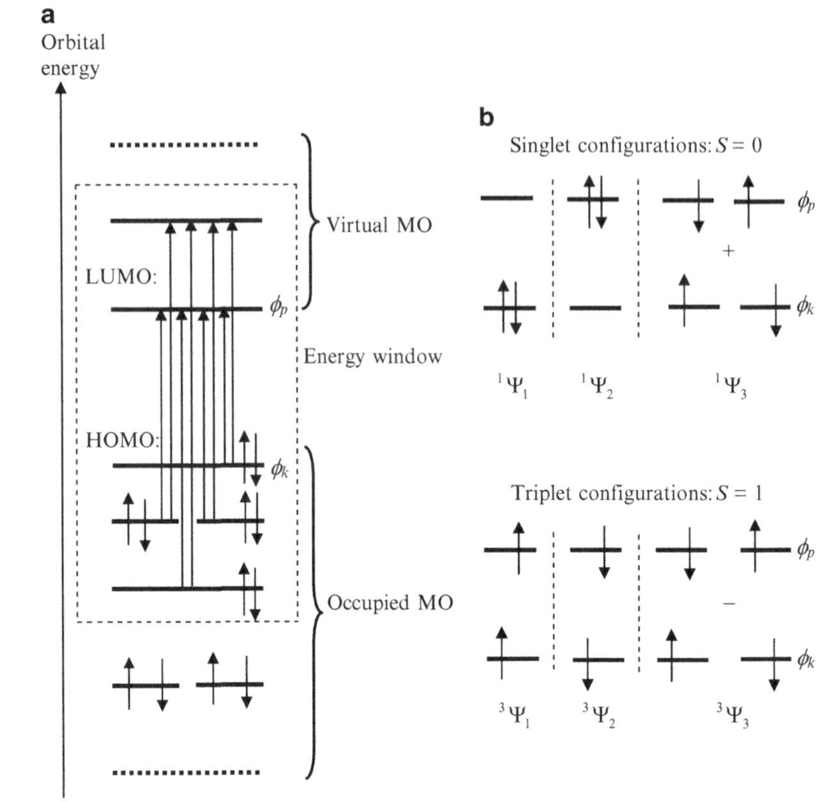

Singlet configurations, $S = M_S = 0$:

$${}^1\Psi_1 = \left|\phi_k^+\phi_k^-\right| = \phi_k(\boldsymbol{r}_1)\phi_k(\boldsymbol{r}_2)\frac{1}{\sqrt{2}}[\alpha(\sigma_1)\beta(\sigma_2) - \beta(\sigma_1)\alpha(\sigma_2)] \equiv \Phi_k^S(\boldsymbol{r}_1,\boldsymbol{r}_2)\, U_{0,0}(\sigma_1,\sigma_2)$$

$${}^1\Psi_2 = \left|\phi_p^+\phi_p^-\right| = \phi_p(\boldsymbol{r}_1)\phi_p(\boldsymbol{r}_2)U_{0,0}(\sigma_1,\sigma_2) \equiv \Phi_p^S(\boldsymbol{r}_1,\boldsymbol{r}_2)U_{0,0}(\sigma_1,\sigma_2)$$

$${}^1\Psi_3 = \frac{1}{\sqrt{2}}\{\left|\phi_k^+\phi_p^-\right| + \left|\phi_p^+\phi_k^-\right|\} = \frac{1}{\sqrt{2}}[\phi_k(\boldsymbol{r}_1)\phi_p(\boldsymbol{r}_2) + \phi_p(\boldsymbol{r}_1)\phi_k(\boldsymbol{r}_2)]U_{0,0}(\sigma_1,\sigma_2)$$
$$\equiv \Phi_{k,p}^S(\boldsymbol{r}_1,\boldsymbol{r}_2)\, U_{0,0}(\sigma_1,\sigma_2)$$

Triplet configurations, $S = 1$, $M_S = (-1, 0, 1)$:

$${}^3\Psi_1 = \left|\phi_k^+\phi_p^+\right| = \frac{1}{\sqrt{2}}[\phi_k(\boldsymbol{r}_1)\phi_p(\boldsymbol{r}_2) - \phi_p(\boldsymbol{r}_1)\phi_k(\boldsymbol{r}_2)]\alpha(\sigma_1)\alpha(\sigma_2) \equiv \Phi_{k,p}^A(\boldsymbol{r}_1,\boldsymbol{r}_2)\, U_{1,1}(\sigma_1,\sigma_2)$$

$${}^3\Psi_1 = \left|\phi_k^-\phi_p^-\right| = \Phi_{k,p}^A(\boldsymbol{r}_1,\boldsymbol{r}_2)\beta(\sigma_1)\beta(\sigma_2) \equiv \Phi_{k,p}^A(\boldsymbol{r}_1,\boldsymbol{r}_2)\, U_{1,-1}(\sigma_1,\sigma_2)$$

$${}^3\Psi_3 = \frac{1}{\sqrt{2}}\{\left|\phi_k^+\phi_p^-\right| - \left|\phi_p^+\phi_k^-\right|\} = \Phi_{k,p}^A(\boldsymbol{r}_1,\boldsymbol{r}_2)\frac{1}{\sqrt{2}}[\alpha(\sigma_1)\beta(\sigma_2) + \beta(\sigma_1)\alpha(\sigma_2)]$$
$$\equiv \Phi_{k,p}^A(\boldsymbol{r}_1,\boldsymbol{r}_2)\, U_{1,0}(\sigma_1,\sigma_2)$$

Fig. 6.2 Energy window determining the range of electron excitations in the limited CI approaches (Panel *a*) and the singlet/triplet configurations involving excitations between the Frontier MO: HOMO → LUMO (Panel *b*)

This additionally limits the number of the spin-adapted configurations, which have to be considered explicitly in order to extract the spectroscopic energy levels.

One similarly uses the spatial symmetry to limit the length of the CI expansion. More specifically, only configurations of the same symmetry type can mix, with the subsets of the (space-spin)-adapted configurations of the same symmetry type determining the nonvanishing diagonal blocks of $\mathbf{H}^{\mathrm{CI}}$, with the *off*-diagonal blocks corresponding to two subsets of different symmetries identically vanishing. Thus, each subset can be considered separately, which radically lowers the number of secular equations to be solved simultaneously by the diagonalization of the associated diagonal block of the CI energy matrix.

Finally, an extra reduction of the configuration number results from using the perturbation theory in determining how effective a given configuration really is in improving the electronic state of interest. This preliminary exploration allows one to remove inefficient excitations and focus solely on those, which strongly couple to the state to be determined. Indeed, in addition to the spin and spatial symmetry restrictions one can directly probe the importance of the current configuration Φ_s in the CI expansion of the ground state wave function, $\Psi_0^{\mathrm{CI}}(N) = \sum_s C_{s,0}^{\mathrm{CI}} \Phi_s(N)$, using the *first*-order estimates from the Ryleigh–Schrödinger PT (5.12):

$$C_{s,0}^{\mathrm{CI}(1)} = H_{s,0}^{\mathrm{CI}} / \left(H_{0,0}^{\mathrm{CI}} - H_{s,s}^{\mathrm{CI}} \right), \tag{6.68}$$

giving rise to the *second*-order (correlation energy) contribution [see (5.16)]:

$$\Delta E_{s,0}^{(2)} = \left| H_{s,0}^{\mathrm{CI}} \right|^2 / \left(H_{0,0}^{\mathrm{CI}} - H_{s,s}^{\mathrm{CI}} \right). \tag{6.69}$$

Both these probes are inversely proportional to the excitation energy measured by the denominator in these expressions. Therefore, by appropriate choice of the energy window for such excitations one can a priori eliminate configurations that most likely generate a minor energy lowering, in advance of the diagonalization step of the $\mathbf{H}^{\mathrm{CI}}$ block grouping only the most important configurations.

It should be also observed that the traditional routines for the diagonalization of $\mathbf{H}^{\mathrm{CI}}$ can be applied only to relatively short and medium CI expansions, e.g., in MC SCF method, while mixing millions of configurations calls for special algorithmic solutions often dictated by the "architectural" designs of the computers themselves (see: Shavitt 1977). For example, the *integral-driven* technique of Roos (Roos 1972; Roos and Siegbahn 1977), called the *direct CI*, maximizes the effects of a single, time-consuming readout of a large number of two-electron integrals from the external memory device to the operational memory of the computer. This is achieved by simultaneously determining the integral contributions to all currently needed elements of the energy matrix. In fact, the full Hamiltonian matrix $\mathbf{H}^{\mathrm{CI}}$, required to determine simultaneously all eigenvectors, is never generated in this procedure. Instead, each eigenvector of interest is determined separately and this reduced problem requires the knowledge of only a small part of $\mathbf{H}^{\mathrm{CI}}$.

6.2.3 Correlation Holes

The essence of the correlation phenomenon lies in an interdependence between the instantaneous, relative positions of electrons. It can be quantified in terms of the corresponding simultaneous (or conditional) density or probability distributions of two electrons, which we have introduced in Sect. 3.1. We first observe that for the *independent* (*ind.*) distributions of the two *distinguishable* electrons, say electron "1" located at $\boldsymbol{r}$ occupying the orbital φ_1, and the electron "2" at $\boldsymbol{r}'$ occupying the orbital φ_2, the simultaneous two-electron probability $p_2^{ind.}(\boldsymbol{r},\boldsymbol{r}') = p_1(\boldsymbol{r})p_2(\boldsymbol{r}')$, where $p_1(\boldsymbol{r}) = |\varphi_1(\boldsymbol{r})|^2$ and $p_2(\boldsymbol{r}) = |\varphi_2(\boldsymbol{r})|^2$. Hence, the conditional probability density of detecting the "*dependent*" electron 2 at $\boldsymbol{r}'$, when the "*reference*" electron 1 is known to be located at $\boldsymbol{r}$, then reads:

$$p^{ind.}(\boldsymbol{r}'|\boldsymbol{r}) = p_2^{ind.}(\boldsymbol{r},\boldsymbol{r}')/p_1(\boldsymbol{r}) = p_2(\boldsymbol{r}'), \qquad \int p^{ind.}(\boldsymbol{r}'|\boldsymbol{r})\,d\boldsymbol{r}' = 1. \tag{6.70a}$$

Similar result follows when one treats two particles in this hypothetical $N = 2$ electron system as *indistinguishable*. The overall one-electron distributions,

$$p(\boldsymbol{r}) = [p_1(\boldsymbol{r}) + p_2(\boldsymbol{r})]/2 = \rho(\boldsymbol{r})/2, \qquad \int p(\boldsymbol{r})\,d\boldsymbol{r} = 1, \quad \int \rho(\boldsymbol{r})\,d\boldsymbol{r} = 2,$$

then generate the simultaneous probability/density of two electrons:

$$p_2^{ind.}(\boldsymbol{r},\boldsymbol{r}') = p(\boldsymbol{r})p(\boldsymbol{r}') = \rho(\boldsymbol{r})\rho(\boldsymbol{r}')/4 = \rho_2^{ind.}(\boldsymbol{r},\boldsymbol{r}')/2,$$
$$\int\int p_2^{ind.}(\boldsymbol{r},\boldsymbol{r}')\,d\boldsymbol{r}\,d\boldsymbol{r}' = 1, \qquad \int\int \rho_2^{ind.}(\boldsymbol{r},\boldsymbol{r}')\,d\boldsymbol{r}\,d\boldsymbol{r}' = 2,$$

and the associated conditional distributions of detecting one (dependent) electron at $\boldsymbol{r}'$ when another (reference) electron is at $\boldsymbol{r}$:

$$p^{ind.}(\boldsymbol{r}'|\boldsymbol{r}) = p_2^{ind.}(\boldsymbol{r},\boldsymbol{r}')/p(\boldsymbol{r}) = p(\boldsymbol{r}'), \qquad \int p^{ind.}(\boldsymbol{r}'|\boldsymbol{r})\,d\boldsymbol{r}' = 1;$$
$$\rho^{ind.}(\boldsymbol{r}'|\boldsymbol{r}) = \rho_2^{ind.}(\boldsymbol{r},\boldsymbol{r}')/\rho(\boldsymbol{r}) = \rho(\boldsymbol{r}')/2, \quad \int \rho^{ind.}(\boldsymbol{r}'|\boldsymbol{r})\,d\boldsymbol{r}' = 1. \tag{6.70b}$$

This independent conditional probability/density provides the reference against which the *exchange*- and/or *Coulomb*-correlated distributions are compared in order to extract the displacements ("*holes*") due to the two-electron correlation.

For a general N-electron system in the Hartree approximation of Sect. 6.1.1, one obtains the following expression for the one- and two-electron densities [see (3.7) and (3.10)]:

$$\begin{aligned}
\rho^{\mathrm{H}}(\boldsymbol{r}) &= \langle \Psi(N)|\hat{\rho}(\boldsymbol{r})|\Psi(N)\rangle = \sum_{k=1}^{N} |\varphi_k(\boldsymbol{r})|^2, \\
\rho_2^{\mathrm{H}}(\boldsymbol{r},\boldsymbol{r}') &= \langle \Psi(N)|\hat{\rho}_2(\boldsymbol{r},\boldsymbol{r}')|\Psi(N)\rangle \\
&= \rho^{\mathrm{H}}(\boldsymbol{r})\left[\rho^{\mathrm{H}}(\boldsymbol{r}') - \frac{1}{\rho^{\mathrm{H}}(\boldsymbol{r})}\sum_{k=1}^{N} |\varphi_k(\boldsymbol{r})|^2 |\varphi_k(\boldsymbol{r}')|^2\right] \\
&\equiv \rho^{\mathrm{H}}(\boldsymbol{r})\rho^{\mathrm{H}}(\boldsymbol{r}'|\boldsymbol{r}) \equiv \rho^{\mathrm{H}}(\boldsymbol{r})\left[\rho^{\mathrm{H}}(\boldsymbol{r}') + h_{\mathrm{H}}^{\mathrm{SI}}(\boldsymbol{r}'|\boldsymbol{r})\right].
\end{aligned} \tag{6.71}$$

It thus follows from the above expression that the Hartree method of Sect. 6.1.1 involves the *self-interaction* (SI) *hole*:

$$h_{\mathrm{H}}^{\mathrm{SI}}(\boldsymbol{r}'|\boldsymbol{r}) = -\frac{1}{\rho^{\mathrm{H}}(\boldsymbol{r})}\sum_{k=1}^{N} |\varphi_k(\boldsymbol{r})|^2 |\varphi_k(\boldsymbol{r}')|^2, \tag{6.72}$$

by which the Hartree conditional probability, $\rho^{\mathrm{H}}(\boldsymbol{r}'|\boldsymbol{r}) = \rho_2^{\mathrm{H}}(\boldsymbol{r},\boldsymbol{r}')/\rho^{\mathrm{H}}(\boldsymbol{r})$, differs from the product reference value $\rho^{ind.}(\boldsymbol{r}'|\boldsymbol{r}) = \rho^{\mathrm{H}}(\boldsymbol{r}')$, which marks the truly independent (SI-contaminated) distributions of two electrons. One also observes that the integration of the SI-hole over positions of the dependent electron gives:

$$\int h_{\mathrm{H}}^{\mathrm{SI}}(\boldsymbol{r}'|\boldsymbol{r})\, d\boldsymbol{r}' = -\frac{1}{\rho^{\mathrm{H}}(\boldsymbol{r})}\sum_{k=1}^{N} |\varphi_k(\boldsymbol{r})|^2 \langle\varphi_k|\varphi_k\rangle = -1. \tag{6.73}$$

Therefore, in this approximation, the SI-hole density effectively eliminates one dependent electron from the surroundings of the reference electron.

Moreover, using (3.15) one expresses the electron repulsion energy in the Hartree method in terms of the classical interaction between independent charge distributions and the self-repulsion correction term involving the SI-hole:

$$\begin{aligned}
V_{e,e}^{\mathrm{H}}(N) &= \frac{1}{2}\iint \frac{\rho_2^{\mathrm{H}}(\boldsymbol{r},\boldsymbol{r}')}{|\boldsymbol{r}-\boldsymbol{r}'|} d\boldsymbol{r}\, d\boldsymbol{r}' = \frac{1}{2}\iint \frac{\rho^{\mathrm{H}}(\boldsymbol{r})\rho^{\mathrm{H}}(\boldsymbol{r}')}{|\boldsymbol{r}-\boldsymbol{r}'|} d\boldsymbol{r}\, d\boldsymbol{r}' + \frac{1}{2}\iint \frac{\rho^{\mathrm{H}}(\boldsymbol{r}) h_{\mathrm{H}}^{\mathrm{SI}}(\boldsymbol{r}'|\boldsymbol{r})}{|\boldsymbol{r}-\boldsymbol{r}'|} d\boldsymbol{r}\, d\boldsymbol{r}' \\
&= J[\rho^{\mathrm{H}}] + E_{\mathrm{SI}}[\rho^{\mathrm{H}}, h^{\mathrm{SI}}] = \frac{1}{2}\sum_{k=1}^{N}\sum_{l=1}^{N} J_{k,l} - \frac{1}{2}\sum_{k=1}^{N} J_{k,k}.
\end{aligned} \tag{6.74}$$

It is seen to be given by the difference between the classical Coulomb repulsion energy of the electronic charge density ρ^{H}, $J[\rho^{\mathrm{H}}]$, and the SI-correction term $E_{\mathrm{SI}}[\rho^{\mathrm{H}}, h^{\mathrm{SI}}]$ representing the interaction between this charge distribution and the SI-hole. It should be also observed that this expression exactly reproduces the two-electron part of the energy expectation value of (6.2).

Next, let us examine the corresponding hole concept in HF theory, in which the state function satisfies Pauli's antisymmetrization postulate. The relevant electron distributions now read:

$$\begin{aligned}
\rho^{\mathrm{HF}}(\boldsymbol{r}) &= \langle \Psi_A(N)|\hat{\rho}(\boldsymbol{r})|\Psi_A(N)\rangle = \sum_{k=1}^{N} |\phi_k(\boldsymbol{r})|^2, \\
\rho_2^{\mathrm{HF}}(\boldsymbol{r},\boldsymbol{r}') &= \langle \Psi_A(N)|\hat{\rho}_2(\boldsymbol{r},\boldsymbol{r}')|\Psi_A(N)\rangle \\
&= \rho^{\mathrm{HF}}(\boldsymbol{r})\left\{ \rho^{\mathrm{HF}}(\boldsymbol{r}') - \frac{1}{\rho^{\mathrm{HF}}(\boldsymbol{r})} \sum_{k=1}^{N}\sum_{l=1}^{N} [\phi_k^*(\boldsymbol{r})\phi_l(\boldsymbol{r})]\,[\phi_l^*(\boldsymbol{r}')\phi_k(\boldsymbol{r}')] \right\} \\
&\equiv \rho^{\mathrm{HF}}(\boldsymbol{r})\rho^{\mathrm{HF}}(\boldsymbol{r}'|\boldsymbol{r}) \equiv \rho^{\mathrm{HF}}(\boldsymbol{r})\left[\rho^{\mathrm{HF}}(\boldsymbol{r}') + h_x(\boldsymbol{r}'|\boldsymbol{r})\right].
\end{aligned} \tag{6.75}$$

Again, the HF conditional density $\rho^{\mathrm{HF}}(\boldsymbol{r}'|\boldsymbol{r}) = \rho_2^{\mathrm{HF}}(\boldsymbol{r},\boldsymbol{r}')/\rho^{\mathrm{HF}}(\boldsymbol{r})$ differs from the associated reference $\rho^{\mathrm{HF}}(\boldsymbol{r}')$ of the independent distribution, which does not eliminate SI, by the *exchange* (Fermi) *hole*:

$$h_x(\boldsymbol{r}'|\boldsymbol{r}) = -\frac{1}{\rho^{\mathrm{HF}}(\boldsymbol{r})} \sum_{k=1}^{N}\sum_{l=1}^{N} [\phi_k^*(\boldsymbol{r})\phi_l(\boldsymbol{r})]\,[\phi_l^*(\boldsymbol{r}')\phi_k(\boldsymbol{r}')]. \tag{6.76}$$

Its integration over positions of the dependent electron again gives:

$$\begin{aligned}
\int h_x(\boldsymbol{r}'|\boldsymbol{r})\,d\boldsymbol{r}' &= -\frac{1}{\rho^{\mathrm{HF}}(\boldsymbol{r})} \sum_{k=1}^{N}\sum_{l=1}^{N} [\phi_k^*(\boldsymbol{r})\phi_l(\boldsymbol{r})]\langle \phi_l \mid \phi_k\rangle \\
&= -\frac{1}{\rho^{\mathrm{HF}}(\boldsymbol{r})} \sum_{k=1}^{N} |\phi_k^*(\boldsymbol{r})|^2 = -1,
\end{aligned} \tag{6.77}$$

where we have recognized the orthonormality relations $\langle \phi_l|\phi_k\rangle = \delta_{l,k}$.

The above expression for the two-electron density in this approximation thus gives the following partition of the electron repulsion energy:

$$\begin{aligned}
V_{e,e}^{\mathrm{HF}}(N) &= \frac{1}{2}\iint \frac{\rho_2^{\mathrm{HF}}(\boldsymbol{r},\boldsymbol{r}')}{|\boldsymbol{r}-\boldsymbol{r}'|}\,d\boldsymbol{r}\,d\boldsymbol{r}' \\
&= \frac{1}{2}\iint \frac{\rho^{\mathrm{HF}}(\boldsymbol{r})\rho^{\mathrm{HF}}(\boldsymbol{r}')}{|\boldsymbol{r}-\boldsymbol{r}'|}\,d\boldsymbol{r}\,d\boldsymbol{r}' + \frac{1}{2}\iint \frac{\rho^{\mathrm{HF}}(\boldsymbol{r})h_x(\boldsymbol{r}'|\boldsymbol{r})}{|\boldsymbol{r}-\boldsymbol{r}'|}\,d\boldsymbol{r}\,d\boldsymbol{r}' \\
&= J[\rho^{\mathrm{HF}}] + E_x[\rho^{\mathrm{HF}}, h_x] = \frac{1}{2}\sum_{k=1}^{N}\sum_{l=1}^{N} J_{k,l} - \frac{1}{2}\sum_{k=1}^{N}\sum_{l=1}^{N} K_{k,l}.
\end{aligned} \tag{6.78}$$

The second term $E_x[\rho^{\mathrm{HF}}, h_x]$, called the *exchange energy*, which corrects the classical energy $J[\rho^{\mathrm{HF}}]$ for the Fermi correlation effects, also removes the *self-interaction* since $J_{k,k} = K_{k,k}$. Indeed, by separating in the exchange hole of (6.76), the orbital-diagonal, SI-part from the remaining interorbital (IOE), *exchange* contribution,

$$h_x(\boldsymbol{r}'|\boldsymbol{r}) = -\frac{1}{\rho^{\mathrm{HF}}(\boldsymbol{r})}\left\{\sum_{k=1}^{N}|\phi_k(\boldsymbol{r})|^2|\phi_k(\boldsymbol{r}')|^2 + \sum_{k,l\neq k}\left[\phi_k^*(\boldsymbol{r})\phi_l(\boldsymbol{r})\right]\left[\phi_l^*(\boldsymbol{r}')\phi_k(\boldsymbol{r}')\right]\right\}$$
$$\equiv h_{\mathrm{HF}}^{\mathrm{SI}}(\boldsymbol{r}'|\boldsymbol{r}) + h_{\mathrm{HF}}^{\mathrm{IOE}}(\boldsymbol{r}'|\boldsymbol{r}), \tag{6.79}$$

the exchange energy in (6.78) can be partitioned into the intraorbital SI- correction, and the IOE energy, respectively:

$$E_x[\rho^{\mathrm{HF}}, h_x] = \frac{1}{2}\iint\frac{\rho^{\mathrm{HF}}(\boldsymbol{r})h_{\mathrm{HF}}^{\mathrm{SI}}(\boldsymbol{r}'|\boldsymbol{r})}{|\boldsymbol{r}-\boldsymbol{r}'|}d\boldsymbol{r}\,d\boldsymbol{r}' + \frac{1}{2}\iint\frac{\rho^{\mathrm{HF}}(\boldsymbol{r})h_{\mathrm{HF}}^{\mathrm{IOE}}(\boldsymbol{r}'|\boldsymbol{r})}{|\boldsymbol{r}-\boldsymbol{r}'|}d\boldsymbol{r}\,d\boldsymbol{r}'$$
$$\equiv E_{\mathrm{SI}}\left[\rho^{\mathrm{HF}}, h_{\mathrm{HF}}^{\mathrm{SI}}\right] + E_{\mathrm{IOE}}\left[\rho^{\mathrm{HF}}, h_{\mathrm{HF}}^{\mathrm{IOE}}\right] = -\frac{1}{2}\sum_{k=1}^{N}J_{k,k}^{\mathrm{HF}} - \sum_{k=1}^{N-1}\sum_{l=k+1}^{N}K_{k,l}. \tag{6.80}$$

One similarly introduces the concept of the resultant *exchange-correlation* (*xc*) hole, combining changes in the conditional two-electron density due to both the Fermi and Coulomb correlations,

$$h_{xc}(\boldsymbol{r}'|\boldsymbol{r}) = h_x(\boldsymbol{r}'|\boldsymbol{r}) + h_c(\boldsymbol{r}'|\boldsymbol{r}). \tag{6.81}$$

It determines the displacements of the fully correlated two-electron density, e.g., that determined from CI calculations, relative to the independent electron reference:

$$\rho_2^{\mathrm{CI}}(\boldsymbol{r},\boldsymbol{r}') = \left\langle\Psi^{\mathrm{CI}}(N)\middle|\hat{\rho}_2(\boldsymbol{r},\boldsymbol{r}')\middle|\Psi^{\mathrm{CI}}(N)\right\rangle$$
$$\equiv \rho^{\mathrm{CI}}(\boldsymbol{r})\rho^{\mathrm{CI}}(\boldsymbol{r}'|\boldsymbol{r}) \equiv \rho^{\mathrm{CI}}(\boldsymbol{r})\left[\rho^{\mathrm{CI}}(\boldsymbol{r}') + h_{xc}^{\mathrm{CI}}(\boldsymbol{r}'|\boldsymbol{r})\right], \tag{6.82}$$

where $\rho^{\mathrm{CI}}(\boldsymbol{r}) = \left\langle\Psi^{\mathrm{CI}}(N)\middle|\hat{\rho}(\boldsymbol{r})\middle|\Psi^{\mathrm{CI}}(N)\right\rangle$ is the electron density in state $\Psi^{\mathrm{CI}}(N)$.

It follows from the normalization of $\rho_2^{\mathrm{CI}}(\boldsymbol{r},\boldsymbol{r}')$[see (3.10)],

$$\iint\rho_2^{\mathrm{CI}}(\boldsymbol{r},\boldsymbol{r}')\,d\boldsymbol{r}\,d\boldsymbol{r}' = N(N-1) = N^2 + \int\rho^{\mathrm{CI}}(\boldsymbol{r})\left[\int h_{xc}^{\mathrm{CI}}(\boldsymbol{r}'|\boldsymbol{r})\,d\boldsymbol{r}'\right]d\boldsymbol{r}, \tag{6.83}$$

that this equation can be satisfied only when the integral in square brackets gives

$$\int h_{xc}^{\mathrm{CI}}(\boldsymbol{r}'|\boldsymbol{r})\,d\boldsymbol{r}' = -1. \tag{6.84}$$

Therefore, since the $h_{xc}(\boldsymbol{r}'|\boldsymbol{r})$ satisfies the same sum rule as its exchange contribution (6.77), the corresponding "normalization" of the Coulomb hole requires:

$$\int h_c(\boldsymbol{r}'|\boldsymbol{r})\,d\boldsymbol{r}' = 0. \tag{6.85}$$

The corresponding expression for the electron repulsion energy then reads:

$$\begin{aligned} V_{e,e}^{\mathrm{CI}}(N) &= \frac{1}{2}\iint \frac{\rho_2^{\mathrm{CI}}(\boldsymbol{r},\boldsymbol{r}')}{|\boldsymbol{r}-\boldsymbol{r}'|}\,d\boldsymbol{r}\,d\boldsymbol{r}' = \frac{1}{2}\iint \frac{\rho^{\mathrm{CI}}(\boldsymbol{r})\rho^{\mathrm{CI}}(\boldsymbol{r}')}{|\boldsymbol{r}-\boldsymbol{r}'|}\,d\boldsymbol{r}\,d\boldsymbol{r}' \\ &+ \frac{1}{2}\iint \frac{\rho^{\mathrm{CI}}(\boldsymbol{r})h_{xc}^{\mathrm{CI}}(\boldsymbol{r}'|\boldsymbol{r})}{|\boldsymbol{r}-\boldsymbol{r}'|}\,d\boldsymbol{r}\,d\boldsymbol{r}' = J[\rho^{\mathrm{CI}}] + E_{xc}^{e,e}\left[\rho^{\mathrm{CI}}, h_{xc}^{\mathrm{CI}}\right], \end{aligned} \tag{6.86}$$

where $E_{xc}^{e,e}\left[\rho^{\mathrm{CI}}, h_{xc}^{\mathrm{CI}}\right]$ stands for the exchange-correlation contribution to the repulsion energy between electrons, expressed as functional of the resultant correlation hole.

It should be observed that this energy component is in fact determined by the *spherically averaged hole*. Define the relative separation between the reference and dependent electrons in the local spherical coordinate system centered at $\boldsymbol{r}$: $\boldsymbol{u} = \boldsymbol{r}' - \boldsymbol{r} = (u, \theta_{\boldsymbol{u}}, \varphi_{\boldsymbol{u}}) \equiv (u, \Omega_{\boldsymbol{u}})$, and introduce the spherically averaged hole:

$$h_{xc}^{\mathrm{CI}}(u|\boldsymbol{r}) = \frac{1}{4\pi}\int h_{xc}^{\mathrm{CI}}(\boldsymbol{r}+\boldsymbol{u}|\boldsymbol{r})\,d\Omega_{\boldsymbol{u}}. \tag{6.87}$$

The functional $E_{xc}^{e,e}\left[\rho^{\mathrm{CI}}, h_{xc}^{\mathrm{CI}}\right]$ of (6.86) can be then transformed into the associated functional of $h_{xc}^{\mathrm{CI}}(u|\boldsymbol{r})$:

$$\begin{aligned} E_{xc}^{e,e}\left[\rho^{\mathrm{CI}}, h_{xc}^{\mathrm{CI}}\right] &= \frac{1}{2}\iint \frac{\rho^{\mathrm{CI}}(\boldsymbol{r})h_{xc}^{\mathrm{CI}}(\boldsymbol{r}+\boldsymbol{u}|\boldsymbol{r})}{|\boldsymbol{u}|}\,d\boldsymbol{r}\,d\boldsymbol{u} \\ &= 2\pi\int_0^{\infty} u\int \rho^{\mathrm{CI}}(\boldsymbol{r})\,h_{xc}^{\mathrm{CI}}(u|\boldsymbol{r})\,d\boldsymbol{r}\,du. \end{aligned} \tag{6.88}$$

The spin-resolved correlation hole can be expressed in terms of the elementary spin-dependent contributions:

$$h_{xc}^{\mathrm{CI}}(\boldsymbol{q}'|\boldsymbol{q}) = h_x^{\mathrm{CI}}(\boldsymbol{q}'|\boldsymbol{q}) + h_c^{\mathrm{CI}}(\boldsymbol{q}'|\boldsymbol{q}) \equiv h_{xc}^{\sigma,\sigma'}(\boldsymbol{r}'|\boldsymbol{r}) = h_x^{\sigma,\sigma'}(\boldsymbol{r}'|\boldsymbol{r}) + h_c^{\sigma,\sigma'}(\boldsymbol{r}'|\boldsymbol{r}), \tag{6.89}$$

where $h_x^{\sigma,\sigma'}(\boldsymbol{r}'|\boldsymbol{r}) = h_x^{\sigma,\sigma}(\boldsymbol{r}'|\boldsymbol{r})\delta_{\sigma,\sigma'}$. They define the corresponding spin contributions to the two-electron density [see (3.11)]:

$$\begin{aligned} \rho_2^{\mathrm{CI}}(\boldsymbol{q},\boldsymbol{q}') &= \left\langle \Psi^{\mathrm{CI}}(N)\middle|\hat{\rho}_2(\boldsymbol{q},\boldsymbol{q}')\middle|\Psi^{\mathrm{CI}}(N)\right\rangle \\ &= \rho^{\mathrm{CI}}(\boldsymbol{q})\rho^{\mathrm{CI}}(\boldsymbol{q}'|\boldsymbol{q}) \equiv \rho^{\mathrm{CI}}(\boldsymbol{q})\left[\rho^{\mathrm{CI}}(\boldsymbol{q}') + h_{xc}^{\mathrm{CI}}(\boldsymbol{q}'|\boldsymbol{q})\right] \\ &\equiv \rho_2^{\sigma,\sigma'}(\boldsymbol{r},\boldsymbol{r}') = \rho^{\sigma}(\boldsymbol{r})\rho^{\sigma,\sigma'}(\boldsymbol{r}'|\boldsymbol{r}) \equiv \rho^{\sigma}(\boldsymbol{r})\left[\rho^{\sigma'}(\boldsymbol{r}') + h_{xc}^{\sigma,\sigma'}(\boldsymbol{r}'|\boldsymbol{r})\right], \end{aligned} \tag{6.90}$$

with the correlated spin density (3.9):

$$\rho^{\mathrm{CI}}(\boldsymbol{q}) = \langle \Psi^{\mathrm{CI}}(N)|\hat{\rho}(\boldsymbol{q})|\Psi^{\mathrm{CI}}(N)\rangle \equiv \rho^{\sigma}(\boldsymbol{r}), \quad \int \rho^{\sigma}(\boldsymbol{r})\,d\boldsymbol{r} = N^{\sigma}, \quad \sum\nolimits_{\sigma} N_{\sigma} = N. \tag{6.91}$$

It then follows from the alternative expressions for $\rho_2^{\mathrm{CI}}(\boldsymbol{r},\boldsymbol{r}')$,

$$\begin{aligned}\rho_2^{\mathrm{CI}}(\boldsymbol{r},\boldsymbol{r}') &= \rho^{\mathrm{CI}}(\boldsymbol{r})\left[\rho^{\mathrm{CI}}(\boldsymbol{r}') + h_{xc}^{\mathrm{CI}}(\boldsymbol{r}'|\boldsymbol{r})\right] = \sum_{\sigma}\sum_{\sigma'} \rho_2^{\sigma,\sigma'}(\boldsymbol{r},\boldsymbol{r}') \\ &= \sum_{\sigma}\sum_{\sigma'} \rho^{\sigma}(\boldsymbol{r})\left[\rho^{\sigma'}(\boldsymbol{r}') + h_{xc}^{\sigma,\sigma'}(\boldsymbol{r}'|\boldsymbol{r})\right] \\ &= \rho^{\mathrm{CI}}(\boldsymbol{r})\rho^{\mathrm{CI}}(\boldsymbol{r}') + \sum_{\sigma}\sum_{\sigma'} \rho^{\sigma}(\boldsymbol{r})h_{xc}^{\sigma,\sigma'}(\boldsymbol{r}'|\boldsymbol{r}),\end{aligned} \tag{6.92}$$

that the spinless hole is given by the following combination of the spin-resolved components:

$$h_{xc}^{\mathrm{CI}}(\boldsymbol{r}'|\boldsymbol{r}) = \sum_{\sigma}\sum_{\sigma'} \frac{\rho^{\sigma}(\boldsymbol{r})}{\rho^{\mathrm{CI}}(\boldsymbol{r})} h_{xc}^{\sigma,\sigma'}(\boldsymbol{r}'|\boldsymbol{r}). \tag{6.93}$$

The relevant sum rules for the *exchange* spin holes read:

$$\begin{aligned}&\int h_{x}^{\sigma,\sigma'}(\boldsymbol{r}'|\boldsymbol{r})\,d\boldsymbol{r}' = \int h_{xc}^{\sigma,\sigma'}(\boldsymbol{r}'|\boldsymbol{r})\,d\boldsymbol{r}' = -\delta_{\sigma,\sigma'} \quad \text{or} \\ &\sum_{\sigma'}\int h_{xc}^{\sigma,\sigma'}(\boldsymbol{r}'|\boldsymbol{r})\,d\boldsymbol{r}' = \int h_{xc}^{\mathrm{CI}}(\boldsymbol{q}'|\boldsymbol{q})\,d\boldsymbol{q}' = \sum_{\sigma'}\int h_{x}^{\sigma,\sigma'}(\boldsymbol{r}'|\boldsymbol{r})\,d\boldsymbol{r}' = \int h_{x}^{\mathrm{CI}}(\boldsymbol{q}'|\boldsymbol{q})\,d\boldsymbol{q}' = -1,\end{aligned} \tag{6.94}$$

while the overall normalization of the *correlation* spin holes requires:

$$\sum_{\sigma'}\int h_{c}^{\sigma,\sigma'}(\boldsymbol{r}'|\boldsymbol{r})\,d\boldsymbol{r}' = \int h_{c}^{\mathrm{CI}}(\boldsymbol{q}'|\boldsymbol{q})\,d\boldsymbol{q}' = 0. \tag{6.95}$$

Moreover, since the *pair* density $\rho_2(\boldsymbol{q},\boldsymbol{q}')$ is symmetrical with respect to exchanging the spin-position coordinates of two (indistinguishable) electrons, $\rho_2(\boldsymbol{q},\boldsymbol{q}') = \rho_2(\boldsymbol{q}',\boldsymbol{q})$ this invariance property must be also reflected by its hole contributions:

$$\rho(\boldsymbol{q})h_{\lambda}(\boldsymbol{q}'|\boldsymbol{q}) = \rho(\boldsymbol{q}')h_{\lambda}(\boldsymbol{q}|\boldsymbol{q}') \quad \text{or} \quad \rho^{\sigma}(\boldsymbol{r})h_{\lambda}^{\sigma,\sigma'}(\boldsymbol{r}'|\boldsymbol{r}) = \rho^{\sigma'}(\boldsymbol{r}')h_{\lambda}^{\sigma',\sigma}(\boldsymbol{r}|\boldsymbol{r}'), \quad \lambda = xc, x, c. \tag{6.96}$$

The hole distribution $h(\boldsymbol{r}'|\boldsymbol{r}) = \rho_2(\boldsymbol{r},\boldsymbol{r}')/\rho(\boldsymbol{r})$ introduces a nonsymmetrical treatment of the "reference" electron at $\boldsymbol{r}$ and the "dependent" electron at $\boldsymbol{r}'$. Alternatively, one can use the *pair correlation function*,

$$f(\boldsymbol{r},\boldsymbol{r}') = \rho_2(\boldsymbol{r},\boldsymbol{r}')/[\rho(\boldsymbol{r})\rho(\boldsymbol{r}')], \tag{6.97}$$

in which both electrons are treated symmetrically. Hence, from the equality $\rho_2(\boldsymbol{r},\boldsymbol{r}') = f(\boldsymbol{r},\boldsymbol{r}')\rho(\boldsymbol{r})\rho(\boldsymbol{r}') = \rho(\boldsymbol{r})[\rho(\boldsymbol{r}') + h(\boldsymbol{r}'|\boldsymbol{r})]$, one finds:

$$h(\boldsymbol{r}'|\boldsymbol{r}) = \rho(\boldsymbol{r}')[f(\boldsymbol{r},\boldsymbol{r}') - 1] \quad \text{or} \quad f(\boldsymbol{r},\boldsymbol{r}') = 1 + h(\boldsymbol{r}'|\boldsymbol{r})/\rho(\boldsymbol{r}'). \tag{6.98}$$

The spherically averaged *pair* correlation function reads:

$$f(\boldsymbol{r},u) = \frac{1}{4\pi}\int f(\boldsymbol{r},\boldsymbol{r}+\boldsymbol{u})\, d\Omega_{\boldsymbol{u}}. \tag{6.99}$$

In terms of the corresponding spin-resolved distributions,

$$\begin{aligned} f_{xc}(\boldsymbol{q},\boldsymbol{q}') &= \frac{\rho_2(\boldsymbol{q},\boldsymbol{q}')}{\rho(\boldsymbol{q})\rho(\boldsymbol{q}')} \equiv f_{xc}^{\sigma,\sigma'}(\boldsymbol{r},\boldsymbol{r}') = \frac{\rho_2^{\sigma,\sigma'}(\boldsymbol{r},\boldsymbol{r}')}{\rho^{\sigma}(\boldsymbol{r})\rho^{\sigma'}(\boldsymbol{r}')} \\ &= f_x(\boldsymbol{q},\boldsymbol{q}') + f_c(\boldsymbol{q},\boldsymbol{q}') \equiv f_x^{\sigma,\sigma'}(\boldsymbol{r},\boldsymbol{r}') + f_c^{\sigma,\sigma'}(\boldsymbol{r},\boldsymbol{r}'), \end{aligned} \tag{6.100}$$

the spinless *pair* correlation function of (6.97) then reads:

$$f(\boldsymbol{r},\boldsymbol{r}') = \sum_{\sigma}\sum_{\sigma'} \frac{1}{\rho(\boldsymbol{r})}\rho_2^{\sigma,\sigma'}(\boldsymbol{r},\boldsymbol{r}')\frac{1}{\rho(\boldsymbol{r}')} = \sum_{\sigma}\sum_{\sigma'} \frac{\rho^{\sigma}(\boldsymbol{r})}{\rho(\boldsymbol{r})} f^{\sigma,\sigma'}(\boldsymbol{r},\boldsymbol{r}')\frac{\rho^{\sigma'}(\boldsymbol{r}')}{\rho(\boldsymbol{r}')}. \tag{6.101}$$

As shown by Kato (1957) the Coulomb interaction between the two electrons implies in the near coalescence limit $\boldsymbol{r}' \to \boldsymbol{r}$, or $u \to 0$, the condition of the so-called *correlation cusp*:

$$f'(\boldsymbol{r},\boldsymbol{r}) \equiv \left.\frac{\partial f(\boldsymbol{r},u)}{\partial u}\right|_{u=0} = f(\boldsymbol{r},\boldsymbol{r}). \tag{6.102}$$

This condition supplements the *nuclear cusp* relation, for the coalescence of electron and the nucleus, when $\boldsymbol{r}_\alpha = \boldsymbol{r} - \boldsymbol{R}_\alpha = (r_\alpha, \Omega_\alpha) \to 0$, where $\boldsymbol{R}_\alpha$ stands for the position of nucleus α, exhibiting the electric charge Z_α (a.u.), formulated in terms of the spherically averaged electron density

$$\rho(r_\alpha) = \frac{1}{4\pi}\int \rho(\boldsymbol{r}_\alpha)\, d\Omega_\alpha, \tag{6.103}$$

$$\rho'(r_\alpha \to 0) \equiv \left.\frac{\partial \rho(r_\alpha)}{\partial r_\alpha}\right|_{r_\alpha=0} = -2Z\rho(r_\alpha = 0). \tag{6.104}$$

Consider, for example, the ground state of the hydrogen-like atom (4.62), for which the spherically symmetrical density $\rho(r) = (Z^3/\pi)\exp(-2Zr)$ and hence

$$\rho'(r \to 0) \equiv \left.\frac{\partial \rho(r)}{\partial r}\right|_{r=0} = -2Z\rho(r=0) = -\frac{2Z^4}{\pi}. \qquad (6.105)$$

In terms of the spin components of the exchange and correlation distribution functions, the correlation cusp reads:

$$f_x{}'[(\boldsymbol{r},\sigma),(\boldsymbol{r},\sigma')] = 0, \quad f_c{}'[(\boldsymbol{r},\sigma),(\boldsymbol{r},\sigma')] = (1-\delta_{\sigma,\sigma'})f_c[(\boldsymbol{r},\sigma),(\boldsymbol{r},\sigma')]. \quad (6.106)$$

Thus, it is the Coulomb correlation between electrons which constitutes the exclusive origin of the correlation cusp of the *pair* correlation function. Indeed, it directly follows from the Pauli antisymmetrization principle that the *near*-coalescence events of two electrons with the parallel spins are excluded, thus giving rise to the vanishing cusp in the exchange contributions to the pair correlation function. Moreover, the nonvanishing cusp is seen to appear only for electrons exhibiting the antiparallel spins. This is also in accord with the above implications of the exclusion principle, since as a result of the requirements of the Fermi statistics the *near*-coalescence of two electrons with identical spins cannot occur due to the exchange correlation.

6.3 Configuration Interaction Techniques

In Sect. 6.2. we have hinted upon a systematic way to include the effects of the Coulomb correlation in the variational N-electron wave function by mixing the HF determinant with the excited configurations obtained by replacing a single or several HF SO by orbitals which remain unoccupied in the ground state. The formal basis for such beyond-HF, SR CI approach was also given in the form of the associated expansion theorem: any N-electron state can be expanded as a combination of all Slater determinants, which can be formed by distributing N electrons among different subsets of the complete basis of SO. Therefore, the FCI scheme in principle offers the exact description of all stationary states of molecular systems, should the set of AO in SCF calculations be large enough to generate a nearly complete basis of MO.

Such *first* principle (ab initio) theories using the CI expansion of *many*-electron wave functions can be developed in both the perturbational and variational formulations. The former methods fall into the category of the *Many-Body Perturbation Theory* (MBPT) (Møller and Plessett 1934; Mattuck 1976; Szabo and Ostlund 1982), while the latter methods are classified as the *Configuration Interaction* (CI) theories (Löwdin 1959; Shavitt 1977). The MBPT variants were shown by Brueckner and Goldstone, using the diagrammatic representation of Feynmann, to be size consistent at any order of the truncation (e.g., Brandow 1967; Freed 1971; Manne 1977; Bartlett 1981, Paldus and Čižek 1975). Both these approaches, at least in principle, are capable to deliver the calculated values of the physical observables to any desired accuracy. We shall briefly summarize in this section some of the most popular variational realizations of the CI techniques, which use a limited CI expansion, truncated due to numerical difficulties involved in the FCI scheme (see Sect. 6.2.2). The necessity to use a limited variety of electron excitations immediately arises, when one intends to apply these methods to large molecules of interest in the contemporary chemistry. For an overview of the wave function techniques of tackling the Coulomb correlation problem the reader is also referred to monographs by Szabo and Ostlund (1982), Christoffersen (1989), and McWeeny (1989).

The SR CI approach relies on the expansion of the current nth electronic state of interest, $\Psi_n^{\mathrm{CI}}(N)$, $n = 0, 1, 2, \ldots$, in terms of the system orthonormal configurations $\{\Phi_s(N), s = 1, 2, \ldots, n_c\}$ [see (6.66)],

$$\Psi_n^{\mathrm{CI}}(N) = \sum_{s=1}^{n_c} \Phi_s(N) C_{s,n}^{\mathrm{CI}}, \quad \langle \Phi_s \mid \Phi_t \rangle = S_{s,t}^{\mathrm{CI}} = \delta_{s,t}, \tag{6.107}$$

grouping in the FCI scheme all n_c antisymmetrized products (Slater determinants) $\{\Phi_s(N) = \det[\boldsymbol{\varphi}_s(N)\}$ of alternative choices of N one-electron functions (SO), $\{\boldsymbol{\varphi}_s(N) = (\phi_{1,s}, \phi_{2,s}, \ldots, \phi_{N,s}\}$, from the whole set of w canonical MO, $\boldsymbol{\varphi} = \{\phi_1, \phi_2, \ldots, \phi_w\}$, determined in the SCF MO calculations in the basis set of the same size, $\boldsymbol{\varphi} = \boldsymbol{\chi}\mathbf{C}$, $\boldsymbol{\chi} = \{\chi_1, \chi_2, \ldots, \chi_w\}$ (6.66).

This expansion may also involve appropriate linear combinations of these determinants, the eigenfunctions of $\hat{\mathrm{S}}^2$ and $\hat{\mathrm{S}}_z$ [see (6.67) and Fig. 6.2] or functions of the specified spatial symmetry type. Indeed, the true configuration functions of the appropriate spatial and spin symmetries in general require linear combinations of several Slater determinants (see, e.g., Fig. 6.2). Therefore, it is only for simplicity of presentation that we call single determinants of (6.109) and (6.110) as "configurations."

The elementary Slater determinants of the excited configurations can be envisaged as resulting from the appropriate electron excitations from the ground state determinant of the HF method,

$$\begin{aligned} \Phi_1(N) &\equiv \Psi_0^{\mathrm{HF}}(N) = |\phi_{1,1}, \phi_{2,1} \ldots, \phi_{N,1}| = \det[\boldsymbol{\varphi}_1(N)] \\ &= |\phi_1, \phi_2, \ldots, \phi_N|, \end{aligned} \tag{6.108}$$

by replacing a subset of the occupied SO in Φ_1, $\boldsymbol{\varphi}^{occd.} = \boldsymbol{\varphi}_1 = \{\phi_k\}$, by an equally numerous subset of the virtual SO, absent in Φ_1, $\boldsymbol{\varphi}^{virt.} = \{\phi_{N+1}, \phi_{N+2}, \ldots, \phi_w\} = \{\phi_p\}$. Thus, in the *singly* excited ($k \to p$) configuration Φ_k^p, the occupied SO ϕ_k has been replaced by the virtual SOϕ_p in the list of the occupied one-electron states:

$$\Phi_k^p = \det|\phi_1, \phi_2, \ldots, \phi_{k-1}, \phi_{k+1}, \ldots, \phi_N, \phi_p| \equiv \Phi[k \to p]. \tag{6.109}$$

Accordingly, in the *doubly* excited configuration $\Phi_{k<l}^{p<q} = \Phi[(k,l) \to (p,q)]$, the indicated pair of SO has been replaced, $(\phi_k,\phi_l) \to \{\phi_p,\phi_q)$,

$$\Phi_{k<l}^{p<q} = \det|\phi_1, \phi_2, \ldots, \phi_{k-1}, \phi_{k+1}, \ldots, \phi_{l-1}, \phi_{l+1}, \ldots, \phi_N, \phi_p, \phi_q|. \quad (6.110)$$

Extension to the *triple*, *quadruple*, and higher excitations is obvious. In this notation the CI expansion of (6.62) and (6.107) reads:

$$\Psi_n^{\mathrm{CI}} = C_0(n)\Psi_0^{\mathrm{HF}} + \sum_k^{occd.}\sum_p^{virt.} C_k^p(n)\,\Phi_k^p + \sum_{k<l}^{occd.}\sum_{p<q}^{virt.} C_{k,l}^{p,q}(n)\,\Phi_{k,l}^{p,q}$$
$$+ \sum_{k<l<m}^{occd.}\ \sum_{p<q<r}^{virt.} C_{k,l,m}^{p,q,r}(n)\,\Phi_{k,l,m}^{p,q,r} + \ldots \quad (6.111)$$

The optimum (linear) variational parameters of the CI method, $\mathbf{C}^{\mathrm{CI}} = \left\{C_{s,t}^{\mathrm{CI}} \equiv C_s(t)\right\}$, which determine the unknown molecular states $\mathbf{\Psi}^{\mathrm{CI}} = \mathbf{\Phi}\mathbf{C}^{\mathrm{CI}} = \{\Psi_n\}$ in the (orthonormal) basis of the N-electron configuration functions $\mathbf{\Phi} = \{\Phi_s(N)\}$, $\mathbf{S}^{\mathrm{CI}} = \left\{S_{s,t}^{\mathrm{CI}}\right\} = \mathbf{I}^{\mathrm{CI}}$, and the associated energies $\mathbf{E}^{\mathrm{CI}} = \left\{E_n^{\mathrm{CI}}\delta_{n,m}\right\}$ are thus determined by the corresponding secular equations of the Ritz method (5.37), $\mathbf{H}^{\mathrm{CI}}\mathbf{C}^{\mathrm{CI}} = \mathbf{C}^{\mathrm{CI}}\mathbf{E}^{\mathrm{CI}}$, or

$$\mathbf{E}^{\mathrm{CI}} = (\mathbf{C}^{\mathrm{CI}})^\dagger\mathbf{H}^{\mathrm{CI}}\mathbf{C}^{\mathrm{CI}}, \qquad (\mathbf{C}^{\mathrm{CI}})^\dagger\mathbf{C}^{\mathrm{CI}} = \mathbf{I}^{\mathrm{CI}}. \quad (6.112)$$

They are solved by the diagonalization of the energy matrix:

$$\mathbf{H}^{\mathrm{CI}} = \left\{\langle\Phi_s(N)|\hat{\mathrm{H}}^e(N)|\Phi_t(N)\rangle = H_{s,t}^{\mathrm{CI}}\right\} \quad (6.113)$$

in the unitary transformation $\mathbf{C}^{\mathrm{CI}}$. Thus, the typical limited-CI calculations in principle consist of determining the set of the orthonormal MO, evaluating integrals in this MO basis, selecting the most important configurations, generating the energy matrix in the adopted configuration basis and solving the secular equations to determine energy levels and the associated wave functions.

The *MacDonald theorem* (see, e.g., Hylleraas and Undheim 1930; MacDonald, 1933; Shull and Löwdin 1958) of the linear variational method states that the ordered (nondegenerate) CI eigenvalues, $E_1^{\mathrm{CI}} < E_2^{\mathrm{CI}} < \ldots < E_{n_c}^{\mathrm{CI}}$, provide the upper bounds to the corresponding exact values of the electronic energy in the molecular ground and excited states, $\{E_n^e\} = \left(E_1^e < E_2^e < \ldots < E_{n_c}^e\right)$:

$$E_n^{\mathrm{CI}} \geq E_n^e, \quad n = 1, \ldots, n_c. \quad (6.114)$$

It also proves that more refined calculations, with successively enlarged basis set of excited configurations, continue to improve these energy estimates:

$$E_n^{\mathrm{CI}}(n_c) \geq E_n^{\mathrm{CI}}(n_c+1) \geq E_n^{\mathrm{CI}}(n_c+2)\ldots \geq E_n^e, \quad (6.115)$$

where $E_n^{\mathrm{CI}}(n_c)$ denotes the energy level from the CI expansion in terms of n_c configurations.

6.3.1 Special Variants of Limited CI

It directly follows from the *closed*-shell HF equations (6.21) and the Slater–Condon rule (5.85) that the matrix elements of $\mathbf{H}^{\mathrm{CI}}$ coupling the ground state (6.108) with the *singly* excited configurations $\{\Phi_k^p\}$ identically vanish:

$$\begin{aligned}\langle\Phi_1|\hat{\mathrm{H}}^e|\Phi_k^p\rangle &= H_{1,k\to p}^{\mathrm{CI}}\\ &= \langle k|\,\hat{\mathrm{h}}\,|p\rangle + \sum_{j=1}^{N}\langle kj|\,g\,|pj - jp\rangle = \langle k|\,\hat{\mathrm{F}}\,|p\rangle = e_p\langle k\mid p\rangle = 0. \quad (6.116)\end{aligned}$$

It indicates that the *singly*-excited configurations alone cannot improve the *closed*-shell HF wave function. This statement, also known as the *Brillouin theorem* (Brillouin 1933, 1934), is no longer satisfied for the *open*-shell ground state.

However, since by the Slater–Condon rules the *single* excitations directly couple to the *double* and *triple* excitations, their presence is generally required and beneficiary, when these higher excitations are also included in the CI expansion [see, e.g., (6.111)]. Nevertheless, the Brillouin theorem indicates the *indirect*, secondary role played by the *single* excitations in representing the ground state Coulomb correlation, when the HF (closed-shell) ground state is used as the reference function, compared with the *direct*, primary role played by the dominating *double* excitations.

Therefore, in determining the Coulomb correlation effects in such molecular ground state, the limited CI approaches must take into account at least double excitations. For example, the CI-*Doubles* (CID) scheme, intended to remedy limitations of HF theory in the ground state applications, uses the variational wave function in the form of the linear combination of the HF determinant and the *doubly* excited configurations:

$$\Psi_0^{\mathrm{CID}} = C_0\Psi_0^{\mathrm{HF}} + \sum_{k<l}^{occd.}\sum_{p<q}^{virt.} C_{k,l}^{p,q}\,\Phi_{k,l}^{p,q}. \quad (6.117)$$

In the CI expansions of excited states, which are dominated by some excited configuration derived from Ψ_0^{HF}, the *singly* excited configurations directly influence the wave function thus being equally important as *double* excitations in the approximate representations of such states. This is the reason why the other popular variant, called CI-(*Singles* and *Doubles*) (CISD) of the limited CI expansion, additionally includes the single excitations in the variational wave function:

$$\Psi_n^{\mathrm{CISD}} = C_0(n)\Psi_0^{\mathrm{HF}} + \sum_{k}^{occd.}\sum_{p}^{virt.} C_k^p(n)\Phi_k^p + \sum_{k<l}^{occd.}\sum_{p<q}^{virt.} C_{k,l}^{p,q}(n)\Phi_{k,l}^{p,q}, \quad n=0,1,\ldots \quad (6.118)$$

In the SR CI approaches, the optimum SO are calculated only once, at the preceding HF stage, with the CI coefficients (linear variational parameters) being subsequently determined for the fixed ("frozen") MO basis, including both the HF occupied and virtual MO. This approach gives rise to a relatively slow convergence of such a *linear* CI representation of the electron correlation effects. It can be improved decisively by allowing the shapes of orbitals, which determine the configuration functions, to adjust to the current form of the CI wave function, with MO being optimized at each iteration simultaneously with the CI coefficients themselves. Such a generalized MC SCF (or MR SCF) technique, which is usually applied to the ground state problems, adopts the CID trial function of (6.117) which guarantees the correct description of the alternative bond-forming/bond-breaking phenomena, with the MO used in all configurations treated as optimized entities in the generalized iteration scheme of the MC SCF process combining the orbital optimization (SCF) and CI stages:

$$\begin{aligned} \dots \rightarrow \left[\mathrm{SCF}_{(k)}\left(\mathbf{C}^{(\mathrm{CI})}_{(k-1)}\right) \rightarrow \boldsymbol{\varphi}_{(k)}\right] \rightarrow \left[\mathrm{CI}_{(k)}(\boldsymbol{\varphi}_{(k)}) \rightarrow \mathbf{C}^{(\mathrm{CI})}_{(k)}\right] \\ \rightarrow \left[\mathrm{SCF}_{(k+1)}\left(\mathbf{C}^{(\mathrm{CI})}_{(k)}\right) \rightarrow \boldsymbol{\varphi}_{(k+1)}\right] \rightarrow \dots \end{aligned}$$

Here $\mathbf{C}^{(\mathrm{CI})}_{(k)}$ and $\boldsymbol{\varphi}_{(k)}$ denote the optimum CI coefficients and MO of the *k*th iteration of such self-consistent optimization process. At its SCF stage, the current shapes of orbitals, i.e., the LCAO MO coefficients, are being determined, using the optimum CI coefficients from the previous iteration, and then the next approximation of $\mathbf{C}^{\mathrm{CI}}$ is generated at the CI stage, using the modified MO obtained from the preceding SCF procedure. The first, SCF stage uses the appropriately generalized HF equations to determine the MO coefficients, while the CI stage involves solving the secular equations (6.112) for the CI coefficients. The Fock operator in the MC SCF method depends on the effective occupancies of orbitals in the CID configurations, thus explicitly depending on the CI coefficients themselves. This full optimization of the CID wave function, which minimizes the electronic energy, is carried out until both the shapes of MO and the CI coefficients in the next iteration agree with those of the preceding iteration, thus determining the doubly self-consistent CI state, in both CI coefficients and MO shapes.

These variational, limited CI approaches suffer from severe *size*-consistency and *size*-extensivity problems. The former concept of the method adequacy deals with the requirement that its estimate of the energy of the system consisting of two infinitely distant (noninteracting) subsystems, say [He + He], must be equal to the sum of energies of the separate subsystems calculated by the same method: $E^e[\mathrm{He} + \mathrm{He}] = 2E^e[\mathrm{He}]$. In other words, the energy of the "dimer" composed of two noninteracting monomers must be twice the energy of the "monomer" and hence the energy of the system being dissociated into subsystems reaches in the limit of their infinite separation the sum of energies of the separate subsystems.

This requirement is satisfied by HF theory: the HF energy of a supermolecule composed of two noninteracting closed-shell subsystems is just the sum of the subsystem HF energies. Unfortunately, at the CI level of theory, this intuitively obvious postulate of the energy additivity is satisfied only by the *full*-CI approach, being violated by all variational methods using a *limited* CI expansion. To illustrate the problem, we observe that the CISD applied to each separate part generates the triple (T) and quadruple (Q) excitations of the system as a whole, since the wave function of the combined system (*supermolecule*) is then given by the product of the subsystem wave functions. Therefore, the configurations appearing at the CISD treatment of fragments are included only at the different CISDTQ levels of the theory for the whole molecule. This indicates that the dimer wave function truncated at CID level does not have sufficient flexibility to generate twice the CID energy of the monomer, thus confirming that this truncated CI scheme does not have the property of size consistency. In principle, no form of limited CI is size consistent. However, CI scheme including quadruple excitations has been shown to be approximately size consistent for small molecules.

To remedy this problem Davidson (Davidson 1974; Langhoff and Davidson 1974) has suggested the following corrected estimate of the ground state correlation energy:

$$\Delta E_{corr.} = \left(1 - C_0^2\right)\Delta E_{corr.}^{\text{CISD}}, \tag{6.119}$$

where $\Delta E_{corr.}^{\text{CISD}}$ stands for its CISD value and C_0^2 determines the participation of Ψ_0^{HF} configuration in the normalized expansion of (6.118). The projected correlation energies per electronic pair were observed to stay remarkably constant, at ~0.042 a.u. (≈1.14 eV).

The related notion of *size* extensivity, which first arose in the nuclear and solid state physics, refers to the method scaling with the number of correlated particles. Indeed, the appropriate scaling of the atomic results to an infinite system is required before the results obtained for an isolated atom can be used in a description of a solid containing an infinite number of atoms. This requirement refers to increasing the size of the "continuous" system, while keeping the particle density constant, e.g., that of the *free* electron gas. When the size of the system is doubled under such constraint, the total energy must be also doubled, thus being proportional to the number of particles N. In other words, the correlation energy per particle should be independent of N.

Thus, in the molecular scenario, the concept of extensivity does not apply to a single atom or molecule. This requirement represents a valid property, however, when the system consisting of many (*weakly* interacting or noninteracting) replicas of a molecule ("monomers"), e.g., H_2, are considered. Then the system energy represents to a good approximation the extensive property, and hence the correlation energy per electron should be expected to be conserved. For example, the HF energy of a crystal is proportional to the number N of constituent molecules, although it is not simply N times the energy of an isolated molecule. We further observe that the MBPT approaches, which we shall discuss in the next section, are both the size consistent and size extensive. The MR-SCF and MR-CI procedures also give rise to an approximate fulfillment of these requirements already at the CISD level.

To conclude this short overview of the limited-CI techniques, we also mention the problem of the most effective orbitals for the CI studies, which will be tackled in the subsequent section. Let us only observe here that the occupied HF orbitals are each determined in an effective field of $N - 1$ electrons, while the corresponding virtual orbitals are determined in an effective field of all N electrons. As a result the *low*-energy virtual MO are generally more diffused, especially in large basis sets, than their filled counterparts. This discrepancy causes a generally low effectiveness of the virtual MO when used as *correlation orbitals*, giving rise to generally slow convergence of the associated CI representation of the correlated wave functions.

Clearly, it would be physically more proper if the correlation virtual orbitals would also "feel" an $N - 1$ electron potential, since the optimum orbitals for correlation purposes should be located in the same regions of space as the electrons whose motion they intend to correlate. One partial solution to this problem is to use the orbitals determined in the MC SCF (small CI) problem, for setting up the MR CI matrix problem, dimension of which is much larger than that of the MC SCF calculation. Indeed, the major orbital relaxation effects would then be already accounted for in the following CI stage. In Sect. 6.3.3 we shall introduce the concept of *natural orbitals*, which give rise to the fastest convergence of the CI expansion and generate a significant physical insight into both the most important correlation mechanisms and the nature of the chemical bond.

6.3.2 Perturbational Theory of Møller and Plesset

The variational bounds for the system total electronic energy are of little value when one calculates the energy differences in which chemistry abounds. For this purpose it is essential to eliminate such nonsystematic errors as a lack of *size* consistency/extensivity and the *Basis Set Superposition Error* (BSSE), which may falsify the predicted structural preferences and reactivity trends, even qualitatively. Since for the solid state physicists, interested in systems of infinite size, the elimination of the former was essential, they have developed MBPT approaches (Brueckner 1955; Brueckner et al. 1955; Brueckner and Levinson 1955; Goldstone 1957; Kelly 1969) or Coupled Cluster (CC) theory (Coester 1958; Coester and Kümmel 1960; Kümmel 1969; Paldus and Čižek 1975), which use the standard Rayleigh–Schrödinger PT coupled with powerful diagrammatic techniques to extract Coulomb correlation corrections from the known HF solutions (Brandow 1967; Freed 1971; Manne 1977; Bartlett 1981; Paldus and Čižek 1975; Čižek and Paldus 1980; Szabo and Ostlund 1982). This theory is wholly satisfactory for molecular structures in the vicinity of the equilibrium geometry, but at present does not provide a useful tool for calculations of the complete PES of chemical reactions.

Such perturbation theory approaches are free from the *size*-consistency problem in all orders of PT. Indeed, let us recall that the FCI scheme, which can be regarded as the *infinite*-order PT, is also free from this problem. In the perturbational treatment this can be the case only when each of n-order corrections also exhibits this property. As also shown by Goldstone, the *finite*-order perturbational estimates of the correlation energy are also size extensive, exhibiting the linear increase with the number of electrons for systems consisting a large number of weakly interacting atoms or molecules: $\Delta E_{corr.}(N) \propto N$, which thus makes them particularly suitable and attractive for applications to infinite systems. This result is due to the exact cancelation in this version of PT of the so-called *unlinked* cluster contributions that are not N-proportional.

We shall now briefly examine the simplest formulation of MBPT proposed by Møller and Plessett (1934), called the MP theory, in which the *zeroth*-order Hamiltonian of N electrons is defined by the sum of the effective Fock operators (6.22) for each electron:

$$\hat{\mathrm{H}}_0^{\mathrm{MP}}(N) = \sum_{i=1}^{N} \hat{\mathrm{F}}(i), \quad \hat{\mathrm{F}}(i) = -\frac{1}{2}\Delta_i + V_{\mathrm{HF}}(\boldsymbol{r}_i), \quad \hat{\mathrm{F}}(\boldsymbol{r})\,\phi_k(\boldsymbol{r}) = e_k\,\phi_k(\boldsymbol{r}). \quad (6.120)$$

This choice of the unperturbed Hamiltonian for the correlation problem is indeed justified by the fact that the configurations appearing in the CI expansion of (6.107) and (6.111) are themselves eigenfunctions of this operator:

$$\hat{\mathrm{H}}_0^{\mathrm{MP}}(N)\Phi_s(N) = \left(\sum_{k=1}^{occd[s]} e_k\right)\Phi_s(N) \equiv E_s^{(0)}\Phi_s(N), \qquad s = 1,\, 2,\, \ldots \quad (6.121)$$

with the eigenvalue $E_s^{(0)}$ given by the sum of orbital energies of all occupied SO in $\Phi_s(N)$. Thus, the lowest eigenvalue corresponding to the HF ground state (6.108) is different from the corresponding HF energy [see (6.14) and (6.23)]:

$$E_0^{\mathrm{HF}} = E_1^{(0)} - \frac{1}{2}\sum_{k=1}^{N}\sum_{l=1}^{N}\{\bar{J}_{k,l} - \bar{K}_{k,l}\}, \quad (6.122)$$

due to a double counting of the electron repulsion terms in $E_1^{(0)}$.

The N-electron perturbation in MP approach, called the correlation (fluctuation) operator, is defined as the difference between the electronic Hamiltonian and this unperturbed effective Hamiltonian of the independent-particle (HF) model:

$$\hat{\mathrm{h}}^{corr.}(N) = \hat{\mathrm{H}}^{e}(N) - \hat{\mathrm{H}}_0^{\mathrm{MP}}(N) = \sum_{i<j}\frac{1}{r_{i,j}} - \sum_{i} V_{\mathrm{HF}}(i). \quad (6.123)$$

Indeed, this correlation operator carries only that part of the interelectronic interactions which has not been included in the averaged (effective) interactions of the resultant potential $V_{\mathrm{HF}}(\boldsymbol{r})$ combining the Coulomb and exchange operators defining the Fock operator.

For simplicity let us assume that we are interested in the Coulomb correlation correction to the (*closed*-shell) ground state of the system, $s = 1$ [see (6.108)].

In accordance with the Rayleigh-Schrödinger theory of Sect. 5.1.1, the consecutive correlation corrections to the unperturbed eigenvalue $E_1^{(0)} = \sum_{k=1}^{N} e_k$, and the associated eigenfunction Φ_1 (6.108), determine the corresponding ground state solutions of the perturbed (correlated) Hamiltonian

$$\hat{\mathrm{H}}_\lambda^{\mathrm{MP}}(N) = \hat{\mathrm{H}}_0^{\mathrm{MP}}(N) + \lambda \hat{\mathrm{h}}^{corr.}(N), \tag{6.124}$$

where $\hat{\mathrm{H}}_{\lambda=0}^{\mathrm{MP}}(N) = \hat{\mathrm{H}}_0^{\mathrm{MP}}(N)$ and $\hat{\mathrm{H}}_{\lambda=1}^{\mathrm{MP}}(N) = \hat{\mathrm{H}}^e(N)$:

$$\begin{aligned} E_\lambda &= \sum_{j=0}^{\infty} \lambda^j E_1^{(j)} \equiv E_1^{(0)} + \Delta E_1^{(1)} + \Delta E_1^{(2)} + \ldots \\ &= E_0^{\mathrm{HF}} + \Delta E_0^{\mathrm{HF}(1)} + \Delta E_0^{\mathrm{HF}(2)} + \ldots \end{aligned} \tag{6.125}$$

$$\Psi_\lambda = \sum_{i=0}^{\infty} \lambda^i \Psi_0^{(i)} = \Phi_1 + \Delta\Phi_1^{(1)} + \Delta\Phi_1^{(2)} + \ldots = \Psi_0^{\mathrm{HF}} + \Delta\Psi_0^{\mathrm{HF}(1)} + \Delta\Psi_0^{\mathrm{HF}(2)} + \ldots \tag{6.126}$$

The *first*-order energy correction [see (5.11)] corrects $E_1^{(0)}$ to E_0^{HF},

$$\begin{aligned} E_1^{(0)} + \Delta E_1^{(1)} &= E_1^{(0)} + \langle \Phi_1 | \hat{\mathrm{h}}^{corr.} | \Phi_1 \rangle = \langle \Psi_0^{\mathrm{HF}} | \hat{\mathrm{H}}_0^{\mathrm{MP}} | \Psi_0^{\mathrm{HF}} \rangle + \langle \Psi_0^{\mathrm{HF}} | \hat{\mathrm{H}}^e - \hat{\mathrm{H}}_0^{\mathrm{MP}} | \Psi_0^{\mathrm{HF}} \rangle \\ &= \langle \Psi_0^{\mathrm{HF}} | \hat{\mathrm{H}}^e | \Psi_0^{\mathrm{HF}} \rangle = E_0^{\mathrm{HF}}, \end{aligned} \tag{6.127}$$

thus removing the double counting of the electron repulsion terms in $E_1^{(0)}$. Therefore, it is the *second*-order term which represents the most important part of the Coulomb correlation energy. A reference to (5.16) indicates that it is determined by the *first*-order correction to $\Phi_1 = \Psi_0^{\mathrm{HF}}$:

$$\Delta\Psi_0^{\mathrm{HF}(1)} = \sum_{s \neq 1} \frac{\langle \Phi_s | \hat{\mathrm{h}}^{corr.} | \Phi_1 \rangle}{E_1^{(0)} - E_s^{(0)}} \Phi_s \equiv \sum_{s \neq 1} \frac{h_{s,1}^{corr.}}{E_1^{(0)} - E_s^{(0)}} \Phi_s = \sum_{s \neq 1} C_{s,1}^{(1)} \Phi_s. \tag{6.128}$$

It also follows from the orthogonality relations between configurations, $\langle \Phi_s \mid \Phi_t \rangle = \delta_{s,t}$, that the *off*-diagonal matrix elements ($s \neq t$) of the correlation operator, coupling different configurations, are identical with the corresponding elements of the energy matrix $\mathbf{H}^{\mathrm{CI}}$ representing the electronic Hamiltonian in the CI method:

$$\langle \Phi_s | \hat{\mathrm{h}}^{corr.} | \Phi_t \rangle = \langle \Phi_s | \hat{\mathrm{H}}^e - \hat{\mathrm{H}}_0^{\mathrm{MP}} | \Phi_t \rangle = H_{s,t}^{\mathrm{CI}} - E_t^{(0)} \langle \Phi_s \mid \Phi_t \rangle = H_{s,t}^{\mathrm{CI}}. \tag{6.129}$$

Therefore, in accordance with the Slater–Condon rules, there is no correlation coupling between this ground state reference configuration and the excited configurations exhibiting higher than double excitations: *Triple* (T), *Quadruple* (Q), etc. Moreover, the Brillouin theorem states that there is no direct coupling between the (closed-shell) ground state and the *Singly* (S) excited configurations, so that the expansion of (6.126) determining the MP2 scheme is then limited only to *Doubly* (D) excited configurations [see (6.117)]:

$$\Delta\Psi_0^{\mathrm{HF}(1)} = \sum_{s\in\mathrm{D}} C_{s,1}^{(1)}\Phi_s = \sum_{k<l}^{occd.}\sum_{p<q}^{virt.} C_{k,l}^{p,q}\Phi_{k,l}^{p,q}, \tag{6.130}$$

with the perturbational expression for the expansion coefficients [see (5.86)]:

$$C_{s,l}^{(1)} = \frac{H_{s,1}^{\mathrm{CI}}}{E_1^{(0)} - E_s^{(0)}} = C_{k,l}^{p,q}\,\frac{\langle kl|g|pq - qp\rangle}{e_k + e_l - e_p - e_q}. \tag{6.131}$$

One thus obtains for the *second*-order correlation energy in this MP2 variant of the perturbational CI method [see (5.16)]:

$$\begin{aligned}\Delta E_0^{\mathrm{HF}(2)} &= \left\langle \Psi_0^{\mathrm{HF}} \middle| \hat{\mathrm{H}}^e - \hat{\mathrm{H}}_0^{\mathrm{MP}} \middle| \Delta\Psi_0^{\mathrm{HF}(1)} \right\rangle = \left\langle \Psi_0^{\mathrm{HF}} \middle| \hat{\mathrm{H}}^e \middle| \Delta\Psi_0^{\mathrm{HF}(1)} \right\rangle - E_1^{(0)}\left\langle \Psi_0^{\mathrm{HF}} \middle| \Delta\Psi_0^{\mathrm{HF}(1)} \right\rangle\\ &= \left\langle \Psi_0^{\mathrm{HF}} \middle| \hat{\mathrm{H}}^e \middle| \Delta\Psi_0^{\mathrm{HF}(1)} \right\rangle,\end{aligned} \tag{6.132}$$

since the correction of (6.130), involving the excited configurations orthogonal to the ground state, must be also orthogonal to the latter. Substituting (6.129)–(6.131) into preceding equation finally gives:

$$\Delta E_0^{\mathrm{HF}(2)} = \sum_{k<l}^{occd.}\left(\sum_{p<q}^{virt.}\frac{|\langle kl|g|pq - qp\rangle|^2}{e_k + e_l - e_p - e_q}\right) \equiv \sum_{k<l}^{occd.} e_{k,l}^{\mathrm{MP2}} \equiv \Delta E_{corr.}^{\mathrm{MP2}}. \tag{6.133}$$

Therefore, this *second*-order correlation energy combines the additive contributions from all electronic pairs of the HF-occupied MO, which define the two-electron *clusters* $\{(k,l)\}$. For this reason, the MP2 method can be classified as corresponding within the PT approach to the *Independent Electron Pair Approximation* (IEPA) of Sect. 6.4.2, since only independent pairs give rise to *additive* energy contributions.

Obviously, by using the standard expressions from PT one could similarly determine the *higher* order contributions to the overall Coulomb correlation energy. Next in importance is the *third*-order correction $\Delta E_0^{\mathrm{HF}(3)}$, determined within the MP3 variant. It results from (5.6) for $p = 3$, projected onto $|n^{(0)}\rangle$, which gives $\Delta E_n^{(3)} = \left\langle n^{(0)} \middle| \hat{\mathrm{h}} \middle| \Delta n^{(2)} \right\rangle$, where the *second*-order correction to the wave function is given by (5.17). For example, for H_2O one finds the following percentages of the correlation energy recovered by different orders of MBPT, $\Delta E_0^{\mathrm{HF}(2)} = 97.7$ and

$\Delta E_0^{\mathrm{HF}(3)} = 1.5$, while in BH_3 the *higher*-order term appears to be relatively more significant: $\Delta E_0^{\mathrm{HF}(2)} = 80.0$ and $\Delta E_0^{\mathrm{HF}(3)} = 16.5$.

6.3.3 Density Matrices and Natural Orbitals

The spin position representation of the Hermitian operator of the projection onto the quantum state of N electrons, $\hat{\mathrm{P}}_{\Psi(N)} = |\Psi(N)\rangle\langle\Psi(N)|$(see Chap. 2), defines the N-particle *density matrix* (Löwdin 1955a, b; Coleman 1963, 1981; Coleman and Erdahl 1968; Davidson 1976):

$$\begin{aligned}\hat{\Gamma}(q_1, q_2, \ldots, q_N; q_1{}', q_2{}', \ldots, q_N{}') &\equiv \hat{\Gamma}(\mathcal{Q}^N; \mathcal{Q}'^N) \\ &= \langle \mathcal{Q}^N | \Psi(N)\rangle\langle\Psi(N)|\mathcal{Q}'^N\rangle = \Psi(\mathcal{Q}^N)\Psi^*(\mathcal{Q}'^N). \end{aligned} \tag{6.134}$$

This continuous matrix represents the kernel of the associated operator $\hat{\Gamma}$ acting in the molecular Hilbert space:

$$\begin{aligned}\langle \mathcal{Q}^N | \hat{\Gamma}\Phi\rangle &= \int \langle \mathcal{Q}^N | \hat{\Gamma} | \mathcal{Q}'^N\rangle\langle\mathcal{Q}'^N|\Phi\rangle \, d\mathcal{Q}'^N = \int \hat{\Gamma}(\mathcal{Q}^N; \mathcal{Q}'^N)\Phi(\mathcal{Q}'^N)\, d\mathcal{Q}'^N \\ &= \langle \mathcal{Q}^N | \Psi\rangle\langle\Psi|\Phi\rangle = \Psi(\mathcal{Q}^N)\langle\Psi \mid \Phi\rangle, \end{aligned} \tag{6.135}$$

and carries the same (complete) information about the system state as does the wave function itself. One observes that its diagonal part, for $\mathcal{Q}^N = \mathcal{Q}'^N$, determines the N-particle probability of (3.4), $\hat{\Gamma}(\mathcal{Q}^N; \mathcal{Q}^N) = p(\mathcal{Q}^N)$, so that

$$\mathrm{tr}\,\hat{\Gamma} = \int p(\mathcal{Q}^N)\, d\mathcal{Q}^N = 1 \tag{6.136}$$

and the *pure* state expectation value of the observable $\hat{\mathrm{A}}$ can be brought into the form similar to the *ensemble* average expression of (3.60a) (3.60b):

$$\langle A\rangle = \langle\Psi|\hat{\mathrm{A}}|\Psi\rangle = \iint \langle\mathcal{Q}^N|\hat{\mathrm{A}}|\mathcal{Q}'^N\rangle\, \Psi(\mathcal{Q}'^N)\Psi^*(\mathcal{Q}^N)\, d\mathcal{Q}^N d\mathcal{Q}'^N = \mathrm{tr}(\hat{\mathrm{A}}\hat{\Gamma}). \tag{6.137}$$

As we have already observed in Sect. 3.3.4, the density operator characterization of molecular systems becomes necessary when the system is in the *mixed* quantum state, which cannot be represented by a single vector in the Hilbert space or the associated wave function; it represents the statistical mixture of several quantum states. For the system to be in the *pure* quantum state, it is necessary and sufficient for its density operator to be idempotent, $\hat{\Gamma}^2 = \hat{\Gamma}$. This condition expresses the idempotency of the associate state projector $\hat{P}_{\Psi(N)}$. In the Schrödinger picture of Quantum Mechanics the time evolution of this pure state density operator is given by (3.101).

The spin-independent electronic Hamiltonian of (5.51) and (5.69) is seen to involve the symmetric combinations of either one- or two-electron terms and the system wave function is antisymmetric with respect to exchanges of the subjective labels attributed to electrons. As we have already observed in Sect. 5.4, one can thus select as representative interactions of the indistinguishable electrons the one-electron energies of electron "1" and two electron repulsion energy between electrons "1" and "2." Therefore, when calculating the expectation values of all one- and two-electron interactions one can take the partial trace (integrate) over the remaining particle variables, equal in the primed and unprimed sets, in the generalized products appearing in the expectation value and matrix element expressions:

$$\begin{aligned}&\Psi(\mathcal{Q}^N)\Psi^*(\mathcal{Q}'^N)\big|_{q_2=q_2', q_3=q_3', \ldots, q_N=q_N'} \quad \text{and}\\ &\Psi(\mathcal{Q}^N)\Psi^*(\mathcal{Q}'^N)\big|_{q_3=q_3', q_4=q_4', \ldots, q_N=q_N'},\end{aligned} \tag{6.138}$$

respectively. This simplification was the basic motivation of Löwdin (1955a, b) when he introduced the concept of the *reduced density matrix* of order p:

$$\begin{aligned}\hat{\gamma}_p(\boldsymbol{q}_1, \boldsymbol{q}_2, \ldots, \boldsymbol{q}_p; \boldsymbol{q}_1', \boldsymbol{q}_2', \ldots, \boldsymbol{q}_p') &= \binom{N}{p} \operatorname{tr} \hat{\Gamma}(\boldsymbol{q}_1, \boldsymbol{q}_2, \ldots, \boldsymbol{q}_N; \boldsymbol{q}_1', \boldsymbol{q}_2', \ldots, \boldsymbol{q}_N')\big|_{\boldsymbol{q}_{p+1}, \boldsymbol{q}_{p+2}, \ldots, \boldsymbol{q}_N}\\ &= \binom{N}{p} \int \Psi(\boldsymbol{q}_1, \ldots, \boldsymbol{q}_p, \boldsymbol{q}_{p+1}, \ldots, \boldsymbol{q}_N)\Psi^*(\boldsymbol{q}_1', \ldots, \boldsymbol{q}_p', \boldsymbol{q}_{p+1}, \ldots, \boldsymbol{q}_N)\\ &\quad \times d\boldsymbol{q}_{p+1} d\boldsymbol{q}_{p+2} \ldots d\boldsymbol{q}_N,\end{aligned} \tag{6.139a}$$

where $\binom{N}{p}$ stands for the binomial coefficient [see (3.12) and (6.66)].

For example, the *first*-order reduced density matrix, or the *1-matrix* for short, reads:

$$\hat{\gamma}_1(\boldsymbol{q}_1; \boldsymbol{q}_1') = N \int \Psi(\boldsymbol{q}_1, \boldsymbol{q}_2, \ldots, \boldsymbol{q}_N)\Psi^*(\boldsymbol{q}_1', \boldsymbol{q}_2, \ldots, \boldsymbol{q}_N)\, d\boldsymbol{q}_2 \ldots d\boldsymbol{q}_N. \tag{6.140a}$$

Its diagonal part thus represents the spin density of the system electrons (3.9): $\rho(\boldsymbol{q}) = \rho(\boldsymbol{r}, \sigma) = \hat{\gamma}_1(\boldsymbol{q}; \boldsymbol{q})$, which implies the associated normalization:

$$\operatorname{tr} \hat{\gamma}_1(\boldsymbol{q}_1; \boldsymbol{q}_1') = \int \hat{\gamma}_1(\boldsymbol{q}_1; \boldsymbol{q}_1)\, d\boldsymbol{q}_1 = N. \tag{6.141a}$$

One similarly obtains the *second*-order reduced density matrix (or *2-matrix*):

$$\hat{\gamma}_2(\boldsymbol{q}_1,\boldsymbol{q}_2;\boldsymbol{q}_1{}',\boldsymbol{q}_2{}') = \binom{N}{2}\int \Psi(\boldsymbol{q}_1,\ \boldsymbol{q}_2,\boldsymbol{q}_3,\ \ldots,\boldsymbol{q}_N) \times \Psi^*(\boldsymbol{q}_1{}',\ \boldsymbol{q}_2{}',\boldsymbol{q}_3,\ \ldots,\boldsymbol{q}_N)\, d\boldsymbol{q}_3 \ldots\ d\boldsymbol{q}_N, \tag{6.142}$$

the *pair* diagonal element of which now represents the *pair* spin density [see (3.12)],

$$\hat{\gamma}_2(\boldsymbol{q},\boldsymbol{q}';\boldsymbol{q},\boldsymbol{q}') = \Gamma(\boldsymbol{q},\boldsymbol{q}') = \int \Psi^*(\mathbb{Q}^N)\hat{\Gamma}(\boldsymbol{q},\boldsymbol{q}')\Psi(\mathbb{Q}^N)\, d\mathbb{Q}^N,$$
$$\hat{\Gamma}(\boldsymbol{q},\boldsymbol{q}') = \sum_{k=1}^{N-1}\sum_{l=k+1}^{N}\delta(\boldsymbol{q}_k-\boldsymbol{q})\delta(\boldsymbol{q}_l-\boldsymbol{q}'). \tag{6.143}$$

This implies the associated normalization of the 2-matrix:

$$\mathrm{tr}\,\hat{\gamma}_2(\boldsymbol{q}_1,\boldsymbol{q}_2;\boldsymbol{q}_1{}',\boldsymbol{q}_2{}') = \iint \hat{\gamma}_2(\boldsymbol{q}_1,\boldsymbol{q}_2;\boldsymbol{q}_1,\boldsymbol{q}_2)\,d\boldsymbol{q}_1 d\boldsymbol{q}_2 = N(N-1)/2 = \binom{N}{2}. \tag{6.144}$$

The expectation value of the sum of all one-electron operators in $\hat{\mathfrak{F}} = \sum_{i=1}^{N}\hat{\mathrm{h}}(i)$ [see (5.69)–(5.71)] now reads:

$$\left\langle \Psi \mid \hat{\mathfrak{F}} \mid \Psi \right\rangle = \mathrm{tr}\,(\hat{\mathfrak{F}}\,\hat{\Gamma}) = \mathrm{tr}\,(\hat{\mathrm{h}}\,\hat{\gamma}_1) = \int \hat{\mathrm{h}}(\boldsymbol{q}_1)\,\hat{\gamma}_1(\boldsymbol{q}_1;\boldsymbol{q}_1{}')|_{\boldsymbol{q}_1=\boldsymbol{q}_1{}'}\, d\boldsymbol{q}_1. \tag{6.145}$$

For the expectation value of the two-electron energy one similarly obtains:

$$\left\langle \Psi \mid \hat{\mathcal{G}} \mid \Psi \right\rangle = \mathrm{tr}\,(\hat{\mathcal{G}}\,\hat{\Gamma}) = \mathrm{tr}(g\,\hat{\gamma}_2) = \iint g(\boldsymbol{q}_1,\boldsymbol{q}_2)\,\hat{\gamma}_2(\boldsymbol{q}_1,\boldsymbol{q}_2;\boldsymbol{q}_1{}',\boldsymbol{q}_2{}')|_{\boldsymbol{q}_1=\boldsymbol{q}_1{}',\boldsymbol{q}_2=\boldsymbol{q}_2{}'}\, d\boldsymbol{q}_1 d\boldsymbol{q}_2. \tag{6.146}$$

Since operators $\hat{\mathrm{h}}$ and g are both spin independent these expressions can be further simplified in terms of the reduced *spinless* density matrices, obtained by summations over the spin variables of the spin-dependent analogs:

$$\begin{aligned}
\hat{\rho}_1(\boldsymbol{r}_1;\boldsymbol{r}_1{}') &= \int \hat{\gamma}_1(\boldsymbol{r}_1,\sigma_1;\boldsymbol{r}_1{}',\sigma_1)\,d\sigma_1 \equiv \sum_{\sigma_1}\hat{\gamma}_1(\boldsymbol{r}_1,\sigma_1;\boldsymbol{r}_1{}',\sigma_1) = \mathrm{tr}\,\hat{\gamma}_1(\boldsymbol{q}_1;\boldsymbol{q}_1{}')|_{\sigma_1}\\
&= \hat{\gamma}_1(\boldsymbol{r}_1,\alpha;\boldsymbol{r}_1{}',\alpha) + \hat{\gamma}_1(\boldsymbol{r}_1,\beta;\boldsymbol{r}_1{}',\beta) \equiv \hat{\rho}_1^{\alpha;\alpha}(\boldsymbol{r};\boldsymbol{r}') + \hat{\rho}_1^{\beta;\beta}(\boldsymbol{r};\boldsymbol{r}'),\\
\hat{\rho}_2(\boldsymbol{r}_1,\boldsymbol{r}_2;\boldsymbol{r}_1{}',\boldsymbol{r}_2{}') &= \iint \hat{\gamma}_2(\boldsymbol{r}_1,\sigma_1,\boldsymbol{r}_2,\sigma_2;\boldsymbol{r}_1{}',\sigma_1,\boldsymbol{r}_2{}',\sigma_2)\,d\sigma_1 d\sigma_2\\
&\equiv \sum_{\sigma_1}\sum_{\sigma_2}\hat{\gamma}_2(\boldsymbol{r}_1,\sigma_1,\boldsymbol{r}_2,\sigma_2;\boldsymbol{r}_1{}',\sigma_1,\boldsymbol{r}_2{}',\sigma_2) = \mathrm{tr}\,\hat{\gamma}_2(\boldsymbol{q}_1,\boldsymbol{q}_2;\boldsymbol{q}_1{}',\boldsymbol{q}_2{}')|_{\sigma_1,\sigma_2}\\
&\equiv \hat{\rho}_2^{\alpha\alpha;\alpha\alpha}(\boldsymbol{r}_1,\boldsymbol{r}_2;\boldsymbol{r}_1{}',\boldsymbol{r}_2{}') + \hat{\rho}_2^{\alpha\beta;\alpha\beta}(\boldsymbol{r}_1,\boldsymbol{r}_2;\boldsymbol{r}_1{}',\boldsymbol{r}_2{}') + \hat{\rho}_2^{\beta\alpha;\beta\alpha}(\boldsymbol{r}_1,\boldsymbol{r}_2;\boldsymbol{r}_1{}',\boldsymbol{r}_2{}')\\
&\quad + \hat{\rho}_2^{\beta\beta;\beta\beta}(\boldsymbol{r}_1,\boldsymbol{r}_2;\boldsymbol{r}_1{}',\boldsymbol{r}_2{}').
\end{aligned} \tag{6.147}$$

Their diagonal elements respectively represent the electron density (3.7) and *pair* density (3.12):

$$\rho(\boldsymbol{r}) = \hat{\rho}_1(\boldsymbol{r};\boldsymbol{r}) \qquad \text{and} \qquad \Gamma(\boldsymbol{r},\boldsymbol{r}') = \hat{\rho}_2(\boldsymbol{r},\boldsymbol{r}';\boldsymbol{r},\boldsymbol{r}'), \tag{6.148}$$

thus giving rise to the associated normalizations:

$$\begin{aligned} \operatorname{tr}\hat{\rho}_1(\boldsymbol{r}_1;\boldsymbol{r}_1{}') &= \int \hat{\rho}_1(\boldsymbol{r};\boldsymbol{r})\,d\boldsymbol{r} = N, \\ \operatorname{tr}\hat{\rho}_2(\boldsymbol{r}_1,\boldsymbol{r}_2;\boldsymbol{r}_1{}',\boldsymbol{r}_2{}') &= \int\int \hat{\rho}_2(\boldsymbol{r},\boldsymbol{r}';\boldsymbol{r},\boldsymbol{r}')\,d\boldsymbol{r}\,d\boldsymbol{r}' = N(N-1)/2. \end{aligned} \tag{6.149}$$

The spinless expression for the expectation value of the system electronic energy thus reads:

$$\begin{aligned} E^e[\hat{\rho}_1,\hat{\rho}_2] &= \int [-\frac{1}{2}\Delta_{\boldsymbol{r}}\,\hat{\rho}_1(\boldsymbol{r};\boldsymbol{r}')|_{\boldsymbol{r}=\boldsymbol{r}'}\,d\boldsymbol{r} + \int v(\boldsymbol{r})\rho(\boldsymbol{r})\,d\boldsymbol{r} \\ &\quad + \int\int \frac{1}{|\boldsymbol{r}-\boldsymbol{r}'|}\Gamma(\boldsymbol{r},\boldsymbol{r}')\,d\boldsymbol{r}\,d\boldsymbol{r}'. \end{aligned} \tag{6.150}$$

The reduced density matrices define kernels of the Hermitian, positive semi-definite operators:

$$\begin{aligned} &\hat{\gamma}_1(\boldsymbol{q}_1;\boldsymbol{q}_1{}') = \hat{\gamma}_1^*(\boldsymbol{q}_1{}';\boldsymbol{q}_1), && \hat{\gamma}_1(\boldsymbol{q}_1;\boldsymbol{q}_1) \geq 0; \\ &\hat{\gamma}_2(\boldsymbol{q}_1,\boldsymbol{q}_2;\boldsymbol{q}_1{}',\boldsymbol{q}_2{}') = \hat{\gamma}_2^*(\boldsymbol{q}_1{}',\boldsymbol{q}_2{}';\boldsymbol{q}_1,\boldsymbol{q}_2), && \hat{\gamma}_2(\boldsymbol{q}_1,\boldsymbol{q}_2;\boldsymbol{q}_1,\boldsymbol{q}_2) \geq 0. \end{aligned} \tag{6.151}$$

The antisymmetry of the wave function also implies that the reduced densities change their sign on exchange of any two primed or two unprimed particle indices:

$$\begin{aligned} \hat{\gamma}_2(\boldsymbol{q}_1,\boldsymbol{q}_2;\boldsymbol{q}_1{}',\boldsymbol{q}_2{}') &= -\hat{\gamma}_2(\boldsymbol{q}_2,\boldsymbol{q}_1;\boldsymbol{q}_1{}',\boldsymbol{q}_2{}') = -\hat{\gamma}_2(\boldsymbol{q}_1,\boldsymbol{q}_2;\boldsymbol{q}_2{}',\boldsymbol{q}_1{}') \\ &= \hat{\gamma}_2(\boldsymbol{q}_2,\boldsymbol{q}_1;\boldsymbol{q}_2{}',\boldsymbol{q}_1{}'). \end{aligned} \tag{6.152}$$

Of particular interest in the electron correlation theory are the eigenvalue problems of $\hat{\gamma}_1$ and $\hat{\gamma}_2$:

$$\int \hat{\gamma}_1(\boldsymbol{q};\boldsymbol{q}')\,\psi_i(\boldsymbol{q}')d\boldsymbol{q}' = n_i^{\text{NO}}\psi_i(\boldsymbol{q}), \tag{6.153}$$

$$\int\int \hat{\gamma}_2(\boldsymbol{q}_1,\boldsymbol{q}_2;\boldsymbol{q}_1{}',\boldsymbol{q}_2{}')\,G_j(\boldsymbol{q}_1{}',\boldsymbol{q}_2{}')\,d\boldsymbol{q}_1{}'d\boldsymbol{q}_2{}' = v_j^{\text{NG}}G_j(\boldsymbol{q}_1,\boldsymbol{q}_2). \tag{6.154}$$

The eigenfunctions $\{\psi_i\}$ of $\hat{\gamma}_1$ are called the *Natural Orbitals* (NO) (Löwdin 1955a, b; Löwdin and Shull 1956; Carlson and Keller 1961; Davidson 1969, 1972a, b, 1976;

Bingel and Kutzelnigg 1970), with the corresponding eigenvalues measuring their effective occupations in Ψ: $\{0 \leq n_i^{\mathrm{NO}} \leq 1\}$. The corresponding two-electron (antisymmetric) eigenfunctions of $\hat{\gamma}_2, \{G_j(\boldsymbol{q}_1, \boldsymbol{q}_2) = -G_j(\boldsymbol{q}_2, \boldsymbol{q}_1)\}$, determine the system *natural geminals* (NG), with the eigenvalues $\{v_j^{\mathrm{NG}}\}$ again reflecting their effective occupations.

It should be observed that these two density matrices assume the most compact (diagonal) representations in terms of their respective eigenfunctions:

$$\begin{aligned}\hat{\gamma}_1 &= \sum_i n_i^{\mathrm{NO}} |\psi_i\rangle\langle\psi_i| \equiv |\boldsymbol{\psi}^{\mathrm{NO}}\rangle \mathbf{n}^{\mathrm{NO}} \langle\boldsymbol{\psi}^{\mathrm{NO}}|,\\ \mathbf{n}^{\mathrm{NO}} &= \{n_i^{\mathrm{NO}}\delta_{i,j}\},\ |\boldsymbol{\psi}^{\mathrm{NO}}\rangle = \{|\psi_i\rangle\},\ \text{or}\\ \hat{\gamma}_1(\boldsymbol{r}_1;\boldsymbol{r}_1') &= \langle\boldsymbol{r}_1|\hat{\gamma}_1|\boldsymbol{r}_1'\rangle = \sum_i n_i^{\mathrm{NO}}\langle\boldsymbol{r}_1|\psi_i\rangle\langle\psi_i|\boldsymbol{r}_1'\rangle = \sum_i n_i^{\mathrm{NO}}\,\psi_i(\boldsymbol{r}_1)\,\psi_i^*(\boldsymbol{r}_1');\end{aligned} \tag{6.155}$$

$$\begin{aligned}\hat{\gamma}_2 &= \sum_j v_j|G_j\rangle\langle G_j| \equiv |\boldsymbol{G}^{\mathrm{NG}}\rangle\boldsymbol{v}^{\mathrm{NG}}\langle\boldsymbol{G}^{\mathrm{NG}}|,\\ \boldsymbol{v}^{\mathrm{NG}} &= \{v_i^{\mathrm{NG}}\delta_{i,j}\},\ |\boldsymbol{G}^{\mathrm{NG}}\rangle = \{|G_j\rangle\},\ \text{or}\\ \hat{\gamma}_2(\boldsymbol{r}_1,\boldsymbol{r}_2;\boldsymbol{r}_1',\boldsymbol{r}_2') &= \langle\boldsymbol{r}_1,\boldsymbol{r}_2|\hat{\gamma}_2|\boldsymbol{r}_1',\boldsymbol{r}_2'\rangle = \sum_j v_j^{\mathrm{NG}}\langle\boldsymbol{r}_1,\boldsymbol{r}_2|G_j\rangle\langle G_j|\boldsymbol{r}_1',\boldsymbol{r}_2'\rangle\\ &= \sum_j v_j^{\mathrm{NG}}\,G_j(\boldsymbol{r}_1,\boldsymbol{r}_2)\,G_j^*(\boldsymbol{r}_1',\boldsymbol{r}_2').\end{aligned} \tag{6.156}$$

Expanding NO in the AO basis functions $\boldsymbol{\chi} = \{\chi_s\}, \boldsymbol{\psi}^{\mathrm{NO}} = \boldsymbol{\chi}\mathbf{D} = \{\psi_i\}$ (the row vector), gives the following expression for the 1-matrix [compare (6.45)]:

$$\begin{aligned}\hat{\gamma}_1(\boldsymbol{r}_1;\boldsymbol{r}_1') &= \langle\boldsymbol{r}_1|\boldsymbol{\psi}^{\mathrm{NO}}\rangle\mathbf{n}^{\mathrm{NO}}\langle\boldsymbol{\psi}^{\mathrm{NO}}|\boldsymbol{r}_1'\rangle = \boldsymbol{\psi}^{\mathrm{NO}}(\boldsymbol{r}_1)\mathbf{n}^{\mathrm{NO}}\boldsymbol{\psi}^{\mathrm{NO}\dagger}(\boldsymbol{r}_1')\\ &= \boldsymbol{\chi}(\boldsymbol{r}_1)[\mathbf{D}\mathbf{n}^{\mathrm{NO}}\mathbf{D}^\dagger]\boldsymbol{\chi}^\dagger(\boldsymbol{r}_1') = \boldsymbol{\chi}(\boldsymbol{r}_1)\mathbf{P}^{\mathrm{NO}}(\mathbf{D})\boldsymbol{\chi}^\dagger(\boldsymbol{r}_1'),\end{aligned} \tag{6.157}$$

where we have grouped the NO occupations as diagonal elements of the square matrix $\mathbf{n}^{\mathrm{NO}} = \{n_i^{\mathrm{NO}}\,\delta_{i,i'}\}$, while $\boldsymbol{\chi}^\dagger = \{\chi_s^*\}^{\mathrm{T}}$ and $\boldsymbol{\psi}^{\mathrm{NO}\dagger} = \{\psi_i^*\}^{\mathrm{T}}$ denote the associated column vectors. Therefore, the NO CBO matrix in the AO representation,

$$\mathbf{P}^{\mathrm{NO}}(\mathbf{D}) = \left\{P_{s,t}^{\mathrm{NO}} = \sum_i^{\mathrm{NO}} D_{s,i}n_i^{\mathrm{NO}}D_{t,i}^*\right\}, \tag{6.158}$$

determines the 1–matrix in terms of basis functions $\boldsymbol{\chi}$.

The corresponding AO expansion of NG,

$$\begin{aligned}\boldsymbol{G}^{\mathrm{NG}}(\boldsymbol{r}_1,\boldsymbol{r}_2) &= \{G_j(\boldsymbol{r}_1,\boldsymbol{r}_2) = \sum_{s,t}^{\mathrm{AO}}\chi_s(\boldsymbol{r}_1)\chi_t(\boldsymbol{r}_2)A_{s,t;j}\\ &\equiv \sum_{s,t}^{\mathrm{AO}}\Omega_{s,t}(\boldsymbol{r}_1,\boldsymbol{r}_2)A_{s,t;j}\} \equiv \boldsymbol{\Omega}(\boldsymbol{r}_1,\boldsymbol{r}_2)\mathbf{A},\end{aligned} \tag{6.159}$$

gives the following expression for the 2-matrix:

$$\hat{\gamma}_2(\boldsymbol{r}_1,\boldsymbol{r}_2;\boldsymbol{r}_1',\boldsymbol{r}_2') = \langle \boldsymbol{r}_1,\boldsymbol{r}_2|\boldsymbol{G}^{\mathrm{NG}}\rangle \boldsymbol{\nu}^{\mathrm{NG}}\langle \boldsymbol{G}^{\mathrm{NG}}|\boldsymbol{r}_1',\boldsymbol{r}_2'\rangle = \boldsymbol{G}^{\mathrm{NG}}(\boldsymbol{r}_1,\boldsymbol{r}_2)\boldsymbol{\nu}^{\mathrm{NG}}\boldsymbol{G}^{\mathrm{NG}\dagger}(\boldsymbol{r}_1',\boldsymbol{r}_2')$$
$$= \boldsymbol{\Omega}(\boldsymbol{r}_1,\boldsymbol{r}_2)[\mathbf{A}\boldsymbol{\nu}^{\mathrm{NG}}\mathbf{A}^\dagger]\boldsymbol{\Omega}^\dagger(\boldsymbol{r}_1',\boldsymbol{r}_2') = \boldsymbol{\Omega}(\boldsymbol{r}_1,\boldsymbol{r}_2)\,\mathbf{P}^{\mathrm{NG}}(\mathbf{A})\,\boldsymbol{\Omega}^\dagger(\boldsymbol{r}_1',\boldsymbol{r}_2'), \tag{6.160}$$

where $\boldsymbol{\nu}^{\mathrm{NG}} = \left\{\nu_j^{\mathrm{NG}}\delta_{j,j'}\right\}$. Again, the corresponding NG CBO matrix

$$\mathbf{P}^{\mathrm{NG}}(\mathbf{A}) = \left\{P^{\mathrm{NG}}_{s,t;u,v} = \sum_j^{\mathrm{NG}} A_{s,t;j}\,\nu_j^{\mathrm{NG}}A^*_{u,v;j}\right\} \tag{6.161}$$

defines the 2-matrix in the AO representation.

The density matrices for the HF wave function, given by a single Slater determinant (5.64), defined by the optimum HF SO $\boldsymbol{\psi} = \{\psi_i\}$, assume particularly simple forms (Löwdin 1955b). The 1-matrix in this approximation reads,

$$\hat{\gamma}_1^{\mathrm{HF}}(\boldsymbol{q}_1;\boldsymbol{q}_1') = \sum_{i=1}^{N}\psi_i(\boldsymbol{q}_1)\,\psi_i^*(\boldsymbol{q}_1'), \tag{6.140b}$$

while the *second*-order reduced density matrix is given by the following determinant of 1-matrices:

$$\hat{\gamma}_2^{\mathrm{HF}}(\boldsymbol{q}_1,\boldsymbol{q}_2;\boldsymbol{q}_1',\boldsymbol{q}_2') = \frac{1}{2}\begin{vmatrix}\hat{\gamma}_1^{\mathrm{HF}}(\boldsymbol{q}_1;\boldsymbol{q}_1') & \hat{\gamma}_1^{\mathrm{HF}}(\boldsymbol{q}_1;\boldsymbol{q}_2') \\ \hat{\gamma}_1^{\mathrm{HF}}(\boldsymbol{q}_2;\boldsymbol{q}_1') & \hat{\gamma}_1^{\mathrm{HF}}(\boldsymbol{q}_2;\boldsymbol{q}_2')\end{vmatrix}$$
$$= \frac{1}{2}[\hat{\gamma}_1^{\mathrm{HF}}(\boldsymbol{q}_1;\boldsymbol{q}_1')\hat{\gamma}_1^{\mathrm{HF}}(\boldsymbol{q}_2;\boldsymbol{q}_2') - \hat{\gamma}_1^{\mathrm{HF}}(\boldsymbol{q}_2;\boldsymbol{q}_1')\hat{\gamma}_1^{\mathrm{HF}}(\boldsymbol{q}_1;\boldsymbol{q}_2')]. \tag{6.141b}$$

As also shown by Löwdin (1955b), this prescription can be extended into a general reduced density matrix of (6.139a) in HF approximation:

$$\hat{\gamma}_p^{\mathrm{HF}}(\boldsymbol{q}_1,\boldsymbol{q}_2,\ldots,\boldsymbol{q}_p;\boldsymbol{q}_1',\boldsymbol{q}_2',\ldots,\boldsymbol{q}_p') = \frac{1}{p!}\begin{vmatrix}\hat{\gamma}_1^{\mathrm{HF}}(\boldsymbol{q}_1;\boldsymbol{q}_1') & \hat{\gamma}_1^{\mathrm{HF}}(\boldsymbol{q}_1;\boldsymbol{q}_2') & \cdots & \hat{\gamma}_1^{\mathrm{HF}}(\boldsymbol{q}_1;\boldsymbol{q}_p') \\ \hat{\gamma}_1^{\mathrm{HF}}(\boldsymbol{q}_2;\boldsymbol{q}_1') & \hat{\gamma}_1^{\mathrm{HF}}(\boldsymbol{q}_2;\boldsymbol{q}_2') & \cdots & \hat{\gamma}_1^{\mathrm{HF}}(\boldsymbol{q}_2;\boldsymbol{q}_p') \\ \cdots & \cdots & \cdots & \cdots \\ \hat{\gamma}_1^{\mathrm{HF}}(\boldsymbol{q}_p;\boldsymbol{q}_1') & \hat{\gamma}_1^{\mathrm{HF}}(\boldsymbol{q}_p;\boldsymbol{q}_2') & \cdots & \hat{\gamma}_1^{\mathrm{HF}}(\boldsymbol{q}_p;\boldsymbol{q}_p')\end{vmatrix}. \tag{6.139b}$$

There is a number of analytical arguments and a growing numerical evidence for small molecules that the convergence of the CI expansion is greatly enhanced by the use of NO. In other words, the correlation energy recovery is greater for the NO-based CI treatment, compared with that using the canonical SCF MO, at the same length of expansion (Shavitt et al. 1976). This analysis of wave functions is basis-set independent and allows one to compare any set of approximate wave functions. There are also conceptual advantages associated with the NO representation. It has been found to provide almost universal patterns of the orbital shapes and occupations for diverse wave functions obtained using widely different basis sets and CI expansions (Davidson 1972b). The observed "invariance" of the NO occupation numbers to choices of the basis set and the length of the CI expansion, respectively, suggests some "universality" of the NO description of atoms and molecules. The NO analysis emphasizes the strongly, near-unity populated HF-like NO and generates the weakly populated *correlation orbitals*, which represent the independent modes of the electron Coulomb correlation in molecular systems, the relative importance of which can be inferred from the corresponding orbital occupations. For example, in He the first correlation NO introduces the angular ("*left–right*")-type correlation, while the second correlation orbital brings in the ("*in–out*")-effect of the radial correlation. The NO analysis has also been widely used to study the nature of the chemical bond and the adequacy of the HF description in the bond-breaking – bond-forming processes (e.g., Shull 1959; Hagstrom and Shull 1963).

Thus, the NO play a significant role as both means for improving the CI convergence and in obtaining the physical insight and understanding of approximate wave functions. However, since NO are determined by "retrospective" analysis of an already known CI wave function, their construction in advance of the CI calculations can be only an approximate one, e.g., in an iterative procedure of determining the so-called *pseudo* NO (PNO). It usually consists of using the SCF virtuals in a limited-CI study, diagonalizing the resulting 1-matrix, which determines the first-approximation PNO, using them as a basis for an improved CI expansion, from which the second-approximation PNO are constructed, etc. Such iterative process is carried out until self-consistency is attained within the adopted level of electron excitations, when the subsequent iterations do not appreciably modify the PNO shapes and occupations.

6.4 Electron Pair Theories

As we have concluded from (6.133), the MP2 theory in fact represents the *perturbational* IEPA approximation (Sinanoĝlu 1964; Nesbet 1965; Szabo and Ostlund 1982) giving rise to the additive contributions to the system Coulomb correlation energy due to each occupied MO-pair of electrons. Clearly, a similar perspective can be also adopted within the *variational* method of determining approximate wave functions of molecular systems. Since the Coulomb correlation effect should be the strongest in pairs of electrons exhibiting opposite spins and occupying the same MO, the simplest formulation of such a variational approach should involve the antisymmetrized product of the electron pair functions, called *geminals*, the name first coined by Shull (1959) to distinguish these two-electron (group) functions from their one-electron analogs – the orbitals. In this *separated pair* approach, each MO-pair of electrons is correlated internally, but the interpair correlation is neglected. This missing part of the Coulomb correlation is accounted for in the more sophisticated *Coupled Electronic Pair Approximation* (CEPA) (Meyer 1977), *Coupled Pair Many Electron Theory* (CPMET) or *Coupled Cluster Approximation* (CCA) (Bartlett 1981, 1989, 1995, 2000; Kucharski and Bartlett 1986; Čižek 1966, 1969; Paldus and Čižek 1973, 1975; Hurley 1976; Kutzelnigg 1977; Čižek and Paldus 1980; Jørgensen and Simons 1981; Szabo and Ostlund 1982), which use appropriate simplifications of the whole hierarchy of the CI equations. As expected, the inclusion of interpair correlation improves accuracy at the expense of the computational complexity and clarity of interpretation. Of similar character is the *Generalized Valence Bond* (GVB) approach of Goddard and collaborators (Goddard III 1967; Goddard III and Ladner 1971; Hunt et al. 1972; Goddard III et al. 1973; Moss and Goddard III 1975; Bobrowicz and Goddard III 1977; Goddard III and Harding 1978), which represents another electron pair generalization of the classical VB theory of Heitler and London.

The *geminal* method can be also viewed as a logical next step in the wave function factorization following the *orbital* approximation. It represents a particular case of the *group* wave functions, describing specific molecular fragments, in which each constituent subsystem is first considered as a separate entity and subsequently their mutual interaction is accounted for, e.g., in the perturbative or self-consistent way. It should be observed that such a general line of thinking also lies behind the separation between the inner and valence shells, the σ and π electrons in aromatic systems, etc. The use of electronic pairs as fundamental structural entities in the limited-CI theories is very much in line with the intuitive chemical thinking. Indeed, the concepts of the bonding pairs in valence shell, as well as the lone pairs of both the valence or inner shells, have been very successful in the early, qualitative theories of electronic structure. Hence, their explicit recognition in the modern beyond-HF calculations brings an additional, chemical insight into the mechanism of the electron Coulomb correlation and represents an element of continuity in the development of such ideas in the theory of chemistry (Pauling 1949; Hurley et al. 1953; Parks and Parr 1958, 1960; Karplus and Grant 1959; McWeeny and Ohno 1960; McWeeny 1989).

One could also include in this category Moffitt's (1951) method of *Atoms-in-Molecules* (AIM) and its subsequent extension involving *deformed* AIM (Arai 1957, 1960; Ellison 1965; Ellison and Wu 1967, 1968; Ellison and Slezak 1969), as well as the *Diatomics-in-Molecules* (DIM) theory (Kuntz 1979; Tully 1977, 1980) and its generalization in the form of the *Molecules-in-Molecules* (MIM) approach, which involves groups of electrons of larger parts of the whole molecular system. All these approaches have played an important role in modeling the molecular PES for the dynamical calculations, e.g., in developing the familiar *London–Eyring–Polanyi–Sato* (LEPS) (e.g.: Hirst 1985; Murrell et al. 1984; Murrell and Bosanac 1989), *Bond-Energy – Bond-Order* (BEBO) (Johnston and Parr 1963) and DIM energy surfaces for the reactive scattering calculations.

6.4.1 Electron Pairs on Strongly Orthogonal Geminals

Following the orbital approximation of the HF theory, marking the separated electrons approach, we now assume the related form of the variational wave function of, say, $N = 2p$ electrons in terms of the separated electronic pairs [compare (5.66)] given by the antisymmetrized product of two-electron functions (geminals) $\{G_j(1, 2)\}$:

$$\Psi_A^G(1, 2, \ldots, N) = \hat{A}'\{G_1(1, 2)\, G_2(3, 4) \ldots G_p(N-1, N)\}; \tag{6.162}$$

here $\hat{A}'$ stands for the partial antisymmetrizer exchanging electrons between different geminals.

These normalized two-electron functions,

$$\iint |G_j(1, 2)|^2 \, d\boldsymbol{q}_1 d\boldsymbol{q}_2 = \langle G_j(1, 2) \mid G_j(1, 2) \rangle_{1,2} = 1, \tag{6.163}$$

have to be also mutually orthogonal:

$$\iint G_i^*(1, 2) G_j(1, 2)\, d\boldsymbol{q}_1 d\boldsymbol{q}_2 = \langle G_i(1, 2) \mid G_j(1, 2) \rangle_{1,2} = 0, \quad i \neq j. \tag{6.164}$$

However, to simplify the expression for the expectation value of the electronic energy and hence also the resulting Euler equations of the variational method using this trial function, the *strong orthogonality* condition is imposed on the optimum geminals:

$$\int G_i^*(1, 2) G_j(1, 2) d\boldsymbol{q}_1 = \langle G_i(1, 2) \mid G_j(1, 2) \rangle_1 = 0, \quad \text{for any } \boldsymbol{q}_2. \tag{6.165}$$

Notice that this requirement automatically implies the ordinary orthogonality of (6.164). The trial wave function of (6.162) is then referred to as the *Antisymmetrized Product of Strongly Orthogonal Geminals* (APSG).

In each (antisymmetric) two-electron state, $G_j(1, 2) = -G_j(2, 1)$, representing the spin-paired electrons in the singlet spin eigenstate of (6.67) (see Fig. 6.2), one can also separate its spatial and spin parts [see (5.103)]:

$$\begin{aligned} G_j(\boldsymbol{q}_1, \boldsymbol{q}_2) \equiv G_j(1, 2) &= \Phi_j(\boldsymbol{r}_1, \boldsymbol{r}_2) \frac{1}{\sqrt{2}} [\alpha(\sigma_1)\beta(\sigma_2) - \beta(\sigma_1)\alpha(\sigma_2)] \\ &\equiv \Phi_j(1, 2)\, U_{0,0}(1, 2). \end{aligned} \tag{6.166}$$

Therefore, since the singlet spin factor of each geminal in APSG of (6.162) is also antisymmetric, $U_{0,0}(1, 2) = -U_{0,0}(2, 1)$, the spatial factors must be symmetric functions of two electrons, $\Phi_j(1, 2) = \Phi_j(2, 1)$, in order to satisfy the Pauli antisymmetrization requirement for these two (spin-paired) fermions.

Let us now examine the corresponding expression for the expectation value of the electronic Hamiltonian in such a trial state [see (5.69)–(5.71)]:

$$\langle E^e\rangle_{\Psi_A^G} = \langle\Psi_A^G|\hat{\mathrm{H}}^e|\Psi_A^G\rangle = \langle\Psi_A^G|\hat{\mathcal{F}}|\Psi_A^G\rangle + \langle\Psi_A^G|\hat{\mathcal{G}}|\Psi_A^G\rangle \equiv \langle\mathcal{F}\rangle_{\Psi_A^G} + \langle\mathcal{G}\rangle_{\Psi_A^G}. \quad (6.167)$$

Due to the strong orthogonality relations between the two-electron functions, the one-electron operator $\hat{\mathrm{h}}(1)$[(5.69) and (5.70)] can only couple the electron in the same geminal of Ψ_A^{G*} and Ψ_A^G. One thus finds the following expression for the overall one-electron contribution to the electronic energy:

$$\langle\mathcal{F}\rangle_{\Psi_A^G} = 2\sum_{j=1}^{p}\langle\Phi_j(1,\,2)|\hat{\mathrm{h}}(1)|\Phi_j(1,\,2)\rangle_{1,2}. \quad (6.168)$$

The two-electron repulsion $g(1,2) = 1/r_{1,2}$ couples both the pairs of particles on the same and different geminals of Ψ_A^{G*} and Ψ_A^G. Thus, using again the orthonormality relations and summing over electron spin orientations gives:

$$\langle\mathcal{G}\rangle_{\Psi_A^G} = \sum_{j=1}^{p}\langle\Phi_j(1,2)|g(1,2)|\Phi_j(1,2)\rangle_{1,2}$$
$$+\sum_{i=1}^{p-1}\sum_{j=i+1}^{p}2\Big(2\langle\Phi_i(1,2)\Phi_j(3,4)|g(1,3)|\Phi_i(1,2)\Phi_j(3,4) - \Phi_i(3,2)\Phi_j(1,4)\rangle_{1,2,3,4}\Big). \quad (6.169)$$

The strong orthogonality condition (6.165) can be automatically satisfied when one uses different (mutually orthogonal) subspaces $\{\psi_{l,j}\}$ of NO to represent different geminals, i.e., when the total (complete) space of the orthonormal orbitals $\boldsymbol{\psi} = \{\psi_s\}$ is partitioned into the exclusive subsets $\{\boldsymbol{\psi}_j = \{\psi_{l,j}\}\}$ for representing jth geminal, $\boldsymbol{\psi} = (\boldsymbol{\psi}_1, \boldsymbol{\psi}_2, \ldots, \boldsymbol{\psi}_p)$,

$$\{\psi_s\} = (\{\psi_{l,1}\},\{\psi_{l,2}\},\ldots,\{\psi_{l,p}\}),\quad \{\psi_{l,j}\} = \left(\psi_{1,j},\psi_{2,j},\ldots,\psi_{w_j,j}\right),$$
$$\langle\psi_{l,j}|\psi_{k,i}\rangle = \delta_{l,k}\delta_{j,i},$$
$$\Phi_j(1,\,2) = \sum_{k=1}^{w_j} b_{k,j}\psi_{k,j}(1)\psi_{k,j}(2),\qquad j = 1,2,\,\ldots,p. \quad (6.170)$$

The normalization of geminals thus implies the following relation to be satisfied by unknown expansion coefficients $\{b_{k,j}\}$ related to NO occupations $\left\{n_{k,j} = b_{k,j}^2\right\}$:

$$\sum_{k=1}^{w_j} b_{k,j}^2 = \sum_{k=1}^{w_j} n_{k,j} = 1. \quad (6.171)$$

In order to justify this NO expansion consider a general spatial two-electron function $\Phi(1, 2)$, symmetric in the antisymmetric (singlet) spin state, e.g., that describing the ground state in He or H_2. It can be expressed in the complete basis set of the orthonormal (real) MO, $\boldsymbol{\varphi} = (\varphi_1, \varphi_2, ..., \varphi_w)$, using the symmetrized CI expansion of (6.60):

$$\begin{aligned}\Phi(1, 2) &= \sum_{k=1}^{w}\sum_{l=1}^{w} d_{k,l}\frac{1}{2}[\varphi_k(1)\varphi_l(2) + \varphi_l(1)\varphi_k(2)] \\ &= \frac{1}{2}[\boldsymbol{\varphi}(1)\,\mathbf{d}\,\boldsymbol{\varphi}^{\mathrm{T}}(2) + \boldsymbol{\varphi}(2)\,\mathbf{d}\,\boldsymbol{\varphi}^{\mathrm{T}}(1)].\end{aligned} \tag{6.172}$$

It gives rise to the following spinless 1-matrix of (6.157):

$$\begin{aligned}\hat{\gamma}_1(\boldsymbol{r}_1;\boldsymbol{r}_1{}') &= 2\sum_{k=1}^{w}\sum_{k'=1}^{w}\varphi_k(\boldsymbol{r}_1)\left[\sum_{l=1}^{w} d_{k,l}d_{k',l}\right]\varphi_{k'}(\boldsymbol{r}_1{}') \\ &\equiv 2\sum_{k=1}^{w}\sum_{k'=1}^{w}\varphi_k(\boldsymbol{r}_1)\,\gamma_{k,k'}^{\mathrm{MO}}\,\varphi_{k'}(\boldsymbol{r}_1{}') \equiv \sum_{k=1}^{w}\sum_{k'=1}^{w}\varphi_k(\boldsymbol{r}_1)\,P_{k,k'}^{\mathrm{MO}}(\mathbf{d})\,\varphi_{k'}^{*}(\boldsymbol{r}_1{}'),\end{aligned} \tag{6.173}$$

where the CBO matrix $\mathbf{P}^{\mathrm{MO}}(\mathbf{d}) = 2\mathbf{d}\mathbf{d}^{\mathrm{T}} = 2\boldsymbol{\gamma}^{\mathrm{MO}}$.

The NO $\boldsymbol{\psi} = \boldsymbol{\varphi}\mathbf{U}$ are then determined by the orthogonal transformation $\mathbf{U}$ which diagonalizes the CBO matrix $\boldsymbol{\gamma}^{\mathrm{MO}}$:

$$\begin{aligned}\mathbf{U}^{\mathrm{T}}\boldsymbol{\gamma}^{\mathrm{MO}}\mathbf{U} &= \mathbf{U}^{\mathrm{T}}\mathbf{d}\mathbf{d}^{\mathrm{T}}\mathbf{U} = (\mathbf{U}^{\mathrm{T}}\mathbf{d}\mathbf{U})(\mathbf{U}^{\mathrm{T}}\mathbf{d}^{\mathrm{T}}\mathbf{U}) = (\mathbf{U}^{\mathrm{T}}\mathbf{d}\mathbf{U})(\mathbf{U}^{\mathrm{T}}\mathbf{d}\mathbf{U})^{\mathrm{T}} \\ &\equiv \mathbf{c}\mathbf{c}^{\mathrm{T}} = \left\{c_\alpha^2\delta_{\alpha,\beta} = n_\alpha\delta_{\alpha,\beta}\right\};\end{aligned} \tag{6.174}$$

and hence $c_\alpha = \pm\sqrt{n_\alpha}$. Finally, substituting the inverse transformation $\boldsymbol{\psi}\mathbf{U}^{\mathrm{T}} = \boldsymbol{\varphi}$ into (6.172) gives the equivalent, more compact natural expansion of the spatial two-electron wave function (geminal):

$$\Phi(1, 2) = \frac{1}{2}[\boldsymbol{\psi}(1)(\mathbf{U}^{\mathrm{T}}\mathbf{d}\mathbf{U})\,\boldsymbol{\psi}^{\mathrm{T}}(2) + \boldsymbol{\psi}(2)(\mathbf{U}^{\mathrm{T}}\mathbf{d}\mathbf{U})\,\boldsymbol{\psi}^{\mathrm{T}}(1)] = \sum_{\alpha=1}^{w} c_\alpha\,\psi_\alpha(1)\,\psi_\alpha(2). \tag{6.175}$$

For example, Shull and colleagues (Shull 1960, 1962, 1964; Hagstrom and Shull 1963; Shull and Prosser 1964; Anex and Shull 1964; Christoffersen and Shull 1968; Christoffersen 1989a, b) have argued that the chemical bond in H_2 (one-geminal system) is already well described by the two-term natural expansion

$$\begin{aligned}\Phi(1, 2) &= c_1\,\psi_1(1)\,\psi_1(2) + c_2\,\psi_2(1)\,\psi_2(2) \\ &= \sqrt{n_1}\,\psi_1(1)\,\psi_1(2) - \sqrt{n_2}\,\psi_2(1)\,\psi_2(2)\,,\end{aligned} \tag{6.176}$$

where $n_1 + n_2 \cong 1$. Indeed, in this highly symmetric (homonuclear) diatomic the two dominating NO basically represent the bonding and antibonding MO, so that the two terms in the preceding equation represent the HF ground and doubly excited MO configurations, respectively, generated in the minimum basis set of two $1s$ orbitals contributed by the two constituent atoms. Therefore, the above wave function in fact corresponds to the minimum CID approximation required to correctly describe the dissociation of the molecule into atoms, which is variationally equivalent to the *Heitler–London* (HL) description of the classical VB theory.

Therefore, in the NO representation of geminals (6.170) the average energy functional of (6.167)–(6.169) becomes the associated function of the NO occupations $\{n_{k,j}\}$ determining the CI expansion coefficients $\left\{b_{k,j} = \pm\sqrt{n_{k,j}}\right\}$, and the functional of the shapes of NO themselves. The latter can be subsequently expressed in terms of the appropriate AO basis, with only the expansion coefficients being optimized, as in the SCF MO method. It should be stressed, however, that the effective energy operators are generally different for each NO, being not invariant to the unitary transformation of the basis set, so that the *off*-diagonal Lagrange multipliers enforcing the orthonormality of NO cannot be easily eliminated. This greatly complicates finding the optimum solutions in practical applications of the method.

The four-electron (heteronuclear) diatomic LiH involves the *inner*-shell (nonbonding) and the *valence*-shell (bonding) geminals. The former represents practically unchanged atomic inner shell of Li, while the dominating NO of the latter exhibits typical effects of the intraatom radial and angular promotion, due to $2s \rightarrow 2p$ and $2s \rightarrow 3s$ effective excitations in Li, as well as the interatom AO mixing generating the covalent (electron delocalization) bond component. The Coulomb correlation energy recovery was about 80%, giving rise to the calculated magnitude of the binding energy, 2.3 eV, which compares favorably with the experimental estimate of 2.52 eV. However, the *strong*-orthogonality constraint has been shown to be a reason for rather poor description of the σ *lone* pair geminal in NH (four- geminal system), since the mainly $2p$ NO has been used to describe the bonding geminal, thus being unavailable for the *lone* pair geminal. Also, the accumulation of four electronic pairs on N gives rise to a strong intergeminal Coulomb correlation, which is missing in this separated pair approach.

One finally observes that the APSG approximation within the separated electron pair method represents a variant of the coupled variational procedure of the MC SCF theory, in which both the shapes of orbitals and CI coefficients are being simultaneously optimized. The quality of the CI expansion determining the reference function is now determined by the length of the NO basis set used to represent the strongly orthogonal geminals.

6.4.2 Independent Electron Pair Approximation

For reasons of simplicity in what follows we shall focus on the Coulomb correlation in the molecular (closed-shell) ground state. The corresponding CI expansion of the exact (unnormalized) ground state function $\bar{\Psi}_0$,

$$\hat{\mathrm{H}}^e\bar{\Psi}_0 = E_0^e\bar{\Psi}_0,$$

where E_0^e denotes the exact electronic energy in the NREL limit (Fig. 6.1), is represented by the CI expansion in the so called *intermediate* normalization representation, for $C_0 \equiv 1$ (e.g., Szabo and Ostlund 1982),

$$\bar{\Psi}_0 = \bar{\Psi}_0^{\text{CI}} \equiv \Psi_0^{\text{HF}} + \sum_k^{occd.} \sum_p^{virt.} \bar{C}_k^p \Phi_k^p + \sum_{k<l}^{occd.} \sum_{p<q}^{virt.} \bar{C}_{k,l}^{p,q} \Phi_{k,l}^{p,q} + \sum_{k<l<m}^{occd.} \sum_{p<q<r}^{virt.} \bar{C}_{k,l,m}^{p,q,r} \Phi_{k,l,m}^{p,q,r} + \ldots; \tag{6.177}$$

here Ψ_0^{HF} stands for the fully optimized Slater determinant of the HF theory, giving rise to the HF energy $E_0^{\text{HF}} = \langle \Psi_0^{\text{HF}} | \hat{\text{H}}^e | \Psi_0^{\text{HF}} \rangle$.

Therefore, it directly follows from the definition of the Coulomb correlation energy, $\Delta E_{corr.} = E_0^e - E_0^{\text{HF}}$, that

$$\left(\hat{\text{H}}^e - E_0^{\text{HF}} \right) \bar{\Psi}_0 = \Delta E_{corr.} \bar{\Psi}_0. \tag{6.178}$$

Multiplying from the left by $\left(\Psi_0^{\text{HF}}\right)^*$ and integrating over the position-spin variables of all electrons then gives the following expression for the correlation energy:

$$\langle \Psi_0^{\text{HF}} | \left(\hat{\text{H}}^e - E_0^{\text{HF}} \right) | \bar{\Psi}_0 \rangle = \Delta E_{corr.}. \tag{6.179}$$

This expression can be further simplified using the Brillouin theorem (6.116) and the Slater–Condon rule of (5.87):

$$\begin{aligned} \langle \Psi_0^{\text{HF}} | \left(\hat{\text{H}}^e - E_0^{\text{HF}} \right) | \bar{\Psi}_0 \rangle &= \sum_{k<l}^{occd.} \sum_{p<q}^{virt.} \bar{C}_{k,l}^{p,q} \langle \Psi_0^{\text{HF}} | \left(\hat{\text{H}}^e - E_0^{\text{HF}} \right) | \Phi_{k,l}^{p,q} \rangle \\ &= \sum_{k<l}^{occd.} \sum_{p<q}^{virt.} \bar{C}_{k,l}^{p,q} \langle \Psi_0^{\text{HF}} | \hat{\text{H}}^e | \Phi_{k,l}^{p,q} \rangle \equiv \sum_{k<l}^{occd.} e_{k,l} = \Delta E_{corr.}. \end{aligned} \tag{6.180}$$

The preceding equation again emphasizes the dominating role of the *doubly* excited configurations in correcting the HF wave function for the Coulomb correlation effects, and it formally expresses the overall correlation energy as the sum of contributions from all electronic pairs identified by labels $\{k < l\}$ of the occupied SO in the (ground-state) HF determinant, which is analogous to the MP2 expression of (6.133).

In the *Independent Electron Pair Approximation* (IEPA) one uses the variational method to determine the correlation energies $\left\{ e_{k,l}^{\text{IEPA}} \right\}$ due to each separate electron pair $k < l$, when the contributions to the correlation energy from the remaining electrons are ignored. For this given electron pair, one thus defines the CID-type correlated wave function in the intermediate normalization representation:

$$\bar{\Psi}_{k,l}^{\mathrm{IEPA}} = \Psi_0^{\mathrm{HF}} + \sum_{p<q}^{virt.} \bar{C}_{p,q}^{\mathrm{IEPA}} \Phi_{k,l}^{p,q}, \tag{6.181}$$

and determines the optimum CI coefficients by minimizing the expectation value of the electronic Hamiltonian:

$$\begin{aligned} E_{k,l}^{\mathrm{IEPA}} &= \left\langle \bar{\Psi}_{k,l}^{\mathrm{IEPA}} \middle| \hat{\mathrm{H}}^e \middle| \bar{\Psi}_{k,l}^{\mathrm{IEPA}} \right\rangle / \left\langle \bar{\Psi}_{k,l}^{\mathrm{IEPA}} \middle| \bar{\Psi}_{k,l}^{\mathrm{IEPA}} \right\rangle \\ &\equiv \bar{E}_{k,l}^{\mathrm{IEPA}} / \left\langle \bar{\Psi}_{k,l}^{\mathrm{IEPA}} \middle| \bar{\Psi}_{k,l}^{\mathrm{IEPA}} \right\rangle \equiv \left\langle \Psi_{k,l}^{\mathrm{IEPA}} \middle| \hat{\mathrm{H}}^e \middle| \Psi_{k,l}^{\mathrm{IEPA}} \right\rangle = E_0^{\mathrm{HF}} + e_{k,l}^{\mathrm{IEPA}}, \end{aligned} \tag{6.182}$$

where the normalized pair wave function $\Psi_{k,l}^{\mathrm{IEPA}} = \bar{\Psi}_{k,l}^{\mathrm{IEPA}} / \left\langle \bar{\Psi}_{k,l}^{\mathrm{IEPA}} \mid \bar{\Psi}_{k,l}^{\mathrm{IEPA}} \right\rangle^{1/2}$.

This variational criterion gives the associated eigenvalue problem:

$$\begin{aligned} \frac{\delta \bar{E}_{k,l}^{\mathrm{IEPA}}}{\delta \bar{\Psi}_{k,l}^{\mathrm{IEPA}*}} &= \frac{\delta E_{k,l}^{\mathrm{IEPA}}}{\delta \bar{\Psi}_{k,l}^{\mathrm{IEPA}*}} \left\langle \bar{\Psi}_{k,l}^{\mathrm{IEPA}} \mid \bar{\Psi}_{k,l}^{\mathrm{IEPA}} \right\rangle + E_{k,l}^{\mathrm{IEPA}} \bar{\Psi}_{k,l}^{\mathrm{IEPA}} = E_{k,l}^{\mathrm{IEPA}} \bar{\Psi}_{k,l}^{\mathrm{IEPA}} \\ &= \hat{\mathrm{H}}^e \bar{\Psi}_{k,l}^{\mathrm{IEPA}}, \end{aligned} \tag{6.183}$$

where we have used the variational principle $\delta E_{k,l}^{\mathrm{IEPA}} / \delta \bar{\Psi}_{k,l}^{\mathrm{IEPA}*} = 0$.

By repeating this procedure for all pairs of electrons and summing over such additive energy terms, one then estimates the system overall Coulomb correlation energy in the IEPA approximation:

$$\Delta E_{corr.}^{\mathrm{IEPA}} \cong \sum_{k<l}^{occd.} e_{k,l}^{\mathrm{IEPA}}. \tag{6.184}$$

This method thus uses different wave functions to estimate different contributions to the total correlation energy between independent pairs of electrons and hence the magnitude of such an estimate may exceed that of the exact correlation energy. Since in this treatment, one neglects the correlation coupling between different pairs of electrons, by neglecting the *off*-diagonal elements of the CI energy matrix between *doubly* excited configurations originating from different pairs of occupied orbitals in the HF reference function, $H_{(k,l)\to(p,q);(m,n)\to(p',q')}^{\mathrm{CI}}$ for $(m, n) \neq (k, l)$, this method is simpler than the CID variant, which violates the size consistency requirement. For the same reason the IEPA approach is size-consistent, i.e., the energy of m noninteracting monomers is m times the energy of a single monomer.

The equations for the pair correlation energy and the optimum (ground-state) CI coefficients $\left\{ \bar{C}_{k,l}^{p,q} \right\} \equiv \bar{\boldsymbol{C}}^{(k,l)}$(the column vector) directly follow from the intermediate normalization projections of the related eigenvalue equations onto the Ψ_0^{HF} and $\left\{ \Phi_{k,l}^{p,q} \right\}$ configurations, respectively [see (6.178)–(6.180)]. It follows from (6.180) that the former gives the expression for the electron pair correlation energy,

$$\left\langle \Psi_0^{\mathrm{HF}} \middle| \hat{\mathrm{H}}^e - E_0^{\mathrm{HF}} \middle| \bar{\Psi}_{k,l}^{\mathrm{IEPA}} \right\rangle = e_{k,l}^{\mathrm{IEPA}} = \sum_{p<q}^{virt.} H_{0,(k,l)\to(p,q)}^{CI} \bar{C}_{k,l}^{p,q}$$
$$\equiv \sum_{p<q}^{virt} (\boldsymbol{A}^{(k,l)})_{p,q} (\boldsymbol{C}^{(k,l)})_{p,q} = \boldsymbol{A}^{(k,l)} \boldsymbol{C}^{(k,l)}, \qquad (6.185)$$

where $\boldsymbol{A}^{(k,l)}$ denotes the row vector. The latter projection and (6.183) generate the additional relation for the expansion coefficients,

$$\left\langle \Phi_{k,l}^{p,q} \middle| \hat{\mathrm{H}}^e - E_0^{\mathrm{HF}} \middle| \bar{\Psi}_{k,l}^{\mathrm{IEPA}} \right\rangle = \left\langle \Phi_{k,l}^{p,q} \middle| \hat{\mathrm{H}}^e \middle| \Psi_0^{\mathrm{HF}} \right\rangle + \sum_{p'<q'}^{virt.} \left\langle \Phi_{k,l}^{p,q} \middle| \hat{\mathrm{H}}^e - E_0^{\mathrm{HF}} \middle| \Phi_{k,l}^{p',q'} \right\rangle \bar{C}_{k,l}^{p',q'}$$
$$= e_{k,l}^{\mathrm{IEPA}} \left\langle \Phi_{k,l}^{p,q} \middle| \bar{\Psi}_{k,l}^{\mathrm{IEPA}} \right\rangle = e_{k,l}^{\mathrm{IEPA}} \bar{C}_{k,l}^{p,q}, \qquad (6.186)$$

or in the matrix notation of (6.185):

$$(\boldsymbol{A}^{(k,l)\dagger})_{p,q} + \sum_{p'<q'}^{virt} (\mathbf{D}^{(k,l)})_{p,q;p',q'} (\boldsymbol{C}^{(k,l)})_{p',q'} = e_{k,l}^{\mathrm{IEPA}} (\boldsymbol{C}^{(k,l)})_{p,q}, \qquad (6.187)$$

where

$$(\mathbf{D}^{(k,l)})_{p,q;p',q'} = \left\langle \Phi_{k,l}^{p,q} \middle| \hat{\mathrm{H}}^e - E_0^{\mathrm{HF}} \middle| \Phi_{k,l}^{p',q'} \right\rangle$$
$$= H_{(k,l)\to(p,q);(k,l)\to(p',q')}^{\mathrm{CI}} - E_0^{\mathrm{HF}} \delta_{p,p'} \delta_{q,q'}. \qquad (6.188)$$

Finally, the coupled matrix equations (6.185) and (6.187),

$$e_{k,l}^{\mathrm{IEPA}} = \boldsymbol{A}^{(k,l)} \bar{\boldsymbol{C}}^{(k,l)} \quad \text{and} \quad \boldsymbol{A}^{(k,l)\dagger} + \mathbf{D}^{(k,l)} \bar{\boldsymbol{C}}^{(k,l)} = e_{k,l}^{\mathrm{IEPA}} \bar{\boldsymbol{C}}^{(k,l)}, \qquad (6.189)$$

which determine the *pair* correlation energies and the associated CI coefficients, can be combined in the following eigenvalue equation:

$$\begin{bmatrix} 0 & \boldsymbol{A}^{(k,l)} \\ \boldsymbol{A}^{(k,l)\dagger} & \mathbf{D}^{(k,l)} \end{bmatrix} \begin{bmatrix} 1 \\ \bar{\boldsymbol{C}}^{(k,l)} \end{bmatrix} \equiv \mathbf{B}^{(k,l)} \begin{bmatrix} 1 \\ \bar{\boldsymbol{C}}^{(k,l)} \end{bmatrix} = e_{k,l}^{\mathrm{IEPA}} \begin{bmatrix} 1 \\ \bar{\boldsymbol{C}}^{(k,l)} \end{bmatrix}. \qquad (6.190)$$

Therefore, the relevant CI coefficients determine the eigenvector of $\mathbf{B}^{(k,l)}$, with the eigenvalue $e_{k,l}^{\mathrm{IEPA}}$ determining the pair correlation energy. In view of the variational principle of (6.183),

$$\min E_{k,l}^{\mathrm{IEPA}} = \min\left(E_0^{\mathrm{HF}} + e_{k,l}^{\mathrm{IEPA}}\right) = \min e_{k,l}^{\mathrm{IEPA}}, \qquad (6.191)$$

one thus diagonalizes the symmetric matrix $\mathbf{B}^{(k,l)}$ and selects the (normalized) eigenvector corresponding to the *lowest* eigenvalue, which determines the optimum value of the *pair* correlation energy. For each pair of the occupied MO in the HF reference state, the associated $\{\mathbf{B}^{(k,l)}\}$ matrices are different, so that there is no variational lower bound to the IEPA overall estimate of the Coulomb (intrapair) correlation energy.

As shown by Ahlrichs et al. (1975a, b) in a series of simple hydrides exhibiting different numbers of the localized bonds and the lone pairs in the valence shell, one finds that the magnitude of the correlation energy of two electrons occupying the nonbonding orbital is lower than the corresponding intrabond contribution. The smaller interorbital correlation energies between bonding and/or nonbonding electrons were found to be of comparable magnitude. In each case, the total estimate of the Coulomb correlation energy exceeded the corresponding CI value, thus confirming the altogether nonvariational aspect of the IEPA method.

6.4.3 *Coupled Electron Pair Approximations*

In fact the internally correlated electronic pairs repel each other and thus are not fully independent. Therefore, several improvements relative to the IEPA approach have been proposed, which approximately take into account the interpair correlation as well. These *size*-consistent *Coupled Electron Pair Approximations* (CEPA) represent a special case of a more general strategy, called the *Coupled Cluster* (CC) approximation, which we shall introduce in Sect. 6.5. Although all these extensions generally give rise to higher accuracy of the predicted effects of the electron correlation, be it at a severe cost of increased computational complexity and hence a reduced capability of an easy interpretation, the need for further improvement still remains. Such more advanced techniques, however, are generally unsuitable for applications to very large supramolecular systems of the contemporary chemistry/biology and in solid state physics.

The interpair correlation is represented by the *quadruply* (Q) excited configurations from the reference HF wave function, involving excitations of both electrons in each pair. Therefore, to cover these effects the CI expansion of (6.177) should be at least of the CIDQ type:

$$\begin{aligned}\bar{\Psi}_0^{\mathrm{CI}} &\cong \Psi_0^{\mathrm{HF}} + \sum_{k<l}^{occd.} \sum_{p<q}^{virt.} \bar{C}_{k,l}^{p,q}\, \Phi_{k,l}^{p,q} + \sum_{k<l<m<n}^{occd.} \sum_{p<q<r<s}^{virt.} \bar{C}_{k,l,m,n}^{p,q,r,s}\, \Phi_{k,l,m,n}^{p,q,r,s} \\ &\equiv \bar{\Psi}_0^{\mathrm{CIDQ}}. \end{aligned} \tag{6.192}$$

Notice that (6.180), obtained from projecting the eigenvalue problem for the correlation energy (6.178) onto the HF reference state, remains valid also for this extended size of the CI expansion,

$$\left\langle\Psi_0^{\mathrm{HF}}\middle|\left(\hat{\mathrm{H}}^e - E_0^{\mathrm{HF}}\right)\middle|\bar{\Psi}_0^{\mathrm{CIDQ}}\right\rangle = \sum_{k<l}^{occd.}\sum_{p<q}^{virt.}\bar{C}_{k,l}^{p,q}\left\langle\Psi_0^{\mathrm{HF}}\middle|\hat{\mathrm{H}}^e\middle|\Phi_{k,l}^{p,q}\right\rangle$$
$$\equiv \sum_{k<l}^{occd.} e_{k,l}^{\mathrm{CIDQ}} = \Delta E_{corr.}^{\mathrm{CIDQ}}. \qquad (6.193)$$

The projection onto the *doubly* excited configuration [compare (6.186)] now gives the following coupled equations for the CI coefficients,

$$\left\langle\Phi_{i,j}^{r,s}\middle|\hat{\mathrm{H}}^e - E_0^{\mathrm{HF}}\middle|\bar{\Psi}_0^{\mathrm{CIDQ}}\right\rangle = \left\langle\Phi_{i,j}^{r,s}\middle|\hat{\mathrm{H}}^e\middle|\Psi_0^{\mathrm{HF}}\right\rangle + \sum_{k<l}^{occd.}\sum_{p<q}^{virt.}\left\langle\Phi_{i,j}^{r,s}\middle|\hat{\mathrm{H}}^e - E_0^{\mathrm{HF}}\middle|\Phi_{k,l}^{p,q}\right\rangle\bar{C}_{k,l}^{p,q}$$
$$+\sum_{m<n}^{occd.}\sum_{t<u}^{virt.}\left\langle\Phi_{i,j}^{r,s}\middle|\hat{\mathrm{H}}^e\middle|\Phi_{i,j,m,n}^{r,s,t,u}\right\rangle\bar{C}_{i,j,m,n}^{r,s,t,u} = \Delta E_{corr.}^{\mathrm{CIDQ}}\bar{C}_{i,j}^{r,s}, \qquad (6.194)$$

which directly follow from the orthogonality relations between the configuration functions and the Slater–Condon rules for the matrix elements of the electronic Hamiltonian. The preceding equation relates the CI coefficients of the *doubly* excited configurations to those corresponding to the quadruple excitations. The latter reflect the interpair correlation, which has been neglected in the IEPA method.

It should be observed that only the assumption $\left\{\bar{C}_{k,l,m,n}^{p,q,r,s} = 0\right\}$ gives the uncoupled equations of the standard CID approach of Sect. 6.3.1 [see also the matrix notation of (6.185), (6.187) and (6.189)]:

$$\left\langle\Psi_0^{\mathrm{HF}}\middle|\left(\hat{\mathrm{H}}^e - E_0^{\mathrm{HF}}\right)\middle|\bar{\Psi}_0^{\mathrm{CID}}\right\rangle = \sum_{k<l}^{occd.}\sum_{p<q}^{virt.}\left\langle\Psi_0^{\mathrm{HF}}\middle|\hat{\mathrm{H}}^e\middle|\Phi_{k,l}^{p,q}\right\rangle\bar{C}_{k,l}^{p,q} \equiv \boldsymbol{A}^{\mathrm{CID}}\bar{\boldsymbol{C}}^{\mathrm{CID}} = \Delta E_{corr.}^{\mathrm{CID}},$$
$$\left\langle\Phi_{k,l}^{p,q}\middle|\hat{\mathrm{H}}^e - E_0^{\mathrm{HF}}\middle|\bar{\Psi}_0^{\mathrm{CID}}\right\rangle = \left\langle\Phi_{k,l}^{p,q}\middle|\hat{\mathrm{H}}^e\middle|\Psi_0^{\mathrm{HF}}\right\rangle + \sum_{m<n}^{occd.}\sum_{r<s}^{virt.}\left\langle\Phi_{k,l}^{p,q}\middle|\hat{\mathrm{H}}^e - E_0^{\mathrm{HF}}\middle|\Phi_{m,n}^{r,s}\right\rangle\bar{C}_{m,n}^{r,s}$$
$$= \Delta E_{corr.}^{\mathrm{CID}}\bar{C}_{k,l}^{p,q}, \qquad (6.195)$$

or in the matrix notation,

$$\boldsymbol{A}^{\mathrm{CID}\dagger} + \mathbf{D}^{\mathrm{CID}}\bar{\boldsymbol{C}}^{\mathrm{CID}} = \Delta E_{corr.}^{\mathrm{CID}}\bar{\boldsymbol{C}}^{\mathrm{CID}}, \qquad (6.196)$$

with the CI energy-difference matrix $\mathbf{D}^{\mathrm{CID}}$ defined in (6.188).

These equations can be again combined into the associated eigenvalue equation for the ground state correlation energy [compare (6.190)]:

$$\begin{bmatrix} 0 & \boldsymbol{A}^{\mathrm{CID}} \\ \boldsymbol{A}^{\mathrm{CID}\dagger} & \mathbf{D}^{\mathrm{CID}} \end{bmatrix}\begin{bmatrix} 1 \\ \bar{\boldsymbol{C}}^{\mathrm{CID}} \end{bmatrix} \equiv \mathbf{B}^{\mathrm{CID}}\begin{bmatrix} 1 \\ \boldsymbol{C}^{\mathrm{CID}} \end{bmatrix} = \Delta E_{corr.}^{\mathrm{CID}}\begin{bmatrix} 1 \\ \bar{\boldsymbol{C}}^{\mathrm{CID}} \end{bmatrix}. \qquad (6.197)$$

Multiplying both sides by the normalization constant of $\bar{\Psi}_0^{\mathrm{CID}}$, equal to C_0^{CID}, and realizing that $\mathbf{B}^{\mathrm{CID}} = \mathbf{H}^{\mathrm{CID}} - E_0^{\mathrm{HF}}\mathbf{I}^{\mathrm{CID}}$ and $\Delta E_{corr.}^{\mathrm{CID}} = E_0^{\mathrm{CID}} - E_0^{\mathrm{HF}}$, one indeed recovers the eigenvalue equation for $\boldsymbol{C}^{\mathrm{CID}\dagger} = C_0^{\mathrm{CID}}[1, \bar{\boldsymbol{C}}^{\mathrm{CID}}]^{\dagger}$ corresponding to E_0^{CID}:

$$\mathbf{H}^{\mathrm{CID}}\boldsymbol{C}^{\mathrm{CID}} = E_0^{\mathrm{CID}}\boldsymbol{C}^{\mathrm{CID}}. \tag{6.198}$$

Should one include the *Hextuple* (H) excitations in the Ψ_0^{CIDQH} expansion of the ground state wave function, which represent the electron correlation between *three* electronic pairs, the projection of the associated (6.178) onto the *doubly* and *quadruply* excited configurations gives rise to yet another set of coupled equations for the underlying CI coefficients. In this case, the coefficients of *Doubles* and *Quadruples* depend upon the *Hextuple* coefficients, etc. In fact, the CI expansion determines the infinite hierarchy of such coupled equations for determining the CI coefficients corresponding to increasing excitation multiplicities.

However, practical calculations call for terminating this hierarchy at some reasonable level of a limited CI, by using an approximate decoupling scheme, which would unable one to obtain the closed set of these equations, say, for the unknowns $\left\{\bar{C}_{k,l}^{p,q}\right\}$, which determine the pair correlations $\{e_{k,l}\}$. This requires an expression of CI coefficients standing before more highly excited configurations in terms of coefficients associated with lower excitations, e.g., for estimating $\left\{\bar{C}_{i,j,m,n}^{r,s,t,u}\right\}$ in terms of $\left\{\bar{C}_{k,l}^{p,q}\right\}$. Existence of such approximate relationships is indeed suggested by the fact that the dominant Coulomb correlation effects originate from the intrapair correlations, between electrons exhibiting the opposite spin orientations. This implies that the most important *quadruply* excited configurations are given by the products of two double excitations, which give the same overall quadruple excitation. This intuition lies behind specific decoupling schemes generating alternative variants of the CEPA or CC approximations.

For example, let us examine the coupled equations (6.194), which introduce the interpair correlation effects neglected in the IEPA scheme. Consider the representative CI coefficient $\bar{C}_{i,j,m,n}^{r,s,t,u}$ multiplying in (6.192) the function $\Phi_{i,j,m,n}^{r,s,t,u}$, which represents the replacements of the HF-occupied SO, identified by the lower indices $(i < j < m < n)$, by the corresponding HF-virtual SO, labeled by the upper indices $(r < s < t < u)$: $(i \to r, j \to s, m \to t, n \to u)$. We want to identify all double (pair) excitations, the product of which generates the same quadruple excitation, e.g., $\Phi_{i,j}^{r,s}\Phi_{m,n}^{t,u}$.

In fact there are 18 distinct products of such pair excitations, which give rise to the same final state as the selected quadruple excitation $\Phi_{i,j,m,n}^{r,s,t,u}$. One first realizes that there are three distinct assignments of the above four occupied SO to two electron pairs in the initial Slater determinants (we disregard the occupied SO which are not replaced in the excitation), in which the second orbital index in the pair is greater than the first one:

$$\{|i<j|m<n|,\ |i<m|j<n|,\ |i<n|j<m|\}. \qquad \text{(I)}$$

These sets of occupied SO thus exhaust the possible lists of the lower indices of the pair excitations, which may be associated with the quadruple excitation $\Phi_{i,j,m,n}^{r,s,t,u}$. The upper indices of such pair excitations, which determine the final configurations in such excitations of two electronic pairs, are similarly determined by the following six Slater determinants, with the intrapair ordered orbital labels:

$$\{|r<s|t<u|,\ |t<u|r<s|,\ |r<t|s<u|,\ |s<u|r<t|,\ |r<u|s<t|,\ |s<t|r<u|\}. \qquad \text{(II)}$$

Each initial assignment of the list (I) can be combined with each final assignment of the list (II), thus giving rise to 18 products of double excitations. For example, the first determinant of (I) combined with the first determinant of (II) generates the canonically ordered product $\Phi_{i,j}^{r,s}\Phi_{m,n}^{t,u}$ corresponding to the associated product of CI coefficients: $\bar{C}_{i,j}^{r,s}\bar{C}_{m,n}^{t,u}$, etc. Each of these 18 *quadruply* excited configurations can be then transformed into the canonically ordered Slater determinant $|r<s|t<u|$, the first in the list (II), by the corresponding number of exchanges of rows in the Slater determinant of the current configuration, with each single exchange changing the sign of the determinant and hence also the sign of the associated product of CI coefficients. The overall number of such exchanges is given by the sum of replacements required to bring the current list (p, q, v, w) of the upper indices to the canonical list (r, s, t, u) and the number of exchanges which transform the current list of lower indices (a, b, c, d) into the ordered set (i, j, m, n).

Therefore, the CI coefficient $\bar{C}_{i,j,m,n}^{r,s,t,u}$ multiplying the quadruple excitation $\Phi_{i,j,m,n}^{r,s,t,u}$ in (6.192) can be associated with the following combination of products of the CI coefficients of double excitations:

$$\begin{aligned}
\bar{C}_{i,j,m,n}^{r,s,t,u} \approx\ & \bar{C}_{i,j}^{r,s}\bar{C}_{m,n}^{t,u} + \bar{C}_{i,j}^{t,u}\bar{C}_{m,n}^{r,s} - \bar{C}_{i,j}^{r,t}\bar{C}_{m,n}^{s,u} - \bar{C}_{i,j}^{s,u}\bar{C}_{m,n}^{r,t} + \bar{C}_{i,j}^{r,u}\bar{C}_{m,n}^{s,t} + \bar{C}_{i,j}^{s,t}\bar{C}_{m,n}^{r,u} \\
& - \bar{C}_{i,m}^{r,s}\bar{C}_{j,n}^{t,u} - \bar{C}_{i,m}^{t,u}\bar{C}_{j,n}^{r,s} + \bar{C}_{i,m}^{r,t}\bar{C}_{j,n}^{s,u} + \bar{C}_{i,m}^{s,u}\bar{C}_{j,n}^{r,t} - \bar{C}_{i,m}^{r,u}\bar{C}_{j,n}^{s,t} - \bar{C}_{i,m}^{s,t}\bar{C}_{j,n}^{r,u} \\
& + \bar{C}_{i,n}^{r,s}\bar{C}_{j,m}^{t,u} + \bar{C}_{i,n}^{t,u}\bar{C}_{j,m}^{r,s} - \bar{C}_{i,n}^{r,t}\bar{C}_{j,m}^{s,u} - \bar{C}_{i,n}^{s,u}\bar{C}_{j,m}^{r,t} + \bar{C}_{i,n}^{r,u}\bar{C}_{j,m}^{s,t} + \bar{C}_{i,n}^{s,t}\bar{C}_{j,m}^{r,u} \\
\equiv\ & \bar{C}_{i,j}^{r,s}\bar{C}_{m,n}^{t,u} - \left\langle \begin{matrix} rs & tu \\ ij & mn \end{matrix} \right\rangle.
\end{aligned} \qquad (6.199)$$

The first three "rows" of the preceding expression are identified by the corresponding determinants in the list (I), while the corresponding "columns" correspond to the associated determinants in the list (II).

Substituting this expression to (6.194) and using (6.193) then gives quadratic equations for the unknowns $\left\{\bar{C}_{k,l}^{p,q}\right\}$:

$$\left\langle \Phi_{i,j}^{r,s} \middle| \hat{\mathrm{H}}^e \middle| \Psi_0^{\mathrm{HF}} \right\rangle + \sum_{k<l}^{occd.} \sum_{p<q}^{virt.} \left\langle \Phi_{i,j}^{r,s} \middle| \hat{\mathrm{H}}^e - E_0^{\mathrm{HF}} \middle| \Phi_{k,l}^{p,q} \right\rangle \bar{C}_{k,l}^{p,q}$$
$$+ \sum_{m<n}^{occd.} \sum_{t<u}^{virt.} \left\langle \Phi_{i,j}^{r,s} \middle| \hat{\mathrm{H}}^e \middle| \Phi_{i,j,m,n}^{r,s,t,u} \right\rangle \bar{C}_{i,j,m,n}^{r,s,t,u}$$
$$= \sum_{m<n}^{occd.} \sum_{t<u}^{virt.} \left\langle \Psi_0^{\mathrm{HF}} \middle| \hat{\mathrm{H}}^e \middle| \Phi_{m,n}^{t,u} \right\rangle \bar{C}_{i,j}^{r,s} \bar{C}_{m,n}^{t,u}$$
$$= \sum_{m<n}^{occd.} \sum_{t<u}^{virt.} \left\langle \Psi_0^{\mathrm{HF}} \middle| \hat{\mathrm{H}}^e \middle| \Phi_{m,n}^{t,u} \right\rangle \left(\bar{C}_{i,j,m,n}^{r,s,t,u} + \left\langle \begin{matrix} rs & tu \\ ij & mn \end{matrix} \right\rangle \right),$$

or

$$\left\langle \Phi_{i,j}^{r,s} \middle| \hat{\mathrm{H}}^e \middle| \Psi_0^{\mathrm{HF}} \right\rangle + \sum_{k<l}^{occd.} \sum_{p<q}^{virt.} \left\langle \Phi_{i,j}^{r,s} \middle| \hat{\mathrm{H}}^e - E_0^{\mathrm{HF}} \middle| \Phi_{k,l}^{p,q} \right\rangle \bar{C}_{k,l}^{p,q}$$
$$- \sum_{m<n}^{occd.} \sum_{t<u}^{virt.} \left\langle \Psi_0^{\mathrm{HF}} \middle| \hat{\mathrm{H}}^e \middle| \Phi_{m,n}^{t,u} \right\rangle \left\langle \begin{matrix} rs & tu \\ ij & mn \end{matrix} \right\rangle = 0, \tag{6.200}$$

where we have also used the identity

$$\left\langle \Phi_{i,j}^{r,s} \middle| \hat{\mathrm{H}}^e \middle| \Phi_{i,j,m,n}^{r,s,t,u} \right\rangle = \left\langle \Psi_0^{\mathrm{HF}} \middle| \hat{\mathrm{H}}^e \middle| \Phi_{m,n}^{t,u} \right\rangle = \langle mn|g|tu - ut \rangle, \tag{6.201}$$

which directly follows from the Slater–Condon rule (5.86).

Such a nonvariational and size-consistent method is known as the *Coupled Pair Many Electron Theory* (CPMET) (Paldus and Čižek 1975; Čižek and Paldus 1980). Determining solutions $\left\{ \bar{C}_{k,l}^{p,q} \right\}$ of (6.200) and using them in (6.193) then gives the correlation energy, which takes into account the effective correlation coupling between different electronic pairs:

$$\Delta E_{corr.}^{\mathrm{CPMET}} \cong \Delta E_{corr.}^{\mathrm{CIDQ}}. \tag{6.202}$$

A number of approximate schemes derived from this full CPMET treatment have also been proposed (see, e.g., Szabo and Ostlund 1982). The simplest variant of such CC equations is obtained, when one neglects in (6.200) the last term $\left\langle \begin{matrix} rs & tu \\ ij & mn \end{matrix} \right\rangle$ altogether:

$$\left\langle \begin{matrix} rs & tu \\ ij & mn \end{matrix} \right\rangle = 0 \quad \text{or} \quad \bar{C}_{i,j,m,n}^{r,s,t,u} \approx \bar{C}_{i,j}^{r,s} \bar{C}_{m,n}^{t,u}. \tag{6.203}$$

This gives rise to linear equations for the unknowns $\left\{ \bar{C}_{k,l}^{p,q} \right\}$:

$$\left\langle \Phi_{i,j}^{r,s} \middle| \hat{\mathrm{H}}^e \middle| \Psi_0^{\mathrm{HF}} \right\rangle + \sum_{k<l}^{occd.} \sum_{p<q}^{virt.} \left\langle \Phi_{i,j}^{r,s} \middle| \hat{\mathrm{H}}^e - E_0^{\mathrm{HF}} \middle| \Phi_{k,l}^{p,q} \right\rangle \bar{C}_{k,l}^{p,q} = 0. \tag{6.204}$$

When substituted in (6.193) they give rise to a nonvariational but size-consistent estimate of the correlation energy: $\Delta E_{corr.}^{\mathrm{CCA}} \approx \Delta E_{corr.}^{\mathrm{CIDQ}}$

In a more accurate CEPA method of Meyer (1977) only the products diagonal in pairs of lower indices in $\left\langle \begin{matrix} rs & tu \\ ij & mn \end{matrix} \right\rangle$ are retained:

$$\left\langle \begin{matrix} rs & tu \\ ij & mn \end{matrix} \right\rangle = \left\langle \begin{matrix} rs & tu \\ ij & ij \end{matrix} \right\rangle \delta_{i,m}\delta_{j,n} = C_{i,j}^{r,s} C_{i,j}^{t,u} \delta_{i,m}\delta_{j,n}. \tag{6.205}$$

This result directly follows from (6.199). Indeed, one observes that all terms in the second row of this equation, corresponding to the lower indices $(i,i,\ j,j)$, must identically vanish since $\bar{C}_{k,k}^{p,q} = 0$ (a given occupied SO can participate only in a single substitution). Moreover, the corresponding terms of the same columns $2 \div 6$ in the first and third row differ in only a single exchange of lower indices, thus exactly canceling each other. The only remaining term is thus given by the first contribution of the third row, $-\bar{C}_{i,j}^{r,s}\bar{C}_{j,i}^{t,u} = \bar{C}_{i,j}^{r,s}\bar{C}_{i,j}^{t,u}$, where we have used the above-mentioned antisymmetry property of the CI coefficients, implied by the same property of the Slater determinants determining the associated configuration functions:

$$\bar{C}_{k,l}^{p,q} = -\bar{C}_{l,k}^{p,q} = -\bar{C}_{k,l}^{q,p}. \tag{6.206}$$

Finally, using the approximation of (6.205) in (6.200) gives the following (quadratic) equations for the unknown CI coefficients $\bar{\boldsymbol{C}}^{\mathrm{CEPA}} = \left\{ \bar{C}_{k,l}^{p,q} \right\}$ of the CEPA variant,

$$\begin{aligned} &\left\langle \Phi_{i,j}^{r,s} \middle| \hat{\mathrm{H}}^e \middle| \Psi_0^{\mathrm{HF}} \right\rangle + \sum_{k<l}^{occd.} \sum_{p<q}^{virt.} \left\langle \Phi_{i,j}^{r,s} \middle| \hat{\mathrm{H}}^e - E_0^{\mathrm{HF}} \middle| \Phi_{k,l}^{p,q} \right\rangle \bar{C}_{k,l}^{p,q} \\ &\qquad - \left\{ \sum_{t<u}^{virt.} \left\langle \Psi_0^{\mathrm{HF}} \middle| \hat{\mathrm{H}}^e \middle| \Phi_{i,j}^{t,u} \right\rangle \bar{C}_{i,j}^{t,u} \right\} \bar{C}_{i,j}^{r,s} = 0, \end{aligned} \tag{6.207}$$

or in terms of the *pair* correlation quantities $\left\{ e_{k,l}^{\mathrm{CIDQ}} \right\}$ defined in (6.193):

$$\left\langle \Phi_{i,j}^{r,s} \middle| \hat{\mathrm{H}}^e \middle| \Psi_0^{\mathrm{HF}} \right\rangle + \sum_{k<l}^{occd.} \sum_{p<q}^{virt.} \left\langle \Phi_{i,j}^{r,s} \middle| \hat{\mathrm{H}}^e - E_0^{\mathrm{HF}} \middle| \Phi_{k,l}^{p,q} \right\rangle \bar{C}_{k,l}^{p,q} = e_{i,j}^{\mathrm{CEPA}} \bar{C}_{i,j}^{r,s}. \tag{6.208}$$

These equations must be solved iteratively until the self-consistency is reached. More specifically, the *pair* correlation contributions $\left\{ e_{k,l}^{\mathrm{CEPA}} \left(\bar{\boldsymbol{C}}_{(n-1)}^{\mathrm{CEPA}} \right) \right\}$ obtained from (6.193), the linear functions of the CI coefficients $\bar{\boldsymbol{C}}_{(n-1)}^{\mathrm{CEPA}}$ obtained in the previous iteration of (6.208), must generate the same expansion coefficients in the next iteration of these equations: $\bar{\boldsymbol{C}}^{\text{CEPA}}_{(n-1)} = \bar{\boldsymbol{C}}^{\text{CEPA}}_{(n)}$. This nonvariational method also gives rise to the size-consistent estimate of the overall Coulomb correlation energy in the molecular ground state: $\Delta E^{\text{CEPA}}_{corr.} = \sum_{k<l}^{occd.} e^{\text{CEPA}}_{k,l}$.

In the next section, we shall demonstrate that the CPMET represents a special case of a more general CC strategy, which is based upon the so-called cluster expansion of the correlated wave function of N electrons. In fact CPMET represents the CC approximation for electronic pairs, i.e., the two-electron *clusters*. This formalism is most elegantly formulated in the so-called second-quantization representation of quantum states of N fermions, by using the *creation* and *annihilation* operators acting in the *Fock space* of state vectors representing all admissible occupations of the one-electron states of the HF (orbital) approximation. In this alternative, occupation number formalism the Pauli exclusion principle is safeguarded by the appropriate for fermions algebraic relations of the anticommutation properties of the creation and annihilation operators. It allows one to treat both the *closed* molecules with the fixed number of electrons, $N = N^0$, and the *open* systems with varying (fluctuating) number of electrons. This more general description of electronic quantum states is the main subject of the next section.

6.5 Second-Quantization Representation

The Hilbert space $\mathcal{H}(N)$ of state vectors (or wave functions) of N-particles (see Chaps. 2 and 3) combines the antisymmetric (A) and symmetric (S) subspaces, $\mathcal{H}_A(N)$ and $\mathcal{H}_S(N)$, corresponding to fermions and bosons, respectively: $\mathcal{H}(N) = \mathcal{H}_A(N) + \mathcal{H}_S(N)$. The expansion theorem of Sect. 6.2 demonstrates that the N-electron subspace $\mathcal{H}_A(N)$ is spanned by determinants constructed from the complete set of SO. These N-electron basis functions are thus uniquely identified by different N-conserving occupation patterns of the one-electron functions (see also Sect. 6.1.3).

However, in many applications of the quantum theory it is convenient to refer to a union of the mutually orthogonal Hilbert spaces for all admissible numbers of particles, $N = 0, 1, 2, \ldots$,

$$\begin{aligned}\mathcal{H} &= \mathcal{H}(0) + \mathcal{H}(1) + \ldots = \{\mathcal{H}_A(0) + \mathcal{H}_A(1) + \ldots\} + \{\mathcal{H}_S(0) + \mathcal{H}_S(1) + \ldots\} \\ &\equiv \mathcal{H}_A + \mathcal{H}_S,\end{aligned} \tag{6.209}$$

with the fermion (antisymmetric) subspace $\mathcal{H}_A$ called the *Fock space*. It combines the state vectors for any number of electrons. Here, the $\mathcal{H}_A(0)$ subspace corresponds to the Hilbert space of no electrons, the electronic *vacuum*. This generalized concept of the vector space has originated in the field theory, with "particles," the photons, being "created" or "annihilated" in the emission or absorption processes, respectively.

The generalized state vector of an electronic system in such an enlarged vector space thus exhibits components corresponding to different numbers of electrons:

$$|\Psi_A\rangle = |\Psi_A(0)\rangle + |\Psi_A(1)\rangle + \ldots + |\Psi_A(N)\rangle + \ldots,$$
$$|\Psi_A(N)\rangle = \hat{\mathrm{P}}_A(N)|\Psi_A\rangle,\ \hat{\mathrm{P}}_A(N)\hat{\mathrm{P}}_A(N') = \delta_{N,N'}; \tag{6.210}$$

here, $\hat{\mathrm{P}}_A(N)$ stands for the projector onto $\mathcal{H}_A(N)$, with the scalar product being determined by the overall projection operator onto $\mathcal{H}_A$, $\hat{\mathrm{P}}_A = \sum_N \hat{\mathrm{P}}_A(N)$, the action of which on an arbitrary fermion state $|\Xi_A\rangle \in \mathcal{H}_A$ amounts to the identity operation, $\hat{\mathrm{P}}_A|\Xi_A\rangle = |\Xi_A\rangle$:

$$\langle\Phi_A \mid \Psi_A\rangle = \sum_N \left(\langle\Phi_A|\hat{\mathrm{P}}_A(N)\right)\left(\hat{\mathrm{P}}_A(N)|\Psi_A\rangle\right) = \sum_N \langle\Phi_A(N) \mid \Psi_A(N)\rangle. \tag{6.211}$$

The state vectors in $\mathcal{H}_A$ exhaust both the N conserving and nonconserving patterns of occupations of the complete set of one-electron states. In order to effect changes in the system number of particles the *creation* and *annihilation* (destruction) operators are being introduced in this *second quantization* (occupation number) *representation* of electronic states (see, e.g.: Jørgensen and Simons 1981; Szabo and Ostlund 1982; McWeeny 1989; Surjan 1989). When acting on an N-electron state the former gives rise to a state of $(N + 1)$ electrons, while the action of the latter generates a state of $(N - 1)$ electrons. In what follows we shall apply this formalism in the specific context of the Coulomb correlation in the N-electron molecular/atomic systems. We shall demonstrate that in this elegant formulation of the generalized CI theory of electronic clusters, the crucial stage of determining the matrix elements of the electronic Hamiltonian in the basis set of Slater determinants defining the excited configurations amounts to straightforward algebraic manipulations on the one- and two-electron integrals calculated in the nearly complete basis set of molecular SO.

In this generalized description the fractional-N states appear as ensembles (statistical mixtures) of the integer-N states. This enables one to treat in quantum theory a variety of classical *chemical* problems involving *open* molecules, dealing with the *Charge Transfer* (CT) phenomena, e.g., the fractional or integral electron attachment or withdrawal processes, and the thermodynamic equilibria of the externally open molecular systems, coupled to electron reservoirs.

6.5.1 Fock Space and Creation/Annihilation Operators

A general Slater determinant

$$\Psi_A(N) = |\psi_i\psi_j\ldots\psi_p| \equiv \langle\mathbf{q}|\psi_i\psi_j\ldots\psi_p\rangle \equiv \langle\mathbf{q}|ij\ldots p\rangle \equiv \langle\mathbf{q}|\Psi_A\rangle \tag{6.212}$$

provides the $\mathbf{q} = \{\boldsymbol{q}_l\} = \mathcal{Q}^N$ (position-spin) representation of the N-electron state vector $|\Psi_A(N)\rangle = |\psi_i\, \psi_j, \ldots, \psi_p\rangle \equiv |i\, j \ldots p\rangle$, which is uniquely identified by its selection of the one-electron state vectors $\{|\psi_k\rangle = |k\rangle\}$ or the singly occupied SO: $\{\psi_k(\boldsymbol{q}) = \langle \boldsymbol{q}|k\rangle\}$. Notice that the order in which SO are listed, although immaterial for the direction of the state vector in the molecular Hilbert space, and hence also for the identity of the quantum state itself, does matter for its sign (phase): any exchange of two SO changes the sign of the state-vector and the associated Slater determinant.

Therefore, such an antisymmetric state of N-electrons is uniquely identified by the occupation numbers $\{n_k\}$ (see Sect. 6.1.3) of the complete set of SO: $n_k = 1$ for the occupied and $n_k = 0$ for the unoccupied SO. For example, the occupation vector $|001010\ldots\rangle$ of the two-electron system identifies the state vector $|\psi_3\, \psi_5\rangle$ and hence also the Slater determinant $|\psi_3\psi_5|$. Similarly, the normalized vacuum state, $\langle vac|vac\rangle = 1$, corresponds to the vanishing occupations of all SO: $|vac\rangle = |00\ldots\rangle$. The Fock ($F$) space thus contains all kets $\{|n_1\, n_2 \ldots n_k \ldots\rangle\}$ for any overall number of electrons $N = \sum_k n_k$:

$$|vac\rangle,\ \{|i\rangle\},\ \{|ij\rangle\}, \{|ijk\rangle\},\ \ldots,\ \{|ij\ldots p\rangle\},\ \ldots \tag{6.213}$$

This enlarged basis of the independent state vectors defines the so called *occupation number* representation. It should be observed that kets corresponding to different permutations of the same set of occupied SO determine the same quantum state of N electrons, which can be uniquely identified by the *ordered* (increasing) labels of SO in $|\Psi_A\rangle$.

As we have already remarked above, the transition from $\Psi_A(N) = |i\, j \ldots p| \rightarrow \Psi_A(N+1) = |i\, j \ldots pr|$ is effected by the action of the *creation operator* $\hat{a}_r^+$, which creates an additional electron in state $|r\rangle$. Its action in the Fock space is defined in the following way:

$$\hat{a}_r^+|i\, j\ldots p\rangle = \begin{cases} |ij\ldots pr\rangle = (-1)^{v_r}|ij\ldots r\ldots p\rangle, & r \notin (i,j,\ \ldots\ ,p); \\ 0, & r \in (i,j,\ \ldots\ ,p); \end{cases} \tag{6.214}$$

here v_r stands for the number of SO exchanges required to shift index r of $|ij\ldots pr\rangle$ to its proper position in the ordered set $|ij\ldots\, r\ldots\, p\rangle$. The zero vector results when the SO to be created is already involved in the initial determinant.

A lowering of the number of electrons is similarly effected by the *annihilation operator* $\hat{a}_r^-$ symbolizing the action opposite to that of $\hat{a}_r^+$, i.e., the destruction of an electron in state $|r\rangle$, $(-1)^{v_r}\hat{a}_r^-|ij\ldots\, r\ldots\, p\rangle = \hat{a}_r^-|ij\ldots pr\rangle = |ij\ldots p\rangle$, or:

$$\hat{a}_r^-|i\, j\ldots p\rangle = \begin{cases} (-1)^{v_r}|ij\ldots\ \not{r}\ldots p\rangle = |ij\ldots p\ \not{r}\rangle, & r \in (i,j,\ \ldots\ ,p); \\ 0, & r \notin (i,j,\ \ldots\ ,p). \end{cases} \tag{6.215}$$

Again, v_r counts the number of orbital exchanges required to move r to the end of the list of SO and $\not{r}$ marks the destroyed SO. The zero vector is seen to result when the SO to be annihilated is not initially present in the determinant.

Therefore, all vectors in the Fock space (6.213) can be derived from the vacuum state by actions of the relevant creation operators,

$$\{|i\rangle = \hat{a}_i^+|vac\rangle\}, \qquad \left\{|ij\rangle = \hat{a}_j^+|i\rangle = \hat{a}_j^+\,\hat{a}_i^+|vac\rangle\right\}, \ \ldots,$$
$$\left\{|ij\ldots p\rangle = \hat{a}_p^+\ldots\hat{a}_j^+\,\hat{a}_i^+|vac\rangle\right\}, \ \ldots, \tag{6.216}$$

or traced back to the vacuum state by actions of the corresponding annihilation operators applied in the reverse order:

$$|vac\rangle = \hat{a}_i^-|i\rangle = \hat{a}_i^-\,\hat{a}_j^-|ij\rangle = \ldots = \hat{a}_i^-\,\hat{a}_j^- \,\ldots\, \hat{a}_p^-|ij\,\ldots\,p\rangle. \tag{6.217}$$

One also observes that $\hat{a}_i^+\hat{a}_i^+|vac\rangle = \hat{a}_i^+|i\rangle = \hat{a}_i^-\hat{a}_i^-|i\rangle = \hat{a}_i^-|vac\rangle = 0$.

These relations imply the anticommutation property of Jordan and Wigner for these operators:

$$\begin{aligned}&\hat{a}_j^+\,\hat{a}_i^+|vac\rangle = |ij\rangle = -|ji\rangle = -\hat{a}_i^+\,\hat{a}_j^+|vac\rangle \quad \text{or}\\ &\hat{a}_j^+\,\hat{a}_i^+ + \hat{a}_i^+\,\hat{a}_j^+ \equiv \left[\hat{a}_i^+, \hat{a}_j^+\right]_+ = 0,\end{aligned} \tag{6.218}$$

$$\hat{a}_i^-\,\hat{a}_j^-|ij\rangle = |vac\rangle = -\hat{a}_j^-\,\hat{a}_i^-|ji\rangle \quad \text{or} \quad \hat{a}_i^-\,\hat{a}_j^- + \hat{a}_j^-\,\hat{a}_i^- = \left[\hat{a}_i^-, \hat{a}_j^-\right]_+ = 0, \tag{6.219}$$

where $[\hat{A}, \hat{B}]_+ = \hat{A}\hat{B} + \hat{B}\hat{A}$ denotes the anticommutator of two operators. It follows from these equations that in the language of the second quantization formalism these relations reflect the Pauli principle of the antisymmetrization of the fermion wave function.

Next, let us examine the products of the creation and annihilation operators. One can easily verify that the nondiagonal ($s \neq r$) product operator $\hat{a}_s^+\hat{a}_r^-$ replaces the occupied state $|r\rangle$ with the occupied state $|s\rangle$, since the right operator in the product destroys an electron in $|r\rangle$ and the left operator creates in its place an electron in state $|s\rangle$:

$$\begin{aligned}\hat{a}_s^+\,\hat{a}_r^-|ij\ldots r\ldots p\rangle &= (-1)^{\nu_r}\hat{a}_s^+\,\hat{a}_r^-|ij\,\ldots\,pr\rangle = (-1)^{\nu_r}\hat{a}_s^+|ij\,\ldots\,p\rangle\\ &= (-1)^{\nu_r}|ij\,\ldots\,ps\rangle = (-1)^{2\nu_r}|ij\,\ldots\,s\,\ldots\,p\rangle\\ &= |ij\,\ldots\,s\,\ldots\,p\rangle \equiv |r \to s\rangle.\end{aligned} \tag{6.220}$$

We have observed above that placing the newly created state $|s\rangle$ in the position originally kept by $|r\rangle$ requires the same number of exchanges as that required to move the latter to the end of the original list of the occupied SO. We therefore conclude that the operator $\hat{a}_s^+\hat{a}_r^-$ indeed amounts to a single $r \to s$ excitation. One similarly verifies the action of the product operator with reverse order of the two factors:

$$\hat{a}_r^- \hat{a}_s^+ |ij\ldots r\ldots p\rangle = \hat{a}_r^- |ij\ldots r\ldots ps\rangle = (-1)^{\nu_r+1} \hat{a}_r^- |ij\ldots psr\rangle$$
$$= (-1)^{\nu_r+1} |ij\ldots ps\rangle. \tag{6.221}$$

The diagonal ($s = r$) products of such creation and annihilation operators similarly give:

$$\hat{a}_r^+ \hat{a}_r^- |ij\ldots p\rangle = \begin{cases} |ij\ldots p\rangle, & r \in (i,j,\ \ldots\ ,p); \\ 0, & r \notin (i,j,\ \ldots\ ,p); \end{cases} \tag{6.222}$$

$$\hat{a}_r^- \hat{a}_r^+ |ij\ldots p\rangle = \begin{cases} |ij\ldots p\rangle, & r \notin (i,j,\ \ldots\ ,p); \\ 0, & r \in (i,j,\ \ldots\ ,p). \end{cases} \tag{6.223}$$

Hence, (6.220)–(6.223) are also compactly summarized by the following anticommutator identity:

$$\hat{a}_s^+ \hat{a}_r^- + \hat{a}_r^- \hat{a}_s^+ = \left[\hat{a}_s^+, \hat{a}_r^-\right]_+ = \delta_{s,r}. \tag{6.224}$$

Indeed, using these four equations gives for the $r \in (i, j, \ldots, p)$ case:

$$\left[\hat{a}_s^+, \hat{a}_r^-\right]_+ |ij\ldots r\ldots p\rangle = |ij\ldots s\ldots p\rangle - |ij\ldots s\ldots p\rangle = 0,$$
$$\left[\hat{a}_r^+, \hat{a}_r^-\right]_+ |ij\ldots r\ldots p\rangle = |ij\ldots r\ldots p\rangle, \text{ etc.}$$

One also observes that the normalization of the state vectors in the Fock space requires:

$$\langle ij\ \ldots\ p|\left(\hat{a}_r^+\right)^\dagger \hat{a}_r^+ |ij\ \ldots\ p\rangle = 1,\ \ r \notin (i,j,\ldots,p), \quad \text{or} \quad \left(\hat{a}_r^+\right)^\dagger \hat{a}_r^+ = 1,$$
$$\langle ij\ \ldots\ r\ \ldots\ p|\left(\hat{a}_r^-\right)^\dagger \hat{a}_r^- |ij\ \ldots\ r\ \ldots\ p\rangle = 1, \quad \text{or} \quad \left(\hat{a}_r^-\right)^\dagger \hat{a}_r^- = 1. \tag{6.225}$$

It thus follows from these operator reciprocity relations that:

$$\left(\hat{a}_r^+\right)^\dagger = \left(\hat{a}_r^+\right)^{-1} = \hat{a}_r^- \equiv \hat{a}_r \qquad \left(\hat{a}_r^-\right)^\dagger = \left(\hat{a}_r^-\right)^{-1} = \hat{a}_r^+ \equiv \hat{a}_r^\dagger. \tag{6.226}$$

Therefore, the creation operators are Hermitian conjugates of the annihilation operators, and vice versa; as such they are not Hermitian themselves. In the new notation introduced in the preceding equation, the anticommutation relations read:

$$\left[\hat{a}_r^\dagger, \hat{a}_s^\dagger\right]_+ = \left[\hat{a}_r, \hat{a}_s\right]_+ = 0 \quad \text{and} \quad \left[\hat{a}_s^\dagger, \hat{a}_r\right]_+ = \delta_{s,r}. \tag{6.227}$$

The creation and destruction operators in the position-spin representation, called the *field operators* $\hat{\psi}^\dagger(\boldsymbol{q})$ and $\hat{\psi}(\boldsymbol{q})$, introduce the *local* aspect of the electron distribution. They respectively create and destroy one-particle state $|\boldsymbol{q}\rangle = |\boldsymbol{r},\sigma_z\rangle$, the eigenfunction of the operator $\hat{\mathbf{q}} = (\hat{\mathbf{r}}, \hat{\sigma}_z)$ associated with the particle position-spin "coordinates" $\boldsymbol{q} = (\boldsymbol{r}, \sigma_z)$, i.e., the state corresponding to the sharply specified position ($\boldsymbol{r}$) and *spin* orientation (σ_z) of an electron,

$$\hat{\mathbf{r}}|\boldsymbol{q}\rangle = \boldsymbol{r}|\boldsymbol{q}\rangle \quad \text{and} \quad \hat{\sigma}_z|\boldsymbol{q}\rangle = \sigma_z|\boldsymbol{q}\rangle, \tag{6.228}$$

or in the combined short notation: $\hat{\mathbf{q}}|\boldsymbol{q}\rangle = \boldsymbol{q}|\boldsymbol{q}\rangle$. This definition of field operators thus implies:

$$\hat{\psi}^{\dagger}(\boldsymbol{q})|vac\rangle = |\boldsymbol{q}\rangle \quad \text{and} \quad \hat{\psi}(\boldsymbol{q})|\boldsymbol{q}\rangle = |vac\rangle.$$

They satisfy the associated anticommutation relations,

$$\begin{aligned} &[\hat{\psi}^{\dagger}(\boldsymbol{q}), \hat{\psi}^{\dagger}(\boldsymbol{q}')]_+ = [\hat{\psi}(\boldsymbol{q}), \hat{\psi}(\boldsymbol{q}')]_+ = 0 \quad \text{and} \\ &[\hat{\psi}^{\dagger}(\boldsymbol{q}), \hat{\psi}(\boldsymbol{q}')]_+ = \delta(\boldsymbol{q} - \boldsymbol{q}'), \end{aligned} \tag{6.229}$$

which represent the local analogs of those summarized in (6.227).

From the relevant identity (completeness) projections,

$$\int d\boldsymbol{q}|\boldsymbol{q}\rangle\langle\boldsymbol{q}| = \sum_r |r\rangle\langle r| = 1, \tag{6.230}$$

and the associated resolutions of SO in the continuous basis $\{|\boldsymbol{q}\rangle\}$ and of $|\boldsymbol{q}\rangle$ in the discrete SO basis $\{|r\rangle\}$,

$$\begin{aligned} |r\rangle &= \int d\boldsymbol{q}|\boldsymbol{q}\rangle\langle\boldsymbol{q}|r\rangle = \int |\boldsymbol{q}\rangle\ \psi_r(\boldsymbol{q})\, d\boldsymbol{q}, \\ |\boldsymbol{q}\rangle &= \sum_r |r\rangle\langle r \mid \boldsymbol{q}\rangle = \sum_r |r\rangle\ \psi_r^*(\boldsymbol{q}), \end{aligned} \tag{6.231}$$

one then arrives at the following relations between the creation/annihilation operators and their (local) field operator analogs:

$$|r\rangle = \hat{a}_r^{\dagger}|vac\rangle = \int d\boldsymbol{q}\, \psi_r(\boldsymbol{q})\, \hat{\psi}^{\dagger}(\boldsymbol{q})|vac\rangle \quad \text{or} \quad \hat{a}_r^{\dagger} = \int d\boldsymbol{q}\ \psi_r(\boldsymbol{q})\, \hat{\psi}^{\dagger}(\boldsymbol{q}), \tag{6.232}$$

$$|\boldsymbol{q}\rangle = \hat{\psi}^{\dagger}(\boldsymbol{q})|vac\rangle = \sum_r \psi_r^*(\boldsymbol{q})\, \hat{a}_r^{\dagger}|vac\rangle \quad \text{or} \quad \hat{\psi}^{\dagger}(\boldsymbol{q}) = \sum_r \psi_r^*(\boldsymbol{q})\, \hat{a}_r^{\dagger}. \tag{6.233}$$

Hence the relations between their adjoints, the associated annihilation operators:

$$\hat{a}_r = \int d\boldsymbol{q}\ \psi_r^*(\boldsymbol{q})\, \hat{\psi}(\boldsymbol{q}) \quad \text{and} \quad \hat{\psi}(\boldsymbol{q}) = \sum_r \psi_r(\boldsymbol{q})\, \hat{a}_r. \tag{6.234}$$

It further follows from (6.222) that kets in the Fock space, $\{|\boldsymbol{n}\rangle\}$, which are identified by the corresponding occupation vector of SO, $\boldsymbol{n} = (n_1, n_2, \ldots, n_r, \ldots)$, are eigenfunctions of the operators $\{\hat{a}_r^+ \hat{a}_r^- = \hat{a}_r^\dagger \hat{a}_r\}$, with the eigenvalues reflecting the SO occupation numbers $\{n_r\}$:

$$\hat{a}_r^\dagger \hat{a}_r |\boldsymbol{n}\rangle \equiv \hat{n}_r^F |\boldsymbol{n}\rangle = n_r |\boldsymbol{n}\rangle, \qquad n_r = \begin{cases} 1, & \text{occupied SO} \\ 0, & \text{unoccupied SO} \end{cases}. \tag{6.235}$$

Therefore, the Hermitian operator $\hat{n}_r^F = \hat{n}_r^{F\dagger}$ represents the occupation number of ψ_r. Hence, the Fock space vectors also satisfy the eigenvalue problem for the overall number of electrons N, represented in the second quantization formalism by the Hermitian operator $\hat{N}^F = \sum_r \hat{n}_r^F = \sum_r \hat{a}_r^\dagger \hat{a}_r$:

$$\hat{N}^F |\boldsymbol{n}\rangle = \left(\sum\nolimits_r n_r\right) |\boldsymbol{n}\rangle = N |\boldsymbol{n}\rangle. \tag{6.236}$$

Let us now examine how the electronic Hamiltonian [(5.51) and (5.69)] and its one- and two-electron contributions $\hat{h}(1)$ and $g(1,2)$ are represented in the Fock space. We first observe that using the discrete projection of (6.230) onto the complete space of SO gives:

$$\hat{h}|i\rangle = \sum_r |r\rangle\langle r|\hat{h}|i\rangle = \sum_r |r\rangle \bar{h}_{r,i}, \tag{6.237}$$

where $\bar{h}_{r,i}$ denotes the matrix element of $\hat{h}(1)$ between the indicated SO (5.73). The one-electron part of the molecular N-electron (Coulomb) Hamiltonian is given by the sum of such one-electron operators:

$$\hat{\mathcal{F}}(N) = \sum_i \hat{h}(i), \tag{6.238}$$

with $\hat{h}(i)$ acting on the one-electron state describing ith electron. Hence, its action on the Slater determinant gives [see (6.237)]:

$$\hat{\mathcal{F}}(N)\det[\psi_{k_1}(1)\psi_{k_2}(2)\ldots\psi_{k_i}(i)\ldots] = \sum_{k_i}\sum_{k_i'} \det[\psi_{k_1}(1)\,\psi_{k_2}(2)\,\ldots\,\psi_{k_i'}(i)\,\ldots]\bar{h}_{k_i',k_i}. \tag{6.239}$$

Here, relative to the original determinant in the l.h.s., the determinant in the r.h.s. sum involves the substitution of the orbital describing ith electron, $\psi_{k_i} \to \psi_{k_i'}$, which in the Fock space can be accomplished by the action of the *single*-excitation operator of (6.220) on the original state:

$$\hat{\mathcal{F}}^F |k_1\,k_2\,\ldots\,k_i\,\ldots\rangle = \sum_{k_i}\sum_{k_i'} \hat{a}_{k_i'}^\dagger \hat{a}_{k_i} |k_1\,k_2\ldots\,k_i\,\ldots\rangle\,\bar{h}_{k_i',\,k_i}. \tag{6.240}$$

Therefore, the *second* quantization form of the (symmetrical) one-electron operator of (6.238) reads:

$$\hat{\mathcal{F}}^F = \sum_r \sum_s \bar{h}_{s,r} \hat{a}_s^\dagger \hat{a}_r. \tag{6.241}$$

It is devoid of any explicit N-dependence, so it applies to both the neutral systems and ions. As we have argued in (6.236), the number of electrons is ultimately recognized by another N-independent operator: $\hat{N}^F = \sum_r \hat{a}_r^\dagger \hat{a}_r$.

One similarly determines the Fock space form of the two-electron operator

$$\hat{\mathcal{G}}(N) = \sum_{i<j} g(i,j). \tag{6.242}$$

We first observe that in the Fock space the *double* excitation $(t,u) \to (r,s)$, consisting of simultaneous replacements of two SO, $\psi_t \to \psi_r$ and $\psi_u \to \psi_s$, is effected by the operator $\hat{a}_r^\dagger \hat{a}_s^\dagger a_u a_t$:

$$\begin{aligned}
\hat{a}_r^\dagger \hat{a}_s^\dagger \hat{a}_u \hat{a}_t |ij \ldots u \ldots t \ldots p\rangle &= (-1)^{v_t} \hat{a}_r^\dagger \hat{a}_s^\dagger \hat{a}_u \hat{a}_t |ij \ldots u \ldots pt\rangle \\
&= (-1)^{v_t} \hat{a}_r^\dagger \hat{a}_s^\dagger \hat{a}_u |ij \ldots u \ldots p\rangle = (-1)^{v_t+v_u} \hat{a}_r^\dagger \hat{a}_s^\dagger \hat{a}_u |ij \ldots pu\rangle \\
&= (-1)^{v_t+v_u} \hat{a}_r^\dagger \hat{a}_s^\dagger |ij \ldots p\rangle = (-1)^{v_t+v_u} \hat{a}_r^\dagger |ij \ldots ps\rangle \\
&= (-1)^{v_t+2v_u} \hat{a}_r^\dagger |ij \ldots s \ldots p\rangle = (-1)^{v_t+2v_u} |ij \ldots s \ldots pr\rangle \\
&= (-1)^{2v_t+2v_u} |ij \ldots s \ldots r \ldots p\rangle = |ij \ldots s \ldots r \ldots p\rangle \equiv |t \to r, u \to s\rangle.
\end{aligned} \tag{6.243}$$

The complete two-electron basis is spanned by all product functions $\{\psi_k(1)\psi_l(2)\}$ [see (6.60)] or the associated Slater determinants $\{|kl\rangle\}$ in the Fock space: $\sum_{k,l} |kl\rangle\langle kl| = 1$. The action of the (multiplicative) operator coupling the two electrons is then determined by the two-electron integrals in the SO basis set [see (5.75)]:

$$g|tu\rangle = \Sigma_{r,s} |rs\rangle\langle rs|g|tu\rangle = \Sigma_{r,s} |rs\rangle\ (rt|su). \tag{6.244}$$

The corresponding action of the symmetrical operator of (6.242) on the representative vector in the Fock space or the associated Slater determinant then reads:

$$\begin{aligned}
\hat{\mathcal{G}}^F |\ldots k_i \ldots k_j \ldots\rangle \equiv \hat{\mathcal{G}}^F |\ldots u \ldots t \ldots\rangle
&= \sum_{t<u} \sum_r \sum_s |\ldots s \ldots r \ldots\rangle\langle sr|g|ut\rangle \\
&\equiv \frac{1}{2} \sum_r \sum_s \sum_t \sum_u |\ldots s \ldots r \ldots\rangle\langle sr|g|ut\rangle \\
&= \frac{1}{2} \sum_r \sum_s \sum_t \sum_u \hat{a}_r^\dagger \hat{a}_s^\dagger \hat{a}_u \hat{a}_t |\ldots u \ldots t \ldots\rangle\langle sr|g|ut\rangle,
\end{aligned} \tag{6.245}$$

and hence

$$\hat{\mathcal{G}}^F = \frac{1}{2}\sum_r \sum_s \sum_t \sum_u \langle sr|g|ut\rangle\, \hat{a}_r^\dagger \hat{a}_s^\dagger \hat{a}_u \hat{a}_t. \tag{6.246}$$

Therefore, the electronic Hamiltonian $\hat{H}^e(N) = \hat{\mathcal{F}}(N) + \hat{\mathcal{G}}(N)$ assumes the following form in the *second* quantization representation of the Fock space:

$$\hat{H}^{e,F} = \hat{\mathcal{F}}^F + \hat{\mathcal{G}}^F = \sum_r \sum_s \bar{h}_{s,r}\, \hat{a}_s^\dagger \hat{a}_r + \frac{1}{2}\sum_r \sum_s \sum_t \sum_u \langle sr|g|ut\rangle\, \hat{a}_r^\dagger \hat{a}_s^\dagger \hat{a}_u \hat{a}_t. \tag{6.247}$$

The equivalence of this formalism to the previous expressions derived in Sect. 5.4 using the position-spin representation can be demonstrated by calculating the expectation value of the electronic energy in state $\Psi_A(N) = |\psi_i\, \psi_j \ldots|$ represented by the associated vector $|i\, j\, \ldots\rangle$ in the Fock space. For the average one-electron energy, one indeed recovers (5.74):

$$\begin{aligned}\langle ij\ldots|\hat{\mathcal{F}}^F\,|ij\ldots\rangle &= \sum_r \sum_s \bar{h}_{s,r}\, \langle ij\ldots r\ldots|\hat{a}_s^\dagger\, \hat{a}_r|ij\ldots r\ldots\rangle \\ &= \sum_r \sum_s \bar{h}_{s,r}\, \langle ij\ldots r\ldots|ij\ldots s\ldots\rangle = \sum_r \sum_s \bar{h}_{s,r}\, \delta_{r,s} = \sum_r \bar{h}_{r,r}.\end{aligned} \tag{6.248}$$

The expectation value of the two-electron operator similarly gives

$$\begin{aligned}\langle ij\ldots|\hat{\mathcal{G}}^F\,|ij\ldots\rangle &= \frac{1}{2}\sum_r \sum_s \sum_t \sum_u \langle sr|g|ut\rangle\, \langle ij\ldots u\ldots t\ldots|\, \hat{a}_r^\dagger \hat{a}_s^\dagger \hat{a}_u \hat{a}_t|ij\ldots u\ldots t\ldots\rangle \\ &= \frac{1}{2}\sum_r \sum_s \sum_t \sum_u \langle sr|g|ut\rangle\, \langle ij\ldots u\ldots t\ldots|ij\ldots s\ldots r\ldots\rangle.\end{aligned} \tag{6.249}$$

The mutual projection of the two vectors in the Fock space does not vanish only in two cases: for $(s = u) \wedge (r = t)$, when $\langle ij\ldots u\ldots t\ldots|ij\ldots u\ldots t\ldots\rangle = 1$, and for $(s = t) \wedge (r = u)$, when $\langle ij\ldots u\ldots t\ldots|ij\ldots t\ldots u\ldots\rangle = -\langle ij\ldots u\ldots t\ldots|ij\ldots u\ldots t\ldots\rangle = -1$. This again gives the familiar result of (5.75):

$$\langle ij\ldots|\hat{\mathcal{G}}^F\,|ij\ldots\rangle = \frac{1}{2}\sum_r \sum_s [\langle sr|g|sr\rangle - \langle sr|g|rs\rangle] = \frac{1}{2}\sum_r \sum_s [\bar{J}_{s,r} - \bar{K}_{s,r}]. \tag{6.250}$$

It should be further emphasized that the creation operator and its field operator analog act as such only when acting on the *ket*, to their right. Indeed, when acting on the *bra*, to their left, they have the opposite, annihilation meaning, e.g.,

$$\begin{aligned}&\hat{a}_r^\dagger|ij\ldots pr\rangle = 0 \quad \text{but} \\ &\langle ij\ldots pr|\hat{a}_r^\dagger = [\hat{a}_r|ij\ldots pr\rangle]^\dagger = |ij\ldots p\rangle^\dagger = \langle ij\ldots p|.\end{aligned} \tag{6.251}$$

In the position-spin representation, one finds analogous expressions for operators depending upon the continuous/discrete variable $\boldsymbol{q}$. For example, the Fock space representation of the spin density operator $\hat{\rho}(\boldsymbol{q})$ (3.9) is given by the product of the creation and annihilation field operators:

$$\hat{\rho}^F(\boldsymbol{q}) = \hat{\psi}^\dagger(\boldsymbol{q})\,\hat{\psi}(\boldsymbol{q}). \tag{6.252}$$

Indeed, using expressions reported in (6.233) and (6.234, one then recovers the known expression for the spin density in terms of the occupied SO:

$$\begin{aligned}\langle ij\ldots|\hat{\rho}^F(\boldsymbol{q})|ij\ldots\rangle &= \sum_r\sum_s \psi_s^*(\boldsymbol{q})\psi_r(\boldsymbol{q})\,\langle ij\ldots r\ldots|\hat{a}_s^\dagger\,\hat{a}_r|ij\ldots r\ldots\rangle \\ &= \sum_r\sum_s \psi_s^*(\boldsymbol{q})\,\psi_r(\boldsymbol{q})\langle ij\ldots r\ldots|ij\ldots s\ldots\rangle \\ &= \sum_r\sum_s \psi_s^*(\boldsymbol{q})\,\psi_r(\boldsymbol{q})\,\delta_{r,s} = \sum_r |\psi_r(\boldsymbol{q})|^2. \end{aligned}\tag{6.253}$$

The kinetic energy operator in the position-spin representation $\hat{\mathrm{T}}_e(\mathbf{q})$ has the following representation in the Fock space (a.u.):

$$\hat{\mathrm{T}}_e^F(\mathbf{q}) = \int d\boldsymbol{q}\,\hat{\psi}^\dagger(\boldsymbol{q})\left(-\frac{1}{2}\Delta\right)\hat{\psi}(\boldsymbol{q}) = \frac{1}{2}\int d\boldsymbol{q}\nabla\hat{\psi}^\dagger(\boldsymbol{q})\cdot\nabla\hat{\psi}(\boldsymbol{q}). \tag{6.254}$$

Notice the similarity of the preceding expressions for the (*many*-electron) quantum mechanical operators to the corresponding one-electron expectation values in the one-electron state $\varphi(\boldsymbol{q})$:

$$\rho(\boldsymbol{q}) = \varphi^*(\boldsymbol{q})\,\varphi(\boldsymbol{q}) \quad \text{and} \quad T_e[\varphi] = \frac{1}{2}\int d\boldsymbol{q}\,\nabla\varphi^*(\boldsymbol{q})\cdot\nabla\varphi(\boldsymbol{q}),$$

with the field operators of the Fock space in the former replacing the wave functions of the latter. This analogy also holds in the Fock space representation of the probability current operator $\hat{\mathbf{j}}(\boldsymbol{q})$[compare (3.128)]:

$$\hat{\mathbf{j}}^F(\boldsymbol{q}) = \frac{1}{2i}[\hat{\psi}^\dagger(\boldsymbol{q})\,\nabla\hat{\psi}(\boldsymbol{q}) - \hat{\psi}(\boldsymbol{q})\,\nabla\hat{\psi}^\dagger(\boldsymbol{q})]. \tag{6.255}$$

The remaining parts of the electronic Hamiltonian in the coordinate-spin representation read:

$$\begin{aligned}\hat{\mathrm{V}}_{ne}^F &= \int d\boldsymbol{q}\;v(\boldsymbol{q})\,\hat{\psi}^\dagger(\boldsymbol{q})\,\hat{\psi}(\boldsymbol{q}), \\ \hat{\mathrm{V}}_{ee}^F &= \frac{1}{2}\int d\boldsymbol{q}_1\int d\boldsymbol{q}_2\frac{1}{|\boldsymbol{r}_1-\boldsymbol{r}_2|}\hat{\psi}^\dagger(\boldsymbol{q}_1)\,\hat{\psi}^\dagger(\boldsymbol{q}_2)\,\hat{\psi}(\boldsymbol{q}_2)\,\hat{\psi}(\boldsymbol{q}_1). \end{aligned}\tag{6.256}$$

In this local *second* quantization formalism, the *first*- and *second*-order reduced density matrices are expectation values of the following products of field operators:

$$\hat{\gamma}_1^F(\boldsymbol{q}_1;\boldsymbol{q}_1{}') = \hat{\psi}^\dagger(\boldsymbol{q}_1)\,\hat{\psi}(\boldsymbol{q}_1{}'),$$
$$\hat{\gamma}_2^F(\boldsymbol{q}_1,\boldsymbol{q}_2;\boldsymbol{q}_1{}',\boldsymbol{q}_2{}') = \frac{1}{2}\hat{\psi}^\dagger(\boldsymbol{q}_1)\,\hat{\psi}^\dagger(\boldsymbol{q}_2)\,\hat{\psi}(\boldsymbol{q}_2{}')\,\hat{\psi}(\boldsymbol{q}_1{}'). \tag{6.257}$$

Indeed, the spin density operator of (6.252) is seen to constitute the diagonal part of $\hat{\gamma}_1^F(\boldsymbol{q}_1;\boldsymbol{q}_1{}')$ thus giving rise to the following position-spin representation of the particle number operator of (6.236):

$$\hat{\mathrm{N}}^F = \int d\boldsymbol{q}\,\hat{\rho}^F(\boldsymbol{q}) = \int d\boldsymbol{q}\,\hat{\psi}^\dagger(\boldsymbol{q})\,\hat{\psi}(\boldsymbol{q}). \tag{6.258}$$

The equivalence of this continuous expression and the previously reported discrete operator can be directly verified using the relevant expansions of the field operators [(6.233) and (6.234)]:

$$\hat{\mathrm{N}}^F = \sum_{r,r'} \langle r\mid r'\rangle\,\hat{\mathrm{a}}_r^\dagger\hat{\mathrm{a}}_{r'} = \sum_{r,r'} \delta_{r,r'}\hat{\mathrm{a}}_r^\dagger\,\hat{\mathrm{a}}_{r'} = \sum\nolimits_r \hat{\mathrm{a}}_r^\dagger\hat{\mathrm{a}}_r = \sum\nolimits_r \hat{\mathrm{n}}_r^F, \tag{6.259}$$

where $\hat{\mathrm{n}}_r^F$ stands for the occupation operator of $|r\rangle$.

6.5.2 *Cluster Expansion of Electronic States*

We have demonstrated in the preceding section that the state vectors of the excited configurations appearing in the CI expansion of the correlated electronic wave functions can be written in the *second* quantization representation as the result of acting on the HF ("vacuum") state $\left|\Psi_0^{\mathrm{HF}}\right\rangle \equiv |vac\rangle = |\ldots,k,l,\ldots,m,\ldots\rangle$ (containing zero excitations) with the corresponding electron excitation operators [see (6.220) and (6.243)]:

$$\left|\Phi_k^p\right\rangle \equiv |k\to p\rangle = \hat{\mathrm{a}}_p^\dagger\hat{\mathrm{a}}_k|vac\rangle, \qquad \left|\Phi_{k,l}^{p,q}\right\rangle \equiv |(k,l)\to(p,q)\rangle = \hat{\mathrm{a}}_p^\dagger\hat{\mathrm{a}}_q^\dagger\hat{\mathrm{a}}_l\hat{\mathrm{a}}_k|vac\rangle,$$
$$\left|\Phi_{k,l,\ldots,m}^{p,q,\ldots,r}\right\rangle \equiv |(k,l,\ldots,m)\to(p,q,\ldots,r)\rangle = \hat{\mathrm{a}}_p^\dagger\hat{\mathrm{a}}_q^\dagger\ldots\hat{\mathrm{a}}_r^\dagger\hat{\mathrm{a}}_m\ldots\hat{\mathrm{a}}_l\hat{\mathrm{a}}_k|vac\rangle,\ \text{etc.} \tag{6.260}$$

Therefore, the FCI expansion of the correlated ground state of electrons can be interpreted as the result of acting on the HF vacuum state with the general electron excitation operator $\hat{\mathrm{T}}$, which combines excitations of all multiplicities:

$$|\Psi_0^{\mathrm{CI}}\rangle \equiv C_0|\bar{\Psi}_0^{\mathrm{CI}}\rangle = C_0\left(1+\sum_k^{occd.}\sum_p^{virt.} \bar{C}_k^p\, \hat{a}_p^\dagger \hat{a}_k + \sum_{k<l}^{occd.}\sum_{p<q}^{virt.} \bar{C}_{k,l}^{p,q}\, \hat{a}_p^\dagger \hat{a}_q^\dagger \hat{a}_l \hat{a}_k + \ldots\right)|vac\rangle$$
$$\equiv C_0\left(1+\sum_k^{occd.} \hat{u}_k + \sum_{k<l}^{occd.} \hat{u}_{k,l} + \ldots + \sum_{k<l<\ldots<m}^{occd.} \hat{u}_{k,l,\ldots,m} + \ldots\right)|vac\rangle$$
$$\equiv C_0(1+\hat{T}_1+\hat{T}_2+\ldots+\hat{T}_n+\ldots)\,|vac\rangle \equiv C_0(1+\hat{T})\,|vac\rangle, \tag{6.261}$$

where $\hat{u}_{k,l,\ldots,m} = \sum_{p<q<\ldots<r}^{virt.} \bar{C}_{k,l,\ldots,m}^{p,q,\ldots,r}\, \hat{a}_p^\dagger \hat{a}_q^\dagger \ldots \hat{a}_r^\dagger \hat{a}_m \ldots \hat{a}_l \hat{a}_k$. The cluster operators

$$\hat{T}_1 = \sum_k^{occd.} \hat{u}_k, \quad \hat{T}_2 = \sum_{k<l}^{occd.} \hat{u}_{k,l}, \quad \ldots, \quad \hat{T}_n = \sum_{k<l<\ldots<m}^{occd.} \hat{u}_{k,l,\ldots,m}, \quad \ldots \tag{6.262}$$

are said to generate the 1-cluster, 2-cluster, . . ., n-cluster, . . ., corrections to the HF (vacuum) state, respectively, through the excitations of single electrons, electronic pairs, . . ., n-electron clusters, etc.

Consider now the action on the HF vacuum state of the *exponential* electron excitation operator of the CC approximation [see (3.92a,b)],

$$\exp(\hat{T}) = 1+\hat{T}+\frac{1}{2}\hat{T}^2+\frac{1}{3!}\hat{T}^3+\ldots, \tag{6.263}$$

$$\exp(\hat{T})\,|\mathrm{vac}\rangle = \left[1+\hat{T}_1+\left(\hat{T}_2+\frac{1}{2}\hat{T}_1^2\right)+\left(\hat{T}_3+\hat{T}_1\hat{T}_2+\frac{1}{3!}\hat{T}_1^3\right)\right.$$
$$\left.+\left(\hat{T}_4+\hat{T}_3\hat{T}_1+\frac{1}{2}\hat{T}_2^2+\frac{1}{2}\hat{T}_2\hat{T}_1^2+\frac{1}{4!}\hat{T}_1^4\right)+\ldots\right]|vac\rangle$$
$$\equiv (1+\hat{C}_1+\hat{C}_2+\hat{C}_3+\hat{C}_4+\ldots)\,|vac\rangle \equiv |\bar{\Psi}_0^{\mathrm{CC}}\rangle. \tag{6.264}$$

It reveals the structure of the n-tuple excitation operator $\hat{C}_n$ in the correlated state $|\bar{\Psi}_0^{\mathrm{CC}}\rangle$ of the CC approximation. Besides $\hat{T}_n$ this operator is seen to contain also all products of the *lower*-order excitations, originating from smaller excitation clusters, which together give rise to the combined multiplicity n of all such partial excitations.

For example, it follows from this expansion that *quadruple* operator $\hat{C}_4$ of the CC expansion contains five different cluster excitations. In accordance with the discussion of Sect. 6.4.3, the most important contribution should be expected to originate from the $\hat{T}_2^2$ term, which represents strong Coulomb correlation interactions between two electronic pairs, e.g., electrons occupying two different MO. The same type of reasoning indicates that the $\hat{T}_4$ term in this operator is much less important, since it represents Coulomb correlation interactions between four electrons, which are already to a large extent *exchange* correlated. Finally, the $\hat{T}_1$

contribution to $\hat{C}_4$ can be strongly limited by using the MC SCF reference function in the associated MR-CC method.

6.5.3 Coupled Cluster Method

As argued above, the most important in the cluster expansion of (6.263) are the 2-clusters (electronic pairs) represented by operator $\hat{T}_2$:

$$\hat{T} \cong \hat{T}_2 \equiv \frac{1}{4} \sum_{k<l}^{occd.} \sum_{p<q}^{virt.} t_{k,l}^{p,q} \, \hat{a}_p^\dagger \hat{a}_q^\dagger \hat{a}_l \hat{a}_k, \tag{6.265}$$

where the unknowns $\left\{ t_{k,l}^{p,q} \right\}$, called the CC *amplitudes*, represent the renormalized CI coefficients $\left\{ \bar{C}_{k,l}^{p,q} \right\}$. This CC-2 approximation is basically equivalent to the CPMET treatment of Sect. 6.4.3.

The relevant equations for determining the coefficients defining the correlated state of the coupled 2-clusters are derived in a way similar to that used to derive the corresponding CPMET equations. One starts with the eigenvalue equation of the electronic Hamiltonian for $|\bar{\Psi}_0^{CC}\rangle \cong \exp(\hat{T}_2)\,|vac\rangle$,

$$\hat{H}^e \exp(\hat{T}_2)\,|vac\rangle = E_0^e \exp(\hat{T}_2)\,|vac\rangle, \tag{6.266}$$

where $E_0^e = E_0^{HF} + \Delta E_{corr.}^{CC}$. Acting from the left on both sides of the preceding equation with the inverse exponential operator $\exp(-\hat{T}_2)$, $\exp(-\hat{T}_2)\exp(\hat{T}_2) = \exp(0) = 1$, one then arrives at the associated eigenvalue problem of the similarity-transformed electronic Hamiltonian:

$$\exp(-\hat{T}_2)\hat{H}^e \exp(\hat{T}_2)\,|vac\rangle \equiv \hat{H}_2^e |vac\rangle = E_0^e\,|vac\rangle. \tag{6.267}$$

Its projection onto the HF vacuum state gives the familiar expression for the correlated energy of the system ground state [see (6.180) and (6.195)]:

$$\begin{aligned}
E_0^e &= E_0^{HF} + \Delta E_{corr.}^{CC} = \langle vac|\hat{H}_2^e|vac\rangle \\
&= \langle vac|(1 - \hat{T}_2 + \frac{1}{2}\hat{T}_2^2 - \ldots)\hat{H}^e(1 + \hat{T}_2 + \frac{1}{2}\hat{T}_2^2 + \ldots)|vac\rangle \\
&= \langle vac|\hat{H}^e|vac\rangle + \langle vac|\hat{H}^e\hat{T}_2|vac\rangle \\
&= E_0^{HF} + \sum_{k<l}^{occd.} \sum_{p<q}^{virt.} t_{k,l}^{p,q} \langle vac|\hat{H}^e|(k,l) \to (p,q)\rangle \\
&= E_0^{HF} + \sum_{k<l}^{occd.} \sum_{p<q}^{virt.} t_{k,l}^{p,q} \langle kl|g|pq - qp\rangle = E_0^{HF} + \sum_{k<l}^{occd.} e_{k,l}^{CC}.
\end{aligned} \tag{6.268}$$

Above, we have observed that

$$\langle vac|\hat{T}_2 = 0, \tag{6.269}$$

since the action of the creation operators of $|p\rangle$ and $|q\rangle$ on the *bra* vacuum vector amounts to annihilation of one-particle states which are not used in the HF determinant [see also (6.251)]. For example,

$$\langle vac|\hat{a}_p^\dagger \hat{a}_q^\dagger \hat{a}_l \hat{a}_k = \left(\hat{a}_k^\dagger \hat{a}_l^\dagger \hat{a}_q \hat{a}_p|vac\rangle\right)^\dagger = 0. \tag{6.270}$$

We have also used in (6.268) the Slater–Condon rules of (5.86) and (5.87), which predict the vanishing matrix element $\langle vac|\hat{H}^e\hat{T}_2\hat{T}_2|vac\rangle$, the ket of which represents the *quadruply* excited configuration relative to the HF vacuum.

As in the electron-pair theories, the equations for the optimum CC amplitudes result from projecting (6.267) into the *doubly* excited states:

$$\langle(kl) \to (p,q)|\hat{H}_2^e|vac\rangle = E_0^e\langle(kl) \to (p,q) \mid vac\rangle = 0, \tag{6.271}$$

by the orthogonality of configurations. Expanding the exponential operators then gives the following nonvanishing contributions to the matrix element on the l.h.s of the preceding equation,

$$\langle(kl) \to (p,q)|\hat{H}^e(1 + \hat{T}_2 + \frac{1}{2}\hat{T}_2^2) - \hat{T}_2\hat{H}^e(1 + \hat{T}_2)|vac\rangle = 0, \tag{6.272}$$

which imply the (coupled) quadratic equations for the CC amplitudes.

Since the number of such unknowns is very large, solving these equations is not an easy task. This severely restricts the range of applications of the CC method to at best medium-size molecular systems. The other problem which still awaits a satisfactory solution is the use of the multireference functions, e.g., the ground-states of the *open*-shell systems or the optimum states from the MC SCF method, as the starting point in the cluster expansion.

6.6 Elements of Valence Bond Approach

The *Valence Bond* (VB) theory originates from the classical Heitler and London (HL) (1927) treatment of the hydrogen molecule (see also: London 1928), the first quantum theory of the chemical bond. It has played an extremely important part in the early history of the quantum treatment of the molecular electronic structure, being strongly advocated by Slater and Pauling, later to be dominated by the MO theory of Mulliken, Hund, and Hückel, and made a strong comeback from 1980s onward. An interesting account of the early VB-MO rivalry has recently been given in an excellent primer by Shaik and Hiberty (2004). This theory connects more directly to the classical chemical concepts and offers important insights into the elementary chemical processes, generates the key paradigms of chemistry, such as the bonding electron pair and octet rule, successfully tackles many classical issues in the theory of molecular structure, and conceptualizes the chemical reactivity (e.g., Shaik 1989, Shaik and Hiberty 1991, 1995; Shaik et al. 2001, 2009). Some of its oversimplified implementations in the past have created notable "failures," which gave a false impression and reputation in some circles of the VB theory as representing an obsolete model. A good example of such a problem is the spin multiplicity of the ground state of O_2. Although it is possible to give the VB explanation of why this molecule has a triplet ground state, this reasoning is rather involved in contrast to the very simple and more natural MO explanation.

However, when properly applied, with all its intrinsic nonorthogonalities and the matrix elements of the Hamiltonian properly accounted for, the VB treatment appears as an alternative approach to molecular electronic structure which is fully equivalent to the MO method. The theory quantitative variants, the ab initio VB methods, e.g., *Generalized* VB (GVB) scheme of Goddard and coworkers (Goddard 1967; Goddard and Ladner 1971; Goddard et al. 1973; Goddard and Harding 1978; Rappé and Goddard 1982; see also: Simonetta 1968; Gerratt 1974), provide very efficient computational tools for determining the molecular PES and predicting the outcomes of chemical reactions.

To obtain from a very large calculation the result which agrees with experiment is only part of science. Of equal importance is a convincing, simple, and elegant model of the molecular phenomena, giving the crucial understanding without elaborate calculations. The VB approach, which closely follows the chemical intuition in coupling atomic states in molecules, has been quite successful in providing such a direct insight into many classical problems of the electronic structure. For example, one of the great merits of the VB theory is its visually intuitive wave function, given by the linear combination of the chemically meaningful "structures." The theory gives rise to our present understanding of the competition between the σ and π electrons in aromatic systems (e.g., Shaik et al. 2001; see also Jug and Köster 1990), implies the new, *charge-shift* bonding mechanism (Shaik et al. 2009), and provides qualitative models of chemical reactions (e.g., Epiotis 1978; Shaik and Hiberty 2004) and the VB ideas have been proven very useful in modeling the Born–Oppenheimer energy surfaces for elementary reactive collisions (e.g., Murrell et al. 1984). Several important developments in the straightforward VB theory have also occurred in recent decades, e.g., the Moffitt (1951) theory of AIM with its subsequent refinements (Balint-Kurti and Karplus 1973), the theory of the separated, strongly orthogonal electron pairs (see Sect. 6.4.1) of Hurley et al. (1953), and the general group function model of McWeeny (1989).

The VB strategy for constructing the molecular wave functions uses the quantum states of constituent atoms (fragments), with a strong emphasis on the spin pairing of electrons on the *singly* occupied valence orbitals of AIM as the source of all chemical bonds in the system. This strategy of determining the antisymmetric basis functions for molecular wave functions is thus multideterminant in character. The VB basis set involves all independent N-electron wave-functions, called the VB structures, constructed directly from the valence shell AO of constituent atoms by using the singlet coupling of the spins in the "bonded" electronic pairs, and admitting all possible distributions of spin factors of the unpaired electrons coupled to the specified length and the projection of the resultant spin (6.67). Such functions, however, which represent the admissible spin-pairing patterns of the valence electrons, are not mutually orthogonal. They may even strongly overlap and in general a selection of the independent set of VB structures is not a trivial mathematical problem, requiring the group-theoretic, antisymmetrizing projections for both the spin and spatial functions, as well as special techniques for constructing the spin and spatial parts of the elementary antisymmetric basis functions (e.g., Gerratt 1974).

The simplest way to generate the desired spin functions is to successively couple together the spins according to the rules for adding angular momenta in quantum mechanics, with the totality of such standard spin functions being conventionally visualized in terms of the familiar "branching" diagrams (e.g., Gerratt 1974). This perspective is thus fundamentally different from that used in the HF theory, which aimed at the best one-determinantal representation in terms of the (delocalized) canonical MO represented by LCAO.

6.6.1 Origins of VB Theory

Let us illustrate the basic proposition of HL theory using the classical case of the hydrogen molecule H_a–H_b, with two atoms contributing a single electron each, and the minimum basis set of the atomic functions, $1s_a \equiv a \in H_a$ and $1s_b \equiv b \in H_b$, respectively. There are two independent Slater determinants, which can be created in this minimum AO basis set:

$$|a^+b^-| = \frac{1}{\sqrt{2}}[a(1)b(2)\alpha(1)\beta(2) - b(1)a(2)\beta(1)\alpha(2)],$$
$$|a^-b^+| = \frac{1}{\sqrt{2}}[a(1)b(2)\beta(1)\alpha(2) - b(1)a(2)\alpha(1)\beta(2)], \tag{6.273}$$

from which two combinations, representing proper spin states of two electrons, can be constructed (see Fig. 6.2),

$$\begin{aligned}\Psi_S^{\mathrm{HL}}(1,2) &= \frac{N_S}{\sqrt{2}}[|a^+b^-| - |a^-b^+|] = \frac{N_S}{\sqrt{2}}[|a^+b^-| + |b^+a^-|]\\ &= \left(\frac{N_S}{\sqrt{2}}[a(1)b(2) + b(1)a(2)]\right)\left(\frac{1}{\sqrt{2}}[\alpha(1)\beta(2) - \beta(1)\alpha(2)]\right)\\ &\equiv \Phi_g(a\text{—}b)U_{0,0}(1,2),\end{aligned} \tag{6.274}$$

$$\begin{aligned}\Psi_T^{\mathrm{HL}}(1,2) &= \frac{N_T}{\sqrt{2}}[|a^+b^-| + |a^-b^+|] = \frac{N_T}{\sqrt{2}}[|a^+b^-| - |b^+a^-|]\\ &= \left(\frac{N_T}{\sqrt{2}}[a(1)b(2) - b(1)a(2)]\right)\left(\frac{1}{\sqrt{2}}[\alpha(1)\beta(2) + \beta(1)\alpha(2)]\right)\\ &\equiv \Phi_u(a\text{—}b)U_{1,0}(1,2),\end{aligned} \tag{6.275}$$

where the normalizing factors $N_s = \left(1 + S_{a,b}^2\right)^{-1/2}$ and $N_T = \left(1 - S_{a,b}^2\right)^{-1/2}$, $S_{a,b} = \langle a|b\rangle$. Here, $U_{0,0}(1, 2)$ and $U_{1,0}(1, 2)$ stand for the $M_S = 0$ singlet (S) and triplet (T) spin states, respectively, while the spatial functions $\Phi_g(a\text{—}b)$ and $\Phi_u(a\text{—}b)$ represent the *even* (g) and *odd* (u) combinations of the two AO product states, which are symmetric and antisymmetric, respectively, with respect to the inversion operation $\hat{\mathrm{i}}$ relative to the bond midpoint, or – equivalently – with respect to the permutation of the two electrons.

It thus follows from the two preceding equations that the two independent orbital products $\{a(1)b(2) \equiv |ab\rangle, b(1)a(2) \equiv |ba\rangle\}$ or their combinations in the bonding (g) and antibonding (u) spatial functions $\{\Phi_g(a\text{—}b), \Phi_u(a\text{—}b)\}$, or equivalently the Slater determinants $\{|a^+b^-|, |a^-b^+|\}$, form the equivalent bases of the two-electron functions in terms of which the singlet and triplet states of the hydrogen molecule can be expressed in the adopted minimum basis set of AO. Let us examine the corresponding matrix elements of the electronic Hamiltonian (a.u.), with $\boldsymbol{r}_{i\mathrm{X}} = |\boldsymbol{r}_i - \boldsymbol{R}_\mathrm{X}|$, $i \in (1, 2)$, $\mathrm{X} \in (\mathrm{H}_a, \mathrm{H}_b)$, $R = |\boldsymbol{R}_a - \boldsymbol{R}_b|$,

$$\begin{aligned}\hat{\mathrm{H}}_e(1,2) &= \hat{\mathrm{h}}(1) + \hat{\mathrm{h}}(2) + g(1,2) + R^{-1} \equiv \hat{\mathrm{H}}^e(1,2) + R^{-1}\\ &= \left(-\frac{1}{2}\Delta_1 - \frac{1}{r_{1a}}\right) + \left(-\frac{1}{2}\Delta_2 - \frac{1}{r_{2b}}\right) + \left(-\frac{1}{r_{1b}} - \frac{1}{r_{2a}} + \frac{1}{r_{1,2}} + \frac{1}{R}\right)\\ &\equiv \hat{\mathrm{H}}_a(1) + \hat{\mathrm{H}}_b(2) + \hat{\mathrm{H}}_{ab}(1,2),\end{aligned} \tag{6.276}$$

where $\{\hat{\mathrm{H}}_X\}$ denote the atomic Hamiltonians satisfying the eigenvalue equations of the separated atoms, $\hat{\mathrm{H}}_a|a\rangle = -1/2\,|a\rangle$ and $\hat{\mathrm{H}}_b|b\rangle = -1/2\,|b\rangle$, $\hat{\mathrm{H}}_{ab}$ groups the interaction terms between the two atoms, all vanishing in the $R \to \infty$ limit, and the one-electron operator of (5.70) reads:

$$\hat{\mathrm{h}}(i) = -\frac{1}{2}\Delta_i - \frac{1}{r_{ia}} - \frac{1}{r_{ib}} = -\frac{1}{2}\Delta_i + v(i). \tag{6.277}$$

The VB-*Coulomb* (diagonal) matrix elements of the electronic Hamiltonian for the elementary VB structures can be then expressed in terms of the orbital one- and two-electron integrals:

$$\begin{aligned}Q &= \langle ab|\hat{\mathrm{H}}^e|ab\rangle = \langle |a^+b^-||\hat{\mathrm{H}}^e||a^+b^-|\rangle = \langle |a^-b^+||\hat{\mathrm{H}}^e||a^-b^+|\rangle\\ &= \langle a|\hat{\mathrm{h}}|a\rangle + \langle b|\hat{\mathrm{h}}|b\rangle + \langle ab|g|ab\rangle = h_{a,a} + h_{b,b} + J_{a,b} \qquad \text{or}\end{aligned} \tag{6.278}$$

$$\begin{aligned}\langle ab|\hat{\mathrm{H}}_e|ab\rangle &= \langle |a^+b^-||\hat{\mathrm{H}}_e||a^+b^-|\rangle = \langle |a^-b^+||\hat{\mathrm{H}}_e||a^-b^+|\rangle = Q + R^{-1} \\ &= -1 + \langle ab|\hat{\mathrm{H}}_{ab}|ab\rangle \equiv -1 + \mathcal{J}(R),\end{aligned} \tag{6.279}$$

where the expectation value of the interatomic hamiltonian

$$\begin{aligned}\mathcal{J}(R) &= \langle ab|\hat{\mathrm{H}}_{ab}|ab\rangle = \langle |a^+b^-||\hat{\mathrm{H}}_{ab}||a^+b^-|\rangle = \langle |a^-b^+||\hat{\mathrm{H}}_{ab}||a^-b^+|\rangle \\ &= \exp(-2R)(R^{-1} + 5/8 - 3R/4 - R^2/6).\end{aligned} \tag{6.280}$$

The VB *exchange* (nondiagonal) matrix element of the electronic Hamiltonian can be similarly expressed in terms of the AO overlap integral,

$$S_{a,b}(R) = \langle a|b\rangle = \exp(-R)(1 + R + R^2/3), \tag{6.281}$$

the coupling one-electron integral $h_{a,b}$, and the exchange two-electron integral $K_{a,b}$:

$$\begin{aligned}K &= \langle ab|\hat{\mathrm{H}}^e|ba\rangle = \langle |a^+b^-||\hat{\mathrm{H}}^e||b^+a^-|\rangle = \langle |a^-b^+||\hat{\mathrm{H}}^e||b^-a^+|\rangle \\ &= 2S_{a,b}\langle a|\hat{\mathrm{h}}|b\rangle + \langle ab|g|ba\rangle = 2S_{a,b}h_{a,b} + K_{a,b} \qquad \text{or}\end{aligned} \tag{6.282}$$

$$\begin{aligned}\langle ab|\hat{\mathrm{H}}_e|ba\rangle &= \langle |a^+b^-||\hat{\mathrm{H}}_e||b^+a^-|\rangle = -S_{a,b}^2(R) + \langle ab|\hat{\mathrm{H}}_{ab}|ba\rangle \\ &\equiv -S_{a,b}^2(R) + \mathcal{K}(R),\end{aligned} \tag{6.283}$$

where $\mathcal{K}(R)$ represents the spin exchange term of the interatomic Hamiltonian $\hat{\mathrm{H}}_{ab}$.

Finally, for the overlap between the two elementary VB product functions one obtains:

$$S(R) = \langle ab \mid ba\rangle = \langle |a^-b^+| \mid |b^-a^+|\rangle = S_{a,b}^2(R). \tag{6.284}$$

For the equilibrium internuclear distance, one thus finds $S_{a,b}(R_e) = 0.75$ and hence $S(R_e) = 0.56$.

In terms of these matrix elements the expectation values of the electronic energy in the singlet (bonding) state (6.274) and its triplet (antibonding) analog (6.275) of H_2 respectively read:

$$\begin{aligned}\langle E^e\rangle_{\Psi_S^{\mathrm{HL}}} &\equiv E^e\left[\Psi_S^{\mathrm{HL}}\right] = \left\langle \Psi_S^{\mathrm{HL}}\middle|\hat{\mathrm{H}}^e\middle|\Psi_S^{\mathrm{HL}}\right\rangle = \frac{Q+K}{1+S} \qquad \text{or} \\ E_e\left[\Psi_S^{\mathrm{HL}}\right] &= \left\langle \Psi_S^{\mathrm{HL}}\middle|\hat{\mathrm{H}}_e\middle|\Psi_S^{\mathrm{HL}}\right\rangle = -1 + \frac{\mathcal{J}+\mathcal{K}}{1+S},\end{aligned} \tag{6.285}$$

$$\begin{aligned}\langle E^e\rangle_{\Psi_T^{\mathrm{HL}}} &\equiv E^e\left[\Psi_T^{\mathrm{HL}}\right] = \left\langle \Psi_T^{\mathrm{HL}}\middle|\hat{\mathrm{H}}^e\middle|\Psi_T^{\mathrm{HL}}\right\rangle = \frac{Q-K}{1-S} \qquad \text{or} \\ E_e\left[\Psi_T^{\mathrm{HL}}\right] &= \left\langle \Psi_T^{\mathrm{HL}}\middle|\hat{\mathrm{H}}_e\middle|\Psi_T^{\mathrm{HL}}\right\rangle = -1 + \frac{\mathcal{J}-\mathcal{K}}{1-S}.\end{aligned} \tag{6.286}$$

In the two preceding equations, the first expression does not include the nuclear repulsion term, while the second expression explicitly separates the dissociation limit, at $R \to \infty$, $E^e\left[\Psi_S^{HL}(\infty)\right] = E^e\left[\Psi_T^{HL}(\infty)\right] = -1$, when $S = \hat{H}_{ab} = 0$ and hence also $\mathcal{J} = \mathcal{K} = 0$.

Should one neglect the overlap, $S \approx 0$, the electronic energies of the singlet and triplet states of H_2 thus read: $E^e\left[\Psi_S^{HL}\right] = Q + K$ and $E^e\left[\Psi_T^{HL}\right] = Q - K$. The energy $Q + R^{-1}$ represents the energy of the separated atoms plus their coulombic interaction and it remains quasiconstant as function of the internuclear distance, from the infinite distance to about the equilibrium distance R_e, showing a shallow minimum near R_e. It corresponds to the energy of the semiclassical state of the two hydrogen atoms, when they are brought together without exchanging their spins. The spin exchange term K describes the nonclassical effect associated with enforcing the proper spin state of two electrons, and it becomes large and negative at the normal interatomic distance, accounting for over 90% of the binding energy of the molecule in the singlet state. In the antibonding triplet state K appears with the opposite sign thus giving rise to an effective repulsion at all distances leading to a spontaneous dissociation of the molecule into two hydrogen atoms.

This explains why in the early quantum theory of the chemical bond the VB energy term associated with the spin exchange between the two AO in the two elementary determinantal states $|a^+b^-|$ and $|a^-b^+|$ has acquired an apparently crucial importance. To quote the Shaik and Hiberty (2004), "*...the physical phenomenon responsible for the bond is the exchange of spins between the two AOs, that is the resonance between the two spin arrangements ....*" However, this diagnosis may be somewhat misleading, since it directly follows from the final expression of (6.282) that only the second term represents the *true* exchange integral $K_{a,b} = \langle a(1)b(2)|g(1,2)|b(1)a(2)\rangle_{1,2} > 0$, involving the *two* electrons and interchange of their variables. Its remaining, dominant contribution, proportional to the overlap of (6.281), $2h_{a,b}S_{a,b}$, is of the one-electron character, $h_{a,b} = \langle a(1)|\hat{h}(1)|b(1)\rangle_1 = t_{a,b} + v_{a,b} < 0$, including the nondiagonal matrix element of the kinetic energy, $t_{a,b} = \langle a(1)| - \frac{1}{2}\Delta_1|b(1)\rangle_1$, representing the nonadditive kinetic energy in AO resolution, and the attraction energy between the nuclei and the overlap charge distribution of the chemical bond, $v_{a,b} = \langle a(1)|v(1)|b(1)\rangle_1$. It is large and negative, completely outweighing the (positive) real exchange integral $K_{a,b}$. Therefore, the VB exchange term K can assume its correct, negative value only when the AO overlap is not neglected. As emphasized by McWeeny (1989), this energy "*... effectively disquises the actual factors involved in chemical bonding by indiscriminately mixing together terms representing kinetic energy, electron–nuclear attraction energy and electron-electron repulsion energy ....*"

Nevertheless, one cannot dismiss altogether the importance of this VB spin exchange term as an important bond increment ("invariant") for a qualitative understanding of the origins of the chemical bond. Indeed, it directly follows from the molecular (BO) virial theorem that formation of the chemical bond involves changes in all energy contributions, both kinetic and potential, attractive and repulsive, of one- and two-electron origin. It thus follows that the above-mentioned mixed character of K, combining different kinetic and potential one- and two-electron energies, cannot be seriously contemplated as a disqualifying feature in a discussion of its potential usefulness in a qualitative understanding/interpretation of the elementary bonding/antibonding effects.

6.6.2 Bond Energies and Ionic Structures

The experimental value of the magnitude of the bonding energy in H_2 (a.u.) (6.65), $-\Delta E_{bond}^{exp.}(R_e^{exp.} = 1.4006) = D_e^{exp.}(R_e^{exp.}) = 0.1745$, is already quite well reproduced in the HL approximation in the minimum basis set of the two 1s orbitals of constituent atoms, $D_e^{\mathrm{HL}}(R_e^{\mathrm{HL}} = 1.64) = 0.115$. It thus gives a better prediction than the corresponding simple MO function, $D_e^{\mathrm{MO}}(R_e^{\mathrm{MO}} = 1.57) = 0.097$, although the predicted equilibrium bond length is much overestimated. The HL function gives the correct atomic dissociation, while the RHF MO approximation predicts the equimixture of the atomic and ionic dissociation products [see (6.63)]. This is because the HL wave function of (6.274) partly accounts for the Coulomb correlation between electrons by representing the state in which the two spin-paired electrons avoid the simultaneous occupancy of the same AO.

Such *ion* pair occupation patterns, of two spin-paired electrons occupying the same AO, generate the elementary *ionic* VB products $\{a(1)a(2) = |aa\rangle \equiv [\mathrm{H}_a^- \mathrm{H}_b^+]$ and $b(1)b(2) = |bb\rangle \equiv [\mathrm{H}_a^+ \mathrm{H}_b^-]\}$ or equivalently their *even* (g) combination, symmetric with respect to the inversion operation $\hat{\mathrm{i}}$ relative to the bond midpoint, compatible with the spatial symmetry type of the singlet wave function Ψ_S^{HL},

$$\Phi_g^{ion.}(a--b) = \frac{1}{\sqrt{2(1+S_{a,b}^2)}}[a(1)a(2) + b(1)b(2)]$$
$$= N^{ion.}\{[\mathrm{H}_a^- \mathrm{H}_b^+] + [\mathrm{H}_a^+ \mathrm{H}_b^-]\}. \qquad (6.287\mathrm{a})$$

It represents the spatial factor of the associated g-type (singlet) ionic VB structure:

$$\Psi_S^{ion.}(1,2) = \frac{N_S}{\sqrt{2}}[|a^+a^-| + |b^+b^-|] = \Phi_g^{ion.}(a--b)\, U_{0,0}(1,2). \qquad (6.287\mathrm{b})$$

Therefore, the Ψ_S^{HL} state represents the *covalent* structure of (6.63), $\Phi_g(a—b) \equiv \Phi_g^{cov.} = \Phi_0^{atom.}$, in which electrons are shared between the two bonded atoms. However, the exact ground state must also partly involve the simultaneous *double* occupancy of the two AO, thus calling for an admixture of the ionic configurations. Indeed, the covalent HL wave function $\Phi_g^{cov.}$ supplemented by the ionic VB structure $\Phi_g^{ion.}$ in the CI-type combination,

$$\Phi_g^{cov.+ion.} \equiv C^{cov.}\Phi_g^{cov.} + C^{ion.}\Phi_g^{ion.}, \qquad (6.288)$$

can be shown to be equivalent to the CID wave function of the MO theory (see the next section).

It should be emphasized, however, that the covalent and ionic combinations are strongly overlapping,

$$\left\langle \Phi_g^{cov.}(R_e) \,\middle|\, \Phi_g^{ion.}(R_e) \right\rangle = 2S_{a,b}(R_e) \Big/ \left[1 + S_{a,b}^2(R_e)\right] = 0.96, \qquad (6.289)$$

so that both states are practically identical and the expansion coefficients in (6.288) have no physical meaning implied by the superposition principle.

In order to extract the really new content $\bar{\Phi}_g^{ion.}$ of $\Phi_g^{ion.}$, which is not already contained in $\Phi_g^{cov.}$, one has to Schmidt orthogonalize the former with respect to the latter [see (3.45a)]:

$$\bar{\Phi}_g^{ion.} = 3.57\Phi_g^{ion.} - 3.43\Phi_g^{cov.}, \qquad (6.290)$$

while the inverse transformation reads:

$$\Phi_g^{ion.} = 0.96\,\Phi_g^{cov.} - 0.28\,\bar{\Phi}_g^{ion.}. \qquad (6.291)$$

The last equation shows directly that the main ingredient of $\Phi_g^{ion.}$ is already present in $\Phi_g^{cov.}$, so that admixture of this ionic function introduces but little new, independent (orthogonal) component to the system ground state wave function. The variationally optimum VB wave function of (6.288) at $R = R_e$,

$$\Phi_g^{cov.+ion.} = 0.454\,\Phi_g^{cov.} + 0.116\,\Phi_g^{ion.} = 0.998\,\Phi_g^{cov.} + 0.058\,\bar{\Phi}_g^{ion.}, \qquad (6.292)$$

thus predicts the 99.7% overall participation of Φ_g^{cov} and only 0.3% of the orthogonal ionic component $\bar{\Phi}_g^{ion.}$, i.e., a practically purely covalent chemical bond. Nevertheless, this slightly (ionically) modified wave function further improves the bonding energy of H_2 but still fails to correct the overestimated equilibrium bond length: $D_e^{\mathrm{HL}/ion.}\left(R_e^{\mathrm{HL}/ion.} = 1.67\right) = 0.119$.

Both MO and VB descriptions can be further improved by scaling the exponents of both $1s$ AO, $\chi_{1s}(r;\ \zeta\ ) = (\zeta^3/\pi)^{1/2}\exp(-\zeta r)$, with the optimum factor ζ, a nonlinear variational parameter, reflecting the effective charge of the atomic nucleus [see (4.62) and (5.98)] in the presence of the unshielded nucleus of the bond partner: $\zeta > 1$. The corresponding predictions then show a decisive improvement in the equilibrium bond length: $D_e^{\mathrm{MO}/\zeta}\left(R_e^{\mathrm{MO}/\zeta} = 1.38\right) = 0.128$, $\zeta^{\mathrm{MO}} = 1.197$; $D_e^{\mathrm{HL}/\zeta}\left(R_e^{\mathrm{HL}/\zeta} = 1.39\right) = 0.139$, $\zeta^{\mathrm{HL}} = 1.166$; $D_e^{\mathrm{HL}/ion./\zeta}\left(R_e^{\mathrm{HL}/ion./\zeta} = 1.43\right) = 0.148$, $\zeta^{\mathrm{HL}/ion.} = 1.193$.

When two atoms approach each other their electron distributions should undergo a cylindrical polarization toward the bonding partner. Therefore, instead of using the spherical AO, one could apply in the VB wave function the optimum (fractional) hybrids $\chi_h = s^x p_z^y$ on both atoms, which are directed along the bond (z) axis:

$$\chi_h(\boldsymbol{r}) = (1 + \mu^2)^{-1/2}[1s(\boldsymbol{r}) + \mu 2p_z(\boldsymbol{r})], \qquad (6.293)$$

where the mixing coefficient μ then constitutes additional variational degree of freedom of the molecular wave function. The corresponding predictions read: $D_e^{HL/h/\zeta}\left(R_e^{HL/h/\zeta}=1.42\right)=0.148, \zeta^{\,HL/h}=1.19$; $D_e^{HL/ion./h/\zeta}\left(R_e^{HL/ion./h/\zeta}=1.41\right)=0.151$, $\zeta^{\,HL/ion./h}=1.19$.

Clearly, such hybrid AO in the MO approach implies an effective extension of the minimum basis set. Let us compare these best VB results with the associated HFL predictions, obtained for the variationally saturated, "complete" basis set, which fully accounts for both the effective contraction and polarization of atomic electron distributions: $D_e^{HFL}\left(R_e^{HFL}=1.40\right)=0.134$. This comparison explicitly shows that a substantial portion of the bond energy due to the Coulomb correlation has already been recovered in these simple VB descriptions of the hydrogen molecule.

6.6.3 Comparison with MO Theory and AO Expansion Theorem

Let us now compare the MO and VB wave functions for H_2, obtained in the minimum basis set of two 1*s* orbitals of constituent atoms. The simple RHF description predicts that the two spin-paired electrons occupy the bonding (even) $1\sigma_g$ MO, symmetric with respect to the inversion $\hat{\mathrm{i}}$ with respect to the bond midpoint,

$$\phi_g(\boldsymbol{r})=N_g[a(\boldsymbol{r})+b(\boldsymbol{r})], \qquad N_g=[2(1+S_{a,b}(R)]^{-1/2}, \qquad \hat{\mathrm{i}}\phi_g=\phi_g, \quad (6.294)$$

with the antibonding (odd) $1\sigma_u$ MO, antisymmetric with respect to inversion,

$$\phi_u(\boldsymbol{r})=N_u[a(\boldsymbol{r})-b(\boldsymbol{r})], \quad N_u=[2(1-S_{a,b}(R)]^{-1/2}, \quad \hat{\mathrm{i}}\phi_u=-\phi_u, \qquad (6.295)$$

remaining unoccupied in the system ground state configuration.

Hence, the spatial part of the singlet (ground state) wave function in the RHF approximation, when expressed in terms of AO, predicts equimixture of the covalent and ionic VB structures:

$$\begin{aligned}\Phi_0^{RHF}(1,2)&=N_g^2[a(1)+b(1)][a(2)+b(2)]\\&=N_g^2\{[a(1)b(2)+b(1)a(2)]+[a(1)a(2)+b(1)b(2)]\}\\&\equiv\bar{N}_g\left[\Phi_g^{cov.}(1,2)+\Phi_g^{ion.}(1,2)\right],\end{aligned} \qquad (6.296)$$

This explains the wrong dissociation limit of this RHF function for $R\to\infty$ [see (6.63)] and its variational inferiority with respect to the VB wave function of (6.288), in which the relative participation of both components is not fixed, with the equilibrium ratio $C^{ion.}(R_e)/C^{cov.}(R_e)=0.256$ approaching zero as $R\to\infty$.

The Pauli principle implies that in the singlet (antisymmetric) spin state of two electrons their spatial function must be symmetric with respect to exchanging the positions of electrons. Moreover, by the Brillouin theorem the *singly* excited configuration $\Phi^S_{g,u}(1,2) = 2^{-1/2}[\phi_g(1)\phi_u(2) + \phi_u(1)\phi_g(2)]$(see Fig. 6.2) does not couple directly to the *closed*-shell RHF function. Therefore, the most important Coulomb correlation contribution originates from including in the CI function the doubly excited configuration $\Phi^S_u(1,2) = \phi_u(1)\phi_u(2)$ in the CID-type trial function

$$\begin{aligned}\Phi^{\mathrm{CID}}(1,2) &= C_0\Phi_0^{\mathrm{RHF}}(1,2) + C_u\Phi_u^S(1,2)\\ &= C_0N_g^2[a(1)+b(1)][a(2)+b(2)] + C_uN_u^2[a(1)-b(1)][a(2)-b(2)]\\ &\equiv \lambda_g\left[\Phi_g^{cov.}(1,2)+\Phi_g^{ion.}(1,2)\right] + \lambda_u\left[\Phi_g^{ion.}(1,2)-\Phi_g^{cov.}(1,2)\right]\\ &= (\lambda_g-\lambda_u)\Phi_g^{cov.}(1,2) + (\lambda_g+\lambda_u)\Phi_g^{ion.}(1,2)\\ &\equiv \lambda_{cov.}\Phi_g^{cov.}(1,2) + \lambda_{ion.}\Phi_g^{ion.}(1,2).\end{aligned} \tag{6.297}$$

This CID trial wave function is variationally equivalent to the $\Phi_g^{cov.+ion.}$ state of (6.288) since it also represents a combination of the independent components $\Phi_g^{cov.}$ and $\Phi_g^{ion.}$. Therefore, the amount of the Coulomb correlation of two electrons in the ground state of H_2 recovered by this VB approximation is the same as that generated in the CID approximation using the same (minimum) basis set of AO.

It should be observed that from the expansion theorem of Sect. 6.2 it directly follows that any *N*-electron function can be expanded in terms of the Slater determinants built from any complete set of the one-electron functions, e.g., the orthonormal SO. Therefore, such state functions should be also represented as combinations of the chemically meaningful elementary VB structures, represented by determinants constructed from the nonorthogonal AO, e.g., the localized canonical or hybridized valence orbitals of constituent atoms, or partly delocalized AO of molecularly promoted AIM, the *group*-MO of molecular fragments, etc.

Let us briefly examine the relevant VB expansion theorem, in which the *atomic* (nonorthogonal) spin orbitals, $\boldsymbol{\chi}(\boldsymbol{r},\sigma) \equiv \boldsymbol{\chi}(\boldsymbol{q}) = \{\boldsymbol{\chi}(\boldsymbol{r})\alpha(\sigma) \equiv \boldsymbol{\chi}^+(\boldsymbol{r}),\ \boldsymbol{\chi}(\boldsymbol{r})\beta(\sigma) \equiv \boldsymbol{\chi}^-(\boldsymbol{r})\}$, constitute the one-electron basis themselves, instead of their *molecular* (orthonormal) SO combinations, $\boldsymbol{\psi}(\boldsymbol{q}) = \boldsymbol{\chi}(\boldsymbol{q})\ \mathbf{C}$, used in the MO expansion of the CI approach. Since AO constitute the normalized, but nonorthogonal basis, $\langle\boldsymbol{\chi}|\boldsymbol{\chi}\rangle = \mathbf{S} \neq \mathbf{I}$, the corresponding completeness projection reads

$$\hat{\mathrm{P}}_{\chi} = |\boldsymbol{\chi}\rangle\mathbf{S}^{-1}\langle\boldsymbol{\chi}| = 1, \tag{6.298}$$

or, in the spin position representation,

$$\begin{aligned}\hat{\mathrm{P}}_{\chi}(\boldsymbol{q},\boldsymbol{q}') &= \langle\boldsymbol{q}|\hat{\mathrm{P}}_{\chi}|\boldsymbol{q}'\rangle = \boldsymbol{\chi}(\boldsymbol{q})\mathbf{S}^{-1}\boldsymbol{\chi}^\dagger(\boldsymbol{q}') = \sum_{k,l}\chi_k(\boldsymbol{q})S_{k,l}^{-1}\chi_l^*(\boldsymbol{q}')\\ &= \langle\boldsymbol{q}\mid\boldsymbol{q}'\rangle = \delta(\boldsymbol{q}'-\boldsymbol{q}).\end{aligned} \tag{6.299}$$

This gives the following AO expansion of any one-electron function $\Psi(\boldsymbol{q})$:

$$\begin{aligned}\Psi(\boldsymbol{q}) \equiv \Psi(1) &= \int d\boldsymbol{q}'\, \hat{\mathrm{P}}_{\boldsymbol{\chi}}(\boldsymbol{q},\boldsymbol{q}')\Psi(\boldsymbol{q}') \\ &= \boldsymbol{\chi}(\boldsymbol{q})\mathbf{S}^{-1}\int d\boldsymbol{q}'\boldsymbol{\chi}^{\dagger}(\boldsymbol{q}')\Psi(\boldsymbol{q}') \equiv \sum_{k}^{\mathrm{AO}} \chi_k(1)\, D_k, \\ D_k &= \sum_{l}^{\mathrm{AO}} S_{k,l}^{-1}\langle \chi_l \mid \Psi\rangle \equiv \sum_{l}^{\mathrm{AO}} S_{k,l}^{-1}\Psi_l. \end{aligned} \tag{6.300}$$

One similarly expands any two-electron wave function $\Psi(\boldsymbol{q}_1,\boldsymbol{q}_2) \equiv \Psi(1,2)$:

$$\Psi(1,2) = \sum_{k}^{\mathrm{AO}} \chi_k(1) \sum_{l}^{\mathrm{AO}} S_{k,l}^{-1}\langle \chi_l(1) \mid \Psi(1,2)\rangle_1 \equiv \sum_{k}^{\mathrm{AO}} \chi_k(1) D_k(2), \tag{6.301}$$

where the subsequent AO expansion of the coefficient function $D_k(2)$ gives:

$$D_k(2) = \sum_{k'}^{\mathrm{AO}} \chi_{k'}(2) \sum_{l'}^{\mathrm{AO}} S_{k',l'}^{-1}\langle \chi_{l'} \mid D_k\rangle \equiv \sum_{k'}^{\mathrm{AO}} D_{k,k'}\chi_{k'}(2). \tag{6.302}$$

It thus follows from the two preceding equations that

$$\Psi(1,2) = \sum_{k}^{\mathrm{AO}}\sum_{k'}^{\mathrm{AO}} D_{k,k'}\chi_k(1)\chi_{k'}(2), \tag{6.303}$$

and hence, by acting with the antisymmetrizer $\hat{\mathrm{A}}$ on both sides of the last equation,

$$\Psi_A(1,2) = \sum_{k}^{\mathrm{AO}}\sum_{k'}^{\mathrm{AO}} D_{k,k'} \det(\chi_k\chi_{k'}). \tag{6.304}$$

Clearly, one can proceed to expand in this way any antisymmetric wave function of N electrons as a linear combination of Slater determinants built directly from AO, which represent the corresponding VB structures or their components,

$$\begin{aligned}\Psi_A(1,\ 2,\ \ldots,\ N) &= \sum_{k_1}^{\mathrm{AO}}\sum_{k_2}^{\mathrm{AO}}\cdots\sum_{k_N}^{\mathrm{AO}} D_{k_1,k_2,\ldots,k_N}\hat{\mathrm{A}}\{\chi_{k_1}(1)\,\chi_{k_2}(2)\ \ldots\ \chi_{k_N}(N)\} \\ &= \sum_{k_1}^{\mathrm{AO}}\sum_{k_2}^{\mathrm{AO}}\cdots\sum_{k_N}^{\mathrm{AO}} D_{k_1,k_2,\ldots,k_N}\det(\chi_{k_1}\chi_{k_2}\ \cdots\ \chi_{k_N}). \end{aligned} \tag{6.305}$$

This AO expansion provides the formal basis for the trial wave functions generated as linear combinations of the N-electron valence structures used in the VB approach.

6.6.4 Semilocalized AO and Extension to Polyatomic Systems

A relatively more complicated combination of the covalent and ionic VB structures of (6.288) can be brought into a simpler HL-like form of (6.274) by replacing the fully localized AO $\{a, b\}$, centered on the corresponding nuclei, by the distorted, strongly overlapping (normalized) orbitals $\{\varphi_a, \varphi_b\}$, both partly delocalized toward the bonding partner, with φ_a strongly resembling a and φ_b representing a b-like orbital:

$$\begin{aligned}\Phi_g^{cov.+ion.}(1,2)\,U_{0,0}(1,2) &\equiv \lambda[a(1)b(2)+b(1)a(2)]+\mu[a(1)a(2)+b(1)b(2)]\}\,U_{0,0}(1,2)\\ &= \lambda\left(\left|a^+b^-\right|-\left|a^-b^+\right|\right)+\mu\left(\left|a^+a^-\right|+\left|b^+b^-\right|\right)\\ &= \mathcal{N}[\varphi_a(1)\varphi_b(2)+\varphi_b(1)\varphi_a(2)]\,U_{0,0}(1,2)\\ &= \mathcal{N}\left[\left|\varphi_a^+\varphi_b^-\right|-\left|\varphi_a^-\varphi_b^+\right|\right]\equiv\Phi^{\mathrm{CF}}(1,2)\,U_{0,0}(1,2)\end{aligned} \tag{6.306}$$

where the normalization factor $\mathcal{N}=[2(1+\langle\varphi_a|\varphi_b\rangle)]^{-1/2}$. These semilocalized, atomic-like one-electron functions of Coulson and Fischer (CF) (1949) are determined by a small AO-mixing parameter ε:

$$\begin{aligned}&\varphi_a=\mathcal{N}_\varphi(a+\varepsilon b),\quad \varphi_b=\mathcal{N}_\varphi(b+\varepsilon a)\},\quad 0<\varepsilon<1,\\ &\mathcal{N}_\varphi=(1+2\varepsilon\, S_{a,b}+\varepsilon^2)^{-1/2}.\end{aligned} \tag{6.307}$$

Substituting these expressions into the effective covalent structure Φ^{CF} indeed gives back the mixture of the covalent and ionic structures in terms of the localized AO:

$$\begin{aligned}\Phi^{\mathrm{CF}}(1,2) &= \mathcal{N}\mathcal{N}_\varphi^2\{(1+\varepsilon^2)[a(1)b(2)+b(1)a(2)]+2\varepsilon[a(1)a(2)+b(1)b(2)]\}\\ &= C^{cov.}\Phi_g^{cov.}+C^{ion.}\Phi_g^{ion.}.\end{aligned} \tag{6.308}$$

This generalized CF representation of the Lewis structure H_a–H_b, embedding the ionic effects in an effective "covalent" structure through the delocalization tails of the semilocalized AO, has been also adopted to polyatomic molecules in the GVB approach, with the semilocalized functions being determined from the variational principle, by minimizing the expectation value of the system electronic energy. These functions are expanded in terms of the localized AO or primitive basis functions centered on the corresponding nuclei, and the optimum AO mixing coefficients are determined by iteratively solving one-electron equations that resemble those of the standard SCF method.

The VB theory is thus closely related to chemist's idea of molecules consisting of AIM held together by localized bonds, each of them described by the combination of two determinantal wave functions of the type given in (6.306), representing the singlet spin-coupled (shared) valence electrons of the covalent bond in question.

This "chemical" quantum theory also views the molecules as composed of atomic "cores," each including the nucleus and the chemically inactive *inner*-shell electrons, and the chemically active *valence* electrons. The spin coupling of the latter is the main objective of the VB treatment of polyatomic systems, which places great emphasis on the spin pairing of electrons.

The chemical intuition can be used in various ways to simplify the VB representation of molecular wave functions. In the so-called *Perfect Pairing Approximation* (PPA) one takes the Lewis structure of the molecule in the assumed singlet ground state, represents each bond by the HL/CF combination of two determinants, and finally expresses the full molecular wave function as product of all such *bond* functions. Therefore, the molecular wave function describing n bonds present in the Lewis structure will be described by 2^n determinants, displaying the possible 2×2 spin permutations between two singlet-coupled orbitals. Consider, as an illustrative example, the Li_a—Li_b molecule, for which the inner-shell $1s$ electrons of both atoms, listed at the beginning of the relevant Slater determinants belong to atomic cores and remain chemically inactive, with only the (unpaired) $2s$ valence electrons participating in the formation of the σ chemical bond. The corresponding VB structure representing this single chemical bond thus reads:

$$\begin{aligned}\mathrm{Li}_a\text{—}\mathrm{Li}_b &= \left|1s_a^+1s_a^-1s_b^+1s_b^-2s_a^+2s_b^-\right| - \left|1s_a^+1s_a^-1s_b^+1s_b^-2s_a^-2s_b^+\right| \\ &\equiv \left|\ldots\ 2s_a^+2s_b^-\right| - \left|\ldots\ 2s_a^-2s_b^+\right| \equiv \left|\ldots\ \overline{2s_a2s_b}\right|, \qquad (6.309a)\end{aligned}$$

or, by retaining only the chemically active part,

$$\begin{aligned}\mathrm{Li}_a\text{—}\mathrm{Li}_b &= [2s_a(1)2s_b(2) + 2s_b(1)2s_a(2)]\ U_{0,0}(1,2) = \left|\overline{2s_a2s_b}\right| \\ &\equiv (2s_a\text{—}2s_b)\ U_{0,0}. \qquad (6.309b)\end{aligned}$$

For the triple bond in $:N_a \equiv N_b:$, with the inactive set of electrons now determined by the *doubly* occupied inner shells ($1s_a$, $1s_b$) and the *lone* pair (lp) $\approx 2sp_z$ hybrids $(2h_a^{lp}, 2h_b^{lp})$, the same approach will involve eight determinants originating from the antisymmetrized product of the covalent/CF valence structures for the three localized bonds: the σ bond resulting from the singlet coupling of two unpaired electrons occupying the bonding $2sp_z$ hybrids of both atoms $(2h_a^{bond.}$ —$2h_b^{bond.})$ and two π bonds resulting from the singlet coupling of the symmetry compatible pairs of $2p_\pi$ orbitals: $(2p_{xa}$—$2p_{xb})$ and $(2p_{ya}$—$2p_{yb})$. The sign of the corresponding determinant in the VB representation of the system electronic wave function is either "+" or "−", according to whether the number of the interchanges required to generate the current determinant from the chosen PPA "*parent*" determinant, say $|\ldots\ 2h_a^{bond.,+}h_a^{bond.,-}2p_{xa}^+2p_{xb}^-2p_{ya}^+2p_{yb}^-|$, is even or odd. For example, the determinant $|\ldots\ 2h_a^{bond.,+}h_a^{bond.,-}2p_{xa}^-2p_{xb}^+2p_{ya}^+2p_{yb}^-|$, representing a single exchange of spins between $2p_x$ AO relative to the reference determinant, will appear with the negative sign in the VB combination representing the ground state of N_2 in PPA.

There are three admissible sets of the localized singlet couplings between the chemically active electrons, determining the joint VB structures of two individual "bonds" in water molecule H_a—O—H_b, in which the oxygen 1s orbital describes the two inner-shell electrons and the two doubly occupied $\approx 2sp^3$ hybrids $\left\{h_O^{lp}\right\}$ determine the state of two *lone* pairs of valence electrons on oxygen,

$$\Phi_A = (2h_a^{bond.}—a,\ 2h_b^{bond.}—b),\quad \Phi_B = (2h_a^{bond.}—2h_b^{bond.},\ a—b),$$
$$\text{and}\quad \Phi_C = (2h_a^{bond.}—b,\ 2h_b^{bond.}—a), \tag{6.310}$$

where $2h_a^{bond.}$ and $2h_b^{bond.}$ stand for the bonding hybrids on oxygen directed toward the corresponding orbitals (a, b) contributed by the two hydrogen atoms. However, by a straightforward multiplication of the spin factors involved in the underlying Slater determinants one can demonstrate that $\Phi_C = (\Phi_A + \Phi_B)$. Therefore, only the first two sets of VB structures exhibiting two localized bonds are linearly independent (see also Fig. 6.3), so that

$$\Phi^{VB} = C_A\Phi_A + C_B\Phi_B. \tag{6.311}$$

This ansatz is still further simplified in GVB approach, which uses only the dominant Lewis structure H_a—O—H_b, with the bond couplings being recognized only between the nearest neighbors, $\Phi^{GVB}(H_2O) = (f_a^{bond.}—g_a, f_b^{bond.}—g_b)$; the generating orbitals on oxygen, $\{f_a^{bond.},\ f_b^{bond.}\}$, and the two hydrogen atoms $\{g_a, g_b\}$ are freely adjusting their shapes in accordance with the variational principle of the minimum electronic energy. The GVB function of methane would similarly involve only the dominant structure of four C—H bonds: $\Phi^{GVB}(CH_4) = (f_a—g_a, f_b—g_b, f_c—g_c, f_d—g_d)$.

The number of admissible spin states of all valence electrons in a general polyelectronic case is very large indeed, but it is effectively limited by selecting the proper eigenfunctions of the resultant spin operators [see (6.67)] and the linearly independent, canonical set of VB structures. The latter task is accomplished by using an appropriate diagrammatic technique, e.g., the Rumer (1932) diagrams (see Figs. 6.3 and 6.4). This technique for selecting the linearly independent set of "canonical" VB structures involves putting the chemically active orbitals on a circle in any a priori order, although interpretation of the results in chemical terms is greatly facilitated by having this ordering as close as possible to the order of orbitals in the molecule. The independent structures are then determined by all valence configurations in which the active orbitals are joint in pairs by noncrossing lines (see Fig. 6.3). The Rumer diagrams for water molecule shown in Fig. 6.3 indeed exclude the Φ_C structure from the canonical set, as being linearly dependent on Φ_A and Φ_B. For the even number n of active orbitals the number of ways of drawing ½n noncrossing lines between n points on a circle is $n!/[(½n)!\ (½n + 1)!]$ (Barriol 1971). For the water molecule, $n = 4$, one indeed finds $4!/(2!3!) = 2$ independent VB structures. For an odd number of orbitals to be paired, one adds a "phantom" orbital, whose contribution is eventually removed.

Next, let us consider the π-bonds in butadiene and benzene. In the former case the first two diagrams of Fig. 6.3 determine the independent valence structures (Φ_a, Φ_b) of Fig. 6.5, while the third diagram determines the linearly dependent structure $\Phi_c = \Phi_b - \Phi_a$. The relevant Rumer diagrams for π-bonds in benzene are shown in Fig. 6.4. The first two diagrams generate the familiar Kekulé structures (Φ_a , Φ_b) of Fig. 6.5, while the remaining diagrams give rise to the Dewar structures (Φ_c, Φ_d, Φ_e) of the same figure, where the associated singly polar structures are also shown.

The chemical intuition can also guide the VB description of chemical changes in terms of the group orbitals, which can be delocalized over the specified molecular fragment(s). Such delocalized one-electron functions can be determined from earlier SCF MO calculations of the separate molecular subsystems, e.g., reactants or their crucial functional groups. This approach allows one to eliminate the explicit description of chemical bonds which remain practically unaffected by the chemical reaction under consideration. For example, in the nucleophilic substitution (S_N2) involving a simultaneous bond-breaking – bond-forming process, $OH^- + CH_3$—Cl → OH—$CH_3 + Cl^-$, the PPA spatial wave function should involve the explicit covalent/CF VB structures of only two bonds, H_3C—Cl and HO—CH_3, i.e., four Slater determinants, with the localized MO of the remaining σ bonds, O—H and C—H, and the lone pairs of electrons in the valence shell of oxygen and chlorine forming an effective "core" for the four truly active orbitals in this chemical reaction.

6.6.5 *Ab Initio VB Calculations*

The *single* structure PPA of molecular wave function, using the strongly overlapping pairs of directed orbitals on individual atoms for each localized chemical bond, invokes the concept of hybrid AO's. In fact, the orbital hybridization arose from efforts to retain the *perfect* pairing picture in interpreting stereochemical situations, which would be ambiguous in terms of the canonical AO's. However, in many situations, in which there are no strong grounds for preferring one valence structure to another, the PPA breaks down, and a mixture

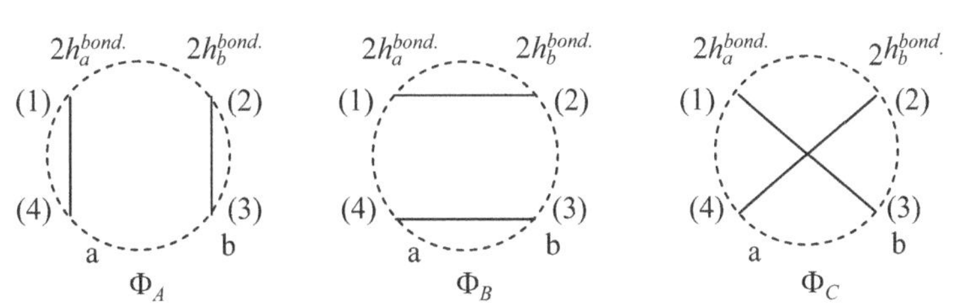

Fig. 6.3 The Rumer diagrams corresponding to the valence structures of water (6.310) and π bonds in butadiene, between $2p_{zi}$, $i = 1, 2, 3, 4$ orbitals (in parentheses) of the carbon atoms (for their consecutive numbering in the π-chain)

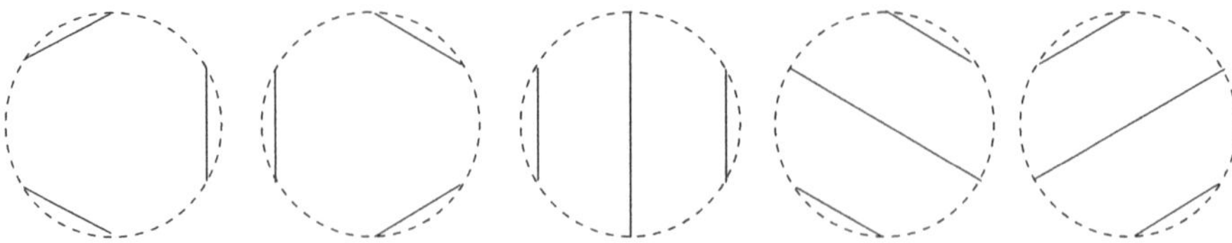

Fig. 6.4 The Rumer diagrams corresponding to the canonical valence structures of π bonds in benzene, between $2p_{zi}$, $i = 1, 2, \ldots, 6$, orbitals of the carbon atoms in the π-ring (distributed consecutively on the perimeter of the circle)

of many VB structures must be used to adequately represent the system wave function. In order to carry out nonempirical VB calculations of comparable accuracy with that of the ab initio SCF MO (CI) theory, it is essential to remove the assumption of the orthonormal orbitals, which is often adopted in qualitative VB considerations, although the calculation of matrix elements is then immensely more difficult. As we have already observed in Sect. 6.6.2, the crude approximation using AO's of the free atoms, without any further adjustment, gives rather poor energies, missing considerable rearrangements these orbitals undergo in a molecule. Therefore, the optimization of the CF orbitals for each chemical bond constitutes another essential requirement of an accurate ab initio VB theory. The perfect-pairing CF representation allows one to determine the HL level of the bond correlation energy in an unbiased variational way. These generalized, orbital-optimized VB methods explore the same variety of *single-* and multiconfigurational approaches, which we have already encountered in the MO theory. For recent surveys of the modern variants of the ab initio VB theories and evaluations of their already remarkable capabilities the reader is referred to the monograph by McWeeny (1989) as well as to reviews by Gerrat (1974) and by Shaik and Hiberty (2004). The following short summary is based upon the latter monographic survey.

The simplest among the *single*-configuration representations is the GVB description of Goddard and coworkers (Goddard 1967; Hunt et al. 1972; Bobrowicz and Goddard 1977; Goddard and Harding 1978) originating from the earlier *separated electron pair* approach of Hurley et al. (1953) (see also Sect. 6.4.1). It is usually based upon the PPA (GVB-PP), by which only a single VB structure of say n valence-pair electron system is generated in the calculation. It is defined in terms of the strongly orthogonal geminals, each describing one particular bond or a single lone electron pair [see also (6.162)], which take the form of the singlet-coupled pair of electrons (in parentheses):

$$G_i(1,2) = \left[\left|\varphi_{i,a}^{+}\varphi_{i,b}^{-}\right| - \left|\varphi_{i,a}^{-}\varphi_{i,b}^{+}\right|\right]$$
$$\equiv N_i[\varphi_{i,a}(1)\varphi_{i,b}(2) + \varphi_{i,a}(2)\varphi_{i,b}(1)]U_{0,0}(1,2), \qquad (6.312)$$

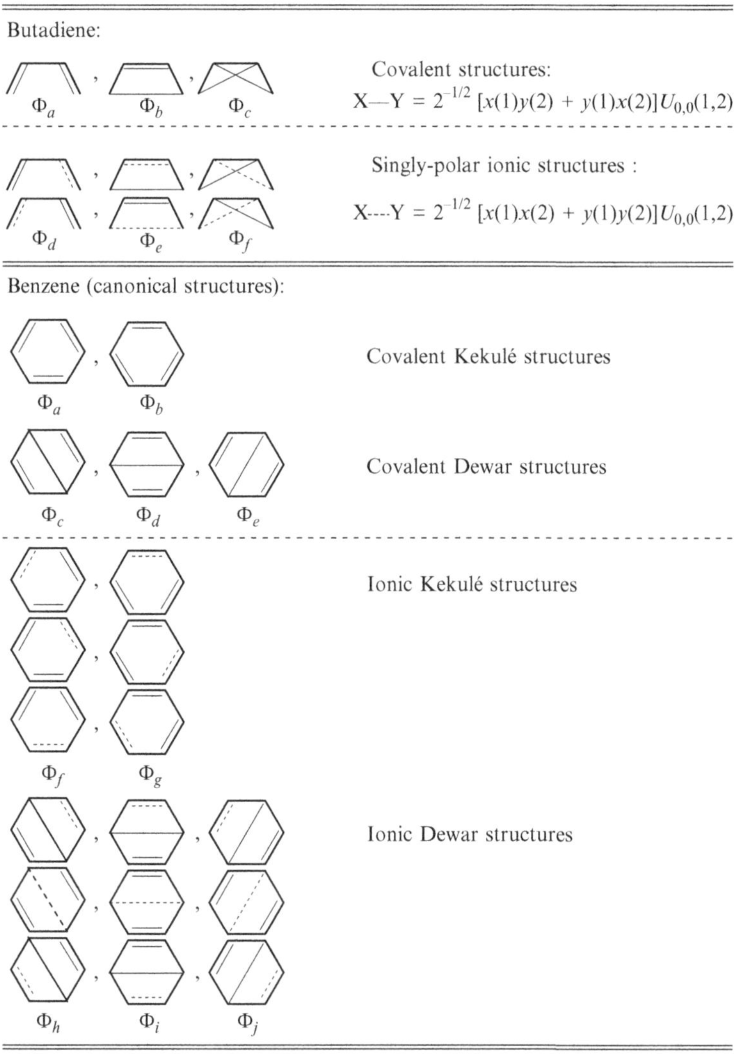

Fig. 6.5 The most important covalent and ionic valence structures of butadiene and benzene; the singly occupied, orthogonalized valence orbitals $2p_\pi$: $x \in$ X and $y \in$ Y are used to form elementary covalent structure, X—Y, and singly polar ionic structure X----Y = $2^{-1/2}(X^+Y^- + X^-Y^+)$ between the constituent carbon atoms X and Y of the π system, with the singlet spin factor (see Fig. 6.2) $U_{0,0}(\sigma_1, \sigma_2) \equiv U_{0,0}(1, 2) = 2^{-1/2}[\alpha(1)\beta(2) - \beta(1)\alpha(2)]$

$$\Psi_A^{\mathrm{GVB}} = \hat{A}'\{G_1(1,2)G_2(3,4)\ldots G_n(2n-1,2n)\} . \tag{6.313}$$

Only orbitals within a given geminal pair display a strong overlap, while those associated with different geminals are required to be mutually orthogonal. This constraint, which greatly simplifies the numerical calculations, is not a serious one, since the orbitals of different geminals are not expected to overlap significantly anyway. For further numerical convenience each *open*-shell geminal product of (6.312) is generated from the two *closed*-shell contributions due to orthonormal natural orbitals (NO) ϕ_i and $\bar{\phi}_i$, bonding and antibonding, respectively, obtained by a simple "rotation" of the original orbitals $\varphi_{i,a}$ and $\varphi_{i,b}$:

$$\varphi_{i,a}(1)\varphi_{i,b}(2) + \varphi_{i,a}(2)\varphi_{i,b}(1) \equiv C_i\,\phi_i(1)\,\phi_i(2) + \bar{C}_i\,\bar{\phi}_i(1)\,\bar{\phi}_i(2). \qquad (6.314)$$

It can be easily verified that the latter are given by the NO combinations

$$\varphi_{i,a} = \mathcal{N}\,[\phi_i + \kappa\bar{\phi}_i], \quad \varphi_{i,b} = \mathcal{N}\,[\phi_i - \kappa\bar{\phi}_i], \quad \mathcal{N} = (1+\kappa^2)^{-1/2}, \qquad (6.315)$$

which give rise to:

$$C_i = 1+\kappa^2, \qquad \bar{C}_i = \frac{-\kappa^2}{1+\kappa^2}, \qquad \langle\varphi_{i,a}|\varphi_{i,b}\rangle = \frac{1-\kappa^2}{1+\kappa^2}. \qquad (6.316)$$

The GVB-PP approach is thus equivalent to the *low*-dimensional MC SCF method, accounting for a part of the nondynamical electron correlation. The PP and strong orthogonality approximations greatly simplify the numerical calculations with no great loss of accuracy in the molecular systems exhibiting clearly separated local bonds. In benzene, however, all electron pairing schemes have to be taken into account to generate a reasonable description of the delocalized π electrons. Inclusion of the full nondynamical correlation would require the CAS SCF calculations, which involve all possible configurations of the *valence*-shell orbitals. As a rule, the GVB method generates a much better agreement with the CAS SCF predictions than the HF theory. However, due to a neglect of the dynamical correlation the GVB bond energies are generally too low. Thus, this *single*-configuration method at best can be regarded as a good starting point for the subsequent Correlation-Consistent CI (CCCI) treatment of Carter and Goddard (1988). For example, this more advanced GVB–CCCI approach has been successfully applied to investigate catalysis by transition metals (Rappé and Goddard 1982) and metallic bonds (McAadon and Goddard 1985, 1987).

A more accurate *single*-configuration VB technique is represented by the Spin-Coupled (SC) approach of Gerratt and coworkers (Cooper et al. 1981, 1987, 1988, 1990a, b; Sironi et al. 2002), which introduces no constraints or preconceptions on the level of the spin coupling, orbital overlaps, and shapes, with all these degrees of freedom of the molecular VB wave function remaining unrestricted in the variational treatment. Therefore, this method represents the ultimate level of accuracy of the *single*-configuration VB wavefunction assuming the fixed orbital occupancies. It provides the unique set of orbitals and measures of a relative importance of alternative spin couplings, which constitute important unbiased chemical descriptors of chemical bonds, thus providing a solid basis for such originally qualitative and intuitive concepts as the valence state of AIM, orbital hybridization, resonance between VB structures, etc.

For example, in the eight spin-coupled orbitals of methane, including four degenerate, symmetry-related (approximately sp^3) hybrids on carbon atom and four degenerate orbitals localized on hydrogen atoms, the PP configuration represents the dominant mode of the spin coupling. The PPA introduces only a minor correlation error to predicted energy, superior to the corresponding HF result, with the SC total energy being very close to the CAS SCF value. In water the method predicts about 20% more p character in the bonding hybrids, compared with those describing the lone pairs on oxygen, in full accord with chemical intuition. In lithium clusters, the interstitial optimum orbitals, localized between two or several nuclei, have been determined. In the π electron system of benzene, the two Kekulé structures have been found to represent the dominant patterns of the spin coupling of electrons occupying slightly distorted p_π AO's, perpendicular to the molecular plane, with the participation of Dewar structures amounting to about 20%.

The single configuration SC description can be further improved by adding the CI involving (nonorthogonal) VB configurations, covalent and ionic, derived from alternative occupations of the optimum (active) orbitals determined at the SC stage. This determines the Spin-Coupled VB (SCVB) variant of the theory. The localized $\sigma \rightarrow \pi$ valence excitations can be then included at the CI stage, with the virtual π orbital being localized in the vicinity of the σ orbital in question.

The most accurate among the ab initio variants are the multiconfigurational VB methods. For example, the VB SCF approach of van Lenthe and Balint-Kurti (1983) can be classified as the MC SCF method in the nonorthogonal AO representation. It uses the trial wave function expressed as the linear combination of the (*fundamental*) VB structures, with both (*occupied*) orbitals and configuration coefficients being simultaneously determined by the variational principle alone. The method is especially suited for studying the resonance stabilization energies, particularly in molecules described by many Lewis structures, and in solving the avoided-crossing problems. The optimized orbitals can be either localized (atomic) or semilocalized (CF) in character, or they may be restricted to specific molecular fragments.

Again the configuration interaction can be used to supplement the VB SCF method with the dynamic correlation effects. Among these post VB SCF treatments the VB CI method of Wu et al. (2001, 2002) using different levels of CI, e.g., VBCIS, VBCISD, VBCISDT, and methods using the *symmetry*-broken wave functions derived from different orbitals for different structures, e.g., Resonating-GVB (RGVB) model of Voter and Goddard (1981a, b) or the Breathing Orbital VB (BOVB and GRVB) schemes of Hiberty and coworkers (Hiberty 1997; Hiberty and Shaik 2002a, b; Hiberty et al. 1992, 1994a, b), appear to be most promising. The virtual orbitals used at the CI stage preserve the interpretability offered by the occupied set defining the fundamental configurations, each having a clear chemical meaning. This is achieved by using the appropriate projection operators to make the *acceptor* virtual orbitals of the electron excitation, receiving electrons from the given subset of the excitation *donor* occupied orbitals of the specified molecular fragment, to be localized in the same molecular fragment, thus describing the same classical VB structure and maximizing the effect of the dynamic correlation correction. Various multiplicities of the electron excitations used at the CI stage of the VB CI treatment can be taken into account, giving rise to the associated limited CI VB variants, VB CIS, VB CISD, etc., which still preserve the interpretability of simple VB methods. The accuracy of VB CISD has been found to be comparable with that of the CCSD and CCSD(T) variants of MO theory.

The MR VB methods are required to adequately describe the molecules, and particularly some radicals, that are naturally described in terms of two or more resonance structures. Care has to be taken to avoid discontinuities observed at a lower level of theory, for which the wave function often exhibits a lower artefactual symmetry of a single VB structure than the nuclear framework itself, due to energy gain resulting from using the optimum orbitals for this structure alone. Therefore, this *symmetry*-breaking effect, resulting from the competition between the resonance and orbital energy stabilization effects, cannot be adequately remedied using a single set of orbitals, optimized within the one-configuration MO or VB theories, since competing resonance structures are associated with the exclusive sets of the optimum orbitals. Within the MO theory Jackels and Davidson (1976) has cured the symmetry-breaking problem in NO_2 by using the *symmetry*-adapted combination of two *symmetry*-broken HF wave functions. The RGVB approach or its generalized (GRVB) version represent the same strategy applied within the VB theory: the symmetry broken GVB wave functions are first determined for each individual resonance structure alone or in the presence of the other structure(s) and then their resonance is determined at the subsequent (nonorthogonal) CI stage. Of similar character is the BOVB method, which focuses on the proper description of both the bond-dissociation phenomena, in which different-orbitals-for-different structures are optimized in the presence of the remaining structures, so that they minimize the energies of each individual valence structure and maximize the stabilization effect due to their mutual resonance (mixing).

References

Ahlrichs R, Lischka H, Staemmler V, Kutzelnigg W (1975a) J Chem Phys 62:1225

Ahlrichs R, Driessler F, Lischka H, Staemmler V, Kutzelnigg W (1975b) J Chem Phys 62:1235

Anex B, Shull H (1964) In: Löwdin P-O, Pullman B (eds) Molecular orbitals in chemistry, physics and biology. Academic, New York, p 227

Arai T (1957) J Chem Phys 26:435; (1960) Rev Mod Phys 32:370

Balint-Kurti GG, Karplus M (1973) Orbital theories of molecules and solids. Clarendon, Oxford

Barriol J (1971) Elements of quantum mechanics with chemical applications. Barnes and Noble, New York

Bartlett RJ (1981) Annu Rev Phys Chem 32:359

Bartlett RJ (1989) J Phys Chem 93:1697

Bartlett RJ (1995) In: Yarkony DR (ed) Modern electronic structure theory, Part II. World Scientific, Singapore, p 1047

Bartlett RJ (2000) In: Keinan E, Schechter I (eds) Chemistry for the 21 Century. Wiley, Weinheim

Bingel WA, Kutzelnigg W (1970) Adv Quant Chem 5:201

Bobrowicz FB, Goddard WA III (1977) In: Schaefer III HF (ed) Modern theoretical chemistry, vol 3. Plenum, New York, p 79
Boys SF (1968) Proc R Soc Lond A 200:542
Boys SF, Bernardi F (1970) Mol Phys 19:553
Brandow BH (1967) Rev Mod Phys 39:771
Brillouin L (1933) Actualities Sci et Ind No 71. Hermann et Cie, Paris
Brillouin L (1934) Actualites Sci et Ind No. 159. Hermann et Cie, Paris
Brueckner KA (1955) Phys Rev 97:1353; 100:36
Brueckner KA, Levinson CA (1955) Phys Rev 97:1344
Brueckner KA, Eden RJ, Francis NC (1955) Phys Rev 93:1445
Carlson BC, Keller JM (1961) Phys Rev 121:659
Carter EA, Goddard WA III (1988) J Chem Phys 88:3132
Christoffersen RE (1989) Basic principles and techniques of molecular quantum mechanics. Springer, Berlin
Christoffersen RE, Shull H (1968) J Chem Phys 48:1790
Čižek J (1966) J Chem Phys 45:4256
Čižek J (1969) Adv Chem Phys 14:35
Čižek J, Paldus J (1980) Phys Script 21:251
Clementi E, Roetti C (1974) At Data Nucl Data Tables 14:177
Coester F (1958) Nucl Phys 1:421
Coester F, Kümmel H (1960) Nucl Phys 17:477
Coleman AJ (1963) Rev Mod Phys 35:668
Coleman AJ (1981) In: Deb BM (ed) The force concept in chemistry. Van Nostrand, Reinhold, New York, p 418
Coleman AJ, Erdahl RM (1968) (Eds) Reduced density matrices with applications to physical and chemical systems. Queen's Papers on Pure and Applied Mathematics, No. 11. Queens University, Kingston
Cooper DL, Gerratt J, Raimondi M (1981) Chem Rev 91:929
Cooper DL, Gerratt J, Raimondi M (1987) Adv Chem Phys 69:319
Cooper DL, Gerratt J, Raimondi M (1988) Int Rev Phys Chem 7:59
Cooper DL, Gerratt J, Raimondi M (1990a) In: Klein DJ, Trinastič N (eds) Valence bond theory and chemical structure. Elsevier, Amsterdam, p 287
Cooper DL, Gerratt J, Raimondi M (1990b) In: Gutman I, Cyvin SJ (eds) Advances in the theory of benzenoid hydrocarbons. Top Curr Chem 153:41
Coulson CA, Fischer I (1949) Philos Mag 40:386
Davidson ER (1969) J Math Phys 10:725
Davidson ER (1972a) Rev Mod Phys 44:451
Davidson ER (1972b) Adv Quant Chem 6:235
Davidson ER (1974) In: Daudel R, Pullman B (eds) The world of quantum chemistry. Reidel, Dordreecht, p 17
Davidson ER (1976) Reduced density matrices in quantum chemistry. Academic, New York
Dunning TH Jr, Hay PJ (1977) In: Schaefer III HF (ed) Methods of electronic structure, vol 3. Plenum, New York, p 1
Ellison FO (1965) J Chem Phys 43:3654
Ellison FO, Slezak JA (1969) J Chem Phys 50:3942
Ellison FO, Wu AA (1967) J Chem Phys 47:4408
Ellison FO, Wu AA (1968) J Chem Phys 48:1103
Epiotis ND (1978) Theory of organic reactions. Springer, Berlin
Fock VA (1930) Z Phys 61:126
Freed KF (1971) Annu Rev Phys Chem 22:313
Froese-Fischer C (1977) The Hartree-Fock method for atoms. Wiley, New York
Gerratt J (1974) A specialist periodical report: theoretical chemistry, vol 1 – Quantum chemistry. The Chemical Society, Burlington House, London, p 60

Goddard WA III (1967) Phys Rev 157:81
Goddard WA III, Harding LB (1978) Annu Rev Phys Chem 29:363
Goddard WA III, Ladner RD (1971) J Am Chem Soc 93:6750
Goddard WA III, Dunning TH Jr, Hunt WJ, Hay PJ (1973) Acc Chem Res 6:368
Goldstone J (1957) Proc R Soc (London) A239:267
Hagstrom S, Shull H (1963) Rev Mod Phys 35:624
Harris FE (1967) Adv Chem Phys 13:205
Hartree DR (1928) Proc Camb Philos Soc 24:89
Heitler W, London F (1927) Z Physik 44:455–472; for an English translation see: Hettema H (2000) Quantum chemistry classic scientific paper. World Scientific, Singapore
Herring C (1940) Phys Rev 57:1169
Hiberty PC (1997) In: Davidson ER (ed) Modern electronic structure theory and applications in organic chemistry. World Scientific, River Edge, NJ, p 289
Hiberty PC, Shaik S (2002a) In: Cooper DL (ed) Valence bond theory. Elsevier, Amsterdam, p 187
Hiberty PC, Shaik S (2002b) Theor Chem Acc 108:225
Hiberty PC, Flament JP, Noizet E (1992) Chem Phys Lett 189:259
Hiberty PC, Humbel S, Archirel P (1994a) J Phys Chem 98:11697
Hiberty PC, Humbel S, Byrman CP, van Lenthe JH (1994b) J Chem Phys 101:5969
Hirst DH (1985) Potential energy surfaces: molecular structure and reaction dynamics. Taylor and Francis, London
Hunt WJ, Hay PJ, Goddard WA III (1972) J Chem Phys 57:738
Hurley AC (1976) Electron correlation in small molecules. Academic Press, New York
Hurley AC, Lennard-Jones JE, Pople JA (1953) Proc R Soc Lond A220:446
Huzinaga S, Andzelm J, Kłobukowski M, Radzio-Andzelm E, Sakar Y, Tatewaki H (1984) Gaussian basis sets for molecular calculations. Elsevier, New York
Hylleraas EA, Undheim B (1930) Z Physik 65:759
Jackels CF, Davidson ER (1976) J Chem Phys 64:2908
Janak JF (1978) Phys Rev B 18:7165
Johnston HS, Parr CJ (1963) J Am Chem Soc 85:2544
Jørgensen P, Simons J (1981) Second quantization–based methods in quantum chemistry. Academic, New York
Jug K, Köster AM (1990) J Am Chem Soc 112:6772
Karplus M, Grant DM (1959) 11:409
Kato T (1957) Commun Pure Appl Math 10:151
Kelly HP (1969) Adv Chem Phys 14:129
Kołos W, Wolniewicz L (1964) J Chem Phys 41:3663
Kołos W, Wolniewicz L (1965) J Chem Phys 43:2429
Kołos W, Wolniewicz L (1968) J Chem Phys 49:404
Kucharski SA, Bartlett RJ (1986) Adv Quant Chem 18:281
Kümmel H (1969) Nucl Phys 22:177
Kuntz PJ (1979) In: Bernstein RB (ed) Atom-molecule collision theory – a guide for the experimentalist. Plenum, New York, p 79
Kutzelnigg W (1977) In: Schaefer III HF (ed) Methods of electronic structure theory, vol 3. Plenum, New York, p 129
Langhoff SR, Davidson ER (1974) Int J Quantum Chem 8:61
London F (1928) Z Phys 46:455
Löwdin P-O (1955a) Phys Rev 97:1474
Löwdin P-O (1955b) Phys Rev 97:1490
Löwdin P-O (1959) In: Advances in chemical physics, vol 2. Prigogine, Ith edn. Interscience, New York, p 207
Löwdin P-O, Shull H (1956) Phys Rev 101:1730
MacDonald JKL (1933) Phys Rev 43:830
Manne R (1977) Int J Quantum Chem S11:175

Mattuck RD (1976) A guide to Feynman diagrams in the many-body problem, 2nd edn. Mc-Graw-Hill, New York
McAadon MH, Goddard WA III (1985) J Phys Chem 91:2607
McAadon MH, Goddard WA III (1987) Phys Rev Lett 55:2563
McWeeny R (1989) Methods of molecular quantum mechanics. Academic, London
McWeeny R, Ohno KA (1960) Proc R Soc (Lond) A255:367
Meyer W (1977) In: Schaefer III HF (ed) Methods of electronic structure theory, vol 3. Plenum, New York, p 413
Moffitt W (1951) Proc R Soc (London) A210:245
Møller C, Plessett MS (1934) Phys Rev 46:618
Moss BJ, Goddard WA III (1975) J Chem Phys 63:3523
Mulliken RS (1935) J Chem Phys 3:573
Mulliken RS (1955) J Chem Phys 23:1833, 1841, 2338, 2343
Mulliken RS (1962) J Chem Phys 36:3428
Murrell JN, Bosanac SD (1989) Introduction to the theory of atomic and molecular collisions. Wiley, New York
Murrell JN, Carter S, Farantos SC, Huxley P, Varandas AJC (1984) Molecular potential energy functions. Wiley, New York
Nesbet RK (1965) Adv Chem Phys 9:321
Paldus J, Čižek J (1973) In: Smith D, McRae WB (eds) Energy, structure and reactivity. Wiley, New York, p 198
Paldus J, Čižek J (1975) Adv Quant Chem 9:105
Parks JM, Parr RG (1958) J Chem Phys 28:335
Parks JM, Parr RG (1960) J Chem Phys 32:1657
Parr RG, Yang W (1984) J Am Chem Soc 106:4049
Pauling L (1949) Proc R Soc Lond A196:343
Phillips JC, Kleinman L (1960) Phys Rev 118:1153
Poirier R, Kari R, Csizmadia IG (1985) Handbook of Gaussian basis sets. Elsevier, New York
Pople JA (1976) In: Schaefer III HF (ed) Modern theoretical chemistry, vol 4. Plenum, New York
Rappé AK, Goddard WA III (1982) J Am Chem Soc 104(297):448
Roos BO (1972) Chem Phys Lett 15:153
Roos BO, Siegbahn EEM (1977) In: Schaefer III HF (ed) Methods of electronic structure theory. Plenum, New York, p 277
Rumer G (1932) Göttinger Nachr 27:337
Shaik S (1989) In: Bertran J, Czismadia IG (eds) New theoretical concepts for understanding organic reactions, NATO ASI Series, vol. C267. Kluwer Academic, Dordrecht, p 165
Shaik S, Hiberty PC (1991) In: Maksić ZB (ed) Theoretical models of chemical bonding, vol 4, pp 269
Shaik S, Hiberty PC (1995) Adv Quant Chem 26:100
Shaik S, Hiberty PC (2004) In: Lipkowitz KB, Larter L, Cundari TR (eds) Reviews in computational chemistry, vol 20, p 1
Shaik S, Shurki A, Danovich D, Hiberty P (2001) Chem Rev 101:1501
Shaik S, Danovich D, Wu W, Hiberty PC (2009) Nat Chem 1:443
Shavitt I (1963) Methods Comput Phys 3:1
Shavitt I (1977) In: Schaefer III HF (ed) Methods of electronic structure theory. Plenum, New York, p 189
Shavitt I, Rosenberg BJ, Palalikit S (1976) Int J Quantum Chem Symp 10:33
Shull H (1959) J Chem Phys 30:1405
Shull H (1960) J Am Chem Soc 82:1287
Shull H (1962) J Phys Chem 66:2320
Shull H (1964) J Am Chem Soc 86:1469
Shull H, Löwdin P-O (1958) Phys Rev 110:1466
Shull H, Prosser F (1964) J Chem Phys 40:233

Simonetta M (1968) In: Rich A, Davidson N (eds) Structural chemistry and molecular biology. Freeman, San Francisco
Sinanoĝlu O (1964) Adv Chem Phys 6:315
Sironi M, Raimondi M, Martinazzo R, Gianturco FA (2002) In: Cooper DL (ed) Valence bond theory. Elsevier, Amsterdam, p 261
Slater JC (1974) Quantum theory of molecules and solids, vol 4: The self-consitent field for molecules and solids. McGraw-Hill, New York
Surjan PR (1989) Second quantized approach to quantum chemistry. Springer, Berlin
Szabo A, Ostlund NS (1982) Modern quantum chemistry: introduction to advanced electronic structure theory. Macmillan, New York
Szasz L (1985) Pseudopotential theory of atoms and molecules. Wiley, New York
Tully JC (1977) In: Segal GA (ed) Semi-empirical methods of electronic structure calculations, Part A. Plenum, New York, p 173
Tully JC (1980) In: Lawley KP (ed) Potential energy surfaces. Wiley, Chichester, p 63
van Lenthe JH, Balint-Kurti GG (1983) J Chem Phys 78:5699
Voter AF, Goddard WA III (1981a) Chem Phys 57:253
Voter AF, Goddard WA III (1981b) J Chem Phys 75:3638
Wu X, Vargas MC, Nayak S, Lotrich V, Scoles G (2001) J Chem Phys 115:8748
Wu W, Song L, Cao Z, Zhang Q, Shaik S (2002) J Phys Chem A 106:2721

Chapter 7
Density Functional Theory

Abstract A short overview is given of fundamentals of the modern *Density Functional Theory* (DFT), an alternative approach to the quantum *many*-body problem. Basic theorems of *Hohenberg* and *Kohn* (HK) are summarized and their extensions for general densities and mixed quantum states are outlined, with the ensemble formulation covering the open molecular systems at finite temperatures. General forms of density functionals for the kinetic and potential energy contributions in both the local density and gradient approximations are then qualitatively examined using the virial theorem and the uniform scaling of the system electron density. The orbital scheme of *Kohn* and *Sham* (KS) for DFT computations is described and its ensemble extension is briefly outlined. The *exchange correlation* (*xc*) energy is introduced and partitioned into the *Fermi* (exchange) and *Coulomb* (correlation) contributions. By using the *Adiabatic Connection* (AC) between the real molecular system of fully interacting electrons and the hypothetical KS system of noninteracting (separable) electrons, both exhibiting the same ground state density, these energy terms are expressed in terms of the effective correlation holes averaged over the *coupling* constant which scales the electron interaction. The Euler equation for the optimum density and the associated KS equations for the optimum orbitals are formulated and the system electronic energy is expressed in terms of the KS eigenvalues.

The importance of the DFT language for elucidating important chemical concepts of the molecular electronic structure and reactivity is illustrated by examining the equilibrium states of the acidic and basic reactants in the *donor–acceptor* (DA) complexes. The *chemical potential* and *hardness* descriptors of atoms and molecules, respectively measuring the first and second derivatives of the system ground state energy with respect to the average number of electrons N, are introduced. Due to the N-convexity of molecular energies, the chemical potential, equal to the system negative electronegativity, is shown to exhibit at integral N values in the limit of zero absolute temperature the N-discontinuity which is responsible for the integer numbers of electrons in the dissociation products. By using the Janak theorem the KS eigenvalues (orbital energies) are linked to energy differences and the orbital interpretations of the chemical potential, hardness, and softness descriptors are given in terms of the *frontier* KS densities.

R.F. Nalewajski, *Perspectives in Electronic Structure Theory*,
DOI 10.1007/978-3-642-20180-6_7, © Springer-Verlag Berlin Heidelberg 2012

An overview of the historical generations of the density functionals for the *exchange correlation* (*xc*) energy is given, covering *Local Density* (LDA) and *Generalized Gradient* (GGA) *Approximations*, as well as the hybrid and the orbital dependent functionals. The basic equations of the *Optimized Potential Method* (OPM), the DFT analog of HF theory, are summarized and rudiments of the density PT, the DFT analog of MP theory, are given and some advantages of the ab initio DFT perspective of Bartlett et al. are commented upon. The variational (Rayleigh–Ritz) principle for ensembles is formulated, which can be used to determine the molecular excitation energies. The HK DFT concepts and relations for ensembles are established and the associated KS scheme is introduced. Illustrative examples of the *density-matrix* functionals are surveyed and special approaches to weak van der Waals (vdW) interactions are summarized. In particular, the *Time-Dependent* DFT (TDFT) is used to generate the adequate nonlocal density functionals for dispersion interactions, via the AC and the zero-temperature *Fluctuation–Dissipation Theorem* (FDT). The HK theorems are extended into the time domain and the linear response functions are examined. The electron excitation energies and the associated oscillator strengths are generated from the poles and residues of the dynamic polarizability, and the rudiments of the orbital-free embedding of molecular subsystems in DFT are summarized. Finally, the physical origins of the correct asymptotic behavior of the dispersion energies in TDDFT are examined. It is argued using the screening approach to dynamic susceptibilities that the *second*-order energy should be only weakly dependent on the particular form of approximate functionals for the KS susceptibility and *xc*-kernel, with even LDA treatment generating the correct *long*-range behavior of vdW interactions.

The modern *Density Functional Theory* (DFT) of Kohn, Hohenberg, and Sham (Hohenberg and Kohn 1964; Kohn and Sham 1965) was born about five decades ago, but the conceptually related Thomas–Fermi (Thomas 1927; Fermi 1928; March 1975) and Hartree–Fock–Slater (HFS) or $X\alpha$ (Slater 1951; Johnson 1973b) theories can be regarded as its historical predecessors. However, while the former is in principle exactly equivalent to the Schrödinger Wave Function Theory (WFT), the latter density models are intrinsically approximate. This alternative, mainly ground state approach to general molecular systems (molecules, clusters, solids, etc.) uses the electron distribution as a vehicle for determining the molecular properties, e.g., the system energy, in the "reduced" mapping problem from the electron density to molecular properties. It has become the physicists' method of choice for describing the electronic structure of solids, and it currently dominates the applications of quantum theory to very large molecular systems, e.g., those encountered in the supramolecular chemistry, catalysis, and molecular biology (e.g., Dreizler and da Providencia 1985; Carr and Parrinello 1985; Parr and Yang 1989; Dreizler and Gross 1990; Kryachko and Ludeña 1990; Gross and Dreizler 1995; Nalewajski 1996a; Geerlings et al. 1999).

The traditional WFT is preferred for small-and medium-size systems, when high accuracy is required. Indeed, given a sufficiently powerful computer, the traditional methods in principle assure a systematic way to obtain any desired level of accuracy.

The present-day DFT calculations, however, which use more and more accurate explicit forms of the density functional for the exchange correlation energy within the *Generalized Gradient Approximation* (GGA) and beyond (meta-GGA), are preferred when a more modest level of accuracy is acceptable. They still suffer from an intrinsic limitation that there is no known systematic way to guarantee an arbitrarily high level of accuracy in the electronic structure computations. Some progress in this direction has recently been made using the exact exchange energy and the orbital-dependent density functionals for the Coulomb correlation energy, within the *Optimized Effective Potential* (OEP) method, which follows the earlier approaches to the HF problem by Sharp and Horton (1953), and by Talman and Shadwick (1976).

The new hope for improvement has thus been generated by the use of the orbitally dependent functionals, which make it possible to harness the experience of the advanced WFT treatments of the electron correlation problem in the modern quantum chemistry in order to construct still better energy functionals. Also, the *Density Matrix Functional Theory* (DMFT), couched in terms of the full *one*-particle density matrix as the system basic state variable instead of only its diagonal part used in DFT, can be expected to bring some improvement into the DFT results. The *Density Functional Perturbation Theory* (DFPT), in the spirit of the MP approach of quantum *many*-body theory, has also been used to design an accurate correlation functional. The ground state limitation of the original DFT has gradually been lifted by a development of several adequate variants of the excited-state approaches. There is also a marked progress in covering the range of weak (hydrogen bond and vdW) interactions, the inclusion of which is crucial for a realistic treatment of large biological systems.

7.1 Hohenberg–Kohn Theory

As we have learned in the preceding chapter the way to determine in quantum mechanics the physical properties of the stationary states of the molecular system described by its electronic Hamiltonian

$$\begin{aligned}\hat{\mathrm{H}}^e(\mathbf{q},\mathbf{Q}) &= V_{ne}(\mathbf{q}) + [\hat{\mathrm{T}}_e(\mathbf{q}) + V_{ee}(\mathbf{q})] = \sum_{i=1}^{N} v(i) + \left[-\frac{1}{2}\sum_{i=1}^{N}\Delta_i + \sum_{i=1}^{N-1}\sum_{j=i+1}^{N} g(i,j)\right]\\ &\equiv V_{ne}(N,v) + \hat{\mathrm{F}}(N) \equiv \hat{\mathrm{H}}^e(N,v),\end{aligned} \tag{7.1}$$

is to solve the Schrödinger equation (SE) (5.57), the eigenvalue problem of $\hat{\mathrm{H}}^e(N,v)$ or $\hat{\mathrm{H}}_e(N,v) = \hat{\mathrm{H}}^e(N,v) + V_{nn}$,

$$\hat{\mathrm{H}}^e\Psi_e = E^e\Psi_e \quad \text{or} \quad \hat{\mathrm{H}}_e\Psi_e = E_e\Psi_e,\ E_e = E^e + V_{nn} \tag{7.2}$$

which directly determines the electronic energy E_e of, say, molecular ground state and the associated electronic wave function Ψ_e. We have indicated above that in the

BO approximation the Hamiltonian is uniquely identified by its overall number of electrons *N* and the external potential $v(\boldsymbol{r})$ due to the nuclei in their assumed (fixed) positions, the shape of which effectively embodies the memory about the charges $\{Z_\alpha\}$ and positions $\{\boldsymbol{R}_\alpha\}$ of nuclear attractors.

It should be also recalled that SE (7.2) represents the Euler equation for the associated wave function variational principle:

$$\delta\{ E_e[\Psi] - E_e\langle\Psi \mid \Psi\rangle \}_{\Psi=\Psi_e} \equiv \delta\,\Omega[\Psi, \Psi^*, E_e]|_{\Psi=\Psi_e} = 0, \tag{7.3}$$

where the exact ground state energy E_e plays the role of the Lagrangian multiplier enforcing the normalization of the optimum wave function: $\langle\Psi|\Psi\rangle|_{\Psi=\Psi_e} = 1$. The complex wave function has two independent components, real and imaginary, or equivalently Ψ^* and Ψ. Thus, by taking the functional derivative of the auxiliary functional of the preceding minimum-energy principle with respect to one of them, say Ψ^*, one immediately arrives at SE (7.2):

$$\left.\frac{\partial\Omega}{\partial\Psi^*(N)}\right|_{\Psi=\Psi_e} = \hat{\mathrm{H}}_e(N)\Psi_e(N) - E_e\Psi_e(N) = 0. \tag{7.4}$$

One also notices that, due to the Hermitian character of the Hamiltonian, the functional differentiation with respect to Ψ gives the complex conjugate of the same equation.

One further observes that the complicated mapping problem of SE, from the variables (N, v) defining the Hamiltonian to the physical observable

$$E_e = \langle\Psi_e|\hat{\mathrm{H}}_e|\Psi_e\rangle = \int v(\boldsymbol{r})\rho(\boldsymbol{r})d\boldsymbol{r} + \langle\Psi_e|\hat{\mathrm{F}}|\Psi_e\rangle \equiv V_{ne}[\rho[\Psi_e]] + F[\Psi_e] \equiv E_e[\Psi_e], \tag{7.5}$$

where the electronic density ρ represents the known functional (3.7) of the wave function, $\rho = \rho[\Psi_e]$, involves 4 *N* position-spin degrees of freedom of electrons in the abstract concept of the system wave function $\Psi_e(N)$, the position-spin representation of yet another abstract entity – the system state vector $|\Psi\rangle_e$ in the molecular Hilbert space, which itself does not represent any physical observable. This WFT mapping also indicates that its final results E_e and Ψ_e are both functionals of the original state parameters $v(\boldsymbol{r})$ and *N* of the Hamiltonian: $E_e = E_e[N, v]$ and $\Psi_e = \Psi_e[N, v]$, with both these relations involving an unknown, complicated dependence upon the local physical observable $v(\boldsymbol{r})$. In these mapping relations the number of degrees of freedom is thus reduced to the minimum of the scalar state parameter *N* and the three coordinates in the physical space, the arguments of the external potential. However, this *existence* statement does not imply any operational knowledge how to realize such a *direct* $(N, v) \to E_e$ mapping without explicitly resorting to SE involving the abstract concept of the wave function. Thus, the known mapping from the Hamiltonian state parameters to the system energy is effected only *indirectly*, by solving (7.2): $(v, N) \to \Psi_e[N, v] \to E_e$.

The notion that the external potential would be the analytically most convenient physical quantity to directly execute such a simplified mapping is by no means obvious either, particularly in view of the divergence of $v(\boldsymbol{r})$ at nuclear positions. Another analytically more suitable candidate for performing such a direct mapping in the physical space is the system *electronic density* $\rho(\boldsymbol{r})$ of (3.7). Its finite *nuclear cusps* [see (6.104)] retain the "memory" about both the charges and positions of the nuclei. We recall that the nuclear-cusp condition expresses the derivative of the spherically averaged density in the vicinity of the nucleus. In accordance with the development of Sect. 6.2.3, one places the origin of the coordinate system at the position of the given nucleus α, with $\boldsymbol{r}_\alpha = \boldsymbol{r} - \boldsymbol{R}_\alpha = (r_\alpha, \vartheta_\alpha, \varphi_\alpha)$ and $\rho(\boldsymbol{r}_\alpha) \equiv \tilde{\rho}(r_\alpha, \vartheta_\alpha, \varphi_\alpha)$, defines the spherically averaged density,

$$\rho(r_\alpha) = \frac{1}{4\pi} \int_0^\pi \sin\vartheta_\alpha \int_0^{2\pi} \tilde{\rho}(r_\alpha, \vartheta_\alpha, \varphi_\alpha)\, d\vartheta_\alpha d\varphi_\alpha, \tag{7.6}$$

and determines its derivative in the limit $r_\alpha \to 0$. By the Kato theorem (6.104):

$$\lim_{r_\alpha \to 0} \left[\left(\frac{\partial}{\partial r_\alpha} + 2Z_\alpha \right) \rho(r_\alpha) \right] = 0. \tag{7.7}$$

For example, in the ground state of the hydrogen-like atom (see Sect. 4.5) $\rho^{av.}(r) = R^2_{1,0}(r) \propto \exp(-2Zr)$, thus fulfilling the nuclear-cusp requirement of (6.105). We also note that another Kato's result, the *correlation-cusp* condition of (6.102) formulated in terms of the *pair* distribution function, predicts an analogous relation for the coalescence of two electrons in the limit of $r_{1,2} \to 0$.

The electron density is also attractive for the following two reasons. On one hand, this physical property represents the partially integrated square of the modulus of the system wave function, $\rho = \rho[\Psi_e]$, which provides the basis for the physical interpretation of quantum states. On the other hand, as the partial functional derivative of the system electronic energy with respect to the external potential, the electron density constitutes the physically matching companion of the latter, i.e., it represents the energy conjugate of $v(\boldsymbol{r})$:

$$\rho(\boldsymbol{r}) = \left(\frac{\partial E_e[v, N]}{\partial v(\boldsymbol{r})} \right)_N. \tag{7.8a}$$

This observation directly follows from the familiar *electrostatic theorem* of Hellmann (1937) and Feynman (1939),

$$\begin{aligned} \frac{\partial E_e[v,N]}{\partial v(\boldsymbol{r})} &= \left\langle \frac{\partial \Psi_e}{\partial v(\boldsymbol{r})} \middle| \hat{\mathrm{H}}_e \middle| \Psi_e \right\rangle + \left\langle \Psi_e \middle| \hat{\mathrm{H}}_e \middle| \frac{\partial \Psi_e}{\partial v(\boldsymbol{r})} \right\rangle + \left\langle \Psi_e \middle| \frac{\partial \hat{\mathrm{H}}_e}{\partial v(\boldsymbol{r})} \middle| \Psi_e \right\rangle \\ &= E_e \frac{\partial \langle \Psi_e \mid \Psi_e \rangle}{\partial v(\boldsymbol{r})} + \left\langle \Psi_e \middle| \sum_{i=1}^N \delta(\boldsymbol{r}_i - \boldsymbol{r}) \middle| \Psi_e \right\rangle = \rho(\boldsymbol{r}), \end{aligned} \tag{7.8b}$$

where the first contribution in the second line of the preceding equation identically vanishes by the normalization condition of the wave function.

In early days of the quantum theory such "classical" density approaches to electronic structure of molecular systems by Thomas (T) (1927) and Fermi (F) (1928), and their extensions due to Dirac (D) (1930) and von Weizsäcker (W) (1935), have indeed been successfully tried giving reasonable results and considerable insight, e.g., in the statistical HFS method (Slater 1951, 1974), also known as X_α model (Johnson 1973b; Connolly 1977), at much reduced computational effort compared with that characterizing the standard methods of the full Schrödinger WFT. In all these density models the nonlocal *xc* effects have been represented by the local, density-dependent potential. Such heuristic approaches had expressed the system energy in terms of the electron density alone. Before the formulation by Hohenberg and Kohn (HK) (1964) of the basic theorems of the modern DFT, which are the main subject of this section, these density "theories" have been treated just as useful approximations of the exact quantum theory and hardly as steps toward the *exact* density treatment which would be fully equivalent to WFT.

Putting such an ultimate goal before these early TFDW density models would indeed imply that the *abstract* wave function $\Psi_e(N)$ defined in 4 N dimensional configuration space can be replaced by the quantum *observable* – the electronic density $\rho(\boldsymbol{r})$ defined in the physical space – as the *equivalent* specification of the system ground state, capable of generating all physical properties of the molecular system, e.g., its energy: $E_e = E_e[\rho]$.

Obviously, the *one*-particle density, the squared modulus of the system electronic wave function $\Psi_e(N)$ integrated over the spin-position variables of the remaining $(N - 1)$ particles [see (3.7)], looses a great deal of information contained in Ψ_e. Therefore, it is not immediately obvious that the remnants of the information still retained after the *cumulant* mapping $\Psi_e \rightarrow \rho$ is still sufficient to uniquely identify back the system quantum state, $\Psi_e = \Psi_e[\rho]$, which would then imply the existence of the reverse mapping: $\rho \rightarrow \Psi_e$. This is the essence of the first HK theorem, which we discuss in the next section. We shall see that the *one-to-one* mapping it establishes between the electron density $\rho(\boldsymbol{r})$ of the nondegenerate ground state Ψ_e and the system external potential $v(\boldsymbol{r})$, $v \leftrightarrow \rho$, involves the intermediate stage of SE and the density wave function mapping, $\rho \leftrightarrow \Psi_e$. Therefore, DFT does not represent any new quantum theory. In fact, it constitutes just another, more compact formulation of the ordinary quantum mechanics, giving a deeper, more causal understanding of molecular processes and in many respects more attractive for interpretation/computation purposes.

7.1.1 First HK Theorem

The "*forward*" (Schrödinger) mapping $\mathcal{M} = \mathcal{A}\mathcal{B}$ from v to ρ,

$$v \xrightarrow{\mathcal{A}} \Psi_e \xrightarrow{\mathcal{B}} \rho, \tag{7.9}$$

or in the operator notation $\mathcal{M}v = \rho$, involves two elementary stages: $\mathcal{A}$, $\mathcal{A}v = \Psi_e$, and $\mathcal{B}$, $\mathcal{B}\Psi_e = \rho$. The first is assured by solving the SE, while the trivial second mapping involves the associated partial integration of the squared modulus of Ψ_e.

For the reverse mapping $\mathcal{M}^{-1} = \mathcal{B}^{-1}\mathcal{A}^{-1}$ from ρ to v,

$$\rho \xrightarrow{\mathcal{B}^{-1}} \Psi_e \xrightarrow{\mathcal{A}^{-1}} v \tag{7.10}$$

or $\mathcal{M}^{-1}\rho = v$, to exist both elementary mappings $\mathcal{A}$ and $\mathcal{B}$ have to be invertible. We first observe that the reverse correspondence $\mathcal{A}^{-1}$ is also assured by SE:

$$\begin{aligned}\frac{[\hat{\mathrm{T}}_e(N) + V_{ee}(N, v)]\Psi_e(N)}{\Psi_e(N)} &= \frac{[\hat{\mathrm{H}}_e(N, v) - V_{ne}(N, v)]\Psi_e(N)}{\Psi_e(N)} \\ &= E_e - V_{ne}(N, v) = const. - \sum_{i=1}^{N} v(i),\end{aligned} \tag{7.11}$$

since $V_{ne}(N, v)$ represents a trivial multiplicative operation.

The existence of the $\mathcal{B}^{-1}\rho = \Psi_e$ mapping follows from the variational argument advanced by Hohenberg and Kohn (1964), which establishes the existence of the overall reverse map $\mathcal{M}^{-1}$. It proves the

First HK theorem. To an additive constant the external potential is uniquely determined by the (nondegenerate) ground state density ρ: $v = v[\rho] + const.$

The proof of this famous theorem goes through the *reductio ad absurdum*. Let us assume that, in contrast to the theorem, there are two different shapes of the external potential, v and $v' \neq v + const.$, associated with the electronic Hamiltonians $\hat{\mathrm{H}} = \hat{\mathrm{H}}_e(N, v)$ and $\hat{\mathrm{H}}' = \hat{\mathrm{H}}_e(N, v')$, respectively, which give rise to the same ground state density of electrons: $\rho = \rho[v] = \rho[v']$. The corresponding SE's,

$$\hat{\mathrm{H}}\Psi = (V_{ne} + \hat{\mathrm{F}})\Psi = E\Psi \quad \text{and} \quad \hat{\mathrm{H}}'\Psi' = (V_{ne}{}' + \hat{\mathrm{F}})\Psi' = E'\Psi', \tag{7.12}$$

where $E = E_e[N, v]$, $V_{ne} = V_{ne}[N,v]$, $E' = E_e[N, v']$, and $V_{ne}{}' = V_{ne}[N, v']$, which determine the associated (nondegenerate) ground states $\Psi = \Psi_e[N, v]$ and $\Psi' = \Psi_e[N, v']$, when combined with the wave function variational principle of Sect. 5.1.2 then directly imply:

$$\begin{aligned} E &= \langle\Psi|\hat{\mathrm{H}}|\Psi\rangle < \langle\Psi'|\hat{\mathrm{H}}|\Psi'\rangle = \langle\Psi'|\hat{\mathrm{H}}'|\Psi'\rangle - \langle\Psi'|\hat{\mathrm{H}}' - \hat{\mathrm{H}}|\Psi'\rangle = E' - \int \rho(\boldsymbol{r})[v'(\boldsymbol{r}) - v(\boldsymbol{r})]d\boldsymbol{r}, \\ E' &= \langle\Psi'|\hat{\mathrm{H}}'|\Psi'\rangle < \langle\Psi|\hat{\mathrm{H}}'|\Psi\rangle = \langle\Psi|\hat{\mathrm{H}}|\Psi\rangle - \langle\Psi|\hat{\mathrm{H}} - \hat{\mathrm{H}}'|\Psi\rangle = E + \int \rho(\boldsymbol{r})[v'(\boldsymbol{r}) - v(\boldsymbol{r})]d\boldsymbol{r}. \end{aligned} \tag{7.13}$$

Adding these two equations thus leads to a contradiction, $E + E' < E' + E$. This completes the proof of the first HK theorem establishing the existence of the reverse mapping $\mathcal{M}^{-1}$.

One similarly proves the map $\mathscr{A}$, $v \rightarrow \Psi_e$, that two different shapes of the external potential have different ground states. Taking the difference between SE's (7.12) with the assumption $\Psi = \Psi'$, which contradicts this theorem, gives:

$$(V_{ne} - V_{ne}{}')\Psi = (E - E')\Psi. \tag{7.14}$$

Again, this contradicts the assumption that the two external potentials differ more than by a trivial additive constant.

It thus confirms the existence of the unique $\mathscr{B}^{-1} = \mathscr{M}^{-1}\mathscr{A}$ transformation, from ρ to Ψ_e, embodied in the functional $\Psi_e = \Psi_e{}^{\mathrm{HK}}[\rho]$, since $\mathscr{A}$ is unique by virtue of SE. Hence, there exists the unique density functional for the system electronic energy (7.5),

$$\begin{aligned} E_v[\rho[v]] &\equiv E_{\mathrm{HK}}[\rho; v] = \int \rho[v; \boldsymbol{r}]v(\boldsymbol{r})\,d\boldsymbol{r} + F_{\mathrm{HK}}[\rho[v]], \\ F_{\mathrm{HK}}[\rho] &= \langle \Psi_e[\rho]|\hat{\mathrm{T}}_e + V_{ee}|\Psi_e[\rho]\rangle \equiv T_e[\rho] + V_{ee}[\rho], \end{aligned} \tag{7.15}$$

where the universal (v-independent) functional $F_{\mathrm{HK}}[\rho]$ generates the sum of the electronic kinetic and repulsion energies. The $\rho = \rho[v]$ notation stresses the fact that ρ is the exact ground state density for the external potential v, i.e., that ρ is v-representable. This $\rho \rightarrow \Psi_e$ mapping, or $\Psi_e = \Psi_e[\rho]$, further implies that the expectation value of any observable $\hat{\mathrm{A}}$ is also given by the unique functional of the ground state density:

$$\langle A\rangle_{\Psi_e} = \langle \Psi_e[\rho]|\hat{\mathrm{A}}|\Psi_e[\rho]\rangle = A[\rho]. \tag{7.16}$$

Now, that we already know from the above HK theorem that the reverse mapping $\rho \rightarrow \Psi_e$ indeed exists for the nondegenerate ground states, its existence for the molecular (Coulombic) systems exhibiting the integer number of electrons can indeed be easily demonstrated and understood. More specifically, by the Kato theorem nuclear cusps of the electronic density identify uniquely the position and charge of each nucleus, and hence also the resultant BO external potential v the nuclei generate for electronic motions. Moreover, by integrating the density over the whole physical space gives N, so that the knowledge the density alone is sufficient to precisely identify the electronic Hamiltionian of (7.1): $\rho \rightarrow \hat{\mathrm{H}}^e(N, v)$. Hence, by solving the stationary SE (7.2) one in principle completes the justification of the existence of the (indirect) density-to-wave-function mapping: $\rho \rightarrow \hat{\mathrm{H}}^e(N, v) \rightarrow \Psi_e$.

7.1.2 Second HK Theorem

The existence of the $\rho \rightarrow \Psi_e$ mapping thus implies that the electronic energy of (7.5) can be regarded as the functional (7.15) $E_{\mathrm{HK}}[\rho[v];v] = E_{v[\rho]}[\rho]$ of the electron density, with the ground state density matching the external potential due to the nuclei as its equilibrium distribution. As we have indicated above, both these local quantities are determined by the respective (exact) functional relationships of the molecular ground state: $\rho = \rho[v]$ and $v = v[\rho]$. Such densities are called *v-representable*. We are now in a position to establish the variational character of this functional, as expressed by the

Second HK theorem. The exact electronic energy $E_e[N,v] = E_v[\rho[N,v]]$ represents the lower bound for energies $E_v[\rho'(N)]$ obtained for any trial (v-representable) density $\rho'(N) = \rho'[N, v']$ associated with $v' \neq v + const.$,

$$E_e[N,v] = E_v[\rho[N,v]] < E_v[\rho'[N,v']].$$

The search for the matching ground state density for the specified external potential v, $\rho = \rho[v]$, thus involves the associated variational principle in terms of trial (v-representable) densities $\rho'[v']$,

$$\delta\left(E_v[\rho'[v']] - \mu \int \rho'[v', \boldsymbol{r}]\, d\boldsymbol{r}\right)_{\rho'[v']=\rho[v]} \equiv \delta\, \Xi\, [\rho'[v'], \mu]\big|_{\rho'[v']=\rho[v]} = 0, \tag{7.17}$$

the density analog of the wave function principle of (7.3).

The identity of the Lagrange multiplier μ, which enforces the "normalization" of the optimum density $\rho' = \rho$, $\int \rho'(\boldsymbol{r})\, d\boldsymbol{r}\big|_{\rho'=\rho} = N[\rho] = N$, is established by the associated Euler equation, $\delta E_e - \mu\, \delta N = 0$, or

$$\mu = \frac{\partial E_e[N,v]}{\partial N} \equiv \mu[N,v]. \tag{7.18}$$

This global quantity of the molecular ground state thus denotes the chemical potential of electrons, the negative of the system electronegativity (Mulliken 1934; Iczkowski and Margrave 1961),

$$\chi[Q,v] \equiv \frac{\partial E_e[Q,v]}{\partial Q} = -\mu[N,v], \tag{7.19}$$

since for the overall net charge of the molecule $Q = \sum_\alpha Z_\alpha - N$, $dQ = -dN$.

The corresponding local Euler equation,

$$\left.\frac{\partial \Xi}{\partial \rho'(\boldsymbol{r})}\right|_{\rho'=\rho} = \left.\frac{\delta E_v[\rho']}{\delta \rho'(\boldsymbol{r})}\right|_{\rho'=\rho} - \mu = 0 \quad \text{or} \quad \frac{\delta E_v[\rho]}{\delta \rho(\boldsymbol{r})} \equiv \mu(\boldsymbol{r}) = \mu = \frac{\partial E_e[N,v]}{\partial N}, \tag{7.20}$$

then indicates that the local value $\mu(\boldsymbol{r})$ of the electronic chemical potential is equalized in the ground state at the global chemical potential level:

$$\mu(\boldsymbol{r}) = v(\boldsymbol{r}) + \frac{\delta F[\rho]}{\delta \rho(\boldsymbol{r})} = \mu[N,v] \quad \text{or} \quad v(\boldsymbol{r}) - \mu \equiv u(\boldsymbol{r}) = -\frac{\delta F[\rho]}{\delta \rho(\boldsymbol{r})}. \tag{7.21}$$

This local Euler equation for the optimum electron density also provides the DFT justification of the Sanderson (1951, 1976) principle of the *Electronegativity Equalization* (EE) throughout the whole molecule in its ground state:

$$-\mu(\boldsymbol{r}) \equiv \chi(\boldsymbol{r}) = \frac{\delta E_v[q]}{\delta q(\boldsymbol{r})} = \chi[Q, v] = \frac{\partial E_e[Q, v]}{\partial Q}, \tag{7.22}$$

where the local electronegativity is defined by the functional derivative of the electronic energy $E_v[q]$ with respect to the local *net* charge distribution

$$q(\boldsymbol{r}) = \sum_\alpha Z_\alpha \delta(\boldsymbol{r} - \boldsymbol{R}_\alpha) - \rho(\boldsymbol{r}) \quad \text{or} \quad \delta q(\boldsymbol{r}) = -\delta\rho(\boldsymbol{r}), \tag{7.23}$$

$$\chi(\boldsymbol{r}) = \frac{\delta E_v[q]}{\delta q(\boldsymbol{r})} = -\mu(\boldsymbol{r}). \tag{7.24}$$

It should be also observed that SE (7.2), after dividing its both sides by $\Psi_e(N)$, implies the equalization of the system *local energy* $\mathscr{E}(N)$ at the eigenvalue level:

$$\mathscr{E}(N) \equiv \Psi_e(N)^{-1}[\hat{\mathrm{H}}_e(N)\Psi_e(N)] = E_e. \tag{7.25}$$

The density principle (7.17), with $E_v[\rho'[v']] = E_{\mathrm{HK}}[\rho';v]$, is limited to the true ground state densities for some (really existing) trial external potentials v' (not necessarily the Coulomb potentials), the so-called v-representable densities of N electrons, $\rho'[N, v']$. Since the a priori conditions for the density to be v-representable are not known and some examples have been identified of electron densities, which are not v-representable, particularly in degenerate systems, a general applicability of the above HK variational principle depends upon replacing the crucial functional $F_{\mathrm{HK}}[\rho]$ with its generalized form $F[\rho]$, which would be applicable to any trial density. Such an extension is provided by Levy's (1979, 1982) constrained-search formulation, which we shall now briefly summarize.

Another requirement is a weaker condition of the so-called N-representability, that a trial density be associated with some antisymmetric wave function of N electrons. As shown by Gilbert (1975), Harrimann (1980), and Lieb (1982, 1983), this condition is always satisfied by any well-behaving, reasonable trial density satisfying the following natural requirements:

$$\rho(\boldsymbol{r}) \geq 0, \int \rho(\boldsymbol{r})\, d\boldsymbol{r} = N, \quad \int \left|\nabla\rho(\boldsymbol{r})^{1/2}\right|^2 d\boldsymbol{r} < \infty. \tag{7.26}$$

As first argued by Gilbert, the corresponding wave function can be then always constructed from N orthonormal SO obtained from the spatial partitioning of the density, thus giving rise to the wave function yielding the specified density: $\Psi(N) \to \rho$. Such an explicit construction has been then proposed by Harriman, using the complex orbitals of Macke (1955a,b) with appropriate density-dependent phase factors, later extended by March (1982), Zumbach and Maschke (1983), and Ciosłowski (1988).

In terms of the energy functional for the fixed external potential of the BO approximation (7.15), $E_v[\rho'] = \int \rho'(\boldsymbol{r})v(\boldsymbol{r})\,d\boldsymbol{r} + F[\rho']$, the generalized variational principle of (7.17), valid for any trial density ρ',

$$\delta\left\{ E_v[\rho'] - \mu \int \rho'(\boldsymbol{r})\,d\boldsymbol{r} \right\}_{\rho'=\rho} \equiv \delta\,\Xi\,[\rho',\mu]|_{\rho'=\rho} = 0, \tag{7.27}$$

will then be satisfied by the exact ground state density $\rho' = \rho[v]$ determined by the Euler equation (7.21) which explicitly testifies that ρ determines the shape of v, i.e., v itself to an additive constant.

The familiar wave function variational principle of Sect. 5.1.2, in terms of the normalized trial functions $\Psi(N)$ of N electrons, can be thus straightforwardly interpreted as the corresponding density principle, involving the search over the trial densities $\rho' = \rho'(N)$ of N electrons:

$$\begin{aligned} E_e[N,v] &= \min_{\Psi(N)} \langle \Psi(N)|\hat{\mathrm{H}}_e[N,v]|\Psi(N)\rangle = E_v[\rho[N,v]] \equiv \min_{\rho'} E_v[\rho'] \\ &= \min_{\rho'} \left[\int \rho'(\boldsymbol{r})v(\boldsymbol{r})\,d\boldsymbol{r} + \inf_{\Psi(N)\to\rho'} \langle \Psi(N)|\hat{\mathrm{F}}(N)|\Psi(N)\rangle \right] \\ &= \int \rho[N,v;\boldsymbol{r}]v(\boldsymbol{r})\,d\boldsymbol{r} + \langle \Psi[\rho[N,v]]|\hat{\mathrm{F}}(N)|\Psi[\rho[N,v]]\,\rangle. \end{aligned} \tag{7.28}$$

Above, we have identified the Levy (1979) (see also: Levy and Perdew 1985) form of the universal (v-independent) functional defined by the constrained search over all antisymmetric functions of N electrons giving rise, after integration, to the specified trial density ρ', the property symbolically denoted by the $\Psi(N) \to \rho'$ notation:

$$F[\rho'] = \inf_{\Psi(N)\to\rho'} \langle \Psi(N)|\hat{\mathrm{F}}(N)|\Psi(N)\rangle. \tag{7.29}$$

It is applicable to any trial density ρ' and by the variational principle of quantum mechanics gives the corresponding HK value, for the v-representable trial density $\rho' = \rho[N, v]$, identified by the lowest value (infimum) of this constrained search:

$$\begin{aligned} F_{\mathrm{HK}}[\rho[N,v]] &= F[\rho[N,v]] = \langle \Psi[\rho[N,v]]|\hat{\mathrm{F}}(N)|\Psi[\rho[N,v]]\,\rangle \\ &= \inf_{\Psi(N)\to\rho} \langle \Psi(N)|\hat{\mathrm{F}}(N)|\Psi(N)\rangle. \end{aligned} \tag{7.30}$$

To summarize, we have established that the electron distribution of the nondegenerate ground state constitutes the *bona fide* state parameter, which in principle can be exactly mapped into the system physical observables, e.g., the system electronic energy, all of them representing some functionals of the density.

Moreover, it has been demonstrated that there exists the associated density variational rule for determining the optimum distribution of electrons $\rho[N, v]$, matching the given external potential v, which replaces the analogous wave function principle of the Schrödinger theory.

These existence theorems, however, do not imply that such maps from the density to physical quantities can be expressed in some closed analytical forms. Nonetheless, the experience of the old density approaches shows that there are quite realistic and in many respects satisfactory approximate algorithms, which realize these functional relations in computational practice, particularly for the $\rho \rightarrow E_e$ map of the energy density functional. When properly harnessed with the variational principle for the electron density within the orbital approximation of Kohn and Sham (KS) (1965) it gives excellent results for both the solid-state and molecular systems.

7.1.3 Refinements

Consider now the degenerate ground state, when there are several independent quantum states $\boldsymbol{\Psi}_v = \{\Psi_{v,k}\}$ corresponding to the ground state energy $E_e = E_e[N,v]$,

$$\hat{\mathrm{H}}_e(N, v)\Psi_{v,k}(N) = E_e[N, v]\Psi_{v,k}(N), \quad k = 1, 2, \ldots, g, \tag{7.31}$$

integrating to the associated densities $\boldsymbol{\rho}_v = \{\rho_{v,k} = \rho_{v,k}[\Psi_{v,k}]\}$. In this case the map $\mathscr{A}$ of (7.9), $v \rightarrow \Psi_{v,k}$, and hence also the product "*forward*" mapping $\mathscr{M} = \mathscr{A}\mathscr{B}$ from the external potential to the electron density, $v \rightarrow \rho_{v,k}$, do not exist: $\Psi_{v,k} \neq \Psi_{v,k}[v]$ and $\rho_{v,k} \neq \rho_{v,k}[v]$. Clearly, the integration of the wave function to the associated density represented by the map $\mathscr{B}$, $\Psi_{v,k} \rightarrow \rho_{v,k}$, always exists. One realizes, however, that in general several degenerate states may give rise to the same electron distribution, say, $\rho_{v,k} = \rho_{v,l}$, so that the reverse map $\mathscr{B}^{-1}$, $\rho_{v,k} \rightarrow \Psi_{v,k}$, does not exist either: $\Psi_{v,k} \neq \Psi_{v,k}[\rho_{v,k}]$. In other words, in the degenerate case, the electron density does not identify uniquely the system quantum state.

However, by using the same argument as in (7.14) it can be straightforwardly demonstrated that the reverse functional relation $\mathscr{A}^{-1}$, $\{\Psi_{v,k} \rightarrow v\}$, constitutes a proper map in the mathematical sense of the word, $v = v[\Psi_{v,k}]$, since the subsets of degenerate wave functions $\boldsymbol{\Psi}_v$ and $\boldsymbol{\Psi}_{v'}$ corresponding to two different shapes of the external potential, $v' \neq v + const.$, are disjoint. More specifically, let Ψ and Ψ' represent any member states of $\boldsymbol{\Psi}_v$ and $\boldsymbol{\Psi}_{v'}$, respectively. The assumption that they may contain the common wave functions, $\Psi = \Psi'$, then implies for the difference of the SE's (7.12) that, contrary to the assumption of the nontrivial difference in *shapes* of two external potentials, $V_{ne} - V_{ne}' = E - E' = const$. The same line of reasoning as in (7.13) then demonstrates the disjoint character of densities associated with these two external potentials: $\boldsymbol{\rho}_v$ and $\boldsymbol{\rho}_{v'}$.

Therefore, any member of the density set $\boldsymbol{\rho}_v$ uniquely identifies the external potential v, $\rho_{v,k} \rightarrow v$, thus proving the existence of the overall reverse map $\mathcal{M}^{-1}$: $v = v[\rho_{v,k}]$. To summarize: also in the degenerate case any degenerate ground state density uniquely specifies the shape of the external potential $v(\boldsymbol{r})$ to an additive constant.

Since in the degenerate case the wave function is no longer the functional of the system electron density, the previous way of introducing the density functionals for the expectation values of physical observables [(7.15), (7.16), and (7.28)] cannot be used. However, one can still uniquely define the nontrivial part of the energy functional of (7.15). This can be done indirectly by using the energy functional defined to an additive constant by the existing $v = v[\rho_{v,k}]$ relation:

$$E_e[N, v[\rho_{v,k}]] = \langle \Psi_{v,k}(N) | \hat{\mathrm{H}}_e(N, v[\rho_{v,k}]) | \Psi_{v,k}(N) \rangle \equiv E_{e,k}, \quad k = 1, 2, \ldots g, \tag{7.32}$$

which fixes the (degenerate) energy level $E_{e,k} = E_e$ to an additive constant. Hence

$$F_{\mathrm{HK}}[\rho_{v,k}] = E_e[N, v[\rho_{v,k}]] - \int \rho_{v,k}(\boldsymbol{r}) v[\rho_{v,k}; \boldsymbol{r}] \, d\boldsymbol{r} = \langle \Psi_{v,k}(N) | \hat{\mathrm{F}}(N) | \Psi_{v,k}(N) \rangle. \tag{7.33}$$

It should be observed that in this definition the ambiguity of the undefined constant contribution introduced by the $v = v[\rho_{v,k}]$ map cancels out. The energy functional for the v-representable densities (7.15) now reads:

$$E_v[\rho_{v,k}[v]] = E_{\mathrm{HK}}[\rho_{v,k}] = \int \rho_{v,k}(\boldsymbol{r}) \, v(\boldsymbol{r}) \, d\boldsymbol{r} + F_{\mathrm{HK}}[\rho_{v,k}] = E_e[N, v], \; k = 1, 2, \ldots g. \tag{7.34}$$

It is now defined for both the degenerate and nondegenerate ground state densities. For all densities $\{\boldsymbol{\rho}_{v'}\}$ corresponding to other shapes of the external potential, $v' \neq v + const.$, one again finds the inequality

$$E_v[\rho_{v',l}] = \int \rho_{v',l}(\boldsymbol{r}) v(\boldsymbol{r}) \, d\boldsymbol{r} + F_{\mathrm{HK}}[\rho_{v',l}] > E_{\mathrm{HK}}[\rho_{v,k}] = E_v[\rho_{v,k}]. \tag{7.35}$$

The previously discussed requirement of the (*pure* state) v-representability for the densities determining the HK functionals, that the trial density be associated with the ground (possibly degenerate) state of the Hamiltonian defined by some local external potential, is obviously violated by the average density ρ_{D} of the ensemble of degenerate ground states, $\boldsymbol{\Psi} = \{\Psi_k\}$, $k = 1, 2, \ldots, g$, corresponding to the statistical density operator (3.60a–c):

$$\hat{\mathrm{D}} = \sum_{k=1}^{g} |\Psi_k\rangle p_k \langle \Psi_k| \equiv \hat{\mathrm{D}}[\boldsymbol{\Psi}], \tag{7.36}$$

with the normalized ensemble probabilities $\{p_k \geq 0\}$, $\sum_{k=1}^{g} p_k = 1$,

$$\rho_{\mathrm{D}}(\boldsymbol{r}) = \mathrm{tr}[\hat{\mathrm{D}}\hat{\mathrm{p}}(\boldsymbol{r})] = \sum_{k=1}^{g} p_k \rho_k(\boldsymbol{r}) = \rho_{ens.}(\boldsymbol{r}), \quad \rho_k(\boldsymbol{r}) = \langle \Psi_k | \, \hat{\mathrm{p}}(\boldsymbol{r}) | \Psi_k \rangle. \tag{7.37}$$

Such densities associated with the mixture of ground states rather than with any single state are called the *ensemble* v-representable.

The domain of the HK functional $F_{\mathrm{HK}}[\rho]$ can be easily extended to handle such average densities as well. It can be demonstrated using a similar variational argument to that used in Sect. 7.1 that the ensemble average densities $\boldsymbol{\rho}_{\mathrm{D}[v]}$ and $\boldsymbol{\rho}_{\mathrm{D}[v']}$ associated with the density operators of (7.36) for two different external potentials, $\hat{\mathrm{D}}[v] = \hat{\mathrm{D}}[\boldsymbol{\Psi}_v]$ and $\hat{\mathrm{D}}[v'] = \hat{\mathrm{D}}[\boldsymbol{\Psi}_{v'}]$, are disjoint when the two external potentials v and v' determining these mixtures of the (degenerate) ground states $\boldsymbol{\Psi}_v$ and $\boldsymbol{\Psi}_{v'}$, respectively, differ by more than a constant. Therefore, ρ_{D} identifies uniquely the external potential $v = v[\rho_{\mathrm{D}}]$ and the associated density operator $\hat{\mathrm{D}}[v] = \hat{\mathrm{D}}[v[\rho_{\mathrm{D}}]] = \hat{\mathrm{D}}[\rho_{\mathrm{D}}]$.

Hence, the extension of the $F_{\mathrm{HK}}[\rho]$ functional to the domain of the ensemble v-representable densities reads:

$$F_{\mathrm{EHK}}[\rho_{\mathrm{D}}] = \mathrm{tr}\{\hat{\mathrm{D}}[\rho_{\mathrm{D}}]\,\hat{\mathrm{F}}\}. \tag{7.38}$$

The corresponding density functional for the ensemble average value of the electronic energy,

$$E_{\mathrm{EHK}}[\rho_{\mathrm{D}}; v] = \int \rho_{\mathrm{D}}(\boldsymbol{r}) v(\boldsymbol{r})\, d\boldsymbol{r} + F_{\mathrm{EHK}}[\rho_{\mathrm{D}}], \tag{7.39}$$

then exhibits the usual variational properties, assuming the minimum value at and only at the correct ground state ensemble average density corresponding to $v(\boldsymbol{r})$.

In a similar way one defines the generalized functionals for the ensemble average value of any physical property A:

$$A_{\mathrm{EHK}}[\rho_{\mathrm{D}}] = \mathrm{tr}\{\hat{\mathrm{D}}[\rho_{\mathrm{D}}]\,\hat{\mathrm{A}}\}. \tag{7.40}$$

These HK-type generalizations, applicable to the ensemble v-representable densities, can be extended to arbitrary nonnegative trial densities integrating to the prescribed number of electrons by using the constrained search construction of Levy (1979, 1982), Lieb (1982), and Valone (1980a,b):

$$F_{\mathrm{D}}[\rho] = \inf_{\hat{\mathrm{D}} \to \rho} \{\mathrm{tr}(\hat{\mathrm{D}}\hat{\mathrm{F}})\}, \tag{7.41}$$

where the lowest value (infimum) is found over all density operators giving rise to the specified ensemble average density of (7.37). This functional is convex and represents the Legendre transform of the system ground state energy $E[N, v]$ [see (7.8a, b)] (Lieb 1982; Nalewajski and Parr 1982):

$$F_{\mathrm{D}}[\rho] = \sup_{v}\left(E[N,v] - \int\left(\frac{\partial E[N,v]}{\partial v(r)}\right)v(\boldsymbol{r})\,d\boldsymbol{r}\right)$$
$$= \sup_{v}\left(E[N,v] - \int \rho(\boldsymbol{r})v(\boldsymbol{r})\,d\boldsymbol{r}\right), \tag{7.42}$$

where the highest value (supremum) is determined for the trial external potentials "searching" for the right ground state density. Therefore, this functional constitutes the "thermodynamic" potential corresponding to a change of the state-variable from v to ρ, with the original state parameter v being associated with the system total energy.

For the *ensemble* v-representable density ρ_{D} this generalized universal functional $F_{\mathrm{D}}[\rho]$ becomes identical with its v-representable analog of (7.38),

$$F_{\mathrm{D}}[\rho_{\mathrm{D}}] = F_{\mathrm{EHK}}[\rho_{\mathrm{D}}]. \tag{7.43}$$

The associated energy functional

$$E_{\mathrm{D}}[\rho; v] = \int \rho(\boldsymbol{r})v(\boldsymbol{r})\,d\boldsymbol{r} + F_{\mathrm{D}}[\rho] \tag{7.44}$$

then recovers the corresponding EHK value,

$$E_{\mathrm{D}}[\rho_{\mathrm{D}}; v] = E_{\mathrm{EHK}}[\rho_{\mathrm{D}}; v], \tag{7.45}$$

thus reaching the exact ensemble-average energy when ρ_{D} corresponds to $v(\boldsymbol{r})$ as its equilibrium distribution.

7.1.4 Finite *Temperature and* Open *System Extensions*

In the generalized variational principle of (7.17), which gives rise to the Euler equation (7.21), the overall number of electrons is not conserved by the trial electron densities ρ. Such fluctuations in the system average (fractional) number of electrons $N[\rho] = \int \rho(\boldsymbol{r})\,d\boldsymbol{r}$ characterize densities of *open* systems, in contact with the external reservoir(s) of electrons, which correspond to the statistical ensemble of molecules defined by Hamiltonians with different (integer) numbers of electrons. In order to cover the atomic plasmas, the *finite* temperature extension of the previous ground state (*zero*-temperature) HK theory is required. The correct theoretical framework for describing the equilibria in the open systems at finite temperatures is that of the *grand canonical ensemble* of thermodynamics

(Gyftopoulos and Hatsopoulos 1965; Mermin 1965; Linderberg 1977; Perdew et al. 1982; Perdew 1985; Parr and Yang 1989).

The original HK formalism has been extended to thermal ensembles by Mermin (1965) who considered the equilibrium states at finite absolute temperature T (see also Sect. 7.3.3). When dealing with the equilibrium states of the externally open molecular systems, in contact with the heat bath of temperature T and the electron reservoir exhibiting the chemical potential μ, the *grand potential*

$$\Omega_{ens.}[\mu, T] = E_{ens.} - \mu N_{ens.} - TS_{ens.}, \tag{7.46}$$

representing the Legendre transform of the system internal energy $E_{ens.}[N_{ens.}, S_{ens.}]$ corresponding to the replacement of the *extensive* state parameters, the ensemble average number of electrons $N_{ens.}$ and its average entropy $S_{ens.}$, with their *intensive* conjugates, $\mu = \partial E_{ens.}/\partial N_{ens.}$ and $T = \partial E_{ens.}/\partial S_{ens.}$, reaches the minimum value at the equilibrium state of the grand canonical ensemble. This thermodynamic potential represents the auxiliary energy function containing the relevant constraint terms, with the Lagrange multipliers μ and T enforcing the prescribed value of the average values of the number of electrons, $N_{ens} = \bar{N}$, and that of the system entropy, $S_{ens} = \bar{S}$.

The equilibrium state for the given external potential v represents the statistical mixture of quantum states $\{\Psi_i^j \equiv \Psi_j[N_i, v]\}$, including the ground ($j = 0$) and excited ($j > 0$) stationary states of molecular Hamiltonians $\{\hat{\mathrm{H}}_i \equiv \hat{\mathrm{H}}^e(N_i, v)\}$ for the given external potential v and specified (integer) numbers of electrons $\{N_i\}$:

$$\hat{\mathrm{H}}_i\Psi_i^j = E_i^j\Psi_i^j, \quad j = 0, 1, 2, \ldots, \tag{7.47}$$

$$\hat{\mathrm{d}}[\mu, T; v] = \sum_i \sum_{j=0}^{\infty} p_i^j[\mu, T; v] \left|\Psi_i^j\right\rangle\left\langle\Psi_i^j\right| \equiv \sum_i \sum_{j=0}^{\infty} p_i^j \hat{\mathrm{P}}_i^j. \tag{7.48}$$

In the ordinary thermodynamics, the equilibrium probabilities of the grand ensemble read:

$$\begin{aligned} p_i^j[\mu, T; v] &= \frac{\exp[-\beta(E_i^j - \mu N_i)]}{\Xi[\mu, T; v]}, \quad \beta = (k_{\mathrm{B}}T)^{-1}, \\ \sum_i \sum_{j=0}^{\infty} p_i^j[\mu, T; v] &= 1, \end{aligned} \tag{7.49}$$

where k_{B} denotes the Boltzmann constant and the grand ensemble partition function

$$\Xi[\mu, T; v] = \sum_i \sum_{j=0}^{\infty} \exp[-\beta(E_i^j - \mu N_i)]. \tag{7.50}$$

This mixture is thus defined by the grand-canonical density operator,

$$\hat{\mathrm{d}}[\mu,T;v]=\sum_i\sum_{j=0}^{\infty}\left|\Psi_i^j\right\rangle p_i^j[\mu,T;v]\left\langle\Psi_i^j\right|, \tag{7.51}$$

which determines the equilibrium values of the ensemble averages of all physical quantities [see (7.40)]:

$$A_{ens.}[\mu,T;v]=\mathrm{tr}\{\hat{\mathrm{d}}[\mu,T;v]\,\hat{\mathrm{A}}\}\equiv A[\hat{\mathrm{d}}[\mu,T;v]]\,. \tag{7.52}$$

The probabilities of (7.49) represent the eigenvalues of the *grand canonical* (*gc*) statistical operator acting in the Fock space (see Sect. 6.5.1),

$$\hat{\mathrm{d}}^F=\frac{\exp[-\beta(\hat{\mathrm{H}}_e^F-\mu\hat{\mathrm{N}}^F)]}{\mathrm{tr}[-\beta(\hat{\mathrm{H}}_e^F-\mu\hat{\mathrm{N}}^F)]}, \tag{7.53}$$

with $\hat{\mathrm{N}}^F$ standing for the electron number operator [see (6.236) and (6.258)], and $\hat{\mathrm{H}}_e^F$ defined in (6.247), (6.254), and (6.256). The quantum mechanical operator associated with the grand potential of (7.46) similarly reads:

$$\hat{\Omega}^F=\hat{\mathrm{H}}_e^F-\mu\hat{\mathrm{N}}^F-T\hat{\mathrm{S}}^F=\hat{\mathrm{H}}_e^F-\mu\hat{\mathrm{N}}^F-\frac{1}{\beta}\ln\hat{\mathrm{d}}^F, \tag{7.54}$$

where we have introduced the *entropy* operator

$$\hat{\mathrm{S}}^F=-k_{\mathrm{B}}\ln\hat{\mathrm{d}}^F. \tag{7.55}$$

To simplify the notation, in what follows we shall drop the Fock space superscript. The equilibrium value of the grand potential and the associated electron density then read [see (7.52)]:

$$\Omega_{ens.}[\mu,T;v]=\mathrm{tr}[\hat{\mathrm{d}}\hat{\Omega}],\quad \rho_{ens.}[\mu,T;v]=\mathrm{tr}[\hat{\mathrm{d}}\hat{\rho}], \tag{7.56}$$

where the density operator in the Fock space is defined in (6.252).

All these operators assume the diagonal form in the energy representation defined by the basis functions $\{\Psi_i^j\}$, the eigenfunctions of Hamiltonians $\{\hat{\mathrm{H}}_i\}$. For example, the statistical operator corresponds to the ensemble probabilities:

$$\mathbf{d}=\{\left\langle\Psi_i^j\middle|\hat{\mathrm{d}}\middle|\Psi_{i'}^{j'}\right\rangle=p_i^j\delta_{i,i'}\delta_{j,j'}\},\quad p_i^j=\left\langle\Psi_i^j\middle|\hat{\mathrm{d}}\middle|\Psi_i^j\right\rangle. \tag{7.57}$$

Since the basis functions correspond to the sharply specified energy of the molecular system in question the associated energy matrix is also diagonal,

$$\mathbf{H}=\{\left\langle\Psi_i^j\middle|\hat{\mathrm{H}}_i\middle|\Psi_{i'}^{j'}\right\rangle=E_i^j\,\delta_{i,i'}\delta_{j,j'}\}, \tag{7.58}$$

and so is the matrix representation of $\hat{\Omega}$:

$$\boldsymbol{\Omega} = \{\left\langle \Psi_i^j \middle| \hat{\Omega} \middle| \Psi_{i'}^{j'} \right\rangle = (E_i^j - \mu N_i + \frac{1}{\beta} \ln p_i^j)\, \delta_{i,i'} \delta_{j,j'} \equiv \Omega_i^j \delta_{i,i'} \delta_{j,j'}\} = \boldsymbol{\Omega}(\mathbf{d}). \quad (7.59)$$

This gives rise to the familiar expression for the equilibrium value of the grand potential:

$$\begin{aligned} \Omega_{ens.}[\mu, T; v] &= \mathrm{tr}(\mathbf{d}\boldsymbol{\Omega}) = \sum_i \sum_{j=0}^{\infty} p_i^j \Omega_i^j \\ &= \sum_i \sum_{j=0}^{\infty} p_i^j E_i^j - \mu \sum_i N_i \sum_{j=0}^{\infty} p_i^j + \frac{1}{\beta} \sum_i \sum_{j=0}^{\infty} p_i^j \ln p_i^j \\ &= E_{ens.}[\mu, T; v] - \mu N_{ens.}[\mu, T; v] - T S_{ens.}[\mu, T; v] \equiv \Omega[\mathbf{d}]. \end{aligned} \quad (7.60)$$

Here, the ensemble average value of the electronic energy,

$$\begin{aligned} E_{ens.}[\mu, T; v] &= \sum_i \sum_{j=0}^{\infty} p_i^j E_i^j = \mathrm{tr}\,(\mathbf{dH}) = \mathrm{tr}\,[\mathbf{d}(\mathbf{V}+\mathbf{F})] \\ &= \sum_i \sum_{j=0}^{\infty} p_i^j V_i^j + \sum_i \sum_{j=0}^{\infty} p_i^j F_i^j \equiv V_{ens.}[\mu, T; v] + F_{ens.}[\mu, T; v], \end{aligned} \quad (7.61)$$

contains the external potential contribution $V_{ens.}[\mu, T; v]$, the ensemble average of its expectation values $\{V_i^j\}$ in individual states,

$$\mathbf{V} = \{\left\langle \Psi_i^j \middle| V_{ne} \middle| \Psi_i^j \right\rangle \delta_{i,i'} \delta_{j,j'} = V_i^j \delta_{i,i'} \delta_{j,j'}\}, \quad (7.62)$$

and the ensemble average term originating from the remaining part $\hat{\mathrm{F}}(N_i)$ of the electronic Hamiltonian over the expectation values $\{F_i^j\}$ in individual states:

$$\mathbf{F} = \{\left\langle \Psi_i^j \middle| \hat{\mathrm{T}}_e + V_{ee} \middle| \Psi_i^j \right\rangle \delta_{i,i'} \delta_{j,j'} = \left\langle \Psi_i^j \middle| \hat{\mathrm{F}} \middle| \Psi_i^j \right\rangle \delta_{i,i'} \delta_{j,j'} = F_i^j \delta_{i,i'} \delta_{j,j'}\}. \quad (7.63)$$

Notice that in this representation, the particle number operator,

$$\hat{\mathrm{N}} = \sum_i N_i \sum_{j=0}^{\infty} \left| \Psi_i^j \right\rangle \left\langle \Psi_i^j \right|, \quad (7.64)$$

is also represented by the diagonal matrix:

$$\mathbf{N} = \{\left\langle \Psi_i^j \middle| \hat{\mathrm{N}} \middle| \Psi_{i'}^{j'} \right\rangle = N_i \delta_{i,i'} \delta_{j,j'}\}, \quad N_i = \left\langle \Psi_i^j \middle| \hat{\mathrm{N}} \middle| \Psi_i^j \right\rangle. \quad (7.65)$$

Hence, the ensemble average number of electrons reads:

$$N_{ens.}[\mu, T; v] = \mathrm{tr}(\mathbf{dN}) = \sum_i N_i \left(\sum_{j=0}^{\infty} p_i^j \right) \equiv \sum_i N_i P_i \equiv N[\mathbf{d}]. \tag{7.66}$$

The electron density operator,

$$\hat{\rho}(\boldsymbol{r}) = \sum_i \sum_{j=0}^{\infty} \left| \Psi_i^j \right\rangle \rho_i^j(\boldsymbol{r}) \left\langle \Psi_i^j \right|, \quad \rho_i^j(\boldsymbol{r}) = \left\langle \Psi_i^j \middle| \hat{\rho}(\boldsymbol{r}) \middle| \Psi_i^j \right\rangle, \tag{7.67}$$

is similarly represented by the diagonal matrix of densities in individual states,

$$\boldsymbol{\rho}(\boldsymbol{r}) = \{ \left\langle \Psi_i^j \middle| \hat{\rho}(\boldsymbol{r}) \middle| \Psi_{i'}^{j'} \right\rangle = \rho_i^j(\boldsymbol{r}) \delta_{i,i'} \delta_{j,j'} \}, \tag{7.68}$$

so that the ensemble average electron density is determined by the mean value expression:

$$\rho_{ens.}[\mu, T; v] \equiv \rho^{gc} = \ \mathrm{tr}\ (\mathbf{d}\boldsymbol{\rho}) = \sum_i \sum_{j=0}^{\infty} p_i^j \rho_i^j \equiv \rho[\mathbf{d}]. \tag{7.69}$$

The entropy operator,

$$\hat{\mathrm{S}} = \sum_i \sum_{j=0}^{\infty} \left| \Psi_i^j \right\rangle S_i^j \left\langle \Psi_i^j \right|, \quad S_i^j = \left\langle \Psi_i^j \middle| \hat{\mathrm{S}} \middle| \Psi_i^j \right\rangle = -k_{\mathrm{B}} \ln p_i^j, \tag{7.70}$$

has also the diagonal representation in eigenstates $\{\Psi_i^j\}$ of $\{\hat{\mathrm{H}}_i\}$,

$$\mathbf{S} = \{ \left\langle \Psi_i^j \middle| \hat{\mathrm{S}} \middle| \Psi_{i'}^{j'} \right\rangle = S_i^j \delta_{i,i'} \delta_{j,j'} \}, \tag{7.71}$$

giving rise to the ensemble average value of the thermodynamic entropy defined in (7.60):

$$S_{ens.}[\mu, T; v] = \mathrm{tr}\ (\mathbf{dS}) = \sum_i \sum_{j=0}^{\infty} p_i^j S_i^j = -k_{\mathrm{B}} \sum_i \sum_{j=0}^{\infty} p_i^j \ln p_i^j \equiv S[\mathbf{d}]. \tag{7.72}$$

It is seen to be proportional to the familiar Shannon (1948) entropy of Information Theory (IT), a measure of the indeterminacy contained in the grand canonical probabilities of this equilibrium ensemble of quantum states.

Consider now the nonequilibrium value of the grand potential functional, defined by some trial (normalized) ensemble probabilities $\mathbf{d}'$,

$$\begin{aligned} \Omega[\mathbf{d}'] &= \mathrm{tr}[\mathbf{d}'\boldsymbol{\Omega}(\mathbf{d}')] = E_v[\mathbf{d}'] - \mu N[\mathbf{d}'] - TS[\mathbf{d}'] \\ &= \sum_i \sum_{j=0}^{\infty} p_i^{j\prime} \{ \int [v(\boldsymbol{r}) - \mu] \rho_i^j(\boldsymbol{r})\, d\boldsymbol{r} \} + (T_e[\mathbf{d}'] + V_{ee}[\mathbf{d}']) - TS[\mathbf{d}'] \\ &= \int u(\boldsymbol{r}) \rho_{ens.}[\mathbf{d}'; \boldsymbol{r}]\, d\boldsymbol{r} + F[\mathbf{d}'] - TS[\mathbf{d}'] \equiv \Omega_u(\mathbf{d}'), \end{aligned} \tag{7.73}$$

where,

$$F[\mathbf{d}'] = \sum_i \sum_{j=0}^{\infty} p_i^{j\prime} \left\langle \Psi_i^j \middle| \hat{\mathrm{T}}_e + V_{ee} \middle| \Psi_i^j \right\rangle \equiv \sum_i \sum_{j=0}^{\infty} p_i^{j\prime} F_i^j = \mathrm{tr}\,(\mathbf{d}'\mathbf{F}). \tag{7.74}$$

It is seen to depend only on the relative external potential $u(\boldsymbol{r}) = v(\boldsymbol{r}) - \mu$.

One can then easily demonstrate using the minimum principle of $\Omega_u(\mathbf{d}')$,

$$\Omega_u(\mathbf{d}') > \Omega_{ens.}[\mathbf{d}[\mu, T; v]] \equiv \Omega_u(\mathbf{d}), \quad \text{for} \quad \mathbf{d}' \neq \mathbf{d}, \tag{7.75}$$

that the optimum (diagonal) matrix **d** contains the grand canonical probabilities (7.49). Indeed, from the stationary character of $\Omega_u[\mathbf{d}']$ at $\mathbf{d}' = \mathbf{d}$, one finds:

$$\delta\Omega_u(\mathbf{d}') = \mathrm{tr}[(\delta\mathbf{d}')\boldsymbol{\Omega}(\mathbf{d}) + \beta^{-1}\delta\mathbf{d}'] = \mathrm{tr}[(\delta\mathbf{d}')\boldsymbol{\Omega}(\mathbf{d})] = 0, \tag{7.76}$$

where we have recognized that for the normalized trial probabilities, $\mathrm{tr}(\mathbf{d}') = 1$ and hence $\mathrm{tr}(\delta\mathbf{d}') = \delta\,\mathrm{tr}(\mathbf{d}') = 0$. One further observes that for arbitrary variations of trial probabilities around **d** this equation can be satisfied only when $\boldsymbol{\Omega}(\mathbf{d}) = \{\Omega\delta_{i,i'}\delta_{j,j'}\}$, since then:

$$\mathrm{tr}[(\delta\mathbf{d}')\boldsymbol{\Omega}(\mathbf{d})] = \Omega\,\mathrm{tr}(\delta\mathbf{d}') = 0. \tag{7.77}$$

Hence, by solving the Euler equation determining the equilibrium density operator,

$$\Omega_i^j = E_i^j - \mu N_i + \beta^{-1} \ln p_i^j = \Omega,$$

for the unknown p_i^j and determining Ω from the probability normalization condition, one finally arrives at the equilibrium ensemble probability summarized in (7.49) and (7.50).

Let us now examine the mapping between the ensemble average electron density [(7.56) and (7.69)] and the relative external potential $u(\boldsymbol{r}) = v(\boldsymbol{r}) - \mu$ [of (7.21) and (7.73)]: $u(\boldsymbol{r}) \leftrightarrow \rho_{ens.}(\boldsymbol{r})$. Following the previous (*reductio ad absurdum*) way of proving the existence of this mapping, we again consider two different external potentials, $u(\boldsymbol{r}) = v(\boldsymbol{r}) - \mu \neq u'(\boldsymbol{r}) = v'(\boldsymbol{r}) - \mu$, which correspond to electronic Hamiltonians,

$$\begin{aligned} \hat{\mathrm{H}} &= \hat{\mathrm{H}}^e(N_i, v) = \hat{\mathrm{H}}' + V_{ne}(N_i, v) - V_{ne}'(N_i, v') \quad \text{and} \\ \hat{\mathrm{H}}' &= \hat{\mathrm{H}}^e(N_i, v') = \hat{\mathrm{H}} + V_{ne}'(N_i, v') - V_{ne}(N_i, v), \end{aligned} \tag{7.78}$$

and generate the same ensemble average density of electrons:

$$\rho_{ens.} = \mathrm{tr}(\mathbf{d}\boldsymbol{\rho}) = \mathrm{tr}(\mathbf{d}'\boldsymbol{\rho}'). \tag{7.79}$$

From the variational principle (7.75) one then finds [compare (7.13)]:

$$\begin{aligned}\Omega_{u'}(\mathbf{d}') &= \mathrm{tr}(\mathbf{\Omega}'\mathbf{d}') = \mathrm{tr}[(\mathbf{\Omega} - \mathbf{V} + \mathbf{V}')\mathbf{d}') = \int [u'(\boldsymbol{r}) - u(\boldsymbol{r})]\rho_{ens.}(\boldsymbol{r})\,d\boldsymbol{r} + \Omega_u(\mathbf{d}') \\ &> \int [u'(\boldsymbol{r}) - u(\boldsymbol{r})]\rho_{ens.}(\boldsymbol{r})\,d\boldsymbol{r} + \Omega_u(\mathbf{d}),\end{aligned} \tag{7.80}$$

$$\begin{aligned}\Omega_u(\mathbf{d}) &= \mathrm{tr}(\mathbf{\Omega}\mathbf{d}) = \mathrm{tr}[(\mathbf{\Omega}' - \mathbf{V}' + \mathbf{V})\mathbf{d}) = \int [u(\boldsymbol{r}) - u'(\boldsymbol{r})]\rho_{ens.}(\boldsymbol{r})\,d\boldsymbol{r} + \Omega_{u'}(\mathbf{d}) \\ &> \int [u(\boldsymbol{r}) - u'(\boldsymbol{r})]\rho_{ens.}(\boldsymbol{r})\,d\boldsymbol{r} + \Omega_{u'}(\mathbf{d}').\end{aligned} \tag{7.81}$$

Summation of these inequalities again leads to contradiction,

$$\Omega_{u'}(\mathbf{d}') + \Omega_u(\mathbf{d}) > \Omega_u(\mathbf{d}) + \Omega_{u'}(\mathbf{d}'),$$

thus proving that different shapes of relative potentials give rise to different equilibrium densities, i.e., $\rho_{ens.}$ determines u uniquely. We also observe that in this ensemble mapping there is no ambiguity with respect to an additive constant in the external potential.

Therefore, the statistical operator can be regarded as functional of $\rho_{ens.}$, $\hat{\mathrm{d}} = \hat{\mathrm{d}}[\rho_{ens.}]$ or $\mathbf{d} = \mathbf{d}[\rho_{ens.}]$, giving rise to the associated density functional for the grand potential:

$$\Omega_u[\rho_{ens.}] = \mathrm{tr}\{\hat{\mathrm{d}}[\rho_{ens.}]\hat{\Omega}\} = \int [v(\boldsymbol{r}) - \mu]\rho_{ens.}(\boldsymbol{r})\,d\boldsymbol{r} + G[\rho_{ens.}], \tag{7.82}$$

where the universal functional

$$G[\rho_{ens.}] = \mathrm{tr}\left\{\hat{\mathrm{d}}[\rho_{ens.}]\left(\hat{\mathrm{F}} + \frac{1}{\beta}\ln\hat{\mathrm{d}}[\rho_{ens.}]\right)\right\} \equiv F[\rho_{ens.}] - TS[\rho_{ens.}], \tag{7.83}$$

and the functional $F[\rho_{ens.}] = \mathrm{tr}\{\hat{\mathrm{d}}[\rho_{ens.}]\hat{\mathrm{F}}\}$ constitutes the ensemble generalization of $F[\rho]$. The limitations of the v-representable densities implicit in these functionals can be again avoided by resorting to the constrained search construction of Lieb and Levy, universal at any given temperature,

$$G[\rho_{ens.}] = \inf_{\hat{\mathrm{d}}\to\rho_{ens.}} \mathrm{tr}\left\{\hat{\mathrm{d}}\left(\hat{\mathrm{F}} + \frac{1}{\beta}\ln\hat{\mathrm{d}}\right)\right\} \equiv F[\rho_{ens.}] - TS[\rho_{ens.}]. \tag{7.84}$$

We finally observe that the above density functional for the grand potential also defines the associated functional $A_v[\rho^c]$ for the *free energy* of Helmholtz, the thermodynamic potential of the *canonical* (c) ensemble consisting of the externally closed molecular systems in thermal equilibrium, characterized by the equilibrium probabilities of eigenstates $\{\Psi^j\}$ of $\hat{\mathrm{H}} \equiv \hat{\mathrm{H}}^e(N, v)$, $\hat{\mathrm{H}}\Psi^j \equiv E^j\Psi^j$,

$$p^j[N,T;v] = \frac{\exp(-\beta E^j[N,T;v])}{\Theta[N,T;v]}, \quad \sum_{j=0}^{\infty} p^j[N,T;v] = 1, \tag{7.85}$$

where the canonical partition function

$$\Theta[N,T;v] = \sum_{j=0}^{\infty} \exp(-\beta E^j[N,T;v]),$$

determining the associated canonical density operator

$$\hat{\mathrm{d}}[N,T;v] = \sum_{j=0}^{\infty} |\Psi^j\rangle p^j[N,T;v] \langle\Psi^j| \tag{7.86}$$

and the system equilibrium density

$$\rho_{ens.}[N,T;v] \equiv \rho^c = \sum_{j=0}^{\infty} p^j[N,T;v]\,\rho^j, \quad \rho^j(\boldsymbol{r}) = \langle\Psi^j|\hat{\rho}(\boldsymbol{r})|\Psi^j\rangle. \tag{7.87}$$

The free energy functional of the equilibrium electron density ρ^c then reads:

$$\begin{aligned} A_v[\rho^c] &= E_v[\rho^c] - TS[\rho^c] = \Omega_u[\rho^c] + \mu N[\rho^c] \\ &= \int v(\boldsymbol{r})\rho^c(\boldsymbol{r})\,d\boldsymbol{r} + (F[\rho^c] - TS[\rho^c]) \\ &= \int [v(\boldsymbol{r})\rho^c(\boldsymbol{r})\,d\boldsymbol{r} + G[\rho^c]. \end{aligned} \tag{7.88}$$

Of interest for the present development also is the multicomponent generalization of the HK theory, e.g., in the non-BO approach to molecular systems (Capitani et al. 1982). In complete analogy to the *one*-component theorems, one finds that in such systems the vector of the component densities $\{\rho_\alpha\}$ implies the existence of the associated density functional for the energy, $E[\{\rho_\alpha\}]$, which yields for the exact ground state densities the exact value of the system energy as its minimum. For example, when treating within the BO theory the electron distribution in molecules as consisting of two spin-components $\{\rho_\sigma, \sigma - \uparrow,\downarrow\}$, as in UHF approach, one uses the two-component energy functional: $E_v[\rho_\uparrow, \rho_\downarrow] = \int[\rho_\uparrow(\boldsymbol{r}) + \rho_\downarrow(\boldsymbol{r})]v(\boldsymbol{r})d\boldsymbol{r} + F[\rho_\uparrow, \rho_\downarrow]$.

7.2 Functionals from the Uniform Scaling of the Electron Density

As demonstrated by Szasz et al. (1975), some useful hints about a general form of density functionals for the kinetic and potential contributions to molecular electronic energy can be gained by considering their underlying overall homogeneities with respect to parameter s of the uniform scaling of electronic positions in the system ground state (normalized) wave function $\Psi(\{\boldsymbol{r}_i\}) = \Psi_{s=1}(\{s\boldsymbol{r}_i\}) \equiv \Psi_{s=1}(\{\boldsymbol{r}_i'\})$: $\int \cdots \int \Psi^*(\{\boldsymbol{r}_i\})\ \Psi(\{\boldsymbol{r}_i\})\ d\boldsymbol{r}_1 \cdots d\boldsymbol{r}_N = 1$. The appropriately renormalized scaled wave function for $s \neq 1$,

$$\begin{aligned}\Psi_s(\{\boldsymbol{r}_i'\}) &= s^{3N/2}\Psi(\{\boldsymbol{r}_i'\}),\\ \langle \Psi_s(\{\boldsymbol{r}_i'\})|\Psi_s(\{\boldsymbol{r}_i'\})\rangle_{\{\boldsymbol{r}_i\}} &= s^{3N}\int \cdots \int \Psi^*(\{\boldsymbol{r}_i'\})\ \Psi(\{\boldsymbol{r}_i'\})\ d\boldsymbol{r}_1 \cdots d\boldsymbol{r}_N\\ &= \int \cdots \int \Psi^*(\{\boldsymbol{r}_i'\})\ \Psi(\{\boldsymbol{r}_i'\})\ d\boldsymbol{r}_1' \cdots d\boldsymbol{r}_N' = 1, \end{aligned} \tag{7.89}$$

then determines the associated electron density:

$$\begin{aligned}\langle \Psi_s(\{\boldsymbol{r}_i'\})|\hat{\rho}(\boldsymbol{r})|\Psi_s(\{\boldsymbol{r}_i'\})\rangle_{\{\boldsymbol{r}_i\}} &= Ns^3 \int \cdots \int |\Psi(s\boldsymbol{r},\ \{\boldsymbol{r}_{j>1}'\})|^2\ d\boldsymbol{r}_2' \cdots d\boldsymbol{r}_N'\\ &= s^3\rho(s\boldsymbol{r}) = s^3\rho(\boldsymbol{r}') \equiv \rho_s(\boldsymbol{r}'). \end{aligned} \tag{7.90}$$

One also observes the trivial scaling relations for the kinetic and electron repulsion (potential) energy operators:

$$\hat{\mathrm{T}}_e(\{\boldsymbol{r}_i\}) = s^2\hat{\mathrm{T}}_e(\{\boldsymbol{r}_i'\}) \quad \text{and} \quad \hat{\mathrm{V}}_{ee}(\{\boldsymbol{r}_i\}) = s\hat{\mathrm{V}}_{ee}(\{\boldsymbol{r}_i'\}). \tag{7.91}$$

They determine the overall (virial) homogeneities of these two energy contributions for the exact ground state wave function:

$$\begin{aligned}T_e[\Psi_s] &= \langle \Psi_s(\{\boldsymbol{r}_i'\})|\hat{\mathrm{T}}_e(\{\boldsymbol{r}_i\})|\Psi_s(\{\boldsymbol{r}_i'\})\rangle_{\{\boldsymbol{r}_i\}} = s^2\langle \Psi(\{\boldsymbol{r}_i'\})|\hat{\mathrm{T}}_e(\{\boldsymbol{r}_i'\})|\Psi(\{\boldsymbol{r}_i'\})\rangle_{\{\boldsymbol{r}_i'\}}\\ &= s^2 T_e[\Psi_{s=1}], \end{aligned} \tag{7.92}$$

$$\begin{aligned}V_{ee}[\Psi_s] &= \langle \Psi_s(\{\boldsymbol{r}_i'\})|\hat{\mathrm{V}}_{ee}(\{\boldsymbol{r}_i\})|\Psi_s(\{\boldsymbol{r}_i'\})\rangle_{\{\boldsymbol{r}_i\}} = s\langle \Psi(\{\boldsymbol{r}_i'\})|\hat{\mathrm{V}}_{ee}(\{\boldsymbol{r}_i'\})|\Psi(\{\boldsymbol{r}_i'\})\rangle_{\{\boldsymbol{r}_i'\}}\\ &= sV_{ee}[\Psi_{s=1}]. \end{aligned} \tag{7.93}$$

Consider now the simplest form of density functionals for these energies, in the so-called *Local Density Approximation* (LDA), in which the densities of these energy contributions depend solely on the local value of the electronic density:

$$\begin{aligned}T_e[\rho] &\cong \int t_e^{\mathrm{LDA}}(\rho(\boldsymbol{r}))\,d\boldsymbol{r} = T_e^{\mathrm{LDA}}[\rho],\\ V_{ee}[\rho] &\equiv J_{ee}[\rho] + V_{xc}^{\mathrm{LDA}}[\rho], \qquad V_{xc}^{\mathrm{LDA}}[\rho] \cong \int v_{xc}^{\mathrm{LDA}}(\rho(\boldsymbol{r}))\,d\boldsymbol{r}, \end{aligned} \tag{7.94}$$

where the classical term

$$J_{ee}[\rho] = \frac{1}{2}\int\int \frac{\rho(\boldsymbol{r}_1)\rho(\boldsymbol{r}_2)}{|\boldsymbol{r}_1 - \boldsymbol{r}_2|} d\boldsymbol{r}_1 d\boldsymbol{r}_2, \tag{7.95}$$

and $V_{xc}^{\mathrm{LDA}}[\rho]$ stands for the remaining (nonclassical) exchange-correlation (xc) part of the electron repulsion (potential) energy. One observes that the former already exhibits the correct overall homogeneity of (7.93):

$$J_{ee}[\rho_s] = \frac{1}{2}\int\int \frac{\rho_s(\boldsymbol{r}_1)\rho_s(\boldsymbol{r}_2)}{|\boldsymbol{r}_1 - \boldsymbol{r}_2|} d\boldsymbol{r}_1 d\boldsymbol{r}_2 = \frac{1}{2}\int\int \frac{s^6\rho_s(\boldsymbol{r}_1{}')\rho_s(\boldsymbol{r}_2{}')}{s^{-1}|\boldsymbol{r}_1{}' - \boldsymbol{r}_2{}'|} \frac{d\boldsymbol{r}_1{}'}{s^3}\frac{d\boldsymbol{r}_2{}'}{s^3} = sJ_{ee}[\rho]. \tag{7.96}$$

For the kinetic energy functional, one similarly finds [see (7.92)]:

$$\begin{aligned} T_e^{\mathrm{LDA}}[\rho_s] &= \int t_e^{\mathrm{LDA}}\left(s^3\rho(s\boldsymbol{r})\right) d\boldsymbol{r} = \int t_e^{\mathrm{LDA}}(\rho_s(\boldsymbol{r}'))\,\frac{d\boldsymbol{r}'}{s^3} \\ &= s^2 T_e^{\mathrm{LDA}}[\rho] = s^2\int t_e^{\mathrm{LDA}}(\rho(\boldsymbol{r}'))\,d\boldsymbol{r}'. \end{aligned} \tag{7.97}$$

One thus predicts the following degree of homogeneity of the LDA functional for the kinetic energy, expressed in terms of the scale factor $\xi = s^3$,

$$t_e^{\mathrm{LDA}}(\xi\rho(\boldsymbol{r})) = \xi^{5/3} t_e^{\mathrm{LDA}}(\rho(\boldsymbol{r})) \quad \text{or} \quad T_e^{\mathrm{LDA}}[\rho] = A\int \rho^{5/3}(\boldsymbol{r})\,d\boldsymbol{r}. \tag{7.98}$$

The same line of reasoning and using (7.93) give for the nonclassical (xc) part of the electron repulsion energy,

$$\begin{aligned} V_{xc}^{\mathrm{LDA}}[\rho_s] &= \int v_{xc}^{\mathrm{LDA}}\left(s^3\rho(s\boldsymbol{r})\right) d\boldsymbol{r} = \int v_{xc}^{\mathrm{LDA}}\left(s^3\rho(\boldsymbol{r}')\right)\frac{d\boldsymbol{r}'}{s^3} \\ &= sV_{xc}^{\mathrm{LDA}}[\rho] = s\int v_{xc}^{\mathrm{LDA}}(\rho(\boldsymbol{r}'))\,d\boldsymbol{r}', \end{aligned} \tag{7.99}$$

and hence

$$v_{xc}^{\mathrm{LDA}}(\xi\rho(\boldsymbol{r})) = \xi^{4/3} v_{xc}^{\mathrm{LDA}}(\rho(\boldsymbol{r})) \quad \text{or} \quad V_{xc}^{\mathrm{LDA}}[\rho] = B\int \rho^{4/3}(\boldsymbol{r})\,d\boldsymbol{r}. \tag{7.100}$$

These functionals have indeed been used in the early density models of the TFD and HFS theories. More specifically, one recognizes in the preceding equations general forms of the Dirac–Slater exchange the TF functional for the kinetic energy.

The density gradient $\nabla\rho(\boldsymbol{r})$ or its Laplacian $\Delta\rho(\boldsymbol{r}) = \nabla^2\rho(\boldsymbol{r})$ explore the nonlocal (NL) aspects of the electron distribution, in the nearest neighborhood of the current monitoring location in space and so do the *higher* order gradient forms of the density. Therefore, in a more advanced NL approximation, called NLDA, the density of the energy functional is expressed as function of $\rho(\boldsymbol{r})$, $\nabla\rho(\boldsymbol{r})$, $\Delta\rho(\boldsymbol{r})$, $[\nabla\rho(\boldsymbol{r})]^4$, $[\nabla\rho(\boldsymbol{r})]^2\Delta\rho(\boldsymbol{r})$, $[\nabla^2\rho(\boldsymbol{r})]^2$, etc., or in terms of the associated gradient forms of the electron spin densities.

Let us now examine general analytical forms of densities of such integral functionals of kinetic and electron repulsion energies, defined by these generalized gradient terms. We again use the above implications of the uniform scaling of electronic positions for these contributions to the system electronic energy. We first observe the following transformations of the scaled density gradient and Laplacian:

$$\nabla_{\boldsymbol{r}}\rho_s(\boldsymbol{r}') = s^3\,\nabla_{\boldsymbol{r}}\rho(s\boldsymbol{r}) = s^4\,\nabla_{\boldsymbol{r}'}\rho(\boldsymbol{r}'),\quad \Delta_{\boldsymbol{r}}\rho_s(\boldsymbol{r}') = s^3\,\Delta_{\boldsymbol{r}}\rho(s\boldsymbol{r}) = s^5\,\Delta_{\boldsymbol{r}'}\rho(\boldsymbol{r}'). \tag{7.101}$$

The simplest form of the gradient correction to the density of the electronic kinetic energy is in the form $t_e^{\text{NLDA}}(\rho(\boldsymbol{r}), |\nabla\rho(\boldsymbol{r})|) = \rho^n(\boldsymbol{r})[\nabla\rho(\boldsymbol{r})]^2$, with the exponent n determined to satisfy the overall degree of homogeneity of (7.92),

$$\begin{aligned} T_e^{\text{NLDA}}[\rho_s] &= \int \rho_s^n(\boldsymbol{r}')[\nabla_{\boldsymbol{r}}\rho_s(\boldsymbol{r}')]^2 d\boldsymbol{r} = \int s^{3n}\rho^n(\boldsymbol{r}_1{}')s^8[\nabla_{\boldsymbol{r}'}\rho(\boldsymbol{r}_1{}')]^2\,\frac{d\boldsymbol{r}'}{s^3} \\ &= s^{3n+5}T_e^{\text{NLDA}}[\rho] = s^2 T_e^{\text{NLDA}}[\rho], \end{aligned} \tag{7.102}$$

and hence $n = -1$. The resulting functional is thus proportional to the von Weizsäcker (vW) (1935) nonhomegeneity correction term of the old density theory:

$$T_e^{\text{NLDA}}[\rho] = \int \frac{|\nabla\rho(\boldsymbol{r})|^2}{\rho(\boldsymbol{r})} d\boldsymbol{r} \propto T_{vW}[\rho]. \tag{7.103}$$

A similar assumption for the gradient correction of the xc-contribution to electron repulsion energy, $v_{xc}^{\text{NLDA}}(\rho(\boldsymbol{r}), |\nabla\rho(\boldsymbol{r})|) = \rho^m(\boldsymbol{r})[\nabla\rho(\boldsymbol{r})]^2$, gives:

$$\begin{aligned} V_{xc}^{\text{NLDA}}[\rho_s] &= \int \rho_s^m(\boldsymbol{r}')[\nabla_{\boldsymbol{r}}\rho_s(\boldsymbol{r}')]^2 d\boldsymbol{r} = \int s^{3m}\rho^m(\boldsymbol{r}_1{}')s^8[\nabla_{\boldsymbol{r}'}\rho(\boldsymbol{r}_1{}')]^2\,\frac{d\boldsymbol{r}'}{s^3} \\ &= s^{3m+5}V_{xc}^{\text{NLDA}}[\rho] = sV_{xc}^{\text{NLDA}}[\rho], \end{aligned} \tag{7.104}$$

and hence $m = -4/3$. This gives rise to the familiar gradient correction to the electron correlation energy:

$$V_{xc}^{\text{NLDA}}[\rho] = \int \frac{|\nabla\rho(\boldsymbol{r})|^2}{\rho^{4/3}(\boldsymbol{r})} d\boldsymbol{r}. \tag{7.105}$$

In the same way, one determines that the gradient corrections to the kinetic energy including $\Delta\rho(\boldsymbol{r})$ or one of the *fourth*-order differentiation terms,

$$[\nabla\rho(\boldsymbol{r})]^4,\quad [\nabla\rho(\boldsymbol{r})]^2\Delta\rho(\boldsymbol{r})\quad \text{or}\quad [\nabla^2\rho(\boldsymbol{r})]^2,$$

are accompanied by the following powers of the density:

$$\int \Delta\rho(\boldsymbol{r})d\boldsymbol{r} = 0, \int \frac{(\nabla\rho(\boldsymbol{r}))^4}{\rho^{11/3}(\boldsymbol{r})}d\boldsymbol{r}, \quad \int \frac{\Delta\rho(\boldsymbol{r})|\nabla\rho(\boldsymbol{r})|^2}{\rho^{8/3}(\boldsymbol{r})}d\boldsymbol{r}, \quad \int \frac{(\Delta\rho(\boldsymbol{r}))^2}{\rho^{5/3}(\boldsymbol{r})}d\boldsymbol{r}, \quad (7.106)$$

where the first identity is satisfied for any well-behaved $\rho(\boldsymbol{r})$. The associated correlation contributions to the electronic repulsion energy read:

$$\int \frac{\Delta\rho(\boldsymbol{r})}{\rho^{1/3}(\boldsymbol{r})}d\boldsymbol{r}, \int \frac{(\nabla\rho(\boldsymbol{r}))^4}{\rho^4(\boldsymbol{r})}d\boldsymbol{r}, \quad \int \frac{\Delta\rho(\boldsymbol{r})|\nabla\rho(\boldsymbol{r})|^2}{\rho^3(\boldsymbol{r})}d\boldsymbol{r}, \quad \int \frac{(\Delta\rho(\boldsymbol{r}))^2}{\rho^2(\boldsymbol{r})}d\boldsymbol{r}. \quad (7.107)$$

7.3 Kohn–Sham Theory

The modern computational DFT originates from the Kohn and Sham (1965) (KS) approach, which determines the density and energy of the real (*interacting*) N-electron system by solving the effective SCF equations for the optimum SO defining the Slater determinant of the hypothetical noninteracting system exhibiting the same electron distribution as the molecular system of interest. In this way, the complex *many*-body effects in the atomic, molecular, and solid-state systems are represented by the effective local potentials, which in principle include all effects due to the Fermi (exchange) and Coulomb electron correlations. The KS scheme uses the orbital approximation to construct the system electron density, and it introduces the crucial concept of the exchange correlation energy $E_{xc}[\rho]$, combining the correlation contributions to the system kinetic and electron repulsion energies, the exact form of which remains unknown. This functional is formally defined by separating the remaining, known contributions to the electronic energy. It satisfies the so-called *adiabatic* relation connecting the hypothetical and real systems (Sect. 7.3.2), which allows for an efficient modeling of this crucial quantity.

7.3.1 Orbital Approximation and Energy Expression

After Kohn and Sham (1965) most DFT calculations for molecular systems employ the exact (orbital dependent) functional for the kinetic energy of the noninteracting system described by the sum of the separable *one*-electron Hamiltonians (a.u.) involving the effective external potential $v_{\mathrm{KS}}(\boldsymbol{r})$ reflecting all correlation effects:

$$\hat{\mathrm{H}}_s(N) = \sum_{i=1}^{N} \hat{\mathrm{H}}_{\mathrm{KS}}(\boldsymbol{r}_i) = \hat{\mathrm{T}}_e(N) + V_s(N), \quad \hat{\mathrm{H}}_{\mathrm{KS}}(\boldsymbol{r}) = -\frac{1}{2}\nabla^2 + v_{\mathrm{KS}}(\boldsymbol{r}). \quad (7.108)$$

The link with the molecular system of the fully interacting electrons is realized through the requirement that this hypothetical (separable) system gives rise to the exact ground state density of the real (nonseparable) system: $\rho_s = \rho = \rho[N, v]$.

Therefore, by the first HK theorem, the effective *one*-body potential $v_{\mathrm{KS}}(\boldsymbol{r})$ is uniquely determined by the true electron density, $v_{\mathrm{KS}}(\boldsymbol{r}) = v_{\mathrm{KS}}[\rho; \boldsymbol{r}]$, and so is the exact ground state wave function of the noninteracting system:

$$\Psi_s[N, v] = \Psi_s[\rho] = \det\{\psi_n\} \equiv |\psi_1\ \psi_2\ \ldots\psi_N|. \tag{7.109}$$

It is exactly given by the KS determinant constructed from N orthonormal, singly occupied molecular SO, $\psi(\boldsymbol{q}) = \{\psi_n(\boldsymbol{q}) = \varphi_n(\boldsymbol{r})\zeta_n(\sigma)\}$, which correspond to N lowest eigenvalues $\{\varepsilon_n\}$ of the KS Hamiltonian $\hat{\mathrm{H}}_{\mathrm{KS}}(\boldsymbol{r})$. These effective *one*-electron Schrödinger equations, called KS equations, determine the optimum spatial parts of SO, i.e., KS MO, $\boldsymbol{\varphi}(\boldsymbol{r}) = \{\varphi_n(\boldsymbol{r}) \equiv \varphi_{n\sigma}(\boldsymbol{r})\}$:

$$\hat{\mathrm{H}}_{\mathrm{KS}}(\boldsymbol{r})\varphi_{n\sigma}(\boldsymbol{r}) = \varepsilon_{n\sigma}\varphi_{n\sigma}(\boldsymbol{r}), \quad n = 1, 2, \ldots, N, \tag{7.110}$$

where $\varphi_{n\sigma}$ stands for nth MO occupied by a single electron with spin orientation σ. The KS eigenvalues $\boldsymbol{\varepsilon} = \{\varepsilon_n\}$ represent the orbital energies of the noninteracting system and the row vector $\boldsymbol{\zeta}(\sigma) = \{\zeta_n(\sigma)\}$ groups the spin functions of the occupied SO, which depend upon the discrete spin variable σ of the two admissible spin orientations of an electron: $\sigma = (\uparrow, \downarrow)$.

Since the *one*-body Hamiltonian $\hat{\mathrm{H}}_{\mathrm{KS}}(\boldsymbol{r})$ of the above effective eigenvalue problem is uniquely specified by the ground state electron density ρ, its solutions are also functionals of this equilibrium distribution of electrons: $\boldsymbol{\varphi} = \boldsymbol{\varphi}[\rho]$ and $\boldsymbol{\varepsilon} = \boldsymbol{\varepsilon}[\rho]$. Notice that the electron density is straightforwardly generated by the sum of orbital densities $\boldsymbol{\rho} = \{\rho_n\}$,

$$\begin{aligned}\rho(\boldsymbol{r}) &= \sum_n |\varphi_n(\boldsymbol{r})|^2 \equiv \sum_n \rho_n(\boldsymbol{r}) \\ &= \sum_{\sigma=\uparrow,\downarrow}\left(\sum_n |\varphi_{n\sigma}(\boldsymbol{r})|^2\right) \equiv \sum_{\sigma=\uparrow,\downarrow}\left[\sum_n \rho_{n\sigma}(\boldsymbol{r})\right] \equiv \sum_{\sigma=\uparrow,\downarrow}\rho_\sigma(\boldsymbol{r}),\end{aligned} \tag{7.111}$$

where $\{\rho_\sigma(\boldsymbol{r})\}$ are the system *spin* densities. The exact kinetic energy of the noninteracting system is similarly given by the sum of the expectation values for the occupied KS MO:

$$T_s[\rho] = -\frac{1}{2}\sum_n \langle\varphi_n[\rho]|\nabla^2|\varphi_n[\rho]\rangle. \tag{7.112}$$

Therefore, in KS theory the density functionals for the electronic energies $E_s[\rho]$ and $E_v[\rho]$ of the noninteracting and real molecular systems, respectively, combine the following contributions:

$$E_s[\rho] = T_s[\rho] + \int \rho(\boldsymbol{r})v_{\mathrm{KS}}(\boldsymbol{r})\,d\boldsymbol{r} = \sum_n \varepsilon_n[\rho], \tag{7.113}$$

$$E_v[\rho] = T_s[\rho] + \int \rho(\boldsymbol{r})v(\boldsymbol{r})\,d\boldsymbol{r} + J_{ee}[\rho] + E_{xc}[\rho], \tag{7.114}$$

where the classical (Hartree) energy of the Coulomb repulsion between electrons has been already defined in (7.95),

$$J_{ee}[\rho] = \frac{1}{2}\iint \frac{\rho(\boldsymbol{r})\rho(\boldsymbol{r}')}{|\boldsymbol{r}-\boldsymbol{r}'|}\,d\boldsymbol{r}\,d\boldsymbol{r}', \tag{7.115}$$

and $E_{xc}[\rho]$ stands for the density functional generating the remaining part of the electronic energy in KS approximation, called the KS *exchange correlation energy*:

$$E_{xc}[\rho] = F[\rho] - T_s[\rho] - J_{ee}[\rho]. \tag{7.116}$$

The exact form of this functional, which now contains all electron correlation contributions of both the potential and kinetic origins, is not known exactly, but its reliable approximations, in both the density and/or orbital-dependent representations, e.g., the sophisticated density-gradient functionals, give remarkably good results for atoms, molecules and solid-state systems.

It should be stressed that the kinetic energy of the interacting electrons includes both the noninteracting (s) and correlation (c) contributions:

$$T[\rho] = T_s[\rho] + T_c[\rho]. \tag{7.117}$$

Therefore, the total electron repulsion energy of the interacting system is given by the sum of the Hartree and exchange correlation terms minus the correlation kinetic energy:

$$V_{ee}[\rho] = J_{ee}[\rho] + (E_{xc}[\rho] - T_c[\rho]) = J_{ee}[\rho] + V_{ee}^{xc}[\rho], \tag{7.118}$$

where $V_{ee}^{xc}[\rho]$ denotes the correlation part of the electron repulsion energy alone [see also (7.94)].

Since, by assumption, the densities of the interacting and noninteracting systems are identical, the chemical potential of (7.21) is equal to the functional derivative of both $E_v[\rho]$ and $E_s[\rho]$:

$$\begin{aligned}\mu[\rho] &= \frac{\delta E_v[\rho]}{\delta\rho(\boldsymbol{r})} = \frac{\delta T_s[\rho]}{\delta\rho(\boldsymbol{r})} + v(\boldsymbol{r}) + \frac{\delta J_{ee}[\rho]}{\delta\rho(\boldsymbol{r})} + \frac{\delta E_{xc}[\rho]}{\delta\rho(\boldsymbol{r})}\\ &= \frac{\delta E_s[\rho]}{\delta\rho(\boldsymbol{r})} = \frac{\delta T_s[\rho]}{\delta\rho(\boldsymbol{r})} + v_{\mathrm{KS}}(\boldsymbol{r}),\end{aligned} \tag{7.119a}$$

or:

$$\frac{\delta T_s[\rho]}{\delta\rho(\boldsymbol{r})} + [v_{\mathrm{KS}}(\boldsymbol{r}) - \mu] \equiv \frac{\delta T_s[\rho]}{\delta\rho(\boldsymbol{r})} + u_{\mathrm{KS}}(\boldsymbol{r}) = 0. \tag{7.119b}$$

Hence, the effective *one*-body potential of the hypothetical noninteracting system (see, e.g., van Leeuwen et al. 1996) includes the external potential due to the nuclei, $v(\boldsymbol{r})$, corrected by two electronic terms: the classical Hartree potential,

$$v_{\rm H}(\boldsymbol{r}) = \frac{\delta J_{ee}[\rho]}{\delta\rho(\boldsymbol{r})} = \int \frac{\rho(\boldsymbol{r}')}{|\boldsymbol{r}-\boldsymbol{r}'|}\,d\boldsymbol{r}' \tag{7.120}$$

and the exchange correlation potential,

$$v_{xc}(\boldsymbol{r}) = \frac{\delta E_{xc}[\rho]}{\delta\rho(\boldsymbol{r})}, \tag{7.121}$$

$$v_{\rm KS}(\boldsymbol{r}) = v(\boldsymbol{r}) + \int \frac{\rho(\boldsymbol{r}')}{|\boldsymbol{r}-\boldsymbol{r}'|}d\boldsymbol{r}' + \frac{\delta E_{xc}[\rho]}{\delta\rho(\boldsymbol{r})} \equiv v(\boldsymbol{r}) + v_{\rm H}(\boldsymbol{r}) + v_{xc}(\boldsymbol{r}). \tag{7.122a}$$

It should be also observed that the total energy of the real molecular system (7.114) can be also expressed in terms of the known KS eigenvalues, which determine the KS orbital (total) energy (7.113):

$$E_s[\rho] = \langle\Psi_s(N)|\hat{\rm H}_s(N)|\Psi_s(N)\rangle = T_s(N) + V_s(N) = \sum_n \varepsilon_n. \tag{7.123}$$

Here,

$$\begin{aligned} V_s(N) &= \langle\Psi_s(N)|\sum_{i=1}^{N} v_{\rm KS}(\boldsymbol{r}_i)|\Psi_s(N)\rangle = \int \rho(\boldsymbol{r})\,v_{\rm KS}(\boldsymbol{r})\,d\boldsymbol{r} \\ &= \int \rho(\boldsymbol{r})\left[v(\boldsymbol{r}) + \int \frac{\rho(\boldsymbol{r}')}{|\boldsymbol{r}-\boldsymbol{r}'|}d\boldsymbol{r}' + v_{xc}(\boldsymbol{r})\right] d\boldsymbol{r} \\ &= \int \rho(\boldsymbol{r})\,v(\boldsymbol{r})\,d\boldsymbol{r} + 2J_{ee}[\rho] + \int \rho(\boldsymbol{r})\,v_{xc}(\boldsymbol{r})\,d\boldsymbol{r}. \end{aligned} \tag{7.124}$$

Hence, by substituting $T_s(N) = E_s[\rho] - V_s(N)$ into (7.114), one obtains the alternative expression for the electronic energy of the interacting system:

$$E_v[\rho] = E_s[\rho] - J_{ee}[\rho] + E_{xc}[\rho] - \int \rho(\boldsymbol{r})v_{xc}(\boldsymbol{r})\,d\boldsymbol{r}. \tag{7.125}$$

The KS equations (7.110) have to be solved iteratively, since the KS effective potential is itself density dependent. Given the explicit functional $E_{xc}[\rho]$, one then calculates for the current variational density, generated by the initial KS orbitals, the effective *one*-body potential of the KS noninteracting system and by solving KS equations determines the next, better approximation to KS orbitals, etc. The resulting scheme, similar to that applied in H or HF eigenvalue problems, is easy to solve, particularly in the analytical representation of a finite basis set.

The computational effort in such calculations is comparable with that required by SCF MO scheme.

These equations determine both the optimum MO of the hypothetical noninteracting system and the electron density and energy of the interacting system. They can be alternatively derived from the associated variational principle for the system electronic energy, subject to the subsidiary conditions of the MO orthogonality and normalization, $\{\langle\varphi_m|\varphi_n\rangle = \delta_{m,n}\}$,

$$\delta\{E_v[\rho] - \sum_n \sum_m \theta_{m,n}\langle\varphi_m|\varphi_n\rangle\} = 0, \qquad (7.126)$$

where $\boldsymbol{\theta} = \{\theta_{m,n}\}$ stands for the matrix of the associated Lagrange multipliers. In the *canonical* representation, which defines the KS MO, this matrix becomes diagonal: $\theta_{m,n} = \varepsilon_n\delta_{m,n}$.

In the spin density variant of KS theory, analogous to the familiar UHF approach, there are separate effective potentials of (7.122b) for the spin-up and spin-down electrons, respectively, due to a dependence of the xc-energy on both spin densities $\{\rho_\sigma(\boldsymbol{r})\}$ of (7.111): $E_{xc} = E_{xc}[\{\rho_\sigma\}]$. In this spin-resolved treatment, one thus separately optimizes the subsets of KS spin orbitals $\{\varphi_{n\sigma}(\boldsymbol{r})\}$ describing the α ($\sigma = ½$) and β ($\sigma = -½$) electrons, which generate the associated spin densities. The corresponding effective (*one*-electron) KS equations for the optimum SO then result from the standard variational principle containing the Lagrange terms associated with the SO orthonormality constraints. They involve the spin-dependent effective potentials [compare (7.122a)]:

$$v_{\mathrm{KS}}^{\sigma}(\boldsymbol{r}) = v(\boldsymbol{r}) + v_{\mathrm{H}}(\boldsymbol{r}) + \frac{\delta E_{xc}[\rho^{\uparrow}, \rho^{\downarrow}]}{\delta\rho^{\sigma}(\boldsymbol{r})}, \quad \sigma = \uparrow(\alpha), \downarrow(\beta). \qquad (7.122b)$$

The spin-polarized DFT builds more physics into approximate models of the exchange correlation functional: it allows treatment of atomic/molecular systems in presence of the magnetic field and it generates the "unrestricted" KS MO, in spirit of UHF, thus correctly accounting for the bond-breaking situations.

7.3.2 Adiabatic Connection

The key functionals $E_{xc}[\rho]$ or $E_{xc}[\{\rho_\sigma\}]$ can be formally expressed in terms of the correlation holes of Sect. 6.2.3 of hypothetical systems exhibiting the exact density of the real system and only a fractional electron repulsion (Harris and Jones 1974; Langreth and Perdew 1975; Gunnarson and Lundqvist 1976). By convention, the strength of the electron repulsion term in such "*scaled*" electronic Hamiltonians is controlled by the so-called *coupling constant* $0 \le \lambda \le 1$, and the external potential is appropriately modified to fulfil the condition of the conserved density:

$$\hat{\mathrm{H}}^{\lambda}(N) = \sum_{i=1}^{N} v^{\lambda}(\boldsymbol{r}_i) + [\hat{\mathrm{T}}_e(N) + \lambda\hat{\mathrm{V}}_{ee}(N)] \equiv \hat{\mathrm{V}}^{\lambda}_{ne}(N) + \hat{\mathrm{F}}^{\lambda}(N),$$

$$\hat{\mathrm{H}}^{\lambda}(N)\,\Psi^{\lambda}(N) = E^{\lambda}(N)\,\Psi^{\lambda}(N),$$
$$\rho^{\lambda}(\boldsymbol{r}) = \langle\Psi^{\lambda}|\hat{\rho}(\boldsymbol{r})|\Psi^{\lambda}\rangle = \rho^{\lambda=1}(\boldsymbol{r}) \equiv \rho(\boldsymbol{r}); \tag{7.128}$$

here, $E^{\lambda}(N) = V^{\lambda}_{ne}(N) + F^{\lambda}(N)$ stands for the expectation value of the Hamiltonian $\hat{\mathrm{H}}^{\lambda}(N)$ for the current interaction strength λ. In this *adiabatic connection* $v^{\lambda=0}(\boldsymbol{r}) = v_{\mathrm{KS}}(\boldsymbol{r})$ corresponds to the noninteracting limit of KS theory and the BO external potential $v^{\lambda=1}(\boldsymbol{r}) = v(\boldsymbol{r})$ identifies the real (fully interacting) molecular system.

The scaled operator $\hat{\mathrm{F}}^{\lambda}(N)$ of the preceding equation is defined by the constrained-search of Levy [see Eq. (7.29)]:

$$F^{\lambda}[\rho] = \inf_{\Psi^{\lambda}(N)\to\rho} \langle\Psi^{\lambda}(N)|\hat{\mathrm{F}}^{\lambda}(N)|\Psi^{\lambda}(N)\rangle \equiv \langle\Psi^{\lambda}[\rho]|\hat{\mathrm{F}}^{\lambda}(N)|\Psi^{\lambda}[\rho]\rangle. \tag{7.129}$$

Its expectation values in the KS and real-system limits then read:

$$F^{\lambda=0}[\rho] = \langle\Psi_s[\rho]|\hat{\mathrm{T}}_e(N)|\Psi_s[\rho]\rangle = T_s[\rho],$$
$$F^{\lambda=1}[\rho] = \langle\Psi[\rho]|\hat{\mathrm{T}}_e(N) + \hat{\mathrm{V}}_{ee}(N)|\Psi[\rho]\rangle = F[\rho] = T_s[\rho] + J_{ee}[\rho] + E_{xc}[\rho], \tag{7.130}$$

where $\Psi_s[\rho]$ denotes the KS determinant of (7.109) and $\Psi[\rho] = \Psi[N, v]$ stands for the exact (correlated) ground state wave function of the interacting system.

Hence, the difference of functionals defined in the preceding equation can be formally expressed as the following integral over the coupling constant:

$$F^{\lambda=1}[\rho] - F^{\lambda=0}[\rho] = J_{ee}[\rho] + E_{xc}[\rho] = \int_0^1 \frac{\partial F^{\lambda}[\rho]}{\partial\lambda}\,d\lambda. \tag{7.131}$$

It then follows from the Hellmann–Feynman theorem that

$$\frac{\partial E^{\lambda}(N)}{\partial\lambda} = \frac{\partial V^{\lambda}_{ne}(N)}{\partial\lambda} + \frac{\partial F^{\lambda}(N)}{\partial\lambda} = \langle\Psi^{\lambda}[\rho]|\frac{\partial\hat{\mathrm{H}}^{\lambda}(N)}{\partial\lambda}|\Psi^{\lambda}[\rho]\rangle$$
$$= \langle\Psi^{\lambda}[\rho]|\frac{\partial\hat{\mathrm{V}}^{\lambda}_{ne}(N)}{\partial\lambda} + \hat{\mathrm{V}}_{ee}(N)|\Psi^{\lambda}[\rho]\rangle = \int \rho(\boldsymbol{r})\frac{\partial v^{\lambda}(\boldsymbol{r})}{\partial\lambda}d\boldsymbol{r} + V_{ee}{}^{\lambda}[\rho], \tag{7.132}$$

and hence

$$\int_0^1 \frac{\partial F^\lambda[\rho]}{\partial\lambda}\,d\lambda = \int_0^1 \left[\frac{\partial E^\lambda(N)}{\partial\lambda} - \frac{\partial V_{ne}^\lambda(N)}{\partial\lambda}\right] d\lambda = \int_0^1 V_{ee}^\lambda[\rho]\,d\lambda. \tag{7.133}$$

The electron repulsion $V_{ee}^\lambda[\rho]$ in the scaled molecular state Ψ^λ can be further expressed in terms of the associated *two*-electron distribution (6.82):

$$\rho_2^\lambda(\boldsymbol{r},\boldsymbol{r}') = \langle\Psi^\lambda|\hat{\rho}_2(\boldsymbol{r},\boldsymbol{r}')|\Psi^\lambda\rangle \equiv \rho(\boldsymbol{r})[\rho(\boldsymbol{r}') + h_{xc}^\lambda(\boldsymbol{r}'|\boldsymbol{r})], \tag{7.134}$$

$$V_{ee}^\lambda[\rho] = \langle\Psi^\lambda|\hat{\mathrm{V}}_{ee}|\Psi^\lambda\rangle = \frac{1}{2}\iint \rho_2^\lambda(\boldsymbol{r},\boldsymbol{r}')\frac{1}{|\boldsymbol{r}-\boldsymbol{r}'|}d\boldsymbol{r}\,d\boldsymbol{r}', \tag{7.135}$$

with $h_{xc}^\lambda(\boldsymbol{r}'|\boldsymbol{r})$ representing the corresponding (resultant) correlation hole of Sect. 6.2.3 for the current interaction strength λ.

It thus follows from (7.131) and (7.133) that the unknown exchange correlation functional is given by the integral over the electron repulsion energies $V_{ee}^\lambda[\rho]$ for the current coupling constant λ:

$$E_{xc}[\rho] = \int_0^1 V_{ee}^\lambda[\rho]\,d\lambda - J_{ee}[\rho]. \tag{7.136}$$

One finally arrives at the following expression for the $E_{xc}[\rho]$ functional of the KS theory in terms of correlation hole $h_{xc}^\lambda(\boldsymbol{r}'|\boldsymbol{r}) = [\rho_2^\lambda(\boldsymbol{r},\boldsymbol{r}')/\rho(\boldsymbol{r})] - \rho(\boldsymbol{r}')$:

$$E_{xc}[\rho] = \frac{1}{2}\iint \rho(\boldsymbol{r})\left[\int_0^1 h_{xc}^\lambda(\boldsymbol{r}'|\boldsymbol{r})\,d\lambda\right]\frac{1}{|\boldsymbol{r}-\boldsymbol{r}'|}d\boldsymbol{r}\,d\boldsymbol{r}' \equiv \frac{1}{2}\iint \rho(\boldsymbol{r})h_{xc}^{av}(\boldsymbol{r}'|\boldsymbol{r})\frac{1}{|\boldsymbol{r}-\boldsymbol{r}'|}d\boldsymbol{r}\,d\boldsymbol{r}'$$
$$\equiv \int_0^1 W_\lambda[\rho]\,d\lambda. \tag{7.137}$$

Therefore, this functional represents the classical interaction between the electron density $\rho(\boldsymbol{r})$ and the distribution $h_{xc}^{av}(\boldsymbol{r}'|\boldsymbol{r})$ of the coupling-constant averaged hole. This expression is fundamental for both the formal theory and modeling of the $E_{xc}[\rho]$ functional in DFT. A similar development for molecular subsystems have also been reported (Nalewajski 2001).

Let us examine more closely the λ-dependent contributions defined in the preceding equation:

$$W_\lambda[\rho] = \frac{1}{2}\iint \rho(\boldsymbol{r})h_{xc}^\lambda(\boldsymbol{r}'|\boldsymbol{r})\frac{1}{|\boldsymbol{r}-\boldsymbol{r}'|}d\boldsymbol{r}\,d\boldsymbol{r}'. \tag{7.138}$$

For $\lambda = 0$, i.e., the noninteracting KS system [see (6.76)],

$$\begin{aligned} h_{xc}^{\lambda=0}(\mathbf{r}'|\mathbf{r}) &= h_x(\mathbf{r}'|\mathbf{r}) = -\frac{1}{\rho(\mathbf{r})}\sum_{\sigma=\uparrow,\downarrow}\sum_{m,n}^{N_\sigma}\varphi_{n\sigma}(\mathbf{r})\varphi_{m\sigma}^*(\mathbf{r})\varphi_{m\sigma}(\mathbf{r}')\varphi_{n\sigma}^*(\mathbf{r}'), \\ &\int h_x(\mathbf{r}'|\mathbf{r})d\mathbf{r}' = -1, \end{aligned} \tag{7.139}$$

since the Coulomb correlation is then completely turned off. Therefore, as in the HF theory,

$$\begin{aligned} W_{\lambda=0}[\rho] &= \frac{1}{2}\iint \rho(\mathbf{r})h_x(\mathbf{r}'|\mathbf{r})\frac{1}{|\mathbf{r}-\mathbf{r}'|}d\mathbf{r}\,d\mathbf{r}' \equiv E_x[\rho] \\ &= -\frac{1}{4}\iint \frac{\left|\hat{\rho}_1^{\mathrm{KS}}(\mathbf{r},\mathbf{r}')\right|^2}{|\mathbf{r}-\mathbf{r}'|}d\mathbf{r}\,d\mathbf{r}', \end{aligned} \tag{7.140}$$

where $\hat{\rho}_1^{\mathrm{KS}}(\mathbf{r},\mathbf{r}')$ stands for corresponding (spinless) reduced density matrix (see Sect. 6.3.3).

It should be emphasized that the correlation contribution of (7.138) is determined solely by the spherically averaged hole of (6.87), over all the spherical angles $\Omega_{\mathbf{u}}$ of $\mathbf{u} = \mathbf{r}' - \mathbf{r} = (u, \Omega_{\mathbf{u}})$:

$$h_{xc}^{\lambda}(u|\mathbf{r}) = \frac{1}{4\pi}\int h_{xc}^{\lambda}(\mathbf{r}+\mathbf{u}|\mathbf{r})d\Omega_{\mathbf{u}}, \tag{7.141}$$

$$\begin{aligned} W_\lambda[\rho] &= \frac{1}{2}\iint \rho(\mathbf{r})h_{xc}^{\lambda}(\mathbf{r}'|\mathbf{r})\frac{1}{|\mathbf{r}-\mathbf{r}'|}d\mathbf{r}\,d\mathbf{r}' \\ &= \int \rho(\mathbf{r})\left(2\pi\int_0^\infty u\,h_{xc}^{\lambda}(u|\mathbf{r})\,du\right)d\mathbf{r} \equiv \int \rho(\mathbf{r})\,\varepsilon_{xc}^{\lambda}(\mathbf{r})\,d\mathbf{r}, \end{aligned} \tag{7.142}$$

where we have introduced the functional density per electron:

$$\varepsilon_{xc}^{\lambda}(\mathbf{r}) = 2\pi\int_0^\infty u\,h_{xc}^{\lambda}(u|\mathbf{r})\,du. \tag{7.143}$$

By explicitly separating the Coulomb hole from the resultant correlation hole,

$$h_c^{\lambda}(\mathbf{r}'|\mathbf{r}) = h_{xc}^{\lambda}(\mathbf{r}'|\mathbf{r}) - h_x(\mathbf{r}'|\mathbf{r}), \qquad \int h_c^{\lambda}(\mathbf{r}'|\mathbf{r})d\mathbf{r}' = 0, \tag{7.144}$$

one obtains the associated expression for the Coulomb correlation functional in KS theory in terms of the average hole:

$$\begin{aligned} E_c[\rho] = E_{xc}[\rho] - E_x[\rho] &= \frac{1}{2}\iint \rho(\mathbf{r})\frac{h_{xc}^{av}(\mathbf{r}'|\mathbf{r}) - h_x(\mathbf{r}'|\mathbf{r})}{|\mathbf{r}'-\mathbf{r}|}d\mathbf{r}'d\mathbf{r} \\ &= \frac{1}{2}\iint \rho(\mathbf{r})\frac{h_c^{av}(\mathbf{r}'|\mathbf{r})}{|\mathbf{r}'-\mathbf{r}|}d\mathbf{r}'d\mathbf{r}, \end{aligned} \tag{7.145}$$

where:

$$h_c^{av}(\mathbf{r}'|\mathbf{r}) = \int_0^1 h_c^\lambda(\mathbf{r}'|\mathbf{r})\,d\lambda. \tag{7.146}$$

A similar partitioning of the *xc*-energy and the underlying correlation holes applies in the spin-resolved KS theory:

$$E_{xc}[\rho_\alpha,\rho_\beta] = E_x[\rho_\alpha,\rho_\beta] + E_c[\rho_\alpha,\rho_\beta].$$

We conclude this section with a short overview of several exact properties of these density functionals. Some of them reflect the homogeneities under uniform scaling of the electron density:

$$\rho(\mathbf{r}) \to s^3\rho(s\mathbf{r}) = s^3\rho(\mathbf{r}') \equiv \rho_s(\mathbf{r}'). \tag{7.147}$$

As observed in Sect. 7.2, this transformation of electronic coordinates, $\mathbf{r} \to s\mathbf{r} \equiv \mathbf{r}'$, preserves the normalization of the scaled wave function and the associated density,

$$\int \rho_s(\mathbf{r}')\,\frac{d\mathbf{r}'}{s^3} = \int \rho(\mathbf{r}')\,d\mathbf{r}' = N, \tag{7.148}$$

while symmetrically contracting ($s > 1$) or expanding ($s < 1$) the electron distribution.

The *kinetic* energy of the noninteracting system has already been diagnosed as the homogeneous function of degree 2 in the scaling factor $s = \lambda$ [(7.92) and (7.97)],

$$T_s[\rho_s] = s^2 T_s[\rho], \tag{7.149}$$

while the exchange (*potential*) energy scales linearly [(7.93) and (7.96)]:

$$E_x[\rho_s] = sE_x[\rho]. \tag{7.150}$$

The KS correlation energy term, which combines the kinetic and potential energy contributions, exhibits a more complicated scaling law (Levy 1987; Perdew 1999):

$$E_c[\rho_s] = s^2 E_c^{1/s}[\rho], \tag{7.151}$$

where [see (7.140)]

$$E_c^{\alpha}[\rho] = \int_0^{\alpha} \{W_\lambda[\rho] - W_0[\rho]\}\,d\lambda = \int_0^{\alpha} W_\lambda[\rho]\,d\lambda - E_x[\rho] = E_{xc}^{\alpha}[\rho] - E_x[\rho] \tag{7.152}$$

stands for the correlation energy for the coupling constant $\alpha = 1/s$.

Therefore, in the *low*-density limit, for $s \to 0$ ($\alpha \to \infty$), where the integrand $W_\lambda[\rho] - W_0[\rho] \equiv \Delta W_\lambda[\rho] \approx \Delta W_\infty[\rho]$ tends to a constant $\Delta W_\infty[\rho]$, independent of the coupling strength λ, the correlation energy scales like the exchange (potential) term (Levy and Perdew 1993, 1997):

$$\lim_{s\to 0} E_c[\rho_s] \approx s^2[\Delta W_\infty[\rho]/s] = s\Delta W_\infty[\rho]. \tag{7.153}$$

Accordingly, in the *high*-density limit, for $s \to \infty$ ($\alpha \to 0$), by the *second*-order PT (Görling and Levy 1993, 1994), ΔW_λ varies like λ as $\lambda \to 0$,

$$\lim_{s\to\infty} E_c[\rho_s] \approx constant. \tag{7.154}$$

This limit is typically observed by electron densities of atoms and molecules, but it does not apply to electron gas of the uniform density, where the *second*-order PT fails (Gell-Mann and Brueckner 1957).

7.3.3 Kohn–Sham–Mermin Theory

We shall now generalize the KS approach to the finite temperature ensembles. We recall that KS method involves the partition of the universal functional, $F[\rho] = T_s[\rho] + J_{ee}[\rho] + E_{xc}[\rho]$, where the first two (dominating) contributions are known, and the adoption of the orbital approximation to generate the system density. We follow these two crucial stages in the following ensemble generalization, called the Kohn–Sham–Mermin (KSM) theory (Perdew 1985).

Let us now decompose the universal part $G[\rho_{ens.}]$ of the grand-potential functional,

$$\begin{aligned}\Omega_u[\rho_{ens.}] &= \int [v(\boldsymbol{r}) - \mu]\rho_{ens.}(\boldsymbol{r})d\boldsymbol{r} + (F[\rho_{ens.}] - TS[\rho_{ens.}])\\ &\equiv \int u(\boldsymbol{r})\rho_{ens.}(\boldsymbol{r})d\boldsymbol{r} + G[\rho_{ens.}],\end{aligned} \tag{7.155}$$

representing the "free" functional $F[\rho_{ens.}]$ [(7.82)–(7.84)] in the Hohenberg–Kohn–Mermin (HKM) theory (see Sect. 7.1.4), into the classical repulsion energy $J_{ee}[\rho_{ens.}]$ for the ensemble average electron distribution $\rho_{ens.}$ (7.69), the free energy of the fictitious noninteracting system exhibiting the same density $\rho_{ens.}(\boldsymbol{r})$,

$$G_s[\rho_{ens.}] = \min_{\hat{\mathrm{d}}\to\rho_{ens.}} \{\mathrm{tr}[\hat{\mathrm{d}}(\hat{\mathrm{T}}_e - T\hat{\mathrm{S}})]\} = T_s[\rho_{ens.}] - TS_s[\rho_{ens.}], \tag{7.156}$$

and the remaining part defining the so called xc-free–energy $F_{xc}[\rho_{ens.}]$:

$$G[\rho_{ens.}] = G_s[\rho_{ens.}] + J_{ee}[\rho_{ens.}] + F_{xc}[\rho_{ens.}]. \tag{7.157}$$

The Euler equation for the equilibrium ensemble density $\rho_{ens.} = \rho_{eq.}[\mu, T; v] \equiv \rho_{eq}$ is then determined by the minimum of the grand potential $\Omega_u[\rho_{ens.}]$:

$$\begin{aligned}\left.\frac{\delta\Omega_u[\rho_{ens.}]}{\delta\rho_{ens.}(\boldsymbol{r})}\right|_{\rho_{eq.}} &= 0 = u(\boldsymbol{r}) + \left.\frac{\delta G[\rho_{ens.}]}{\delta\rho_{ens.}(\boldsymbol{r})}\right|_{\rho_{eq.}} \\ &= (v(\boldsymbol{r}) - \mu) + \left(v_{\mathrm{H}}[\rho_{ens.}; \boldsymbol{r}] + \left.\frac{\delta G_s[\rho_{ens.}]}{\delta\rho_{ens.}(\boldsymbol{r})}\right|_{\rho_{eq.}} + \left.\frac{\delta F_{xc}[\rho_{ens.}]}{\delta\rho_{ens.}(\boldsymbol{r})}\right|_{\rho_{eq.}}\right) \quad \text{or}\end{aligned} \tag{7.158}$$

$$\begin{aligned}&\left.\frac{\delta G_s[\rho_{ens.}]}{\delta\rho_{ens.}(\boldsymbol{r})}\right|_{\rho_{eq.}} + \left(v(\boldsymbol{r}) + v_{\mathrm{H}}[\rho_{eq.}[\mu, T; v]; \boldsymbol{r}] + \left.\frac{\delta F_{xc}[\rho_{ens.}]}{\delta\rho_{ens.}(\boldsymbol{r})}\right|_{\rho_{eq.}}\right) - \mu \\ &\quad\equiv \left.\frac{\delta G_s[\rho_{ens.}]}{\delta\rho_{ens.}(\boldsymbol{r})}\right|_{\rho_{eq.}} + \left(v_{eff.}[\rho_{eq.}; \boldsymbol{r}] - \mu\right) \\ &\quad\equiv \left.\frac{\delta G_s[\rho_{ens.}]}{\delta\rho_{ens.}(\boldsymbol{r})}\right|_{\rho_{eq.}} + u_{eff.}[\rho_{eq.}; \boldsymbol{r}] = 0.\end{aligned} \tag{7.159}$$

This Euler equation in the KSM theory for the statistical ensemble at finite temperature complements (7.119a, b) of the original, *zero*-temperature KS theory. It corresponds to the hypothetical system of noninteracting electrons exhibiting the chemical potential μ and temperature T, moving in the effective external potential $v_{eff.}[\rho_{ens.}; \boldsymbol{r}]$.

In order to derive the associated *one*-electron Schrödinger equations [see (7.110)], one again adopts the orbital representation of the ensemble average density:

$$\rho_{ens.}(\boldsymbol{r}) = \sum_i d_i |\phi_i(\boldsymbol{r})|^2, \tag{7.160}$$

with the Slater determinant $\det\{\varphi_i\}$, constructed from the set of the occupied SO $\{\varphi_i(\boldsymbol{q}) = \phi_i(\boldsymbol{r})\zeta_i(\sigma)\}$, describing the fictitious system of noninteracting electrons. The vector $\boldsymbol{d}[\mu, T] = \{d_i = d_i[\mu, T]\}$ groups the equilibrium (fermion) occupations of these ensemble SO of the KSM theory in the prescribed thermodynamical conditions $[\mu, T]$:

$$d_i[\mu, T] = \{\exp[\beta(\varepsilon_i - \mu)] + 1\}^{-1}, \quad i = 1, 2, \ldots, \tag{7.161}$$

with ε_i standing for the orbital energy of φ_i. The ensemble average density of (7.160) is thus determined by both the shapes of orbitals and their effective equilibrium occupations, with the latter being explicitly dependent on the orbital energies $\{\varepsilon_i\}$. From the Euler equations (7.159), one then arrives at the associated Schrödinger equations determining the optimum (orthonormal) KSM orbitals.

One first observes that the noninteracting part of the free energy is readily calculated for the assumed form of the ensemble average electron density:

$$\begin{aligned} G_s[\rho_{ens.}] &= T_s[\rho_{ens.}] - TS_s[\rho_{ens.}], \\ T_s[\rho_{ens.}] &= \sum_i d_i \langle \phi_i | -\frac{1}{2}\Delta | \phi_i \rangle, \\ S_s[\rho_{ens.}] &= -k_B \sum_i \{ d_i \ln d_i + (1-d_i)\ln(1-d_i) \}, \end{aligned} \tag{7.162}$$

and hence:

$$d_i^{-1}\delta G_s/\delta\phi_i^*(\boldsymbol{r}) = d_i^{-1}\delta T_s/\delta\phi_i^*(\boldsymbol{r}) = -\frac{1}{2}\Delta\phi_i(\boldsymbol{r}). \tag{7.163}$$

The effective Schrödinger equations determining the optimum *canonical* KSM orbitals $\{\phi_i\}$ are obtained by supplementing the grand potential with the relevant constraints of the orbital normalization:

$$\delta\{\Omega_u[\rho_{ens.}[\{\phi_i\}]] - \sum_i \varepsilon_i \langle \phi_i | \phi_i \rangle\} \equiv \delta\Theta[\{\phi_i\}] = 0. \tag{7.164}$$

By functionally differentiating this subsidiary functional, $d_i^{-1}\delta\Theta[\{\phi_i\}]/\delta\phi_i^*(\boldsymbol{r})$, and using (7.159) and (7.163), one finally arrives at the associated *one*-particle SE:

$$\begin{aligned} &\left\{-\frac{1}{2}\Delta + u_{eff}[\rho_{ens.};\boldsymbol{r}]\right\}\phi_i(\boldsymbol{r}) = (\varepsilon_i - \mu)\phi_i(\boldsymbol{r}) \quad \text{or} \\ &\left\{-\frac{1}{2}\Delta + v_{eff}[\rho_{ens.};\boldsymbol{r}]\right\}\phi_i(\boldsymbol{r}) = \varepsilon_i\phi_i(\boldsymbol{r}), \quad i = 1, 2, \ldots \end{aligned} \tag{7.165}$$

These *one*-electron Schrödinger equations have to be solved iteratively, since the effective potential depends upon the ensemble average density itself, until the self-consistent orbitals and their occupations are reached. The optimum solutions finally determine the equilibrium ensemble average electron density of the molecular system under the specified thermodynamic conditions.

It should be emphasized that in the grand ensemble, which represents the open system at finite temperature T, the instantaneous number of electrons N exhibits fluctuations relative to the average value $N_{ens.}$, $\delta N = N - N_{ens.}$, due to the instantaneous flows of electrons between the molecule and its hypothetical reservoir. The ensemble average number can thus assume both the fractional and integral values. In the following sections, we shall briefly examine the properties of the ensemble average energy as function of $N_{ens.}$, $E_{ens.}(N_{ens.})$, and its derivatives in the $T \to 0$ limit.

7.3.4 Donor–Acceptor Complexes

As we have already emphasized in Sect. 7.1.2, the first derivative of the ensemble average energy function $E_{ens.}(N_{ens.})$ of an open system,

$$\mu_{ens.}(N_{ens.}) = \partial E_{ens.}(N_{ens.})/\partial N_{ens.}, \tag{7.166}$$

determines the system chemical potential of (7.18) and (7.19), the equilibrium level of the system local chemical potentials of (7.21) and (7.22). In the ground state ensemble, for $T = 0$, it can be estimated using the finite difference approach from the known energies (E_X, E_{X^+}, E_{X^-}) corresponding to the neutral system X(N_0), where $N_0 = \sum_\alpha Z_\alpha$ (integer), its cation $X^+(N_0 - 1)$, and anion $X^-(N_0 + 1)$, respectively (see Fig. 7.1).

When the atomic or molecular system X acts as the Lewis acid (electron acceptor) relative to the (infinite, macroscopic) reservoir $N_{ens.} > N_0$, thus exhibiting on average the negative net charge $\delta N_X = N_{ens.} - N_0 > 0$, the energy slope can be estimated by taking the finite displacement $\Delta N_X = 1$,

$$\mu_X^{(-)} \approx \Delta E_X(\Delta N_X)/\Delta N_X = (E_{X^-} - E_X)/1 \equiv -A_X, \tag{7.167}$$

where A_X denotes the electron affinity of X.

Accordingly, when X acts as the Lewis basis (electron donor) relative to the reservoir $N_{ens.} < N_0$, thus exhibiting on average the positive net charge $\delta N_X = N_{ens.} - N_0 < 0$, which can be represented by the finite displacement $\Delta N_X = -1$,

$$\mu_X^{(+)} \approx \Delta E_X(\Delta N_X)/\Delta N_X = (E_{X^+} - E_X)/(-1) \equiv -I_X, \tag{7.168}$$

with I_X standing for the (first) ionization potential of X.

In accordance with the intuitive electronegativity estimates [Mulliken (M) 1934; Iczkowski and Margrave 1961] for the neutral system itself, i.e., for $N_{ens.} = N_0$, one takes the arithmetic average of these two *finite* difference estimates,

$$\mu_M = \frac{\mu_X^{(-)} + \mu_X^{(+)}}{2} = -\frac{(I_X + A_X)}{2} = \left.\frac{\partial E^{(2)}(N_{ens.})}{\partial N_{ens.}}\right|_{N_0}. \tag{7.169}$$

It represents (see Figs. 7.1 and 7.2) the derivative of the parabola $E^{(2)}(N_{ens.})$ passing through the points (E_X, E_{X^+}, E_{X^-}). Mulliken's ("radical") estimate $\mu_X \approx \mu_M$ should be applied as the unbiased measure of the N-gradient of the system average energy, when we have no advance information about the donor/acceptor character of the system under consideration relative to its molecular environment.

The Mulliken parabolic interpolation of the energy displacements relative to E_X (see Fig. 7.2),

$$\Delta E^{(2)}(\Delta N) = \mu_M \Delta N + \frac{1}{2}\eta_M(\Delta N)^2, \quad \Delta N = N - N_0, \tag{7.170}$$

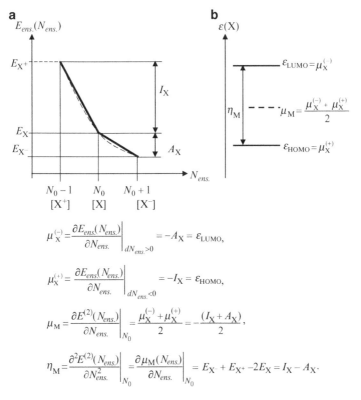

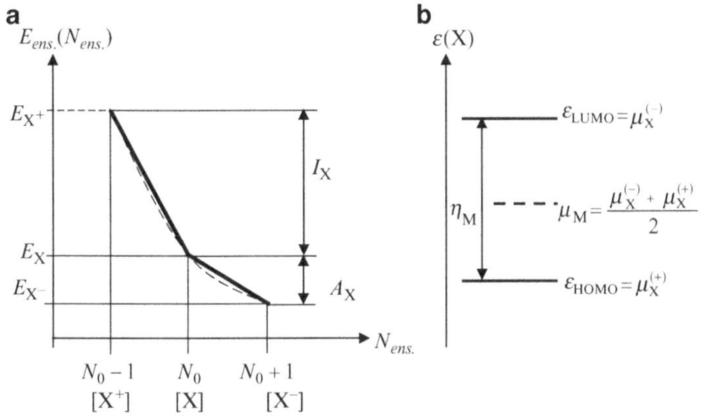

$$\mu_X^{(-)} = \left.\frac{\partial E_{ens.}(N_{ens.})}{\partial N_{ens.}}\right|_{dN_{ens.}>0} = -A_X = \varepsilon_{LUMO},$$

$$\mu_X^{(+)} = \left.\frac{\partial E_{ens.}(N_{ens.})}{\partial N_{ens.}}\right|_{dN_{ens.}<0} = -I_X = \varepsilon_{HOMO},$$

$$\mu_M = \left.\frac{\partial E^{(2)}(N_{ens.})}{\partial N_{ens.}}\right|_{N_0} = \frac{\mu_X^{(-)} + \mu_X^{(+)}}{2} = -\frac{(I_X + A_X)}{2},$$

$$\eta_M = \left.\frac{\partial^2 E^{(2)}(N_{ens.})}{\partial N_{ens.}^2}\right|_{N_0} = \left.\frac{\partial \mu_M(N_{ens.})}{\partial N_{ens.}}\right|_{N_0} = E_{X^-} + E_{X^+} - 2E_X = I_X - A_X.$$

Fig. 7.1 The average electronic energy $E_{ens.}(N_{ens.})$ of the equilibrium (ground) state of the *open* molecular system X at $T = 0$ K as function of the grand canonical average number of electrons $N_{ens.}$ (Panel *a*) and the chemical potential (μ) and chemical hardness (η) estimates it implies. It is seen to consist of the straight line segments connecting the exact energies for the neighboring integer values of $N_{ens.}$. In Panel *b* the associated KS orbital interpretations are given. The Mulliken estimate, $\mu_M = \partial E^{(2)}(N_{ens})/\partial N_{ens}$, measures the derivative of the quadratic function $E^{(2)}(N_{ens})$ obtained by interpolation of the ground state energies of the neutral system X, its anion X^- and cation X^+, shown as the broken line in diagram *a*, while $\eta_M = \partial^2 E^{(2)}(N_{ens.})/\partial N_{ens.}^2$ measures its curvature. It should be observed that the "left" and "right" derivatives $\mu_X^{(+)}$ and $\mu_X^{(-)}$ of $E_{ens.}(N_{ens.})$, for $\delta N_{ens.} \equiv N_{ens.} - N_0 < 0$ and $\delta N_{ens.} > 0$, respectively, are not equal at $T = 0$ K, thus giving rise to the $N_{ens.}$-discontinuity of the system chemical potential $\mu_{ens.}(N_{ens.}) = \partial E_{ens.}(N_{ens.})/\partial N_{ens}$ at the integer value $N_{ens.} = N_0$ which identifies the neutral system X. As indicated in the Panel *b* of the figure, these chemical potential derivatives for the system acting as the Lewis acid or base, respectively, can be identified via the Janak theorem with the corresponding *frontier* KS orbitals, while the system chemical hardness measures the energy gap between them

around the ground-state energy of the neutral atom/molecule X, for which $\Delta E^{(2)}(0) = 0$, fits the energy displacements corresponding to the system cation, $\Delta E^{(2)}(-1) = I_X$, and anion, $\Delta E^{(2)}(1) = -A_X$. As shown in Fig. 7.2, this smooth energy function exactly reproduces the biased chemical potentials of (7.167) and (7.168) at the ionic *transition states*, corresponding to $|\Delta N| = ½$ (Nalewajski 2010e).

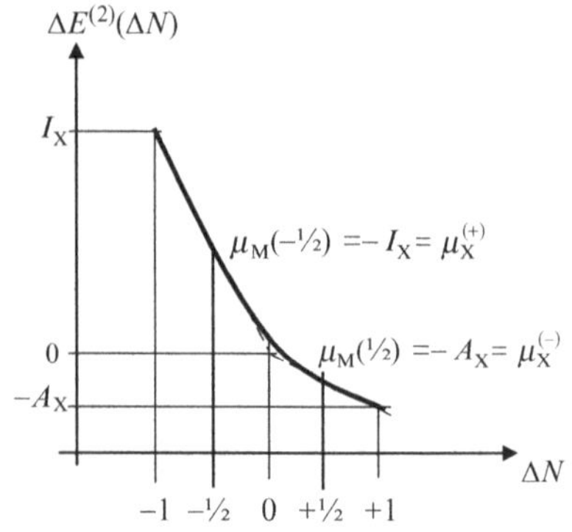

Fig. 7.2 The parabolic interpolation of Mulliken reproduces the *biased* chemical potentials at the ionic transition states corresponding to $|\Delta N| = \pm\,{}^1\!/_2$

Following the familiar ideas of Parr and Pearson (1983), the second derivative of $E_{ens.}(N_{ens.})$,

$$\eta_{ens.}(N_{ens.}) = \partial^2 E_{ens.}(N_{ens.})/\partial N_{ens.}^2, \tag{7.171}$$

is used to measure the "hardness" of the system electron distribution. Such concept is used in several intuitive rules of chemistry and provides an important reactivity descriptor together with its "softness" and Fukui function companions (Parr and Yang 1989; Sen 1993; Nalewajski and Korchowiec 1997; Nalewajski 1996b; Geerlings et al. 2003; Chattaraj 2009). For example, the *Hard/Soft Acids and Bases* (HSAB) principle of Pearson (1973, 1988) states that acids (bases) prefer in interactions with their basic (acidic) partners of the *Donor–Acceptor* (DA) complexes the comparable degree of the reactant hardnesses: hard acids form relatively stable compounds with hard bases and so do the soft acids and soft bases. Another familiar example comes from the coordination chemistry. The *symbiosis* rule of Jørgensen (1964) predicts that hard (soft) ligands enhance the tendency of the central (transition metal) ion to bind more hard (soft) ligands in its coordination sphere.

The hardness derivative of the neutral system X can be again estimated by finite differences (see Fig. 7.1) as the derivative of the Mulliken estimate of the neutral system chemical potential:

$$\eta_M \approx \left.\frac{\partial^2 E^{(2)}(N_{ens.})}{\partial N_{ens.}^2}\right|_{N_0} = \left.\frac{\partial \mu^{(2)}(N_{ens.})}{\partial N_{ens.}}\right|_{N_0} = E_{X^-} + E_{X^+} - 2E_X = I_X - A_X. \tag{7.172}$$

As indicated in the preceding equation, η_M measures the energy difference corresponding to the *disproportionation* reaction, $2X \to X^+ + X^-$, of the simultaneous reduction and oxidation of X.

The Mulliken parabolic energy curve exhibits the minimum value

$$\Delta E^{(2)}(\Delta N_{min.}) = -\frac{\mu_M^2}{2\eta_M} = -\frac{(I_X + A_X)^2}{2(I_X - A_X)} < 0, \tag{7.173}$$

at the optimum amount of CT,

$$\Delta N_{min.} = -\mu_{\mathrm{M}}/\eta_{\mathrm{M}} = \frac{I_{\mathrm{X}} + A_{\mathrm{X}}}{2(I_{\mathrm{X}} - A_{\mathrm{X}})} > 0, \tag{7.174}$$

determined from the associated equilibrium condition:

$$\frac{d\Delta E^{(2)}(\Delta N)}{d\Delta N} = \mu_{\mathrm{M}} + \eta_{\mathrm{M}}\Delta N \equiv \mu^{(2)}(\Delta N) = 0. \tag{7.175}$$

Therefore, when $3A_{\mathrm{X}} < I_{\mathrm{X}}$, which is more a rule than an exception, this formula gives unphysical result predicting the positive chemical potential (negative electronegativity) of stable anions (Nalewajski 2010e):

$$\mu^{(2)}(1) = \frac{1}{2}(I_{\mathrm{X}} - 3A_{\mathrm{X}}) = \mu_{\mathrm{X}^-}^{(2)} > 0, \tag{7.176}$$

due to the artificial energy minimum at the fractional negative charge: $0<\Delta N_{min.}<1$. Hence, in the Mulliken interpolation scheme, such anions are erroneously diagnosed as being unstable relative to the optimum CT intermediate of (7.173) and (7.174).

An alternative continuous interpolation of the N-dependence of the system energy is provided by the exponential function (Nalewajski 2010e):

$$\Delta E^{(e)}(\Delta N) = \alpha + \beta \exp(\gamma \Delta N), \tag{7.177}$$

where the condition $\Delta E^{(e)}(0) = 0$ implies: $\alpha = -\beta$. Fitting the remaining two parameters to satisfy the conditions $\Delta E^{(e)}(-1) = I_{\mathrm{X}}$ and $\Delta E^{(e)}(1) = -A_{\mathrm{X}}$ (see Fig. 7.2) finally gives:

$$\Delta E^{(e)}(\Delta N) = \frac{I_{\mathrm{X}} A_{\mathrm{X}}}{A_{\mathrm{X}} - I_{\mathrm{X}}}\left[1 - \left(\frac{A_{\mathrm{X}}}{I_{\mathrm{X}}}\right)^{\Delta N}\right]. \tag{7.178}$$

This formula predicts the following expressions for the system chemical potential and hardness:

$$\begin{aligned}
\mu^{(e)}(\Delta N) &= \frac{I_{\mathrm{X}} A_{\mathrm{X}}}{I_{\mathrm{X}} - A_{\mathrm{X}}}\left[\ln\left(\frac{A_{\mathrm{X}}}{I_{\mathrm{X}}}\right)\right]\left(\frac{A_{\mathrm{X}}}{I_{\mathrm{X}}}\right)^{\Delta N} < 0, \\
\eta^{(e)}(\Delta N) &= \frac{I_{\mathrm{X}} A_{\mathrm{X}}}{I_{\mathrm{X}} - A_{\mathrm{X}}}\left[\ln\left(\frac{A_{\mathrm{X}}}{I_{\mathrm{X}}}\right)\right]^2\left(\frac{A_{\mathrm{X}}}{I_{\mathrm{X}}}\right)^{\Delta N} > 0.
\end{aligned} \tag{7.179}$$

Therefore, this continuous interpolation always generates the physical (negative) values of the chemical potential for both the neutral and ionic species in the whole region $-1 < \Delta N < 1$:

$$\mu^{(e)}(-1) = \tfrac{I_X^2}{I_X - A_X} \ln\left(\tfrac{A_X}{I_X}\right) = \mu_{X^+}^{(e)} < \mu^{(e)}(0) = \tfrac{I_X A_X}{I_X - A_X} \ln\left(\tfrac{A_X}{I_X}\right) \\ < \mu^{(e)}(1) = \tfrac{A_X^2}{I_X - A_X} \ln\left(\tfrac{A_X}{I_X}\right) = \mu_{X^-}^{(e)}. \tag{7.180}$$

Equation (7.179) also correctly predicts that chemical species become harder with increasing ionization:

$$\eta^{(e)}(-1) = \tfrac{I_X^2}{I_X - A_X} \left[\ln\left(\tfrac{A_X}{I_X}\right)\right]^2 = \eta_{X^+}^{(e)} > \eta^{(e)}(0) = \tfrac{I_X A_X}{I_X - A_X} \left[\ln\left(\tfrac{A_X}{I_X}\right)\right]^2 = \eta_X^{(e)} \\ > \eta^{(e)}(1) = \tfrac{A_X^2}{I_X - A_X} \left[\ln\left(\tfrac{A_X}{I_X}\right)\right]^2 = \eta_{X^-}^{(e)}. \tag{7.181}$$

These relations from the exponential interpolation can be compactly summarized in terms of the I_X/A_X ratio:

$$\mu_{X^+}^{(e)}/\mu_X^{(e)} = \mu_X^{(e)}/\mu_{X^-}^{(e)} = \eta_{X^+}^{(e)}/\eta_X^{(e)} = \eta_X^{(e)}/\eta_{X^-}^{(e)} = I_X/A_X > 1. \tag{7.182}$$

The relative levels of the chemical potentials exhibited by two molecular reactants A and B in the given DA system M = A----B, where A and B stand for its acidic (acceptor) and basic (donor) subsystems, respectively, establishes the *direction* of a spontaneous *charge transfer* (CT) A ← B in M, from the donor reactant B, exhibiting higher chemical potential (lower electronegativity), to the acceptor reactant A, characterized by the lower chemical potential (higher electronegativity).

However, to determine the optimum amount of CT,

$$N_{CT} \equiv N_A - N_A^0 \equiv \Delta N_A = N_B^0 - N_B \equiv -\Delta N_B, \tag{7.183}$$

the reactant hardnesses are also needed. The initial state of the DA system, before CT, invokes the separated (infinitely distant, mutually noninteracting) *free* reactants A^0 and B^0 in $M^0 = (A^0$----$B^0)$, containing N_A^0 and N_B^0 electrons, respectively, for which (see Fig. 7.3)

$$\mu_A^0 < \mu_B^0 \quad \text{or} \quad \chi_A^0 > \chi_B^0. \tag{7.184}$$

The acidic reactant contains the "harder" and less polarizable valence electrons, compared with its basic partner in DA system, which represent the system "softer" part, containing more polarizable valence electrons. Therefore, one observes in the separate or very distant (noninteracting) subsystems of DA complexes:

$$\eta_A^0 = \left.\frac{\partial \mu_A(N_A)}{\partial N_A}\right|_{N_A^0} > \eta_B^0 = \left.\frac{\partial \mu_B(N_B)}{\partial N_B}\right|_{N_B^0}. \tag{7.185}$$

For the (interacting) reactants, at their finite mutual separations, the average energy of M becomes a function of the effective numbers of electrons on both reactants, $E_{ens.}(\mathrm{M}) = E_{ens.}[N_{ens.}(\mathrm{A}), N_{ens.}(\mathrm{B}); v_{\mathrm{M}}] \equiv E_{\mathrm{M}}(N_{\mathrm{A}}, N_{\mathrm{B}})$, where v_{M} stands for the resultant external potential due to the nuclei of both subsystems at their fixed positions in M. In the molecular complex containing the closed reactants at their finite separation and the specified mutual orientation between the two subsystems, $\mathrm{M}^+ = (\mathrm{A}^+|\mathrm{B}^+)$, the two fragments are closed both *mutually* and *externally*, so that their chemical potentials are not equalized: $\mu_{\mathrm{A}}{}^+ < \mu_{\mathrm{B}}{}^+$.

After the barrier for the electron flow between them has been lifted in $\mathrm{M}^{\mathrm{CT}} = (\mathrm{A}^*{\vdots}\mathrm{B}^*) \equiv \mathrm{M}^*$ one observes the net (spontaneous) CT $\mathrm{B} \rightarrow \mathrm{A}$, to the amount N_{CT}, when the chemical potentials of electrons in two reactants equalize: $\mu_{\mathrm{A}}{}^* = \mu_{\mathrm{B}}{}^*$. Here, the solid vertical line separating the two reactants signifies that they are mutually closed, while the broken vertical line separating these molecular fragments implies their freedom to exchange electrons.

Hence, the initial state of M^+ involves the integer numbers of electrons $N_{\mathrm{A}}^0 = N_{\mathrm{A}}^+$ and $N_{\mathrm{B}}^0 = N_{\mathrm{B}}^+$ on each *polarized* subsystem. Then the electron flow between them becomes allowed in M^{CT}, giving rise to the fractional (average) numbers of electrons on both subsystems in the ensuing equilibrium state of M as a whole:

$$N_{\mathrm{A}}^* = N_{\mathrm{A}}^0 + N_{\mathrm{CT}} \quad \text{and} \quad N_{\mathrm{B}}^* = N_{\mathrm{B}}^0 - N_{\mathrm{CT}}. \tag{7.185a}$$

In order to exert the independent control over these two population variables one could envisage the *mutually* closed but *externally* open reactants in the composite system,

$$\mathcal{M}^+ = (\mathcal{R}_{\mathrm{A}}{\vdots}\mathrm{A}^+\,|\,\mathrm{B}^+{\vdots}\mathcal{R}_{\mathrm{B}}) \equiv (\mathcal{M}_{\mathrm{A}}{}^+\,|\,\mathcal{M}_{\mathrm{B}}{}^+), \tag{7.186}$$

in which the acidic and basic fragments of M are hypothetically coupled to the *separate* (macroscopic) electron reservoirs $\mathcal{R}_{\mathrm{A}}$ and $\mathcal{R}_{\mathrm{B}}$, respectively, with $\mu_{\mathrm{A}} = \mu_{\mathcal{R}_{\mathrm{A}}}$ and $\mu_{\mathrm{B}} = \mu_{\mathcal{R}_{\mathrm{B}}}$. Depending on the current values of chemical potentials of these reservoirs each reactant can then acquire arbitrary fractional charge reflected by their average electron populations:

$$N_{\mathrm{A}}^+ = N_{\mathrm{A}}^+(\mathcal{M}_{\mathrm{A}}^+) = N_{\mathrm{A}}^0 + \Delta N_{\mathrm{A}}^+ \quad \text{and} \quad N_{\mathrm{B}}^+ = N_{\mathrm{B}}^+(\mathcal{M}_{\mathrm{B}}^+) = N_{\mathrm{B}}^0 + \Delta N_{\mathrm{B}}^+. \tag{7.185b}$$

These spontaneous exchanges between subsystems and their reservoirs give rise to the partial equalization of the chemical potentials involved: $\mu_{\mathrm{A}}{}^+ = \mu_{\mathcal{R}_{\mathrm{A}}} \neq \mu_{\mathrm{B}}{}^+ = \mu_{\mathcal{R}_{\mathrm{B}}}$.

Thus, by appropriately changing the chemical potentials of such independent reservoirs one can realize arbitrary partial flows $\{(\mathrm{X}^+ \leftrightarrow \mathcal{R}_{\mathrm{X}})\}$. For example, one could then effect the net transfer of N_{CT} electrons between the mutually open reactants, for

$$\mu_{\mathrm{A}}{}^* = \mu_{\mathcal{R}_{\mathrm{A}}} = \mu_{\mathrm{B}}{}^* = \mu_{\mathcal{R}_{\mathrm{B}}} = \mu_{\mathcal{R}_{\mathrm{M}}}, \tag{7.187}$$

as a result of the synchronized simultaneous exchanges $B \xrightarrow{N_{CT}} \mathcal{R}_B$ and $\mathcal{R}_A \xrightarrow{N_{CT}} A$, which indeed amount to the net $B \xrightarrow{N_{CT}} A$ flow of N_{CT} electrons between the *mutually* and *externally* open reactants, now coupled to a common electron reservoir $\mathcal{R}_M$:

$$\begin{aligned} \mathcal{M}^* &\equiv (\mathcal{M}_A{}^* \vdots \mathcal{M}_B{}^*) = (\mathcal{R}_M \vdots A^* \mid B^* \vdots \mathcal{R}_M) \equiv \mathcal{M}^+(\mathcal{R}_M) \\ &= (\mathcal{R}_M \vdots A^* \vdots B^* \vdots \mathcal{R}_M) \equiv \mathcal{M}^{CT}(\mathcal{R}_M) \equiv (A^* \vdots B^* \vdots \mathcal{R}_M) \equiv (M^* \vdots \mathcal{R}_M); \end{aligned} \tag{7.188}$$

here, $\mathcal{R}_M$ denotes the hypothetical molecular reservoir with $\mu_M{}^* = \mu_{\mathcal{R}_M}$. In this composite supersystem scenario the average numbers of electrons on the two reactants, $N_A{}^+(\mathcal{R}_M) = N_A^*(\mathcal{M}_A^*)$ and $N_B{}^+(\mathcal{R}_M) = N_B^*(\mathcal{M}_B^*)$, are in fact the equilibrium average values of each of the two mutually closed combined systems $\mathcal{M}_A{}^+(\mathcal{R}_M) = (\mathcal{R}_M \vdots A^*) = \mathcal{M}_A{}^*$ and $\mathcal{M}_B{}^+(\mathcal{R}_M) = (B^* \vdots \mathcal{R}_M) = \mathcal{M}_B{}^*$, respectively.

Obviously, these two CT scenarios, involving the *externally* open and *externally* closed subsystems, respectively, are phenomenologically equivalent. Indeed, the global equilibrium $(A^* \vdots B^*) = (\mathcal{R}_M \vdots A^* \mid B^* \vdots \mathcal{R}_M) \equiv (M^* \vdots \mathcal{R}_M) \equiv (A^* \vdots B^* \vdots \mathcal{R}_M)$ calls for the total equalization of the chemical potentials involved, $\mu_A{}^* = \mu_{\mathcal{R}_A} = \mu_B{}^* = \mu_{\mathcal{R}_B} = \mu_{\mathcal{R}_M} = \mu_M{}^*$, since then subsequent opening of the two combined subsystems in $\mathcal{M}^{CT}(\mathcal{R}_M) = (\mathcal{R}_M \vdots A^* \vdots B^* \vdots \mathcal{R}_M)$ would not affect the electron populations of the two subsystems already in their mutual-equilibrium state. Indeed, this chemical potential equalization criterion implies that subsystems have been brought to their mutual equilibrium state already in $\mathcal{M}^+(\mathcal{R}_M) = (\mathcal{R}_M \vdots A^* \mid B^* \vdots \mathcal{R}_M)$, before lifting the barrier for their direct electron exchanges.

In the interacting DA complex both diagonal and *off*-diagonal elements of the subsystem chemical potential derivatives are required to determine the equilibrium amount of CT. These *second*-order partial derivatives of $E_M(N_A, N_B)$ determine the *hardness tensor* of this reactive system:

$$\boldsymbol{\eta}_M = \left\{ \eta_{X,Y} = \frac{\partial^2 E_M(N_A,N_B)}{\partial N_X \partial N_Y} = \left.\frac{\partial \mu_X(N_A,N_B)}{\partial N_Y}\right|_{N_X} = \left.\frac{\partial \mu_Y(N_A,N_B)}{\partial N_X}\right|_{N_Y}, (X,Y) \in \{A,B\} \right\}. \tag{7.189}$$

Thus, due to nonvanishing *off*-diagonal elements in this matrix, the change in the effective number of electrons on one subsystem affects the chemical potentials of both subsystems:

$$\Delta\mu_A = \eta_{A,A}\,\Delta N_A + \eta_{A,B}\,\Delta N_B \quad \text{and} \quad \Delta\mu_B = \eta_{B,A}\,\Delta N_A + \eta_{B,B}\,\Delta N_B. \tag{7.190a}$$

Let us now examine some implications of the global equilibrium, which determines the optimum amount of CT between the two reactants,

$$N_{CT} = \Delta N_A = -\Delta N_B, \tag{7.190b}$$

$$N_{CT} = -[\mu_A(M) - \mu_B(M)]/[\eta_A(M) + \eta_B(M)] \equiv -\mu_{CT}(M)/\eta_{CT}(M) \equiv \chi_{CT}(M)/\eta_{CT}(M),$$
$$\mu_A(M) = \mu_A[N_A{}^0, N_B{}^0; v_M] \equiv \mu_A^+, \qquad \mu_B(M) = \mu_B[N_A{}^0, N_B{}^0; v_M] \equiv \mu_B^+,$$
$$\eta_A(M) = \eta_{A,A} - \eta_{A,B}, \qquad \eta_B(M) = \eta_{B,B} - \eta_{B,A}. \tag{7.191}$$

Therefore, the resultant (in situ) characteristics of the DA complex, the initial chemical potential difference $\mu_{CT}(M)$ of the polarized reactants and their effective hardness $\eta_{CT}(M)$ in M read:

$$\mu_{CT}(M) = \frac{\partial \Delta E_M(N_{CT})}{\partial N_{CT}} = \mu_A(M) - \mu_B(M) < 0, \tag{7.192}$$

$$\eta_{CT}(M) = \frac{\partial^2 \Delta E_M(N_{CT})}{\partial N_{CT}^2} = \eta_A(M) + \eta_B(M) = \eta_{A,A} + \eta_{B,B} - 2\eta_{A,B}, \tag{7.193}$$

where to *first*-order

$$\begin{aligned} \Delta E_M(N_{CT}) &= E_M(N_A{}^0 + N_{CT}, N_B{}^0 - N_{CT}) - E_M(N_A{}^0, N_B{}^0) \\ &\approx \Delta E_M{}^{(1)}(N_{CT}) \equiv \mu_{CT}(M) N_{CT} = -[\mu_{CT}(M)]^2/\eta_{CT}(M). \end{aligned} \tag{7.194}$$

To summarize, the in situ chemical potential difference $\mu_{CT}(M)$ determines the resultant driving force behind the interreactant flow of electrons, the in situ hardness $\eta_{CT}(M)$ similarly represents the associated effective "stiffness" modulus for the net A ← B CT, and $\Delta E_M(N_{CT})$ defines the dominating energy change in this process. It should be observed that the *second*-order CT energy reads:

$$\begin{aligned} \Delta E_M(N_{CT}) &\cong \Delta E_M{}^{(2)}(N_{CT}) = \frac{\partial \Delta E_M(N_{CT})}{\partial N_{CT}} N_{CT} + \frac{1}{2}\frac{\partial^2 \Delta E_M(N_{CT})}{\partial N_{CT}^2} N_{CT}^2 \\ &= \Delta E_M{}^{(1)}(N_{CT}) + \frac{1}{2}\eta_{CT}(M) N_{CT}^2 = -\frac{1}{2}[\mu_{CT}(M)]^2/\eta_{CT}(M). \end{aligned} \tag{7.195}$$

It should be realized that the spontaneous CT from the basic to the acidic reactant represents the net effect of all interorbital flows involving the occupied MO of one subsystem and the virtual orbitals of the other subsystem. As schematically shown in Fig. 7.3, these orbital excitations generate within the CI approach both the partial [A ← B] and [A → B] flows of electrons, which add up to the resultant (net) CT, A ← B = [A ← B] + [A → B], which accords with the electronegativity difference of the initially polarized, mutually closed reactants: $\chi_{CT}(M) > 0$ or $\mu_{CT}(M) < 0$. As also shown in this qualitative diagram the initial "promotion" of the two reactants of the DA complex in presence of each other can be linked to the electron excitations within each reactant. It should be emphasized, that CT between reactants induces further polarizations of then already bonded (mutually open) fragments. The fact that the basic fragment of DA complex is softer, relative to its acidic partner (7.185), is reflected in this diagram by the corresponding magnitudes of their HOMO–LUMO gaps (see also Fig. 7.1).

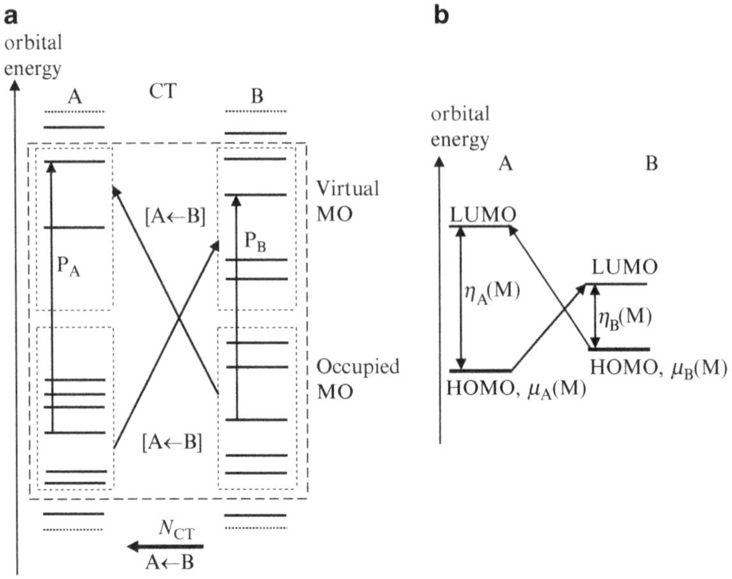

Fig. 7.3 The polarization (P) and CT electron excitations within the adopted CI energy-window (Panel *a*) in the DA complex M = A----B, consisting of the basic (B) and acidic (A) reactants, and partial flows of electrons between the *frontier* MO (HOMO, LUMO) of both subsystems (Panel *b*)

7.3.5 Zero-Temperature Limit

Let us now examine the limit $T \to 0$, or $\beta = (k_B T)^{-1} \to \infty$, of the ensemble-average energy for the equilibrium state of the grand-ensemble exhibiting the continuously varying average number of electrons $N_{ens.}$, which can assume both the fractional and integral values covering both the positive ($N_{ens.} < N_0$), negative ($N_{ens.} > N_0$), and neutral ($N_{ens.} = N_0$) net charges of the system in question (see Fig. 7.1). At the absolute zero temperature only the ground states $\{\Psi_i^0\}$ for different (integer) values $\{N_i\}$ of the system number of electrons can mix in the equilibrium state of the ensemble, with the associated probabilities, $\bar{\boldsymbol{p}} = \{\bar{p}_i \equiv p_i^0\}$, defined by (7.49):

$$\bar{p}_i = \lim_{T \to 0} \left\{ \exp[-\beta(E_i^0 - \mu N_i)] / \sum_k \exp[-\beta(E_k^0 - \mu N_k)] \right\}, \quad \sum_i \bar{p}_i = 1. \quad (7.196)$$

The ensemble averages of physical quantities are then given by the corresponding probability-weighted mean values, e.g.,

$$N_{ens.} = \sum_i \bar{p}_i N_i, \quad E_{ens.} = \sum_i \bar{p}_i E_i^0, \quad \text{etc.} \quad (7.197)$$

This section summarizes the thermodynamic analysis of electronegativity by Gyftopoulos and Hatsopoulos (1965) and the grand-ensemble study of the chemical potential by Perdew et al. (1982) [see also: Perdew and Norman 1982; Perdew and Levy 1983, 1984; Perdew 1985; Parr and Yang 1989]. We first observe that the finite values of probabilities $\bar{p}$ in the $\beta \to \infty$ limit imply the finite values of their inverses:

$$\bar{p}_i^{-1} = \lim_{T\to 0} \sum_k \exp\left(\beta[(E_i^0 - E_k^0) + \mu(N_k - N_i)]\right). \tag{7.198}$$

For $\beta \to \infty$ this is the case only when the remaining factor in the exponent of the preceding equation vanishes, $(E_i^0 - E_k^0) + \mu(N_k - N_i) \to 0$, for at least a single $k \neq i$, when

$$\mu = (E_i^0 - E_k^0)/(N_i - N_k). \tag{7.199}$$

Consider now the mixed states involving the ground state for $N_i = N_0$. Excluding the $\bar{p}_i = 1$ case, which implies the pure quantum state for N_0, we now examine the statistical mixtures corresponding to the negative and positive net charge of the system, when $N_{ens.} > N_0$ and $N_{ens.} < N_0$, respectively. The former case implies that the mixture must involve Ψ_i^0 and at least one ground state of the *anion* systems, with the ground state $\Psi_k^0 = \Psi_{i+1}^0$ of X^{-1} for $N_k = N_i + 1 = N_0 + 1$ being energetically most accessible (see Fig. 7.1). The finite participation of this state in the equilibrium grand ensemble then implies [see (7.199)]:

$$\mu = -(E_i^0 - E_{i+1}^0) = -A_X \equiv \mu_i^{(-)}, \tag{7.200}$$

and $\bar{p}_k^{-1} \to \infty$, i.e., $\bar{p}_k \to 0$, for $k \neq (i, i + 1)$.

This is due to the observed *convexity* of the ground state eigenvalues for the molecular (Coulombic) systems:

$$\ldots < E_{i-2}^0 < E_{i-1}^0 < E_i^0 < E_{i+1}^0 < \ldots \tag{7.201}$$

It should be observed that this property further implies

$$\ldots > (I_{i-2} = A_{i-3}) > (I_{i-1} = A_{i-2}) > (I_i = A_{i-1}) > (I_{i+1} = A_i) > \ldots \tag{7.202}$$

where I_i and A_i denote the ionization potential and electron affinity of the molecular system containing N_i electrons. Above we have also recognized that the energy difference defining I_i at the same time determines A_{i-1}.

Therefore, the mixture of Ψ_i^0 and Ψ_{i+1}^0 excludes participation of any other ground state in the ensemble. In other words the approximate, *finite*-difference estimate of the chemical potential of the negatively charged molecular/atomic system, when it acts as the Lewis acid (7.167), reflects the *exact* chemical potential of such an anionic system, for $N_0 < N_{ens} < N_0 + 1$, as indeed indicated in Fig. 7.1.

One can repeat the above argument for the *left*-derivative of the ensemble energy function (7.168). The state energetically most accessible from among all the admissible *cation* states is then the neighboring $\Psi_k^0 = \Psi_{i-1}^0$ ground state of X^{+1}, for $N_k = N_i - 1 = N_0 - 1$. Its finite participation in the grand-canonical mixture then implies

$$\mu = E_i^0 - E_{i-1}^0 = -I_X \equiv \mu_i^{(+)}, \tag{7.203}$$

and $\bar{p}_k^{-1} \to \infty$, i.e., $\bar{p}_k \to 0$, for $k \neq (i, i-1)$, so that the finite participation in the ensemble of Ψ_{i-1}^0 and Ψ_i^0 excludes the participation of any other state in the mixture. Again, this indicates that the approximation of (7.168) is in fact the *exact* relation, as also shown in Fig. 7.1.

This result remains valid for any fractional value of $(N_k \equiv k) < N_{ens.} < (N_{k+1} \equiv k+1)$: only the neighboring ground states Ψ_k^0 and Ψ_{k+1}^0 participate in such an equilibrium mixed state at $T \to 0$ K, with probabilities of the remaining states exactly vanishing. The finite probabilities of these "*bracketing*" states are then uniquely determined by the assumed value of $N_{ens.}$ (7.197) and the normalization condition of (7.196):

$$N_{ens.} = \bar{p}_k k + \bar{p}_{k+1}(k+1) \quad \text{and} \quad \bar{p}_k + \bar{p}_{k+1} = 1, \tag{7.204}$$

Solving these equations for the ensemble probabilities gives:

$$\begin{aligned} &\bar{p}_k = k + 1 - N_{ens.} \equiv 1 - \omega_k > 0 \quad \text{and} \quad \bar{p}_{k+1} = N_{ens.} - k \equiv \omega_k > 0; \\ &\quad k < N_{ens.} < k+1 \quad \text{or} \quad 0 < \omega_k < 1. \end{aligned} \tag{7.205}$$

Therefore, at $T \to 0$ K the energy function $E_{ens.}(N_{ens.})$ consists of the straight-line segments $\{E_k(\omega_k)\}$, interpolating between the $E_k^0 = E[N_k, v]$ and $E_{k+1}^0 = E[N_{k+1}, v]$ eigenvalues of the electronic Hamiltonian (see Fig. 7.1):

$$E_{ens.}(N_{ens.}) = (1 - \omega_k)E_k^0 + \omega_k E_{k+1}^0 \equiv E_k(\omega_k). \tag{7.206}$$

In the vicinity of each integer value $N_{ens.} = k \geq 1$ only the "left" (7.203) and "right" (7.200) chemical-potential derivatives $\mu_k^{(+)}$ and $\mu_k^{(-)}$, respectively, determining the so called "*biased*" chemical potentials, are defined:

$$\begin{aligned} \mu_k^{(+)} &= \frac{\partial E_{k-1}(\omega_{k-1})}{-\partial \omega_{k-1}} = -(E_k^0 - E_{k-1}^0) = -I_k, \\ \mu_k^{(-)} &= \frac{\partial E_k(\omega_k)}{\partial \omega_k} = E_{k+1}^0 - E_k^0 = -A_k. \end{aligned} \tag{7.207}$$

As schematically shown in Fig. 7.2, the parabolic interpolation of Mulliken reproduces these derivatives as the transition-state slopes for the fractional displacements $\Delta N_{ens.} = \pm ½$ relative to the neutral system.

Hence, at the zero temperature the chemical potential exhibits the discontinuity at each integral value of the average number of electrons. This discovery of Perdew et al. (1982) has important implications for the dissociation products. Consider the molecular system consisting of the A and B fragments, M = A–B, at very large value of their mutual separation, $R_{A\text{–}B} \to \infty$, when both fragments no longer interact with one another. For definiteness, we again assume that among the two products of the dissociation reaction, $M \to A^0 + B^0$, A^0 represents the system acidic part while B^0 denotes its basic fragment. This nearly separated state of the two subsystems also implies the vanishing values of the coupling hardnesses, $\eta_{A,B} = \eta_{B,A} = 0$, and hence

$$\mu_A(M) = \mu_A{}^0 < \mu_B(M) = \mu_B{}^0, \quad \eta_A(M) = \eta_A{}^0 > \eta_B(M) = \eta_B{}^0. \tag{7.208}$$

The following question then naturally arises: why do the dissociation products always exhibit the integral values N_A and N_B of electrons, in spite of the above differences in their electronegativities? The answer to this intriguing question directly results from the chemical potential discontinuity by examining the energy changes for this dissociation system, shown in Fig. 7.4, which accompany the transfer of an infinitesimal amount of electrons $dN > 0$ between these two (separate, noninteracting) initially neutral molecular fragments, which contain the integer values of electrons N_A^0 and N_B^0, respectively.

Consider first the $B \xrightarrow{dN} A$ CT, $dN = N_{CT} = N_A - N_A^0 = N_B^0 - N_B > 0$, globally isoelectronic, $N_A^0 + N_B^0 = N_A + N_B = N$, in the direction consistent with the assumed chemical-potential difference of the preceding equation:

$$\Delta E_{B\to A}(dN) = [\mu_A^{(-)} - \mu_B^{(+)}]\, dN = [-A_A + I_B]\, dN > 0. \tag{7.209}$$

The above inequality follows from the experimental observation that for Coulombic systems the ionization potential is always higher that the electron affinity:

$$I_A > I_B > A_A > A_B. \tag{7.210}$$

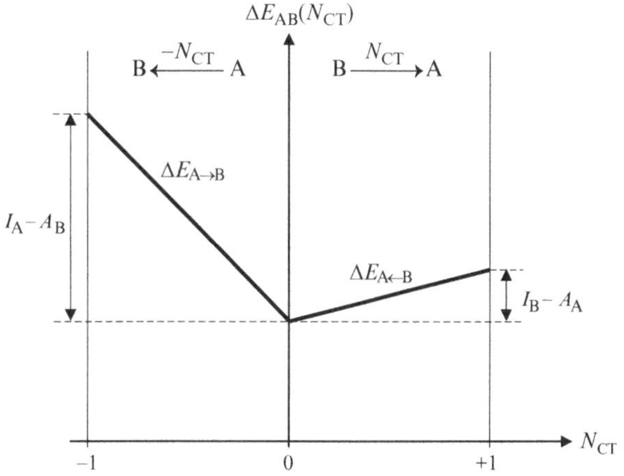

Fig. 7.4 Variations $\Delta E_{AB}(N_{CT})$ of the ensemble-average electronic energy of the noninteracting DA system A·····B, consisting of the practically separated subsystems A (acid) and B (base), with the amount of CT between reactants, $-1 \le N_{CT} \le 1$: $N_{CT} \equiv \Delta N_A = N_A - N_A^0 \equiv -\Delta N_B = N_B^0 - N_B$

The reverse direction of this elementary CT, $A \xrightarrow{dN} B$, contrary to the net spontaneous flow of electrons, $dN = -N_{CT} = N_B - N_B^0 = N_A^0 - N_A > 0$ implies even faster increase in the system energy:

$$\Delta E_{A\to B}(dN) = [\mu_B^{(-)} - \mu_A^{(+)}]\, dN = [-A_B + I_A]\, dN > \Delta E_{B\to A}(dN) > 0. \quad (7.211)$$

Therefore, any deviation from neutrality of the dissociation products costs energy and this is why the neutral dissociation products preserve their integral electron populations.

It should be also observed that the constant chemical potential $\mu_k^{(-)}$ in kth fractional region of $N_{ens.}$, representing the slope of $E_k(\omega_k)$ (7.207), formally implies the vanishing value of the corresponding hardness descriptor:

$$\eta_k(\omega_k) = \frac{\partial^2 E_k(\omega_k)}{\partial \omega_k^2} = \frac{\partial \mu_k^{(-)}(\omega_k)}{\partial \omega_k} = 0. \quad (7.212)$$

This somewhat surprising, *ensemble* result is different from the *molecular*, finite-difference (Mulliken-type) estimate η_M of Fig. 7.1 and (7.172) and corresponding estimate from the exponential interpolation [(7.179), (7.181)]. The latter refers to the externally *closed* molecule X, in which the interpolated (fractional) values of N_X are assumed to be *fixed*, thus exhibiting zero fluctuations. In the ensemble scenario the equilibrium state of the molecular fragment X is represented by the grand ensemble, so it corresponds to X coupled to the *macroscopic* electron reservoir $\mathcal{R}_X$ in $\mathcal{M}_X = (X|\mathcal{R}_X)$ (see Sect. 7.3.4). Therefore, as a part of such an infinitely soft combined system, in which a small displacement in the number of electrons does not affect the chemical potential of the infinitely large $\mathcal{M}_X$, equal to that of X itself, this molecular fragment must exhibit the zero hardness (infinite softness) predicted in the preceding equation. This result simply states that in this range of the electron population of subsystem X, its chemical potential remains insensitive to an inflow/outflow of electrons due to its being a part of the macroscopic combined system.

7.3.6 Physical Interpretation of KS Eigenvalues

The optimum SO $\{\varphi_k(\boldsymbol{n})\}$ of the KS method for their given electron occupation numbers $\boldsymbol{n} = \{n_k = 1,\ k = 1, 2, \ldots, N;\ n_k = 0,\ k > N\}$ (see also Sect. 6.1.3) determine the system ground state density $\rho[\boldsymbol{n},\{\varphi_k(\boldsymbol{n})\}]$ of (7.111),

$$\rho(\boldsymbol{r}) \equiv \rho[\boldsymbol{n}, \{\varphi_k(\boldsymbol{n};\boldsymbol{r})\}] \equiv \rho(\boldsymbol{n};\boldsymbol{r}) = \sum_k n_k |\varphi_k(\boldsymbol{n};\boldsymbol{r})|^2, \quad (7.213)$$

and the associated minimum of the system energy:

$$E_v[\rho] = E_v[\boldsymbol{n}, \rho[\boldsymbol{n}, \{\varphi_k(\boldsymbol{n};\boldsymbol{r})\}]] \equiv E_v[\boldsymbol{n}, \rho(\boldsymbol{n})] \equiv E_v(\boldsymbol{n}). \quad (7.214)$$

Therefore, as we have already indicated in the preceding equation, the electronic energy can be regarded as a function of the orbital occupations. It should be recalled (see Sect. 6.1.2) that we have treated in a similar manner the HF energy in the context of the Koopmans theorem [(6.24b) and (6.25)] and Slater's transition state (Sect. 6.1.3). This function satisfies the Janak (1978) theorem,

$$\frac{\partial E_v(\boldsymbol{n})}{\partial n_k} = \varepsilon_k, \quad k = 1, 2, \ldots, \tag{7.215}$$

where ε_k denotes the eigenvalue (orbital energy) of the KS equation (7.110).

A comparison between (6.25) and (7.215) indicates that Janak's theorem represents the exact analog of the approximate Koopmans' relation. The proof of the former is straightforward. It adopts the KS partition of the system electronic energy (7.114),

$$\begin{aligned} E_v(\boldsymbol{n}) = E_v[\boldsymbol{n}, \rho(\boldsymbol{n})] &= -\sum_k n_k \langle \varphi_k[\rho(\boldsymbol{n})] | \frac{1}{2}\Delta | \varphi_k[\rho(\boldsymbol{n})] \rangle + \int \rho(\boldsymbol{n}; \boldsymbol{r}) v(\boldsymbol{r})\, d\boldsymbol{r} \\ &+ J_{ee}[\rho(\boldsymbol{n})] + E_{xc}[\rho(\boldsymbol{n})], \end{aligned} \tag{7.216}$$

and uses the relevant chain rule for the differentiation with respect to the occupation-number n_k. This directly gives [see (7.108), (7.110) and (7.122a)]:

$$\begin{aligned} \frac{\partial E_v(\boldsymbol{n})}{\partial n_k} &= \left(\frac{\partial E_v[\boldsymbol{n}, \rho(\boldsymbol{n})]}{\partial n_k} \right)_\rho + \int \left(\frac{\partial E_v[\boldsymbol{n}, \rho(\boldsymbol{n})]}{\partial \rho(\boldsymbol{r})} \right)_{\boldsymbol{n}} \left(\frac{\partial \rho(\boldsymbol{r})}{\partial n_k} \right) d\boldsymbol{r} \\ &= \langle \varphi_k | -\frac{1}{2}\Delta + v_{\mathrm{KS}} | \varphi_k \rangle = \varepsilon_k. \end{aligned} \tag{7.217}$$

As we have already remarked above, the KS equations defining the optimum orbitals can be derived from the minimum energy principle of (7.126). Its extension including both the SO $\boldsymbol{\varphi} = \{\varphi_k\}$ and the occupation "variables" of the electronic energy $E_v(\boldsymbol{n}) \equiv E_v[\boldsymbol{\varphi}(\boldsymbol{n}), \boldsymbol{n}]$ then reads:

$$\delta\{E_v[\boldsymbol{\varphi}, \boldsymbol{n}] - \sum_k \sum_l \theta_{k,l}(\langle \varphi_k | \varphi_l \rangle - \delta_{k,l}) - \mu[N(\boldsymbol{n}) - N^0]\} = 0, \tag{7.218}$$

where $N(\boldsymbol{n}) = \sum_k n_k$. In the canonical representation the matrix of the Lagrange multipliers associated with the SO orthonormality constraints becomes diagonal: $\{\theta_{k,l} = n_k \varepsilon_k \delta_{k,l}\}$. It can be then straightforwardly verified that independent variations of $\boldsymbol{\varphi}^*$ for the fixed $\boldsymbol{n}$, when the last term in the preceding equation can be eliminated as the redundant constraint, then gives the KS equations,

$$\sum_k n_k \langle \delta\varphi_k | \hat{\mathrm{H}}_{\mathrm{KS}} - \varepsilon_k | \varphi_k \rangle = 0 \quad \text{or} \quad \hat{\mathrm{H}}_{\mathrm{KS}} \varphi_k = \varepsilon_k \varphi_k, \quad k = 1, 2, \ldots. \tag{7.219}$$

Let us now focus on the occupation part of the generalized variational principle of (7.218), for determining the optimum occupations $\boldsymbol{n}$. As in (7.213) and (7.214)], the optimum KS orbitals $\boldsymbol{\varphi} = \{\varphi_k(\boldsymbol{n})\} \equiv \boldsymbol{\varphi}(\boldsymbol{n})$ are assumed, which already satisfy the preceding Euler equations. In order to automatically obey the Pauli restrictions $\{0 \le n_k \le 1\}$, we express after Gilbert (1975) the orbital occupations in terms of the associated *angle*-variables $\boldsymbol{a} = \{a_k\}$: $n_k \equiv \cos^2 a_k$. Using the Janak theorem then gives the following Euler equations for the unknowns $\boldsymbol{a}$:

$$\left[\frac{\partial E_v(\boldsymbol{n})}{\partial n_k} - \mu\right]\left(\frac{dn_k}{da_k}\right) = (\varepsilon_k - \mu)\sin 2a_k = 0, \quad k = 1, 2, \ldots \tag{7.220}$$

There are three cases of their solutions:

(1) $\varepsilon_k = \mu$, for arbitrary a_k, i.e., for generally fractional occupation $0 \le n_k \le 1$.
(2) $\varepsilon_k \ne \mu$, for $a_k = 0$, i.e., for the maximum occupation $n_k = 1$.
(3) $\varepsilon_k \ne \mu$, for $a_k = \pi/2$, i.e., for the vanishing occupation $n_k = 0$.

For $T \to 0$, the first case identifies the chemical potential of an open molecular system as the energy level of the KS HOMO, which determines the system *Fermi level*. In the equilibrium state at this temperature, all KS SO below this critical energy, $\varepsilon_k < \mu$, must be fully occupied (the second case), while the virtual KS SO, $\varepsilon_k > \mu$, remain empty (the third case).

In other words, the fractional occupations at $T \to 0$ can be observed only at the Fermi level, with the KS eigenvalue for HOMO determining the system chemical potential, when it acts as the electron donor (Lewis base) in the electron *outflow* to the system *electrophilic* environment. Indeed, the most energetically accessible HOMO electrons then determine $dN = dn_{\mathrm{HOMO}} < 0$, and hence by Janak theorem

$$\mu_{\mathrm{X}} = \frac{\partial E_{\mathrm{X}}[N, v]}{\partial N} = \frac{\partial E_{\mathrm{X}}(\boldsymbol{n})}{\partial n_{\mathrm{HOMO}}} = \varepsilon_{\mathrm{HOMO}} = \mu_{\mathrm{X}}^{(+)}. \tag{7.221}$$

Similarly, in the electron-*inflow* processes, when electrons are accepted from the system *nucleophilic* environment, the energetically most favorable KS LUMO level becomes populated, $dN = dn_{\mathrm{LUMO}} > 0$, and hence the LUMO eigenvalue determines the system chemical potential when it accepts an electron from its environment (Lewis acid):

$$\mu_{\mathrm{X}} = \frac{\partial E_{\mathrm{X}}[N, v]}{\partial N} = \frac{\partial E_{\mathrm{X}}(\boldsymbol{n})}{\partial n_{\mathrm{LUMO}}} = \varepsilon_{\mathrm{LUMO}} = \mu_{\mathrm{X}}^{(-)}. \tag{7.222}$$

The Mulliken electronegativity formula, which represents the arithmetic average of these two estimates, thus provides the unbiased (radical) measure of the chemical potential, valid in processes exhibiting negligible external CT:

$$\begin{aligned}\mu_{\mathrm{M}} &= \frac{\partial E_{\mathrm{X}}^{(2)}[N, v]}{\partial N} = \frac{1}{2}\left(\frac{\partial E_{\mathrm{X}}(\boldsymbol{n})}{\partial n_{\mathrm{LUMO}}} + \frac{\partial E_{\mathrm{X}}(\boldsymbol{n})}{\partial n_{\mathrm{HOMO}}}\right) \\ &= \frac{1}{2}(\varepsilon_{\mathrm{LUMO}} + \varepsilon_{\mathrm{HOMO}}) = \frac{1}{2}[\mu_{\mathrm{X}}^{(-)} + \mu_{\mathrm{X}}^{(+)}].\end{aligned} \tag{7.223}$$

It should be used in chemical reactions in which the system chemical environment exhibits the "radical" character or when its acidic/basic character cannot be established in advance. All these frontier KS MO identifications are shown in the qualitative diagram of Fig. 7.1b.

These interpretations of the chemical potential as the frontier orbital energies attribute the physical and chemical significance to KS FO. Thus, Janak's theorem makes a connection between the physical energy differences and Kohn–Sham orbital energies. The latter provide a solid and attractive concepts for interpretations of several chemical reactivity phenomena, and particularly, for diagnosing the CT effects in both the gas phase and catalytic reactions (e.g., Nalewajski and Korchowiec 1997). These eigenvalues are also important in the solid state physics by providing the foundation of most band-structure calculations for crystals.

The Janak theorem also admits the approximate orbital interpretation of the chemical hardness descriptor, measuring the system chemical potential response per unit CT, as the energy gap between the KS FO (Fig. 7.1b). Indeed, using the orbital identifications of (7.221) and (7.222) in the *finite*-difference estimate of (7.172) gives:

$$\eta_X \approx -A_X + I_X = \mu_X^{(-)} - \mu_X^{(+)} = \varepsilon_{LUMO} - \varepsilon_{HOMO}. \tag{7.224}$$

Therefore, the "hard" systems, e.g., the *acidic* reactants, should exhibit a relatively large gap between the frontier KS eigenvalues, while the "soft" systems, e.g., the *basic* reactants, are characterized by a relatively narrow gap between these critical KS FO (see Fig. 7.3). It should be also observed that the *finite*-difference expression (7.172) measures in the Pariser (1953) approximation of the semiempirical *Pariser–Parr–Pople* (PPP) SCF MO theory (Pariser and Parr 1953; Pople 1953; see also: Parr 1963) the representative electron-repulsion in the system valence-shell (Nalewajski et al. 1988).

Thus, the fictitious *single*-particle orbitals $\{\varphi_k\}$ and eigenvalues $\{\varepsilon_k\}$ of KS theory, although lacking the strict physical significance, are nonetheless quite appropriate for defining and elucidating important universal concepts of the molecular structure and chemical reactivity. In WFT it is often impossible to conceptualize how the *many*-body wave functions are related to structure and behavior of chemical species. This task is quite natural in DFT, where both the electron density and electron number have the central place in the theory. Thus, the KS theoretical framework is of great semiquantitative value, even more so than the corresponding HF reference system, because it also reflects the electron correlation. We finally observe that for isolated systems with $v(r \to \infty) = 0$, the highest eigenvalue, ε_N, controls the asymptotic decay of the associated orbital density $|\varphi_N|^2$, which then dominates ρ and hence can be shown to be the negative of the exact (*many*-body) ionization potential.

7.3.7 Chemical Reactivity Concepts

In probing changes in the equilibrium energy $E[N, v] = E_v[\rho]$ of an open molecular system in the DFT-based theory of chemical reactivity (e.g., Nalewajski and

Korchowiec 1997; Chattaraj 2009; Johnson et al. 2011), one uses the associated Taylor expansion in powers of displacements of these canonical variables, ΔN and $\Delta v(\boldsymbol{r})$, or of changes in their energy conjugates, $\Delta\mu$ and $\Delta\rho(\boldsymbol{r})$ [see (7.8a) and (7.18)].

Alternatively, the associated expansion of the system grand potential at T $= 0$ K, $\Omega[\mu, v] = E - N\mu = \Omega[u]$ [see (7.82)], where $\mu = \partial E[N, v]/\partial N$ stands for the system chemical potential and the relative potential $u(\boldsymbol{r}) = v(\boldsymbol{r}) - \mu$ (7.21), can be used to explore the equilibrium system responses to external perturbations.

We recall that at zero temperature the corresponding differentials of these thermodynamic potentials read:

$$\Delta E[N, v] = \mu \Delta N + \int \rho(\boldsymbol{r})\, \Delta v(\boldsymbol{r})\, d\boldsymbol{r}, \tag{7.225}$$

$$\begin{aligned} \Delta\Omega[\mu, v] &= \Delta E - \mu\Delta N - N\Delta\mu = -N\Delta\mu + \int \rho(\boldsymbol{r})\, \Delta v(\boldsymbol{r})\, d\boldsymbol{r} \\ &= \int \rho(\boldsymbol{r})\, \Delta u(\boldsymbol{r})\, d\boldsymbol{r} = \Delta\Omega[u]. \end{aligned} \tag{7.226}$$

The *second*-order estimate of a displacement in the system electronic energy $E[N, v]$ due to changes $\{\Delta N, \Delta v(\boldsymbol{r})\}$ in the (canonical) state parameters is similarly determined by the associated *charge sensitivities* (CS) defined by the second partials of the energy:

$$\begin{aligned} \Delta^{(1+2)}E[N, v] = {} & \left(\frac{\partial E}{\partial N}\right)_v \Delta N + \int \left(\frac{\partial E}{\partial v(\boldsymbol{r})}\right)_N \Delta v(\boldsymbol{r})\, d\boldsymbol{r} \\ & + \frac{1}{2}\left[\left(\frac{\partial^2 E}{\partial N^2}\right)_v (\Delta N)^2 + 2\Delta N \int \frac{\partial^2 E}{\partial N \partial v(\boldsymbol{r})} \Delta v(\boldsymbol{r})\, d\boldsymbol{r} \right. \\ & \left. + \iint \left(\frac{\partial^2 E}{\partial v(\boldsymbol{r})\, \partial v(\boldsymbol{r}')}\right)_N \Delta v(\boldsymbol{r}) \Delta v(\boldsymbol{r}')\, d\boldsymbol{r}\, d\boldsymbol{r}'\right] \\ \equiv {} & \mu\Delta N + \int \rho(\boldsymbol{r})\, \Delta v(\boldsymbol{r})\, d\boldsymbol{r} + \frac{1}{2}\left(\eta (\Delta N)^2 + 2\Delta N \int f(\boldsymbol{r})\, \Delta v(\boldsymbol{r})\, d\boldsymbol{r} \right. \\ & \left. + \iint \beta(\boldsymbol{r}, \boldsymbol{r}')\, \Delta v(\boldsymbol{r})\, \Delta v(\boldsymbol{r}')\, d\boldsymbol{r}\, d\boldsymbol{r}' \right). \end{aligned} \tag{7.227}$$

Above, we have used the Hellmann–Feynman theorem (7.8a) as well as the definitions of the system chemical potential (7.18) and its chemical hardness (7.172). The inverse of the latter generates the overall softness descriptor:

$$S = \frac{1}{\eta} = -\left(\frac{\partial^2 \Omega[\mu, v]}{\partial \mu^2}\right)_v = \left(\frac{\partial N}{\partial \mu}\right)_v. \tag{7.228}$$

Therefore, while the hardness measure can be thought of as a *resistance* to CT, the softness descriptor measures the *ease* of such an electron flow. The hard/soft chemical species are thus associated with the low/high polarizability of the valence electrons.

Moreover, two additional (v-related) *second*-order differential descriptors of molecules have been introduced: the system electronic *Fukui function* (FF) of Parr and Yang (1984, 1989),

$$f(\boldsymbol{r}) = \frac{\partial^2 E}{\partial N \partial v(\boldsymbol{r})} = \left(\frac{\partial \rho(\boldsymbol{r})}{\partial N}\right)_v = \left(\frac{\partial \mu}{\partial v(\boldsymbol{r})}\right)_N, \quad \int f(\boldsymbol{r})\, d\boldsymbol{r} = 1, \tag{7.229}$$

and the *linear-response* (density polarization) kernel:

$$\beta(\boldsymbol{r}, \boldsymbol{r}') = \left(\frac{\partial^2 E}{\partial v(\boldsymbol{r})\, \partial v(\boldsymbol{r}')}\right)_N = \left(\frac{\partial \rho(\boldsymbol{r}')}{\partial v(\boldsymbol{r})}\right)_N. \tag{7.230}$$

In CT processes the associated changes in the system density are indeed governed by densities of FO (Fukui 1975, 1987) (see also: Fujimoto and Fukui 1974), which justifies the name of this important reactivity concept. This local index can be interpreted as dimensionless measure $f(\boldsymbol{r}) = s(\boldsymbol{r})/S$ of the *local softness* $s(\boldsymbol{r})$ defined by the mixed second derivative of the grand potential:

$$\begin{aligned} s(\boldsymbol{r}) = \frac{\partial^2 \Omega[\mu, v]}{\partial \mu\, \partial v(\boldsymbol{r})} = \left(\frac{\partial \rho(\boldsymbol{r})}{\partial \mu}\right)_v &= \left(\frac{\partial \rho(\boldsymbol{r})}{\partial N}\right)_v \left(\frac{\partial N}{\partial \mu}\right)_v = f(\boldsymbol{r}) S \\ &= -\frac{\delta N}{\delta u(\boldsymbol{r})} = -\left(\frac{\partial N}{\partial v(\boldsymbol{r})}\right)_\mu. \end{aligned} \tag{7.231}$$

All these descriptors can be expressed in terms of the canonical, mutually inverse *hardness* and *softness kernels* represented by the corresponding *two*-point functional derivatives (e.g., Berkowitz and Parr 1988; Nalewajski 2002d, 2003a, 2006f, 2009a; Nalewajski and Korchowiec 1997; Nalewajski et al. 1996, 2008). The hardness kernel is defined by the (partial) second derivative of the energy:

$$\begin{aligned} \eta(\boldsymbol{r}, \boldsymbol{r}') &= \frac{\delta^2 E_v[\rho]}{\delta \rho(\boldsymbol{r})\, \delta \rho(\boldsymbol{r}')} = \frac{\delta^2 F[\rho]}{\delta \rho(\boldsymbol{r})\, \delta \rho(\boldsymbol{r}')} = -\frac{\delta u(\boldsymbol{r}')}{\delta \rho(\boldsymbol{r})} \\ &= \left(\frac{\partial^2 E[\rho, v]}{\partial \rho(\boldsymbol{r})\, \partial \rho(\boldsymbol{r}')}\right)_v = \left(\frac{\partial \mu(\boldsymbol{r}')}{\partial \rho(\boldsymbol{r})}\right)_v, \end{aligned} \tag{7.232}$$

while the softness kernel represents the associated second derivative of the open-system grand potential:

$$\begin{aligned} s(\boldsymbol{r}', \boldsymbol{r}) &= \frac{\delta^2 \Omega[u]}{\delta u(\boldsymbol{r})\, \delta u(\boldsymbol{r}')} = \frac{\delta^2 G[\rho]}{\delta \rho(\boldsymbol{r})\, \delta \rho(\boldsymbol{r}')} = -\frac{\delta \rho(\boldsymbol{r}')}{\delta u(\boldsymbol{r})} \\ &= \left(\frac{\partial^2 \Omega[\mu, v]}{\partial v(\boldsymbol{r})\, \partial v(\boldsymbol{r}')}\right)_\mu = -\left(\frac{\partial \rho(\boldsymbol{r}')}{\partial v(\boldsymbol{r})}\right)_\mu. \end{aligned} \tag{7.233}$$

They satisfy the following reciprocity relations:

$$\int \eta(\boldsymbol{r},\boldsymbol{r}'')s(\boldsymbol{r}'',\boldsymbol{r}')d\boldsymbol{r}'' = \int \frac{\delta u(\boldsymbol{r}'')}{\delta \rho(\boldsymbol{r})}\frac{\delta \rho(\boldsymbol{r}')}{\delta u(\boldsymbol{r}'')}d\boldsymbol{r}'' = \frac{\delta \rho(\boldsymbol{r}')}{\delta \rho(\boldsymbol{r})} = \delta(\boldsymbol{r}'-\boldsymbol{r}),$$
$$\int s(\boldsymbol{r},\boldsymbol{r}'')\eta(\boldsymbol{r}'',\boldsymbol{r}')d\boldsymbol{r}'' = \int \frac{\delta \rho(\boldsymbol{r}'')}{\delta u(\boldsymbol{r})}\frac{\delta u(\boldsymbol{r}')}{\delta \rho(\boldsymbol{r}'')}d\boldsymbol{r}'' = \frac{\delta u(\boldsymbol{r}')}{\delta u(\boldsymbol{r})} = \delta(\boldsymbol{r}'-\boldsymbol{r}), \qquad (7.234)$$

thus identifying each kernel as the inverse of the other.

By using the appropriate chain-rule transformations of the defining density derivative the softness kernel can be expressed in terms of the softness quantities (Berkowitz and Parr 1988):

$$s(\boldsymbol{r},\boldsymbol{r}') = -\beta(\boldsymbol{r},\boldsymbol{r}') + s(\boldsymbol{r})f(\boldsymbol{r}') = -\beta(\boldsymbol{r},\boldsymbol{r}') + f(\boldsymbol{r})Sf(\boldsymbol{r}'). \qquad (7.235)$$

The first term in the preceding equation measures the "internal" density response of the closed system, while the second contribution reflects the effect of an external CT.

These derivatives determine the associated *second*-order expansion of the system grand potential $\Omega[\mu, v] = \Omega[u]$ in terms of powers of displacements in their arguments (state variables), $\{\Delta\mu, \Delta v(\boldsymbol{r})\}$ (or $\Delta u(\boldsymbol{r}) = \Delta[v(\boldsymbol{r}) - \mu]$),

$$\begin{aligned}
\Delta^{(1+2)}\Omega[\mu,v] &= \left(\frac{\partial\Omega}{\partial\mu}\right)_v \Delta\mu + \int \left(\frac{\partial\Omega}{\partial v(\boldsymbol{r})}\right)_\mu \Delta v(\boldsymbol{r})\,d\boldsymbol{r} \\
&\quad + \frac{1}{2}\left[\left(\frac{\partial^2\Omega}{\partial\mu^2}\right)_v (\Delta\mu)^2 + 2\Delta\mu \int \frac{\partial^2\Omega}{\partial\mu\,\partial v(\boldsymbol{r})}\Delta v(\boldsymbol{r})\,d\boldsymbol{r}\right. \\
&\quad \left. + \iint \left(\frac{\partial^2\Omega}{\partial v(\boldsymbol{r})\,\partial v(\boldsymbol{r}')}\right)_\mu \Delta v(\boldsymbol{r})\Delta v(\boldsymbol{r}')\,d\boldsymbol{r}\,d\boldsymbol{r}'\right] \\
&\equiv -N\Delta\mu + \int \rho(\boldsymbol{r})\,\Delta v(\boldsymbol{r})\,d\boldsymbol{r} \\
&\quad + \frac{1}{2}\left[-S(\Delta\mu)^2 + 2\Delta\mu \int s(\boldsymbol{r})\,\Delta v(\boldsymbol{r})\,d\boldsymbol{r}\right. \\
&\quad \left. + \iint s(\boldsymbol{r},\boldsymbol{r}')\,\Delta v(\boldsymbol{r})\,\Delta v(\boldsymbol{r}')\,d\boldsymbol{r}\,d\boldsymbol{r}'\right] \\
&= \int \rho(\boldsymbol{r})\,\Delta u(\boldsymbol{r})\,d\boldsymbol{r} - \frac{1}{2}\iint s(\boldsymbol{r},\boldsymbol{r}')\,\Delta u(\boldsymbol{r})\,\Delta u(\boldsymbol{r}')\,d\boldsymbol{r}\,d\boldsymbol{r}'. \qquad (7.236)
\end{aligned}$$

Above, we have used the following chain-rule expressions for the local and global softnesses in terms of the softness kernel:

$$\begin{aligned}
S = \left(\frac{\partial N}{\partial\mu}\right)_v &= \iint \frac{\delta\rho(\boldsymbol{r}')}{\delta u(\boldsymbol{r})}\left(\frac{\partial u(\boldsymbol{r})}{\partial\mu}\right)_v d\boldsymbol{r}\,d\boldsymbol{r}' = \iint s(\boldsymbol{r},\boldsymbol{r}')d\boldsymbol{r}\,d\boldsymbol{r}' \\
&= \int \left(\frac{\partial\rho(\boldsymbol{r})}{\partial\mu}\right)_v d\boldsymbol{r} = \int s(\boldsymbol{r})d\boldsymbol{r}, \qquad (7.237)
\end{aligned}$$

where the local softness

$$s(\boldsymbol{r}) = \int s(\boldsymbol{r}, \boldsymbol{r}')d\boldsymbol{r}'. \tag{7.238}$$

One similarly expresses the global hardness in terms of its kernel:

$$\begin{aligned}\eta = \left(\frac{\partial^2 E}{\partial N^2}\right)_v &= \iint \left(\frac{\partial \rho(\boldsymbol{r})}{\partial N}\right)_v \frac{\delta^2 E_v[\rho]}{\delta\rho(\boldsymbol{r})\delta\rho(\boldsymbol{r}')} \left(\frac{\partial \rho(\boldsymbol{r}')}{\partial N}\right)_v d\boldsymbol{r}\, d\boldsymbol{r}' \\ &= \iint f(\boldsymbol{r})\, \eta(\boldsymbol{r}, \boldsymbol{r}') f(\boldsymbol{r}') d\boldsymbol{r}\, d\boldsymbol{r}'. \end{aligned} \tag{7.239}$$

In the DA complexes of Sect. 7.3.4 the two reactants A (acid) and B (base) experience the coupled changes in their electron numbers, due to the CT between them, $\Delta \boldsymbol{N} = (\Delta N_{\mathrm{A}} = N_{\mathrm{CT}}, \Delta N_{\mathrm{B}} = -N_{\mathrm{CT}})$ (row vector), and in the external potentials felt by electrons due to the presence of the reaction partner, $\Delta \boldsymbol{v} = (\Delta v_{\mathrm{A}} \approx \phi_{\mathrm{B}}, \Delta v_{\mathrm{B}} \approx \phi_{\mathrm{A}})$ (row vector), where

$$\phi_{\mathrm{X}}(\boldsymbol{r}) = -\sum_{\alpha \in \mathrm{X}} Z_\alpha / |\boldsymbol{r} - \boldsymbol{R}_\alpha| + \int \rho_{\mathrm{X}}(\boldsymbol{r}')/|\boldsymbol{r}' - \boldsymbol{r}| d\boldsymbol{r}'$$

denotes the negative electrostatic potential (ESP) due to the nuclei and electrons of X. In this reactant resolution the *second*-order expansion for the DA complex will involve the matrices of CS introduced in (7.227):

$$\begin{aligned}\Delta^{(1+2)} E_{\mathrm{DA}}[\boldsymbol{N}, \boldsymbol{v}] &= \left(\frac{\partial E_{\mathrm{DA}}}{\partial \boldsymbol{N}}\right)_{\boldsymbol{v}} \Delta \boldsymbol{N}^{\mathrm{T}} + \int \left(\frac{\partial E_{\mathrm{DA}}}{\partial \boldsymbol{v}(\boldsymbol{r})}\right)_{\boldsymbol{N}} \Delta \boldsymbol{v}(\boldsymbol{r})^{\mathrm{T}} d\boldsymbol{r} \\ &\quad + \frac{1}{2}\left[\Delta \boldsymbol{N} \left(\frac{\partial^2 E_{\mathrm{DA}}}{\partial \boldsymbol{N}\, \partial \boldsymbol{N}}\right)_{\boldsymbol{v}} \Delta \boldsymbol{N}^{\mathrm{T}} + 2\Delta \boldsymbol{N} \int \frac{\partial^2 E_{\mathrm{DA}}}{\partial \boldsymbol{N}\, \partial \boldsymbol{v}(\boldsymbol{r})} \Delta \boldsymbol{v}(\boldsymbol{r})^{\mathrm{T}} d\boldsymbol{r}\right. \\ &\quad \left. + \iint \Delta \boldsymbol{v}(\boldsymbol{r}) \left(\frac{\partial^2 E_{\mathrm{DA}}}{\partial \boldsymbol{v}(\boldsymbol{r})\, \partial \boldsymbol{v}(\boldsymbol{r}')}\right)_{\boldsymbol{N}} \Delta \boldsymbol{v}(\boldsymbol{r}')^{\mathrm{T}} d\boldsymbol{r}\, d\boldsymbol{r}'\right] \\ &\equiv \boldsymbol{\mu}_{\mathrm{DA}} \Delta \boldsymbol{N}^{\mathrm{T}} + \int \boldsymbol{\rho}_{\mathrm{DA}}(\boldsymbol{r})\, \Delta \boldsymbol{v}(\boldsymbol{r})^{\mathrm{T}} d\boldsymbol{r} \\ &\quad + \frac{1}{2}\left(\Delta \boldsymbol{N}\, \boldsymbol{\eta}_{\mathrm{DA}} \Delta \boldsymbol{N}^{\mathrm{T}} + 2\Delta \boldsymbol{N} \int \mathbf{f}_{\mathrm{DA}}(\boldsymbol{r})\, \Delta \boldsymbol{v}(\boldsymbol{r})^{\mathrm{T}} d\boldsymbol{r}\right. \\ &\quad \left. + \iint \Delta \boldsymbol{v}(\boldsymbol{r})\, \boldsymbol{\beta}_{\mathrm{DA}}(\boldsymbol{r}, \boldsymbol{r}')\, \Delta \boldsymbol{v}(\boldsymbol{r}')^{\mathrm{T}} d\boldsymbol{r}\, d\boldsymbol{r}'\right). \end{aligned} \tag{7.240}$$

Here, the (row) vectors of the first partial derivatives of the system electronic energy $\boldsymbol{\mu}_{\mathrm{DA}} = (\mu_{\mathrm{A}}, \mu_{\mathrm{B}})$ and $\boldsymbol{\rho}_{\mathrm{DA}} = (\rho_{\mathrm{A}}, \rho_{\mathrm{B}})$ group the chemical potentials and electron densities of reactants, respectively, while the hardness, FF, and linear response matrices combine the corresponding second partial derivatives [see (7.189)]:

$$\boldsymbol{\eta}_{\mathrm{DA}} = \left(\frac{\partial^2 E_{\mathrm{DA}}[\boldsymbol{N}, \boldsymbol{v}]}{\partial \boldsymbol{N}\, \partial \boldsymbol{N}}\right)_{\boldsymbol{v}} = \left(\frac{\partial^2 \boldsymbol{\mu}_{\mathrm{DA}}}{\partial \boldsymbol{N}}\right)_{\boldsymbol{v}} = \left\{\left(\frac{\partial \mu_{\mathrm{Y}}}{\partial N_{\mathrm{X}}}\right)_{\boldsymbol{v}}\right\}, \tag{7.241}$$

$$\begin{aligned} \mathbf{f}_{\mathrm{DA}}(\boldsymbol{r}) &= \frac{\partial^2 E_{\mathrm{DA}}[\boldsymbol{N}, \boldsymbol{v}]}{\partial \boldsymbol{N} \partial \boldsymbol{v}(\boldsymbol{r})} = \frac{\partial \boldsymbol{\rho}_{\mathrm{DA}}(\boldsymbol{r})}{\partial \boldsymbol{N}} = \left(\frac{\partial \boldsymbol{\mu}_{\mathrm{DA}}}{\partial \boldsymbol{v}(\boldsymbol{r})}\right)^{\mathrm{T}} \\ &= \left\{\frac{\partial^2 E_{\mathrm{DA}}}{\partial N_{\mathrm{X}} \partial v_{\mathrm{Y}}(\boldsymbol{r})} = \left(\frac{\partial \rho_{\mathrm{Y}}(\boldsymbol{r})}{\partial N_{\mathrm{X}}}\right)_{\boldsymbol{v}, N_{\mathrm{Y}}} = \left(\frac{\partial \mu_{\mathrm{X}}}{\partial v_{\mathrm{Y}}(\boldsymbol{r})}\right)_{N, v_{\mathrm{X}}}\right\}, \end{aligned} \tag{7.242}$$

$$\boldsymbol{\beta}_{\mathrm{DA}}(\boldsymbol{r}, \boldsymbol{r}') = \left(\frac{\partial^2 E_{\mathrm{DA}}[\boldsymbol{N}, \boldsymbol{v}]}{\partial \boldsymbol{v}(\boldsymbol{r})\, \partial \boldsymbol{v}(\boldsymbol{r}')}\right)_{N} = \left(\frac{\partial \boldsymbol{\rho}_{\mathrm{DA}}(\boldsymbol{r}')}{\partial \boldsymbol{v}(\boldsymbol{r})}\right)_{N} = \left\{\left(\frac{\partial \rho_{\mathrm{Y}}(\boldsymbol{r}')}{\partial v_{\mathrm{X}}(\boldsymbol{r})}\right)_{N, v_{\mathrm{Y}}}\right\}. \tag{7.243}$$

It should be emphasized, that this quadratic Taylor expansion of the interaction energy between two molecular reactants remains meaningful only at an early stage of their mutual approach, when they are nearly separable thus preserving their chemical identity. Only then the state parameters of the isolated reactants can be used to determine the gross effects of their presence at the finite separation between these two subsystems. Comparing trends in these representative and simplified interactions for a series of the alternative mutual orientations of the two reactants then admits valid qualitative conjectures about the preferred course of the reaction at smaller separations.

7.4 Local Density and Gradient Approximations

It follows from (7.140) and (7.145) that densities of the exchange and correlation energies are nonlocal objects, reflecting the fact that the associated density-hole interactions for the reference electron at $\boldsymbol{r}$ depend on the presence of the remaining electrons at all their locations, through the nonlocal exchange and correlation holes. However, as already indicated in Sect. 7.2, the historically first applications of modern DFT of Kohn and Sham have adopted the kinetic energy contribution of the noninteracting system, $T_s[\rho]$, and the LDA to the unknown density functional for the exchange-correlation energy,

$$E_{xc}^{\mathrm{LDA}}[\rho] = \int \rho(\boldsymbol{r}) \varepsilon_{xc}(\rho(\boldsymbol{r})) d\boldsymbol{r}, \tag{7.244}$$

which makes a reference to the homogeneous electron gas by regarding the inhomogeneous electron distribution as locally homogeneous. This quite reliable approach makes use of the known *Quantum Monte Carlo* (QMC) results of Ceperley and Alder (1980) for the homogeneous electron gas. Its spin-resolved version, known as the *Local Spin Density Approximation* (LSDA) (von Barth and Hedin 1972; Oliver and Perdew 1979),

$$E_{xc}^{\mathrm{LSDA}}[\rho_\alpha, \rho_\beta] = \int \rho(\boldsymbol{r}) \varepsilon_{xc}(\rho_\alpha(\boldsymbol{r}), \rho_\beta(\boldsymbol{r})) d\boldsymbol{r}, \tag{7.245}$$

which is required for the magnetic and open-shell systems, similarly uses the appropriate interpolation between the fully spin-polarized and unpolarized data for the homogeneous system of the specified electron density (Vosko et al. 1980; Perdew and Zunger 1981). These parametrizations are local, i.e., the value of the xc-potential at position $\boldsymbol{r}$ depends solely on the values of the density arguments at this point.

Similar ideas have been used within the historic predecessor of DFT, known as the *Thomas–Fermi–Dirac* (TFD) theory, which uses the following LDA-type functional for the system electronic energy (see Sect. 7.2):

$$E_{\mathrm{TFD}}[\rho] = C_k \int \rho(\boldsymbol{r})^{5/3} d\boldsymbol{r} + \int \rho(\boldsymbol{r}) v(\boldsymbol{r})\, d\boldsymbol{r} + J_{ee}[\rho] - C_x \int \rho(\boldsymbol{r})^{4/3} d\boldsymbol{r}, \tag{7.246}$$

where the proportionality constants $C_k = 3(3\pi^2)^{2/3}/10$ and $C_x = 3(3\pi)^{1/3}/4$. The (Coulomb) correlation can be added using Wigner's (1934) functional

$$E_c{}^{\mathrm{W}} = -0.056 \int \rho(\boldsymbol{r})^{4/3} [0.079 + \rho(\boldsymbol{r})^{1/3}] d\boldsymbol{r}, \tag{7.247}$$

and the nonhomogeneity correction to the local kinetic energy functional, proportional to the density-gradient integral $\int \rho(\boldsymbol{r})^{-1} |\nabla \rho(\boldsymbol{r})|^2 d\boldsymbol{r}$, has been proposed by von Weizsäcker (1935). Indeed, through the density gradient this correction probes the local nonhomogeneities of the electronic gas.

To summarize, the coupling-strength-averaged *pair*-correlation function of (6.97),

$$\begin{aligned} f_{xc}^{av}(\boldsymbol{r}, \boldsymbol{r}') &= \int_0^1 f_{xc}^{\lambda}(\boldsymbol{r}, \boldsymbol{r}')\, d\lambda = 1 + h_{xc}^{av}(\boldsymbol{r}'|\boldsymbol{r})/\rho(\boldsymbol{r}'), \\ f_{xc}^{\lambda}(\boldsymbol{r}, \boldsymbol{r}') &= 1 + h_{xc}^{\lambda}(\boldsymbol{r}'|\boldsymbol{r})/\rho(\boldsymbol{r}'), \end{aligned} \tag{7.248}$$

and the associated average hole $h_{xc}^{av}(\boldsymbol{r}'|\boldsymbol{r})$, which ultimately determine the exchange correlation energy functional of (7.137),

$$\begin{aligned} E_{xc}[\rho] &= \frac{1}{2} \iint \frac{\rho(\boldsymbol{r})\rho(\boldsymbol{r}')}{|\boldsymbol{r} - \boldsymbol{r}'|} [f_{xc}^{av}(\boldsymbol{r}, \boldsymbol{r}') - 1]\, d\boldsymbol{r}\, d\boldsymbol{r}' \\ &= \frac{1}{2} \iint \frac{\rho(\boldsymbol{r}) h_{xc}^{av}(\boldsymbol{r}'|\boldsymbol{r})}{|\boldsymbol{r} - \boldsymbol{r}'|} d\boldsymbol{r}\, d\boldsymbol{r}' \equiv \int \rho(\boldsymbol{r}) \varepsilon_{xc}(\boldsymbol{r})\, d\boldsymbol{r}, \end{aligned} \tag{7.249}$$

are approximated in LDA by the appropriate distributions for the homogeneous electron gas density ρ^h which locally equals $\rho(\boldsymbol{r})$. They depend only on the distance $u = |\boldsymbol{r} - \boldsymbol{r}'|$ between the two electrons:

$$\begin{aligned} h_{xc}^{\mathrm{LDA}}(\boldsymbol{r}'|\boldsymbol{r}) &= \rho(\boldsymbol{r}')[f_{xc}^{\mathrm{LDA}}(\boldsymbol{r}, \boldsymbol{r}') - 1] \\ &\approx \rho(\boldsymbol{r})\{f^h[u, \rho^h = \rho(\boldsymbol{r})] - 1\}, \end{aligned} \tag{7.250}$$

and reduce the interaction of the hole at $\boldsymbol{r}$ with the density in the same position. Therefore, in LDA the exchange-correlation hole centered at $\boldsymbol{r}$ is in fact interacting only locally, with the electronic density in the same position, which amounts to approximating the *pair*-correlation kernel

$$f_{xc}^{\mathrm{LDA}}(\boldsymbol{r},\boldsymbol{r}') \approx f^h[u,\rho(\boldsymbol{r})]\rho(\boldsymbol{r})/\rho(\boldsymbol{r}'),$$

with the density ratio formally effecting this local LDA constraint.

In the LSDA of the spin-resolved DFT the two spin densities $\rho_\alpha(\boldsymbol{r}) = \rho_\uparrow(\boldsymbol{r})$ and $\rho_\beta(\boldsymbol{r}) = \rho_\downarrow(\boldsymbol{r})$, or equivalently the overall density, $\rho(\boldsymbol{r}) = \rho_\downarrow(\boldsymbol{r}) + \rho_\uparrow(\boldsymbol{r})$, and the magnetization density, $\zeta(\boldsymbol{r}) = [\rho_\uparrow(\boldsymbol{r}) - \rho_\downarrow(\boldsymbol{r})]/\rho(\boldsymbol{r})$, form the independent local state-parameters and hence:

$$\begin{aligned} E_{xc}^{\mathrm{LSDA}}[\rho_\alpha,\rho_\beta] &= \int \rho(\boldsymbol{r})\varepsilon_{xc}(\rho_\uparrow(\boldsymbol{r}),\rho_\downarrow(\boldsymbol{r}))\,d\boldsymbol{r} \\ &= E_{xc}^{\mathrm{LSDA}}[\rho,\zeta] = \int \rho(\boldsymbol{r})\varepsilon_{xc}(\rho(\boldsymbol{r}),\zeta(\boldsymbol{r}))\,d\boldsymbol{r}. \end{aligned} \tag{7.251}$$

The density per electron of the LSDA xc-energy $\varepsilon_{xc}(\rho_\uparrow(\boldsymbol{r}),\rho_\downarrow(\boldsymbol{r}))$ (7.245) replaces the LDA density $\varepsilon_{xc}(\rho(\boldsymbol{r}))$ of (7.244), with the locally interpolated expression for the homogeneous electronic distributions in the paramagnetic ($\zeta = 0$) and ferromagnetic ($\zeta = 1$) limits. This can be accomplished using von Barth and Hedin's (1972) formula:

$$\begin{aligned} &\varepsilon_{xc}(\rho_\uparrow(\boldsymbol{r}),\rho_\downarrow(\boldsymbol{r})) \approx \varepsilon_{xc}^h(\rho(\boldsymbol{r}),\zeta(\boldsymbol{r})) = x(\zeta)\ \varepsilon_{xc}^{polarized}[\rho;\boldsymbol{r}] + [1-x(\zeta)]\varepsilon_{xc}^{unpolarized}[\rho;\boldsymbol{r}], \\ &\quad \varepsilon_{xc}^{polarized}[\rho;\boldsymbol{r}] = \varepsilon_{xc}^h(\rho(\boldsymbol{r}),\zeta(\boldsymbol{r})=1), \\ &\quad \varepsilon_{xc}^{unpolarized}[\rho;\boldsymbol{r}] = \varepsilon_{xc}^h(\rho(\boldsymbol{r}),\zeta(\boldsymbol{r})=0), \\ &\quad x(\zeta) = [(1+\zeta)^{4/3} + (1-\zeta)^{4/3} - 2]/(2^{4/3}-2), \end{aligned} \tag{7.252}$$

or a more advanced parameterization by Vosko et al. (1980) of the known QMC results and the *Random Phase Approximation* (RPA) calculations on the uniform electron gas.

These local, spherical-hole approximations, being derived from the real physical system (homogeneous electron gas), satisfy all relevant sum rules for the correlation holes and provide a decent representation of the spherically averaged correlation holes (6.87), which ultimately determine $E_{xc}[\rho]$ (6.88) (Ernzerhof et al. 1996; Perdew 1999). This explains a remarkable performance of the LDA/LSDA calculations, particularly in numerous solid-state applications. However, these variants of KS theory have been shown to favor homogeneous systems and they overbind molecules/solids. In the chemically bonded molecular systems, this approximation generally gives good equilibrium geometries, but $v_{xc}(\boldsymbol{r})$ potential does not exhibit the correct $-e^2/r$ decay, a consequence of LDA and LSDA failing at canceling the *self*-interaction included in the Hartree term $J_{ee}[\rho]$, thus affecting both the dissociation limits and ionization energies.

Since the electron correlation, of both the Fermi and Coulomb origins, represents an inherently nonlocal phenomenon, the local inhomogeneity of the electron gas reflected by the density gradient should affect $E_{xc}[\rho]$. Indeed, for densities that vary slowly in space, the following *Gradient Expansion Approximation* (GEA) of the exchange-correlation energy [(7.100) and (7.105)] (Hohenberg and Kohn 1964; Kohn and Sham 1965) applies,

$$E_{xc}[\rho] = E_{xc}^{\mathrm{GEA}}[\rho] = E_{xc}^{\mathrm{LDA}}[\rho] + \int C_{xc}[\rho(\boldsymbol{r})]\frac{|\nabla\rho(\boldsymbol{r})|^2}{\rho(\boldsymbol{r})^{4/3}}d\boldsymbol{r} + \ldots, \tag{7.253}$$

with the leading LDA term. However, a straightforward evaluation of this expansion is ill-behaved, with the *first*-order correction worsening the quality of the LDA predictions. The reasons for that are by now fairly well understood: the expansion to order $|\nabla\rho(\boldsymbol{r})|^2$ of the exchange-correlation hole does not correspond to any real physical system since it violates the relevant sum rules. To remedy this, one has to retain all relevant contributions to the desired order, from all terms in this expansion, and one must then enforce the exact key conditions satisfied by the real correlation holes, which may be violated by truncated expansions. Perdew and collaborators have used this sum-rule explanation in their first-principles construction of $\varepsilon_{xc}(\rho, |\nabla\rho|)$ within the *Generalized Gradient Approximation* (GGA) (Perdew and Wang 1986, 1989; Perdew et al. 1996a, b, c, 1998; Perdew 1991; Ernzerhof et al. 1996; Perdew 1999), through the cut-off holes, missing the spurious *long*-rage part, to satisfy the sum rules. Essentially the same GGA has been derived by Perdew, Burke and Ernzerhoff (PBE) (1996a, b, c, 1997), who imposed exact conditions directly on $E_{xc}[\rho]$, without appeal to the exchange-correlation hole. The GGA and PBE functionals have shown that imposing these exact constraints on functionals that originally do not verify them results in a remarkable improvement of the functional quality.

To summarize, in the GGA,

$$\begin{aligned} E_{xc}^{\mathrm{GGA}}[\rho] &= \int \rho(\boldsymbol{r})\,\varepsilon_{xc}(\,\rho(\boldsymbol{r}), |\nabla\rho(\boldsymbol{r})|)\,d\boldsymbol{r} \quad \text{or} \\ E_{xc}^{\mathrm{GGA}}[\rho_\alpha, \rho_\beta] &= \int \rho(\boldsymbol{r})\,\varepsilon_{xc}\big(\,\rho_\alpha(\boldsymbol{r}), \rho_\beta(\boldsymbol{r}), |\nabla\rho_\alpha(\boldsymbol{r})|, |\nabla\rho_\beta(\boldsymbol{r})|\big)\,d\boldsymbol{r}, \end{aligned} \tag{7.254}$$

generating at present the functionals of choice in quantum chemistry, one can construct a parameterized analytic form of the functional density and then fit the parameters to a data set, e.g., energies of atoms and molecules, or approach the problem using the quantum mechanical constraints imposed on the functional density through general properties of the correlation holes. A convenient way to visualize the nonlocality of GGA is through the "*enhancement*" factor $F_{xc}[\boldsymbol{r}_S(\boldsymbol{r}), s(\boldsymbol{r})]$ over the *local* exchange $\varepsilon_x[\rho(\boldsymbol{r})]$:

$$E_{xc}^{\mathrm{GGA}}[\rho] = \int \rho(\boldsymbol{r})\,\varepsilon_x[\rho(\boldsymbol{r})]F_{xc}[r_S(\boldsymbol{r}), s(\boldsymbol{r})]\;d\boldsymbol{r}. \tag{7.255a}$$

Here, the local radius of the sphere enclosing a single electron in the uniform electron gas, $r_S(\boldsymbol{r}) = \{3/[4\pi\rho(\boldsymbol{r})]\}^{1/3}$, and the dimensionless density gradient

$$s(\boldsymbol{r}) = |\nabla\rho(\boldsymbol{r})|/[2k_F(\boldsymbol{r})\rho(\boldsymbol{r})], \quad k_F = [3\pi^2\rho(\boldsymbol{r})]^{1/3}. \tag{7.255b}$$

The LDA for exchange and correlation is recovered for $F_{xc}(r_S, 0)$, while the GGA for the exchange alone corresponds to $F_{xc}(0, s)$.

What is thus needed in this generalized nonlocal approach is the analytical form that mimics the resummation to infinite order. Clearly, there is no unique recipe for constructing such gradient-dependent functionals through some heuristic resummation of the gradient expansion, since not all formal properties can be enforced simultaneously. This nonuniqueness has resulted in a number of GGA's, e.g., functionals due to Langreth and Mehl (LM) (1981), Perdew and Wang (PW86 and PW91), Becke's (1985, 1988) exchange (B88), and Lee–Yang–Parr (1988) (LYP) correlation energy. Currently, the GGA of choice appears to be the Perdew–Burke–Ernzerhof (1996a, b, c, 1997) (PBE) functional,

$$E_{xc}^{\text{PBE}}[\rho] = \int \rho(\boldsymbol{r})\,\varepsilon_x[\rho(\boldsymbol{r})]F_{xc}[\rho(\boldsymbol{r}),\ \zeta(\boldsymbol{r}),\ s(\boldsymbol{r})]\,d\boldsymbol{r}, \tag{7.256}$$

very satisfactory from the theoretical point of view, as obeying many of the exact conditions verified by exact correlation holes and devoid of any fitting parameters, i.e., expressed only in terms of fundamental physical constants.

In meta-GGA's (MGGA) one goes beyond the *second*-order gradient expansion of the exchange-energy, which introduces the term proportional to the squared gradient of the density in $\varepsilon_{xc}(\rho, |\nabla\rho|^2)$. The *fourth*-order expansion similarly gives the contributions proportional to the squared Laplacian of the electron density and other fourth-order combinations of the gradient operator: $\varepsilon_{xc}(\rho, |\nabla\rho|^2, |\nabla^2\rho|^2, |\nabla\rho|^2\nabla^2\rho, |\nabla\rho|^4)$ (see also Sect. 7.2). However, the fully *first*-principle construction of this function is not available, so that all meta-GGA functionals include some fitted parameters.

Another way to introduce the Laplacian-type nonlocality is to introduce the KS kinetic energy density to *xc*-functional (e.g., Becke and Roussel 1989; Van Voorhis and Scuseria 1998),

$$\begin{aligned} \tau(\boldsymbol{r}) &= \frac{1}{2}\sum_{n}^{occupied} |\nabla\varphi_n(\boldsymbol{r})|^2 \\ &= \frac{1}{2}\sum_{\sigma}^{\alpha,\beta}\sum_{i=1}^{N_\sigma} |\nabla\varphi_{i\sigma}(\boldsymbol{r})|^2 = \sum_{\sigma}^{\alpha,\beta}\tau_\sigma(\boldsymbol{r}). \end{aligned} \tag{7.257}$$

The general form of the spin-resolved density functional for the KS exchange-correlation energy then reads:

$$E_{xc}^{\text{MGGA}}[\rho_\alpha, \rho_\beta] = \int \rho\,\varepsilon_{xc}\big(\rho_\alpha, \rho_\beta, |\nabla\rho_\alpha(\boldsymbol{r})|, |\nabla\rho_\beta(\boldsymbol{r})|, \Delta\rho_\alpha, \Delta\rho_\beta, \tau_\alpha, \tau_\beta\big)\,d\boldsymbol{r}. \tag{7.258}$$

Contrary to LDA, all these nonlocal gradient corrections favor density inhomogeneity thus reducing the LDA tendency to "homogenize" the system, overbind atoms and molecules, and thus to overly contract the bond lengths. The GGA decisively improves the predicted binding and atomization energies, bond lengths and angles. The GGA energetics, geometries, and dynamics of the hydrogen-bonded system also exhibit a marked improvement over the LDA results.

However, the GGA functionals still do not satisfy the correct asymptotic behavior, since they do not compensate completely for the *self*-interaction, and they contain insufficient degree of nonlocality in the exchange part. This motivated the development of the new generation of *hybrid* functionals, called hyper-GGAs (HGGA), which additionally depend on the exact exchange densities or related quantities (Becke 1993, 1996, 1997; Ernzerhof 1996; Perdew et al. 1996a, b, c). For example, one may adopt the mixture

$$E_{xc}^{hybrid} = E_{xc}^{\mathrm{GGA}} + a(E_x^{exact} - E_{xc}^{\mathrm{GGA}}), \tag{7.259a}$$

where a is a mixing parameter, with $a = 1$ recovering the exact exchange energy of the Kohn–Sham orbitals (7.140),

$$E_x^{exact} = \langle \Psi_s | \hat{\mathrm{K}}^{\mathrm{KS}} | \Psi_s \rangle = -\sum_{\sigma=\uparrow,\downarrow} \frac{1}{4} \iint \frac{\left| \sum_n^{occupied} \varphi_{n\sigma}^*(\boldsymbol{r}) \varphi_{n\sigma}(\boldsymbol{r}') \right|^2}{|\boldsymbol{r} - \boldsymbol{r}'|} d\boldsymbol{r}\, d\boldsymbol{r}', \tag{7.259b}$$

and Becke's choice of $a = ¼$ giving the best reproductions of the MP2 and CI geometries and binding energies of molecular systems. The B3LYP and B3PW91 functionals (Becke 1993, 1996, 1997) are examples of most popular and remarkably successful hybrid functionals in contemporary computational quantum chemistry.

The spin-resolved HGGA functional for $E_{xc}[\rho_\alpha, \rho_\beta]$ thus assumes the following general form:

$$E_{xc}^{\mathrm{HGGA}}[\rho_\alpha, \rho_\beta] = \int \rho\, \varepsilon_{xc}\left(\rho_\alpha, \rho_\beta, |\nabla\rho_\alpha(\boldsymbol{r})|, \left|\nabla\rho_\beta(\boldsymbol{r})\right|, \Delta\rho_\alpha, \Delta\rho_\beta, \tau_\alpha, \tau_\beta, \varepsilon_{x\alpha}, \varepsilon_{x\beta}\right) d\boldsymbol{r},$$
$$\varepsilon_{x\sigma}(\boldsymbol{r}) = -\frac{1}{2\rho_\sigma(\boldsymbol{r})} \int \left| \sum_{i=1}^{N_\sigma} \varphi_{i\sigma}^*(\boldsymbol{r}) \varphi_{i\sigma}(\boldsymbol{r}') \right|^2 |\boldsymbol{r} - \boldsymbol{r}'|^{-1}\, d\boldsymbol{r}'. \tag{7.260}$$

These functional approximations climb the first four rungs of the "Jacob's ladder" of the DFT accuracy (Perdew et al. 2005) shown in Fig. (7.5). It also establishes the hierarchy of the complexity/advancement in the "functional ZOO" (Perdew 1999), between the relative simplicity of the ordinary Hartree approach and the upper, FCI or CC world representing the "heaven of chemical accuracy," with the lowest LDA/LSDA rung and the currently most advanced generalized RPA functionals, which also take into account the unoccupied KS orbitals. Predictably,

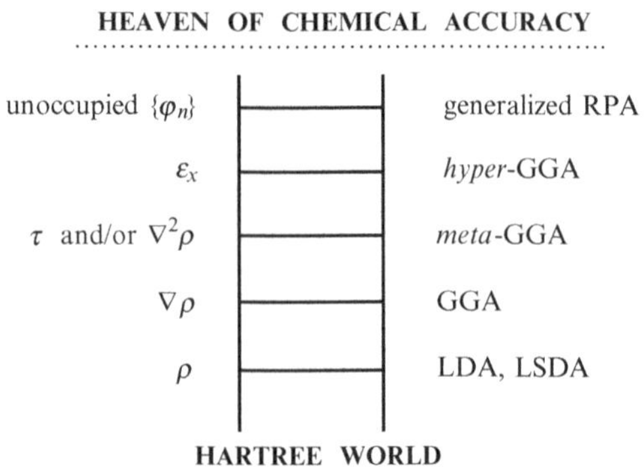

Fig. 7.5 Functional Jacob's ladder of Perdew et al. (2005)

the higher one climbs this functional ladder the further one gets from Kohn's ideal of a simplified and affordable treatment of the electron correlation problem allowing calculations for systems of very many (10^3–10^5) atoms.

The *third*-generation (orbitally dependent) functionals use the exact exchange plus some approximation for the (Coulomb) correlation, thus replacing the KS partition of the electronic energy (7.114) with its *full*-exchange analog:

$$E_v[\rho] = T_s[\rho] + \int \rho(\boldsymbol{r})v(\boldsymbol{r})\,d\boldsymbol{r} + J_{ee}[\rho] + E_x^{exact}[\rho] + E_c[\rho]. \tag{7.261}$$

These functionals will be discussed in the next section. For the asymptotic behavior care has to be taken to ensure that the correlation hole cancels the *long*-range component of the exact exchange hole.

By neglecting the last, Coulomb correlation term in the preceding equation, one can formulate the x-only KS-type variational procedure giving rise to effective *one*-body equations for the optimum orbitals $\{\varphi_n\}$ of the hypothetical noninteracting system of the same density as the real interacting system exhibiting the Fermi correlation. This reverse HF problem of finding the variationally best local exchange potential $v_{x-only}(\boldsymbol{r})$, which minimizes the HF energy $E_{\rm HF}[\{\varphi_n\}]$ (6.16) and gives rise to the exact HF density $\rho_{\rm HF}(\boldsymbol{r})$, is known as the *Optimized Potential Method* (OPM) (Sharp and Horton 1953; Talman and Shadwick 1976). It is thus summarized by the following KS-type equations:

$$\begin{aligned}
&\left(-\frac{\nabla^2}{2} + v(\boldsymbol{r}) + \int \frac{\rho(\boldsymbol{r}')}{|\boldsymbol{r}'-\boldsymbol{r}|}d\boldsymbol{r}' + \frac{\delta E_x^{exact}[\rho]}{\delta\rho(\boldsymbol{r})}\right)\varphi_n(\boldsymbol{r}) \\
&\quad\equiv \left(-\frac{\nabla^2}{2} + v(\boldsymbol{r}) + v_{\rm H}(\boldsymbol{r}) + v_{x-only}(\boldsymbol{r})\right)\varphi_n(\boldsymbol{r}) = \varepsilon_n\varphi_n(\boldsymbol{r}), \\
&\rho(\boldsymbol{r}) = \rho_{\rm HF}(\boldsymbol{r}) = \sum_n^{occupied} |\varphi_n(\boldsymbol{r})|^2.
\end{aligned} \tag{7.262}$$

Clearly, the KS scheme can be viewed as a special case of this OPM procedure. It has been first addressed by Sharp and Horton, long before the birth of the modern

DFT, with Talman and Shadwick performing first numerical calculations for small atoms. This approach leads to an integral equation for the associated effective potential $v_{x-only}(\boldsymbol{r})$, which has to be solved self-consistently with the preceding KS-like equations. The optimum OPM orbitals were found to be practically identical with the corresponding HF analogs (Grabo et al. 1999; Ivanov et al. 1999; Görling 1999; Hirata et al. 2001; Krieger et al. 1990; Kümmel and Perdew 2003; Kümmel and Kronik 2008). This practically demonstrates the fundamental concept of the KS method, that nonlocal interactions can be treated exactly with local potentials. The exact exchange method introduces the explicit dependence of the exchange-correlation energy on orbitals, now the complex functional of the density, in the same way as the KS scheme has introduced such a dependence of the kinetic energy.

7.5 Orbital-Dependent Functionals

The modern density functionals for the exchange-correlation energy are constructed in the explicitly *orbital*-dependent forms (Grabo et al. 1999), with only the implicit dependence on the electronic distribution. In the spirit of (7.261) the exact exchange energy of the KS system can be then directly expressed in terms of KS orbitals using the HF-like expression of (7.140), with only the Coulomb correlation energy of the interacting system to be realistically approximated using the eigensolutions for the noninteracting Hamiltonian.

The interplay between the specific orbital $\varphi(\boldsymbol{r})$, solution of the effective *one*-body Schrödinger equation

$$\left[-\frac{\hbar^2}{2m_e}\nabla^2 + V(\boldsymbol{r})\right]\varphi(\boldsymbol{r}) = \varepsilon\varphi(\boldsymbol{r}) \quad \text{or} \quad [\nabla^2 + k^2]\varphi(\boldsymbol{r}) = U(\boldsymbol{r})\varphi(\boldsymbol{r}), \qquad (7.263)$$

where $k^2 = 2m_e\varepsilon/\hbar^2$ and $U(r) = 2m_eV(\boldsymbol{r})/\hbar^2$, and the potential $U(\boldsymbol{r})$ is conveniently described using the concept of the *Green function* of the operator $\nabla^2 + k^2$. It emerges in the associated integral (scattering) problem, which is formally equivalent to the preceding differential equation.

This transformation is performed most efficiently by regarding the product $U\varphi$ in r.h.s of the preceding equation as an inhomogeneity term. We recall that the general solution of the inhomogeneous Schrödinger equation is obtained by adding to the general solution $\varphi_0(\boldsymbol{r})$ of the homogeneous equation,

$$[\nabla^2 + k^2]\varphi_0(\boldsymbol{r}) = 0, \qquad (7.264)$$

a particular solution of the complete equation. The latter is formally constructed by introducing the "inverse" of the homogenous operator $\nabla^2 + k^2$, called *Green's function* (kernel) $G(\boldsymbol{r} - \boldsymbol{r}')$, such that

$$[\nabla^2 + k^2]G(\boldsymbol{r} - \boldsymbol{r}') = \delta(\boldsymbol{r} - \boldsymbol{r}') \quad \text{and} \quad G(\boldsymbol{r}' - \boldsymbol{r})[\nabla^2 + k^2] = \delta(\boldsymbol{r}' - \boldsymbol{r}). \qquad (7.265)$$

The general solution of (7.263) is then conveniently given by the following combination,

$$\varphi(\boldsymbol{r}) = \varphi_0(\boldsymbol{r}) + \int G(\boldsymbol{r}-\boldsymbol{r}')\, U(\boldsymbol{r}')\varphi(\boldsymbol{r}')d\boldsymbol{r}', \tag{7.266}$$

with the integral involving Green's function generating the particular solution of the inhomogeneous equation.

The equivalence of the above integral equation and the original differential Schrödinger equation can be demonstrated by applying the operator $\nabla^2 + k^2$ to both sides of the preceding equation:

$$\begin{aligned}[\nabla^2 + k^2]\varphi(\boldsymbol{r}) &= [\nabla^2 + k^2]\int G(\boldsymbol{r}-\boldsymbol{r}')\, U(\boldsymbol{r}')\varphi(\boldsymbol{r}')d\boldsymbol{r}' \\ &= \int \delta(\boldsymbol{r}-\boldsymbol{r}')\, U(\boldsymbol{r}')\varphi(\boldsymbol{r}')d\boldsymbol{r}' = U(\boldsymbol{r})\varphi(\boldsymbol{r}).\end{aligned} \tag{7.267}$$

Inversely, any solution of the stationary Schrödinger equation can be shown to satisfy the integral equation (7.266). Therefore, the differential equation (7.263) can be replaced by its equivalent integral form (7.266). It also follows from the latter that the Green function can be expressed as the functional derivative,

$$G(\boldsymbol{r}-\boldsymbol{r}') = \frac{1}{\varphi(\boldsymbol{r}')}\frac{\delta\varphi(\boldsymbol{r})}{\delta U(\boldsymbol{r}')}, \tag{7.268}$$

which is seen to measure the dependence of the orbital φ on the external potential U.

7.5.1 Optimized Potential Method

Let us now examine the OEP, $v_{xc,\sigma}^{\text{OEP}}[\{\varphi_{n\sigma'}[\{\rho_{\sigma''}\}]\};\boldsymbol{r}] \equiv v_{xc,\sigma}^{\text{OEP}}(\boldsymbol{r})$, the *explicit* functional of KS SO $\{\varphi_{n\sigma}[\{\rho_{\sigma'}\}]\}$, all *implicit* functionals of the system spin-densities $\{\rho_{\sigma'}\}$. One also realizes that these *one*-electron functions are connected with spin-densities through the effective potentials of the associated KS-like equations:

$$\varphi_{n\sigma}[\{\rho_{\sigma''}\}] = \varphi_{n\sigma}\left[\left\{ v_{\text{KS}}^{\sigma'}[\{\rho_{\sigma''}\}] \right\}\right]. \tag{7.269}$$

This complex functional dependence thus calls for the *double* chain-rule transformation of the defining derivative,

$$\begin{aligned} v_{xc,\sigma}^{\text{OEP}}(\boldsymbol{r}) &= \frac{\delta E_{xc}^{\text{OEP}}[\{\varphi_{n\sigma}[\{v_{\text{KS}}^{\sigma'}[\{\rho_{\sigma''}\}]\}]\}]}{\delta\rho_\sigma(\boldsymbol{r})} \\ &\equiv \sum_{\sigma',\sigma''}^{\alpha,\beta}\sum_{p}^{occupied}\iint\left\{\frac{\delta E_{xc}^{\text{OEP}}[\{\varphi_{n\sigma}\}]}{\delta\varphi_{p\sigma'}(\boldsymbol{r}')}\frac{\delta\varphi_{p\sigma'}(\boldsymbol{r}')}{\delta v_{\text{KS}}^{\sigma''}(\boldsymbol{r}'')}\frac{\delta v_{\text{KS}}^{\sigma''}(\boldsymbol{r}'')}{\delta\rho_\sigma(\boldsymbol{r})} + c.c.\right\} d\boldsymbol{r}'d\boldsymbol{r}''. \end{aligned} \tag{7.270}$$

where $c.c.$ stands for the complex conjugate of the preceding term.

The first factor in the above product of derivatives is determined by the given form of the orbital functional $E_{xc}^{\mathrm{OEP}}[\{\varphi_{n\sigma}\}]$, while the second factor, the KS Green-function [see (7.268)], can be estimated using the *first*-order PT (see Sect. 5.5.1) by considering the influence on the orbital $\varphi_{p\sigma'}$ at $\boldsymbol{r}'$ of an infinitesimal, local displacement in the effective potential $v_{\mathrm{KS}}^{\sigma''}$ at $\boldsymbol{r}''$, through the spin-resolved KS equation:

$$\left(-\frac{\nabla^2}{2}+v_{\mathrm{KS}}^{\sigma}[\rho_\alpha,\rho_\beta;\boldsymbol{r}]\right)\varphi_{n\sigma}(\boldsymbol{r})=\varepsilon_{n\sigma}\varphi_{n\sigma}(\boldsymbol{r}),\ \rho_\sigma(\boldsymbol{r})=\sum_{n}^{N_\sigma}|\varphi_{n\sigma}(\boldsymbol{r})|^2,\ \sigma=\alpha,\beta. \quad (7.271)$$

The resulting derivative then reads:

$$\frac{\delta\varphi_{p\sigma'}(\boldsymbol{r}')}{\delta v_{\mathrm{KS}}^{\sigma''}(\boldsymbol{r}'')}=\delta_{\sigma',\sigma''}\sum_{q\neq p}\frac{\varphi_{q\sigma''}^*(\boldsymbol{r}')\varphi_{q\sigma''}(\boldsymbol{r}'')}{\varepsilon_{p\sigma''}-\varepsilon_{q\sigma''}}\varphi_{p\sigma''}(\boldsymbol{r}'')=\delta_{\sigma',\sigma''}G_{p\sigma''}^{\mathrm{KS}}(\boldsymbol{r}',\boldsymbol{r}'')\varphi_{p\sigma''}(\boldsymbol{r}''). \quad (7.272)$$

It also follows from this expression that $G_{p\sigma}^{\mathrm{KS}}$ is orthogonal to $\varphi_{p\sigma}$, since this Green function constitutes the linear combination of KS SO orthogonal to $\varphi_{p\sigma}$.

Finally, the inverse of the third factor in the product of (7.270) determines the KS linear-response function, measuring the density response to a perturbation in the effective potential, also known from PT:

$$\begin{aligned} X_{\mathrm{KS}}^{\sigma,\sigma''}(\boldsymbol{r},\boldsymbol{r}'') &= \frac{\delta\rho_\sigma(\boldsymbol{r})}{\delta v_{\mathrm{KS}}^{\sigma''}(\boldsymbol{r}'')} = \delta_{\sigma,\sigma''}\sum_{k,l}\frac{\varphi_{k\sigma}^*(\boldsymbol{r})\varphi_{l\sigma}(\boldsymbol{r})\varphi_{l\sigma}^*(\boldsymbol{r}'')\varphi_{k\sigma}(\boldsymbol{r}'')}{\varepsilon_{k\sigma}-\varepsilon_{l\sigma}}+c.c. \\ &= \delta_{\sigma,\sigma''}X_{\mathrm{KS}}^{\sigma}(\boldsymbol{r},\boldsymbol{r}'') = \delta_{\sigma,\sigma''}\left[\sum_k \varphi_{k\sigma}(\boldsymbol{r}'')G_{k\sigma}^{\mathrm{KS}}(\boldsymbol{r}'',\boldsymbol{r})\varphi_{k\sigma}^*(\boldsymbol{r})+c.c.\right]. \end{aligned} \quad (7.273)$$

Hence the effective *xc*-potential of the OEP method is defined by the following integral equations in terms of KS orbitals satisfying (7.271):

$$v_{xc,\sigma}^{\mathrm{OEP}}(\boldsymbol{r})=\sum_{p=1}^{N_\sigma}\iint\left\{\frac{\delta E_{xc}^{\mathrm{OEP}}[\{\varphi_{n\sigma'}\}]}{\delta\varphi_{p\sigma}(\boldsymbol{r}')}G_{p\sigma}^{\mathrm{KS}}(\boldsymbol{r}',\boldsymbol{r}'')\varphi_{p\sigma}(\boldsymbol{r}'')+c.c.\right\}X_{\mathrm{KS}}^{\sigma,-1}(\boldsymbol{r}'',\boldsymbol{r})\,d\boldsymbol{r}'d\boldsymbol{r}''. \quad (7.274)$$

The first derivative in (7.270) follows from the adopted form of the orbital-dependent functional for the *xc*-energy. It can be taken from the MBPT or CCSD(T) (Grabowski 2008; Grabowski et al. 2002, 2007; Bartlett et al. 2005a, b; Grabowski and Lotrich 2005; Lotrich et al. 2005), e.g., at the MP2 level. Alternatively, the Colle–Salvetti (1975) ansatz for the correlation energy, which also provides the basis of the successful LYP density functional, can be used to supplement the exact exchange energy of (7.259b) (Grabo et al. 1999),

$$E_x^{\mathrm{OEP}}[\{\varphi_{p\sigma}\}] = \langle\Psi_s|\hat{\mathrm{K}}^{\mathrm{KS}}|\Psi_s\rangle$$
$$= -\frac{1}{2}\sum_{\sigma}\sum_{n,m}^{N_\sigma}\int\int\frac{\varphi_{n\sigma}^*(\boldsymbol{r})\varphi_{m\sigma}(\boldsymbol{r})\varphi_{m\sigma}^*(\boldsymbol{r}')\varphi_{n\sigma}(\boldsymbol{r}')}{|\boldsymbol{r}-\boldsymbol{r}'|}\,d\boldsymbol{r}\,d\boldsymbol{r}', \tag{7.275}$$

which approximates $E_{xc}^{\mathrm{OEP}}[\{\varphi_{n\sigma'}\}]$ in the x-only OEP approach or OEP$_x$–KS. In the latter case, one finds:

$$\frac{\delta E_x^{\mathrm{OEP}}[\{\varphi_{n\sigma'}\}]}{\delta\varphi_{p\sigma}(\boldsymbol{r})} = -\sum_{q=1}^{N_\sigma}\varphi_{q\sigma}^*(\boldsymbol{r})\int\frac{\varphi_{p\sigma}^*(\boldsymbol{r}')\varphi_{q\sigma}(\boldsymbol{r}')}{|\boldsymbol{r}-\boldsymbol{r}'|}\,d\boldsymbol{r}'. \tag{7.276}$$

An equivalent and numerically more convenient form of these integral equations for the effective *single*-particle spin-potentials is obtained by multiplying both sides of (7.274) with $X_{\mathrm{KS}}^{\sigma}(\boldsymbol{r},\boldsymbol{r}''')$, integrating over $\boldsymbol{r}$, and using (7.273):

$$\sum_{p=1}^{N_\sigma}\int\varphi_{p\sigma}^*(\boldsymbol{r}')[v_{xc,\sigma}^{\mathrm{OEP}}(\boldsymbol{r}') - u_{p\sigma}^{xc}(\boldsymbol{r}')]G_{p\sigma}^{\mathrm{KS}}(\boldsymbol{r}',\boldsymbol{r})\varphi_{p\sigma}(\boldsymbol{r}')\,d\boldsymbol{r} + c.c. = 0, \tag{7.277}$$

$$u_{p\sigma}^{xc}(\boldsymbol{r}) = \frac{1}{\varphi_{p\sigma}^*(\boldsymbol{r})}\frac{\delta E_{xc}^{\mathrm{OEP}}[\{\varphi_{n\sigma'}\}]}{\delta\varphi_{p\sigma}(\boldsymbol{r})}. \tag{7.278}$$

These equations have to be solved in each KS iteration, for the current set of orbitals. The OEP scheme can be applied rigorously by projecting the integral equations on the appropriately chosen basis set of Gaussian orbitals, usually those approximating the KS MO themselves, or it can be recast in an approximate version, e.g., in the Krieger–Li–Yafrate (KLI) (1990, 1992, 1995) method, which correctly reproduces the N-discontinuity feature of Fig. (7.4).

Finally, we would like to emphasize that the OEP procedure searching for the optimum KS spin potentials $\{v_{\mathrm{KS},\sigma}^{\mathrm{OEP}}(\boldsymbol{r}')\}$ which minimize the energy functional

$$E_v^{\mathrm{OEP}}[\{\varphi_{n\sigma}[\{v_{\mathrm{KS},\sigma'}^{\mathrm{OEP}}\}]\}] \equiv E_v^{\mathrm{OEP}}[\{v_{\mathrm{KS},\sigma'}^{\mathrm{OEP}}\}], \tag{7.279}$$

$$\left.\frac{\delta E_v^{\mathrm{OEP}}[\{v_{\mathrm{KS}}^{\sigma'}\}]}{\delta v_{\mathrm{KS}}^{\sigma}(\boldsymbol{r})}\right|_{v_{\mathrm{KS}}^{\sigma}=v_{\mathrm{KS},\sigma}^{\mathrm{OEP}}} = 0, \tag{7.280}$$

is equivalent to the HK variational principle of Sect. 7.1.2. It also explains the origin of method's name. This can be explicitly demonstrated using the appropriate chain-rule transformation of the preceding derivative:

$$\frac{\delta E_v^{\mathrm{OEP}}[\{v_{\mathrm{KS}}^{\sigma'}\}]}{\delta v_{\mathrm{KS}}^{\sigma}(\boldsymbol{r})} = \sum_{\sigma''}^{\alpha,\beta}\int\frac{\delta E_v^{\mathrm{OEP}}[\{\rho_{\sigma'}\}]}{\delta\rho_{\sigma''}(\boldsymbol{r}')}\frac{\delta\rho_{\sigma''}(\boldsymbol{r}')}{\delta v_{\mathrm{KS}}^{\sigma}(\boldsymbol{r})}\,d\boldsymbol{r}' = 0. \tag{7.281}$$

Multiplying both sides of this equation with the inverse of the linear-response function $\delta v_{\mathrm{KS}}^{\sigma}(\boldsymbol{r})/\delta\rho_{\sigma''}(\boldsymbol{r}'')$ and integrating over $\boldsymbol{r}$ then gives the HK variational rule:

$$\int \frac{\delta E_v^{\mathrm{OEP}}[\{v_{\mathrm{KS}}^{\sigma'}\}]}{\delta v_{\mathrm{KS}}^{\sigma}(\boldsymbol{r})} \frac{\delta v_{\mathrm{KS}}^{\sigma}(\boldsymbol{r})}{\delta\rho_{\sigma''}(\boldsymbol{r}'')} d\boldsymbol{r} = \frac{\delta E_v^{\mathrm{OEP}}[\{\rho_{\sigma'}\}]}{\delta\rho_{\sigma''}(\boldsymbol{r}'')} = 0. \qquad (7.282)$$

A good summary of the performance of OPM and OEP methods in predicting various physical properties of atoms, molecules, and solids is given in the review by Kohanoff and Gidopoulos (2003). To conclude this section, we briefly summarize some formal properties of the exact-exchange (EXX, OEP_x) approach and its KLI approximation. Like in HF treatment the *self*-interaction is exactly removed, so that the asymptotics of the exchange potentials is now correct, with the exchange potential decaying as $-1/r$ at long distances, for all orbital states, irrespectively of whether they are occupied or empty. This is contrary to HF theory, where this Coulomb decay characterizes only the occupied orbitals, but for all virtual states it decays exponentially. As a result, many negatively charged ions are not even bound at HF level of theory. The N-discontinuities of the energy and effective *one*-body potential, which give rise to the preference of the integer number of electrons in molecular systems and their dissociation products, are correctly verified in both EXX and KLI. Another difference is that all occupied HF orbitals decay exponentially, with the same exponent, while in OEP each orbital decays with its own exponent, as it should be. The OPM energies (for local x-potential) are only marginally larger, virtually identical with the UHF energy (for nonlocal x-potential), which determines the lower bound for all x-only orbital schemes, with the KLI energies being still marginally above the corresponding EXX values.

7.5.2 Density-Functional Perturbation Theory

In Sect. 6.3.2, we have described the MP approach to the electron correlation problem in which one starts from the HF solution and introduces in a perturbative way the Coulomb electron correlation on top of the HF theory. In DFT an analogous treatment is known as the *Görling–Levy* (GL) (1993, 1994) theory or DFPT. In this approach the separable Hamiltonian $\hat{\mathrm{H}}^{\lambda=0}(N)$ of the noninteracting ($\lambda = 0$) system defines the associated 0th-order (unperturbed) eigenvalue problem of the KS theory, while the difference

$$\hat{\mathrm{h}}^{\lambda}(N) = \hat{\mathrm{H}}^{\lambda}(N) - \hat{\mathrm{H}}^{\lambda=0}(N) = \sum_{i=1}^{N} [v^{\lambda}(\boldsymbol{r}_i) - v_{\mathrm{KS}}(\boldsymbol{r}_i)] + \lambda\hat{\mathrm{V}}_{ee}(N), \qquad (7.283)$$

represents the correlation perturbation due to the scaled λ-interaction between N electrons and the change in the external potential required by the adiabaticity condition (7.128):

$$\hat{\mathrm{h}}^{\lambda}(N) = \hat{\mathrm{V}}^{\lambda}_{ne}(N) - V_s(N) + \lambda \hat{\mathrm{V}}_{ee}(N) \equiv \lambda \hat{\mathrm{h}}'_{\lambda}(N). \tag{7.284}$$

The total electronic energy $E^{\lambda}(N) = E^{\lambda}[\rho]$ for the current interaction strength λ can be then expressed as the *xc*-corrected energy of the noninteracting system, $E^{\lambda=0}[\rho] = E^{0}(N)$,

$$E^{\lambda}[\rho] = E^{\lambda=0}[\rho] + E^{(1)}[\rho] + E^{\lambda}_{c}[\rho]; \tag{7.285}$$

here the first-order correction $E^{(1)}[\rho] = E^{\lambda=0}_{xc}[\rho] = E_x[\rho]$ is the exact-exchange energy, given by the familiar Fock expression in terms of KS orbitals [(7.259b) and (7.275)], while $E^{\lambda}_{c}[\rho]$ denotes the contribution due to the Coulomb correlation. Indeed, using the Levy constrained-search construction of (7.29) the latter can be formally expressed as the difference between the expectation values of the operator $\hat{\mathrm{F}}^{\lambda}(N)$ [see (7.131)], calculated for the wave functions $\Psi^{\lambda}(N)$ and $\Psi^{0}(N)$, which minimize the expectation values of $\hat{\mathrm{F}}^{\lambda}(N) = \hat{\mathrm{T}}_e(N) + \lambda \hat{\mathrm{V}}_{ee}(N)$ and $\hat{\mathrm{F}}^{\lambda=0}(N) = \hat{\mathrm{T}}_e(N)$, respectively,

$$\begin{aligned}
E^{\lambda}_{c}[\rho] &= \langle \Psi^{\lambda}(N)|\hat{\mathrm{F}}^{\lambda}(N)|\Psi^{\lambda}(N)\rangle - \langle \Psi^{\lambda=0}(N)|\hat{\mathrm{F}}^{\lambda}(N)|\Psi^{\lambda=0}(N)\rangle \\
&= (T_s[\rho] + \lambda J_{ee}[\rho] + E^{\lambda}_{xc}[\rho]) - (T_s[\rho] + \lambda J_{ee}[\rho] + E_x[\rho]) \\
&= \langle \Psi^{\lambda}(N)|\hat{\mathrm{H}}^{\lambda}(N) - \hat{\mathrm{V}}^{\lambda}_{ne}(N)|\Psi^{\lambda}(N)\rangle \\
&\quad - \langle \Psi^{\lambda=0}(N)|\hat{\mathrm{H}}^{\lambda=0}(N) + \{\hat{\mathrm{H}}^{\lambda}(N) - \hat{\mathrm{H}}^{\lambda=0}(N) - \hat{\mathrm{V}}^{\lambda}_{ne}(N)\}|\Psi^{\lambda=0}(N)\rangle \\
&= \langle \Psi^{\lambda}(N)|\hat{\mathrm{H}}^{\lambda}(N)|\Psi^{\lambda}(N)\rangle - \langle \Psi^{\lambda=0}(N)|\hat{\mathrm{H}}^{\lambda=0}(N)|\Psi^{\lambda=0}(N)\rangle \\
&\quad - \langle \Psi^{\lambda=0}(N)|\hat{\mathrm{h}}^{\lambda}(N)|\Psi^{\lambda=0}(N)\rangle \\
&= (E^{\lambda}[\rho] - E^{\lambda=0}[\rho]) - \langle \Psi^{\lambda=0}(N)|\hat{\mathrm{h}}^{\lambda}(N)|\Psi^{\lambda=0}(N)\rangle = E^{\lambda}_{xc}[\rho] - E_x[\rho].
\end{aligned} \tag{7.286}$$

Here, we have used the DFT adiabaticity condition of (7.128), that both $\Psi^{\lambda}(N)$ and $\Psi^{\lambda=0}(N)$ give rise to the same electron density, which directly gives

$$\langle \Psi^{\lambda}(N)|\hat{\mathrm{V}}^{\lambda}_{ne}(N)|\Psi^{\lambda}(N)\rangle = \langle \Psi^{\lambda=0}(N)|\hat{\mathrm{V}}^{\lambda}_{ne}(N)|\Psi^{\lambda=0}(N)\rangle = \int v^{\lambda}(\boldsymbol{r})\,\rho(\boldsymbol{r})\,d\boldsymbol{r}.$$

It should be observed that (7.285) and (7.286) then indeed confirm that the exchange energy constitutes the first-order correction in DFPT, given by the expectation value of the perturbation operator of (7.284) calculated for the KS determinant $\Psi^{\lambda=0}(N)$ describing the noninteracting system:

$$\begin{aligned}
E_x[\rho] &= \langle \Psi^{\lambda=0}(N)|\hat{\mathrm{h}}^{\lambda}(N)|\Psi^{\lambda=0}(N)\rangle \equiv \Delta E^{(1)}[\rho] \\
&= \langle \Psi^{\lambda=0}(N)|\lambda \hat{\mathrm{h}}'_{\lambda}(N)|\Psi^{\lambda=0}(N)\rangle \equiv \lambda E^{(1)}[\rho].
\end{aligned} \tag{7.287}$$

We further observe that the exact scaling relation of (7.151) for $s = 1/\lambda$ allows one to express the correlation energy $E_c^\lambda[\rho]$ for the current interaction strength λ in terms of the correlation energy of the fully interacting system ($\lambda = 1$), $E_c[\rho_{1/\lambda}]$, calculated for the scaled density $\rho_s = \rho_{1/\lambda}$ (7.147):

$$E_c[\rho_{1/\lambda}] = \frac{1}{\lambda^2} E_c^\lambda[\rho]. \tag{7.288}$$

Hence, the Taylor expansion in powers of λ around the known noninteracting KS limit ($\lambda = 0$) then gives the following expansion of $E_c[\rho_{1/\lambda}]$:

$$E_c[\rho_{1/\lambda}] = \sum_{j=2}^{\infty} \lambda^{j-2} E_c^{(j)}[\rho], \tag{7.289}$$

with contributions $E_c^{(j)}[\rho]$ being determined from the standard expressions of PT (see Sect. 5.1.1) using the known KS eigensolutions.

Finally, the total interaction energy for the fully interacting system is obtained by the (adiabatic) integration over the coupling constant λ (see Sect. 7.3.2):

$$E_c[\rho] = \int_0^1 E_c[\rho_{1/\lambda}]\, d\lambda = \sum_{j=2}^{\infty} \frac{1}{j-1} E_c^{(j)}[\rho]. \tag{7.290}$$

The explicit formulas for calculating the complicated and computationally demanding nth-order contribution $E_c^{(n)}[\rho]$ to the Coulomb correlation energy have been derived by Görling and Levy (1993, 1994). As in the MP theory the *lowest*-order of the perturbative correction in GL approach, which represents the true Coulomb correlation in the molecular ground state, is the second-order, MP2-like term:

$$E_c^{(2)} = -\sum_{t=1}^{\infty} \frac{\left|\langle \Phi_0^0(N)|\hat{V}_{ee}(N) - \hat{V}_{\mathrm{H}}(N) - \hat{V}_x(N)|\Phi_t^0(N)\rangle\right|^2}{E_t^0 - E_0^0}; \tag{7.291}$$

where the summation is over all excited states of the noninteracting system $\{\Phi_{t>0}^0(N)\}$, i.e., the configurations defined by the KS determinantal states, with the ground state wave function $\Phi_0^0(N) = \Psi_s(N)$ (7.109), $\{E_t^0 = \sum_k^{occd[t]} \varepsilon_k\}$ stand for the associated energies, the overall Hartree-potential of N electrons $\hat{V}_{\mathrm{H}}(N) = \sum_{i=1}^{N} v_{\mathrm{H}}(\boldsymbol{r}_i)$ [(see 7.122a)] and the *one*-electron exchange operator of N electrons $\hat{V}_x(N) = \sum_{i=1}^{N} v_x(\boldsymbol{r}_i)$ groups the (*local*) exchange potentials, in the sense of the OPM of Sharp and Horton (see the preceding section):

$$v_x(\boldsymbol{r}) = \frac{\delta E_x[\rho]}{\delta \rho(\boldsymbol{r})}. \tag{7.292}$$

Above, the EX functional $E_x[\rho] = E_x[\{\varphi_k[\rho]\}]$ is again given by the Fock expression in terms of KS orbitals $\{\varphi_i\}$ [(7.259b) and (7.275)].

In terms of KS eigensolutions, this *second*-order DFPT (GL2) gives the MP2-like expression [see (6.133)] in terms of the KS *single*-particle states $\{\varphi_i\}$ and their associated eigenvalues $\{\varepsilon_i\}$:

$$E_c^{\mathrm{GL2}} = -\frac{1}{4}\sum_{k,l}^{occd.}\sum_{p,q}^{virt.}\frac{\left|\langle\varphi_k\varphi_l|g|\varphi_p\varphi_q\rangle\right|^2}{\varepsilon_p+\varepsilon_q-\varepsilon_k-\varepsilon_l} - \sum_{k}^{occd.}\sum_{p}^{virt.}\frac{\left|\langle\varphi_k|v_x-\hat{\mathrm{K}}^{\mathrm{KS}}|\varphi_p\rangle\right|^2}{\varepsilon_p-\varepsilon_k}, \quad (7.293)$$

where $\hat{\mathrm{K}}^{\mathrm{KS}}$ denotes the Fock-like (nonlocal) exchange operator expressed in terms of the KS *single*-particle states [(7.259b) and (7.275)].

For practical reason of a prohibitively expensive character of the *higher*-order DFPT, one can normally afford only its GL2 level. As demonstrated by Ernzerhof (1996), the energetics of molecular atomization resulting from this truncation is worsened compared to predictions from the traditional DFT approaches and quantum-chemical calculations. This is typical of truncated perturbational approaches: unless the perturbation is really week, the *second*-order treatments are not really successful, since the *higher*-order terms can be comparable to the *second*-order contribution.

It has been shown by Engel et al. (2000) that this is particularly true in the vdW systems, e.g., Ne_2, where MP2 approach underbinds (too long bond length) and the combined EXX, x-only KLI-GL2 treatment overbinds (too short bond length). A favorable comparison with the experiment and a marked improvement over MP2 results can be achieved only after an empirical resummation of the perturbative expansion, in the spirit of GGA, after which some exact limits not satisfied in the bare GL2 approach, e.g., of a strongly interacting system (for $\lambda \to \infty$), are finally fulfilled.

Indeed, for some values of the coupling constant λ the perturbative series might even be divergent. It has also been demonstrated by Seidl et al. (2000), who generalized the resummation ideas of Ernzerhof to verify both the *strong*- and *weak*-interaction limits in the correlation functional and achieved good predictions of atomization energies, that in some cases the radius of convergence can be very small indeed. The expansion was shown to most likely fail in the limit $\lambda \to \infty$, where electronic positions become strictly correlated giving rise to the Wigner crystallization. It should be recalled that the uniform-electron-gas limit, marking a weak interaction between electrons, also gives rise to divergent terms in the perturbational series and has to be treated using the RPA theory.

7.5.3 *Ab Initio DFT*

The cost of standard KS GGA-method is of the order of the ordinary HF (SCF) calculations in quantum chemistry. In fact, the formal DFT can be regarded as an exactification of the Hartree and Thomas–Fermi theories, by reducing the *many*-electron problem to in principle exact, but explicitly unknown *single*-electron equations. This scheme, however, suffers from several drawbacks such as the *self*-interaction contamination, inability to describe the weak interactions, and wrong asymptotics. The latter affect predictions of the polarization quantities and excitation energies, which strongly depend on the KS eigenvalues. Neither can this approximation be systematically improved with high confidence that a higher level of theory leads to better results, as is the case in treating the electron correlation in the standard variational methods of quantum chemistry, such as the advanced MBPT and CC approaches.

In a search for the "*right answers for the right reason*" Bartlett and coworkers have combined the OEP approach with the correlation treatment in the standard CI methods developed by quantum chemists, coining the name of the *ab initio* DFT (Bartlett 2000) to emphasize the *first*-principle character of this correlation treatment in the orbitally dependent DFT. This procedure allows one to systematically improve over the standard DFT methods. The cost of such rather advanced ab-initio DFT treatments, however, soon becomes comparable to that of the original MBPT/CC calculations, thus prohibiting a wider use of the accurate orbital-dependent density functionals for the electron correlation, e.g., those developed by Grabowski et al. (Grabowski 2008; Grabowski and Lotrich 2005, Grabowski et al. 2002, 2007), and by Bartlett et al. (2005a,b), in calculations on large molecular systems of interest in contemporary chemistry and biology.

In this approach the addition of the electron correlation on top of the OEP_x (OPM) scheme uses the *density* (adiabaticity) *condition* of the KS method, that the densities of the hypothetical noninteracting KS system and of the real interacting system are identical, in an efficient determination of the effective *single*-particle potentials corresponding to a variety of the orbitally dependent correlation functionals taken from the CC theory. This approach avoids the troublesome functional differentiation of the energy and often uses the self-consistent HF orbitals or those determined at a simpler OEP_x stage. Such variants effectively cover the quasidegenerate systems and vdW interactions, giving predictions comparable with the CCSD(T) method of WFT and systematically converge into the FCI solutions with the improvement of the correlation coverage. Another key problem was the choice of the basis functions, which determine the quality of the analytical potentials in the OEP–KS scheme. The often observed overestimation of the correlation energies in the OEP–MBPT(2) approach has been remedied by the appropriate choice of the *zero*-th order Hamiltonian in defining the correlation functionals and the associated effective potentials in the *second* order of PT.

It has been also demonstrated that, contrary to the standard DFT scheme, the effective potentials resulting from the *ab initio* DFT approach reproduce the exact potentials in model systems, giving rise to an excellent HOMO predictions and ionization potentials. Although the cost of such approaches is by now still prohibitively high, this line of DFT research has demonstrated that one can profit by using in the OEP–DFT the experience gained in advanced *ab initio* treatments of the electron correlation within WFT.

The *ab initio* DFT thus constitutes a way to a systematic improvement of the quality of predictions toward the benchmark MBPT–CC and FCI results. These advanced calculations probe the subtle relations between the DFT and WFT perspectives on molecular electronic structure and provide an explicit construction and a practical demonstration of the basic equivalence of these two complementary viewpoints in contemporary quantum chemistry.

7.6 Rudiments of Ensemble Theory for Excited States

In this section we shall briefly outline the Rayleigh–Ritz variational principle for statistical ensembles. This approach of Theophilou (1979) and of Kohn and collaborators (Kohn 1986; Gross et al. 1988; Oliveira et al. 1988; Gross and Kurth 1993) can be used to determine the excitation energies of atoms and molecules, following the Slater TS approach of Sect. 6.1.3. For clarity, only the simplest case will be considered explicitly, of the statistical mixture of two (nondegenerate) molecular states, $|\Psi_1\rangle = |1\rangle$ and $|\Psi_2\rangle = |2\rangle$, e.g., the system ground state and the first excited state, respectively:

$$\hat{\mathrm{H}}\,|i\rangle = E_i|i\rangle, \quad i = 1, 2; \tag{7.294}$$

here $\hat{\mathrm{H}} = \hat{\mathrm{H}}(N,v)$ denotes the molecular electronic Hamiltonian and E_i stands for its ith eigenvalue. This mixed quantum state is defined by the density operator,

$$\hat{\mathrm{D}} = p_1|1\rangle\langle 1| + p_2|2\rangle\langle 2|, \quad p_1 + p_2 = 1, \tag{7.295}$$

where the ensemble weight p_i stands for the probability of the pure state $|i\rangle$ in the statistical mixture.

We recall (see Sect. 3.3.4) that the ensemble averages of physical quantities are given by the corresponding mean values determined by the ensemble probability weights and the corresponding properties of the individual pure states (3.60a–c). For example, the ensemble averages of the system energy and electron density read:

$$E_{ens.} = \mathrm{tr}(\hat{\mathrm{D}}\hat{\mathrm{H}}) = p_1E_1 + p_2E_2, \quad E_i = \langle i|\hat{\mathrm{H}}|i\rangle, \quad i = 1, 2; \tag{7.296}$$

$$\rho_{ens.}(\boldsymbol{r}) = \mathrm{tr}[\hat{\mathrm{D}}\hat{\rho}(\boldsymbol{r})] = p_1\rho_1(\boldsymbol{r}) + p_2\rho_2(\boldsymbol{r}), \quad \rho_i(\boldsymbol{r}) = \langle i|\hat{\rho}(\boldsymbol{r})|i\rangle, \quad i = 1, 2. \tag{7.297}$$

Both pure-state densities are assumed to integrate to the same particle number of the (*closed*) molecular system, $N = \int\rho_i(\boldsymbol{r})\, d\boldsymbol{r}$, $i = 1, 2$, and the weight of the ground state is assumed to be greater, or equal to that of the first excited state,

$$0 \le p_2 \equiv \lambda \le p_1 = 1 - \lambda \le 1, \quad 0 \le \lambda \le \frac{1}{2}, \tag{7.298}$$

where $\lambda = ½$ case defines the equiensemble formalism of Theophilou (1979). The nondegenerate, *two*-state limitation is by no means essential; for a general case of

M-state ensemble, including degeneracies, the reader is referred to the original paper by Gross et al. (1988).

The ensemble of (7.295) is thus characterized by a single parameter $\lambda = p_2$,

$$\hat{\mathrm{D}}(\{\Psi_i\};\ \lambda) = (1-\lambda)\,|1\rangle\langle 1| + \lambda\,|2\rangle\langle 2| \equiv \hat{\mathrm{D}}(\lambda). \tag{7.299}$$

It corresponds to the molecular electronic Hamiltonian $\hat{\mathrm{H}}(N, v)$, identified by the fixed number of electrons N and the external potential v due to the nuclei in their assumed fixed positions, and gives the following ensemble averages of (7.296) and (7.297):

$$E_{ens.}(\{\Psi_i\};\lambda) = (1-\lambda)E_1 + \lambda E_2, \quad \rho_{ens.}(\boldsymbol{r};\lambda) = (1-\lambda)\rho_1(\boldsymbol{r}) + \lambda\rho_2(\boldsymbol{r}). \tag{7.300}$$

The familiar Rayleigh–Ritz principle of (5.20) can be then extended into the following *variational principle for ensembles*:

For any pair of trial states $(|\Phi_1\rangle, |\Phi_2\rangle)$, which approximate the exact eigenstates $(|1\rangle, |2\rangle)$, respectively, $|\Phi_1\rangle \approx |\Psi_1\rangle$ and $|\Phi_2\rangle \approx |\Psi_2\rangle$, thus defining the trial ensemble with the same probability weights

$$\hat{\mathrm{D}}'(\{\Phi_i\};\ \lambda) = (1-\lambda)\,|\Phi_1\rangle\langle\Phi_1| + \lambda\,|\Phi_2\rangle\langle\Phi_2| \equiv \hat{\mathrm{D}}'(\lambda), \tag{7.301}$$

the following inequality holds:

$$\begin{aligned} E_{ens.}(\{\Phi_i\};\lambda) &= \mathrm{tr}[\hat{\mathrm{D}}'(\lambda)\hat{\mathrm{H}}] = (1-\lambda)\langle\Phi_1|\hat{\mathrm{H}}|\Phi_1\rangle + \lambda\langle\Phi_2|\hat{\mathrm{H}}|\Phi_2\rangle \\ &\geq E_{ens.}(\{\Psi_i\};\lambda) = \mathrm{tr}[\hat{\mathrm{D}}(\lambda)\hat{\mathrm{H}}] = (1-\lambda)E_1 + \lambda E_2 = E_1 + \lambda(E_2 - E_1). \end{aligned} \tag{7.302}$$

For $\lambda < ½$, i.e., $p_1 > p_2$, the equality sign holds only if and only if both trial states are equal to the corresponding exact eigenstates, $|\Phi_1\rangle = |1\rangle,\ |\Phi_2\rangle = |2\rangle$, while for $\lambda = ½$ the equality sign in (7.302) holds if and only if the trial states lie in the subspace spanned by the two exact eigenstates: $|\Phi_1\rangle, |\Phi_2\rangle \in \{|1\rangle,\ |2\rangle\}$.

This principle can be justified by the expansion of the two trial states in the complete set of eigenstates of $\hat{\mathrm{H}}(N, v)$, $\sum_k |k\rangle\langle k| = 1$,

$$|\Phi_1\rangle = \sum_k |k\rangle\langle k|\Phi_1\rangle \equiv \sum_k |k\rangle C_{k,1} \quad \text{and} \quad |\Phi_2\rangle = \sum_l |l\rangle\langle l|\Phi_2\rangle \equiv \sum_i |l\rangle C_{l,2}. \tag{7.303}$$

Hence, the expectation value of the system energy in $|\Phi_i\rangle$,

$$\langle\Phi_i|\hat{\mathrm{H}}|\Phi_i\rangle = \sum_{k,l} C^*_{k,i} C_{l,i}\langle k|\hat{\mathrm{H}}|l\rangle = \sum_k |C_{k,i}|^2 E_k \equiv \sum_k P(k|i)E_k, \quad i = 1,2, \tag{7.304}$$

where the conditional probabilities $\{P(k|i)\}$ of the quantum-mechanical superposition principle satisfy the usual normalization condition:

$$\sum_k P(k|i) = 1. \tag{7.305}$$

Thus, the l.h.s. of the inequality of (7.302),

$$\begin{aligned} E_{ens.}(\{\Phi_i\};\lambda) &= (1-\lambda)\langle\Phi_1|\hat{\mathrm{H}}|\Phi_1\rangle + \lambda\langle\Phi_2|\hat{\mathrm{H}}|\Phi_2\rangle \\ &= (1-\lambda)(P(1|1)E_1 + P(2|1)E_2 + \ldots) + \lambda(P(1|2)E_1 + P(2|2)E_2 + \ldots) \\ &\equiv \sum_k p_k^{ens.} E_k, \end{aligned} \tag{7.306}$$

is equal to $E_{ens.}(\{\Psi_i\};\ \lambda)$ for the assumed weight factor $\lambda < ½$ if and only if simultaneously $P(1|1) = P(2|2) = 1$, i.e., for $\{P(k \neq 1|1) = 0\}$ and $\{P(l \neq 2|2) = 0\}$, when the two trial functions happen to be the respective exact eigenfunctions of the molecular electronic Hamiltonian. Otherwise, the resultant (positive) factors, which in the preceding equation multiply E_1 and E_2, respectively, satisfy the following inequalities:

$$\begin{aligned} p_1^{ens.} &\equiv (1-\lambda)P(1|1) + \lambda P(1|2) < 1-\lambda, \\ p_2^{ens.} &\equiv (1-\lambda)P(2|1) + \lambda P(2|2) > \lambda, \\ p_{k>2}^{ens.} &\equiv (1-\lambda)P(k|1) + \lambda P(k|2) > 0. \end{aligned} \tag{7.307}$$

Indeed, relative to the exact ground state solution, $P(1|1) = 1 \wedge \{P(k \neq 1|1) = 0\}$, the *weighted* probability $p_1^{ens.}$ represents a diminished fraction of the higher ensemble probability $p_1 = 1 - \lambda > ½$ and an increased contribution of the lower ensemble probability $p_2 = \lambda < ½$, thus justifying the first inequality. Similarly, relative to the exact solution for the first excited state, $P(2|2) = 1$ and $\{P(l \neq 2|2) = 0\}$, $p_2^{ens.}$ involves a diminished participation of $p_2 = \lambda < ½$ and an increased fraction of $p_1 = 1 - \lambda > ½$ in the ensemble average probabilities $\{p_i^{ens.}\}$, which explains the reversed character of the second inequality in the preceding equation. The third inequality results from the finite conditional probabilities of higher excited states in the two trial states: $P(k|1) > 0$ and $P(k|2) > 0$.

To summarize, any approximate estimate of the ensemble-average energy of (7.306) for $\hat{\mathrm{D}}'(\lambda) \neq \hat{\mathrm{D}}(\lambda)$ implies a lower contribution of the lowest, ground state energy and increased contributions of all excited states, relative to the exact solution case $\{|\Phi_1\rangle = |1\rangle,\ |\Phi_2\rangle = |2\rangle\}$, thus justifying the inequality of (7.302):

$$\mathrm{tr}\,[\hat{\mathrm{D}}'(\lambda)\,\hat{\mathrm{H}}] > \mathrm{tr}\,[\hat{\mathrm{D}}(\lambda)\,\hat{\mathrm{H}}]. \tag{7.308}$$

It finally follows from (7.302) that the excitation energy $E_2 - E_1$ is given by one of the alternative expressions involving either the ensemble energy itself,

$$\begin{aligned} E_2 - E_1 &= \lambda^{-1}[E_{ens.}(\{\Psi_i\};\lambda) - E_{ens.}(\{\Psi_i\};\lambda = 0)] \\ &= \lambda^{-1}[E_{ens.}(\{\Psi_i\};\lambda) - E_1], \end{aligned} \tag{7.309}$$

or its derivative reflecting the explicit dependence on λ of the functional for the ensemble electronic energy:

$$E_2 - E_1 = \frac{dE_{ens.}(\{\Psi_i\};\lambda)}{d\lambda}. \tag{7.310}$$

The associated DFT for ensembles can be established following the standard HK argument of (7.13), (7.80) and (7.81). More specifically, one compares two ensemble densities for the fixed value of λ, $\rho_{ens.}(\lambda) \equiv \rho(\lambda)$ and $\rho'_{ens.}(\lambda) \equiv \rho'(\lambda)$, which are generated by the statistical mixtures of the two lowest eigenfunctions $\{|1\rangle, |2\rangle\}$ and $\{|1'\rangle, |2'\rangle\}$ of Hamiltonians $\hat{\mathrm{H}} \equiv \hat{\mathrm{H}}(N, v)$ and $\hat{\mathrm{H}}' \equiv \hat{\mathrm{H}}(N, v')$, respectively, exhibiting the essentially different external potentials, $v(\boldsymbol{r}) \neq v'(\boldsymbol{r}) + const.$ These two ensembles are respectively defined by the density operator $\hat{\mathrm{D}}(\lambda)$ of (7.299) and the corresponding "primed" statistical operator

$$\hat{\mathrm{d}}(\lambda) = (1 - \lambda)\,|1'\rangle\langle 1'| + \lambda\,|2'\rangle\langle 2'|, \tag{7.311}$$

which give rise to the associated ensemble average energies:

$$E(\lambda) \equiv \mathrm{tr}[\hat{\mathrm{D}}(\lambda)\,\hat{\mathrm{H}}] \quad \text{and} \quad E'(\lambda) \equiv \mathrm{tr}[\hat{\mathrm{d}}(\lambda)\,\hat{\mathrm{H}}']. \tag{7.312}$$

Next, we examine the unique mapping from the ensemble density to the density operator, which parallels the previous mapping in the *pure*-state case, between the electron density and the system wave function. Following the same *reductio ad absurdum* argument we first assume that the two Hamiltonians $\hat{\mathrm{H}}$ and $\hat{\mathrm{H}}'$ can give rise to the same ensemble average density: $\rho(\lambda) = \rho'(\lambda) \equiv \rho(\boldsymbol{r})$. The variational principle of (7.308) then implies that the following inequalities have to be observed:

$$\begin{aligned} E(\lambda) &< \mathrm{tr}[\hat{\mathrm{d}}(\lambda)\,\hat{\mathrm{H}}] = \mathrm{tr}[\hat{\mathrm{d}}(\lambda)\,\hat{\mathrm{H}}'] + \mathrm{tr}[\hat{\mathrm{D}}(\lambda)(\hat{\mathrm{H}} - \hat{\mathrm{H}}')] = E'(\lambda) + \int \rho(\boldsymbol{r})[v(\boldsymbol{r}) - v'(\boldsymbol{r})]\,d\boldsymbol{r}, \\ E'(\lambda) &< \mathrm{tr}[\hat{\mathrm{D}}(\lambda)\,\hat{\mathrm{H}}'] = \mathrm{tr}[\hat{\mathrm{D}}(\lambda)\,\hat{\mathrm{H}}] + \mathrm{tr}[\hat{\mathrm{D}}(\lambda)(\hat{\mathrm{H}}' - \hat{\mathrm{H}})] = E(\lambda) + \int \rho(\boldsymbol{r})[v'(\boldsymbol{r}) - v(\boldsymbol{r})]\,d\boldsymbol{r}. \end{aligned} \tag{7.313}$$

Since their summation leads to contradiction, $E(\lambda) + E'(\lambda) < E'(\lambda) + E(\lambda)$, we again conclude that the map between the external potential and the ensemble average density is invertible. Thus, $\rho(\lambda)$ uniquely identifies the Hamiltonian, and hence also its eigenstates, which define the density operator:

$$\rho(\lambda) \to \hat{\mathrm{D}}(\lambda) \quad \text{or} \quad \hat{\mathrm{D}}(\lambda) = \hat{\mathrm{D}}[\rho(\lambda)]. \tag{7.314}$$

Therefore, the ensemble average of any observable $\hat{\mathrm{A}}$ also becomes the functional of the ensemble density (see also Sect. 7.1.3):

$$A_{ens.}(\lambda) = \mathrm{tr}[\hat{\mathrm{D}}(\lambda)\,\hat{\mathrm{A}}] = A_{ens.}[\rho(\lambda)]. \tag{7.315}$$

In particular, the average energy of the ensemble can be expressed in the familiar form of (7.15):

$$E_v^{(\lambda)}[\rho] = \mathrm{tr}[\hat{\mathrm{D}}(\lambda)\,\hat{\mathrm{H}}] = \int \rho(\boldsymbol{r})\,v(\boldsymbol{r})\,d\boldsymbol{r} + F^{(\lambda)}[\rho], \tag{7.316}$$

with the universal (v-independent) functional

$$F^{(\lambda)}[\rho] = \mathrm{tr}[\hat{\mathrm{D}}(\lambda)\,\hat{\mathrm{F}}] = \mathrm{tr}\{\hat{\mathrm{D}}(\lambda)[\hat{\mathrm{T}}_e + V_{ee}]\}\ . \tag{7.317}$$

The latter can be also extended to cover the non v-representable densities by using Levy's constrained search construction [(7.41) and (7.84)].

It then directly follows from the above Rayleigh–Ritz principle for ensembles that this energy functional has the following variational properties for a trial ensemble density ρ:

$$\begin{aligned} E_v^{(\lambda)}[\rho] &> E_v^{(\lambda)}\left[\rho^{(\lambda)}[v]\right] \quad \text{if} \quad \rho(\boldsymbol{r}) \neq \rho^{(\lambda)}[v;\boldsymbol{r}], \\ E_v^{(\lambda)}[\rho] &= E_v^{(\lambda)}\left[\rho^{(\lambda)}[v]\right] \quad \text{if} \quad \rho(\boldsymbol{r}) = \rho^{(\lambda)}[v;\boldsymbol{r}], \end{aligned} \tag{7.318}$$

where $\rho^{(\lambda)}[v;\boldsymbol{r}] \equiv \rho^{(\lambda)}[v]$ denotes the ensemble density matching the fixed external potential $v(\boldsymbol{r})$. Thus, the exact ensemble energy and density can be calculated by minimizing $E_v^{(\lambda)}[\rho]$ for any assumed value of λ in (7.298).

The associated KS scheme for ensembles is again build upon the fundamental assumption that the density of the real interacting system can be calculated as the ensemble density of the hypothetical, noninteracting (separable) N-electron system, obtained from its ground state Slater determinant $|1_s\rangle = |\varphi_1 \ldots \varphi_N|$ and that of the system first excited state $|2_s\rangle = |\varphi_1 \ldots \varphi_{N-1}\varphi_{N+1}|$:

$$\begin{aligned} \rho(\boldsymbol{r}) &= (1-\lambda)\langle 1_s|\hat{\rho}(\boldsymbol{r})|1_s\rangle + \lambda\langle 2_s|\hat{\rho}(\boldsymbol{r})|2_s\rangle \\ &= \sum_{k=1}^{N-1} |\varphi_k(\boldsymbol{r})|^2 + (1-\lambda)|\varphi_N(\boldsymbol{r})|^2 + \lambda|\varphi_{N+1}(\boldsymbol{r})|^2. \end{aligned} \tag{7.319}$$

It determines the ensemble electronic energy [see (7.125) and (7.306)]:

$$\begin{aligned} E_v^{(\lambda)}[\rho] &= \int \rho(\boldsymbol{r})\, v(\boldsymbol{r})\, d\boldsymbol{r} + F^{(\lambda)}[\rho] \\ &= \int \rho(\boldsymbol{r})\, v(\boldsymbol{r})\, d\boldsymbol{r} + \left(T_s^{(\lambda)}[\rho] + \frac{1}{2}\iint \frac{\rho(\boldsymbol{r})\rho(\boldsymbol{r}')}{|\boldsymbol{r}-\boldsymbol{r}'|} d\boldsymbol{r}d\boldsymbol{r}' + E_{xc}^{(\lambda)}[\rho]\right) \\ &= \sum_{k=1}^{N-1} \varepsilon_k^{(\lambda)} + (1-\lambda)\,\varepsilon_N^{(\lambda)} + \lambda\varepsilon_{N+1}^{(\lambda)} - \frac{1}{2}\iint \frac{\rho(\boldsymbol{r})\rho(\boldsymbol{r}')}{|\boldsymbol{r}-\boldsymbol{r}'|} d\boldsymbol{r}d\boldsymbol{r}' + E_{xc}^{(\lambda)}[\rho] \\ &\quad - \int \rho(\boldsymbol{r}) v_{xc}^{(\lambda)}[\rho;\boldsymbol{r}]\, d\boldsymbol{r}, \end{aligned} \tag{7.320}$$

where $\{\varepsilon_k^{(\lambda)}\}$ groups the KS eigenvalues, $v_{xc}^{(\lambda)}[\rho;\boldsymbol{r}]$ stands for the effective KS potential, and the kinetic energy of the noninteracting ensemble [compare (7.112)] is given by the following mean-value expression:

$$T_s^{(\lambda)}[\rho] = -\frac{1}{2}\left(\sum_{k=1}^{N-1} \langle\varphi_k|\Delta|\varphi_k\rangle + (1-\lambda)\langle\varphi_N|\Delta|\varphi_N\rangle + \lambda\langle\varphi_{N+1}|\Delta|\varphi_{N+1}\rangle\right). \tag{7.321}$$

The relevant KS equations representing the Euler equations determining the optimum orbitals of the ensemble, to be solved self-consistently for each value of the state-mixing probability λ, then read:

$$\left(-\frac{1}{2}\Delta + v(\boldsymbol{r}) + \int \frac{\rho(\boldsymbol{r}')}{|\boldsymbol{r}-\boldsymbol{r}'|}d\boldsymbol{r}' + v_{xc}^{(\lambda)}[\rho;\boldsymbol{r}]\right)\varphi_k(\boldsymbol{r}) = \varepsilon_k^{(\lambda)}\varphi_k(\boldsymbol{r}), \tag{7.322}$$

where:

$$v_{xc}^{(\lambda)}[\rho;\boldsymbol{r}] = \frac{\delta E_{xc}^{(\lambda)}[\rho]}{\delta\rho(\boldsymbol{r})}. \tag{7.323}$$

Notice that for $\lambda = 0$ these equations and the above expression for the ensemble energy reproduce the previously reported ground state analogs.

Finally, it follows from (7.310) that the excitation energy can be expressed in terms of the corresponding ensemble-KS eigenvalues,

$$E_2 - E_1 = \frac{\partial E_v^{(\lambda)}[\rho]}{\partial\lambda} = \varepsilon_{N+1}^{(\lambda)} - \varepsilon_N^{(\lambda)} + \frac{\partial E_{xc}^{(\lambda)}[\rho]}{\partial\lambda}, \tag{7.324}$$

which can be regarded as an exact analog of the approximate relation (6.41) formulated in Slater's TS theory. In the preceding equation, the partial derivative of $E_{xc}^{(\lambda)}[\rho]$ denotes the derivative of the functional only, which neglects its dependence on λ through $\rho = \rho(\lambda)$. Practical calculations using this ensemble approach require an adequate approximation for $E_{xc}^{(\lambda)}[\rho]$. For the equiensemble of M states, with entropy $S = k_B \ln M$, Kohn (1986) has successfully used the LDA functional of a thermal ensemble.

7.7 Density-Matrix Functional Theory

The reduced density matrices of Sect. 6.3.3 can also serve as alternative basic variables in quantum-mechanical calculations. Attempts to use the 2-matrix, for which the exact functional for the electronic energy is known (6.150), are hampered by the unknown sufficient conditions for its (*pure*-state) N-representability. This prompted a development of the DMFT (Gilbert 1975; Berrondo and Gościnski 1975; Donnelly and Parr 1978; Donnelly 1979; Levy 1979; Valone 1980a,b; Nguyen-Dang et al. 1985; Zumbach and Maschke 1985), which uses as the basic state-variable the system 1-matrix

$$\hat{\gamma}_1(\boldsymbol{q};\boldsymbol{q}') = N\int\ldots\int\Psi^*(\boldsymbol{q}',\boldsymbol{q}_2,\ldots,\boldsymbol{q}_N)\Psi(\boldsymbol{q},\boldsymbol{q}_2,\ldots,\boldsymbol{q}_N)\,d\boldsymbol{q}_2\ldots d\boldsymbol{q}_N \equiv \gamma(\boldsymbol{q};\boldsymbol{q}'), \tag{7.325}$$

in terms of which the exact functional for the kinetic energy of the interacting system is known. Notice that its diagonal part (6.148) defines the electron (spin) density itself: $\gamma_1(\boldsymbol{q},\boldsymbol{q}) = \rho(\boldsymbol{q}) \equiv \rho_\sigma(\boldsymbol{r})$. The necessary and sufficient conditions for the N-representability of 1-matrix are known, and there are several exact conditions to be satisfied by the γ-functional for the system electronic energy. This is very much in spirit of the KS treatment in DFT, in which the kinetic energy is approximated by the known functional of orbitals in the noninteracting limit and the adequate approximation for the electron-repulsion term is searched for.

The 1-matrix is hermitian and positive semidefinite and its eigenfunctions, the natural spin-orbitals (NSO) (6.153) $\{\psi_i\}$ constitute the orthonormal basis with fractional occupations $\{0 \leq n_i \leq 1\}$ in its (diagonal) spectral representation

$$\gamma(\boldsymbol{q};\boldsymbol{q}') = \sum_i n_i \psi_i^*(\boldsymbol{q}')\psi_i(\boldsymbol{q}). \tag{7.326}$$

As indicated in (6.150), the expectation value of the electronic Hamiltonian

$$\hat{\mathrm{H}}^e(N) = \sum_{i=1}^{N} \hat{\mathrm{h}}(i) + \frac{1}{2}\sum_{i \neq j}^{N} g(i,j), \quad \hat{\mathrm{h}}(i) = -\frac{1}{2}\nabla^2 + v(i),$$

involves only the 1-matrix and the *pair*-density (3.12), the diagonal part of the 2-matrix (6.148), $\Gamma(\boldsymbol{q},\boldsymbol{q}') = \hat{\gamma}_2(\boldsymbol{q},\boldsymbol{q}';\boldsymbol{q},\boldsymbol{q}')$,

$$\begin{aligned} E^e[\Psi(N)] &= \langle \Psi(N)|\hat{\mathrm{H}}^e(N)|\Psi(N)\rangle \\ &= \int \hat{\mathrm{h}}(\boldsymbol{q})\gamma(\boldsymbol{q},\boldsymbol{q}')|_{\boldsymbol{q}=\boldsymbol{q}'} d\boldsymbol{q} + \iint \Gamma(\boldsymbol{q},\boldsymbol{q}')g(\boldsymbol{q},\boldsymbol{q}')\, d\boldsymbol{q}\, d\boldsymbol{q}', \end{aligned} \tag{7.327}$$

In the spirit of the HK theory, Gilbert (1975) has shown that the v-representable 1-matrix uniquely determines the shape of the (local or nonlocal) external potential $v = v[\gamma]$, thus giving rise to the following mapping relations:

$$\rho \leftarrow \gamma \rightarrow v \rightarrow \Psi_0 \rightarrow E_0, \tag{7.328}$$

where ρ, Ψ_0, and E_0 stand for the ground state electron density, wave function and energy, respectively. Since the $\Psi_0 \rightarrow \gamma$ map $\gamma = \gamma[\Psi_0]$ follows directly from the very definition of γ, this theorem also implies the reversible map between γ and Ψ_0, $\gamma \leftrightarrow \Psi_0$, and hence the existence of the energy functional

$$E_v[\gamma] = \int \hat{\mathrm{h}}(\boldsymbol{q})\gamma(\boldsymbol{q},\boldsymbol{q}')|_{\boldsymbol{q}=\boldsymbol{q}'} d\boldsymbol{q} + E_{ee}[\gamma] \equiv \mathcal{F}_v[\gamma] + \mathcal{G}[\gamma], \tag{7.329}$$

satisfying the variational principle for any trial, N-representable density matrix γ':

$$E_v[\gamma'] \geq E_v[\gamma] = E_0. \tag{7.330}$$

The relevant sufficient conditions of the (ensemble) N-representability of the variational density matrix γ' are well known: the underlying NO have to be orthonormal,

$$\int \psi_i^*(\boldsymbol{q})\,\psi_j(\boldsymbol{q})d\boldsymbol{q} = \delta_{i,j}, \quad i,j = 1,2,\ldots \tag{7.331}$$

and their normalized occupations, $\sum_i n_i = N$, should not exceed 1: $0 \leq n_i \leq 1$.

The preceding equations determine the formal basis of DMFT. As in the DFT, the main effort has been devoted to establishing the workable approximation to the universal functional $E_{ee}[\gamma]$ for the electron repulsion energy. It can be given a precise definition in terms of the associated Levy constrained search [compare (7.41)]:

$$E_{ee}[\gamma] = \inf_{\Psi\to\gamma_2\to\gamma} \iint \gamma_2(\boldsymbol{q},\boldsymbol{q}';\boldsymbol{q},\boldsymbol{q}')|\boldsymbol{r}-\boldsymbol{r}'|^{-1}\,d\boldsymbol{q}\,d\boldsymbol{q}', \tag{7.332}$$

where $\boldsymbol{q} = (\boldsymbol{r},\sigma), \boldsymbol{q}' = (\boldsymbol{r}',\sigma')$, and $\Psi \to \gamma_2 \to \gamma$ stand for any trial wave function of N electrons giving rise to the 2-matrix γ_2, the partial contraction of which yields the current 1-matrix γ:

$$\gamma(\boldsymbol{q};\boldsymbol{q}') = \frac{2}{N-1}\int \gamma_2(\boldsymbol{q},\boldsymbol{q}_2;\boldsymbol{q}',\boldsymbol{q}_2)\,d\boldsymbol{q}_2. \tag{7.333}$$

This in principle exact but impractical prescription has to be replaced by some adequate, workable approximation using the KS-like partition of $E_{ee}[\gamma]$, into the classical, Hartree term of (7.115),

$$\begin{aligned} J_{ee}[\gamma] &= \frac{1}{2}\iint \frac{\gamma(\boldsymbol{q},\boldsymbol{q})\gamma(\boldsymbol{q}',\boldsymbol{q}')}{|\boldsymbol{r}-\boldsymbol{r}'|}d\boldsymbol{q}d\boldsymbol{q}' = \frac{1}{2}\iint \frac{\rho(\boldsymbol{q})\rho(\boldsymbol{q}')}{|\boldsymbol{r}-\boldsymbol{r}'|}d\boldsymbol{q}d\boldsymbol{q}' = J_{ee}[\rho], \\ &= \frac{1}{2}\iint \frac{\left(\sum_\sigma \rho_\sigma(\boldsymbol{r})\right)\left(\sum_{\sigma'} \rho_{\sigma'}(\boldsymbol{r}')\right)}{|\boldsymbol{r}-\boldsymbol{r}'|}d\boldsymbol{r}d\boldsymbol{r}' = J_{ee}[\{\rho_\sigma\}] \end{aligned} \tag{7.334}$$

and the remaining xc-energy $E_{xc}^{ee}[\gamma]$ of the electron repulsion:

$$E_{ee}[\gamma] = J_{ee}[\gamma] + E_{xc}^{ee}[\gamma] = J_{ee}[\gamma] + (E_x[\gamma] + E_c^{ee}[\gamma]). \tag{7.335}$$

As also indicated in the preceding equation, the latter term is conventionally separated into the exact exchange energy of (7.275),

$$E_x[\gamma] = -\frac{1}{2}\iint \frac{\gamma(\boldsymbol{q},\boldsymbol{q}')\gamma(\boldsymbol{q}',\boldsymbol{q})}{|\boldsymbol{r}-\boldsymbol{r}'|}d\boldsymbol{q}d\boldsymbol{q}', \tag{7.336}$$

and the Coulomb correlation contribution to the electron repulsion energy, $E_c^{ee}[\gamma]$, for which several exact analytical properties are known (e.g.: Goedecker and Umrigar 1998; Yasuda 2001; Buijse and Baerends 2002).

More specifically, the Coulomb correlation energy $E_c^{ee}[\gamma]$ should vanish for the idempotent 1-matrix of the HF approximation, which exhibits the integer eigenvalues of (6.153): $\{n_i = 0 \text{ or } n_i = 1\}$. Moreover, under the uniform scaling of (7.89) and (7.147), which generates the scaled density matrix

$$\gamma_s(\boldsymbol{q}; \boldsymbol{q}') = s^3 \gamma_s(s\boldsymbol{r}, \sigma; s\boldsymbol{r}', \sigma'), \tag{7.337}$$

this contribution now scales exactly as each purely potential-energy term should [compare (7.93), (7.96) and (7.150)]:

$$E_c^{ee}[\gamma] = sE_c^{ee}[\gamma]. \tag{7.338}$$

This correlation energy should additionally exhibit the particle-hole symmetry:

$$E_c^{ee}[\gamma] = E_c^{ee}[\delta - \gamma], \tag{7.339}$$

where in the position representation δ stands for the Dirac delta.

Yet another exact constraint is provided by the known DMF of the electron repulsion in the singlet state of *two*-electron systems (Kutzelnigg 1963):

$$E_{ee}[\gamma] = \frac{1}{2} \min_{\{f_i\}} \left[\sum_{i,j} f_i f_j \sqrt{n_i n_j} \bar{K}_{i,j} \right], \tag{7.340}$$

where $\{n_i\}$ and $\{f_i\}$ denote the occupations and phases of NSO $\{\psi_i\}$, respectively, with the minimum repulsion energy corresponding to the opposite phases of the *strongly*- and *weakly*-occupied NSO, and $\bar{K}_{i,j}$ stands for the exchange integral of (5.75) and (5.77):

$$\bar{K}_{i,j} = \int\int \psi_i^*(\boldsymbol{q})\, \psi_j^*(\boldsymbol{q}') |\boldsymbol{r} - \boldsymbol{r}'|^{-1} \psi_j(\boldsymbol{q})\, \psi_i(\boldsymbol{q}')\, d\boldsymbol{q}\, d\boldsymbol{q}' \equiv \langle ij|g|ji\rangle \equiv (ij|ji). \tag{7.341}$$

To summarize, in DMFT the *one*-electron energy $\mathcal{F}_v[\gamma]$ is known exactly, with only the *two*-electron energy $\mathcal{G}[\gamma] = E_{ee}[\gamma]$ remaining to be approximated. The first DMF, e.g., those proposed by Müller (1984), Goedecker and Umrigar (1998), as well as by Buijse and Baerends (BB) (2002), represent a generalization of the HF expression (*JK*-type functionals) for the electron-repulsion energy as function of the Coulomb $\bar{J}_{i,j} = \langle ij|g|ij\rangle = (ii|jj)$ and exchange integrals (7.341) in terms of NSO, as well as their occupations $\boldsymbol{n} = \{n_i\}$:

$$E_{ee}^{\mathrm{HF}}[\gamma] = \frac{1}{2} \sum_{i,j} n_i n_j (\bar{J}_{i,j} - \bar{K}_{i,j}). \tag{7.342}$$

A systematic development of such functionals has been later undertaken by Ciosłowski, Pernal and collaborators (Ciosłowski et al. 2003; Ciosłowski and Pernal 2004a,b; Pernal and Ciosłowski 2004, 2005; Pernal 2005; Gritsenko et al.

2005; Pernal and Baerends 2006). In particular, for the closed-shell states the following generalization of the preceding expression for the correlated 1-matrix has been examined (Ciosłowski et al. 2003; Pernal and Ciosłowski 2004):

$$E_{ee}[\gamma] = \frac{1}{2}\sum_{i,j} A_{i,j}(\boldsymbol{n})\,\bar{J}_{i,j} - \frac{1}{2}\sum_{i,j} B_{i,j}(\boldsymbol{n})\,\bar{K}_{i,j}, \tag{7.343}$$

where the occupation functions satisfy the exact condition

$$\sum_j A_{i,j}(\boldsymbol{n}) = (N-1)n_i + B_{i,i}(\boldsymbol{n}), \quad i = 1, 2, \ldots \tag{7.344}$$

Several heuristic approaches to design such generalized *JK* functionals have been proposed. For example, the limited CI wave functions can be used for this purpose and the correlated APSG wave functions (see Sect. 6.4.1) and their generalizations describing the intergeminal correlations can both be used to construct the DMF for electron repulsion energy. Such functionals were shown to be capable of describing the dispersion interactions.

Alternatively, the adequate representations of the *two*-electron density $\rho_2(\boldsymbol{r}, \boldsymbol{r}')$ (3.11), in terms of which [see (3.15)]

$$E_{ee} = \frac{1}{2}\int\int \rho_2(\boldsymbol{r}, \boldsymbol{r}')|\boldsymbol{r} - \boldsymbol{r}'|^{-1} d\boldsymbol{r}\, d\boldsymbol{r}', \tag{7.345}$$

can be used in designing the DMF. For example, the BB functional for the closed-shell states results from

$$\rho_2^{\mathrm{BB}}(\boldsymbol{r}, \boldsymbol{r}') = \rho(\boldsymbol{r})\rho(\boldsymbol{r}') - \sum_{i,j} \sqrt{n_i n_j}\, \varphi_i^*(\boldsymbol{r})\varphi_j(\boldsymbol{r})\varphi_i(\boldsymbol{r}')\varphi_j^*(\boldsymbol{r}'), \tag{7.346}$$

where NO $\{\varphi_i\}$ group the spatial parts of the corresponding NSO and exhibit occupations in the range [0, 2]. The BB functional,

$$E_{ee}^{\mathrm{BB}} = \frac{1}{2}\int\int \rho_2^{\mathrm{BB}}(\boldsymbol{r}, \boldsymbol{r}')|\boldsymbol{r} - \boldsymbol{r}'|^{-1} d\boldsymbol{r}\, d\boldsymbol{r}' = J_{ee}[\rho] - \frac{1}{2}\sum_{i,j} \sqrt{n_i n_j} K_{i,j}, \tag{7.347}$$

where $K_{i,j}$ stands for the exchange integral in terms of NO, was found to overestimate the correlation energy in diatomic molecules.

To correct the NO-phase problem, which was diagnosed as responsible for this shortcoming, has motivated Gritsenko et al. (2005) to add in the closed-shell systems to BB distribution of (7.346) the correction:

$$\Delta^{(1)}\rho_2(\boldsymbol{r}, \boldsymbol{r}') = 2\sum_{\alpha,\beta\neq\alpha>N/2} \sqrt{n_\alpha n_\beta}\, \varphi_\alpha^*(\boldsymbol{r})\varphi_\beta(\boldsymbol{r})\varphi_\alpha(\boldsymbol{r}')\varphi_\beta^*(\boldsymbol{r}'), \tag{7.348}$$

where NO are assumed to be ordered in accordance with their decreasing occupations, with the first $N/2$ orbitals grouping the strongly occupied states. The corrected pair distribution function then gives the modified BB functional

$$E_{ee}^{\mathrm{BB}+(1)} = E_{ee}^{\mathrm{BB}} + \sum_{\alpha,\beta\neq\alpha>N/2} \sqrt{n_\alpha n_\beta}\, K_{\alpha,\beta}, \tag{7.349}$$

which improves predictions for H_2 but requires additional physically motivated corrections for larger systems. This analysis has uncovered the important role played by the mixed terms in $\rho_2(\boldsymbol{r}, \boldsymbol{r}')$, simultaneously containing the strongly- and weakly-occupied NO, in the static correlation in diatomic molecules. Such corrected BB functional was shown to dramatically improve the correlation treatment in such systems.

The local KS-type DMF has also been proposed by Ciosłowski and Pernal (2005) and the effective Hamiltonian for NO (see Gilbert 1975, Donnelly and Parr 1978),

$$\hat{\mathrm{H}}_{eff.} = \hat{\mathrm{h}} + \hat{\mathrm{v}}_{ee}, \tag{7.350}$$

with the nonlocal effective potential originating from the functional derivative of the electron repulsion energy,

$$v_{ee}(\boldsymbol{q}, \boldsymbol{q}') = \frac{\delta E_{ee}[\gamma]}{\delta\gamma(\boldsymbol{q},\boldsymbol{q}')}, \tag{7.351}$$

defining the associated operator

$$\hat{\mathrm{v}}_{ee}\psi_i(\boldsymbol{q}) = \int v_{ee}(\boldsymbol{q}, \boldsymbol{q}')\psi_i(\boldsymbol{q}')\, d\boldsymbol{q}', \tag{7.352}$$

has been derived by Pernal (2005). The responses in NO shapes and occupations to static perturbations have been examined by Pernal and Baerends (2006).

This effective Hamiltonian formulation allows for a new algorithm for determining NO via successive diagonalizations of the generalized Fock matrix containing the elements $\{\langle\psi_i|\hat{\mathrm{H}}_{eff.}|\psi_j\rangle\}$. The DMFT has also been successfully applied by Pernal and Ciosłowski (2005) in determining the ionization potentials in context of the *Extended Koopmans' Theorem* (EKT).

7.8 Weak Molecular Interactions in DFT

Although the functionals corresponding to first four rungs in Perdew's "Jacob's ladder" of Fig. 7.5 decently account for the *long*-range electrostatic and induction interactions, they fail to properly describe the asymptotic behavior of the vdW dispersion energy, $E_{disp.}$, i.e., the *long*-range electron correlation effect, at very large separations between two atoms/molecules, the densities of which do not overlap, e.g., in dimmers of the closed-shell systems such as He_2. The origin of this interaction between two chemically nonbonded fragments is the coupling between the electric field generated by the spontaneous fluctuations in the electronic density of one subsystem with the density of the other subsystem. At large separations R between two fragments this interaction approaches the classical dipole–dipole interaction which decays as R^{-6}.

The weak interactions between atomic/molecular systems, which so decisively depend on the correct asymptotics of the correlation tails, have thus acquired a status of the crucial test of the applicability of DFT calculations in molecular biology and to macromolecular systems, where an adequate representation of vdW interactions is critical. In preparing this short summary of the basics, recent trends, and applications of the DFT treatment of weak interactions between atoms and molecules the author was greatly helped by the instructive overview of the subject contained in the lecture notes by Jansen (2009).

For rare gas dimmers the LDA dramatically overbinds, but correctly reproduces trends in interactions, while the GGA functionals improve on average but fail for trends (e.g., van Mourik and Gdanitz 2002). These functionals may still do well for *short*-range part of $E_{disp.}$ and for the overlapping molecular subsystems, especially when supplemented by the DFT embedding (Cortona 1991; Wesołowski and Warshel 1993), and in cases where $E_{disp.}$ is not dominant, e.g., in the hydrogen-bonded systems. In the latter case, the hybrid functionals give predictions closest to MP2 which performs reasonably well, with the BLYP functional representing the "best" GGA (Boese et al. 2007).

A simple way to effectively introduce the dispersion is by fitting parameters in the semiempirical functionals, fitting databases of weak interactions in the appropriate training sets. This approach has been used on all higher rungs of Jacob's ladder. For example, Adamo and Barone (1998) have used the differential exchange energies of rare gas dimmers in the training set, while the interaction energies and gradients of hydrogen-bonded systems have been used to fit the empirical functionals developed by Boese et al. (2000) and by Boese and Handy (2001). Representative examples of the higher rung functionals are the X3LYP hyper-GGA (hybrid) functional of Xu and Goddard III (2004), again trained on the rare-gas dimmers, and a series of functionals developed by Zhao and colleagues (Zhao et al. 2005, 2006; Zhao and Truhlar 2006, 2008): M05 hyper-GGA (hybrid meta), trained on vdW and H-bridged complexes, M06-L meta-GGA, fitting the interaction energies of 31 noncovalently bonded systems, and M06, M06-2X, *hyper*-GGA functionals, fitted on the interaction energies of π-stacking systems and noncovalent complexes exhibiting H-bonds as well as the vdW, CT, and dipole interactions.

These empirical functionals improve the equilibrium geometries and the associated energetics but still suffer from wrong *long*-range behavior. A simple DFT + D correction scheme, suggested by the familiar dispersion part of the *second*-order expression from the Rayleigh–Schrödinger PT, in the spirit of an earlier HF + D approach (Hepburn et al. 1975; Ahlrichs et al. 1977), involve an addition of the *long*-distance dispersion via C_6/R_6-terms:

$$E_{int.}(\text{DFT + D}) = E_{int.}(\text{DFT}) - \sum_{i,j}^{atoms} C_{i,j}/R_{i,j}^6, \tag{7.353}$$

with the empirical atom–atom dispersion coefficients $\{C_{i,j}\}$, as in the force fields. In order to avoid a double counting of the electron correlation, this vdW correction must be damped for smaller separations, and the standard E_{xc} functionals have to be readjusted accordingly (e.g., Elstner et al. 2001; Wu et al. 2001; Wu and Yang 2002; Zimmerli et al. 2004; Grimme 2004; McNamara and Hillier 2007; Tuttle and Thiel 2008). These DFT + D functionals have been successfully applied to biologically relevant complexes (Morgado et al. 2008) of the Jurečka et al. (2006) database of benchmark CCSD(T) energies for over 100 nucleobase and amino acid complexes.

Alternatively, within the effective *core*-potential approximation (see Sect. 6.1.5), the nonlocal part of *ab initio* AIM pseudopotentials (in the plane-wave basis) can be adjusted to simulate dispersion interactions (von Lilienfeld et al. 2004, 2005). In this *Dispersion-Corrected Atom-Centered Potentials* (DCACP) approach one obtains quite good energetics of the H-bonded DNA base and amino acid pairs, while BLYP severely under-binds (Lin and Röthlisberger 2008). It also improves the DNA stacking and interstrand base-pairs energies, as well as the binding energies in aminoacids, but still fails to reproduce the correct *long*-range behavior (von Lilienfeld et al. 2005; Lin et al. 2007).

In a search for the nonempirical functional covering the vdW interactions, represented by the fifth rung in Fig. 7.5, one goes back to the scaled electron interaction idea of Sect. 7.3.2. In the relevant *Adiabatic-Connection Fluctuation-Dissipation Theorem* (ACFDT) DFT one uses the *fluctuation-dissipation theorem* (FDT) of quantum statistical thermodynamics (Nyquist 1928; Callen and Welton 1951). It establishes a connection between the response of the system and the magnitude of fluctuations in the quantum-mechanical observable $\hat{O}$ of the unperturbed system:

$$\langle\Psi|(\hat{O}-O)^2|\Psi\rangle = -\mathrm{Im}\left[\frac{1}{\pi}\int_0^\infty\int O(\boldsymbol{r})\chi(\boldsymbol{r},\boldsymbol{r};\omega)\,O(\boldsymbol{r})d\boldsymbol{r}d\omega\right]. \tag{7.354}$$

Here, the frequency-dependent linear density-density response function $\chi(\boldsymbol{r},\boldsymbol{r}';\omega)$ determines the equilibrium fluctuations $\delta\rho(\boldsymbol{r};\omega)$ in the system ground state electron density,

$$\delta\rho(\boldsymbol{r};\omega) = \int\chi(\boldsymbol{r},\boldsymbol{r}';\omega)\delta V_{pert.}(\boldsymbol{r}';\omega)\,d\boldsymbol{r}' \equiv \delta\rho(\boldsymbol{r})e^{-i\omega t},$$

due to the time-dependent perturbation $\delta V_{pert.}(\boldsymbol{r}';\omega)=\delta(\boldsymbol{r}-\boldsymbol{r}')\delta Ve^{-i\omega t}\equiv\delta V(\boldsymbol{r}')e^{-i\omega t}$.

For example, even in molecules with no permanent dipole moment, i.e., the zero value of the expectation value of the associated quantum-mechanical observable, the expectation value of its square does not vanish, and so every molecule exhibits nonzero dipole fluctuations. The molecular polarizability is related to the magnitude of these fluctuations with large dipole fluctuations implying high polarizability. The connection between the system response (polarizability) and the magnitude of the charge density fluctuations of the unperturbed system represents an example of FDT. The use of this theorem has been introduced in WFT by McLachlan and

Ball (1964) and then extended to DFT (Harris and Jones 1974; Langreth and Perdew 1975; Gunnarson and Lundqvist 1976).

Expressing the *xc*-energy of (7.137) in terms of the linear density–density response function $\chi_\lambda(\boldsymbol{r}, \boldsymbol{r}'; \omega)$ for the current coupling constant λ gives:

$$\begin{aligned} E_{xc}[\rho] &= \frac{1}{2}\int_0^1 d\lambda \int d\boldsymbol{r} \int d\boldsymbol{r}' \, |\boldsymbol{r}-\boldsymbol{r}'|^{-1} \\ &\quad \times \left[\langle \Psi^\lambda[\rho] | (\hat{\rho}(\boldsymbol{r}) - \rho(\boldsymbol{r}))(\hat{\rho}(\boldsymbol{r}') - \rho(\boldsymbol{r}')) | \Psi^\lambda[\rho] \rangle - \delta(\boldsymbol{r}-\boldsymbol{r}')\rho(\boldsymbol{r})\right] \\ &= \frac{1}{2}\int_0^1 d\lambda \int d\boldsymbol{r} \int d\boldsymbol{r}' \, |\boldsymbol{r}-\boldsymbol{r}'|^{-1} \left[-\frac{1}{\pi}\int_0^\infty \chi_\lambda(\boldsymbol{r}, \boldsymbol{r}'; iu)\, du - \delta(\boldsymbol{r}-\boldsymbol{r}')\rho(\boldsymbol{r}) \right]. \end{aligned} \tag{7.355}$$

One can further separate in $\chi_\lambda(\boldsymbol{r}, \boldsymbol{r}'; \omega)$ the KS-reference contribution (see Sect. 7.9.2),

$$\chi_{\mathrm{KS}}(\boldsymbol{r}, \boldsymbol{r}'; \omega) = \sum_{i,j} \frac{f_i - f_j}{\omega - (\varepsilon_j - \varepsilon_i) + i\eta} \varphi_i^*(\boldsymbol{r}) \varphi_j^*(\boldsymbol{r}') \varphi_j(\boldsymbol{r}) \varphi_i(\boldsymbol{r}'), \tag{7.356}$$

where η stands for a positive infinitesimal, f_i and ε_i respectively denote the usual Fermi occupation (0 or 1) and orbital energy of φ_i, and the summations are ranging over both occupied and unoccupied KS MO, including the continuum states. This separation gives

$$\begin{aligned} \chi_\lambda(\boldsymbol{r}, \boldsymbol{r}'; \omega) &= \chi_{\mathrm{KS}}(\boldsymbol{r}, \boldsymbol{r}'; \omega) \\ &\quad + \iint d\boldsymbol{r}_1 d\boldsymbol{r}_2 \chi_{\mathrm{KS}}(\boldsymbol{r}, \boldsymbol{r}_1; \omega) \left(\frac{\lambda}{|\boldsymbol{r}_1 - \boldsymbol{r}_2|} + f_{xc,\lambda}(\boldsymbol{r}_1, \boldsymbol{r}_2; \omega) \right) \chi_\lambda(\boldsymbol{r}_2, \boldsymbol{r}'; \omega), \end{aligned} \tag{7.357}$$

where $f_{xc,\lambda}(\boldsymbol{r}_1, \boldsymbol{r}_2; \omega)$ denotes the *xc*-kernel for the interaction strength λ. One can then separate the exchange from Coulomb correlation energies [(7.140) and (7.145)], $E_{xc}[\rho] = E_x[\rho] + E_c[\rho]$, finding the following expression for the latter:

$$E_c[\rho] = -\frac{1}{2\pi}\int_0^1 d\lambda \int_0^\infty du \int d\boldsymbol{r} \int d\boldsymbol{r}' \, |\boldsymbol{r}-\boldsymbol{r}'|^{-1} [\chi_\lambda(\boldsymbol{r}, \boldsymbol{r}'; iu) - \chi_{\mathrm{KS}}(\boldsymbol{r}, \boldsymbol{r}'; iu)]. \tag{7.358}$$

The full ACFDT DFT requires model *xc*-kernel $f_{xc,\lambda}(\boldsymbol{r}_1, \boldsymbol{r}_2; \omega)$ to solve (7.357) for each value of the coupling constant λ and yields "seamless" vdW DFT (Andersson et al. 1996; Dion et al. 2004). It uses both the occupied and virtual KS orbitals and accounts for the dispersion interactions (Lundqvist et al. 1995; Dobson and Dinte 1996; Chakarova-Käck et al. 2006).

In the RPA variant one neglects the exchange-correlation kernel $f_{xc,\lambda}(\boldsymbol{r}_1, \boldsymbol{r}_2; \omega)$ in $\chi_\lambda(\boldsymbol{r}, \boldsymbol{r}'; \omega)$. This RPA–ACFDT, or in short RPA DFT method, which has been used for model extended systems like Jellium, has recently been applied to molecules as well (Furche 2001; Aryasetiawan et al. 2002; Fuchs and Gonze 2002). On the positive side, it improves description of the bond dissociation and gives good reproduction of cohesive energies and lattice constants in solids (Marini et al. 2006). Moreover, it recovers the right dependence of the cohesive energy upon the system volume. This RPA–ACFDT approach also appears to be much less dependent on the functional used to evaluate orbitals, than the corresponding "pure" LDA or GGA schemes (Harl and Kresse 2008). It scales like N^5–N^4 (Furche 2008; Scuseria et al. 2008), so that it may be computationally efficient, and provides correct description of the dispersion forces at large separations (Dobson et al. 2005). On the negative side, this approximation provides rather poor representation of the *short*-range correlations, giving rise to exceedingly negative correlation energies (Yan et al. 2000), exhibits a slow basis-set convergence, and may produce unphysical maxima in the dissociation energy curves (Furche 2001).

Several attempts to cure these deficiencies of the RPA–ACFDT DFT have been undertaken (e.g., Yan et al. 2000; Dobson and Wang 2000). One way is to apply the *range-separation* (RS) of the electron–electron interaction $g(1,2) = 1/|r_1 - r_2| \equiv 1/r$,

$$g(r) = 1/r = [1 - \mathrm{erf}(\mu r/2)]/r + \mathrm{erf}(\mu r/2)/r \equiv g^{sr}(r) + g^{lr}(r), \qquad (7.359)$$

for a suitably chosen exponent μ, with different treatments of the *short*-range (*sr*) and *long*-range (*lr*) correlation effects by the PBE functional, using the RS–hybrid-MO, and the RPA*x* functional, with the HF exchange kernel $f_{x,\lambda}$, respectively. The RS–ACFDT scheme is much less basis set dependent, produces no artificial maxima in dissociation curves, and it performs better for rare-gas dimers than the MP2 theory.

Of similar character is the ACFDT DFT approach of Kohn et al. (1998), in which the RS of the preceding equation is effected using the exponential weighting factor for $g^{sr}(r) \approx \exp(-\kappa r)/r$ and $g^{lr}(r) \approx [1 - \exp(-\kappa r)]/r$, with the coupling constant $0 \le \lambda \le 1$ being applied to "turn on" only the $g^{lr}(r)$ component, which solely contributes to the vdW energy:

$$\begin{aligned} &\hat{\mathrm{H}}(\lambda) = \hat{\mathrm{T}}_e + V_{ne}^{\lambda} + V_{ee}^{\lambda}, \\ &V_{ee}^{\lambda} = V_{ee}^{sr} + \lambda V_{ee}^{lr}; \quad V_{ee}^{sr} = \frac{1}{2}\sum_{i \neq j}^{N} g^{sr}(r_{i,j}), \quad V_{ee}^{lr} = \frac{1}{2}\sum_{i \neq j}^{N} g^{lr}(r_{i,j}). \end{aligned} \qquad (7.360)$$

Therefore, the full (Coulombic) Hamiltonian operator again corresponds to $\lambda = 1$, with the $\lambda = 0$ system exhibiting only *short*-range interactions. The external potential $v^{\lambda}(\boldsymbol{r})$ in V_{ne}^{λ} is again chosen to keep the electron density fixed, $\rho_\lambda(\boldsymbol{r}) = \rho_{\lambda=1}(\boldsymbol{r}) = \rho(\boldsymbol{r})$, and the molecular electronic energy, $E(\lambda = 1) = \langle\Psi^\lambda|\hat{\mathrm{H}}(\lambda)|\Psi^\lambda\rangle|_{\lambda=1} = E$, for the bare-nuclei external potential $v^{\lambda=1}(\boldsymbol{r}) = v(\boldsymbol{r})$, now includes the vdW contribution:

$$\begin{aligned} E &= E(\lambda = 0) + \int_0^1 \frac{dE(\lambda)}{d\lambda} d\lambda \\ &= E(0) + \int \left(\int_0^1 \frac{dv^\lambda(\boldsymbol{r})}{d\lambda} d\lambda \right) \rho(\boldsymbol{r})\, d\boldsymbol{r} + \int_0^1 \langle\Psi^\lambda|V_{ee}^{lr}|\Psi^\lambda\rangle d\lambda \\ &= E(0) + \int [v(\boldsymbol{r}) - v_0(\boldsymbol{r})]\rho(\boldsymbol{r})\, d\boldsymbol{r} \\ &\quad + \frac{1}{2} \iint d\boldsymbol{r}\, d\boldsymbol{r}'\, V_{ee}^{lr}(|\boldsymbol{r} - \boldsymbol{r}'|) \left[\int_0^1 \langle\Psi^\lambda|\hat{\rho}(\boldsymbol{r})\hat{\rho}(\boldsymbol{r}')|\Psi^\lambda\rangle d\lambda - \rho(\boldsymbol{r})\delta(\boldsymbol{r} - \boldsymbol{r}') \right]. \end{aligned} \tag{7.361}$$

Here, $E(0)$ is the DFT energy for $V_{ee} = V_{ee}^{sr}$,

$$E(0) = T_s[\rho] + \int v_0(\boldsymbol{r})\rho(\boldsymbol{r})\, d\boldsymbol{r} + \frac{1}{2} \iint d\boldsymbol{r} d\boldsymbol{r}'\, V_{ee}^{sr}(|\boldsymbol{r} - \boldsymbol{r}'|)\rho(\boldsymbol{r})\rho(\boldsymbol{r}') + E_{xc}^{sr}[\rho], \tag{7.362}$$

and $E_{xc}^{sr}[\rho]$ stands for the *xc*-energy for the molecular ground state density and the full interaction V_{ee}^{sr}. Combining the two preceding equations finally gives:

$$E = T_s[\rho] + \int v(\boldsymbol{r})\rho(\boldsymbol{r})\, d\boldsymbol{r} + J_{ee}[\rho] + E_{xc}^{sr}[\rho] - \frac{1}{2} V_{ee}^{lr}(0)N + E_{pol.}^{lr}[\rho], \tag{7.363}$$

with the last term, the *long*-range polarization energy now including the vdW energy:

$$E_{pol.}^{lr}[\rho] = \frac{1}{2} \iint V_{ee}^{lr}(|\boldsymbol{r} - \boldsymbol{r}'|) \int_0^1 \langle\Psi^\lambda|[\hat{\rho}(\boldsymbol{r}) - \rho(\boldsymbol{r})][\hat{\rho}(\boldsymbol{r}') - \rho(\boldsymbol{r}')]|\Psi^\lambda\rangle d\lambda\, d\boldsymbol{r} d\boldsymbol{r}'. \tag{7.364}$$

The first four terms in (7.363) are calculated using traditional DFT methods, e.g., by calculating the electron density using LDA with the full $V_{ee}(N)$. This calculation also gives $T_s[\rho]$ and $E_{xc}^{sr}[\rho]$. Finally, using the FDT allows one to express the expectation value of the preceding equation in terms of the imaginary part of the retarded linear susceptibility $\chi_\lambda(\boldsymbol{r}_2, \boldsymbol{r}'; \omega)$ (e.g., Doniach and Sondheim 1982),

$$\langle\Psi^\lambda|[\hat{\rho}(\boldsymbol{r}) - \rho(\boldsymbol{r})][\hat{\rho}(\boldsymbol{r}') - \rho(\boldsymbol{r}')]|\Psi^\lambda\rangle = -\frac{1}{\pi} \int_0^\infty \mathrm{Im}\, \chi_\lambda(\boldsymbol{r}, \boldsymbol{r}'; \omega)\, d\omega, \tag{7.365}$$

which gives the *long*-range polarization energy:

$$E^{lr}_{pol.}[\rho] = -\frac{1}{2\pi}\iint drdr'\, V^{lr}_{ee}(|\boldsymbol{r}-\boldsymbol{r}'|)\int_0^1 d\lambda\int_0^\infty d\omega\,\mathrm{Im}\,\chi_\lambda(\boldsymbol{r},\boldsymbol{r}';\omega). \tag{7.366}$$

The $\chi_\lambda(\boldsymbol{r}_2,\boldsymbol{r}';\omega)$ is defined as before: a small perturbing potential $\delta V_{pert.}(\boldsymbol{r},\omega)= \delta V(\boldsymbol{r})e^{-i\omega t}$ acting on the ground state Ψ^λ of $\hat{\mathrm{H}}(\lambda)$ produces the electron density response $\delta\rho_\lambda(\boldsymbol{r},\omega) = \delta\rho_\lambda(\boldsymbol{r})e^{-i\omega t}$ with the amplitude

$$\delta\rho_\lambda(\boldsymbol{r},\omega) = \int \chi_\lambda(\boldsymbol{r},\boldsymbol{r}';\omega)\,\delta V(\boldsymbol{r}';\omega)\,d\boldsymbol{r}'. \tag{7.367}$$

The response kernel results from the integral *screening equation* (7.357):

$$\chi_\lambda(\boldsymbol{r},\boldsymbol{r}';\omega) = \chi_{\mathrm{KS}}(\boldsymbol{r},\boldsymbol{r}';\omega) + \iint d\boldsymbol{r}_1 d\boldsymbol{r}_2 \chi_{\mathrm{KS}}(\boldsymbol{r},\boldsymbol{r}_1;\omega)\left[g(1,\,2) + f_{xc,\lambda}(\boldsymbol{r}_1,\boldsymbol{r}_2;\omega)\right]\chi_\lambda(\boldsymbol{r}_2,\boldsymbol{r}';\omega). \tag{7.368}$$

The Hartree (or better) approximations for $\chi_\lambda(\boldsymbol{r}_2,\boldsymbol{r}';\omega)$ then generate the adequate functionals for the vdW energies.

To avoid the *self*-consistent solution of these computationally forbidding equations for each value of λ and ω, the $\chi_\lambda(\boldsymbol{r}_2,\boldsymbol{r}';\omega)$ can be expressed as the Fourier transform

$$\chi_\lambda(\boldsymbol{r}_2,\boldsymbol{r}';\omega) = \int dt\,\chi_\lambda(\boldsymbol{r}_2,\boldsymbol{r}';t)e^{i\omega t}$$

of the *time-dependent* response function $\chi_\lambda(\boldsymbol{r}_2,\boldsymbol{r}';t)$, which generates the density response to perturbation $\delta V(\boldsymbol{r}';t')$:

$$\delta\rho_\lambda(\boldsymbol{r},t) = \iint \chi_\lambda(\boldsymbol{r},\boldsymbol{r}';t-t')\,\delta V(\boldsymbol{r}';t')\,d\boldsymbol{r}'dt'. \tag{7.369}$$

The corresponding (time-domain) expression for the *long*-range polarization energy then reads:

$$E^{lr}_{pol.}[\rho] = -\frac{1}{2\pi}\iint drdr'\, V^{lr}_{ee}(|\boldsymbol{r}-\boldsymbol{r}'|)\int_0^1 d\lambda\int_0^\infty \frac{dt}{t}\chi_\lambda(\boldsymbol{r},\boldsymbol{r}';t). \tag{7.370}$$

This approach has been applied with excellent results to determine the asymptotic vdW interaction between two helium atoms and between hydrogen and helium.

Gordon and Kim (1972) have proposed a simple model to estimate within Thomas–Fermi (TF) theory the interaction energies in noble-gas diatomics A----B,

$$E_{int}(\mathrm{A}\ \text{----}\ \mathrm{B}) = E_v^{\mathrm{TF}}[\rho_{\mathrm{A}}^0 + \rho_{\mathrm{B}}^0] - E_v^{\mathrm{TF}}[\rho_{\mathrm{A}}^0] - E_v^{\mathrm{TF}}[\rho_{\mathrm{B}}^0], \qquad (7.371)$$

where $E_v^{\mathrm{TF}}[\rho]$ stands for the (orbital-free) density functional for the electronic energy in TF method, calculated for the molecular external potential due to both atomic nuclei, $v(\boldsymbol{r}) = v_{\mathrm{A}}(\boldsymbol{r} - \boldsymbol{R}_{\mathrm{A}}) + v_{\mathrm{B}}(\boldsymbol{r} - \boldsymbol{R}_{\mathrm{B}})$, where $(\boldsymbol{R}_{\mathrm{A}}, \boldsymbol{R}_{\mathrm{B}})$ denote the fixed atomic positions in the BO approximation, and $(\rho_{\mathrm{A}}^0 = \rho_{\mathrm{A}}[v_{\mathrm{A}}], \rho_{\mathrm{B}}^0 = \rho_{\mathrm{B}}[v_{\mathrm{B}}])$ are the electron densities of the separated (isolated) free atoms. Astonishingly, this nonvariational approach, requiring the prior knowledge of the free-atom densities, gives surprisingly good interaction energies, with the energy curves identifying a bit too short equilibrium distances and failing to reflect the known asymptotics of vdW interactions.

This approach has introduced the idea of the energy *bifunctional*,

$$E_v^{\mathrm{TF}}[\rho_{\mathrm{A}}^0 + \rho_{\mathrm{B}}^0] \equiv E_v^{\mathrm{TF}}[\rho_{\mathrm{A}}^0, \rho_{\mathrm{B}}^0] \equiv E_{total}^{\mathrm{TF}}[\boldsymbol{\rho}^0], \qquad (7.372a)$$

with the interaction energy of (7.371) representing its nonadditive part in the isolated-atom (promolecule) resolution, given by difference between this total energy and its additive component

$$E_{v_{\mathrm{A}}}^{\mathrm{TF}}[\rho_{\mathrm{A}}^0] + E_{v_{\mathrm{B}}}^{\mathrm{TF}}[\rho_{\mathrm{B}}^0] \equiv E_{add.}^{\mathrm{TF}}[\boldsymbol{\rho}^0], \qquad (7.372b)$$

$$E_{int}(\mathrm{A}\ \text{----}\ \mathrm{B}) = E_{total}^{\mathrm{TF}}[\boldsymbol{\rho}^0] - E_{add.}^{\mathrm{TF}}[\boldsymbol{\rho}^0] \equiv E_{nonadd.}^{\mathrm{TF}}[\boldsymbol{\rho}^0]. \qquad (7.372c)$$

This bifunctional approach has been further developed in the DFT treatments by Cortona (1991) as well as by Wesołowski and collaborators (Wesołowski and Warshel 1993; Wesołowski et al. 1995; Wesołowski and Weber 1998; Wesołowski and Tran 2003; Wesołowski 2004a,b). In this (variational) *orbital-free embedding* scheme for intermolecular complexes one can adequately describe the selected fragment A, not covalently bonded to its environment B in a complex material A----B, without the need to construct KS orbitals representing the entire system.

Assuming the integer number of electrons in each subsystem,

$$\int \rho_{\mathrm{A}}(\boldsymbol{r})d\boldsymbol{r} = N_{\mathrm{A}}^0, \quad \int \rho_{\mathrm{B}}(\boldsymbol{r})d\boldsymbol{r} = N_{\mathrm{B}}^0, \quad N_{\mathrm{A}}^0 + N_{\mathrm{B}}^0 = N,$$

and partitioning the kinetic energy bifunctional for the overall density $\rho(\boldsymbol{r}) = \rho_{\mathrm{A}}(\boldsymbol{r}) + \rho_{\mathrm{B}}(\boldsymbol{r})$ of the noninteracting KS system,

$$\begin{aligned} T_s[\rho_{\mathrm{A}} + \rho_{\mathrm{B}}] &\equiv T_s^{total}[\rho_{\mathrm{A}}, \rho_{\mathrm{B}}] = (T_s[\rho_{\mathrm{A}}] + T_s[\rho_{\mathrm{B}}]) + T_s^{nadd.}[\rho_{\mathrm{A}}, \rho_{\mathrm{B}}] \\ &\equiv T_s^{add.}[\rho_{\mathrm{A}}, \rho_{\mathrm{B}}] + T_s^{nadd.}[\rho_{\mathrm{A}}, \rho_{\mathrm{B}}], \end{aligned} \qquad (7.373)$$

one can then variationally optimize the densities of the embedded subsystems, feeling the presence of their complementary environment, by using the Euler–Lagrange type minimization of the energy. The gradient-dependent bifunctionals $T_s^{nadd.}[\rho_A, \rho_B]$ for the small overlap between the two densities have been developed, which together with PW91 functional for *xc*-energy give decent predictions of vdW energies close to the equilibrium geometries of molecular complexes.

This development allows one to study various intermolecular systems by performing either the *joint*-minimization of both densities in $E_v[\rho_A, \rho_B]$ or the *constrained*-optimization of the density of the fragment of interest, say A, for the "frozen" ρ_B. The latter calculations determine the electronic structure of the embedded molecule A from the KS-like calculations in which the effect of the presence of the surrounding B is expressed exactly by means of the orbital-free effective potential simulating the presence of the molecular environment of A:

$$\begin{aligned} V_A^{emb.}(\boldsymbol{r};B) &\equiv V_A^{emb.}[\boldsymbol{r},\rho_A;\rho_B] \\ &= \left(v_B(\boldsymbol{r}) + \int \frac{\rho_B(\boldsymbol{r}')}{|\boldsymbol{r}-\boldsymbol{r}'|} d\boldsymbol{r}' \right) + \left(\frac{\delta E_{xc}[\rho]}{\delta\rho(\boldsymbol{r})}\Big|_{\rho=\rho_A+\rho_B} - \frac{\delta E_{xc}[\rho]}{\delta\rho(\boldsymbol{r})}\Big|_{\rho=\rho_A} \right) + \frac{\delta T_s^{nadd.}[\rho_A,\rho_B]}{\delta\rho_A(\boldsymbol{r})} \\ &\equiv v_A^{H}(\boldsymbol{r};B) + v_A^{xc}(\boldsymbol{r};B) + v_A^{kin.}(\boldsymbol{r};B). \end{aligned} \tag{7.374}$$

Here, $v_A^H(\boldsymbol{r};\ B)$ is the effective Hartree potential due to the nuclei and electrons of the subsystem B, $v_A^{xc}(\boldsymbol{r};\ B)$ stands for corresponding exchange-correlation contribution, and $v_A^{kin.}(\boldsymbol{r};\ B)$ is the extra term due to the nonadditive kinetic energy. Such formalism is particularly suitable for describing the interactions between two subsystems bonded by the electrostatic and/or vdW forces, e.g., in solvation (Wesołowski et al. 1995) and biological systems (Olsson et al. 2004). In combination with the linear-response DFT it can be also applied to study excited states of embedded molecules (Casida and Wesołowski 2004).

In the ab initio vdW-corrected DFT scheme of (7.353) one requires both the accurate nonempirical description of the dispersion energy between DFT monomers and an adequate (dispersion-free) treatment of the supermolecular interaction energy within DFT. For the simplest scenario of the above noncovalently bonded "dimer" complex A----B, Rajchel et al. (2009a, b) have suggested a promising approach using the energy bifunctional of densities of the embedded monomers to determine the effects of the noncovalent interactions via the monomer polarization and the Pauli Blockade of Gutowski and Piela (1988). The latter is used to enforce the orthogonality between KS orbitals determining the electron densities of the two polarized monomers. It ensures that the intersubsystem Pauli exclusion principle is exactly fulfilled in the optimum solutions.

The a posteriori vdW correction from the Coupled-Perturbed Kohn–Sham (CPKS) formulation within the *Symmetry Adapted Perturbation Theory* (SAPT) of molecular interactions (Jeziorski et al. 1994; Jeziorski and Szalewicz 2002; Bukowski et al. 2008) use the DFT description of monomers (DFT–SAPT) (Williams and Chabalowski 2001; Jansen and Heßelmann 2001; Heßelmann and

Jansen 2003; Heßelmann et al. 2005; Misquitta and Szalewicz 2002, 2005; Misquitta et al. 2003, 2005; Bukowski et al. 2005; Szalewicz et al. 2005; Podeszwa et al. 2006). This treatment of long-range interactions is based on the generalized Casimir–Polder expression for the dispersion energy (e.g., Zaremba and Kohn 1976), using the time-dependent DFT response theory (see the next section), which is added to the electronic energy of the molecular complex consisting of the polarized monomers.

This promising avenue represents a systematic *ab initio* Many-Body (MB) perspective on molecular interactions, giving results of CCSD(T) quality. It expresses the interaction energy between molecular subsystems as a hierarchy of the physically meaningful terms in MBPT:

$$E_{int}(\mathrm{A} - - - - \mathrm{B}) = E^{(1)}_{el.} + E^{(1)}_{exch.} + E^{(2)}_{ind.} + E^{(2)}_{exch.-ind.} + E^{(2)}_{disp.} + E^{(2)}_{exch.-disp.} + \ldots \quad (7.375)$$

It is also well suited for fitting the molecular force fields and it gives a deeper understanding of the interaction mechanism. At the same time the MB SAPT, using the correlated *ab initio* MBPT wave functions for monomers, provides efficient means for an explicit calculation of all these electrostatic (*el.*), exchange (*exch.*), inductive (*ind.*), and dispersion (*disp.*) contributions to the interaction energy. Extensions to the *three*-body effects and open-shell cases are also available (Podeszwa and Szalewicz 2007).

In the combined DFT–SAPT approach, the *first*-order electrostatic and exchange terms, $E^{(1)}_{el.}$ and $E^{(1)}_{exch.}$, are calculated using the monomer DFT densities and density matrices. The associated *second*-order induction terms, $E^{(2)}_{ind.}$ and $E^{(2)}_{exch.-ind.}$, are determined from the CPKS response functions, while the *second*-order dispersion contributions, $E^{(2)}_{disp.}$ and $E^{(2)}_{exch.-disp.}$, result from the *Time-Dependent* DFT (TDDFT) response functions. This computational scheme is relatively "cheap," giving at best the MP2-like N^5 scaling (Heßelmann and Schütz 2006). It is capable of generating potentially exact $E^{(1)}_{el.}$, $E^{(2)}_{ind.}$, and $E^{(2)}_{disp.}$ terms, but the remaining exchange corrections have been found to be strongly dependent on a quality of the *xc*-potential at the DFT stage, and less – on the *xc*-kernel f_{xc}, used in CPKS and TDDFT stages (Jansen and Heßelmann 2001). For the acetylene–benzene (Tekin and Jansen 2007) and acetylene–furan (Sánchez-Garcia et al. 2008) complexes, the resulting energy curves are very close to CCSD(T) results. Another successful benchmark application of this approach (Heßelmann and Schütz 2006; Fiethen et al. 2008) deals with the stacking interactions in DNA, which are of the utmost importance for the DNA structure. The goal was to compare them to other intermolecular interactions and to determine their variations with changing geometries of the complex (Jurečka and Hobza 2003).

Other approaches also combine the DFT and WFT perspectives, e.g., within the *ab initio* DFT (Sect. 7.5.3), or attempt to model the dispersion directly via the exchange hole (Becke and Johnson 2007). This general goal has been pursued vigorously along many lines, e.g., via the range separation (Stoll and Savin 1985; Savin and Flad 1995; Leininger et al. 1997; Gerber and Ángyán 2007; Goll et al. 2008) of the "*double hybrid*" method (Grimme 2006a,b).

7.9 Time-Dependent DFT

Ordinary (time-*independent*) DFT uses exclusively the ground state electron density to completely describe the N-electron system, with each observable quantity being in principle given as the corresponding functional of this state-variable alone. The density variational principle of DFT or the associated Euler equation for the optimum electron density then replace the stationary Schrödinger equation of WFT. The associated KS scheme allows one to determine the density of the interacting system as the density of an auxiliary system of noninteracting electrons moving in an effective (local) *single*-particle potential. We recall that this original formulation is based on the existence of an exact mapping between densities and external potentials.

The *Time-Dependent* DFT (TDDFT), covering the time-dependent external and effective potentials, represents a more recent development in the theory (Pueckert 1978, Zangwill and Soven 1980; Deb and Ghosh 1982; Ghosh and Deb 1982, 1983a,b; Bartolotti 1981, 1982, 1984, 1987; Runge and Gross 1984; Gross and Kurth 1993; Gross et al. 1996; Marques and Gross 2004; Ghosh 2009), which effectively replaces the time-dependent Schrödinger equation of WFT in determining the evolution in time of the densities and currents of *many*-particle systems, through the associated time-dependent KS scheme. This formal development, the rudiments of which are summarized in the following subsections, again establishes the mapping between the time-dependent densities and external potentials, expresses each observable as the unique functional of the time-dependent density, and replaces the energy variational principle of the time-independent theory with the associated principle of the stationary quantum-mechanical action, which ultimately allows one to establish the time-dependent KS scheme. The need for this replacement arises since in time-dependent systems the energy is no longer conserved. This important extension of the original (static) formulation of the theory also generates the computationally efficient scheme for determining the excitation energies. For the recent reviews by physicists see Gross et al. (1996), Gross and Kurth (1993), Grabo et al. (1999), Marques and Gross (2004), and for the survey written from the point of view of quantum chemistry see Casida (1996).

7.9.1 *Extensions of HK Theorems into Time Domain*

Let us again examine the system of N electrons moving in an explicitly time dependent external potential $V_{ext.}(N,t) = \sum_{i=1}^{N} v_{ext.}(\boldsymbol{r}_i, t)$. Its electronic Hamiltonian [see (7.1)],

$$\hat{\mathrm{H}}_e(N,t) = V_{ext.}(N,t) + \hat{\mathrm{F}}(N) \equiv \hat{\mathrm{H}}(t),$$

shapes the time evolution of quantum states (3.94) (a.u. are used throughout):

$$\hat{\mathrm{H}}(N,t)\Psi(N,t) = i\frac{\partial\Psi(N,t)}{\partial t}, \tag{7.376}$$

thus determining the TD analog of the stationary map $\mathscr{A}$ (7.9), between the time-dependent external potentials $v_{ext.}(\boldsymbol{r},t) \equiv v(t)$ and quantum states of the N-electron system in question: $v_{ext.}(r,t) \rightarrow \Psi(N,t) \equiv \Psi(t)$ or $\mathscr{A}v(t) = \Psi(t)$. The trivial map $\mathscr{B}$, from TD wave function to the associated electron density, $\mathscr{B}\Psi(t) = \rho(t)$ is again given by the expectation value of the density operator:

$$\Psi(N,t) \rightarrow \rho(\boldsymbol{r},t) = \langle \Psi(N,t)|\hat{\rho}(\boldsymbol{r})|\Psi(N,t)\rangle \equiv \rho(t). \qquad (7.377)$$

Together they define the unique forward mapping $\mathscr{M} = \mathscr{A}\mathscr{B}$, from the TD external potentials to the associated TD densities: $\mathscr{M}v(t) = \rho(t)$. The basic aim of TDDFT was to prove the invertability of this product-mapping, i.e., the existence of the unique relation $\mathscr{M}^{-1}\rho(t) = v(t)$, which then establishes the 1–1 correspondence between TD densities and potentials.

In TDDFT the external potential is assumed to be time-independent for $t < t_0$, $v_{ext.}(\boldsymbol{r},t_0) = v_0(\boldsymbol{r})$, so that the time-dependent field is switched on exactly at (finite) time t_0. The Runge–Gross (1984) scenario deals with the quantum states evolving from the fixed initial state

$$\Psi(N,t_0) = \Psi_0 \equiv \Psi_0[N,v_0], \qquad (7.378)$$

giving rise to the initial density $\rho(\boldsymbol{r},t_0) = \rho_0(\boldsymbol{r}) = \langle \Psi_0|\hat{\rho}(\boldsymbol{r})|\Psi_0\rangle$, which is not required to be the ground state or some other eigenstate of the initial potential $v_0(\boldsymbol{r})$. The basic HK-like theorem of TD theory of Gross and collaborators then states:

the densities $\rho(t)$ and $\rho'(t)$ evolving from the common initial state Ψ_0 under the influence of two TD potentials $v(t)$ and $v'(t)$, respectively, both Taylor-expandable around t_0, are always different provided that these potentials differ by more than the purely t-dependent ($\boldsymbol{r}$-independent) function:

$$v(\boldsymbol{r},t) \neq v'(\boldsymbol{r},t) + c(t), \qquad (7.379)$$

Notice that otherwise the resulting wave functions $\Psi(t)$ and $\Psi'(t)$ evolving from the same initial state would differ solely by a purely time-dependent phase factor, thus giving rise to identical densities: $\rho(t) = \rho'(t)$.

The proof (Gross and Kurth 1993; Gross et al. 1996; Grabo et al. 1999) goes in two steps. One first demonstrates, by using the quantum mechanical equation of motion (3.117) for the (paramagnetic) current density operator [see (3.128) and (3.129)],

$$\hat{\mathbf{j}}(\boldsymbol{r}) = \frac{1}{2i}\sum_{k=1}^{N}[\nabla_{\boldsymbol{r}_k}\delta(\boldsymbol{r}_k - \boldsymbol{r}) + \delta(\boldsymbol{r}_k - \boldsymbol{r})\nabla_{\boldsymbol{r}_k}], \qquad (7.380)$$

and the Taylor expandability of the two potentials that the current densities $\boldsymbol{j}(\boldsymbol{r},t) = \langle\Psi(t)|\hat{\mathbf{j}}(\boldsymbol{r})|\Psi(t)\rangle$ and $\boldsymbol{j}'(\boldsymbol{r},t) = \langle\Psi'(t)|\hat{\mathbf{j}}(\boldsymbol{r})|\Psi'(t)\rangle$ are then different for the essentially different potentials of (7.379): $\boldsymbol{j}(\boldsymbol{r}, t) \neq \boldsymbol{j}'(\boldsymbol{r}, t)$. In the second step one uses the continuity equation (3.127) to prove, again by the *reductio ad absurdum*, that the two densities will become different infinitesimally later than t_0: $\rho(\boldsymbol{r}, t) \neq \rho'(\boldsymbol{r}, t)$. It can be demonstrated that the difference $\rho(\boldsymbol{r}, t) - \rho'(\boldsymbol{r}, t)$ is nonvanishing already in the *first* order of $v(\boldsymbol{r}, t) - v'(\boldsymbol{r}, t)$, thus ensuring the invertibility of the linear response operators.

By virtue of this established 1–1 correspondence, for the given Ψ_0, the time-dependent density determines the external potential uniquely up to within an additive, purely time-dependent function. The potential in turn determines the time-dependent wave function, which can therefore be considered as a functional of the time-dependent density, unique up to within a purely time-dependent phase $\alpha(t)$,

$$\Psi(t) = e^{-i\alpha(t)}\tilde{\Psi}[\rho; t]. \tag{7.381}$$

Therefore, the expectation value of any quantum mechanical operator $\hat{\mathrm{O}}(t)$ is the unique functional of the density, with the phase ambiguity canceling out:

$$\langle O \rangle_{\Psi(t)} = \langle \Psi(t)|\hat{\mathrm{O}}(t)|\Psi(t)\rangle = \langle \tilde{\Psi}[\rho; t]|\hat{\mathrm{O}}(t)|\tilde{\Psi}[\rho; t]\rangle = O[\rho; t]. \tag{7.382}$$

It should be recalled (Gross et al. 1996) that the Schrödinger equation (7.376) with the initial condition (7.378) corresponds to a stationary point of the quantum-mechanical *action* integral, itself the unique density functional

$$\begin{aligned}
\mathcal{A}[\Psi] &= \int_{t_0}^{t_1} \langle \Psi(t)|i\frac{\partial}{\partial t} - \hat{\mathrm{H}}(t)|\Psi(t)\rangle dt = \int_{t_0}^{t_1} \langle \tilde{\Psi}[\rho; t]|i\frac{\partial}{\partial t} - \hat{\mathrm{H}}(t)|\tilde{\Psi}[\rho; t]\rangle dt \equiv \mathcal{A}[\rho] \\
&= \int_{t_0}^{t_1} \langle \tilde{\Psi}[\rho; t]|i\frac{\partial}{\partial t} - \hat{\mathrm{F}}|\tilde{\Psi}[\rho; t]\rangle dt - \iint_{t_0}^{t_1} v(\boldsymbol{r}, t)\rho(\boldsymbol{r}, t)\, d\boldsymbol{r}\, dt \\
&\equiv \mathcal{B}[\rho] - \iint_{t_0}^{t_1} v(\boldsymbol{r}, t)\rho(\boldsymbol{r}, t)\, d\boldsymbol{r}\, dt.
\end{aligned} \tag{7.383}$$

The exact time-dependent density marks the stationary point of this functional, thus enabling one to compute $\rho(\boldsymbol{r}, t)$ from the associated Euler equation:

$$\frac{\delta \mathcal{A}[\rho]}{\delta \rho(\boldsymbol{r}, t)} = 0. \tag{7.384}$$

This *stationary-action principle* also facilitates the introduction of the time-dependent KS scheme, in which the exact time-dependent density of the system of interacting particles is calculated as the density of the hypothetical system of the noninteracting electrons occupying the time-dependent KS orbitals $\{\varphi_n(\boldsymbol{r},t)\}$:

$$\rho(\boldsymbol{r},t) = \sum_{n=1}^{N} \varphi_n^*(\boldsymbol{r},t)\varphi_n(\boldsymbol{r},t). \tag{7.385}$$

They define the associated time-dependent determinant $\Psi_s(N,t) = \det(\{\varphi_n\})$ and satisfy the time-dependent KS equation,

$$i\frac{\partial}{\partial t}\phi_n(r,t) = \left(-\frac{\nabla^2}{2} + v_{\mathrm{KS}}[\rho;\boldsymbol{r},t]\right)\varphi_n(\boldsymbol{r},t), \tag{7.386}$$

for the effective time-dependent *one*-body potential

$$v_{\mathrm{KS}}[\rho;\boldsymbol{r},t] = v(\boldsymbol{r},t) + \int \frac{\rho(\boldsymbol{r}',t)}{|\boldsymbol{r}'-\boldsymbol{r}|}d\boldsymbol{r}' + v_{xc}[\rho;\boldsymbol{r},t] \tag{7.387}$$

with the effective (time-dependent) exchange-correlation potential $v_{xc}[\rho;\boldsymbol{r},t]$ defined by the functional derivative

$$v_{xc}[\rho;\boldsymbol{r},t] = \frac{\delta \mathcal{A}_{xc}[\rho]}{\delta\rho(\boldsymbol{r},t)}. \tag{7.388}$$

Here, the *xc*-part of the action integral of (7.383) reads

$$\mathcal{A}_{xc}[\rho] = \mathcal{B}_s[\rho] - \mathcal{B}[\rho] - \frac{1}{2}\int_{t_0}^{t_1} dt \iint \frac{\rho(\boldsymbol{r}',t)\rho(\boldsymbol{r},t)}{|\boldsymbol{r}'-\boldsymbol{r}|}d\boldsymbol{r}'\,d\boldsymbol{r}, \tag{7.389}$$

with the noninteracting action integral $\mathcal{B}_s[\rho]$ being similarly defined in terms of the KS Slater determinant $\Psi_s(N,t) = \tilde{\Psi}_s[\rho,t]$:

$$\mathcal{B}_s[\rho] = \int_{t_0}^{t_1} \langle\tilde{\Psi}_s[\rho;t]|i\frac{\partial}{\partial t} - \hat{\mathrm{T}}_e(t)|\tilde{\Psi}_s[\rho;t]\rangle dt. \tag{7.390}$$

In practice $v_{xc}[\rho;\boldsymbol{r},t]$ has to be approximated, as in the static KS theory. Compared to the time-dependent HF and CI theories, the TDKS scheme is computationally simpler since $v_{\mathrm{KS}}[\rho;\boldsymbol{r},t]$ is a *local* potential, i.e., the multiplicative operator in the position representation.

It should be emphasized that in TDDFT, the above mapping relations are established only for the fixed initial state, so that the above density functionals parametrically depend on Ψ_0. However, when the initial state is the nondegenerate ground state, Ψ_0 is then the unique functional of ρ_0, by the ordinary HK theorem, so that in this case the TD functionals are also functionals of this density alone.

The OEP/OPM method of Sect. 7.5.1 has been generalized to the time-dependent case by Ullrich et al. (1995). The relevant integral equations determining the time-dependent x-potential $v_{x\sigma}^{\mathrm{EXX}}(\boldsymbol{r},t)$ for spin σ in the EXX approximation, again derived via a series of chain rules involving derivatives of the Fock action,

$$\mathcal{A}_x^{\mathrm{EXX}}[\boldsymbol{\varphi}[\rho]] = -\frac{1}{2}\sum_{\sigma}\sum_{j,k}^{occd.}\int_{t_0}^{t_1} dt \iint \frac{\varphi_{j\sigma}^*(\boldsymbol{r}',t)\varphi_{k\sigma}(\boldsymbol{r}',t)\varphi_{j\sigma}(\boldsymbol{r},t)\varphi_{k\sigma}^*(\boldsymbol{r},t)}{|\boldsymbol{r}'-\boldsymbol{r}|} d\boldsymbol{r}'d\boldsymbol{r}, \quad (7.391)$$

now read (Marques and Gross 2004):

$$\sum_{j}^{occd.}\int dt' \int d\boldsymbol{r}'[v_{x\sigma}^{\mathrm{EXX}}(\boldsymbol{r}',t') - u_{xj\sigma}(\boldsymbol{r}',t')]\ \varphi_{j\sigma}(\boldsymbol{r},t)G_{R\sigma}(\boldsymbol{r},t;\boldsymbol{r}',t')\varphi_{j\sigma}^*(\boldsymbol{r}',t') + c.c. = 0, \quad (7.392)$$

where $G_{R\sigma}(\boldsymbol{r},t;\boldsymbol{r}',t')$ stands for the KS (retarded) *one*-particle Green's function and

$$u_{xj\sigma}(\boldsymbol{r},t) = \frac{1}{\varphi_{j\sigma}^*(\boldsymbol{r},t)}\frac{\delta\mathcal{A}_{xc}^{\mathrm{EXX}}}{\delta\varphi_{j\sigma}(\boldsymbol{r},\tau)}\bigg|_{\varphi_{j\sigma}=\varphi_{j\sigma}(\boldsymbol{r},t)}. \quad (7.393)$$

7.9.2 Linear-Response Functions

Next, let us examine the linear response $\delta^{(1)}\rho(\boldsymbol{r},t) \equiv \delta\rho(\boldsymbol{r},t)$ in the electron density of an N-electron molecular system, initially in its ground state at $t = t_0$, to the external time-dependent perturbation $\delta v(\boldsymbol{r}, t)$, with $\delta v(\boldsymbol{r}, t) = 0$ for $t \le t_0$. Therefore, the initial density $\rho_0(\boldsymbol{r})$ can be calculated from the ordinary ground state (static) KS equations of (7.110),

$$\left(-\frac{1}{2}\Delta + v_0(\boldsymbol{r}) + \int \frac{\rho_0(\boldsymbol{r}')}{|\boldsymbol{r}'-\boldsymbol{r}|}d\boldsymbol{r}' + v_{xc}[\rho_0;\boldsymbol{r}]\right)\varphi_n(\boldsymbol{r}) = \varepsilon_n\varphi_n(\boldsymbol{r}), \quad (7.394)$$

with the density given by the sum of densities of the lowest N (singly occupied) KS orbitals:

$$\rho_0(\boldsymbol{r}) = \sum_{n=1}^{N}\varphi_n^*(\boldsymbol{r})\varphi_n(\boldsymbol{r}). \quad (7.395)$$

In this case the time-dependent density of the preceding section is uniquely determined by the perturbing external potential $v_{ext.}(\boldsymbol{r},t) = v_0(\boldsymbol{r}) + \delta v(\boldsymbol{r},t)$:

$\rho(\mathbf{r},t) = \rho[v_{ext.};\mathbf{r},t]$. This functional relation can be then inverted by virtue of the Runge–Gross theorem of TDDFT: $v_{ext.}(\mathbf{r}, t) = v_{ext.}[\rho; \mathbf{r}, t]$. Since the theorem also holds for the time-dependent KS system with $\rho_s(\mathbf{r}, t) = \rho(\mathbf{r}, t)$, the associated functional relations exist in this noninteracting limit as well: $\rho(\mathbf{r},t) = \rho[v_{KS};\mathbf{r},t]$ and $v_{KS}(\mathbf{r},t) = v_{KS}[\rho;\mathbf{r},t]$.

Turning now to the density PT, for a sufficiently small $\delta v(\mathbf{r},t)$ the functional for the density response $\rho^{resp.}(\mathbf{r},t) = \rho(\mathbf{r},t) - \rho_0(\mathbf{r})$ can be expanded in power series of the perturbation [compare (5.4)] with coefficients $\{\delta^{(k)}\rho(\mathbf{r}, t)\}$ at the kth power of $\delta v(\mathbf{r}, t)$:

$$\rho^{resp.}(\mathbf{r},t) = \delta\rho(\mathbf{r},t) + \delta^{(2)}\rho(\mathbf{r},t) + \ldots \tag{7.396}$$

The *first*-order, linear response in the density is customarily written as the integral

$$\delta\rho(\mathbf{r},t) = \iint \chi(\mathbf{r},t;\mathbf{r}',t')\,\delta v(\mathbf{r}',t')\,dt'\,d\mathbf{r}', \tag{7.397}$$

where the full (density–density) linear-response function

$$\chi(\mathbf{r},t;\mathbf{r}',t') = \left.\frac{\delta\rho[v_{ext.};\mathbf{r},t]}{\delta v_{ext.}(\mathbf{r}',t')}\right|_{v_0}. \tag{7.398}$$

This functional derivative depends also on ρ_0, which uniquely determines v_0. The associated KS response function, i.e., the density–density response function of the noninteracting system with the unperturbed density ρ_0 similarly reads:

$$\chi_{KS}(\mathbf{r},t;\mathbf{r}',t') = \left.\frac{\delta\rho[v_{KS};\mathbf{r},t]}{\delta v_{KS}(\mathbf{r}',t')}\right|_{v_{KS}[\rho_0]}. \tag{7.399}$$

One then formally establishes the time-dependent KS scheme by inserting the functional $\rho[v_{ext.}; \mathbf{r}, t]$ into $v_{KS}[\rho; \mathbf{r}, t]$. It gives the effective potential $v_{KS}[\rho;\mathbf{r},t]= v_{KS}[v_{ext.};\mathbf{r},t] \equiv v_{KS}(\mathbf{r},t)$, such that the noninteracting electrons moving in this field have the time-dependent density identical with that produced by the fully interacting electrons moving in $v_{ext.}(\mathbf{r}, t)$:

$$v_{KS}(\mathbf{r},t) = v_{ext.}(\mathbf{r},t) + \int \frac{\rho(\mathbf{r}',t)}{|\mathbf{r}'-\mathbf{r}|}d\mathbf{r}' + v_{xc}(\mathbf{r},t). \tag{7.400}$$

Therefore, the functional dependence between ρ and $v_{ext.}$ can be chain-rule transformed using the dependence of v_{KS} on $v_{ext.}$:

$$\chi(\mathbf{r},t;\mathbf{r}',t') = \int d\mathbf{r}'' \int dt'' \left.\frac{\delta\rho[v_{ext.};\mathbf{r},t]}{\delta v_{KS}(\mathbf{r}'',t'')}\frac{\delta v_{KS}(\mathbf{r}'',t'')}{\delta v_{ext.}(\mathbf{r}',t')}\right|_{\rho_0}. \tag{7.401}$$

The complex functional dependence $v_{\mathrm{KS}} = v_{\mathrm{KS}}[v_{ext.}, \rho[v_{ext.}]]$ calls for the second chain-rule transformation in calculating the derivative:

$$\left.\frac{\delta v_{\mathrm{KS}}(\boldsymbol{r},t)}{\delta v_{ext.}(\boldsymbol{r}',t')}\right|_{\rho_0} = \delta(\boldsymbol{r}-\boldsymbol{r}')\delta(t-t') + \int d\boldsymbol{r}'' \int dt'' \left[\frac{\delta(t-t'')}{|\boldsymbol{r}-\boldsymbol{r}''|} + \frac{\delta v_{xc}(\boldsymbol{r},t)}{\delta\rho(\boldsymbol{r}'',t'')}\right] \frac{\delta\rho(\boldsymbol{r}'',t'')}{\delta v_{ext.}\boldsymbol{r}',t')}. \quad (7.402)$$

Finally, combining (7.398), (7.399), (7.401), and (7.402) gives:

$$\chi(\boldsymbol{r},t;\boldsymbol{r}',t') = \chi_{\mathrm{KS}}(\boldsymbol{r},t;\boldsymbol{r}',t') + \int d\boldsymbol{r}_1 \int d\tau \int d\boldsymbol{r}_2 \int d\tau' \chi_{\mathrm{KS}}(\boldsymbol{r},t;\boldsymbol{r}_1,\tau) \left[\frac{\delta(\tau-\tau')}{|\boldsymbol{r}_1-\boldsymbol{r}_2|} + f_{xc}(\boldsymbol{r}_1,\tau;\boldsymbol{r}_2,\tau')|_{\rho_0}\right] \chi(\boldsymbol{r}_2,\tau';\boldsymbol{r}',t'), \quad (7.403)$$

where the time-dependent *xc*-kernel, implicitly defined in (7.402),

$$f_{xc}(\boldsymbol{r},t;\boldsymbol{r}',t')|_{\rho_0} = \left.\frac{\delta v_{xc}[\rho;\boldsymbol{r},t]}{\delta\rho(\boldsymbol{r}',t')}\right|_{\rho_0}, \quad (7.404)$$

is the functional of the initial density ρ_0.

Multiplying both sides of (7.403) by the perturbation $\delta v(\boldsymbol{r}',t')$ and integrating over $\boldsymbol{r}'$and t' gives the alternative exact relation between the linear responses in the density and effective KS potential, respectively (compare (7.397):

$$\delta\rho(\boldsymbol{r},t) = \int\int \chi_{\mathrm{KS}}(\boldsymbol{r},t;\boldsymbol{r}',t')\, \delta v_{\mathrm{KS}}(\boldsymbol{r}',t')\, dt'\, d\boldsymbol{r}', \quad (7.405)$$

where

$$\delta v_{\mathrm{KS}}(\boldsymbol{r},t) = \delta v(\boldsymbol{r},t) + \int \frac{\delta\rho(\boldsymbol{r}',t)}{|\boldsymbol{r}'-\boldsymbol{r}|} d\boldsymbol{r}' + \int\int f_{xc}(\boldsymbol{r},t;\boldsymbol{r}',t')|_{\rho_0} \delta\rho(\boldsymbol{r}',t') d\boldsymbol{r}' dt'. \quad (7.406)$$

The Fourier transform with respect to $\tau = t - t'$ of these KS relations gives the exact *frequency-dependent* linear density response:

$$\delta\rho(\boldsymbol{r},\omega) = \int \chi_{\mathrm{KS}}(\boldsymbol{r},\boldsymbol{r}_1;\omega)\, \delta v(\boldsymbol{r}_1,\omega)\, d\boldsymbol{r}_1 + \int d\boldsymbol{r}_1 \int d\boldsymbol{r}_2 \chi_{\mathrm{KS}}(\boldsymbol{r},\boldsymbol{r}_1;\omega) \left[\frac{1}{|\boldsymbol{r}_1-\boldsymbol{r}_2|} + f_{xc}(\boldsymbol{r}_1,\boldsymbol{r}_2;\omega)|_{\rho_0}\right] \delta\rho(\boldsymbol{r}_2,\omega), \quad (7.407)$$

where KS response function $\chi_{\mathrm{KS}}(\boldsymbol{r},\boldsymbol{r}_1;\omega)$ is given by (7.356). Here, we have used the usual Fourier-transform convention:

$$f(\omega) = \frac{1}{2\pi}\int_{-\infty}^{+\infty} e^{+i\omega t} f(t)\, dt \quad \text{and} \quad f(t) = \frac{1}{2\pi}\int_{-\infty}^{+\infty} e^{-i\omega t} f(\omega)\, d\omega. \quad (7.408)$$

The same operation performed on (7.403) gives:

$$\chi(\boldsymbol{r},\boldsymbol{r}';\omega) = \chi_{\mathrm{KS}}(\boldsymbol{r},\boldsymbol{r}';\omega) + \int d\boldsymbol{r}_1 \int d\boldsymbol{r}_2 \chi_{\mathrm{KS}}(\boldsymbol{r},\boldsymbol{r}_1;\omega)\left[\frac{1}{|\boldsymbol{r}_1-\boldsymbol{r}_2|} + f_{xc}(\boldsymbol{r}_1,\boldsymbol{r}_2;\omega)\big|_{\rho_0}\right]\chi(\boldsymbol{r}_2,\boldsymbol{r}';\omega). \tag{7.409}$$

The static (*stat.*) analog of (7.403), within the time-independent KS theory which examines density responses to static perturbations, then reads:

$$\chi^{stat.}(\boldsymbol{r},\boldsymbol{r}') = \chi_{\mathrm{KS}}^{stat.}(\boldsymbol{r},\boldsymbol{r}') + \int d\boldsymbol{r}_1 \int d\boldsymbol{r}_2 \chi_{\mathrm{KS}}^{stat.}(\boldsymbol{r},\boldsymbol{r}_1)\left[\frac{1}{|\boldsymbol{r}_1-\boldsymbol{r}_2|} + f_{xc}^{stat.}(\boldsymbol{r}_1,\boldsymbol{r}_2)\big|_{\rho_0}\right]\chi^{stat.}(\boldsymbol{r}_2,\boldsymbol{r}'), \tag{7.410}$$

where $\chi^{stat.}(\boldsymbol{r},\boldsymbol{r}') = \chi(\boldsymbol{r},\boldsymbol{r}';\omega=0)$ and $\chi_{\mathrm{KS}}^{stat.}(\boldsymbol{r},\boldsymbol{r}') = \chi_{\mathrm{KS}}(\boldsymbol{r},\boldsymbol{r}';\omega=0)$ are the molecular and KS response functions to static perturbation, respectively, while the xc-kernel

$$f_{xc}^{stat.}(\boldsymbol{r},\boldsymbol{r}')\big|_{\rho_0} = \left.\frac{\delta v_{xc}[\rho;\boldsymbol{r}]}{\delta\rho(\boldsymbol{r}')}\right|_{\rho_0} = \left.\frac{\delta^2 E_{xc}[\rho]}{\delta\rho(\boldsymbol{r})\,\delta\rho(\boldsymbol{r}')}\right|_{\rho_0}. \tag{7.411}$$

As also demonstrated by Gross et al. (1996)

$$\lim_{\omega\to 0} f_{xc}(\boldsymbol{r},\boldsymbol{r}';\omega)\Big|_{\rho_0} = f_{xc}^{stat.}(\boldsymbol{r},\boldsymbol{r}')\big|_{\rho_0}. \tag{7.412}$$

7.9.3 Excitation Energies

Applications of the ensemble theory to excited states (Sect. 7.4) are hampered by the fundamental difficulty that the xc-functional in this approach depends on the particular ensemble used and on the symmetries of the two states involved in the transition. Very little is known on how this functional might differ from the ordinary ground state analog. The TDDFT offers an alternative way to determine the excitation energies and the associated oscillator strengths. To extract these quantities one exploits the fact that the frequency-dependent linear density response of a finite system has discrete poles at the excitation energies of the unperturbed system. The differences of the KS orbital energies, which similarly determine the poles of the associated noninteracting system, can be then systematically shifted toward the true excitation energies (Casida 1996; Gross et al. 1996; Marques and Gross 2004).

For finite systems, e.g., atoms and molecules, which exhibit the discrete spectrum of electron excitations, the Fourier transform (7.409) can be recast in the Lehmann (spectral) form, involving the sum-over-states:

$$
\chi_{\sigma,\sigma'}(\boldsymbol{r},\boldsymbol{r}';\omega) = \lim_{\eta\to 0^+}\sum_m\left(\frac{\langle\Psi_0|\hat{\rho}_\sigma(\boldsymbol{r})|\Psi_m\rangle\langle\Psi_m|\hat{\rho}_{\sigma'}(\boldsymbol{r}')|\Psi_0\rangle}{\omega-(E_m-E_0)+i\eta}-\frac{\langle\Psi_0|\hat{\rho}_{\sigma'}(\boldsymbol{r}')|\Psi_m\rangle\langle\Psi_m|\hat{\rho}_\sigma(\boldsymbol{r})|\Psi_0\rangle}{\omega+(E_m-E_0)+i\eta}\right), \tag{7.413}
$$

where $\hat{\rho}_\sigma(\boldsymbol{r}) \equiv \hat{\rho}(\boldsymbol{q})$ is the density operator for the σ-spin electrons at position $\boldsymbol{r}$, $\{\Psi_m(N)\}$ form the complete set of the N-electron states with energies E_m, and η is a positive infinitesimal. It then directly follows from this expression that the density response function of the interacting system has poles at frequencies that correspond to the system excitation energies: $\Omega_m = E_m - E_0$. In the same way the KS response function of the noninteracting system $\chi_{\mathrm{KS};\sigma,\sigma'}(\boldsymbol{r},\boldsymbol{r}';\omega)$ [see (7.356)],

$$
\chi_{\mathrm{KS},\sigma,\sigma'}(\boldsymbol{r},\boldsymbol{r}';\omega) = \delta_{\sigma,\sigma'}\sum_{j,k}\frac{f_{k\sigma}-f_{j\sigma}}{\omega-(\varepsilon_{j\sigma}-\varepsilon_{k\sigma})+i\eta}\varphi^*_{k\sigma}(\boldsymbol{r})\varphi^*_{j\sigma}(\boldsymbol{r}')\varphi_{j\sigma}(\boldsymbol{r})\varphi_{k\sigma}(\boldsymbol{r}'), \tag{7.414}
$$

diagonal in the spin variable, exhibits poles at the differences $\{\omega_{j,k;\sigma} = \varepsilon_{j\sigma} - \varepsilon_{k\sigma}\}$ of KS eigenvalues corresponding to single-particle excitations for the electron spin σ.

The spin-resolved analog of (7.407) reads:

$$
\begin{aligned}
\delta\rho_\sigma(\boldsymbol{r},\omega) =& \sum_{\sigma'}\int\chi_{\mathrm{KS};\sigma,\sigma'}(\boldsymbol{r},\boldsymbol{r}_1;\omega)\delta v_{\sigma'}(\boldsymbol{r}_1,\omega)d\boldsymbol{r}_1\\
&+\sum_{\sigma',\sigma''}\int d\boldsymbol{r}_1\int d\boldsymbol{r}_2\chi_{\mathrm{KS};\sigma,\sigma'}(\boldsymbol{r},\boldsymbol{r}_1;\omega)\left[\frac{1}{|\boldsymbol{r}_1-\boldsymbol{r}_2|}+f_{xc;\sigma',\sigma''}(\boldsymbol{r}_1,\boldsymbol{r}_2;\omega)\Big|_{\rho_{0\uparrow},\rho_{0\downarrow}}\right]\\
&\times\delta\rho_{\sigma''}(\boldsymbol{r}_2,\omega),
\end{aligned} \tag{7.415}
$$

where the spin-dependent xc-kernel is given by the Fourier transform of [compare (7.404)]

$$
f_{xc;\sigma,\sigma'}(\boldsymbol{r},t;\boldsymbol{r}',t')\Big|_{\rho_{0\uparrow},\rho_{0\downarrow}} = \left.\frac{\delta v_{xc\sigma}[\rho_{0\uparrow},\rho_{0\downarrow};\boldsymbol{r},t]}{\delta\rho_{\sigma'}(\boldsymbol{r}',t')}\right|_{\rho_{0\uparrow},\rho_{0\downarrow}}. \tag{7.416}
$$

One further observes that formally $\delta\rho_\sigma(\boldsymbol{r},\omega) = \sum_{\sigma''}\int d\boldsymbol{r}_2\delta_{\sigma,\sigma''}\ \delta(\boldsymbol{r}_2-\boldsymbol{r})\ \delta\rho_{\sigma''}(\boldsymbol{r}_2,\omega)$ so that (7.415) can be rewritten in an equivalent operator form:

$$
\begin{aligned}
&\sum_{\sigma''}\int d\boldsymbol{r}_2\{\delta(\boldsymbol{r}_2-\boldsymbol{r})\delta_{\sigma,\sigma''}\\
&-\sum_{\sigma'}\int d\boldsymbol{r}_1\left[|\boldsymbol{r}_1-\boldsymbol{r}_2|^{-1}+f_{xc;\sigma',\sigma''}(\boldsymbol{r}_1,\boldsymbol{r}_2;\omega)\Big|_{\rho_{0\uparrow},\rho_{0\downarrow}}\right]\chi_{\mathrm{KS};\sigma,\sigma'}(\boldsymbol{r},\boldsymbol{r}_1;\omega)\}\delta\rho_{\sigma''}(\boldsymbol{r}_2,\omega)\\
&=\sum_{\sigma'}\int\chi_{\mathrm{KS};\sigma,\sigma'}(\boldsymbol{r},\boldsymbol{r}_1;\omega)\,\delta v_{\sigma'}(\boldsymbol{r}_1,\omega)\,d\boldsymbol{r}_1.
\end{aligned} \tag{7.417}
$$

This equation represents in its l.h.s the action of a complicated integral operator on density responses giving rise in its r.h.s to the perturbation containing term. A simple *reductio ad absurdum* argument shows that it cannot be inverted to determine the density response directly. One observes that since the excitation energies $\mathbf{\Omega} = \{\Omega_m\}$ are not identical with the KS excitation energies $\boldsymbol{\omega} = \{\omega_{j,k;\sigma}\}$, in the limit $\boldsymbol{\omega} \rightarrow \mathbf{\Omega}$ the r.h.s. of this equation remains finite while the density response $\delta\rho_\sigma(\boldsymbol{r},\omega)$ exhibit poles for exact excitation energies. Indeed, should the inverse integral operator exist for $\boldsymbol{\omega} \rightarrow \mathbf{\Omega}$, its action on both sides of this equation would produce a finite result on r.h.s. and a pole in the l.h.s. of the resulting equation.

One thus concludes, by integrating over the delta-function, that for the unperturbed system, when $\delta v_\sigma(\boldsymbol{r},\omega) = 0$, the following equation determines the excitation frequencies:

$$\sum_{\sigma',\sigma''}\iint d\boldsymbol{r}_1 d\boldsymbol{r}_2 \chi_{\mathrm{KS};\sigma,\sigma'}(\boldsymbol{r},\boldsymbol{r}_1;\omega)\left[|\boldsymbol{r}_1-\boldsymbol{r}_2|^{-1}+f_{xc;\sigma',\sigma''}(\boldsymbol{r}_1,\boldsymbol{r}_2;\omega)\Big|_{\rho_{0\uparrow},\rho_{0\downarrow}}\right]\varsigma_{\sigma''}(\boldsymbol{r}_2,\omega)$$
$$= \lambda(\omega)\varsigma_\sigma(\boldsymbol{r},\omega), \tag{7.418}$$

marking the vanishing result of the integral operation of the l.h.s. of (7.417), where

$$\lambda(\omega \in \mathbf{\Omega}) = 1. \tag{7.419}$$

This condition rigorously determines the true excitation spectrum of the real interacting system (Gross et al. 1996). After a series of algebraic manipulations one arrives at the pseudoeigenvalue equation for the excitation energy Ω (Marques and Gross 2004):

$$\sum_{j',k'}\sum_{\sigma'}\left[\delta_{j,j'}\delta_{k,k'}\delta_{\sigma,\sigma'}\omega_{j,k;\sigma}+(f_{k'\sigma'}-f_{j'\sigma'})K_{jk,\sigma;j'k',\sigma'}(\Omega)\right]\beta_{j'k',\sigma'} = \Omega\beta_{jk,\sigma}, \tag{7.420}$$

where

$$K_{jk,\sigma;j'k',\sigma'}(\omega) = \iint \varphi^*_{j\sigma}(\boldsymbol{r})\varphi_{k\sigma}(\boldsymbol{r})[|\boldsymbol{r}'-\boldsymbol{r}|^{-1}+f_{xc;\sigma,\sigma'}(\boldsymbol{r},\boldsymbol{r}';\omega)]\,\varphi_{j'\sigma'}(\boldsymbol{r}')\varphi^*_{k'\sigma'}(\boldsymbol{r}')d\boldsymbol{r}'d\boldsymbol{r}. \tag{7.421}$$

When the excitation of interest is well described by a *single*-electron excitation from the occupied to unoccupied KSMO, one can neglect the *off*-diagonal coupling coefficients of the preceding equation. In this *Single-Pole Approximation* (SPA), the KS excitation energy is corrected by the additive Coulomb and f_{xc} terms and the spin-multiplet structure is preserved through the spin-dependence of f_{xc} (Marques and Gross 2004).

To simplify the pseudoeigenvalue equations for the excitation energies Ω, it is possible to derive the equivalent equation for their squares, defined in the product space $\mathscr{S}_i \times \mathscr{S}_a$ of subspaces of the *occupied* ($\mathscr{S}_i = \{\varphi_{i\sigma}\}$) and *virtual* ($\mathscr{S}_a = \{\varphi_{a\sigma}\}$) KS orbitals, which determine the linear change in the electron density:

$$\delta\rho_\sigma(\boldsymbol{r},\omega) = \sum_i^{occ.}\sum_a^{virt.}\left[\xi_{ia,\sigma}(\omega)\,\varphi_{a\sigma}^*(\boldsymbol{r})\varphi_{i\sigma}(\boldsymbol{r}) + \xi_{ai,\sigma}(\omega)\,\varphi_{a\sigma}(\boldsymbol{r})\varphi_{i\sigma}^*(\boldsymbol{r})\right]. \quad (7.422)$$

The matrix pseudoeigenvalue problem for squares Ω^2 of excitation energies reads:

$$\sum_{a',i'}\sum_{\sigma'}\left[\delta_{a,a'}\delta_{i,i'}\delta_{\sigma,\sigma'}(\varepsilon_{a\sigma}-\varepsilon_{i\sigma})^2 + 2(\varepsilon_{a\sigma}-\varepsilon_{i\sigma})K_{ai,\sigma;a'i',\sigma'}(\Omega)(\varepsilon_{a'\sigma'}-\varepsilon_{i'\sigma'})\right]\beta_{a'i',\sigma'} = \Omega^2\beta_{ai,\sigma}. \quad (7.423)$$

The eigenvectors of this reduced matrix problem then determine the associated oscillator strengths of the electronic transitions. For interpretative purposes it is then necessary to assign these excitations to the specific excited states of the system. In order to make such an assignment, it is necessary to introduce some approximate assumptions about the nature of the ground and excited states involved (Casida 1996). This approach has been also implemented to subsystems in molecular complexes within the orbital-free embedding of Cortona and Wesołowski (Casida and Wesołowski 2004).

7.9.4 Van der Waals Interactions Revisited

As we have already indicated in the overview of Sect. 7.8, the weak-interaction aspect of the ground state energy calculations is also assisted by TDDFT. Indeed, the dispersion interactions in vdW complexes, which arise from correlations between dynamic fluctuations in the electron density, cannot be adequately represented within the standard LDA and GGA calculations using the local or near-local approximations (Perez-Jorda and Becke 1995). For widely separated fragments, e.g., in the dispersion interaction between a pair of neutral spherical atoms separated by a distance R, the tail of the well-known Lennard–Jones potential falls off as R^{-6} (e.g., Mahanty and Ninham 1976). This term, in addition to some polarization contributions relating to any static electric moments, is readily derived for nonoverlapping electronic systems by regarding electrons on one subsystem as distinguishable from those on the other subsystem and by using the *second*-order Rayleigh–Schrödinger perturbation theory in treating the Coulomb interaction between the two groups of electrons as the perturbation supplement to the sum of Hamiltonians of separated fragments (e.g., Jeziorski et al. 1994). For very large separations, when the retardation of the electromagnetic interactions between the two subsystems cannot be ignored, the R^{-6} law is replaced by R^{-7} (Mahanty and Ninham 1976).

In the ACFDT approach, based on the *adiabatic connection* (AC) formula and the (*zero*-temperature) FDT, vdW functionals have been generated via the frequency integration over the *dynamic* susceptibilities, thus emphasizing the fundamental part played by TDDFT in generating the heavily nonlocal functional for the dispersion energy. Both the Coulomb kernel and the dynamic *xc*-kernel $f_{xc}(\boldsymbol{r},\boldsymbol{r}';\omega)$ for inhomogeneous systems appear in the Dyson-type screening equation (7.409) of Petersilka et al. (1996), which relates $\chi(\boldsymbol{r},\boldsymbol{r}';\omega)$ and $\chi_{\mathrm{KS}}(\boldsymbol{r},\boldsymbol{r}';\omega)$. They are essential for generating the right *long*-range behavior of the resulting RPA functionals. A closer look into the origins of this effect indicates that the *long*-range vdW interaction comes from the *long*-range of the Coulomb interaction in the screening equation (7.409), since χ_{KS} is not normally *long*-ranged.

In other words, even the simplest LDA approximations for χ or f_{xc} can be quite successful in generating the correct *long*-range vdW interaction via the screening equation. Therefore, if one makes the *short*-ranged local or gradient approximation for χ_{KS} or f_{xc} in (7.357) and (7.358), one obtains an approximate but highly nonlocal, *long*-ranged prescription for the ground state correlation energy, with the ground-state density as the only input argument. Should the *short*-ranged aspects of the ground state correlation be safeguarded by the appropriate set of conditions imposed on f_{xc}, the *xc*-energy resulting from (7.355) and (7.358) should cover molecules, metals, and subdivided systems at all separations, thus providing a truly universal density functional facilitating calculations on very large systems, too complicated to handle by the CI methods of WFT.

It follows from the work of Casimir, Lifshitz, London, and many others (see, e.g., Mahanty and Ninham 1976) that the dispersion energy between such widely separated systems can be related to their electric polarizabilities. As shown by Zaremba and Kohn (1976), the *second*-order interaction energy is given by the functional derived directly from the *second*-order PT:

$$E^{(2)} = -\frac{1}{2\pi}\int d\boldsymbol{r}_1 \int d\boldsymbol{r}_2 \int d\boldsymbol{r}_1{}' \int d\boldsymbol{r}_2{}' \frac{1}{|\boldsymbol{r}_1-\boldsymbol{r}_2|}\frac{1}{|\boldsymbol{r}_1{}'-\boldsymbol{r}_2{}'|}\int_0^\infty du\,\chi_{\mathrm{A}}(\boldsymbol{r}_1,\boldsymbol{r}_1{}';iu)\chi_{\mathrm{B}}(\boldsymbol{r}_2,\boldsymbol{r}_2{}';iu), \tag{7.424}$$

where $\chi_{\mathrm{X}}(\boldsymbol{r}_2,\boldsymbol{r}_2{}';\omega)$ stands for the density–density response function of the separate subsystem X (in the absence of the other system). It is defined as the linear response in electron density of X,

$$\delta\rho_{\mathrm{X}}(\boldsymbol{r},t) = \delta\rho_{\mathrm{X}}(\boldsymbol{r})\exp(ut),$$

to an externally applied perturbation

$$\delta v(\boldsymbol{r},t) = \delta v_{ext.}(\boldsymbol{r})\exp(ut),$$

with $\delta\rho_{\mathrm{X}}(\boldsymbol{r}) = \int d\boldsymbol{r}'\chi_{\mathrm{X}}(\boldsymbol{r},\boldsymbol{r}';iu)\delta v_{ext.}(\boldsymbol{r}')$.

This exact *second*-order formula have been used (Andersson et al. 1996) to derive approximate vdW functionals as explicit, highly nonlocal functionals of the ground state density for widely separated subsystems A and B. Consider, e.g., the LDA-type hydrodynamic approximation for χ,

$$\chi(\boldsymbol{r},\boldsymbol{r}';\omega) = \nabla_{\boldsymbol{r}}\nabla_{\boldsymbol{r}'}\left(\frac{\rho(\boldsymbol{r})\delta(\boldsymbol{r}-\boldsymbol{r}')}{\omega^2 - \omega_P^2(\rho(\boldsymbol{r}))}\right), \tag{7.425}$$

where $\omega_P(\rho(\boldsymbol{r})) = (4\pi\rho(\boldsymbol{r})/m)^{1/2}$ is the local plasma frequency at point $\boldsymbol{r}$. It does not violate the charge conservation,

$$\int \chi(\boldsymbol{r},\boldsymbol{r}';\omega)\, d\boldsymbol{r} = 0, \tag{7.426}$$

as well as the reciprocity condition, $\chi(\boldsymbol{r},\boldsymbol{r}';iu) = \chi(\boldsymbol{r},\boldsymbol{r}';-iu)$ for real u, and generates via (7.424)

$$E^{(2)} = -\frac{3}{32\pi^2}\iint d\boldsymbol{r}\, d\boldsymbol{r}' \frac{1}{|\boldsymbol{r}-\boldsymbol{r}'|^6} \frac{\omega_P(\rho_{\mathrm{A}}(\boldsymbol{r}))\omega_P(\rho_{\mathrm{B}}(\boldsymbol{r}'))}{[\omega_P(\rho_{\mathrm{A}}(\boldsymbol{r})) + \omega_P(\rho_{\mathrm{B}}(\boldsymbol{r}'))]}; \tag{7.427}$$

here $\omega_P(\rho_{\mathrm{A}}(\boldsymbol{r}))$ is the plasma frequency at $\boldsymbol{r}$ in A, while $\omega_P(\rho_{\mathrm{B}}(\boldsymbol{r}'))$ similarly denotes the plasma frequency at $\boldsymbol{r}'$ in B. One observes the presence in the integrand of this functional the harmonic mean of the local plasma frequencies on the two interacting subsystems.

A related functional has been derived on different grounds by Andersson et al. (1996) who generalized a somewhat similar formula by Rapcewicz and Ashcroft (1991) postulated on the basis of a diagrammatic analysis. These approximate functionals validate the pairwise addition of R^{-6} vdW corrections to GGA energies. The numerical experience for the isotropic dispersion coefficients for various atomic pairs shows that this simple approach gives good answers provided that one uses an appropriate cutoff in the *low*-density tails of electron distributions.

7.10 Conclusion

The basic premise of DFT, that the ground state electron density uniquely determines the system energy and its remaining physical properties, can be also rephrased in terms of the density per particle (probability) state variable, which provides the shape factor of the electron distribution (Ayers 2000a; Ayers and Cedillo 2009). It can also serve as a fundamental descriptor in problems of the atomic/molecular similarity (Ayers and Cedillo 2009). The above nonrelativistic development can be extended to cover the relativistic treatment (Rajagopal and Callaway 1973; Engel and Dreizler 1996; Ramana and Rajagopal 1983;

Engel et al. 1995). The DFT perspective can be also recast as the local thermodynamics (Ghosh et al. 1984; Ghosh and Berkowitz 1985; Nalewajski 2002c; 2003b, 2004a, 2006h).

The HK theorems of Sect. 7.1 establish the formal (existence) link between the electron density and the physical properties of a molecule, e.g., its energy, without invoking the concept of orbitals. Therefore, the orbitals are in principle dispensable in the "true" (HK) DFT. This prompted theoreticians to perform DFT calculations on both materials and molecules using the orbital-free density functionals for the electronic energy (see, e.g., Yang 1987, 1988; Wang and Carter 2000; Wesołowski 2004b). It should be recalled, however, that the historic predecessor of such an approach, the TF theory (Thomas 1927; Fermi 1928), has been shown by Teller (1962) to fail to predict the very stability of molecules.

Despite this original failure of the TF approach in predicting the electronic energy of molecular systems the search for better approximations of the energy density functional has been an object of continuous interest for theoreticians, resulting in the orbital-free studies and the computer modeling at the quantum-mechanical level of complex materials. However, the applications to molecules, although now already predicting stable systems, are still no match for the more accurate conventional approaches to very large systems in DFT which do invoke orbitals (Carr and Parrinello 1985; Galli and Parinello 1992; Yang 1992; Kohn 1993, 1995). Besides, the orbitals provide a solid basis for interpretation and hence an "anchor" with the contemporary chemical concepts. As stressed by Kohn (1993, 1995), the KS orbitals are the "far-sighted" concepts, depending on the potential everywhere, while the well-localized Wannier functions are "near-sighted," depending only on their neighborhood. The latter property enables one to construct the linear scaling algorithms for determining the electron density and energy of a truly large systems. Of particular interest are the "divide-and-conquer" strategy of Yang (1992) and the generalized Wannier Function approach of Kohn (1993).

This interest in the orbital-free DFT has brought back the issue of an explicit density functional for the kinetic energy (e.g.: Murphy and Parr 1979; Murphy 1981; Acharya et al. 1980; Yang and Harriman 1986; Parr and Yang 1989). Its modeling can be also enhanced using some information-theoretic ideas (Sears 1980; Sears et al. 1980; Nalewajski 2003f; Nagy 2003; Ghirinhelli et al. 2010). It should be recalled that IT has also been used to analyze the electron correlation and other properties for which the electron localization, measured by the Fisher information, and electron delocalization, reflected by the Shannon entropy (see Chap. 8) are important (e.g., Hõ et al. 1995; Yáñez et al. 1995; Ziesche 1995; Romera and Dehesa 2004; Sen et al. 2007).

References

Acharya PK, Bartolotti LJ, Sears LB, Parr, RG (1980) Proc Natl Acad Sci USA 77:6978
Adamo C, Barone V (1998) J Chem Phys 108:664

Ahlrichs R, Penco R, Scoles G (1977) Chem Phys Lett 19:119
Andersson Y, Langreth DC, Lundqvist BI (1996) Phys Rev Lett 76:102
Aryasetiawan F, Miyake T, Terakura K (2002) Phys Rev Lett 88:166401
Ayers PW (2000a) Proc Natl Acad Sci USA 97:1959
Ayers PW, Cedillo A (2009) In: Chattaraj PK (ed) Chemical reactivity theory: a density functional view. CRC, Boca Raton, p 269
Bartlett RJ (2000) In: Keinan E, Schechter I (eds) Chemistry for the 21 Century. Wiley, Weinheim
Bartlett RJ, Grabowski I, Hirata S, Ivanov S (2005a) J Chem Phys 122:034104
Bartlett RJ, Lotrich VF, Schweigert IV (2005b) J Chem Phys 123:062205
Bartolotti LJ (1981) Phys Rev A 24:1661
Bartolotti LJ (1982) Phys Rev A 26:2243
Bartolotti LJ (1984) J Chem Phys 80:5687
Bartolotti LJ (1987) Phys Rev A 36:4492
Becke AD (1985) Int J Quantum Chem 27:585
Becke AD (1988) Phys Rev A 38:3098
Becke AD (1993) J Chem Phys 98:5648
Becke AD (1996) J Chem Phys 104:1040
Becke AD (1997) J Chem Phys 107:8554
Becke AD, Johnson ER (2007) J Chem Phys 127:124108
Becke AD, Roussel MR (1989) Phys Rev A 39:3761
Berkowitz M, Parr RG (1988) J Chem Phys 88:2554
Berrondo M, Gościnski O (1975) Int J Quantum Chem Symp 9:67
Boese AD, Handy NC (2001) J Chem Phys 114:5487
Boese AD, Doltsinis NL, Handy NC, Sprik M (2000) J Chem Phys 112:1670
Boese AD, Martin JML, Klopper W (2007) J Phys Chem A 111:11122
Buijse MA, Baerends EJ (2002) Mol Phys 100:401
Bukowski R, Podeszwa R, Szalewicz K (2005) Chem Phys Lett 414:111
Bukowski R, Cencek W, Jankowski P, Jeziorska M, Jeziorski B, Kucharski SA, Lotrich VF, Misquitta AJ, Moszyński R, Patkowski K, Podeszwa R, Rybak S, Szalewicz K, Williams HL, Wheatley RJ, Wormer PES, Żuchowski PS (2008) SAPT2008: an *ab initio* program for *many*-body symmetry-adapted perturbation theory calculations of intermolecular interaction energies. University of Delaware and University of Warsaw, Delaware, Warsaw
Callen HB, Welton TA (1951) Phys Rev 83:34
Capitani JF, Nalewajski RF, Parr RG (1982) J Chem Phys 76:568
Carr R, Parrinello M (1985) Phys Rev Lett 55:2471
Casida ML (1996) In: Seminario JM (ed) Recent developments and applications in modern density functional theory. Elsevier, Amsterdam, p 391
Casida ML, Wesołowski TA (2004) Int J Quantum Chem 96:577
Ceperley DM, Alder BJ (1980) Phys Rev Lett 45:566
Chakarova-Käck SD, Schröder E, Lundqvist BI, Langreth DC (2006) Phys Rev Lett 96:146107
Chattaraj PK (ed) (2009) Chemical reactivity theory – a density functional view. CRC, Boca Raton
Ciosłowski J (1988) Phys Rev Lett 60:2141
Ciosłowski J, Pernal K (2004a) J Chem Phys 120:10364
Ciosłowski J, Pernal K (2004b) Phys Rev B 71:113103
Ciosłowski J, Pernal K (2005) Phys Rev B 71:113103
Ciosłowski J, Pernal K, Buchowiecki M (2003) J Chem Phys 119:6443
Colle R, Salvetti O (1975) Theor Chim Acta (Berl) 37:329
Connolly JWD (1977) In: Segal GA (ed) Semiempirical methods of electronic structure calculations, Part A: Techniques. Plenum, New York, p 105
Cortona P (1991) Phys Rev B 44:8454
Deb BM, Ghosh SK (1982) J Chem Phys 77:342
Dion M, Rydberg H, Schröder E, Langreth DC, Lundqvist BI (2004) Phys Rev Lett 92:246401
Dirac PAM (1930) Proc Cambr Philos Soc 26:376

Dobson JF, Dinte BP (1996) Phys Rev Lett 76:1780
Dobson JF, Wang J (2000) Phys Rev B 62:10038
Dobson JF, Wang J, Dinte BP, McLennan K, Lee HM (2005) Int J Quantum Chem 101:579
Doniach S, Sondheim EH (1982) Green's functions for solid state physicists. Addison-Wesley, Redwood City
Donnelly RA (1979) J Chem Phys 71:2874
Donnelly RA, Parr RG (1978) J Chem Phys 69:4431
Dreizler RM, da Providencia J (1985) (eds) Density functional methods in physics. NATO ASI vol 123. Plenum, New York
Dreizler RM, Gross EKU (1990) Density functional theory: an approach to the quantum many-body problem. Springer, Berlin
Elstner M, Hobza P, Frauenheim T, Suhai S, Kaxiras E (2001) J Chem Phys 114:5149
Engel E, Dreizler RM (1996) Density functional theory ii: relativistic and time dependent extensions. In: Nalewajski RF (ed) Topics in current chemistry, vol. 181. Springer, Berlin, p 1
Engel E, Müller H, Speicher C, Dreizler RM (1995) Density functional theory, NATO ASI, vol 337. Plenum, New York, p 65
Engel E, Höck A, Dreizler RM (2000) Phys Rev A 61:032502
Ernzerhof M (1996) Chem Phys Lett 263:499
Ernzerhof M, Perdew JP, Burke K (1996) Density functional theory I: functionals and effective potentials. In: Nalewajski RF (ed) Topics in current chemistry, vol 180. Springer, Berlin, p 1
Fermi E (1928) Z Phys 48:73
Feynman RP (1939) Phys Rev 56:340
Fiethen A, Heßelmann A, Schütz M (2008) J Am Chem Soc 130:1802
Fuchs M, Gonze X (2002) Phys Rev B 65:235109
Fujimoto H, Fukui K (1974) In: Klopman G (ed) Chemical reactivity and reaction paths. Wiley-Interscience, New York, p 23
Fukui K (1975) Theory of orientation and stereoselection. Springer, Berlin
Fukui K (1987) Science 218:747
Furche F (2001) Phys Rev B 64:195120
Furche F (2008) J Chem Phys 129:114105
Galli G, Parinello M (1992) Phys Rev Lett 69:3547
Geerlings P, De Proft F, Langenaeker W (eds) (1999) Density functional theory: a bridge between chemistry and physics. VUB University Press, Brussels
Geerlings P, De Proft F, Langenaeker W (2003) Chem Rev 103:1793
Gell-Mann M, Brueckner KA (1957) Phys Rev 106:364
Gerber IC, Ángyán JG (2007) J Chem Phys 126:044103
Ghirinhelli LM, Delle Site L, Mosna RA, Hamilton IP (2010) Information theoretic approach to kinetic-energy functionals: the nearly uniform electron gas. J Math Chem (in press)
Ghosh SK (2009) In: Chattaraj PK (ed) Chemical reactivity theory: a density functional view. CRC, Boca Raton, p 71
Ghosh SK, Berkowitz M (1985) J Chem Phys 83:2976
Ghosh SK, Deb BM (1982) Chem Phys 71:295
Ghosh SK, Deb BM (1983a) Theor Chim Acta (Berl.) 62:209
Ghosh SK, Deb BM (1983b) J Mol Struct 103:163
Ghosh SK, Berkowitz M, Parr RG (1984) Proc Natl Acad Sci USA 81:8028
Gilbert TL (1975) Phys Rev B 12:2111
Goedecker S, Umrigar CJ (1998) Phys Rev Lett 81:866
Goll E, Leininger T, Manby FR, Mitrushchenkov A, Werner H-J, Stoll H (2008) Phys Chem Chem Phys 10:3353
Gordon RG, Kim YS (1972) J Chem Phys 56:3122
Görling A (1999) Phys Rev Lett 83:5459
Görling A, Levy M (1993) Phys Rev B 47:13101
Görling A, Levy M (1994) Phys Rev A 50:196

Grabo T, Kreibich T, Kurth S, Gross EKU (1999) In: Anisimov VI (ed) Strong Coulomb correlations in electronic structure: beyond the local density approximation. Gordon & Breach, Amsterdam, p 1
Grabowski I (2008) Int J Quantum Chem 108:2076
Grabowski I, Lotrich V (2005) Mol Phys 103:2085
Grabowski I, Hirata S, Ivanov S, Bartlett RJ (2002) J Chem Phys 116:4415
Grabowski I, Lotrich V, Bartlett RJ (2007) J Chem Phys 127:154111
Grimme S (2004) J Comput Chem 25:1463
Grimme S (2006a) JComput Chem 27:1787
Grimme S (2006b) J Chem Phys 124:034108
Gritsenko O, Pernal K, Baerends EJ (2005) J Chem Phys 122:204102
Gross EKU, Dreizler RM (1995) Density functional theory, NATO ASI, vol 337. Plenum, New York
Gross EKU, Kurth S (1993) In: Malli GL (ed) Relativistic and electron correlation effects in molecules and solids, NATO ASI series. Plenum, New York
Gross EKU, Oliveira LN, Kohn W (1988) Phys Rev A 37:2805
Gross EKU, Dobson JF, Petersilka M (1996) Density functional theory II: relativistic and time dependent extensions. In: Nalewajski RF (ed) Topics in current chemistry, vol 181. Springer, Berlin, p 81
Gunnarson O, Lundqvist BI (1976) Phys Rev B 13:4274
Gutowski M, Piela L (1988) Mol Phys 64:337
Gyftopoulos EP, Hatsopoulos GN (1965) Proc Natl Acad Sci USA 60:786
Harl J, Kresse G (2008) Phys Rev B 77:045136
Harrimann JE (1980) Phys Rev A 24:680
Harris J, Jones RO (1974) J Phys F 4:1170
Hellmann H (1937) Einführung in die quantenchemie. Deuticke, Leipzig
Hepburn J, Scoles G, Penco R (1975) Chem Phys Lett 36:451
Heßelmann A, Jansen G (2003) Chem Phys Lett 367:778
Heßelmann A, Schütz M (2006) J Am Chem Soc 128:11730
Heßelmann A, Jansen G, Schütz M (2005) J Chem Phys 122:014103
Hirata S, Ivanov S, Grabowski I, Bartlett RJ, Burke K, Talman JD (2001) J Chem Phys 115:1635
Hõ M, Sagar RB, Schmier H, Weaver DF, Smith VH Jr (1995) Int J Quantum Chem 53:627
Hohenberg P, Kohn W (1964) Phys Rev 136B:864
Iczkowski RP, Margrave JL (1961) J Am Chem Soc 83:3547
Ivanov S, Hirata S, Bartlett RJ (1999) Phys Rev Lett 83:5455
Janak JF (1978) Phys Rev B 18:7165
Jansen G (2009) Density functional theory approach to intermolecular interactions. Lecture Notes
Jansen G, Heßelmann A (2001) J Phys Chem A 105:11156
Jeziorski B, Szalewicz K (2002) In: Wilson S (ed) Handbook of molecular physics and quantum chemistry, vol 3, Part 2. Wiley, New York, p 232
Jeziorski B, Moszyński R, Szalewicz K (1994) Chem Rev 94:1887
Johnson KH (1973b) Adv Quant Chem 7:143
Johnson PA, Bartolotti LJ, Ayers P, Fievez T, Geerlings P (2011) Charge density and chemical reactions: a unified view from conceptual DFT. In: Gatti C, Macchi P (eds) Modern charge-density analysis. Springer, Berlin (in press)
Jørgensen CK (1964) Inorg Chem 3:1201
Jurečka P, Hobza P (2003) J Am Chem Soc 125:15608
Jurečka P, Sponer J, Cerny J, Hobza P (2006) Phys Chem Chem Phys 8:1985
Kohanoff J, Gidopoulos NI (2003) In: Wilson S (ed) Handbook of molecular physics and quantum chemistry, vol 2: Molecular electronic structure. Wiley, New York, p 532
Kohn W (1986) Phys Rev A 34:737
Kohn W (1993) Chem Phys Lett 208:167

Kohn W (1995) In: Gross EKU, Dreizler RM (eds) Density functional theory. Pleum, New York, p 3
Kohn W, Sham LJ (1965) Phys Rev 140A:1133
Kohn W, Meir Y, Makarov DE (1998) Phys Rev Lett 80:4153
Krieger JB, Li Y, Iafrate GJ (1990) Phys Lett A 146:256
Krieger JB, Li Y, Iafrate GJ (1992) Phys Rev A 45:101
Krieger JB, Li Y, Iafrate GJ (1995) In: Gross EKU and Dreizler RM (eds) Density functional theory. Pleum, New York, p 191
Kryachko ES, Ludeña EV (1890) Energy density functional theory of many-electron systems. Kluwer, Dordrecht
Kümmel S, Kronik L (2008) Rev Mod Phys 80:1
Kümmel S, Perdew JP (2003) Phys Rev B 68:035103
Kutzelnigg W (1963) Theor Chim Acta (Berl.) 1:327
Langreth DC, Mehl MJ (1981) Phys Rev Lett 47:446
Langreth DC, Perdew JP (1975) Solid State Commun 17:1425
Lee C, Yang W, Parr RG (1988) Phys Rev B 37:785
Leininger T, Stoll H, Werner H-J, Savin A (1997) Chem Phys Lett 275:151
Levy M (1979) Proc Natl Acad Sci USA 76:6062
Levy M (1982) Phys Rev A 26:1200
Levy M (1987) In: March NH, Deb BM (eds) Single particle density in physics and chemistry. Academic, London, p 45
Levy M, Perdew JP (1985) In: Dreizler RM, da Providencia J (eds) Density functional methods in physics. Plenum, New York, p 11
Levy M, Perdew JP (1993) Phys Rev B 48:11638
Levy M, Perdew JP (1997) Phys Rev B 55:13321
Lieb EH (1982) In: Feshbach M, Shimony A (eds) Physics as natural philosophy; Essays in Honor of Laszlo Tisza on His 75th birthday. MIT, Cambridge, p 111. For a revised version see: Lieb EH (1983) Int J Quantum Chem 24:243; In: Dreizler RM and da Providencia J (eds) Density functional methods in physics. Plenum, New York, p 31
Lin I-C, Röthlisberger U (2008) Phys Chem Chem Phys 10:2730
Lin I-C, Continho-Neto MD, Felsenheimer C, von Lilienfeld OA, Tavernelli IA, Röthlisberger U (2007) Phys Rev B 75:205131
Linderberg J (1977) Int J Quantum Chem 12(suppl 1):267
Lotrich V, Bartlett RJ, Grabowski I (2005) Chem Phys Lett 405:43
Lundqvist BI, Andersson Y, Shao H, Chan S, Langreth DC (1995) Int J Quantum Chem 57:247
Macke W (1955a) Ann Phys (Leipzig) 17:1
Macke W (1955b) Phys Rev 100:992
Mahanty J, Ninham BW (1976) Dispersion forces. Academic, London
March NH (1975) Self-consistent fields in atoms. Pergamon, Oxford
March NH (1982) Phys Rev A 26:1845
Marini A, Gárcia-González P, Rubio A (2006) Phys Rev Lett 96:136404
Marques MAL, Gross EKU (2004) Annu Rev Phys Chem 55:427
McLachlan AD, Ball MA (1964) Rev Mod Phys 36:844
McNamara JP, Hillier IH (2007) Phys Chem Chem Phys 9:2362
Mermin ND (1965) Phys Rev A 137:1441
Misquitta AJ, Szalewicz K (2002) Chem Phys Lett 357:301
Misquitta AJ, Szalewicz K (2005) J Chem Phys 122:214109
Misquitta AJ, Jeziorski R, Szalewicz K (2003) Phys Rev Lett 91:033201
Misquitta AJ, Podeszwa R, Jeziorski B, Szalewicz K (2005) J Chem Phys 123:214103
Morgado C, Vincent MA, Hillier IH, Shan X (2008) Phys Chem Chem Phys 9:448
Müller AMK (1984) Phys Lett A 105:446
Mulliken RS (1934) J Chem Phys 2:782
Murphy DR (1981) Phys Rev A 24:1682

Murphy DR, Parr RG (1979) Chem Phys Lett 60:377
Nagy Á (2003) J Chem Phys 119:9401
Nalewajski RF (1996a) (ed) Density functional theory I–IV. Topics in current chemistry, vols 180–183. Springer, Berlin
Nalewajski RF (1996b) (ed) Density functional theory IV: theory of chemical reactivity. Topics in current chemistry, vol 183. Springer, Berlin
Nalewajski RF (2001) Adv Quant Chem 38:217
Nalewajski RF (2002c) Acta Chim Phys Debr 34–35:131
Nalewajski RF (2002d) In: Sen KD (ed) Reviews of modern quantum chemistry: a celebration of the contributions of Robert G. Parr, vol 2. World Scientific, Singapore, p 1071
Nalewajski RF (2003a) Adv Quant Chem 43:119
Nalewajski RF (2003b) J Phys Chem A 107:3792
Nalewajski RF (2003f) Chem Phys Lett 367:414
Nalewajski RF (2004a) Ann Phys (Leipzig) 13:201
Nalewajski RF (2006f) Adv Quant Chem 51:235
Nalewajski RF (2006h) Mol Phys 104:255
Nalewajski RF (2009a) In: Chattaraj PK (ed) Chemical reactivity theory: a density functional view. Taylor and Francis, London, p 453
Nalewajski RF (2010e) J Math Chem 47:1068
Nalewajski RF, Korchowiec J (1997) Charge sensitivity approach to electronic structure and chemical reactivity. World-Scientific, Singapore
Nalewajski RF, Parr RG (1982) J Chem Phys 77:399; the extremum principle in Eqs. (69) and (70) of this paper is a maximum principle, not a minimum principle
Nalewajski RF, Korchowiec J, Zhou Z (1988) Int J Quantum Chem Symp 22:349
Nalewajski RF, Korchowiec J, Michalak A (1996) Top Curr Chem 183:25
Nalewajski RF, Błażewicz D, Mrozek J (2008) J Math Chem 44:325
Nguyen-Dang TT, Ludeña EV, Tall Y (1985) J Mol Struct THEOCHEM 120:247
Nyquist H (1928) Phys Rev 32:110
Oliveira LN, Gross EKU, Kohn W (1988) Phys Rev A 37:2821
Oliver GL, Perdew JP (1979) Phys Rev A 20:397
Olsson MHM, Hong GY, Warshel A (2004) J Am Chem Soc 96:577
Pariser R (1953) J Chem Phys 21:568
Pariser R, Parr RG (1953) J Chem Phys 21:767
Parr RG (1963) The quantum theory of molecular electronic structure. Benjamin, New York
Parr RG, Pearson RG (1983) J Am Chem Soc 105:7512
Parr RG, Yang W (1984) J Am Chem Soc 106:4049
Parr RG, Yang W (1989) Density-functional theory of atoms and molecules. Oxford University Press, New York
Pearson RG (1973) Hard and soft acids and bases. Dowden, Hutchinson, and Ross, Stroudsburg
Pearson RG (1988) Inorg Chem 27:734
Perdew JP (1985) In: Dreizler RM, da Providencia J (eds) Density functional methods in physics. Plenum, New York, p 265
Perdew JP (1991) In: Ziesche P, Eschrig H (eds) Electronic structure of solids '91. Akademie, Berlin, p 11
Perdew JP (1999) In: Geerlings P, De Proft F, Langenaeker W (eds) Density functional theory – a bridge between chemistry and physics. VUB University Press, Brussels, p 87
Perdew JP, Levy M (1983) Phys Rev Lett 51:1884
Perdew JP, Levy M (1984) In: Langreth D, Suhl H (eds) Many-body phenomena at surfaces. Academic, Orlando, p 71
Perdew JP, Norman MR (1982) Phys Rev B 26:5445
Perdew JP, Wang Y (1986) Phys Rev B 33:8800
Perdew JP, Wang Y (1989) Phys Rev B 40:3399
Perdew JP, Zunger A (1981) Phys Rev B 23:5048

Perdew JP, Parr RG, Levy M, Balduz JL (1982) Phys Rev Lett 49:1691

Perdew JP, Burke K, Ernzerhoh M (1996a) Phys RevLett 77:3865

Perdew JP, Burke K, Wang Y (1996b) Phys Rev B 54:16533

Perdew JP, Ernzerhof M, Burke K (1996c) J Chem Phys 105:9982

Perdew JP, Burke K, Ernzerhoh M (1997) Phys Rev Lett 78:1396

Perdew JP, Burke K, Wang Y (1998) Phys Rev B 57:14999

Perdew JP, Ruzsinsky A, Tao J, Staroverov VN, Scuseria GE, Csonka GI (2005) J Chem Phys 123:062201

Perez-Jorda JM, Becke AD (1995) Chem Phys Lett 233:134

Pernal K (2005) Phys Rev Lett 94:233002

Pernal K, Baerends EJ (2006) J Chem Phys 124:014102

Pernal K, Ciosłowski J (2004) J Chem Phys 120:5987

Pernal K, Ciosłowski J (2005) Chem Phys Lett 412:71

Petersilka M, Gossmann UJ, Gross EKU (1996) Phys Rev Lett 76:1212

Podeszwa R, Szalewicz K (2007) J Chem Phys 126:194101

Podeszwa R, Bukowski R, Szalewicz K (2006) J Chem Theo Comp 2:400

Pople JA (1953) Trans Faraday Soc 49:1375

Pueckert V (1978) J Phys C11:4945

Rajagopal AK, Callaway J (1973) Phys Rev B7:1912

Rajchel Ł, Żuchowski PS, Szczęśniak MM, Chałasiński G (2009a) A DFT approach to non-covalent interactions *via* monomer polarization and Pauli blockade. *arXiv*, 0908.0798v [physics.chem.-ph] (in press)

Rajchel Ł, Żuchowski PS, Szczęśniak MM, Chałasiński G (2009b) Derivation of the supermolecular interaction energy from the monomer densities in the density functional theory. *arXiv*:0908.0798v [physics.chem.-ph] (in press)

Ramana MV, Rajagopal AK (1983) Adv Chem Phys 54:231

Rapcewicz K, Ashcroft NW (1991) Phys Rev B 44:4032

Romera E, Dehesa JS (2004) J Chem Phys 120:8906

Runge E, Gross EKU (1984) Phys Rev Lett 52:997

Sánchez-Garcia E, Mardyukov A, Tekin A, Crespo-Otero R, Montero LA, Sander W, Jansen G (2008) J Chem Phys 343:168

Sanderson RT (1951) Science 114:670

Sanderson RT (1976) Chemical bonds and bond energy, 2nd edn. Academic, New York

Savin A, Flad H-J (1995) Int J Quantum Chem 56:327

Scuseria GE, Henderson TM, Sorensen DC (2008) J Chem Phys 129:231101

Sears SB (1980) Ph.D Thesis, The University of North Carolina at Chapel Hill, North Carolina

Sears SB, Parr RG, Dinur U (1980) Israel J Chem 19:165

Seidl M, Perdew JP, Kurth S (2000) Phys Rev A 62:012502

Sen KD (ed) (1993) Chemical hardness: structure and bonding, vol 80. Springer, Berlin

Sen KD, Antolín J, Angulo JC (2007) Phys Rev A 76:032502

Shannon CF (1948) Bell Syst Technol J 27(379):623

Sharp RT, Horton GK (1953) Phys Rev 90:317

Slater JC (1951) Phys Rev 81:385

Slater JC (1974) Quantum theory of molecules and solids, vol 4: The self-consitent field for molecules and solids. McGraw-Hill, New York

Stoll H, Savin A (1985) In: Dreizler RM, da Providencia J (eds) Density functional methods in physics. Plenum, New York, p 177

Szalewicz K, Bukowski R, Jeziorski B (2005) In: Dykstra CE, Frenking G, Kim KS, Scuseria GE (eds) Theory and applications of computational chemistry: the first 40 years. A volume of technical and historical perspectives. Elsevier, Amsterdam, p 919

Szasz L, Berrios-Pagan I, McGinn G (1975) Z Naturforsch 30a:1516

Talman JD, Shadwick WF (1976) Phys Rev A 14:36

Tekin A, Jansen G (2007) Phys Chem Chem Phys 9:1680

Teller E (1962) Rev Mod Phys 34:627
Theophilou A (1979) J Phys C 12:5419
Thomas LH (1927) Proc Camb Philos Soc 23:542
Tuttle T, Thiel W (2008) Phys Chem Chem Phys 10:2159
Ullrich CA, Grossmann UJ, Gross EKU (1995) Phys Rev Lett 74:872
Valone SM (1980a) J Chem Phys 73:1344
Valone SM (1980b) J Chem Phys 73:4653
van Leeuwen R, Gritsenko OV, Baerends EJ (1996) Density functional theory I: functionals and effective potentials. In: Nalewajski RF (ed) Topics in current chemistry, vol 180. Springer, Berlin, p 107
van Mourik T, Gdanitz RJ (2002) J Chem Phys 116:9620
van Voorhis T, Scuseria GE (1998) J Chem Phys 109:400
von Barth U, Hedin L (1972) J Phys C 5:1629
von Lilienfeld OA, Tavernelli I, Röthlisberger U, Sebastianini D (2004) Phys Rev Lett 93:15300490
von Lilienfeld OA, Tavernelli I, Röthlisberger U, Sebastianini D (2005) Phys Rev B 71:195119
von Weizsäcker CF (1935) Z Phys 96:431
Vosko SH, Wilk L, Nusair M (1980) Can J Phys 58:1200
Wang YA, Carter EA (2000) In: Schwartz (ed) Theoretical methods in condensed phase chemistry. Kluwer, Dordrecht, p 117
Wesołowski TA (2004a) J Am Chem Soc 126:11444
Wesołowski TA (2004b) Chimia 58:311
Wesołowski TA, Tran F (2003) J Chem Phys 118:2072
Wesołowski TA, Warshel A (1993) J Phys Chem 97:8050
Wesołowski TA, Weber J (1998) Chem Phys Lett 248:71
Wesołowski TA, Muller RP, Warshel A (1995) J Phys Chem 100:15444
Wigner EP (1934) Phys Rev 46:1002
Williams HL, Chabalowski CF (2001) J Phys Chem A 105:646
Wu Q, Yang W (2002) J Chem Phys 116:515
Wu X, Vargas MC, Nayak S, Lotrich V, Scoles G (2001) J Chem Phys 115:8748
Xu X, Goddard WA III (2004) Proc Natl Acad Sci USA 101:2673
Yan Z, Perew JP, Kurth S (2000) Phys Rev B 61:16430
Yáñez RJ, Angulo JC, Dehesa SJ (1995) Int J Quantum Chem 56:489
Yang W (1987) Phys Rev Lett 59:1569
Yang W (1988) Phys Rev A 38(5494):5504
Yang W (1992) Phys Rev Lett 66:1438
Yang W, Harriman JE (1986) J Chem Phys 84:3323
Yasuda K (2001) Phys Rev A 63:032517
Zangwill A, Soven P (1980) Phys Rev A21:1561
Zaremba E, Kohn W (1976) Phys Rev B 13:2270
Zhao Y, Truhlar DG (2006) J Chem Phys 125:194101
Zhao Y, Truhlar DG (2008) Theor Chem Acc 120:215
Zhao Y, Schultz NE, Truhlar DG (2005) J Chem Phys 123:161103
Zhao Y, Schultz NE, Truhlar DG (2006) J Chem Theor Comput 2:364
Ziesche P (1995) Int J Quantum Chem 56:363
Zimmerli U, Parrinello M, Koumoutsakos P (2004) J Chem Phys 120:2693
Zumbach G, Maschke K (1983) Phys Rev A 28:544; Errata (1984) Phys Rev A 29:1585
Zumbach G, Maschke K (1985) J Chem Phys 82:5604

Part III
Insights from Information Theory

Chapter 8
Elements of Information Theory

Abstract The *Information Theory* (IT) of Fisher and Shannon provides convenient tools for the systematic and unbiased extraction of the *chemical* interpretation of the known (experimental or calculated) electron distribution in a molecule. A short overview of the basic concepts, relations, and techniques of IT is presented. The Shannon (S) entropy, reflecting the amount of the uncertainty (spread, disorder) contained in the given probability distribution, and the complementary Fisher (F) (intrinsic-accuracy) measure, focusing on the distribution narrowness (order), are introduced. The relative ("*cross*") entropy (entropy deficiency, missing information, directed-divergence) concept of Kullback and Leibler (KL), probing the information distance between the compared probability distributions, is presented. Rudiments of the IT descriptors of the communication channels are outlined and applied to the illustrative symmetric binary channel (SBC). The average conditional-entropy (communication noise) and mutual-information (information flow) quantities of information networks are then discussed in a more detail in view of their importance for interpreting the covalent and ionic bond components within the "communication" theory of the chemical bond. The information characteristics or several dependent probability schemes are then briefly summarized and the variational principle for the constrained extremum of the adopted measure of information, called the extreme physical information (EPI) principle, is advocated as a powerful tool for an unbiased assimilation in the optimum probability distribution of the information contained in the relevant constraints and/or references.

8.1 Introduction

The *Information Theory* (IT) (Fisher 1922, 1925, 1959; Hartley 1928; Shannon 1948; Shannon and Weaver 1949; Kullback and Leibler 1951; Khinchin 1957; Kullback 1959; Abramson, 1963; Ash 1965; Mathai and Rathie 1975; Pfeifer 1978) is one of the youngest branches of the applied probability theory, in which the probability ideas have been introduced into the field of communication, control, and data

R.F. Nalewajski, *Perspectives in Electronic Structure Theory*,
DOI 10.1007/978-3-642-20180-6_8,

processing. It has originated from the needs of practice, to create a theoretical model for a transmission of information, and evolved into an important chapter of the general theory of probability. Its foundations have been laid in 1920s by Sir R.A. Fisher (1922, 1925), in his classical measurement theory, and in 1940s by C.E. Shannon (1948), in his mathematical theory of communication. The theory provides the unifying principle of science (Brillouin 1956; Frieden 2000; Jaynes 1957a, b, 1985), and it has proved its utility in diverse areas of chemical physics (e.g., Aslangul et al. 1972; Sears 1980; Bernstein 1982; Gadre et al. 1985a, b; Gadre 2002; Esquivel et al. 1996; Nagy and Parr 1994, 1996, 2000) including a novel outlook on the origins of the chemical bond (Nalewajski 2006g, 2010f).

The distributions of electrons in molecular or reactive systems carry the associated information content, which changes during the chemical bond formation or in the course of a chemical reaction, i.e., a concerted bond-forming–bond-breaking process. Atoms-in-molecules (AIM) and larger molecular fragments (open subsystems) constantly exchange electrons, and hence also the information. In a sense, they "talk" to each other, both at the equilibrium (ground) state and during a transition to the displaced equilibrium state, in response to the applied perturbation. It is thus a challenging task to describe and understand the information content of electronic probability distributions in molecules, reactive systems, and their subsystems. Such an IT approach provides the alternative perspective on the molecular electronic structure, similarity, and reactivity. In fact, an insight into the entropic origins of chemical bonds and their couplings in diverse chemical phenomena is central to many branches of chemistry, particularly to the reactivity theory.

The quantum mechanical state of a molecule is determined by the system electronic wave function, the (complex) amplitude of the associated probability distribution, which carries the information. It is intriguing to explore the information content of the electronic probability distribution in a molecule or in its amplitude, and to extract the pattern of chemical bonds, trends in chemical reactivity, and other molecular descriptors from it, e.g., the bond multiplicities ("orders") and their covalent and ionic components. It is also of great interest to examine the identity of bonded atoms in molecules, the exchanges of information (communications) between them, as well as the information representation of subtle electron redistributions in chemical processes, e.g., those accompanying formation or breaking of bonds in a molecular or reactive system. It has been already amply demonstrated that many classical problems in theoretical chemistry can be approached anew using this novel IT perspective (e.g., Gadre 1984, 2002; Gadre et al. 1985a, b; Gadre and Sears 1979; Sears 1980; Sears et al. 1980; Gadre and Bendale 1985; Nalewajski 2002b, 2006g, 2010f).

The concepts and techniques of IT have been successfully used to explore the chemical properties of molecules or their fragments, and to examine the bonding patterns in both the molecular and reactive systems. For example, the displacements in the information distribution in molecules, relative to the "promolecular" reference consisting of the nonbonded constituent atoms in molecular positions, have been investigated and the least biased partition of the molecular electron distributions into the subsystem contributions, e.g., densities of bonded AIM, has been examined

(Nalewajski 2002a, 2003b, 2004a, 2006g, 2010f; Nalewajski and Broniatowska 2003a, 2005, 2007; Nalewajski and Parr 2000, 2001; Nalewajski et al. 2002; Parr et al. 2005; Ayers 2000b). The IT approach has been shown to lead to the "stockholder" molecular fragments of Hirshfeld (1977). These pieces of the molecular electron density have been derived from alternative global and local variational principles of IT, thus providing a solid theoretical basis for these formerly intuitive constructs. The concept of electronic loges (Daudel 1969, 1974) can also be tackled anew using IT techniques (Aslangul et al. 1972; Nalewajski 2003d).

The spatial localization of specific bonds, not to mention some qualitative questions about the very existence of some controversial chemical bonds, e.g., between the bridgehead carbon atoms in small propellanes, presents another challenging problem for this novel IT treatment of molecular systems. Another important aspect of the molecular electronic structure deals with the shell structure and the electron localization in atoms and molecules. The nonadditive Fisher information in the AO resolution has been recently used as the *contra-gradience* (CG) criterion for localizing the bonding regions in molecules (Nalewajski 2008e, 2010a, f; Nalewajski et al. 2010a, b), while the related information density in the MO resolution has been shown (Nalewajski et al. 2005) to determine the vital ingredient of the *electron-localization function* (ELF) (Becke and Edgecombe 1990; see also: Silvi and Savin 1994; Savin et al. 1997).

The *Communication Theory of the Chemical Bond* (CTCB) has been developed using the basic entropy/information descriptors of the molecular information (communication) channels in the AIM, orbital, and local resolution levels of the electron probability distributions (Nalewajski 2000c, 2004b, c, d, e, 2005a, b, c, 2006a, b, c, d, f, 2008a, b, 2009a, b, c, d, e, f, g, 2010b, c, d, h, i; Nalewajski et al. 2010c). The same bond descriptors have been used to provide the information scattering perspective on the intermediate stages of the electron redistribution processes (Nalewajski 2008b), including the atom promotion via the orbital hybridization (Nalewajski 2007), and the elements of the communication theory for excited electron configurations have been established (Nalewajski 2006c, e). Moreover, a phenomenological description of equilibria in molecular subsystems has been developed (Nalewajski 2002c, 2003b, 2004a, 2006h), which closely resembles that in ordinary thermodynamics (e.g., Callen 1962; Tisza 1977).

In the next chapter the information roots of quantum mechanics (see, e.g., Frieden 2000) will also be stressed by demonstrating that the Schrödinger equations of quantum mechanics, which determine the equilibrium distribution of electrons and its time dependence for the fixed external potential due to the system "frozen" nuclei, result from the constrained extremum principles of the Fisher information functional, related to the system average kinetic energy (Nalewajski 2008e, 2010f). The importance of the nonadditive effects in the chemical bond phenomena will also be emphasized, and the use of alternative IT probes of the electronic structure and of chemical bonds in molecules will be illustrated.

The electron redistribution accompanying the bond formation, from the (molecularly placed) free atoms $\{X^0\}$ of the "promolecule," exhibiting the ground-state densities $\{\rho_X^0\}$, to the bonded AIM with densities $\{\rho_X\}$, is marked by the difference

$\Delta\rho = \rho - \rho^0$ between the molecular ($\rho = \sum_X \rho_X$) and promolecular ($\rho^0 = \sum_X \rho_X^0$) electron densities, which mark the process final and initial stages, respectively. The entropy/information content of these distributions provides a basis for a novel IT perspective on the molecular electronic structure. The densities of displacements in the Shannon entropy and missing information (cross entropy), relative to the promolecular reference, will be used as sensitive diagnostic tools for detecting the chemical bonds, and to monitor the promotion/hybridization changes that the bonded atoms undergo in the molecular environment. As we have already mentioned above, when applied to a classical problem of the density partition into AIM pieces, the IT approach gives the familiar "stockholder" division scheme of Hirshfeld (1977). The nonadditive Fisher information density in the MO and AO resolutions, respectively, generate the ELF and CG information probes for localizing electrons and detecting the bonding regions in molecules, respectively. Representative examples of such an exploration of the molecular ground-state probability distributions will be presented.

The *Orbital Communication Theory* (OCT) of the chemical bond uses the standard entropy/information descriptors of the Shannon theory of communication to characterize the scattering of AO electron probabilities, throughout the network of chemical bonds generated by the system-occupied MO. The molecule is thus treated as an information system which propagates, from the channel AO "inputs" to AO "outputs," the "signals" of the electron allocations to these basis functions of the molecular SCF LCAO MO calculations. The underlying conditional probabilities, generated from the (bond-projected) superposition principle of quantum mechanics (Dirac 1967), are shown to be proportional to the squares of the corresponding elements of the *first*-order density matrix in the AO representation (Nalewajski 2009e), thus being also related to the Wiberg (1968) quadratic index of the chemical bond multiplicity (see also: Gopinathan and Jug 1983; Mayer 1983, 1985; Jug and Gopinathan 1990; Nalewajski and Jug 2002; Nalewajski et al. 1993, 1994a, b, 1996a, b, 1997; Nalewajski and Mrozek 1994, 1996; Nalewajski 2004b).

Such an information propagation in molecules exhibits the communication "noise" due to electron delocalization via the system chemical bonds, which effectively lowers the information content in the *output*-signal distribution, compared with that contained in probabilities determining the channel *input*-signal, molecular or promolecular. The orbital information systems are now being used to generate the entropic measures of the chemical bond multiplicity and their covalent/ionic composition for both the molecule as a whole and its diatomic fragments. The average *conditional-entropy*, which measures the channel communication noise due to electron delocalization (communication-indeterminacy), measures the IT *covalency* in the molecule, while the complementary descriptor of the network *mutual-information* (information-flow, communication-determinacy) reflects the electron localization effects and measures the system IT *ionic* component. The illustrative examples of applying these novel IT tools for exploring the electronic and bonding structures of representative molecules will be reported and discussed.

To summarize, the entropic probes of the molecular electronic structure have provided novel, attractive tools for describing the chemical bond phenomenona in information terms. It is the main purpose of this chapter to introduce the key concepts and techniques of IT which will be used in subsequent chapters of this part of the book, intended to review some of the recent developments in alternative local entropy/information probes of the molecular electronic structure and in the orbital formulation of CTCB (OCT).

8.2 Shannon and Fisher Measures of Information

The *Shannon* (S) (Shannon 1948) entropy content $S[p]$ in the (normalized) spatial probability distribution $p(\boldsymbol{r})$,

$$S[p] \equiv I^{S}[p] = -\int p(\boldsymbol{r}) \log p(\boldsymbol{r}) d\boldsymbol{r}, \quad \int p(\boldsymbol{r})\, d\boldsymbol{r} = 1, \tag{8.1}$$

where the definite integration is over the whole range of the random (position) variable $\boldsymbol{r}$, provides a measure of the average *indeterminacy* in $p(\boldsymbol{r})$ for the locality events $\{\boldsymbol{r}\}$. As indicated above, it also measures the average amount of information $I^{S}[p]$ obtained when this spatial uncertainty is removed by an appropriate localization measurement (experiment). Here the logarithm is taken to an arbitrary but fixed base: when taken to base 2, $\log = \log_2$, the information is measured in *bits*, while selecting $\log = \ln$ expresses the information in *nats*: 1 nat = 1.44 bits.

The *Fisher* (F) (Fisher 1922, 1925, 1959) information measure historically predates the Shannon entropy by about 25 years, being proposed in about the same time when the final form of the quantum mechanics was shaped. It emerges as an expected error in a "smart" measurement, in the context of the efficient estimators of a parameter. Fisher was the first to suggest that data samples in an experiment together with a given parametric distribution model contain the statistical information about the parameter(s). Let $p(\boldsymbol{r}|\theta)$ be the probability distribution function depending upon the parameter θ. The Fisher measure of information contained in this probability density is then defined as follows:

$$I(\theta) \equiv \int p(\boldsymbol{r}|\theta)\left(\frac{\partial \ln p(\boldsymbol{r}|\theta)}{\partial \theta}\right)^{2} d\boldsymbol{r} = \int \frac{[p'(\boldsymbol{r}|\theta)]^{2}}{p(\boldsymbol{r}|\theta)} d\boldsymbol{r}, \quad p'(\boldsymbol{r}|\theta) = \frac{\partial p(\boldsymbol{r}|\theta)}{\partial \theta}. \tag{8.2}$$

The *intrinsic accuracy* is a special case of this parametric measure, for the locality parameter $\boldsymbol{\theta}$, when $p(\boldsymbol{r}|\boldsymbol{\theta}) = p(\boldsymbol{r} + \boldsymbol{\theta}) = p(\boldsymbol{r}')$. In this case the parametric Fisher measure provides the information about the probability distribution itself:

$$\frac{\partial p(\boldsymbol{r}|\boldsymbol{\theta})}{\partial \boldsymbol{\theta}} = \frac{\partial \boldsymbol{r}'}{\partial \boldsymbol{\theta}} \frac{\partial p(\boldsymbol{r}')}{\partial \boldsymbol{r}'} = \frac{\partial p(\boldsymbol{r}')}{\partial \boldsymbol{r}'} = \nabla p(\boldsymbol{r}'). \tag{8.3}$$

Hence, for the *single*-component probability distribution

$$I(\boldsymbol{\theta}) = I[p] = \int p(\boldsymbol{r}')[\nabla \ln p(\boldsymbol{r}')]^2 d\boldsymbol{r}' = \int [\nabla p(\boldsymbol{r})]^2 / p(\boldsymbol{r})\, d\boldsymbol{r} \equiv I^{\mathrm{F}}[p]. \tag{8.4}$$

This Fisher information functional, reminiscent of von Weizsäcker's (1935) inhomogeneity correction to the electronic kinetic energy in TF theory, characterizes the compactness (order) of the probability density $p(\boldsymbol{r})$. For example, the Fisher information in the familiar *normal* distribution measures the inverse of its variance, called the *invariance*, while the complementary Shannon entropy is proportional to the logarithm of variance, thus monotonically increasing with the spread of the Gaussian distribution.

The Shannon entropy and the Fisher information for locality thus describe the complementary facets of the probability density: the former reflects distribution's "spread" (a measure of uncertainty, "disorder"), while the latter measures its "narrowness" ("order"). The analytical properties of the Shannon and Fisher information functionals are quite different (e.g., Frieden 2000; Frieden and Soffer 2010). When extremized through variation of the probability distribution the Shannon entropy gives an exponential solution, while the Fisher information generates the differential equation, and hence multiple solutions specified by the appropriate boundary conditions.

The form of the intrinsic accuracy functional can be simplified by expressing it through the associated classical (real) amplitude $A(\boldsymbol{r}) = \sqrt{p(\boldsymbol{r})}$ of the probability distribution $p(\boldsymbol{r})$:

$$I[p] = 4\int [\nabla A(\boldsymbol{r})]^2\, d\boldsymbol{r} \equiv I[A]. \tag{8.5}$$

It is naturally generalized into the case of complex probability amplitudes encountered in quantum mechanics, i.e., the system wave functions (e.g., Nalewajski 2008e). For the simplest case of the spinless *one*-particle system, when $A(\boldsymbol{r}) = \psi(\boldsymbol{r})$ and $p(\boldsymbol{r}) = \psi^*(\boldsymbol{r})\psi(\boldsymbol{r}) = |\psi(\boldsymbol{r})|^2$,

$$I[\psi] = 4\int |\nabla\psi(\boldsymbol{r})|^2\, d\boldsymbol{r} = 4\int \nabla\psi^*(\boldsymbol{r})\cdot\nabla\psi(\boldsymbol{r})\, d\boldsymbol{r} \equiv \int f(\boldsymbol{r})\, d\boldsymbol{r}. \tag{8.6}$$

Therefore, $I[A]$ or $I[\psi]$ measures the gradient content in the amplitude of the probability density. Its extension to the multicomponent (vector) probabilities $\boldsymbol{p}(\boldsymbol{r}) \equiv \{p_n(\boldsymbol{r})\}$, expressed in terms of the associated amplitudes, classical $\boldsymbol{A}(\boldsymbol{r}) \equiv \{A_n(\boldsymbol{r}) \equiv p_n(\boldsymbol{r})^{1/2}\}$, or quantum mechanical $\boldsymbol{\psi} \equiv \{\psi_n(\boldsymbol{r})\}$, in terms of which $p_n(\boldsymbol{r}) = [A_n(\boldsymbol{r})]^2$ or $p_n(\boldsymbol{r}) = |\psi_n(\boldsymbol{r})|^2$, reads:

$$I[\boldsymbol{p}] \equiv \sum_n I[p_n] = 4\sum_n \int [\nabla A_n(\boldsymbol{r})]^2 d\boldsymbol{r} = \sum_n I[A_n] \equiv I[\boldsymbol{A}]$$

or

$$I[\boldsymbol{p}] = 4\sum_n \int |\nabla\psi_n(\boldsymbol{r})|^2 d\boldsymbol{r} = \sum_n I[\psi_n] \equiv I[\boldsymbol{\psi}]. \tag{8.7}$$

8.3 Entropy Deficiency

An important generalization of Shannon's entropy, called the *relative* (*cross*) *entropy*, also known as the *entropy deficiency*, *missing information*, or the *directed divergence*, has been proposed by Kullback and Leibler (KL) (Kullback and Leibler 1951). It measures the information "distance" between the two (normalized) probability distributions for the same set of events. For example, in the discrete probability scheme $\mathbf{A} = [\boldsymbol{a}, \boldsymbol{P}(\boldsymbol{a})]$, involving events identified by the admissible values of the discrete random variable $\boldsymbol{a} = \{a_i\}$, and their probabilities $\boldsymbol{P}(\boldsymbol{a}) = \{P(a_i) = p_i\} \equiv \boldsymbol{p}$, this discrimination information in $\boldsymbol{p}$ with respect to the *reference* distribution $\boldsymbol{p}^0 = \{P(a_i^0) = p_i^0\}$ reads:

$$\Delta S(\boldsymbol{p}|\boldsymbol{p}^0) \equiv I^{\mathrm{KL}}(\boldsymbol{p}|\boldsymbol{p}^0) = \sum\nolimits_i p_i \log(p_i/p_i^0) \geq 0. \tag{8.8}$$

In the continuous distribution case, e.g., $\mathbf{A} = \{\boldsymbol{r}, p(\boldsymbol{r})\}$, this directed-divergence measure of the entropy deficiency in the probability density $p(\boldsymbol{r})$, relative to the prior distribution $p^0(\boldsymbol{r})$, is defined by the related functional

$$\Delta S[p|p^0] \equiv I^{\mathrm{KL}}[p|p^0] = \int p(\boldsymbol{r}) \log[p(\boldsymbol{r})/p^0(\boldsymbol{r})]\, d\boldsymbol{r} \geq 0. \tag{8.9}$$

For individual events the logarithm of the probability ratio, $I_i = \log(p_i/p_i^0)$, $I(\boldsymbol{r}) = \log[p(\boldsymbol{r})/p^0(\boldsymbol{r})]$, or for the *one*-dimensional continuous distributions $p(x)$ and $p^0(x)$, $I(x) = \log[p(x)/p^0(x)]$, called the *surprisal*, provides a measure of the information contained in the current distribition relative to the reference distribution. The equality in the last two equations takes place only when the surprisal identically vanishes for all events, i.e., when the two compared probability schemes are identical.

As we have observed above, the Shannon entropy measures a "*disorder*" (uncertainty, indeterminacy, "smoothness") of the probability distribution. On a finite interval the distribution possessing the highest entropy is the uniform distribution, and any deviation from uniformity indicates the "perturbing" presence of "*order*." The Kullback–Leibler measure, i.e., the *referenced* Shannon's entropy, generates a similar description, but in reference to some prior distribution. The entropy deficiency thus provides a measure of an information resemblance between the two compared probability schemes. The more the two probability distributions differ from one another, the larger the information distance.

The nonnegative character of the entropy deficiency in $p(x)$ relative to $p^0(x)$,

$$\Delta S[p|p^0] = \int p(x) \log[p(x)/p^0(x)]\, dx \equiv \int p(x) I(x)\, dx, \tag{8.10}$$

directly follows from the observation that the line $y = z - 1$ lies above the curve $y = \log z$, with two functions having equal (zero) value only for $z = 1$. Taking $z(x) = p^0(x)/p(x)$ and using the condition of the probability normalization then give:

$$\begin{aligned}\Delta S[p|p^0] &= -\int p(x) \log z(x)\, dx \geq \int p(x)[z(x) - 1]\, dx \\ &= \int [p^0(x) - p(x)]\, dx = 0.\end{aligned} \tag{8.11}$$

Notice, however, that the surprisal itself becomes negative, when the current probability is smaller than its reference value. Therefore, the directed divergence functional is not symmetrical with respect to the two probability distributions involved and exhibits negative values of the integrand. To avoid this limitation, Kullback (K) (Kullback 1959) has proposed an alternative measure, called the *divergence*, defined by the symmetrized combination of the two admissible entropy deficiencies,

$$\begin{aligned}\Delta S(\boldsymbol{p}, \boldsymbol{p}^0) &\equiv I^{\mathrm{K}}(\boldsymbol{p}, \boldsymbol{p}^0) = \Delta S(\boldsymbol{p}|\boldsymbol{p}^0) + \Delta S(\boldsymbol{p}^0|\boldsymbol{p}) \\ &= \sum\nolimits_i (p_i - p_i^0) I[p_i/p_i^0] \equiv \sum\nolimits_i \Delta p_i I_i, \qquad \Delta p_i I_i \geq 0, \\ \Delta S[p, p^0] &\equiv I^{\mathrm{K}}[p, p^0] = \Delta S[p|p^0] + \Delta S[p^0|p] \\ &= \int [p(\boldsymbol{r}) - p^0(\boldsymbol{r})] I(\boldsymbol{r})\, d\boldsymbol{r} \equiv \int \Delta p(\boldsymbol{r}) I(\boldsymbol{r})\, d\boldsymbol{r}, \qquad \Delta p(\boldsymbol{r}) I(\boldsymbol{r}) \geq 0,\end{aligned} \tag{8.12}$$

which in the continuous case gives rise to a nonnegative integrand.

8.4 Dependent Probability Distributions

For two mutually dependent (discrete) probability distributions,

$$\boldsymbol{P}(\boldsymbol{a}) = \{P(a_i) = p_i\} \equiv \boldsymbol{p} \quad \text{and} \quad \boldsymbol{P}(\boldsymbol{b}) = \{P(b_j) = q_j\} \equiv \boldsymbol{q}, \tag{8.13}$$

determining the associated probability schemes $\mathbf{A} = [\boldsymbol{a}, \boldsymbol{P}(\boldsymbol{a})]$ and $\mathbf{B} = [\boldsymbol{b}, \boldsymbol{P}(\boldsymbol{b})]$, respectively, we decompose the *joint*-probabilities $\mathbf{P}(\boldsymbol{a}\wedge\boldsymbol{b}) = \{P(a_i b_j) = \pi_{i,j}\} \equiv \boldsymbol{\pi}$ of the simultaneous events $\boldsymbol{a}\wedge\boldsymbol{b} = \{a_i \wedge b_j \equiv a_i b_j\}$, as products $\{\pi_{i,j} = p_i\, P(j|i)\}$ of the *marginal* probability $p_i = P(a_i)$ of ith event in $\boldsymbol{a}$ and the *conditional* probability $P(j|i) = P(a_i b_j)/P(a_i)$ of the event b_j of $\boldsymbol{b}$, given that the event a_i has already occurred. The relevant normalization conditions for the joint probabilities $\boldsymbol{\pi}$ and the conditional probabilities $\mathbf{P}(\boldsymbol{b}|\boldsymbol{a}) = \{P(j|i)\}$ then read:

$$\sum\nolimits_j \pi_{i,j} = p_i, \quad \sum\nolimits_i \pi_{i,j} = q_j, \quad \sum\nolimits_i \sum\nolimits_j \pi_{i,j} = 1; \quad \sum\nolimits_j P(j|i) = 1, \\ i = 1, 2, \ldots, n. \tag{8.14}$$

In this *two*-scheme scenario the Shannon entropy of the product distribution $\boldsymbol{\pi}$,

$$\begin{aligned} S(\boldsymbol{\pi}) &= -\sum\nolimits_i \sum\nolimits_j \pi_{i,j} \log \pi_{i,j} = -\sum\nolimits_i \sum\nolimits_j p_i P(j|i)[\log p_i + \log P(j|i)] \\ &= -\left[\sum\nolimits_j P(j|i)\right] \sum\nolimits_i p_i \log p_i - \sum\nolimits_i p_i \left[\sum\nolimits_j P(j|i) \log P(j|i)\right] \\ &\equiv S(\boldsymbol{p}) + \sum\nolimits_i p_i S(\boldsymbol{q}|i) \equiv S(\boldsymbol{p}) + S(\boldsymbol{q}|\boldsymbol{p}), \end{aligned} \tag{8.15}$$

is seen to be given by the sum of the average entropy in the marginal probability distribution, $S(\boldsymbol{p})$, and the average *conditional entropy* in $\boldsymbol{q}$ given $\boldsymbol{p}$:

$$\begin{aligned} S(\boldsymbol{q}|\boldsymbol{p}) &= -\sum\nolimits_i \sum\nolimits_j \pi_{i,j} \log P(j|i) = -\sum\nolimits_i p_i \left[\sum\nolimits_j P(j|i) \log P(j|i)\right] \\ &\equiv \sum\nolimits_i P_i S(\boldsymbol{q}|i). \end{aligned} \tag{8.16}$$

The latter represents the extra amount of uncertainty about the occurrence of outcomes $\boldsymbol{b}$, given that the events $\boldsymbol{a}$ are known to have occurred. In other words: the amount of information obtained as a result of simultaneously observing the events $\boldsymbol{a}$ and $\boldsymbol{b}$ of the two discrete probability distributions $\boldsymbol{p} = \boldsymbol{P}(\boldsymbol{a})$ and $\boldsymbol{q} = \boldsymbol{P}(\boldsymbol{b})$, respectively, equals the amount of information observed in one set, say $\boldsymbol{a}$, supplemented by the extra information provided by the occurrence of events in the other set $\boldsymbol{b}$, when $\boldsymbol{a}$ are known to have occurred already. These information quantities are shown in Fig. 8.1.

Clearly, by using the other probability distribution $\boldsymbol{q} = \{P(b_i)\}$ as the marginal one, one arrives at the alternative expression for $S(\boldsymbol{\pi})$:

$$S(\boldsymbol{\pi}) = S(\boldsymbol{q}) + \sum\nolimits_j q_j S(\boldsymbol{p}|\, j) \equiv S(\boldsymbol{q}) + S(\boldsymbol{p}|\boldsymbol{q}), \tag{8.17a}$$

where

$$S(\boldsymbol{p}|\boldsymbol{q}) = -\sum\nolimits_j \sum\nolimits_i \pi_{j,i} \log P(i|\, j)] = -\sum\nolimits_j q_j \left[\sum\nolimits_i P(i|\, j) \log P(i|\, j)\right] = \sum\nolimits_j q_j S(\boldsymbol{p}|\, j), \tag{8.17b}$$

and probabilities $\boldsymbol{P}(\boldsymbol{a}|\boldsymbol{b}) = \{P(i|j)\}$ satisfy the normalization conditions: $\sum_i P(i|j) = 1, j = 1, 2, \ldots, m$. Here, the average conditional entropy $S(\boldsymbol{p}|\boldsymbol{q})$ represents the residual uncertainty about the occurrence of events $\boldsymbol{a}$, when $\boldsymbol{b}$ are known to have

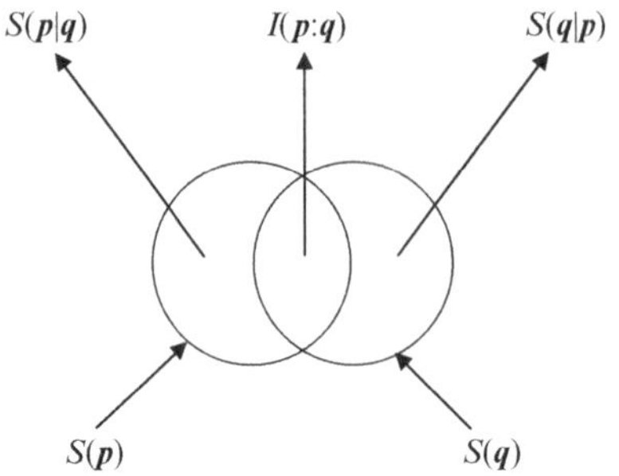

Fig. 8.1 A qualitative diagram of the conditional entropy and mutual information quantities of two dependent probability distributions $\boldsymbol{p}$ and $\boldsymbol{q}$. Two circles enclose the areas representing the Shannon entropies $S(\boldsymbol{p})$ and $S(\boldsymbol{q})$ of the two *separate* distributions, while the common (*overlap*) area of the two circles represents the mutual information $I(\boldsymbol{p}{:}\boldsymbol{q})$ in two distributions. The remaining parts of two circles correspond to conditional entropies $S(\boldsymbol{p}|\boldsymbol{q})$ and $S(\boldsymbol{q}|\boldsymbol{p})$, measuring the residual uncertainty about events in one set, when one has the full knowledge of the occurrence of events in the other set of outcomes. The area enclosed by the envelope of the two overlapping circles then represents the entropy of the "product" (joint) distribution: $S(\boldsymbol{\pi}) = S(\mathbf{P}(\boldsymbol{a} \wedge \boldsymbol{b})) = S(\boldsymbol{p}) + S(\boldsymbol{q}) - I(\boldsymbol{p}{:}\boldsymbol{q}) = S(\boldsymbol{p}) + S(\boldsymbol{q}|\boldsymbol{p}) = S(\boldsymbol{q}) + S(\boldsymbol{p}|\boldsymbol{q})$

been observed. Since the conditional probability is a probability measure the properties of the conditional entropy are similar to those of the information entropy itself.

The common amount of the information content in two events a_i and b_j, $I(i{:}j)$, measuring the information about a_i provided by the occurrence of b_j, or the information about b_j provided by the occurrence of a_i,

$$\begin{aligned} I(i{:}j) &= \log[P(a_i b_j)/P(a_i)P(b_j)] = \log[\pi_{i,j}/(p_i q_j)] \\ &\equiv \log[P(i|\,j)/p_i] \equiv \log[P(j|i)/q_j] = I(j{:}i), \end{aligned} \tag{8.18a}$$

is called the *mutual information* in two events. This quantity may take on any real value, positive, negative, or zero. It vanishes, when both events are independent,

$$P(a_i b_j) = P(a_i)P(b_j), \qquad P(i|\,j) = P(a_i), \qquad P(j|i) = P(b_j),$$

i.e., when the occurrence of one event does not influence (or condition) the probability of the occurrence of the other event, and it is negative, when the occurrence of one event makes the nonoccurrence of the other event more likely. It also follows from the preceding equation that

$$\begin{aligned} &I(i{:}j) = I(i) - I(i|\,j) = I(j) - I(j|i) = I(i) + I(j) - I(i \wedge j) \qquad \text{or} \\ &I(i \wedge j) = I(i) + I(j) - I(i : j), \end{aligned} \tag{8.18b}$$

where the information of the joint event $I(i \wedge j) = -\log \pi_{i,j}$. Thus, the information in the joint occurrence of two events a_i and b_j is the information in the occurrence of

a_i plus that in the occurrence of b_j minus their mutual information. Clearly, for the independent events $I(i \wedge j) = I(i) + I(j)$, since $I(i{:}j) = \log 1 = 0$.

The mutual information of an event with itself defines its *self-information*:

$$I(i{:}i) \equiv I(i) = \log[P(i|i)/p_i] = -\log p_i, \tag{8.19a}$$

since $P(i|i) = 1$. It vanishes when $p_i = 1$, i.e., when there is no uncertainty about the occurrence of a_i so that the occurrence of this event removes no uncertainty, and hence conveys no information. This quantity provides a measure of the uncertainty about the occurrence of the event, i.e., the information received when the event actually occurs. The Shannon entropy of (8.1) can be thus interpreted as the mean value of the *self*-information in all individual events defining the probability distribution, e.g.,

$$S(\boldsymbol{p}) = -\sum\nolimits_i p_i \log p_i \equiv \sum\nolimits_i p_i I(i). \tag{8.19b}$$

One similarly defines the average mutual information in two probability distributions, $I(\boldsymbol{p}{:}\boldsymbol{q})$, as the $\boldsymbol{\pi}$-weighted mean value of the mutual information quantities for the individual *joint*-events:

$$\begin{aligned} I(\boldsymbol{p}{:}\boldsymbol{q}) &= \sum\nolimits_i \sum\nolimits_j \pi_{i,j} I(i{:}j) = \sum\nolimits_i \sum\nolimits_j \pi_{i,j} \log(\pi_{i,j}/\pi_{i,j}^0) \\ &= S(\boldsymbol{p}) + S(\boldsymbol{q}) - S(\boldsymbol{\pi}) = S(\boldsymbol{p}) - S(\boldsymbol{p}|\boldsymbol{q}) = S(\boldsymbol{q}) - S(\boldsymbol{q}|\boldsymbol{p}) \geq 0, \end{aligned} \tag{8.20}$$

where the equality holds only for the independent distributions, when $\{\pi_{i,j} = \pi_{i,j}^0 \equiv p_i q_j\}$. These average entropy/information relations are also shown in Fig. 8.1. Indeed, the amount of uncertainty in $\boldsymbol{q}$ can only decrease, when $\boldsymbol{p}$ has been known beforehand, $S(\boldsymbol{q}) \geq S(\boldsymbol{q}|\boldsymbol{p}) = S(\boldsymbol{q}) - I(\boldsymbol{p}{:}\boldsymbol{q})$, as indeed seen in Fig. 8.1, with equality being observed only when the two sets of events are independent giving rise to nonoverlapping entropy circles.

It should also be observed that the average mutual information is an example of the entropy deficiency, which measures the missing information between the joint probabilities $\mathbf{P}(\boldsymbol{a} \wedge \boldsymbol{b}) = \boldsymbol{\pi}$ of the *dependent* events $\boldsymbol{a}$ and $\boldsymbol{b}$ and the joint probabilities $\mathbf{P}^{ind.}(\boldsymbol{a} \wedge \boldsymbol{b}) = \boldsymbol{\pi}^0 \equiv \boldsymbol{p} \otimes \boldsymbol{q}$ for the independent events: $I(\boldsymbol{p}{:}\boldsymbol{q}) = \Delta S(\boldsymbol{\pi}|\boldsymbol{\pi}^0)$. The average mutual information thus measures a degree of a dependence between events defining the two probability schemes. A similar information distance interpretation can be attributed to the conditional entropy: $S(\boldsymbol{p}|\boldsymbol{q}) = S(\boldsymbol{p}) - \Delta S(\boldsymbol{\pi}|\boldsymbol{\pi}^0)$.

Again, the nonnegative character of this average information measure can be demonstrated using the inequality $\log z \leq z - 1$ for $z = \pi_{i,j}^0/\pi_{i,j}$ and the normalization conditions of $\boldsymbol{\pi}$ and $\boldsymbol{\pi}^0$:

$$I(\boldsymbol{p}{:}\boldsymbol{q}) = \sum\nolimits_i \sum\nolimits_j \pi_{i,j} \log(\pi_{i,j}/\pi_{i,j}^0) \geq \sum\nolimits_i \sum\nolimits_j (\pi_{i,j}{}^0 - \pi_{i,j}) = 0. \tag{8.21}$$

8.5 Information Propagation in Communication Systems

We continue this short overview with the key entropy/information descriptors of a transmission of signals in the communication systems (Abramson 1963). The basic elements of such a "device" are shown in Fig. 8.2. The *input signal* emitted from n "inputs" $\boldsymbol{a} = (a_1, a_2, \ldots, a_n)$ of the channel *source* (**A**) is characterized by the input probability distribution $\boldsymbol{P}(\boldsymbol{a}) = \boldsymbol{p} = (p_1, p_2, \ldots, p_n) \equiv \boldsymbol{P}(\mathbf{A})$. It can be received at m "outputs" $\boldsymbol{b} = (b_1, b_2, \ldots, b_m)$ of the system *receiver* (**B**). The distribution of the output signal among the detection "events" $\boldsymbol{b}$ gives rise to the *output* probability distribution $\boldsymbol{P}(\boldsymbol{b}) = \boldsymbol{q} = (q_1, q_2, \ldots, q_m) \equiv \boldsymbol{P}(\mathbf{B})$. The transmission of signals is randomly disturbed within the communication system, thus exhibiting a typical communication *noise*, since the signal sent at the given input is in general received with a nonzero probability at several outputs. This feature of communication systems is described by the spread in the conditional probabilities of the *outputs-given-inputs*,

$$\mathbf{P}(\mathbf{B}|\mathbf{A}) = \{P(b_j|a_i) = P(a_i \wedge b_j)/P(a_i) \equiv P(j|i)\},$$

or the conditional probabilities of *inputs-given-outputs*,

$$\mathbf{P}(\mathbf{A}|\mathbf{B}) = \{P(a_i|b_j) = P(a_i \wedge b_j)/P(b_j) \equiv P(i|j)\},$$

where $P(a_i \wedge b_j) \equiv \pi_{i,j}$ stands for the probability of the *joint*-occurrence of the specified pair of the output and input events.

The Shannon entropy of the input (source) probabilities $\boldsymbol{p}$, $H(\mathbf{A}) \equiv S(\boldsymbol{p})$, determines the channel a priori entropy. The average *conditional entropy* $H(\mathbf{B}|\mathbf{A}) \equiv S(\mathbf{B}|\mathbf{A}) \equiv S(\boldsymbol{q}|\boldsymbol{p})$, of the outputs given inputs, is thus determined by the scattering probabilities $\mathbf{P}(\mathbf{B}|\mathbf{A}) = \{P(b_j|a_i) \equiv P(j|i)\} \equiv \mathbf{P}(\boldsymbol{b}|\boldsymbol{a})$. This entropy measures the average noise in the "*forward*" transmission of signals, from $\boldsymbol{a}$ to $\boldsymbol{b}$.

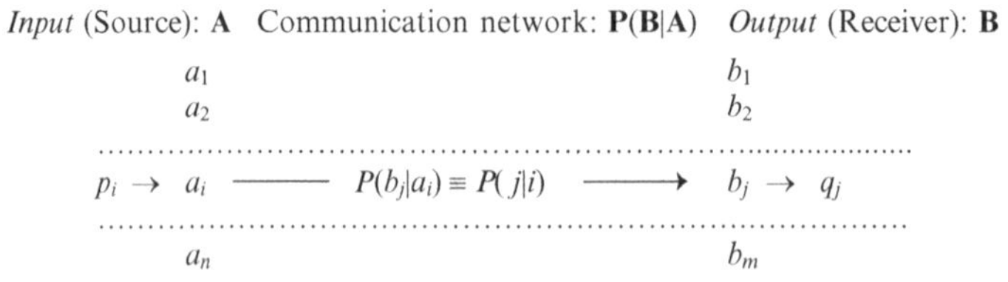

Fig. 8.2 The communication system is characterized by the probability vectors: $\boldsymbol{P}(\boldsymbol{a}) = \{P(a_i)\} = \boldsymbol{p} = (p_1, \ldots, p_n) \equiv \boldsymbol{P}(\mathbf{A})$, of the channel "*input*" events $\boldsymbol{a} = (a_1, \ldots, a_n)$ in the system *source* **A**, and $\boldsymbol{P}(\boldsymbol{b}) = \{P(b_j)\} = \boldsymbol{q} = (q_1, \ldots, q_m) \equiv \boldsymbol{P}(\mathbf{B})$, of the "*output*" events $\boldsymbol{b} = (b_1, \ldots, b_m)$ in the system *receiver* **B**. The transmission of signals *via* a network of this communication channel is described by the $(n \times m)$ matrix of the conditional probabilities $\mathbf{P}(\mathbf{B}|\mathbf{A}) = \{P(b_j|a_i) \equiv P(j|i)\} \equiv \mathbf{P}(\boldsymbol{b}|\boldsymbol{a})$, of observing different "outputs" (*columns*, $j = 1, 2, \ldots, m$), given the specified "inputs" (*rows*, $i = 1, 2, \ldots, n$). For clarity, only the single "forward" scattering $a_i \rightarrow b_j$ is shown in the diagram

The so-called a posteriori entropy, of the inputs given outputs, $H(\mathbf{A}|\mathbf{B}) \equiv S(\boldsymbol{p}|\boldsymbol{q})$, is similarly defined by the conditional probabilities of the "*reverse*" probability scattering, from $\boldsymbol{b}$ to $\boldsymbol{a}$: $\mathbf{P}(\mathbf{A}|\mathbf{B}) = \{P(a_i|b_j) = P(i|j)\} \equiv \mathbf{P}(\boldsymbol{a}|\boldsymbol{b})$. It reflects the residual indeterminacy about the input signal, when the output signal has already been received.

The average conditional entropy $H(\mathbf{A}|\mathbf{B}) \equiv S(\boldsymbol{p}|\boldsymbol{q})$ thus measures the indeterminacy of the source with respect to the receiver, while the conditional entropy $H(\mathbf{B}|\mathbf{A}) \equiv S(\boldsymbol{q}|\boldsymbol{p})$ reflects the uncertainty of the receiver relative to the source. Hence, an observation of an output signal provides on average the amount of information given by the difference between the a priori and a posteriori uncertainties, $S(\boldsymbol{p}) - S(\boldsymbol{p}|\boldsymbol{q}) = I(\boldsymbol{p}:\boldsymbol{q}) \equiv I(\mathbf{A}:\mathbf{B})$, which defines the *mutual information* in the source and receiver. In other words, the mutual information $I(\boldsymbol{p}:\boldsymbol{q})$ measures the net amount of information transmitted through the communication channel, while the conditional information $S(\boldsymbol{p}|\boldsymbol{q})$ reflects a portion of the initial information content $S(\boldsymbol{p})$ transformed into the communication "noise" as a result of the input signal being scattered in the information channel. Accordingly, $S(\boldsymbol{q}|\boldsymbol{p})$ reflects the noise part of $S(\boldsymbol{q})$: $S(\boldsymbol{q}) = S(\boldsymbol{q}|\boldsymbol{p}) + I(\boldsymbol{p}:\boldsymbol{q})$.

As an illustrative example consider the *Symmetric Binary Channel* (SBC) shown in Fig. 8.3, consisting of two inputs and two outputs, with the input probabilities $\boldsymbol{p} = (x, 1 - x)$ and the symmetric conditional probability matrix

$$\mathbf{P}^{\mathrm{SBC}}(\mathbf{B}|\mathbf{A}) = \begin{bmatrix} 1-\omega & \omega \\ \omega & 1-\omega \end{bmatrix}.$$

Its input entropy is determined by the *Binary Entropy Function* (BEF) shown in Fig. 8.4,

$$H(\mathbf{A}) = -\,x\log x - (1-x)\log(1-x) \equiv H(x),$$

which defines the channel a priori entropy. The system output entropy is also generated by BEF, $H(\mathbf{B}) = H(z(x,\ \omega))$, where $z(x,\ \omega) \equiv q_2 = x\omega + (1 - x)(1 - \omega)$. One also finds that the channel conditional entropy $H(\mathbf{B}|\mathbf{A}) = H(\omega)$ measures its average communication noise in the forward transmission of signals and hence the mutual information between the system inputs and outputs, $I(\mathbf{A}:\mathbf{B}) = S(\mathbf{B}) - S(\mathbf{B}|\mathbf{A}) = H[z(x,\ \omega)] - H(\omega)$, reflects the net information flow in SBC. These relations are illustrated in Fig. 8.4. It should be observed that z always lies between ω and $1 - \omega$, and hence:

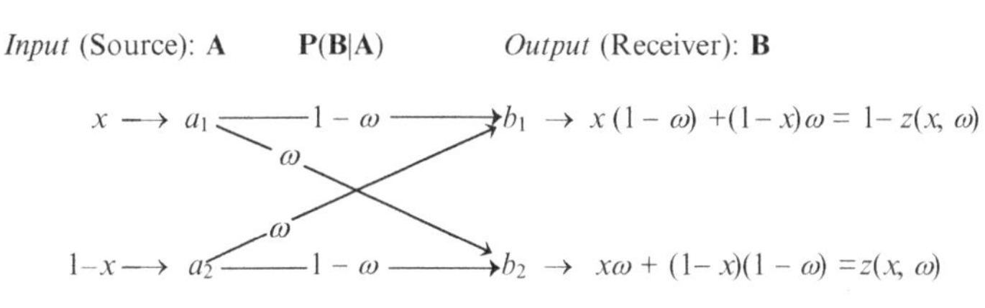

Fig. 8.3 The symmetric binary channel (SBC)

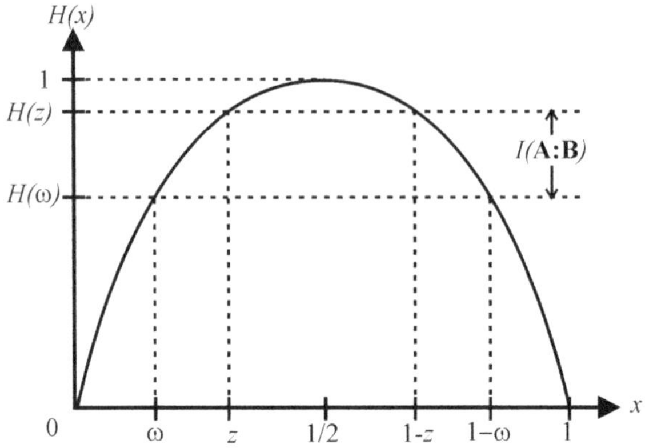

Fig. 8.4 The binary entropy function (BEF) $H(x) = -x\log_2 x - (1-x)\log_2(1-x)$ and a geometric interpretation of the conditional entropy $H(\mathbf{B}|\mathbf{A}) = H(\omega)$ and mutual information $I(\mathbf{A}:\mathbf{B}) = H(z) - H(\omega)$ in SBC shown in Fig. 8.3

$$H(z) = H(1-z) \geq H(\omega) = H(1-\omega).$$

This demonstrates a nonnegative character of the mutual information, represented by the overlap area between the two entropy circles in a qualitative diagram shown in Fig. 8.1.

The amount of information $I(\mathbf{A}:\mathbf{B})$ flowing through SBC thus depends on both the conditional probability parameter ω, characterizing the communication system itself, and on the input probability parameter x, which determines the way the channel is used (or probed). For $x = 0$ (or 1) $H(z) = H(\omega)$ and thus $I(\mathbf{A}:\mathbf{B}) = 0$, i.e., there is no net flow of information from the source to the receiver whatever. For $x = ½$, when the two inputs are equally probable, one finds $H(z) = 1$ bit, thus giving rise to the maximum value of the channel mutual information determining the system transmission *capacity*:

$$C(\omega) \equiv \max_{\mathbf{A}} I(\mathbf{A}:\mathbf{B}) = \max_x \{H[z(x,\omega)] - H(\omega)\} = 1 - H(\omega)(\text{bits}).$$

Hence, for $\omega = ½$, the information capacity of SBC identically vanishes.

We have already remarked in Sect. 8.1 that in molecular communication systems the conditional entropy $S(\boldsymbol{q}|\boldsymbol{p})$, where $\boldsymbol{p}$ and $\boldsymbol{q}$ stand for the molecular input and output probabilities, measures the entropy *covalency* of all bonds in the molecular system, while the complementary mutual information relative to the reference ("promolecular") distribution $\boldsymbol{p}^0$ reflects the system overall IT ionicity. One can similarly regard the SBC for $\omega = ½$ as a prototype of the symmetric chemical bond, e.g., the σ bond in H_2 or π bond in ethylene. The channel IT covalency $S(\boldsymbol{q}|\boldsymbol{p}) = H(\omega = ½) = 1$ bit and the complementary IT ionicity relative to the "promolecular" input probabilities $\boldsymbol{p}^0 = (½, ½)$, $I(\boldsymbol{p}^0:\boldsymbol{q}) = C(\omega = ½) = 0$, reflecting the maximum flow of information in this channel, then generate the overall IT "bond" index of this prototype information network of the purely covalent "single" chemical bond:

$$N(\mathbf{A}^0;\mathbf{B}) = N(\boldsymbol{p}^0;\boldsymbol{q}) \equiv S(\boldsymbol{q}|\boldsymbol{p}) + I(\boldsymbol{p}^0{:}\boldsymbol{q}) = 1 \text{ bit.} \tag{8.22}$$

8.6 Several Probability Schemes

Let us now examine the entropic quantities characterizing three probability distributions, represented by overlapping entropy circles in the lower part of Fig. 8.5; the diagram upper part again summarizes the *two*-distributions case also shown in Fig. 8.1. This variety of the entropy/information descriptors of three dependent probability schemes, $\mathbf{A} = [\boldsymbol{a}, \boldsymbol{P}(\boldsymbol{a})]$, $\mathbf{B} = [\boldsymbol{b}, \boldsymbol{P}(\boldsymbol{b})]$, and $\mathbf{C} = [\boldsymbol{c}, \boldsymbol{P}(\boldsymbol{c})]$, now involves the conditional (relative) entropies with respect to a single or two probability schemes, the mutual information contained in two or three probability distributions, and the mutual information characteristics of two probability distributions conditional on the third one. In what follows we denote by a, b, and c, the representative single events in the three probability schemes involved: $a \in \boldsymbol{a}$, $b \in \boldsymbol{b}$ and $c \in \boldsymbol{c}$.

The diagrams shown in Fig. 8.5 demonstrate the *additivity* of the information quantities in the three-probability scenario, which may now involve two *output* signals resulting from a single *input* signal in the communication channel, or the repeated, two *input* signals transmitted in the noisy information system. For example, the information about the input signal **A** obtained by observing two output signals **B** and **C** is measured by the difference of the Shannon (a priori) entropy $H(\mathbf{A})$ and the conditional (double a posteriori) entropy in **A** given (**B** and **C**), i.e., the indeterminacy of **A** relative to (**B** and **C**), which defines the *three*-scheme mutual information quantity

$$\begin{aligned} I(\mathbf{A}:\mathbf{BC}) &= \sum_{a\in\boldsymbol{a}}\sum_{b\in\boldsymbol{b}}\sum_{c\in\boldsymbol{c}} P(a\wedge b\wedge c)\log\frac{P(a\wedge b\wedge c)}{P(a)P(b\wedge c)} \\ &= \sum_{a\in\boldsymbol{a}}\sum_{b\in\boldsymbol{b}}\sum_{c\in\boldsymbol{c}} P(a\wedge b\wedge c)\log\frac{P(a|b\wedge c)}{P(a)} \\ &= H(\mathbf{A}) - H(\mathbf{A}|\mathbf{BC}) = S(\boldsymbol{P}(\boldsymbol{a})) - S(\boldsymbol{P}(\boldsymbol{a})|\mathbf{P}(\boldsymbol{b}\wedge\boldsymbol{c})) \\ &\equiv I(\boldsymbol{P}(\boldsymbol{a}):\mathbf{P}(\boldsymbol{b}\wedge\boldsymbol{c})), \end{aligned} \tag{8.23}$$

where

$$\begin{aligned} H(\mathbf{A}|\mathbf{BC}) &= -\sum_{a\in\boldsymbol{a}}\sum_{b\in\boldsymbol{b}}\sum_{c\in\boldsymbol{c}} P(a\wedge b\wedge c)\log P(a|b\wedge c) \\ &\equiv S(\boldsymbol{P}(\boldsymbol{a})|\mathbf{P}(\boldsymbol{b}\wedge\boldsymbol{c})). \end{aligned} \tag{8.24}$$

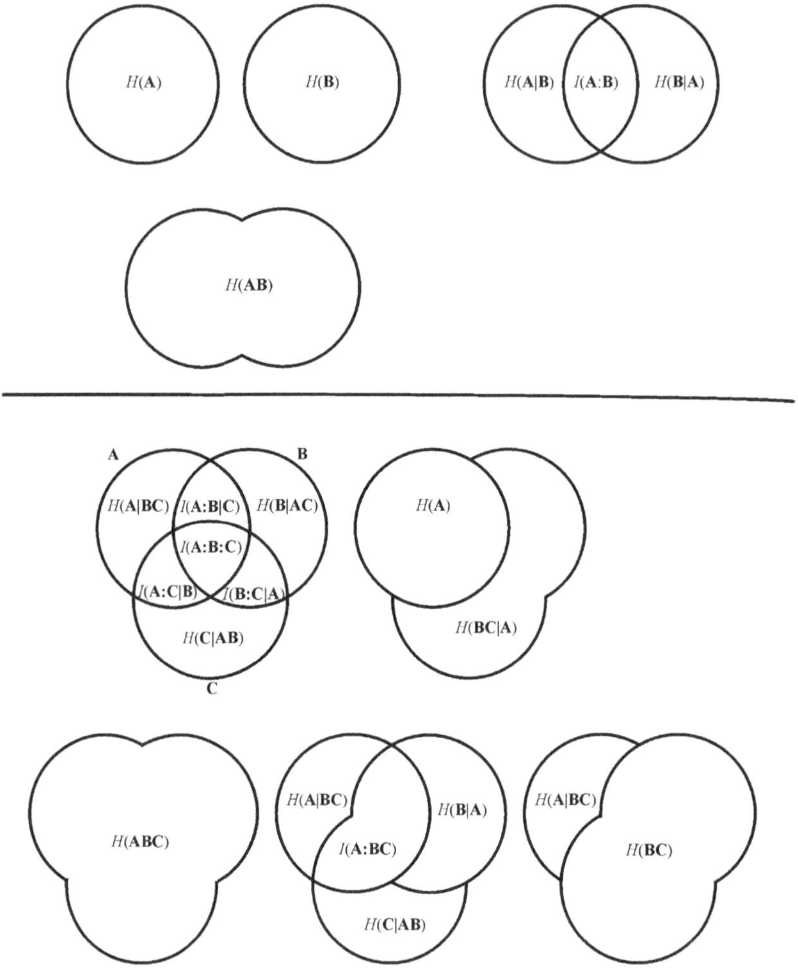

Fig. 8.5 A variety of the conditional-entropy and mutual-information descriptors of two (upper part) and three (lower part) dependent probability schemes. Here, the individual circles $H(\mathbf{A}) = S(\boldsymbol{P}(\boldsymbol{a}))$, $H(\mathbf{B}) = S(\boldsymbol{P}(\boldsymbol{b}))$, and $H(\mathbf{C}) = S(\boldsymbol{P}(\boldsymbol{c}))$ represent the Shannon entropies of the separate probability distributions, $H(\mathbf{AB}) = S(\mathbf{P}(\boldsymbol{a} \wedge \boldsymbol{b}))$, $H(\mathbf{B}|\mathbf{A}) = S(\boldsymbol{P}(\boldsymbol{b})|\boldsymbol{P}(\boldsymbol{a}))$, $I(\mathbf{A}:\mathbf{B}) = I(\boldsymbol{P}(\boldsymbol{a}):\boldsymbol{P}(\boldsymbol{b}))$, etc.

The mutual information quantity of (8.23) can be alternatively expressed in terms of other entropic quantities shown in the qualitative diagrams of Fig. 8.5. For example, it is seen to measure the sum of elementary amounts of the mutual information in three or two probability schemes, or it can be equivalently expressed in terms of the relevant entropies:

$$\begin{aligned} I(\mathbf{A}:\mathbf{BC}) &= I(\mathbf{A}:\mathbf{B}|\mathbf{C}) + I(\mathbf{A}:\mathbf{C}|\mathbf{B}) + I(\mathbf{A}:\mathbf{B}:\mathbf{C}) = I(\mathbf{A}:\mathbf{B}) + I(\mathbf{A}:\mathbf{C}|\mathbf{B}) \\ &= H(\mathbf{AB}) - H(\mathbf{A}|\mathbf{BC}) - H(\mathbf{B}|\mathbf{A}). \end{aligned} \tag{8.25}$$

Here, the mutual information in two probability distributions, given the third distribution, reads:

$$
\begin{aligned}
I(\mathbf{A}{:}\mathbf{B}|\mathbf{C}) = H(\mathbf{A}|\mathbf{C}) - H(\mathbf{A}|\mathbf{BC}) &= \sum_{a\in\boldsymbol{a}}\sum_{b\in\boldsymbol{b}}\sum_{c\in\boldsymbol{c}} P(a\wedge b\wedge c)\log\frac{P(a|b\wedge c)}{P(a|c)} \\
&= \sum_{a\in\boldsymbol{a}}\sum_{b\in\boldsymbol{b}}\sum_{c\in\boldsymbol{c}} P(a\wedge b\wedge c)\log\frac{P(a\wedge b|c)}{P(a|c)P(b|c)} \\
&= \sum_{a\in\boldsymbol{a}}\sum_{b\in\boldsymbol{b}}\sum_{c\in\boldsymbol{c}} P(a\wedge b\wedge c)\log\frac{P(a\wedge b\wedge c)p(c)}{P(a\wedge c)P(b\wedge c)}.
\end{aligned}
\tag{8.26}
$$

The mutual information in three probability schemes,

$$
\begin{aligned}
I(\mathbf{A}{:}\mathbf{B}{:}\mathbf{C}) &= I(\mathbf{A}{:}\mathbf{B}) - I(\mathbf{A}{:}\mathbf{B}|\mathbf{C}) = I(\mathbf{A}{:}\mathbf{C}) - I(\mathbf{A}{:}\mathbf{C}|\mathbf{B}) = I(\mathbf{B}{:}\mathbf{C}) - I(\mathbf{B}{:}\mathbf{C}|\mathbf{A}) \\
&= \sum_{a\in\boldsymbol{a}}\sum_{b\in\boldsymbol{b}}\sum_{c\in\boldsymbol{c}} P(a\wedge b\wedge c)\log\frac{P(a\wedge b)P(a\wedge c)P(b\wedge c)}{p(a)p(b)p(c)P(a\wedge b\wedge c)} \\
&= H(\mathbf{A}) + H(\mathbf{B}) + H(\mathbf{C}) - H(\mathbf{AB}) - H(\mathbf{AC}) - H(\mathbf{BC}) + H(\mathbf{ABC}),
\end{aligned}
\tag{8.27}
$$

which may assume negative values, is represented in Fig. 8.5 by the common area of three entropy circles. The above expression can be straightforwardly generalized for larger numbers of probability schemes. For example, in the *four*-scheme case as shown in Fig. 8.6, one finds (Abramson 1963):

$$
\begin{aligned}
I(\mathbf{A}{:}\mathbf{B}{:}\mathbf{C}{:}\mathbf{D}) &= I(\mathbf{A}{:}\mathbf{B}{:}\mathbf{C}) - I(\mathbf{A}{:}\mathbf{B}{:}\mathbf{C}|\mathbf{D}) \\
&= H(\mathbf{A}) + H(\mathbf{B}) + H(\mathbf{C}) + H(\mathbf{D}) - H(\mathbf{AB}) - H(\mathbf{AC}) \\
&\quad - H(\mathbf{BC}) - H(\mathbf{AD}) - H(\mathbf{BD}) - H(\mathbf{CD}) + H(\mathbf{ABC}) \\
&\quad + H(\mathbf{ABD}) + H(\mathbf{ACD}) + H(\mathbf{BCD}) - H(\mathbf{ABCD}).
\end{aligned}
\tag{8.28}
$$

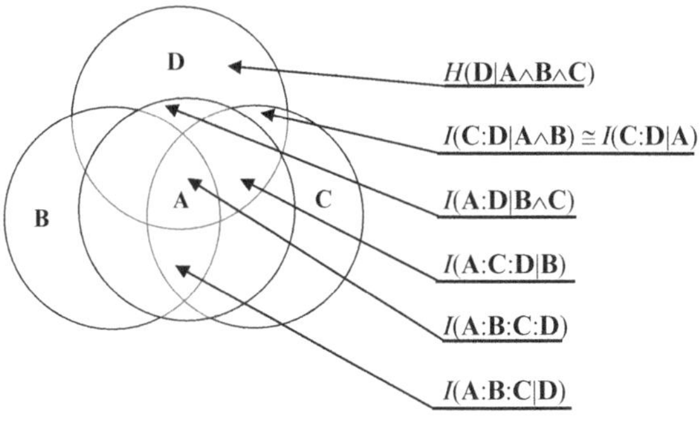

Fig. 8.6 General entropy/information diagrams of four dependent probability schemes (**A**, **B**, **C**, **D**). The entropies of the separate probability distributions are represented by circles, the circle overlap area denotes the mutual-information quantity, while the circle remainder, after removal of the overlap(s) with other circle(s), signifies the corresponding conditional entropy descriptor

The *three*-scheme conditional entropy of (8.24),

$$H(\mathbf{A}|\mathbf{BC}) \equiv H(\mathbf{A}|\mathbf{D}) = H(\mathbf{AD}) - H(\mathbf{D}), \tag{8.29}$$

where the "product" probability scheme $\mathbf{D} = \mathbf{BC} \equiv \mathbf{B}\wedge\mathbf{C} = [\boldsymbol{d}, \boldsymbol{P}(\boldsymbol{d})] \equiv [\boldsymbol{b}\wedge\boldsymbol{c}, \mathbf{P}(\boldsymbol{b}\wedge\boldsymbol{c})]$, measures the average noise in the $\boldsymbol{d} \rightarrow \boldsymbol{a}$ propagation of information in the underlying communication system. This entropy can be compared (Fig. 8.5) with the *two*-scheme relative entropy $H(\mathbf{A}|\mathbf{B}) = H(\mathbf{A}) - I(\mathbf{A}{:}\mathbf{B})$, the communication noise in the $\boldsymbol{b} \rightarrow \boldsymbol{a}$ probability scattering,

$$\begin{aligned} H(\mathbf{A}|\mathbf{BC}) &= -\sum_{a\in\boldsymbol{a}}\sum_{b\in\boldsymbol{b}}\sum_{c\in\boldsymbol{c}} P(a\wedge b\wedge c)\log\frac{P(a\wedge b\wedge c)}{P(b\wedge c)} \\ &= -\sum_{a\in\boldsymbol{a}}\sum_{b\in\boldsymbol{b}}\sum_{c\in\boldsymbol{c}} P(a\wedge b\wedge c)\log\frac{P(a\wedge b\wedge c)p(a)}{p(a)P(b\wedge c)} \\ &= H(\mathbf{A}) - I(\mathbf{A}{:}\mathbf{BC}) < H(\mathbf{A}|\mathbf{B}), \end{aligned} \tag{8.30}$$

since $I(\mathbf{A}{:}\mathbf{D}) > I(\mathbf{A}{:}\mathbf{B})$.

The information propagation from the product events in the $\boldsymbol{d} \rightarrow \boldsymbol{a}$ channel thus results in less noise (IT covalency) compared with that characterizing the $\boldsymbol{b} \rightarrow \boldsymbol{a}$ communication system. Accordingly, the opposite trend must be detected in the complementary mutual information quantity:

$$I(\mathbf{A}{:}\mathbf{BC}) = H(\mathbf{A}) - H(\mathbf{A}|\mathbf{D}) > I(\mathbf{A}{:}\mathbf{B}) = H(\mathbf{A}) - H(\mathbf{A}|\mathbf{B}). \tag{8.31}$$

Therefore, the *product* event scattering preserves in **D** a larger portion of the information content of **A**, compared with that conserved in scheme **B** alone, because of the extra information in **C** about **A** given **B**, as measured by $I(\mathbf{A}{:}\mathbf{C}|\mathbf{B})$. In other words, the $\boldsymbol{d} \rightarrow \boldsymbol{a}$ probability propagation results in a larger IT ionic (deterministic) component, compared with the $\boldsymbol{b} \rightarrow \boldsymbol{a}$ scattering.

To summarize, the $(\boldsymbol{d} = \boldsymbol{b}\wedge\boldsymbol{c}) \rightarrow \boldsymbol{a}$ communications appear less noisy (more deterministic) compared with the $\boldsymbol{b} \rightarrow \boldsymbol{a}$ probability propagation. Therefore, the a posteriori indeterminacy of **A** relative to **B**, $H(\mathbf{A}|\mathbf{B})$, exceeds the *double* a posteriori entropy $H(\mathbf{A}|\mathbf{BC})$, which measures the indeterminacy of **A** relative to (**B** and **C**), while the information about **A** provided by **B** alone, $I(\mathbf{A}{:}\mathbf{B})$, is lower than that provided by (**B** and **C**), $I(\mathbf{A}{:}\mathbf{BC})$, by the extra amount of information $I(\mathbf{A}{:}\mathbf{C}|\mathbf{B})$ about **A** provided by **C**, when **B** is known beforehand.

Let us finally examine the remaining *four*-scheme entropy/information descriptors shown in Fig. 8.6. In this qualitative diagram we have delineated areas representing various indices reflecting the entropy/information couplings between the four dependent probability schemes, $\{\mathbf{X} = (\boldsymbol{x}, \boldsymbol{P}(\boldsymbol{x}))\} = (\mathbf{A}, \mathbf{B}, \mathbf{C}, \mathbf{D})$, which combine the relevant sets of events, $\{\boldsymbol{x}\} = (\boldsymbol{a}, \boldsymbol{b}, \boldsymbol{c}, \boldsymbol{d})$, and their associated probabilities, $\{\boldsymbol{P}(\boldsymbol{x})\} = [\boldsymbol{P}(\boldsymbol{a}), \boldsymbol{P}(\boldsymbol{b}), \boldsymbol{P}(\boldsymbol{c}), \boldsymbol{P}(\boldsymbol{d})]$. For example, these partial events may refer to finding in the bond system of the molecule an electron on the basis functions $\{\boldsymbol{x} = \boldsymbol{\chi}_X\}$ contributed by molecular fragments $\boldsymbol{X} = \{X\}$. In these diagrams the Shannon entropies $\{H(\mathbf{X}) = H[\boldsymbol{P}(\boldsymbol{x})]\}$, are depicted by circles, the overlap areas reflect the average mutual information quantities, measuring the associated IT ionicities, while the corresponding circle remainders represent the complementary average conditional entropies, i.e., the corresponding IT covalencies.

Consider, e.g., the conditional entropy of **D**, given the simultaneous occurrence of events in **A**, **B**, and **C**,

$$\begin{aligned} H(\mathbf{D}|\mathbf{A}\wedge\mathbf{B}\wedge\mathbf{C}) &= S[\boldsymbol{P}(\boldsymbol{d})|\boldsymbol{P}(\boldsymbol{a})\wedge\boldsymbol{P}(\boldsymbol{b})\wedge\boldsymbol{P}(\boldsymbol{c})] \\ &= -\sum_{a\in\boldsymbol{a}}\sum_{b\in\boldsymbol{b}}\sum_{c\in\boldsymbol{c}}\sum_{d\in\boldsymbol{d}} P(a\wedge b\wedge c\wedge d)\log P(d|a\wedge b\wedge c). \end{aligned} \tag{8.32}$$

It represents the residual uncertainty in **D**, when the events of the three remaining schemes are known to have occurred already. In the molecular fragment scenario it can measure the overall IT covalency components of all bonds in fragment D, due to the basis functions of the remaining subsystems A, B, and C. Together with the complementary IT ionicity index,

$$\begin{aligned} I(\mathbf{A}\wedge\mathbf{B}\wedge\mathbf{C}{:}\mathbf{D}) &= I[\boldsymbol{P}(\boldsymbol{a})\wedge\boldsymbol{P}(\boldsymbol{b})\wedge\boldsymbol{P}(\boldsymbol{c}){:}\boldsymbol{P}(\boldsymbol{d})] \\ &= H(\mathbf{D}) - H(\mathbf{D}|\mathbf{A}\wedge\mathbf{B}\wedge\mathbf{C}) \\ &= \sum_{a\in\boldsymbol{a}}\sum_{b\in\boldsymbol{b}}\sum_{c\in\boldsymbol{c}}\sum_{d\in\boldsymbol{d}} P(a\wedge b\wedge c\wedge d)\log\frac{P(a\wedge b\wedge c\wedge d)}{P(d)P(a\wedge b\wedge c)} \\ &= \sum_{a\in\boldsymbol{a}}\sum_{b\in\boldsymbol{b}}\sum_{c\in\boldsymbol{c}}\sum_{d\in\boldsymbol{d}} P(a\wedge b\wedge c\wedge d)\log\frac{P(d|a\wedge b\wedge c)}{P(d)}, \end{aligned} \tag{8.33}$$

it gives rise to the overall (conditional) bond index of D due to interactions with A, B and C,

$$N(\mathbf{A}\wedge\mathbf{B}\wedge\mathbf{C};\mathbf{D}) = H(\mathbf{D}|\mathbf{A}\wedge\mathbf{B}\wedge\mathbf{C}) + I(\mathbf{A}\wedge\mathbf{B}\wedge\mathbf{C}{:}\mathbf{D}) = H(\mathbf{D}). \tag{8.34}$$

Using the property of the information *additivity* (Abramson 1963) one can alternatively express the amount of information of (8.33) in terms of more elementary mutual information quantities, which generate the underlying "overlap" area between the envelope of three circles (**A**, **B**, **C**) and the entropy circle of **D** (Fig. 8.6):

$$\begin{aligned} I(\mathbf{A}\wedge\mathbf{B}\wedge\mathbf{C}{:}\mathbf{D}) &= I(\mathbf{A}{:}\mathbf{D}) + I(\mathbf{B}{:}\mathbf{D}|\mathbf{A}\wedge\mathbf{C}) + I(\mathbf{C}{:}\mathbf{D}|\mathbf{A}\wedge\mathbf{B}) \\ &= I(\mathbf{A}{:}\mathbf{D}|\mathbf{B}\wedge\mathbf{C}) + I(\mathbf{B}{:}\mathbf{D}|\mathbf{A}\wedge\mathbf{C}) + I(\mathbf{C}{:}\mathbf{D}|\mathbf{A}\wedge\mathbf{B}) \\ &\quad + I(\mathbf{A}{:}\mathbf{C}{:}\mathbf{D}|\mathbf{B}) + I(\mathbf{A}{:}\mathbf{B}{:}\mathbf{D}|\mathbf{C}) + I(\mathbf{A}{:}\mathbf{B}{:}\mathbf{C}{:}\mathbf{D}). \end{aligned} \tag{8.35}$$

To conclude this section, we summarize the key expressions for the complementary communication-*noise* (IT covalency) and information-*flow* (IT ionicity) descriptors involving several probability distributions. For *two* dependent probability schemes, these basic quantities read:

$$H(\mathbf{B}|\mathbf{A}) = -\sum_{a\in\boldsymbol{a}}\sum_{b\in\boldsymbol{b}} P(a\wedge b)\log P(b|a) = H(\mathbf{A}\wedge\mathbf{B}) - H(\mathbf{A}), \tag{8.36}$$

$$\begin{aligned} I(\mathbf{A}{:}\mathbf{B}) &= \sum_{a\in\boldsymbol{a}}\sum_{b\in\boldsymbol{b}} P(a\wedge b)\log\frac{P(b|a)}{P(b)} = H(\mathbf{B}) - H(\mathbf{B}|\mathbf{A}) \\ &= \sum_{a\in\boldsymbol{a}}\sum_{b\in\boldsymbol{b}} P(a\wedge b)\log\frac{P(a|b)}{P(a)} = H(\mathbf{A}) - H(\mathbf{A}|\mathbf{B}) \\ &= H(\mathbf{A}) + H(\mathbf{B}) - H(\mathbf{A}\wedge\mathbf{B}). \end{aligned} \tag{8.37}$$

The associated *three*-scheme quantities have been similarly expressed by the relevant entropies and probabilities:

$$\begin{aligned} I(\mathbf{A}{:}\mathbf{B}{:}\mathbf{C}) &= I(\mathbf{A}{:}\mathbf{B}) - I(\mathbf{A}{:}\mathbf{B}|\mathbf{C}) \\ &= H(\mathbf{A}) + H(\mathbf{B}) + H(\mathbf{C}) - H(\mathbf{A}\wedge\mathbf{B}) - H(\mathbf{A}\wedge\mathbf{C}) \\ &\quad - H(\mathbf{B}\wedge\mathbf{C}) + H(\mathbf{A}\wedge\mathbf{B}\wedge\mathbf{C}) \\ &= \sum_{a\in\boldsymbol{a}}\sum_{b\in\boldsymbol{b}}\sum_{c\in\boldsymbol{c}} P(a\wedge b\wedge c)\log\frac{P(c|a)P(c|b)}{P(c)P(c|a\wedge b)}, \end{aligned} \tag{8.38}$$

where:

$$\begin{aligned} I(\mathbf{A}{:}\mathbf{B}|\mathbf{C}) &= H(\mathbf{A}|\mathbf{C}) - H(\mathbf{A}|\mathbf{B}\wedge\mathbf{C}) \\ &= \sum_{a\in\boldsymbol{a}}\sum_{b\in\boldsymbol{b}}\sum_{c\in\boldsymbol{c}} P(a\wedge b\wedge c)\log\frac{P(a|b\wedge c)}{P(a|c)}. \end{aligned} \tag{8.39}$$

Finally, for *four* dependent probability distributions one finds:

$$\begin{aligned} I(\mathbf{A}{:}\mathbf{B}{:}\mathbf{C}{:}\mathbf{D}) &= I(\mathbf{A}{:}\mathbf{B}{:}\mathbf{C}) - I(\mathbf{A}{:}\mathbf{B}{:}\mathbf{C}|\mathbf{D}) \\ &= \sum_{a\in\boldsymbol{a}}\sum_{b\in\boldsymbol{b}}\sum_{c\in\boldsymbol{c}}\sum_{d\in\boldsymbol{d}} P(a\wedge b\wedge c\wedge d)\log\frac{P(d|a\wedge b\wedge c)P(d|a)P(d|b)P(d|c)}{P(d)P(d|a\wedge b)P(d|b\wedge c)P(d|a\wedge c)}, \end{aligned} \tag{8.40}$$

$$I(\mathbf{A}{:}\mathbf{B}{:}\mathbf{C}|\mathbf{D}) = I(\mathbf{A}{:}\mathbf{B}|\mathbf{D}) - I(\mathbf{A}{:}\mathbf{B}|\mathbf{C}\wedge\mathbf{D}), \tag{8.41}$$

where the conditional mutual information measure

$$\begin{aligned} I(\mathbf{A}{:}\mathbf{B}|\mathbf{C}\wedge\mathbf{D}) &= H(\mathbf{A}|\mathbf{C}\wedge\mathbf{D}) - H(\mathbf{A}|\mathbf{B}\wedge\mathbf{C}\wedge\mathbf{D}) \\ &= \sum_{a\in\boldsymbol{a}}\sum_{b\in\boldsymbol{b}}\sum_{c\in\boldsymbol{c}}\sum_{d\in\boldsymbol{d}} P(a\wedge b\wedge c\wedge d)\log\frac{P(a|b\wedge c\wedge d)}{P(a|c\wedge d)}. \end{aligned} \tag{8.42}$$

8.7 Variational Principles

Let us briefly summarize some general properties of the Shannon entropy, which might be expected to characterize a reasonable measure of uncertainty/information. For the simplest case of a finite (discrete) probability distribution $\boldsymbol{p} = \{p_i\}$, $S(\boldsymbol{p}) = -\sum_i p_i \log p_i = 0$ if and only if one of the probabilities in $\boldsymbol{p}$ equals one (sure event) and all the others exactly vanish. Indeed, in this case the outcome of the experiment carries no uncertainty, since the result can be predicted beforehand with complete certainty. In all other cases $S(\boldsymbol{p}) > 0$. For the fixed number of n outcomes the probability scheme exhibiting the most uncertainty, i.e., the largest value of the entropy, is the one with equally probable outcomes: $p_i = 1/n$.

This "maximum smoothness" property of the optimum probability distribution is summarized by the (constrained) variational principle of Shannon and Jaynes (1957a, b, 1985), called the *maximum entropy* (ME) rule: given data $F^0 = \{F_i^0\}$, the probability distribution $\boldsymbol{p}$ [or $p(\boldsymbol{r})$], which describes these constraints most objectively, must maximize the entropy $S(\boldsymbol{p})$ (or $S[p]$) with respect to all $\boldsymbol{p}$s (ps) satisfying $\boldsymbol{F}^0$. It implies the associated Euler–Lagrange variational principles containing the relevant Lagrange multipliers $\{\lambda_i\}$ enforcing the auxiliary conditions, which should be eventually determined from the constraint values themselves:

$$\delta\left\{S(\boldsymbol{p}) - \sum_i \lambda_i F_i(\boldsymbol{p})\right\} = 0 \quad \text{or} \quad \delta\left\{S[p] - \sum_i \lambda_i F_i[p]\right\} = 0. \tag{8.43}$$

Thus, the entropy maximization results in the most "evenly spread" of all probability distributions consistent with the imposed constraints. This principle represents a device allowing one to assimilate in the optimum distribution the physical information contained in the constraints in the most unbiased manner possible.

The ME principle represents a powerful method for determining the equilibrium probability distributions of physical systems or processes, given some information about them. It has been successfully used to reconstruct the equilibrium thermodynamics, asserting at the same time the principle applicability to far more general problems. The ME rule involves the maximization of Shannon's entropy subject to the imposed constraints and provides a unifying principle in statistical physics, allowing a construction of thermodynamic laws based upon statistical inference and the unbiased assimilation of available data. In other words, the entropy becomes the starting point in the construction of the statistical mechanics, instead of being identified in the end as a byproduct of the energy-centered arguments. The Shannon–Jaynes principle is the uniquely correct method for inductive inference, when new information is given in the form of the statistical expectation values.

Clearly, this optimum assimilation of the physical information contained in the constraints can also be accomplished using any other admissible information measure, e.g., the local Fisher information or the information distance (entropy deficiency) quantities introduced in this chapter. For example, Kullback has proposed the generalization of the ME rule to problems involving the reference probability distribution(s), which can be called the *Maximum-Resemblance* principle or the *Minimum Entropy Deficiency* (MED) rule, as another unbiased procedure for assimilating in the optimum probability distribution the constraints of available experimental data, and thus its highest similarity to the reference distribution afforded by the constraints.

Suppose that an experiment has been performed, which yields the expectation values of several functions of the discrete probability distribution $\boldsymbol{p}$, $\{F_i(p) = F_i^0\}$, or functionals of the continuous probability density $p(\boldsymbol{r})$, $\{F_i[p] = F_i^0\}$. The Kullback principle asserts that the task of assimilating this information in the optimum probability distribution, which is to resemble the reference distribution as much as possible, can be accomplished by minimizing the entropy deficiency subject to these constraints:

$$\delta\left\{\Delta S(\boldsymbol{p}|\boldsymbol{p}^0) - \sum_i \lambda_i F_i(\boldsymbol{p})\right\} = 0 \quad \text{or} \quad \delta\left\{\Delta S[p|p^0] - \sum_i \lambda_i F_i[p]\right\} = 0. \tag{8.44}$$

Alternatively, the Fisher measure of information, which reflects the overall "*order*" ("sharpness") of the probability distribution, can be used in the variational procedure of assimilating the available data in the probability distribution, e.g.,

$$\delta\left\{I[p] - \sum_i \lambda_i F_i[p]\right\} = 0. \tag{8.45}$$

These variational procedures can thus be regarded as examples of the *Extreme Physical Information* (EPI) principle (Frieden 2000), which is closely related to the theory of measurement (Fisher 1959). The maximum Shannon entropy and the minimum Fisher information principles may have coincident solutions for the given set of constraints. The Maxwell–Boltzmann velocity dispersion law of the equilibrium statistical thermodynamics results from both information principles, with the Fisher information principle generating additional nonequilibrium solutions as subsidiary minima, with the absolute minimum being attained by the Maxwell–Boltzmann solution. Thus, the coincidence is observed at the equilibrium level of statistical mechanics, with the Shannon ME rule being unable to cover the nonequilibrium phenomena.

References

Abramson N (1963) Information theory and coding. McGraw-Hill, New York
Ash RB (1965) Information theory. Interscience, New York
Aslangul C, Constanciel R, Daudel R, Kottis P (1972) Adv Quant Chem 6:94
Ayers PW (2000b) J Chem Phys 113:10886
Becke AD, Edgecombe KE (1990) J Chem Phys 92:5397

Bernstein RB (1982) Chemical dynamics via molecular beam and laser techniques. Clarendon, Oxford
Brillouin L (1956) Science and information theory. Academic, New York
Callen HB (1962) Thermodynamics: an introduction to the physical theories of equilibrium thermostatics and irreversible thermodynamics. Wiley, New York
Daudel R (1969) The fundamentals of theoretical chemistry. Pergamon, Oxford
Daudel R (1974) The quantum theory of the chemical bond. D. Reidel, Dordrecht
Dirac PAM (1967) The principles of quantum mechanics. Clarendon, Oxford
Esquivel RO, Rodriquez AL, Sagar RP, Hõ M, Smith VH Jr (1996) Phys Rev A 54:259
Fisher RA (1922) Philos Trans R Soc A (Lond) 222:309
Fisher RA (1925) Proc Camb Philos Soc 22:700
Fisher RA (1959) Statistical methods and scientific inference, 2nd edn. Oliver and Boyd, London
Frieden BR (2000) Physics from the Fisher information – a unification. Cambridge University Press, Cambridge
Frieden BR, Soffer BH (2010) Weighted fisher informations, their derivation and use. Phys Lett A (in press)
Gadre SR (1984) Phys Rev A 30:620
Gadre SR (2002) In: Sen KD (ed) Reviews of modern quantum chemistry: a celebration of the contributions of Robert G. Parr, vol 1. World Scientific, Singapore, p 108
Gadre SR, Bendale RD (1985) Int J Quantum Chem 28:311
Gadre SR, Sears SB (1979) J Chem Phys 71:4321
Gadre SR, Bendale RD, Gejii SP (1985a) Chem Phys Lett 117:138
Gadre SR, Sears SB, Chakravorty SJ, Bendale RD (1985b) Phys Rev A 32:2602
Gopinathan MS, Jug K (1983) Theor Chim Acta (Berl.) 63:497, 511
Hartley RVL (1928) Bell Syst Tech J 7:535
Hirshfeld FL (1977) Theor Chim Acta (Berl) 44:129
Jaynes ET (1957a) Phys Rev 106:620
Jaynes ET (1957b) Phys Rev 108:171
Jaynes ET (1985) In: Smith CR, Grandy WT (eds) Maximum entropy and Bayesian methods in inverse problems. Reidel, Dordrecht
Jug K, Gopinathan MS (1990) In: Maksić ZB (ed) Theoretical models of chemical bonding, vol 2. Springer, Heidelberg, p 77
Khinchin AI (1957) Mathematical foundations of the information theory. Dover, New York
Kullback S (1959) Information theory and statistics. Wiley, New York
Kullback S, Leibler RA (1951) Ann Math Stat 22:79
Mathai AM, Rathie PM (1975) Basic concepts in information theory and statistics: axiomatic foundations and applications. Wiley, New York
Mayer I (1983) Chem Phys Lett 97:270
Mayer I (1985) Theor Chim Acta (Berl.) 67:315
Nagy Á, Parr RG (2000) J Mol Struct THEOCHEM 501:101
Nagy Á, Parr RG (1994) Proc Indian Acad Sci Chem Sci 106:217
Nagy Á, Parr RG (1996) Int J Quantum Chem 58:323
Nalewajski RF (2000c) J Phys Chem A 104:11940
Nalewajski RF (2002a) Phys Chem Chem Phys 4:1710
Nalewajski RF (2002b) Int J Mol Sci 3:237
Nalewajski RF (2002c) Acta Chim Phys Debr 34–35:131
Nalewajski RF (2003b) J Phys Chem A 107:3792
Nalewajski RF (2003d) Chem Phys Lett 375:196
Nalewajski RF (2004a) Ann Phys (Leipzig) 13:201
Nalewajski RF (2004b) Chem Phys Lett 386:265
Nalewajski RF (2004c) Mol Phys 102:531
Nalewajski RF (2004d) Mol Phys 102:547
Nalewajski RF (2004e) Struct Chem 15:395

Nalewajski RF (2005a) Theor Chem Acc 114:4
Nalewajski RF (2005b) Mol Phys 103:451
Nalewajski RF (2005c) J Math Chem 38:43
Nalewajski RF (2006a) Mol Phys 104:365
Nalewajski RF (2006b) Mol Phys 104:493
Nalewajski RF (2006c) Mol Phys 104:1977
Nalewajski RF (2006d) Mol Phys 104:2533
Nalewajski RF (2006e) Mol Phys 104:3339
Nalewajski RF (2006f) Adv Quant Chem 51:235
Nalewajski RF (2006g) Information theory of molecular systems. Elsevier, Amsterdam
Nalewajski RF (2006h) Mol Phys 104:255
Nalewajski RF (2007) J Phys Chem A 111:4855
Nalewajski RF (2008a) J Math Chem 43:265
Nalewajski RF (2008b) J Math Chem 43:780
Nalewajski RF (2008e) Int J Quantum Chem 108:2230
Nalewajski RF (2009a) In: Chattaraj PK (ed) Chemical reactivity theory: a density functional view. Taylor and Francis, London, p 453
Nalewajski RF (2009b) J Math Chem 45:709
Nalewajski RF (2009c) J Math Chem 45:776
Nalewajski RF (2009d) J Math Chem 45:1041
Nalewajski RF (2009e) Int J Quantum Chem 109:425
Nalewajski RF (2009f) Int J Quantum Chem 109:2495
Nalewajski RF (2009g) Adv Quant Chem 56:217
Nalewajski RF (2010a) J Math Chem 47:667
Nalewajski RF (2010b) J Math Chem 47:692
Nalewajski RF (2010c) J Math Chem 47:709
Nalewajski RF (2010d) J Math Chem 47:808
Nalewajski RF (2010f) Information origins of the chemical bond. Nova Science, New York
Nalewajski RF (2010h) Information perspective on molecular electronic structure. In: Mathematical chemistry. Nova Science, New York (in press)
Nalewajski RF (2010i) Information tools for probing chemical bonds. In: Putz M (ed) Chemical information and computation challenges in 21st: a celebration of 2011 international year of chemistry. Nova Science, New York (in press)
Nalewajski RF, Broniatowska E (2003a) J Phys Chem A 107:6270
Nalewajski RF, Broniatowska E (2005) Int J Quantum Chem 101:349
Nalewajski RF, Broniatowska E (2007) Theor Chem Acc 117:7
Nalewajski RF, Jug K (2002) In: Sen KD (ed) Reviews of modern quantum chemistry: a celebration of the contributions of Robert G. Parr, vol 1. World Scientific, Singapore, p 148
Nalewajski RF, Mrozek J (1994) Int J Quantum Chem 51:187
Nalewajski RF, Mrozek J (1996) Int J Quantum Chem 57:377
Nalewajski RF, Parr RG (2000) Proc Natl Acad Sci USA 97:8879
Nalewajski RF, Parr RG (2001) J Phys Chem A 105:7391
Nalewajski RF, Köster AM, Jug K (1993) Theor Chim Acta (Berl.) 85:463
Nalewajski RF, Formosinho SJ, Varandas AJC, Mrozek J (1994a) Int J Quantum Chem 52:1153
Nalewajski RF, Korchowiec J, Michalak A (1994b) Proc Indian Acad Sci Chem Sci 106:353
Nalewajski RF, Korchowiec J, Michalak A (1996a) Top Curr Chem 183:25
Nalewajski RF, Mrozek J, Mazur G (1996b) Can J Chem 100:1121
Nalewajski RF, Mrozek J, Michalak A (1997) Int J Quantum Chem 61:589
Nalewajski RF, Świtka E, Michalak A (2002) Int J Quantum Chem 87:198
Nalewajski RF, Köster AM, Escalante S (2005) J Phys Chem A 109:10038
Nalewajski RF, de Silva P, Mrozek J (2010a) Kinetic-energy/Fisher-information indicators of chemical bonds. In: Wang A, Wesołowski TA (eds) Kinetic energy functional. World Scientific, Singapore (in press)

Nalewajski RF, de Silva P, Mrozek J (2010b) J Mol Struct THEOCHEM 954:57
Nalewajski RF, Szczepanik D, Mrozek J (2011) Adv Quant Chem 68:1
Parr RG, Ayers PW, Nalewajski RF (2005) J Phys Chem A 109:3957
Pfeiffer PE (1978) Concepts of probability theory. Dover, New York
Savin A, Nesper R, Wengert S, Fässler TF (1997) Angew Chem Int Ed Engl 36:1808
Sears SB (1980) PhD Thesis, The University of North Carolina, Chapel Hill
Sears SB, Parr RG, Dinur U (1980) Israel J Chem 19:165
Shannon CE (1948) Bell Syst Technol J 27:379, 623
Shannon CE, Weaver W (1949) The mathematical theory of communication. University of Illinois, Urbana
Silvi B, Savin A (1994) Nature 371:683
Tisza L (1977) Generalized thermodynamics. MIT Press, Cambridge
von Weizsäcker CF (1935) Z Phys 96:431
Wiberg KB (1968) Tetrahedron 24:1083

Chapter 9
Schrödinger Equations from Information Principles

Abstract The amplitude and phase contributions to the Fisher-information density and its current are examined. Its local balance is probed and the source term in the associated continuity equation is established. The physically constrained information principles generating the stationary and time-dependent Schrödinger equations are summarized and IT variational rule for the adiabatic separation of the fast (electronic) and slow (nuclear) motions in molecular systems is given. The Kohn–Sham equations of the computational DFT are derived from the relevant EPI principle using the Fisher measure of the information content.

9.1 Fisher-Information Density and Its Current

For simplicity, let us consider a single, spinless particle of mass $\mu = m_e$, characterized by the Hamiltonian (energy) operator

$$\hat{\mathrm{H}}(\boldsymbol{r}) = -(\hbar^2/2\mu)\nabla^2 + v(\boldsymbol{r}) = \hat{\mathrm{T}}(\boldsymbol{r}) + \hat{\mathrm{V}}(\boldsymbol{r}), \tag{9.1}$$

with $\hat{\mathrm{V}}(\boldsymbol{r}) \equiv v(\boldsymbol{r})$ standing for the (multiplicative) potential energy operator, e.g., of the nuclear–electron attraction energy, $v = v_{ne}$, and the Laplacian operator $\Delta = \nabla^2$. In the Schrödinger mechanics, the system quantum state is specified by the complex wave-function in the position representation,

$$\psi(\boldsymbol{r}, t) = R(\boldsymbol{r}, t)\ \exp[i\Phi(\boldsymbol{r}, t)], \tag{9.2}$$

where the real functions $R(\boldsymbol{r}, t)$ and $\Phi(\boldsymbol{r}, t)$ describe the probability-amplitude and the phase of ψ, respectively. Its physical interpretation has been established through the particle-probability density $p(\boldsymbol{r}, t)$ defined by the amplitude $R(\boldsymbol{r}, t)$ alone,

$$p(\boldsymbol{r}, t) = |\psi(\boldsymbol{r}, t)|^2 = \psi^*(\boldsymbol{r}, t)\ \psi(\boldsymbol{r}, t) = R^2(\boldsymbol{r}, t), \qquad \int p(\boldsymbol{r}, t)\ d\boldsymbol{r} = 1, \tag{9.3}$$

R.F. Nalewajski, *Perspectives in Electronic Structure Theory*,
DOI 10.1007/978-3-642-20180-6_9,

and the *probability-current* density:

$$\begin{aligned} \boldsymbol{j} &= \frac{\hbar}{2\mu i}[\psi^*\nabla\psi - \psi\nabla\psi^*] = \frac{\hbar}{\mu}\mathrm{Im}(\psi^*\nabla\psi) \equiv \boldsymbol{j}[\psi] \equiv p\boldsymbol{V} \\ &= \frac{\hbar}{\mu}p\nabla\Phi = p\nabla\left[\frac{\hbar\Phi}{\mu}\right]. \end{aligned} \tag{9.4a}$$

The preceding equation expresses the local speed of the probability "fluid," $\boldsymbol{V}(\boldsymbol{r}, t) = \boldsymbol{j}(\boldsymbol{r}, t)/p(\boldsymbol{r}, t)$, in terms of the gradient of the phase part of the system wave-function. The probability-current distribution is seen to explore the gradient of the phase function weighted by the local probability density. The latter is determined solely by the density-amplitude, while the speed of the probability fluid is proportional to the gradient of the phase part of the wave-function. We recall that the probability-current density can be regarded as the expectation value of the associated (Hermitian) quantum-mechanical operator (3.129 and 7.380):

$$\boldsymbol{j}(\boldsymbol{r}) = \langle\psi|\hat{\mathbf{j}}(\boldsymbol{r})|\psi\rangle, \quad \hat{\mathbf{j}}(\boldsymbol{r}) = \frac{\hbar}{2\mu i}\sum_{k=1}^{N}[\nabla_k\delta(\boldsymbol{r}_k - \boldsymbol{r}) + \delta(\boldsymbol{r}_k - \boldsymbol{r})\nabla_k]. \tag{9.4b}$$

For such complex probability-amplitudes of quantum mechanics the classical Fisher information of (8.5), $I[p] = I[R]$, is generalized into the functional of (8.6) with its integrand then defining the associated density $f(\boldsymbol{r}, t)$ of the intrinsic accuracy measure:

$$I[\psi] = 4\int|\nabla\psi(\boldsymbol{r},t)|^2 d\boldsymbol{r} = 4\int\nabla\psi^*(\boldsymbol{r},t)\cdot\nabla\psi(\boldsymbol{r},t)\,d\boldsymbol{r} \equiv \int f(\boldsymbol{r},t)\,d\boldsymbol{r}. \tag{9.5}$$

Its multicomponent analog, described by the vector of the component wave-functions $\boldsymbol{\psi} = \{\psi_n\}$, similarly reads:

$$I[\boldsymbol{\psi}] = 4\sum\nolimits_n\int|\nabla\psi_n(\boldsymbol{r},t)|^2\,d\boldsymbol{r} = \sum\nolimits_n I[\psi_n] \equiv \sum\nolimits_n\int f_n(\boldsymbol{r},t)\,d\boldsymbol{r}. \tag{9.6}$$

In what follows we shall examine some properties of this generalized information measure (Nalewajski 2008e, 2010f). A somewhat different, dimensionless approach has recently been proposed by Frieden and Soffer (2010).

We first observe that, by a straightforward integration by parts, the Schrödinger functional for the expectation value of the kinetic energy operator can be interpreted as being proportional to the average Fisher information contained in the wave-function ψ :

$$\langle T\rangle = \int\psi^*\hat{\mathrm{T}}\,\psi\,d\boldsymbol{r} \equiv \langle\psi|\hat{\mathrm{T}}|\psi\rangle = \frac{\hbar^2}{2\mu}\int\nabla\psi^*\cdot\nabla\psi\,d\boldsymbol{r} = \frac{\hbar^2}{8\mu}I[\psi]. \tag{9.7}$$

In terms of the probability-amplitude and phase parts of the system wave-function, this Fisher-information functional reads:

$$I[\psi] = 4\int \{(\nabla R)^2 + p(\nabla\Phi)^2\}\, d\boldsymbol{r} \equiv I[R] + I[R,\Phi]$$
$$= \int \frac{(\nabla p)^2}{p} d\boldsymbol{r} + \frac{4\mu^2}{\hbar^2}\int \frac{\boldsymbol{j}^2}{p} d\boldsymbol{r} \equiv I[p] + I[\boldsymbol{j}] \equiv I[p,\boldsymbol{j}]. \tag{9.8}$$

Therefore, this generalized measure of information becomes identical with the classical Fisher functional $I[p]$ for the *stationary* quantum states characterized by the time-independent probability amplitude $R = \varphi(\boldsymbol{r})$ and the position-independent phase $\Phi = -\omega t$:

$$\psi(\boldsymbol{r},t) = \varphi(\boldsymbol{r})\ \exp[-iEt/\hbar] \equiv \varphi(\boldsymbol{r})\ \exp(-i\omega t). \tag{9.9}$$

These eigenfunctions of the Hamiltonian operator,

$$\hat{\mathrm{H}}\psi = E\psi \qquad \text{or} \qquad \hat{\mathrm{H}}\varphi = E\varphi, \tag{9.10}$$

correspond to the sharply specified system energy E and frequency ω. Indeed, such states imply the vanishing current, $\boldsymbol{j} = 0$, and hence also $I[\boldsymbol{j}] = 0$.

The generalized functional $I[\psi]$ is seen to symmetrically probe the gradient content of both parts of the complex wave-function, with the gradient of the probability-amplitude determining the classical Fisher measure of information contained in the probability distribution p, with the latter also providing the weight in the local contribution from the phase gradient, due to the probability-current density,

$$I[p] \equiv I^F[R] = \int \left(\frac{\nabla p}{\sqrt{p}}\right)^2 d\boldsymbol{r} \equiv \int (\bar{\nabla} p)^2 d\boldsymbol{r}, \tag{9.11}$$

$$I[\boldsymbol{j}] = \int (2R\nabla\Phi)^2 d\boldsymbol{r} \equiv I[R,\Phi] = \int \left(\frac{2\mu\boldsymbol{j}}{\hbar\sqrt{p}}\right)^2 d\boldsymbol{r} \equiv \int \bar{\boldsymbol{j}}^2 d\boldsymbol{r}. \tag{9.12}$$

It follows from these expressions that the classical Fisher information measures the "length" of the "reduced" gradient of the probability density, $\bar{\nabla} p(\boldsymbol{r})$, while the other contribution represents the length of the reduced vector of the probability-current density $\bar{\boldsymbol{j}}(\boldsymbol{r})$.

These two-information contributions can be alternatively expressed in terms of the real and imaginary parts of the gradient of the wave-function logarithm, $\nabla \ln\psi = (\nabla\psi)/\psi$:

$$I[R] = 4\int p[\mathrm{Re}(\nabla\ln\psi)]^2 d\boldsymbol{r} \quad \text{and} \quad I[R,\Phi] = 4\int p[\mathrm{Im}(\nabla\ln\psi)]^2 d\boldsymbol{r}. \tag{9.13}$$

Thus, these complementary components of the generalized Fisher information have a common interpretation in quantum mechanics as the p-weighted averages of the gradient content of the real and imaginary parts of the logarithmic gradient of the system wave-function. As such they represent a natural (complex) generalization of the classical (*real*) information concept of (8.5).

Of interest also is the information density per electron:

$$\tilde{f} \equiv \frac{f}{p} = \left(\frac{\nabla p}{p}\right)^2 + \left(\frac{2\mu \boldsymbol{j}}{\hbar p}\right)^2 \equiv (\tilde{\nabla} p)^2 + (\tilde{\boldsymbol{j}})^2 \geq 0. \tag{9.14}$$

It is seen to be generated by squares of the local values of the related quantities per electron: the probability gradient $(\tilde{\nabla} p)^2$ and the current density $(\tilde{\boldsymbol{j}})^2$. This expression emphasizes the basic equivalence of the roles played by the probability density and its current in shaping the resultant value of the generalized Fisher-information density in quantum mechanics.

9.2 Continuity Equations and Information Source

For the IT interpretation of electron redistributions in molecules the concept of the information *flow* is paramount. The Hamiltonian operator $\hat{\mathrm{H}}$ determines the time-evolution of the system wave-function through the Schrödinger equation

$$i\hbar \frac{\partial \psi}{\partial t} = \hat{\mathrm{H}} \psi, \tag{9.15}$$

which implies the conservation of the wave-function normalization in time:

$$\frac{d}{dt} \int \psi^*(\boldsymbol{r}, t)\, \psi(\boldsymbol{r}, t)\, d\boldsymbol{r} = \frac{d}{dt} \int p(\boldsymbol{r}, t) d\boldsymbol{r} = 0. \tag{9.16}$$

The probability density and its current together determine the local balance of the probability distribution in quantum mechanics, which is summarized by the familiar *continuity equation*:

$$\frac{\partial p}{\partial t} = -\nabla \cdot \boldsymbol{j} \quad \text{or} \quad \frac{dp}{dt} \equiv \dot{p} = \frac{\partial p}{\partial t} + \nabla \cdot \boldsymbol{j} = 0. \tag{9.17}$$

The first form of the preceding equation expresses the fact that for the norm-conserving evolution of the system wave-function the local rate of change of the probability density is determined by the probability density leaving that location, so that the local net production (source) $\dot{p}$ of the probability density identically vanishes, as expressed by the second form of this equation. Indeed, the particles are neither created nor destroyed in (closed) molecular systems.

Moreover, by Green's theorem, the volume integral representing the overall outflow of the particle-probability from the volume V enclosed by the surface S can be expressed as the surface integral measuring the global flux through S:

$$\int_V \nabla \cdot \boldsymbol{j}\, d\boldsymbol{r} = \int_S \boldsymbol{j} \cdot d\boldsymbol{S}, \tag{9.18}$$

with $d\boldsymbol{S} = dS\boldsymbol{n}$ standing for the normal vector along the unit vector $\boldsymbol{n}$ of the surface element dS. By using (9.18), the generalized Fisher information of (9.5) can be thus expressed as the following difference between the relevant surface and volume integrals,

$$I[\psi] = 4\int \nabla\psi^* \cdot \nabla\psi\, d\boldsymbol{r} = 4\left[\int_S \psi^*(\nabla\psi)\cdot d\boldsymbol{S} - \int_V \psi^*\Delta\psi d\boldsymbol{r}\right], \tag{9.19}$$

with $V \to \infty$ corresponding to a very large volume defined by the closed surface $S \to \infty$. Subtracting from this equation its complex conjugate and taking into account the Hermitian property of the Laplacian then gives the conservation of the overall probability in V:

$$\int_S \boldsymbol{j} \cdot d\boldsymbol{S} = 0. \tag{9.20}$$

In the *Time-Dependent Density Functional Theory* (TDDFT) (see Sect. 7.9), for the given initial state $\psi(\boldsymbol{r}, t_0)$ the time-dependent density determines the external potential uniquely up to within an additive, purely time-dependent function. The latter in turn determines the time-dependent wave-function up to within a purely time-dependent phase: $\Psi(t) = \exp[-i\alpha(t)]\,\tilde{\Psi}[\rho; t]$. This phase ambiguity cancels out in the expectation values of physical quantities, which for the given initial state can be thus regarded as unique functionals of the system time-dependent probability distribution $p(\boldsymbol{r}, t)$, or of the associated electron density $\rho(\boldsymbol{r}, t) = Np(\boldsymbol{r}, t)$, e.g.,

$$\boldsymbol{j}(\boldsymbol{r}, t) = \langle\Psi(t)|\hat{\mathbf{j}}(\boldsymbol{r})|\Psi(t)\rangle = \boldsymbol{j}[\rho; \boldsymbol{r}, t]. \tag{9.21}$$

From the expression for the time-dependence of the quantum-mechanical expectation values (3.117),

$$\frac{\partial}{\partial t}\langle\Psi(t)|\hat{\mathrm{Q}}(t)|\Psi(t)\rangle = \langle\Psi(t)|\frac{\partial\hat{\mathrm{Q}}(t)}{\partial t} - \frac{i}{\hbar}[\hat{\mathrm{Q}}(t), \hat{\mathrm{H}}(t)]|\Psi(t)\rangle, \tag{9.22}$$

one then obtains for $\hat{\mathrm{Q}}(\boldsymbol{r}, t) = \hat{\mathbf{j}}(\boldsymbol{r})$:

$$\begin{aligned} \partial\boldsymbol{j}(\boldsymbol{r}, t)/\partial t &= \partial\langle\Psi(t)|\hat{\mathbf{j}}(\boldsymbol{r})|\Psi(t)\rangle/\partial t = \frac{i}{\hbar}\langle\Psi(t)|[\hat{\mathrm{H}}(t), \hat{\mathbf{j}}(\boldsymbol{r})]|\Psi(t)\rangle \\ &= -\frac{p}{\mu}\nabla v \equiv \frac{p}{\mu}\boldsymbol{F} = \boldsymbol{G}[\rho; \boldsymbol{r}, t], \end{aligned} \tag{9.23}$$

where the force acting on the particle $\boldsymbol{F}(\boldsymbol{r}) = -\nabla v(\boldsymbol{r})$.

The two "hydrodynamical" equations (9.17) and (9.23) are equivalent to the TDDFT stationary-action principle of (7.384).

It should be stressed, however, that – with the exception of the stationary quantum states – the overall amount of information in the system is not conserved, when the particle-probability density evolves in time. Therefore, the continuity equation expressing the local balance of the Fisher information in quantum mechanics has to include a nonvanishing "source" term $df/dt = \dot{f} \neq 0$, since the overall information content, proportional to the system average kinetic energy, changes for different shapes of the electronic probability density:

$$\frac{\partial f}{\partial t} = \dot{f} - \nabla \cdot \boldsymbol{J} \quad \text{or} \quad \frac{df}{dt} = \dot{f} = \frac{\partial f}{\partial t} + \nabla \cdot \boldsymbol{J} \neq 0, \quad \boldsymbol{J} = f\boldsymbol{V} = \left(\frac{f}{p}\right)\boldsymbol{j} = \tilde{f}\boldsymbol{j}. \quad (9.24)$$

Above, the information-current density $\boldsymbol{J}$ exhibits the same local velocity $\boldsymbol{V}$ as the probability "fluid" in (9.4a, 9.4b), since the information current in the system is effected through the probability flow $\boldsymbol{j}$.

It should be observed that in the continuity equations (9.17) and (9.24) the partial derivatives $\partial p/\partial t$ and $\partial f/\partial t$ measure the rate of change at the *fixed* point in space, inside an infinitesimal volume element at rest. Alternatively, these continuity equations have been interpreted in terms of the total derivatives (sources) $\dot{p}$ and $\dot{f}$, which represent the time rate of change of these densities in a volume element of the particle-probability fluid as it moves in space with the speed $\boldsymbol{V}$.

The full time-dependence of the Fisher information contained in state ψ can be determined directly using (9.8) and (9.17):

$$\begin{aligned} \frac{dI[\psi]}{dt} &= \int \frac{\partial I[p,\boldsymbol{j}]}{\partial p(\boldsymbol{r})} \frac{dp(\boldsymbol{r})}{dt} d\boldsymbol{r} + \int \frac{\partial I[p,\boldsymbol{j}]}{\partial \boldsymbol{j}(\boldsymbol{r})} \cdot \frac{d\boldsymbol{j}(\boldsymbol{r})}{dt} d\boldsymbol{r} \\ &= \int \frac{\partial I[\boldsymbol{j}]}{\partial \boldsymbol{j}(\boldsymbol{r})} \cdot \frac{d\boldsymbol{j}(\boldsymbol{r})}{dt} d\boldsymbol{r} = \frac{8\mu}{\hbar^2} \int \boldsymbol{j}(\boldsymbol{r}) \cdot \boldsymbol{F}(\boldsymbol{r}) d\boldsymbol{r} \equiv \int \frac{df(\boldsymbol{r})}{dt} d\boldsymbol{r}. \end{aligned} \quad (9.25)$$

Hence, the local production of the Fisher information density is proportional to the product of the local force $[\boldsymbol{F}(\boldsymbol{r})]$ and flow $[\boldsymbol{j}(\boldsymbol{r})]$ vectors, in perfect analogy to the entropy "source" in irreversible thermodynamics:

$$\dot{f}(\boldsymbol{r}) = \frac{8\mu}{\hbar^2} \boldsymbol{j}(\boldsymbol{r}) \cdot \boldsymbol{F}(\boldsymbol{r}). \quad (9.26)$$

Indeed, since by the probability continuity $dp(\boldsymbol{r})/dt = 0$, the only source of variations of $I[\psi] = \int f(\boldsymbol{r})\, d\boldsymbol{r} \equiv \int p(\boldsymbol{r})\tilde{f}(\boldsymbol{r}) d\boldsymbol{r}$ in time is the dependence of the information density per electron, $\tilde{f}(\boldsymbol{r})$ (9.14), on the probability-current density $\boldsymbol{j}(\boldsymbol{r})$:

$$\begin{aligned} \frac{dI[\psi]}{dt} &= \int p(\boldsymbol{r}) \left(\frac{\partial \tilde{f}(\boldsymbol{r})}{\partial \boldsymbol{j}(\boldsymbol{r})}\right)_p \cdot \frac{d\boldsymbol{j}(\boldsymbol{r})}{dt} d\boldsymbol{r} \\ &= \int p(\boldsymbol{r}) \left[2\left(\frac{2\mu}{\hbar p(\boldsymbol{r})}\right)^2 \boldsymbol{j}(\boldsymbol{r})\right] \cdot \left(\frac{p(\boldsymbol{r})}{\mu} F(\boldsymbol{r})\right) d\boldsymbol{r} \\ &= \frac{8\mu}{\hbar^2} \int \boldsymbol{j}(\boldsymbol{r}) \cdot F(\boldsymbol{r})\, d\boldsymbol{r}, \end{aligned}$$

where we have used the hydrodynamical equation (9.23).

Another, equivalent way to explicitly identify this information density is by using the time derivative of the expectation value of the associated observable [see (3.117) and (3.118)]. We first observe that the quantum-mechanical operator $\hat{\mathrm{I}}(\boldsymbol{r})$ (linear and Hermitian) of the generalized Fisher information, which defines its expectation value of (9.5),

$$\begin{aligned} I \equiv \langle I \rangle &= \int \psi^*(\boldsymbol{r})\hat{\mathrm{I}}(\boldsymbol{r})\,\psi(\boldsymbol{r})\,d\boldsymbol{r}, \\ \hat{\mathrm{I}}(\boldsymbol{r}) &= -4\Delta = 4(\hat{\mathbf{p}}(\boldsymbol{r})/\hbar)^2 \equiv 4\hat{\mathbf{k}}(\boldsymbol{r})^2, \end{aligned} \tag{9.27}$$

where $\hat{\mathbf{p}}(\boldsymbol{r}) = -i\hbar\nabla \equiv \hbar\hat{\mathbf{k}}(\boldsymbol{r})$ stands for the quantum-mechanical operator of the particle momentum with $\hat{\mathbf{k}}(\boldsymbol{r})$ standing for the particle wave-vector operator. The average information $\langle I \rangle$ thus measures the average length of the particle momentum vector. Indeed, for the wave-functions and their gradients vanishing at infinity the identity of (9.25) follows directly by a straightforward integration by parts.

Since $\hat{\mathrm{I}}(\boldsymbol{r})$ does not depend explicitly on time, the quantum-mechanical expression for the time rate of change of $\langle I \rangle$ is determined by the expectation value of the commutator $[\hat{\mathrm{H}}, \hat{\mathrm{I}}]$ [see (3.118)]:

$$\frac{d\langle I \rangle}{dt} \equiv \int \frac{df(\boldsymbol{r},t)}{dt} d\boldsymbol{r} = \frac{i}{\hbar}\left\langle [\hat{\mathrm{H}}, \hat{\mathrm{I}}] \right\rangle = \frac{i}{\hbar}\left\langle [\hat{\mathrm{V}}, \hat{\mathrm{I}}] \right\rangle. \tag{9.28}$$

The net time rate of change of the Fisher information density is thus seen to be determined by the commutator

$$[\hat{\mathrm{V}}, \hat{\mathrm{I}}] = 4[\Delta, v] = 8(\nabla v)\cdot\nabla + 4(\Delta v). \tag{9.29}$$

Using the identity from the integration by parts, $\int p\Delta v\, d\boldsymbol{r} = -\int \nabla p.\nabla v\, d\boldsymbol{r}$, and (9.4a) again recovers (9.25) defining the total time derivative (9.26) of the Fisher information density, which determines the source term in the associated continuity equation. A general form of this expression for the local production of the Fisher information resembles the familiar terms of the local entropy source in the irreversible thermodynamics, which are expressed as products of thermodynamic forces and the conjugated fluxes (e.g., Callen 1962).

We again recall that in a general case of the nonstationary quantum states, when the system average energy is not conserved, it is the stationary property of the quantum-mechanical action integral

$$\mathcal{a} = \int_{t_0}^{t_1} \langle \psi(t)|\hat{\mathrm{A}}(t)|\psi(t)\rangle dt, \tag{9.30}$$

with the expectation value $A(t) = \langle \psi(t)|\hat{\mathrm{A}}(t)|\psi(t)\rangle$ defined by the action operator $\hat{\mathrm{A}}(t) = i\hbar\partial/\partial t - \hat{\mathrm{H}}$, which generates the time-dependent Schrödinger equation.

Indeed, (9.15) is seen to result directly from the *stationary-action principle* $\delta\mathcal{A} = 0$, which can be equivalently expressed by the vanishing functional derivative:

$$\frac{\delta\mathcal{A}}{\delta\psi^*(\boldsymbol{r},t)} = 0. \tag{9.31}$$

9.3 Physical Information Principles Generating Schrödinger Equations

Consider again the stationary wave-equation of quantum mechanics for a general N-electron molecular system,

$$\hat{\mathrm{H}}(N)\ \Psi(N) = E\ \Psi(N). \tag{9.32}$$

It marks the eigenvalue equation of the electronic (fixed-nuclei) Hamiltonian in the familiar adiabatic (BO) approximation,

$$\hat{\mathrm{H}}(N) = -(\hbar^2/2m_e)\sum_{i=1}^{N}\Delta_i + V(N) = \hat{\mathrm{T}}(N) + \hat{\mathrm{V}}(N), \tag{9.33}$$

where the (multiplicative) potential-energy operator $\hat{V}(N) = V_{ne}(N) + V_{ee}(N) = V(N)$ includes the nuclear–electron (ne) attraction as well as the electron–electron (ee) repulsion terms, and E stands for the system ground-state electronic energy. This equation directly follows from the Schrödinger variational principle for the system minimum average energy $\langle E(N)\rangle$ subject to the normalization constraint of the electronic ground-state wave-function:

$$\delta\{\langle E(N)\rangle - E(N)\langle\Psi(N)\mid\Psi(N)\rangle\} = 0, \tag{9.34}$$

where the exact value of the electronic energy $E(N)$ plays the role of the Lagrange multiplier enforcing the constraint.

The average electronic energy, the expectation value of the Hamiltonian (9.33), contains the kinetic and potential contributions,

$$\begin{aligned}\langle E(N)\rangle &= \langle\Psi(N)|\hat{\mathrm{H}}(N)|\Psi(N)\rangle = \langle\Psi(N)|\hat{\mathrm{T}}(N)|\Psi(N)\rangle + \langle\Psi(N)|\hat{\mathrm{V}}(N)|\Psi(N)\rangle\\ &= \langle T(N)\rangle + \langle V(N)\rangle,\end{aligned} \tag{9.35}$$

with the kinetic-energy $\langle T\rangle$ component being proportional to the N-electron Fisher information:

$$\langle I(N)\rangle = 4\sum_{i=1}^{N}\langle\nabla_i\Psi^*(N)\cdot\nabla_i\Psi(N)\rangle = -4\sum_{i=1}^{N}\langle\Psi(N)|\Delta_i|\Psi(N)\rangle = \frac{8m_e}{\hbar^2}\langle T(N)\rangle. \tag{9.36}$$

One also observes that the joint-probability distribution of N electrons,

$$P(N) = \Psi^*(N)\Psi(N) = |\Psi(N)|^2, \quad \int P(N)d\tau^N = 1, \tag{9.37}$$

determines the average value of the system potential energy,

$$\langle V(N)\rangle = \int P(N)\, V(N)\, d\tau^N. \tag{9.38}$$

The wave-function principle (9.34) thus expresses the stationary character of the constrained Fisher information in the associated EPI rule (Nalewajski 2008e, 2010f):

$$\begin{aligned} &\delta\left\{\langle I(N)\rangle + \frac{8m_e}{\hbar^2}\left[\int P(N)\, V(N)\, d\tau^N - E(N)\int P(N)\, d\tau^N\right]\right\} \\ &\quad = \delta\left(\langle I[P(N)]\rangle - \lambda_{\langle V\rangle}\langle V[P(N)]\rangle - \lambda_{Norm.}\int P(N)\, d\tau^N\right) = 0. \end{aligned} \tag{9.39}$$

Here, the average Fisher-information term $\langle I(N)\rangle$ represents the EPI *intrinsic*-information term and the remaining, constraint part, stands for the EPI *bound*-information contribution (Frieden 2000). The latter consists of two terms: the physical, potential-energy part, effected by the Lagrange multiplier $\lambda_{\langle V\rangle} = -8m_e/\hbar^2$, and the "geometric" condition of the system probability normalization enforced by $\lambda_{Norm.} = 8m_e E/\hbar^2$.

Therefore, the time-independent Schrödinger equation can be regarded as having the information origins, by resulting from the above EPI rule. This Schrödinger equation for stationary states thus determines the optimum wave-function (probability-amplitude) which marks the extremum of the N-electron Fisher information subject to the probability-normalization and the average potential energy constraints. It is satisfied by the optimum probability distribution $P(N)$ of the electronic ground state. The two Lagrange multipliers in this information principle can be also interpreted as corresponding derivatives of the optimum Fisher information with respect to the associated constraints (Nalewajski 2008e).

Next, let us examine sufficient physical constraints in the relevant Euler–Lagrange information principle giving rise to the time-dependent Schrödinger equation. For simplicity, we again assume a single, spin-less particle of mass $\mu = m_e$, with the energy operator (9.1), described by the wave-finction (9.2). We have demonstrated above that the time-independent (stationary) Schrödinger equation results from the Euler–Lagrange EPI principle for the trial probability-amplitude $R'(\boldsymbol{r}) = \varphi'(\boldsymbol{r})$ (9.9), which determines the variational probability density $p'(\boldsymbol{r}) = |\varphi'(\boldsymbol{r})|^2$, including the constraint terms due to the probability/wave-function normalization,

$$\|\psi'\|^2 = \int \psi'^*\psi'\, d\boldsymbol{r} = \int p'(\boldsymbol{r})\, d\boldsymbol{r} = 1, \tag{9.40}$$

and the fixed value of the average potential energy, of the interaction between the electron density and the external potential $v(\boldsymbol{r})$ due to the fixed nuclei:

$$\langle V\rangle = \int \psi^*(\boldsymbol{r})\, v(\boldsymbol{r})\, \psi(\boldsymbol{r})\, d\boldsymbol{r} = \int p(\boldsymbol{r})\, v(\boldsymbol{r})\, d\boldsymbol{r} = V^0, \tag{9.41}$$

These constraints have been found to be enforced by the global Lagrange multipliers $\alpha = 8E\mu/\hbar^2$ and $\beta = -8\mu/\hbar^2$, respectively, in the underlying EPI rule:

$$\delta\left\{I[\psi'] - \alpha\int \psi'^*(\boldsymbol{r})\psi'(\boldsymbol{r})\, d\boldsymbol{r} - \beta\int \psi'^*(\boldsymbol{r})v(\boldsymbol{r})\psi'(\boldsymbol{r})\, d\boldsymbol{r}\right\} = 0, \tag{9.42}$$

where E stands for the fixed ground-state energy of this *one*-particle system.

The nonstationary wave-equation additionally requires the time-dependent constraint. The stationary action principle (Gross et al. 1996) of (9.31) for the trial variations $\delta\psi^*(\boldsymbol{r},t) = \langle\delta\psi(t)|\boldsymbol{r}\rangle$, with

$$\begin{aligned}\delta\mathcal{A} &= \int_{t_0}^{t_1} dt\langle\delta\psi(t)|\hat{\mathrm{A}}(t)|\psi(t)\rangle \\ &= i\hbar\int_{t_0}^{t_1} dt\langle\delta\psi(t)|\partial\psi(t)/\partial t\rangle - \int_{t_0}^{t_1} dt\,\langle\delta\psi(t)|\hat{\mathrm{H}}|\psi(t)\rangle = 0,\end{aligned} \tag{9.43}$$

suggests that the associated nonstationary EPI rule has to include the time derivative of the squared norm of (9.40),

$$\frac{\partial\|\psi'\|^2}{\partial t} = \int\left(\frac{\partial\psi'^*}{\partial t}\psi' + \psi'^*\frac{\partial\psi'}{\partial t}\right) d\boldsymbol{r} = \int\frac{\partial p'(\boldsymbol{r})}{\partial t} d\boldsymbol{r}, \tag{9.44}$$

besides the variation of the system average electronic energy. Therefore, the conservation of the probability normalization in time (9.16),

$$\frac{\partial\langle\psi\mid\psi\rangle}{\partial t} = \frac{\partial}{\partial t}\int p(\boldsymbol{r})d\boldsymbol{r} = 0, \tag{9.45}$$

provides the relevant time-dependent constraint in the EPI principle generating the Schrödinger dynamics of quantum states.

For the already normalized wave-functions, when $\alpha = 0$ (redundant constraint), the corresponding Fisher-information EPI rule reads (Nalewajski 2008e):

$$\delta\left\{I[\psi'] - \beta\int \psi'^*(\boldsymbol{r})v(\boldsymbol{r})\psi'(\boldsymbol{r})d\boldsymbol{r} - \gamma\frac{\partial}{\partial t}\int \psi'^*(\boldsymbol{r})\psi'(\boldsymbol{r})\, d\boldsymbol{r}\right\} \equiv \delta I[\psi'^*,\psi'] = 0, \tag{9.46}$$

where β has already been identified before and $\gamma = 8\mu i/\hbar$. This identification follows from the functional differentiation of the above auxiliary functional $I[\psi'^*,\psi']$ with respect to ψ'^*. These Lagrange multipliers can be also interpreted as derivatives of the optimum Fisher information with respect to the associated constraints.

9.4 Information Principle for Adiabatic Approximation

As further application of the physically constrained information principle let us examine the quantum mechanical system with particles differing in mass, in the familiar scenario of the BO separation of the (stationary) electronic and nuclear distributions in molecules. Since the Fisher-information principle generating the electronic Schrödinger equation, for the fixed (parametric) positions of nuclei, has already been discussed in the preceding section, here we shall focus on the effective Schrödinger equation for nuclear motions, in which the influence of the *fast*-moving (light) electrons is averaged out in the resultant potential determining forces acting on the *slowly* moving (heavy) nuclei. To simplify the notation a.u. will be used throughout this section.

A reference to (9.26) indicates that the Fisher-information operator probes the length of the electron velocity, i.e., of its momentum per unit particle mass. Therefore, in order to bring on equal footing the Fisher intrinsic-information terms for electrons and nuclei in the molecular EPI principle, which generates the adiabatic separation of the electronic and nuclear motions, one has to combine the electronic information term with the corresponding nuclear contribution per unit mass.

The molecular wave-function $\Psi(\mathbf{q}, \mathbf{Q})$ of N electrons at positions $\mathbf{r} = \{\boldsymbol{r}_i\}$ exhibiting spins $\boldsymbol{\sigma} = \{\sigma_i\}$, or in the combined notation

$$\mathbf{q} = \{\boldsymbol{r}_i, \sigma_i\} = \{\boldsymbol{q}_i\} = \{\boldsymbol{r}, \boldsymbol{\sigma}\},$$

and of m nuclei with masses $\{M_\alpha\}$, charges $\{Z_\alpha\}$, and spins $\boldsymbol{\Sigma} = \{\Sigma_\alpha\}$, in positions $\mathbf{R} = \{\boldsymbol{R}_\alpha\}$,

$$\mathbf{Q} = \{\boldsymbol{R}_\alpha, \Sigma_\alpha\} = \{\boldsymbol{Q}_\alpha\} = \{\mathbf{R}, \boldsymbol{\Sigma}\},$$

generates the probability distribution of the joint electronic–nuclear events, $P(\mathbf{q}, \mathbf{Q}) = |\Psi(\mathbf{q}, \mathbf{Q})|^2$, satisfying the relevant overall and partial normalizations:

$$\iint P(\mathbf{q}, \mathbf{Q})\, d\mathbf{q}\, d\mathbf{Q} = \int \Pi(\mathbf{Q})\, d\mathbf{Q} = \int \pi(\mathbf{q})\, d\mathbf{q} = 1; \tag{9.47}$$

here $\Pi(\mathbf{Q})$ and $\pi(\mathbf{q})$ denote the partially integrated nuclear and electronic probability distributions, respectively.

The essence of the adiabatic approximation (see Sect. 5.2) lies in extracting the probability density of the heavy (slow) nuclei as the *marginal* (reference, parameter) distribution (see also Sect. 8.4):

$$P(\mathbf{q}, \mathbf{Q}) = \Pi(\mathbf{Q}) \frac{P(\mathbf{q}, \mathbf{Q})}{\Pi(\mathbf{Q})} \equiv \Pi(\mathbf{Q}) p(\mathbf{q}|\mathbf{Q}), \quad \int p(\mathbf{q}|\mathbf{Q})\, d\mathbf{q} = 1, \tag{9.48a}$$

where in the (conditional) electronic probability density $p(\mathbf{q}|\mathbf{Q})$ the nuclear variables $\mathbf{Q}$ constitute parameters, as indeed reflected by the above normalization condition of this parametric probability density. This further implies the familiar factorization of the system wave-function (probability amplitude) in terms of the nuclear and electronic functions $\chi(\mathbf{Q})$ and $\phi(\mathbf{q}|\mathbf{Q})$, respectively [see (5.48)],

$$\Psi(\mathbf{q},\mathbf{Q}) = \chi(\mathbf{Q})\,\phi(\mathbf{q}|\mathbf{Q}), \tag{9.49}$$

representing the amplitudes of the associated probability distributions:

$$\Pi(\mathbf{Q}) = |\chi(\mathbf{Q})|^2 \qquad \text{and} \qquad p(\mathbf{q}|\mathbf{Q}) = |\phi(\mathbf{q}|\mathbf{Q})|^2. \tag{9.48b}$$

The "intrinsic"-information functional of the EPI principle for this molecular scenario of N electrons and m nuclei now combines the average Fisher-information terms per unit mass for all constituent particles (Nalewajski 2008e):

$$I[\Psi] = \langle I(N)\rangle + \sum_{\alpha=1}^{m} \frac{1}{M_\alpha}\langle I_\alpha\rangle. \tag{9.50}$$

Here, the electronic contribution $\langle I(N)\rangle$ represents the nuclear probability-weighted mean-value of (equal) contributions due to the indistinguishable N electrons:

$$\begin{aligned}\langle I(N)\rangle &= 4\sum_{i=1}^{N}\int\!\!\int \nabla_i\Psi^*\cdot\nabla_i\Psi\,d\mathbf{q}\,d\mathbf{Q}\\ &= 4\sum_{i=1}^{N}\int \Pi(\mathbf{Q})\Big[\int \nabla_i\phi^*\cdot\nabla_i\phi\,d\mathbf{q}\Big]\,d\mathbf{Q}\\ &= \sum_{i=1}^{N}\int \Pi(\mathbf{Q})\,\langle I_i(\mathbf{Q})\rangle\,d\mathbf{Q} = N\int \Pi(\mathbf{Q})\langle I_1(\mathbf{Q})\rangle d\mathbf{Q}.\end{aligned} \tag{9.51}$$

The average Fisher information due to nucleus α in (9.50) similarly reads:

$$\begin{aligned}\langle I_\alpha\rangle &= 4\int\!\!\int \nabla_\alpha\Psi^*\cdot\nabla_\alpha\Psi\,d\mathbf{q}\,d\mathbf{Q}\\ &= 4\Big(\int |\nabla_\alpha\chi|^2 d\mathbf{Q} + \int \Pi\int |\nabla_\alpha\phi|^2 d\mathbf{q}\,d\mathbf{Q} + \int \chi(\nabla_\alpha\chi^*)\cdot\int \phi^*\nabla_\alpha\phi\,d\mathbf{q}\,d\mathbf{Q}\\ &\quad + \int \chi^*(\nabla_\alpha\chi)\cdot\int \phi\,\nabla_\alpha\phi^* d\mathbf{q}\,d\mathbf{Q}\Big).\end{aligned} \tag{9.52}$$

The molecular EPI principle should now include, besides the usual subsidiary conditions of the probability normalizations, the constraint of the fixed value of the overall Coulombic potential energy:

$$\begin{aligned}\langle W\rangle &= \int\int P(\mathbf{q},\mathbf{Q})\,W(\mathbf{q},\mathbf{Q})\,d\mathbf{q}\,d\mathbf{Q}\\ &= \int \Pi(\mathbf{Q})\int p(\mathbf{q}|\mathbf{Q})[V_{ne}(\mathbf{q},\mathbf{Q})+V_{ee}(\mathbf{q})]\,d\mathbf{q}\,d\mathbf{Q} + \int \Pi(\mathbf{Q})\,V_{nn}(\mathbf{Q})\,d\mathbf{Q}\\ &\equiv \int \Pi(\mathbf{Q})\,\langle W(\mathbf{Q})\rangle_{\mathbf{q}}\,d\mathbf{Q} = \langle W\rangle^0.\end{aligned} \tag{9.53}$$

Here, $\langle W(\mathbf{Q})\rangle_{\mathbf{q}}$ denotes the overall BO potential energy, for the specified nuclear positions, including the nuclear repulsion, averaged over the electronic coordinates $\mathbf{q}$, defined by the multiplicative operator

$$\begin{aligned}W(\mathbf{q},\mathbf{Q}) &= -\sum_{\alpha=1}^{m}\sum_{i=1}^{N}\frac{Z_\alpha}{|\boldsymbol{r}_i-\boldsymbol{R}_\alpha|}+\sum_{i=1}^{N-1}\sum_{j=i+1}^{N}\frac{1}{|\boldsymbol{r}_i-\boldsymbol{r}_j|}+\sum_{\alpha=1}^{m-1}\sum_{\beta=\alpha+1}^{m}\frac{Z_\alpha Z_\beta}{|\boldsymbol{R}_\alpha-\boldsymbol{R}_\beta|}\\ &\equiv V_{ne}(\mathbf{q},\mathbf{Q})+V_{ee}(\mathbf{q})+V_{nn}(\mathbf{Q}).\end{aligned} \tag{9.54}$$

The adiabatic EPI principle using the Fisher-information contributions per unit mass,

$$\begin{aligned}\delta\{I[\Psi] &- \lambda_1\int \Pi(\mathbf{Q})d\mathbf{Q} - \int \lambda_2(\mathbf{Q})\int p(\mathbf{q}|\mathbf{Q})d\mathbf{q}\,d\mathbf{Q}\\ &- \lambda_3\int \Pi(\mathbf{Q})\int p(\mathbf{q}|\mathbf{Q})\,W(\mathbf{q},\mathbf{Q})d\mathbf{q}\,d\mathbf{Q}\} = 0,\end{aligned} \tag{9.55}$$

includes the global (λ_1, λ_3) and local [$\lambda_2(\mathbf{Q})$] Lagrange multipliers, which enforce the subsidiary conditions of the normalization of $\Pi(\mathbf{Q})$, the fixed value of the average potential energy, and the normalization of $p(\mathbf{q}|\mathbf{Q})$ for any given $\mathbf{Q}$, respectively.

The functional differentiation of the overall Fisher information of (9.50) with respect to $\chi^*(\mathbf{Q})$ then gives:

$$\begin{aligned}\frac{\delta I[\Psi]}{\delta\chi^*(\mathbf{Q})} = 4\Bigg\{&\left(\sum_{i=1}^{N}\int|\nabla_i\phi|^2 d\mathbf{q}\right)\chi(\mathbf{Q})\\ &-\sum_{\alpha=1}^{m}\frac{1}{M_\alpha}\left[\Delta_\alpha\chi(\mathbf{Q})+\left(\int\phi^*\Delta_\alpha\phi\,d\mathbf{q}\right)\chi(\mathbf{Q})+2\left(\int\phi^*\nabla_\alpha\phi\,d\mathbf{q}\right)\cdot\nabla_\alpha\chi(\mathbf{Q})\right]\Bigg\},\end{aligned} \tag{9.56}$$

where we have taken into account the integration by parts and the conservation of the normalization of the electronic (conditional) probability, which further implies

$$\nabla_\alpha\int p(\mathbf{q}|\mathbf{Q})\,d\mathbf{q} = \int(\phi^*\nabla_\alpha\phi+\phi\,\nabla_\alpha\phi^*)\,d\mathbf{q} = 0. \tag{9.57}$$

A similar differentiation of the constraint part of the auxiliary functional of (9.55) generates the remaining terms of the Euler equation for the optimum nuclear wave-function of (9.49):

$$\frac{\delta I[\Psi]}{\delta\chi^*(\mathbf{Q})} - [\lambda_1 + \lambda_3 \langle W(\mathbf{Q})\rangle_{\mathbf{q}}]\chi(\mathbf{Q}) = 0. \tag{9.58}$$

Let us interpret the contributions in (9.56) in terms of the electronic and nuclear kinetic-energy operators:

$$\hat{\mathrm{T}}(\mathbf{q}) = -1/2\sum_{i=1}^{N}\Delta_i \qquad \text{and} \qquad \hat{\mathrm{T}}_n(\mathbf{Q}) = -\sum_{\alpha=1}^{m}\frac{1}{2M_\alpha}\Delta_\alpha. \tag{9.59}$$

The nuclear Euler equation then reads:

$$\left\{\hat{\mathrm{T}}_n(\mathbf{Q}) + [\langle T(\mathbf{Q})\rangle_{\mathbf{q}} - \frac{\lambda_3}{8}\langle W(\mathbf{Q})\rangle_{\mathbf{q}}] + \langle T_n^{\phi}(\mathbf{Q})\rangle_{\mathbf{q}} - \frac{\lambda_1}{8}\right\}\chi(\mathbf{Q})$$
$$-\sum_{\alpha=1}^{m}\frac{1}{M_\alpha}\langle\phi \mid \nabla_\alpha\phi\rangle_{\mathbf{q}}\cdot\nabla_\alpha\chi(\mathbf{Q}) = 0. \tag{9.60}$$

Here [see (5.54) and (5.56)],

$$\langle T_n^{\phi}(\mathbf{Q})\rangle_{\mathbf{q}} = \langle\phi|\hat{\mathrm{T}}_n \mid \phi\rangle_{\mathbf{q}} \tag{9.61}$$

stands for the ("diagonal") kinetic-energy correction to the electronic PES of (5.55),

$$\langle E(\mathbf{Q})\rangle_{\mathbf{q}} = \langle T(\mathbf{Q})\rangle_{\mathbf{q}} + \langle W(\mathbf{Q})\rangle_{\mathbf{q}} = E_e(\mathbf{Q}), \tag{9.62}$$

in the resultant effective potential for nuclear motions in the adiabatic approximation (5.54),

$$U(\mathbf{Q}) = E_e(\mathbf{Q}) + \langle T_n^{\phi}(\mathbf{Q})\rangle_{\mathbf{q}}. \tag{9.63}$$

It complements $\hat{\mathrm{T}}_n(\mathbf{Q})$ in the effective Hamiltonian of (5.54),

$$\hat{\mathrm{H}}_n^{eff.}(\mathbf{Q}) = \hat{\mathrm{T}}_n(\mathbf{Q}) + U(\mathbf{Q}), \tag{9.64}$$

which accounts for the averaged (integrated) influence, due to fast electronic motions, of the Schrödinger equation determining the nuclear distributions:

$$\{\hat{\mathrm{T}}_n(\mathbf{Q}) + U(\mathbf{Q}) - E_{mol.}\}\chi(\mathbf{Q}) \equiv [\hat{\mathrm{H}}_n^{eff.}(\mathbf{Q}) - E_{mol.}]\chi(\mathbf{Q}) = 0. \tag{9.65}$$

Here, $E_{mol.}$ denotes the full Coulomb molecular energy, the eigenvalue of the effective molecular Hamiltonian $\hat{\mathrm{H}}_n^{eff.}(\mathbf{Q})$.

In this approximation one neglects the last, nonadiabatic term in (9.60), which involves the nuclear gradients of both the electronic and nuclear wave-functions. This equation is seen to assume the form of the preceding eigenvalue equation of the effective nuclear Hamiltonian for $\lambda_3 = -8$ and $\lambda_1 = 8E_{mol.}$. It should be finally recalled that by additionally neglecting the small diagonal correction $\langle T_n^{\phi}(\mathbf{Q})\rangle_{\mathbf{q}}$ in the effective potential for nuclear motions, $U(\mathbf{Q}) \approx E_e(\mathbf{Q})$, one arrives at the original (crude-adiabatic) approximation of Born and Oppenheimer.

9.5 Kohn–Sham Equations from Information Rule

Next, we shall briefly demonstrate that the KS equations (7.110) of the computational DFT also result from the associated EPI principle (Nalewajski 2003c, 2006g). To simplify notation a.u. will be used again. The relevant intrinsic-data information functional for this orbital approximation is now in the multicomponent Fisher form $I[\boldsymbol{\psi}]$ (8.7), where $\boldsymbol{\psi} = \{\psi_n\}$ groups the N-lowest (*singly* occupied) KS SO. Indeed, in the spin-resolved KS theory, with $\{\psi_n \equiv \psi_{n\sigma} = \varphi_{n\sigma}\xi_{n\sigma}\}$ defined by the spatial functions (KS MO) $\boldsymbol{\varphi} = \{\varphi_{n\sigma}\}$ and the corresponding spin functions $\boldsymbol{\xi} = \{\xi_{n\sigma}\}$, each orbital (probability) density $\rho_{n\sigma}(\boldsymbol{r}) = |\varphi_{n\sigma}(\boldsymbol{r})|^2$ constitutes a distinct, independent component of the overall *one*-electron probability distribution $p(\boldsymbol{r}) = \rho(\boldsymbol{r})/N$. Here, the ground-state electron density $\rho(\boldsymbol{r})$ of the fictitious noninteracting system of N particles is by hypothesis equal to that of the real system of the fully interacting electrons:

$$\rho(\boldsymbol{r}) = \sum_{\sigma=\uparrow,\downarrow}\sum_n |\varphi_{n\sigma}(\boldsymbol{r})|^2 \equiv \sum_n \rho_n(\boldsymbol{r}) \equiv \sum_{\sigma=\uparrow,\downarrow}\rho_\sigma(\boldsymbol{r}),$$
$$\int \rho(\boldsymbol{r})\, d\boldsymbol{r} = N, \quad \int \rho_\sigma(\boldsymbol{r})\, d\boldsymbol{r} = N_\sigma; \tag{9.66}$$

above, $\{\rho_\sigma(\boldsymbol{r})\}$ groups the two spin densities and N_σ denotes the number of electrons exhibiting the spin-up ($\uparrow$, $\sigma = ½$) or the spin-down ($\downarrow$,$\sigma = -½$) states.

The physical information functional $K[\boldsymbol{\psi}] = I[\boldsymbol{\psi}] - J[\boldsymbol{\psi}]$ of the associated EPI principle, where $I[\boldsymbol{\psi}]$ denotes the usual (*intrinsic*) Fisher-information term and $J[\boldsymbol{\psi}]$ stands for the multicomponent constrained (*bound*) information part, then uniquely defines the variational KS problem for the molecular system in question, which constitutes the basis of traditional DFT computations.

Let us now consider in some detail this key *one*-body problem of KS theory of N noninteracting electrons moving in an effective (local) external potential $v_{\mathrm{KS}}(\boldsymbol{r})$, which also determines the exact ground-state density of the real molecular system of N interacting electrons moving in the external potential $v(\boldsymbol{r})$ due to the system nuclei in their fixed positions in space. The ground-state of the hypothetical noninteracting system is exactly described by the KS determinant $\Psi_s(N) \equiv |\boldsymbol{\psi}| = \det\{\varphi_{n\sigma}\xi_{n\sigma}\}$ constructed from the N lowest (orthonormal), *singly* occupied KS SO.

We begin with a brief reminder of the theory basic assumptions and concepts (see Sect. 7.3). By the Hohenberg–Kohn theorem both the effective potential $v_{KS}(\boldsymbol{r})$ and the system exact electronic energy are unique functionals of the ground-state electronic density:

$$v_{KS}(\boldsymbol{r}) = v_{KS}[\boldsymbol{r};\rho] \qquad \text{and} \qquad E[N,v] = E_v[\rho]. \tag{9.67}$$

The Hohenberg–Kohn variational principle also implies that the optimum KS orbitals, which also mark the solution point of the underlying KS–EPI principle, are the unique functionals of the system electron density: $\boldsymbol{\varphi} = \boldsymbol{\varphi}[\rho]$. It also follows from Sect. 7.3 that the density functional for the ground-state electronic energy, $E_v[\rho]$, consists of the trivial external potential energy functional, $V_{ne}[\rho] = \int \rho(\boldsymbol{r})v(\boldsymbol{r})\, d\boldsymbol{r}$, and the v-independent Hohenberg–Kohn–Levy functional for the sum of the expectation values of the electron kinetic and repulsion energies, $F[\rho] = T[\rho] + V_{ee}[\rho]$:

$$E_v[\rho] = \int \rho(\boldsymbol{r})v(\boldsymbol{r})\, d\boldsymbol{r} + F[\rho]. \tag{9.68}$$

We also recall that the total kinetic energy $T[\rho]$ of N electrons includes both the noninteracting (s) and (Coulomb) correlation (c) parts:

$$T[\rho] = T_s[\rho] + T_c[\rho], \tag{9.69}$$

where the separable (additive) contribution

$$T_s[\rho] = -\frac{1}{2}\sum_{\sigma}\sum_{n}\langle\varphi_{n\sigma}[\rho]|\Delta|\varphi_{n\sigma}[\rho]\rangle. \tag{9.70}$$

In KS theory one further extracts from the electron-repulsion energy of the real (interacting) system, $V_{ee}[\rho]$, the classical (Hartree) energy,

$$J_{ee}[\rho] = \frac{1}{2}\int\int\frac{\rho(\boldsymbol{r})\rho(\boldsymbol{r}')}{|\boldsymbol{r}-\boldsymbol{r}'|}\, d\boldsymbol{r}\, d\boldsymbol{r}', \tag{9.71}$$

and the (potential) correlation contribution $E_{xc}[\rho] - T_c[\rho]$,

$$V_{ee}[\rho] = J_{ee}[\rho] + (E_{xc}[\rho] - T_c[\rho]),$$

so that

$$F[\rho] = T_s[\rho] + J_{ee}[\rho] + E_{xc}[\rho]. \tag{9.72}$$

Here $E_{xc}[\rho]$ stands for the *xc*-energy of KS theory, including the $T_c[\rho]$ contribution to $T[\rho]$, which determines the correlation part $v_{xc}(r)$ of the effective KS potential,

$$v_{\mathrm{KS}}(\boldsymbol{r}) = v(\boldsymbol{r}) + \int \frac{\rho(\boldsymbol{r}')}{|\boldsymbol{r}' - \boldsymbol{r}|} d\boldsymbol{r}' + \frac{\delta E_{xc}[\rho]}{\delta\rho(\boldsymbol{r})} \equiv v(\boldsymbol{r}) + v_{\mathrm{H}}(\boldsymbol{r}) + v_{xc}(\boldsymbol{r}), \tag{9.73}$$

in the effective *one*-body Hamiltonian of (7.108),

$$\hat{\mathrm{H}}_{\mathrm{KS}}(\boldsymbol{r}) = -{}^{1}\!/\!{}_{2}\nabla^2 + v_{\mathrm{KS}}(\boldsymbol{r}), \tag{9.74}$$

determining the optimum KS MO of the N lowest eigenvalues (orbital energies) $\{\varepsilon_{n\sigma}\}$ (7.110):

$$\hat{\mathrm{H}}_{\mathrm{KS}}(\boldsymbol{r})\varphi_{n\sigma}(\boldsymbol{r}) = \varepsilon_{n\sigma}\varphi_{n\sigma}(\boldsymbol{r}), \quad n = 1, 2, \ldots, N. \tag{9.75}$$

In the KS theory, the electronic energy of the interacting system is thus given by the density functional:

$$E_v[\rho] = T_s[\rho] + (V_{ne}[\rho] + J_{ee}[\rho] + E_{xc}[\rho]) \equiv T_s[\boldsymbol{\varphi}[\rho]] + V_e^{\mathrm{KS}}[\rho], \tag{9.76}$$

where the KS functional for the electronic potential energy $V_e^{\mathrm{KS}}[\rho]$ contains the correlation part of the kinetic energy of interacting electrons:

$$V_e^{\mathrm{KS}}[\rho] = (V_{ne}[\rho] + V_{ee}[\rho]) + T_c[\rho] \equiv V_e[\rho] + T_c[\rho]. \tag{9.77}$$

The KS equations (9.75), which determine the optimum shapes of orbitals of the hypothetical noninteracting system and hence also the electron density and energy of the interacting system, follow from the HK variational principle for the system electronic energy,

$$\begin{aligned} &\delta\Big\{E_v[\rho] - \sum\nolimits_\sigma \sum\nolimits_{\sigma'} \sum\nolimits_n \sum\nolimits_m \Theta_{m\sigma,n\sigma'} \langle\varphi_{m\sigma}|\varphi_{n\sigma'}\rangle\Big\} \equiv \delta K_s[\boldsymbol{\varphi}[\rho]] = 0 \quad \text{or} \\ &8\delta T_{\mathrm{S}}[\boldsymbol{\varphi}] \equiv \delta I_s[\boldsymbol{\varphi}[\rho]] = 8\Big(\sum\nolimits_n \sum\nolimits_m \Theta_{m\sigma,n\sigma'} \delta\langle\varphi_{m\sigma}|\varphi_{n\sigma'}\rangle - \delta V_e^{\mathrm{KS}}[\rho]\Big) \equiv \delta J_s[\boldsymbol{\varphi}[\rho]], \end{aligned} \tag{9.78}$$

where, in the canonical representation the matrix of Lagrange multipliers, which enforce the constraints of MO normalization/orthogonality, becomes diagonal: $\{\Theta_{m\sigma,n\sigma'} = \varepsilon_{n\sigma}\delta_{m,n}\delta_{\sigma,\sigma'}\}$. The preceding equation identifies the intrinsic ($I_s[\boldsymbol{\varphi}[\rho]]$) and bound ($J_s[\boldsymbol{\varphi}[\rho]]$) information terms in the KS EPI problem. The former is related to the kinetic energy of noninteracting system. Indeed, a straightforward integration by parts shows that the expectation value of the kinetic energy of noninteracting electrons of the separable KS problem,

$$\begin{aligned} T_s[\boldsymbol{\varphi}] &= -\frac{1}{2}\sum\nolimits_\sigma \sum\nolimits_n \int \varphi_{n\sigma}(\boldsymbol{r})\nabla^2\varphi_{n\sigma}(\boldsymbol{r})\, d\boldsymbol{r} \\ &= \frac{1}{2}\sum\nolimits_\sigma \sum\nolimits_n \int [\nabla\varphi_{n\sigma}(\boldsymbol{r})]^2\, d\boldsymbol{r}, \end{aligned} \tag{9.79}$$

is proportional to the multicomponent Fisher information $I_s[\boldsymbol{\psi}] = I_s[\boldsymbol{\varphi}]$:

$$I_s[\boldsymbol{\varphi}] = 4\sum\nolimits_\sigma \sum\nolimits_n \int [\nabla\varphi_{n\sigma}(\boldsymbol{r})]^2\, d\boldsymbol{r} = 8T_s[\boldsymbol{\varphi}]. \tag{9.80}$$

Thus, the KS variational rule of (9.78) can indeed be interpreted as another example of the EPI principle (8.45) using the Fisher measure of information:

$$\delta\{I_s[\boldsymbol{\varphi}[\rho]] - J_s[\boldsymbol{\varphi}[\rho]]\} = \delta K_s[\boldsymbol{\varphi}[\rho]] = 0. \tag{9.81}$$

In a similar way one derives the EPI problem for the orbital approximation in the wave-function theory, which defines the HF approach. Again, the kinetic energy functional of HF orbitals determines the intrinsic information of the variational problem, while the combined electronic attraction and repulsion terms of the potential energy, supplemented by the constraints of the orbital orthonormality define the associated bound-information part of the underlying HF EPI principle.

References

Callen HB (1962) Thermodynamics: an introduction to the physical theories of equilibrium thermostatics and irreversible thermodynamics. Wiley, New York

Frieden BR (2000) Physics from the Fisher information – a unification. Cambridge University Press, Cambridge

Frieden BR, Soffer BH (2010) Weighted Fisher informations, their derivation and use. Phys Lett A (in press)

Gross EKU, Dobson JF, Petersilka M (1966) In: Nalewajski RF (ed) Density functional theory II: relativistic and time dependent extensions, Topics in current chemistry, vol 181. Springer, Berlin, p 81

Nalewajski RF (2003c) Chem Phys Lett 372:28

Nalewajski RF (2006g) Information theory of molecular systems. Elsevier, Amsterdam

Nalewajski RF (2008e) Int J Quantum Chem 108:2230

Nalewajski RF (2010f) Information origins of the chemical bond. Nova Science, New York

Chapter 10
Electron Density as Carrier of Information

Abstract The information densities of electronic ground-state distributions are used as local probes of the chemical bonds in molecules. The entropy-deficiency and entropy-displacement relative to the promolecular electron density identify effects due to formation of the chemical bond. They reflect typical atom-promotion (polarization) and charge-transfer processes. The nonadditive information contributions to the density of molecular Fisher information (kinetic energy) of electrons are investigated in the MO and AO resolution levels. The former gives rise to the *Electron Localization Function*, while the latter is used to define the *Contra-gradience* criterion for locating the bonding regions in molecules. Illustrative applications to typical atomic/molecular systems are presented and discussed. They are seen to reflect quite adequately the system bonding pattern and the valence state of bonded atoms in the molecular environment, in accordance with intuitive chemical expectations.

10.1 Local Entropy-Deficiency (Surprisal) Analysis

As we have already remarked before, the electron densities $\{\rho_i^0\}$ of the separated atoms define the molecular (isoelectronic, $N = N^0$) prototype called the atomic "promolecule" (e.g., Hirshfeld 1977; Nalewajski 2006g), given by the sum of the atomic ground-state electron densities shifted to the actual locations of *Atoms-in-Molecules* (AIM). The resulting electron density $\rho^0 = \sum_i \rho_i^0$ of this collection of the "frozen" *free*-atom distributions defines the initial stage in the bond-formation process and determines a natural reference for extracting changes due to the chemical bonds in the familiar *density difference function* $\Delta\rho = \rho - \rho^0$. This deformation density has been widely used to probe the electronic structure of molecular systems. In this section we shall compare these plots with some local IT probes introduced in Chap. 8, to explore the molecular electron distributions $\rho(\boldsymbol{r})$ or their shape(probability) factors $p(\boldsymbol{r}) = \rho(\boldsymbol{r})/N$ generated using the KS LDA calculations.

R.F. Nalewajski, *Perspectives in Electronic Structure Theory*,
DOI 10.1007/978-3-642-20180-6_10,

Consider first the density $\Delta s(\boldsymbol{r})$ (in nats per unit volume) of the molecular KL entropy-deficiency (directed-divergence) reflecting the average information distance between the molecular and promolecular electron distributions:

$$\Delta S[\rho|\rho^0] = I^{KL}[\rho|\rho^0] = \int \rho(\boldsymbol{r}) \ln[\rho(\boldsymbol{r})/\rho^0(\boldsymbol{r})]\, d\boldsymbol{r} = \int \rho(\boldsymbol{r}) I[w(\boldsymbol{r})]\, d\boldsymbol{r} \equiv \int \Delta s(\boldsymbol{r})\, d\boldsymbol{r}, \tag{10.1}$$

where $w(\boldsymbol{r}) = \rho(\boldsymbol{r})/\rho^0(\boldsymbol{r}) = p(\boldsymbol{r})/p^0(\boldsymbol{r})$ stands for the local *enhancement* factor of the density/probability distribution and $I[w(\boldsymbol{r})]$ denotes the associated *surprisal* function. The functional density $\Delta s(\boldsymbol{r})$ also represents the local (renormalized) missing information between the shape(probability) factors $p(\boldsymbol{r})$ and $p^0(\boldsymbol{r}) = \rho^0(\boldsymbol{r})/N^0$ of the two compared electron densities:

$$\Delta s(\boldsymbol{r}) \equiv \Delta s[\rho(\boldsymbol{r})|\rho^0(\boldsymbol{r})] = Np(\boldsymbol{r})\ \ln[p(\boldsymbol{r})/p^0(\boldsymbol{r})] \equiv N\Delta s[p(\boldsymbol{r})|p^0(\boldsymbol{r})]. \tag{10.2}$$

The KL-information density $\Delta s(\boldsymbol{r})$ thus measures the local value of the electron-density/probability-weighted surprisal of the electronic distribution relative to the promolecular reference. The related divergence measure of Kullback's (K) symmetrized missing information reads:

$$\begin{aligned}\Delta S[\rho, \rho^0] = I^{K}[\rho, \rho^0] &= \int [\rho(\boldsymbol{r}) - \rho^0(\boldsymbol{r})] \ln[\rho(\boldsymbol{r})/\rho^0(\boldsymbol{r})] d\boldsymbol{r} \\ &\equiv \int \Delta\rho(\boldsymbol{r}) I[w(\boldsymbol{r})]\, d\boldsymbol{r} \equiv \int \Delta D(\boldsymbol{r})\, d\boldsymbol{r}.\end{aligned} \tag{10.3}$$

Therefore, its density $\Delta D(\boldsymbol{r})$ represents the local surprisal "weighted" by the density difference $\Delta\rho(\boldsymbol{r}) = \rho(\boldsymbol{r}) - \rho^0(\boldsymbol{r})$ or the probability difference $\Delta p(\boldsymbol{r}) = p(\boldsymbol{r}) - p^0(\boldsymbol{r}) = \Delta\rho(\boldsymbol{r})/N$:

$$\begin{aligned}\Delta D(\boldsymbol{r}) &\equiv \Delta s[\rho(\boldsymbol{r}), \rho^0(\boldsymbol{r})] = \Delta\rho(\boldsymbol{r}) I[w(\boldsymbol{r})] \\ &= N\Delta p(\boldsymbol{r}) I[w(\boldsymbol{r})] \equiv N\Delta s[p(\boldsymbol{r}), p^0(\boldsymbol{r})].\end{aligned} \tag{10.4}$$

The molecular surprisal $I[w(\boldsymbol{r})]$ thus measures the density-per-electron of the local entropy-deficiency relative to the promolecular reference,

$$I[w(\boldsymbol{r})] = \Delta s(\boldsymbol{r})/\rho(\boldsymbol{r}), \tag{10.5}$$

or the density-per-electron–displacement of the molecular divergence,

$$I[w(\boldsymbol{r})] = \Delta D(\boldsymbol{r})/\Delta\rho(\boldsymbol{r}). \tag{10.6}$$

We further recall that the molecular electron density $\rho(\boldsymbol{r})$ is on average only slightly modified relative to the promolecular distribution $\rho^0(\boldsymbol{r})$, $\rho(\boldsymbol{r}) \approx \rho^0(\boldsymbol{r})$ or

$w(\boldsymbol{r}) \approx 1$. Indeed, the formation of chemical bonds involves only a minor reconstruction of the electronic structure, mainly in the *valence*-shells of the constituent atoms, so that $|\Delta\rho(\boldsymbol{r})| \equiv |\rho(\boldsymbol{r}) - \rho^0(\boldsymbol{r})| << \rho(\boldsymbol{r}) \approx \rho^0(\boldsymbol{r})$ and hence the ratio $\Delta\rho(\boldsymbol{r})/\rho(\boldsymbol{r}) \approx \Delta\rho(\boldsymbol{r})/\rho^0(\boldsymbol{r})$ is generally small in the energetically important regions of the large density values, near the atomic nuclei. As explicitly shown in the first column of Fig. 10.1, the largest values of the density difference $\Delta\rho(\boldsymbol{r})$ are observed mainly in the bond region, between the nuclei of chemically bonded atoms; the reconstruction of atomic lone pairs can also lead to an appreciable displacement in the molecular electron density in the outer regions of atomic densities.

By expanding the logarithm of the molecular surprisal $I[w(\boldsymbol{r})]$ around $w(\boldsymbol{r}) = 1$, to *first*-order in the relative displacement of the electron density, one obtains the following approximate relations between the local values of the molecular surprisal and the density-difference function:

$$I[w(\boldsymbol{r})] = \ln[\rho(\boldsymbol{r})/\rho^0(\boldsymbol{r})] = \ln\{[\rho^0(\boldsymbol{r}) + \Delta\rho(\boldsymbol{r})]/\rho^0(\boldsymbol{r})\} \\ \cong \Delta\rho(\boldsymbol{r})/\rho^0(\boldsymbol{r}) \approx \Delta\rho(\boldsymbol{r})/\rho(\boldsymbol{r}). \tag{10.7}$$

This relation provides a semiquantitative information-theoretic interpretation of the relative density difference diagrams and links the local surprisal of IT to the density difference function of quantum chemistry (Nalewajski et al. 2002; Nalewajski and Świtka 2002). It also relates the integrands of the alternative information-distance functionals to the corresponding functions of displacements in the electron density:

$$\Delta s(\boldsymbol{r}) = \rho(\boldsymbol{r})I[w(\boldsymbol{r})] \cong \Delta\rho(\boldsymbol{r})w(\boldsymbol{r}) \approx \Delta\rho(\boldsymbol{r}), \tag{10.8}$$

$$\Delta D(\boldsymbol{r}) = \Delta\rho(\boldsymbol{r})I[w(\boldsymbol{r})] \cong [\Delta\rho(\boldsymbol{r})]^2/\rho^0(\boldsymbol{r}) \geq 0. \tag{10.9}$$

The first of these relations qualitatively explains a remarkable similarity between the density difference and the KL information density plots observed in Fig. 10.1, where the contour maps of these quantities are reported for selected linear diatomics and triatomics.

The approximate equalities of (10.8) and (10.9) are numerically verified in Figs. 10.1 and 10.2, respectively. In the former, the contour diagram of the directed divergence density $\Delta s(\boldsymbol{r})$ is compared with the corresponding map of its *first*-order approximation, $\Delta\rho(\boldsymbol{r})w(\boldsymbol{r})$, and the density difference function $\Delta\rho(\boldsymbol{r})$ itself. A general similarity between these diagrams in each row of the figure confirms the semiquantitative character of the *first*-order expansions of the directed divergence densities. The corresponding numerical validation of (10.9) is shown in Fig. 10.2, where the contour maps of Kullback's divergence density $\Delta D(\boldsymbol{r})$ are compared with the corresponding diagrams of its *first*-order approximation $[\Delta\rho(\boldsymbol{r})]^2/\rho^0(\boldsymbol{r})$. Again, a remarkable similarity between the two diagrams in each row numerically validates this approximate relation thus also testifying to the overall smallness of the density adjustments due to the bond-formation processes in molecules.

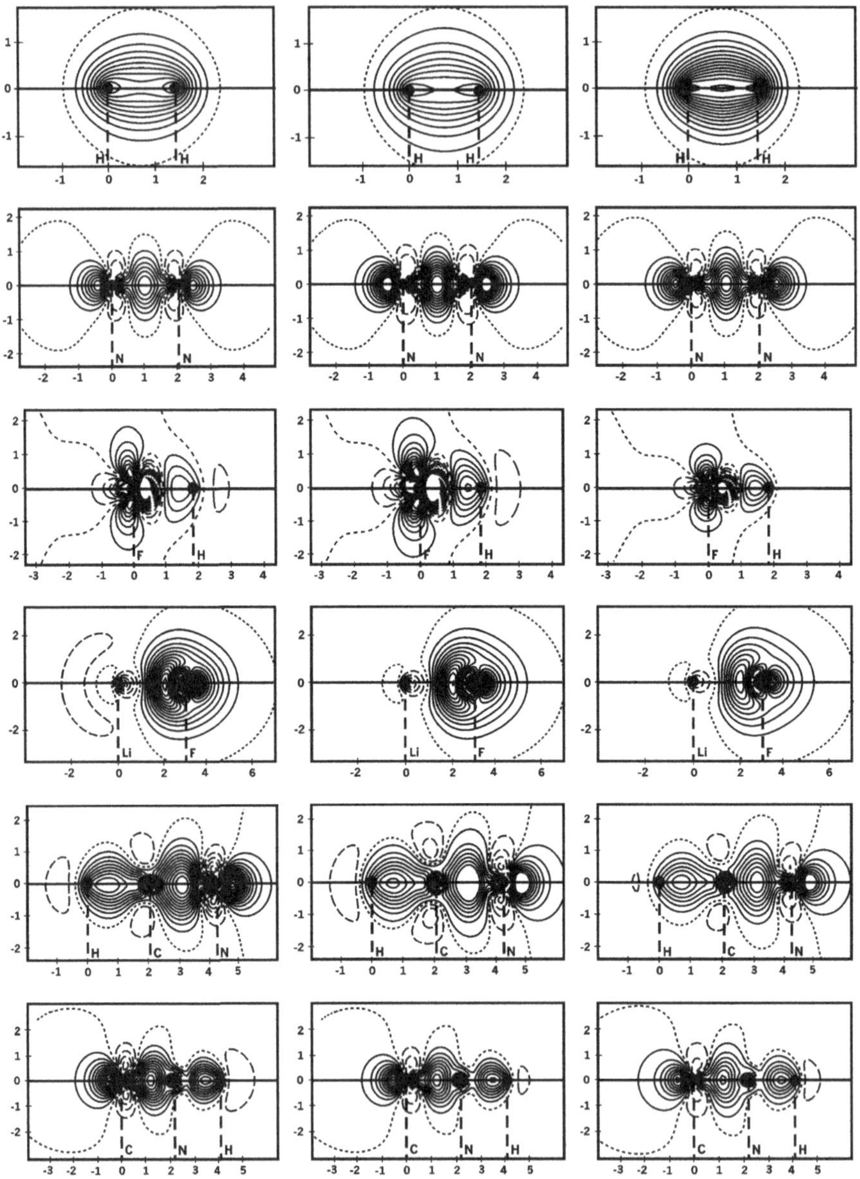

Fig. 10.1 Contour diagrams of the molecular density difference function, $\Delta\rho(\boldsymbol{r}) = \rho(\boldsymbol{r}) - \rho^0(\boldsymbol{r})$ (first column), the information-distance density, $\Delta s(\boldsymbol{r}) = \rho(\boldsymbol{r}) I[w(\boldsymbol{r})]$ (second column) and its approximate, *first*-order expansion $\Delta s(\boldsymbol{r}) \cong \Delta\rho(\boldsymbol{r}) w(\boldsymbol{r})$ (10.8) (third column), for selected diatomic and linear triatomic molecules: H_2, HF, LiF, HCN, and HNC. The solid, pointed, and broken lines denote the positive, zero, and negative values, respectively, of the equally spaced contours; the same convention is applied in Fig. 10.2

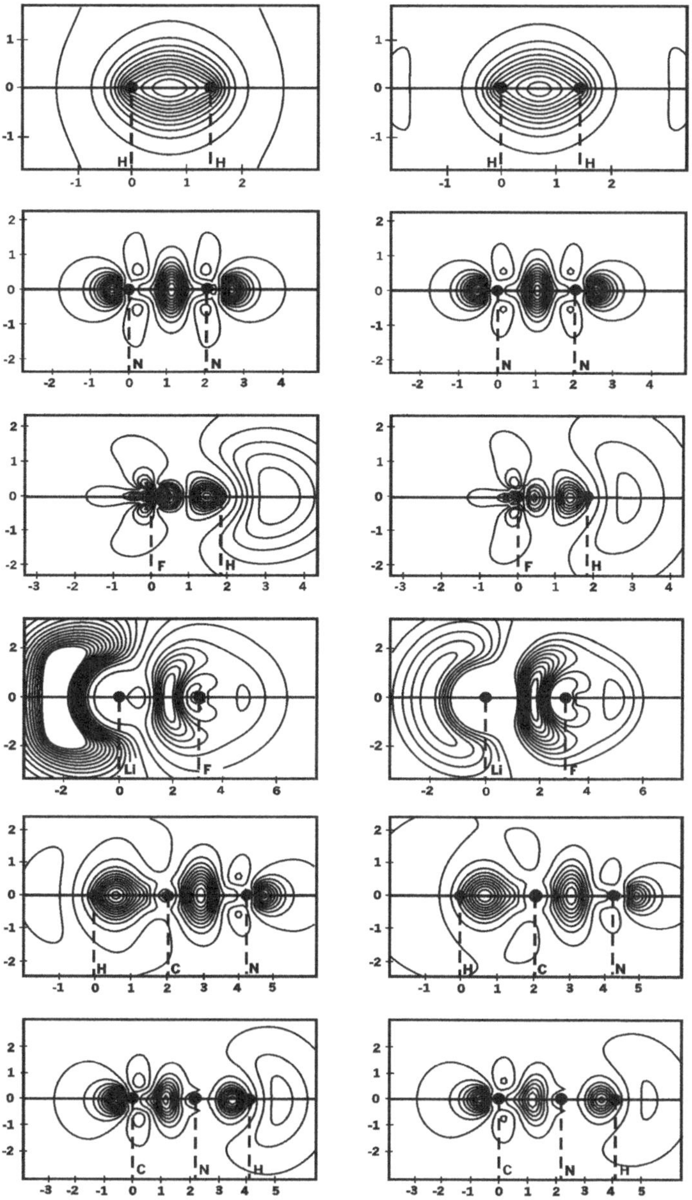

Fig. 10.2 A comparison between contour maps of the molecular divergence density, $\Delta D(\boldsymbol{r}) = \Delta\rho(\boldsymbol{r})I[w(\boldsymbol{r})]$ (first column) and its *first*-order term $[\Delta\rho(\boldsymbol{r})]^2/\rho^0(\boldsymbol{r})$ (second column), for molecules of Fig. 10.1, which validates the approximate relation (10.9)

In Fig. 10.1 the density difference function $\Delta\rho(\boldsymbol{r})$ for representative linear diatomic and triatomic molecules exhibits typical aspects of the equilibrium reconstructions of the *free* atoms during formation of the single and multiple chemical bonds, which exhibit varying degree of the bond covalency (electron-sharing) and ionicity (electron-transfer) components. Let us first examine the contour maps for the two homonuclear diatomics. The single covalent bond in H_2 gives rise to a relative accumulation of electrons in the bond region, between the two nuclei, at the expense of the outer, nonbonding regions of space. The *triple*-bond pattern for N_2 is seen to be more complex, reflecting the density accumulations in the bonding region, due to both the σ and π bonds, and the accompanying increase in the density of the lone pairs on both nitrogen atoms, due to their expected *sp*-hybridization in the promoted valence state of each atom. One also observes a density decrease in the vicinity of the nuclei and an outflow of electrons from the $2p_\pi$ AO to their overlap area, a clear sign of these orbitals involvement in the formation of the double π bond.

Both heteronuclear diatomics, HF and LiF, represent partially ionic bonds between the two atoms exhibiting small and large differences in their electronegativity and chemical hardness descriptors, respectively. A pattern of the density displacement in HF reflects a weakly ionic (strongly covalent) bond, while in LiF the two AIM are seen to be connected by the strongly ionic (weakly covalent) bond. Indeed, in HF one detects a relatively high degree of a "common possession" of the valence electrons by the two atoms, which significantly contribute to the shared bond-charge located between them, and a comparatively weak H $\rightarrow$ F polarization. In LiF, a substantial Li $\rightarrow$ F CT can be detected so that an ion-pair picture indeed provides more adequate *zeroth*-order description of the chemical bond in this diatomic.

Finally, in the two triatomic molecules shown in Fig. 10.1, one identifies a strongly covalent pattern of the electron density displacements in the regions of the single N–H and C–H bonds. A typical buildup of the bond charge due to the multiple CN bonds in the two isomers HCN and HNC can be also observed. The increase in the lone-pair electron density on the terminal heavy atom, N in HCN and or C in HNC, can be also detected, thus again confirming the expected *sp*-hybridization of these bonded atoms in their promoted, valence states in the molecular environment.

A comparison between the corresponding panels of the first two columns in Fig. 10.1 shows that the two displacement maps so strongly resemble one another that they are hardly distinguishable. This confirms a close relation between the local density and entropy-deficiency relaxation patterns, thus attributing to the former the complementary IT interpretation of the latter. This strong resemblance between the two types of molecular diagrams also indicates that the local inflow of electrons increases the relative entropy, while the outflow of electrons gives rise to a diminished level of this relative-uncertainty content of the electron distribution in the molecule. The density displacement and the missing-information distribution can be thus viewed as equivalent probes of the system chemical bonds.

Similar diagnostic conclusions follow from the divergence density plots of Fig. 10.2, where all crucial bonding and nonbonding regions of space are now identified by the *positive* values of Kullback's symmetrized information-distance density. Therefore, all information-distance densities can indeed be regarded as efficient tools for diagnosing the presence of chemical bonds and monitoring the effective valence states of the bonded atoms in molecules.

10.2 Displacements in Entropy Density

We shall now examine the molecular displacements in the average Shannon entropy, relative to the promolecular reference value,

$$\mathscr{H}[\rho] \equiv S[\rho] - S[\rho^0] = -\int \rho(\boldsymbol{r}) \ln \rho(\boldsymbol{r})\, d\boldsymbol{r} + \int \rho^0(\boldsymbol{r}) \ln \rho^0(\boldsymbol{r})\, d\boldsymbol{r} \equiv \int h_\rho(\boldsymbol{r})\, d\boldsymbol{r}, \tag{10.10}$$

and its density $h_\rho(\boldsymbol{r})$ as alternative candidates for the global/local probes of the electron distributions in molecules. The corresponding entropy shifts in terms of the unity-normalized probability distributions,

$$\mathscr{H}[p] \equiv S[p] - S[p^0] = -\int p(\boldsymbol{r}) \ln p(\boldsymbol{r})\, d\boldsymbol{r} + \int p^0(\boldsymbol{r}) \ln p^0(\boldsymbol{r})\, d\boldsymbol{r} \equiv \int h_p(\boldsymbol{r})\, d\boldsymbol{r}, \tag{10.11}$$

can also be used as alternative tools to explore the local relaxations in electron uncertainties, which accompany the chemical bond formation in molecules.

In Fig. 10.3, the contour maps of the entropy-displacement density $h_\rho(\boldsymbol{r})$ are compared with the corresponding density-difference diagrams $\Delta\rho(\boldsymbol{r})$ for the representative linear molecules of Fig. 10.1. To better visualize details of the two functions and to facilitate a qualitative comparison between their topographies, nonequidistant contour values have been selected. Therefore, only the profile of $h_\rho(\boldsymbol{r})$, shown in the third column of the figure, reflects the relative importance of each feature. When interpreting these plots one should realize that a negative (positive) value of $h_\rho(\boldsymbol{r})$ [or $h_p(\boldsymbol{r})$] signifies a decrease (increase) in the local electron *uncertainty* in the molecule, relative to the associated promolecular reference value.

Again, the $\Delta\rho$ and h_ρ diagrams for H_2 are seen to qualitatively resemble one another and the corresponding Δs map shown in Fig. 10.1. The main feature of the h_ρ diagram, an increase in the electron uncertainty in the bonding region between the two nuclei, is due to the inflow of electrons to this region. This manifests the bond-covalency phenomenon, which can be attributed to the electron-sharing effect and a delocalization of the bonding electrons, now effectively moving in the field of both nuclei. One detects in all these maps a similar nodal structure and finds that the nonbonding regions exhibit a decreased uncertainty, due to a transfer of the electron density from this area to the vicinity of the two nuclei and the region between them or as a result of the orbital hybridization (cylindrical polarization).

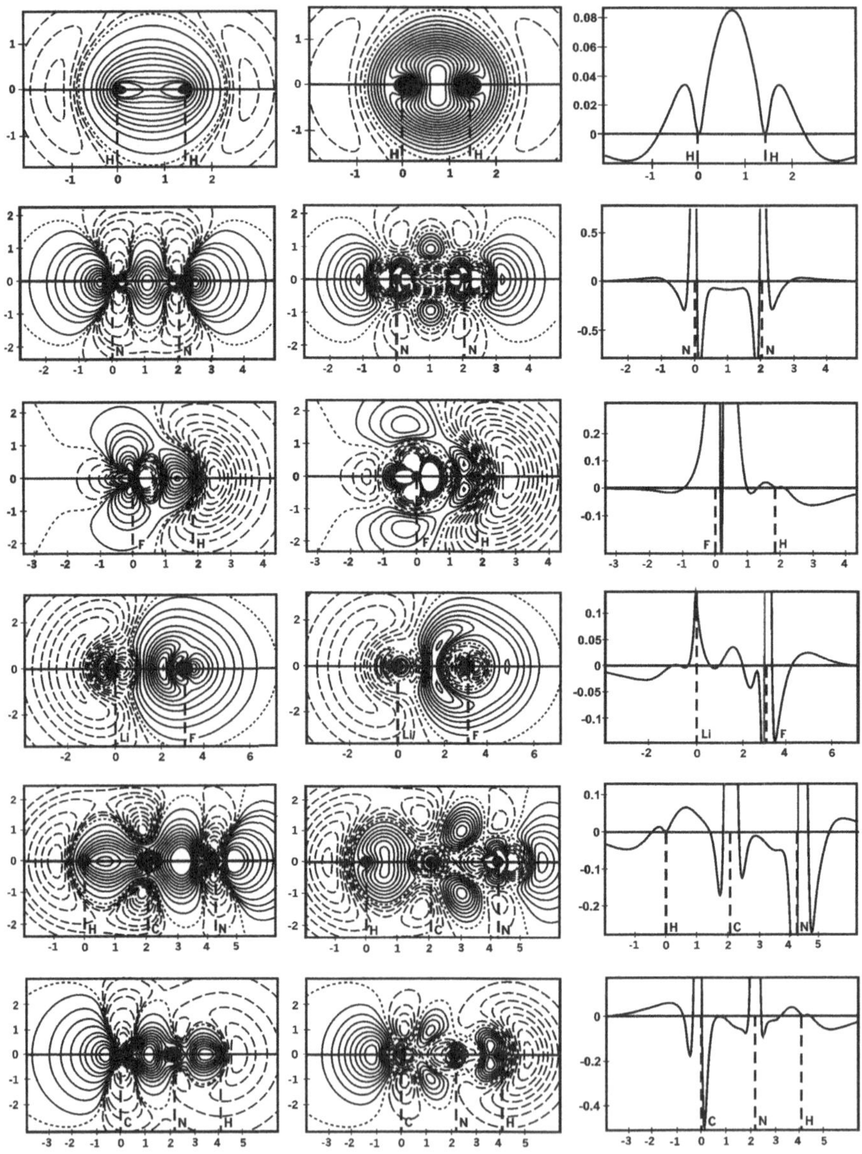

Fig. 10.3 A comparison between the (nonequidistant) contour diagrams of the density difference $\Delta\rho(\boldsymbol{r})$ (first column) and entropy-difference $h_\rho(\boldsymbol{r})$ (second column) functions for the linear molecules of Fig. 10.1. The corresponding profiles of $h_\rho(\boldsymbol{r})$ for the cuts along the bond axis are shown in the third column of the figure

Therefore, the molecular entropy difference function displays all typical features in the reconstruction of electron distributions in a molecule, relative to those of the corresponding free atoms. Its diagrams thus provide an alternative information tool

Table 10.1 Displacements of Shannon entropies (in bits) for molecules of Fig. 10.1

Molecule	$\mathcal{H}[\rho] \equiv S[\rho]-S[\rho^0]$	$S[\rho]$	$S[\rho^0]$
H_2	−0.84	6.61	7.45
N_2	−0.68	8.95	9.63
HF	−1.00	3.00	4.00
LiF	−3.16	5.12	8.28
HCN	−1.44	12.99	14.45
HNC	−1.39	13.06	14.45

for diagnosing the presence of chemical bonds through displacements in the entropy/information content of the molecular electron densities.

In fact, the comparison of Fig. 10.3 also demonstrates that, compared to the corresponding density difference diagrams, the entropy difference plots provide in many respects a more detailed account of the reorganization of the electronic structure relative to the free atoms in the promolecule, particularly in the *inner*-shell regions of heavy atoms.

In Table 10.1 we have listed the representative values of the molecular entropy difference (10.10) together with the associated Shannon entropies for the molecular and promolecular electron densities (Nalewajski and Broniatowska 2003a). These results show that in general the molecular distribution gives rise to a lower level of the information-entropy (less electron uncertainty) compared to the promolecule. This confirms an expected higher degree of compactness exhibited by electron distributions of the *bonded* atoms, which experience the presence of the remaining atoms, compared to their free (separated) analogs.

Thus, the degree of uncertainty contained in the electron distribution on average decreases when the constituent free atoms form the chemical bonds. Indeed, the dominating overall contraction of atomic electron distributions in the field of all nuclear attractors in the molecule should imply a higher degree of "order" (less uncertainty) in the molecular electron density in comparison to that present in the promolecular distribution. The largest magnitude of this relative decrease in the entropy content of the molecular electron density is observed for LiF, which exhibits the most ionic bond (largest amount of CT) among all molecules included in the table.

There is no apparent correlation in the table between the global entropy-displacement and the chemical bond multiplicity. For example, a triple covalent bond in N_2 generates less overall entropy loss than does a single bond in H_2. The reason for a low magnitude of the entropy-displacement in N_2 is the result of a mutual cancellation of the negative and positive contributions due to valence electrons. Indeed, the orbital hybridization and AO contraction should lower the entropy of the atomic electron distribution, since they increase charge inhomogeneity in the molecule, relative to the atomic promolecule. By the same criterion, the effective expansion of atomic densities due to the AIM promotion, as well as the electron delocalization *via* the system chemical bonds and the charge-transfer between AIM should have the opposite effect, of relatively increasing the uncertainty content of the electron distribution in the molecule. Notice that, should one assume a similar entropy-displacement of about −0.7 bits for all triple bonds in a series of isoelectronic molecules N_2, HCN and CNH, one obtains a contribution due to a single C–H or N–H bond of about −0.7, a result close to that found for the H–H bond.

10.3 Illustrative Application to Propellanes

As an additional illustration we now present the combined density-difference, entropy-displacement, and the information-distance analysis of the central bond in small propellane systems (Nalewajski and Broniatowska 2003a). The main purpose of this study was to examine the effect on the central C′–C′ bond, between the (primed) "bridgehead" carbon atoms, of a successive increase in the size of the carbon bridges in the series of the [1.1.1], [2.1.1], [2.2.1], and [2.2.2] propellanes shown in Fig. 10.4. Figure 10.5 reports the contour maps of the molecular density difference function $\Delta\rho(\boldsymbol{r})$, the KL integrand $\Delta s(\boldsymbol{r})$, and the entropy-displacement density $h_\rho(\boldsymbol{r})$, for the planes of sections displayed in Fig. 10.4. The corresponding central-bond profiles of the density- and entropy-difference functions are shown in Fig. 10.6. The corresponding ground-state densities have been generated using the DFT-LDA calculations in the extended (DZVP) basis set.

The density difference plots show that in the small [1.1.1] and [2.1.1] propellanes there is on average a depletion of the electron density between the bridgehead carbon atoms, relative to the atomic promolecule, while the [2.2.1] and

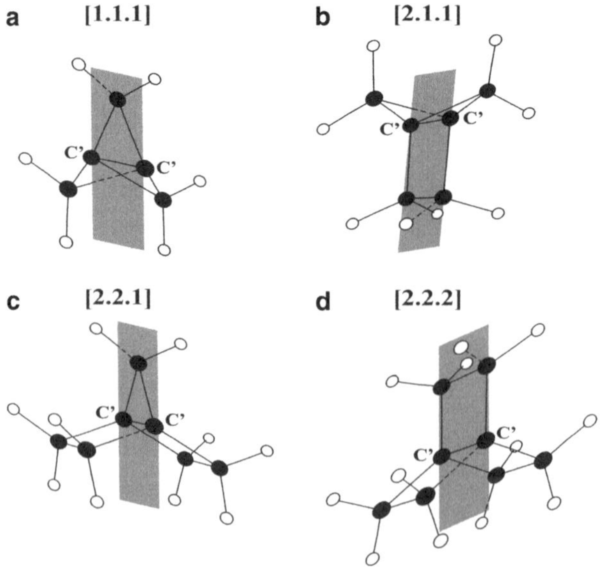

Fig. 10.4 The propellane structures and the planes of sections containing the bridge and bridgehead (C′) carbon atoms, identified by black circles

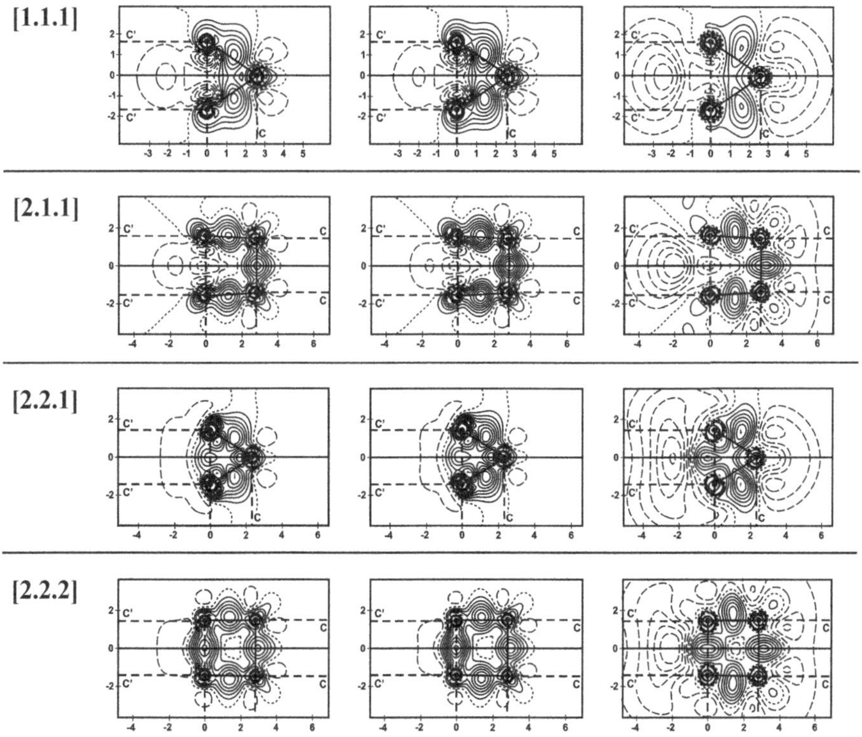

Fig. 10.5 A comparison between the equidistant contour maps of the density-difference function $\Delta\rho(\boldsymbol{r})$ (first column), the information-distance density $\Delta s(\boldsymbol{r})$ (second column), and the entropy-displacement density $h_{\rho}(\boldsymbol{r})$ (third column), for the four propellanes of Fig. 10.4

[2.2.2] systems exhibit a net density buildup in this region. A similar conclusion follows from the entropy-displacement and entropy-deficiency plots of these diagrams. The two entropic maps are again seen to be qualitatively similar to the corresponding density-difference plots. This resemblance is seen to be particularly strong between $\Delta\rho(\boldsymbol{r})$ and $\Delta s(\boldsymbol{r})$ diagrams shown in first two columns of Fig. 10.5.

In the generalized outlook on the bond-order concept (Nalewajski 2011a, b, c; Nalewajski and Gurdek 2010), emerging from extensions of both the Wiberg-type (Wiberg 1968; Gopinathan and Jug 1983; Mayer 1983, 1985; Jug and Gopinathan 1990) or the quadratic-difference approaches (Nalewajski et al. 1993, 1994a, 1996b, 1997; Nalewajski and Mrozek 1994, 1996; Mrozek et al. 1998; Nalewajski 2004b) to bond multiplicities in MO theory, and from the entropic bond-multiplicity concept in the *Orbital Communication Theory* (OCT) of the chemical bond (Nalewajski 2009e, f, g, 2010b, c, d, f; see also Chap. 12), one identifies the overall chemical bond multiplicity as a measure of the statistical "*dependence*" (nonadditivity) between orbitals on different atomic centers.

On one hand, this dependence between a given pair of AO on different atoms, the basis functions contributed to the bond system of the molecule, can be realized *directly* (through space), by the constructive interference of these two orbitals, which increases the electron density between them. For the positive AO overlap it is then manifested by the positive value of the *off*-diagonal CBO (1-matrix) element coupling the two AO in the molecule. In OCT this through-space component, which signifies the absence of any AO intermediaries in the chemical interaction, originates from the *direct* probability/information scattering between the two AO under consideration.

On the other hand, such an interdependence between the two orbitals can also originate from *indirect* sources, through their chemical coupling to the remaining basis functions used to construct the system-occupied MO. Indeed, the orthonormality relations of the occupied MO, which determine the framework of chemical bonds in the molecule, introduce the implicit dependencies between orbitals, which generate the associated "through-bridge" bond-order contributions (Nalewajski 2011a,b,c,e,f; Nalewajski and Gurdek 2011a, b). In OCT this "cascade" bond component is effected through the *indirect* probability/information propagation between the specified pair of AO, through the specified subsets of the remaining basis functions (Nalewajski 2011e,f; Nalewajski and Gurdek 2011b). This component is thus realized through several AO-intermediates, which form an effective "bridge" for the probability/information scattering from one basis function to another.

Therefore, each pair of AO or AIM exhibits the partial through-*space* and through-*bridge* bond components. The bond-order of the former quickly decays with an increase in the interatomic separation. It is also small, when interacting orbitals are heavily engaged in forming chemical bonds with other atoms. The through-*bridge* bond-order, however, can still assume appreciable values, when the remaining atoms form an effective bridge of neighboring (bonded) atoms, which links the atomic pair in question. We shall discuss the basic concepts of this novel IT outlook on statistical origins of bond-orders in Chap. 12.

The numerical bond-orders reported in Fig. 10.6 originate from the *two*-electron difference approach (Nalewajski et al. 1993, 1994a, 1996a, b, 1997; Nalewajski and Mrozek 1994, 1996; Mrozek et al. 1998; Nalewajski 2004b), an extension of the original bond multiplicity index of Wiberg (1968) and its subsequent generalizations (Gopinathan and Jug 1983; Mayer 1983, 1985; Jug and Gopinathan 1990). Together with the corresponding density profiles shown in this figure they reveal a changing nature of the central bond in the four propellanes included in this analysis. The central "bonds" in the smallest systems, lacking the accumulation of the electron density or the entropy/entropy-deficiency density between the bridgehead atoms, are seen to be mostly of the *indirect* character, being realized through-bridges rather than directly through-space (Nalewajski 2011a,b,e,f). Clearly, the most important atomic intermediates for this indirect bond mechanism in propellanes are the *bridge* carbons, which strongly overlap (communicate) with the *bridgehead* carbons.

The missing through-space component in the smallest [1.1.1] system is due to nearly tetrahedral (sp^3) hybridization on the bridgehead carbons (see Sect. 12.10.6), optimum to form strong bonds with the bridging carbon atoms, with the three hybrids on each of these central atoms being used to form the chemical bonds with the bridge carbons and the fourth (nonbonding, *singly*-occupied) hybrid being

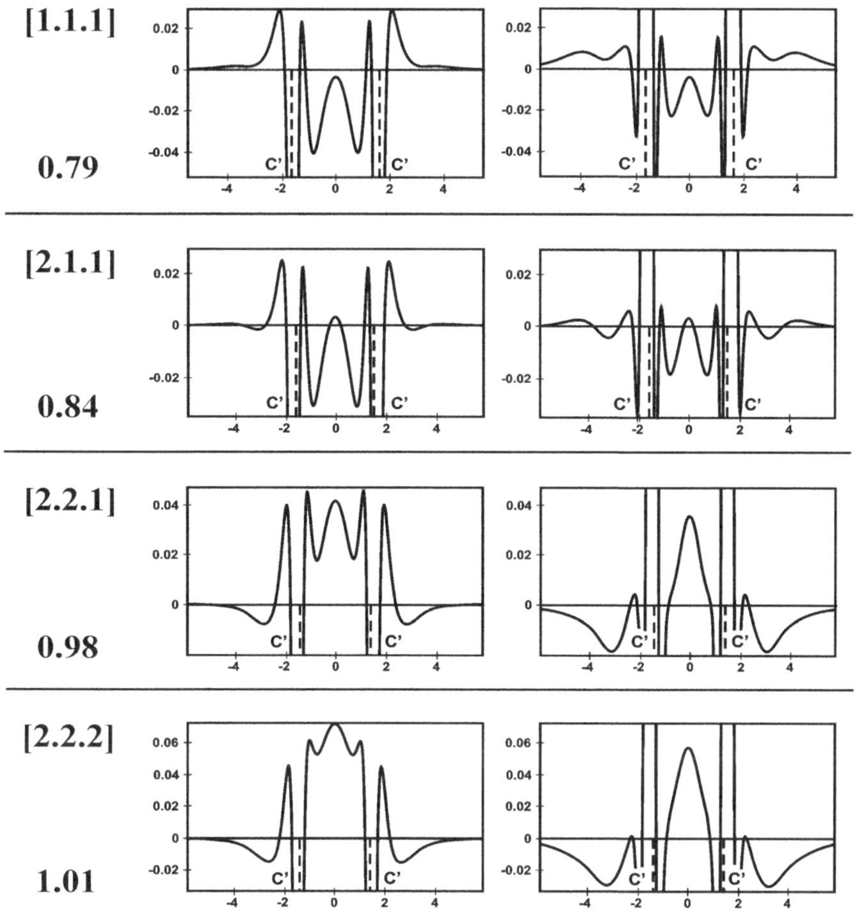

Fig. 10.6 The bridgehead bond profiles of the density difference function (*left panel*) and the molecular entropy-displacement (*right panel*) for the four propellanes shown in Fig. 10.4. For comparison the numerical values of the bond-multiplicities from the *two*-electron difference approach (Nalewajski et al. 1996b) are also reported

directed away from the central-bond region. In the largest [2.2.2] propellane these central carbons acquire a nearly trigonal (sp^2) hybridization, to form bonds with the bridge neighbors, with one 2*p* orbital, not used in this hybridization scheme, being now directed along the central-bond axis thus being capable of forming a strong through-space component of the overall multiplicity of the C′–C′ bond.

A gradual emergence of the direct, through-space component of the central bond, due to accumulation of the electron density and the entropy (entropy-deficiency) density between the bridgehead carbons, is observed when the bridges are enlarged in the two largest propellanes. Using the above *two*-electron difference approach one roughly estimates a full (single) central C′–C′ bond in the [2.2.1] and [2.2.2] propellanes, and approximately 0.8 (through-bridge) bond-order in the [1.1.1] propellane. A more realistic orbital model of these interactions (Nalewajski 2011b) predicts for the largest [2.2.2] system the 0.62 of the direct Wiberg-type component and 0.14 of the indirect bond-order *via* the three double-carbon bridges, giving rise to the total measure of 0.76 bond multiplicity between the bridge-head carbons. The corresponding model estimate of the through-bridges chemical interaction in the smallest [1.1.1] propellane predict a weaker 0.40 bond-order. Therefore, the smaller system exhibits a higher through-bridge component, compensating for the lack of the direct central bond in this system, while the larger system generates almost twice as large overall bond multiplicity, mainly due to the direct component.

10.4 Nonadditive Information Measures

Each scheme $\rho = \sum_\alpha \rho_\alpha$ of exhaustively resolving the molecular ground-state electron density ρ into the corresponding pieces $\boldsymbol{\rho} = \{\rho_\alpha\}$ attributed to molecular fragments of interest, e.g., AIM, MO or AO, molecular fragments, etc., implies the associated division of the physical quantity $A \equiv A[\rho]$ into its subsystem additive and nonadditive components [Nalewajski 2003e, 2008e, 2010a, c, f; Nalewajski et al. 2010a, b; see also (7.372a)–(7.372c) in Sect. 7.8]:

$$A = A[\rho] \equiv A^{total}[\boldsymbol{\rho}] = A^{add.}[\boldsymbol{\rho}] + A^{nadd.}[\boldsymbol{\rho}]. \tag{10.12}$$

Indeed, by the Hohenberg–Kohn theorem *A* correspond to the unique functional of the ground-state density ρ, which defines the overall (*total*) multicomponent functional in the adopted subsystem resolution: $A^{total}[\boldsymbol{\rho}] \equiv A[\rho]$. This partition of the electron density also determines the associated *additive* contribution,

$$A^{add.}[\boldsymbol{\rho}] = \sum\nolimits_\alpha A[\rho_\alpha], \tag{10.13}$$

and hence also (indirectly) the *nonadditive* component:

$$A^{nadd.}[\boldsymbol{\rho}] = A^{total}[\boldsymbol{\rho}] - A^{add.}[\boldsymbol{\rho}] = A[\rho] - \sum\nolimits_\alpha A[\rho_\alpha]. \tag{10.14}$$

For example, this Gordon and Kim (1972) type division (see Sect. 7.8) of the kinetic energy functional defines its nonadditive contribution, which constitutes the basis of the DFT-embedding concept of Cortona (1991) and Wesołowski et al. (Wesołowski and Warshel 1993; Wesołowski et al. 1995; Wesołowski and Weber 1998; Nalewajski 2002f, 2003e; Wesołowski and Tran 2003; Wesołowski 2004a, b; Casida and Wesołowski 2004). It has also been demonstrated (Nalewajski et al. 2005) that the inverse of the nonadditive Fisher information in the MO resolution defines the IT-ELF concept, in spirit of the original ELF formulation by Becke and Edgecombe (1990), while the related quantity in the AO resolution of the SCF MO theory offers the key *Contra-gradience* (CG) criterion for localizing the chemical bonds in molecular systems (Nalewajski 2008e, 2010f; Nalewajski et al. 2010a, b).

Such a division can also be used to partition the information quantities themselves (Nalewajski 2006g, 2010a, c). As an illustration consider the *local* partitioning of the molecular electron density into the AIM components, e.g., in the Hirshfeld ("stockholder") division, in which the molecular probability density at point $\boldsymbol{r}$ is divided into atomic contributions in accordance with the conditional probabilities (*share* factors) (Nalewajski and Parr 2000, 2001; Nalewajski 2002a, 2006g):

$$\begin{aligned} \boldsymbol{d}(\boldsymbol{r}) &= \{P(\mathrm{X}|\boldsymbol{r}) = p_{\mathrm{X}}(\boldsymbol{r})/p(\boldsymbol{r}) = \rho_{\mathrm{X}}(\boldsymbol{r})/\rho(\boldsymbol{r})\} \quad \text{or} \\ \boldsymbol{\rho}(\boldsymbol{r}) &= \{\rho_{\mathrm{X}}(\boldsymbol{r}) = Np_{\mathrm{X}}(\boldsymbol{r}) = \rho(\boldsymbol{r})P(\mathrm{X}|\boldsymbol{r})\}. \end{aligned} \tag{10.15a}$$

The promolecular reference similarly determines the initial conditional probabilities of the free atoms:

$$\begin{aligned} \boldsymbol{d}^0(\boldsymbol{r}) &= \{P^0(\mathrm{X}|\boldsymbol{r}) = p^0_{\mathrm{X}}(\boldsymbol{r})/p^0(\boldsymbol{r}) = \rho^0_{\mathrm{X}}(\boldsymbol{r})/\rho^0(\boldsymbol{r})\} \quad \text{or} \\ \boldsymbol{\rho}^0(\boldsymbol{r}) &= \{\rho^0_{\mathrm{X}}(\boldsymbol{r}) = Np^0_{\mathrm{X}}(\boldsymbol{r}) = \rho^0(\boldsymbol{r})P^0(\mathrm{X}|\boldsymbol{r})\}. \end{aligned} \tag{10.15b}$$

In this local partition scheme the overall distributions $\rho(\boldsymbol{r})$ [or $p(\boldsymbol{r}) = \rho(\boldsymbol{r})/N$] as well as $\rho^0(\boldsymbol{r})$ [or $p^0(\boldsymbol{r}) = \rho^0(\boldsymbol{r})/N$] and the associated promolecular conditional probabilities $\boldsymbol{d}^0(\boldsymbol{r})$ are assumed to be known from earlier molecular and atomic calculations, respectively. The "stockholder" division rule (Hirshfeld 1977) then assumes $\boldsymbol{d}(\boldsymbol{r}) = \boldsymbol{d}^0(\boldsymbol{r})$.

The overall KL missing-information density in this local partition of the electron density,

$$\Delta s[\rho; \boldsymbol{r}] = \rho(\boldsymbol{r}) \log[\rho(\boldsymbol{r})/\rho^0(\boldsymbol{r})\}] = \rho(\boldsymbol{r})I(\boldsymbol{r}) \equiv \Delta s^{total}[\boldsymbol{\rho}(\boldsymbol{r})|\boldsymbol{\rho}^0(\boldsymbol{r})], \tag{10.16}$$

where $w(\boldsymbol{r})$ stands for the molecular enhancement factor relative to the promolecule and $I(\boldsymbol{r}) \equiv I[w(\boldsymbol{r})]$ denotes the associated surprisal, can be subsequently divided into its additive and nonadditive contributions in this atomic resolution:

$$\begin{aligned} &\Delta s^{add.}[\boldsymbol{\rho}(\boldsymbol{r})|\boldsymbol{\rho}^0(\boldsymbol{r})] = \textstyle\sum_{\mathrm{X}} \Delta s[\rho_{\mathrm{X}}; \boldsymbol{r}], \\ &\qquad \Delta s[\rho_{\mathrm{X}}; \boldsymbol{r}] = \rho_{\mathrm{X}}(\boldsymbol{r}) \log[\rho_{\mathrm{X}}(\boldsymbol{r})/\rho^0_{\mathrm{X}}(\boldsymbol{r})] \equiv \rho_{\mathrm{X}}(\boldsymbol{r}) \log w_{\mathrm{X}}(\boldsymbol{r}) \equiv \rho_{\mathrm{X}}(\boldsymbol{r})I_{\mathrm{X}}(\boldsymbol{r}); \end{aligned} \tag{10.17}$$

here $w_{\mathrm{X}}(\boldsymbol{r})$ again denotes the atomic *enhancement* factor and $I_{\mathrm{X}}(\boldsymbol{r}) \equiv I[w_{\mathrm{X}}(\boldsymbol{r})]$ stands for the associated local value of the atomic surprisal. Finally, the difference of the two entropy-deficiency densities determines the associated nonadditive contribution:

$$\Delta s^{nadd.}[\boldsymbol{\rho}(\boldsymbol{r})|\boldsymbol{\rho}^0(\boldsymbol{r})] = \Delta s^{total}[\boldsymbol{\rho}(\boldsymbol{r})|\boldsymbol{\rho}^0(\boldsymbol{r})] - \Delta s^{add.}[\boldsymbol{\rho}(\boldsymbol{r})|\boldsymbol{\rho}^0(\boldsymbol{r})]. \tag{10.18}$$

10.5 Electron Localization Function

Consider now the nonadditivities of the Fisher measure of information in the MO resolution, say, defined by the occupied KS MO from DFT calculations. In the spin-resolved approach one uses the energy functional of spin densities $\{\rho_\sigma(\boldsymbol{r})\}$, each defined by the spin-like electrons occupying the corresponding spin-subsets of MO $\boldsymbol{\psi} = \{\boldsymbol{\psi}_\sigma = \{\psi_i^\sigma\}\}$,

$$\rho_\sigma(\boldsymbol{r}) = \sum_{i\in\sigma}^{occd.} |\psi_i^\sigma(\boldsymbol{r})|^2. \tag{10.19}$$

The (double) noninteracting kinetic energy density (a.u.),

$$2t_{s\sigma}[\boldsymbol{r}, \rho_\sigma] \equiv \tau_\sigma[\boldsymbol{r}; \rho_\sigma] = \sum_{i\in\sigma}^{occd.} |\nabla\psi_i^\sigma(\boldsymbol{r})|^2, \tag{10.20}$$

is then proportional to the distribution of the multicomponent (additive) Fisher information functional in the amplitude representation [see (8.7)],

$$I^{add.}[\boldsymbol{\psi}_\sigma] = 4\sum_{i\in\sigma}^{occd.} \int |\nabla\psi_i^\sigma(\boldsymbol{r})|^2 d\boldsymbol{r} \equiv \int f_\sigma^{add.}(\boldsymbol{r})\, d\boldsymbol{r}, \tag{10.21}$$

$$\tau_\sigma(\boldsymbol{r}) = \frac{1}{4} f_\sigma^{add.}(\boldsymbol{r}). \tag{10.22}$$

The leading term of the Taylor expansion of the spherically averaged HF (conditional) *pair*-probability of finding in distance s from the reference electron of spin σ at position $\boldsymbol{r}$ the other spin-like electron then reads (Becke and Edgecombe 1990; see also: Fuentalba et al. 2009):

$$P_c^{\sigma\sigma}(s|\boldsymbol{r}) = \frac{1}{3} D_\sigma(\boldsymbol{r}) s^2 + \ldots,$$
$$D_\sigma(\boldsymbol{r}) = \tau_\sigma(\boldsymbol{r}) - \frac{|\nabla\rho_\sigma(\boldsymbol{r})|^2}{4\rho_\sigma(\boldsymbol{r})} = -\frac{1}{4}[f_\sigma^{total}(\boldsymbol{r}) - f_\sigma^{add.}(\boldsymbol{r})] = -\frac{1}{4} f_\sigma^{nadd.}(\boldsymbol{r}) \geq 0. \tag{10.23}$$

Above, we have explicitly indicated that the function $D_\sigma(\boldsymbol{r})$, a key concept for the definition of ELF, is proportional to the negative nonadditive component of the Fisher information in the spin-resolved MO resolution.

The appropriately calibrated square of its inverse, which determines the original ELF, has been successfully used as a probe of the electron localization patterns in atoms (Fig. 10.7) and molecules (Nalewajski et al. 2005). Indeed, the magnitude of

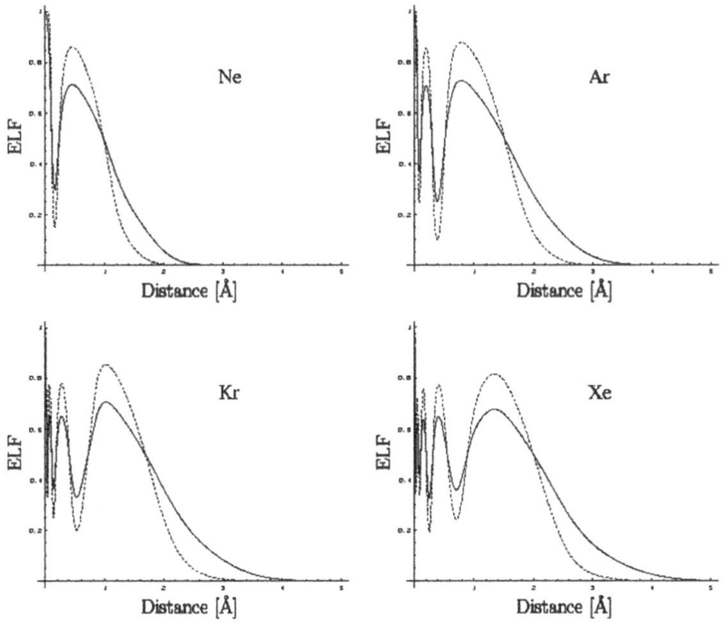

Fig. 10.7 Plots of ELF (*dashed line*) and IT-ELF (*solid line*) for Ne, Ar, Kr, and Xe

the (negative) nonadditive information component increases with the electron *delocalization*; therefore, its inverse reflects the complementary aspect of the electron *localization*. When supplemented with elements of the distribution topological analysis, the ELF concept provides a powerful tool for approaching many problems in structural chemistry (Silvi and Savin 1994; Savin et al. 1997).

A somewhat improved behavior at long distances is observed in the DFT-tailored simple inverse of this function, known as IT-ELF (Nalewajski et al. 2005). In Fig. 10.7, representative graphs of both functions are presented for Ne, Ar, Kr, and Xe. A qualitative behavior of the two curves is seen to be very similar, emphasizing the shell structure of these noble gas atoms. One also observes that ELF exhibits a faster decay at large distances from the nucleus and thus gives rise to smaller spatial extension of the valence basins compared to those in IT-ELF.

In Fig. 10.8, we have compared the perspective views of these functions for N_2, NH_3, PH_3, and B_2H_6. These ELF plots convincingly validate the use of this local probe as an indicator of the localization of the valence electrons in the bonding and nonbonding (lone-pair) regions of these illustrative molecular systems. Indeed, in the homonuclear diatomic N_2 one detects in both plots the typical accumulation of electrons between the nuclei, due to the formation of the triple covalent bond, and the accompanying increase in the localization of the lone-pair electrons in the nonbonding regions of both atoms, a clear manifestation of the accompanying *sp*-hybridization. The three localized N–H bonds are also clearly visualized in both NH_3 panels of the figure and a similar ELF features are detected in its PH_3 part. The final B_2H_6 plots are also seen to successfully locate the bonding electrons of the four terminal B–H bonds.

The final application, shown in Fig. 10.9, again investigates the direct central and bridge bonds between the carbon atoms in the smallest [1.1.1] and largest [2.2.2] propellanes of Fig. 10.4. In these contour maps the absence (presence) of the through-space component of the central bond, between the bridgehead carbon atoms, is clearly seen in the upper (lower) panel for both versions of ELF, while the structure of the C–C bonds in the bridges is also transparently revealed. An interesting feature of the bridge bonds in the upper diagrams, also seen in Fig. 10.5, is a slight displacement of the bonding electrons away from the line connecting the bridge and bridgehead carbons, due to the near sp^3-hybridization on the bridge-carbon, which is required to additionally accommodate the two hydrogen atoms (Nalewajski 2010l).

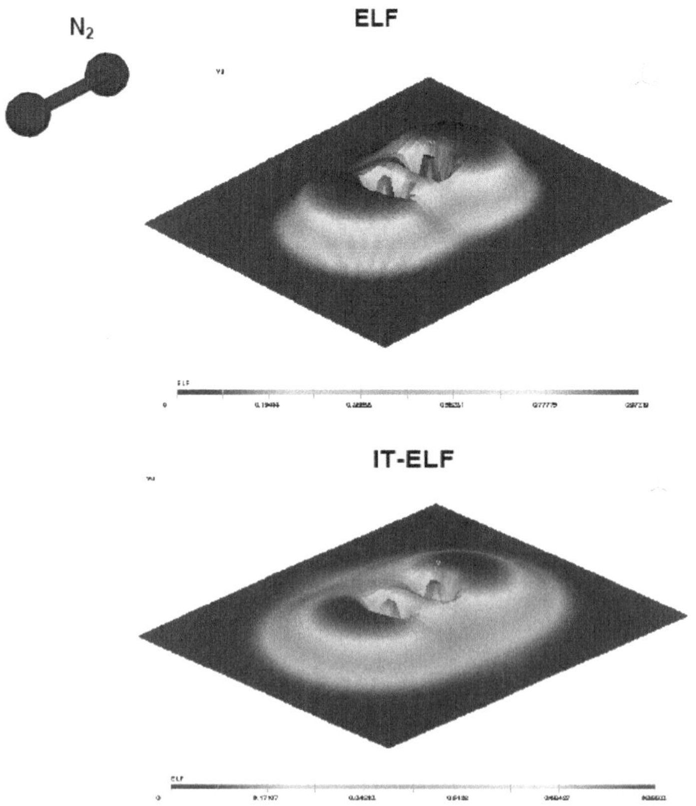

Fig. 10.8 (continued)

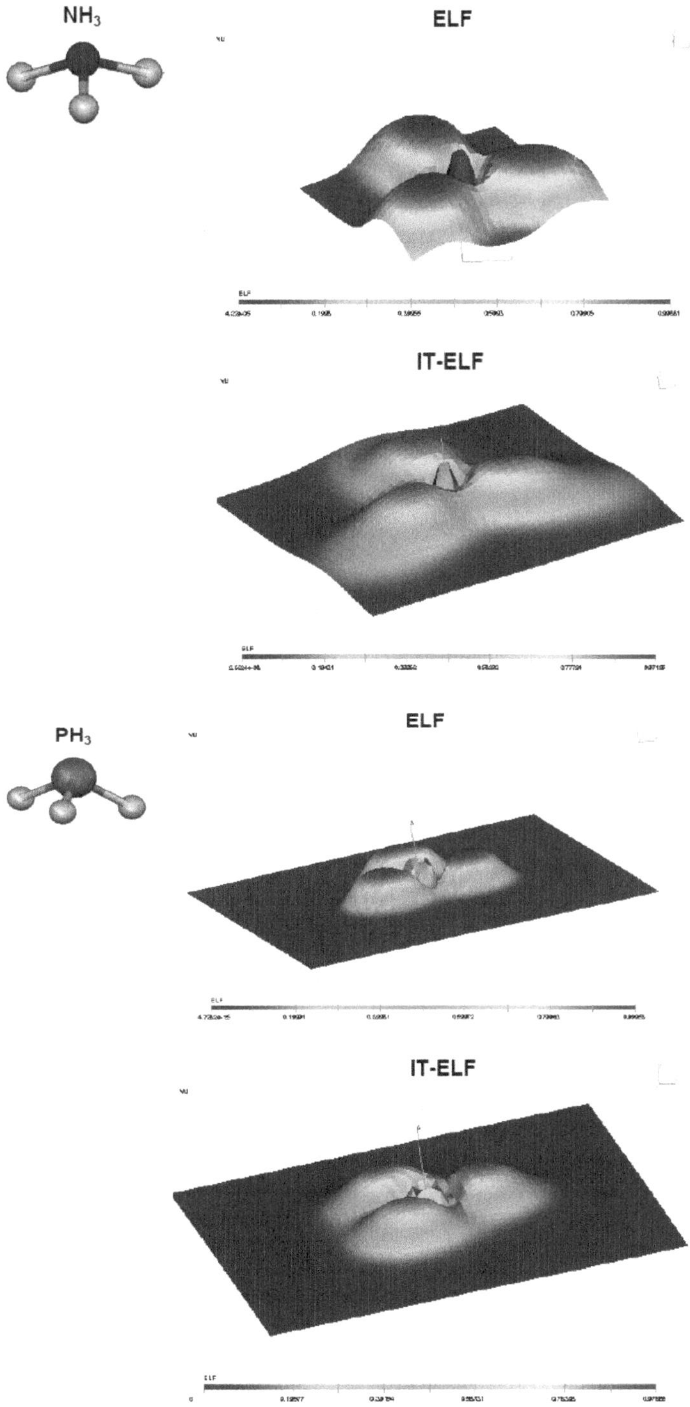

Fig. 10.8 (continued)

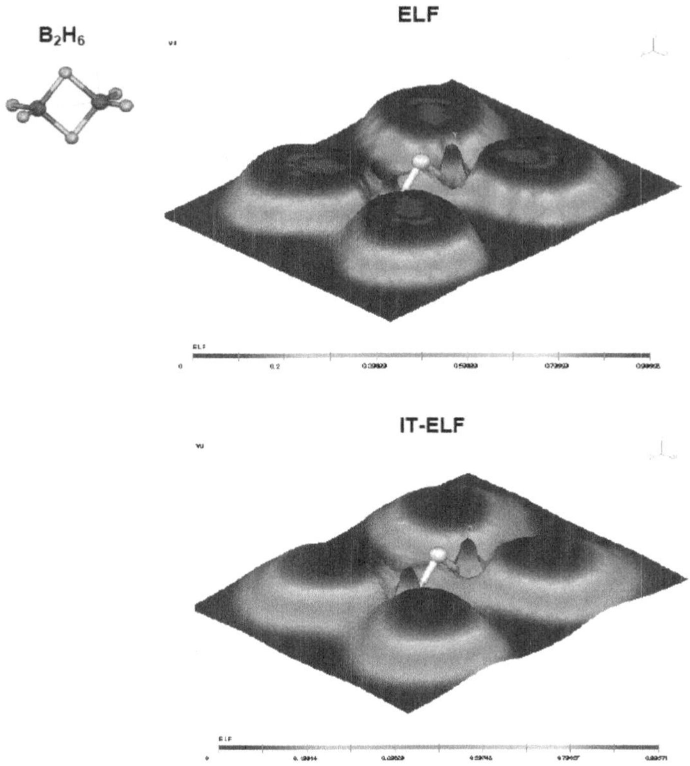

Fig. 10.8 Comparison between perspective views of the ELF and IT-ELF surfaces for N_2, NH_3, PH_3, and B_2H_6, on the corresponding planes of section: along the bond axis (N_2), in the plane determined by three hydrogen atoms (NH_3 and PH_3), and in the plane passing through both terminal BH_2 groups (B_2H_6), respectively

10.6 Contra-gradience Criterion for Locating Bonding Regions in Molecules

Consider next the familiar problem of combining the two (Löwdin-orthogonalized) AO (OAO), $A(\boldsymbol{r})$ and $B(\boldsymbol{r})$, say, two 1*s* orbitals centered on nuclei A and B, respectively, which contribute a single electron each, $N = 2$, to form the chemical bond A—B. The two basis functions $\boldsymbol{\chi} = (A, B)$ then form the bonding (φ_b) and antibonding (φ_a) MO combinations, $\boldsymbol{\varphi} = (\varphi_b, \varphi_a) = \boldsymbol{\chi}\mathbf{C}$:

$$\varphi_b = \sqrt{P}A + \sqrt{Q}B \equiv \sum\nolimits_{k=A,B} \chi_k C_{k,b}, \quad \varphi_a = -\sqrt{Q}A + \sqrt{P}B \equiv \sum\nolimits_{k=A,B} \chi_k C_{k,a}, \quad P + Q = 1, \tag{10.24}$$

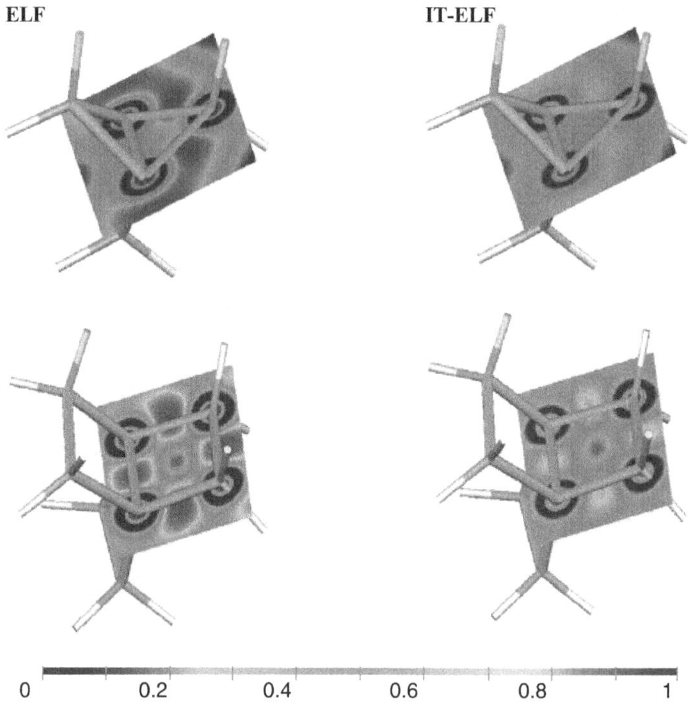

Fig. 10.9 Plots of the ELF(first column) and IT-ELF (second column) for the [1.1.1] (top row) and [2.2.2] (bottom row) propellanes of Fig. 10.4, on the indicated planes of section. The color scale for the ELF values is given in the bottom of the figure

where the square matrix $\mathbf{C} = [\boldsymbol{C}_b, \boldsymbol{C}_a]$ groups the LCAO MO expansion coefficients expressed in terms of the complementary probabilities: P and $Q = 1-P$. The former marks the conditional probabilities $P(A|\varphi_b) = |C_{A,b}|^2 = P(B|\varphi_a) = |C_{B,a}|^2 \equiv P$, and the latter measures the remaining elements of the conditional-probability matrix, of observing AO in MO: $P(B|\varphi_b) = |C_{B,b}|^2 = P(A|\varphi_a) = |C_{A,a}|^2 = Q$.

We now examine the (bonding) ground-state, $\Psi_0 = [\varphi_b^2]$, the (nonbonding) *singly* excited configuration $\Psi_1 = [\varphi_b^1 \varphi_a^1]$, and the (antibonding) *doubly* excited state $\Psi_2 = [\varphi_a^2]$. Consider the *charge-and-bond-order* (CBO), density matrix of the SCF LCAO MO theory for each of these model configurations,

$$\begin{aligned}\boldsymbol{\gamma}_i &= \mathbf{C}\,\mathbf{n}(\Psi_i)\mathbf{C}^{\mathrm{T}} = \langle \boldsymbol{\chi} \mid \boldsymbol{\psi}^{occd.}(\Psi_i)\rangle \mathbf{n}(\Psi_i)\langle \boldsymbol{\psi}^{occd.}(\Psi_i) \mid \boldsymbol{\chi}\rangle \\ &= N\langle \boldsymbol{\chi} \mid \boldsymbol{\psi}^{occd.}(\Psi_i)\rangle \mathbf{p}(\Psi_i)\langle \boldsymbol{\psi}^{occd.}(\Psi_i) \mid \boldsymbol{\chi}\rangle = N\langle \boldsymbol{\chi}|\hat{\mathrm{d}}(\Psi_i)|\boldsymbol{\chi}\rangle, \quad i = 0, 1, 2,\end{aligned} \tag{10.25}$$

with the diagonal matrix of the MO electron occupations $\{n_s(\Psi_i)\}$, $\mathbf{n}(\Psi_i) = \{n_s(\Psi_i)\delta_{s,s'}\} = N\mathbf{p}(\Psi_i) = \{Np_s(\Psi_i)\delta_{s,s'}\}$, reflecting the MO probabilities $\{p_s(\Psi_i) = n_s(\Psi_i)/N\}$ in the electron configuration under consideration. As shown in the preceding equation, this matrix constitutes the OAO representation of the density operator $\hat{d}(\Psi_i)$ determined by the appropriate statistical mixture of the subspace of the *singly* occupied spin-MO in the given electron configuration Ψ_i, $\boldsymbol{\psi}^{occd.}(\Psi_i)$. The three CBO matrices of the preceding equation read:

$$\boldsymbol{\gamma}_0 = 2\begin{bmatrix} P & \sqrt{PQ} \\ \sqrt{PQ} & Q \end{bmatrix}, \quad \boldsymbol{\gamma}_1 = \begin{bmatrix} 1 & 0 \\ 0 & 1 \end{bmatrix}, \quad \boldsymbol{\gamma}_2 = 2\begin{bmatrix} Q & -\sqrt{PQ} \\ -\sqrt{PQ} & P \end{bmatrix}. \tag{10.26}$$

They are seen to reflect the configuration bonding status. Indeed, the (bonding) ground state exhibits the positive *off*-diagonal bond-order, $\gamma_{A,B} = 2(PQ)^{1/2}$, which vanishes in the nonbonding configuration Ψ_1, $\gamma_{A,B} = 0$, while its negative value in Ψ_2, $\gamma_{A,B} = -2(PQ)^{1/2}$, correctly reflects the configuration antibonding character.

In this (real) 2-OAO model the AO-partition of the Fisher information densities of the bonding and antibonding MO gives the following total MO information densities:

$$\begin{aligned} f_b = 4(\nabla\varphi_b)^2 \equiv f_{\boldsymbol{\chi}}^{total}[\varphi_b] &= 4[P(\nabla A)^2 + Q(\nabla B)^2] + 8\sqrt{PQ}\nabla A\cdot\nabla B \\ &\equiv f_{\boldsymbol{\chi}}^{add.}[\varphi_b] + f_{\boldsymbol{\chi}}^{nadd.}[\varphi_b], \end{aligned} \tag{10.27}$$

$$\begin{aligned} f_a = 4(\nabla\varphi_a)^2 \equiv f_{\boldsymbol{\chi}}^{total}[\varphi_a] &= 4[Q(\nabla A)^2 + P(\nabla B)^2] - 8\sqrt{PQ}\nabla A\cdot\nabla B \\ &\equiv f_{\boldsymbol{\chi}}^{add.}[\varphi_a] + f_{\boldsymbol{\chi}}^{nadd.}[\varphi_a]. \end{aligned} \tag{10.28}$$

Along the bond-axis, in the AO-overlap region between the two atoms, which is decisive for the (direct) bonding or antibonding character of MO, the CG density defined by the product of gradients of the two interacting basis functions is negative: $i^{c-g} = \nabla A\cdot\nabla B < 0$ (see Fig. 10.10). Therefore, for these crucial locations

$$\begin{aligned} f_{\boldsymbol{\chi}}^{nadd.}[\varphi_b] &= 8\sqrt{PQ}\,\nabla A\cdot\nabla B \equiv 8\sqrt{PQ}\,i^{c-g} < 0, \\ f_{\boldsymbol{\chi}}^{nadd.}[\varphi_a] &= -8\sqrt{PQ}\,\nabla A\cdot\nabla B = -8\sqrt{PQ}\,i^{c-g} > 0. \end{aligned} \tag{10.29}$$

One similarly defines the average densities of the nonadditive and additive Fisher information contributions per electron for the configuration in question:

$$\begin{aligned} f_{\boldsymbol{\chi}}^{nadd.}[\Psi_i] &= \sum\nolimits_s [n_s(\Psi_i)/N] f_{\boldsymbol{\chi}}^{nadd.}[\varphi_s] \equiv \sum\nolimits_s p_s(\Psi_i) f_{\boldsymbol{\chi}}^{nadd.}[\varphi_s], \\ f_{\boldsymbol{\chi}}^{add.}[\Psi_i] &= \sum\nolimits_s p_s(\Psi_i) f_{\boldsymbol{\chi}}^{add.}[\varphi_s]. \end{aligned} \tag{10.30}$$

In particular, for the three model electron configurations one finds:

$$f_{\boldsymbol{\chi}}^{nadd.}[\Psi_0] = 8\sqrt{PQ}\,i^{c-g}, \quad f_{\boldsymbol{\chi}}^{nadd.}[\Psi_1] = 0, \quad f_{\boldsymbol{\chi}}^{nadd.}[\Psi_2] = -8\sqrt{PQ}\,i^{c-g}. \tag{10.31}$$

Therefore, the AO-phase-dependent, nonadditive contribution to the MO Fisher information density, proportional to the CG integral

$$I^{c-g} = \int i^{c-g}(\boldsymbol{r})\, d\boldsymbol{r} = \int \nabla A(\boldsymbol{r}) \cdot \nabla B(\boldsymbol{r})\, d\boldsymbol{r} = -\int A(\boldsymbol{r})\, \Delta B(\boldsymbol{r})\, d\boldsymbol{r}$$
$$= \frac{2m_e}{\hbar^2} \langle A|\, \hat{\mathrm{T}} |B\rangle \equiv \frac{2m_e}{\hbar^2} T_{A,B}, \tag{10.32}$$

reflects the bonding and antibonding characters of both MO as well as the bonding, nonbonding, and antibonding nature of the model configurations Ψ_0, Ψ_1, and Ψ_2, respectively. It thus provides an attractive concept for both locating and indexing the chemical bonds in molecules (Nalewajski 2010a, g; Nalewajski et al. 2010a, b).

As also indicated in the preceding equation, the CG integral measures the coupling (*off*-diagonal) element $T_{A,B}$ of the electronic kinetic energy operator $\hat{\mathrm{T}}$ between the two basis functions. Such integrals are routinely calculated in typical quantum-chemical packages for determining the electronic structure of molecular systems. This observation also emphasizes the crucial role of the kinetic energy terms in the IT interpretations of the entropic origins of the chemical bonding, using the Fisher measure of information.

A reference to Fig. 10.10 indicates that the $i^{c-g}(\boldsymbol{r}) = 0$ contour is defined by the equation $\boldsymbol{r}_{\mathrm{A}} \cdot \boldsymbol{r}_{\mathrm{B}} = 0$. It separates the region of positive contributions $i^{c-g}(\boldsymbol{r}) > 0$, outside this contour, from the region of negative CG density, $i^{c-g}(\boldsymbol{r}) < 0$, inside the contour. Consider, e.g., its section in xz-plane, for $y = 0$. For the Cartesian reference frame located at the bond midpoint and the two nuclei in positions $\boldsymbol{R}_{\mathrm{A}} = (0, 0, -½R)$ and $\boldsymbol{R}_{\mathrm{B}} = (0, 0, ½R)$, where the internuclear separation $R_{\mathrm{AB}} = R$, the electron-position vectors $\{\boldsymbol{r}_{\mathrm{X}} = \boldsymbol{r} - \boldsymbol{R}_{\mathrm{X}}\}$ in this plane of section are: $\boldsymbol{r}_{\mathrm{A}} = (x, 0, z + ½R)$ and $\boldsymbol{r}_{\mathrm{B}} = (x, 0, z - ½R)$. The equation determining the $i^{c-g}(\boldsymbol{r}) = 0$ contour

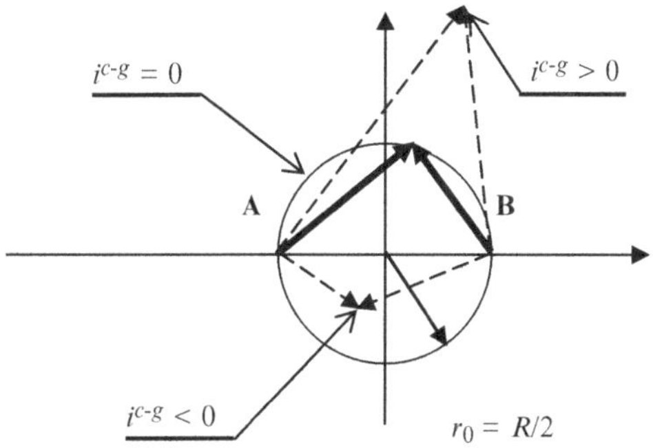

Fig. 10.10 The circular contour of the vanishing CG integrand for two 1s orbitals on atoms A and B, $i^{c-g}(\boldsymbol{r}) = 0$, passing through both nuclei, which separates the *bonding region* (inside the circle), where $i^{c-g}(\boldsymbol{r}) < 0$, from the region of positive contributions $i^{c-g}(\boldsymbol{r}) > 0$ (outside the circle). At each location $\boldsymbol{r}$ the sign of $i^{c-g}(\boldsymbol{r})$ is determined by the scalar product of the electron-position vectors $\boldsymbol{r}_\alpha = \boldsymbol{r} - \boldsymbol{R}_\alpha$, $\alpha = $ A, B, which are mutually perpendicular on the zero contour passing through both nuclei

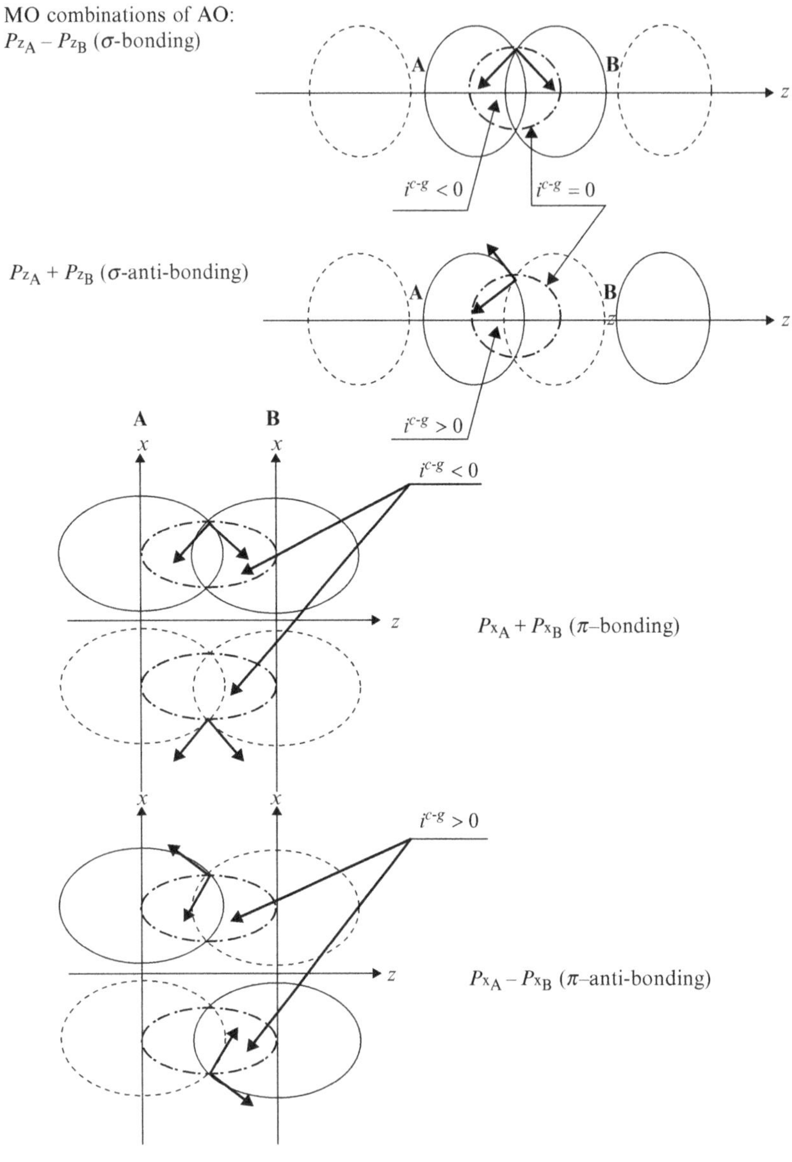

Fig. 10.11 Schematic representation of the bonding $[i^{c-g}(\boldsymbol{r}) < 0]$ and antibonding $[i^{c-g}(\boldsymbol{r}) > 0]$ regions of the strong AO overlap in the chemical interaction between the valence orbitals p_σ and p_π on atoms A and B, enclosed by the vanishing CG contours, $i^{c-g}(\boldsymbol{r}) = 0$, represented by the pointed–broken line. The AO gradients, each in the direction perpendicular to the orbital contour, remain mutually perpendicular on the $i^{c-g}(\boldsymbol{r}) = 0$ surface (see also Fig. 10.10). It again follows from these qualitative diagrams that the negative density of the nonadditive Fisher information accompanies the electron delocalization via the *constructive*-combination of AO in the bonding MO, while the positive density of this information contribution is associated with an effective electron localization, due to the *destructive*-combination of AO in the antibonding MO.

then reads: $x^2 + z^2 = (^1/_2 R)^2 \equiv r_0^2$. It thus determines the circle shown in the figure, centered at the bond midpoint and passing through both nuclei.

As also argued in the qualitative diagram of Fig. 10.11, the negative CG density is essential for any bonding interaction between basis functions, e.g., the two *p*-type orbitals. Therefore, it is vital for the direct *bonding* interaction between a given pair of basis functions that the gradient of one orbital exhibits a nonvanishing, *negative* component along the direction of the gradient of the other orbital, which justifies the name of the CG criterion itself.

10.7 Illustrative Applications of CG Probe

In the ground-state of N-electron system and the OAO basis set $\boldsymbol{\chi} = (\chi_1, \chi_2, \ldots, \chi_m)$ the nonadditive Fisher information density in the AO resolution for the electron configuration defined by N lowest (*singly*-occupied) molecular *spin*-orbitals $\boldsymbol{\psi} = \{\psi_k\}$, with the spatial MO parts $\boldsymbol{\varphi} = \boldsymbol{\chi}\mathbf{C} = \{\varphi_k,\ k = 1, 2, \ldots, N\}$, e.g., those from the SCF MO or KS calculations, reads:

$$I^{nadd.}[\boldsymbol{\chi}] = 4\sum_{k=1}^{m}\sum_{l=1}^{m}(1-\delta_{k,l})\int \gamma_{k,l}\nabla\chi_l^*(\boldsymbol{r})\cdot\nabla\chi_k(\boldsymbol{r})\,d\boldsymbol{r} \equiv 2\int f^{nadd.}(\boldsymbol{r})\,d\boldsymbol{r}$$
$$= 8T^{nadd.}[\boldsymbol{\chi}], \qquad (10.33)$$

where the CBO (density) matrix now provides the AO representation of the projection operator onto the occupied subspace of SO, $\hat{\mathrm{d}}(\Psi_0) = \hat{\mathrm{P}}_{\psi} = |\boldsymbol{\psi}\rangle\langle\boldsymbol{\psi}|$,

$$\boldsymbol{\gamma} = \langle\boldsymbol{\chi}|\left(\sum_{k=1}^{N}|\psi_k\rangle\langle\psi_k|\right)|\boldsymbol{\chi}\rangle = \langle\boldsymbol{\chi}|\boldsymbol{\psi}\rangle\langle\boldsymbol{\psi}|\boldsymbol{\chi}\rangle = \mathbf{CC}^{\dagger} = \{\gamma_{u,w}\}. \qquad (10.34)$$

It is proportional to the nonadditive component $T^{nadd.}[\boldsymbol{\chi}]$ of the system average kinetic energy: $T^{\mathrm{total}}[\boldsymbol{\chi}] = \mathrm{tr}(\boldsymbol{\gamma}\mathbf{T})$, where the kinetic-energy matrix in AO representation $\mathbf{T} = \{T_{k,l} = \langle\chi_k|\hat{\mathrm{T}}|\chi_l\rangle\}$.

In this general molecular scenario of the orbital approximation one uses the most extended (valence) basins of the negative CG density, $f^{nadd.}(\boldsymbol{r}) < 0$, enclosed by the corresponding $f^{nadd.}(\boldsymbol{r}) = 0$ surfaces, as locations of the system chemical bonds. This proposition has been recently validated numerically (Nalewajski et al. 2010a, b). These results have been obtained using the standard SCF MO calculations [GAMESS (1993) software] in the minimum AO basis; the STO-3G basis has been selected to facilitate a comparison with intuitive chemical considerations. In the remaining part of this section we present in Figs. 10.12–10.21 representative results of this extensive study (see also: Nalewajski 2010a, f). The contour maps, for the optimized geometries, will be reported in a.u. The negative CG basins, also shown in the perspective views, are identified by the broken-line contours. For the visualization purposes the Matpack and DISLIN graphic libraries have been used.

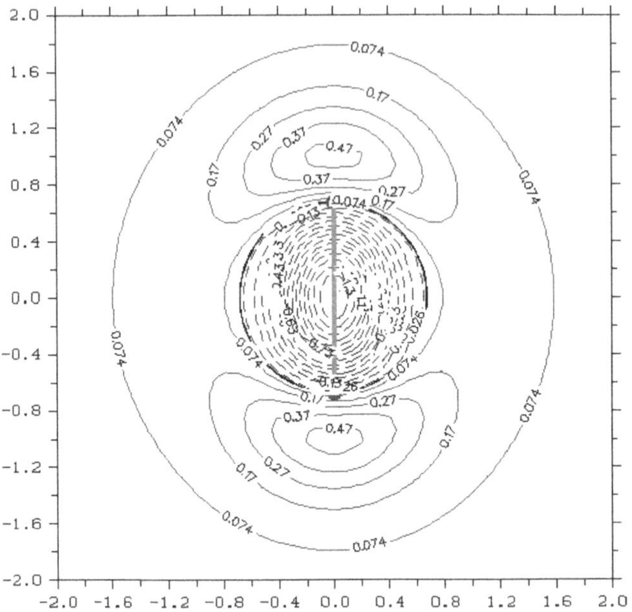

Fig. 10.12 The contour map of the CG density $f^{nadd.}(\boldsymbol{r})$ for H_2 (see also Fig. 10.10)

The contour map of Fig. 10.12 confirms qualitative predictions of Fig. 10.10. In this axial cut of $f^{\,nadd.}(\boldsymbol{r})$ for H_2 the nonadditive Fisher information is seen to be lowered in the spherical *bonding* region between the two nuclei. At the same time the accompanying increases in this quantity are observed in the nonbonding regions of each hydrogen atom, signifying the increased localization (structure) in this homonuclear diatomic due to the axial polarization of the initially spherical atomic densities. It should be stressed that the molecular CG integral over the whole space must be positive, since by the *virial theorem* for the equilibrium geometry the shift in kinetic component of the BO potential, relative to the separated atom (dissociation) limit, must be positive, thus giving rise to the overall *"production"* of the nonadditive Fisher information in the molecular hydrogen.

A similar analysis for HF is presented in Fig. 10.13. The perspective view of $f^{nadd.}(\boldsymbol{r}) < 0$ volumes (upper panel) and the contour map of the axial cut of $f^{nadd.}(\boldsymbol{r})$ (lower panel) indicate the existence of three basins of a decreased nonadditive Fisher information: a large, dominating (bonding) region Ω_1 located in the valence shells of two atoms, and the axially centered two small volumes detected in the inner-shell of fluorine. The shape of the bonding volume is seen to exhibit a polarization toward the $2p_{\pi}$ orbitals of a more electronegative fluorine atom.

In HCl (Fig. 10.14) one again observes two smaller (inner-shell) and a large (valence-shell) basins of the negative CG density. The softer heavy atom is now seen to undergo a more substantial inner-shell reconstruction in the nonadditive Fisher information. There is an axial build-up of $f^{nadd.}(\boldsymbol{r})$ seen in the nonbonding regions of two atoms, particularly on the hydrogen atom. It should be realized that

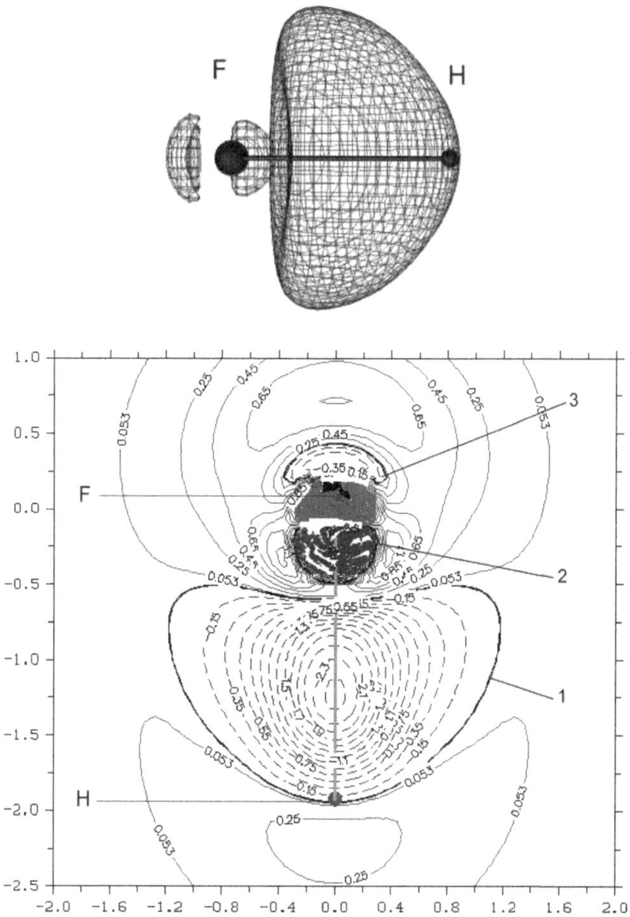

Fig. 10.13 The perspective view of the negative basins of CG (upper panel) and the contour map of the CG density (lower panel) for HF

compared to HF, where both atoms exhibit the "hard" (difficult to polarize) electron distributions in their valence shells, the (soft) chlorine atom combined with the (hard) hydrogen generates a stronger ionic (electron-transfer) component H $\rightarrow$ Cl, ultimately giving rise to the ionic pair H^+Cl^- in the dissociation limit, and hence a smaller covalent (electron-sharing) component H—Cl in the resultant chemical bond.

Consider next the triple chemical bond in N_2 (Fig. 10.15), where the bonding (valence) basin is now distinctly extended away from the bond axis, due to the presence of two π bonds accompanying the central σ bond. Small core-polarization basins, now symmetrically distributed near each constituent atom along the bond axis, are observed in the perspective view, while the *sp*-hybridization reconstruction of the nonbonding regions on both atoms is again much in evidence in the

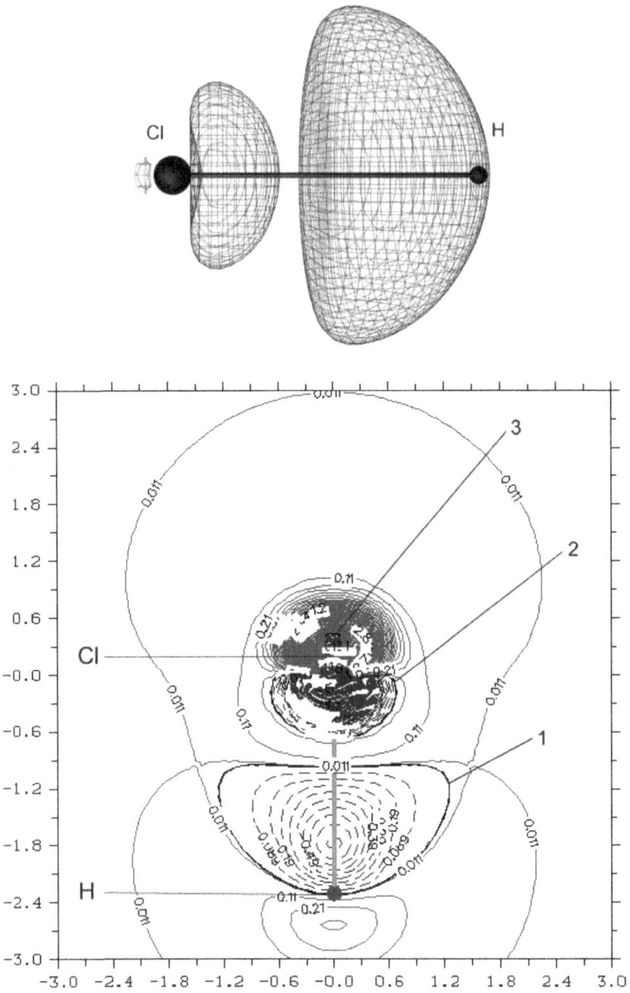

Fig. 10.14 The same as in Fig. 10.13 for HCl

accompanying contour map. The dominating (bonding) region around the bond middle-point is now "squeezed" between the two cores of nitrogen atoms. The small, axially placed core regions of the depleted contra-gradience are seen to be surrounded by the volumes of the positive values of this information density in transverse directions. They reflect the charge displacements accompanying the π-bond formation, which is also seen in the corresponding density difference diagram of Figs. 10.1 and 10.3.

In the water molecule (Fig. 10.16) one detects two slightly overlapping outer-basins of the negative nonadditive Fisher information in the O–H bonding regions, and two small inner-shell basins of the negative CG density on oxygen atom. The bonding basins, which define the two localized single bonds, are located between

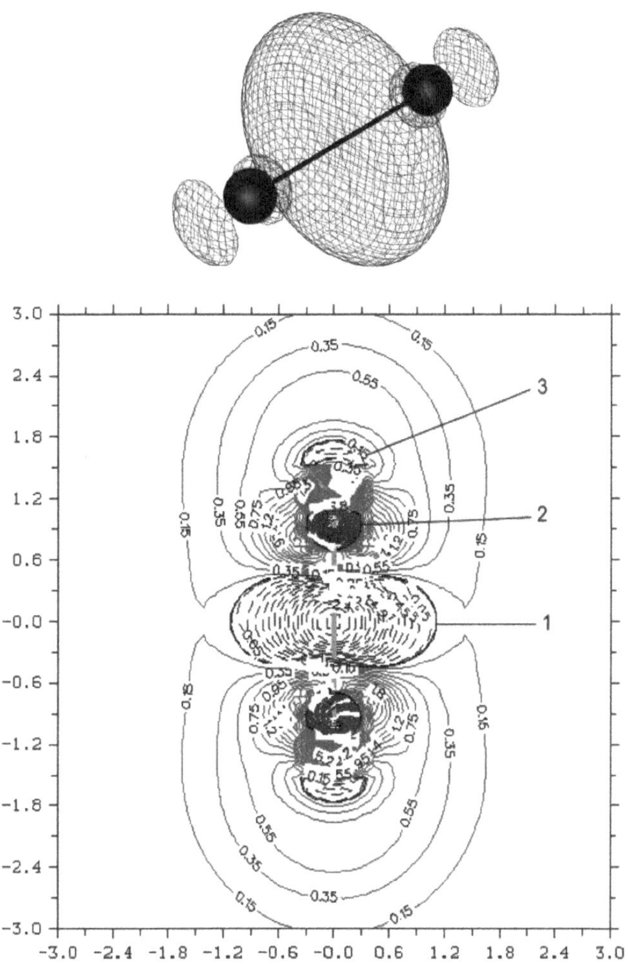

Fig. 10.15 The same as in Fig. 10.13 for N_2

the corresponding pairs of nuclei and the lowering of the CG density in each bond is seen to be the strongest in the direction linking the two O and H nuclear attractors. The overlapping character of these two regions of the negative nonadditive kinetic-energy, reflected by the present nonadditive Fisher-information probe, indicates a partial delocalization of the bonding electrons of one O–H bond into the bonding region of the other chemical bond, as indeed implied by the delocalized character of the occupied canonical MO. The contour map for the cut in the molecular plane also reveals a strong buildup of this information/kinetic-energy quantity in the lone-pair region of oxygen, and – to a lesser degree – in the nonbonding regions of two hydrogens. This effect on the heavy atom should indeed be expected as a result of its promotion due to the nearly tetrahedral sp^3-hybridization.

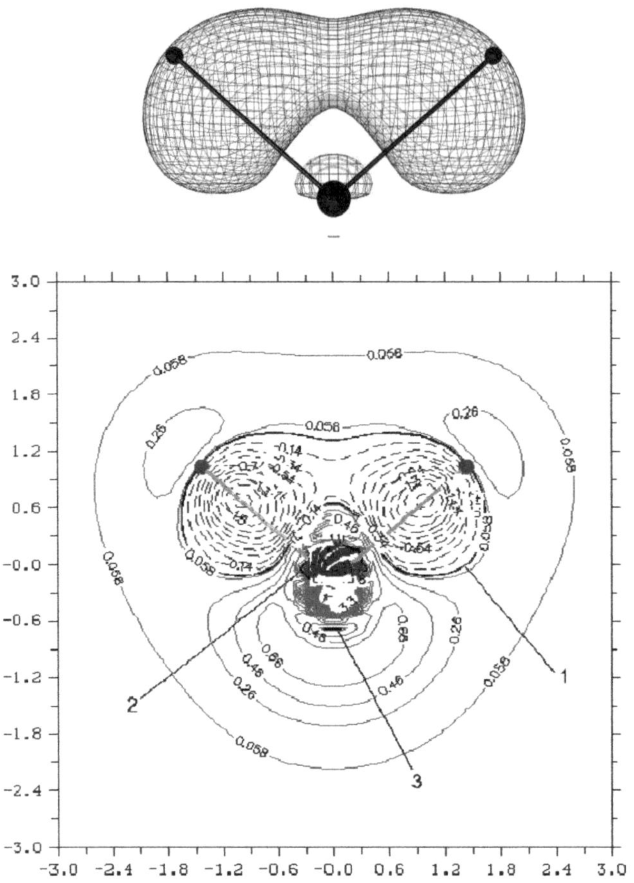

Fig. 10.16 The same as in Fig. 10.13 for H_2O

One thus concludes on the basis of this numerical evidence that CG criterion for detecting the valence basins of a diminished nonadditive Fisher information in AO-resolution indeed provides an efficient tool for locating the bonding regions in representative diatomics. It diagnoses all typical displacements of the bonded atoms in the bond formation process, relative to the corresponding free atoms of the promolecule, which have already been diagnosed from the density difference diagrams, e.g., the AIM polarization *via* the promotion/hybridization mechanism, the interatomic CT, and the constructive interference of AO in the bonding region, which is responsible for the electron accumulation between the covalently bonded atoms.

The chemical bonds in small hydrocarbons are investigated in contour maps of Figs. 10.17–10.20. These diagrams testify to the efficiency of the CG criterion in localizing all C–C and C–H bonding regions in ethane, ethylene, acetylene, butadiene, and benzene. The CG pattern of the triple bond between the carbon

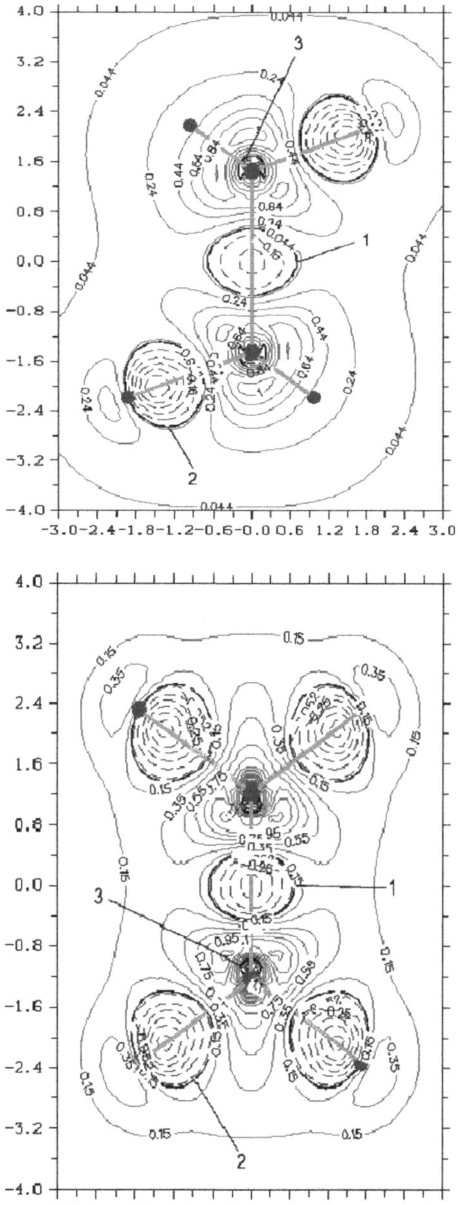

Fig. 10.17 The same as in Fig. 10.12 for ethane (upper panel) and ethylene (lower panel)

atoms in acetylene (upper panel of Fig. 10.18) strongly resembles that observed in N_2 (Fig. 10.15). In acetylene the two cylindrical bonding regions of the C–H bonds, axially extended due to a strong linear promotion of both carbon atoms *via* the *sp*-hybridization, and the central bonding basin due to the triple C–C bond, now transversely extended in the directions perpendicular to the bond axis, can be

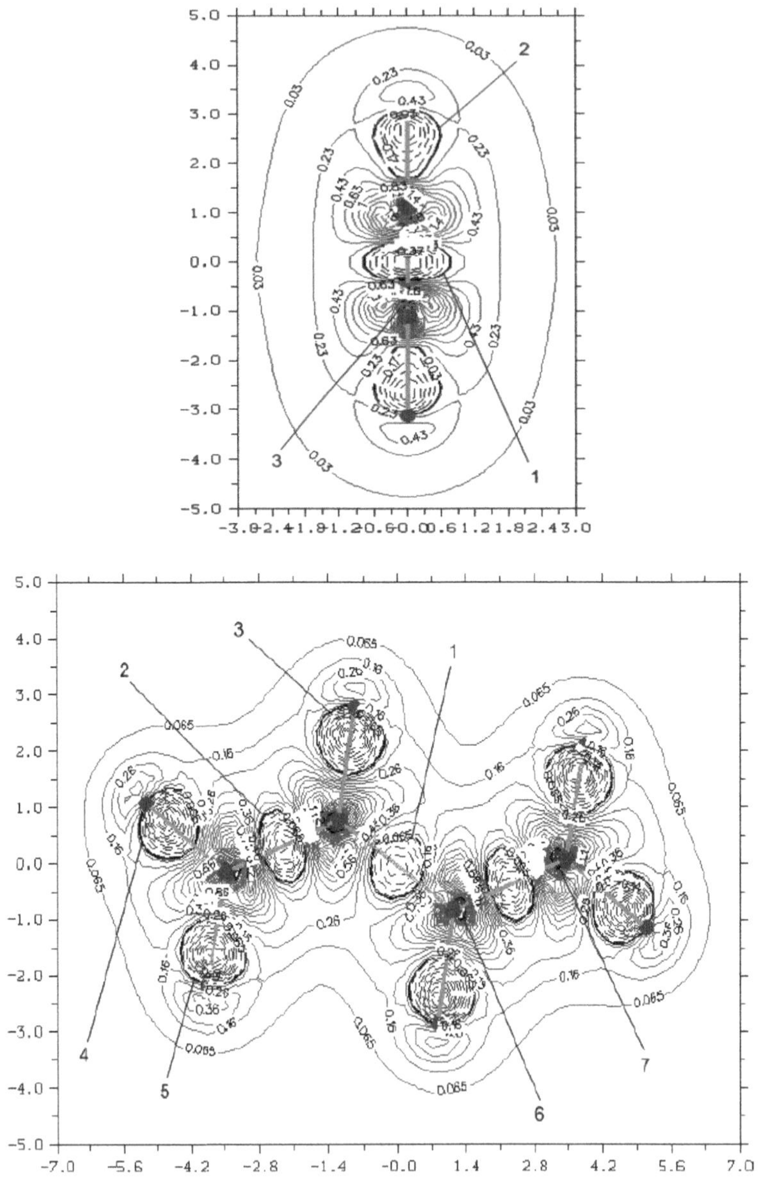

Fig. 10.18 The same as in Fig. 10.12 for acetylene (upper panel) and butadiene (lower panel)

clearly seen in the contour map. The $f^{nadd.}(\boldsymbol{r}) > 0$ regions on each carbon atom, very much resembling the atomic $2p_\pi$ distributions, reflect the presence of the bond π component. The depletion of the $2p_\pi$ electron density near the carbon nuclei generates more structure in the electron π-donating (nonbonding) regions of both

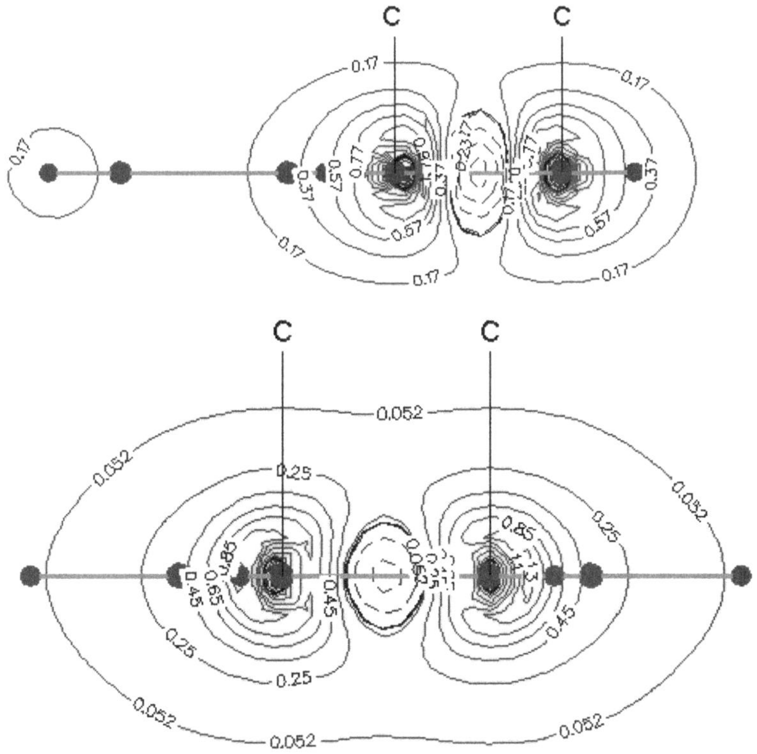

Fig. 10.19 The same as in Fig. 10.18 for the planes of section perpendicular to the molecular plane in butadiene passing through the peripheral (upper diagram) and middle (lower diagram) C—C bond axis

carbon atoms, and hence less structure (more delocalization) in the π-accepting (bonding) volume between the two nuclei.

The butadiene contour map in the molecular plane containing all nuclei is shown in the lower panel in Fig. 10.18. As seen in this diagram all bonds are properly accounted for by the IT CG probe. The same conclusion follows from examining Fig. 10.20, where the CG contour maps for benzene are shown, in the molecular plane (upper diagram) and in the perpendicular section containing the C–C bond (lower panel). It also follows from Fig. 10.19, where the additional CG cuts for butadiene are shown, in the planes of sections perpendicular to the molecular plane, along the peripheral and middle C–C bonds, respectively, that the π bond between the neighboring peripheral carbons in butadiene is indeed stronger than its central counterpart, in full accord with familiar quantum chemical predictions from the SCF LCAO MO theory.

Finally, the bonding patterns in a series of four small propellanes of Fig. 10.4 are examined in CG contour maps of Fig. 10.21. Each row of the figure is devoted to a different propellane, arranged from the smallest [1.1.1] molecule, exhibiting three single-carbon bridges, to the largest [2.2.2] system, consisting of three double-

Fig. 10.20 The same as in Fig. 10.12 for benzene. The upper panel shows the contour map in the molecular plane, while the lower panel corresponds to the perpendicular plane of section passing through one of the C—C bonds

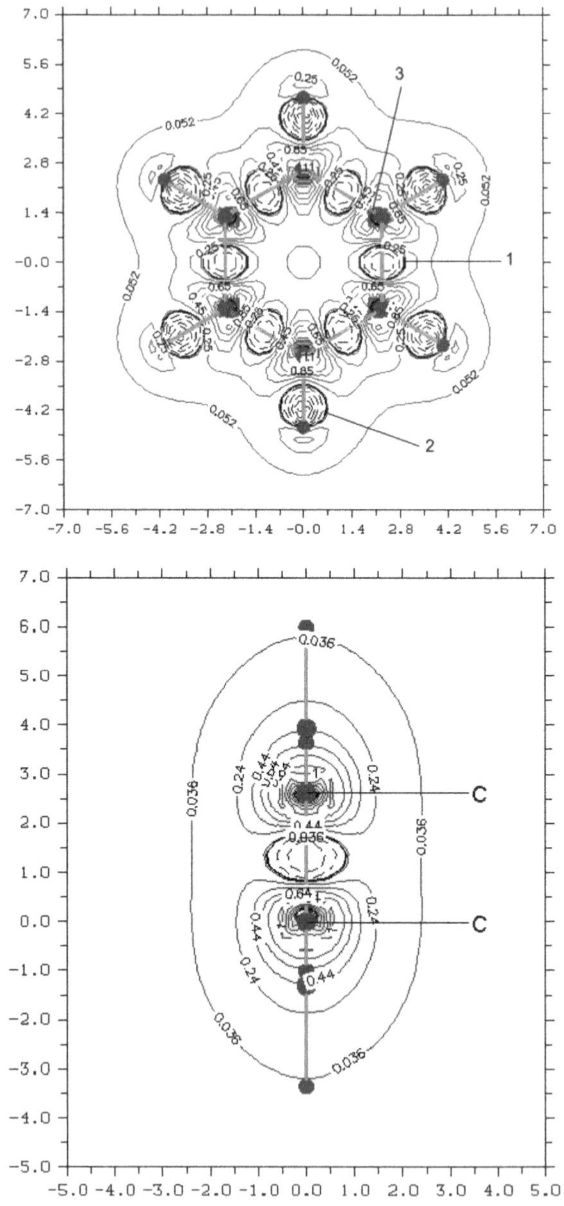

carbon bridges; the left panel of each row corresponds to the plane of section perpendicular to the central bond between the bridgehead carbons, at the bond midpoint, while the axial cut of the right panel involves one of the system carbon bridges.

The main result of the previous IT and density-difference analyses, the apparent lack of the *direct* (through-space) bond between the carbon bridgeheads in the

[1.1.1] and [2.1.1] systems, and a presence of practically single bond in the [2.2.1] and [2.2.2] propellanes, remain generally confirmed by the new CG probe, but this transition is now seen to be less sharp, with very small bonding basins between bridgeheads being also observed in the two smallest molecules. Thus, in accordance with the CG criterion the transition from the missing direct bonding in [1.1.1] system to the full, direct central bond in [2.2.2] propellane appears to be less abrupt: a very small bonding basin identified in the former case is steadily evolving into that attributed to the full bond in both [2.2.1] and [2.2.2] propellanes. The C–C and C–H chemical bonds in the bridges are again perfectly delineated by the valence surfaces of the vanishing CG density. The observed patterns of the nonadditive

[1.1.1]:

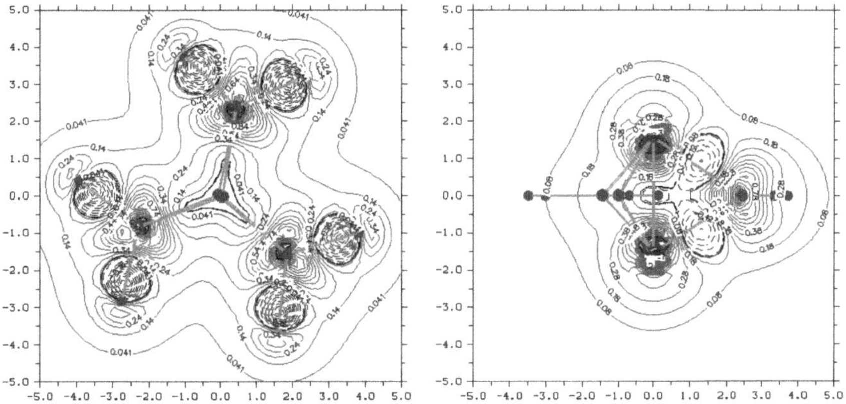

[2.1.1]:

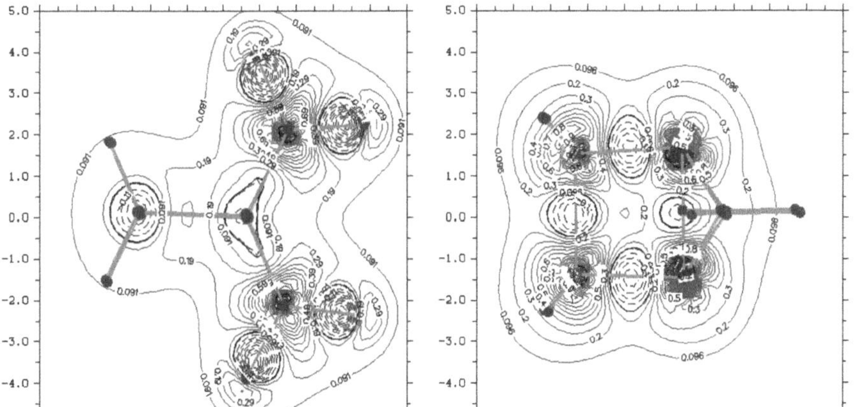

Fig. 10.21 (continued)

[2.2.1]:

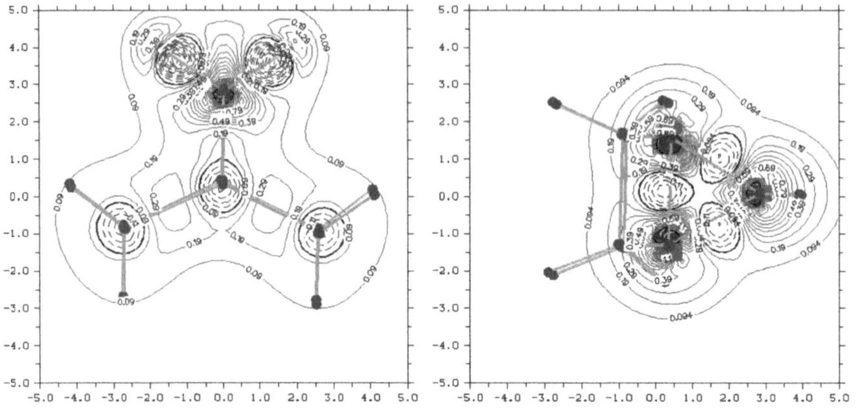

[2.2.2]:

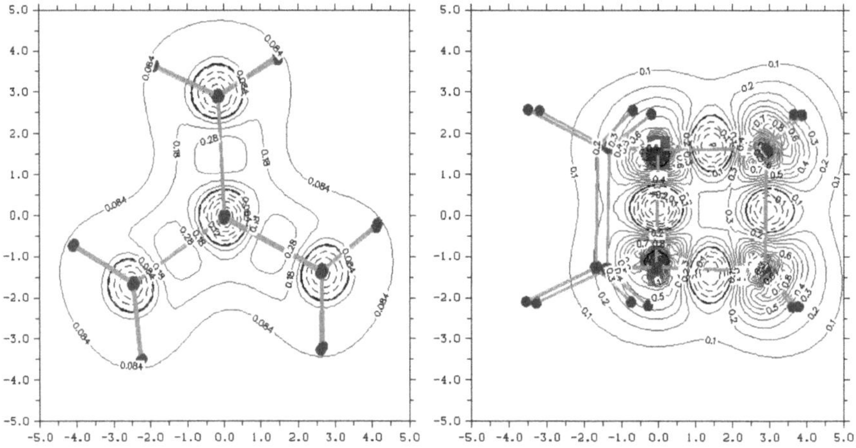

Fig. 10.21 The same as in Fig. 10.12 for the four propellanes of Fig. 10.4

Fisher information density always appear to be very much polarizational in character, with the closed bonding-regions of the negative CG being separated by the molecular environment of the positive values of this quantity, marking the system nonbonding regions.

References

Becke AD, Edgecombe KE (1990) J Chem Phys 92:5397
Casida ML, Wesołowski TA (2004) Int J Quantum Chem 96:577
Cortona P (1991) Phys Rev B 44:8454

Fuentalba P, Guerra D, Savin A (2009) In: Chattaraj PK (ed) Chemical reactivity theory: a density functional view. CRC, Boca Raton, p 281
Gopinathan MS, Jug K (1983) Theor Chim Acta (Berl.) 63:497, 511
Gordon RG, Kim YS (1972) J Chem Phys 56:3122
Hirshfeld FL (1977) Theor Chim Acta (Berl) 44:129
Jug K, Gopinathan MS (1990) In: Maksić ZB (ed) Theoretical models of chemical bonding, vol 2. Springer, Heidelberg, p 77
Mayer I (1983) Chem Phys Lett 97:270
Mayer I (1985) Theor Chim Acta (Berl.) 67:315
Mrozek J, Nalewajski RF, Michalak A (1998) Polish J Chem 72:1779
Nalewajski RF (2002a) Phys Chem Chem Phys 4:1710
Nalewajski RF (2002f) In: Barone V, Bencini A, Fantucci P (eds) Recent advances in density functional methods, part III. World Scientific, Singapore, p 257
Nalewajski RF (2003e) Mol Phys 101:2369
Nalewajski RF (2004b) Chem Phys Lett 386:265
Nalewajski RF (2006g) Information theory of molecular systems. Elsevier, Amsterdam
Nalewajski RF (2008e) Int J Quantum Chem 108:2230
Nalewajski RF (2009e) Int J Quantum Chem 109:425
Nalewajski RF (2009f) Int J Quantum Chem 109:2495
Nalewajski RF (2009g) Adv Quant Chem 56:217
Nalewajski RF (2010a) J Math Chem 47:667
Nalewajski RF (2010b) J Math Chem 47:692
Nalewajski RF (2010c) J Math Chem 47:709
Nalewajski RF (2010d) J Math Chem 47:808
Nalewajski RF (2010f) Information origins of the chemical bond. Nova Science, New York
Nalewajski RF (2010g) J Math Chem 48:752
Nalewajski RF (2011a) J Math Chem 49:371
Nalewajski RF (2011b) J Math Chem 49:546
Nalewajski RF (2011c) J Math Chem 49:806
Nalewajski RF (2011d) J Math Chem 49:592
Nalewajski RF (2011e) J Math Chem 49:2308
Nalewajski RF (2011f) Int J Quantum Chem (in press)
Nalewajski RF, Broniatowska E (2003a) J Phys Chem A 107:6270
Nalewajski RF, Gurdek P (2011a) J Math Chem 49:1226
Nalewajski RF, Gurdek P (2011b) Struct Chem (in press)
Nalewajski RF, Mrozek J (1994) Int J Quantum Chem 51:187
Nalewajski RF, Mrozek J (1996) Int J Quantum Chem 57:377
Nalewajski RF, Parr RG (2000) Proc Natl Acad Sci USA 97:8879
Nalewajski RF, Parr RG (2001) J Phys Chem A 105:7391
Nalewajski RF, Świtka E (2002) Phys Chem Chem Phys 4:4952
Nalewajski RF, Köster AM, Jug K (1993) Theor Chim Acta (Berl.) 85:463
Nalewajski RF, Formosinho SJ, Varandas AJC, Mrozek J (1994a) Int J Quantum Chem 52:1153
Nalewajski RF, Mrozek J, Mazur G (1996b) Can J Chem 100:1121
Nalewajski RF, Mrozek J, Michalak A (1997) Int J Quantum Chem 61:589
Nalewajski RF, Świtka E, Michalak A (2002) Int J Quantum Chem 87:198
Nalewajski RF, Köster AM, Escalante S (2005) J Phys Chem A 109:10038
Nalewajski RF, de Silva P, Mrozek J (2010a) Kinetic-energy/Fisher-information indicators of chemical bonds. In: Wang A, Wesołowski TA (eds) Kinetic energy functional. World Scientific, Singapore (in press)
Nalewajski RF, de Silva P, Mrozek J (2010b) J Mol Struct THEOCHEM 954:57
Savin A, Nesper R, Wengert S, Fässler TF (1997) Angew Chem Int Ed Engl 36:1808
Silvi B, Savin A (1994) Nature 371:683

Wesołowski TA (2004a) J Am Chem Soc 126:11444
Wesołowski TA (2004b) Chimia 58:311
Wesołowski TA, Tran F (2003) J Chem Phys 118:2072
Wesołowski TA, Warshel A (1993) J Phys Chem 97:8050
Wesołowski TA, Weber J (1998) Chem Phys Lett 248:71
Wesołowski TA, Muller RP, Warshel A (1995) J Phys Chem 100:15444
Wiberg KB (1968) Tetrahedron 24:1083

Chapter 11
Bonded Atoms from Information Theory

Abstract The chemical concept of bonded atoms, the building blocks of molecules, is approached using the variational principles of IT. The Hirshfeld stockholder rule of partitioning the molecular density into pieces attributed to Atoms in Molecules (AIM) is derived from the principle of maximum information resemblance of AIM to the free constituent atoms defining the promolecular prototype. This principle can be straightforwardly extended into the related schemes for partitioning *many*-electron distributions, which give slightly different effective *one*-electron distributions for light AIM, compared with their Hirshfeld analogs. Representative information densities of such bonded atoms are presented and the combination relations between their charge sensitivities and the associated properties of the molecule as a whole are derived.

11.1 Chemical Concepts

In chemistry an understanding of the electronic structure of molecules and of their preferences in reactions comes from interpreting the computed (or experimental) electron densities as collections of bonded atoms, and from locating/characterizing the chemical bonds, which represent the molecular "connectivities" between AIM. Of interest also are some larger fragments of the whole molecular system, e.g., the functional groups or reactants, or such chemically meaningful subsystems as the σ and π electrons in benzene. Indeed, an understanding of molecules as combinations of atoms is fundamental to chemistry. It is not surprising, then, that the classical concept of AIM and efficient probes of the chemical bonds have been recently much discussed in scientific literature (e.g., Bader 1990; Nalewajski and Parr 2000, 2001; Parr et al. 2005; Nalewajski 2002a, 2006g, 2010f; Nalewajski and Broniatowska 2007; Savin et al. 1997; Silvi and Savin 1994).

In fact, chemistry deals mainly with rather small changes in bonded atoms and/or larger molecular fragments, with reasonably well-understood and transferable molecular invariants, such as AIM, functional groups, molecular subsystems,

R.F. Nalewajski, *Perspectives in Electronic Structure Theory*,
DOI 10.1007/978-3-642-20180-6_11, © Springer-Verlag Berlin Heidelberg 2012

e.g., reactants and products of an elementary chemical reaction, etc., which tend to maintain their identity in different molecular environments. Molecular systems do indeed consist of only slightly perturbed atoms (or atomic ions), deformed by the presence of their molecular environment and exhibiting somewhat modified net electric charges. These displacements in the electronic structure of AIM, relative to the corresponding states of the isolated atoms, are due to the coupled processes of the intraatomic *Polarization* (P), responsible for the "promotion" of bonded atoms to their effective *valence states* in the molecule, and the interatomic *Charge Transfer* (CT), which accompany the formation of the system chemical bonds.

An important part of chemical science is based on such intuitive notions, which ultimately escape the rigorous definition in molecular quantum mechanics, thus representing the Kantian *noumenons* of chemistry (Parr et al. 2005). The natural question to be addressed in the novel IT approach to the bonded fragment problem (e.g., Nalewajski 2006g) is whether this theory can actually help in making these concepts more precisely defined in terms of the molecular and/or promolecular electron distributions. Other questions of interest are about the location of chemical bonds, an extent of the information redistribution accompanying the bond formation processes, the information probes of bond multiplicity, and the entropic descriptors of their covalent/ionic composition.

The bonded fragment of a given molecular system represents the externally *open* subsystem capable of exchanging electrons with the molecular reminder. One would hope to find that a given AIM, like its free (nonbonded) analog, would possess a single cusp at the nucleus in its electron density, linked to the effective atomic number of the nucleus. Their electron densities $\{\rho_i\}$ must sum up to the molecular electron density $\rho = \sum_i \rho_i$. However, since each bonded atom preserves to a remarkably high degree the free atom identity, the AIM distributions should be also closely related to electron densities $\{\rho_i^0\}$ of the separated atoms, which define the atomic promolecule consisting of such *free* atom densities shifted to the actual AIM locations in the molecule. We also recall that the resultant electron density $\rho^0 = \sum_i \rho_i^0$ of this collection of the "frozen" atomic electron distributions defines the initial stage in the bond formation process, a natural reference for extracting changes due to chemical bonds. Indeed, the density difference function $\Delta\rho = \rho - \rho^0$ has been widely used to probe the chemical bonds in molecules (see the preceding chapter).

The electronic structure of molecular systems is effectively characterized by their *one-*, *two-*, and *many*-electron probability distributions in the continuous ("fine-grained") description. To obtain its *chemical* interpretation, e.g., in terms of AIM, functional groups, reactants, or other type of chemically significant subsystems, e.g., the σ and π electrons in aromatic compounds, these overall distributions have to be "discretized" in terms of the relevant pieces of the overall density attributed to the constituent parts of the molecular system under consideration, e.g., the bonded atoms or molecular fragments (clusters of AIM). The densities of molecular subsystems constitute their fine-grained description. By an appropriate integration of the electron/probability densities, one subsequently obtains the corresponding *condensed* descriptors of the electronic structure of molecules and their fragments, providing the associated discrete ("coarse-grained") indices (occupations or populations) of the underlying continuous electron distributions. Additional resolution levels are provided by the AO and MO representations resulting from the quantum mechanical calculations of the molecular electronic/geometrical structure.

It should be emphasized, however, that an exhaustive partitioning of the given molecular electron density between constituent (bonded) atoms, which determines the AIM effective net charges (oxidation states) in a given molecular environment, is not unique, since it depends on the adopted criterion for such a division. The imposing variety of published theoretical methods for partitioning the molecular density into "*best*" AIM contributions testifies to the importance of this theme in chemistry. Different methods are based on different principles, some to a degree arbitrary or heuristic, which can produce conflicting trends in the associated AIM charges. Methods differ in theoretical techniques used, e.g., topological analysis of the molecular density, wave function description, or the density functional approach. They also differ in the physical/heuristic principles invoked, e.g., electronegativity equalization, zero flux, the minimum promotion energy rules, and the minimum entropy deficiency (information distance, missing information) criteria of the IT approach.

The historically first scheme of the Mulliken/Löwdin *population analysis* (Mulliken 1955; Löwdin 1950, 1956) has used the *function* space partitioning, in which one distributes electrons between AO which form the basis set for expanding MO of the Hartree–Fock (SCF LCAO MO) theory and its *Configuration Interaction* (CI) extension, nonorthogonal in the Mulliken approach and the symmetrically orthogonalized in the Löwdin variant. Another popular approach of Bader (1990), with a solid topological and quantum mechanical basis, uses the *physical* space partitioning, i.e., a division of space into the exclusive atomic domains (basins), with the boundaries determined by the *zero* flux surfaces, on which the flow of electrons between subsystems vanishes. In the latter approach, the spatially nonconfined bonded atoms of the population analysis are replaced by the topological, nonspherical pieces of the molecular density, obtained as cuts along the *zero* flux surfaces. As a result, the topological AIM represent the spatially confined (nonoverlapping) and strongly nonsymmetrical atoms.

11.2 Stockholder Atoms in Molecules

Yet another *stockholder* division scheme of Hirshfeld (1977), which has been widely exploited in crystallography, uses the commonsense *local* partitioning principle, which parallels the familiar stockmarket rule: in forming a molecule each bonded atom locally partakes of the molecular density gain or loss ("profit") in proportion to its share in the promolecular density ("investment"). Thus, by construction, the (overlapping) stockholder AIM are infinitely extending (spatially nonconfined) and known to be only slightly polarized relative to the *free* atom reference. Both the topological and stockholder atoms are derived from the molecular electron density, and as such they preserve a "memory" of their molecular origin.

This *one*-electron stockholder division scheme has recently been shown to have a strong basis in IT (e.g., Nalewajski and Parr 2000, 2001; Nalewajski 2002a, 2006g). However, in molecules involving light atoms, e.g., hydrogen or lithium, a generalization of the stockholder principle applied to partition the *two*-electron density (Nalewajski and Broniatowska 2007) generates slightly different effective *one*-electron distributions of the associated bonded hydrogens, compared to the corresponding stockholder atoms originating from partitioning the *one*-electron density, which more strongly emphasize the bonding (overlap) regions between the bond partners.

It has been shown by Hirshfeld (1977) that the electron density $\rho(\boldsymbol{r})$ of the molecular system $\mathrm{M} = (\mathrm{A}^{\mathrm{H}}\!\!\mid \mathrm{B}^{\mathrm{H}}\!\!\mid \ldots)$, consisting of the mutually open atoms $\mathrm{X}^{\mathrm{H}} = (\mathrm{A}^{\mathrm{H}}, \mathrm{B}^{\mathrm{H}}, \ldots)$, as marked by the perpendicular *broken* lines separating the AIM symbols in M, can be exhaustively partitioned into the "stockholder" AIM densities $\{\rho_{\mathrm{X}}^{\mathrm{H}}(\boldsymbol{r})\} \equiv \boldsymbol{\rho}^{\mathrm{H}}(\boldsymbol{r})$:

$$\begin{aligned} \rho_{\mathrm{X}}^{\mathrm{H}}(\boldsymbol{r}) &= \rho_{\mathrm{X}}^{0}(\boldsymbol{r})[\rho(\boldsymbol{r})/\rho^{0}(\boldsymbol{r})] \equiv \rho_{\mathrm{X}}^{0}(\boldsymbol{r})\, w(\boldsymbol{r}) \\ &= \rho(\boldsymbol{r})[\rho_{\mathrm{X}}^{0}(\boldsymbol{r})/\rho^{0}(\boldsymbol{r})] \equiv \rho(\boldsymbol{r})\, d_{\mathrm{X}}^{\mathrm{H}}(\boldsymbol{r}), \qquad \mathrm{X} = \mathrm{A}, \mathrm{B}, \ldots \\ \sum\nolimits_{\mathrm{X}} d_{\mathrm{X}}^{\mathrm{H}}(\boldsymbol{r}) &= 1, \qquad \rho(\boldsymbol{r}) = \sum\nolimits_{\mathrm{X}} \rho_{\mathrm{X}}^{\mathrm{H}}(\boldsymbol{r}). \end{aligned} \tag{11.1}$$

Here, $\boldsymbol{\rho}^{0}(\boldsymbol{r}) = \{\rho_{\mathrm{X}}^{0}(\boldsymbol{r})\}$ groups the densities of the free constituent atoms, giving rise to the reference electron density $\rho^{0}(\boldsymbol{r}) = \sum_{\mathrm{X}} \rho_{\mathrm{X}}^{0}(\boldsymbol{r})$ of the (isoelectronic) atomic promolecule $\mathrm{M}^{0} = (\mathrm{A}^{0}|\ \mathrm{B}^{0}|\ \ldots)$, consisting of the nonbonded (mutually closed) atoms $\mathrm{X}^{0} = (\mathrm{A}^{0}, \mathrm{B}^{0}, \ldots)$, as marked by the perpendicular *solid* lines separating the atomic symbols in M^{0}:

$$\begin{aligned} \int \rho^{0}(\boldsymbol{r})\, d\boldsymbol{r} &= \sum\nolimits_{\mathrm{X}} \int \rho_{\mathrm{X}}^{0}(\boldsymbol{r})\, d\boldsymbol{r} = \sum\nolimits_{\mathrm{X}} N_{\mathrm{X}}^{0} = N^{0} \\ &= \int \rho(\boldsymbol{r})\, d\boldsymbol{r} = \sum\nolimits_{\mathrm{X}} \int \rho_{\mathrm{X}}^{\mathrm{H}}(\boldsymbol{r})\, d\boldsymbol{r} = \sum\nolimits_{\mathrm{X}} N_{\mathrm{X}}^{\mathrm{H}} = N. \end{aligned} \tag{11.2}$$

Above, the free atom densities $\boldsymbol{\rho}^{0}$ in M^{0} are shifted to the respective atomic positions in the molecule, and the vectors $\boldsymbol{N}^{\mathrm{H}} = \{N_{\mathrm{X}}^{\mathrm{H}}\}$ and $\boldsymbol{N}^{0} = \{N_{\mathrm{X}}^{0}\}$ group the atomic average numbers of electrons of the bonded and free atoms, respectively. As we have already observed in the preceding chapter, the same promolecular reference is used to determine the density difference function $\Delta\rho(\boldsymbol{r}) = \rho(\boldsymbol{r}) - \rho^{0}(\boldsymbol{r})$.

A reference to (11.1) shows that the Hirshfeld AIM densities satisfy the local principle of the *one*-electron "stockholder" division, which can be stated as the following equality between the local molecular and promolecular *conditional* probabilities [see also (10.15a, b)]:

$$d_X^H(\boldsymbol{r}) = \rho_X^H(\boldsymbol{r})/\rho(\boldsymbol{r}) \equiv P^H(X|\boldsymbol{r}) = d_X^0(\boldsymbol{r}) = \rho_X^0(\boldsymbol{r})/\rho^0(\boldsymbol{r}) \equiv P^0(X|\boldsymbol{r}),$$
$$\sum_X P^H(X|\boldsymbol{r}) = \sum_X P^0(X|\boldsymbol{r}) = 1. \tag{11.3}$$

As we have also remarked above, this relation has been interpreted by Hirshfeld using the stockmarket analogy: each atom participates locally in the molecular "profit" $\rho(\boldsymbol{r})$ in proportion to its "share" $d_X^0(\boldsymbol{r}) = P^0(X|\boldsymbol{r})$ in the promolecular "investment" $\rho^0(\boldsymbol{r})$. In this section, we shall demonstrate that this commonsense division rule has a solid basis in IT.

By extracting the overall number of electrons $N = N^0$ from the molecular and subsystem densities, one introduces the associated (molecularly normalized) probability distributions:

$$\rho(\boldsymbol{r}) = Np(\boldsymbol{r}) = N\sum_X p_X^H(\boldsymbol{r}) \quad \text{and} \quad \boldsymbol{\rho}^H(\boldsymbol{r}) = N\boldsymbol{p}^H(\boldsymbol{r}) = N\{p_X^H(\boldsymbol{r})\}. \tag{11.4}$$

Here, $p(\boldsymbol{r})$ and $\boldsymbol{p}^H(\boldsymbol{r}) = \{p_X^H(\boldsymbol{r})\}$ stand for the *shape* factors of the system as a whole and of its Hirshfeld fragments, respectively,

$$\sum_X \int p_X^H(\boldsymbol{r})\, d\boldsymbol{r} = \sum_X (N_X^H/N) \equiv \sum_X P_X^H = 1, \tag{11.5}$$

while $\boldsymbol{P}^H = \{P_X^H\}$ groups the condensed probabilities of finding an electron of M on the specified stockholder AIM.

This molecular normalization reflects the important fact that bonded atoms are constituent parts of the molecule, so that the full normalization condition has to involve the summation/integration over the complete set of *one*-electron events, consisting of all possible "values" of the discrete argument X (atomic label) and all spatial locations of an electron, identified by the continuous coordinates $\boldsymbol{r} = (x, y, z)$ in the subsystem probability distributions:

$$\boldsymbol{p}^H(\boldsymbol{r}) = \{p_X^H(\boldsymbol{r}) \equiv P^H(X \wedge \boldsymbol{r}) = p(\boldsymbol{r})P^H(X|\boldsymbol{r})\}. \tag{11.6}$$

The same, global normalization has to be adopted for the *free*-atom pieces of the *one*-electron probability distribution in the isoelectronic promolecule and for its *free*-atom components, respectively,

$$p^0(\boldsymbol{r}) = \rho^0(\boldsymbol{r})/N^0 = \sum_X p_X^0(\boldsymbol{r}), \quad \boldsymbol{p}^0(\boldsymbol{r}) = \boldsymbol{\rho}^0(\boldsymbol{r})/N^0 = \{p_X^0(\boldsymbol{r}) = \rho_X^0(\boldsymbol{r})/N^0\};$$
$$\sum_X \int p_X^0(\boldsymbol{r})\, d\boldsymbol{r} = \sum_X (N_X^0/N^0) \equiv \sum_X P_X^0 = 1, \tag{11.7}$$

where $\boldsymbol{P}^0 = \{P_X^0\}$ collects the condensed probabilities of observing an electron of M^0 on the specified free atom. Again, the full normalization of the shape (probability) factors $\boldsymbol{p}^0(\boldsymbol{r}) = \{p_X^0(\boldsymbol{r}) \equiv P^0(X \wedge \boldsymbol{r})\}$ of the nonbonded atoms in the promolecular system involves summation over the discrete atomic "variable"

X and integration over all positions $\boldsymbol{r}$ of an electron, the latter representing the continuous event label of the probability distributions of atomic fragments in M^0.

It also follows from (11.1) that in Hirshfeld's (*one*-electron) stockholder division scheme each free subsystem density (or its shape factor) is locally modified in accordance with the *molecular* (subsystem-independent) enhancement factor $w(\boldsymbol{r})$:

$$\begin{aligned} w_X^H(\boldsymbol{r}) &\equiv \rho_X^H(\boldsymbol{r})/\rho_X^0(\boldsymbol{r}) = p_X^H(\boldsymbol{r})/p_X^0(\boldsymbol{r}) \\ &= \rho(\boldsymbol{r})/\rho^0(\boldsymbol{r}) = p(\boldsymbol{r})/p^0(\boldsymbol{r}) \equiv w(\boldsymbol{r}). \end{aligned} \quad (11.8)$$

Therefore, this procedure is devoid of any subsystem bias and as such appears to be fully objective.

Representative plots of the overlapping electron densities of the stockholder hydrogens in H_2 are shown in Fig. 11.1. They are seen to be distributed all over the physical space, decaying exponentially at large distances from the molecule and exhibiting in the bond density profile a single cusp at the atomic nucleus. They also display the expected molecular polarization toward the bonding partner. These subsystem densities are highly transferable and their overlap in the molecule accords with the classical interpretation of the origin of the (direct) chemical bonding. One also observes a higher AIM density at the atomic nucleus, in comparison with the *free* hydrogen density, i.e., a contraction of the AIM

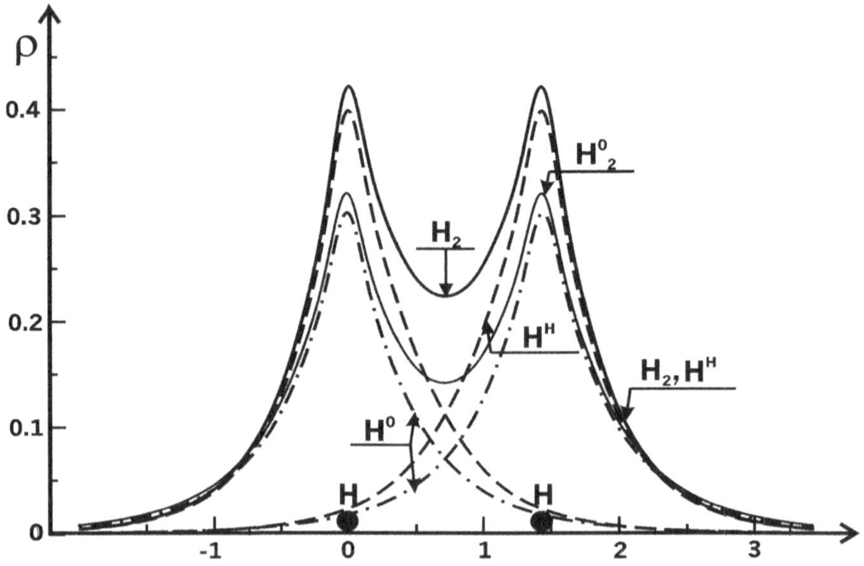

Fig. 11.1 Profiles of the Hirshfeld electron densities of the bonded Hirshfeld hydrogen atoms (H^H) obtained from the molecular density (H_2). The free hydrogen densities (H^0) and the resulting electron density of the promolecule (H_2^0) are also shown for comparison. The density and internuclear distance are in a.u. The zero cusps at nuclear positions are the artifacts of the Gaussian basis set used in DFT calculations

distribution at the expense of the nonbonding, outer part of the free atom density. This is due to the presence of the other atom causing an effective lowering of the molecular external potential relative to the external potential of the separated atom.

Let us examine the asymptotic properties of the stockholder atomic densities. For simplicity, we consider a diatomic system $M = (A^H|B^H)$, $\rho = \rho_A^H + \rho_B^H$, consisting of two Hirshfeld atoms A^H and B^H, the free analogs of which, A^0 and B^0, are assumed to exhibit relative electron *acceptor* (acidic) and *donor* (basic) properties, respectively. This further implies $I_A^0 > I_B^0$, where I_X^0 denotes the *ionization potential* of X^0. Rewriting (11.1) in terms of the local density ratio $x = \rho_B^0/\rho_A^0$ gives:

$$\rho_A^H = (1+x)^{-1}\rho \quad \text{and} \quad \rho_B^H = (1+x^{-1})^{-1}\rho. \tag{11.9}$$

Hence, $x \to \infty$ for $r_X = |\boldsymbol{r} - \boldsymbol{R}_X| \to \infty$, when distances from both nuclei become large compared with the interatomic separation R_{AB}, since the asymptotic behavior of the free subsystems of the promolecule is determined by their electronegativities, measured by the negative energies of the highest occupied KS orbitals, i.e., their ionization potentials:

$$\rho_X^0 \to \exp[-2(2I_X^0)r_X] \quad (r_X \to \infty), \qquad X = A, B. \tag{11.10}$$

Therefore, for $r_X \to \infty$ $\rho_A^H \to 0$ and $\rho_B^H \to \rho$, so that the density of the softer (donor) atom B has a dominant contribution to the molecular density at distances from the molecule large compared with R_{AB}.

In Table 11.1, we have listed the net charges and average entropy deficiencies of the Hirshfeld AIM for a series of illustrative linear molecules discussed in Sect. 10.1. It follows from these numerical results that the Hirshfeld charges represent the chemical intuition quite well. For example, in the series (HF, LiH, LiCl, and LiF) of heteronuclear diatomics of increasing bond ionicity due to a growing electronegativity difference between constituent atoms, the amount of CT monotonically increases, as intuitively expected.

The atomic missing information,

$$\Delta S_X[p_X^H|p_X^0] = \int p_X^H(\boldsymbol{r}) \, \log[p_X^H(\boldsymbol{r})/p_X^0(\boldsymbol{r})] \, d\boldsymbol{r}, \tag{11.11}$$

reflects the information distance between the atomic *shape* factors $p_X^H(\boldsymbol{r})$ and $p_X^0(\boldsymbol{r})$. The reported values of these quantities are quite small, thus numerically confirming that the bonded atoms do indeed strongly resemble their *free* atom analogs. The same general conclusion follows from examining the reported global entropy deficiencies:

$$\Delta S[p|p^0] = \sum_X \Delta S_X[p_X^H|p_X^0] = \int p(\boldsymbol{r}) \log[p(\boldsymbol{r})/p^0(\boldsymbol{r})] \, d\boldsymbol{r} \equiv \int p(\boldsymbol{r}) \, I[w(\boldsymbol{r})] \, d\boldsymbol{r}. \tag{11.12}$$

Table 11.1 Representative net charges $\{q_X^H = N_X^H - Z_X\}$ (a.u.) of the stockholder AIM, where Z_X denotes the charge of nucleus X, the AIM entropy deficiencies $\{\Delta S_X[p_X^H|p_X^0]\}$, and the global entropy deficiency $\Delta S[p|p^0]$ (in bits), for the linear molecules of Fig. 10.1

Molecule	X	q_X^H	$\Delta S_X[p_X^H\|p_X^0]$	$\Delta S[p\|p^0]$
H_2	H	0.00	0.056	0.056
N_2	N	0.00	0.006	0.006
HF	H	0.24	0.144	0.020
	F	−0.24	0.005	
LiH	Li	0.35	0.157	0.136
	H	−0.35	0.012	
LiF	Li	0.58	0.244	0.063
	F	−0.58	0.007	
LiCl	Li	0.53	0.212	0.033
	Cl	−0.53	0.003	
HCN	H	0.14	0.104	0.017
	C	0.03	0.015	
	N	−0.17	0.005	
HNC	H	0.20	0.110	0.018
	N	−0.10	0.008	
	C	−0.10	0.011	
HNCS	H	0.19	0.114	0.008
	N	−0.13	0.007	
	C	0.05	0.008	
	S	−0.11	0.002	
HSCN	H	0.22	0.088	0.008
	S	−0.07	0.002	
	C	0.04	0.008	
	N	−0.19	0.004	

This strong similarity between the molecular and promolecular electron distributions is also seen in Fig. 11.1: the appreciable changes of the free atom densities in the molecule are observed only around the nuclei (a contraction of the *free* atom density) and in the bond region between the two nuclei (a polarization of the free atoms toward the bonding partner). As expected, in heavier atoms only a slight distortion of the valence (external) electrons is observed in the stockholder (bonded) atoms, with the inner-shell structure being left practically intact.

11.3 Information Theoretic Justification

The problem of the optimum *local* partition of the molecular density can be best formulated in terms of the unknown *conditional* probabilities $\boldsymbol{d}(\boldsymbol{r}) = \{d_X(\boldsymbol{r}) \equiv P(X|\boldsymbol{r})\}$, which uniquely determine the AIM pieces of $p(\boldsymbol{r})$, $\{p_X(\boldsymbol{r}) = d_X(\boldsymbol{r})p(\boldsymbol{r})\}$ [see (10.15a, b)]. The relevant (local, multicomponent) KL function of the unknown share factors $\boldsymbol{d}(\boldsymbol{r})$, which measures the local information distance relative to the promolecular reference values $\boldsymbol{d}^0(\boldsymbol{r}) = \{P^0(\mathrm{X}|\boldsymbol{r})\}$, is then given (in nats) by the sum of AIM contributions [see also (11.12)] (Nalewajski and Parr 2000; Nalewajski 2002a, 2003c, 2006g):

$$\Delta s^{add.}[\boldsymbol{d}(\boldsymbol{r})|\boldsymbol{d}^0(\boldsymbol{r})] = \sum\nolimits_{\mathrm{X}} P(\mathrm{X}|\boldsymbol{r}) \ln[P(\mathrm{X}|\boldsymbol{r})/P^0(\mathrm{X}|\boldsymbol{r})]. \tag{11.13}$$

The best (unbiased) share factors of subsystems, giving rise to the maximum similarity of the bonded fragment densities to their nonbonded (free) analogs, must minimize this missing information function subject to the normalization constraint of the local conditional probabilities, $\sum_{\mathrm{X}} P(\mathrm{X}|\boldsymbol{r}) = 1$,

$$\delta\left\{\Delta s^{add.}[\boldsymbol{d}(\boldsymbol{r})|\boldsymbol{d}^0(\boldsymbol{r})] - \lambda(\boldsymbol{r}) \sum\nolimits_{\mathrm{X}} P(\mathrm{X}|\boldsymbol{r})\right\} = 0, \tag{11.14}$$

where $\lambda(\boldsymbol{r})$ stands for the Lagrange multiplier associated with this auxiliary condition at the given location $\boldsymbol{r}$. The resulting Euler equation for the optimum local conditional probability $P^{opt.}(\mathrm{X}|\boldsymbol{r})$,

$$\ln[P^{opt.}(\mathrm{X}|\boldsymbol{r})/P^0(\mathrm{X}|\boldsymbol{r})] + [1 - \lambda(\boldsymbol{r})] \equiv \ln\{P^{opt.}(\mathrm{X}|\boldsymbol{r})/[C(\boldsymbol{r})\, P^0(\mathrm{X}|\boldsymbol{r})]\} = 0, \tag{11.15}$$

or $P^{opt.}(\mathrm{X}|\boldsymbol{r}) = C(\boldsymbol{r})P^0(\mathrm{X}|\boldsymbol{r})$, when combined with the normalization constraint, $\sum_{\mathrm{X}} P^{opt.}(\mathrm{X}|\boldsymbol{r}) = C(\boldsymbol{r}) \sum_X P^0(\mathrm{X}|\boldsymbol{r}) = C(\boldsymbol{r}) = 1$, is then seen to give the Hirshfeld solution of (11.1) and (11.8): $\boldsymbol{d}^{opt.}(\boldsymbol{r}) = \boldsymbol{d}^0(\boldsymbol{r}) = \boldsymbol{d}^{\mathrm{H}}(\boldsymbol{r})$. Therefore, the Hirshfeld, promolecular choice of the local share factors minimizes the local information distance to the lowest value possible: $\Delta s^{add.}[\boldsymbol{d}^H(\boldsymbol{r})|\boldsymbol{d}^0(\boldsymbol{r})] = 0$.

The same answer follows from the alternative, *global* information principles formulated in terms of either the electron densities or their shape factors (e.g., Nalewajski and Parr 2000), in which one seeks the optimum overall atomic (or fragment) distributions exhibiting the strongest resemblance to the corresponding nonbonded, reference distributions of the associated promolecular system. Let us define the KL information distance functional between the trial *one*-electron densities of atomic fragments $\{\rho_{\mathrm{X}}\} \equiv \boldsymbol{\rho}$ (or the associated probability distributions $\{p_{\mathrm{X}}\} \equiv \boldsymbol{p} = \boldsymbol{\rho}/N$) of the bonded atoms and the corresponding reference densities $\{\rho^0_{\mathrm{X}}\} \equiv \boldsymbol{\rho}^0$ (or $\{p^0_{\mathrm{X}}\} \equiv \boldsymbol{p}^0 = \boldsymbol{\rho}^0/N$) of the free atoms:

$$\begin{aligned}\Delta S^{add.}[\boldsymbol{\rho}|\boldsymbol{\rho}^0] &= \sum\nolimits_{\mathrm{X}} \int \rho_{\mathrm{X}}(\boldsymbol{r}) \ln\left[\rho_{\mathrm{X}}(\boldsymbol{r})/\rho^0_{\mathrm{X}}(\boldsymbol{r})\right] d\boldsymbol{r} \equiv \sum\nolimits_{\mathrm{X}} \int \rho_{\mathrm{X}}(\boldsymbol{r}) I_{\mathrm{X}}[w_{\mathrm{X}}(\boldsymbol{r})]\, d\boldsymbol{r}\\ &\equiv \sum\nolimits_{\mathrm{X}} \Delta S_{\mathrm{X}}\left[\rho_{\mathrm{X}}|\rho^0_{\mathrm{X}}\right]\\ &= N \sum\nolimits_{\mathrm{X}} \int p_{\mathrm{X}}(\boldsymbol{r}) \ln\left[p_{\mathrm{X}}(\boldsymbol{r})/p^0_{\mathrm{X}}(\boldsymbol{r})\right] d\boldsymbol{r} \equiv N \sum\nolimits_{\mathrm{X}} \int p_{\mathrm{X}}(\boldsymbol{r})\, I_{\mathrm{X}}[w_{\mathrm{X}}(\boldsymbol{r})]\, d\boldsymbol{r}\\ &\equiv N \Delta S^{add.}[\boldsymbol{p}|\boldsymbol{p}^0] \equiv N \sum\nolimits_{\mathrm{X}} \Delta S_{\mathrm{X}}\left[p_{\mathrm{X}}|p^0_{\mathrm{X}}\right].\end{aligned} \tag{11.16}$$

Here, $I_X[w_X(\boldsymbol{r})] = \ln[p_X(\boldsymbol{r})/p_X^0(\boldsymbol{r})] \equiv I_X(\boldsymbol{r})$ stands for (variational) surprisal of atom X, for the current value of the local enhancement factor relative to the *free* subsystem reference: $w_X(\boldsymbol{r}) = \rho_X(\boldsymbol{r})/\rho_X^0(\boldsymbol{r}) = p_X(\boldsymbol{r})/p_X^0(\boldsymbol{r})$.

In the preceding equation we have also indicated that the directed divergence of $\boldsymbol{\rho}$ relative to $\boldsymbol{\rho}^0$, $\Delta S^{add.}[\boldsymbol{\rho}|\boldsymbol{\rho}^0]$ is just N times the entropy deficiency $\Delta S^{add.}[\boldsymbol{p}|\boldsymbol{p}^0]$. The same relation holds between the subsystem missing information: $\Delta S_X[\rho_X|\rho_X^0] = N\Delta S_X[p_X|p_X^0]$. Therefore, for the isoelectronic molecular and promolecular systems, the problem of normalization of the compared electronic densities does not influence the corresponding constrained variational principle for determining the optimum densities of atomic pieces:

$$\begin{aligned}&\delta\left\{\Delta S^{add.}[\boldsymbol{\rho}|\boldsymbol{\rho}^0] - \int \lambda(\boldsymbol{r})\sum\nolimits_X \rho_X(\boldsymbol{r})d\boldsymbol{r}\right\}\\ &= N\delta\left\{\Delta S^{add.}[\boldsymbol{p}|\boldsymbol{p}^0] - \int \lambda(\boldsymbol{r})\sum\nolimits_X p_X(\boldsymbol{r})d\boldsymbol{r}\right\} = 0,\end{aligned} \tag{11.17}$$

where the local Lagrange multiplier $\lambda(\boldsymbol{r})$ enforces the corresponding condition of the exhaustive division at point $\boldsymbol{r}$: $\sum_X \rho_X(\boldsymbol{r}) = \rho(\boldsymbol{r})$. The above variational problem in terms of electron densities is thus equivalent to the associated principle in terms of the probability distributions (shape factors):

$$\delta\left\{\Delta S^{add.}[\boldsymbol{p}|\boldsymbol{p}^0] - \int \lambda(\boldsymbol{r})\sum\nolimits_X p_X(\boldsymbol{r})\, d\boldsymbol{r}\right\} = 0, \tag{11.18}$$

with $\lambda(\boldsymbol{r})$ now multiplying the local exhaustive division constraint $\sum_X p_X(\boldsymbol{r}) = p(\boldsymbol{r})$.

Finally, it can be directly verified by a straightforward functional differentiation that both these variational principles give the same answer of the local stockholder division: $p_X^{opt.}(\boldsymbol{r}) = p_X^H(\boldsymbol{r}) = \rho_X^H(\boldsymbol{r})/N$ or $w_X^H(\boldsymbol{r}) = w(\boldsymbol{r})$. In these global minimum entropy deficiency principles the missing information term provides an entropy "penalty" for the AIM densities deviating from the corresponding free atom densities.

This Hirshfeld solution of the density partitioning problem in IT is thus independent of the adopted entropy deficiency (missing information) measure. The symmetrized divergence of Kullback,

$$\begin{aligned}\Delta S^{add.}[\boldsymbol{\rho}^0, \boldsymbol{\rho}] &= \Delta S^{add.}[\boldsymbol{\rho}|\boldsymbol{\rho}^0] + \Delta S^{add.}[\boldsymbol{\rho}^0|\boldsymbol{\rho}] = \sum\nolimits_X \int \Delta\rho_X(\boldsymbol{r}) \ln[\rho_X(\boldsymbol{r})/\rho_X^0(\boldsymbol{r})]\, d\boldsymbol{r}\\ &= N\Delta S^{add.}[\boldsymbol{p}^0, \boldsymbol{p}],\end{aligned} \tag{11.19}$$

where $\Delta\rho_X(\boldsymbol{r}) = \rho_X(\boldsymbol{r}) - \rho_X^0(\boldsymbol{r}) = N\Delta p_X(\boldsymbol{r})$, gives rise to the same optimum division of the molecular electron distribution into the AIM fragments.

It follows from (11.1), (11.2), (11.3), and (11.8) that the molecular missing information

$$\begin{aligned}\Delta S[\rho|\rho^0] &= \int \rho(\boldsymbol{r}) \log[\rho(\boldsymbol{r})/\rho^0(\boldsymbol{r})]\, d\boldsymbol{r} = \int \rho(\boldsymbol{r}) \log[w(\boldsymbol{r})]\, d\boldsymbol{r} \\ &= \int \rho(\boldsymbol{r}) I(\boldsymbol{r})\, d\boldsymbol{r},\end{aligned} \tag{11.20}$$

which determines the total multicomponent entropy deficiency functional in the Hirshfeld partition of the molecular electron density $\rho(\boldsymbol{r})$ (see Sect. 10.4),

$$\Delta S[\rho|\rho^0] \equiv \Delta S^{total}[\boldsymbol{\rho}^{\mathrm{H}}|\boldsymbol{\rho}^0], \tag{11.21}$$

is exactly equal to its *additive* component:

$$\begin{aligned}\Delta S^{add.}[\boldsymbol{\rho}^{\mathrm{H}}|\boldsymbol{\rho}^0] &= \sum\nolimits_{\mathrm{X}} \Delta S_{\mathrm{X}}[\rho_{\mathrm{X}}^{\mathrm{H}}|\rho_{\mathrm{X}}^0] = \Delta S[\rho|\rho^0] \equiv \Delta S^{total}[\boldsymbol{\rho}^{\mathrm{H}}|\boldsymbol{\rho}^0], \\ \Delta S_{\mathrm{X}}[\rho_{\mathrm{X}}^{\mathrm{H}}|\rho_{\mathrm{X}}^0] &= \int \rho_{\mathrm{X}}^{\mathrm{H}}(\boldsymbol{r}) \log[\rho_{\mathrm{X}}^{\mathrm{H}}(\boldsymbol{r})/\rho_{\mathrm{X}}^0(\boldsymbol{r})]\, d\boldsymbol{r} = \int d_{\mathrm{X}}^{\mathrm{H}}(\boldsymbol{r})\rho(\boldsymbol{r})I(\boldsymbol{r})\, d\boldsymbol{r}.\end{aligned} \tag{11.22}$$

Therefore, this particular division scheme marks the exactly vanishing nonadditive information contribution (Nalewajski 2006g, 2010c):

$$\Delta S^{nadd.}[\boldsymbol{\rho}^{\mathrm{H}}|\boldsymbol{\rho}^0] = \Delta S^{total}[\boldsymbol{\rho}^{\mathrm{H}}|\boldsymbol{\rho}^0] - \Delta S^{add.}[\boldsymbol{\rho}^{\mathrm{H}}|\boldsymbol{\rho}^0] = 0. \tag{11.23}$$

The Fisher intrinsic accuracy similarly generates the differential Euler equation for the optimum subsystem distributions. One of its particular solutions is also the Hirshfeld prescription for the bonded atom pieces of the molecular electron density (Nalewajski and Parr 2001).

The minimum entropy deficiency principle can be similarly applied to a division of the molecular *joint* distribution of k electrons into the corresponding pieces describing the AIM k-clusters (Nalewajski 2002a, 2003a). This approach gives rise to the associated *many*-electron stockholder principle (Nalewajski 2006g; Nalewajski and Broniatowska 2007). By an appropriate partial summation and integration these cluster distributions can be used to determine the associated effective *one*-particle densities $\left\{\rho_{\mathrm{X}}^{eff.}(\boldsymbol{r};k)\right\}$ of these k-electron "stockholder" (k-S) atoms, which in general slightly differ from their *one*-electron (1-S) analogs of Hirshfeld. The largest differences (see Fig. 11.2) between densities of 2-S and 1-S bonded atoms have been observed for the lightest hydrogen and lithium atoms (Nalewajski and Broniatowska 2007). As seen in the figure, the 2-S H and Li AIM exhibit more bonding (overlapping) character compared with their 1-S analogs.

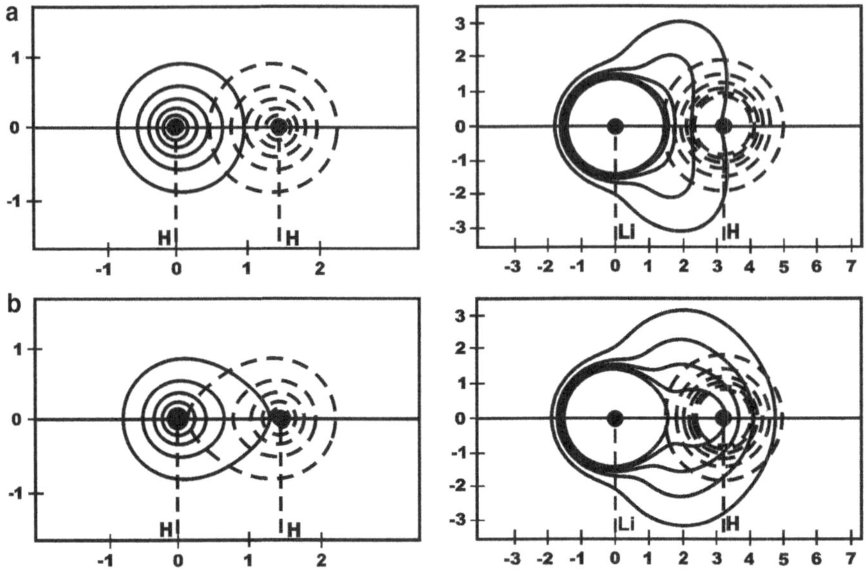

Fig. 11.2 A comparison of the contour diagrams of the $\rho_X^H(\boldsymbol{r})$ (1-*S*, Panel *a*) and $\rho_X^{eff}(\boldsymbol{r})$ (2-*S*, Panel *b*) distributions of bonded atoms in H_2 (*left column*) and LiH (*right column*)

11.4 Representative Information Densities

The same entropic descriptors, which have been used in Sections 10.1-10.3 to probe the molecular electron densities/probabilities, can be applied to diagnose changes the bonded atoms undergo in molecules relative to the free atom reference (e.g., Nalewajski et al. 2002; Nalewajski and Świtka 2002; Nalewajski and Broniatowska 2003a; Nalewajski 2006g). The entropy deficiency of the stockholder AIM,

$$\Delta S\left[\rho_X^H|\rho_X^0\right] = \int \rho_X^H(\boldsymbol{r}) \log[\rho_X^H(\boldsymbol{r})/\rho_X^0(\boldsymbol{r})]\, d\boldsymbol{r} \equiv \int \rho_X^H(\boldsymbol{r}) I_X^H(\boldsymbol{r})\, d\boldsymbol{r}$$
$$= \int P^H(X|\boldsymbol{r})\rho(\boldsymbol{r})I(\boldsymbol{r})\, d\boldsymbol{r} \equiv \int P^H(X|\boldsymbol{r})\Delta s(\boldsymbol{r})\, d\boldsymbol{r} \equiv \int \Delta s_X^H(\boldsymbol{r})\, d\boldsymbol{r}, \tag{11.24}$$

defines the density of atomic *cross*-entropy (shown in Figs. 11.3 and 11.4), $\Delta s_X^H(\boldsymbol{r}) = P^H(X|\boldsymbol{r})\Delta s(\boldsymbol{r})$, where the conditional probability $P^H(X|\boldsymbol{r})$ is defined by the stockholder share factor $d_X^H(\boldsymbol{r})$ of (11.3), and the atomic surprisals $\{I_X^H(\boldsymbol{r})\}$ are all equalized at the global (molecular) value $I(\boldsymbol{r}) = I(w(\boldsymbol{r}))$ (11.8).

The global values of the entropy displacement of the Hirshfeld AIM,

$$\mathcal{H}_X^H = S[\rho_X^H] - S[\rho_X^0] = -\int \rho_X^H(\boldsymbol{r}) \log \rho_X^H(\boldsymbol{r})\, d\boldsymbol{r} + \int \rho_X^0(\boldsymbol{r}) \log \rho_X^0(\boldsymbol{r})\, d\boldsymbol{r}$$
$$\equiv \int h_X^H(\boldsymbol{r})\, d\boldsymbol{r}, \tag{11.25}$$

for selected linear molecules of Table 10.1 are listed in Table 11.2 together with the corresponding stockholder and *free* atom values of the Shannon entropy. The preceding equation also defines the atomic entropy displacement density $h_X^H(\boldsymbol{r})$, shown in Figs. 11.3 and 11.4.

A reference to H_2 and N_2 entries of Tables 10.1 and 11.2 shows that the atomic entropy displacements for bonded hydrogen atoms are approximately additive: $2\mathcal{H}_X^H \cong \mathcal{H}[\rho(H_2)]$. A similar *near*-additivity is observed for most of the remaining molecules, with the largest deviation from such an uncoupled (independent) behavior of changes in the AIM entropy being observed for the most ionic Li–F bond, which exhibits the largest amount of CT. In general, the bonded atom exhibits a lower degree of uncertainty compared with the *free* atom value, a clear sign of the dominating effect of a relatively more compact electron distribution in a molecule. In strong electron acceptors, e.g., in F of LiF, F[LiF], and in N of HCN, N[HCN], one detects positive displacements due to the dominating CT contribution, which should result in a softer atomic distribution of electrons, thus exhibiting more "disorder" (uncertainty).

The atomic entropy displacements for HF and LiF indicate that the *donor* atom exhibits the dominating (negative) displacement, while the *acceptor* AIM only slightly increases its entropy (see also the net AIM charges in Table 11.1). The triatomic data in Table 11.2 provide an additional confirmation of this rule, with an exception of N in HNC. A reference to the atomic charges reported in Table 11.1 again shows a strong sensitivity of the atomic entropy displacements to a magnitude of the interatomic CT. As expected, a degree of this sensitivity to a change in the atomic overall electron population decreases with a growing overall number of electrons on the atom in question. Indeed, a given displacement in the AIM charge is seen to produce relatively larger reconstructions of the free atom electron distributions in H or Li, compared with N or F.

In Fig. 11.3 we have compared the contour maps of the density difference function for the Hirshfeld AIM,

$$\Delta\rho_X^H(\boldsymbol{r}) = \rho_X^H(\boldsymbol{r}) - \rho_X^0(\boldsymbol{r}), \tag{11.26}$$

with the corresponding entropy displacement density $h_X^H(\boldsymbol{r})$, for the constituent AIM of the representative diatomics of Fig. 10.1. The density difference plot (left panel) for the "stockholder" hydrogen in H_2, $H[H_2]$, exhibits changes typical for the covalently bonded atom, which have been already observed in the density profile of Fig. 11.1: the electron density buildup around the nucleus and in the bond region between nuclei, at the expense of the outer, nonbonding regions of the atomic density distribution. This is due to the contraction of the bonded hydrogen and its

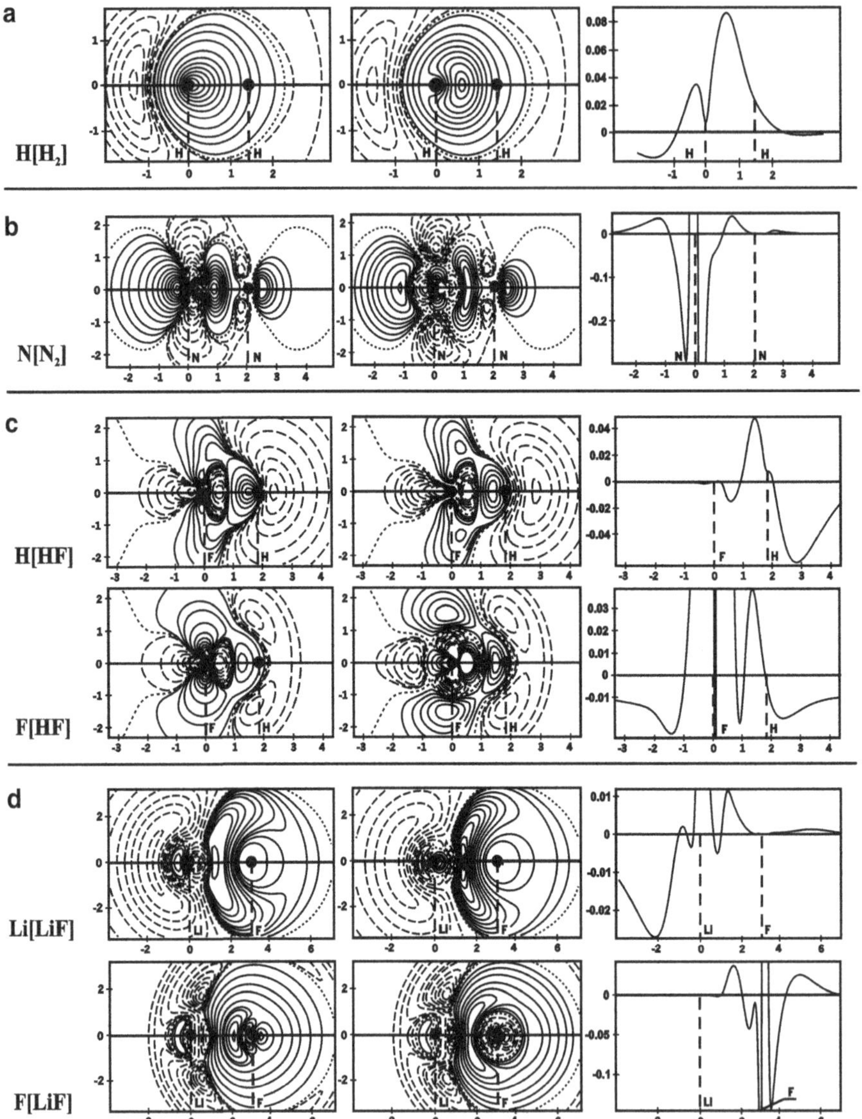

Fig. 11.3 Representative contour maps of the electron density difference function of bonded stockholder atoms, $\Delta\rho_X^H(\boldsymbol{r})$ (first column), and the associated entropy displacement density $h_X^H(\boldsymbol{r})$ (second column), for the constituent atoms of representative diatomic molecules (see also Fig. 10.1): H_2 (**a**), N_2 (**b**), HF (**c**), and LiF (**d**). The contour values are not equidistant, having been selected only for the purpose of revealing the topographic features of the quantities compared. In the third column the associated bond axis profiles of $h_X^H(\boldsymbol{r})$ (a.u.) are reported

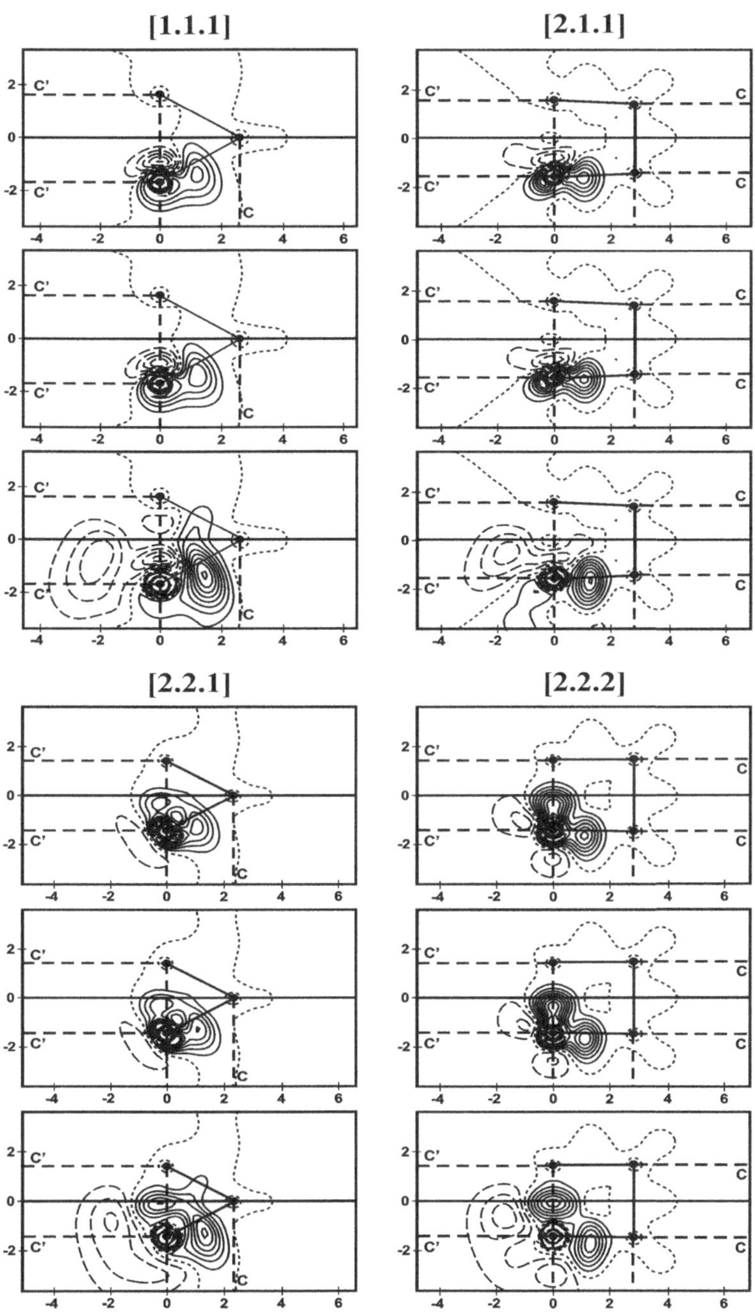

Fig. 11.4 Contour diagrams of displacements in the electron density (upper panel), information distance density (medium panel) and of the atomic Shannon entropy (lowest panel), for the stockholder bridgehead carbon atoms in the four propellanes of Fig. 10.4

Table 11.2 Displacements of the Hirshfeld AIM entropies (in bits) for representative linear molecules of Table 10.1. The molecular and promolecular entropy data are also listed. The corresponding molecular entropy shifts are reported in Table 10.1

Molecule	X	$\mathcal{H}_X^H$	$S[\rho_X{}^H]$	$S[\rho_X{}^0]$
H_2	H	−0.41	3.77	4.18
N_2	N	−0.34	5.86	6.20
HF	H	−1.09	3.09	4.18
	F	0.03	1.22	1.19
LiF	Li	−4.02	3.87	7.89
	F	0.97	2.14	1.17
HCN	H	−0.87	3.31	4.18
	C	−0.73	7.29	8.03
	N	0.15	6.35	6.20
CNH	C	0.01	8.04	8.03
	N	−0.44	5.76	6.20
	H	−0.98	3.20	4.18

polarization toward the other atom. This directional (cylindrical) polarization of the atomic electron density is clearly seen in the left, electron density panel. A similar pattern, with a somewhat more emphasized bond polarization, is seen in the entropy difference plot of the right contour map. The bonding part of the AIM entropy displacement is positive, thus marking an increase in the local uncertainty due to the electron delocalization toward the bonding partner.

A reference to part *b* of the figure, devoted to N[N_2], again shows a general similarity between the two compared contour diagrams. The displacements observed in the valence shell of both panels accord with the known density deformation changes in the triply bonded nitrogen: the ($2s$, $2p_\sigma$)-hybridization along the molecular axis, a transfer of the $2p_\pi$ electrons to the bond charge region between two nuclei, with the buildup (lowering) of the electron density of the left panel giving rise to the associated increase (decrease) in the local electron uncertainty of the right panel. Both these diagrams confirm the molecular character of the Hirshfeld atoms, with the atom displacement "tails" extending all over the molecule.

In both the density and entropy panels for H[HF], again strongly resembling one another, the hydrogen is seen to be strongly polarized toward the fluorine atom, with the transferred electron density being channeled to both the σ bond region and to the $2p_\pi$ (lone pair) regions on fluorine. The valence shell part of the F[HF] panels reveals a similar polarization of the fluorine toward the hydrogen, with the accompanying increase in the lone pair ($2p_\pi$) density, in the direction perpendicular to the bond axis. These observations confirm a relatively strong covalency of H–F bond. The localization of the fluorine bond charge close to the proton position provides an additional support to this observation.

A different CT pattern is found in the strongly ionic Li–F bond, for which the difference diagrams of the AIM density and entropy consistently show a transfer of electrons from the peripheral part of the Li[LiF] electron distribution toward F[LiF] as a whole, giving rise to a much lower degree of localization of the density displacement. This reflects much lower covalent (electron-sharing) contribution, compared with that observed in HF. This observation accords with the relative hardnesses of the two atoms in these two molecules: the two hard atoms in HF give rise to a strongly covalent bond, while the soft (Li) and hard (F) atoms in LiF generate a relatively more ionic (CT) bond character.

As a final illustration of this section we examine the bridgehead carbon atoms of the four propellanes of Fig. 10.4. The contour maps of Fig. 11.4 report the density difference function, $\Delta\rho_X^H$, the KL missing information density, Δs_X^H, and the entropy displacement density, h_X^H. They fully support the conclusions already drawn from the molecular plots of Figs. 10.5 and 10.6. The three contour diagrams for the given bridgehead carbon atom in Fig. 11.4 are seen to be qualitatively similar, thus further validating the usefulness of all these quantities as sensitive probes into changes the free atoms undergo in the molecule. These alternative diagnostic tools are seen to be to a large extent equivalent.

For the small [1.1.1] and [2.1.1] propellanes, which are lacking a large portion of the *direct* ("through-space") component of the central chemical bond between the bridgehead carbons, the atomic density/information buildups observed in the three panels indeed reflect a presence of only the bridge chemical bonds. It should be emphasized that the latter also imply the associated *indirect* ("through-bridge") component of the central bond linking the two bridgehead carbons (see Sect. 12.10). In the contour maps of Fig. 11.4 one indeed observes a distinct lowering of the AIM densities and information densities in the direction of the other bridgehead atom, which is indicative of the antibonding direct interaction between these central carbons in the [1.1.1] and [2.1.1] propellanes. For two larger systems of the [2.2.1] and [2.2.2] propellanes, in which the presence of the "through-space" component of the central bond has already been inferred from both the molecular density difference and the direct bond order measures, the displacement in the AIM electron density and the corresponding information-distance/entropy densities all exhibit increases in directions of both the neighboring bridgehead and bridge carbon atoms.

To summarize, the observed features of displacements in the entropy/cross-entropy distributions indicate that they can indeed serve as an alternative, sensitive diagnostic tools for monitoring the valence state of bonded atoms.

11.5 Charge Sensitivities of Stockholder AIM

Several attractive features of the stockholder pieces of the molecular electron distribution make them attractive concepts for chemical applications (e.g., Nalewajski 2006g). In particular, these entropy equilibrium densities were shown to give rise to the local equalization of the fragment surprisals. The role of the entropy penalty term in the variational principle of the entropy deficiency in the atomic resolution, which is responsible for the Hirshfeld density localization around the atomic nucleus, has been examined and the additivity of the information distances of Hirshfeld AIM has been established. In addition to the extremum principle of the additive entropy deficiency of atomic components of the molecular electron density the complementary variational rule has been formulated for the extremum of the nonadditive missing information in atomic densities relative to the *free*-atom distributions (Nalewajski 2006g, 2010c). The latter has also been shown to generate the stockholder partition of the molecular electron distribution into atomic fragments. In the preceding section the displacements of the electron density of bonded atoms from the corresponding *free*-atom references and the associated missing information densities and changes in the Shannon entropy density have all been used to diagnose the atomic promotion to their respective valence states in molecules. All these probes were shown to give a consistent diagnosis of the main changes the atoms undergo, when they form the chemical bonds.

Thus, these novel entropic concepts attribute to the atomic density difference function the complementary information theoretic interpretation. The reported illustrative applications demonstrate the potential of IT in extracting the chemical interpretation from the known molecular electron distributions and reflect the information origins of the chemical bonds (Nalewajski 2010f). The *charge sensitivities* (CS) of the Hirshfeld atoms, e.g., their hardness, softness, and Fukui Function descriptors, have been expressed in terms of the share factors of atomic fragments as fractions of the corresponding molecular properties. It has also been argued that the Hirshfeld atoms equalize their chemical potentials and are in principle the (effective) external potential representable (Nalewajski 2006g).

It should be realized that irrespective of the adopted density partition scheme the chemical potentials of the mutually open AIM pieces of the molecular ground state density are always equalized at the global value of the chemical potential, for the system as a whole. Indeed, for any exhaustive partitioning $\rho = \sum_X \rho_X$ of the ground state density ρ one obtains, via a straightforward *chain*-rule transformation of the functional derivative of the density functional for the system electronic energy,

$$\begin{aligned} E_v[\rho] &= E_v\left[\sum\nolimits_X \rho_X\right] = E_v^{total}[\boldsymbol{\rho}] \equiv E_v[\boldsymbol{\rho}], \\ \mu_X[\boldsymbol{\rho}] &= \{\partial E_v[\boldsymbol{\rho}]/\partial \rho_X(\boldsymbol{r})\}_v = \{\delta E_v[\rho]/\delta \rho(\boldsymbol{r})\}[\partial \rho(\boldsymbol{r})/\partial \rho_X(\boldsymbol{r})] \\ &= \delta E_v[\rho]/\delta \rho(\boldsymbol{r}) = \mu. \end{aligned} \tag{11.27}$$

This chemical potential equalization should indeed be expected, by analogy to the ordinary thermodynamics, for any set of the mutually open subsystems, i.e., for arbitrary shapes of the atomic pieces of the molecular electron density. Therefore, the fragment chemical potential, representing in the energy representation the subsystem "intensity" associated with the fragment electron density, does not discriminate between alternative divisions of the molecular electron density. In other words, such quantities are insensitive to all admissible "*vertical*" displacements in the electronic structure, which preserve the specified ground state density of the molecule as a whole.

Therefore, in order to determine the *equilibrium* vertical partition, one needs the *entropic* principle of IT to uniquely characterize the states of such embedded open AIM. As we have demonstrated above the alternative extremum principles of the entropy deficiency (cross-entropy), which give rise to the intersubsystem equalization of the AIM local surprisals, are indeed required to identify the Hirshfeld AIM as the optimum (equilibrium) atomic fragments, which exhibit the least information distance from (i.e., the maximum resemblance to) the corresponding free atoms.

This failure of the energetic criterion to identify the equilibrium partition among all admissible, exhaustive divisions of a given molecular density is because the overall electronic energy in the subsystem resolution is the same for all vertical partitioning schemes:

$$E_v[\boldsymbol{\rho}^H] = E_v[\boldsymbol{\rho}] = E_v[\rho],$$
$$\sum\nolimits_X \rho_X^H(\boldsymbol{r}) = \sum\nolimits_X \rho_X[\rho;\boldsymbol{r}] = \rho(\boldsymbol{r}).$$

This invariance of the overall electronic energy with respect to the density partitioning schemes further confirms a need for the subsidiary entropy/information principles in determining the "best" bonded atomic fragments of the molecule.

One further observes that any "*horizontal*" displacement $d\rho \neq 0$ of the molecular ground state density is uniquely partitioned into the corresponding displacements of the "stockholder" AIM,

$$d\boldsymbol{\rho}^H[d\rho] = \{d\rho_X^H(\boldsymbol{r}) = d_X^H(\boldsymbol{r})d\rho(\boldsymbol{r})\}.$$

This partitioning results from the Hirshfeld division of both the initial distribution,

$$\rho(\boldsymbol{r}) = \sum\nolimits_X \rho_X^H[\rho;\boldsymbol{r}],$$
$$\rho_X^H[\rho;\boldsymbol{r}] = \rho(\boldsymbol{r})\, d_X^H(\boldsymbol{r}) = \rho_X^0(\boldsymbol{r})\, w(\boldsymbol{r}), \quad w(\boldsymbol{r}) = \rho(\boldsymbol{r})/\rho^0(\boldsymbol{r}),$$

and the displaced molecular density $\rho' = \rho + d\rho$:

$$\rho'(\boldsymbol{r}) = \sum\nolimits_X \rho_X^H[\rho';\boldsymbol{r}],$$
$$\rho_X^H[\rho';\boldsymbol{r}] = \rho'(\boldsymbol{r})\, d_X^H(\boldsymbol{r}) = \rho_X^0(\boldsymbol{r})w'(\boldsymbol{r}), \;\; w'(\boldsymbol{r}) = \rho'(\boldsymbol{r})/\rho^0(\boldsymbol{r}).$$

Hence, the associated AIM partitioning of the horizontal density displacement reads:

$$d\rho(\boldsymbol{r}) = \sum\nolimits_X \{\rho_X^H[\rho';\boldsymbol{r}] - \rho_X^H[\rho;\boldsymbol{r}]\} = \sum\nolimits_X d_X^H(\boldsymbol{r})d\rho(\boldsymbol{r})$$
$$= \sum\nolimits_X d\rho_X^H(\boldsymbol{r}) = \sum\nolimits_X \rho_X^0(\boldsymbol{r})\, dw(\boldsymbol{r}),$$

where the displacement in the molecular enhancement factor $dw = w' - w$.

This allows one to interpret the local "share" factors (conditional probabilities) $\{d_X^H(\boldsymbol{r}) = P^H(X|\boldsymbol{r}) = d_X^0(\boldsymbol{r})\}$ and their inverses as the corresponding local derivatives:

$$d\rho_X^H(\boldsymbol{r})/d\rho(\boldsymbol{r}) = d_X^H(\boldsymbol{r}) \quad \text{and} \quad d\rho(\boldsymbol{r})/d\rho_X^H(\boldsymbol{r}) = 1/d_X^H(\boldsymbol{r}), \quad X = A, B. \quad (11.28)$$

Moreover, since the molecular density can be expressed in terms of any *single* Hirshfeld component,

$$\rho(\boldsymbol{r}) = \rho[\rho_X^H(\boldsymbol{r})] = \rho_X^H(\boldsymbol{r})/d_X^H(\boldsymbol{r}) = \rho_X^H(\boldsymbol{r})/d_X^0(\boldsymbol{r}), \quad (11.29)$$

the overall electronic energy also represents the unique functional of the density of any single Hirshfeld atom:

$$\begin{aligned}\bar{E}_v[\rho_X^H] = E_v[\rho(\rho_X^H)] &\equiv \int \rho_X^H(\boldsymbol{r})\, V_X^H(\boldsymbol{r})\, d\boldsymbol{r} + \bar{F}[\rho_X^H] \equiv \bar{E}[\rho_X^H, V_X^H] \\ \equiv E_v[\rho] &= \int \rho(\boldsymbol{r})\, v(\boldsymbol{r})\, d\boldsymbol{r} + F[\rho],\end{aligned} \quad (11.30)$$

where $\bar{F}[\rho_X^H] = F[\rho(\rho_X^H)]$ and the Hirshfeld external potential

$$V_X^H(\boldsymbol{r}) = v(\boldsymbol{r})/d_X^H(\boldsymbol{r}). \quad (11.31)$$

Equation (11.30) expresses the *full* dependence of the molecular energy on the selected stockholder density, including a dependence due to $\rho_{Y(\neq X)}^H = \rho_Y^H(\rho_X^H)$. Therefore, the functional derivative of $\bar{E}_v[\rho_X^H]$ with respect to its density argument does not represent the *partial* derivative of the molecular energy with respect to ρ_X^H [(11.27)], for the fixed electron densities of the remaining constituent atoms, and as such is not equalized throughout the space [see (7.21)]. Indeed, using the relevant local *chain* rule transformation gives for the *total* energy conjugate of the Hirshfeld subsystem density, called the Hirshfeld potential,

$$\begin{aligned}\delta\bar{E}_v[\rho_X^H]/\delta\rho_X^H(\boldsymbol{r}) &= \{\delta E_v[\rho]/\delta\rho(\boldsymbol{r})\}[d\rho(\boldsymbol{r})/d\rho_X^H(\boldsymbol{r})] \\ &= \mu/d_X^H(\boldsymbol{r}) \equiv m_X^H(\boldsymbol{r}) \neq \mu/d_X^H(\boldsymbol{r}') \equiv m_X^H(\boldsymbol{r}')\ldots\end{aligned} \quad (11.32)$$

In Sect. 7.3.7 we have introduced several charge sensitivities characterizing the equilibrium distribution of electrons in the molecule as a whole. In what follows we examine specific examples of CS of the *one*-electron "stockholder" (1-*S*) AIM, resulting from the Hirshfeld division scheme, which describe the mutually closed and open subsystems. Following the previous molecular development, we shall investigate the *second*-derivative properties emerging in both the *chemical softness representation*, within the *Electron Following* (EF) perspective of the BO approximation, and the *chemical hardness representation*, in the *Electron Preceding* (EP) perspective of the Hellmann–Feynmann theorem. An overview of the *chemical softness* (density response) and *hardness* (potential response) properties of the Hirshfeld atoms has been given in (Nalewajski 2006g).

Consider first the EF outlook, in which the *chemical softness* descriptors of molecular systems are defined. In this representation the external potential due to the nuclei or the *effective* external potentials of molecular fragments define the controlled *local* state variable(s) of the molecular system in the fragment resolution. For simplicity, we assume a diatomic molecule consisting of two Hirshfeld AIM. They are considered to be either mutually closed, when densities of each AIM can be independently modified in the hypothetical system $M_c = (A^H|B^H)$, or open, when the densities of these molecular fragments are determined by the electron distribution of the molecule as a whole, as in the real molecule $M = (A^H \,\vdots\, B^H)$.

It follows from the strict preservation of the stockholder proportions of the displaced molecular densities [see (11.28)] that

$$\frac{d\rho(\boldsymbol{r})}{d\rho_X^H(\boldsymbol{r})} = \frac{1}{d_X^H(\boldsymbol{r})}, \qquad X = A, B. \tag{11.33}$$

Therefore, any single density of the Hirshfeld AIM uniquely specifies the current (horizontally displaced) molecular ground state density: $\rho = \rho(\rho_X^H)$. The molecular Euler equation (7.21),

$$u(\boldsymbol{r}) \equiv v(\boldsymbol{r}) - \mu = -\delta F[\rho]/\delta\rho(\boldsymbol{r}), \tag{11.34}$$

can be thus transcribed as the equivalent equation for its single Hirshfeld component. The $F[\rho(\rho_X^H)] \equiv \bar{F}[\rho_X^H]$ conjugates of the stockholder AIM densities $\{\rho_X^H\}$ derived from the ground state density ρ, for which $\boldsymbol{\mu} = \mu\boldsymbol{I}$ (11.27), then read:

$$\begin{aligned} -\boldsymbol{U}^H(\boldsymbol{r}) \equiv \boldsymbol{m}^H(\boldsymbol{r}) - \boldsymbol{V}^H(\boldsymbol{r}) &= -\left\{U_X^H(\boldsymbol{r}) \equiv m_X^H(\boldsymbol{r}) - V_X^H(\boldsymbol{r}) = \mu/d_X^H(\boldsymbol{r}) - v(\boldsymbol{r})/d_X^H(\boldsymbol{r})\right\} \\ &= \left\{\frac{\delta\bar{F}[\rho_X^H]}{\delta\rho_X^H(\boldsymbol{r})} = \frac{\delta F[\rho]}{\delta\rho(\boldsymbol{r})}\frac{d\rho(\boldsymbol{r})}{d\rho_X^H(\boldsymbol{r})} = -\frac{u(\boldsymbol{r})}{d_X^H(\boldsymbol{r})}\right\}. \end{aligned} \tag{11.35}$$

This Euler equation for the single Hirshfeld AIM density component further implies the associated functional relations:

$$\rho_X^H = \rho_X^H[U_X^H], \qquad U_X^H = U_X^H[\rho_X^H] \quad \text{and} \quad \bar{F}[\rho_X^H] = \bar{F}[\rho_X^H[U_X^H]] \equiv \tilde{F}[U_X^H].$$

Equation (11.35) is thus fully equivalent to the molecular Euler equation (11.34): the latter is recovered by multiplying both sides of the former by the atomic share factor $d_X^H(\boldsymbol{r})$. This indeed should be expected, since for the fixed promolecular reference in the Hirshfeld scheme the subsystem density uniquely specifies the molecular distribution of electrons and, conversely, the overall density uniquely determines the stockholder densities of subsystems.

The ground state electronic energy $E_v[\rho]$ of the whole molecular system has been similarly expressed in Eq. (11.30) as the functional $\bar{E}_v[\rho_X^H]$ of the single Hirshfeld AIM component. It gives rise to the associated *first* differential:

$$\begin{aligned} d\bar{E}[\rho_X^H, V_X^H] &= \int \left(\frac{\partial \bar{E}[\rho_X^H, V_X^H]}{\partial \rho_X^H(\boldsymbol{r})} \right)_{V_X^H} d\rho_X^H(\boldsymbol{r})\, d\boldsymbol{r} + \int \left(\frac{\partial \bar{E}[\rho_X^H, V_X^H]}{\partial V_X^H(\boldsymbol{r})} \right)_{\rho_X^H} dV_X^H(\boldsymbol{r})\, d\boldsymbol{r} \\ &= \int m_X^H(\boldsymbol{r})\, d\rho_X^H(\boldsymbol{r})\, d\boldsymbol{r} + \int \rho_X^H(\boldsymbol{r})\, dV_X^H(\boldsymbol{r})\, d\boldsymbol{r} \\ &= \mu dN + \int \rho(\boldsymbol{r})\, dv(\boldsymbol{r})\, d\boldsymbol{r} = dE[N, v]. \end{aligned} \tag{11.36}$$

In the foregoing equation, we have identified the local intensity $m_X^H(\boldsymbol{r})$ as the $\bar{E}[\rho_X^H, V_X^H]$-conjugate of $\rho_X^H(\boldsymbol{r})$ with the latter representing the $\bar{E}[\rho_X^H, V_X^H]$-conjugate of $V_X^H(\boldsymbol{r})$ (the Hellmann–Feynman theorem for the Hirshfeld AIM component).

One similarly interprets the Legendre transforms of the system energy as the associated functionals of any single Hirshfeld component of the molecular electron density. For example, for the *grand*-potential functional at zero temperature [see (7.82)] one finds:

$$\begin{aligned} \Omega_u[\rho] = \Omega[\mu, v] &= \int \rho(\boldsymbol{r})\, u(\boldsymbol{r})\, d\boldsymbol{r} + F[\rho] = \Omega[u] \\ &= \int \rho_X^H(\boldsymbol{r})\, U_X^H(\boldsymbol{r})\, d\boldsymbol{r} + \bar{F}[\rho_X^H] \\ &\equiv \bar{\Omega}_{U_X^H}[\rho_X^H] \equiv \bar{\Omega}[m_X^H, V_X^H] \equiv \tilde{\Omega}[U_X^H], \end{aligned} \tag{11.37}$$

Again, as in (11.36), the associated differential expressions are equivalent to those of the corresponding molecular density functionals:

$$\begin{aligned} d\bar{\Omega}[m_X^H, V_X^H] &= \int \left(\frac{\partial \bar{\Omega}[m_X^H, V_X^H]}{\partial m_X^H(\boldsymbol{r})} \right)_{V_X^H} dm_X^H(\boldsymbol{r})\, d\boldsymbol{r} + \int \left(\frac{\partial \bar{\Omega}[m_X^H, V_X^H]}{\partial V_X^H(\boldsymbol{r})} \right)_{m_X^H} dV_X^H(\boldsymbol{r})\, d\boldsymbol{r} \\ &= -\int \rho_X^H(\boldsymbol{r})\, dm_X^H(\boldsymbol{r})\, d\boldsymbol{r} + \int \rho_X^H(\boldsymbol{r})\, dV_X^H(\boldsymbol{r})\, d\boldsymbol{r} \\ &= \int \rho_X^H(\boldsymbol{r})\, dU_X^H(\boldsymbol{r})\, d\boldsymbol{r} \\ &= -N d\mu + \int \rho(\boldsymbol{r})\, dv(\boldsymbol{r})\, d\boldsymbol{r} = d\Omega[\mu, v] \\ &= \int \rho(\boldsymbol{r})\, du(\boldsymbol{r})\, d\boldsymbol{r} = d\Omega[u], \end{aligned} \tag{11.38}$$

$$d\bar{F}[\rho_X^H] = \int \frac{\delta \bar{F}[\rho_X^H]}{\delta \rho_X^H(\boldsymbol{r})} d\rho_X^H(\boldsymbol{r})\, d\boldsymbol{r} = -\int U_X^H(\boldsymbol{r})\, d\rho_X^H(\boldsymbol{r})\, d\boldsymbol{r}$$
$$= -\int u(\boldsymbol{r})\, d\rho(\boldsymbol{r})\, d\boldsymbol{r} = dF[\rho]. \quad (11.39)$$

The (AIM-diagonal) second functional derivatives of $\bar{E}[\rho_X^H, V_X^H]$ and $\bar{\Omega}[m_X^H, U_X^H] \equiv \bar{\Omega}_{U_X^H}[\rho_X^H]$ with respect to the chosen stockholder density argument ρ_X^H, calculated for the fixed atomic external potential V_X^H in the molecule, then define the hardness kernel of the bonded (embedded) atom X^H:

$$\left(\frac{\delta^2 \bar{E}[\rho_X^H, V_X^H]}{\delta\rho_X^H(\boldsymbol{r})\,\delta\rho_X^H(\boldsymbol{r}')}\right)_{V_X^H} = \frac{\delta^2 \Omega_{U_X^H}[\rho_X^H]}{\delta\rho_X^H(\boldsymbol{r})\,\delta\rho_X^H(\boldsymbol{r}')} = \frac{\delta^2 \bar{F}[\rho_X^H]}{\delta\rho_X^H(\boldsymbol{r})\,\delta\rho_X^H(\boldsymbol{r}')}$$
$$\equiv \eta_{X,X}^H(\boldsymbol{r},\boldsymbol{r}') = -\frac{\delta U_X^H(\boldsymbol{r}')}{\delta \rho_X^H(\boldsymbol{r})} = \frac{\eta(\boldsymbol{r},\boldsymbol{r}')}{d_X^H(\boldsymbol{r})\, d_X^H(\boldsymbol{r}')}. \quad (11.40)$$

Since the specification of any Hirshfeld component is equivalent to the specification of the molecular ground state density itself, $\bar{\Omega}_{U_X^H}[\rho_X^H] = \bar{\Omega}_{U_Y^H}[\rho_Y^H]$, one could similarly define the AIM *off*-diagonal hardness kernels. Together, these descriptors determine the corresponding *matrix* of the hardness kernels in the Hirshfeld AIM resolution:

$$\frac{\delta^2 \tilde{F}[\boldsymbol{\rho}^H]}{\delta\boldsymbol{\rho}^H(\boldsymbol{r})\,\delta\boldsymbol{\rho}^H(\boldsymbol{r}')} = -\frac{\delta \boldsymbol{U}^H(\boldsymbol{r}')}{\delta\boldsymbol{\rho}^H(\boldsymbol{r})} \equiv \boldsymbol{\eta}^H(\boldsymbol{r},\boldsymbol{r}')$$
$$= \left\{\eta_{X,Y}^H(\boldsymbol{r},\boldsymbol{r}') = -\frac{\delta U_Y^H(\boldsymbol{r}')}{\delta\rho_X^H(\boldsymbol{r})} = \frac{\eta(\boldsymbol{r},\boldsymbol{r}')}{d_X^H(\boldsymbol{r})\, d_Y^H(\boldsymbol{r}')}\right\}, \quad (11.41)$$

with each element representing the corresponding fraction of the molecular kernel determined by the inverse product of atomic share factors. Its "inverse" determines the corresponding matrix of the softness kernels in the stockholder AIM resolution:

$$\boldsymbol{\sigma}^H(\boldsymbol{r},\boldsymbol{r}') = -\frac{\delta\boldsymbol{\rho}^H(\boldsymbol{r}')}{\delta\boldsymbol{U}^H(\boldsymbol{r})} \equiv \boldsymbol{\eta}^H(\boldsymbol{r},\boldsymbol{r}')^{-1}$$
$$= \left\{\sigma_{X,Y}^H(\boldsymbol{r},\boldsymbol{r}') = -\frac{\delta\rho_Y^H(\boldsymbol{r}')}{\delta U_X^H(\boldsymbol{r})} = d_X^H(\boldsymbol{r})\,\sigma(\boldsymbol{r},\boldsymbol{r}')\, d_Y^H(\boldsymbol{r}')\right\}. \quad (11.42)$$

They satisfy the relevant reciprocity relation,

$$\int \eta_{X,Z}^H(\boldsymbol{r},\boldsymbol{r}'')\sigma_{Z,Y}^H(\boldsymbol{r}'',\boldsymbol{r}')\, d\boldsymbol{r}'' = \delta(\boldsymbol{r}'-\boldsymbol{r})\frac{d_Y^H(\boldsymbol{r}')}{d_X^H(\boldsymbol{r})} = \frac{\delta\rho_Y^H(\boldsymbol{r}')}{\delta\rho_X^H(\boldsymbol{r})},$$
$$X, Y, Z \in (A, B). \quad (11.43)$$

Similar interpretations of other linear response properties of the Hirshfeld AIM as stockholder fractions of the corresponding molecular quantities follow from the relevant *chain* rule manipulations of the defining derivatives. We recall that the *two*-point molecular kernel defined by the second (partial) functional derivative of the ground state energy $E_v[\rho] = E[N, v]$ with respect to the molecular external potential represents the molecular *Linear Response* (LR) function (7.230):

$$\beta(\boldsymbol{r}, \boldsymbol{r}') \equiv \{\partial^2 E[N, v]/\partial v(\boldsymbol{r})\partial v(\boldsymbol{r}')\}_N = [\partial\rho(\boldsymbol{r}')/\partial v(\boldsymbol{r})]_N. \tag{11.44}$$

In the externally *closed* molecular systems, for the fixed overall number of electrons N, it transforms the external potential perturbation Δv into the associated linear response of the system density, $[\Delta\rho]_N$:

$$[\Delta\rho(\boldsymbol{r})]_N = \int \Delta v(\boldsymbol{r}')\beta(\boldsymbol{r}', \boldsymbol{r})d\boldsymbol{r}'. \tag{11.45}$$

One similarly introduces the square matrix of the *doubly* AIM-resolved kernels of the second partial derivatives of the electronic energy $E[N, v] = \mathcal{E}[N, \boldsymbol{V}^{\mathrm{H}}(v)]$ with respect to the Hirshfeld external potentials:

$$\begin{aligned}\boldsymbol{\beta}^{\mathrm{H}}(\boldsymbol{r}, \boldsymbol{r}') &= \left(\frac{\partial^2\mathcal{E}[N, \boldsymbol{V}^{\mathrm{H}}]}{\partial\boldsymbol{V}^{\mathrm{H}}(\boldsymbol{r})\ \partial\boldsymbol{V}^{\mathrm{H}}(\boldsymbol{r}')}\right)_N = \left(\frac{\partial\boldsymbol{\rho}^{\mathrm{H}}(\boldsymbol{r}')}{\partial\boldsymbol{V}^{\mathrm{H}}(\boldsymbol{r})}\right)_N \\ &= \left\{\beta^{\mathrm{H}}_{\mathrm{X,Y}}(\boldsymbol{r}, \boldsymbol{r}') = \left(\frac{\partial\rho^{\mathrm{H}}_{\mathrm{Y}}(\boldsymbol{r}')}{\partial V^{\mathrm{H}}_{\mathrm{X}}(\boldsymbol{r})}\right)_N\right\}\end{aligned} \tag{11.46}$$

This matrix of the Hirshfeld responses transforms the given displacements of the atomic effective external potentials $\Delta\boldsymbol{V}^{\mathrm{H}}$ into the associated linear responses of electron densities of the externally closed AIM:

$$[\Delta\rho^{\mathrm{H}}(\boldsymbol{r})]_{N^{\mathrm{H}}} = \int \Delta\boldsymbol{V}^{\mathrm{H}}(\boldsymbol{r}')\,\boldsymbol{\beta}^{\mathrm{H}}(\boldsymbol{r}', \boldsymbol{r})\,d\boldsymbol{r}'. \tag{11.47}$$

The double *chain*-rule transformation of the $\beta^{\mathrm{H}}_{\mathrm{X,Y}}(\boldsymbol{r}, \boldsymbol{r}')$ derivative then expresses this kernel as the corresponding fraction of the molecular kernel:

$$\begin{aligned}\beta^{\mathrm{H}}_{\mathrm{X,Y}}(\boldsymbol{r}, \boldsymbol{r}') &= \left(\frac{\partial\rho^{\mathrm{H}}_{\mathrm{Y}}(\boldsymbol{r}')}{\partial V^{\mathrm{H}}_{\mathrm{X}}(\boldsymbol{r})}\right)_{N^{\mathrm{H}}} = \frac{dv(\boldsymbol{r})}{dV^{\mathrm{H}}_{\mathrm{X}}(\boldsymbol{r})}\left(\frac{\partial\rho(\boldsymbol{r}')}{\partial v(\boldsymbol{r})}\right)_N \frac{d\rho^{\mathrm{H}}_{\mathrm{Y}}(\boldsymbol{r}')}{d\rho(\boldsymbol{r}')} \\ &= d^{\mathrm{H}}_{\mathrm{X}}(\boldsymbol{r})\beta(\boldsymbol{r}, \boldsymbol{r}')d^{\mathrm{H}}_{\mathrm{Y}}(\boldsymbol{r}').\end{aligned} \tag{11.48}$$

Consider next the FF indices of the Hirshfeld atoms, the bonded atom analogs of the molecular FF (7.229):

$$f(\boldsymbol{r}) = \partial^2 E[N, v]/\partial N \partial v(\boldsymbol{r}) = [\partial \rho(\boldsymbol{r})/\partial N]_v = [\partial \mu/\partial v(\boldsymbol{r})]_N. \tag{11.49}$$

In the atomic resolution one introduces the row vector of the stockholder FF:

$$\begin{aligned} \boldsymbol{f}^{\mathrm{H}}(\boldsymbol{r}) &= \frac{\partial^2 \mathcal{E}[N, \boldsymbol{V}^{\mathrm{H}}]}{\partial N\, \partial \boldsymbol{V}^{\mathrm{H}}(\boldsymbol{r})} = \left(\frac{\partial \boldsymbol{\rho}^{\mathrm{H}}(\boldsymbol{r})}{\partial N}\right)_{\boldsymbol{V}^{\mathrm{H}}} = \left(\frac{\partial \mu}{\partial \boldsymbol{V}^{\mathrm{H}}(\boldsymbol{r})}\right)_N^{\mathrm{T}} \\ &= \left\{ f_{\mathrm{X}}^{\mathrm{H}}(\boldsymbol{r}) = \left(\frac{\partial \rho_{\mathrm{X}}^{\mathrm{H}}(\boldsymbol{r})}{\partial N}\right)_{\boldsymbol{V}^{\mathrm{H}}} = \left(\frac{\partial \mu}{\partial V_{\mathrm{X}}^{\mathrm{H}}(\boldsymbol{r})}\right)_N \right\}. \end{aligned} \tag{11.50}$$

They can be also expressed as the associated fractions of the molecular FF by the appropriate *chain*-rule transformation of the defining derivative:

$$\boldsymbol{f}^{\mathrm{H}}(\boldsymbol{r}) = \{ f_{\mathrm{X}}^{\mathrm{H}}(\boldsymbol{r}) = \left(\frac{\partial v(\boldsymbol{r})}{\partial V_{\mathrm{X}}^{\mathrm{H}}(\boldsymbol{r})}\right)\left(\frac{\partial^2 E[N, v]}{\partial v(\boldsymbol{r})\, \partial N}\right) = d_{\mathrm{X}}^{\mathrm{H}}(\boldsymbol{r}) f(\boldsymbol{r}) \}. \tag{11.51}$$

These contributions thus define the additive atomic contributions to the molecular FF: $\sum_{\mathrm{X}} f_{\mathrm{X}}^{\mathrm{H}}(\boldsymbol{r}) = f(\boldsymbol{r})$. By integrating these atomic FF one finds the associated condensed FF index of the stockholder subsystems:

$$F_{\mathrm{X}}^{\mathrm{H}} \equiv [\partial N_{\mathrm{X}}^{\mathrm{H}}/\partial N]_v = \int f_{\mathrm{X}}^{\mathrm{H}}(\boldsymbol{r}) d\boldsymbol{r} = \int f(\boldsymbol{r}) d_{\mathrm{X}}^{\mathrm{H}}(\boldsymbol{r}) d\boldsymbol{r}, \tag{11.52}$$

measuring the linear response in atomic electron population $N_{\mathrm{X}}^{\mathrm{H}}$ per unit shift in the global number of electrons N.

As we have already indicated in (11.36)–(11.39), the first partials of the single-AIM density functionals reproduce the corresponding first differential of the associated molecular functional. One can similarly demonstrate such consistency of the second differential, e.g., of the energy functional $\bar{E}[\rho_{\mathrm{X}}^{\mathrm{H}}, V_{\mathrm{X}}^{\mathrm{H}}]$. In order to show that the above CS of AIM recover the second differential of the molecular electronic energy in (7.227), we observe that for the "frozen" molecular external potential v of the BO approximation, the shift of the Hirshfeld AIM density is determined by its FF [Eq. (11.51):

$$[d\rho_{\mathrm{X}}^{\mathrm{H}}(\boldsymbol{r})]_v = dN[\partial \rho_{\mathrm{X}}^{\mathrm{H}}(\boldsymbol{r})/\partial N]_v = dN\, f_{\mathrm{X}}^{\mathrm{H}}(\boldsymbol{r}) = dN\, d_{\mathrm{X}}^{\mathrm{H}}(\boldsymbol{r}) f(\boldsymbol{r}). \tag{11.53}$$

The second differential of $\bar{E}[\rho_X^H, V_X^H]$ then gives:

$$
\begin{aligned}
d^2\bar{E}[\rho_X^H, V_X^H] &= \frac{1}{2}\left[\iint [d\rho_X^H(\boldsymbol{r}')]_v \left(\frac{\partial m_X^H(\boldsymbol{r})}{\partial \rho_X^H(\boldsymbol{r}')}\right)_v [d\rho_X^H(\boldsymbol{r})]_v \, d\boldsymbol{r}\, d\boldsymbol{r}'\right. \\
&+ \iint dV_X^H(\boldsymbol{r}')\left(\frac{\partial m_X^H(\boldsymbol{r})}{\partial V_X^H(\boldsymbol{r}')}\right)_N [d\rho_X^H(\boldsymbol{r})]_v \, d\boldsymbol{r}\, d\boldsymbol{r}' \\
&+ \iint dV_X^H(\boldsymbol{r}')\left(\frac{\partial \rho_X^H(\boldsymbol{r})}{\partial V_X^H(\boldsymbol{r}')}\right)_N dV_X^H(\boldsymbol{r}) \, d\boldsymbol{r}\, d\boldsymbol{r}' \\
&+ \left.\iint [d\rho_X^H(\boldsymbol{r}')]_v \left(\frac{\partial \rho_X^H(\boldsymbol{r})}{\partial \rho_X^H(\boldsymbol{r}')}\right)_v dV_X^H(\boldsymbol{r}) \, d\boldsymbol{r}\, d\boldsymbol{r}'\right] \\
&= \frac{1}{2}[(dN)^2 \iint f_X^H(\boldsymbol{r}')\eta_{X,X}^H(\boldsymbol{r}',\boldsymbol{r}) f_X^H(\boldsymbol{r}) \, d\boldsymbol{r}\, d\boldsymbol{r}' \\
&+ dN \iint \frac{d\rho_X^H(\boldsymbol{r})}{d_X^H(\boldsymbol{r})} f_X^H(\boldsymbol{r}')\, dV_X^H(\boldsymbol{r}') f_X^H(\boldsymbol{r})\, d\boldsymbol{r}\, d\boldsymbol{r}' \\
&+ \iint dV_X^H(\boldsymbol{r}')\beta_{X,X}^H(\boldsymbol{r}',\boldsymbol{r}) dV_X^H(\boldsymbol{r}) \, d\boldsymbol{r}\, d\boldsymbol{r}' \\
&+ dN \iint f_X^H(\boldsymbol{r}') \frac{d_X^H(\boldsymbol{r})}{d_X^H(\boldsymbol{r}')} \delta(\boldsymbol{r}-\boldsymbol{r}')\, dV_X^H(\boldsymbol{r}) \, d\boldsymbol{r}\, d\boldsymbol{r}'] \\
&= \frac{1}{2}[(dN)^2\eta + 2dN \int f(\boldsymbol{r})\, dv(\boldsymbol{r}) d\boldsymbol{r} + \iint dv(\boldsymbol{r}')\beta(\boldsymbol{r}',\boldsymbol{r}) dv(\boldsymbol{r}) \, d\boldsymbol{r}\, d\boldsymbol{r}''] \\
&= d^2E[N, v].
\end{aligned}
\tag{11.54}
$$

It should be emphasized that ultimately the "*vertical*," submolecular reality of the system resolution into atomic fragments, which is so important for both the language and understanding in chemistry, cannot be verified by any direct experiment, since descriptors of bonded atoms cannot be formulated as the unique quantum mechanical "observables." Therefore, the bonded atoms, functional groups, and chemical bonds have to be ultimately classified as Kantian *noumenons* (Parr et al. 2005). Nonetheless, the partial understanding and indirect probes of these important chemical concepts are available from several different perspectives (e.g., Bader 1990; Nalewajski 2006g). Moreover, the close analogy between the phenomenological description of molecules/fragments in IT and the ordinary thermodynamics introduces the thermodynamic-like causality into relations between perturbations and responses of molecular subsystems thus bringing more consistency into chemical theories (Nalewajski 2006g).

References

Bader RFW (1990) Atoms in molecules. Oxford University Press, New York (and references therein)

Hirshfeld FL (1977) Theor Chim Acta (Berl) 44:129

Löwdin P-O (1950) J Chem Phys 16:365
Löwdin P-O (1956) Adv Phys 5:111
Mulliken RS (1955) J Chem Phys 23:1833, 1841, 2338, 2343
Nalewajski RF (2002a) Phys Chem Chem Phys 4:1710
Nalewajski RF (2003a) Adv Quant Chem 43:119
Nalewajski RF (2003c) Chem Phys Lett 372:28
Nalewajski RF (2006g) Information theory of molecular systems. Elsevier, Amsterdam
Nalewajski RF (2010c) J Math Chem 47:709
Nalewajski RF (2010f) Information origins of the chemical bond. Nova Science, New York
Nalewajski RF, Broniatowska E (2003a) J Phys Chem A 107:6270
Nalewajski RF, Broniatowska E (2007) Theor Chem Acc 117:7
Nalewajski RF, Parr RG (2000) Proc Natl Acad Sci USA 97:8879
Nalewajski RF, Parr RG (2001) J Phys Chem A 105:7391
Nalewajski RF, Świtka E (2002) Phys Chem Chem Phys 4:4952
Nalewajski RF, Świtka E, Michalak A (2002) Int J Quantum Chem 87:198
Parr RG, Ayers PW, Nalewajski RF (2005) J Phys Chem A 109:3957
Savin A, Nesper R, Wengert S, Fässler TF (1997) Angew Chem Int Ed Engl 36:1808
Silvi B, Savin A (1994) Nature 371:683

Chapter 12
Orbital Communication Theory of the Chemical Bond

Abstract The rudiments of the Orbital Communication Theory (OCT) of the chemical bond are presented. Molecules are interpreted as communication systems in which the electron probability (information) is scattered between AO of the basis set of molecular calculations. They are defined by the conditional probabilities derived from the bond-projected superposition principle. The IT multiplicity of all chemical bonds in a molecule is introduced, originating from the orbital interactions between AO of all constituent atoms, and its covalent and ionic components are linked to the channel average communication noise and information flow descriptors, respectively. In the illustrative *two*-AO model of the single chemical bond this approach conserves the overall bond descriptor for all admissible MO polarizations, from the purely covalent structure to the *ion* pair configuration. This bond order preservation signifies the competition between the covalent and ionic bond components. The orbital communications are partitioned into the internal (intra-atomic) and external (interatomic) subchannels.

The multiconditional probabilities of orbital events in the chemical bond system of the molecule, required for determining the information theoretic indices of the bond couplings between molecular fragments, are established within the theoretical framework of the *one*-determinantal orbital representation of molecular electronic structure. They are again derived from an appropriate generalization of the bond-projected superposition principle. The *triply* conditional probabilities, relating one conditional event to another, are shown to satisfy the relevant nonnegativity and symmetry requirements. The probability/information scattering perspective on the localized diatomic interactions between AO originating from a given pair of AIM is presented. It uses the ensemble averaging, known as the *flexible input* approach, with the weights provided by the *joint* (bond) *two*-orbital probabilities of the interacting AO. This procedure is first applied to the *two*-orbital model, where it is shown to reproduce (in bits) the corresponding Wiberg measure of the bond order. Its generalization to atoms contributing several AO to the chemical bond system is shown to exactly reproduce the corresponding Wiberg index in diatomic molecules, while closely approximating the latter in larger systems. The coupling effects between chemical bonds, which require conditional probabilities of several

R.F. Nalewajski, *Perspectives in Electronic Structure Theory*,
DOI 10.1007/978-3-642-20180-6_12,

AO on molecular subsystems, are examined and the effect of the IT-ionic activation of adsorbates is predicted.

The direct (through-*space*) and indirect (through-*bridge*) components of chemical interactions between atomic orbitals are identified in both the Wiberg bond order formalism and in OCT. The illustrative examples using the Hückel description of the conjugated π-bonds in benzene and butadiene are given and the existence of the through-bridge bond between bridgehead carbons in small propellanes is conjectured. The amplitude channels of probability scattering in molecules are introduced and the operator representation of the direct and multiple probability propagations is developed. The independent (principal) AO communications are defined and the stationary probability distribution is shown to be conserved in the multiple bridge propagations of the scattering amplitudes. The bridge amplitudes are expressed in terms of the bond overlap (density) matrix elements using the chain rule expressions for the implicit derivatives between AO in the molecular bond system.

12.1 Molecular Communication Systems

The key concept of CTCB is the molecular communication (information) channel (see Sect. 8.5), which can be constructed at alternative levels of resolving the electron probabilities into the underlying elementary "events" determining the channel inputs $\boldsymbol{a} = \{a_i\}$ and outputs $\boldsymbol{b} = \{b_j\}$. For example, they may involve finding an electron on the basis set orbitals (AO), MO, AIM, molecular fragment, etc. Such communication channels can be generated within both the *local* and *condensed* descriptions of electronic probabilities. These networks describe the probability/information propagation in the molecule and are characterized using standard entropic quantities developed in IT for real communication devices (e.g., Shannon 1948; Shannon and Weaver 1949; Abramson 1963), describing the covalent (communication noise) and ionic (information flow) components of the chemical bonds, which together determine the IT measure of the overall bond multiplicity, expressed in the entropy/information units (Nalewajski 2006g, 2010f and references therein).

Due to the electron delocalization throughout the network of chemical bonds in a molecule the transmission of "signals" about the electron assignment to the underlying elementary events of the resolution in question becomes randomly disturbed, thus exhibiting typical communication "noise." Indeed, an electron initially attributed to the given atom/orbital in the channel "input" $\boldsymbol{a}$ (molecular, promolecular) can be later found with nonzero probabilities at several locations in the molecular "output" $\boldsymbol{b}$. Therefore, the input signal (molecular, promolecular, or ensemble "tailored") becomes scattered in the molecule in accordance with its communication connections determined by the conditional probabilities of finding the channel outputs given inputs in the specific resolution under consideration.

The input signal propagation due to electron delocalization in the molecular bond system is thus embodied in the conditional probabilities of the *outputs-given-inputs*, $\mathbf{P}(\boldsymbol{b}|\boldsymbol{a}) = \{P(b_j|a_i) \equiv P(j|i)\}$, which define the molecular information network. Both the *one*- and *two*-electron approaches have been devised to construct this matrix (Nalewajski 2006g, 2010f, and references therein). The latter have used the joint probabilities of simultaneous, joint events involving two electrons in a molecule, in the AIM input and output, respectively, to determine the network conditional probabilities (Nalewajski 2006g and references therein), while the former constructs the relevant orbital probabilities using the bond-projected superposition principle of quantum mechanics (Nalewajski 2010f and references therein).

In OCT of the chemical bond, which explores the information propagation in the AO-resolved molecular communication channels (Nalewajski 2009e, f, g, 2010b, f, 2011d, e, f, g, h; Nalewajski et al. 2011a), the overall IT bond orders and their entropic covalent (communication "noise") and ionic (information "flow") components result from the appropriate conditional probabilities between AO contributed by the system constituent atoms. The latter are generated by the *bond-projected* superposition principle of quantum mechanics. The basis functions (AO) of the standard SCF LCAO MO calculations indeed determine a natural set of the elementary electron occupation events within the orbital representation of the molecular electronic structure. The ultimate goal of these orbital (Hartree–Fock or Kohn–Sham) theories is to determine the subspace of the occupied *Molecular Orbitals* (MO), which determine the equilibrium (ground state) distribution of electrons in a molecule and thus its system of chemical bonds.

The *two*-electron CTCB treatment (Nalewajski 2000c, 2004b, c, d, e, 2005a, b, c, 2006a, b, c, d, e) has been found to give rise to rather poor representation of the bond differentiation patterns in molecules (Nalewajski 2006g). These bonding patterns are decisively improved within OCT using the AO resolution (Nalewajski 2009f, g, 2010c, f, 2011g, h; Nalewajski et al. 2011a). The latter scheme complements its earlier orbital implementation using the effective AO promotion channel generated from the sequential cascade of the intermediate orbital transformation stages (Nalewajski 2008a, b, c). Such consecutive cascades of elementary information systems have been used to represent the orbital transformations and the electron excitations in the resultant propagations of the electron probabilities in molecules. The information cascade approach also provides the probability scattering perspective on the atomic promotion due to the orbital hybridization (Nalewajski 2007).

In OCT the conditional probabilities determining the molecular communication channel in the basis function resolution follow from the quantum mechanical superposition principle (Dirac 1967) supplemented by the "physical" projection onto the subspace of the system occupied MO, which determine the molecular network of chemical bonds (Nalewajski 2009e, 2010b, 2011g, h; Nalewajski et al. 2011a). Both the molecule as a whole and its constituent subsystems have been shown to be adequately described by the CTCB/OCT bond indices (Nalewajski 2005b, c, 2009g, 2010d; Nalewajski et al. 2010c). The internal and external indices of molecular fragments (groups of AO) can be efficiently generated using the appropriate *reduction* of the molecular channel by combining several inputs/outputs into a single molecular fragment (Nalewajski 2005a).

In the OCT formulation of CTCB, the *off*-diagonal orbital communications have been shown (Nalewajski 2009e) to be proportional to the corresponding Wiberg (1968) or related quadratic indices of the chemical bond (Gopinathan and Jug 1983; Mayer 1983; 1985; Jug and Gopinathan 1990; Nalewajski and Mrozek 1994, 1996; Nalewajski et al. 1993, 1994a, 1996b, 1997; Nalewajski 2004b). The Wiberg-calibrated IT indices of diatomic interactions in molecules, generated using the *input*-weighted approach, which adopts the flexible ("ensemble") input probabilities to probe the localized bonds in the molecule, have been successfully implemented in the spin-*restricted Hartree–Fock* (RHF) theory (Nalewajski et al. 2011a). The resulting IT descriptors have been shown to account for the chemical intuition quite well, at the same time providing the resolution of the overall bond multiplicities into the complementary IT-covalent and IT-ionic components. In the same study, the need for recognizing the signs of the *off*-diagonal matrix elements of the CBO matrix has been stressed, in order to properly account for the so-called "occupation" decoupling, when the antibonding MO become successively populated in the excited electron configurations.

In this chapter we shall summarize the theoretical basis and representative ground state applications of OCT to the localized chemical bonds and the bond coupling phenomena in molecular and/or reactive systems. The novel, generalized perspective on the bond multiplicity origins, recognizing both the *direct* (through-space) and *indirect* (through-bridge) contributions to chemical interactions in molecules will be presented and appropriate generalization of the orbital communication contributions will be introduced. This generalized outlook on orbital interactions in molecules rectifies some artifacts of the π-interactions in benzene and butadiene, and it gives an additional insight into the origins of the central bond in small propellanes.

12.2 Information Channels in Atomic Orbital Resolution

We begin with a short overview of the molecular communication systems in the AO/basis-function resolution for the propagation of the condensed electron probabilities of AO in the molecular bond system (Nalewajski 2009e, f, g). The underlying conditional probabilities generate the entropy/information descriptors of both the overall pattern of chemical bonds and their covalent and ionic composition in the molecular system under consideration. The conditional entropy (communication noise) and mutual information (information flow) descriptors of the molecular channel then provide the IT measures of the system covalency and ionicity, respectively (Nalewajski 2000c, 2004b, c, d, e, 2006g).

In MO theory, the network of chemical bonds is determined by the occupied MO in the system ground state. Let us assume, for reasons of simplicity, the closed-shell (*cs*) electron configuration of $N = 2n$ electronic system, within the standard RHF description, which involves the n lowest (*doubly* occupied, orthonormal) MO. In the familiar SCF LCAO MO approach, they are generated as linear combinations $\boldsymbol{\varphi} = (\varphi_1, \varphi_2, \ldots, \varphi_m) = \{\varphi_s\} = \boldsymbol{\chi}\mathbf{C}$ of the adopted basis functions of the generally overlapping AO contributed by the constituent AIM, $\boldsymbol{\chi} = (\chi_1, \chi_2, \ldots, \chi_m) = \{\chi_i\}$, for which the overlap matrix $\mathbf{S} = \langle\boldsymbol{\chi}|\boldsymbol{\chi}\rangle = \{S_{i,j} \neq \delta_{i,j}\}$. Alternatively, the Löwdin (symmetrically) orthogonalized AO (OAO) can be used, for which $\mathbf{S} = \langle\boldsymbol{\chi}|\boldsymbol{\chi}\rangle = \{S_{i,j} = \delta_{i,j}\} \equiv \mathbf{I}$; unless specified otherwise the OAO basis is assumed throughout this chapter. The square matrix $\mathbf{C} = \{C_{i,s}\} = \langle\boldsymbol{\chi}|\boldsymbol{\varphi}\rangle = (\mathbf{C}^{o}|\mathbf{C}^{v})$ then groups the relevant LCAO MO coefficients to be determined using the iterative SCF procedure, with the rectangular submatrices $\mathbf{C}^{o}$ and $\mathbf{C}^{v}$ grouping the expansion coefficients of the *occupied* (o) and *virtual* (v) MO:

$$\boldsymbol{\varphi}^{o} = (\varphi_1, \varphi_2, \ldots, \varphi_n) = \boldsymbol{\chi}\mathbf{C}^{o} \quad \text{and} \quad \boldsymbol{\varphi}^{v} = (\varphi_{n+1}, \varphi_{n+2}, \ldots, \varphi_m) = \boldsymbol{\chi}\mathbf{C}^{v}.$$

The system electron density $\rho(\boldsymbol{r})$,

$$\rho(\boldsymbol{r}) = 2\boldsymbol{\varphi}(\boldsymbol{r})\boldsymbol{\varphi}^{\dagger}(\boldsymbol{r}) = \boldsymbol{\chi}(\boldsymbol{r})[2\mathbf{C}\mathbf{C}^{\dagger}]\boldsymbol{\chi}^{\dagger}(\boldsymbol{r}) \equiv \boldsymbol{\chi}(\boldsymbol{r})\boldsymbol{\gamma}\boldsymbol{\chi}^{\dagger}(\boldsymbol{r}) = Np(\boldsymbol{r}), \tag{12.1}$$

and hence also the *one*-electron probability distribution $p(\boldsymbol{r}) = \rho(\boldsymbol{r})/N$, the *shape-*factor of ρ, are then determined by the (CBO) density matrix $\boldsymbol{\gamma}$ of (6.45), (10.25), and (10.34),

$$\begin{aligned}\boldsymbol{\gamma} &= 2\langle\boldsymbol{\chi}|\boldsymbol{\varphi}^{o}\rangle\langle\boldsymbol{\varphi}^{o}|\boldsymbol{\chi}\rangle = 2\mathbf{C}^{o}\mathbf{C}^{o\dagger} \equiv 2\langle\boldsymbol{\chi}|\hat{\mathrm{P}}^{o}_{\varphi}|\boldsymbol{\chi}\rangle = 2\left(\langle\boldsymbol{\chi}|\hat{\mathrm{P}}^{o}_{\varphi}\right)\left(\hat{\mathrm{P}}^{o}_{\varphi}|\boldsymbol{\chi}\rangle\right) \equiv 2\langle\boldsymbol{\chi}^{b}|\boldsymbol{\chi}^{b}\rangle \\ &= \left\{\gamma_{i,j} = 2\langle\chi_i|\hat{\mathrm{P}}^{o}_{\varphi}|\chi_j\rangle \equiv 2\langle i|\hat{\mathrm{P}}^{o}_{\varphi}|j\rangle = 2\langle i^{b}\mid j^{b}\rangle\right\}.\end{aligned} \tag{12.2}$$

The latter thus constitutes the AO representation of the projection operator onto the subspace of all *doubly* occupied MO, $\hat{\mathrm{P}}^{o}_{\varphi} = |\boldsymbol{\varphi}^{o}\rangle\langle\boldsymbol{\varphi}^{o}| = \sum_{s}^{occd.} |s\rangle\langle s| \equiv \Sigma_s \hat{\mathrm{P}}^{o}_{s}$. Above, we have also introduced the AO projections onto the bond subspace $\boldsymbol{\varphi}^{o} : |\boldsymbol{\chi}^{b}\rangle = |\boldsymbol{\varphi}^{o}\rangle\langle\boldsymbol{\varphi}^{o}|\boldsymbol{\chi}\rangle = \{|i^{b}\rangle\}$. In orbital approximation of the single Slater determinant the density matrix $\boldsymbol{\gamma}$ of the closed-shell system then satisfies the following idempotency relation

$$(\boldsymbol{\gamma})^{2} = 4\langle\boldsymbol{\chi}|\hat{\mathrm{P}}^{o}_{\varphi}|\boldsymbol{\chi}\rangle\langle\boldsymbol{\chi}|\hat{\mathrm{P}}^{o}_{\varphi}|\boldsymbol{\chi}\rangle = 4\langle\boldsymbol{\chi}|\left(\hat{\mathrm{P}}^{o}_{\varphi}\right)^{2}|\boldsymbol{\chi}\rangle = 4\langle\boldsymbol{\chi}|\hat{\mathrm{P}}^{o}_{\varphi}|\boldsymbol{\chi}\rangle = 2\boldsymbol{\gamma}. \tag{12.3}$$

This CBO (density) matrix reflects the promoted, *valence* state of AO in the molecule, with the diagonal elements measuring the effective electron occupations of basis functions, $\{N_i = \gamma_{i,i} = Np_i\}$, and hence also the net charges of AIM, with probabilities $\boldsymbol{p} = \{p_i = \gamma_{i,i}/N\}$ of the AO being occupied in the molecule, $\sum_i p_i = 1$. The signs of its *off*-diagonal (interatomic) elements, called *bond orders*, reflect the character of the effective chemical interaction between the given pair (χ_i,χ_j) of AO: for their positive overlap $S_{i,j} > 0$, $\gamma_{i,j} > 0$ then signifies their bonding combination and $\gamma_{i,j} < 0$ implies the resultant antibonding coupling in all occupied MO, with $\gamma_{i,j} = 0$ identifying the mutually nonbonding status of the two basis functions [see (10.26)].

The molecular information channel in the (condensed) orbital resolution thus involves the AO events $\boldsymbol{\chi}$ in its input $\boldsymbol{a} = \{\chi_i\}$ and output $\boldsymbol{b} = \{\chi_j\}$. It represents the effective communication promotion of these basis functions in the molecule, via the probability/information scattering described by the conditional probabilities of the AO outputs given the AO inputs, identified by the *column* (output) and *row* (input) indices, respectively. In this *one*-electron description, the AO → AO communication network is determined by the conditional probabilities of the output AO events, given the input AO events,

$$\mathbf{P}(\boldsymbol{b}|\boldsymbol{a}) = \{P(\chi_j|\chi_i) \equiv P(j|i) = P(i \wedge j)/p_i\}, \quad \sum_j P(j|i) = 1, \tag{12.4}$$

where the associated joint probabilities of simultaneously observing two AO in the system chemical bonds $\mathbf{P}(\boldsymbol{a}\wedge\boldsymbol{b}) = \{P(i\wedge j)\}$ satisfy the usual partial and total normalization relations:

$$\sum_i P(i \wedge j) = p_j, \quad \sum_j P(i \wedge j) = p_i, \quad \sum_i \sum_j P(i \wedge j) = 1. \tag{12.5}$$

The conditional probabilities $\mathbf{P}(\boldsymbol{b}|\boldsymbol{a})$ define the information scattering in the AO promotion channel of the molecule, in which the "signals" of the molecular electron allocations to basis functions are transmitted between the AO inputs and outputs. Such communication system constitutes the basis of the OCT.

In quantum mechanics, the "*geometric*" (g) conditional probability $P^g(\phi|\psi)$ of observing the normalized (*variable*) state ϕ, given another (*parameter*, reference) state ψ, emerges in the context of the superposition principle (Dirac 1967):

$$P^g(\phi|\psi) = |\langle\psi|\phi\rangle|^2 = \langle\phi|\psi\rangle\langle\psi|\phi\rangle = \langle\phi|\hat{\mathrm{P}}_\psi|\phi\rangle = \langle\psi|\hat{\mathrm{P}}_\phi|\psi\rangle = P^g(\psi|\phi). \tag{12.6}$$

It should be observed that the quantum state conditioned upon itself in the molecular Hilbert space gives the probability of the sure event: $P^g(\psi|\psi) = 1$.

Since we are interested in the simultaneous AO events occurring in the bond system of the molecule, the two scalar products in the preceding equations have to be calculated between the AO projections into the occupied subspace $\boldsymbol{\varphi}^o$ of MO (Nalewajski 2009e). Such "*physical*" conditional probabilities between AO are obtained by inserting the projector $\hat{\mathrm{P}}^o_{\boldsymbol{\varphi}}$ between the two states, say $\phi = \chi_j$ and $\psi = \chi_i$, involved in the two scalar products of the preceding geometrical expression:

$$\begin{aligned} P(j|i) &= \mathcal{N}_i|\langle i|\hat{\mathrm{P}}^o_{\boldsymbol{\varphi}}|j\rangle|^2 = \mathcal{N}_i\langle j|\hat{\mathrm{P}}^o_{\boldsymbol{\varphi}}|i\rangle\langle i|\hat{\mathrm{P}}^o_{\boldsymbol{\varphi}}|j\rangle = \mathcal{N}_i\langle j|\hat{\mathrm{P}}^o_{\boldsymbol{\varphi}}\hat{\mathrm{P}}_i\hat{\mathrm{P}}^o_{\boldsymbol{\varphi}}|j\rangle \equiv \mathcal{N}_i\langle j|\hat{\mathrm{S}}_i|j\rangle \\ &= (2\gamma_{i,i})^{-1}\gamma_{i,j}\gamma_{j,i}. \end{aligned} \tag{12.7}$$

This probability is thus determined as expectation value in the *final* (variable, output) state χ_j of the molecular scattering operator $\hat{S}_i$ from the *initial* (reference, input) state χ_i (Nalewajski 2011c):

$$\hat{S}_i = \hat{P}^o_{\varphi}|i\rangle\langle i|\hat{P}^o_{\varphi} = \hat{P}^o_{\varphi}\hat{P}^o_i\hat{P}^o_{\varphi} = |i^b\rangle\langle i^b| \equiv \hat{P}^b_i. \tag{12.8}$$

The proportionality constant $\mathcal{N}_i = (2\gamma_{i,i})^{-1}$ satisfies the required normalization condition [see (12.4)],

$$\sum_j P(j|i) = \mathcal{N}_i \sum_j \gamma_{i,j}\gamma_{j,i} = 2\mathcal{N}_i\gamma_{i,i} = 1. \tag{12.9}$$

For the open-shell generalization, see Nalewajski (2009g).

To summarize, the generalized, *bond*-projected superposition principle of quantum mechanics generates the (*physical*) conditional probabilities as renormalized squares of corresponding elements of the CBO matrix:

$$\mathbf{P}(\boldsymbol{b}|\boldsymbol{a}) = \left\{P(j|i) = \mathcal{N}_i\left|\langle i|\hat{P}^o_{\varphi}|j\rangle\right|^2 = (2\gamma_{i,i})^{-1}\gamma_{i,j}\gamma_{j,i} = (2\gamma_{i,i})^{-1}{\gamma_{i,j}}^2\right\}. \tag{12.10}$$

These probabilities explore the dependencies between AO resulting from their simultaneous participation in the framework of all occupied MO, i.e., their involvement in the entire network of chemical bonds in the molecule. This molecular channel can be probed using both the promolecular $(\boldsymbol{p}^0 = \{p^0_i\})$, molecular $(\boldsymbol{p} = \{p_i\})$, or arbitrary, e.g., the ensemble input probabilities, in order to extract the desired IT descriptors of the system/fragment bond multiplicities and their *ionic* and *covalent* components (Nalewajski 2009e, f, g, 2010b, c, f, 2011g, h; Nalewajski et al. 2010c).

In this approach, the *off*-diagonal conditional probability of *j*th AO output given *i*th AO input is thus proportional to the squared element of the CBO matrix linking the specified pair of AO, $\gamma_{j,i} = \gamma_{i,j}$, thus being also proportional to the corresponding AO contribution $\mathcal{M}_{i,j} = \gamma^2_{i,j}$ to the Wiberg index of the overall chemical bond order between two atoms A and B in the molecule (Wiberg 1968),

$$\mathcal{M}_{\mathrm{A,B}} = \sum_{i\in \mathrm{A}}\sum_{j\in \mathrm{B}}\mathcal{M}_{i,j}, \tag{12.11}$$

or to related generalized quadratic descriptors of the molecular bond multiplicities (Gopinathan and Jug 1983; Mayer 1983; Jug and Gopinathan 1990; Nalewajski and Mrozek 1994, 1996; Nalewajski et al. 1993, 1994a, 1996a, b, 1997).

It can be straightforwardly verified using the idempotency relation of (12.3) that the associated matrix of the joint *two*-AO probabilities,

$$\mathbf{P}(\boldsymbol{a}\wedge\boldsymbol{b}) = \left\{P(i\wedge j) = p_i P(j|i) = (2N)^{-1}\gamma_{i,j}\gamma_{j,i} = (2N)^{-1}\langle i|\hat{P}^o_{\varphi}|j\rangle\langle j|\hat{P}^o_{\varphi}|i\rangle\right\}, \tag{12.12}$$

indeed satisfies the normalization conditions of (12.5):

$$\sum_i P(i \wedge j) = (2N)^{-1} \sum_i \gamma_{j,i}\gamma_{i,j} = (2N)^{-1} 2\gamma_{j,j} = p_j. \tag{12.13}$$

12.3 Entropy/Information Descriptors of Bond Components

In OCT the entropy/information indices of the covalent/ionic components of all chemical bonds in the given molecular system as a whole represent the complementary descriptors of the average communication *noise* and the average amount of the information *flow* in the molecular channel (e.g., Nalewajski 2000c, 2004b, c, d, e, 2005a, b, c, 2006g, 2010f; 2011e; Nalewajski et al. 2010c). The molecular input signal $\boldsymbol{P(a)} \equiv \boldsymbol{p}$ then generates the same distribution in the channel output,

$$\boldsymbol{p}\,\mathbf{P}(\boldsymbol{b}|\boldsymbol{a}) = \left\{\sum_i p_i P(j|i) \equiv \sum_i P(i \wedge j) = p_j\right\} = \boldsymbol{p}, \tag{12.14}$$

thus identifying $\boldsymbol{p}$ as the *stationary* probability vector of AO in the molecular ground state, while the promolecular input $\boldsymbol{P(a^0)} \equiv \boldsymbol{p}^0$ in general produces slightly different output probability.

The purely *molecular* communication channel, with $\boldsymbol{p}$ defining its input signal, is devoid of any reference (history) of the chemical bond formation and generates the average noise index of the *molecular* IT bond *covalency*, measured by the *conditional entropy* of the molecular outputs given molecular inputs (see Sects. 8.4 and 8.5):

$$\begin{aligned} S(\boldsymbol{P(b)}|\boldsymbol{P(a)}) &\equiv H(\mathbf{B}|\mathbf{A}) = -\textstyle\sum_i \sum_j P(i \wedge j) \log[P(i \wedge j)/p_i] \\ &= \textstyle\sum_i p_i \left[-\sum_j P(j|i) \log P(j|i)\right] \equiv \sum_i p_i S_i \equiv S(\boldsymbol{p}|\boldsymbol{p}) \equiv S. \end{aligned} \tag{12.15}$$

This average noise descriptor expresses the difference between the Shannon entropies of the molecular *one*- and *two*-orbital probabilities (see Sect. 8.4),

$$\begin{aligned} S &= H(\mathbf{AB}) - H(\mathbf{A}), \\ H(\mathbf{A}) &= -\sum_i p_i \log p_i \equiv S(\boldsymbol{P(a)}) = H(\mathbf{B}) \equiv S(\boldsymbol{P(b)}), \\ H(\mathbf{AB}) &\equiv S(\mathbf{P}(\boldsymbol{a} \wedge \boldsymbol{b})) \equiv H(\mathbf{A} \wedge \mathbf{B}) = H(\mathbf{A}) + H(\mathbf{B}) - I(\mathbf{A}{:}\mathbf{B}) \\ &= -\sum_i \sum_j P(i \wedge j) \log P(i \wedge j). \end{aligned} \tag{12.16}$$

Hence, for the *independent* input and output events, when $\mathbf{P}^{ind.}(\boldsymbol{a}\wedge\boldsymbol{b}) = \{p_i p_j\}$, $S(\mathbf{P}^{ind.}(\boldsymbol{a}\wedge\boldsymbol{b})) = 2S(\boldsymbol{p})$ and hence $S^{ind.} = S(\boldsymbol{p})$.

The AO channel with the *promolecular* AO probabilities $\boldsymbol{P(a^0)} = \boldsymbol{p}^0$ as its input "signal" refers to the initial state in the bond formation process. It corresponds to the ground state (fractional) occupations of the AO contributed by the system constituent (free) atoms, before their mixing into MO. This input signal gives rise to the average information flow descriptor of the system overall IT bond *ionicity*, given by the *mutual information* in the channel inputs and outputs (Nalewajski 2011e):

$$\begin{aligned} I(\boldsymbol{P}(\boldsymbol{a}^0){:}\boldsymbol{P}(\boldsymbol{b})) &\equiv I(\mathbf{A}^0{:}\mathbf{B}) = \sum_i \sum_j P(i \wedge j) \log\left[P(i \wedge j)/\left(p_j p_i^0\right)\right] \\ &= \sum_i p_i \left\{\sum_j P(j|i) \log\left[P(i|j)/p_i^0\right]\right\} \equiv \sum_i p_i I_i^0 \equiv I(\boldsymbol{p}^0{:}\boldsymbol{p}) \equiv I^0 \\ &= S(\boldsymbol{P}(\boldsymbol{b})) + S(\boldsymbol{P}(\boldsymbol{a}^0)) - S(\boldsymbol{P}(\boldsymbol{a} \wedge \boldsymbol{b})) = \Delta S(\boldsymbol{p}|\boldsymbol{p}^0) + S(\boldsymbol{p}) - S, \end{aligned} \quad (12.17)$$

where (see Section 8.3) the entropy deficiency in the molecular output distribution $\boldsymbol{p}$ relative to the promolecular reference $\boldsymbol{p}^0$ reads: $\Delta S(\boldsymbol{p}|\boldsymbol{p}^0) = \sum_i p_i \log(p_i|p_i^0)$. This *amount of information* reflects the fraction of the initial (promolecular) information content $S(\boldsymbol{p}^0)$ received in the channel output, which has not been dissipated as noise in the molecular communication system. In particular, for the molecular input, when $\boldsymbol{p}^0 = \boldsymbol{p}$ and $\Delta S(\boldsymbol{p}|\boldsymbol{p}) = 0$,

$$I(\boldsymbol{P}(\boldsymbol{a}){:}\boldsymbol{P}(\boldsymbol{b})) = \sum_i \sum_j P(i,j) \log[P(i,j)/(p_j p_i)] = S(\boldsymbol{p}) - S \equiv I(\boldsymbol{p}{:}\boldsymbol{p}). \quad (12.18)$$

Hence, for the independent input and output events $I^{ind.}(\boldsymbol{P}(\boldsymbol{a}) : \boldsymbol{P}(\boldsymbol{b})) = 0$.

Finally, the sum of these two bond components,

$$\begin{aligned} \mathcal{N}(\boldsymbol{P}(\boldsymbol{a}^0); \boldsymbol{P}(\boldsymbol{b})) &= S + I^0 \equiv \mathcal{N}(\boldsymbol{p}^0; \boldsymbol{p}) \equiv \mathcal{N}^0 = \Delta S(\boldsymbol{p}|\boldsymbol{p}^0) + S(\boldsymbol{p}) \\ &= \sum_i p_i \left(S_i + I_i^0\right) \equiv \sum_i p_i \mathcal{N}_i^0, \end{aligned} \quad (12.19)$$

where $\mathcal{N}_i^0 = -\log p_i^0$ stands for the *self*-information in the promolecular AO input event χ_i, measures the overall IT multiplicity of all bonds in the molecular system under consideration. Alternatively, for the molecular input, when $\boldsymbol{P}(\boldsymbol{a}) = \boldsymbol{p}$, this quantity preserves the Shannon entropy of the molecular input probabilities:

$$\mathcal{N}(\boldsymbol{P}(\boldsymbol{a}); \boldsymbol{P}(\boldsymbol{b})) = S(\boldsymbol{P}(\boldsymbol{b})|\boldsymbol{P}(\boldsymbol{a})) + I(\boldsymbol{P}(\boldsymbol{a}){:}\boldsymbol{P}(\boldsymbol{b})) = S(\boldsymbol{P}(\boldsymbol{a})) = S(\boldsymbol{p}). \quad (12.20)$$

We recall (see diagrams of Figs. 8.1 and 8.5) that for two dependent probability schemes the common (overlap) area of the associated entropy circles corresponds to the mutual information $I(\boldsymbol{P}(\boldsymbol{a}){:}\boldsymbol{P}(\boldsymbol{b}))$ in both distributions, while the remaining parts of individual circles represent the corresponding conditional entropies $S(\boldsymbol{P}(\boldsymbol{b})|\boldsymbol{P}(\boldsymbol{a}))$ and $S(\boldsymbol{P}(\boldsymbol{a})|\boldsymbol{P}(\boldsymbol{b}))$. The latter measure the residual uncertainty about events in one set, when one has the full knowledge of the occurrence of the events in the other set of outcomes. Accordingly, the area enclosed by the envelope of these two overlapping circles represents the entropy in the joint distribution of these two sets of events [see also (12.16)]:

$$\begin{aligned} S(\boldsymbol{P}(\boldsymbol{a}) \wedge \boldsymbol{P}(\boldsymbol{b})) &= S(\boldsymbol{P}(\boldsymbol{a})) + S(\boldsymbol{P}(\boldsymbol{b})) - I(\boldsymbol{P}(\boldsymbol{a}) : \boldsymbol{P}(\boldsymbol{b})) \\ &= S(\boldsymbol{P}(\boldsymbol{a})) + S(\boldsymbol{P}(\boldsymbol{b})|\boldsymbol{P}(\boldsymbol{a})) = S(\boldsymbol{P}(\boldsymbol{b})) + S(\boldsymbol{P}(\boldsymbol{a})|\boldsymbol{P}(\boldsymbol{b})). \end{aligned} \quad (12.21)$$

12.4 *Two*-Orbital Model of Chemical Bond

To illustrate these IT concepts, let us again examine the 2-AO model of the chemical bond (see Sect. 10.6). The ground state density matrix $\boldsymbol{\gamma}_0$ for the doubly occupied bonding MO φ_b (10.26) generates the following conditional probability matrix $\mathbf{P}(\boldsymbol{b}|\boldsymbol{a}) = \mathbf{P}(\boldsymbol{\chi}|\boldsymbol{\chi}) = \{P(j|i)\}$ (12.10):

$$\mathbf{P}(\boldsymbol{\chi}|\boldsymbol{\chi}) = \begin{bmatrix} P & Q \\ P & Q \end{bmatrix}, \quad (12.22)$$

which determines the relevant AO → AO communication network for this model bond system, shown in Fig. 12.1. In this nonsymmetrical binary channel, one adopts the molecular input signal, $\boldsymbol{p} = (P, Q = 1 - P)$, to extract the bond IT-covalency index, which measures the channel average communication noise, and the promolecular input signal, $\boldsymbol{p}^0 = (½, ½)$, to calculate the IT-ionicity index measuring the channel information capacity relative to this *covalent* promolecule, in which the two basis functions contribute a single electron each to form the chemical bond.

The bond IT-covalency $S(P)$ is thus determined by the binary entropy function $H(P) = -P\log_2 P - Q\log_2 Q = H(\boldsymbol{p})$ of Fig. 8.4. It reaches the maximum value $H(P = ½) = 1$ bit for the symmetric bond $P = Q = ½$, e.g., the σ bond in H_2 or π-bond in ethylene, and vanishes for the *lone* pair molecular configurations, when $P = (0, 1)$, $H(P = 0) = H(P = 1) = 0$, marking the alternative *ion* pair configurations A^+B^- and A^-B^+, respectively, relative to the initial AO occupations $N^0 = (1, 1)$ in the assumed (atomic) promolecular reference, in which both atoms contribute a single electron each to form the chemical bond.

The complementary descriptor of IT-ionicity, determining the channel mutual information (*capacity*), $I^0(P) = \Delta S(\boldsymbol{p}|\boldsymbol{p}^0) = H[\boldsymbol{p}^0] - H(P) = 1 - H(P)$, reaches the highest value for these two limiting electron-transfer configurations $P = (0, 1)$:

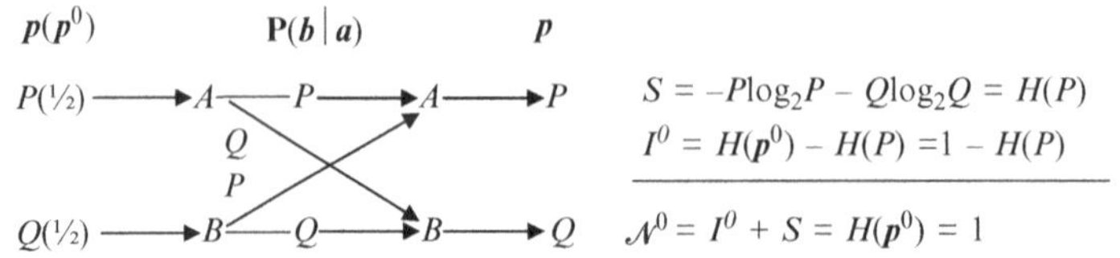

Fig. 12.1 Communication channel of the 2-OAO model of the chemical bond and its entropy/information descriptors (in bits)

$I^0(P = 0) = I^0(P = 1) = H(½) = 1$ bit. This ionicity descriptor identically vanishes for the purely covalent, symmetric bond: $I^0(P = ½) = 0$.

Therefore, these two components of the chemical bond multiplicity compete with one another, yielding the conserved overall IT bond index $\mathcal{N}^0(P) = S(P) + I^0(P) = 1$ bit., marking a single bond in IT, in the whole range of admissible bond polarizations $P \in [0, 1]$. This simple model thus properly accounts for the competition between the bond covalency and ionicity, while preserving the single bond order measure reflected by the conserved overall IT multiplicity of the chemical bond. Similar effects transpire from this model description in the (*two*-electron) CTCB and the quadratic bond indices formulated in the MO theory.

12.5 Additive and Nonadditive Components of Information Channels

Let us combine the molecular basis functions of typical SCF LCAO MO calculations into the corresponding atomic subsets:

$$\boldsymbol{\chi} = \{\boldsymbol{\chi}_X\} = (\boldsymbol{\chi}_A, \boldsymbol{\chi}_B, \boldsymbol{\chi}_C, \ldots) \equiv \boldsymbol{\chi}^{AIM}. \tag{12.23}$$

This arrangement determines the associated block structure of the AO conditional probability matrix:

$$\mathbf{P}(\boldsymbol{\chi}^{AIM}|\boldsymbol{\chi}^{AIM}) = \{\mathbf{P}(\boldsymbol{\chi}_X|\boldsymbol{\chi}_Y)\}, \quad (X, Y) \in A, B, C, \ldots \tag{12.24}$$

As schematically shown in Fig. 12.2, each diagonal block $\mathbf{P}(\boldsymbol{\chi}_X|\boldsymbol{\chi}_X)$ then determines the internal (*one*-center) communications $X \to X$, within atom X alone, which are responsible for the AIM *promotion* to its bonding (*valence*) state in the molecule. The *off*-diagonal blocks $\mathbf{P}(\boldsymbol{\chi}_X|\boldsymbol{\chi}_Y)$ and $\mathbf{P}(\boldsymbol{\chi}_Y|\boldsymbol{\chi}_X)$, $X \neq Y$, similarly generate the external (*two*-center) communications $Y \to X$ and $X \to Y$, respectively, between AO contributed by both atoms, which are ultimately responsible for the truly *bonding* contributions to the overall IT multiplicities of the localized chemical bonds between the specified pair of AIM. It should be emphasized, however, that the chemical values of diatomic bond multiplicities combine both the *one*- and *two*-center effects of the intra-atom polarization (promotion) and interatomic delocalization and CT effects, respectively, since in bonding phenomena both these processes are synergetically combined with one another.

The interatomic communications in the molecular channel reflect the covalent interactions between the given pair of atoms, so that nonbonded (separated) atoms of the promolecule exhibit only the intra-atom probability propagations. The same principle can be used to naturally partition the molecular AO communication system of the AIM-arranged basis set $\boldsymbol{\chi}^{AIM}$ into its AIM *additive* and nonadditive subchannels (Fig. 12.2) (Nalewajski 2010c):

$$
\begin{bmatrix}
P(\chi_A|\chi_A) & P(\chi_B|\chi_A) & P(\chi_C|\chi_A) & \ldots \\
P(\chi_A|\chi_B) & P(\chi_B|\chi_B) & P(\chi_C|\chi_B) & \ldots \\
P(\chi_A|\chi_C) & P(\chi_B|\chi_C) & P(\chi_C|\chi_C) & \ldots \\
\ldots & \ldots & \ldots & \ldots
\end{bmatrix}
$$

$$
= \begin{bmatrix}
P(\chi_A|\chi_A) & \mathbf{0} & \mathbf{0} & \ldots \\
\mathbf{0} & P(\chi_B|\chi_B) & \mathbf{0} & \ldots \\
\mathbf{0} & \mathbf{0} & P(\chi_C|\chi_C) & \ldots \\
\ldots & \ldots & \ldots & \ldots.
\end{bmatrix}
+
\begin{bmatrix}
\mathbf{0} & P(\chi_B|\chi_A) & P(\chi_C|\chi_A) & \ldots \\
P(\chi_A|\chi_B) & \mathbf{0} & P(\chi_C|\chi_B) & \ldots \\
P(\chi_A|\chi_C) & P(\chi_B|\chi_C) & \mathbf{0} & \ldots \\
\ldots & \ldots & \ldots & \ldots
\end{bmatrix}
$$

$$
= \begin{bmatrix}
A \to A & \mathbf{0} & \mathbf{0} & \ldots \\
\mathbf{0} & B \to B & \mathbf{0} & \ldots \\
\mathbf{0} & \mathbf{0} & C \to C & \ldots \\
\ldots & \ldots & \ldots & \ldots
\end{bmatrix}
+
\begin{bmatrix}
\mathbf{0} & A \to B & A \to C & \ldots \\
B \to A & \mathbf{0} & B \to B & \ldots \\
C \to A & C \to B & \mathbf{0} & \ldots \\
\ldots & \ldots & \ldots & \ldots
\end{bmatrix}
$$

total = internal (additive) + external (*non*-additive)

Fig. 12.2 Partitioning of the conditional AO probabilities defining the molecular information system into the *one*-center (AIM-internal, additive) and *two*-center (AIM-external, nonadditive) subchannels, and the underlying communications between the constituent bonded atoms

$$
\begin{aligned}
\mathbf{P}(\boldsymbol{\chi}^{\mathrm{AIM}}|\boldsymbol{\chi}^{\mathrm{AIM}}) &\equiv \mathbf{P}^{total}(\boldsymbol{\chi}^{\mathrm{AIM}}|\boldsymbol{\chi}^{\mathrm{AIM}}) \\
&= \mathbf{P}^{add.}(\boldsymbol{\chi}^{\mathrm{AIM}}|\boldsymbol{\chi}^{\mathrm{AIM}}) + \mathbf{P}^{nadd.}(\boldsymbol{\chi}^{\mathrm{AIM}}|\boldsymbol{\chi}^{\mathrm{AIM}}).
\end{aligned}
\tag{12.25}
$$

The former combines all internal (intra-atomic) communications within each (externally decoupled) AIM, thus being solely determined by the diagonal, atomic blocks of the molecular conditional probabilities $\mathbf{P}(\boldsymbol{\chi}^{\mathrm{AIM}}|\boldsymbol{\chi}^{\mathrm{AIM}})$:

$$
\mathbf{P}^{int.}(\boldsymbol{\chi}^{\mathrm{AIM}}|\boldsymbol{\chi}^{\mathrm{AIM}}) = \{\mathbf{P}(\boldsymbol{\chi}_{\mathrm{X}}|\boldsymbol{\chi}_{\mathrm{X}})\delta_{\mathrm{X,Y}}\} \equiv \mathbf{P}^{add.}(\boldsymbol{\chi}^{\mathrm{AIM}}|\boldsymbol{\chi}^{\mathrm{AIM}}). \tag{12.26}
$$

The latter groups all complementary, external (interatomic) probability propagations between the (externally coupled) pairs of bonded atoms in the molecular system under consideration:

$$
\mathbf{P}^{ext.}(\boldsymbol{\chi}^{\mathrm{AIM}}|\boldsymbol{\chi}^{\mathrm{AIM}}) = \{\mathbf{P}(\boldsymbol{\chi}_{\mathrm{X}}|\boldsymbol{\chi}_{\mathrm{Y}})(1 - \delta_{\mathrm{X,Y}})\} \equiv \mathbf{P}^{nadd.}(\boldsymbol{\chi}^{\mathrm{AIM}}|\boldsymbol{\chi}^{\mathrm{AIM}}). \tag{12.27}
$$

It should be stressed, however, that these scattering probabilities of atomic subchannels, which originate from the given input, do no longer sum up to 1, since this normalization condition applies only to the total list of outputs involving both the AIM diagonal and *off*-diagonal communications. In OCT only the full list of the AO inputs determines the complete *origins* (sources) of all chemical bonds in the molecule. Accordingly, the full list of such outputs signifies that all chemical bonds have been counted in the resulting resultant IT bond multiplicities. Thus, should one focus on the effective chemical bonds between the specified pair of atoms A and B, only the $\boldsymbol{\chi}_{AB} = (\boldsymbol{\chi}_A, \boldsymbol{\chi}_B)$ outputs should be included in the relevant communication network. Again, the full list $\boldsymbol{\chi} = \boldsymbol{\chi}^{AIM}$ of the AO inputs generates the resultant chemical connectivity between the two bonded atoms in the molecule under consideration, while limiting this list to $\boldsymbol{\chi}_{AB}$ generates the corresponding entropy/information measures solely due to internal communications (bonds) in this diatomic fragment.

Therefore, both the inter-atom and interatom communications ultimately contribute to the overall IT bond index in the molecular system in question. Indeed, the chemical bond concept combines both the intra-atomic promotion (polarization) and the interatomic delocalization/CT phenomena. As we have already argued in the preceding section, the promoted (valence) state of each AIM is determined mainly by the associated atomic (diagonal) block of molecular conditional probabilities. Important though it is for the full characterization of the AIM valence preparation in the molecule and the resultant, chemical values of bond multiplicities reflected by the overall IT multiplicities and their covalent/ionic components, it has no direct relevance for the pattern of diatomic "connectivities" between bonded atoms. Therefore, the partition of (12.25)–(12.27) again emphasizes the importance of separating the additive and nonadditive subchannels for distinguishing the chemical *one*-center promotion of AIM from the molecular *two*-center interaction phenomena in OCT of the chemical bond.

It should be stressed, however, that the *exact* AO additivity calls for the strictly deterministic (unit matrix) character of the intra-atomic communications. Therefore, the presence of the *off*-diagonal probability scattering introduces the atomic information nonadditivities, which are synonymous with the presence of some chemical bonds associated with this *one*-center covalency.

One also realizes that the external/internal proportions of the chemical bond contributions depend on the adopted orbital representation. Clearly, in the canonical AO representation, there are always some *internal* covalency (communication noise) and ionicity (information flow) contributions involved in the atomic promotion processes. One also observes that in the *Natural Hybrid Orbital* (NHO) framework on constituent atoms, in which the atomic (diagonal) blocks of the *first*-order density matrix become diagonal, the intra-atomic communications become exactly deterministic in character, so that *one*-center (additive) IT-covalency then identically vanishes. As in the CG probe of the chemical bond localization, the entropic descriptors of the bond IT-covalency are then determined solely by the nonadditive subchannel, which combines the communications between AO originating from *different* atoms.

The partitioning of the conditional probabilities in (12.25)–(12.27) is in the spirit of the related division of the AO representations of quantum mechanical operators in the context of their *internal* and *external* eigenvalue problems (Nalewajski et al. 1996a; Nalewajski 2008d). The latter approach has been successfully applied in identifying the partially decoupled channels of the collective electron displacements in reactants (Nalewajski and Korchowiec 1997; Nalewajski 2008d), and it has recently been used to determine the interatomic flows of electrons in molecules (Mitoraj 2007; Mitoraj and Michalak 2005, 2007; Mitoraj et al. 2006,

2007). In the next chapter, we shall demonstrate that, when additionally supplemented by the atom promotion communications and the ensemble bond weighting in the channel input, this separation exactly reproduces the Wiberg bond orders in diatomic molecules, thus fully accounting for the chemical bond differentiation patterns in diatomic fragments of typical molecules.

12.6 Quantum Conditioning of Orbital Subspaces and Conditional AO Events

It follows from (12.6) that the essence of the quantum conditioning of two states in $P^g(\phi|\psi)$, measuring the expectation value in the variable state of the projector onto the parameter state, is the projection operator onto the reference state. This prescription is then supplemented in the physical probability of (12.7) by the additional projections of OAO onto the occupied subspace $\boldsymbol{\varphi}^o$ of MO, which reflects the actual participation of the basis functions in the whole system of chemical bonds in a molecule. It is the main goal of this section to devise a similar procedure for conditioning the OAO subspaces involving several basis functions and/or two conditional events on different molecular fragments. Again, the appropriately generalized bond-projected superposition principle will be used to derive these *singly* and *triply* conditional probabilities, respectively, which simultaneously involve four basis functions in the molecular system of chemical bonds (Nalewajski 2010b, 2011d).

Let us first examine the quantum conditioning of two subspaces in $\boldsymbol{\chi}$, spanned by the disjoint pairs of OAO, $\varepsilon_1 \equiv \{u(1)\} = \{\chi_i, \chi_j\}$ and $\varepsilon_2 \equiv \{w(2)\} = \{\chi_k, \chi_l\}$, which define the associated subspace projectors:

$$\hat{\mathrm{P}}_{\varepsilon_1} = |i\rangle\langle i| + |j\rangle\langle j| = \hat{\mathrm{P}}_i + \hat{\mathrm{P}}_j, \quad \hat{\mathrm{P}}_{\varepsilon_2} = |k\rangle\langle k| + |l\rangle\langle l| = \hat{\mathrm{P}}_k + \hat{\mathrm{P}}_l. \tag{12.28}$$

A natural generalization (Nalewajski 2011d) of the procedure adopted in the *two*-AO development calls for the subtrace over the variable subspace ε_1 of the projector onto the reference subspace ε_2. This gives the following *geometric* (g) expression, corresponding to the whole molecular Hilbert space,

$$\begin{aligned} P^g(\varepsilon_1|\varepsilon_2) &= \mathcal{N}^g \sum\nolimits_{u(1)} \langle u(1)|\hat{\mathrm{P}}_{\varepsilon_2}|u(1)\rangle \\ &= \mathcal{N}^g[\langle i|k\rangle\langle k|i\rangle + \langle i|l\rangle\langle l|i\rangle + \langle j|k\rangle\langle k|j\rangle + \langle j|l\rangle\langle l|j\rangle] \\ &= \mathcal{N}^g(\delta_{i,k} + \delta_{i,l} + \delta_{j,k} + \delta_{j,l}) \\ &= \mathcal{N}^g \sum\nolimits_{w(2)} \langle w(2)|\hat{\mathrm{P}}_{\varepsilon_1}|w(2)\rangle = P^g(\varepsilon_2|\varepsilon_1) = 0. \end{aligned} \tag{12.29}$$

Its *physical*, bond-projected analog then reads:

$$P(\varepsilon_1|\varepsilon_2) \equiv P(i,j|k,l) = \mathcal{N}_2 \sum\nolimits_{u(1)} \left\langle u(1) \middle| \hat{\mathrm{P}}^0_\varphi \hat{\mathrm{P}}_{\varepsilon_2} \hat{\mathrm{P}}^0_\varphi \middle| u(1) \right\rangle \equiv \mathcal{N}_2 \mathrm{tr}_{\varepsilon_1} \hat{\mathrm{S}}_{\varepsilon_2}$$
$$= (\mathcal{N}_2/4)(\gamma_{i,k}\gamma_{k,i} + \gamma_{i,l}\gamma_{l,i} + \gamma_{j,k}\gamma_{k,j} + \gamma_{j,l}\gamma_{l,j})$$
$$= (\mathcal{N}_2/4)\left[(\gamma_{i,k})^2 + (\gamma_{i,l})^2 + (\gamma_{j,k})^2 + (\gamma_{j,l})^2\right] \geq 0, \qquad (12.30)$$

where the probability scattering operator from the reference subspace ε_2

$$\hat{\mathrm{S}}_{\varepsilon_2} = \hat{\mathrm{P}}^0_\varphi \, \hat{\mathrm{P}}_{\varepsilon_2} \, \hat{\mathrm{P}}^0_\varphi \equiv \hat{\mathrm{P}}_{\varepsilon_2^b} \equiv \sum_{w \in \varepsilon_2} |w^b\rangle\langle w^b|. \qquad (12.31)$$

As required, up to the normalization constant the resulting expression for the (nonnegative) conditional probability $P(\varepsilon_1|\varepsilon_2)$ is seen to be symmetrical with respect to exchanges of orbitals inside and between the two subspaces. It combines the (renormalized) additive *two*-orbital contributions of (12.7) for all selections of single orbitals in the variable and parameter subspaces, respectively. It is straightforward to calculate once the self-consistent density matrix $\boldsymbol{\gamma}$ is known. This result can be extended to cover any number of basis functions in each subspace. Such conditional probabilities reflect the communications between all members of the two spaces involved and generate the associated IT-covalent and -ionic descriptors of the collective chemical interactions between these groups of orbitals in the molecular ground state.

As illustrated in Fig. 12.3, this subspace information scattering involving pairs of AO actually calls for the enlarged channel $\mathbf{P}(\boldsymbol{\chi}, \boldsymbol{\chi}'|\boldsymbol{\chi}, \boldsymbol{\chi}')$ containing *doubled* OAO inputs and outputs, where: $\varepsilon_1 = \{i \in \boldsymbol{\chi}, j \in \boldsymbol{\chi}'\}$ and $\varepsilon_2 = \{k \in \boldsymbol{\chi}, l \in \boldsymbol{\chi}'\}$.

Thus, the proportionality constant $\mathcal{N}_2$ has to satisfy the modified normalization condition involving summation over the union $\cup$ of all *double*-AO (output) subspaces $\varepsilon_1 = (|i\rangle, |j\rangle)$ amounting to the double trace over of the entire list $(\boldsymbol{\chi}, \boldsymbol{\chi}')$ of the output events:

$$\mathrm{tr}_{\cup\{\varepsilon_1=(i,j)\}} P(i,j|k,l) = 1. \qquad (12.32)$$

Therefore, the overall normalization of (12.32) can be expressed in terms of the double AO trace:

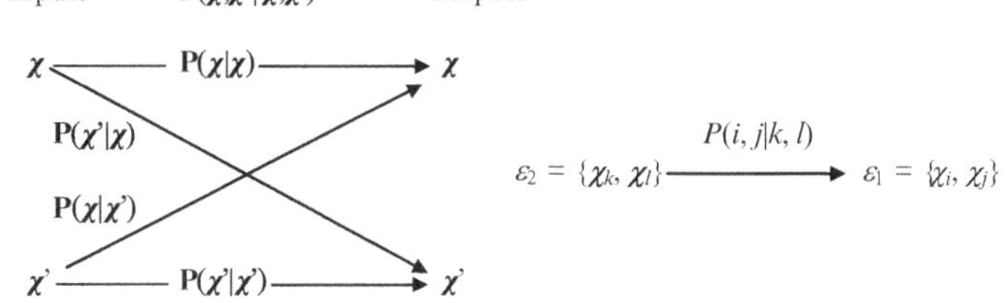

Fig. 12.3 The *double* AO information channel $\mathbf{P}(\boldsymbol{\chi}, \boldsymbol{\chi}'|\boldsymbol{\chi}, \boldsymbol{\chi}')$ and its elementary conditional probability $P(\varepsilon_1|\varepsilon_2) \equiv P(i, j|k, l)$

$$\begin{aligned}\mathcal{N}_2\left[\mathrm{tr}_{\boldsymbol{\chi}}\hat{\mathrm{S}}_{\varepsilon_2}+\mathrm{tr}_{\boldsymbol{\chi}'}\hat{\mathrm{S}}_{\varepsilon_2}\right]&=\mathcal{N}_2\left[\sum\nolimits_{i\in\boldsymbol{\chi}}|\langle i|\hat{\mathrm{P}}^0_{\varphi}\,\hat{\mathrm{P}}_{\varepsilon_2}\,\hat{\mathrm{P}}^0_{\varphi}|i\rangle+\sum\nolimits_{j\in\boldsymbol{\chi}'}\langle j|\hat{\mathrm{P}}^0_{\varphi}\,\hat{\mathrm{P}}_{\varepsilon_2}\,\hat{\mathrm{P}}^0_{\varphi}|j\rangle\right]\\&=(\mathcal{N}_2/4)\left[\sum\nolimits_{i\in\boldsymbol{\chi}}(\gamma_{k,i}\,\gamma_{i,k}+\gamma_{l,i}\gamma_{i,l})+\sum\nolimits_{j\in\boldsymbol{\chi}'}(\gamma_{k,j}\,\gamma_{j,k}+\gamma_{l,j}\gamma_{j,l})\right]\\&=\mathcal{N}_2(\gamma_{k,k}+\gamma_{l,l})=1,\end{aligned}\tag{12.33}$$

where we have used the idempotency relation of (12.3). Hence,

$$\mathcal{N}_2=1/(\gamma_{k,k}+\gamma_{l,l}).\tag{12.34}$$

Therefore, the normalized expression for the *singly* conditional probability, relating in the bond space one pair of OAO to another, reads:

$$P(i,j|k,l)=(\gamma_{k,i}\gamma_{i,k}+\gamma_{l,i}\gamma_{i,l}+\gamma_{k,j}\gamma_{j,k}+\gamma_{l,j}\gamma_{j,l})/(\gamma_{k,k}+\gamma_{l,l}).\tag{12.35}$$

It constitutes a natural generalization of the *two*-AO expression (12.10). As expected, this probability vanishes when there is no chemical coupling between the two AO subsets, $\gamma_{i,k}=\gamma_{i,l}=\gamma_{j,k}=\gamma_{j,l}=0$, either because of the symmetry requirements or due to large spatial separation $R_{1,2}$ between two sets of AO,

$$P(i,j|k,l)\to 0(R_{1,2}\to\infty).\tag{12.36}$$

This probability is seen to assume a general form of the known *two*-orbital expression (12.7),

$$P(\varepsilon_1|\varepsilon_2)=\gamma^2_{\varepsilon_1,\varepsilon_2}/(2\gamma_{\varepsilon_2,\varepsilon_2}),\tag{12.37}$$

when expressed in terms of the square of the chemical coupling $\gamma_{\varepsilon_1,\varepsilon_2}$ between the conditioned subspaces,

$$\gamma^2_{\varepsilon_1,\varepsilon_2}=\gamma^2_{i,k}+\gamma^2_{i,l}+\gamma^2_{j,k}+\gamma^2_{j,l},\tag{12.38}$$

and the average AO electron occupation in the parameter (reference) subspace:

$$\gamma_{\varepsilon_1,\varepsilon_2}=\frac{1}{2}(\gamma_{k,k}+\gamma_{l,l}).\tag{12.39}$$

As an alternative measure of an effective information coupling between the two subspaces one could also adopt the average of the four intersubspace (*two*-AO) conditional probabilities of Fig. 12.4:

$$\begin{aligned}\bar{P}(\varepsilon_1|\varepsilon_2)&=\bar{P}(i,j|k,l)=1/4[P(i|k)+P(j|k)+P(i|l)+P(j|l)]\\&=\frac{1}{8}[(\gamma_{k,i}\gamma_{i,k}+\gamma_{k,j}\gamma_{j,k})/\gamma_{k,k}+(\gamma_{l,i}\gamma_{i,l}+\gamma_{l,j}\gamma_{j,l})/\gamma_{l,l}].\end{aligned}\tag{12.40}$$

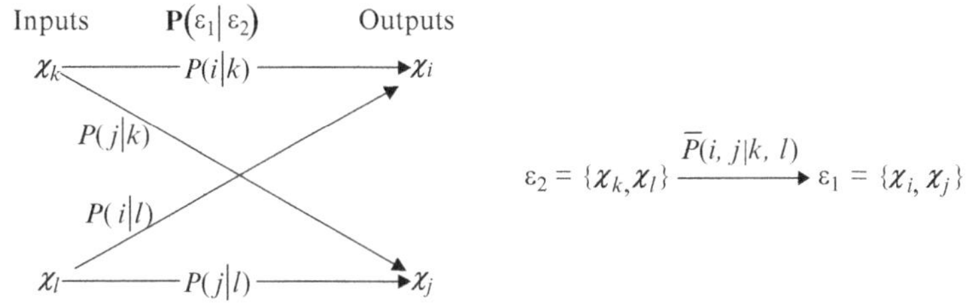

Fig. 12.4 The elementary *two*-AO scatterings between subspaces ε_2 and ε_1 and the associated *average* conditional probability $\bar{P}(\varepsilon_1|\varepsilon_2) = \bar{P}(i,j|k,l)$

It also satisfies the normalization of (12.32) and (12.33):

$$\begin{aligned}\mathrm{tr}_{\cup\{\varepsilon_1=(i,j)\}}\bar{P}(i,j|k,l) = &\left(\sum\nolimits_{i\in\boldsymbol{\chi}}\gamma_{k,i}\gamma_{i,k} + \sum\nolimits_{j\in\boldsymbol{\chi}'}\gamma_{k,j}\gamma_{j,k}\right)/(8\gamma_{k,k})\\ &+\left(\sum\nolimits_{i\in\boldsymbol{\chi}}\gamma_{l,i}\gamma_{i,l} + \sum\nolimits_{j\in\boldsymbol{\chi}'}\gamma_{l,j}\gamma_{j,l}\right)/(8\gamma_{l,l}) = 1,\end{aligned}$$

where we have again used the idempotency property of the closed-shell CBO matrix.

The conditional probability (12.35) subsequently defines the associated *joint* probability of the two subspaces in the bonding space of the molecule [see (12.12)]:

$$\begin{aligned}P(\varepsilon_1 \wedge \varepsilon_2) &= P[(i,j) \wedge (k,l)] \equiv P(i,j,k,l) = P(k \wedge l)P(i,j|k,l)\\ &= [2N(\gamma_{k,k} + \gamma_{l,l})]^{-1}(\gamma_{k,l})^2[(\gamma_{i,k})^2 + (\gamma_{i,l})^2 + (\gamma_{j,k})^2 + (\gamma_{j,l})^2],\end{aligned} \tag{12.41a}$$

which also conforms to the adopted normalization:

$$\begin{aligned}\mathrm{tr}_{\cup\{\varepsilon_1=(i,j)\}}P[(i,j) \wedge (k,l)] &= P(k \wedge l)\mathrm{tr}_{\cup\{\varepsilon_1=(i,j)\}}P(i,j|k,l) = P(k \wedge l)\\ &\equiv P(k,l).\end{aligned} \tag{12.41b}$$

These joint probabilities of the simultaneous *two* subspace (*four*-orbital) events in the molecular bond space originate from the actual participation of these orbitals in the system chemical bonds described by the occupied MO. The corresponding average measure associated with $\bar{P}(\varepsilon_1|\varepsilon_2)$ reads:

$$\begin{aligned}\bar{P}(\varepsilon_1 \wedge \varepsilon_2) &= P(k \wedge l)\bar{P}(i,j|k,l)\\ &= \frac{\gamma_{k,l}^2}{16N}\left[\left(\gamma_{i,k}^2 + \gamma_{j,k}^2\right)/\gamma_{k,k} + \left(\gamma_{i,l}^2 + \gamma_{j,l}^2\right)/\gamma_{l,l}\right].\end{aligned} \tag{12.41c}$$

Let us now address the associated problem of conditioning the probability propagations in molecular fragments. In the communication theory the localized chemical interaction between the given AO pair $\{\chi_i, \chi_j\}$ reflects the probability scattering between these two basis functions. It is embodied in the respective elements $P(j|i)$ and $P(i|j)$ of the conditional probability matrix $\mathbf{P}(\boldsymbol{b}|\boldsymbol{a})$. In order to quantify in IT the coupling effects between such localized chemical bonds, say between χ_i and χ_j in one molecular fragment, and between χ_k and χ_l in another part of the molecule, when $\varepsilon_1 = (\chi_i, \chi_j)$ and $\varepsilon_2 = (\chi_k, \chi_l)$ are disjoint, one has to relate two conditional events $\alpha = (j|i)$ and $\beta = (l|k)$ in these subsystems. One then requires the *triply* conditional probability $P[(j|i)|(l|k)] = P(\alpha|\beta)$, of the *variable* (conditional) *two*-orbital event α in one fragment, of the $i \to j$ scattering, conditional on the *parameter* (conditional) *two*-orbital event β in another fragment, of the $k \to l$ probability propagation.

Earlier attempt (Nalewajski 2010b) to solve this nonunique problem of generating such coupling, *triply* conditional probabilities $P[(j|i)|(l|k)]$ and the associated joint probabilities,

$$\begin{aligned} P[(j|i) \wedge (l|k)] &\equiv P[(j|i), (l|k)] = P(\alpha \wedge \beta) \\ &= P(\alpha)P(\beta|\alpha) = P(j|i)P[(k|l)|(j|i)] \\ &= P(\beta)P(\alpha|\beta) = P(l|k)P[(j|i)|(l|k)], \end{aligned} \tag{12.42}$$

was shown to violate the requirements of the nonnegativity and of the relevant symmetries with respect to exchange of orbitals. However, for equal AO occupations, these model probabilities were shown to satisfy Bayes' rule,

$$P(\beta|\alpha) = P(\beta)P(\alpha|\beta)/P(\alpha) = P(\beta)P(\alpha|\beta)/\left[\sum\nolimits_\beta P(\beta)P(\alpha|\beta)\right], \tag{12.43a}$$

about "hypotheses" $\{\beta\}$ accounting for the occurrence of α. Its first part implies that the probability ratio of individual propagations can be expressed as the corresponding ratio of the mutually reverse conditional probabilities:

$$P(\beta)/P(\alpha) = P(\beta|\alpha)/P(\alpha|\beta). \tag{12.43b}$$

The simplest way to automatically satisfy the preceding Bayes relation is to use the subspace conditioning of (12.35) to identity the probability of the joint conditional events,

$$\begin{aligned} P(\alpha \wedge \beta) &= P(\beta \wedge \alpha) = P[(j|i), (l|k)] \\ &\equiv P[j, l|i, k] = P(\varepsilon_{\mathrm{I}}|\varepsilon_{\mathrm{II}}) = \frac{(\gamma_{j,i}^2 + \gamma_{j,k}^2 + \gamma_{l,i}^2 + \gamma_{l,k}^2)}{(\gamma_{i,i} + \gamma_{k,k})}; \end{aligned}$$

here, the AO labels (j, l) combine the *variable* labels of two conditional events and span the subspace ε_{I}, while the remaining labels (i, k), including the reference AO labels of two conditional events, determine the parameter subspace $\varepsilon_{\mathrm{II}}$. In accordance with the standard practice of the probability conditioning, one then defines the conditional probabilities were seek as the ratios:

$$P(\alpha|\beta) = P(\alpha \wedge \beta)/P(\beta) \quad \text{and} \quad P(\beta|\alpha) = P(\alpha \wedge \beta)/P(\alpha).$$

This procedure gives:

$$P(\alpha|\beta) = P[(j|i)|(l|k)] = \frac{2\gamma_{k,k}(\gamma_{i,j}^2 + \gamma_{j,k}^2 + \gamma_{i,l}^2 + \gamma_{k,l}^2)}{(\gamma_{i,i} + \gamma_{k,k})\gamma_{k,l}^2},$$
$$P(\beta|\alpha) = P[(l|k)|(j|i)] = \frac{2\gamma_{i,i}(\gamma_{i,j}^2 + \gamma_{j,k}^2 + \gamma_{i,l}^2 + \gamma_{k,l}^2)}{(\gamma_{i,i} + \gamma_{k,k})\gamma_{i,j}^2}. \tag{12.44a}$$

The same conditioning prescription adopted to the average probability measure of (12.40) gives:

$$\bar{P}(\alpha \wedge \beta) \equiv \bar{P}[j,l|i,k] = \bar{P}(\varepsilon_{\mathrm{I}}|\varepsilon_{\mathrm{II}}) = \frac{1}{8}\left[(\gamma_{j,i}^2 + \gamma_{l,i}^2)/\gamma_{i,i} + (\gamma_{j,k}^2 + \gamma_{l,k}^2)/\gamma_{k,k}\right],$$
$$\bar{P}(\alpha|\beta) = \frac{\gamma_{k,k}}{4\gamma_{k,l}^2}\left[(\gamma_{j,i}^2 + \gamma_{l,i}^2)/\gamma_{i,i} + (\gamma_{j,k}^2 + \gamma_{l,k}^2)/\gamma_{k,k}\right],$$
$$P(\beta|\alpha) = \frac{\gamma_{i,i}}{4\gamma_{i,j}^2}\left[(\gamma_{j,i}^2 + \gamma_{l,i}^2)/\gamma_{i,i} + (\gamma_{j,k}^2 + \gamma_{l,k}^2)/\gamma_{k,k}\right]. \tag{12.44b}$$

In designing the adequate measure of the *triply* conditional probabilities of AO events in the molecular bond system, one could also follow the heuristic approach of (12.37)–(12.39):

$$\tilde{P}(\alpha|\beta) = \gamma_{\alpha,\beta}^2/(2\gamma_{\beta,\beta}), \tag{12.45a}$$

where $\gamma_{\alpha,\beta}$ denotes the appropriate average chemical interaction attributed to two conditional AO events α and β, and $\gamma_{\beta,\beta}$ stands for the effective electron occupation associated with the reference event. In defining these mean quantities, we use as weighting factors the known probabilities of the separate conditional events,

$$P(\alpha) = \frac{\gamma_{i,j}^2}{2\gamma_{i,i}} \quad \text{and} \quad P(\beta) = \frac{\gamma_{k,l}^2}{2\gamma_{k,k}},$$
$$\gamma_{\alpha,\beta}^2 = P(\alpha)\gamma_{\varepsilon_1,\varepsilon_2}^2 P(\beta) = (\gamma_{i,k}^2 + \gamma_{i,l}^2 + \gamma_{j,k}^2 + \gamma_{j,l}^2)\frac{\gamma_{i,j}^2\gamma_{k,l}^2}{4\gamma_{i,i}\gamma_{k,k}},$$
$$\gamma_{\beta,\beta} = P(\beta)\gamma_{\varepsilon_2,\varepsilon_2} = \frac{\gamma_{k,l}^2}{4\gamma_{k,k}}(\gamma_{k,k} + \gamma_{l,l}),$$
$$\gamma_{\alpha,\alpha} = P(\alpha)\gamma_{\varepsilon_1,\varepsilon_1} = \frac{\gamma_{i,j}^2}{4\gamma_{i,i}}(\gamma_{i,i} + \gamma_{j,j}). \tag{12.45b}$$

The emerging heuristic expressions for the triply conditional probabilities then read:

$$\tilde{P}(\alpha|\beta) = P(\alpha)\gamma_{\varepsilon_1,\varepsilon_2}^2/(2\gamma_{\varepsilon_2,\varepsilon_2}), \quad \tilde{P}(\beta|\alpha) = P(\beta)\gamma_{\varepsilon_1,\varepsilon_2}^2/(2\gamma_{\varepsilon_1,\varepsilon_1}). \tag{12.46}$$

Hence, the Bayes' ratio of (12.43b),

$$\tilde{P}(\beta|\alpha)/\tilde{P}(\alpha|\beta) = [P(\beta)/P(\alpha)](\gamma_{\varepsilon_1,\varepsilon_1}/\gamma_{\varepsilon_2,\varepsilon_2}), \tag{12.47}$$

indeed reproduces the probability ratio of individual conditional events,

$$P(\beta)/P(\alpha) = \gamma_{k,l}^2 \gamma_{i,i} / (\gamma_{i,j}^2 \gamma_{k,k}), \tag{12.48}$$

for equal occupations of the two-orbital subspaces:

$$N_{\varepsilon_1} = \gamma_{i,i} + \gamma_{j,j} = 2\gamma_{\varepsilon_1,\varepsilon_1} = N_{\varepsilon_2} = \gamma_{k,k} + \gamma_{l,l} = 2\gamma_{\varepsilon_2,\varepsilon_2}.$$

The additional test comes from the symmetry requirement for the joint probabilities of the two conditional events:

$$\tilde{P}(\alpha \wedge \beta) = P(\beta)\tilde{P}(\alpha|\beta) = P(\alpha)\tilde{P}(\beta|\alpha). \tag{12.49}$$

The two probability products give:

$$\begin{aligned} P(\beta)\tilde{P}(\alpha|\beta) &= P(\alpha)\gamma_{\varepsilon_1,\varepsilon_2}^2 P(\beta)/(2\gamma_{\varepsilon_2,\varepsilon_2}) = \gamma_{\alpha,\beta}^2/N_{\varepsilon_2}, \\ P(\alpha)\tilde{P}(\beta|\alpha) &= P(\alpha)\gamma_{\varepsilon_1,\varepsilon_2}^2 P(\beta)/(2\gamma_{\varepsilon_1,\varepsilon_1}) = \gamma_{\alpha,\beta}^2/N_{\varepsilon_1}. \end{aligned} \tag{12.50}$$

Therefore, these probabilities are again equal to one another for identical occupations of the two-orbital subspaces.

In practical molecular calculations, for general AO occupations, the average of these two estimates can be used as the joint probability measure,

$$\begin{aligned} \tilde{P}(\alpha \wedge \beta) &\cong \tfrac{1}{2}\left[P(\beta)\tilde{P}(\alpha|\beta) + P(\alpha)\tilde{P}(\beta|\alpha)\right] \\ &= \tfrac{1}{2}\gamma_{\alpha,\beta}^2[1/N_{\varepsilon_1} + 1/N_{\varepsilon_2}] \equiv \frac{\gamma_{\alpha,\beta}^2}{2N_{(\varepsilon_1,\varepsilon_2)}^h}. \end{aligned} \tag{12.51}$$

It then assumes a general form of the *two*-orbital probability of (12.12) when expressed in terms of the *harmonic* average of the two subspace occupations:

$$N_{(\varepsilon_1,\varepsilon_2)}^h = N_{\varepsilon_1}N_{\varepsilon_2}/(N_{\varepsilon_1} + N_{\varepsilon_2}). \tag{12.52}$$

One thus concludes that the heuristic probabilities $\tilde{P}(\beta|\alpha)$ and $\tilde{P}(\alpha|\beta)$ can indeed serve as adequate measures of the *triply* conditional probabilities emerging in the *bond* coupling phenomena of OCT.

12.7 Flexible Input Approach

In typical SCF LCAO MO calculations the lone pairs of the *valence*- and/or *inner*-shell electrons can strongly affect the overall IT descriptors of the chemical bonds (see Sect. 12.3). Elimination of such *lone* pair contributions to the resultant IT bond indices of diatomic fragments in molecules requires an *ensemble* approach, in which the input probabilities are derived from the *joint* (bond) probabilities of two AO centered on different atoms (Nalewajski 2010c, f; Nalewajski et al. 2011a). Indeed, the contributions due to each AO input on atom A to chemical bond(s) with AO on atom B should be weighted using the corresponding joint (*two*-orbital) probabilities, which reflect the actual, *simultaneous* participation of the given pair of basis functions in the A—B chemical bonds. Such an approach effectively projects out the spurious contributions due to the *inner*- and *outer*-shell AO, which are excluded from mixing into the delocalized, bonding MO combinations. This probability weighting procedure, known as the flexible input approach, is capable of reproducing the Wiberg bond order in diatomics, at the same time providing the IT-covalent/ionic resolution of this overall bond index (Nalewajski et al. 2011a).

The localized (diatomic) bond multiplicities in molecules are mainly determined by the constituent AO of both atoms, $\boldsymbol{\chi}_{\mathrm{AB}} = (\boldsymbol{\chi}_{\mathrm{A}}, \boldsymbol{\chi}_{\mathrm{B}})$. This partial basis corresponds to the diatomic block $\boldsymbol{\gamma}_{\mathrm{AB}} = \{\boldsymbol{\gamma}_{\mathrm{X,Y}};\ \mathrm{X, Y} \in (\mathrm{A, B})\}$ of the molecular density matrix, and to the associated block of molecular conditional probabilities between the AO contributed by both atoms: $\mathbf{P}_{\mathrm{AB}}(\boldsymbol{\chi}_{\mathrm{AB}}|\boldsymbol{\chi}_{\mathrm{AB}}) = \{\mathbf{P}(\boldsymbol{\chi}_{\mathrm{Y}}|\boldsymbol{\chi}_{\mathrm{X}});\ \mathrm{X,Y} \in (\mathrm{A,B})\}$. The former also determines the effective number of electrons on AB in the molecule, given by the partial trace $N_{\mathrm{AB}} = \sum_{i \in \mathrm{AB}} \gamma_{i,i}$.

We begin this section by applying this weighting procedure to the 2-OAO model of Sect. 12.4. In the bond-weighted approach one distinguishes in the molecular channel of Fig. 12.1 the elementary (*row*) subchannels (Nalewajski 2005b, 2006g) due to each AO input (see Fig. 12.5). The conditional entropy and mutual information quantities for these partial communication systems, $\{S_{\mathrm{AB}}(\boldsymbol{\chi}_{\mathrm{AB}}|i), I^0_{\mathrm{AB}}(i : \boldsymbol{\chi}_{\mathrm{AB}});\ i = A, B\}$, respectively, with the latter being determined for the *covalent*-reference probabilities $\boldsymbol{p}^0 = (½, ½)$ marking the single electrons contributed by each AO to the diatomic chemical bond, are also listed in the diagram. Since the row descriptors represent the IT indices *per electron*, these contributions have to be multiplied by $N_{\mathrm{AB}} = 2$ in the corresponding resultant covalent/ionic and overall measures.

Therefore, using the *off*-diagonal joint probability $P(A \wedge B) = P(B \wedge A) = PQ = \gamma_{A,B}\gamma_{B,A}/4$ as the ensemble probability for both OAO inputs gives the following average quantities for this model diatomic bond:

$$\begin{aligned} S_{\mathrm{AB}} &= N_{\mathrm{AB}}[P(A \wedge B)S_{\mathrm{AB}}(\boldsymbol{\chi}_{\mathrm{AB}}|A) + P(B \wedge A)S_{\mathrm{AB}}(\boldsymbol{\chi}_{\mathrm{AB}}|B)] \\ &= 4PQH(P) = \mathcal{M}_{A,B}H(P), \\ I^0_{\mathrm{AB}} &= N_{\mathrm{AB}}\left[P(A \wedge B)I^0_{\mathrm{AB}}(A{:}\boldsymbol{\chi}_{\mathrm{AB}}) + P(B \wedge A)I^0_{\mathrm{AB}}(B{:}\boldsymbol{\chi}_{\mathrm{AB}})\right] \\ &= 4PQ[1 - H(P)] = \mathcal{M}_{A,B}[1 - H(P)], \\ \mathcal{N}^0_{\mathrm{AB}} &= S_{\mathrm{AB}} + I^0_{\mathrm{AB}} = 4PQ = (\gamma_{A,B})^2 = \mathcal{M}_{A,B}. \end{aligned} \tag{12.53}$$

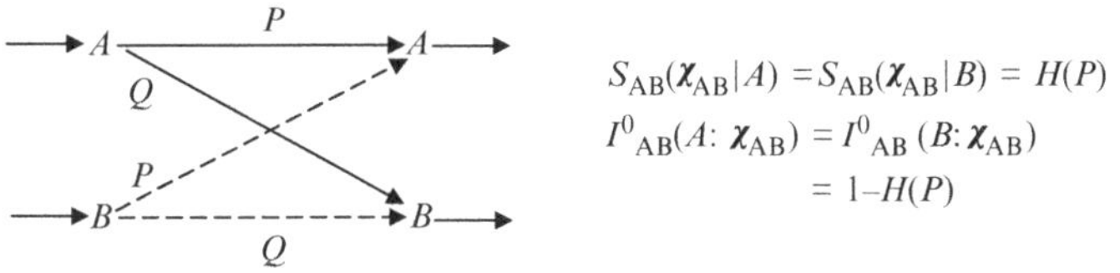

Fig. 12.5 The elementary (*row*) subchannels due to inputs *A* (*solid lines*) and *B* (*broken lines*) in the 2-OAO model of the chemical bond of Fig. 12.1

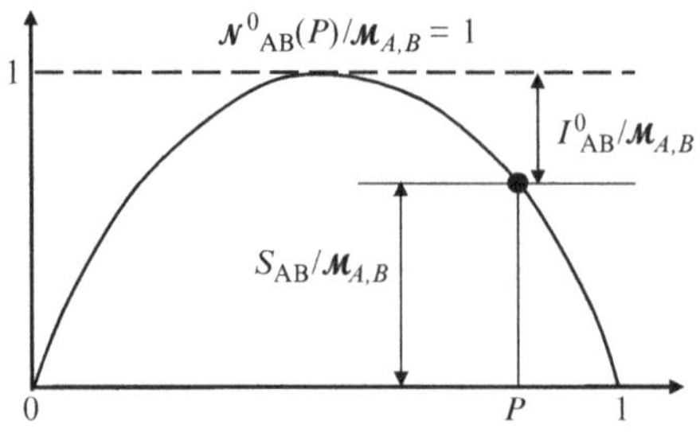

Fig. 12.6 Variations of the IT-covalent $[S_{AB}(P)]$ and IT-ionic $\left[I^0_{AB}(P)\right]$ components (in the Wiberg $\mathcal{M}_{A,B}$ units) of the chemical bond in the 2-OAO model (see Figs. 12.1 and 12.5) with changing MO polarization P, and the conservation of the relative bond order $\mathcal{N}^0_{AB}(P)/\mathcal{M}_{A,B} = \left[S_{AB}(P) + I^0_{AB}(P)\right]/\mathcal{M}_{A,B} = 1$

We have thus recovered the Wiberg index as the overall IT descriptor of the chemical bond in 2-OAO model, $\mathcal{N}^0_{AB} = \mathcal{M}_{A,B}$, at the same time establishing its covalent, $S_{AB} = \mathcal{M}_{A,B}H(P)$, and ionic, $I^0_{AB} = \mathcal{M}_{A,B}[1 - H(P)]$, contributions. It follows from Fig. 12.6 that these IT-covalency and IT-ionicity components compete with one another while conserving the Wiberg bond order of this model as the overall information measure of the bond multiplicity (in bits).

12.8 Localized Bonds in Diatomic Fragments

This development can be straightforwardly generalized to a general case of several basis functions contributed by each bonded atom (Nalewajski et al. 2011a). The molecular probability scattering in the specified diatomic fragment (A, B) involving the basis functions $\boldsymbol{\chi}_{AB} = (\boldsymbol{\chi}_A, \boldsymbol{\chi}_B)$ contributed by these two atoms to the overall set of AO, $\boldsymbol{\chi} = \{\boldsymbol{\chi}_X\}$, is fully characterized by the corresponding block $\mathbf{P}_{AB}(\boldsymbol{\chi}_{AB}|\boldsymbol{\chi}_{AB})$ of the molecular conditional probability matrix $\mathbf{P}(\boldsymbol{\chi}|\boldsymbol{\chi}) = \{\mathbf{P}_{XY}(\boldsymbol{\chi}_{XY}|\boldsymbol{\chi}_{XY})\}$. This diatomic block contains only the intradiatomic communications, missing the probability propagations to (A, B) fragment originating from AO of the remaining constituent atoms $\boldsymbol{\chi}_Z \notin \boldsymbol{\chi}_{AB}$, thus neglecting *indirect* A—B bond components, due to the presence of the system remainder. However, in the spirit of the Wiberg approach, such diatomic basis should be perfectly capable of describing the *direct* (localized) chemical interactions, between A and B alone (see Sect. 12.10).

The atomic *output* reduction $\boldsymbol{X}_{AB} = (A, B)$ (Nalewajski 2005a, 2006g) of $\mathbf{P}(\boldsymbol{\chi}_{AB}|\boldsymbol{\chi}_{AB})$, carried out by combining the output AO events $\boldsymbol{\chi}_X$ into a single AIM event X in the output of the molecular channel, gives the associated *condensed* conditional probabilities of such (partially) reduced information system of the diatomic fragment in question:

$$\boldsymbol{P}_{\mathrm{AB}}(\boldsymbol{X}_{\mathrm{AB}}|\boldsymbol{\chi}_{\mathrm{AB}}) = [\boldsymbol{P}(\mathrm{A}|\boldsymbol{\chi}_{\mathrm{AB}}), \boldsymbol{P}(\mathrm{B}|\boldsymbol{\chi}_{\mathrm{AB}})] = \{P(\mathrm{X}|i)\} = \sum_{j\in\mathrm{X}} \boldsymbol{P}(j|\boldsymbol{\chi}_{\mathrm{AB}});$$
$$\chi_i \in \boldsymbol{\chi}_{\mathrm{AB}}, \quad \boldsymbol{X}_{\mathrm{AB}} = \{\mathrm{X} = \mathrm{A}, \mathrm{B}\}. \qquad (12.54)$$

Here, $P(\mathrm{X}|i)$ measures the conditional probability that an electron originating from χ_i will be found on atom X in the diatomic AB of the molecule.

The sum of these conditional probabilities over all AO contributed by the selected two atoms then determines the communication connections $\{P(\mathrm{AB}|i)\}$ linking the totally condensed (reduced) diatomic output AB and the given AO input χ_i in the communication system of the diatomic fragment under consideration:

$$\boldsymbol{P}(\mathrm{A}|\boldsymbol{\chi}_{\mathrm{AB}}) + \boldsymbol{P}(\mathrm{B}|\boldsymbol{\chi}_{\mathrm{AB}}) = \boldsymbol{P}(\mathrm{AB}|\boldsymbol{\chi}_{\mathrm{AB}})$$
$$= \left\{ P(\mathrm{AB}|i) = P(\mathrm{A}|i) + P(\mathrm{B}|i) = \sum_{j\in(\mathrm{A,B})} P(j|i) \le 1 \right\}. \qquad (12.55)$$

In other words, $P(\mathrm{AB}|i)$ measures the probability that an electron occupying χ_i will be detected in the diatomic fragment AB of the molecule. The inequality in the preceding equation reflects the fact that the atomic functions participate in chemical bonds with *all* constituent atoms, with the equality sign thus corresponding only to the diatomic molecule, when $\boldsymbol{\chi}_{\mathrm{AB}} = \boldsymbol{\chi}$.

The associated *fragment*-normalized AO probabilities,

$$\tilde{\boldsymbol{p}}(\mathrm{AB}) = \{\tilde{p}_i(\mathrm{AB}) = \gamma_{i,i}/N_{\mathrm{AB}}, \quad \chi_i \in \boldsymbol{\chi}_{\mathrm{AB}}\}, \quad \sum_{i\in(\mathrm{A,B})} \tilde{p}_i(\mathrm{AB}) = 1, \qquad (12.56)$$

where $N_{\mathrm{AB}} = \sum_{i\in(\mathrm{A,B})} \gamma_{i,i}$ stands for the number of electrons found in the molecule on this diatomic fragment and $\tilde{p}_i(\mathrm{AB})$ denotes the probability that one of them occupies $\chi_{i\in(\mathrm{A,B})}$, then determine the simultaneous probabilities of the joint *two*-orbital events:

$$\mathbf{P}_{\mathrm{AB}}(\boldsymbol{\chi}_{\mathrm{AB}} \wedge \boldsymbol{\chi}_{\mathrm{AB}}) = \{P_{\mathrm{AB}}(i \wedge j) = \tilde{p}_i(\mathrm{AB})P(j|i) = \gamma_{i,j}\gamma_{j,i}/(2N_{\mathrm{AB}})\}. \qquad (12.57)$$

These probabilities in turn generate via the relevant partial summations the joint atom-orbital probabilities in AB, $\{P_{\mathrm{AB}}(\mathrm{X}, i)\}$:

$$\boldsymbol{P}_{\mathrm{AB}}(\boldsymbol{X}_{\mathrm{AB}} \wedge \boldsymbol{\chi}_{\mathrm{AB}}) = [\boldsymbol{P}_{\mathrm{AB}}(\mathrm{A} \wedge \boldsymbol{\chi}_{\mathrm{AB}}), \ \boldsymbol{P}_{\mathrm{AB}}(\mathrm{B} \wedge \boldsymbol{\chi}_{\mathrm{AB}})]$$
$$= \{P_{\mathrm{AB}}(\mathrm{X} \wedge i) = \sum_{j\in\mathrm{X}} P_{\mathrm{AB}}(i \wedge j) \equiv \tilde{p}_i(\mathrm{AB})P(\mathrm{X}|i), \ \mathrm{X} = \mathrm{A}, \mathrm{B}\}. \qquad (12.58)$$

For the closed-shell molecular ground state, one thus finds:

$$\boldsymbol{P}_{AB}(X \wedge \boldsymbol{\chi}_{AB}) = \left\{ P_{AB}(X \wedge i) = \tilde{p}_i(AB) \sum_{j \in X} P(j|i) = \sum_{j \in X} \frac{\gamma_{i,j}\gamma_{j,i}}{2N_{AB}} \right\}$$
$$\equiv \boldsymbol{P}_{AB}(\boldsymbol{\chi}_{AB} \wedge X)^{T},$$
$$X = A, B. \tag{12.59}$$

These vectors of AO probabilities in the diatomic fragment AB subsequently define the condensed probabilities $\{P_X(AB)\}$ of both bonded atoms in this diatomic subsystem:

$$P_X(AB) = \frac{N_X(AB)}{N_{AB}} = \sum_{i \in (A,B)} P_{AB}(X \wedge i) = \sum_{i \in (A,B)} \sum_{j \in X} \frac{\gamma_{i,j}\gamma_{j,i}}{2N_{AB}},$$
$$X = A, B, \tag{12.60}$$

where the effective number of electrons $N_X(AB) = N_{AB}P_X(AB)$ found in fragment AB on atom X $\in$(A, B) now reads:

$$N_X(AB) = \sum_{i \in (A,B)} \sum_{j \in X} \frac{\gamma_{i,j}\gamma_{j,i}}{2}. \tag{12.61}$$

Therefore, in diatomic molecules, when $\boldsymbol{\chi}_{AB} = \boldsymbol{\chi}$, one finds using the idempotency relation of (12.3),

$$P_X(AB) = \sum_{j \in X} \left(\sum_{i} \frac{\gamma_{j,i}\gamma_{i,j}}{2N_{AB}} \right) = \sum_{j \in X} \frac{\gamma_{j,j}}{N_{AB}} = \sum_{j \in X} \tilde{p}_j(AB) \quad X = A,\ B, \tag{12.62}$$

and hence: $P_A(AB) + P_B(AB) = 1$.

One further observes that the relative importance of the basis functions contributed by one atom in forming the chemical bonds with the other atom of the specified diatomic fragment is reflected by the joint *bond* (*b*) probabilities of the two atoms, defined only by the diatomic components of their simultaneous probabilities:

$$P_b(A \wedge B) \equiv \sum_{j \in B} P_{AB}(A \wedge j) \equiv \sum_{i \in A} P_{AB}(i \wedge B) = P_b(B \wedge A)$$
$$= \sum_{i \in A} \sum_{j \in B} \frac{\gamma_{i,j}\gamma_{j,i}}{2N_{AB}}. \tag{12.63}$$

Indeed, the joint atom-orbital bond probabilities, $\{P_{AB}(A \wedge j),\ j \in B\}$ and $\{P_{AB}(i \wedge B,),\ i \in A\}$, to be used as weighting factors in determining the average conditional entropy (covalency) and mutual information (ionicity) descriptors of the chemical bond(s) between A and B, assume appreciable magnitudes only when the electron occupying the atomic orbital χ_i of one atom is simultaneously found with a significant probability on the other atom, thus effectively excluding the contributions to the entropy/information bond descriptors due to the *lone* pair electrons.

The *reference* bond probabilities of AO, to be used to calculate the mutual information (IT-ionicity) bond index of the diatomic channel, have to be normalized to the corresponding sums $\boldsymbol{P}(\mathrm{AB}|\boldsymbol{\chi}_{\mathrm{AB}}) = \{P(\mathrm{AB}|i)\}$ of (12.55). Since the bond probability concept of the preceding equation involves symmetrically the two bonded atoms, one applies the same symmetry requirement in determining the associated reference bond probabilities of AO:

$$\{p_b(i) = P(\mathrm{AB}|i)/2; \quad i \in (\mathrm{A}, \mathrm{B})\}, \tag{12.64}$$

where $P(\mathrm{AB}|i)$ denotes the probability that an electron originating from orbital χ_i will be found on atom A or B in the molecule.

In OCT the complementary quantities characterizing the average *noise* (conditional entropy of the channel output given input) and the information *flow* (mutual information in the channel output and the reference input) in the diatomic communication system defined by the AO conditional probabilities provide the overall descriptors of the fragment bond covalency and ionicity, respectively. Both molecular and promolecular reference (input) probability distributions have been used in the past to determine the information index characterizing the displacement (ionicity) aspect of the system chemical bonds. In the bond-weighted diatomic development, the equal bond probabilities of (12.64) will be used as the input reference values for this purpose.

In the fragment development one defines the following ("ensemble") average contributions of both constituent atoms to the diatomic covalency (delocalization) entropy:

$$\begin{aligned} S_{\mathrm{AB}}(\mathrm{B}|\boldsymbol{\chi}_{\mathrm{A}}) &= \sum_{i\in\mathrm{A}} P_{\mathrm{AB}}(i \wedge \mathrm{B}) S_{\mathrm{AB}}(\boldsymbol{\chi}_{\mathrm{AB}}|i) \quad \text{and} \\ S_{\mathrm{AB}}(\mathrm{A}|\boldsymbol{\chi}_{\mathrm{B}}) &= \sum_{i\in\mathrm{B}} P_{\mathrm{AB}}(i \wedge \mathrm{A}) S_{\mathrm{AB}}(\boldsymbol{\chi}_{\mathrm{AB}}|i), \end{aligned} \tag{12.65}$$

where the Shannon entropy (in bits) of the conditional probabilities for the given AO input $\chi_i \in \boldsymbol{\chi}_{\mathrm{AB}} = (\boldsymbol{\chi}_{\mathrm{A}}, \boldsymbol{\chi}_{\mathrm{B}})$ in the diatomic channel:

$$S_{\mathrm{AB}}(\boldsymbol{\chi}_{\mathrm{AB}}|i) = -\sum_{j\in(\mathrm{A},\mathrm{B})} P(j|i)\log_2 P(j|i). \tag{12.66}$$

Finally, since in (12.65) the conditional entropy $S_{\mathrm{AB}}(\mathrm{Y}|\boldsymbol{\chi}_{\mathrm{X}})$ quantifies (in bits) the $\mathrm{X} \rightarrow \mathrm{Y}$ delocalization per electron, the absolute IT-covalency in the diatomic fragment A—B reads:

$$\mathcal{S}_{AB} = N_{AB}[S_{AB}(B|\boldsymbol{\chi}_A) + S_{AB}(A|\boldsymbol{\chi}_B)]. \tag{12.67}$$

The bond-weighted contributions to the average *mutual* information quantities (in bits) of the two bonded atoms are similarly defined in reference to the unbiased bond probabilities of AO (12.64):

$$\begin{aligned} I_{AB}(\boldsymbol{\chi}_A{:}B) &= \sum_{i\in A} P_{AB}(i \wedge B) I(i{:}\boldsymbol{\chi}_{AB}), \\ I_{AB}(A{:}\boldsymbol{\chi}_B) &= \sum_{i\in B} P_{AB}(i \wedge A) I(i{:}\boldsymbol{\chi}_{AB}), \end{aligned} \tag{12.68}$$

where:

$$I(i{:}\boldsymbol{\chi}_{AB}) = \sum_{j\in(A,B)} P(j|i)\log_2\left(\frac{P(j|i)}{p_b(j)}\right). \tag{12.69}$$

They generate the total IT-ionicity of all chemical bonds in the diatomic fragment:

$$\mathcal{I}_{AB} = N_{AB}[I_{AB}(\boldsymbol{\chi}_A{:}B) + I_{AB}(A{:}\boldsymbol{\chi}_B)]. \tag{12.70}$$

Hence, the sum of the above total (diatomic) entropy-covalency and information ionicity indices determines the overall information-theoretic bond multiplicity for the diatomic fragment in question:

$$\mathcal{N}_{AB} = \mathcal{S}_{AB} + \mathcal{I}_{AB}. \tag{12.71}$$

For diatomic molecules, for which $\boldsymbol{\chi}_{AB} = \boldsymbol{\chi}$ and the reference probabilities $\{p_b(k) = P(AB|k)/2 = ½\}$, the identity [see (12.11)]

$$\mathcal{N}_{AB} = \mathcal{S}_{AB} + \mathcal{I}_{AB} = \mathcal{M}_{A,B} \tag{12.72}$$

can be readily demonstrated:

$$\begin{aligned} \mathcal{N}_{AB} &= \mathcal{S}_{AB} + \mathcal{I}_{AB} \\ &= N_{AB}\left\{\sum_{i\in A} P_{AB}(i\wedge B)[S_{AB}(\boldsymbol{\chi}|i) + I(i{:}\boldsymbol{\chi})] + \sum_{i\in B} P_{AB}(i\wedge A)[S_{AB}(\boldsymbol{\chi}|i) + I(i{:}\boldsymbol{\chi})]\right\} \\ &\equiv N_{AB}\left\{\sum_{i\in A} P_{AB}(i\wedge B) N(\boldsymbol{\chi};i) + \sum_{i\in B} P_{AB}(i\wedge A) N(\boldsymbol{\chi};i)\right\} \\ &= N_{AB}\left\{\sum_{i\in A} P_{AB}(i\wedge B) + \sum_{i\in B} P_{AB}(i\wedge A)\right\} = 2N_{AB}P_b(A\wedge B) = \mathcal{M}_{A,B}. \end{aligned} \tag{12.73}$$

Table 12.1 Comparison of the diatomic Wiberg index $\mathscr{M}_{A,B}$ and the entropy/information bond-multiplicities $\mathscr{N}_{A,B}$, $\mathscr{S}_{A,B}$, and $\mathscr{I}_{A,B}$ (in bits) from the *bond*-weighted AO communication channels of selected diatomic fragments A—B in representative molecules M; RHF results for equilibrium geometries in the minimum (Min., STO-3G) and extended (Ext., 6-31G*) basis sets (Nalewajski et al. 2011a)

M	A—B	$\mathscr{M}_{A,B}$		$\mathscr{N}_{A,B}$		$\mathscr{S}_{A,B}$		$\mathscr{I}_{A,B}$	
		Min.	Ext.	Min.	Ext.	Min.	Ext.	Min.	Ext.
F_2	F—F	1.00	1.23	1.00	1.23	0.95	1.01	0.05	0.27
HF	H—F	0.98	0.82	0.98	0.82	0.89	0.60	0.09	0.22
LiH	Li—H	1.00	1.00	1.00	1.00	1.00	1.00	0.00	0.00
LiF	Li—F	1.59	1.12	1.59	1.12	0.97	0.49	0.62	0.63
CO	C—O	2.60	2.90	2.60	2.90	2.09	2.37	0.51	0.53
H_2O	O—H	0.99	0.88	1.01	0.90	0.86	0.66	0.15	0.23
AlF_3	Al—F	1.07	1.15	1.09	1.15	0.78	0.75	0.31	0.41
CH_4	C—H	1.00	0.98	1.02	1.00	0.93	0.92	0.09	0.08
C_2H_6	C—C	1.02	1.13	1.07	1.18	1.00	1.08	0.07	0.11
	C—H	0.99	0.95	1.02	0.98	0.94	0.88	0.08	0.11
C_2H_4	C—C	2.03	2.16	2.09	2.23	2.00	2.12	0.09	0.11
	C—H	0.98	0.93	1.01	0.97	0.95	0.88	0.07	0.09
C_2H_2	C—C	3.00	3.13	3.06	3.19	2.98	3.09	0.06	0.10
	C—H	0.99	0.91	1.02	0.94	0.98	0.88	0.04	0.06
C_6H_6[a]	C_1—C_2	1.44	1.51	1.53	1.59	1.41	1.47	0.14	0.12
	C_1—C_3	0.00	0.06	0.00	0.06	0.00	0.03	0.00	0.02
	C_1—C_4	0.12	0.11	0.12	0.12	0.08	0.08	0.03	0.03

[a]For the sequential numbering of carbon atoms in the benzene ring

We have observed above that the conditional IT bond multiplicity due to the input χ_k (per single electron)

$$N(\boldsymbol{\chi};k) = \sum_{l\in\boldsymbol{\chi}} \{-P(l|k)\log_2 P(l|k) + P(l|k)\log_2[P(l|k)/p_b(l)]\}$$
$$= \left[\sum_{l\in\boldsymbol{\chi}} P(l|k)\right]\log_2 2 = 1. \qquad (12.74)$$

In Table 12.1, we have compared the illustrative numerical RHF results (Nalewajski et al. 2011a) of the IT bond multiplicities for the localized (diatomic) interactions in representative diatomic and polyatomic molecules, calculated for their equilibrium geometries. These predictions have been obtained using two choices of the Gaussian basis set: the minimum basis set (STO-3G), combining the ground state occupied Slater-type AO of constituent atoms expanded into three GTO each, and the extended basis (6-31G*), involving a larger GTO expansion of the inner shells and the Gaussian polarization functions.

In diatomic systems the trends exhibited by the entropic covalent and ionic components of the exactly conserved Wiberg overall bond order generally agree with intuitive chemical expectations. For example, in the minimum basis set description the roughly "single" chemical bond in F_2, HF, and LiH is seen to be almost purely covalent, although a more substantial IT-ionicity is diagnosed for the fluorine compounds in the extended basis set calculations. For the most ionic LiF, which exhibits in the minimum basis set roughly 3/2 bond, consisting of approximately 1 covalent and ½ ionic bond multiplicities, the extended basis set gives approximately a "single" bond order estimate, with the information theory again predicting the ionic dominance over the covalent component of the resultant bond index. In CO, for which the extended basis set calculations have diagnosed approximately a "triple" bond, this chemical interaction is seen to be predominantly covalent.

The basis set dependence of the predicted IT bond descriptors is seen to be relatively weak with the extended basis calculations often giving rise to predictions exhibiting slightly better agreement with intuitive chemical estimates. One also finds that in polyatomic systems the Wiberg bond orders are very well reproduced by the overall IT descriptors. The (direct) carbon–carbon interactions in the benzene ring are seen to be properly differentiated and the intuitive multiplicities of the carbon–carbon chemical bonds in ethane, ethylene, and acetylene are correctly accounted for.

The IT bond descriptors thus provide the covalent/ionic resolution of the Wiberg bond order $\mathcal{M}_{A,B}$, which has been customarily regarded as being of purely "covalent" origin. However, the LCAO MO coefficients carry the information about both the electron-sharing (covalent) and electron-separation/transfer (ionic) phenomena in the chemical bond. Therefore, this overall index in fact combines the covalent and ionic contributions, which remain to be separated (Nalewajski et al. 1993, 1994a, 1996a, b, 1997; Nalewajski and Jug 2002; Nalewajski and Mrozek 1994, 1996; Nalewajski 2004b). The present IT approach provides a novel resolution scheme for this in fact *resultant* bond order.

The significant information ionicity contribution is also detected for all metal halides in the upper part of Table 12.2, where additional predictions from the

Table 12.2 Additional RHF predictions (Nalewajski et al. 2011a) obtained using the extended 6-31G* basis set for illustrative diatomic and polyatomic molecules. The reported IT bond orders are in bits

Molecule	A—B	$\mathcal{M}_{A,B}$	$\mathcal{N}_{A,B}$	$\mathcal{S}_{A,B}$	$\mathcal{I}_{A,B}$
LiCl	Li—Cl	1.39	1.39	0.73	0.66
LiBr	Li—Br	1.39	1.39	0.73	0.66
NaF	Na—F	0.91	0.91	0.43	0.48
KF	K—F	0.83	0.83	0.37	0.46
SF_2	S—F	1.06	1.08	0.68	0.40
SF_4	S—F_a	1.05	1.06	0.67	0.39
	S—F_b	0.91	0.93	0.60	0.32
SF_6	S—F	0.98	0.98	0.73	0.25
B_2H_6[a]	B—B	0.82	0.85	0.79	0.06
	B—H_t	0.97	0.99	0.94	0.06
	B—H_b	0.48	0.49	0.46	0.03
Propellanes[b]:					
[1.1.1]	C_b—C_b	0.80	0.83	0.76	0.07
[2.1.1]	C_b—C_b	0.83	0.86	0.79	0.07
[2.2.1]	C_b—C_b	0.95	0.99	0.87	0.11
[2.2.2]	C_b—C_b	1.01	1.05	0.99	0.06

[a] H_t and H_b denote the *terminal* and *bridge* hydrogen atoms, respectively
[b] Central bonds between the *bridgehead* carbon atoms C_b

extended basis set RHF calculations are reported. The subtle bond differentiation of the "equatorial" and "axial" S—F bonds in the irregular tetrahedron of SF_4 is correctly reproduced, and an increase in the strength of the direct central bond in propellanes with the increase in size of the bridges is correctly predicted. Moreover, as intuitively expected, the C—H bonds are seen to slightly increase their information ionicity, when the number of these terminal bonds increases in a series: acetylene, ethylene, ethane. In B_2H_6 the correct, around ½ bond order of the bridging B—H bond is found and approximately single terminal bond multiplicity is detected. For the alkali metal fluorides, the increase in the bond entropy-covalency (decrease in information ionicity) with increasing size (softness) of the metal is also observed. For the fixed alkali metal in halides, e.g., in a series LiF, LiCl, and LiBr, the overall IT bond multiplicity is increased for larger (softer) halogen atoms, mainly due to a higher entropy-covalency (delocalization, noise) component of the molecular information channel in AO resolution.

12.9 *Many*-Orbital Effects

The entropy/information quantities for several (dependent) probability schemes (see Sect. 8.6), which relate *two* conditional events in the molecular bond system, each involving a pair of basis functions (see also Sect. 12.6), have recently been used to describe within the OCT the interfragment couplings between internal orbital communications (chemical bonds) in molecular subsystems (Nalewajski 2009g, 2010b, d, f). In this section, we shall briefly summarize potential applications of these generalized IT bond descriptors in probing such subtle bonding effects in molecular and/or reactive systems. For simplicity, we shall limit our discussion to IT quantities involving three or four probability schemes in the AO resolution.

The molecular scenarios invoking the IT bond multiplicities of three probability distributions and their covalent and ionic components may involve three separate species A, B, and C, e.g., two reactants A and B and the catalyst/surface C, with the corresponding sets of the AO events ($\boldsymbol{a}$, $\boldsymbol{b}$, $\boldsymbol{c}$) of the associated probability schemes (**A**, **B**, **C**) then referring to the basis functions provided by the constituent atoms of these subsystems. Alternatively, three molecular fragments can be involved. The *three*-orbital development then enables one to discuss the influence of one reactant/fragment, say C, on the bond structure (or reactivity) of two remaining fragments A and B.

For example, one could address in such an IT framework a natural question about the influence of the catalyst on the structure/reactivity of two adsorbed species, and ultimately assess the cooperation effects between the catalyst–adsorbate bonds (A—C, B—C) and the A—B bond linking the two adsorbates. For this catalytic scenario, the conditional *three*-scheme entropy

$$S(\boldsymbol{P}(\boldsymbol{b})|\mathbf{P}(\boldsymbol{a} \wedge \boldsymbol{c})) = H(\mathbf{B}|\mathbf{AC}) = H(\mathbf{B}|\mathbf{A}) - I(\mathbf{B}{:}\mathbf{C}|\mathbf{A}), \qquad (12.75)$$

reflects the molecular indeterminacy of the AO output events $\boldsymbol{b}$ in B with respect to the input *product* events $\boldsymbol{a} \wedge \boldsymbol{c}$ in the combined subsystem (A¦C). As seen in Fig. 8.5, this IT-covalency accounts for only a *part* of the overall noise component in the $\boldsymbol{a} \rightarrow \boldsymbol{b}$ communications, measured by the indeterminacy of $\boldsymbol{b}$ with respect to $\boldsymbol{a}$ alone, $H(\mathbf{B}|\mathbf{A}) = S(\boldsymbol{P}(\boldsymbol{b})|\boldsymbol{P}(\boldsymbol{a}))$, which reflects the *overall* information loss in the $\boldsymbol{a} \rightarrow \boldsymbol{b}$ probability scattering between A and B. The remaining part, the mutual information $I(\mathbf{B}{:}\mathbf{C}|\mathbf{A})$, by which $H(\mathbf{B}|\mathbf{AC})$ is decreased relative to $H(\mathbf{B}|\mathbf{A})$ and by which $I(\mathbf{B}{:}\mathbf{AC}) = I(\mathbf{A}{:}\mathbf{B}) + I(\mathbf{B}{:}\mathbf{C}|\mathbf{A})$ is increased relative to $I(\mathbf{A}{:}\mathbf{B})$, should now be attributed to the influence of the catalyst C on interactions between B and (A¦C). Similarly, $H(\mathbf{A}|\mathbf{BC}) = H(\mathbf{A}|\mathbf{B}) - I(\mathbf{A}{:}\mathbf{C}|\mathbf{B})$ and $I(\mathbf{A}{:}\mathbf{BC}) = I(\mathbf{A}{:}\mathbf{B}) + I(\mathbf{A}{:}\mathbf{C}|\mathbf{B})$, so that the presence of the catalyst C is again felt by an effective increase in IT-ionic character of the interaction of A with (B¦C), compared with the separate interaction of A with B alone.

Thus, the mutual information in **A** and **C**, given **B**, $I(\mathbf{A}{:}\mathbf{C}|\mathbf{B})$, and in **B** and **C**, given **A**, $I(\mathbf{B}{:}\mathbf{C}|\mathbf{A})$, together account for the effect of an increased ionic (deterministic) character of the chemical interactions (communications) between the chemisorbed reactants A and B, compared with the AO communications involving separate species:

$$I(\mathbf{A}{:}\mathbf{BC}) + I(\mathbf{B}{:}\mathbf{AC}) - 2I(\mathbf{A}{:}\mathbf{B}) = I(\mathbf{A}{:}\mathbf{C}|\mathbf{B}) + I(\mathbf{B}{:}\mathbf{C}|\mathbf{A}). \tag{12.76}$$

This increase in the IT-ionicity of A–B interaction (decrease in its IT-covalency) results from the fact that due to the electron delocalization Y → C the chemisorbed reactant Y, now representing the modified interaction partner of X in the combined subsystem (Y¦C), effectively increases its direct information "overlap" with the other adsorbate X in (X¦C), by extra areas (ionicities) due to the overlaps between their entropy circles and that of the catalyst, in the (Y, C) and (X, C) entropy "envelopes," respectively.

Such an IT-ionic "activation" of adsorbates, as a result of forming the partial A—C and B—C bonds on active sites of the catalytic surface, also manifests the *competition effect* between these surface bonds and the interadsorbate bond A—B in the catalytic system: the more heavily are the valence orbitals/electrons of A and B involved in chemical bonds with C, the less noisy (more deterministic) are their mutual communications, thus giving rise to the less IT-covalent (more IT-ionic) interactions between the *chemisorbed* species.

The surface *chemical* bond between a given adsorbate and the catalyst should be strongly felt at the position of the other adsorbate, and hence the information "coupling" between probability distributions of the chemisorbed species and that of the catalyst should be relatively strong. The *physical* adsorption of these reactants should be marked by a relatively small value of $I(\mathbf{B}{:}\mathbf{C}|\mathbf{A})$, since the dependencies ("overlaps") between entropy circles of schemes (**B**, **A**) and **C** should be relatively small, thus grossly diminishing the above IT-ionic activation effect generated by catalyst's presence (Nalewajski 2010b, d).

Another molecular scenario, in which the *three*-scheme entropy/information descriptors are expected to be useful, is the influence of one reactive site in a molecule upon another, e.g., in the contexts of subtle reactivity preferences in the DA reactive systems implied by the HSAB rule (Nalewajski 2010d), and particularly its regional formulation for predicting the regioselectivity trends in cyclization reactions (Chandra et al. 1998), the *Maximum Complementarity* principle (Nalewajski 2000d), and the *Bond Length Variation Rules* of Gutmann (1978). Such cooperative interaction between different sites in a molecule is also responsible for the directing *trans/cis* influence of ligands in transition metal complexes and the familiar *substituent effect* in aromatic systems. The adequate IT description of the AIM cooperation in *many*-center bonds, e.g., in boron hydrides or propellanes, may also require the entropy/information indices involving several probability schemes (see Sect. 12.10).

The four probability schemes generate a diversity of the conditional entropy and mutual information descriptors introduced in Sect. 8.6. They have been delineated in a qualitative entropy diagram of Fig. 8.6. Figures 12.7 and 12.8 correspond to the weakly and strongly (chemically) interacting subsystems M_1 and M_2 in the molecular complex $M = (M_1 \vdots M_2)$, which involves smaller molecular fragments of interest, $M_1 = $ A—B and $M_2 = $ C—D, to which the probability schemes **A**, **B**, **C**, **D** are ascribed.

Alternatively, the pairs of the input and output distributions in the communication systems describing M_1 and M_2, respectively, may constitute the four probability schemes under consideration. Let us examine the IT-covalent and IT-ionic couplings between the orbital information systems of these complementary subsystems in M (Fig. 12.9). Let the probability schemes **A** and **B,** respectively, denote the AO input and outputs in M_1 with the remaining schemes **C** and **D** having a similar meaning for M_2. It should be stressed that each scheme in M_1 now combines the AO contributed by its both constituent fragments A and B, while

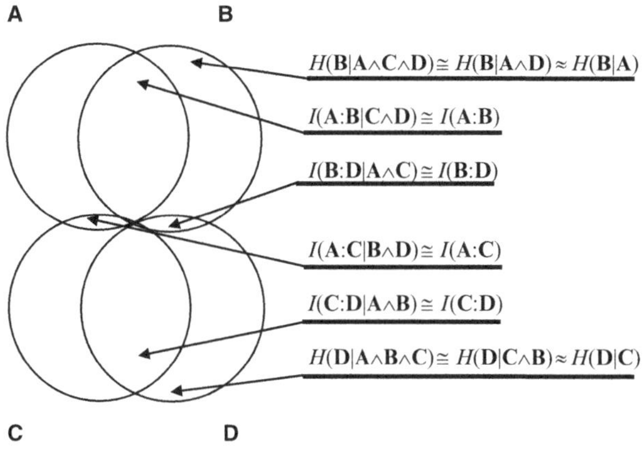

Fig. 12.7 General entropy/information diagrams of four probability schemes (**A**, **B**, **C**, **D**) corresponding to *weakly* interacting, *internally* bonded subsystems A—B and C—D in the reaction complex $\begin{bmatrix} \text{A}\text{—}\text{B} \\ \vdots \quad \vdots \\ \text{C}\text{—}\text{D} \end{bmatrix}$

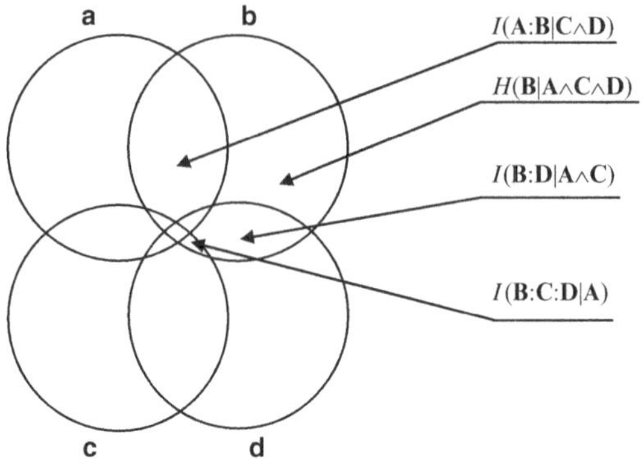

Fig. 12.8 Entropy/information diagram of the interaction between the internal communications $\boldsymbol{a} \rightarrow \boldsymbol{b}$ and $\boldsymbol{c} \rightarrow \boldsymbol{d}$ in the strongly interacting complementary fragments $M_1 = (A\vert B)$ and $M_2 = (C\vert D)$ of the molecular complex $M = (M_1\vert M_2)$

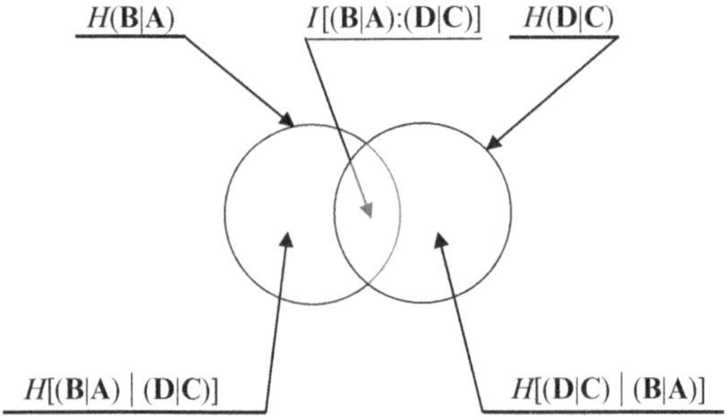

Fig. 12.9 Alternative entropy diagram for the complementary fragments M_1 and M_2 of $M = (M_1\vert M_2)$

each scheme in M_2 combines the AO of its fragments C and D, with the fragment AO events extending over all basis functions contributed by its constituent atoms.

Our aim now is to identify prospective candidates for the entropy/information description, of the mutual influence of chemical bond(s) in M_1, originating from communications $\boldsymbol{a} \rightarrow \boldsymbol{b}$, on bonds in M_2, generated by communications $\boldsymbol{c} \rightarrow \boldsymbol{d}$. These intrafragment communications are characterized by the conditional probabilities $\mathbf{P}(\boldsymbol{b}|\boldsymbol{a}) = \{P(b|a) \equiv P(\alpha)\}$ and $\mathbf{P}(\boldsymbol{d}|\boldsymbol{c}) = \{P(d|c) \equiv P(\beta)\}$, respectively, which define the associated *singly* conditional probability schemes (**B**|**A**) and (**D**|**C**) (see Sect. 12.2). The mutual dependencies of the internal communications in the complementary subsystems of M are then described by the *triply* conditional probabilities of the probability propagations $(\boldsymbol{a} \rightarrow \boldsymbol{b}) \equiv \boldsymbol{\alpha}$ in M_1 conditional on the information scattering $(\boldsymbol{c} \rightarrow \boldsymbol{d}) \equiv \boldsymbol{\beta}$ in M_2 (see Sect. 12.6):

$$\mathbf{P}[(\boldsymbol{b}|\boldsymbol{a})|(\boldsymbol{d}|\boldsymbol{c})] = \{P[(b|a)|(d|c)] \equiv P(\alpha|\beta) \equiv P(\alpha \wedge \beta)/P(\beta)\}. \tag{12.77}$$

Here, the joint probabilities of two conditional AO events, $P(\alpha \wedge \beta)$, are assumed to satisfy the usual normalization:

$$\sum_{\alpha}^{M_1} \sum_{\beta}^{M_2} P(\alpha \wedge \beta) = \sum_{\beta}^{M_2} P(\beta) = 1. \tag{12.78}$$

with summations ranging over all internal communications in each subsystem.

The average conditional entropy indices of this interfragment covalent coupling then read:

$$H[(\mathbf{B}|\mathbf{A})|(\mathbf{D}|\mathbf{C})] = S(\boldsymbol{P}(\boldsymbol{\alpha})|\boldsymbol{P}(\boldsymbol{\beta})) = -\sum_{\alpha}^{M_1}\sum_{\beta}^{M_2} P(\alpha \wedge \beta) \log P(\alpha|\beta),$$

$$H[(\mathbf{D}|\mathbf{C})|(\mathbf{B}|\mathbf{A})] = S(\boldsymbol{P}(\boldsymbol{\beta})|\boldsymbol{P}(\boldsymbol{\alpha})) = -\sum_{\alpha}^{M_1}\sum_{\beta}^{M_2} P(\alpha \wedge \beta) \log P(\beta|\alpha), \qquad (12.79)$$

while the IT-ionic coupling between the two fragments is embodied in the associated average mutual information quantity:

$$I[(\mathbf{B}|\mathbf{A}):(\mathbf{D}|\mathbf{C})] = I(\boldsymbol{P}(\boldsymbol{\alpha}):\boldsymbol{P}(\boldsymbol{\beta})) = \sum_{\alpha}^{M_1}\sum_{\beta}^{M_2} P(\alpha \wedge \beta) \log \frac{P(\alpha \wedge \beta)}{P(\alpha)P(\beta)}$$
$$= H(\mathbf{B}|\mathbf{A}) - H[(\mathbf{B}|\mathbf{A})|(\mathbf{D}|\mathbf{C})] = H(\mathbf{D}|\mathbf{C}) - H[(\mathbf{D}|\mathbf{C})|(\mathbf{B}|\mathbf{A})]. \qquad (12.80)$$

These IT-covalency and -ionicity descriptors of the communication interaction between the two subsystems thus conserve the internal uncertainties in each molecular fragment (Fig. 12.9) measured by the corresponding Shannon entropies of the internal communications in each subsystem:

$$N[(\mathbf{B}|\mathbf{A});(\mathbf{D}|\mathbf{C})] = H[(\mathbf{B}|\mathbf{A})|(\mathbf{D}|\mathbf{C})] + I[(\mathbf{B}|\mathbf{A}):(\mathbf{D}|\mathbf{C})] = H(\mathbf{B}|\mathbf{A}),$$
$$N[(\mathbf{D}|\mathbf{C});(\mathbf{B}|\mathbf{A})] = H[(\mathbf{D}|\mathbf{C})|(\mathbf{B}|\mathbf{A})] + I[(\mathbf{B}|\mathbf{A}):(\mathbf{D}|\mathbf{C})] = H(\mathbf{D}|\mathbf{C}). \qquad (12.81)$$

Next, let us attribute the four probability schemes (**A**, **B**, **C**, **D**) to AO events in fragments A, B (of M_1) and C, D (of M_2), respectively (Fig. 12.8). Other entropy/information descriptors discussed in Sect. 8.6 should also reflect specific ionic and covalent influences between the chemically interacting subsystems in $M = (M_1 \vdots M_2)$. For example, the mutual information $I(\mathbf{A}:\mathbf{B}|\mathbf{C}\wedge\mathbf{D})$ (8.42) accounts for the influence of M_2 on the IT-ionicity in M_1. The associated conditional entropy,

$$H[(\mathbf{B}|\mathbf{A})|\mathbf{C} \wedge \mathbf{D}] = H(\mathbf{B}|\mathbf{A}) - I(\mathbf{B}:\mathbf{D}|\mathbf{A} \wedge \mathbf{C}) - I(\mathbf{B}:\mathbf{C}:\mathbf{D}|\mathbf{A}), \qquad (12.82)$$

similarly reflects the effect of M_2 on the IT-covalency in M_1.

Finally, by attributing the four schemes **A**, **B**, **C**, and **D** to the corresponding fragments in the weakly interacting bimolecular reactive system (Fig. 12.7),

$$M_1\text{----}M_2 = \begin{bmatrix} A\text{—}B \\ \vdots \quad \vdots \\ C\text{—}D \end{bmatrix},$$

one predicts $I(\mathbf{A}:\mathbf{B}|\mathbf{C}\wedge\mathbf{D}) \cong I(\mathbf{A}:\mathbf{B})$, $I(\mathbf{B}:\mathbf{C}:\mathbf{D}|\mathbf{A}) \cong 0$, and hence $H[(\mathbf{B}|\mathbf{A})|\mathbf{C}\wedge\mathbf{D}] \cong H(\mathbf{B}|\mathbf{A}) - I(\mathbf{B}:\mathbf{D}|\mathbf{A}\wedge\mathbf{C}) \approx H(\mathbf{B}|\mathbf{A})$. Therefore, weak M_1----M_2 interactions have practically vanishing effect on the internal ionicities of reactants, with only their covalencies being slightly affected.

12.10 Through-Space and Through-Bridge Bond Components

In MO theory the chemical interaction between, say, two (valence) AO or general basis functions originating from different atoms is strongly influenced by their direct overlap/interaction, which conditions the bonding effect experienced by electrons occupying their bonding combination in the molecule, compared with the nonbonding reference of electrons on separated AO. This "through-*space*" bonding mechanism is then associated with typical accumulation of the valence electrons in region between the two nuclei, due to the constructive interference between the two AO functions, which can also exhibit some polarization reflecting the initial electronegativity difference of the two atoms involved.

Indeed, such "shared" bond charge is synonymous with the presence of the bond *covalency* in the direct interaction between the two AO. It is also reflected by the associated *covalent* VB structure. Similar effect of the bonding accumulation of the information densities relative to the promolecular distribution has been detected in maps of alternative information densities relative to the promolecule, e.g., of the entropy deficiency and the displacement in Shannon's entropy (Sects. 10.1 and 10.2). Accordingly, the complementary bond *ionicity* aspect is manifested by the MO polarization and CT or – alternatively – by the participation of the orthogonal part of the *ionic* VB structure in the ground state wave function. Let us recall that on the elementary CID level in the minimum basis set both MO and VB descriptions of the chemical bond in H_2 are exactly equivalent, differing only in specific routes of arriving at the same *two*-electron (ground state) wave function describing the *singlet*-paired electrons. Finally, in OCT the bond ionicity descriptor reflects a degree of "localization" (determinicity) in communications between bonded atoms, while the bond covalency is manifested by the "delocalization" (noise) aspect of the molecular (direct) information channel.

Therefore, the *direct* ("through-space") bonding interaction between neighboring atoms, as reflected by the associated Wiberg bond orders, is usually associated with the presence of the bond charge between the two nuclei. However, for more distant atomic partners such an accumulation of valence electrons can be absent, e.g., in the cross-ring π-interactions in benzene or between the bridgehead carbon atoms in small propellanes (see Sect. 10.3 and Table 12.2). It has been already argued above that the bonding interaction lacking such an accumulation of the bond charge (information) can be also realized *indirectly*, through the neighboring AO intermediaries forming a "bridge" for an effective interaction between more distant (terminal) AO, e.g., in the cross-ring interactions between two *meta*- or *para*-carbons in benzene (Nalewajski 2011a, b, d, f). This indirect ("through-bridges") mechanism reflects the implicit dependencies between AO resulting from their participation in all chemical bonds in the molecular system under consideration, determined by the subspace of the occupied MO (Nalewajski and Gurdek 2011a, b).

Indeed, the orthonormality constraints imposed on the occupied MO imply the *implicit* dependencies between AO on different (terminal) atoms (Nalewajski and Gurdek 2011a), which can generate the indirect contributions to their chemical interactions in molecules also when they are widely separated in a molecule, provided they directly couple to the chemically interacting orbitals of real bridges connecting the terminal atoms (Nalewajski 2011a, b).

In such a generalized outlook on the bond order concept, emerging from both the Wiberg and quadratic difference measures formulated in the MO theory, as well as from the IT bond multiplicity of OCT, one identifies the chemical bond multiplicity as a measure of the statistical "*dependence*" (nonadditivity) between orbitals on different atomic centers. On one hand, this dependence between basis functions of different atoms can be realized *directly* (through space), by the constructive interference of orbitals (probability amplitudes) on two atoms, which generally increases the electron density between them. On the other hand, it can also have an *indirect* origin, through the dependence on orbitals of the remaining AIM used to construct the system occupied MO. As we have already remarked above, the latter component is due to the orthonormality relations between the occupied MO, which determine the entire framework of chemical bonds in the molecule.

In OCT the *internal* A—B bonds result from the $\boldsymbol{\chi}_{AB} \to \boldsymbol{\chi}_{AB}$ probability scatterings, while the communications $\boldsymbol{\chi}_{C\neq(A,B)} \to \boldsymbol{\chi}_{AB}$ originating from the molecular complement of this diatomic fragment generate the *external* bond components. In this IT approach the *indirect* interactions between A and B result from effective communications between AB basis functions $\boldsymbol{\chi}_{AB}$ via the AO intermediaries $\boldsymbol{\chi}_{C\neq(A,B)}$ in the molecular communication cascade: $\boldsymbol{\chi}_{AB} \to \boldsymbol{\chi}_{C\neq(A,B)} \to \boldsymbol{\chi}_{AB}$. Admissible also are the multiple cascade propagations of information between basis functions (Nalewajski 2011a, c, f; Nalewajski and Gurdek 2011a, b).

Therefore, each pair of AO or AIM exhibits the partial through-*space* and through-*bridge* components: the bond order of the former quickly vanishes with an increase of the interatomic separation or when the interacting AO are heavily engaged in forming chemical bonds with other atoms, while the latter can still assume appreciable values, when the remaining atoms form an effective bridge of the neighboring, chemically interacting atoms, which links the specified AO/AIM in question (Nalewajski 2011a, b, c, f; Nalewajski and Gurdek 2011a, b). In this section, we shall identify these components of general chemical interactions between AO using the Wiberg measure of bond multiplicities. The corresponding IT-covalencies originating from the indirect (cascade) communications will be also examined within the recently proposed OCT of the chemical bond.

12.10.1 Bond Projections and Density Matrix

In standard SCF MO theory (see Sect. 12.2) the network of chemical bonds is determined by the occupied MO in the system ground state. Let us again assume the *closed*-shell (*cs*) configuration of $N = 2n$ electrons in the standard spin-restricted Hartree–Fock (RHF) description, which involves n lowest, *doubly* occupied (orthonormal) MO. In the familiar LCAO MO approach, they are generated as linear combinations of the (Löwdin-orthogonalized) OAO (basis functions)

$\boldsymbol{\chi} = (\chi_1, \chi_2, \ldots, \chi_m) = \{\chi_i\}$, $\langle\boldsymbol{\chi}|\boldsymbol{\chi}\rangle = \{\delta_{i,j}\} \equiv \mathbf{I}$, contributed by the system constituent atoms:

$$\boldsymbol{\varphi} = [(\varphi_1, \varphi_2, \ldots \varphi_n), (\varphi_{n+1}, \ldots \varphi_m) = (\boldsymbol{\varphi}^o, \boldsymbol{\varphi}^v) = \{\varphi_s\} = \boldsymbol{\chi}\mathbf{C} = \boldsymbol{\chi}(\mathbf{C}^0|\mathbf{C}^v). \quad (12.84)$$

Here, the rectangular matrices $\mathbf{C}^o = \langle\boldsymbol{\chi}|\boldsymbol{\varphi}^o\rangle$ and $\mathbf{C}^v = \langle\boldsymbol{\chi}|\boldsymbol{\varphi}^v\rangle$ group the relevant expansion coefficients of n (*doubly* occupied) and $m - n$ virtual (empty) MO, respectively, to be determined using the iterative self-consistent field (SCF) procedure. The full LCAO MO matrix $\mathbf{C}$ is unitary, $\mathbf{C}^\dagger = \mathbf{C}^{-1}$, since it "rotates" m orthonormal AO into the m orthonormal MO; hence, the inverse transformation reads: $\boldsymbol{\chi} = \boldsymbol{\varphi}\mathbf{C}^\dagger$.

The 1-density (CBO) matrix $\boldsymbol{\gamma}$ of (12.2) now reads

$$\begin{aligned}\boldsymbol{\gamma} &= 2\langle\boldsymbol{\chi}|\boldsymbol{\varphi}^o\rangle\langle\boldsymbol{\varphi}^o|\boldsymbol{\chi}\rangle = 2\mathbf{C}^o\mathbf{C}^{o\dagger} = \mathbf{C}\mathbf{d}\mathbf{C}^\dagger \\ &\equiv 2\langle\boldsymbol{\chi}|\hat{\mathrm{P}}^o_\varphi|\boldsymbol{\chi}\rangle = 2\left(\langle\boldsymbol{\chi}|\hat{\mathrm{P}}^o_\varphi\right)\left(\hat{\mathrm{P}}^o_\varphi|\boldsymbol{\chi}\rangle\right) \equiv 2\langle\boldsymbol{\chi}^b|\boldsymbol{\chi}^b\rangle \\ &= \left\{\gamma_{i,j} = 2\langle i|\hat{\mathrm{P}}^o_\varphi|j\rangle = 2\langle i|\left(\hat{\mathrm{P}}^o_\varphi\right)^2|j\rangle \equiv 2\langle i^b \mid j^b\rangle\right\},\end{aligned} \quad (12.85)$$

where the diagonal matrix $\mathbf{d}$ groups the MO occupations, $\mathbf{d} = \{\delta_{s,s'}(2, s \leq n;\ 0, s{>}n)\}$, and the basis set projections onto the occupied (*bond*) subspace $\boldsymbol{\varphi}^o$,

$$|\boldsymbol{\chi}^b\rangle = \hat{\mathrm{P}}^o_\varphi|\boldsymbol{\chi}\rangle = |\boldsymbol{\varphi}^o\rangle\langle\boldsymbol{\varphi}^o|\boldsymbol{\chi}\rangle = |\boldsymbol{\varphi}^o\rangle\mathbf{C}^{o\dagger} = \left\{\hat{\mathrm{P}}^o_\varphi|i\rangle = |i^b\rangle\right\}, \quad (12.86)$$

determine the bond projections of AO. The closed-shell CBO matrix thus constitutes the AO representation of the projection operator onto the subspace of all *doubly* occupied MO,

$$\begin{aligned}\hat{\mathrm{P}}^o_\varphi &= |\boldsymbol{\varphi}^o\rangle\langle\boldsymbol{\varphi}^o| = \sum\nolimits_s^{occd.} |\varphi_s\rangle\langle\varphi_s| \\ &\equiv \sum\nolimits_s^{occd.} \hat{\mathrm{P}}_s = \tfrac{1}{2}|\boldsymbol{\chi}\rangle\boldsymbol{\gamma}\langle\boldsymbol{\chi}|, \quad \left(\hat{\mathrm{P}}^o_\varphi\right)^2 = \hat{\mathrm{P}}^o_\varphi,\end{aligned} \quad (12.87)$$

and satisfies the idempotency relation of (12.3).

It also follows from (12.85) that MO determine the eigenvectors of $\boldsymbol{\gamma}$ corresponding to the eigenvalues (occupations) $\mathbf{d}$: $\boldsymbol{\gamma}\mathbf{C} = \mathbf{C}\mathbf{d}$. Thus, m bond projections $|\boldsymbol{\chi}^b\rangle$ of AO effectively span the n-dimensional subspace of the occupied MO, $|\boldsymbol{\varphi}^o\rangle = \{|s^o\rangle\}$. Indeed, the occupied MO determine the complete orthonormal basis in the bonding vector space $|\boldsymbol{\chi}^b\rangle$, so that any bond projection of AO can be exactly expanded in terms of $|\boldsymbol{\varphi}^o\rangle$ [see (12.86)].

We again recall that the 1-matrix reflects the promoted, *valence* state of AO in the molecule, with the diagonal elements measuring the effective electron occupations of basis functions, $\{N_i = \gamma_{i,i} = Np_i\}$, with the normalized probabilities $\boldsymbol{p} = \{p_i = \gamma_{i,i}/N\}$ of the basis functions occupancy in molecule: $(1/N)\ \mathrm{tr}\boldsymbol{\gamma} = \sum_i p_i = 1$. The *off*-diagonal CBO elements $\{\gamma_{i,j}\}$ between AO on different atoms similarly reflect the bonding status of the resultant interaction of the specified AO pair in the molecule. For the standard (positive) AO overlap, the positive (negative) values of $\gamma_{i,j}$ signify the resultant bonding (antibonding) coupling between basis functions, and the vanishing bond order $\gamma_{i,j} = 0$ identifies the net nonbonding chemical interaction, i.e., $|i^b\rangle = 0$ or $|j^b\rangle = 0$ [see (10.26)]. Thus, the "constructive" (bonding) interference requires that the two AO exhibit the positive product of their bond projections, while the negative product value identifies their resultant "destructive" interference in the molecular bond system.

Therefore, the (nonorthonormal) bond-projected AO basis contains $m - n$ linearly dependent vectors, with 1-matrix determining the singular overlap matrix:

$$\mathbf{S}^b = \langle \boldsymbol{\chi}^b | \boldsymbol{\chi}^b \rangle = \boldsymbol{\gamma}/2 = \mathbf{C}^o\mathbf{C}^{o\dagger} = \{S_{i,j}{}^b = \langle i^b \mid j^b \rangle\}. \tag{12.88}$$

The $|i_b\rangle$ projector can be expressed in terms of the renormalized bond projection of AO onto $|\bar{\chi}_i^b\rangle = |i^b\rangle\sqrt{2/\gamma_{i,i}}, \quad \langle \bar{\chi}_i^b | \bar{\chi}_i^b \rangle = 1$,

$$\hat{\bar{\mathrm{P}}}_i^b = |\bar{\chi}_i^b\rangle\langle\bar{\chi}_i^b| = |i^b\rangle 2\gamma_{i,i}^{-1}\langle i^b| = 2\gamma_{i,i}^{-1}\hat{\mathrm{P}}_i^b, \quad \left(\hat{\bar{\mathrm{P}}}_i^b\right)^2 = \hat{\bar{\mathrm{P}}}_i^b, \tag{12.89}$$

while the idempotent projection onto the whole bonding subspace $|\boldsymbol{\chi}^b\rangle = |\boldsymbol{\varphi}^o\rangle$ amounts to $\hat{\mathrm{P}}_{\varphi}^o$ projection:

$$\begin{aligned}\sum_{i=1}^{m} \hat{\bar{\mathrm{P}}}_i^b &= \sum_{s,s'}^{occd.} |s^o\rangle \left(\sum_{i=1}^{m} \langle s^o | i \rangle \langle i | s'^o \rangle \right) \langle s'^o | = \sum_{s,s'}^{occd.} |s^o\rangle \left(\sum_{i=1}^{m} C_{s,i}^{\dagger} C_{i,s'} \right) \langle s'^o | \\ &= \sum_{s,s'}^{occd.} |s^o\rangle\, \delta_{s,s'} \langle s'^o | = \sum_{s}^{occd.} |s^o\rangle\langle s^o| = \hat{\mathrm{P}}_{\varphi}^o,\end{aligned} \tag{12.90}$$

where we have recognized the unitary character of the LCAO MO matrix: $\mathbf{C}^{\dagger}\mathbf{C} = \mathbf{I}$.

12.10.2 *Through*-Space *and Through*-Bridge *Bond Orders*

The square of the *off*-diagonal CBO matrix element $\gamma_{i,j}$ linking two different AO χ_i and χ_j, contributed by atoms A and B, respectively, determines the ground state index proposed by Wiberg (1968) (12.11) for the (ground state) chemical *bond* order between these two basis functions:

$$\begin{aligned}\mathcal{M}_{i,j} &= \gamma_{i,j}\gamma_{j,i} = 4\langle j^b \mid i^b \rangle\langle i^b \mid j^b \rangle = 4\langle j^b|\hat{\mathrm{P}}_i^b|j^b\rangle \\ &= 2\gamma_{i,i}\langle j^b|\hat{\bar{\mathrm{P}}}_i^b|j^b\rangle = 4|\langle i^b \mid j^b \rangle|^2 \equiv 4|S_{i,j}^b|^2.\end{aligned} \tag{12.91}$$

It constitutes the additive contribution to the overall index of the molecular *bond* multiplicity between these two atoms,

$$\mathcal{M}_{A,B} = \sum_{i \in A} \sum_{j \in B} \mathcal{M}_{i,j}. \qquad (12.92)$$

This bond multiplicity concept has been subsequently extended (Gopinathan and Jug 1983; Mayer 1983; Jug and Gopinathan 1990) and generalized in terms of the *quadratic* bond orders from the *two*-electron difference approach (Nalewajski et al. 1993, 1994a, 1996a, b, 1997; Nalewajski and Mrozek 1994, 1996; Nalewajski 2004b). It follows from (12.91) that this "through-*space*" dependence between two AO located on different atoms originates from the *direct* "overlap" $S^b_{i,j}$ between the bond projections $|i^b\rangle$ and $|j^b\rangle$ of the two interacting orbitals:

$$S_{i,j}{}^b = \left(\langle i|\hat{\mathrm{P}}^o_{\varphi}\right)\left(\hat{\mathrm{P}}^o_{\varphi}|j\rangle\right) = \langle i^b \mid j^b\rangle = \gamma_{i,j}/2, \qquad (12.93)$$

which reflect the overall involvement of these two basis functions in all chemical bonds in the molecular system under consideration.

However, the overall dependence between two AO, say $\chi_i \in A$ and $\chi_j \in B$, in the molecular bond subspace combining all occupied MO has also an *indirect* ("through-*bridge*") origins, as represented by the associated amplitude $S_{i,j}(bridges)$ (Nalewajski 2011a). It originates from the implicit dependencies between two specified AO through the remaining AO in the molecular bonding subspace (Nalewajski and Gurdek 2011a). As shown in Fig. 12.10, these intermediate AO interactions can be classified as originating from the AO α-bridges, $\alpha = 1, 2, \ldots, t, \ldots, m-2$, including α AO intermediaries in communications between the specified pair of basis functions:

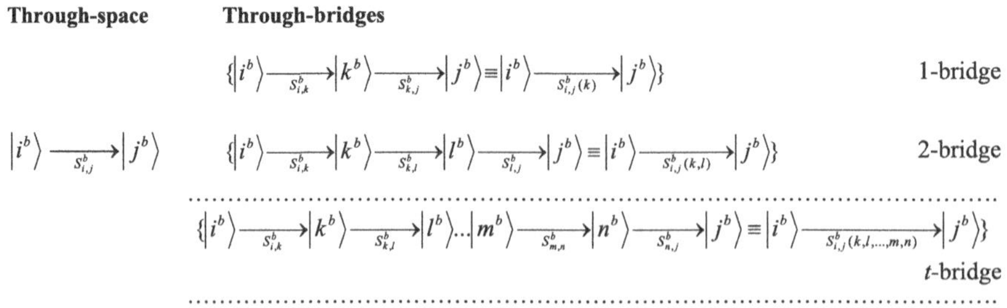

Fig. 12.10 Direct (through-space) and indirect (through-bridges) information propagations between orbitals χ_i and χ_j. The latter involve communications through the $\alpha = 1, 2, \ldots, t, \ldots m-2$ AO intermediaries defining α-bridges for the *implicit* probability scattering in molecules

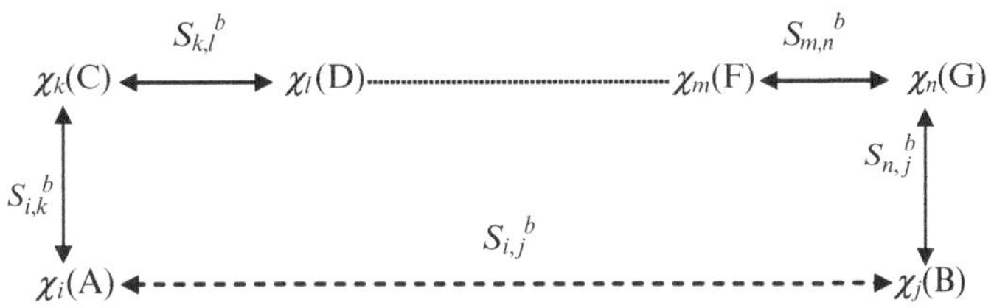

Fig. 12.11 Direct (through-space) chemical interaction (*broken line*) between orbitals χ_i and χ_j contributed by atoms A and B, respectively, and the indirect (through-*bridge*) interaction (*solid lines*), through t-AO intermediaries $(\chi_k, \chi_l, \ldots, \chi_m, \chi_n) = \{\chi_r\ , r = 1, 2, \ldots, t\}$ contributed by the neighboring bonded atoms (C, D, …, F, G), respectively

$$\begin{aligned} S_{i,j}(bridges) &= \sum_{k\neq(i,j)} S^b_{i,j}(k) + \sum_{k,l\neq(i,j)} S^b_{i,j}(k,l) + \ldots + \sum_{k,l,\ldots,m,n\neq(i,j)} S^b_{i,j}(k,l,\ldots,m,n) + \ldots \\ &\equiv S^{(1)}_{i,j} + S^{(2)}_{i,j} + \ldots + S^{(t)}_{i,j} + \ldots = \sum_{\alpha} S^{(\alpha)}_{i,j}. \end{aligned} \tag{12.94}$$

Let us examine the representative bond overlap $S^b_{i,j}(k,l,\ldots,m,n)$ between χ_i and χ_j due to t-bridge, originating from the bond projections of t strongly overlapping intermediate AO $\left(|k^b\rangle, |l^b\rangle, \ldots, |m^b\rangle, |n^b\rangle\right)$ (see Fig. 12.11), contributed by a cluster of the neighboring, bridging atoms {C, D, …, F, G}, which connect the specified AIM pair A and B in the molecule. This indirect overlap is proportional to the associated product of the CBO matrix elements:

$$\begin{aligned} S^b_{i,j}(k,l,..,m,n) &= \langle i^b \mid k^b\rangle\langle k^b \mid l^b\rangle\langle l^b \mid m^b\rangle\langle m^b|\ldots|n^b\rangle\langle n^b|j^b\rangle \\ &= 2^{-t}\gamma_{i,k}\gamma_{k,l}\gamma_{l,m}\cdots\gamma_{m,n}\gamma_{n,j}. \end{aligned} \tag{12.95}$$

Clearly, the most important 1-bridge bonding overlaps $\{S_{i,j}{}^b(k)\}$ are realized through orbitals $\chi_k \in$C contributed by the atomic neighbor(s) C, chemically bonded to both atoms A and B, which contribute the specified pair of communicating AO, thus forming the real atomic bridge A—C—B. Similarly, in a general t-bridge case, the intermediate AO which contribute the most to the overlap $S^b_{i,j}(k,l,\ldots,m,n)$ between $\chi^b_i \in$ A and $\chi^b_j \in$ B, through t AO intermediaries of Fig. 12.11 $\left(\chi^b_k \in \mathrm{C}, \chi^b_l \in \mathrm{D}, \ldots, \chi^b_m \in \mathrm{F}, \chi^b_n \in \mathrm{G}\right)$, are orbitals contributed by AIM forming the real bridge of chemical bonds: A—[C—D—…—F—G]—B. Thus, such indirect interactions between A and B can indeed be *long* ranged in character, provided there exist real chemical bridge connecting the two terminal atoms A and B.

This representative indirect overlap through t-bridge, $S^b_{i,j}(k,\ l,\ ..,\ m,\ n)$ constitutes a natural generalization of its direct, through-space analog of (12.91) and (12.93), by additionally including the product of (nonidempotent) *bond* projections onto the indicated intermediate AO,

$$S^b_{i,j}(k,l,\ldots,m,n) = \left\langle i^b \middle| \prod_{r=1}^{t} \hat{\mathrm{P}}^b_r \middle| j^b \right\rangle \equiv \left\langle i^b \middle| \hat{\mathrm{P}}^b_{t-bridge} \middle| j^b \right\rangle. \tag{12.96}$$

For example, for specific bridges of Fig. 12.10, this indirect bond overlap reads:

$$S^b_{i,j}(k) = S^b_{i,k}S^b_{k,j}, \quad S^b_{i,j}(k,l) = S^b_{i,k}S^b_{k,l}S^b_{l,j}, \quad \ldots,$$
$$S^b_{i,j}(k,l,..,m,n) = S^b_{i,k}S^b_{k,l}\ldots S^b_{m,n}S^b_{n,j}, \quad \ldots \tag{12.97}$$

The square of such an overlap defines the associated Wiberg-type bond order of the implicit interaction between orbitals χ_i and χ_j originating from the bridge in question:

$$\mathcal{M}_{i,j}(k,l,...,m,n) = 2^{2t}|S^b_{i,j}(k,l,..,m,n)|^2 = \gamma_{i,k}\{\gamma_{k,l}\cdots[\gamma_{m,n}(\gamma_{n,j}\gamma_{j,n})\gamma_{n,m}]\cdots\gamma_{l,k}\}\gamma_{k,i}$$
$$= \mathcal{M}_{i,k}\mathcal{M}_{k,l}\mathcal{M}_{l,m}\ldots\mathcal{M}_{m,n}\mathcal{M}_{n,j}. \tag{12.98}$$

This indirect bond multiplicity is thus given by the product of the partial (direct) bond orders of Wiberg, which involve the specified pair of orbitals and a sequence of AO intermediaries of the bridge under consideration. These orbital contributions in turn define the overall interaction between two atoms through the specified AO bridge:

$$\mathcal{M}_{\mathrm{A,B}}(k,l,...,m,n) = \sum\nolimits_{i\in\mathrm{A}}\sum\nolimits_{j\in\mathrm{B}}\mathcal{M}_{i,j}(k,l,...,m,n), \tag{12.99}$$

and hence also the overall indirect bond order due to the given t-AIM bridge:

$$\mathcal{M}_{\mathrm{A,B}}(\mathrm{C,D},...,\mathrm{F,G}) = \sum\nolimits_{k\in\mathrm{C}}\sum\nolimits_{l\in\mathrm{D}}\cdots\sum\nolimits_{m\in\mathrm{F}}\sum\nolimits_{n\in\mathrm{G}}\mathcal{M}_{\mathrm{A,B}}(k,l,...,m,n)$$
$$= \sum\nolimits_{i\in\mathrm{A}}\sum\nolimits_{j\in\mathrm{B}}\mathcal{M}^{(t)}_{i,j}(\mathrm{C,D},...,\mathrm{F,G}). \tag{12.100}$$

This bond index explores all implicit dependencies between the bonded atoms A and B, which originate from the basis functions of the t-AIM intermediaries defining the atomic bridge in question.

Thus, the overall implicit bond order between atoms A and B due to all admissible bridges:

$$\mathcal{M}_{\mathrm{A,B}}(bridges) = \sum_{k\neq(i,j)}\mathcal{M}_{i,j}(k) + \sum_{k,l\neq(i,j)}\mathcal{M}_{i,j}(k,l) + \ldots + \sum_{k,l,...,m,n\neq(i,j)}\mathcal{M}_{i,j}(k,l,...,m,n) + \ldots$$
$$\equiv \mathcal{M}^{(1)}_{i,j} + \mathcal{M}^{(2)}_{i,j} + \ldots + \mathcal{M}^{(t)}_{i,j} + \ldots = \sum_{\alpha}\mathcal{M}^{(\alpha)}_{i,j}, \tag{12.101}$$

where $\mathcal{M}^{(t)}_{i,j}$ stands for the bond order generated by all t-AIM bridges. Together with the direct component of (12.92), this bridge contribution thus determines the full quadratic bond multiplicitiy between atoms A and B:

$$\mathcal{M}(\mathrm{A},\ \mathrm{B}) = \mathcal{M}_{\mathrm{A,B}} + \mathcal{M}_{\mathrm{A,B}}(bridges). \tag{12.102}$$

The indirect bond component due to AO bridges can be alternatively viewed in the AIM resolution defined by the atomic bond projectors:

$$\hat{\mathrm{P}}^{b}_{\mathrm{X}} = \sum_{x \in \mathrm{X}} |x^b\rangle\langle x^b|, \quad \mathrm{X} = \mathrm{A}, \mathrm{B}, \mathrm{C}, \ldots, \tag{12.103}$$

which define the associated AIM bridge projections:

$$\begin{aligned}\hat{\mathrm{P}}^{b}_{bridges} &= \sum_{\mathrm{C} \neq (\mathrm{A,B})} \hat{\mathrm{P}}^{b}_{\mathrm{C}} + \sum_{\mathrm{C,D} \neq (\mathrm{A,B})} \hat{\mathrm{P}}^{b}_{\mathrm{C}} \hat{\mathrm{P}}^{b}_{\mathrm{D}} + \ldots \sum_{\mathrm{C,D,\ldots,F,G} \neq (\mathrm{A,B})} \hat{\mathrm{P}}^{b}_{\mathrm{C}} \hat{\mathrm{P}}^{b}_{\mathrm{D}} \ldots \hat{\mathrm{P}}^{b}_{\mathrm{F}} \hat{\mathrm{P}}^{b}_{\mathrm{G}} + \ldots \\ &\equiv \sum_{\beta}^{\mathrm{AIM}} \hat{\mathrm{P}}^{(b)}_{\beta-\mathrm{AIM}\ bridge},\end{aligned} \tag{12.104}$$

and the corresponding A–B bond order contributions:

$$\begin{aligned}\mathcal{M}_{\mathrm{A,B}}(bridges) &= \sum_{\mathrm{C} \neq (\mathrm{A,B})} \mathcal{M}_{\mathrm{A,B}}(\mathrm{C}) + \sum_{\mathrm{C,D} \neq (\mathrm{A,B})} \mathcal{M}_{\mathrm{A,B}}(\mathrm{C,D}) + \ldots + \sum_{\mathrm{C,D,\ldots,F,G} \neq (\mathrm{A,B})} \mathcal{M}_{\mathrm{A,B}}(\mathrm{C,D,\ldots,F,G}) \\ &\equiv \mathcal{M}^{(\mathrm{I})}_{\mathrm{A,B}} + \mathcal{M}^{(\mathrm{II})}_{\mathrm{A,B}} + \ldots + \mathcal{M}^{(t-\mathrm{AIM})}_{\mathrm{A,B}} + \ldots = \sum_{\beta}^{\mathrm{AIM}} \mathcal{M}^{(\beta-\mathrm{AIM})}_{\mathrm{A,B}},\end{aligned} \tag{12.105}$$

where:

$$\begin{aligned}\mathcal{M}^{(t-\mathrm{AIM})}_{\mathrm{A,B}} &= |S^{b}_{\mathrm{A,B}}(\mathrm{C,D,\ldots,F,G})|^2 = \sum_{i \in \mathrm{A}} \sum_{j \in \mathrm{B}} \left| \left\langle i^b \middle| \prod_{\mathrm{X}=1}^{t} \hat{\mathrm{P}}^{b}_{\mathrm{X}} \middle| j^b \right\rangle \right|^2 \\ &\equiv \sum_{i \in \mathrm{A}} \sum_{j \in \mathrm{B}} \left| \left\langle i^b \middle| \hat{\mathrm{P}}^{b}_{t-\mathrm{AIM}\ bridge} \middle| j^b \right\rangle \right|^2.\end{aligned} \tag{12.106}$$

This atomic resolution is better suited for interpretations in chemistry by expressing the overall bridge bond order $\mathcal{M}_{\mathrm{A,B}}(bridges)$ in terms of all admissible clusters of AIM, with the dominant contributions again being expected from the real t-AIM bridges of the chemically bonded atoms connecting A and B.

As an illustration let us examine the resultant bond order due to 1-AIM bridges. One first observes that $\sum_{\mathrm{C}} \hat{\mathrm{P}}^{b}_{\mathrm{C}} = \hat{\mathrm{P}}^{o}_{\varphi}$ and hence $\sum_{\mathrm{C} \neq (\mathrm{A,B})} \hat{\mathrm{P}}^{b}_{\mathrm{C}} = \hat{\mathrm{P}}^{o}_{\varphi} - \hat{\mathrm{P}}^{b}_{\mathrm{A}} - \hat{\mathrm{P}}^{b}_{\mathrm{B}}$:

$$\begin{aligned}\mathcal{M}^{(\mathrm{I})}_{\mathrm{A,B}} &= \sum_{\mathrm{C} \neq (\mathrm{A,B})} \mathcal{M}_{\mathrm{A,B}}(\mathrm{C}) = 4 \sum_{i \in \mathrm{A}} \sum_{j \in \mathrm{B}} \left| \left\langle i^b \middle| \sum_{\mathrm{C} \neq (\mathrm{A,B})} \hat{\mathrm{P}}^{b}_{\mathrm{C}} \middle| j^b \right\rangle \right|^2 \\ &= 4 \sum_{i \in \mathrm{A}} \sum_{j \in \mathrm{B}} \left| \left\langle i^b \middle| \hat{\mathrm{P}}^{o}_{\varphi} - \hat{\mathrm{P}}^{b}_{\mathrm{A}} - \hat{\mathrm{P}}^{b}_{\mathrm{B}} \middle| j^b \right\rangle \right|^2.\end{aligned} \tag{12.107}$$

Recognizing next that $\hat{P}^o_\varphi|\boldsymbol{\chi}^b\rangle = (\hat{P}^o_\varphi)^2|\boldsymbol{\chi}\rangle = \hat{P}^o_\varphi|\boldsymbol{\chi}\rangle = |\boldsymbol{\chi}^b\rangle$ and taking into account the AO orthogonality,

$$(\hat{P}^b_A + \hat{P}^b_B)|j^b\rangle = \hat{P}^b_B|j^b\rangle = |j^b\rangle\gamma_{j,j}/2 \quad \text{and} \quad (\hat{P}^b_A + \hat{P}^b_B)|i^b\rangle = \hat{P}^b_A|i^b\rangle = |i^b\rangle\gamma_{i,i}/2,$$

finally gives:

$$\begin{aligned}\mathcal{M}^{(I)}_{A,B} &= 4\sum_{i\in A}\sum_{j\in B}\left|\left\langle i^b\middle|\hat{P}^o_\varphi - \hat{P}^b_A - \hat{P}^b_B\middle|j^b\right\rangle\right|^2 \\ &= 4\sum_{i\in A}\sum_{j\in B}\left\langle j^b\middle|\hat{P}^o_\varphi - \hat{P}^b_A - \hat{P}^b_B\middle|i^b\right\rangle\left\langle i^b\middle|\hat{P}^o_\varphi - \hat{P}^b_A - \hat{P}^b_B\middle|j^b\right\rangle \\ &= 4\sum_{i\in A}\sum_{j\in B}\left|S^b_{i,j}\right|^2\left[1 - \frac{1}{2}(\gamma_{i,i} + \gamma_{j,j}) + \frac{1}{4}\gamma_{i,i}\gamma_{j,j}\right] \\ &= \mathcal{M}_{A,B} - \frac{1}{4}\sum_{i\in A}\sum_{j\in B}\mathcal{M}_{i,j}\left[2(\gamma_{i,i} + \gamma_{j,j}) - \gamma_{i,i}\gamma_{j,j}\right].\end{aligned} \tag{12.108}$$

This equation thus determines the exact relation between the indirect A—B bond order $\mathcal{M}^{(I)}_{A,B}$, realized through *single*-AIM bridges of all remaining atoms, and the direct through-space Wiberg component $\mathcal{M}_{A,B}$ of the overall bond multiplicity of (12.102).

Let us now examine two limiting occupations of the valence AO on atoms A and B. For the *full* occupations in the molecule of all chemically active AO contributed by these two atoms, $\{\gamma_{i,i} = \gamma_{j,j} = 2\}$, i.e., $\{2(\gamma_{i,i} + \gamma_{j,j}) - \gamma_{i,i}\gamma_{j,j} = 4\}$, when they remain effectively nonbonded, exhibiting only the lone pairs in their inner and valence shells, (12.108) gives: $\mathcal{M}^{(I)}_{A,B} = \mathcal{M}_{A,B} - \mathcal{M}_{A,B} = 0$. Thus, as intuitively expected, the nonbonded AIM do not generate the implicit bond component through the *single*-AIM bridges. Indeed, since the bridge projectors probe the common part of the diatomic bonding subspace $\{|\boldsymbol{\chi}_A{}^b\rangle, |\boldsymbol{\chi}_B{}^b\rangle\}$ and the remaining basis functions $\{|\chi_{C\neq(A,B)}{}^b\rangle\}$, for the nonbonded A and B, which do not exhibit the common part of their AO subspace with any single remaining atom and hence also with AO bases of any cluster of such bridging atoms, one predicts the vanishing bond order contribution generated through any bridges: atomic, diatomic, triatomic, etc.

It should be stressed that the common bonding subspaces for a larger number of AIM should steadily decrease with increasing order of the AIM-cluster in bridges, thus implying their expected diminished contribution to the overall bridge component of the chemical bond under consideration: $\mathcal{M}^{(I)}_{A,B} > \mathcal{M}^{(II)}_{A,B} > \ldots$ The bridging atoms must be also mutually bonded to generate the appreciable through bridge overlap of the interacting AO, so that significant *hypothetical* bridges are in fact limited to *real* chemical bridges of atoms in the molecular structural formula.

The next configuration of interest is the promoted state in which all valence orbitals from A and B are *half* occupied in the molecule: $\{\gamma_{i,i} = \gamma_{j,j} = 1\}$, i.e.,

$\{2(\gamma_{i,i}+\gamma_{j,j})-\gamma_{i,i}\gamma_{j,j}=3\}$, e.g., on hydrogens and the carbon atoms of the π-systems in hydrocarbons, in their promoted, valence state in the molecule. This implies that $|\boldsymbol{\chi}_A\rangle$ and $|\boldsymbol{\chi}_B\rangle$ are heavily engaged in forming the chemical bonds and gives: $\mathcal{M}^{(I)}_{A,B}=\mathcal{M}_{A,B}-\frac{3}{4}\mathcal{M}_{A,B}=\frac{1}{4}\mathcal{M}_{A,B}$. Therefore, quite a substantial indirect bond order, amounting to a quarter of the through-space component, is then realized already through all single-atom bridges in the molecule. It should be also realized that in this case one also expects significant, nonvanishing contributions from higher orders of the AIM bridges.

12.10.3 Conditional Probabilities for Information Propagation

The key concept of CTCB is the molecular information system (see the opening two sections of this chapter). It can be constructed at alternative levels of resolving the electron probabilities into the underlying electron localization "events," which determine the channel inputs $\boldsymbol{a}=\{a_i\}$ and outputs $\boldsymbol{b}=\{b_j\}$. In OCT the AO basis functions of SCF MO calculations determine a natural resolution level for discussing the information contributions to the multiplicity (order) of the system chemical bonds: $\boldsymbol{a}=\{\chi_i\}$ and $\boldsymbol{b}=\{\chi_j\}$. These AO-resolved networks describe the *direct* probability/information propagation $\boldsymbol{a}\rightarrow\boldsymbol{b}$ in the molecule, which can be described by the standard quantities developed in IT for real communication devices. Due to electron delocalization throughout the network of chemical bonds, the transmission of "signals" about the electron assignments to AO becomes randomly disturbed, thus exhibiting typical communication "noise." Indeed, an electron initially attributed to the given AO in the channel "input" $\boldsymbol{a}$ can be later found with a nonzero probability at several locations in the molecular "output" $\boldsymbol{b}$. This feature of the electron delocalization is embodied in the conditional probabilities of the "outputs-given-inputs," $\mathrm{P}(\boldsymbol{b}|\boldsymbol{a})=\{\,P(\chi_j|\chi_i)\equiv P(j|i)\}$, which define the molecular information network. As we have argued in the preceding section, the two basis functions on atoms A and B, respectively, $\chi_i\in$A and $\chi_j\in$B, can also "communicate" indirectly via the AO intermediaries $\{\chi_k\neq(\chi_i,\chi_j)\}$, e.g., in the *single*-AO information cascade: $\{i\rightarrow\{\chi_k\}\rightarrow j\}$. Determining the conditional probabilities for such bridge propagation of the electronic information in molecules is the basic goal of this section.

As we have seen in Sect. 12.2, in OCT one constructs the orbital pair probabilities (12.7) using the superposition principle of quantum mechanics (Dirac 1967) supplemented by the "physical" projection onto the subspace of the system occupied MO, which determine the molecular network of all chemical bonds. The AIM *off*-diagonal orbital communications are thus related to Wiberg's (1968) bond order contributions or to related generalized "quadratic" bond multiplicities formulated in MO theory (Gopinathan and Jug 1983; Mayer 1983; Jug and Gopinathan 1990; Nalewajski and Mrozek 1994, 1996; Nalewajski et al. 1993, 1994a, 1996b, 1997; Nalewajski 2004b). The IT descriptors have been shown to account for the chemical intuition quite well, at the same time providing the resolution of the diatomic bond multiplicities into their complementary IT-covalent and IT-ionic components (e.g., Nalewajski 2010f).

The orbital information system thus involves the AO events in the channel input $\boldsymbol{a} = \{\chi_i\}$ and output $\boldsymbol{b} = \{\chi_j\}$. In this description the AO $\rightarrow$ AO communication network is determined by the conditional probabilities of (12.7) for the (direct) communications between AO, involving squares of corresponding *off*-diagonal elements of the system CBO matrix. We recall that for the closed-shell systems

$$\mathbf{P}(\boldsymbol{b}|\boldsymbol{a}) = \left\{P(j|i) = (2\gamma_{i,i})^{-1}\gamma_{i,j}\gamma_{j,i} = (2\gamma_{i,i})^{-1}(\gamma_{i,j})^2 \equiv (\bar{N}_{i\rightarrow j}\gamma_{i,j})^2\right\}, \qquad (12.109)$$

where the amplitude normalization constant $\bar{N}_{i\rightarrow j}$ satisfies the requirement: $\sum_j P(j|i) = 1$. One observes that this conditional probability has the following interpretation in terms of the AO projectors of (12.89):

$$P(j|i) = \frac{2}{\gamma_{i,i}}\langle j^b|\hat{\mathrm{P}}_i^b|j^b\rangle = \langle j^b|\hat{\bar{\mathrm{P}}}_i^b|j^b\rangle. \qquad (12.110)$$

In order to estimate the IT bond contributions due to the through-bridge interactions between the specified pair of basis functions, in the molecular input and output, respectively, one requires the associated conditional probabilities realized through the indicated AO intermediaries. A straightforward projection generalization of the preceding expression gives:

$$\begin{aligned} P[(j|i)|k,l,...,m,n] &= \langle j^b|\left(\hat{\bar{\mathrm{P}}}_n^b\hat{\bar{\mathrm{P}}}_m^b...\hat{\bar{\mathrm{P}}}_l^b\hat{\bar{\mathrm{P}}}_k^b\right)\hat{\bar{\mathrm{P}}}_i^b\left(\hat{\bar{\mathrm{P}}}_k^b\hat{\bar{\mathrm{P}}}_l^b...\hat{\bar{\mathrm{P}}}_m^b\hat{\bar{\mathrm{P}}}_n^b\right)|j^b\rangle \\ &\equiv \langle j^b|\hat{\bar{\mathrm{P}}}_{bridge}^{b\dagger}\hat{\bar{\mathrm{P}}}_i^b\hat{\bar{\mathrm{P}}}_{bridge}^b|j^b\rangle \\ &= \frac{1}{2\gamma_{i,i}\gamma_{k,k}^2\gamma_{l,l}^2...\gamma_{m,m}^2\gamma_{n,n}^2}\gamma_{i,k}\{\gamma_{k,l}\cdots[\gamma_{m,n}(\gamma_{n,j}\gamma_{j,n})\gamma_{n,m}]\cdots\gamma_{l,k}\}\gamma_{k,i} \\ &= \frac{1}{2\gamma_{i,i}}\left(\frac{\gamma_{i,k}\gamma_{k,l}...\gamma_{m,n}\gamma_{n,j}}{\gamma_{k,k}\gamma_{l,l}...\gamma_{m,m}\gamma_{n,n}}\right)^2 \equiv P[(j|i)|t-\mathrm{AO}]. \end{aligned} \qquad (12.111a)$$

or

$$\begin{aligned} &P(k)\ P(l)\dots P(m)\ P(n)\ P[(j|i)|k,l,\dots,m,n] \\ &\quad = P(i\rightarrow k)P(k\rightarrow l)\dots P(m\rightarrow n)P(n\rightarrow j) \\ &\quad = P[(j|i)\wedge(k,l,\dots,m,n)], \end{aligned} \qquad (12.111b)$$

For example, for the *single*-AO bridge χ_k one obtains

$$P[(j|i)|k] = \frac{1}{2\gamma_{i,i}}\left(\frac{\gamma_{i,k}\gamma_{k,j}}{\gamma_{k,k}}\right)^2, \qquad (12.112)$$

while *two*-AO intermediaries (χ_k, χ_l) give

$$P[(j|i)|k,l] = \frac{1}{2\gamma_{i,i}}\left(\frac{\gamma_{i,k}\gamma_{k,l}\gamma_{l,j}}{\gamma_{k,k}\gamma_{l,l}}\right)^2. \quad (12.113)$$

where $P(r) = \gamma_{r,r}/2$ is the probability of χ_r being occupied in the molecule. These conditional probabilities determine the effective through-bridge (cascade) communications between the given AO *input* (χ_i) and AO *output* (χ_j), thus defining the associated information system for each order of the specified AO or AIM bridges.

The above bridge probabilities of information propagation in molecules render the associated *information cascade* interpretation. Consider first the *single*-AO bridge case of (12.112). This probability expression can be recast in terms of the conditional probabilities linking corresponding orbital pairs in the bonding subspace of the molecule (12.7):

$$P[(j|i)|k] = P(j|k)P(k)^{-1}P(k|i), \quad (12.114)$$

where $P(k) = \gamma_{k,k}/2$ denotes the probability that the intermediate AO is occupied in the molecular ground state. Hence, the associated *joint* probability of the indirect probability scattering ($i \to j$) through k^{th} AO bridge,

$$P[(j|i) \wedge k] \equiv P(k)P[(j|i)|k] = P(j|k)P(k|i), \quad (12.115)$$

is given by the product of elementary *two*-AO scatterings in the sequential (*double*) information cascade: ($i \to [k] \to j$). By the probability continuity the summation over all such AO intermediaries should amount to probability of the direct scattering event: $(i \to j) = \sum_k (i \to [k] \to j)$, or

$$\sum\nolimits_k P[(j|i) \wedge k] = P(j|i). \quad (12.116)$$

This partial normalization can be explicitly demonstrated using the quantum-mechanical (amplitude) interference of orbital communications [(12.206) in Sect. 12.11.3)]. Its overall normalization for the specified initial input can be verified by taking the sum of the preceding equations for $j = 1, 2, \ldots, m$:

$$\sum\nolimits_j \left\{\sum\nolimits_k P[(j|i) \wedge k]\right\} = \sum\nolimits_k \left[\sum\nolimits_j P(j|k)\right] P(k|i) = \sum\nolimits_k P(k|i) = 1, \quad (12.117)$$

where we have twice used the familiar sum rule for conditional probabilities: $\sum_l P(l|n) = 1$.

The same property can be demonstrated for any higher order of the bridge propagations of AO probabilities in a molecule. For example, for the case of two AO intermediaries one finds [see (12.109) and (12.113)],

$$P[(j|i)|k,l] = P(j|l)P(l)^{-1}P(l|k)P(k)^{-1}P(k|i), \quad (12.118)$$

and hence

$$P[(j|i) \wedge (k,l)] \equiv P(l)P(k)P[(j|i)|k,l] = P(j|l)P(l|k)P(k|i). \quad (12.119)$$

This product of conditional probabilities again represents elementary *two*-AO scatterings in the sequential (*triple*) information cascade: $(i \to [k \to l] \to j)$. By the probability continuity, the summation over all such *two*-AO intermediaries should amount to probability of the direct scattering event: $(i \to j) = \sum_k \sum_l (i \to [k \to l] \to j)$, or [see (12.208) in Sect. 12.11.3]

$$\sum_k \sum_l P[(j|i) \wedge (k,l)] = P(j|i), \tag{12.120}$$

The overall normalization for the given i can be again verified by taking the sum of these partial normalizations over the final output events $j = 1, 2, \ldots, m$:

$$\begin{aligned}\sum_j \left\{\sum_k \sum_l P[(j|i) \wedge (k,l)]\right\} &= \sum_k \sum_l \left[\sum_j P(j|l)\right] P(l|k)P(k|i)\\ &= \sum_k P(k|i)\left[\sum_l P(l|k)\right] = \sum_k P(k|i) = 1.\end{aligned} \tag{12.121}$$

To summarize, the through-bridge probability scattering gives rise to the associated information cascade for the joint probabilities. This additionally validates the claim that the appreciable through-bridges bonding should result mainly from the truly bonded bridging AO chain between the given pair of terminal orbitals.

This cascade development can be straightforwardly generalized into the AIM resolution, with the single conditional probabilities being then replaced by the corresponding AIM-resolved blocks of (12.24). For example, the joint probabilities of the conditional scatterings between $\boldsymbol{\chi}_A$ and $\boldsymbol{\chi}_B$ through the general t-AIM bridge (C, D, ..., F, G), i.e., all basis functions $\{\boldsymbol{\chi}_C, \boldsymbol{\chi}_D, \ldots, \boldsymbol{\chi}_F, \boldsymbol{\chi}_G\}$ contributed by these atoms to form MO (see also Fig. 12.11), are then given by the products of the diatomic blocks of conditional probabilities:

$$\begin{aligned}&P[(\boldsymbol{\chi}_B|\boldsymbol{\chi}_A) \wedge (\boldsymbol{\chi}_C, \boldsymbol{\chi}_D, \cdots, \boldsymbol{\chi}_F, \boldsymbol{\chi}_G)]\\ &\quad = \left\{\sum_{k\in C} \sum_{l\in D} \cdots \sum_{m\in F} \sum_{n\in G} P[(j|i) \wedge (k,l,\ldots,m,n)]\right.\\ &\quad \equiv \sum_{k\in C} \sum_{l\in D} \cdots \sum_{m\in F} \sum_{n\in G} P[(j|i)|k,l,\ldots,m,n]P(k)P(l)\ldots P(m)P(n)\\ &\quad = \left.\sum_{k\in C} \sum_{l\in D} \cdots \sum_{m\in F} \sum_{n\in G} P(j|n)P(n|m)\ldots P(l|k)P(k|i)\right\}\\ &\quad = P[(\boldsymbol{\chi}_B|\boldsymbol{\chi}_G)P[(\boldsymbol{\chi}_G|\boldsymbol{\chi}_F)\ldots P[(\boldsymbol{\chi}_D|\boldsymbol{\chi}_C)P[(\boldsymbol{\chi}_C|\boldsymbol{\chi}_A).\end{aligned} \tag{12.122}$$

They represent the sequential t-AIM cascade,

$$\begin{aligned}&(\boldsymbol{\chi}_A \to [\boldsymbol{\chi}_C \to \boldsymbol{\chi}_D \to \ldots \to \boldsymbol{\chi}_F \to \boldsymbol{\chi}_G] \to \boldsymbol{\chi}_B)\\ &\quad \equiv (A \to [C \to D \to \ldots \to F \to G] \to B),\end{aligned}$$

where the independent sums over *all* AIM bridges $\{\sum_X \sum_{x\in X} = \sum_x\}$ amount to the direct probability scattering:

$$\begin{aligned}&\sum_C \sum_D \cdots \sum_F \sum_G (A \to [C \to D \to \ldots \to F \to G] \to B) = (A \to B) \quad \text{or}\\ &\sum_C \sum_D \cdots \sum_F \sum_G P[(\boldsymbol{\chi}_B|\boldsymbol{\chi}_A) \wedge (\boldsymbol{\chi}_C, \boldsymbol{\chi}_D, \cdots, \boldsymbol{\chi}_F, \boldsymbol{\chi}_G)] = P(\boldsymbol{\chi}_B|\boldsymbol{\chi}_A).\end{aligned} \tag{12.123}$$

We further recall (see Sect. 12.3) that in OCT the entropy/information indices of the covalent/ionic components of chemical bonds represent the complementary descriptors of the average amounts of the information scattered (communication *noise*) and the information conserved (information *flow*) in the molecular communication channel of interest. The purely molecular communication channel, devoid of any reference (history) of the chemical bond formation process, uses the molecular AO probabilities $\boldsymbol{p} = \{p_i = \gamma_{i,i}/N\}$ in the channel input, while the promolecular signal $\boldsymbol{p}^0 = \left\{p_i^0 = \gamma_{i,i}^0/N\right\}$, reflecting the ground state electron configurations of the collection of the constituent free atoms, is used to extract the ionic IT-component of the resultant IT bond order. Thus, the average (through-space) noise index of the molecular IT bond *covalency* is measured by the conditional entropy of (12.15), while the average (through-space) information flow descriptor of the system IT bond *ionicity* is given by the mutual information in the channel promolecular inputs and molecular outputs (12.17). The latter "amount of information" reflects the fraction of the initial (promolecular) information content $S(\boldsymbol{p}^0)$, which has not been dissipated as noise in the molecular communication system. The sum of these two bond components (12.19) then measures the overall *direct* IT multiplicity of all bonds in the molecular system under consideration.

The IT-covalent and IT-ionic components of the *implicit* chemical bonds, realized via intermediate propagations of AO probabilities through bridges, can be defined in a similar way for any order of the AO/AIM bridge. For example, for the given t-AO bridge $(k, l, \ldots, m, n)$ of Fig. 12.11, the conditional probabilities $\{P[(j|i)|k, l, \ldots, m, n]\}$ determine the associated through-bridge communications between the terminal input (i) and output (j) AO. The corresponding average noise contribution,

$$S[(\boldsymbol{b}|\boldsymbol{a})|k,l,\ldots,m,n] = -\sum\nolimits_i \sum\nolimits_j P[(i \wedge j)|k,l,\ldots,m,n] \log P[(j|i)|k,l,\ldots,m,n] \\ \equiv S(k,l,\ldots,m,n). \tag{12.124}$$

where $P[(i \wedge j)|k, l, \ldots, m, n] = p_i\, P[(j|i)|k, l, \ldots, m, n]$, then reflects the system indirect IT-covalency generated by this particular AO bridge.

12.10.4 Illustrative Application to π-Electron Systems in Benzene and Butadiene

Next, let us examine the indirect π-bonds between carbon atoms in benzene and butadiene in the familiar Hückel approximation (Nalewajski 2011a). For the consecutive numbering of carbons in the ring, the relevant CBO matrix elements in benzene, the "amplitudes" of the corresponding *direct* Wiberg bond orders, read:

$$\gamma_{i,i} = 1, \quad \gamma_{i,i+1} = 2/3, \quad \gamma_{i,i+2} = 0, \quad \gamma_{i,i+3} = -1/3. \tag{12.125}$$

They generate the associated direct (through-space) bond components:

$$\mathcal{M}_{i,i+1} = 0.44\ (ortho), \quad \mathcal{M}_{i,i+2} = 0\ (meta), \quad \mathcal{M}_{i,i+3} = 0.11\ (para). \tag{12.126}$$

The vanishing direct bond order between the two *meta*-carbons is then complemented by the following most important indirect interactions in the π-system:

$$\mathcal{M}_{i,i+2}(i+1) = \mathcal{M}_{i,i+1}\mathcal{M}_{i+1,i+2} = 0.20, \tag{12.127}$$

$$\mathcal{M}_{i,i+2}(i+5, i+4, i+3) = \mathcal{M}_{i,i+5}\mathcal{M}_{i+5,i+4}\mathcal{M}_{i+4,i+3}\mathcal{M}_{i+3,i+2} = 0.04, \tag{12.128}$$

which together amount to the overall 0.24 indirect π-bond multiplicity between *meta*-carbons realized through the remaining (neighboring) C—C bonds in benzene. There also are two small Wiberg-type cross-ring contributions, due to bridges involving nonneighbors in the ring:

$$\mathcal{M}_{i,i+2}(i+3) = \mathcal{M}_{i,i+3}\mathcal{M}_{i+3,i+2} = \mathcal{M}_{i,i+2}(i+5) = \mathcal{M}_{i,i+5}\mathcal{M}_{i+5,i+2} \\ = 0.05, \tag{12.129}$$

$$\mathcal{M}_{i,i+2}(i+1,\ i+4,\ i+3) = \mathcal{M}_{i,i+1}\mathcal{M}_{i+1,i+4}\mathcal{M}_{i+4,i+3}\mathcal{M}_{i+3,i+2} \\ = \mathcal{M}_{i,i+2}(i+1,\ i+4,\ i+5) = \mathcal{M}_{i,i+1}\mathcal{M}_{i+1,i+4}\mathcal{M}_{i+4,i+5}\mathcal{M}_{i+5,i+2} = 0.01. \tag{12.130}$$

Therefore, the *meta*-carbons in benzene, which in this approximation exhibit the vanishing direct (through-space) π-bond, are linked by the indirect bond multiplicity $\mathcal{M}_{i,i+2}(bridges) \cong 0.3$. In this rough estimate, we have neglected some very small contributions due to bridges involving several cross-ring links, e.g.,

$$\mathcal{M}_{i,i+2}(i+3,\ i+4,\ i+5) = \mathcal{M}_{i,i+3}\mathcal{M}_{i+3,i+4}\mathcal{M}_{i+4,i+5}\mathcal{M}_{i+5,i+2} = 0.002. \tag{12.131}$$

The neighboring, *ortho*-carbons, which exhibit the highest direct π-bond order, generate relatively small overall bridge contribution. More specifically, all *single*- and *triple*-AO bridges generate the vanishing indirect components, while the only nonvanishing contributions result from the following *double*- and *quadruple*-AO bridges:

$$\mathcal{M}_{i,i+1}(i+5, i+4) = \mathcal{M}_{i,i+5}\mathcal{M}_{i+5,i+4}\mathcal{M}_{i+4,i+1} \\ = \mathcal{M}_{i,i+1}(i+2,\ i+3) = \mathcal{M}_{i+1,i+2}\mathcal{M}_{i+2,i+3}\mathcal{M}_{i+3,i} = 0.02, \tag{12.132}$$

$$\mathcal{M}_{i,i+1}(i+2,\ i+3,\ i+4,\ i+5)$$
$$= \mathcal{M}_{i+1,i+2}\mathcal{M}_{i+2,i+3}\mathcal{M}_{i+3,i+4}\mathcal{M}_{i+4,i+5}\mathcal{M}_{i+5,i} = 0.02, \quad (12.133)$$

generating altogether $\mathcal{M}_{i,i+1}(bridges) = 0.06$. This implicit contribution and the direct part of (12.126) give rise to a *half* total bond order between neighboring carbons in the benzene ring, as intuitively expected. In this estimate, we have again neglected very small contributions generated by bridges involving several *para*-links in the ring, e.g.,

$$\mathcal{M}_{i,i+1}(i+5,\ i+2,\ i+3,\ i+4)$$
$$= \mathcal{M}_{i,i+5}\mathcal{M}_{i+5,i+2}\mathcal{M}_{i+2,i+3}\mathcal{M}_{i+3,i+4}\mathcal{M}_{i+4,i+1} = 0.001, \quad (12.134)$$

$$\mathcal{M}_{i,i+1}(i+3,\ i+2,\ i+5,\ i+4)$$
$$= \mathcal{M}_{i,i+3}\mathcal{M}_{i+3,i+2}\mathcal{M}_{i+2,i+5}\mathcal{M}_{i+5,i+4}\mathcal{M}_{i+4,i+1} = 0.0004. \quad (12.135)$$

Finally, let us consider the bridge components of π-interactions between two *para*-carbons in the ring, which exhibit a relatively small direct bond order reported in (12.126). Again, all *single*- and *triple*-AO bridges generate the vanishing indirect components, while the most important *double*-AO bridges realized through neighboring bonds give:

$$\mathcal{M}_{i,i+3}(i+5,\ i+4) = \mathcal{M}_{i,i+5}\mathcal{M}_{i+5,i+4}\mathcal{M}_{i+4,i+3}$$
$$= \mathcal{M}_{i,i+3}(i+1, i+2) = \mathcal{M}_{i,i+1}\mathcal{M}_{i+1,i+2}\mathcal{M}_{i+2,i+3} = 0.09, \quad (12.136)$$

i.e., $\mathcal{M}_{i,i+3}(bridges) \cong 0.18$. There are also two *quadruple*-AO bridges involving two cross-ring links, which contribute a very small indirect bond order between *para*-carbons in benzene:

$$\mathcal{M}_{i,i+3}(i+1,\ i+4,\ i+5,\ i+2) =$$
$$\mathcal{M}_{i,i+1}\mathcal{M}_{i+1,i+4}\mathcal{M}_{i+4,i+5}\mathcal{M}_{i+5,i+2}\mathcal{M}_{i+2,i+3} = 0.001. \quad (12.137)$$

Therefore, two *para*-carbons in benzene ring also exhibit the resultant π-bond order of (12.102) of about 0.3, similar to that characterizing two *meta*-carbons.

The artificial distinction of the cross-ring π-interactions in Wiberg's (direct) multiplicities, with the vanishing bond order for *meta*-carbons, is thus effectively removed when the through-bridges components are also taken into account:

$$\mathcal{M}(para) \cong \mathcal{M}(meta) = 0.3 < \mathcal{M}(ortho) = 0.5. \quad (12.138)$$

We again emphasize the differences in their compositions: the *para* interactions exhibit comparable through-space and through-bridge components, the *meta* multiplicities are realized exclusively through bridges, while the strongest *ortho* bond orders have practically direct, through-space origin.

Of interest also is a comparison of the bond order contributions realized through the ring bridges of increasing length:

$$\begin{aligned}\mathcal{M}_{i,i+2}(i+1) &= (\mathcal{M}_{i,i+1})^2 = 0.20;\\ \mathcal{M}_{i,i+3}(i+1,\ i+2) &= (\mathcal{M}_{i,i+1})^3 = 0.09;\\ \mathcal{M}_{i,i+4}(i+1,\ i+2,\ i+3) &= (\mathcal{M}_{i,i+1})^4 = 0.04;\\ \mathcal{M}_{i,i+5}(i+1,\ i+2,\ i+3,\ i+4) &= (\mathcal{M}_{i,i+1})^5 = 0.02.\end{aligned} \tag{12.139}$$

Thus, the longer the bridge, the smaller indirect bond order it contributes.

Next, let us examine the π-interactions in butadiene. For the consecutive numbering of carbon atoms, the *off*-diagonal part of the CBO matrix in the Hückel approximation is fully characterized by the following elements:

$$\begin{aligned}&\gamma_{1,2} = \gamma_{3,4} = 2/\sqrt{5}, \quad \gamma_{1,3} = \gamma_{2,4} = 0, \quad \gamma_{1,4} = -1/\sqrt{5},\\ &\gamma_{2,3} = 1/\sqrt{5},\end{aligned} \tag{12.140}$$

which determine the associated through-space bond orders:

$$\mathcal{M}_{1,2} = \mathcal{M}_{3,4} = 0.80, \quad \mathcal{M}_{1,3} = \mathcal{M}_{2,4} = 0, \quad \mathcal{M}_{1,4} = \mathcal{M}_{2,3} = 0.20. \tag{12.141}$$

This somewhat artificial distinction of the (1–3) and (2–4) interactions as nonbonding can be again expected to be remedied by the inclusion of the indirect bond components.

For the strongest, terminal bond (1–2) the only nonvanishing contribution is due to the *two*-AO bridge,

$$\mathcal{M}_{1,2}(4,3) = \mathcal{M}_{1,4}\mathcal{M}_{4,3}\mathcal{M}_{3,2} = 0.03 = \mathcal{M}_{1,2}(bridges), \tag{12.142}$$

and hence the resultant bond order of the terminal bonds reads:

$$\mathcal{M}(1-2) = \mathcal{M}(3-4) = 0.83. \tag{12.143}$$

The bridge contributions to the *second*-neighbor interactions (1–3) are:

$$\mathcal{M}_{1,3}(4) = \mathcal{M}_{1,4}\mathcal{M}_{4,3} = \mathcal{M}_{1,3}(2) = \mathcal{M}_{1,2}\mathcal{M}_{2,3} = 0.16, \tag{12.144}$$

and hence

$$\mathcal{M}_{1,3}(bridges) = 0.32 = \mathcal{M}(1-3). \tag{12.145}$$

The bridge contribution to π-interactions between terminal carbons,

$$\mathcal{M}_{1,4}(2,3) = \mathcal{M}_{1,2}\mathcal{M}_{2,3}\mathcal{M}_{3,4} = 0.13 = \mathcal{M}_{1,4}(bridges), \quad (12.146)$$

and between the two carbons of the middle bond,

$$\mathcal{M}_{2,3}(1,4) = \mathcal{M}_{2,1}\mathcal{M}_{1,4}\mathcal{M}_{4,3} = 0.13 = \mathcal{M}_{1,3}(bridges), \quad (12.147)$$

similarly give:

$$\mathcal{M}(1,4) = \mathcal{M}(2,3) = 0.33. \quad (12.148)$$

Therefore, this novel perspective on the π-bond multiplicities in butadiene, more rational than that following from the simple direct bond orders of Wiberg, predicts:

$$\begin{aligned}\mathcal{M}(1-2) &= \mathcal{M}(3-4) = 0.83\\ &> \{\mathcal{M}(1-4) = \mathcal{M}(2-3) = 0.33 \cong \mathcal{M}(1-3) = \mathcal{M}(2-4) = 0.32\}.\end{aligned} \quad (12.149)$$

Again, the strongest terminal bonds (1–2) and (3–4) are almost exclusively of the through-space origin, the π-bonds (1–3) and (2–4) connecting the *second*-neighbors exhibit the pure through-bridge character, while the remaining bonds (1–4) and (2–3) include comparable direct and indirect components.

It has been argued in Sect. 12.10.1 that the CBO matrix elements reflect overlaps between projections of the basis functions onto the subspace of the occupied MO. They quantify the mutual dependencies between these bonded projections. For the given pair $\chi_i \neq \chi_j$ of the chemically interacting orbitals from atoms A and B, respectively, all elements $\{\gamma_{i,k} = \gamma_{k,i},\ k \neq i,$ and $\gamma_{j,l} = \gamma_{l,j},\ l \neq j\}$ embody the resultant interdependencies between these two AO and the remaining bonded parts of the basis set functions. The Wiberg bond order concept makes use of only the single $k = j$ (or $l = i$) projection constraint, corresponding to a single *off*-diagonal matrix element $\gamma_{i,j} = \gamma_{j,i}$ while neglecting all the remaining effective conditions between the AO bond projections involving this pair of basis functions. They are effectively taken into account only in the through-bridge bond multiplicity concept (Nalewajski and Gurdek 2011a), which in principle is capable to include all these additional constraints represented by the specified values of the remaining elements of the density matrix in rows/columns i and j (see also Sect. 12.11.3).

12.10.5 Indirect Orbital Communications

Next, let us briefly examine the through-bridge orbital communications in benzene. We focus on the indirect probabilities determining the communication system for the directly noninteracting *meta*-carbons in the ring. We recall that the symmetry-unrelated, direct conditional probabilities in benzene,

$$\begin{aligned} P(i|i) = 1/2, \quad P(i+1|i) = 2/9, \quad P(i+2|i) = 0, \\ P(i+3|i) = 1/18, \end{aligned} \tag{12.150}$$

define the through-space π-communications in this molecule (Nalewajski 2010d).

For the most important *meta*-bridges of (12.127)–(12.129), one finds the following *off*-diagonal probabilities linking these two carbon atoms through the corresponding bridges:

$$\begin{aligned} P[(i+2|i)|i+1] &= 0.0988, \\ P[(i+2|i)|i+3] &= 0.0247, \\ P[(i+2|i)|i+5,\ i+4,\ i+3] &= 0.0195, \end{aligned} \tag{12.151}$$

and hence the overall *off*-diagonal (switching) probability

$$P[(i+2|i)|bridges] = P[(i|i+2)|bridges] \cong 0.1430. \tag{12.152}$$

One similarly estimates the corresponding bridge contributions for the diagonal probabilities realized through these three bridges,

$$\begin{aligned} &P[(i|i)|i+1] = P[(i+2|i+2)|i+1] = P[(i+2|i+2)|i+3] = 0.0988, \\ &P[(i|i)|i+3] = 0.0062; \\ &P[(i|i)|i+5,\ i+4,\ i+3] = P[(i+2|i+2)|i+5,\ i+4,\ i+3] = 0.0049, \end{aligned} \tag{12.153}$$

which give:

$$P[(i|i)|bridges] \cong 0.110 \quad \text{and} \quad P[(i+2|i+2)|bridges] \cong 0.202. \tag{12.154}$$

Resultant probabilities of (12.152) and (12.154) then define the communication channel for the indirect π-interactions between the *meta*-carbons in benzene ring shown in Fig. 12.12. Its average conditional entropy (communication noise), which measures the implicit IT-covalency of the π-bond order between the *meta*-carbons

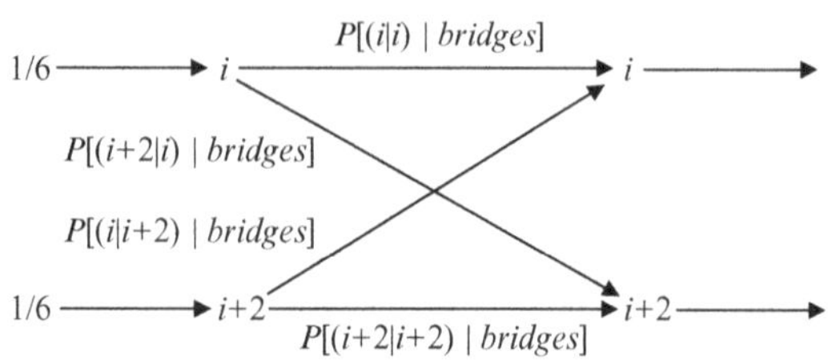

Fig. 12.12 Indirect (through-bridges) communication system for the indirect π-interactions between two *meta*-carbons in the benzene ring

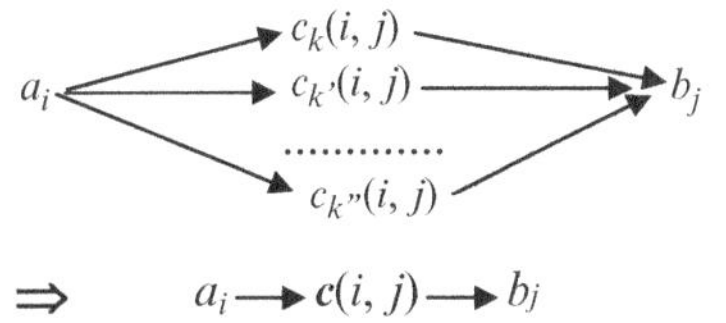

Fig. 12.13 The intermediate probability scattering between the specified AO events a_i and b_j, in the channel input and output, respectively, through the single-AO intermediates $\boldsymbol{c}(i, j) = \{c_k(i, j), k \neq (i, j)\}$

in the benzene ring, realized through these three bridges: $S[(i + 2|i)|bridges] \cong 0.27$ (bits). This IT estimate is close to the associated Wiberg-type indirect bond multiplicity $\mathcal{M}_{i,i+2}(bridges) \cong 0.3$ reported in the preceding section.

It is also of interest to examine the conditional probabilities between the specified pairs of AO originating from communications through the *parallel* set of the *single*-AO bridges containing all the remaining basis functions (Nalewajski 2011b). The indirect probability scattering through the remaining basis functions $\boldsymbol{\chi}' = \{\chi_{k \neq (i,j)}\}$, which constitute the effective (parallel) bridge for the specified AO pair χ_i (input) and χ_j (output), can be then determined as conditional probabilities of the underlying *information cascade* of Fig. 12.13, in which the signal emitted at the input $a_i = \chi_i$ is propagated into the specified output $b_j = \chi_j$ through all admissible single orbital bridges including all remaining AO: $\boldsymbol{c}(i, j) = \{c_k(i, j) = \chi_{k \neq (i,j)}\}$. The associated conditional probability for such a through-bridges propagation reads:

$$\begin{aligned} P[(j|i)|\boldsymbol{c}(i,j)] &= \sum_{k \neq (i,j)} P(k|i)P(j|k) = \sum_l P(l|i)P(j|l) - P(j|i)[P(i|i) + P(j|j)] \\ &= P^2(j|i) - \frac{1}{2}P(j|i)[\gamma_{i,i} + \gamma_{j,j}] \equiv P^2(j|i) - P(j|i)\gamma^{av.}(i,j), \end{aligned} \tag{12.155}$$

where the square of the AO conditional probabilities,

$$\begin{aligned} \mathbf{P}^2(\boldsymbol{b}|\boldsymbol{a}) \equiv \Big\{ P^2(j|i) &\equiv [\mathbf{P}^2(\boldsymbol{\chi}|\boldsymbol{\chi})]_{i,j} \equiv [\mathbf{P}(\boldsymbol{c}|\boldsymbol{a})\mathbf{P}(\boldsymbol{b}|\boldsymbol{c})]_{i,j} \\ &= \sum_l P(l|i)P(j|l) = \sum_l \frac{\gamma_{i,l}^2 \gamma_{j,l}^2}{4\gamma_{i,i}\gamma_{l,l}} \Big\}, \end{aligned} \tag{12.156}$$

characterizes the *complete* sequential probability cascade (Fig. 12.14), involving *all* orbital intermediaries. It follows from (12.155) that the intermediate probability propagation via the full "bridge" consisting of all the remaining AO, $\chi_i \rightarrow \boldsymbol{c}(i,j) \rightarrow \chi_j$, is determined by the corresponding "squared" probability $P^2(j|i)$ of the sequential AO cascade corrected by the product of the direct scattering probability $P(j|i)$ and the average occupation $\gamma^{av.}(i,j)$ of the specified input and output AO in the molecule.

It follows from (12.109) that the single AO χ_k in this sequential approach gives:

$$P[(j|i)|k] = P(k|i)P(j|k) = \frac{\gamma_{i,k}^2\,\gamma_{j,k}^2}{4\gamma_{i,i}\,\gamma_{k,k}}, \quad \Sigma_j P[(j|i)|k] = P(k|i). \tag{12.157}$$

For the *parallel* two AO intermediaries of Fig. 12.15a, one similarly finds:

$$P[(j|i)|(k,l)] = P(k|i)P(j|k) + P[(l|i)P(j|l) = \frac{1}{4\gamma_{i,i}}\left(\frac{\gamma_{i,k}^2\,\gamma_{j,k}^2}{\gamma_{k,k}} + \frac{\gamma_{i,l}^2\,\gamma_{j,l}^2}{\gamma_{l,l}}\right), \tag{12.158}$$

while the *sequential two*-AO bridge of Fig. 12.15b gives:

$$P[(j|i)|(k \to l)] = P(k|i)P(l|k)P(j|l) = \frac{\gamma_{i,k}^2\,\gamma_{l,k}^2\gamma_{j,l}^2}{8\gamma_{i,i}\gamma_{k,k}\gamma_{l,l}}. \tag{12.159}$$

The additional degrees of freedom of Fig. 12.13, for the indirect communications between different π-AO through the single member of the AO subset $\boldsymbol{c}(i, j)$ combining all remaining basis functions, are then given by the associated probabilities of (12.157). All these intermediate π-AO communications in benzene are illustrated in Fig. 12.16.

Fig. 12.14 Effective sequential probability cascade of two information systems in AO resolution

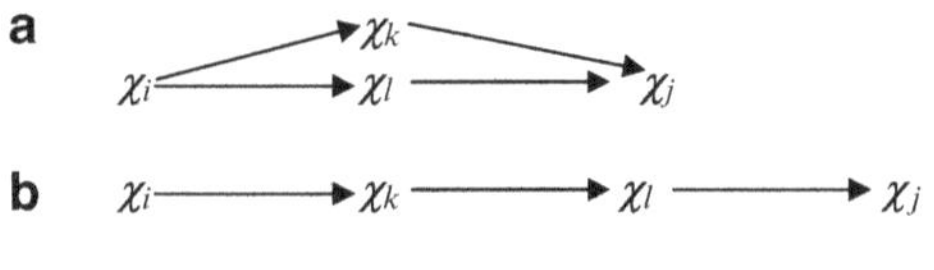

Fig. 12.15 The intermediate probability scattering from χ_i to χ_j *via* the *two*-AO bridges $\{\chi_k, \chi_l\}$: parallel (Panel *a*) and sequential (Panel *b*)

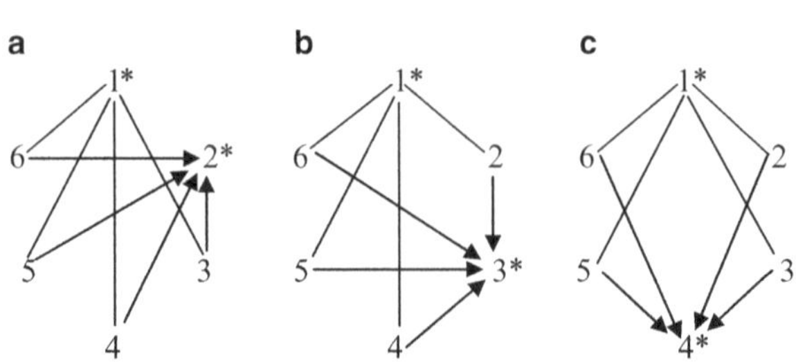

Fig. 12.16 Indirect communications between the *terminal* (identified by an *asterisk*) *ortho* (Panel *a*), *meta* (Panel *b*) and *para* (Panel *c*) carbons in benzene, through the single π–AO intermediate

For benzene, one finds,

$$P[(i+1|i)|\boldsymbol{c}(i,i+1)] = P[(i+3|i)|\boldsymbol{c}(i,i+3)] = 0,$$
$$P[(i+2|i)|\boldsymbol{c}(i,i+2)] = 2/27, \quad (12.160)$$

while these implicit communications in butadiene read:

$$P[(2|1)|(3,4)] = P[(4|3)|(1,2)] = P[(4|1)|(2,3)] = P[(3|2)|(1,4)] = 0,$$
$$P[(3|1)|(2,4)] = P[(4|2)|(1,3)] = 2/25, \quad (12.161)$$

Therefore, these results for the complete bridge of the sequential probability propagation between two π-AO through the *single* orbital of the set combining the remaining basis functions predicts the vanishing *single*-AO bridge communications between the two *ortho*- and *para*-carbons, since all combined communications of Fig. 12.16a, c involve at least one direct (*meta*) step in the signal bridge propagation, which exactly vanishes in the Hückel approximation. Only the *meta*-carbons exhibit nonvanishing indirect probability scatterings of Fig. 12.16b, since these bridge communications involve the direct *ortho* and *para* steps. A similar trend is observed for butadiene, with only the $1 \rightarrow 3$ and $2 \rightarrow 4$ communications, which exhibit the vanishing *direct* component, now acquiring the nonvanishing *indirect* communication links.

The additional, indirect IT-covalency (in bits) between two *meta*-carbons in benzene is thus reflected by the associated conditional entropy (noise) descriptor:

$$\mathscr{S}_{i,i+2}[\boldsymbol{c}(i,i+2)] = -(2/27)\log_2(2/27) = 0.28. \quad (12.162)$$

The indirect entropic covalency of the 1–3 and 2–4 IT bond orders in butadiene similarly reads:

$$\mathscr{S}_{1,3}(2,4) = \mathscr{S}_{2,4}(1,3) = -(2/25)\log_2(2/25) = 0.29. \quad (12.163)$$

These implicit (total) bond IT-covalencies can be compared with the direct (total) entropies of the remaining two-orbital interactions in benzene,

$$\mathscr{S}_{i,i+1} = -(2/9)\log_2(2/9) = 0.48,$$
$$\mathscr{S}_{i,i+3} = -(1/18)\log_2(1/18) = 0.23, \quad (12.164)$$

and in butadiene:

$$\mathscr{S}_{1,2} = \mathscr{S}_{3,4} = -(2/5)\log_2(2/5) = 0.53,$$
$$\mathscr{S}_{1,4} = \mathscr{S}_{2,3} = -(1/10)\log_2(1/10) = 0.33. \quad (12.165)$$

Therefore, this perspective combining the entropy-covalencies due to the direct AO communications and indirect probability propagations via the *single*-AO bridges of all remaining basis functions gives even more dichotomous distinction of diatomic π-interactions in these two molecules compared with that resulting from all admissible bridges in the AO information system. The bridge contributions now correct only the atomic pairs, which do not interact directly: in benzene $\mathscr{S}_{i,i+2}[\boldsymbol{c}(i, i+2)] \cong 0.3$ bits and in butadiene $\mathscr{S}_{1,3}(2,4) = \mathscr{S}_{2,4}(1,3) \cong 0.3$ bits. One again observes that this *indirect* correction is of the order of the weaker *direct* bonds in these two prototype molecules.

12.10.6 Qualitative Model of Bonds in Propellanes

Let us again interpret the patterns of chemical bonds in the representative [1.1.1] and [2.2.2] propellanes (Nalewajski 2011b). As already remarked in Sect. 10.3, in the minimum basis set approach the bond structure in these two systems can be qualitatively understood in terms of the localized MO resulting from interactions between the *directed* orbitals on neighboring atoms and the nonbonding electrons occupying nonoverlapping atomic hybrids.

In the smallest [1.1.1] system, the nearly tetrahedral ($h = sp^3$) hybridization on both bridgehead and bridging carbons is required to form the chemical bonds of the three carbon bridges and to accommodate two hydrogens on each bridge carbons. Thus, three sp^3 hybrids on each of the bridgehead atoms are used to form the chemical bonds with the bridge carbons and the fourth hybrid, directed away from the central bond region, remains nonbonding and *singly* occupied (Fig. 12.17a).

In the largest [2.2.2] propellane the two central carbons acquire a nearly trigonal ($h' = sp^2$) hybridization to form bonds with the bridge neighbors, each with a single $2p$ orbital directed along the central bond axis, which has not been used in this hybridization scheme, now being available to form a strong through-space component of the overall multiplicity of the C′—C′ bond (see Figs. 10.4 and 12.17b). This explains the missing strong through-space component in the (diradical) smaller [1.1.1] propellane and its presence in the larger [2.2.2] system.

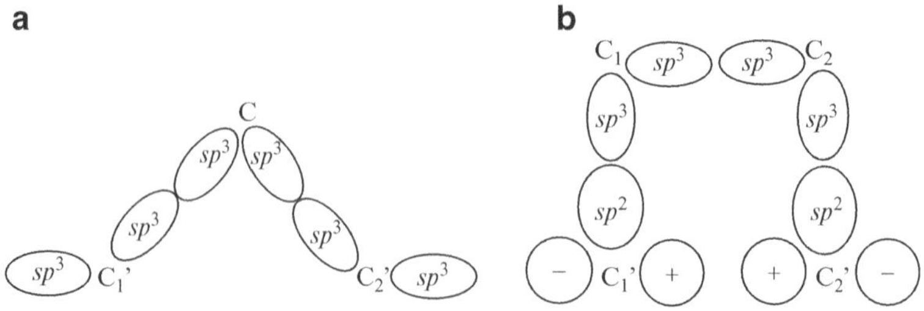

Fig. 12.17 Schematic diagrams rationalizing the patterns of the localized bonds in [1.1.1] (Panel *a*) and [2.2.2] (Panel *b*) propellanes; the bridgehead carbon atoms are primed

In this qualitative picture each directed AO participates in a single localized, *two*-center (*doubly* occupied) bonding MO, which allows one to estimate the diatomic CBO matrix elements determining the direct and indirect components of the central bonds in these two propellanes:

$$\gamma_{h,h} = \left(1 + S^{\sigma}_{p,p}\right)^{-1}, \quad \gamma_{h,h} = (1 + S_{h,h})^{-1} \quad \text{and} \quad \gamma_{h,h'} = \left(1 + S_{h,h'}\right)^{-1},$$

where $S^{\sigma}_{p,p}$, $S_{h,h}$ and $S_{h,h'}$ stand for the overlap integrals between two $2p_{\sigma}$ orbitals and between the indicated hybrid orbitals, respectively. These overlaps can be realistically estimated using the standard overlap integrals between valence orbitals on carbon atoms in ethane (single C—C bond): $S_{s,s} = 0.36$, $S^{\sigma}_{s,p} = 0.42$, $S^{\sigma}_{p,p} = 0.28$, giving rise to the associated standard overlaps between hybrid AO: $S^{\sigma}_{h,h} = 0.66$, $S^{\sigma}_{h,h'} = 0.67$.

Hence, the estimate of the *direct*, Wiberg component of the central bond in the [2.2.2] system,

$$\mathcal{M}_{1',2'} \approx \left(1 + S^{\sigma}_{p,p}\right)^{-2} = 0.62,$$

and the *indirect* contribution due to three (*double*-carbon) bridges,

$$\mathcal{M}_{1',2'}(bridges) = 3\mathcal{M}_{C'_1,C_1}\mathcal{M}_{C_1,C_2}\mathcal{M}_{C_2,C'_2} \approx 3\left(1 + S^{\sigma}_{h,h'}\right)^{-4}\left(1 + S^{\sigma}_{h,h}\right)^{-2} = 0.14,$$

which give rise to the *total* bond multiplicity:

$$\mathcal{M}(1' — 2') = \mathcal{M}_{1',2'} + \mathcal{M}_{1',2'}(bridges) \approx 0.76.$$

The corresponding *indirect* (*total*) component for the [1.1.1] system gives:

$$\mathcal{M}_{1',2'}(bridges) = 3\mathcal{M}_{C'_1,C}\mathcal{M}_{C,C'_2} \approx 3\left(1 + S^{\sigma}_{h,h}\right)^{-4} = 0.40 \cong \mathcal{M}(1' — 2').$$

Therefore, the smaller system is predicted to exhibit higher through-bridge component, compared with larger propellane, with the latter generating greater overall bond order. This trend is also reflected by numerical SCF and DFT calculations (see Sect. 10.3).

Finally, let us briefly examine the associated conditional entropy (communication noise) contributions, reflecting the associated IT bond orders due to through-bridge covalencies. The conditional probabilities of the information scattering via a single bridge in the [1.1.1] and [2.2.2] propellanes, respectively, read:

$$P[(C_2'|C_1')|C] \approx \frac{1}{4}(\gamma_{h,h})^4 = 0.0333 \quad \text{and}$$

$$P[(C_2'|C_1')|C_1 \to C_2] \approx \frac{1}{8}(\gamma_{h,h})^2(\gamma_{h,h'})^4 = 0.0058.$$

They again reflect a higher through-bridge propagation of the electron probability in the single-carbon bridge. These probabilities generate the associated entropies due to three identical (parallel) bridges, which measure the bridge IT-covalencies (in bits) of the central bond in these two molecular systems:

$$\mathcal{S}_{1',2'}(bridges) \approx 3(-0.0333\ \log_2 0.0333) = 0.49 \quad \text{and}$$
$$\mathcal{S}_{1',2'}(bridges) \approx 3(-0.0058\ \log_2 0.0058) = 0.13.$$

These entropies are seen to compare favorably with the corresponding Wiberg estimates for this model, reported at the beginning of this section.

We again conclude that the entropic and Wiberg measures of the through-bridge component of the central bond covalency in this simple model of the electronic structure in the representative [1.1.1] and [2.2.2] propellanes are in general agreement with one another thus providing consistent insights into the novel indirect bond components in these two prototype molecular systems.

12.11 Amplitude Channels and Interference of Orbital Communications

The atomic orbitals contributed by bonded atoms of molecular systems emit or receive "signals" of electronic allocations to basis functions thus acting as both the signal "*source*" (input) and "*receiver*" (output), respectively, in the associated communication network. Each orbital simultaneously participates in both the through-space and through-bridge probability propagations: the former involve direct communications between two AO, while the latter are realized indirectly via the orbital intermediates. For the internal consistency of OCT it is vital that the stationary, ground state of electrons and the AO probabilities it implies are both recovered from the general, multiple probability scatterings between these basis functions in the molecule (Nalewajski 2011c; Nalewajski and Gurdek 2011a).

In OCT the chemical bonds originate from molecular communications between AO events determining the channel input and output of Fig. 12.18a, in accordance with the associated conditional probability matrix $\mathbf{P}(\boldsymbol{b}|\boldsymbol{a}) = \mathbf{P}(\boldsymbol{\chi}|\boldsymbol{\chi}) = \{P(j|i)\}$. The quantum amplitudes $\mathbf{A}(\boldsymbol{b}|\boldsymbol{a}) = \{A(j|i) \equiv A_{i\to j}\} \equiv \mathbf{A}_{a\to b}$ of such conditional probabilities, $\{P(j|i) \equiv |A_{i\to j}|^2\}$, which define the associated *amplitude channel* of Fig. 12.18b, have been shown to be proportional to the corresponding elements of the system CBO (density) matrix $\boldsymbol{\gamma}$ (Nalewajski 2010c, 2011c). For the positive

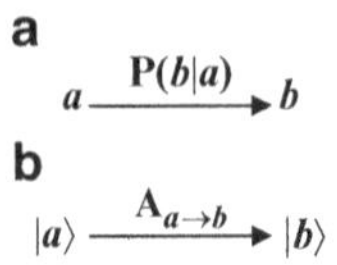

Fig. 12.18 The orbital channels in the probability (Panel *a*) and amplitude (Panel *b*) representations

overlap between AO the "constructive" (bonding) interference between two AO implies the positive (*in phase*) product of their bond projections, while its negative (*out of phase*) value identifies their resultant "destructive" interference in the molecular bond system.

The probability and amplitude channels of Fig. 12.18 thus summarize communications between AO events, $\boldsymbol{a} \to \boldsymbol{b}$, and corresponding state vectors, $|\boldsymbol{a}\rangle \to |\boldsymbol{b}\rangle$, respectively. It should be emphasized that only the information scattering *states*, defined by the communication amplitudes, are capable of "interference" effects. One should also envisage the generalized (*consecutive*) bridges (Fig. 12.19) for the AO probability propagation in molecules, which involve the (*parallel*) set of all basis functions at each scattering step. Each propagation stage then involves all AO inputs communicating with all AO outputs.

We recall that in the Born (statistical) interpretation of quantum mechanics the state probability distribution is given by the squared modulus of the corresponding (complex-valued) amplitude, the system wave function. This wave function "power" is then interpreted as the probability density over the representation elementary events. It is the superposition of the wave functions (quantum states) that gives rise to the interference of *micro*objects. Therefore, it is both natural and indeed compulsory that in the reconstruction of the underlying AO communications behind the stationary (ground state) probability distribution via the multiple probability propagations between the basis functions one must insist on the wave character, capable of the interference, of such elementary molecular communications.

Indeed, the classical combination rules for conditional probabilities determining the molecular communications between AO [see, e.g., (12.156) and Fig. 12.14] would fail to reconstruct this molecular probability distribution and the underlying wave function (probability amplitude) of the system as a whole. Instead, one must combine the AO probability amplitudes in order to recover the stationary conditional probabilities. The amplitudes of such elementary steps of the information propagations between AO have been shown to be proportional to the corresponding elements of the CBO matrix, which exhibit both positive and negative values, thus being perfectly capable of both the constructive and destructive interference in the relevant scattering states.

12.11.1 Probability Scattering States and Stationary Communication Modes

The 1-density matrix $\mathbf{D} = \langle\boldsymbol{\chi}|\boldsymbol{\varphi}^o\rangle\langle\boldsymbol{\varphi}^o|\boldsymbol{\chi}\rangle = \mathbf{C}^o\mathbf{C}^{o\dagger} = \langle\boldsymbol{\chi}^b|\boldsymbol{\chi}^b\rangle = \mathbf{S}^b$, which measures the bond overlaps $\mathbf{S}^b$ between AO $|\boldsymbol{\chi}\rangle$, constitutes the AO representation of the projection operator $\hat{\mathrm{P}}^o_{\varphi}$ onto the bond subspace $\boldsymbol{\varphi}^o$ of the (doubly) occupied MO, thus satisfying the associated idempotency relations,

$$\mathbf{D}^2 = \langle \boldsymbol{\chi}^b | \boldsymbol{\chi}^b \rangle \langle \boldsymbol{\chi}^b | \boldsymbol{\chi}^b \rangle = \langle \boldsymbol{\chi}^b | \boldsymbol{\chi}^b \rangle = \mathbf{D} \quad \text{and} \quad \mathbf{D}^n = \mathbf{D}. \tag{12.166}$$

The quantum mechanical amplitudes $\mathbf{A}(\boldsymbol{b}|\boldsymbol{a}) \equiv \mathbf{A}_{\boldsymbol{a}\to\boldsymbol{b}} = \{A(j|i) \equiv A_{i\to j}\}$ of the AO conditional probabilities (12.109),

$$\mathbf{P}(\boldsymbol{b}|\boldsymbol{a}) = \left\{ P(j|i) = (\bar{N}_{i\to j})^2 \gamma_{i,j}\gamma_{j,i} = (2\gamma_{i,i})^{-1} |\gamma_{i,j}|^2 \equiv A_{i\to j} A^*_{i\to j} = |A_{i\to j}|^2 \right\}, \tag{12.167}$$

are seen to be determined by the corresponding (renormalized) elements of the density matrix,

$$A_{i\to j} = \gamma_{i,j} / \sqrt{2\gamma_{i,i}} \equiv \gamma_{i,j} \bar{N}_{i\to j} = D_{i,j} / \sqrt{D_{i,i}} \equiv D_{i,j} N_{i\to j}, \tag{12.168}$$

and hence $|A_{i\to j}| \leq 1$ since $|\langle i_b | j_b \rangle| \leq \langle i_b | i_b \rangle$.

The (*non*-Hermitian) matrix $\mathbf{A}_{\boldsymbol{a}\to\boldsymbol{b}} = \{A_{i\to j}\}$ of all "forward" (*input* $\to$ *output*) communications then determines the AO scattering states of the amplitude channel of Fig. 12.18b. It can be viewed as the AO representation of the associated communication operator $\hat{\mathrm{A}} \equiv \hat{\mathrm{A}}(\boldsymbol{a} \to \boldsymbol{b}) : \mathbf{A}_{\boldsymbol{a}\to\boldsymbol{b}} \equiv \langle \boldsymbol{b} | \hat{\mathrm{A}} | \boldsymbol{a} \rangle$. Therefore, the scattering of the input states in the forward channel reads:

$$\hat{\mathrm{A}} \, |\boldsymbol{a}\rangle = A_{\boldsymbol{a}\to\boldsymbol{b}} |\boldsymbol{a}\rangle. \tag{12.169}$$

The Hermitian conjugate matrix $(\mathbf{A}_{\boldsymbol{a}\to\boldsymbol{b}})^\dagger = \langle \boldsymbol{a} | \hat{\mathrm{A}}^\dagger | \boldsymbol{b} \rangle \equiv \mathbf{A}_{\boldsymbol{b}\to\boldsymbol{a}} = \{A_{j\to i}\}$ then combines the AO amplitudes of the "reverse" (*output* $\to$ *input*) probability propagations, with the conjugate operator $\hat{\mathrm{A}}^\dagger \equiv \hat{\mathrm{A}}(\boldsymbol{b} \to \boldsymbol{a})$ standing for the scattering of the output states in the reverse channel:

$$\hat{\mathrm{A}}^\dagger |\boldsymbol{b}\rangle = \mathbf{A}_{\boldsymbol{b}\to\boldsymbol{a}} |\boldsymbol{b}\rangle. \tag{12.170}$$

These *directed*-scattering operators can be combined into the symmetrized (Hermitian) *communication operator* of AO amplitudes:

$$\hat{\mathrm{B}} = \tfrac{1}{2}(\hat{\mathrm{A}} + \hat{\mathrm{A}}^\dagger) = \tfrac{1}{2}[\hat{\mathrm{A}}(\boldsymbol{a} \to \boldsymbol{b}) + \hat{\mathrm{A}}(\boldsymbol{b} \to \boldsymbol{a})] \quad \text{or}$$
$$\mathbf{B} = \langle \boldsymbol{\chi} | \hat{\mathrm{B}} | \boldsymbol{\chi} \rangle = \tfrac{1}{2}(\mathbf{A}_{\boldsymbol{a}\to\boldsymbol{b}} + \mathbf{A}_{\boldsymbol{b}\to\boldsymbol{a}}) = \tfrac{1}{2}\left(\mathbf{A}_{\boldsymbol{a}\to\boldsymbol{b}} + \mathbf{A}_{\boldsymbol{a}\to\boldsymbol{b}}{}^\dagger\right). \tag{12.171}$$

Its eigenvalue (diagonalization) problem then determines the decoupled (principal, normal) modes $|\boldsymbol{\psi}\rangle = |\boldsymbol{\chi}\rangle \mathbf{G}$ of the information propagation between AO:

$$\hat{\mathrm{B}} |\boldsymbol{\psi}\rangle = \boldsymbol{\lambda} |\boldsymbol{\psi}\rangle \quad \text{or} \quad \mathbf{G}^\dagger \mathbf{B} \, \mathbf{G} = \boldsymbol{\lambda};$$
$$\mathbf{G} = \langle \boldsymbol{\chi} | \boldsymbol{\psi} \rangle, \quad |\boldsymbol{\psi}\rangle = \{|\psi_\alpha\rangle\}, \quad \langle \boldsymbol{\psi} | \hat{\mathrm{B}} | \boldsymbol{\psi} \rangle = \boldsymbol{\lambda} = \{\lambda_\alpha \delta_{\alpha,\beta}\}. \tag{12.172}$$

Here, the stationary states $\{|\psi_\alpha\rangle\}$ combine the forward and reverse probability scatterings into the corresponding (delocalized) "standing waves" of AO communications in molecules, with the eigenvalues $\{\lambda_\alpha\}$ providing the *principal* propagation amplitudes of the associated (diagonal) spectral resolution of the AO communication operator:

$$\hat{\mathrm{B}} = |\boldsymbol{\psi}\rangle\boldsymbol{\lambda}\langle\boldsymbol{\psi}| = \Sigma_\alpha|\psi_\alpha\rangle\lambda_\alpha\langle\psi_\alpha|. \tag{12.173}$$

The conditional probability amplitude of (12.168) determines the associated *forward*-scattered state,

$$|i \to j\rangle = |i\rangle + A_{i\to j}|j\rangle, \tag{12.174}$$

for the unit probability of the incident input state $|i\rangle$. Accordingly, the *reverse*-scattered state originating from the output state $|j\rangle$ reads:

$$|j \to i\rangle = |j\rangle + A_{j\to i}|i\rangle. \tag{12.175}$$

These equations identify the forward and reverse propagation amplitudes as the scalar products of the corresponding scattered states and the relevant final ("detection") AO state:

$$A_{i\to j} = \langle j|i \to j\rangle \quad \text{and} \quad A_{j\to i} = \langle i|j \to i\rangle. \tag{12.176}$$

In this vector interpretation the elementary conditional probabilities $P(j|i) = P_{i\to j}$ and $P(i|j) = P_{j\to i}$ of the specified forward and reverse conditional events represent the expectation value in the detection state of the corresponding projection operators onto the scattering state:

$$\begin{aligned} P(j|i) &= |A_{i\to j}|^2 = \langle j|i \to j\rangle\langle i \to j|j\rangle \equiv \langle j|\hat{\mathrm{P}}_{i\to j}|j\rangle, \\ P(i|j) &= |A_{j\to i}|^2 = \langle i|j \to i\rangle\langle j \to i|i\rangle \equiv \langle i|\hat{\mathrm{P}}_{j\to i}|i\rangle. \end{aligned} \tag{12.177a}$$

Alternatively, these probabilities are seen to determine the expectation value in the relevant scattered state of the projection operator onto the detection AO state in question:

$$\begin{aligned} P(j|i) &= |A_{i\to j}|^2 = \langle i \to j|j\rangle\langle j|i \to j\rangle \equiv \langle i \to j|\hat{\mathrm{P}}_j|i \to j\rangle, \\ P(i|j) &= |A_{j\to i}|^2 = \langle j \to i|i\rangle\langle i|j \to i\rangle \equiv \langle j \to i|\hat{\mathrm{P}}_i|j \to i\rangle. \end{aligned} \tag{12.177b}$$

12.11.2 Cascade Probability Scatterings and Their Interference

The conditional probabilities of (12.167) and the associated amplitudes of (12.168) describe the delocalized electrons in the molecular bond system and originate from the standard (Slater determinant) wave function of the molecule in the adopted MO

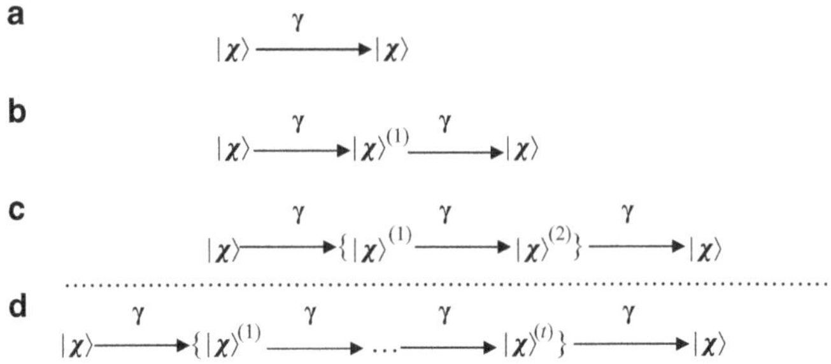

Fig. 12.19 Hierarchy of the amplitude channels for the multiple scatterings between AO states $|\boldsymbol{\chi}\rangle$: direct communications (Panel *a*); *single*(*S*)-AO bridges $|\boldsymbol{\chi}\rangle_{(1)} \equiv \boldsymbol{\chi}^{(1)}$ (Panel *b*); *double*(*D*)-AO bridges $|\boldsymbol{\chi}\rangle_{(2)} \equiv \{|\boldsymbol{\chi}\rangle^{(1)} \to |\boldsymbol{\chi}\rangle^{(2)}\}$ (Panel *c*); *t*-AO bridges $|\boldsymbol{\chi}\rangle_{(t)} \equiv \{|\boldsymbol{\chi}\rangle^{(1)} \to |\boldsymbol{\chi}\rangle^{(2)} \to \ldots \to |\boldsymbol{\chi}\rangle^{(t)}\}$ (Panel *d*). The resultant amplitudes of such multistep propagations are proportional to consecutive powers of the CBO (density) matrix $\boldsymbol{\gamma}$: $\mathbf{A}_{(t)} = \{A^{(t)}_{i \to j}\} \propto \boldsymbol{\gamma}^{t+1}$

approximation. This stationary distribution must effectively combine contributions from both the direct and generalized indirect probability propagations between AO. It is vital for the internal consistency of OCT to demonstrate that an inclusion of these generalized, *multiple* through-bridge communications preserves this stationary conditional probability distribution.

Consider the generalized through-bridge (forward) probability propagations $\{i \to j\}$ of Fig. 12.19. Each output in the *direct* information network of Panel *a* can subsequently emit the signal to any of the system basis functions in the sequential cascade of Panel *b*, in which the events of the secondary output are linked to the original input *via* all *single*(*S*)-AO intermediates $\boldsymbol{\chi}^{(1)}$. This intermediate scattering can be repeated still further, e.g., by adding the *double*(*D*)-AO bridges $\boldsymbol{\chi}_{(2)} \equiv \{\boldsymbol{\chi}^{(1)} \to \boldsymbol{\chi}^{(2)}\}$, additionally including the AO intermediates $\boldsymbol{\chi}^{(2)}$ in the sequential cascade of Panel *c*, or in general *t*-AO bridges $\boldsymbol{\chi}_{(t)} \equiv \{\boldsymbol{\chi}^{(1)} \to \boldsymbol{\chi}^{(2)} \to \ldots \to \boldsymbol{\chi}^{(t)}\}$ shown in Panel *d* of the figure. In principle this order of the intermediate probability propagation can be extended to infinity, $t \to \infty$, with the combined effect of all propagation orders $t = 0, 1, 2, \ldots$, where $t = 0$ corresponds to the direct molecular communications, eventually establishing the stationary molecular communications between AO. In this multiscattering perspective the stationary distribution of the electronic probabilities in molecular systems is thus seen as the net result of all such elementary (multiple) probability propagations, in which the communications $i \to j$ between two specified basis functions effectively involve bridges of all orders.

In the *classical* sequential cascade of several information channels the resultant conditional probability matrix between the initial input and the final output is given by the product of conditional probabilities of all constituent subchannels (see, e.g., Fig. 12.14). As we have argued above, this classical combination rule would fail to reconstruct the molecular distributions of electrons, the quantum particles capable of interference, since such products do not conserve the elementary molecular probabilities of the direct AO communications. For such microobjects one has to propagate the probability amplitudes at each bridge order and then determine the effective amplitude due to interference of amplitudes from all orders $\{t\}$. In determining the overall effect of the superposition of the scattering states corresponding to all bridge orders one thus has to add the resultant amplitudes for all orders.

The bridge generalization of the probability scattering amplitude, *via* several sequential sets of all AO used in SCF MO calculations (see Fig. 12.19), involves the corresponding powers of the projection operator onto the bonding subspace. For example, the amplitude for the communication $i \to j$ through the bridge order t (see Fig. 12.19d) reads:

$$\begin{aligned} A_{i\to j}^{(t)} &= \bar{N}_{i\to j}^{(t)}(\langle i|\hat{\mathrm{P}}_{\varphi}^{o})(\hat{\mathrm{P}}_{\varphi}^{o})^{t}(\hat{\mathrm{P}}_{\varphi}^{o}|j\rangle) = [\bar{N}_{i\to j}^{(t)}/2^{t+1}](\gamma^{t+1})_{i,j} \\ &= \bar{N}_{i\to j}^{(t)}\langle i^{b}|(\hat{\mathrm{P}}_{\varphi}^{o})^{t}|j^{b}\rangle \equiv \bar{N}_{i\to j}\langle i^{b}|j^{b}\rangle = A_{i\to j}, \end{aligned} \tag{12.178}$$

since by the idempotency of the bond projection $(\hat{\mathrm{P}}_{\varphi}^{o})^{m} = \hat{\mathrm{P}}_{\varphi}^{o}$ and $\hat{\mathrm{P}}_{\varphi}^{o}|j^{b}\rangle = |j^{b}\rangle$, $\hat{\mathrm{P}}_{\varphi}^{o}|i^{b}\rangle = |i^{b}\rangle$. Thus, at each order of the through-bridge scattering involving the entire set of AO the amplitude for the representative propagation $i \to j$ stays the same as in the direct propagation.

Therefore, the resultant (*res.*) amplitude from the interference of all bridge orders, the normalized linear combination of amplitudes generated at each bridge order, is identical with the direct amplitude, which generates the stationary conditional probability distribution in the molecule:

$$A_{i\to j}^{res.} = \sum_{m=0}^{\infty} C^{(m)} A_{i\to j}^{(m)} = A_{i\to j}, \quad \sum_{m=0}^{\infty} \left|C^{(m)}\right|^{2} = 1. \tag{12.179}$$

Hence, the square of the modulus of the amplitudes thus obtained, reflecting the net effect of the amplitude superposition at all bridge orders, generates the stationary molecular probability. Indeed, in accordance with the quantum superposition principle any combination of the state with itself represents the same state of the system.

We thus conclude, that the interference of the quantum amplitudes for the generalized conditional probabilities of AO, resulting from the multiple through-bridge scatterings at each stage involving *all* basis functions, do not alter the stationary communications resulting from the direct information scattering in the molecular ground state. This explicitly demonstrates the internal consistency of OCT as the *quantum* theory of the molecular electronic structure.

12.11.3 Implicit Dependency Origins of Through-Bridge Interactions

The through-bridge mechanism was conjectured to result from the *implicit* dependencies between the (nonorthogonal) AO projections into the bonding subspace of the occupied MO (Nalewajski and Broniatowska 2003a; Nalewajski

2006g), as reflected by the bond overlap (1-density) matrix $\mathbf{S}^b = \mathbf{D}$ (12.88) [see (12.166)]. These AO components reflect the resultant participation of the basis functions in the whole system of chemical bonds. It is the main purpose of this section to explore the CBO framework of these indirect dependencies in a more detail and to demonstrate that the novel through-bridge mechanism represents a natural extension of the direct dependencies already manifested in the through-space bond mechanism (Nalewajski and Gurdek 2011a).

It follows from (12.86) that the bond projections $|\boldsymbol{\chi}^b\rangle$ of basis functions can be also expressed in terms of the basis functions $|\boldsymbol{\chi}\rangle$,

$$|\boldsymbol{\chi}^b\rangle = |\boldsymbol{\varphi}^o\rangle \mathbf{C}^{o\dagger} = |\boldsymbol{\chi}\rangle(\mathbf{C}^o\mathbf{C}^{o\dagger}) = |\boldsymbol{\chi}\rangle\mathbf{D}, \tag{12.180}$$

and hence the bond-overlap (density) matrix $\mathbf{S}^b = \mathbf{D} = \{D_{i,j}\}$ can be interpreted as matrix combining the derivatives

$$\mathbf{D} = \frac{\partial \boldsymbol{\chi}^b}{\partial \boldsymbol{\chi}} = \left\{ D_{i,j} = \langle i^b | j^b \rangle = \left(\frac{\partial \chi_j^b}{\partial \chi_i} \right) \right\}. \tag{12.181}$$

Moreover, using the idempotency relation of (12.166) allows one to interpret (12.180) as the linear transformation of the AO projections themselves:

$$|\boldsymbol{\chi}^b\rangle = (|\boldsymbol{\chi}\rangle\mathbf{D})\mathbf{D} = |\boldsymbol{\chi}^b\rangle\mathbf{D}. \tag{12.182}$$

Therefore, the partial derivatives of (12.181) can be also interpreted as reflecting the linear dependencies between the bond projections of the basis functions:

$$\mathbf{D} = \frac{\partial \boldsymbol{\chi}^b}{\partial \boldsymbol{\chi}^b} = \left\{ D_{i,j} = \left(\frac{\partial \chi_j^b}{\partial \chi_i^b} \right) \right\}. \tag{12.183}$$

This allows one to interpret the idempotency relation of the density matrix as the following chain rule identity:

$$(\mathbf{D}^2)_{i,j} = \sum_k \left(\frac{\partial \chi_k^b}{\partial \chi_i^b} \right) \left(\frac{\partial \chi_j^b}{\partial \chi_k^b} \right) = \left(\frac{\partial \chi_j^b}{\partial \chi_i^b} \right) = D_{i,j}. \tag{12.184}$$

In fact, using the resolution of the identity operator combining the complementary projections onto the mutually orthogonal subspaces of the occupied and virtual MO,

$$1 = |\boldsymbol{\varphi}^o\rangle\langle\boldsymbol{\varphi}^o| + |\boldsymbol{\varphi}^v\rangle\langle\boldsymbol{\varphi}^v| \equiv \hat{\mathrm{P}}_{\varphi}^o + \hat{\mathrm{P}}_{\varphi}^v, \quad \hat{\mathrm{P}}_{\varphi}^o \hat{\mathrm{P}}_{\varphi}^v = 0, \tag{12.185}$$

gives the associated resolution of the basis functions:

$$|\boldsymbol{\chi}\rangle = \hat{\mathrm{P}}^{o}_{\varphi}|\boldsymbol{\chi}\rangle + \hat{\mathrm{P}}^{v}_{\varphi}|\boldsymbol{\chi}\rangle = |\boldsymbol{\chi}^{b}\rangle + |\boldsymbol{\chi}^{v}\rangle. \tag{12.186}$$

This allows one to partition the unit matrix of the OAO-overlap into the complementary contributions originating from these two subspaces of MO:

$$\begin{aligned}\langle\boldsymbol{\chi}|\boldsymbol{\chi}\rangle &= \frac{\partial\boldsymbol{\chi}}{\partial\boldsymbol{\chi}} = \mathbf{I} = (\langle\boldsymbol{\chi}^{b}| + \langle\boldsymbol{\chi}^{v}|)\,(|\boldsymbol{\chi}^{b}\rangle + |\boldsymbol{\chi}^{v}\rangle) = \langle\boldsymbol{\chi}^{b}|\boldsymbol{\chi}^{b}\rangle + \langle\boldsymbol{\chi}^{v}|\boldsymbol{\chi}^{v}\rangle = \frac{\partial\boldsymbol{\chi}^{b}}{\partial\boldsymbol{\chi}^{b}} + \frac{\partial\boldsymbol{\chi}^{v}}{\partial\boldsymbol{\chi}^{v}}\\ &= (\langle\boldsymbol{\chi}^{b}| + \langle\boldsymbol{\chi}^{v}|)|\boldsymbol{\chi}\rangle = \frac{\partial\boldsymbol{\chi}}{\partial\boldsymbol{\chi}^{b}} + \frac{\partial\boldsymbol{\chi}}{\partial\boldsymbol{\chi}^{v}} = \langle\boldsymbol{\chi}|(|\boldsymbol{\chi}^{b}\rangle + |\boldsymbol{\chi}^{v}\rangle) = \frac{\partial\boldsymbol{\chi}^{b}}{\partial\boldsymbol{\chi}} + \frac{\partial\boldsymbol{\chi}^{v}}{\partial\boldsymbol{\chi}}.\end{aligned} \tag{12.187}$$

These subspace-overlaps thus define the partial orbital dependencies in these two subsets of MO:

$$\begin{aligned}&\left(\frac{\partial\chi_j^b}{\partial\chi_i^b}\right) = \left(\frac{\partial\chi_j^b}{\partial\chi_i}\right) = \left(\frac{\partial\chi_j}{\partial\chi_i^b}\right) \quad \text{and}\\ &\left(\frac{\partial\chi_j^v}{\partial\chi_i^v}\right) = \left(\frac{\partial\chi_j^v}{\partial\chi_i}\right) = \left(\frac{\partial\chi_j}{\partial\chi_i^v}\right).\end{aligned} \tag{12.188}$$

We recall that the density matrix also determines the conditional probabilities for the *direct* information propagation in the AO information system (12.109):

$$\begin{aligned}\mathbf{P}(\boldsymbol{b}|\boldsymbol{a}) &= \Big\{P(j|i) = (2\gamma_{i,i})^{-1}\gamma_{i,j}\gamma_{j,i} = (2\gamma_{i,i})^{-1}|\gamma_{i,j}|^2\\ &= (D_{i,i})^{-1}|D_{i,j}|^2 = |N_{i\to j}D_{i,j}|^2\\ &\equiv A(j|i)A(j|i)^* \equiv |A_{i\to j}|^2 \equiv P(i\to j)\Big\},\end{aligned} \tag{12.189}$$

where the normalization constant results from the requirement $\sum_j P(j|i) = 1$. They have been determined (Nalewajski 2009e) from the superposition principle of quantum mechanics (Dirac 1967) supplemented by the "physical" projection onto the bond subspace of the occupied MO. The preceding equation also introduces the quantum mechanical *amplitude* $A(j|i) \equiv A_{i\to j}$ associated with the conditional probability $P(j|i) \equiv P_{i\to j}$. The latter is seen to be determined by the corresponding (occupation renormalized) element of the CBO/density matrix (12.168):

$$A_{i\to j} = \gamma_{i,j}/\sqrt{2\gamma_{i,i}} = D_{i,j}/\sqrt{D_{i,i}} \equiv N_{i\to j}D_{i,j}. \tag{12.190}$$

It also follows from (12.189) that this (direct) conditional probability is related to the Wiberg bond order contribution of (12.91), $\mathcal{M}_{i,j} = (\gamma_{i,j})^2$,

$$P(i\to j) = \mathcal{M}_{i,j}(N_{i\to j})^2/4. \tag{12.191}$$

This Wiberg bond order measure between two basis functions $|i\rangle$ and $|j\rangle$ makes use of only their *explicit* dependency in the molecular bond system, reflected by the coupling CBO matrix element $\gamma_{i,j} = 2\langle i^b | j^b\rangle = 2D_{i,j}$. It neglects all the remaining constraints, embodied by other CBO matrix elements involving these two AO, which introduce the *implicit* dependencies between the two AO in question through the remaining orbitals participating in the bond subspace of MO. These indirect relations are responsible for the bridge contributions to the overall bond multiplicity between the specified pair of AO, since the mutually bonding status of two basis functions can be felt even at large distances due to their coupling to the chain of the chemically interacting AO intermediaries.

We have interpreted in (12.181), (12.183) and (12.187) the bond-overlaps as derivatives between AO projections in the bond system of the molecule, $D_{i,j} = \gamma_{i,j}/2 = \partial\chi_j^b/\partial\chi_i^b$, satisfying the associated chain (idempotency) rule of (12.184), $\sum_k D_{i,k}D_{k,j} = D_{i,j}$. Therefore, one can express a displacement in one bond-projection in terms of displacements of all basis set projections:

$$\delta\chi_k^b = \Sigma_l \delta\chi_l^b D_{l,k}. \tag{12.192}$$

In probing the bond dependencies between the given pair (i, j) of AO the Wiberg approach makes use of only the *direct* terms $D_{j,i} = D_{i,j}$ in the expansions of bond components of both these basis functions, while neglecting the *implicit* dependencies reflected by the remaining derivatives involving these two AO. The latter are taken into account only in the *indirect* bond components, due to the bond projections of all remaining orbitals.

The amplitudes of the through-bridge probability propagations can be explicitly expressed in terms of these implicit derivatives by using the chain rules of (12.184). Consider the simplest case of a single-AO bridge in the information scattering between $|i\rangle$ and $|j\rangle$ through $|k\rangle$, $i \to k \to j$, $k \neq (i, j)$, reflected by the associated conditional bridge probability

$$\begin{aligned} P(i \to j|k) = & \ |A_{(i\to j|k)}|^2 = (D_{i,k}D_{k,j})^2/(D_{i,i}D_{k,k}) = \left(D_{i,k}/\sqrt{D_{i,i}}\right)^2\left(D_{k,j}/\sqrt{D_{k,k}}\right)^2 \\ = & \ |A_{i\to k}|^2|A_{k\to j}|^2 = P(i\to k)P(k\to j). \end{aligned} \tag{12.193}$$

It is defined by the associated amplitude, the renormalized implicit derivative of χ_j^b on χ_i^b through χ_k^b,

$$A_{(i\to j|k)} = N_{(i\to j|k)}\left(\frac{\partial\chi_k^b}{\partial\chi_i^b}\right)\left(\frac{\partial\chi_j^b}{\partial\chi_k^b}\right) = N_{(i\to j|k)}D_{i,k}D_{k,j}, \tag{12.194}$$

with the normalization constant,

$$N_{(i\to j|k)} = 1/(D_{i,i}D_{k,k})^{1/2}, \tag{12.195}$$

fixed to satisfy the relevant sum rule:

$$\Sigma_j P(i \to j|k) = P(k|i) \equiv P(i \to k). \tag{12.196}$$

Indeed, the summation of the conditional probabilities over all possible final outputs $\{j\}$ in the sequential scatterings $i \to k \to \{j\}$ must reproduce the conditional probability of its first step $i \to k$.

Therefore, the amplitude for the bridge scattering $i \to (k) \to j$ is given by the product of amplitudes of the elementary *two*-AO scatterings through the bridge: $i \to k$ and $k \to j$. Since the Wiberg-type bond order $\mathcal{M}(i \to j|k)$ contribution due to this bridge scattering is proportional to the conditional probability of (12.157), it is also seen to be related to the product of the Wiberg bond orders of the associated *two*-AO propagation stages:

$$\mathcal{M}(i \to j|k) = \mathcal{M}_{i,k}\mathcal{M}_{k,j}, \quad \mathcal{M}_{i,k} = (\gamma_{i,k})^2, \quad \mathcal{M}_{k,j} = (\gamma_{k,j})^2,$$
$$P(i \to j|k) = \mathcal{M}(i \to j|k)(N_{(i\to j|k)})^2/4^2. \tag{12.197}$$

This development for the probability propagation *via* a *single*-AO bridge can be also extended to cover several AO in the bridge. Consider, e.g., the *two*-AO bridge in the sequential scattering $i \to (k \to l) \to j$ described by the bridge conditional probability

$$\begin{aligned} P(i \to j|k,l) &= |A_{(i\to j|k,l)}|^2 = (D_{i,k}D_{k,l}D_{l,j})^2/(D_{i,i}D_{k,k}D_{l,l}) \\ &= \left(D_{i,k}/\sqrt{D_{i,i}}\right)^2\left(D_{k,l}/\sqrt{D_{k,k}}\right)^2\left(D_{l,j}/\sqrt{D_{l,l}}\right)^2 \\ &= |A_{i\to k}|^2|A_{k\to l}|^2|A_{l\to j}|^2 = P(i \to k)P(k \to l)P(l \to j). \end{aligned} \tag{12.198}$$

It is defined by the associated amplitude, the renormalized implicit derivative of χ_j^b on χ_i^b through χ_k^b and χ_l^b:

$$A_{(i\to j|k,l)} = N_{(i\to j|k,l)}\left(\frac{\partial\chi_k^b}{\partial\chi_i^b}\right)\left(\frac{\partial\chi_l^b}{\partial\chi_k^b}\right)\left(\frac{\partial\chi_j^b}{\partial\chi_l^b}\right) = N_{(i\to j|k,l)}D_{i,k}D_{k,l}D_{l,j}. \tag{12.199}$$

Here, the normalization constant,

$$N_{(i\to j|k,l)} = 1/(D_{i,i}D_{k,k}D_{l,l})^{1/2}, \tag{12.200}$$

satisfies the sum rule

$$\Sigma_j P(i \to j|k,l) = P(i \to l|k). \tag{12.201}$$

It again expresses the fact that the summation of the conditional probabilities over all possible final outputs $\{j\}$ in the sequential scattering events $i \to (k \to l) \to \{j\}$

must reproduce the conditional probability of the preceding step of the *single*-AO bridge propagation $i \to (k) \to l$. These probabilities are related to the corresponding Wiberg-type bond order

$$\mathcal{M}(i \to j|k,l) = \mathcal{M}_{i,k}\mathcal{M}_{k,l}\mathcal{M}_{l,j}, \quad \mathcal{M}_{l,j} = (\gamma_{l,j})^2,$$
$$P(i \to j|k,l) = \mathcal{M}(i \to j|k,l)(N_{(i\to j|k,l)})^2/4^3. \tag{12.202}$$

Of interest also is the probability scattering *via* the *parallel* single-AO bridges (Nalewajski 2011c), consisting of all basis functions $\boldsymbol{\chi}$, which determine the AO cascade between orbitals χ_i and χ_j in the bond system of the molecule,

$$i \to \{k\} \to j \equiv i \to \boldsymbol{\chi} \to j, \quad P(i \to j|\boldsymbol{\chi}) = |A_{(i\to j|\boldsymbol{\chi})}|^2, \tag{12.203}$$

defined by the amplitude

$$A_{(i\to j|\boldsymbol{\chi})} = N_{(i\to j|\boldsymbol{\chi})}\sum_k \left(\frac{\partial \chi_k^b}{\partial \chi_i^b}\right)\left(\frac{\partial \chi_j^b}{\partial \chi_k^b}\right) = N_{(i\to j|\boldsymbol{\chi})}\Sigma_k D_{i,k}D_{k,j}$$
$$= N_{(i\to j|\boldsymbol{\chi})}D_{i,j}, \tag{12.204}$$

where we have used the idempotency relation of (12.166). The normalization constant

$$N_{(i\to j|\boldsymbol{\chi})} = 1/D_{i,i} \tag{12.205}$$

then indeed assures that

$$P(i \to j|\boldsymbol{\chi}) \equiv \Sigma_k P(i \to j|k) = P(i \to j). \tag{12.206}$$

This *single*-cascade development can be extended to probe the *multiple* cascade bridges (Nalewajski 2011c). Consider for example the probability scattering through the *double*-cascade

$$i \to \{k\} \to \{l\} \to j \equiv i \to \boldsymbol{\chi} \to \boldsymbol{\chi}' \to j, \quad P(i \to j|\boldsymbol{\chi},\boldsymbol{\chi}') = |A_{(i\to j|\boldsymbol{\chi},\boldsymbol{\chi}')}|^2,$$

$$A_{(i\to j|\boldsymbol{\chi},\boldsymbol{\chi}')} = N_{(i\to j|\boldsymbol{\chi},\boldsymbol{\chi}')}\sum_k\sum_l \left(\frac{\partial \chi_k^b}{\partial \chi_i^b}\right)\left(\frac{\partial \chi_l^b}{\partial \chi_k^b}\right)\left(\frac{\partial \chi_j^b}{\partial \chi_l^b}\right)$$
$$= N_{(i\to j|\boldsymbol{\chi},\boldsymbol{\chi}')}\Sigma_k\Sigma_l D_{i,k}D_{k,l}D_{l,j} = N_{(i\to j|\boldsymbol{\chi},\boldsymbol{\chi}')}D_{i,j}, \tag{12.207}$$

where again the normalization of (12.205) gives:

$$P(i \to j|\boldsymbol{\chi},\boldsymbol{\chi}') \equiv \Sigma_k\Sigma_l P(i \to j|k,l) = P(i \to j). \tag{12.208}$$

Therefore, the multiple cascades, with each step involving all basis functions, indeed conserve the stationary direct probability scattering (12.167) (Nalewajski 2011c; see also the preceding section).

12.12 Conclusion

Communication theory has been shown to provide a novel perspective on the entropic origins of the chemical bond. In this short overview of OCT we have introduced its key concepts and techniques, which have been subsequently used to explore the electronic structure of prototype molecules in terms of both the overall bond multiplicity and its ionic/covalent components, as well as through the corresponding entropy/information descriptors of the localized (diatomic) chemical interactions. Illustrative numerical results have been presented to validate the claim that these communication *noise* (covalency) and information *flow* (ionic) measures (in bits) of the overall IT bond order indeed reflect the chemical intuition quite well.

The localized bond multiplicities were shown to approximate the quadratic Wiberg index of quantum chemistry in typical polyatomic molecules, at the same time providing its IT-covalent(ionic) resolution. It should be also emphasized, that the extra computational effort of this complementary IT analysis of the molecular bonding patterns is negligible, compared to the cost of standard SCF LCAO MO calculations of the molecular electronic structure, since practically all computations using MO approximation already determine the CBO data required in the OCT probe of bond multiplicities. This diatomic development extends our understanding of the chemical bond from the complementary viewpoint of the Communication Theory. We have also briefly outlined the *many*-orbital generalization of OCT, which allows one to describe the interbond coupling phenomena, origins of catalytical activity, multibond reactivity, polycenter bonds, etc.

The OCT also introduces the communication (entropy/information) perspective on several classical issues in the electronic structure theory (Nalewajski 2006g, 2010f). Until recently, a wider use of CTCB in probing the molecular electronic structure has been hindered by the use of the *two*-electron conditional probabilities, which blur the chemical bond differentiation (Nalewajski 2006g). The AO-resolved OCT using the flexible input probabilities and recognizing the bonding/antibonding character of the orbital interactions in the molecule, reflected by the signs of the underlying CBO matrix elements, to a large extent remedies this problem (Nalewajski et al. 2011a).

We have also explored in this chapter a novel through-bridge mechanism of bonding interactions in molecular systems, which has been first conjectured to explain the bonding patterns in small propellane systems. Thermodynamical data for these systems, CG analysis of Fig. 10.21, and quadratic bond order indices reported in Fig. 10.6 and in Table 12.2, both confirm the existence of some central bond, even in the smallest systems, where no accumulation of the electron or information density between the bridgehead carbons is observed. This prompted the alternative propositions of the VB-inspired *charge-shift* (Shaik et al. 2009) and the *through-bridge* (Nalewajski 2006g, 2010f, 2011a, b, c, f; Nalewajski and Gurdek 2011a, b) mechanisms, with the former attributing this bonding effect to the instantaneous charge fluctuations, to explain the apparent existence of some chemical bonding between the central carbons even in the smallest propellanes, despite the absence of the charge accumulation between the bridgehead carbons.

We have also demonstrated using both the Wiberg and OCT bond multiplicities how atoms exhibiting the vanishing direct chemical interaction can be still bonded indirectly, *via* the AO/AIM bridges. This novel mechanism has been shown to have important implications for the bonding patterns of π-interactions in hydrocarbons: in the π-system of benzene the *ortho*-carbons exhibit a strong Wiberg bond multiplicity measure of almost exclusively through-space origin, the *cross*-ring interactions between the *meta*- and *para*-carbons where shown to be described by much smaller but practically equalized overall resultant bond orders, being distinguished solely by the direct/indirect composition of these resultant chemical interactions: the *meta* bonds have been shown to be realized exclusively through bridges, while the *para* bonds exhibit comparable direct and indirect components.

We thus conclude that the chemical interaction between the specified pair of AO/AIM has both the through-*space* and, hitherto neglected, through-*bridge* components. The former reflects the *direct* interactions between bonded atoms while the latter is realized *indirectly*, through the remaining atoms, which constitute an effective bridge for the chemical coupling between more distant AIM. The most efficient bridges for such an implicit bonding mechanism *via* atomic intermediaries are the real chemical bridges, originating from the basis functions contributed by the chemically bonded atoms connecting such "terminal" atoms of the AIM chain in question. Therefore, the bonded status of the given pair of orbitals/atoms can be felt even at large separations provided there exist real bridge(s) of direct chemical bonds connecting them. The effective range of bridge interactions in representative linear polymers (Nalewajski 2011f; Nalewajski and Gurdek 2011b) has been shown to extend to up to three AIM intermediates in the polymer chain.

In OCT the *direct* bond component due to the specified pair of interacting AO originates from the mutual probability scattering between this two basis functions, which constructively mix into the bonding MO. Its covalency originates from the finite conditional probability of their communications in the molecule, related to the square of the corresponding element of the system density matrix, coupling the two basis functions, and hence also to the associated Wiberg bond order contribution. The direct AO communications are in accordance with the electron delocalization pattern implied by the system occupied (bonding) subspace of MO. The "*implicit*" (through-bridge) bond component can be similarly viewed as resulting from the indirect (cascade) information propagation *via* the bridging AO. Therefore, while the through-space bonding reflects the direct "conversation" between AO, the through-bridge channel(s) can be compared to a chatty talk reporting "hearsay," the "*rumor*" spread between the two AO in question *via* the connecting chain of the AO intermediaries involved in the effective chemical bridge under consideration.

One thus distinguishes in OCT the direct ("dialogue") and indirect ("gossip") origins of the chemical bond, which together contribute to the resultant IT bond multiplicity between the given pair of AO or AIM. We have demonstrated that a similar description follows from the Wiberg-type bond multiplicities formulated in the MO theory. The direct (explicit) bonding interaction between neighboring atoms, reflected by the Wiberg bond orders, is generally associated with the presence of the bond charge or the increase in information density between the two nuclei. However, for more distant atomic partners such an accumulation of valence electrons can be absent, e.g., in the cross-ring π-interactions in benzene or between the bridgehead carbon atoms in small propellanes. For the latter the "charge-shift" bonding mechanism has also been proposed, involving instantaneous charge fluctuations due to a strong resonance between covalent and ionic VB structures. As we have argued in this chapter, such an effectively bonding interaction lacking an accumulation of the bond charge (information) can be also realized *indirectly*, through the neighboring AO intermediaries forming a "bridge" for an effective interaction (communication) between more distant ("terminal") AO.

To summarize, each pair of AO exhibits partial through-*space* and through-*bridge* bond components. The "order" of the former quickly vanishes with increasing interatomic separation. It is also small when the interacting AO are heavily engaged in forming chemical bonds with other atoms or remain nonbonding, thus describing the lone electron pairs. In these cases the chemical interaction multiplicity can still assume appreciable values, when the remaining atoms form an effective bridge of the neighboring, chemically bonded atoms, which links the two AO in question. Thus, a nonvanishing density matrix element coupling the two AO in the molecule, which in MO theory reflects their directly–bonding status, is not essential for the existence of their through-bridge interaction. The latter may exist even when the direct interaction vanishes provided the two AO strongly couple to the chemically bonded chain of orbitals connecting them [see (12.98), (12.111) and (12.157–12.159)].

The Wiberg measure of bond multiplicities has been used to explicitly identify both these components in chemical interactions between AO, by using appropriate projections of basis functions onto the bonding subspace of MO, the scalar products of which determine in the SCF MO theory the associated elements of the system CBO (density) matrix. We have also explored the through-bridge mechanism of bonding interactions in molecular systems using the OCT analysis. Both these approaches independently confirm a presence of the explicit and implicit chemical interactions in molecules giving consistent predictions for model and prototype systems.

As also mentioned above, the direct, more familiar bonding mechanism is associated with an accumulation of the electronic charge between bonded atoms. Being conditioned by the direct overlap and coupling between the interacting orbitals it is possible only at relatively short distances between AIM. The indirect bonding does not require a presence of such a bond charge and depends on the existence of the bridge of chemically bonded atoms between the interacting AO. As such it can be effected at larger separations between atoms, thus having profound implications for biological, *supra*-molecular and solid-state systems. The bottom line of this analysis is that chemical bonding between two AO can be realized despite the vanishing CBO matrix element coupling directly these basis functions in the molecule, provided that they both exhibit the nonvanishing elements with the bridge basis functions. In other words, the two AO may exhibit the indirect chemical bonding when they strongly couple to other directly bonded basis functions.

The simple orbital model of such direct and indirect interactions in small propellanes further confirms the apparent existence of the through-bridge bond even in the smallest [1.1.1] system lacking a strong direct bond component, thus offering an alternative explanation of the experimentally conjectured central bonding in this molecule despite the absence of the charge/information accumulation between central (bridgehead) carbons. This constitutes an additional insight into the bond pattern in these molecules, alternative to the VB-inspired *charge-shift* mechanisms of the instantaneous charge fluctuations between the central carbon atoms.

The through-bridge mechanism adds to the complexity/diversity of chemical interactions in molecular systems and it offers an alternative explanation of some controversial issues in structural chemistry, e.g., of the central bond problem in propellanes. We recall that both the Shannon-type information densities (Sect. 10.3) and the *contragradience* criterion (Sect. 10.7), related to the *Electron Localization Function*, fail to detect the presence of an appreciable direct chemical bond in the smallest propellane. It has also been shown to remove some artifacts of the over-simplified approach based solely upon the through-space mechanism. Within both the generalized Wiberg-type bond order description and the cascade information propagation approaches this indirect bonding interaction, realized through the orbital/AIM intermediaries, has been shown to give an additional insight into the π-bonding in benzene and butadiene, by supplementing the traditional Wiberg bond order description with complementary bridge contributions. This extension of the chemical bond concept has been shown to be crucial for the benzene *cross*-ring interactions. Similar extra bonding multiplicity follows from the OCT treatment using the conditional probability corrections due to chemical bridges.

The quadratic indices of the chemical bond multiplicity and the IT descriptors of the bond order, as well as their through-bridge generalizations are related to molecular *probabilities*, squares of the corresponding quantum amplitudes. We have reconstructed the molecular (stationary) probability distribution *via* the multiple probability scattering between the elementary AO states. The operator representation of such scattering processes has facilitated the vector interpretation of the probability amplitudes as projections of the forward- and reverse-scattered states, eventually leading to the establishment of the independent modes (standing waves) of the molecular conditional probability propagation between basis functions, linked to the eigenvalue problem of the associated AO communication operator.

The interference of amplitudes of the generalized, multiple scatterings through all basis functions at arbitrary bridge-order, has also resolved the apparent paradox that in quantum mechanics the resultant effect of the superposition of such scattering states must ultimately conserve the initial (stationary) conditional probability distribution of the molecular ground state. This consistency requirement is not satisfied, when the bridge conditional probabilities are determined classically, as products of probabilities of each consecutive subchannel in the information cascade. Only the *wave*-like superposition of the scattering amplitudes was explicitly shown to satisfy this stationary condition at any bridge order. The idempotency of the molecular density matrix in *one*-determinantal approximation was shown to be vital for the fulfillment of this conditional probability preservation principle. This demonstrates that OCT provides the internally consistent *quantum* description of the molecular electronic structure and of the associated information communications, provided that the elementary scattering amplitudes are superimposed, rather than probabilities.

The IT approach to chemical bonding is very much in spirit of the Eugene Wigner's observation, often quoted by Walter Kohn, that the understanding in science requires insights from *several* different points of view. The kinetic energy probe of the bonding regions in molecules and the communication perspective on the information genesis of the chemical bonds provide such an alternative. Only together these complementary tools constitute what one would call a more "complete" theory of the complex bond phenomenon, which – to paraphrase yet another famous citation from Samuel Beckett – is one of old, good problems that never die out.

References

Abramson N (1963) Information theory and coding. McGraw-Hill, New York
Chandra AK, Michalak A, Nguyen MT, Nalewajski RF (1998) J Phys Chem A 102:10182
Dirac PAM (1967) The principles of quantum mechanics. Clarendon, Oxford
Gopinathan MS, Jug K (1983) Theor Chim Acta (Berl.) 63:497, 511
Gutmann V (1978) The donor-acceptor approach to molecular interactions. Plenum, New York
Jug K, Gopinathan MS (1990) In: Maksić ZB (ed) Theoretical models of chemical bonding, vol 2. Springer, Heidelberg, p 77
Mayer I (1983) Chem Phys Lett 97:270
Mayer I (1895) Theor Chim Acta (Berl.) 67:315
Mitoraj M (2007) Ph.D. Thesis, Jagiellonian University, Krakow, Poland
Mitoraj M, Michalak A (2005) J Mol Model 11:341
Mitoraj M, Michalak A (2007) J Mol Model 13:347
Mitoraj M, Zhu H, Michalak A, Ziegler T (2006) J Org Chem 71:9208
Mitoraj M, Zhu H, Michalak A, Ziegler T (2007) Organometallics 16:1627
Nalewajski RF (2000c) J Phys Chem A 104:11940
Nalewajski RF (2000d) Top Catal 11–12:469
Nalewajski RF (2004b) Chem Phys Lett 386:265
Nalewajski RF (2004c) Mol Phys 102:531
Nalewajski RF (2004d) Mol Phys 102:547
Nalewajski RF (2004e) Struct Chem 15:395
Nalewajski RF (2005a) Theor Chem Acc 114:4
Nalewajski RF (2005b) Mol Phys 103:451
Nalewajski RF (2005c) J Math Chem 38:43
Nalewajski RF (2006a) Mol Phys 104:365

Nalewajski RF (2006b) Mol Phys 104:493
Nalewajski RF (2006c) Mol Phys 104:1977
Nalewajski RF (2006d) Mol Phys 104:2533
Nalewajski RF (2006e) Mol Phys 104:3339
Nalewajski RF (2006g) Information theory of molecular systems. Elsevier, Amsterdam
Nalewajski RF (2007) J Phys Chem A 111:4855
Nalewajski RF (2008a) J Math Chem 43:265
Nalewajski RF (2008b) J Math Chem 43:780
Nalewajski RF (2008c) J Math Chem 44:414
Nalewajski RF (2008d) J Math Chem 44:802
Nalewajski RF (2009e) Int J Quantum Chem 109:425
Nalewajski RF (2009f) Int J Quantum Chem 109:2495
Nalewajski RF (2009g) Adv Quant Chem 56:217
Nalewajski RF (2010b) J Math Chem 47:692
Nalewajski RF (2010c) J Math Chem 47:709
Nalewajski RF (2010d) J Math Chem 47:808
Nalewajski RF (2010f) Information origins of the chemical bond. Nova Science, New York
Nalewajski RF (2011h) Information perspective on molecular electronic structure. In: Hong WI (ed) Mathematical Chemistry, Nova Science, New York, p 247
Nalewajski RF (2011g) Information tools for probing chemical bonds. In: Putz M (ed) Chemical information and computation challenges in 21st: a celebration of 2011 international year of chemistry. Nova Science, New York (in press)
Nalewajski RF (2011a) J Math Chem 49:371
Nalewajski RF (2011b) J Math Chem 49:546
Nalewajski RF (2011c) J Math Chem 49:806
Nalewajski RF (2011d) J Math Chem 49:592
Nalewajski RF (2011e) J Math Chem 49:2308
Nalewajski RF (2011f) Int J Quantum Chem (in press)
Nalewajski RF, Broniatowska E (2003a) J Phys Chem A 107:6270
Nalewajski RF, Gurdek P (2011a) J Math Chem 49:1226
Nalewajski RF, Gurdek P (2011b) Struct Chem (in press)
Nalewajski RF, Jug K (2002) In: Sen KD (ed) Reviews of modern quantum chemistry: a celebration of the contributions of Robert G. Parr, vol 1. World Scientific, Singapore, p 148
Nalewajski RF, Korchowiec J (1997) Charge sensitivity approach to electronic structure and chemical reactivity. World-Scientific, Singapore
Nalewajski RF, Mrozek J (1994) Int J Quantum Chem 51:187
Nalewajski RF, Mrozek J (1996) Int J Quantum Chem 57:377
Nalewajski RF, Köster AM, Jug K (1993) Theor Chim Acta (Berl.) 85:463
Nalewajski RF, Formosinho SJ, Varandas AJC, Mrozek J (1994a) Int J Quantum Chem 52:1153
Nalewajski RF, Korchowiec J, Michalak A (1996a) Top Curr Chem 183:25
Nalewajski RF, Mrozek J, Mazur G (1996b) Can J Chem 100:1121
Nalewajski RF, Mrozek J, Michalak A (1997) Int J Quantum Chem 61:589
Nalewajski RF, Szczepanik D, Mrozek J (2011a) Adv Quant Chem 61:1
Shaik S, Danovich D, Wu W, Hiberty PC (2009) Nat Chem 1:443
Shannon CF (1948) Bell Syst Technol J 27(379):623
Shannon CE, Weaver W (1949) The mathematical theory of communication. University of Illinois, Urbana
Wiberg KB (1968) Tetrahedron 24:1083

Part IV
Chemical Concepts for Molecular Structure and Reactivity

Chapter 13
Alternative Perspectives in Chemical Theories

Abstract A general outlook on concepts and principles of the electronic structure and chemical reactivity is presented with an emphasis on their importance for *understanding* the molecular behavior. It is argued that *chemical* interpretation of molecular processes in terms of electronic pairs, chemical bonds, AIM, functional groups, reactants, reactivity sites, etc., calls for the conceptual approaches in theoretical chemistry. The classical rules of the molecular structure and reactivity are briefly summarized and a need for the electronically and geometrically initiated perspectives on molecular changes, called the *electron-preceding* (chemical hardness) and *electron-following* (chemical softness) representations, is stressed. The quadratic Taylor expansion of the electronic energy of molecular systems in powers of displacements (perturbations) of the system *state* parameters provides theoretical framework capable of accounting for the main couplings between state parameters of both molecules and reactive systems. It introduces the theory generalized responses: "*potentials*" – corresponding to the *first* partials of the energy, and *charge sensitivities* (CS) – defined by the *second* partials of the electronic energy with respect to the system canonical parameters of state. This power series also constitutes an adequate framework for describing reactant subsystems in the bimolecular reactive system.

The role of electronic density as the source and carrier of the complete information about the ground state equilibrium is stressed. A distinction is made between transitions from one ground state density to another, called the "*horizontal*" displacements of the system electronic structure, and hypothetical flows of electrons between molecular subsystems for the fixed density (energy) of the molecule as a whole, called the "*vertical*" displacements. The internal equilibria in the *mutually* closed molecular subsystems are examined, when they are *externally* open or closed relative to their associated electron reservoirs. The optimum densities of molecular fragments then appear as solutions of the Euler equations for the *embedded* subsystems, which are briefly examined. The alternative sets of the subsystem *state* parameters are identified and the corresponding "thermodynamic" potentials, Legendre transforms of the system electronic energy, are introduced. Their CS in the subsystem resolution, including the chemical *hardness*,

R.F. Nalewajski, *Perspectives in Electronic Structure Theory*,
DOI 10.1007/978-3-642-20180-6_13,

softness, and *Fukui function* quantities of molecular fragments, are discussed and the corresponding *second*-order Taylor expansion of the electronic energy is examined. The equilibrium transformations between perturbations and linear responses of the system as a whole and its constituent fragments are summarized. These subsystem descriptors, for both the externally closed and open molecular fragments, are applied to the illustrative bimolecular system, at both the polarizational and charge-transfer stages of the chemical reaction. The associated in situ quantities for the externally closed reactants are examined and the internal/external stability criteria are linked to the structure of the condensed hardness matrix in the reactant resolution. Finally, the implications of the equilibrium and stability criteria are summarized.

13.1 Survey of Reactivity Phenomena and Need for Conceptual Approaches

The last decades have witnessed a dramatic growth of modern quantum chemistry, both in its conceptual ideas and computational techniques. The conceptual theory generates means for understanding the structure and chemical behavior of molecular systems and for interpreting results of theoretical calculations. The ab initio data, often of an admirable accuracy, are now generated using both the wave function and DFT methods, with a strong tendency of the latter to dominate calculations on very large systems. These qualitative and quantitative theoretical results are often synergetically combined with laboratory techniques, verifying experimental data and guiding the researchers in their planning of future experiments. However, the wave functions resulting from the modern high-level methods of computational quantum chemistry are so immensely complex that they cannot be immediately understood in simple and physically or chemically meaningful terms. The categorization and interpretation objectives in theoretical chemistry call for the well-founded principles and conceptual models, which are transparent, intuitively appealing, and useful in qualitative and semiquantitative applications to chemical systems.

The classical perspective on the molecular electronic structure in terms of electron pairs, bonds, AIM, functional groups, etc., represents a central and most fruitful theme in chemistry. A knowledge of the electronic and geometric structure parameters of isolated molecules already gives important clues for understanding the behavior of chemical compounds in different reactive environments. It constitutes a starting point for a subsequent, perturbative studies of molecular interactions. This *Separated Reactant Limit* (SRL) provides a natural and convenient reference state, at an early stage of the reactant mutual approach. The structure of separated reactants qualitatively reveals the expected main features of the preferred *Minimum Energy Path* (MEP), thus already determining gross features of the easiest ascent in the reactant valley of PES toward the transition state (TS) complex, the exact location of which ultimately determines the activation barrier for the chemical process in question.

Chemistry is concerned with properties and reactions of an enormous number of different compounds which for the purpose of expediency are classified into similarity groups, e.g., those with the same functional group(s), so that the physical and chemical properties of a particular compound may be inferred from the behavior of any other member. A number of qualitative and quantitative approaches have been formulated to relate properties of members belonging to the same and different similarity groups. Representative examples in the area of chemical reactivity are provided by the familiar directing influences of the electron-*withdrawing* and electron-*donating* substituents in benzene derivatives, as well as the related (experimental) correlations of Hammett (Hammett 1935, 1937; Johnson 1973a). The "*free energy*" relationships (Marcus 1968, 1969; Chapman and Shorter 1972; Johnson 1975) have been extremely valuable in helping chemists to predict the reactivity of chemical compounds and to understand a subtle interrelationship between *reactivity* and *selectivity* in chemical processes.

Trends in chemical reactivity are the main objectives of the so-called reactivity "theories." Their basic aim is to qualitatively *predict* reactivity patterns and to find an *explanation*, in chemical terms, of the experimentally or theoretically determined course of specific reactions. Such theories have to provide means of systematization, recognition of regularities, and rationalization of the myriads of established experimental and computational facts to disclose the fundamental causes governing the reactivity phenomena. The most general of them are formulated in terms of the appropriate variational principles or the most favorable "matching" rules for the crucial physical properties of reactants (global or regional), which uncover the decisive factors responsible for the preferred direction of the given chemical reaction.

Investigations into the primary sources of the observed chemical behavior of molecules cover both the thermodynamic/statistical and quantum mechanical laws of chemical change. For example, the concept of the activation energy in the bimolecular reaction is statistical in character, but the actual value of this critical energy of reactants, which is required for the reactive outcome of their collision, cannot be understood without the quantum mechanical description of the elementary changes in the electronic structure of reacting species. In predicting/understanding the electronic structure of chemical compounds and their reactions theoretical chemistry uses concepts and techniques of both the *static* and *dynamical* approaches. The basic objective of the dynamical treatment is to calculate/explain the *rates* of chemical reactions from first principles. Given the interaction potential for the nuclear motion in the specified system of reactants, one should in general be able to determine the probabilities, cross sections, and rate constants for fundamental elementary reaction processes by solving the quantum mechanical equation of motion for the system. This dynamical goal, however, has so far been realized only for very simple systems involving three or four atoms, due to the computationally immense task in the theoretic determination of the complete PES and in solving the Schrödinger equation for motions of the system nuclei.

Therefore, much of the present understanding of the chemical reaction dynamics at the molecular level has come about by using quite limited information about the multidimensional PES. For example, the model (analytical) PES, reproducing a network of selected ab initio points, and some approximate methods, e.g., the classical trajectories, have been used to probe the dynamics of elementary reactive collisions. Another familiar example is the statistical *Transition State Theory* (TST), in which only the geometry and frequencies of the separated reactants and the TS complex are required to convert this limited information about the interaction between reactants into the rate quantities. The DFT-based molecular *charge sensitivities* (CS) also constitute attractive (static) concepts, in terms of which useful reactivity criteria can be formulated (e.g., Nalewajski and Koniński 1988; Gázquez 1993, 2009; Gázquez et al. 1987; Nalewajski and Korchowiec 1997; Mortier and Schoonheydt 1997; Nalewajski et al. 1996a; Nalewajski 1988, 1989a, 2002e, 2005d, e; Geerlings et al. 2003; Chattaraj 2009; Johnson et al. 2011).

In principle, the rates and mechanisms of chemical reactions can be predicted by the standard methods of statistical thermodynamics in terms of the partition functions of reactants and of the TS complex. However, the range of applicability of the absolute rate theory is severely limited by the fact that an evaluation of the vibrational partition function for the TS complex of the elementary process of interest requires a detailed consideration of the whole PES for the reactive system. The calculation of the absolute rate constants is thus possible only for relatively simple systems. This indicates a need for a more approximate theoretical treatment of chemical reactions, i.e., the *conceptual* reactivity "theory," which would allow chemists to go further in their predictions and understanding of properties of new compounds and outcomes of chemical interactions, particularly in large reactive systems of interest in the contemporary organic chemistry and biology.

Due to diversity and ever increasing complexity of molecules and reactions, relatively crude assumptions have to be made in such simplified approaches to elementary molecular processes, and empirical factors are often introduced into theoretical expressions. Thus, from the purist point of view, such "theories" should be more appropriately classified as mere theoretical *models* of reactivity. A classical example of such a heuristic approach is the celebrated Hammond (1955) postulate of a relative similarity of the TS complex to reactants (products) in the exothermic (endothermic) reactions (see also: Dunning 1984; Ciosłowski 1991; Nalewajski and Broniatowska 2003b).

On one hand, such general, conceptual tools a posteriori reduce the overwhelming amount of information embodied in the ab initio wave functions to a more manageable, qualitative level by extracting common roots of seemingly unrelated data. On the other hand, they provide a valuable means for the *chemical* understanding of the molecular structure and reactivity, enabling a subsequent informed "guess work" about the system behavior in a changed molecular environment and a more precise planning of future experiments. Such adequate theoretical models thus offer a rationale for *trends* within families of related compounds, and they bridge a gap between the rigorous quantum mechanics and empirical concepts of the intuitive, phenomenological chemistry. Therefore, qualitative models and quantitative theories of the electronic structure and chemical reactivity constitute inevitable and necessary ingredients of the scientific method of chemistry. Only a parallel advancement of both these branches marks the harmonious development of theoretical chemistry. The qualitative concepts determine the scientific *vocabulary* of the interpretative chemistry, while the approximate model relations allow for a semiquantitative prediction of trends implied by changing structural and experimental conditions.

The IT approach allows one to treat in an unbiased way the molecular fragments in both the "*horizontal*" and "*vertical*" rearrangements of the molecular electronic structure, for the displaced and fixed molecular electron density, respectively, in a thermodynamic-like fashion (e.g., Nalewajski 2006g, 2010f). The vertical problem is vital for extracting the chemical interpretation from the known molecular electron density in terms of such chemical concepts as bonded atoms, functional groups, reactants, lone electron pairs, and bonds, which connect the constituent subsystems in the molecule. For example, in Chap. 10 we have shown how alternative entropy/information densities can be used to probe chemical bonds in molecular systems. We have also demonstrated in Chap. 11 that IT can be successfully used to tackle the definition of AIM by searching for atomic densities, which reproduce the density of the system as a whole and exhibit the least information distance relative to the corresponding free atoms of the "promolecule."

The resulting effective distributions of electrons in such chemical atoms can be monitored at different stages of their reconstruction in a molecular environment, e.g., for the optimum *polarization* (P) of the mutually closed subsystems and after the *charge transfer* (CT) between the system constituent parts. In chemistry these infinite (overlapping) AIM, referenced to the corresponding free atoms of the promolecule and immersed in the molecular environment composed of the remaining atoms, constitute natural building units of molecules. Indeed, they conform to several classical ideas in chemistry, which strongly emphasize the atomic density/orbital overlap as the primary source of the chemical bond. This entropic definition of bonded atoms complements Bader's (1990, 1991) concept of the nonoverlapping *topological* atoms, defined by partitioning of the physical space into exclusive atomic "basins," separated by the "zero-flux" surfaces of the molecular electron density. These quantum mechanically defined boundaries effectively partition the molecular electron densities into the exclusive atomic pieces, which are solely referenced to the *molecular* state. This is in contrast to the stockholder AIM, which are defined with respect to both the molecular and promolecular (free atom) references, with the latter customarily marking the "starting point" in the bond-formation process.

This IT treatment of the submolecular reality of bonded molecular fragments also gives rise to a "thermodynamic" description of molecules and their constituent fragments in terms of the *entropy* equilibrium molecular subsystems (Nalewajski 2003b, 2004a, 2006g, h). However, since molecular fragments cannot be related to specific quantum mechanical "observables", they cannot be verified experimentally. Thus, the bonded atoms of chemistry ultimately represent the *noumenons* of Kant (Parr et al. 2005). Nonetheless, they can be partially validated either by their ability to conform to the established chemical concepts or by the extra causality they offer in describing the molecular phenomena, e.g., via the demonstrated parallelism to the ordinary thermodynamics. It has been argued elsewhere (Nalewajski 2003a, b, 2004a, 2006g) that by using the IT approach to define molecular subsystems one indeed generates a "chemical" interpretation with thermodynamic-like causal relations between perturbations and responses of molecular subsystems. Within such an IT outlook on the molecular and submolecular electronic structure the whole experience of the ordinary thermodynamics can be employed in treating a variety of subtle processes in chemistry.

Besides generating the entropic tools for probing chemical bonds and providing the justification of the "stockholder" AIM, IT has been shown to give rise to fresh outlook on classical VB and "loge" theories (Nalewajski 2003d, 2006a, g, 2009b), new criteria of molecular similarity (Nalewajski 2009h), the electron and bond localization probes (Chap. 10), and to thermodynamic-like description of molecular systems and their fragments (Nalewajski 2003b, 2004a, 2006g, h). In particular, the density fluctuations and flows of electrons between subsystems have also been tackled in the local "thermodynamic" description, which closely follows the ordinary irreversible thermodynamics (2003b, 2004a), and the system chemical potential has been interpreted as the information "temperature" (Nalewajski 2006h).

The first task confronting the chemist is to identify the compound reactive sites as functions of the changing molecular structure and to determine their relative reactivity trends. A complex organic molecule may contain several alternative *Nucleophilic* (N), *Electrophilic* (E), and/or *Radical* (R) sites, and hence the competition of the attacking agents for these reaction centers constitutes a very important problem in reactivity theory. To meet this challenge, one has to understand the intersite coupling mechanism, i.e., how the molecular structure and the presence of the catalyst affect the reactivities at various active centers of the molecule. The *relative* reactivity of an active site may vary with the nature of the attacking agent (*ambident reactivity*). Ambidency may also be exhibited as a result of changing experimental conditions. Any *bona-fide* theory of chemical reactivity must provide a framework which in principle is capable of accounting for all these diverse reactivity phenomena.

A distinction between the thermodynamic and kinetic controls of competing reaction paths is also essential for a satisfactory explanation of all such processes. Simplified, *single*-reactant approaches to chemical reactivity problems, with the underlying assumption that the reactivities of molecules are intrinsically embedded in their own electronic/geometrical structure, can only serve as a starting point in tackling finer reactivity phenomena shaped by the interaction between both reactants of the given bimolecular system. An understanding of subtle reactivity trends in terms of the static reactivity criteria thus calls for the truly *two*-reactant theoretical treatment (e.g., Nalewajski 1995a, 1997b; Nalewajski and Korchowiec 1997; Nalewajski et al. 1996a; Nalewajski and Michalak 1995, 1996, 1998), which combines both the molecule and the perturbation due to the attacking agent. Indeed, only such approaches provide an adequate basis for describing variations in reactivity of one reactant, and/or its particular site(s), with a changing position/character of the other reactant and of its reactive sites.

Moreover, the functional groups in the specific molecular reactant are mutually coupled via the connecting atoms and bonds. Therefore, the chemical reaction taking place at one site is not without an influence on the reactivities of the remaining sites. Therefore, the adequate reactivity theory must be sufficiently rich in its conceptual basis and flexible in its theoretical framework to fully account for all such *inductive* (coupled) reactivity effects.

When two large species orient themselves relative to one another at an early stage of the chemical reaction, an even more subtle challenge for the reactivity theory emerges. It is related to the fact that the very classification of chemical species as the *electrophilic* (electron deficient, acceptor, acidic) or *nucleophilic* (electron rich, donor, basic) is only a relative one. Indeed, the mutual acidic/basic properties of reactants or their respective active sites depend on the current identity and state of the reaction partner, since reactants represent strongly coupled parts of a single reactive supersystem. Thus, a given molecular site may simultaneously act as a base toward one (relatively acidic) site of the other reactant, or as an acid toward another (relatively basic) site of the reaction partner.

A satisfactory reactivity theory must be also able to cover the issues of a subtle interplay between the electronic and geometrical coordinates of the molecular and reactive systems (e.g., Nalewajski 1995b, 2000a, 2006f, 2010g; Nalewajski et al. 2008). The so-called *"mapping"* transformations (Baekelandt et al. 1995; Nalewajski and Korchowiec 1997; Nalewajski et al. 1996a; Nalewajski and Sikora 2000; Nalewajski 2006f) between these two aspects of the molecular structure provide such unifying concepts for both the qualitative understanding and semi-quantitative characterization of such couplings in molecules and between reactants. Both the EP and EF transformations can be approached in this way. The former, Hellmann–Feynman viewpoint envisages a shift in the electron distribution as preceding, accelerating the associated motion of the system nuclei, while the latter, BO perspective views the nuclear displacements as the driving force of changes in the molecular electron density.

This dual development has decisively extended the range of applications of the theory of chemical reactivity, in comparison with the (empirical) structural rules of Gutmann (1978). The EF mapping relations, of the BO perspective, in principle allow one to diagnose trends in the electron redistribution, in response to a given (hypothetical or real) displacement of the system geometry. The "inverse" EP relations, within the Hellmann–Feynman-type perspective, are closer to an intuitive chemical thinking. They are required to solve another typical reactivity problem: how to manipulate the system electronic structure, e.g., the charge distribution of the *fine-grained*, local description, or the effective oxidation states (net charges) of AIM in the *coarse-grained*, atomic resolution, to bring about a desired change in the system geometry and/or breaking the specific bond(s). Through the mapping relations any shift in the nuclear position space can be "translated" into the conjugate displacement in the electron distribution, and vice versa. These transformations enrich a variety of diagnostic and interpretative tools of theoretical chemistry and provide a semiquantitative characterization of couplings between the nuclear and electronic molecular structures in chemical processes.

Alternatively, the electronic parameters can be related to the *Minimum Energy Coordinates* (MEC) of nuclear displacements, which are formulated in the *compliant* (geometrical softness) approach (Nalewajski 1995b, 1998b, 2000a, 2006f, 2009a;

Nalewajski and Korchowiec 1997; Nalewajski et al. 1996a, 2008), in the combined treatment of the electronic-nuclear displacements (see Sect. 14.4).

13.2 Chemical Understanding of Molecular Processes

Accurate theoretical calculations of the energy profile along the MEP on molecular PES, which use the standard computational methods of quantum chemistry, cannot be itself regarded as *theory* either, but rather as mere computer *experiments*. In fact, most of the activity of theoretical chemists has been directed not toward an *understanding* of the rules governing reactions, but rather to numerical determination of the physical properties of molecules and reactants at the given stage of their mutual approach. In the past, the use of the simplified models or principles, which permits a useful chemical information to be derived without such elaborate calculations, e.g., relative rates of admissible reaction channels, has provided a valuable insight into our understanding of molecular reactivity preferences. It has also provided means to analyze molecules and calculations on molecules in theoretically consistent framework that allows chemists to *understand* the results of calculations in terms of the intuitive concepts which dominate the *language* of chemistry.

The electronic ground state energy of a molecule can be regarded as functional of alternative sets of the system "variables," both global and local in character, which uniquely specify the system equilibrium state. Chemistry is fundamentally the science about transformations (reactions) and responses of molecules due to some displacements in their environment. The chemical understanding of the molecular electronic structure is not limited to properties of isolated species, but it also covers a behavior of molecular reactants, when they are in contact with other agents, which create a perturbation in the system environment. These displacements of either electronic or geometrical (nuclear) origin can ultimately lead to a change in the pattern of bonds between the system constituent atoms/subsystems, i.e., to an elementary chemical reaction. Such an understanding of the molecular behavior calls for CS, which measure the system responses to such hypothetical or real changes in its physical degrees of freedom, both electronic and nuclear.

When considering a behavior of a single molecule or the family of chemically similar molecules in the given type of chemical reaction, e.g., during the electrophilic, nucleophilic, or radical attacks by small agents, various *single*-reactant reactivity concepts have proven their utility in predicting the most reactive site. Such criteria are based upon the underlying notion of an inherent chemical reactivity of a molecule or a hierarchy of relative reactivities of its parts, for the fixed reaction *stimulus* at each of the compared locations, due to the perturbation created by the same attacking agent. This notion implies that the way the molecule reacts is somehow predetermined by its own structure.

This approximate point of view neglects the mutual influence of one reactant upon another. A more subtle *two*-reactant description of chemical reactivity is required to probe alternative arrangements of two large molecules, e.g., in the cyclization reactions, when several chemical bonds are being simultaneously formed or broken. In order to account for the mutual influence of both molecular subsystems in a given bimolecular reactive system, the adequate reactivity criteria have to include the relevant *embedding* (nonadditive) energy terms. For each part of the reactive system, they involve the appropriate reaction "stimulus," i.e., the subsystem perturbation created by a presence of the complementary subsystem at the given, say, early stage of the reactant mutual approach, and the conjugate response of the perturbed species. The normalized response quantities, per unit displacement in the system state parameters, determine the generalized "polarizabilities" (CS) of the reactants.

Such embedding energy contributions are in principle included in all second-order perturbational approaches to reactive systems. The responses of reactants can be classified as "diagonal" (intrareactant), when both displacements in the defining second derivative of the system electronic energy refer to the same subsystem, or "*off*-diagonal" (interreactant), when the two perturbations in the energy derivative correspond to different subsystems. In particular, the diagonal CS representing the normalized nonlocal responses reflect the influence of an attack (perturbation) on one site of the subsystem under consideration on its reactivity at the other location. Accordingly, the *off*-diagonal (*two*-point) CS account for the influence of an attack in one reactant on the reactivity of its reaction partner.

The *Molecular Orbital* (MO) and *Valence Bond* (VB) theories have facilitated a deeper understanding of a wide range of physical properties of molecules, the relative reactivity trends, and the preferred pathways of chemical reactions. They provide the standard frameworks to understand, at qualitative or semiquantitative levels, what happens to the electronic structure, when the molecule is placed in a changed environment, e.g., in the presence of the catalyst or the reaction partner. These theories also warrant predictions of changes in the molecular geometry accompanying a given, real or hypothetical change in the system electronic structure. Such reactivity "theories" deliberately de-emphasize the computational aspect, aiming instead at a more qualitative understanding of both the electronic structure and chemical reactions. They have proven to be very useful to an experimentalist, who requires an understanding of *why* molecules react the way they do, *what* determines their electronic structure, and *how* this influences reactivity.

In other words, the MO and VB reactivity ideas provide the basis for an understanding of the development of reactions along specific routes, even without an assistance of rather complex computer calculations. Such simple models invoke the classical concepts of the orbital symmetry, electronic pairs, overlap of the electron distributions, as well as the electrostatic potential, electronegativity, hardness/softness (Fukui function) descriptors of molecules and their constituent subsystems. The MO reactivity theories are not limited to the *one*-determinantal description of the standard Hartree–Fock (SCF MO) and Kohn–Sham (DFT) methods. Various *Configuration Interaction* (CI) and *Valence Bond* (VB) ideas have often been invoked to improve the qualitative models of chemical reactions (e.g., Epiotis 1978; Shaik and Hiberty 2004).

The concept of a conservation of the orbital symmetry in both organic and inorganic chemistry has proved to be a major advance in the theory of chemical reactivity (Woodward and Hoffmann 1969, 1970, 1971). It has succeeded in bringing together and rationalizing many diverse areas of the subject. This concept has also provided a basis for the unified mechanistic approach to the cycloaddition reactions and various molecular rearrangements. Nowadays criteria of the orbital symmetry conservation and the related correlation diagram approaches constitute the standard part of the qualitative vocabulary of the modern organic chemistry (e.g., Gilchrist and Storr 1972; Gill and Willis 1974; Stone 1978; Jones 1979; Halevi 1992). The celebrated *Woodward–Hoffmann Rules* have correlated a great number of existing chemical facts and stimulated further widespread experimentation. Similar ways of comprehending the geometry and reactivity of inorganic systems have also been proposed (Albright et al. 1985), thus demonstrating that simple concepts of symmetry and bonding are applicable to the chemical understanding of all molecules. General symmetry rules of chemical reactions have been formulated by considering the symmetry restrictions on the excited state contributions to the perturbed ground state wave functions of reactants. Such terms describe the polarization and charge transfer due to the perturbation created by the normal-mode displacements of the nuclei along the reaction coordinate (Pearson 1976; Bader 1960; Bader and Bandrauk 1968).

It should be stressed, however, that the very characterization of reactions as *allowed* or *forbidden* by the symmetry criteria carries no quantitative information. In many cases, there are several allowed reaction paths and it becomes necessary to distinguish between them in order to determine which one is the most probable and – ideally – to estimate yield ratios and relative reaction rates. For this purpose, a number of perturbative methods have emerged. Early work on organic reactivity concentrated on the conjugated π-electron systems, described by the Hückel theory, and this analysis has produced several reactivity indices, e.g., the *π-electron density*, *free valence* or *self-polarizability* of atoms, with the high value of either index assumed to imply high reactivity (e.g., Coulson and Longuet-Higgins 1947a, b).

The localization energy method has assumed a model for the transition state complex, the so-called Wheland intermediate, in which both the attacking reagent and the substituted atom are bonded to a roughly tetrahedrally promoted carbon, unable to form π-bonds with the remainder of the original π-electron system. The associated change in the π-electron energy, the so-called *localization energy*, is then taken as reactivity index by assuming that other contributions to the activation energy are likely to be approximately constant for the given type of the approaching agent. The Perturbational MO (PMO) theory of Dewar (1969) has proposed an approximate way to estimate the localization energy, the so-called *reactivity number*.

The MO theory has been quite successful in interpreting and predicting the molecular orientations and stereo-selections in a large variety of chemical reactions (Dewar 1969; Fukui 1975, 1981, 1987; Klopman 1974a,b; Dewar and Dougherty 1975; McWeeny 1979). The *Frontier Orbital* (FO) theory of Fukui and coworkers (Fukui 1975, 1981, 1987; Fujimoto and Fukui 1974) uses the FO density and the related *superdelocalizability* index to predict the reactivity preferences and most suitable orientations of the molecular reactants. It has been found that the *electrophilic* (E) aromatic substitutions take place predominantly at the carbon position, where the π-electron density of the HOMO reaches the maximum value. Accordingly, the atomic sites exhibiting the highest value of the π-electron density in the LUMO were confirmed as the preferred locations for the *nucleophilic* (N) aromatic substitution. These two crucial MO of each molecular system determine its FO. They are expected to dominate the chemical interaction between reactants. The *superdelocalizability* index has been derived taking into account the hyperconjugation in the transition state complex of an aromatic substitution, between the aromatic π-electron system of the attacked molecule and the pseudo π-orbital of the subsystem consisting of the reagent and the hydrogen to be replaced in the reaction product.

The PMO methods were also widely invoked after the formulation of the Woodward–Hoffmann rules to treat chemical reactions more comprehensively (Klopman 1974a). In these approaches the reactivity trends are not linked to a single term in the corresponding Taylor expansion of the interaction energy, but rather to the combined sum of contributions due to the steric interactions, electrostatic, polarization and electron-transfer effects, the solvation energy, etc. All these more elaborate treatments, using the semiempirical formulation of the MO theory, take account of the charge distribution and the overlap between orbitals of both reactants. In the familiar Klopman–Salem energy expression (Salem 1968a, b, 1969; Klopman 1968, 1974b) a domination of the electrostatic interaction between the substrates marks the so-called *Charge Control* of the reaction, when reactants are both highly charged and relatively difficult to polarize. This is characteristic of the "*hard*" species in Pearson's (1973, 1988, 1997) terminology. One encounters the other extreme case of the dominating CT contribution to the interaction energy, due to the mixing of the filled orbitals on one molecule with the empty orbitals on the other molecule, when both reactants are uncharged and highly polarizable ("*soft*"). This category of chemical reactions is called to exhibit the *FO Control.*

In this more general perturbational framework Klopman was able to rationalize the Pearson's (1973) HSAB principle that hard acids form stable complexes with hard bases, and soft acids with soft bases, respectively, whereas the complexes of hard acids with soft bases (or of soft acids with hard bases) remain relatively less stable. He was also able to take into account the nature of the attacking reagent in the electrophilic or nucleophilic aromatic substitution, and to show that the ratios of yields of the *ortho*, *meta*, and *para* products of the substitution of the benzene derivatives depend on the competition between the charge and frontier orbital controls.

This perturbation theory of chemical reactivity focuses on an early stage of the reactant approach, when the molecules are still distinct though close enough for the molecular orbital description of the combined reactive system to be valid, say separated by a distance of the order of 5–10 a.u. The implicit assumption is that the reaction profiles for the compared reaction paths are of similar shape, so that the trends of the predicted energy differences at an early point on the reaction coordinate reflect the differences in the activation energies.

The Fukui FO approximation recognizes the interaction between the HOMO and LUMO on both reactants as the crucial factor controlling the course of a chemical process. In many cases an additional approximation is introduced by considering only a single HOMO–LUMO pair for the bimolecular system, for which the orbital energy separation is the smallest, e.g., the HOMO of the *donor* (basic) reactant and the LUMO of the *acceptor* (acidic) reactant. The argument against such a drastic approximation is that it neglects many contributions to the Klopman–Salem equation from other molecular orbitals, the combined effect of which may outweigh the selected FO interaction. Further uncertainties arise in the unique determination of the orbital energies. To remedy this shortcoming, the orbital energies have been substituted by the ionization potentials and electron affinities, by virtue of the Koopmans' and Slater–Janak theorems. Nevertheless, the frontier orbital theory undoubtedly works in most cases, though it may not be as universally successful as are the symmetry rules of Woodward and Hoffmann.

In order to further justify the *frontier* electron model, Fukui has formulated the three supplementary principles (Fukui 1975; Stone 1978): of the "positional parallelism between the Charge Transfer and Bond Interchange," of "narrowing the frontier orbital separation," and of "growing the frontier electron density" along the reaction path. These principles have correctly recognized the need to include the relaxation effects, of both the electronic and geometrical structures of reactants, with the progress along the reaction path. Indeed, a chemical reaction always involves a subtle coupling between the equilibrium electron (chemical bond) distribution on one side and the molecular geometry on the other side, with the latter determining the external potential for the fast movements of electrons within the BO approximation.

As we have already remarked before, changes in the distribution of electrons due to the substrate interaction create extra forces acting on the nuclei. This Hellmann–Feynman (EP) perspective (Nakatsuji 1973, 1974a, b; Nalewajski 1999, 2000a, 2006f, g, 2009a, 2010g; Nalewajski and Korchowiec 1997; Nalewajski and Sikora 2000; Nalewajski et al. 1996a, b, 2008) is close to the intuitive chemical thinking, in which manipulations of the electronic structure are considered as preceding and ultimately accelerating the subsequent changes in the molecular geometry. The BO, EF perspective provides the complementary description, in which displacements in the nuclear positions precede the concomitant electronic relaxation. Clearly, the complete understanding of molecular mechanisms of chemical reactions must ultimately involve reactivity criteria related to both these representations. In the former the displacements in the reactant *electronic* degrees of freedom, e.g., electron densities or the condensed electron populations (or net charges) of AIM, are considered as the independent state parameters of the reactive system, with the nuclear/geometric parameters responding to this electronic *perturbation*. In the latter the displacements in the nuclear coordinates and the associated shifts in the reactant external potentials are viewed as the system state variables, with the electronic parameters responding to this perturbation in the system geometry.

The geometrical relaxations, in response to displacements in the electronic structure in the acid (acceptor)–base (donor) DA reactive system are also the subject of the intuitive bond variation rules of Gutmann (1978). They also follow the Hellmann–Feynman (EP) perspective of Nakatsuji (1973, 1974a, b), who obtained interesting interrelations between changes in the electron density and nuclear configuration in a variety of contexts associated with chemical reactions. For example, it was observed that the centroid of a change in the electron density tends to lag the change in the nuclear coordinates in a movement away from a stable configuration, and it tends to lead the geometrical change in a movement away from an unstable geometrical structure toward the equilibrium one, thus always creating the force acting toward restoring the system equilibrium geometry.

The interaction between FO is the strongest, when the mutual orientation of both reactants, at a given intermolecular separation, gives the maximum overlap between the electron-donating (HOMO) orbitals of one reactant and the electron-accepting (LUMO) orbitals of the reaction partner. The FO theory identifies such a maximum overlap direction as the preferred one. It thus represents the overlap matching rule of chemical reactivity, valid for the *frontier*-controlled reactions in the Klopman classification. The corresponding rule for the *charge*-controlled reactions follows from the corresponding matching of the ESP of both reactants. Indeed, it can be expected that the largest charge stabilization should result, when the electron-deficient, positive regions of the ESP of one reactant overlap the most with the electron-rich, negative regions of ESP of the other reactant. This electrostatic analysis has now become firmly established as an effective guide to molecular interactions (e.g., Politzer and Truhlar 1981; Murray and Sen 1996; Politzer and Murray 2009). It is being applied to a variety of chemical and biological systems, covering the preferred sites for the N and E attacks, solvent effects, catalysis, as well as the molecular cluster and crystal behavior.

The ESP is the unique functional of the molecular electron density and it exhibits interesting critical points, which reflect the opposing contributions from the nuclei and electrons. Its topological analysis (Gadre and Shirsat 2000) supplements the related treatment of the electron density (Bader 1990, 1991) and ELF (Silvi and Savin 1994). Bader's investigations of the local features of the electron density distribution and the associated Laplace field lead to the unique topological definition of AIM, chemical bonds, molecular structure, and structural changes. It can also be used to gain a valuable insight into the bonding mechanism, to identify the sites that are prone to the electrophilic or nucleophilic attacks, and the bonds that can be easily broken in a molecule (Bader 1990, 1991; Kraka and Cremer 1990). The maps of local features of the Laplacian of the molecular density can be also used to predict the best matching of the electron-depleted regions of the acidic reactant with the electron-rich regions of the basic reactant, which can qualitatively determine their preferred mutual orientation in a chemical reaction.

To conclude this short outline of the main ideas of modern theories of the molecular electronic structure and chemical reactivity, one should also recognize the important insights from the VB theory (Heitler and London 1927; London 1928). This historic rival of the MO approach has dominated chemistry until the mid-1950s and made a strong comeback from 1980s onward (e.g., Cooper 2002;

Shaik 1989; Shaik and Hiberty 1991, 1995, 2004). Its roots can be traced to the classical paper of Lewis (1916). The VB theory introduces into chemistry the concept of the bonding electron pair and the octet rule. The qualitative VB theory gives rise to a lucid insight into the elementary chemical processes and produces key paradigms of chemical bonding and reactivity. It allows one to successfully tackle various issues in the molecular theory, including the aromaticity–antiaromaticity and the VB diagrams conceptualizing the chemical reactivity and the barrier formation by avoided crossing (resonance mixing) of the VB states that describe reactants and products. This qualitative theory also offers complementary insights into the factors that control the barrier heights and into the competition between the σ and π electrons in determining the regular structure of the benzene ring. The quantitative variants, the ab initio VB methods, e.g., Generalized VB (GVB) scheme of Goddard and Harding (1978), provide efficient computational tools for determining the outcomes of chemical reactions. One should also mention the use of the VB ideas in modeling the PES of elementary chemical reactions for subsequent dynamical simulations, e.g., the familiar LEPS (London–Eyring–Polanyi–Sato), DIM (Diatomics in Molecules), and BEBO (Bond Energy–Bond Order) approaches. These analytical PES are routinely used in numerical studies of the reactive scattering (e.g., Murrell et al. 1984).

For the bimolecular reactive system the *second*-order Taylor expansion of the electronic energy provides a consistent *two*-reactant framework in which all couplings between the familiar descriptors of reactants can be adequately accounted for. It should be realized, however, that in this approximation one probes the reactant behavior only through their leading responses, which are linear functions of the applied perturbations. In other words, such a quadratic approach amounts to the *Linear Response* (LR) approximation in theory of chemical reactivity. In order to include the nonlinear effects an expansion to *higher* orders, e.g., including cubic terms, should be applied (Senet 1996, 1997, 2009). However, the extra energy contributions in such more elaborate Taylor series should generally be small compared with those already present in the quadratic, LR approximation of the shift in electronic energy relative to the SRL, $\Delta E^{(1+2)}$, while grossly hampering the interpretation. These higher order terms can be thus thought of as only slightly modifying CS of the quadratic approach into those representing the "dressed" fragments of the reactive system.

The presence of the mutual coupling between reactants indicates that for the fixed *reaction stimuli* (perturbations) in the reactive system the trends in the electronic energy (reactivity) have to be indexed by vectors of generalized potentials and matrices of CS, since no single response quantity can fully reflect the net effect of a complicated pattern of all such couplings present in the bimolecular system. Moreover, changes in the electronic energy have to be supplemented in the full (*second*-order) interaction potential in BO approximation, $\Delta W^{(1+2)}$, with the trivial nuclear repulsion between reactants, V_{nn},

$$\Delta W^{(1+2)} \equiv \Delta E^{(1+2)} + V_{nn}, \tag{13.1}$$

for the current geometry of the whole reactive system. At the finite interreactant separation the V_{nn} term has to be taken into account also in approximate, semiquantitative treatments which use the separated reactant responses to approximate the Taylor expansion of the electronic energy.

For a series of similar reactions, e.g., attacks at alternative sites of the same molecule by the atomic agents of a similar chemical character, one can assume that at comparable stages of the reactant approach the sum of the first-order energy $\Delta W^{(1)}$ remains approximately constant for all compared locations, so that trends in the second-order energy (see Sect. 7.3.7) for alternative reaction pathways determine the preferred course of the reaction in question. Only for comparable values of the reaction perturbations (stimuli) at alternative sites the trends in the quadratic terms of the electronic energy are then reflected by CS themselves.

Notice, however, that such a simplified treatment cannot be used in a general reactive system involving two large reactants, reactivity of which changes in a series of compared mutual orientations, often leading to different reaction paths. This is because at each site alternative orientations of reactants imply a different matching between the perturbation/potential and responses quantities. In such a general case the whole set of molecular potentials and CS has to be combined with the appropriate reaction stimuli to produce the overall interaction energy, to be eventually compared for a series of the probed geometries or pathways of the whole reactive system under consideration. This is important for predicting a direction of the energetically preferred approach of reactants at the crucial, early stage of the reaction, which sets the least activation course of the process and thus ultimately selects the preferred reaction event itself.

A remarkable progress in the DFT of Kohn, Hohenberg, and Sham besides offering efficient schemes for the electronic structure computations has also provided an attractive framework for formulating novel concepts and rules describing behavior of molecular systems in different chemical environments. In chemistry, and particularly in reactivity theory, this conceptual development has had a distinctly unifying character. For example, some of the originally intuitive, but remarkably successful tools of chemistry, such as the electronegativity and hardness, which have long been part of the chemical vocabulary, have been shown to be fundamental and well defined.

According to HK theorems the ground state electron density $\rho(\boldsymbol{r})$, or the density per electron $p(\boldsymbol{r}) = \rho(\boldsymbol{r})/N$ (one-electron probability distribution called the *shape* factor), carries the complete "information" about the nondegenerate quantum mechanical state of a molecule. It uniquely identifies the shape of the system external potential due to the nuclei and the overall number of electrons and hence also the Coulombic molecular Hamiltonian. This exact result has given a new impetus toward the density-based chemical interpretations of the electronic structure of molecular systems and their diverse reactivity preferences in terms of AIM and bond multiplicities.

Such density-related characteristics of reactants have already been discussed in Sect. 7.3.7. In this chapter, we shall further elaborate on the use of these CS in the full *two*-reactant description of the bimolecular reactive systems, in which all electronic–geometrical interactions are explicitly taken into account (e.g., Nalewajski 2006f, g, 2009a, 2010g; Nalewajski and Korchowiec 1997; Nalewajski et al. 1996a, 2008).

13.3 *Horizontal* and *Vertical* Displacements of Molecular Electronic Structure

In a thermodynamic-like treatment of molecular systems one aims at determining changes in the electronic structure accompanying the displacements from one equilibrium (ground state) density to another: $\rho_1 = \rho[N_1, v_1] \rightarrow \rho_2 = \rho[N_2, v_2]$, where for the externally closed systems $N_1 = N_2 = N$. As indicated above, these electron distributions are uniquely determined by the two state parameters determining the corresponding electronic Hamiltonians: the overall number of electrons N_i and the external potential due to the nuclei, $v_i(\mathbf{r})$, $i = 1, 2$ (see Sect. 7.1). We call such shifts the "*horizontal*" displacements in the system electronic structure, along the ground state energy surface $E[N, v] = E_v[\rho[N, v]]$, for the current ground state density $\rho = \rho[N, v]$.

The given horizontal displacement $\Delta\rho = \rho_2 - \rho_1$, from one equilibrium ($v$-representable) electron density to another, due to $\Delta N = N_2 - N_1$ and $\Delta v = v_2 - v_1$, gives rise to the associated *first*-order change,

$$\Delta^{(1)}E[N,v] = \mu_1\Delta N + \int \Delta v(\mathbf{r})\rho_1(\mathbf{r})d\mathbf{r}, \quad \mu_1 = \mu[N_1, v_1], \tag{13.2}$$

in the EF density functional for the system ground state energy:

$$E_v[\rho[N,v]] \equiv \int v(\mathbf{r})\rho[N,v;\mathbf{r}]d\mathbf{r} + F[\rho[N,v]].$$

The latter can be expressed in the equivalent EP form,

$$E[\rho] \equiv E_{v[\rho]}[\rho] = E[N[\rho], v[\rho]] = \int v[\rho;\mathbf{r}]\rho(\mathbf{r})d\mathbf{r} + F[\rho],$$

which also gives [see (7.21)]:

$$\begin{aligned}\Delta^{(1)}E[\rho] &= \int \{v_1(\mathbf{r}') - u_1(\mathbf{r}') + \int \left.\frac{\delta v(\mathbf{r})}{\delta\rho(\mathbf{r}')}\right|_1 \rho_1(\mathbf{r})\,d\mathbf{r}\}\,\Delta\rho(\mathbf{r}')\,d\mathbf{r}' \\ &= \mu_1\Delta N + \int \Delta v(\mathbf{r})\rho_1(\mathbf{r})d\mathbf{r}.\end{aligned}$$

Here, the external potential changes with the current electron density in such a way that the current electron distribution ρ matches v as its ground state (equilibrium) density: $\rho = \rho[v]$ and $v = v[\rho]$. It should be emphasized that this density functional for the *ground state* energy differs from the HK functional for the *variational* density $E_v[\rho']$, in which the external potential is fixed [not related to the current (trial) density ρ']. Only for the exact ground state density, satisfying the HK minimum principle (7.21),

$$E_v[\rho[N,v]] = E_{v[\rho]}[\rho] \equiv E[\rho] = \bar{E}_u[\rho[u]] = \bar{E}_{u[\rho]}[\rho] \equiv \bar{E}[u]. \qquad (13.3)$$

We have indicated in the preceding equation that the energy functional $E[\rho]$ can be alternatively interpreted as the functional $\bar{E}[u]$ of the relative external potential $u(\boldsymbol{r}) = v(\boldsymbol{r}) - \mu$. Indeed, the Euler equation (7.21) implies that the density and relative external potential are the unique functionals of each other.

In what follows we shall use the term *Softness Representation* to denote the EF description of molecular equilibrium states, which uses the (nuclear) external potential as the *independent* (local) state parameter. In the associated variational principle the given external potential u then determines its companion (equilibrium) density $\rho[u]$. In this approach, to which both the variational functional, $E_v[\rho']$, of a trial ρ', and the ground state functional $\bar{E}_u[\rho[u]] = \bar{E}[u]$ belong, the electron density represents the *dependent* state variable. Therefore, this *softness* "picture" of DFT indeed adopts the EF perspective of the BO approximation.

Selecting the electron density as an *independent* state variable, e.g., in the variational functional $E[\rho] = E_{v[\rho]}[\rho] = \bar{E}_{u[\rho]}[\rho]$, for the *dependent* v or u, gives rise to the complementary EP description. In the associated variational procedure, the given (v-representable) electron density ρ determines the matching external potential $u[\rho]$. In this *Hardness Representation* of DFT, in which one defines the ground state energy functional $E[\rho] = \bar{E}_{u[\rho]}[\rho] \equiv E_\rho[u[\rho]]$, one thus adopts the EP perspective of the Hellmann–Feynman theorem.

The Euler equation (7.21) also implies that the external potential $v(\boldsymbol{r})$ in the open molecular systems is determined by ρ only up to a constant μ, which can be related to the chemical potential of an external *electron reservoir* (r): $\mu = \mu_r = \mu[\rho]$. Therefore, the unique specification of the external potential as the functional of ρ additionally requires the knowledge of this equalized state parameter, $\mu = \mu(\boldsymbol{r})$, the "*intensive*" conjugate of N or $\rho(\boldsymbol{r})$:

$$v(\boldsymbol{r}) = u(\boldsymbol{r}) + \mu = \mu - \delta F/\delta\rho(\boldsymbol{r}) \equiv v[\mu[\rho], \rho; \boldsymbol{r}] = v[\rho; \boldsymbol{r}]. \qquad (13.4)$$

This "horizontal" character of changes in the ground state electronic structure is in contrast to a search for the equilibrium (exhaustive) partition of the molecular ground state density $\rho(\boldsymbol{r})$ into densities $\boldsymbol{\rho}(\boldsymbol{r}) \equiv \{\rho_\alpha(\boldsymbol{r}) = \rho_\alpha[\rho; \boldsymbol{r}]\}$ (a *row* vector) of the constituent subsystems, e.g., the density pieces attributed to AIM, which at each point sum up to this given molecular density: $\rho(\boldsymbol{r}) = \sum_\alpha \rho_\alpha(\boldsymbol{r})$ (see Chap. 11). This density division problem is "*vertical*" (entropic) in character, with the search being performed for the fixed molecular density and hence also for the fixed value of the electronic energy.

This is also the case in Levy's constrained search construction of the universal functional $F[\rho]$ (7.29):

$$F[\rho] = \inf_{\Psi \to \rho} \langle \Psi | \hat{\mathrm{F}} | \Psi \rangle = \langle \Psi[\rho] | \hat{\mathrm{F}} | \Psi[\rho] \rangle, \tag{13.5}$$

in which one searches over all wave functions (or density operators) yielding a given electron density and calculates the density functional $F[\rho]$ as the lowest value (infimum) of the expectation (or ensemble-average) values of the sum of the electron kinetic and repulsion energy operators. Since this search is performed for the fixed (ground state) density, it also implies the fixed value of the system electronic energy. Therefore, by analogy to the maximum principle of the thermodynamic entropy for constant internal energy in the ordinary phenomenological thermodynamics, this DFT minimum principle can be regarded as being also "entropic" in character.

Moreover, since $u = u[\rho]$ and $\rho = \rho[u]$, the universal density functional $F[\rho]$ can be alternatively regarded as the associated functional $\bar{F}[u]$ of the relative external potential:

$$F[\rho] = F[\rho[u]] \equiv \bar{F}[u]. \tag{13.6}$$

The corresponding density-constrained search for the external potential matching the given (v-representable) density ρ thus reads:

$$F[\rho] = \sup_{v' \to \rho} \{E[N, v'] - \int v'(\boldsymbol{r})\, \rho(\boldsymbol{r})\, d\boldsymbol{r}\}. \tag{13.7}$$

In this extremum principle, one searches over external potentials $v' \to \rho$, which give rise to the specified ground state density ρ, and determines the maximum (suprimum) of the Legendre transform of the system energy, which replaces the external potential v' by the ground state density ρ in the list of the system independent state parameters. At the solution point this variational principle yields the optimum external potential $v = v[\rho]$, which identifies the specified (v-representable) density as its ground state, equilibrium distribution. This constrained search construction can be generalized to any trial density ρ':

$$F[\rho'] = \sup_{v'} \{E_{v'}[\rho'] - \int v'(\boldsymbol{r})\, \rho'(\boldsymbol{r})\, d\boldsymbol{r}\}. \tag{13.8}$$

As argued by Nalewajski and Parr (2001), the AIM division problem of the fixed molecular density ρ also represents a search for the optimum *effective* external potentials $v^{eff} = \{v^{eff}_{\alpha}\}$ of atomic subsystems:

$$v_\alpha^{eff.}(\boldsymbol{r}) = v_\alpha^{eff.}[\boldsymbol{\rho}[\rho]; \boldsymbol{r}] = v(\boldsymbol{r}) + \left(\frac{\partial F^{nadd.}[\boldsymbol{\rho}]}{\partial \rho_\alpha(\boldsymbol{r})}\right)_{\beta \neq \alpha}, \quad \alpha = 1, 2, \ldots \tag{13.9}$$

Here, the partial differentiation with respect to $\rho_\alpha(\boldsymbol{r})$ of the nonadditive part $F^{nadd.}[\boldsymbol{\rho}]$ of the *total* Hohenberg–Kohn–Levy functional in AIM resolution, $F[\rho] \equiv F^{total}[\boldsymbol{\rho}]$,

$$F^{nadd.}[\boldsymbol{\rho}] \equiv F[\rho] - \Sigma_\gamma F[\rho_\gamma] \equiv F^{total}[\boldsymbol{\rho}] - F^{add.}[\boldsymbol{\rho}], \tag{13.10}$$

where $F^{add.}[\boldsymbol{\rho}]$ denotes the *additive* part of the overall functional $F^{total}[\boldsymbol{\rho}]$, is carried out for the fixed densities of the remaining subsystems $\{\rho_{\beta \neq \alpha}\}$.

These effective external potentials of the (*embedded*) bonded atoms in the molecule are then related to their respective densities through the Euler equation for the subsystem densities:

$$v_\alpha^{eff.}(\boldsymbol{r}) - \mu_\alpha = u_\alpha^{eff}[\boldsymbol{\rho}[\rho]; \boldsymbol{r}] \equiv -\frac{\delta F[\rho_\alpha]}{\delta \rho_\alpha(\boldsymbol{r})}, \tag{13.11}$$

where the equalized subsystem chemical potential $\mu_\alpha = \mu_\alpha(\boldsymbol{r})$ is given by the partial derivative,

$$\mu_\alpha(\boldsymbol{r}) = \frac{\partial E_v^{total}[\boldsymbol{\rho}]}{\partial \rho_\alpha(\boldsymbol{r})} = \left(\frac{\partial E^{total}[\boldsymbol{N}, v]}{\partial N_\alpha}\right)_v, \tag{13.12}$$

of the system electronic energy in atomic resolution,

$$E_v[\rho] \equiv E_v^{total}[\boldsymbol{\rho}[\rho]] = E[N, v] \equiv E^{total}[\boldsymbol{N}, v]. \tag{13.13}$$

Here, the vector $\boldsymbol{N} \equiv \{N_\alpha = \int \rho_\alpha(\boldsymbol{r}) d\boldsymbol{r} \equiv N[\rho_\alpha]\}$ groups the average numbers of electrons in atomic subsystems.

Clearly, for the mutually open atomic subsystems, with no barriers preventing the flow of electrons between them, the AIM chemical potentials are equalized at the global level characterizing the molecular system as a whole [see (7.18)]:

$$\boldsymbol{\mu}(\boldsymbol{r}) \equiv \{\mu_\alpha(\boldsymbol{r})\} = \boldsymbol{\mu} \equiv \{\mu_\alpha\} = \mu \boldsymbol{1}, \tag{13.14}$$

where the unit row vector $\boldsymbol{1} = (1, 1, \ldots)$. The same (equalized) chemical potentials of AIM thus follow from the density functional for the electronic energy of the embedded atom α:

$$\mathcal{E}_{v_\alpha}[\boldsymbol{\rho}] = \left\{\int v(\boldsymbol{r}) \rho_\alpha(\boldsymbol{r}) d\boldsymbol{r} + F[\rho_\alpha]\right\} + F^{nadd.}[\boldsymbol{\rho}] \equiv E_v[\rho_\alpha] + F^{nadd.}[\boldsymbol{\rho}], \tag{13.15}$$

where $E_v[\rho_\alpha]$ stands for the electronic energy of ρ_α alone in the molecular external potential v, and $F^{nadd.}[\boldsymbol{\rho}]$ represents the *embedding energy* due to the presence of electrons on the remaining subsystems:

$$\mu_\alpha(\boldsymbol{r}) = \partial\mathcal{E}_{v_\alpha}[\boldsymbol{\rho}]/\partial\rho_\alpha(\boldsymbol{r}) = \mu_\alpha = \mu, \quad \alpha = 1, 2, \ldots \tag{13.16}$$

We therefore conclude that the DFT description of the equilibrium states in molecules and of their mutually open subsystems are isomorphic, since the subsystem effective (relative) external potentials are related to their electron densities through the same "horizontal"-type Euler equation linking the complementary EF and EP representations of the molecular electronic structure.

13.4 Constrained Equilibria in Molecular Subsystems and Charge Sensitivities of Reactants

The chemical potentials $\boldsymbol{\mu} = \{\mu_\alpha\}$ of the mutually-open subsystems have been defined by the *partial* functional derivatives:

$$\mu_\alpha(\boldsymbol{r}) = \left\{\partial E_v^{total}[\boldsymbol{\rho}]/\partial\rho_\alpha(\boldsymbol{r})\right\}_{v,\beta\neq\alpha} = \mu_\alpha = \mu, \quad \alpha = 1, 2, \ldots, m. \tag{13.17}$$

Here the row vector of subsystem densities $\boldsymbol{\rho} = (\rho_\alpha, \rho_\beta, \ldots)$ gives rise to the overall density $\rho = \sum_\gamma \rho_\gamma = \boldsymbol{1}\boldsymbol{\rho}^{\mathrm{T}}$. One similarly introduces the row vector the molecular external potential, due to all nuclei in the molecule: $\boldsymbol{v}(\boldsymbol{r}) = v(\boldsymbol{r})\boldsymbol{1}$. The chemical potentials of molecular fragments are thus determined for the fixed external potential $v(\boldsymbol{r})$ due to the system nuclei and the "frozen" embedding densities $\{\rho_{\beta\neq\alpha}(\boldsymbol{r})\}$ of all remaining subsystems. They are equalized at the molecular chemical potential level μ, $\boldsymbol{\mu} = \mu\boldsymbol{1}$, when these fragments are mutually *opened*, in the *global* equilibrium state. We shall denote such relative condition of molecular subsystems by the vertical *broken* lines in the symbolic representation of the composite (molecular) system in the *global* (g) (intersubsystem) equilibrium of the ground state of an externally open molecule: $M_g = (\alpha\mathord{\vdots}\beta\mathord{\vdots}\gamma\mathord{\vdots}\ldots)$. This equalization of the subsystem chemical potentials, $\boldsymbol{\mu} = (\mu_\alpha, \mu_\beta, \mu_\gamma, \ldots) = \mu\boldsymbol{1}$, can be also attributed to a single external reservoir (r) of electrons, *common* to all constituent fragments, exhibiting the chemical potential

$$\mu_r = \mu_\alpha = \mu[\rho], \quad \alpha = 1, 2, \ldots, m, \tag{13.18}$$

to which these molecular subsystems are coupled in the hypothetical combined system $(r\mathord{\vdots}M_g) = (r\mathord{\vdots}\alpha\mathord{\vdots}\beta\mathord{\vdots}\gamma\mathord{\vdots}\ldots)$.

In this subsystem resolution, one also considers the *constrained* (intrasubsystem) equilibrium states, when all subsystems are mutually *closed*. This is accordingly symbolized by the vertical *solid* lines in the symbolic representation of a collection of the mutually closed fragments of $M_c = (\alpha \mid \beta \mid \ldots)$. In order to probe properties of the externally open subsystems in M_c, when each molecular fragment is characterized by the intrasubsystem equalized chemical potential, at generally different level for each subsystem,

$$\mu_\alpha(\boldsymbol{r}) = \mu_\alpha = \mu_r^\alpha \neq \mu_\beta(\boldsymbol{r}) = \mu_\beta = \mu_r^\beta \neq \ldots \neq \mu, \tag{13.19}$$

one envisages the *separate* electron reservoirs $\{r_\alpha\}$ for each subsystem, characterized by the independently controlled chemical potentials $\{\mu_r^\alpha\}$, in the combined system $(r_\alpha \mid \alpha \mid r_\beta \mid \beta \mid \ldots)$. Notice that only for the global equilibrium, when $r_\alpha = r_\beta = \ldots = r$, the fragment chemical potentials are equalized: $\boldsymbol{\mu} = \mu \boldsymbol{1}$.

When only a single subsystem or a subset of molecular fragments is considered externally open, while the remaining subsystems are externally closed, one similarly envisages the coupling of a specified single or selection of subsystems to their corresponding reservoirs, e.g., in $(r_\alpha \mid \alpha | \beta | \gamma | \ldots)$. The equilibrium state of the subsystem α in contact with r_α is then characterized by the equalization of this subsystem chemical potential and that of its reservoir: $\mu_r^\alpha = \mu_\alpha$. Therefore, the hypothetical, independent displacements of the subsystem chemical potentials reflect changes performed on the corresponding subsystem reservoirs, e.g., $d\mu_\alpha = d\mu_r^\alpha$.

In this subsystem resolution one defines the row vector of the AIM relative external potentials:

$$\begin{aligned} \boldsymbol{u}(\boldsymbol{r}) &\equiv -\partial F^{total}[\boldsymbol{\rho}]/\partial \boldsymbol{\rho}(\boldsymbol{r}) = v(\boldsymbol{r})\boldsymbol{1} - \boldsymbol{\mu} \equiv \boldsymbol{v}(\boldsymbol{r}) - \boldsymbol{\mu} \\ &= \{u_\alpha(r) \equiv v(r) - \mu_\alpha = -\,\partial F^{total}[\boldsymbol{\rho}]/\partial \rho_\alpha(\boldsymbol{r}) \\ &\qquad = -\,\partial F^{add.}[\boldsymbol{\rho}]/\partial \rho_\alpha(\boldsymbol{r}) - \partial F^{nadd.}[\boldsymbol{\rho}]/\partial \rho_\alpha(\boldsymbol{r})\}, \end{aligned} \tag{13.20}$$

They are defined by the negative partial derivatives of the functional $F^{total}[\boldsymbol{\rho}]$, which has been partitioned in (13.10) into the additive (*add.*) and nonadditive (*nadd.*) parts in the subsystem resolution in question. It provides the universal (v-independent) part of the energy density functional $E_v[\boldsymbol{\rho}]$ in the subsystem resolution:

$$\begin{aligned} E_v^{total}[\boldsymbol{\rho}] &= \int \boldsymbol{v}(\boldsymbol{r})\boldsymbol{\rho}(\boldsymbol{r})^{\mathrm{T}} d\boldsymbol{r} + F^{total}[\boldsymbol{\rho}] \\ &= \int v(\boldsymbol{r})\rho(\boldsymbol{r}) d\boldsymbol{r} + F^{add}[\boldsymbol{\rho}] + F^{nadd}[\boldsymbol{\rho}]. \end{aligned} \tag{13.21}$$

The optimum (open) subsystem densities in such constrained-equilibrium state then follow from the variational principle,

$$\delta\{E_v^{total}[\boldsymbol{\rho}] - \Sigma_\gamma \mu_\gamma N[\rho_\gamma]\} = 0,$$

giving rise to the associated Euler equations

$$\begin{aligned}\{\partial E_v^{total}[\boldsymbol{\rho}]/\partial\rho_\alpha(\boldsymbol{r})\}_{\beta\neq\alpha} &\equiv \mu_\alpha(\boldsymbol{r}) = v(\boldsymbol{r}) + \{\partial F^{total}[\boldsymbol{\rho}]/\partial\rho_\alpha(\boldsymbol{r})\}_{\beta\neq\alpha}\\ &= v_\alpha^{eff.}(\boldsymbol{r}) + \delta F[\rho_\alpha]/\delta\rho_\alpha(\boldsymbol{r}) = \mu_\alpha, \quad \alpha = 1, 2, \ldots, m,\end{aligned} \tag{13.22}$$

and hence:

$$u_\alpha(\boldsymbol{r}) = v(\boldsymbol{r}) - \mu_\alpha = -\{\partial F^{total}[\boldsymbol{\rho}]/\partial\rho_\alpha(\boldsymbol{r})\}_{\beta\neq\alpha} \quad \text{or} \quad u_\alpha^{eff}(\boldsymbol{r}) \equiv -\delta F[\rho_\alpha]/\delta\rho_\alpha(\boldsymbol{r}).$$

Therefore, the equilibrium densities of the open subsystems are unique functionals of the molecular external potential and the reservoir chemical potentials,

$$\boldsymbol{\rho} = \boldsymbol{\rho}[\boldsymbol{u}[\boldsymbol{\mu}, v]] = \boldsymbol{\rho}[\boldsymbol{\mu}, v],$$

and so is the associated row vector of the average (fractional) numbers of electrons:

$$\boldsymbol{N}[\boldsymbol{\rho}] = \int \boldsymbol{\rho}[\boldsymbol{\mu}, v; \boldsymbol{r}]d\boldsymbol{r}.$$

Similarly, when all subsystems are both mutually and externally closed in $M \equiv (\alpha|\beta| \ldots)$, the external potential and the numbers of electrons in subsystems uniquely determine the equilibrium densities $\boldsymbol{\rho} = \boldsymbol{\rho}[\boldsymbol{N},v]$ of molecular fragments, the system energy, $E_v^{total}[\boldsymbol{\rho}] = E_v^{total}[\boldsymbol{\rho}[\boldsymbol{N}, v]] = E_v^{total}[\boldsymbol{N}, v]$, and all its physical properties, e.g., the subsystem chemical potentials $\boldsymbol{\mu} = \boldsymbol{\mu}[\boldsymbol{N},v]$.

One introduces the "thermodynamic" potentials corresponding to the four alternative sets of the subsystem *state* parameters, which define the corresponding constrained equilibrium states,

$$\{\boldsymbol{N}, v\}, \quad \{\boldsymbol{\mu}, v\} \equiv \{\boldsymbol{u}\}, \quad \{\boldsymbol{N}, \boldsymbol{\rho}\} \equiv \{\boldsymbol{\rho}\}, \quad \{\boldsymbol{\mu}, \boldsymbol{\rho}\}, \tag{13.23}$$

as the associated Legendre transforms of the electronic energy $E^{total}[\boldsymbol{N}, v]$:

$$\Omega^{total}[\boldsymbol{\mu}, v] = E - \boldsymbol{N}(\partial E/\partial \boldsymbol{N})_v^{\mathrm{T}} = E[\boldsymbol{\rho}[\boldsymbol{u}]] - \boldsymbol{N}[\boldsymbol{\rho}[\boldsymbol{u}]]\boldsymbol{\mu}[\boldsymbol{\rho}[\boldsymbol{u}]]^{\mathrm{T}} = \Omega^{total}[\boldsymbol{u}], \tag{13.24}$$

$$F^{total}[\boldsymbol{\rho}] = E - \int v(\boldsymbol{r})\{\partial E/\partial v(\boldsymbol{r})\}_{\boldsymbol{N}}^{\mathrm{T}} d\boldsymbol{r} = E - \int v(\boldsymbol{r})\boldsymbol{\rho}(\boldsymbol{r})^{\mathrm{T}} d\boldsymbol{r}, \tag{13.25}$$

$$\begin{aligned}R^{total}[\boldsymbol{\mu}, \boldsymbol{\rho}] &= E - \boldsymbol{N}(\partial E/\partial \boldsymbol{N})_v^{\mathrm{T}} - \int v(\boldsymbol{r})\{\partial E/\partial v(\boldsymbol{r})\}_{\boldsymbol{N}}^{\mathrm{T}} d\boldsymbol{r}\\ &= F^{total}[\boldsymbol{\rho}] - \boldsymbol{N}[\boldsymbol{\rho}]\boldsymbol{\mu}[\boldsymbol{\rho}]^{\mathrm{T}}.\end{aligned} \tag{13.26}$$

In this subsystem resolution one could also consider all intermediate specifications of the molecular constrained equilibria, when only a part of the subsystems remains externally open (characterized by the fixed chemical potentials of a common or separate reservoirs) with the remaining, complementary set of molecular fragments being closed, characterized by the fixed (integer) numbers of electrons. We would like to observe that in the theory of chemical reactivity these partially open situations do indeed arise, e.g., in the surface reactions, when one adsorbate reactant is closed (physisorbed) and the other remains open (chemisorbed) with respect to the catalytic surface, with the latter then acting as the external electron reservoir.

The corresponding differentials of the system electronic energy and its Legendre transforms in this subsystem resolution then read (e.g., Nalewajski 1983, 1993a, b, 2000b, 2002d, 2003a, 2006f, g; Nalewajski and Korchowiec 1997; Nalewajski et al. 1996a):

$$dE^{total}[\boldsymbol{N}, \boldsymbol{v}] = \boldsymbol{\mu}[\boldsymbol{N}, \boldsymbol{v}]\, d\boldsymbol{N}^{\mathrm{T}} + \int \boldsymbol{\rho}[\boldsymbol{N}, \boldsymbol{v}; \boldsymbol{r}] d\boldsymbol{v}(\boldsymbol{r})^{\mathrm{T}} d\boldsymbol{r}. \tag{13.27}$$

$$\begin{aligned} d\Omega^{total}[\boldsymbol{\mu}, \boldsymbol{v}] &= -\boldsymbol{N}[\boldsymbol{\mu}, \boldsymbol{v}] d\boldsymbol{\mu}^{\mathrm{T}} + \int \boldsymbol{\rho}[\boldsymbol{\mu}, \boldsymbol{v}; \boldsymbol{r}] d\boldsymbol{v}(\boldsymbol{r})^{\mathrm{T}} d\boldsymbol{r} \\ &= \int \boldsymbol{\rho}[\boldsymbol{u}; \boldsymbol{r}] d\boldsymbol{u}(\boldsymbol{r})^{\mathrm{T}} d\boldsymbol{r} = d\Omega^{total}[\boldsymbol{u}], \end{aligned} \tag{13.28}$$

$$dF^{total}[\boldsymbol{\rho}] = \boldsymbol{\mu}[\boldsymbol{\rho}] d\boldsymbol{N}^{\mathrm{T}} - \int \boldsymbol{v}[\boldsymbol{\rho}; \boldsymbol{r}] d\boldsymbol{\rho}(\boldsymbol{r})^{\mathrm{T}} d\boldsymbol{r} = -\int \boldsymbol{u}[\boldsymbol{\rho}; \boldsymbol{r}] d\boldsymbol{\rho}(\boldsymbol{r})^{\mathrm{T}} d\boldsymbol{r}, \tag{13.29}$$

$$dR^{total}[\boldsymbol{\mu}, \boldsymbol{\rho}] = -\boldsymbol{N}[\boldsymbol{\mu}, \boldsymbol{\rho}] d\boldsymbol{\mu}^{\mathrm{T}} - \int \boldsymbol{v}[\boldsymbol{\mu}, \boldsymbol{\rho}; \boldsymbol{r}]\, d\boldsymbol{\rho}(\boldsymbol{r})^{\mathrm{T}} d\boldsymbol{r}, \tag{13.30}$$

where displacements $d\boldsymbol{v}(\boldsymbol{r})$ allow for the independent changes of the external potential of each subsystem. These differential expressions identify the associated conjugate "intensities" to state-variables defining the representation under consideration:

$$\boldsymbol{\mu}[\boldsymbol{N}, \boldsymbol{v}] = (\partial E^{total}[\boldsymbol{N}, \boldsymbol{v}]/\partial \boldsymbol{N})_{\boldsymbol{v}}, \qquad \boldsymbol{\rho}[\boldsymbol{N}, \boldsymbol{v}; \boldsymbol{r}] = \{\partial E^{total}[\boldsymbol{N}, \boldsymbol{v}]/\partial \boldsymbol{v}(\boldsymbol{r})\}_{\boldsymbol{N}}; \tag{13.31}$$

$$\begin{aligned} \boldsymbol{N}[\boldsymbol{\mu}, \boldsymbol{v}] = -(\partial \Omega^{total}[\boldsymbol{\mu}, \boldsymbol{v}]/\partial \boldsymbol{\mu})_{\boldsymbol{v}}, \quad \boldsymbol{\rho}[\boldsymbol{\mu}, \boldsymbol{v}; \boldsymbol{r}] &= \{\partial \Omega^{total}[\boldsymbol{\mu}, \boldsymbol{v}]/\partial \boldsymbol{v}(\boldsymbol{r})\}_{\boldsymbol{\mu}} \\ &= \delta \Omega^{total}[\boldsymbol{u}]/\delta \boldsymbol{u}(\boldsymbol{r}); \end{aligned} \tag{13.32}$$

$$-\boldsymbol{u}[\boldsymbol{\rho}; \boldsymbol{r}] = \boldsymbol{\mu} - \boldsymbol{v}(\boldsymbol{r}) = \delta F^{total}[\boldsymbol{\rho}]/\delta \boldsymbol{\rho}(\boldsymbol{r}); \tag{13.33}$$

$$-\boldsymbol{N}[\boldsymbol{\mu}, \boldsymbol{\rho}] = (\partial R^{total}[\boldsymbol{\mu}, \boldsymbol{\rho}]/\partial \boldsymbol{\mu})_{\boldsymbol{\rho}}, \quad -\boldsymbol{v}[\boldsymbol{\mu}, \boldsymbol{\rho}; \boldsymbol{r}] = \{\partial R^{total}[\boldsymbol{\mu}, \boldsymbol{\rho}]/\partial \boldsymbol{\rho}(\boldsymbol{r})\}_{\boldsymbol{\mu}}. \tag{13.34}$$

In this subsystem resolution the corresponding *second*-order Taylor expansion of the molecular electronic energy in powers of displacements [$d\boldsymbol{N}$, $d\boldsymbol{v}(\boldsymbol{r})$] of the *canonical* state parameters involves the relevant principal derivatives, potentials and canonical CS of the energy representation:

$$\begin{aligned}\Delta^{(1+2)}E^{total}[\boldsymbol{N},\boldsymbol{v}] &= (\partial E^{total}[\boldsymbol{N},\boldsymbol{v}]/\partial \boldsymbol{N})_{\boldsymbol{v}} d\boldsymbol{N}^{\mathrm{T}} + \int \{\partial E^{total}[\boldsymbol{N},\boldsymbol{v}]/\partial \boldsymbol{v}(\boldsymbol{r})\}_{\boldsymbol{N}} d\boldsymbol{v}(\boldsymbol{r})^{\mathrm{T}} d\boldsymbol{r} \\ &+ {}^{1}\!/_{2}\{d\boldsymbol{N}(\partial^2 E^{total}[\boldsymbol{N},\boldsymbol{v}]/\partial \boldsymbol{N}\partial \boldsymbol{N})_{\boldsymbol{v}} d\boldsymbol{N}^{\mathrm{T}} \\ &+ 2d\boldsymbol{N}[\partial/\partial \boldsymbol{N}\int \{\partial E^{total}[\boldsymbol{N},\boldsymbol{v}]/\partial \boldsymbol{v}(\boldsymbol{r})\}_{\boldsymbol{N}}]_{\boldsymbol{v}} d\boldsymbol{v}(\boldsymbol{r})^{\mathrm{T}} d\boldsymbol{r} \\ &+ \iint d\boldsymbol{v}(\boldsymbol{r})\{\partial^2 E^{total}[\boldsymbol{N},\boldsymbol{v}]/\partial \boldsymbol{v}(\boldsymbol{r})\partial \boldsymbol{v}(\boldsymbol{r}')\}_{\boldsymbol{N}} d\boldsymbol{v}(\boldsymbol{r}')^{\mathrm{T}} d\boldsymbol{r}\, d\boldsymbol{r}'\} \\ &\equiv \boldsymbol{\mu}\, d\boldsymbol{N}^{\mathrm{T}} + \int \boldsymbol{\rho}(\boldsymbol{r}) d\boldsymbol{v}(\boldsymbol{r})^{\mathrm{T}} d\boldsymbol{r} + {}^{1}\!/_{2}\{d\boldsymbol{N}\mathbf{H}\, d\boldsymbol{N}^{\mathrm{T}} + 2d\boldsymbol{N}\int \mathbf{f}(\boldsymbol{r}) d\boldsymbol{v}(\boldsymbol{r})^{\mathrm{T}} d\boldsymbol{r} \\ &+ \iint d\boldsymbol{v}(\boldsymbol{r})\mathbf{B}(\boldsymbol{r},\boldsymbol{r}') d\boldsymbol{v}(\boldsymbol{r}')^{\mathrm{T}} d\boldsymbol{r}\, d\boldsymbol{r}'\}.\end{aligned} \tag{13.35}$$

The second differential in the preceding equation is determined by the following matrices of the principal CS in the subsystem resolution:

(a) *Hardness* matrix,

$$\mathbf{H} = (\partial \boldsymbol{\mu}/\partial \boldsymbol{N})_{\boldsymbol{v}} = \{H_{\alpha,\beta} = (\partial \mu_\beta/\partial N_\alpha)_{\boldsymbol{v}}\};$$

(b) *Fukui function* matrix,

$$\begin{aligned}\mathbf{f}(\boldsymbol{r}) = [\partial \boldsymbol{\rho}(\boldsymbol{r})/\partial \boldsymbol{N}]_{\boldsymbol{v}} &= [\partial \boldsymbol{\mu}/\partial \boldsymbol{v}(\boldsymbol{r})]_{\boldsymbol{N}}^{\mathrm{T}} \\ &= \{f_{\alpha,\beta}(\boldsymbol{r}) = [\partial \rho_\beta(\boldsymbol{r})/\partial N_\alpha]_{\boldsymbol{v}} = [\partial \mu_\alpha/\partial v_\beta(\boldsymbol{r})]_{\boldsymbol{N}}\};\end{aligned}$$

(c) *Linear response* matrix,

$$\mathbf{B}(\boldsymbol{r},\boldsymbol{r}') = [\partial \boldsymbol{\rho}(\boldsymbol{r}')/\partial \boldsymbol{v}(\boldsymbol{r})]_{\boldsymbol{N}} = \{B_{\alpha,\beta}(\boldsymbol{r},\boldsymbol{r}') = [\partial \rho_\beta(\boldsymbol{r}')/\partial v_\alpha(\boldsymbol{r})]_{\boldsymbol{N}}\}.$$

One also defines the corresponding matrices of the *softness* quantities at this resolution level:

(a) *Softness* matrix,

$$\mathbf{S} = (\partial \boldsymbol{N}/\partial \boldsymbol{\mu})_{\boldsymbol{v}} = \mathbf{H}^{-1} = \{S_{\alpha,\beta} = (\partial N_\beta/\partial \mu_\alpha)_{\boldsymbol{v}}\} = \int \mathbf{s}(\boldsymbol{r}) d\boldsymbol{r};$$

(b) *Local softness* matrix,

$$\mathbf{s}(\boldsymbol{r}) = -\partial \boldsymbol{N}/\partial \boldsymbol{u}(\boldsymbol{r}) = [\partial \boldsymbol{\rho}(\boldsymbol{r})/\partial \boldsymbol{\mu}]_{\boldsymbol{v}} = \{s_{\alpha,\beta}(\boldsymbol{r}) = -\partial N_\beta/\partial u_\alpha(\boldsymbol{r}) = [\partial \rho_\beta(\boldsymbol{r})/\partial \mu_\alpha]_{\boldsymbol{v}}\};$$

(c) *Softness-kernel* matrix,

$$\boldsymbol{\sigma}(\boldsymbol{r},\boldsymbol{r}') = -\partial\boldsymbol{\rho}(\boldsymbol{r}')/\partial\boldsymbol{u}(\boldsymbol{r}) = -[\partial\boldsymbol{\rho}(\boldsymbol{r}')/\partial\boldsymbol{v}(\boldsymbol{r})]_{\boldsymbol{\mu}} = \{\sigma_{\alpha,\beta}(\boldsymbol{r},\boldsymbol{r}') \\ = -[\partial\rho_\beta(\boldsymbol{r}')/\partial v_\alpha(\boldsymbol{r})]_{\boldsymbol{\mu}}\}.$$

A straightforward chain rule transformation then gives the following expression for $\mathbf{s}(\boldsymbol{r})$ in terms of $\boldsymbol{\sigma}(\boldsymbol{r}, \boldsymbol{r}')$:

$$\mathbf{s}(\boldsymbol{r}) = -\int [\partial\boldsymbol{u}(\boldsymbol{r}')/\partial\boldsymbol{\mu}]_v[\partial\boldsymbol{\rho}(\boldsymbol{r})/\partial\boldsymbol{u}(\boldsymbol{r}')]d\boldsymbol{r}' = \int \mathbf{I}\boldsymbol{\sigma}(\boldsymbol{r},\boldsymbol{r}')d\boldsymbol{r}', \tag{13.36}$$

where the identity matrix $\mathbf{I} = \{\delta_{\alpha,\beta}\}$. A similar transformation gives the expression for the Fukui function matrix in terms of the local softnesses of subsystems:

$$\mathbf{f}(\boldsymbol{r}) = \int [\partial\boldsymbol{u}(\boldsymbol{r}')/\partial N]_v[\partial\boldsymbol{\rho}(\boldsymbol{r})/\partial\boldsymbol{u}(\boldsymbol{r}')]\, d\boldsymbol{r}' = [\partial\boldsymbol{\mu}/\partial N]_v \int \boldsymbol{\sigma}(\boldsymbol{r},\boldsymbol{r}')d\boldsymbol{r}' \\ = \mathbf{H}\mathbf{s}(\boldsymbol{r}) = \mathbf{S}^{-1}\mathbf{s}(\boldsymbol{r}). \tag{13.37}$$

Multiplying both sides of the last equation from the left by $\mathbf{S} = \mathbf{H}^{-1}$ also gives:

$$\mathbf{s}(\boldsymbol{r}) = \mathbf{S}\,\mathbf{f}(\boldsymbol{r}). \tag{13.38}$$

The matrix of hardness kernels in the subsystem resolution is the inverse matrix of the softness kernels:

$$\boldsymbol{\eta}(\boldsymbol{r},\boldsymbol{r}') = \{\partial^2 E_v^{total}[\boldsymbol{\rho}]/\partial\boldsymbol{\rho}(\boldsymbol{r})\partial\boldsymbol{\rho}(\boldsymbol{r}')\}_v = \partial^2 F^{total}[\boldsymbol{\rho}]/\partial\boldsymbol{\rho}(\boldsymbol{r})\partial\boldsymbol{\rho}(\boldsymbol{r}') \\ = -\,\partial\boldsymbol{u}(\boldsymbol{r}')/\partial\boldsymbol{\rho}(\boldsymbol{r}) = -[\partial\boldsymbol{v}(\boldsymbol{r}')/\partial\boldsymbol{\rho}(\boldsymbol{r})]_{\boldsymbol{\mu}} \\ = \{\eta_{\alpha,\beta}(\boldsymbol{r},\boldsymbol{r}') = -\partial u_\beta(\boldsymbol{r}')/\partial\rho_\alpha(\boldsymbol{r}) = -[\partial v_\beta(\boldsymbol{r}')/\partial\rho_\alpha(\boldsymbol{r})]_{\boldsymbol{\mu}}\} = \boldsymbol{\sigma}^{-1}(\boldsymbol{r}',\boldsymbol{r}). \tag{13.39}$$

These two sets of kernels thus satisfy the following reciprocity relation:

$$\int \boldsymbol{\eta}(\boldsymbol{r}',\boldsymbol{r})\boldsymbol{\sigma}(\boldsymbol{r},\boldsymbol{r}'')d\boldsymbol{r} = \left\{\Sigma_\gamma \int \eta_{\alpha,\gamma}(\boldsymbol{r}',\boldsymbol{r})\sigma_{\gamma,\beta}(\boldsymbol{r},\boldsymbol{r}'')d\boldsymbol{r} \right. \\ \left. = \partial\rho_\beta(\boldsymbol{r}'')/\partial\rho_\alpha(\boldsymbol{r}') = \delta_{\alpha,\beta}\delta(\boldsymbol{r}'-\boldsymbol{r}'')\right\}, \tag{13.40}$$

since in the fragment resolution the subsystem densities are the independent *function* variables of the molecular state.

The hardness matrix of subsystems can be similarly expressed in terms of the hardness kernel matrix using the double chain rule transformation:

$$\mathbf{H} = (\partial\boldsymbol{\mu}/\partial N)_v = \iint [\partial\boldsymbol{\rho}(\boldsymbol{r})/\partial N]_v[\partial^2 E_v[\boldsymbol{\rho}]/\partial\boldsymbol{\rho}(\boldsymbol{r})\,\partial\boldsymbol{\rho}(\boldsymbol{r}')]_v[\partial\boldsymbol{\rho}(\boldsymbol{r}')/\partial N]_v^{\mathrm{T}} d\boldsymbol{r}\, d\boldsymbol{r}' \\ = \iint \mathbf{f}(\boldsymbol{r})\boldsymbol{\eta}(\boldsymbol{r},\boldsymbol{r}')\,\mathbf{f}(\boldsymbol{r}')^{\mathrm{T}} d\boldsymbol{r}\, d\boldsymbol{r}'. \tag{13.41}$$

The differential of the subsystem densities of the externally closed subsystems, $\boldsymbol{\rho} = \boldsymbol{\rho}[N, v; \boldsymbol{r}]$, now reads:

$$d\boldsymbol{\rho}(\boldsymbol{r}) = d\boldsymbol{\rho}[N, v; \boldsymbol{r}] = dN\mathbf{f}(\boldsymbol{r}) + \int dv(\boldsymbol{r}')\, \mathbf{B}(\boldsymbol{r}', \boldsymbol{r}) d\boldsymbol{r}'. \tag{13.42}$$

It can be alternatively expressed as $d\boldsymbol{\rho}[\boldsymbol{u}; \boldsymbol{r}]$,

$$d\boldsymbol{\rho}(\boldsymbol{r}) = d\boldsymbol{\rho}[\boldsymbol{u}; \boldsymbol{r}] = \int [d\boldsymbol{\mu} - dv(\boldsymbol{r}')]\, \boldsymbol{\sigma}(\boldsymbol{r}', \boldsymbol{r})\, d\boldsymbol{r}', \tag{13.43}$$

where

$$d\boldsymbol{\mu} = d\boldsymbol{\mu}[N, v] = dN\mathbf{H} + \int dv(\boldsymbol{r})\, \mathbf{f}(\boldsymbol{r})^{\mathrm{T}} d\boldsymbol{r}. \tag{13.44}$$

This gives equivalent expression for $d\boldsymbol{\rho}[N, v; \boldsymbol{r}]$ in terms of CS for the externally open molecular subsystems:

$$\begin{aligned} d\boldsymbol{\rho}[N, v; \boldsymbol{r}] &= \int \left[dN\mathbf{H} + \int dv(\boldsymbol{r}'')\, \mathbf{f}(\boldsymbol{r}'')^{\mathrm{T}} d\boldsymbol{r}'' - dv(\boldsymbol{r}')\right] \boldsymbol{\sigma}(\boldsymbol{r}', \boldsymbol{r})\, d\boldsymbol{r}' \\ &= dN[\mathbf{H}\, \mathbf{s}(\boldsymbol{r})] + \int dv(\boldsymbol{r}')\, [\mathbf{f}(\boldsymbol{r}')^{\mathrm{T}} \mathbf{s}(\boldsymbol{r}) - \boldsymbol{\sigma}(\boldsymbol{r}', \boldsymbol{r})] d\boldsymbol{r}'. \end{aligned} \tag{13.45}$$

A comparison between coefficients at displacements dN and $dv(\boldsymbol{r})$ in these alternative expressions for $d\boldsymbol{\rho}(\boldsymbol{r})$ finally gives [see (13.37)]:

$$\begin{aligned} \mathbf{f}(\boldsymbol{r}) &= \mathbf{H} \int \boldsymbol{\sigma}(\boldsymbol{r}, \boldsymbol{r}') d\boldsymbol{r}' = \mathbf{H}\mathbf{s}(\boldsymbol{r}) \quad \text{and} \\ \mathbf{B}(\boldsymbol{r}', \boldsymbol{r}) &= \mathbf{f}(\boldsymbol{r}')^{\mathrm{T}} \mathbf{s}(\boldsymbol{r}) - \boldsymbol{\sigma}(\boldsymbol{r}', \boldsymbol{r}) = \mathbf{f}(\boldsymbol{r}')^{\mathrm{T}} \mathbf{S}\, \mathbf{f}(\boldsymbol{r}) - \boldsymbol{\sigma}(\boldsymbol{r}', \boldsymbol{r}). \end{aligned} \tag{13.46}$$

Hence, the matrix $\boldsymbol{\sigma}(\boldsymbol{r}', \boldsymbol{r})$ of the subsystem softness kernels can be expressed in terms of the softness matrix of molecular fragments, their Fukui functions, and linear response kernels:

$$\boldsymbol{\sigma}(\boldsymbol{r}', \boldsymbol{r}) = \mathbf{f}(\boldsymbol{r}')^{\mathrm{T}} \mathbf{S}\, \mathbf{f}(\boldsymbol{r}) - \mathbf{B}(\boldsymbol{r}', \boldsymbol{r}). \tag{13.47}$$

13.5 Transformations of Perturbations into Responses

In the global equilibrium state of the molecular system in question, when all constituent subsystems are mutually open, i.e., free to exchange electrons between themselves, the linear responses of the unconstrained (dependent) *state* variables can be expressed as transformations of the corresponding perturbations, i.e., displacements in the representation independent state parameters. In this section, we shall briefly summarize these relations for the four admissible choices of the independent parameters of state:

$$[N, v], \quad [v - \mu] \equiv [u], \quad [\rho], \quad [\mu, \rho]. \tag{13.48}$$

The associated "thermodynamic" potentials are then defined by the energy functional, $E[N, v] = E_v[\rho] = \int v(\boldsymbol{r})\rho(\boldsymbol{r})\, d\boldsymbol{r} + F[\rho]$, and its Legendre transforms:

$$\Omega[v - \mu] = E[N, v] - N\mu = \int u(\boldsymbol{r})\rho(\boldsymbol{r})\, d\boldsymbol{r} + F[\rho] \equiv \Omega_u[\rho], \tag{13.49}$$

$$F[\rho] = E_v[\rho] - \int v(\boldsymbol{r})\rho(\boldsymbol{r})\, d\boldsymbol{r}, \tag{13.50}$$

$$R[\mu, \rho] = E[N, v] - N\mu - \int v(\boldsymbol{r})\rho(\boldsymbol{r})\, d\boldsymbol{r} = F[\rho] - N\mu. \tag{13.51}$$

In the $[N,v]$-representation of the EF perspective the displacements $[\Delta N, \Delta v]$ of the independent state parameters generate the following equilibrium linear responses of the dependent energy conjugates $[\mu, \rho]$ [see (7.227)–(7.230)]:

$$\Delta\mu = \left(\frac{\partial\mu}{\partial N}\right)_v \Delta N + \int \left(\frac{\partial\mu}{\partial v(\boldsymbol{r})}\right)_v \Delta v(\boldsymbol{r})\, d\boldsymbol{r} = \eta\,\Delta N + \int f(\boldsymbol{r})\,\Delta v(\boldsymbol{r})\, d\boldsymbol{r}, \tag{13.52}$$

$$\begin{aligned} \Delta\rho(\boldsymbol{r}) &= \left(\frac{\partial\rho(\boldsymbol{r})}{\partial N}\right)_v \Delta N + \int \Delta v(\boldsymbol{r}') \left(\frac{\partial\rho(\boldsymbol{r})}{\partial v(\boldsymbol{r}')}\right)_N d\boldsymbol{r}' \\ &= f(\boldsymbol{r})\Delta N + \int \Delta v(\boldsymbol{r}')\, \beta(\boldsymbol{r}', \boldsymbol{r})\, d\boldsymbol{r}'. \end{aligned} \tag{13.53}$$

These two relations can be jointly expressed in the form of the following integral (matrix) transformation:

$$[\Delta\mu, \Delta\rho(\boldsymbol{r})] = [\Delta N, \int d\boldsymbol{r}'\, \Delta v(\boldsymbol{r}')] \begin{bmatrix} \eta & f(\boldsymbol{r}) \\ f(\boldsymbol{r}') & \beta(\boldsymbol{r}', \boldsymbol{r}) \end{bmatrix}. \tag{13.54}$$

Consider next the "inverse" representation $[\mu, \rho]$ of the EP perspective, in which the independent perturbations $[\Delta\mu, \Delta\rho]$ determine the linear responses in the conjugate (unconstrained) variables of state: $\Delta N = \int \Delta\rho(\boldsymbol{r})\, d\boldsymbol{r}$ and

$$\begin{aligned} \Delta v(\boldsymbol{r}) &= \left(\frac{\partial v(\boldsymbol{r})}{\partial\mu}\right)_\rho \Delta\mu + \int \Delta\rho(\boldsymbol{r}') \left(\frac{\partial v(\boldsymbol{r})}{\partial\rho(\boldsymbol{r}')}\right)_\mu d\boldsymbol{r}' \\ &= \Delta\mu - \int \Delta\rho(\boldsymbol{r}')\eta(\boldsymbol{r}', \boldsymbol{r})\, d\boldsymbol{r}'. \end{aligned} \tag{13.55}$$

Again, these linear transformations can be combined into a single integral (matrix) transformation:

$$[\Delta N, \Delta v(\boldsymbol{r})] = \left[\Delta\mu, \int d\boldsymbol{r}'\, \Delta\rho(\boldsymbol{r}')\right] \begin{bmatrix} 0 & 1 \\ 1 & -\eta(\boldsymbol{r}', \boldsymbol{r}) \end{bmatrix}. \tag{13.56}$$

The corresponding transformations in the mutually reverse $[\rho]$ (EP) and $[v - \mu] = [u]$ (EF) representations are determined by the respective *total* functional derivatives [see (7.232) and (7.233)]:

$$\begin{aligned} \Delta u(\boldsymbol{r}) &= -\int \Delta\rho(\boldsymbol{r}')\, \eta(\boldsymbol{r}', \boldsymbol{r})\, d\boldsymbol{r}' = \Delta u[\Delta\rho; \boldsymbol{r}], \\ \Delta\rho(\boldsymbol{r}) &= -\int \Delta u(\boldsymbol{r}')\, \sigma(\boldsymbol{r}', \boldsymbol{r})\, d\boldsymbol{r}' = \Delta\rho[\Delta u; \boldsymbol{r}]. \end{aligned} \tag{13.57}$$

These transformations can be also given a more resolved form in terms of the corresponding *partial* derivatives. Consider, e.g., the transformations of perturbations $[\Delta\mu, \Delta v] = \Delta u$ of the $[v - \mu] = u$ representation into the conjugate responses $[\Delta N, \Delta\rho]$:

$$\Delta N = \left(\frac{\partial N}{\partial \mu}\right)_v \Delta\mu + \int \left(\frac{\partial N}{\partial v(\boldsymbol{r})}\right)_\mu \Delta v(\boldsymbol{r})\, d\boldsymbol{r} = S\, \Delta\mu - \int s(\boldsymbol{r})\, \Delta v(\boldsymbol{r})\, d\boldsymbol{r}, \tag{13.58}$$

$$\begin{aligned} \Delta\rho(\boldsymbol{r}) &= \left(\frac{\partial\rho(\boldsymbol{r})}{\partial\mu}\right)_v \Delta\mu + \int \Delta v(\boldsymbol{r}') \left(\frac{\partial\rho(\boldsymbol{r})}{\partial v(\boldsymbol{r}')}\right)_\mu d\boldsymbol{r}' \\ &= s(\boldsymbol{r})\Delta\mu - \int \Delta v(\boldsymbol{r}')\, \sigma(\boldsymbol{r}', \boldsymbol{r})\, d\boldsymbol{r}'. \end{aligned} \tag{13.59}$$

Combining these differential expressions into the joint matrix form gives the following integral transformation:

$$[\Delta N, \Delta\rho(\boldsymbol{r})] = \left[\Delta\mu, \int d\boldsymbol{r}'\, \Delta v(\boldsymbol{r}')\right] \begin{bmatrix} S & s(\boldsymbol{r}) \\ -s(\boldsymbol{r}') & -\sigma(\boldsymbol{r}', \boldsymbol{r}) \end{bmatrix}. \tag{13.60}$$

When the constituent subsystems are mutually closed (see the fragment development of the preceding section), the corresponding matrices of charge sensitivities replace the global quantities in the corresponding transformations of perturbations of molecular fragments into the conjugate responses. In the canonical $[\boldsymbol{N}, \boldsymbol{v}]$ representation this integral matrix transformation then reads:

$$[\Delta\boldsymbol{\mu}, \Delta\boldsymbol{\rho}(\boldsymbol{r})] = \left[\Delta\boldsymbol{N}, \int d\boldsymbol{r}'\, \Delta\boldsymbol{v}(\boldsymbol{r}')\right] \begin{bmatrix} \mathbf{H} & \mathbf{f}(\boldsymbol{r}) \\ \mathbf{f}(\boldsymbol{r}')^{\mathrm{T}} & \mathbf{B}(\boldsymbol{r}, \boldsymbol{r}') \end{bmatrix}, \tag{13.61}$$

while the inverse $[\boldsymbol{\mu}, \boldsymbol{\rho}]$ representation generates the following matrix *linear* response relationship:

$$[\Delta N, \Delta \boldsymbol{v}(\boldsymbol{r})] = \left[\Delta \boldsymbol{\mu}, \int dr' \, \Delta \boldsymbol{\rho}(\boldsymbol{r}')\right] \begin{bmatrix} \mathbf{0} & \mathbf{I} \\ \mathbf{I} & -\boldsymbol{\eta}(\boldsymbol{r}', \boldsymbol{r}) \end{bmatrix}. \tag{13.62}$$

Finally, in the $[\boldsymbol{\rho}]$ and $[\boldsymbol{v}-\boldsymbol{\mu}] = [\boldsymbol{u}]$ representations, one similarly finds:

$$\Delta \boldsymbol{u}(\boldsymbol{r}) = -\int \Delta \boldsymbol{\rho}(\boldsymbol{r}') \, \boldsymbol{\eta}(\boldsymbol{r}', \boldsymbol{r}) d\boldsymbol{r}', \tag{13.63}$$

$$\Delta \boldsymbol{\rho}(\boldsymbol{r}) = -\int \Delta \boldsymbol{u}(\boldsymbol{r}') \, \boldsymbol{\sigma}(\boldsymbol{r}', \boldsymbol{r}) d\boldsymbol{r}', \quad \Delta N[\boldsymbol{\rho}] = \int \Delta \boldsymbol{\rho}(\boldsymbol{r}) d\boldsymbol{r}. \tag{13.64}$$

13.6 Illustrative Description of Bimolecular Reactive Systems

13.6.1 Equilibria and Charge Sensitivities of Reactants

The most important division scheme in the theory of chemical reactivity is the reactant partitioning of the bimolecular reactive system $R = A—B$, with A and B denoting the *acidic* (electron acceptor) and *basic* (electron donor) partners in the DA complex, respectively. This A or B classification of interacting species can be carried out by examining the chemical potentials (negative electronegativities) of the separate (infinitely distant) reactants at $R^0(\infty) = A^0 + B^0$, $\boldsymbol{\mu}_R^0 = (\mu_A^0, \mu_B^0)$, satisfying the inequality $\mu_A^0 < \mu_B^0$. The electron densities of the geometrically and electronically "frozen" free subsystems, $\boldsymbol{\rho}_R^0 = (\rho_A^0, \rho_B^0)$, where $\int \rho_A^0(\boldsymbol{r}) d\boldsymbol{r} = N_A^0$ and $\int \rho_B^0(\boldsymbol{r}) d\boldsymbol{r} = N_B^0$, when shifted to their (finite) separation and the assumed mutual orientation specified by the nuclear coordinates $\mathbf{R}_{AB}$, then determine the associated "promolecular" distribution $\rho_R^0 = \rho_A^0 + \rho_B^0$ of the bimolecular complex. Against this reference, $R^0(\mathbf{R}_{AB}) = (A^0|B^0)$, consisting of the mutually closed (nonbonded) reactants, one then extracts the effects due to the chemical bonds between the two molecular fragments.

It is the accepted convention in the reactivity theory to view the interaction between reactants in two stages. At the intermediate, *Polarization* (P) stage of the chemical reaction the mutually closed subsystems in $R^{\mathrm{P}}(\mathbf{R}_{AB}) = (A^+|B^+)$ modify their electron distributions $\boldsymbol{\rho}_R^{\mathrm{P}} = (\rho_A^+, \rho_B^+)$ due to the presence of the other reactant, thus determining the associated overall density distribution of the polarized (promoted) reactants $\rho_R^{\mathrm{P}} = \rho_A^+ + \rho_B^+$. The subsequent internal CT between the mutually open reactants in $R^{\mathrm{CT}}(\mathbf{R}_{AB}) = (A^*|B^*)$, which preserves the overall number of electrons in R as a whole, $N_R = N_A + N_B = N_A^0 + N_B^0$, then establishes the final equilibrium electron distributions of both subsystems, $\boldsymbol{\rho}_R^{\mathrm{CT}} = (\rho_A^*, \rho_B^*)$, in such an externally closed and internally open reactive system, which give rise to the molecular density: $\rho_R^{CT} = \rho_A^* + \rho_B^* = \rho_R$ (Nalewajski 2006f; Korchowiec and Uchimaru 1998).

This final, CT-stage of the interaction gives rise to the equilibrium distribution of electrons ρ_R^{CT} in the whole molecular complex $R^{CT}(\mathbf{R}_{AB}) = (A^*|B^*)$, for the specified external potential

$$v_R(\boldsymbol{r}; \mathbf{R}_{AB}) = v_A(\boldsymbol{r}; \mathbf{R}_{AB}) + v_B(\boldsymbol{r}; \mathbf{R}_{AB}),$$

where $v_\alpha(\boldsymbol{r}; \mathbf{R}_{AB})$ denotes the external potential due to the nuclei of subsystem α alone. In the reactive system one can separately modify geometries of individual reactants, thus independently changing their contributions to the external potential of the whole system. Therefore, these two components of the overall external potential can be regarded as *independent* nuclear (geometric) degrees of freedom of the reactive system. Thus, the bonded, mutually open reactants in R^{CT} reach their equilibrium distributions $\boldsymbol{\rho}_R^* = (\rho_A^*, \rho_B^*)$ for the CT-displaced average numbers of electrons $N_A^* = \int \rho_A^*(\boldsymbol{r})d\boldsymbol{r} = N_A^0 + N_{CT}$ and $N_B^* = \int \rho_B^*(\boldsymbol{r})d\boldsymbol{r} = N_B^0 - N_{CT}$ preserving the overall number of electrons: $N_R = N_A^* + N_B^* = N_A^0 + N_B^0 = N_R^0$.

Of interest also are, e.g., in the chemisorption systems, the *externally* open bimolecular systems $\mathcal{R}^{CT}(\mathbf{R}_{AB}) = (\mathfrak{r}|A^*|B^*)$ and $\mathcal{R}^{P}(\mathbf{R}_{AB}) = (\mathfrak{r}_A|A^+|B^+|\mathfrak{r}_B)$, consisting of reactants coupled in $\mathcal{R}^{CT}$ to a common electron reservoir $\mathfrak{r}$, and in $\mathcal{R}^{P}$ – to separate reservoirs $\mathfrak{r}_A$ and $\mathfrak{r}_B$, respectively. In the former case, the system global average number of electrons N_R determines a single populational parameter of state, as reflected by the *global* FF of both subsystems in $\mathcal{R}^{CT}$ combined into FF (row) *vector*

$$\begin{aligned} \boldsymbol{f}_{\mathcal{R}}^{CT}(\boldsymbol{r}) = \left(\frac{\partial \boldsymbol{\rho}_{\mathcal{R}}^{CT}(\boldsymbol{r})}{\partial N_R}\right)_{v_R} &= \left[\left(\frac{\partial \rho_A^*(\boldsymbol{r})}{\partial N_R}\right)_{v_R}, \left(\frac{\partial \rho_B^*(\boldsymbol{r})}{\partial N_R}\right)_{v_R}\right] \\ &\equiv \{f_A^{CT}(\boldsymbol{r}), f_B^{CT}(\boldsymbol{r})\}. \end{aligned} \tag{13.65}$$

In latter case the overall electron populations of both reactants, $\boldsymbol{N}_{\mathcal{R}}^{P} = (N_A^+, N_B^+)$, can be independently varied by exchanges (external CT) with their separate reservoirs. This gives rise to the FF *matrix* of reactants in $\mathcal{R}^{P}$:

$$\mathbf{f}_{\mathcal{R}}^{P}(\boldsymbol{r}) = \left(\frac{\partial \boldsymbol{\rho}_{\mathcal{R}}^{P}(\boldsymbol{r})}{\partial \boldsymbol{N}_{\mathcal{R}}^{P}}\right)_{v_R} = \begin{bmatrix} \left(\frac{\partial \rho_A^+(\boldsymbol{r})}{\partial N_A}\right)_{v_R,N_B} & \left(\frac{\partial \rho_B^+(\boldsymbol{r})}{\partial N_A}\right)_{v_R,N_B} \\ \left(\frac{\partial \rho_A^+(\boldsymbol{r})}{\partial N_B}\right)_{v_R,N_A} & \left(\frac{\partial \rho_B^+(\boldsymbol{r})}{\partial N_B}\right)_{v_R,N_A} \end{bmatrix} \equiv \{f_{\alpha,\beta}(\boldsymbol{r})\}. \tag{13.66}$$

The ground state electron density of $\mathcal{R}^{CT}$, $\rho_{\mathcal{R}}^{CT} = \rho_A^* + \rho_B^* = \rho_{\mathcal{R}}$ and the associated energy $E_{v_R}[\rho_{\mathcal{R}}^{CT}]$ are uniquely determined by the system external potential v_R and the overall number of electrons N_R:

$$\rho_{\mathcal{R}}^{CT} = \rho_{\mathcal{R}}^{CT}[N_R, v_R], \quad E_{v_R}[\rho_{\mathcal{R}}^{CT}] \equiv E_{\mathcal{R}}^{CT}[N_R, v_R] \tag{13.67}$$

The corresponding functional relations for the externally open reactants in $\mathcal{R}^{\mathrm{P}}$ read:

$$\boldsymbol{\rho}_{\mathcal{R}}^{\mathrm{P}} = \boldsymbol{\rho}_{\mathcal{R}}^{\mathrm{P}}[\boldsymbol{N}_{\mathcal{R}}^{\mathrm{P}}, v_R], \quad E_{v_R}[\boldsymbol{\rho}_{\mathcal{R}}^{\mathrm{P}}] \equiv E_{\mathcal{R}}^{\mathrm{P}}[\boldsymbol{N}_{\mathcal{R}}^{\mathrm{P}}, v_R], \tag{13.68}$$

where the row vectors of electron densities of the polarized reactants $\boldsymbol{\rho}_{\mathcal{R}}^{\mathrm{P}} = (\rho_A^+, \rho_B^+)$ and $\boldsymbol{N}_{\mathcal{R}}^{\mathrm{P}} = (N_A^+, N_B^+) = \boldsymbol{N}_{\mathcal{R}}^{\mathrm{P}}[\boldsymbol{\rho}_{\mathcal{R}}^{\mathrm{P}}]$.

The populational partial derivative of $E_{v_R}[\rho_{\mathcal{R}}^{\mathrm{CT}}] \equiv E_{\mathcal{R}}^{\mathrm{CT}}[N_R, v_R]$ determines the system global chemical potential, equal to that of the common reservoir of electrons,

$$\mu_{\mathcal{R}}^{\mathrm{CT}} = \left(\frac{\partial E_{v_R}[\rho_{\mathcal{R}}^{\mathrm{CT}}]}{\partial \rho_{\mathcal{R}}^{\mathrm{CT}}(\boldsymbol{r})}\right)_{v_R} = \left(\frac{\partial E_{\mathcal{R}}^{\mathrm{CT}}[N_R, v_R]}{\partial N_R}\right)_{v_R}, \tag{13.69}$$

while the functional derivative with respect to v_R gives, by the Hellmann–Feynman theorem, the system overall electron density:

$$\rho_{\mathcal{R}}^{\mathrm{CT}}(\boldsymbol{r}) = \left(\frac{\partial E_{\mathcal{R}}^{\mathrm{CT}}[N_R, v_R]}{\partial v_R(\boldsymbol{r})}\right)_{N_R}. \tag{13.70}$$

These derivatives determine the corresponding *first* differential of $E_{\mathcal{R}}^{\mathrm{CT}}[N_R, v_R]$:

$$dE_{\mathcal{R}}^{\mathrm{CT}}[N_R, v_R] = \mu_{\mathcal{R}}^{\mathrm{CT}} dN_R + \int \rho_{\mathcal{R}}^{\mathrm{CT}}(\boldsymbol{r})\, dv_R(\boldsymbol{r})\, d\boldsymbol{r}. \tag{13.71}$$

The second partial derivatives of this energy functional define CS of the externally open reactive system $\mathcal{R}^{\mathrm{CT}}$:

hardness:

$$\eta_{\mathcal{R}}^{\mathrm{CT}} = \left(\frac{\partial^2 E_{\mathcal{R}}^{\mathrm{CT}}[N_R, v_R]}{\partial N_R^2}\right)_{v_R} = \left(\frac{\partial \mu_{\mathcal{R}}^{\mathrm{CT}}[N_R, v_R]}{\partial N_R}\right)_{v_R}, \tag{13.72}$$

Fukui function:

$$\begin{aligned} f_{\mathcal{R}}^{\mathrm{CT}}(\boldsymbol{r}) &= \frac{\partial^2 E_{\mathcal{R}}^{\mathrm{CT}}[N_R, v_R]}{\partial N_R\, \partial v_R(\boldsymbol{r})} = \left(\frac{\partial \rho_{\mathcal{R}}^{\mathrm{CT}}(\boldsymbol{r})}{\partial N_R}\right)_{v_R} = \left(\frac{\partial \mu_{\mathcal{R}}^{\mathrm{CT}}}{\partial v_R(\boldsymbol{r})}\right)_{N_R} \\ &= f_A^{\mathrm{CT}}(\boldsymbol{r}) + f_B^{\mathrm{CT}}(\boldsymbol{r}), \end{aligned} \tag{13.73}$$

and the density-response kernel:

$$\beta_{\mathcal{R}}^{\mathrm{CT}}(\boldsymbol{r}, \boldsymbol{r}') = \left(\frac{\partial^2 E_{\mathcal{R}}^{\mathrm{CT}}[N_R, v_R]}{\partial v_R(\boldsymbol{r})\, \partial v_R(\boldsymbol{r}')}\right)_{N_R} = \left(\frac{\partial \rho_{\mathcal{R}}^{\mathrm{CT}}(\boldsymbol{r}')}{\partial v_R(\boldsymbol{r})}\right)_{N_R} = \beta_{\mathcal{R}}^{\mathrm{CT}}(\boldsymbol{r}', \boldsymbol{r}), \tag{13.74}$$

for which the closure relation implies:

$$\int \beta_{\mathcal{R}}^{\mathrm{CT}}(\boldsymbol{r},\boldsymbol{r}')\, d\boldsymbol{r}' = [\partial N_R/\partial v_R(\boldsymbol{r})]_{N_R} = 0. \tag{13.75}$$

These canonical quantities determine the second differential of the system energy $E_{\mathcal{R}}^{\mathrm{CT}}[N_R, v_R]$:

$$\begin{aligned} d^2E_{\mathcal{R}}^{\mathrm{CT}}[N_R, v_R] = \frac{1}{2}\Big(\eta_{\mathcal{R}}^{\mathrm{CT}}(dN_R)^2 + 2dN_R \int f_{\mathcal{R}}^{\mathrm{CT}}(\boldsymbol{r})\, dv_R(\boldsymbol{r})\, d\boldsymbol{r} \\ + \iint dv_R(\boldsymbol{r})\beta_{\mathcal{R}}^{\mathrm{CT}}(\boldsymbol{r},\boldsymbol{r}')dv_R(\boldsymbol{r}')d\boldsymbol{r}\, d\boldsymbol{r}'\Big). \end{aligned} \tag{13.76}$$

The corresponding *first* derivatives for reactants in $\mathcal{R}^P$, which define the first differential of $E_{v_R}[\boldsymbol{\rho}_{\mathcal{R}}^{\mathrm{P}}] \equiv E_{\mathcal{R}}^{\mathrm{P}}[\boldsymbol{N}_{\mathcal{R}}^{\mathrm{P}}, v_R]$,

$$dE_{\mathcal{R}}^{\mathrm{P}}[\boldsymbol{N}_{\mathcal{R}}^{\mathrm{P}}, v_R] = \boldsymbol{\mu}_{\mathcal{R}}^{\mathrm{P}} d\boldsymbol{N}_{\mathcal{R}}^{\mathrm{P,T}} + \int \rho_{\mathcal{R}}^{\mathrm{P}}(\boldsymbol{r})\, dv_R(\boldsymbol{r})\, d\boldsymbol{r}, \tag{13.77}$$

include the row vectors of the reactant chemical potentials, equal to those of their respective hypothetical reservoirs,

$$\boldsymbol{\mu}_{\mathcal{R}}^{\mathrm{P}} = \left(\frac{\partial E_{v_R}[\boldsymbol{\rho}_{\mathcal{R}}^{\mathrm{P}}]}{\partial \boldsymbol{\rho}_{\mathcal{R}}^{\mathrm{P}}(\boldsymbol{r})}\right)_{v_R} = \left(\frac{\partial E[\boldsymbol{N}_{\mathcal{R}}^{\mathrm{P}}, v_R]}{\partial \boldsymbol{N}_{\mathcal{R}}^{\mathrm{P}}}\right)_{v_R} = (\mu_A^+, \mu_B^+), \tag{13.78}$$

and the sum of the promoted densities of both (mutually closed) subsystems:

$$\begin{aligned} \rho_{\mathcal{R}}^{\mathrm{P}}(\boldsymbol{r}) &= \left(\frac{\partial E[\boldsymbol{N}_{\mathcal{R}}^{\mathrm{P}}, v_R]}{\partial v_R(\boldsymbol{r})}\right)_{\boldsymbol{N}_{\mathcal{R}}^{\mathrm{P}}} \\ &= \rho_A^+(\boldsymbol{r}) + \rho_B^+(\boldsymbol{r}), \int \rho_{\mathcal{R}}^{\mathrm{P}}(\boldsymbol{r})\, d\boldsymbol{r} = N_{\mathcal{R}}^{\mathrm{P}} = N_A^+ + N_B^+. \end{aligned} \tag{13.79}$$

The *second* derivatives of this electronic energy define the associated CS of the *mutually* closed but *externally* open reactants in $\mathcal{R}^P$:

hardness matrix,

$$\boldsymbol{\eta}_{\mathcal{R}}^{\mathrm{P}} = \left(\frac{\partial^2 E[\boldsymbol{N}_{\mathcal{R}}^{\mathrm{P}}, v_R]}{\partial \boldsymbol{N}_{\mathcal{R}}^{\mathrm{P}}\, \partial \boldsymbol{N}_{\mathcal{R}}^{\mathrm{P}}}\right)_{v_R} = \left(\frac{\partial \boldsymbol{\mu}_{\mathcal{R}}^{\mathrm{P}}}{\partial \boldsymbol{N}_{\mathcal{R}}^{\mathrm{P}}}\right)_{v_R} = \{\eta_{\alpha,\beta}\}, \tag{13.80}$$

FF vector,

$$\boldsymbol{f}_{\mathcal{R}}^{\mathrm{P}}(\boldsymbol{r}) = \frac{\partial^2 E[\boldsymbol{N}_{\mathcal{R}}^{\mathrm{P}}, v_R]}{\partial \boldsymbol{N}_{\mathcal{R}}^{\mathrm{P}} \partial v_R(\boldsymbol{r})} = \left(\frac{\partial \rho_{\mathcal{R}}^{\mathrm{P}}(\boldsymbol{r})}{\partial \boldsymbol{N}_{\mathcal{R}}^{\mathrm{P}}}\right)_{v_R} = \left(\frac{\partial \boldsymbol{\mu}_{\mathcal{R}}^{\mathrm{P}}}{\partial v_R(\boldsymbol{r})}\right)_{\boldsymbol{N}_{\mathcal{R}}^{\mathrm{P}}}^{\mathrm{T}} = \{f_\alpha^{\mathrm{P}}(\boldsymbol{r})\}, \tag{13.81}$$

and the density-response kernel,

$$\beta_{\mathfrak{R}}^{\mathrm{P}}(\boldsymbol{r},\boldsymbol{r}') = \left(\frac{\partial^2 E_{\mathfrak{R}}^{\mathrm{P}}[\boldsymbol{N}_{\mathfrak{R}}^{\mathrm{P}}, v_R]}{\partial v_R(\boldsymbol{r})\,\partial v_R(\boldsymbol{r}')}\right)_{\boldsymbol{N}_{\mathfrak{R}}^{\mathrm{P}}} = \left(\frac{\partial \rho_{\mathfrak{R}}^{\mathrm{P}}(\boldsymbol{r}')}{\partial v_R(\boldsymbol{r})}\right)_{\boldsymbol{N}_{\mathfrak{R}}^{\mathrm{P}}} = \beta_{\mathfrak{R}}^{\mathrm{P}}(\boldsymbol{r}',\boldsymbol{r}). \tag{13.82}$$

Again, by the closure relation [see (13.79)],

$$\int \beta_{\mathfrak{R}}^{\mathrm{P}}(\boldsymbol{r},\boldsymbol{r}')\,d\boldsymbol{r}' = \left[\partial \boldsymbol{N}_{\mathfrak{R}}^{\mathrm{P}}/\partial v_R(\boldsymbol{r})\right]_{\boldsymbol{N}_{\mathfrak{R}}^{\mathrm{P}}} = 0. \tag{13.83}$$

The FF indices of (13.81) can be expressed in terms of the corresponding descriptors of (13.66) using the following *chain* rule transformation:

$$\begin{aligned} f_{\alpha}^{\mathrm{P}}(\boldsymbol{r}) = \left(\frac{\partial \rho_{\mathfrak{R}}^{\mathrm{P}}(\boldsymbol{r})}{\partial N_{\alpha}}\right)_{v_R} &= \sum_{\beta}\sum_{\gamma}\left(\frac{\partial \rho_{\beta}^{+}(\boldsymbol{r})}{\partial N_{\gamma}}\right)_{v_R}\left(\frac{\partial N_{\gamma}}{\partial N_{\alpha}}\right)_{v_R} \\ &= \sum_{\beta}\sum_{\gamma} f_{\gamma,\beta}(\boldsymbol{r})\delta_{\gamma,\alpha} = \sum_{\beta} f_{\alpha,\beta}(\boldsymbol{r}). \end{aligned} \tag{13.84}$$

Finally, the CS (13.80)–(13.82) generate the associated second differential of $E_{\mathfrak{R}}^{\mathrm{P}}[\boldsymbol{N}_{\mathfrak{R}}^{\mathrm{P}}, v_R]$:

$$\begin{aligned} d^2E_{\mathfrak{R}}^{\mathrm{P}}[\boldsymbol{N}_{\mathfrak{R}}^{\mathrm{P}}, v_R] = \frac{1}{2}\Bigg(& d\boldsymbol{N}_{\mathfrak{R}}^{\mathrm{P}}\boldsymbol{\eta}_{\mathfrak{R}}^{\mathrm{P}} d\boldsymbol{N}_{\mathfrak{R}}^{\mathrm{P,T}} + 2d\boldsymbol{N}_{\mathfrak{R}}^{\mathrm{P}}\int \boldsymbol{f}_{\mathfrak{R}}^{\mathrm{P}}(\boldsymbol{r})^{\mathrm{T}}\,dv_R(\boldsymbol{r})\,d\boldsymbol{r} \\ & + \iint dv_R(\boldsymbol{r})\beta_{\mathfrak{R}}^{\mathrm{P}}(\boldsymbol{r},\boldsymbol{r}')dv_R(\boldsymbol{r}')\,d\boldsymbol{r}\,d\boldsymbol{r}'\Bigg). \end{aligned} \tag{13.85}$$

Integration of the local FF indices of reactants (13.65) and (13.66) gives the condensed indices in the reactant resolution, measuring responses in the overall numbers of electrons on each subsystem per unit external inflows of electrons from the corresponding reservoirs:

$$\boldsymbol{F}_{\mathfrak{R}}^{\mathrm{CT}} = \int \boldsymbol{f}_{\mathfrak{R}}^{\mathrm{CT}}(\boldsymbol{r})d\boldsymbol{r} = \left(\frac{\partial \boldsymbol{N}_{\mathfrak{R}}^{\mathrm{CT}}}{\partial N_R}\right)_{v_R} = \left\{\left(\frac{\partial N_A}{\partial N_R}\right)_{v_R}, \left(\frac{\partial N_B}{\partial N_R}\right)_{v_R}\right\} \equiv \{F_A, F_B\}, \tag{13.86}$$

$$\mathbf{F}_{\mathfrak{R}}^{\mathrm{P}} = \int \mathbf{f}_{\mathfrak{R}}^{\mathrm{P}}(\boldsymbol{r})d\boldsymbol{r} = \left\{\left(\frac{\partial N_{\beta}}{\partial N_{\alpha}}\right)_{v_R} = \delta_{\alpha,\beta}\right\}, \quad \alpha, \beta \in \{A, B\}. \tag{13.87}$$

The FF of the whole bimolecular system, reflected by the normalized responses of the overall density $\rho_R = \rho_A + \rho_B$ per unit shift in the global number of electrons N_R, are then given by the relevant combinations of these elementary responses of the externally open subsystems:

$$f_{\mathfrak{R}}^{\mathrm{CT}}(\boldsymbol{r}) = \left(\frac{\partial \rho_{\mathfrak{R}}^{\mathrm{CT}}(\boldsymbol{r})}{\partial N_R}\right)_{v_R} = f_A^{\mathrm{CT}}(\boldsymbol{r}) + f_B^{\mathrm{CT}}(\boldsymbol{r}), \tag{13.88}$$

$$
\begin{aligned}
f_{\mathfrak{R}}^{\mathrm{P}}(\boldsymbol{r}) &= \left(\frac{\partial \rho_{\mathfrak{R}}^{\mathrm{P}}(\boldsymbol{r})}{\partial N_R}\right)_{v_R} = \sum_{\alpha,\beta} f_{\alpha,\beta}(\boldsymbol{r}) \\
&= [f_{A,A}(\boldsymbol{r}) + f_{B,A}(\boldsymbol{r})] + [f_{A,B}(\boldsymbol{r}) + f_{B,B}(\boldsymbol{r})] \\
&\equiv f_A^{\mathrm{P}}(\boldsymbol{r}) + f_B^{\mathrm{P}}(\boldsymbol{r}).
\end{aligned}
\tag{13.89}
$$

where $\rho_{\mathfrak{R}}^{\mathrm{CT}}$ and $\rho_{\mathfrak{R}}^{\mathrm{P}}$ denote the overall electron densities of the externally open systems of reactants in $\mathfrak{R}^{\mathrm{CT}}$ and $\mathfrak{R}^{P}$, respectively. The FF descriptors $f_{\mathfrak{R}}^{\mathrm{CT}}$ and f_{α}^{P} (or the "diagonal" contributions $f_{\alpha,\alpha}$) satisfy the usual normalization:

$$
\int f_{\mathfrak{R}}^{\mathrm{CT}}(\boldsymbol{r})\, d\boldsymbol{r} = \left(\frac{\partial N_R}{\partial N_R}\right)_{v_R} = \int f_{\alpha}^{\mathrm{P}}(\boldsymbol{r})\, d\boldsymbol{r} = \int f_{\alpha,\alpha}(\boldsymbol{r})\, d\boldsymbol{r} = \left(\frac{\partial N_\alpha}{\partial N_\alpha}\right)_{v_R} = 1, \tag{13.90}
$$

while the "*off*-diagonal" terms in the subsystem resolution, reflecting the polarization of one (closed) reactant induced by the unit CT to/from the other reactant, must integrate to zero, e.g.,

$$
\int f_{A,B}(\boldsymbol{r})\, d\boldsymbol{r} = \left(\frac{\partial N_B}{\partial N_A}\right)_{v_R, N_B} = 0. \tag{13.91}
$$

The reactant partitioning of the system overall electron density implies the *two*-component approach. The electrons of the mutually *open* reactants in $\mathfrak{R}^{\mathrm{CT}}$ exhibit identical chemical potentials, equalized at the common reservoir level, $\mu_A^* = \mu_B^* = \mu_r$, or in equivalent matrix notation $\boldsymbol{\mu}_{\mathfrak{R}}^{\mathrm{CT}} = (\mu_A^*, \mu_B^*) = \mu_r \boldsymbol{1}$. In $\mathfrak{R}^{\mathrm{P}}$ the mutually *closed* subsystems in general exhibit different chemical potentials: $\mu_{r_A} = \mu_A^+ \neq \mu_B^+ = \mu_{r_B}$, giving rise to the chemical potential vector $\boldsymbol{\mu}_{\mathfrak{R}}^{\mathrm{P}} = (\mu_A^+, \mu_B^+)$, which reflect the populational derivatives of the *two*-component density functional (bifunctional) for the system electronic energy:

$$
E_{v_R}[\rho_{\mathfrak{R}}^{\mathrm{P}}] = E_{v_R}[\boldsymbol{\rho}_{\mathfrak{R}}^{\mathrm{P}}] \equiv E_{v_R}^{total}[\boldsymbol{\rho}_{\mathfrak{R}}^{\mathrm{P}}] = \int \rho_{\mathfrak{R}}^{\mathrm{P}}(\boldsymbol{r}) v_R(\boldsymbol{r}) d\boldsymbol{r} + F^{total}[\boldsymbol{\rho}_{\mathfrak{R}}^{\mathrm{P}}], \tag{13.92}
$$

with the repulsive (universal, v_R – independent) part $F^{total}[\boldsymbol{\rho}_{\mathfrak{R}}^{\mathrm{P}}]$ generating the sum of the electron kinetic and repulsion energies of the mutually closed reactants. It is defined by the Levy-type construction:

$$
F^{total}[\boldsymbol{\rho}_{\mathfrak{R}}^{\mathrm{P}}] = \inf_{\Psi_A \to \rho_A^+, \Psi_B \to \rho_B^+} \langle \Psi_{\mathfrak{R}}^+ \mid \hat{\mathrm{T}}_e + \hat{\mathrm{V}}_{ee} | \Psi_{\mathfrak{R}}^+ \rangle, \quad \Psi_{\mathfrak{R}}^+ \in \{\hat{\mathrm{A}}(\Psi_A \Psi_B)\}, \tag{13.93}
$$

where the constrained search is over all antisymmetrized products $\Psi_{\mathfrak{R}}^+$ of the subsystem wave functions Ψ_A and Ψ_B, which integrate to the prescribed densities of the mutually closed, polarized reactants in $\mathfrak{R}^{\mathrm{P}}$.

From the variational principle for the optimum (ground state) densities of both reactants,

$$\delta\{E_{v_R}^{total}[\boldsymbol{\rho}_{\mathfrak{R}}^{\mathrm{P}}] - \boldsymbol{\mu}_{\mathfrak{R}}^{\mathrm{P}}\boldsymbol{N}_{\mathfrak{R}}^{\mathrm{P}}[\boldsymbol{\rho}_{\mathfrak{R}}^{\mathrm{P}}]^{\mathrm{T}}\} = 0, \tag{13.94}$$

one derives the associated Euler equations:

$$v_R(\boldsymbol{r}) - \mu_\alpha^+ \equiv u_\alpha^+(\boldsymbol{r}) = -\frac{\partial F^{total}[\boldsymbol{\rho}_{\mathfrak{R}}^{\mathrm{P}}]}{\partial \rho_\alpha^+(\boldsymbol{r})} = -\frac{\partial F^{nadd.}[\boldsymbol{\rho}_{\mathfrak{R}}^{\mathrm{P}}]}{\partial \rho_\alpha^+(\boldsymbol{r})} - \frac{\partial F^{add.}[\boldsymbol{\rho}_{\mathfrak{R}}^{\mathrm{P}}]}{\partial \rho_\alpha^+(\boldsymbol{r})} \quad \text{or}$$

$$\left(v_R(\boldsymbol{r}) + \frac{\delta F^{nadd.}[\boldsymbol{\rho}_{\mathfrak{R}}^{\mathrm{P}}]}{\delta \rho_\alpha^+(\boldsymbol{r})}\right) - \mu_\alpha^+ \equiv v_\alpha^{eff.}(\boldsymbol{r}) - \mu_\alpha^+ \equiv u_\alpha^{eff.+}(\boldsymbol{r}) = -\frac{\delta F[\rho_\alpha^+]}{\delta \rho_\alpha^+(\boldsymbol{r})}, \quad \alpha = A, B. \tag{13.95}$$

Therefore, the relative external potentials of the polarized reactants,

$$\begin{aligned} \boldsymbol{u}_{\mathfrak{R}}^{\mathrm{P}}(\boldsymbol{r}) &= \boldsymbol{v}_{\mathfrak{R}}^{\mathrm{P}}(\boldsymbol{r}) - \boldsymbol{\mu}_{\mathfrak{R}}^{\mathrm{P}} = \{u_A^+(\boldsymbol{r}), u_B^+(\boldsymbol{r})\}, \\ \boldsymbol{v}_{\mathfrak{R}}^{\mathrm{P}}(\boldsymbol{r}) &= \{v_A^{\mathrm{P}}(\boldsymbol{r}), v_B^{\mathrm{P}}(\boldsymbol{r})\} \equiv v_R(\boldsymbol{r})\boldsymbol{1}, \end{aligned} \tag{13.96}$$

are functionals of the subsystem densities: $\boldsymbol{u}_{\mathfrak{R}}^{\mathrm{P}} = \boldsymbol{u}_{\mathfrak{R}}^{\mathrm{P}}[\boldsymbol{\rho}_{\mathfrak{R}}^{\mathrm{P}}]$. This functional dependence defines the hardness (EP) representation in the reactant resolution. It also follows from (13.95) that in the complementary, softness (EF) representation the reactant densities can be regarded as functionals of the relative external potentials of both subsystems: $\boldsymbol{\rho}_{\mathfrak{R}}^{\mathrm{P}} = \boldsymbol{\rho}_{\mathfrak{R}}^{\mathrm{P}}[\boldsymbol{u}_{\mathfrak{R}}^{\mathrm{P}}]$.

In the hardness representation one thus obtains the following expression for the first differential of $\boldsymbol{u}_{\mathfrak{R}}^{\mathrm{P}}[\boldsymbol{\rho}_{\mathfrak{R}}^{\mathrm{P}}]$ in terms of the relevant hardness kernels in the reactant resolution:

$$\begin{aligned} du_\alpha^+(\boldsymbol{r}) = d[v_R(\boldsymbol{r}) - \mu_\alpha^+] &= -\sum_\beta \int d\rho_\beta^+(\boldsymbol{r}') \frac{\partial^2 F^{total}[\boldsymbol{\rho}_{\mathfrak{R}}^{\mathrm{P}}]}{\partial \rho_\beta^+(\boldsymbol{r}')\partial \rho_\alpha^+(\boldsymbol{r})} d\boldsymbol{r}' \\ &= \sum_\beta \int d\rho_\beta^+(\boldsymbol{r}') \frac{\delta u_\alpha^+(\boldsymbol{r})}{\delta \rho_\beta^+(\boldsymbol{r}')} d\boldsymbol{r}' \\ &\equiv -\sum_\beta \int d\rho_\beta^+(\boldsymbol{r}')\ \eta_{\beta,\alpha}(\boldsymbol{r}', \boldsymbol{r})\ d\boldsymbol{r}', \quad \alpha = A, B, \end{aligned}$$

or in a short matrix notation:

$$\Delta \boldsymbol{u}_{\mathfrak{R}}^{\mathrm{P}}(\boldsymbol{r}) = -\int \Delta \boldsymbol{\rho}_{\mathfrak{R}}^{\mathrm{P}}(\boldsymbol{r}')\ \boldsymbol{\eta}_{\mathfrak{R}}^{\mathrm{P}}(\boldsymbol{r}', \boldsymbol{r}) d\boldsymbol{r}'. \tag{13.97}$$

Here, the elements of the square matrix of the hardness kernels in reactant resolution, $\boldsymbol{\eta}_{\mathfrak{R}}^{\mathrm{P}}(\boldsymbol{r}', \boldsymbol{r}) = \{\eta_{\alpha,\beta}(\boldsymbol{r}', \boldsymbol{r})\}$, measure the negative responses of the relative external potential of subsystem β, per unit shift in the electron density of subsystem α. They are seen to transform displacements in the reactant electron densities $\Delta\boldsymbol{\rho}_{\mathfrak{R}}^{\mathrm{P}}$ into the concomitant responses $\Delta\boldsymbol{u}_{\mathfrak{R}}^{\mathrm{P}}$ of the reactant relative potentials.

The inverse of $\boldsymbol{\eta}_{\mathfrak{R}}^{\mathrm{P}}(\boldsymbol{r}',\boldsymbol{r})$ defines the square matrix of the corresponding softness kernels of these two complementary molecular fragments:

$$\boldsymbol{\sigma}_{\mathfrak{R}}^{\mathrm{P}}(\boldsymbol{r},\boldsymbol{r}') = -\frac{\delta\boldsymbol{\rho}_{\mathfrak{R}}^{\mathrm{P}}(\boldsymbol{r}')}{\delta\boldsymbol{u}_{\mathfrak{R}}^{\mathrm{P}}(\boldsymbol{r})} = \left\{\sigma_{\alpha,\beta}(\boldsymbol{r},\boldsymbol{r}') = -\frac{\delta\rho_{\beta}^{+}(\boldsymbol{r}')}{\delta u_{\alpha}^{+}(\boldsymbol{r})}\right\} = \boldsymbol{\eta}_{\mathfrak{R}}^{\mathrm{P}}(\boldsymbol{r}',\boldsymbol{r})^{-1}. \tag{13.98}$$

The two matrix kernels satisfy the following reciprocity relation:

$$\int \boldsymbol{\sigma}_{\mathfrak{R}}^{\mathrm{P}}(\boldsymbol{r},\boldsymbol{r}')\boldsymbol{\eta}_{\mathfrak{R}}^{\mathrm{P}}(\boldsymbol{r}',\boldsymbol{r}'')d\boldsymbol{r}'' = \delta(\boldsymbol{r}-\boldsymbol{r}'')\mathbf{I}^{\mathrm{R}}, \tag{13.99}$$

where the 2×2 identity matrix $\mathbf{I}^{R} = \{\delta_{\alpha,\beta}\}$. These softness kernels transform the displacements $\Delta\boldsymbol{u}_{\mathfrak{R}}^{\mathrm{P}}$ in the relative external potentials of reactants into the conjugate shifts $\Delta\boldsymbol{\rho}_{\mathfrak{R}}^{\mathrm{P}}$ of their electron densities,

$$\Delta\rho_{\beta}^{+}(\boldsymbol{r}') = -\sum_{\alpha}\int \Delta u_{\alpha}^{+}(\boldsymbol{r})\sigma_{\alpha,\beta}(\boldsymbol{r},\boldsymbol{r}')d\boldsymbol{r}, \quad \beta = A,B,$$

or in the compact matrix notation

$$\begin{aligned}\Delta\boldsymbol{\rho}_{\mathfrak{R}}^{\mathrm{P}}(\boldsymbol{r}') &= -\int \Delta\boldsymbol{u}_{\mathfrak{R}}^{\mathrm{P}}(\boldsymbol{r})\ \boldsymbol{\sigma}_{\mathfrak{R}}^{\mathrm{P}}(\boldsymbol{r},\boldsymbol{r}')\ d\boldsymbol{r}\\ &= -\int \Delta\boldsymbol{v}_{\mathfrak{R}}^{\mathrm{P}}(\boldsymbol{r})\ \boldsymbol{\sigma}_{\mathfrak{R}}^{\mathrm{P}}(\boldsymbol{r},\boldsymbol{r}')\ d\boldsymbol{r} + \Delta\boldsymbol{\mu}_{\mathfrak{R}}^{\mathrm{P}}\int \boldsymbol{\sigma}_{\mathfrak{R}}^{\mathrm{P}}(\boldsymbol{r},\boldsymbol{r}')\ d\boldsymbol{r}\\ &\equiv -\int \Delta\boldsymbol{v}_{\mathfrak{R}}^{\mathrm{P}}(\boldsymbol{r})\ \boldsymbol{\sigma}_{\mathfrak{R}}^{\mathrm{P}}(\boldsymbol{r},\boldsymbol{r}')\ d\boldsymbol{r} + \Delta\boldsymbol{\mu}_{\mathfrak{R}}^{\mathrm{P}}\ \mathbf{s}_{\mathfrak{R}}^{\mathrm{P}}(\boldsymbol{r}').\end{aligned} \tag{13.100}$$

Here, the local softness matrix in reactant resolution is defined as follows:

$$\mathbf{s}_{\mathfrak{R}}^{\mathrm{P}}(\boldsymbol{r}) = \left(\frac{\partial\boldsymbol{\rho}_{\mathfrak{R}}^{\mathrm{P}}(\boldsymbol{r})}{\partial\boldsymbol{\mu}_{\mathfrak{R}}^{\mathrm{P}}}\right)_{v_R} = \int \boldsymbol{\sigma}_{\mathfrak{R}}^{\mathrm{P}}(\boldsymbol{r}',\boldsymbol{r})\ d\boldsymbol{r}' = \left\{s_{\alpha,\beta}(\boldsymbol{r}) = \left(\frac{\partial\rho_{\beta}^{+}(\boldsymbol{r})}{\partial\mu_{\alpha}^{+}}\right)_{v_R}\right\}. \tag{13.101}$$

Writing $d\boldsymbol{\mu}_{\mathfrak{R}}^{\mathrm{P}}$ [of $d\boldsymbol{u}_{\mathfrak{R}}^{\mathrm{P}}(\boldsymbol{r})$] as differential $d\boldsymbol{\mu}_{\mathfrak{R}}^{\mathrm{P}}[\boldsymbol{N}_{\mathfrak{R}}^{\mathrm{P}}, v_R]$ then gives [see (13.44)]:

$$d\boldsymbol{\mu}_{\mathfrak{R}}^{\mathrm{P}}[\boldsymbol{N}_{\mathfrak{R}}^{\mathrm{P}}, v_R] = d\boldsymbol{N}_{\mathfrak{R}}^{\mathrm{P}}\boldsymbol{\eta}_{\mathfrak{R}}^{\mathrm{P}} + \int dv_R(\boldsymbol{r})\ \boldsymbol{f}_{\mathfrak{R}}^{\mathrm{P}}(\boldsymbol{r})\ d\boldsymbol{r}. \tag{13.102}$$

It should be further observed that the condensed hardness matrix $\boldsymbol{\eta}_{\mathfrak{R}}^{\mathrm{P}}$ in this equation can be expressed in terms of the hardness kernels $\boldsymbol{\eta}_{\mathfrak{R}}^{\mathrm{P}}(\boldsymbol{r}',\boldsymbol{r})$ of (13.97) using the relevant *chain* rule transformation:

$$\begin{aligned}\boldsymbol{\eta}^{\mathrm{P}}_{\mathfrak{R}} &= \iint \left(\frac{\partial \boldsymbol{\rho}^{\mathrm{P}}_{\mathfrak{R}}(\boldsymbol{r}')}{\partial \boldsymbol{N}^{\mathrm{P}}_{\mathfrak{R}}}\right)_{v_R} \frac{\partial^2 F^{total}[\boldsymbol{\rho}^{\mathrm{P}}_{\mathfrak{R}}]}{\partial \boldsymbol{\rho}^{\mathrm{P}}_{\mathfrak{R}}(\boldsymbol{r}')\partial \boldsymbol{\rho}^{\mathrm{P}}_{\mathfrak{R}}(\boldsymbol{r})} \left(\frac{\partial \boldsymbol{\rho}^{\mathrm{P}}_{\mathfrak{R}}(\boldsymbol{r})}{\partial \boldsymbol{N}^{\mathrm{P}}_{\mathfrak{R}}}\right)_{v_R} d\boldsymbol{r}\, d\boldsymbol{r}' \\ &= \iint \mathbf{f}^{\mathrm{P}}_{\mathfrak{R}}(\boldsymbol{r}')\ \boldsymbol{\eta}^{\mathrm{P}}_{\mathfrak{R}}(\boldsymbol{r}',\boldsymbol{r})\ \mathbf{f}^{\mathrm{P}}_{\mathfrak{R}}(\boldsymbol{r}')^{\mathrm{T}} d\boldsymbol{r}\, d\boldsymbol{r}'.\end{aligned} \tag{13.103}$$

Combining these equations finally gives:

$$\begin{aligned}d\boldsymbol{\rho}^{\mathrm{P}}_{\mathfrak{R}}(\boldsymbol{r}) = d\boldsymbol{N}^{\mathrm{P}}_{\mathfrak{R}}\,\boldsymbol{\eta}^{\mathrm{P}}_{\mathfrak{R}}\ \mathbf{s}^{\mathrm{P}}_{\mathfrak{R}}(\boldsymbol{r}) - \int d\boldsymbol{v}^{\mathrm{P}}_{\mathfrak{R}}(\boldsymbol{r}')\ \boldsymbol{\sigma}^{\mathrm{P}}_{\mathfrak{R}}(\boldsymbol{r}',\boldsymbol{r})\, d\boldsymbol{r}' \\ + \left[\int d\boldsymbol{v}^{\mathrm{P}}_{\mathfrak{R}}(\boldsymbol{r}')\mathbf{f}^{\mathrm{P}}_{\mathfrak{R}}(\boldsymbol{r}')^{\mathrm{T}} d\boldsymbol{r}'\right] \mathbf{s}^{\mathrm{P}}_{\mathfrak{R}}(\boldsymbol{r}).\end{aligned} \tag{13.104}$$

One can independently write this differential using the alternative functional dependence $\boldsymbol{\rho}^{\mathrm{P}}_{\mathfrak{R}}[\boldsymbol{u}^{\mathrm{P}}_{\mathfrak{R}}] = \boldsymbol{\rho}^{\mathrm{P}}_{\mathfrak{R}}[\boldsymbol{N}^{\mathrm{P}}_{\mathfrak{R}}, v_R]$:

$$d\boldsymbol{\rho}^{\mathrm{P}}_{\mathfrak{R}}(\boldsymbol{r}) = d\boldsymbol{N}^{\mathrm{P}}_{\mathfrak{R}}\ \mathbf{f}^{\mathrm{P}}_{\mathfrak{R}}(\boldsymbol{r}) + \int d\boldsymbol{v}^{\mathrm{P}}_{\mathfrak{R}}(\boldsymbol{r}')\ \boldsymbol{\beta}^{\mathrm{P}}_{\mathfrak{R}}(\boldsymbol{r}',\boldsymbol{r})d\boldsymbol{r}'. \tag{13.105}$$

A comparison between the preceding two equations gives the reactant-resolved generalization of the Berkowitz–Parr (1988) relations [see also (13.47)],

$$\mathbf{f}^{\mathrm{P}}_{\mathfrak{R}}(\boldsymbol{r}) = \boldsymbol{\eta}^{\mathrm{P}}_{\mathfrak{R}}\ \mathbf{s}^{\mathrm{P}}_{\mathfrak{R}}(\boldsymbol{r}) \quad \text{or} \quad \mathbf{s}^{\mathrm{P}}_{\mathfrak{R}}(\boldsymbol{r}) = \mathbf{S}^{\mathrm{P}}_{\mathfrak{R}}\ \mathbf{f}^{\mathrm{P}}_{\mathfrak{R}}(\boldsymbol{r}), \tag{13.106}$$

$$\begin{aligned}\boldsymbol{\beta}^{\mathrm{P}}_{\mathfrak{R}}(\boldsymbol{r}',\boldsymbol{r}) &= -\boldsymbol{\sigma}^{\mathrm{P}}_{\mathfrak{R}}(\boldsymbol{r}',\boldsymbol{r}) + \mathbf{f}^{\mathrm{P}}_{\mathfrak{R}}(\boldsymbol{r}')^{\mathrm{T}}\mathbf{s}^{\mathrm{P}}_{\mathfrak{R}}(\boldsymbol{r}) \\ &= -\boldsymbol{\sigma}^{\mathrm{P}}_{\mathfrak{R}}(\boldsymbol{r}',\boldsymbol{r}) + \mathbf{f}^{\mathrm{P}}_{\mathfrak{R}}(\boldsymbol{r}')^{\mathrm{T}}\mathbf{S}^{\mathrm{P}}_{\mathfrak{R}}\,\mathbf{f}^{\mathrm{P}}_{\mathfrak{R}}(\boldsymbol{r}),\end{aligned} \tag{13.107}$$

where the condensed softness matrix is the inverse of the hardness matrix of (13.80) and (13.103):

$$\mathbf{S}^{\mathrm{P}}_{\mathfrak{R}} = (\boldsymbol{\eta}^{\mathrm{P}}_{\mathfrak{R}})^{-1} = \{S_{\alpha,\beta} = (\partial N^{+}_{\beta}/\partial \mu^{+}_{\alpha})_{v_R}\}. \tag{13.108}$$

The explicit integral transformation of "perturbations,"

$$[\Delta\boldsymbol{N}^{\mathrm{P}}_{\mathfrak{R}},\ \Delta\boldsymbol{v}^{\mathrm{P}}_{\mathfrak{R}}(\boldsymbol{r}')],$$

into the conjugate responses in the chemical softness (EF) representation,

$$[\Delta\boldsymbol{\mu}^{\mathrm{P}}_{\mathfrak{R}},\ \Delta\boldsymbol{\rho}^{\mathrm{P}}_{\mathfrak{R}}(\boldsymbol{r})],$$

can be thus summarized by the following matrix equation:

$$[\Delta\boldsymbol{\mu}^{\mathrm{P}}_{\mathfrak{R}},\ \Delta\boldsymbol{\rho}^{\mathrm{P}}_{\mathfrak{R}}(\boldsymbol{r})] = \left[\Delta\boldsymbol{N}^{\mathrm{P}}_{\mathfrak{R}}, \int d\boldsymbol{r}'\ \Delta\boldsymbol{v}^{\mathrm{P}}_{\mathfrak{R}}(\boldsymbol{r}')\right] \begin{bmatrix} \boldsymbol{\eta}^{\mathrm{P}}_{\mathfrak{R}} & \mathbf{f}^{\mathrm{P}}_{\mathfrak{R}}(\boldsymbol{r}) \\ \mathbf{f}^{\mathrm{P}}_{\mathfrak{R}}(\boldsymbol{r}')^{\mathrm{T}} & \boldsymbol{\beta}^{\mathrm{P}}_{\mathfrak{R}}(\boldsymbol{r}',\boldsymbol{r}) \end{bmatrix}. \tag{13.109}$$

The inverse transformation, in the chemical hardness (EP) representation, reads:

$$[\Delta N_{\mathfrak{R}}^{\mathrm{P}},\ \Delta v_{\mathfrak{R}}^{\mathrm{P}}(\boldsymbol{r})] = \left[\Delta \boldsymbol{\mu}_{\mathfrak{R}}^{\mathrm{P}},\ \int d\boldsymbol{r}'\ \Delta \boldsymbol{\rho}_{\mathfrak{R}}^{\mathrm{P}}(\boldsymbol{r}')\right]\begin{bmatrix} \mathbf{0} & \mathbf{I} \\ \mathbf{I} & -\boldsymbol{\eta}_{\mathfrak{R}}^{\mathrm{P}}(\boldsymbol{r}',\boldsymbol{r}) \end{bmatrix}. \tag{13.110}$$

where $\mathbf{0}$ denotes the (2×2) square matrix containing zero elements.

To conclude this section, we summarize the explicit expressions for the condensed charge sensitivities in reactant resolution in terms of the canonical matrix of the subsystem hardnesses,

$$\boldsymbol{\eta}_{\mathfrak{R}}^{\mathrm{P}} = \begin{bmatrix} \eta_{A,A} & \eta_{A,B} \\ \eta_{B,A} & \eta_{B,B} \end{bmatrix}, \tag{13.111}$$

where, by the Maxwell (*cross* differentiation) identity, $\eta_{A,B} = \eta_{B,A}$. The associated softness matrix is provided by the inverse of $\boldsymbol{\eta}_{\mathfrak{R}}^{\mathrm{P}}$:

$$\mathbf{S}_{\mathfrak{R}}^{\mathrm{P}} = (\boldsymbol{\eta}_{\mathfrak{R}}^{\mathrm{P}})^{-1} = \begin{bmatrix} \eta_{B,B}/D & -\eta_{A,B}/D \\ -\eta_{A,B}/D & \eta_{A,A}/D \end{bmatrix}, \tag{13.112}$$

where the hardness determinant

$$D = \det \boldsymbol{\eta}_{\mathfrak{R}}^{\mathrm{P}} = \eta_{A,A}\,\eta_{B,B} - (\eta_{A,B})^2.$$

The softness matrix subsequently generates the respective fragment and global softness descriptors,

$$\begin{aligned} \boldsymbol{S}_{\mathfrak{R}}^{\mathrm{CT}} &= (S_A = \eta_B^R/D,\ S_B = \eta_A^R/D), \quad \eta_A^R = \eta_{A,A} - \eta_{A,B}, \quad \eta_B^R = \eta_{B,B} - \eta_{A,B}, \\ S_{\mathfrak{R}}^{\mathrm{CT}} &= 1/\eta_{\mathfrak{R}}^{\mathrm{CT}} = S_A + S_B, \end{aligned} \tag{13.113}$$

and hence also the condensed Fukui functions of reactants:

$$\boldsymbol{F}_{\mathfrak{R}}^{\mathrm{CT}} = \{F_A = S_A \eta_{\mathfrak{R}}^{\mathrm{CT}} = \eta_B^R/(\eta_A^R + \eta_B^R),\quad F_B = S_B \eta_{\mathfrak{R}}^{\mathrm{CT}} = \eta_A^R/(\eta_A^R + \eta_B^R)\}. \tag{13.114}$$

13.6.2 In Situ Quantities in Donor–Acceptor Systems

In this section, we shall examine selected in situ derivatives characterizing the internal (N_R-restricted) $B \to A$ CT process in the donor–acceptor (DA) reactive system $R = A{-}{-}{-}{-}B \equiv (A|B)$, i.e., a transfer of electrons from the basic reactant B to its acidic partner A in the externally closed R. We further assume the fixed external potential $v_R(\boldsymbol{r}) = v_A(\boldsymbol{r}) + v_B(\boldsymbol{r})$ due to the nuclei of the constituent atoms of both reactants, i.e., the "frozen" geometry of the reactive system as a whole. Here, the vertical *broken* line again symbolizes the freedom of the bonded, mutually open but externally closed reactants to exchange electrons. For simplicity, we shall further suppose that the internal geometries of the isolated reactants A^0 and B^0 are held frozen in the DA complex R, so that there exists the unique "promolecular" reference of the reactive system, $R^0 \equiv (A^0|B^0)$, consisting of the free, "frozen" geometry reactants shifted to their mutual orientation and separation in R. Above, the vertical solid line separating the nonbonded reactants in R^0 accordingly implies that they are both mutually and externally closed.

For such an *internal* electron transfer in R, for which the overall number of electrons $N_A + N_B = N_R$ is held fixed, the equilibrium electron populations on reactants, $\boldsymbol{N}_R = (N_A, N_B)$, which result from the integration of the subsystem densities at the internal CT equilibrium,

$$\boldsymbol{\rho}_R = (\rho_A, \rho_B), \quad N_\alpha = \int \rho_\alpha(\boldsymbol{r})d\boldsymbol{r}, \quad \alpha = A, B,$$

determine the current amount of the interreactant CT in R:

$$\begin{aligned} N_{\mathrm{CT}} = N_A - N_A^0 = N_B^0 - N_B > 0 \quad \text{or} \\ \boldsymbol{N}_R(N_{\mathrm{CT}}) = (N_A^0 + N_{\mathrm{CT}}, N_B^0 - N_{\mathrm{CT}}) \equiv \boldsymbol{N}_R^0 + \Delta\boldsymbol{N}_R(N_{\mathrm{CT}}). \end{aligned} \quad (13.115)$$

The row vector $\boldsymbol{N}_R^0 = (N_A^0, N_B^0)$ in the preceding equation groups the reference electron populations of the mutually closed reactants defining the promolecular system R^0 consisting of the separate (isolated) reactants.

The current amount of such an internal CT in R represents the independent "reaction coordinate" for the *intra-R* (interreactant) displacement of the system electronic structure. It should be observed that for the equilibrium geometries of the isolated reactants, which determine the fragment (fixed) external potentials, the electron densities of the separated reactants are functions of their overall electron populations, $\boldsymbol{\rho}_R^0 = \{\rho_\alpha^0(N_\alpha^0)\}$, since N_α^0 and $v_\alpha(\boldsymbol{r})$ uniquely identify the electronic Hamiltonian of the separated reactant α. A similar $\boldsymbol{N}_R$-dependence characterizes the electron densities of the bonded reactants in $R, \boldsymbol{\rho}_R = \{\rho_\alpha[\boldsymbol{N}_R(N_{\mathrm{CT}})] = \rho_\alpha(N_{\mathrm{CT}})\}$, and the system electronic energy in the reactant resolution for the given amount of CT: $E_R = E_R[\boldsymbol{N}_R(N_{\mathrm{CT}})] = E_R(N_{\mathrm{CT}})$.

The in situ population derivatives (Ciosłowski and Mixon 1993; Nalewajski 1994) involve the differentiation with respect to the amount of internal CT. The following FF-like populational derivatives along the internal CT-coordinate,

$$\boldsymbol{f}_A = (dN_A/dN_{\mathrm{CT}})_{N_R} = 1 \quad \text{and} \quad \boldsymbol{f}_B = (dN_B/dN_{\mathrm{CT}})_{N_R} = -1, \quad (13.116)$$

then enter the *chain* rule expressions for the reactant in situ derivatives. For example, the energy conjugate of N_{CT},

$$\mu_{\mathrm{CT}}(N_{\mathrm{CT}}) = \frac{\partial E_R(N_{\mathrm{CT}})}{\partial N_{\mathrm{CT}}} = \frac{\partial E_R}{\partial N_A}\boldsymbol{f}_A + \frac{\partial E_R}{\partial N_B}\boldsymbol{f}_B = \mu_A(N_{\mathrm{CT}}) - \mu_B(N_{\mathrm{CT}}), \quad (13.117)$$

measures the *relative* chemical potential of two (polarized) subsystems in R. The condition of the vanishing CT-gradient $\mu_{\mathrm{CT}}(N^*_{\mathrm{CT}}) = 0$ yields the equilibrium value of $N_{\mathrm{CT}} = N^*_{\mathrm{CT}}$, for which the chemical potentials of reactants in R are equalized: $\boldsymbol{\mu}^*_A(N^*_{\mathrm{CT}}) = \boldsymbol{\mu}^*_B(N^*_{\mathrm{CT}})$. This equation represents the *global* equilibrium criterion for such (geometry-"frozen") DA reactive system.

One similarly obtains the following expression for the in situ global hardness of the DA system,

$$\begin{aligned}\eta_{\mathrm{CT}} &= \frac{\partial^2 E_R(N_{\mathrm{CT}})}{\partial {N_{\mathrm{CT}}}^2} = \frac{\partial \mu_{\mathrm{CT}}}{\partial N_{\mathrm{CT}}}\\ &= \frac{\partial \mu_A}{\partial N_A}\boldsymbol{f}_A + \frac{\partial \mu_A}{\partial N_B}\boldsymbol{f}_B - \frac{\partial \mu_B}{\partial N_A}\boldsymbol{f}_A - \frac{\partial \mu_B}{\partial N_B}\boldsymbol{f}_B\\ &= (\eta_{A,A} - \eta_{B,A}) - (\eta_{A,B} - \eta_{B,B})\\ &\equiv \eta^R_A + \eta^R_B = \eta_{A,A} + \eta_{B,B} - 2\eta_{A,B},\end{aligned} \quad (13.118)$$

where η^R_α stands for the hardness of reactant α in R (13.113). It should be emphasized that for the chemically interacting reactants at finite mutual separations these descriptors include the corresponding diagonal hardness,

$$\eta_{\alpha,\alpha} = \partial^2 E_R/\partial N^2_\alpha = \partial\mu_\alpha/\partial N_\alpha \cong \eta^0_\alpha = \partial^2 E^0_\alpha/(\partial N^0_\alpha)^2 = \partial\mu^0_\alpha/\partial N^0_\alpha,$$

modified by the finite coupling hardness

$$\eta_{\beta,\alpha} = \partial^2 E_R/\partial N_\beta \partial N_\alpha = \partial\mu_\alpha/\partial N_\beta$$

due to the populational displacement of the complementary subsystem $\beta \neq \alpha$ (a finite electron "reservoir"), which effectively softens the fragment α in R. This coupling hardness term vanishes at very large separations between reactants, when $\eta^0_{\mathrm{CT}} = \eta^0_A + \eta^0_B$.

Next, let us consider the in situ softness quantities in R and the related FF indices of reactants. One first observes that the derivatives of (13.116) represent the condensed in situ FF indices of reactants. Their vanishing sum,

$$\boldsymbol{f}_A + \boldsymbol{f}_B = (\partial N_R/\partial N_{\mathrm{CT}})_{N_R} = 0,$$

reflects the conservation of the overall number of electrons in the externally closed R. The relevant FF of reactants in R then follow from the following chain rules:

$$\begin{aligned}f^{\mathrm{CT}}_A(\boldsymbol{r}) &= \frac{\partial \rho_A(\boldsymbol{r})}{\partial N_{\mathrm{CT}}} = \frac{\partial \rho_A(\boldsymbol{r})}{\partial N_A}\boldsymbol{f}_A + \frac{\partial \rho_A(\boldsymbol{r})}{\partial N_B}\boldsymbol{f}_B = f_{A,A}(\boldsymbol{r}) - f_{B,A}(\boldsymbol{r}),\\ f^{\mathrm{CT}}_B(\boldsymbol{r}) &= \frac{\partial \rho_B(\boldsymbol{r})}{\partial N_{\mathrm{CT}}} = \frac{\partial \rho_B(\boldsymbol{r})}{\partial N_A}\boldsymbol{f}_A + \frac{\partial \rho_B(\boldsymbol{r})}{\partial N_B}\boldsymbol{f}_B = f_{A,B}(\boldsymbol{r}) - f_{B,B}(\boldsymbol{r}),\end{aligned} \quad (13.119)$$

They give rise to the overall in situ FF of R as a whole:

$$f_{\mathrm{CT}}(\boldsymbol{r}) = \frac{\partial \rho_R(\boldsymbol{r})}{\partial N_{\mathrm{CT}}} = f_A^{\mathrm{CT}}(\boldsymbol{r}) + f_B^{\mathrm{CT}}(\boldsymbol{r}). \tag{13.120}$$

The global CT softness is defined by the inverse of CT hardness of (13.118):

$$\begin{aligned} S_{\mathrm{CT}} &= \frac{1}{\eta_{\mathrm{CT}}} = \frac{\partial N_{\mathrm{CT}}}{\partial \mu_{\mathrm{CT}}} = \left(\frac{\partial N_A}{\partial \mu_{\mathrm{CT}}}\right)_N = -\left(\frac{\partial N_B}{\partial \mu_{\mathrm{CT}}}\right)_N \\ &= (\eta_{A,A} + \eta_{B,B} - 2\eta_{A,B})^{-1}. \end{aligned} \tag{13.121}$$

Finally, by multiplying the subsystem FF by S_{CT} gives the in situ local softnesses of reactants in R:

$$s_{\mathrm{CT}}(\boldsymbol{r}) = \{s_\alpha^{\mathrm{CT}}(\boldsymbol{r}) = f_\alpha^{\mathrm{CT}}(\boldsymbol{r})\ S_{\mathrm{CT}}\}. \tag{13.122}$$

13.6.3 Implications of Equilibrium and Stability Criteria

The equalization of reactant chemical potentials, when $\mu_{\mathrm{CT}} = 0$, constitutes the electronic equilibrium criterion for the DA reactive system. Thus, the initial in situ chemical potential of the ground state electron distributions of the mutually closed, polarized reactants in R^{P},

$$\mu_{\mathrm{CT}}^{+} = \mu_{\mathrm{CT}}(N_{\mathrm{CT}} = 0) = \mu_A^{+} - \mu_B^{+}. \tag{13.123}$$

constitutes the driving force for the interreactant CT. The subsequent equalization of the reactant chemical potentials in R^{CT},

$$\begin{aligned} \mu_{\mathrm{CT}}^{*}(N_{\mathrm{CT}}^{*}) &= \mu_A^{*}(N_{\mathrm{CT}}^{*}) - \mu_B^{*}(N_{\mathrm{CT}}^{*}) = (\mu_A^{+} + \eta_A^R N_{\mathrm{CT}}^{*}) - (\mu_B^{+} - \eta_B^R N_{\mathrm{CT}}^{*}) \\ &= \mu_{\mathrm{CT}}^{+} + \eta_{\mathrm{CT}} N_{\mathrm{CT}}^{*} = 0, \end{aligned} \tag{13.124}$$

then gives the following expression for the equilibrium amount of CT [see (7.191)]:

$$N_{\mathrm{CT}}^{*}(\mu_{\mathrm{CT}}^{+}) = -\mu_{\mathrm{CT}}^{+}/\eta_{\mathrm{CT}} = -\mu_{\mathrm{CT}}^{+} S_{\mathrm{CT}}. \tag{13.125}$$

It should be emphasized that both the initial chemical potential difference and the effective in situ hardness include the "embedding" terms due to the presence of the other reactant. Only at the very early stage of a reaction, at large interreactant separation, when there is little charge coupling between the two species, can they be approximated by the separate reactant quantities:

$$\mu_{\mathrm{CT}}^{+} \cong \mu_{\mathrm{CT}}^{0} = \mu_{A}^{0} - \mu_{B}^{0}, \quad \eta_{\mathrm{CT}} \cong \eta_{\mathrm{CT}}^{0} = \eta_{A}^{0} + \eta_{B}^{0}, \tag{13.126}$$

where η_{α}^{0} stands for the chemical hardness of the free reactant α. The coupling hardnesses $\eta_{A,B} = \eta_{B,A}$ are crucial for a realistic *two*-reactant description of the chemical reactivity trends in both the gas-phase and chemisorption processes.

The related in situ expression for the sum of the *first*- and *second*-order CT energy terms thus reads [compare (7.195)]:

$$E_{\mathrm{CT}}^{*} = \mu_{\mathrm{CT}}^{+} N_{\mathrm{CT}}^{*} + \frac{1}{2}\eta_{\mathrm{CT}}(N_{\mathrm{CT}}^{*})^{2} = \frac{1}{2}\mu_{\mathrm{CT}}^{+} N_{\mathrm{CT}}^{*} = -\frac{1}{2}(\mu_{\mathrm{CT}}^{+})^{2} S_{\mathrm{CT}}. \tag{13.127}$$

It is proportional to the chemical potential difference for the polarized and (initially) mutually closed reactants, marking the in situ chemical potential of R, and the amount of CT taking place when the barrier for the flow of electrons between these two subsystems is removed. The last term of the preceding equation indicates that the overall dependence of the magnitude of the CT stabilization energy on the chemical potential difference is quadratic, with a half of the negative in situ softness determining the relevant proportionality constant.

Next, let us summarize the *stability* requirements for the molecular/reactive systems (Nalewajski 1994, 2003a; Nalewajski and Korchowiec 1997; Nalewajski et al. 1996a). For simplicity we again envisage a general partitioning of the system under consideration into two complementary subsystems A and B, e.g., the two reactants in DA reactive complex. We shall then examine the *external* and *internal* perspectives on the electronic stability problem, associated with the charge stability in the externally *open* bimolecular system $\mathcal{R}^{\mathrm{CT}} = (\mathfrak{r}|A^{*}|B^{*})$ and the externally *closed* analog $R^{\mathrm{CT}} = R^{*} = (A^{*} \mid B^{*})$, respectively.

Let us first assume the initial *internal* equilibrium in R^{*}, marked by the equalized chemical potentials of both subsystems: $\mu_{A}^{*} = \mu_{B}^{*}$. The internal stability problem then refers to a hypothetical displacement of electron populations of reactants in R^{*}, $\delta\boldsymbol{N}^{*}(\Delta) = (\delta N_{A}^{*}, \delta N_{B}^{*}) = (\Delta, -\Delta)$, generated by the population-shift Δ relative to the initial (equilibrium) electron populations $\boldsymbol{N}^{*} = (N_{A}^{*}, N_{B}^{*})$, which automatically preserves the overall number of electrons $N_{R} = N_{A} + N_{B}$. The charge distribution in R^{*} is stable when any finite $\delta\boldsymbol{N}^{*}(\Delta)$ implies an increase in the system electronic energy:

$$\delta E_{\mathrm{CT}}(\Delta) = \frac{1}{2}\eta_{\mathrm{CT}}\Delta^{2} > 0 \quad \text{or} \quad \eta_{\mathrm{CT}} > 0. \tag{13.128}$$

This stability criterion can be transcribed into the following inequality in terms of the condensed hardnesses of reactants [see (13.118)]:

$$\boldsymbol{a} \equiv (\eta_{A,A} + \eta_{B,B})/2 > \eta_{A,B} > 0. \tag{13.129}$$

Hence, the charge distribution of the reactive system is *internally* stable when the *arithmetic* average $\boldsymbol{a}$ of the diagonal hardnesses $\{\eta_{\alpha,\alpha} > 0\}$ of both subsystems exceeds the magnitude of the coupling (*off*-diagonal) hardness between these two complementary fragments of R^*.

We next assume the initial equilibrium between R^* and an external electron reservoir $\mathfrak{r}$ in $\mathfrak{R}^{\mathrm{CT}} = (\mathfrak{r}|A^*|B^*)$, e.g., the surface of the heterogeneous catalyst. It implies the chemical potential equalization $\mu_A^* = \mu_B^* = \mu_{\mathfrak{r}}$ and thus the vanishing *first*-order change in the system electronic energy due to the hypothetical electron exchange between the reactive system R^* and its reservoir $\mathfrak{r}$. In order to diagnose the *external* stability, of the reactive system relative to its reservoir, one examines a virtual flow δN_R between R^* and $\mathfrak{r}$, which generates the associated *second*-order change in the electronic energy of $\mathfrak{R}^{\mathrm{CT}}$,

$$\delta E_{\mathrm{CT}}(\delta N_R) = \frac{1}{2}\eta_R(\delta N_R)^2 > 0 \quad \text{or} \quad \eta_R > 0, \tag{13.130}$$

where $\eta_R = \partial^2 E_R[N_R, v_R]/\partial N_R^2 = \partial \mu_R[N_R, v_R]/\partial N_R$ denotes the chemical hardness of R^* as a whole. We have indicated above that in the *externally* stable reactive system this charge displacement must increase the energy of the combined system $\mathfrak{R}^{\mathrm{CT}}$. Expressing η_R in terms of the condensed hardnesses of reactants (13.113),

$$\eta_R = 1/S_R = 1/(S_A + S_B) = [\eta_{A,A}\eta_{B,B} - (\eta_{A,B})^2]/\eta_{\mathrm{CT}}, \tag{13.131}$$

and realizing that $\eta_{\mathrm{CT}} > 0$, by the internal stability (13.128), then allows one to transcribe the external stability requirement in terms of the *geometric* average $\boldsymbol{g}$ of the diagonal hardnesses of both reactants:

$$\boldsymbol{g} \equiv (\eta_{A,A}\ \eta_{B,B})^{1/2} > \eta_{A,B} > 0, \tag{13.132}$$

Therefore, for the *internally* stable reactive system to be also *externally* stable the geometric average of the diagonal hardnesses of the mutually closed, polarized reactants must be higher than the coupling (*off*-diagonal) hardness.

In further examining the stability regimes as functions of $\eta_{A,B} > 0$, one recalls that $\boldsymbol{g}^2 = \boldsymbol{a}^2 - [(\eta_{A,A} - \eta_{B,B})/2]^2$, and hence $\boldsymbol{g} < \boldsymbol{a}$. In a typical region of the coupling hardness between the two reactants, $0 < \eta_{A,B} < \eta_{A,A}$, where we have recognized the acidic fragment as the electronically harder subsystem, $\eta_{A,A} > \eta_{B,B}$, one therefore obtains the following stability predictions for the externally open reactive system R^* in $\mathfrak{R}^{\mathrm{CT}}$:

$0 < \eta_{A,B} < \boldsymbol{g}$, internally and externally stable;

$\boldsymbol{g} \le \eta_{A,B} < \boldsymbol{a}$, internally stable and externally unstable;

$\boldsymbol{a} \le \eta_{A,B} < \eta_{A,A}$, internally unstable and externally stable.

Thus, for the weakly coupled reactants, when $\eta_{A,B} < \boldsymbol{g}$, the system exhibits both the internal and external stability. With an increase of the charge coupling between these two complementary parts of the bimolecular reactive system, as measured by the intersubsystem hardness, one first observes the onset of the external instability of R^*, relative to the electron reservoir, when the *off*-diagonal hardness reaches the critical value $\eta_{A,B} = \boldsymbol{g}$. The second critical point for the stability predictions is $\eta_{A,B} = \boldsymbol{a}$, above which the internal instability sets in, while the system regains the external stability. This precedence of the external instability before the internal one, when coupling between the two reactant increases, has qualitative implications for catalysis (Nalewajski 1993b). Indeed, one requires a smaller coupling between reactants to effect the charge instability involving the catalyst (external electron reservoir), compared with the interreactant coupling required to generate the internal instability in the gas phase. This external instability subsequently triggers the internal instability, which effects chemical reaction between the chemisorbed reactants. Specific examples of such stability analysis in exploring reactivity trends of model chemisorption complexes and oxide clusters have been reported elsewhere (Nalewajski 1994, 1995a; Nalewajski and Michalak 1995; Nalewajski and Korchowiec 1997).

References

Albright TA, Burdett JK, Whangbo M-H (1985) Orbital interactions in chemistry. Wiley-Interscience, New York

Bader RF (1960) Mol Phys 3:137

Bader RFW (1990) Atoms in molecules. Oxford University Press, New York (and references therein)

Bader RFW (1991) Chem Rev 91:893

Bader RF, Bandrauk AD (1968) J Chem Phys 49:1666

Baekelandt BG, Janssens GOA, Toufar H, Mortier WJ, Schoonheydt RA, Nalewajski RF (1995) J Phys Chem 99:9784

Berkowitz M, Parr RG (1988) J Chem Phys 88:2554

Chapman NB, Shorter J (eds) (1972) Advances in linear free energy relationships. Plenum, New York

Chattaraj PK (ed) (2009) Chemical reactivity theory – a density functional view. CRC, Boca Raton

Ciosłowski J (1991) J Am Chem Soc 113:6756

Ciosłowski J, Mixon ST (1993) J Am Chem Soc 115:1084

Cooper DL (ed) (2002) Valence bond theory. Elsevier, Amsterdam

Coulson CA, Longuet-Higgins HC (1947a) Proc R Soc A Lond 191:39

Coulson CA, Longuet-Higgins HC (1947b) Proc R Soc A Lond 192:16

Dewar MJS (1969) Molecular orbital theory of organic chemistry. McGraw Hill, New York

Dewar MJS, Dougherty RC (1975) The PMO theory of organic chemistry. Plenum, New York

Dunning TH Jr (1984) J Phys Chem 88:2469

Epiotis ND (1978) Theory of organic reactions. Springer, Berlin

Fujimoto H, Fukui K (1974) In: Klopman G (ed) Chemical reactivity and reaction paths. Wiley-Interscience, New York, p 23

Fukui K (1975) Theory of orientation and stereoselection. Springer, Berlin

Fukui K (1981) Acc Chem Res 14:363

Fukui K (1987) Science 218:747
Gadre SR, Shirsat RN (2000) Electrostatics of atoms and molecules. Universities Press, Hyderabad
Gázquez JL (1993) In: Sen KD (ed) Hardness; Struct Bond 80:27
Gázquez JL (2009) In: Chattaraj PK (ed) Chemical reactivity theory: a density functional view. CRC, Boca Raton, p 7
Gázquez JL, Vela A, Galvàn M (1987) In: Sen KD and Jørgensen, C.K (eds) Electronegativity; Struct Bond 66:79
Geerlings P, De Proft F, Langenaeker W (2003) Chem Rev 103:1793
Gilchrist TL, Storr RC (1972) Organic reactions and orbital symmetry. Cambridge University, Cambridge
Gill GB, Willis MR (1974) Pericyclic reactions. Chapman and Hall, New York
Goddard WA III, Harding LB (1978) Annu Rev Phys Chem 29:363
Gutmann V (1978) The donor-acceptor approach to molecular interactions. Plenum, New York
Halevi EA (1992) Orbital symmetry and reaction mechanism – the orbital correspondence in maximum symmetry view. Springer, Berlin
Hammett LP (1935) Chem Rev 17:125
Hammett LP (1937) J Am Chem Soc 59:96
Hammond GS (1955) J Am Chem Soc 77:334
Heitler W, London F (1927) Z Physik 44:455–472; for an English translation see: Hettema H (2000) Quantum chemistry classic scientific paper. World Scientific, Singapore
Johnson CD (1973a) The Hammett equation. Cambridge University, London
Johnson CD (1975) Chem Rev 75:755
Johnson PA, Bartolotti LJ, Ayers P, Fievez T, Geerlings P (2011) Charge density and chemical reactions: a unified view from conceptual DFT. In: Gatti C, Macchi P (eds) Modern charge-density analysis. Springer, Berlin (in press)
Jones RAY (1979) Physical and mechanistic organic chemistry. Cambridge University, Cambridge
Klopman G (1968) J Am Chem Soc 90:223
Klopman G (ed) (1974a) Chemical reactivity and reaction paths. Wiley-Interscience, New York
Klopman G (1974b) In: Klopman G (ed) Chemical reactivity and reaction paths. Wiley-Interscience, New York, p 55
Korchowiec J, Uchimaru T (1998) J Phys Chem A 102:10167
Kraka E, Cremer D (1990) In: Maksić ZB (ed) Theoretical models of chemical bonding, Part 3 – The concept of the chemical bond. Springer, Berlin, p 453
Lewis GN (1916) J Am Chem Soc 38:762
London F (1928) Z Phys 46:455
Marcus RA (1968) J Phys Chem 72:891
Marcus RA (1969) J Am Chem Soc 91:7224
McWeeny R (1979) Coulson's valence. Oxford University Press, Oxford
Mortier WM, Schoonheydt RA (eds) (1997) Developments in the theory of chemical reactivity and heterogeneous catalysis. Research Signpost, Trivandrum
Murray JS, Sen K (eds) (1996) Molecular electrostatic potentials: concepts and applications. Elsevier, Amsterdam
Murrell JN, Carter S, Farantos SC, Huxley P, Varandas AJC (1984) Molecular potential energy functions. Wiley, New York
Nakatsuji H (1973) J Am Chem Soc 95:345
Nakatsuji H (1974a) J Am Chem Soc 96:24
Nakatsuji H (1974b) J Am Chem Soc 96:30
Nalewajski RF (1983) J Chem Phys 78:6112
Nalewajski RF (1988) Z Naturforsch 43a:65
Nalewajski RF (1989a) J Phys Chem 93:2658
Nalewajski RF (1993a) Struct Bond 80:115

Nalewajski RF (1993b) J Mol Catal 82:371

Nalewajski RF (1994) Int J Quantum Chem 49:675

Nalewajski RF (1995a) Int J Quantum Chem 56:453

Nalewajski RF (1995b) In: Gross EKU and Dreizler RM (eds) Proceedings of the NATO ASI on density functional theory. Plenum, New York, p 339

Nalewajski RF (1997b) In: Mortier WM, Schoonheydt RA (eds) Developments in the theory of chemical reactivity and heterogeneous catalysis. Research Signpost, Trivandrum, p 135

Nalewajski RF (1998b) Int J Quantum Chem 69:591

Nalewajski RF (1999) Phys Chem Chem Phys 1:1037

Nalewajski RF (2000a) Comput Chem 24:243

Nalewajski RF (2000b) Int J Quantum Chem 78:168

Nalewajski RF (2002d) In: Sen KD (ed) Reviews of modern quantum chemistry: a celebration of the contributions of Robert G. Parr, vol 2. World Scientific, Singapore, p 1071

Nalewajski RF (2002e) Chem Phys Lett 353:143

Nalewajski RF (2003a) Adv Quant Chem 43:119

Nalewajski RF (2003b) J Phys Chem A 107:3792

Nalewajski RF (2003d) Chem Phys Lett 375:196

Nalewajski RF (2004a) Ann Phys (Leipzig) 13:201

Nalewajski RF (2005d) Chem Phys Lett 410:335

Nalewajski RF (2005e) J Chem Sci 117:455

Nalewajski RF (2006a) Mol Phys 104:365

Nalewajski RF (2006f) Adv Quant Chem 51:235

Nalewajski RF (2006g) Information theory of molecular systems. Elsevier, Amsterdam

Nalewajski RF (2006h) Mol Phys 104:255

Nalewajski RF (2009a) In: Chattaraj PK (ed) Chemical reactivity theory: a density functional view. Taylor and Francis, London, p 453

Nalewajski RF (2009b) J Math Chem 45:709

Nalewajski RF (2009h) J Math Chem 45:607

Nalewajski RF (2010f) Information origins of the chemical bond. Nova Science, New York

Nalewajski RF (2010g) J Math Chem 48:752

Nalewajski RF, Broniatowska E (2003b) Chem Phys Lett 376:33

Nalewajski RF, Koniński M (1988) Acta Phys Polon A 74:255

Nalewajski RF, Korchowiec J (1997) Charge sensitivity approach to electronic structure and chemical reactivity. World-Scientific, Singapore

Nalewajski RF, Michalak A (1995) Int J Quantum Chem 56:603

Nalewajski RF, Michalak A (1996) J Phys Chem 100:20076

Nalewajski RF, Michalak A (1998) J Phys Chem 102:636

Nalewajski RF, Parr RG (2001) J Phys Chem A 105:7391

Nalewajski RF, Sikora O (2000) J Phys Chem A 104:5638

Nalewajski RF, Korchowiec J, Michalak A (1996a) Top Curr Chem 183:25

Nalewajski RF, Mrozek J, Mazur G (1996b) Can J Chem 100:1121

Nalewajski RF, Błażewicz D, Mrozek J (2008) J Math Chem 44:325

Parr RG, Ayers PW, Nalewajski RF (2005) J Phys Chem A 109:3957

Pearson RG (1973) Hard and soft acids and bases. Dowden, Hutchinson, and Ross, Stroudsburg

Pearson RG (1976) Symmetry rules for chemical reactions: orbital topology and elementary processes. Wiley, New York

Pearson RG (1988) Inorg Chem 27:734

Pearson RG (1997) Chemical hardness, applications from molecules to solids. Wiley-VCH, Weindheim

Politzer P, Murray JS (2009) In: Chattaraj PK (ed) Chemical reactivity theory: a density functional view. CRC, Boca Raton, p 243

Politzer P, Truhlar D (eds) (1981) Chemical applications of atomic and molecular electrostatic potentials. Plenum, New York

Salem L (1968a) J Am Chem Soc 90:543
Salem L (1968b) J Am Chem Soc 90:553
Salem L (1969) Chem Br 5:449
Senet P (1996) J Chem Phys 105:6471
Senet P (1997) J Chem Phys 107:2516
Senet P (2009) In: Chattaraj PK (ed) Chemical reactivity theory: a density functional view. CRC, Boca Raton, p 331, vol 4, p 269
Shaik S (1989) In: Bertran J, Czismadia IG (eds) New theoretical concepts for understanding organic reactions, NATO ASI Series, vol. C267. Kluwer Academic, Dordrecht, p 165
Shaik S, Hiberty PC (1991) In: Maksić ZB (ed) Theoretical models of chemical bonding, vol 4, p 269
Shaik S, Hiberty PC (1995) Adv Quant Chem 26:100
Shaik S, Hiberty PC (2004) In: Lipkowitz KB, Larter L, Cundari TR (eds) Reviews in computational chemistry, vol 20, p 1
Silvi B, Savin A (1994) Nature 371:683
Stone AJ (1978) In: Dixon RN, Thomson C (eds) Specialist periodical reports: theoretical chemistry, vol. 3. Bartholomew, Dorking, p 39
Woodward RB, Hoffmann R (1969) Angew Chem Int Ed Engl 8:781
Woodward RB, Hoffmann R (1970) The conservation of orbital symmetry. Academic, London
Woodward RB, Hoffmann R (1971) The conservation of orbital symmetry. Verlag Chemie, Weinheim

Chapter 14
Coupling Between Electronic and Geometrical Structures

Abstract The equilibria in both the geometrically rigid and relaxed molecules, externally closed or open, are examined using the combined treatment of the electronic and nuclear degrees of freedom, which fully accounts for their mutual coupling. New CS descriptors of the molecular states are explored in all admissible geometrical representations, which use the explicit dependence of the Legendre transforms of the system BO potential on the geometric coordinates $\boldsymbol{Q}$ or the conjugate forces $\boldsymbol{F}$. In the canonical geometric representation the principal CS of the BO potential define the system electronic–nuclear Hessian, including the diagonal blocks of the electronic hardness and geometric force constants, with the nuclear Fukui function indices determining the coupling between these two aspects of the molecular structure. Its partial or complete inversion determines the associated compliant matrices in alternative Legendre-transformed representations, in which these principal *state* variables are partly or totally replaced by their respective energy conjugates [generalized "forces" (intensities)]: the system chemical potential μ, characterizing the *electronically* relaxed (open) system, and the vanishing forces on nuclei, $\boldsymbol{F} = \boldsymbol{0}$, identifying the *geometrically* relaxed molecular system.

These geometric representations provide more complete theoretical frameworks for describing diverse reactivity phenomena, carried out under alternative experimental conditions. They may involve the *externally* closed molecular systems: geometrically rigid at the principal (canonical) level of the $(N, \boldsymbol{Q})$-representation, with N standing for the system overall number of electrons, or the geometrically relaxed systems in the geometrically inverted $(N, \boldsymbol{F})$-representation. The *externally* open systems, coupled to an external electron reservoir exhibiting the chemical potential μ, geometrically rigid or relaxed, are described by the electronically inverted $(\mu, \boldsymbol{Q})$-representation or the completely inverted $(\mu, \boldsymbol{F})$-representation, respectively, with the latter defining the system generalized softness representation.

A possible use of the derivative properties generated in this generalized framework as reactivity indices is discussed. A special attention is paid to the *compliance constants* reflecting the interaction between the electronic and nuclear degrees of freedom in the electronically and/or geometrically relaxed systems. Specific

R.F. Nalewajski, *Perspectives in Electronic Structure Theory*,
DOI 10.1007/978-3-642-20180-6_14,

coupling constants describing a subtle interplay between electronic and nuclear state parameters in the geometrically rigid or relaxed systems are identified and discussed. The *minimum-energy coordinates* (MEC) and other compliant descriptors of molecules in such a combined electronic–nuclear treatment of molecular systems and reactants are introduced and discussed within the complementary EF and EP perspectives. These compliant descriptors provide attractive reactivity criteria, since they directly measure the system linear responses, electronic or geometric, and reflect conditions required for the molecule to undergo specific displacements, e.g., those responsible for the chemical reaction (breaking/forming of bonds) of interest. The MEC similarly generate semiquantitative indicators of how specific electronic and/or nuclear displacements affect the remaining state parameters, thus facilitating a qualitative analysis of molecular responses due to the environmental or reactivity perturbations.

Finally, this formalism is applied to the atom–diatom collisions using the semiempirical representation of the canonical hardness tensor in atomic resolution. The nuclear Fukui function indices, which determine the coupling between the molecular electronic and geometrical structures, are explicitly modeled using this analytical representation of the canonical hardness tensor in AIM resolution. Sensitivities for collective charge displacements are derived and the couplings along the minimum-energy reaction path are examined.

14.1 Electronic–Geometric Representations of Molecular States

The internal degrees of freedom of molecular systems are of either electronic or nuclear (geometric) origins. In the BO approximation the equilibrium (ground) state of the externally closed, rigid molecule is specified by the system overall number of electrons N (integer) and the external potential $v(\boldsymbol{r}; \boldsymbol{Q})$ due to the nuclei at their fixed locations, uniquely identified by the internal geometric coordinates $\boldsymbol{Q}$. Alternatively, the state parameters N and $\boldsymbol{Q}$ uniquely identify the system (Coulombic) Hamiltonian $\hat{\mathrm{H}}(N, v) = \hat{\mathrm{H}}(N, \boldsymbol{Q})$, its ground state $\Psi[N, v] = \Psi(N, \boldsymbol{Q})$, the electronic energy $E[N, v] = \langle \Psi[N, v] | \hat{\mathrm{H}}(N, v) | \Psi[N, v] \rangle = E(N, Q)$, and the BO potential of (13.1),

$$W(N, \boldsymbol{Q}) = E(N, \boldsymbol{Q}) + V_{nn}(\boldsymbol{Q}), \tag{14.1}$$

where $V_{nn}(\boldsymbol{Q})$ stands for the repulsion energy between the system nuclei in positions defined by the geometrical parameters $\boldsymbol{Q}$.

One similarly describes the equilibrium state of an externally open system characterized by the (fractional) average number of electrons N, which is coupled to an external electron reservoir $\mathfrak{r}$ controlling the system chemical potential $\mu_{\mathfrak{r}} = \mu = \partial E(N, \boldsymbol{Q}) / \partial N$. In the geometrically rigid molecule, this equilibrium state is identified by μ and $\boldsymbol{Q}$, while the corresponding equilibria of the flexible (geometrically relaxed) molecular systems are identified by the vanishing forces $\boldsymbol{F} = -[\partial W(N, \boldsymbol{Q})/\partial \boldsymbol{Q}]^{\mathrm{T}} = \boldsymbol{0}$, giving rise to alternative sets of state parameters, $(N, \boldsymbol{F})$ or $(\mu, \boldsymbol{F})$, in the electronically closed and open systems, respectively.

As we have already argued in Sect. 13.2, each molecular process involves the mutually coupled displacements in the distribution of the system electrons and positions of its nuclei. In chemistry the mutual interaction between the electronic and geometric structures of molecules or reactive systems plays a vital role in diagnosing their behavior in different chemical environments. Therefore, designing the adequate descriptors of this coupling and establishing principles for qualitative predictions of its structural and reactivity manifestations constitute challenging problems in theoretical chemistry. Indeed, the rules governing this subtle interplay between the electronic and geometric degrees of freedom in molecular systems constitute an important part of the structural chemistry and reactivity theory. They reflect effects of the mutual interaction between an internal polarization (P) and/or external charge transfer (CT) on one side, and the concomitant geometrical relaxation on the other side, e.g., in molecular subsystems of the DA complexes. The qualitative Gutmann (1978) rules of structural chemistry and their semiquantitative extension provided by the Mapping Relations formulated within CSA (Baekelandt et al. 1995; Nalewajski 1995b, 1999, 2006f; Nalewajski and Sikora 2000; Nalewajski and Korchowiec 1997; Nalewajski et al. 1996a) allow for predictions of such general relaxational effects. Another example is provided by the Minimum-Energy Coordinates (MEC) of the *compliant* approach in CSA (Nalewajski 1995b, 2006f, 2009a; Nalewajski and Michalak 1995, Nalewajski and Korchowiec 1997; Nalewajski et al. 1996a, 2008), a development in the spirit of the related treatment of nuclear vibrations (Decius 1963; Jones and Ryan 1970; Swanson 1976; Swanson and Satija 1977).

All these approaches aim at diagnosing the molecular electronic and/or geometrical responses to hypothetical electronic or nuclear displacements (perturbations). The "thermodynamical" Legendre-transformed approach provides a versatile theoretical framework for describing all such alternative equilibrium states of molecules in different chemical environments. The essence of the combined *linear* response treatment of the electronic and geometric state variables is that all their mutual interactions are then explicitly taken into account in the generalized electronic–nuclear Hessian. The relevant coupling terms are represented by the *off*-diagonal elements of such generalized "force constant" tensors for all admissible selections of the system state parameters, which specify the equilibria in the externally closed or open molecules, for their rigid or relaxed geometries. The overall number of electrons, N, and its energy conjugate, the chemical potential μ, determine the *electronic* state variables in the externally closed and open molecular systems, respectively. Accordingly, the internal coordinates $\boldsymbol{Q}$ or their energy conjugates – the (vanishing) forces $\boldsymbol{F}$ of the geometrically optimized systems – similarly describe the rigid or relaxed molecular geometries, respectively.

This theoretical framework unites the EF and EP perspectives on molecular changes, in the spirit of the BO approximation and the Hellmann-Feynman theorem or DFT, respectively. We recall that in the former the electron distribution responds to the geometrical (nuclear) perturbation, i.e., a given displacement in nuclear positions, while the latter implies the system geometrical relaxation following a given test displacement in the system electronic state parameters. All such generalized "polarizabilities" of molecules can be generated within CSA of molecular systems. They provide reliable criteria in the reactivity theories based upon the modern DFT concepts (e.g., Nalewajski and Korchowiec 1997; Nalewajski et al. 1996a).

In the EF outlook the adjustment $\Delta\rho$ in the electron distribution represents the unconstrained (dependent) local state variable of the molecular system in question: $\Delta\rho = \Delta\rho[N, \Delta v] = \Delta\rho(N, \Delta\boldsymbol{Q})$ or $\Delta\rho = \Delta\rho[\mu, \Delta v] = \Delta\rho(\mu, \Delta\boldsymbol{Q})$. In other words, the electron density responds to ("follows") the displacements $\Delta\boldsymbol{Q}$ in nuclear positions, which generate the associated displacement in the system external potential: $\Delta v = \Delta v(\Delta\boldsymbol{Q})$. This selection of the dependent (ρ) and independent (v) state variables generates the softness kernel $\sigma(\boldsymbol{r}, \boldsymbol{r}')$ and the linear response function of the reactivity theory; it has been previously classified as the *chemical softness representation* of molecular states.

These roles are reversed in the EP perspective of DFT, which can be also referred to as the *chemical hardness representation*, since it defines another key concept of the electronic structure and reactivity theories – the hardness kernel $\eta(\boldsymbol{r}, \boldsymbol{r}')$, the inverse of $\sigma(\boldsymbol{r}, \boldsymbol{r}')$. In this EP approach the displacement in electron density of the geometrically relaxed molecule, effected either by the controlled change in the system number of electrons ΔN or by a shift $\Delta\mu$ in the chemical potential of the external reservoir, $\Delta\rho = \Delta\rho(\Delta N, \boldsymbol{F} = \boldsymbol{0})$ or $\Delta\rho = \Delta\rho(\Delta\mu, \boldsymbol{F} = \boldsymbol{0})$, is now regarded as the controlling, independent parameter of state, while the external potential (system geometry) responds to this redistribution of electrons, thus representing a dependent (unconstrained) state variable $\Delta v = \Delta v[\Delta\rho]$. The shift in electron distribution thus "precedes" the movements of nuclei, $\Delta\boldsymbol{Q} = \Delta\boldsymbol{Q}(\Delta N, \boldsymbol{F} = \boldsymbol{0})$ or $\Delta\boldsymbol{Q} = \Delta\boldsymbol{Q}(\Delta\mu, \boldsymbol{F} = \boldsymbol{0})$, in the spirit of the Hellmann–Feynman (force) theorem of quantum chemistry and in accordance with a general philosophy of DFT. This way of approaching the molecular displacements is quite common in qualitative chemical thinking and in reactivity theory. Indeed, chemists often envisage the key manipulation of the system electronic structure as the primary cause of the desired reconstruction of the molecular geometry, e.g., that leading to breaking/forming of the specified bonds in the molecular/reactive system.

As we have already emphasized above, one requires both these perspectives to tackle all issues in the theory of electronic structure of molecules and their chemical reactivity. The wave function and density functional formulations of the quantum theory thus emerge as the complementary descriptions which *together* provide a more complete theoretical framework for *thinking* about molecular systems and discussing diverse issues in chemical reactivity. The emergence of the modern DFT has provided the EP perspective a solid theoretical basis and generated new approaches to many classical problems in chemistry (e.g., Chattaraj 2009; Geerlings et al. 2003; Johnson et al. 2011; Nalewajski 2003a, 2006f, 2009a; Nalewajski and Korchowiec 1997; Nalewajski et al. 1996a). It has offered an alternative point of view from which one can approach the diverse physical/chemical properties and processes involving atomic, molecular, and reactive systems. This novel perspective is much in the spirit of Sanderson's (1951, 1976) Electronegativity Equalization (EE) description of the equilibrium distributions of electrons in molecular systems (see also: Nalewajski 1985, 1998a). Examples of the reactivity indices quantifying the electronic–geometric coupling are provided by the electronic FF (Parr and Yang 1984, 1989) and its nuclear analog (Cohen et al. 1994, 1995; Cohen 1996).

In this short overview of the problem we shall explore diverse coupling constants and MEC components, which can be formulated in the combined electronic–geometrical approaches to the equilibrium states in molecules and reactants. The reactivity implications of the derivative descriptors emerging in this generalized framework, of the interaction between the electronic and geometric aspects of the molecular structure in both the EP and EF perspectives, will be also commented upon. We begin this analysis with a brief survey of the basic concepts and relations of the generalized compliant description of molecular systems, which simultaneously involve the electronic and nuclear degrees of freedom. Illustrative numerical data of these generalized derivative properties for selected polyatomic molecules (Nalewajski et al. 2008) will be discussed and their use as possible reactivity criteria will be advocated. The trends exhibited by such molecular descriptors will be interpreted as manifestations of the familiar LeChâtelier–Braun principle of thermodynamics (Callen 1962; Tisza 1977).

14.2 Perturbation–Response Relations in Geometric Representations

We first examine coupling relations within the canonical geometric representation, which corresponds to the BO potential (14.1) combining the electronic energy and nuclear repulsion term. In this description the BO potential $W(N, \boldsymbol{Q})$ is regarded as function of the adopted geometric coordinates $\boldsymbol{Q}$, e.g., bond lengths and angles.

The canonical *geometrical* derivatives of $W(N, \boldsymbol{Q})$ then determine the *forces* acting on nuclei along these internal coordinates,

$$\boldsymbol{F}(N,\boldsymbol{Q}) = -[\partial W(N,\boldsymbol{Q})/\partial \boldsymbol{Q}]^{\mathrm{T}} = \{F_s\}, \tag{14.2}$$

and the second partials define the geometric Hessian of the molecular system in question, the tensor of geometric force constants in the adopted reference frame $\boldsymbol{Q}$,

$$\mathbf{H} = \frac{\partial^2 W(N,\boldsymbol{Q})}{\partial \boldsymbol{Q}\,\partial \boldsymbol{Q}} = -\frac{\partial \boldsymbol{F}(N,\boldsymbol{Q})}{\partial \boldsymbol{Q}} = \left\{H_{s,s'} = -\frac{\partial F_{s'}}{\partial Q_s}\right\}. \tag{14.3}$$

Accordingly, the corresponding canonical *electronic* derivatives, calculated for the rigid molecular geometry, determine the system chemical potential,

$$\mu = \partial E(N,\boldsymbol{Q})/\partial N = \partial W(N,\boldsymbol{Q})/\partial N, \tag{14.4}$$

and its chemical hardness:

$$\eta = \frac{\partial^2 E(N,\boldsymbol{Q})}{\partial N^2} = \frac{\partial^2 W(N,\boldsymbol{Q})}{\partial N^2} = \frac{\partial \mu(N,\boldsymbol{Q})}{\partial N}. \tag{14.5}$$

It should be recalled that this *rigid*-geometry measure the system global hardness also determines its inverse, the associated global softness $S = (\partial N/\partial \mu)_{\boldsymbol{Q}}$:

$$\eta = \left(\frac{\partial \mu}{\partial N}\right)_{\boldsymbol{Q}} = S^{-1}. \tag{14.6}$$

Finally, the canonical *mixed* second derivatives, coupling the electronic state parameter N with the geometric coordinates $\boldsymbol{Q}$, define the *nuclear Fukui Function* (NFF) indices:

$$\boldsymbol{\varphi}(N,\boldsymbol{Q}) = -\left(\frac{\partial^2 W(N,\boldsymbol{Q})}{\partial N \partial \boldsymbol{Q}}\right)^{\mathrm{T}} = \left(\frac{\partial \boldsymbol{F}(N,\boldsymbol{Q})}{\partial N}\right)_{\boldsymbol{Q}} = -\left(\frac{\partial \mu(N,\boldsymbol{Q})}{\partial \boldsymbol{Q}}\right)_N^{\mathrm{T}}, \tag{14.7}$$

where we have recognized the Maxwell cross-differentiation identity.

These first partials give rise to the first differential of $W(N, \boldsymbol{Q})$:

$$\begin{aligned} dW(N,\boldsymbol{Q}) &= (\partial W/\partial N)_{\boldsymbol{Q}} dN + d\boldsymbol{Q}(\partial W/\partial \boldsymbol{Q})_N \\ &= \mu dN - d\boldsymbol{Q}\,\boldsymbol{F}^{\mathrm{T}} \equiv \mu dN - \Sigma_s F_s dQ_s. \end{aligned} \tag{14.8}$$

In this canonical $(N,\boldsymbol{Q})$-representation the given displacements $\Delta \boldsymbol{p} = (\Delta N, \Delta \boldsymbol{Q})$ in the system independent state parameters determine its "perturbations." The associated linear responses in the energy conjugate quantities, i.e., the unconstrained electronic and geometric state variables, are grouped in the associated gradient vector $\Delta \boldsymbol{g} = (\Delta \mu, -\Delta \boldsymbol{F})$. It then follows from the preceding definitions that the combined Hessian $\mathbf{H}$, which effects the transformation of perturbations $\Delta \boldsymbol{p}$ into the responses $\Delta \boldsymbol{g}$,

$$(\Delta \mu, -\Delta \boldsymbol{F}) = \Delta \boldsymbol{g} = (\Delta N, \Delta \boldsymbol{Q})\mathbf{H} = \Delta \boldsymbol{p}\mathbf{H}, \tag{14.9}$$

includes the following blocks of the system generalized "force" constants:

$$\mathbf{H} = \begin{bmatrix} \left(\frac{\partial \mu}{\partial N}\right)_{\boldsymbol{Q}} & -\left(\frac{\partial \boldsymbol{F}}{\partial N}\right)_{\boldsymbol{Q}} \\ \left(\frac{\partial \mu}{\partial \boldsymbol{Q}}\right)_N & -\left(\frac{\partial \boldsymbol{F}}{\partial \boldsymbol{Q}}\right)_N \end{bmatrix} \equiv \begin{bmatrix} H_{N,N} & \boldsymbol{H}_{N,\boldsymbol{Q}} \\ \boldsymbol{H}_{\boldsymbol{Q},N} & \mathbf{H}_{\boldsymbol{Q},\boldsymbol{Q}} \end{bmatrix} = \begin{bmatrix} \eta & -\boldsymbol{\varphi} \\ -\boldsymbol{\varphi}^{\mathrm{T}} & \mathbf{H} \end{bmatrix}. \tag{14.10}$$

This overall transformation thus combines the following partial (electronic and geometric) ground state relations:

$$\Delta\mu = \Delta N\,\eta - \Delta \boldsymbol{Q}\,\boldsymbol{\varphi}^{\mathrm{T}} \quad \text{and} \quad -\Delta \boldsymbol{F} = -\Delta N \boldsymbol{\varphi} + \Delta \boldsymbol{Q}\,\mathbf{H}. \tag{14.11}$$

The inverse of $\mathbf{H}$ determines the corresponding *compliance matrix* (Nalewajski 2000a, 2006f) describing the open system in the (μ, $\boldsymbol{F}$)-representation. The relevant thermodynamic potential is thus defined by the total Legendre transform of the system BO potential, which replaces the state parameters (N, $\boldsymbol{Q}$) with their energy conjugates (μ, $\boldsymbol{F}$), respectively:

$$\Sigma(\mu, \boldsymbol{F}) = W - N(\partial W/\partial N)_{\boldsymbol{Q}} - \boldsymbol{Q}(\partial W/\partial \boldsymbol{Q})_N = W - N\mu + \boldsymbol{Q}\,\boldsymbol{F}^{\mathrm{T}}. \tag{14.12}$$

Its first differential,

$$d\Sigma = -N\,d\mu + \boldsymbol{Q}\,d\boldsymbol{F}^{\mathrm{T}}, \tag{14.13}$$

then expresses the initial set of state variables as intensities in this inverse, compliant representation:

$$-N = (\partial\Sigma/\partial\mu)_{\boldsymbol{F}} \quad \text{and} \quad \boldsymbol{Q} = (\partial\Sigma/\partial \boldsymbol{F})_{\mu}^{\mathrm{T}}. \tag{14.14}$$

Setting $\boldsymbol{F} = \boldsymbol{0}$ then identifies properties for the equilibrium (relaxed) molecular geometry.

The generalized compliance matrix, combining the relevant blocks corresponding to the electronic and geometric degrees of freedom,

$$\mathbf{S} = \mathbf{H}^{-1} = \begin{bmatrix} -\left(\frac{\partial N}{\partial \mu}\right)_{\boldsymbol{F}} & \left(\frac{\partial \boldsymbol{Q}}{\partial \mu}\right)_{\boldsymbol{F}} \\ -\left(\frac{\partial N}{\partial \boldsymbol{F}}\right)_{\mu} & \left(\frac{\partial \boldsymbol{Q}}{\partial \boldsymbol{F}}\right)_{\mu} \end{bmatrix} \equiv \begin{bmatrix} S_{\mu,\mu} & \mathbf{S}_{\mu,\boldsymbol{F}} \\ \mathbf{S}_{\boldsymbol{F},\mu} & \mathbf{S}_{\boldsymbol{F},\boldsymbol{F}} \end{bmatrix}, \tag{14.15}$$

relates displacements of the representation independent variables ($\Delta\mu$, $\Delta\boldsymbol{F}$) with the conjugate responses in the unconstrained quantities ($-\Delta N$, $\Delta\boldsymbol{Q}$):

$$(-\Delta N, \Delta \boldsymbol{Q}) = (\Delta\mu, \Delta \boldsymbol{F})\mathbf{S}. \tag{14.16}$$

It thus summarizes the coupled responses in the system average number of electrons and its geometry:

$$-\Delta N = \Delta\mu\, S_{\mu,\mu} + \Delta\boldsymbol{Q}\,\mathbf{S}_{\boldsymbol{F},\mu}, \quad \Delta\boldsymbol{Q} = \Delta\mu\,\mathbf{S}_{\mu,\boldsymbol{F}} + \Delta\boldsymbol{F}\,\mathbf{S}_{\boldsymbol{F},\boldsymbol{F}}. \tag{14.17}$$

A reference to (14.15) shows that the diagonal element $S_{\mu,\mu}$ represents the relaxed-geometry analog of the negative rigid-geometry softness of (14.6).

It follows from the second of the preceding equations that a change in the chemical potential of an open system induces an extra relaxation of the geometrical frame. This geometric "softness" effect is described by the derivatives of the row vector $\mathbf{S}_{\mu,F} = \{S_{\mu,s}\} = \mathbf{S}^{\mathrm{T}}_{F,\mu} \equiv \mathbf{S}$.

One can express the compliance matrix in terms of the elements of the principal CS defining the generalized electronic–nuclear "hardness" matrix $\mathbf{H}$ of (14.10), by eliminating ΔN and $\Delta \boldsymbol{Q}$ from (14.11):

$$\begin{aligned} -\Delta N &= -\Delta\mu(\eta - B)^{-1} + \Delta \boldsymbol{F}\,\mathbf{H}^{-1}\boldsymbol{\varphi}^{\mathrm{T}}(\eta - B)^{-1}, \qquad & B &= \boldsymbol{\varphi}\,\mathbf{H}^{-1}\boldsymbol{\varphi}^{\mathrm{T}}; \\ \Delta \boldsymbol{Q} &= \Delta\mu\,\boldsymbol{\varphi}\,\mathbf{H}^{-1}\mathbf{C} - \Delta \boldsymbol{F}\,\mathbf{H}^{-1}\mathbf{C}\eta, & \mathbf{C} &= (\eta\mathbf{I} - \boldsymbol{\varphi}^{\mathrm{T}}\boldsymbol{\varphi}\,\mathbf{H}^{-1})^{-1}. \end{aligned} \tag{14.18}$$

or in the combined matrix form:

$$\mathbf{S} = \begin{bmatrix} -\left(\frac{\partial N}{\partial \mu}\right)_{F} = -(\eta - B)^{-1} \equiv -S^{rel} & \left(\frac{\partial \boldsymbol{Q}}{\partial \mu}\right)_{F} = \boldsymbol{\varphi}\mathbf{H}^{-1}\mathbf{C} \\ -\left(\frac{\partial N}{\partial \boldsymbol{F}}\right)_{\mu} = \mathbf{H}^{-1}\boldsymbol{\varphi}^{\mathrm{T}}(\eta - B)^{-1} & \left(\frac{\partial \boldsymbol{Q}}{\partial \boldsymbol{F}}\right)_{\mu} = -\mathbf{H}^{-1}\mathbf{C}\eta \equiv \mathbf{G}^{rel} \end{bmatrix}, \tag{14.19}$$

where $S^{rel} = (\eta^{rel})^{-1}$ stands for the geometrically relaxed softness, inverse of the associated relaxed hardness $\eta^{rel} = (\partial\mu/\partial N)_{\boldsymbol{F}}$, and $\mathbf{G}^{rel}$ denotes the electronically relaxed geometrical compliant matrix, which differs from its closed system analog $\mathbf{G} = -\mathbf{H}^{-1} = (\partial \boldsymbol{Q}/\partial \boldsymbol{F})_N$.

Let us now turn to the partly inverted (N, $\boldsymbol{F}$)-representation describing the geometrically relaxed but externally closed molecular system. The relevant thermodynamic potential is now defined by the Legendre transform of $W(N,\boldsymbol{Q})$ which replaces $\boldsymbol{Q}$ by $\boldsymbol{F}$ in the list of the system parameters of state:

$$\Theta(N, \boldsymbol{F}) = W - \boldsymbol{Q}(\partial W/\partial \boldsymbol{Q})_N = W + \boldsymbol{Q}\boldsymbol{F}^{\mathrm{T}}. \tag{14.20}$$

Its differential,

$$d\Theta = \mu dN + \boldsymbol{Q}\,d\boldsymbol{F}^{\mathrm{T}}, \tag{14.21}$$

defines the corresponding conjugate (unconstrained) variables in this representation:

$$\mu = (\partial\Theta/\partial N)_{\boldsymbol{F}} \quad \text{and} \quad \boldsymbol{Q} = (\partial\Theta/\partial \boldsymbol{F})_N^{\mathrm{T}}. \tag{14.22}$$

Eliminating $\Delta\mu$ from (14.17) and inserting it into the second of these equations then give the following transformation of the representation independent displacements (ΔN, $\Delta \boldsymbol{F}$) into the linear responses of their conjugates ($\Delta\mu$, $\Delta \boldsymbol{Q}$),

$$(\Delta\mu, \Delta \boldsymbol{Q}) = (\Delta N, \Delta \boldsymbol{F})\mathbf{V}, \tag{14.23}$$

where the relevant Hessian $\mathbf{V}$ expressed in terms of the principal compliance coefficients of (14.15) reads:

$$\mathbf{V} = \begin{bmatrix} \left(\frac{\partial\mu}{\partial N}\right)_{\boldsymbol{F}} & \left(\frac{\partial \boldsymbol{Q}}{\partial N}\right)_{\boldsymbol{F}} \\ \left(\frac{\partial\mu}{\partial \boldsymbol{F}}\right)_N & \left(\frac{\partial \boldsymbol{Q}}{\partial \boldsymbol{F}}\right)_N \end{bmatrix} \equiv \begin{bmatrix} V_{N,N} & \boldsymbol{V}_{N,\boldsymbol{F}} \\ \boldsymbol{V}_{\boldsymbol{F},N} & \mathbf{V}_{\boldsymbol{F},\boldsymbol{F}} \end{bmatrix} = \begin{bmatrix} -S_{\mu,\mu}^{-1} & -\boldsymbol{S}_{\mu,\boldsymbol{F}} S_{\mu,\mu}^{-1} \\ -\boldsymbol{S}_{\boldsymbol{F},\mu} S_{\mu,\mu}^{-1} & \mathbf{S}_{\boldsymbol{F},\boldsymbol{F}} - \boldsymbol{S}_{\boldsymbol{F},\mu}\boldsymbol{S}_{\mu,\boldsymbol{F}} S_{\mu,\mu}^{-1} \end{bmatrix}. \tag{14.24}$$

Again, the diagonal element $V_{N,N}$ represents the molecular hardness estimated for the relaxed geometry of the molecule, a companion parameter of the rigid geometry hardness measure of (14.5). The two partial relations for the electronic and geometric responses in (14.23),

$$\Delta\mu = \Delta N\ V_{N,N} + \Delta \boldsymbol{F}\ \boldsymbol{V}_{\boldsymbol{F},N} \quad \text{and} \quad \Delta \boldsymbol{Q} = \Delta N\ \boldsymbol{V}_{N,\boldsymbol{F}} + \Delta \boldsymbol{F}\ \mathbf{V}_{\boldsymbol{F},\boldsymbol{F}}, \tag{14.25}$$

imply that there is an additional geometry relaxation due to a finite external CT between the open molecule and its electron reservoir, besides the usual term for constant N, due to the forces acting on the system nuclei. This extra relaxation of the molecular frame is described by the coupling vectors $\boldsymbol{V}_{N,\boldsymbol{F}} = \boldsymbol{V}_{\boldsymbol{F},N}^{\mathrm{T}}$.

The blocks of $\mathbf{V}$ can be alternatively expressed in terms of the principal geometric derivatives defining the generalized Hessian of (14.10). This can be accomplished by first expressing $\Delta\boldsymbol{Q}$ as function of ΔN and $\Delta\boldsymbol{F}$, using (14.11), and then inserting the result into the first of these equations:

$$\Delta\mu = \Delta N(\eta - B) + \Delta \boldsymbol{F}\ \mathbf{H}^{-1}\boldsymbol{\varphi}^{\mathrm{T}} \qquad \text{and} \qquad \Delta\boldsymbol{Q} = \Delta N\ \boldsymbol{\varphi}\mathbf{H}^{-1} - \Delta\boldsymbol{F}\ \mathbf{H}^{-1}. \tag{14.26}$$

A comparison between (14.25) and (14.26) then gives:

$$\mathbf{V} = \begin{bmatrix} \left(\frac{\partial\mu}{\partial N}\right)_{\boldsymbol{F}} = (\eta - B) = \eta^{rel} & \left(\frac{\partial \boldsymbol{Q}}{\partial N}\right)_{\boldsymbol{F}} = \boldsymbol{\varphi}\mathbf{H}^{-1} \equiv \boldsymbol{f} \\ \left(\frac{\partial\mu}{\partial \boldsymbol{F}}\right)_N = \mathbf{H}^{-1}\boldsymbol{\varphi}^{\mathrm{T}} \equiv \boldsymbol{f}^{\mathrm{T}} & \left(\frac{\partial \boldsymbol{Q}}{\partial \boldsymbol{F}}\right)_N = -\mathbf{H}^{-1} \equiv \mathbf{G} \end{bmatrix}, \tag{14.27}$$

where the *row* vector $\boldsymbol{f} = \{f_s\}$ groups the *Geometric Fukui Function* (GFF) indices.

Finally, let us examine the remaining (μ,$\boldsymbol{Q}$)-representation describing the equilibrium state of an open molecular system with the "frozen" nuclear framework. The relevant partial Legendre transform of the BO potential energy surface, which replaces N by μ in the list of independent state parameters, represents the BO grand potential:

$$\Xi(\mu, \boldsymbol{Q}) = W - N(\partial W/\partial N)_{\boldsymbol{Q}} = W - N\mu. \tag{14.28}$$

Its differential,

$$d\Xi = -N d\mu - \boldsymbol{F}\, d\boldsymbol{Q}^{\mathrm{T}}, \tag{14.29}$$

defines the representation "intensities,"

$$-N = (\partial\Xi/\partial\mu)_{\boldsymbol{Q}} \quad \text{and} \quad -\boldsymbol{F} = (\partial\Xi/\partial\boldsymbol{Q})_{\mu}^{\mathrm{T}}, \tag{14.30}$$

which respond to displacements in the system-independent state parameters μ and $\boldsymbol{Q}$.

Eliminating $\Delta\boldsymbol{F}$ from (14.17) and inserting the result into the first of these two equations then give the following transformation of the representation independent perturbations ($\Delta\mu$, $\Delta\boldsymbol{Q}$) into the linear responses of their conjugates ($-\Delta N$, $-\Delta\boldsymbol{F}$), expressed in terms of the matrix elements of the compliance matrix $\mathbf{S}$ of (14.15):

$$-(\Delta N, \Delta\boldsymbol{F}) = (\Delta\mu, \Delta\boldsymbol{Q})\mathbf{G}, \tag{14.31}$$

$$\mathbf{G} = \begin{bmatrix} -\left(\frac{\partial N}{\partial\mu}\right)_{\boldsymbol{Q}} & -\left(\frac{\partial \boldsymbol{F}}{\partial\mu}\right)_{\boldsymbol{Q}} \\ -\left(\frac{\partial N}{\partial\boldsymbol{Q}}\right)_{\mu} & -\left(\frac{\partial \boldsymbol{F}}{\partial\boldsymbol{Q}}\right)_{\mu} \end{bmatrix} \equiv \begin{bmatrix} G_{\mu,\mu} & \mathbf{G}_{\mu,Q} \\ \mathbf{G}_{Q,\mu} & \mathbf{G}_{Q,Q} \end{bmatrix} = \begin{bmatrix} S_{\mu,\mu} - \mathbf{S}_{\mu,\boldsymbol{F}}\mathbf{S}_{\boldsymbol{F},\boldsymbol{F}}^{-1}\mathbf{S}_{\boldsymbol{F},\mu} & \mathbf{S}_{\mu,\boldsymbol{F}}\mathbf{S}_{\boldsymbol{F},\boldsymbol{F}}^{-1} \\ \mathbf{S}_{\boldsymbol{F},\boldsymbol{F}}^{-1}\mathbf{S}_{\boldsymbol{F},\mu} & -\mathbf{S}_{\boldsymbol{F},\boldsymbol{F}}^{-1} \end{bmatrix}. \tag{14.32}$$

This matrix transformation combines the electronic and geometric relations:

$$-\Delta N = \Delta\mu\, G_{\mu,\mu} + \Delta\boldsymbol{Q}\, \mathbf{G}_{Q,\mu} \quad \text{and} \quad -\Delta\boldsymbol{F} = \Delta\mu\, \mathbf{G}_{\mu,Q} + \Delta\boldsymbol{Q}\, \mathbf{G}_{Q,Q}. \tag{14.33}$$

The elements of $\mathbf{G}$ can be alternatively expressed in terms of the generalized hardness matrix of (14.10), by eliminating ΔN from (14.11) and inserting the result into the second of these equations:

$$\begin{aligned} -\Delta N &= -\Delta\mu\, \eta^{-1} - \Delta\boldsymbol{Q}\, \boldsymbol{\varphi}^{\mathrm{T}}\eta^{-1} \equiv -\Delta\mu\, S - \Delta\boldsymbol{Q}\, \mathbf{s}^{\mathrm{T}}, \\ -\Delta\boldsymbol{F} &= -\Delta\mu\, \mathbf{s} + \Delta\boldsymbol{Q}(\mathbf{H} - \boldsymbol{\varphi}^{\mathrm{T}} S\, \boldsymbol{\varphi}) = -\Delta\mu\, \mathbf{s} + \Delta\boldsymbol{Q}(\mathbf{H} - \boldsymbol{\varphi}^{\mathrm{T}}\mathbf{s}). \end{aligned} \tag{14.34}$$

Here S denotes the rigid geometry measure of the system global softness and the *row* vector $\mathbf{s}$ of the geometric softnesses is defined as product of the global softness and the NFF vector:

$$\mathbf{s} = S\boldsymbol{\varphi} = (\partial\boldsymbol{F}/\partial\mu)_{\boldsymbol{Q}} = (\partial N/\partial\boldsymbol{Q})_{\mu}^{\mathrm{T}} = (\partial\boldsymbol{F}/\partial N)_{\boldsymbol{Q}}/(\partial\mu/\partial N)_{\boldsymbol{Q}} = \{s_s \equiv (F_s)_{\mu}\}. \tag{14.35}$$

A reference to (14.33) shows that the effective geometrical Hessian of an open molecular system differs from that for the closed system (14.10) by the extra CT-contribution involving the geometrical softnesses and NFF.

One finally identifies the corresponding blocks of **G** by comparing the partial relations of (14.33) with the explicit transformations of (14.34):

$$\mathbf{G} = \begin{bmatrix} -\left(\frac{\partial N}{\partial \mu}\right)_{\boldsymbol{Q}} = -S & -\left(\frac{\partial \boldsymbol{F}}{\partial \mu}\right)_{\boldsymbol{Q}} = -\boldsymbol{s} \\ -\left(\frac{\partial N}{\partial \boldsymbol{Q}}\right)_{\mu} = -\boldsymbol{s}^{\mathrm{T}} & -\left(\frac{\partial \boldsymbol{F}}{\partial \boldsymbol{Q}}\right)_{\mu} = \mathbf{H} - \boldsymbol{\varphi}^{\mathrm{T}}\boldsymbol{s} \equiv \mathbf{H}^{rel} \end{bmatrix}. \tag{14.36}$$

Here, $\mathbf{H}^{rel}$ denotes the electronically relaxed geometrical Hessian, which differs from its closed-system analog $\mathbf{H} = -(\partial \boldsymbol{F}/\partial \boldsymbol{Q})_N$. The **G** matrix thus involves the negative, rigid geometry electronic softness as diagonal element associated with the electronic state variable μ, the *off*-diagonal elements representing the geometric softnesses, and the open-system (electronically relaxed) geometrical Hessian as the nuclear diagonal block. The latter differs from the closed system (electronically rigid) Hessian **H** by the softening contribution implied by the LeChâtelier–Braun principle of the ordinary thermodynamics:

$$\begin{aligned} \boldsymbol{\varphi}^{\mathrm{T}}\boldsymbol{s} = \boldsymbol{\varphi}^{\mathrm{T}} S\, \boldsymbol{\varphi} &= (\partial \mu/\partial \boldsymbol{Q})_N (\partial \boldsymbol{F}/\partial \mu)_{\boldsymbol{Q}} = [(\partial N/\partial \boldsymbol{Q})_\mu (\partial \boldsymbol{F}/\partial N)_{\boldsymbol{Q}}]^{\mathrm{T}} \\ &= (\partial \boldsymbol{F}/\partial \boldsymbol{Q})_N - (\partial \boldsymbol{F}/\partial \boldsymbol{Q})_\mu. \end{aligned} \tag{14.37}$$

14.3 Descriptors of Electronic–Geometric Interaction

Several geometrical quantities introduced in the preceding section provide indices reflecting a strength of the mutual coupling between the molecular electronic and geometrical structures. In the canonical geometrical (N,$\boldsymbol{Q}$)-representation the diagonal blocks of the generalized Hessian **H**, measuring the system (rigid geometry) hardness η and its force constants **H**, describe the decoupled aspects of the electronic and geometric structures, respectively. The NFF $\boldsymbol{\varphi}$, defining the *off*-diagonal blocks in **H**, reflects the coupling between the electronic and nuclear aspects of the molecular structure. They describe the influence of geometrical displacements in the externally closed system on the molecular chemical potential or the effect of an external CT on the geometrical forces. The geometric softnesses $\boldsymbol{S}$ [see (14.17 and (14.19)] or $\boldsymbol{s}$ (14.35) reflect similar couplings in the externally open molecular systems. It should be stressed, however, that in the geometric compliance matrix **S** the interaction between these two facets of molecular structure enters the diagonal blocks as well, as explicitly indicated in (14.18) and (14.19).

A similar effect of the system electronic or nuclear "softening," due to its opening relative to the reservoir or the relaxation of its geometry, is seen in the diagonal blocks of the partial compliant matrices **V** and **G**. This spontaneous relaxation of the system electronic–nuclear structure reflects the LeChâtelier–Braun principle (of "moderation") in the ordinary thermodynamics (Callen 1962). Indeed, the extra electronic relaxation $\delta N(\Delta \boldsymbol{Q})$ induced by the primary nuclear perturbation $\Delta \boldsymbol{Q}$ in the externally *open* system, in which a spontaneous CT between the molecule and its reservoir is allowed, effectively lowers the increases in the magnitude of forces on the system nuclei, compared with those in the externally *closed* system: $|\Delta \boldsymbol{F}(\Delta \boldsymbol{Q})|_N > |\Delta \boldsymbol{F}(\Delta \boldsymbol{Q})|_\mu$ [see (14.37)]. The effect of the spontaneous geometry relaxation $\delta \boldsymbol{Q}(\Delta N)$ induced by the primary electronic perturbation ΔN similarly lowers the increase in the system chemical potential, compared to that in the rigid system: $|\Delta\mu(\Delta N)|_{\boldsymbol{Q}} > |\Delta\mu(\Delta N)|_{\boldsymbol{F}=\boldsymbol{0}}$.

It should be also realized that the generalized softness matrix represents the full compliant description of the electronic "coordinate" N coupled to the system geometric relaxation. Indeed, the global softness for the relaxed geometry,

$$\begin{aligned} -S_{\mu,\mu} &= (\partial N/\partial\mu)_{\boldsymbol{F}} = (\eta - B)^{-1} = (\eta - \boldsymbol{\varphi}\mathbf{H}^{-1}\boldsymbol{\varphi}^{\mathrm{T}})^{-1} \equiv S^{rel} \equiv (\eta^{rel})^{-1} \\ &> \eta^{-1} = S = (\partial N/\partial\mu)_{\boldsymbol{Q}} > 0, \end{aligned} \tag{14.38}$$

where the last inequality states the familiar LeChâtelier (stability) requirement, differs from the conventional definition of the electronic global softness S, which invokes the *rigid* geometry constraint. The geometric hardness contribution B in the preceding equation effectively softens the electronic distribution via the relaxation in nuclear positions, represented by term including the purely geometric compliant $\mathbf{H}^{-1}$, with the "weighting" factors provided by the NFF $\boldsymbol{\varphi}$ reflecting components of the *relative* geometric softness of the molecule.

The other diagonal block of the generalized geometrical compliants, which contains the contributions due to electron–nuclear couplings,

$$\mathbf{S}_{\boldsymbol{F},\boldsymbol{F}} = (\partial \boldsymbol{Q}/\partial \boldsymbol{F})_\mu = -\mathbf{H}^{-1}\mathbf{C}\eta = -\mathbf{H}^{-1}(\eta\mathbf{I} - \boldsymbol{\varphi}^{\mathrm{T}}\boldsymbol{\varphi}\mathbf{H}^{-1})^{-1}\eta \neq -\mathbf{H}^{-1} = \mathbf{V}_{\boldsymbol{F},\boldsymbol{F}}, \tag{14.39}$$

is also seen to differ from the purely geometrical compliant $\mathbf{V}_{F,F}$ by the additional factor exhibiting both the electronic and nuclear origins. The mixture of the electronic and nuclear inputs is also seen to determine the *off*-diagonal blocks $\boldsymbol{S}_{\mu,\boldsymbol{F}}$ and $\boldsymbol{S}_{\boldsymbol{F},\mu}$ of the geometric compliant matrix, respectively, measuring the effect of the chemical potential on the relaxed nuclear positions ($\boldsymbol{S}_{\mu,\boldsymbol{F}}$) or the influence of forces on the effective charge of an open molecule ($\boldsymbol{S}_{\boldsymbol{F},\mu}$).

Next, let us examine the compliance descriptors of the externally *closed* system in the $\Theta(N,\boldsymbol{F})$ representation, defined by the corresponding blocks of the geometric CS in **V** (14.24). Again, the first diagonal derivative in this matrix, $V_{N,N} = (\partial\mu/\partial N)_{\boldsymbol{F}} = \eta - B$, allows the geometry of the system to relax, after an addition/removal of an electron, until the forces on nuclei exactly vanish. The electronic–geometric interaction is also detected in the coupling blocks

$\boldsymbol{V}_{N,\boldsymbol{F}} = (\partial \boldsymbol{Q}/\partial N)_{\boldsymbol{F}}$ and $\boldsymbol{V}_{\boldsymbol{F},N} = (\partial \mu/\partial \boldsymbol{F})_N$. A reference to (14.26) (14.27) indicates that they are determined by the purely nuclear compliants $\mathbf{V}_{\boldsymbol{F},\boldsymbol{F}} = -\mathbf{H}^{-1}$ and NFF.

The electronic–nuclear coupling in molecules is also detected in the other partial Legendre-transformed representation $\Xi(\mu, \boldsymbol{Q})$, which defines the combined Hessian $\mathbf{G}$ of (14.32). Its first diagonal derivative,

$$G_{\mu,\mu} = -(\partial N/\partial \mu)_{\boldsymbol{Q}} = -(\partial N/\partial \mu)_v = -S, \tag{14.40}$$

represents the purely electronic, global compliant reflecting the negative softness of the geometry rigid system. The *off*-diagonal blocks $\boldsymbol{G}_{\mu,\boldsymbol{Q}} = -(\partial \boldsymbol{F}/\partial \mu)_{\boldsymbol{Q}}$ and $\boldsymbol{G}_{\boldsymbol{Q},\mu} = -(\partial N/\partial \boldsymbol{Q})_\mu$ represent the geometric softnesses. For the rigid nuclear frame, they measure the effect of the system chemical potential on forces on nuclei ($\boldsymbol{G}_{\mu,\boldsymbol{Q}}$) or the influence of nuclear displacements on the effective molecular charge ($\boldsymbol{G}_{\boldsymbol{Q},\mu}$). Since in this representation the molecular system is externally open, one detects in the geometrical Hessian of this representation the contribution due to external CT, triggered by nuclear displacements:

$$\mathbf{G}_{\boldsymbol{Q},\boldsymbol{Q}} = -(\partial \boldsymbol{F}/\partial \boldsymbol{Q})_\mu = \mathbf{H} - \boldsymbol{\varphi}^{\mathrm{T}}\boldsymbol{s} = \mathbf{H} - \boldsymbol{\varphi}^{\mathrm{T}} S\, \boldsymbol{\varphi} \neq \mathbf{H} = -(\partial \boldsymbol{F}/\partial \boldsymbol{Q})_N. \tag{14.41}$$

This block thus contains the electronically relaxed force constants along the system internal coordinates.

14.4 Compliance Formalism and Minimum-Energy Coordinates

We start with a brief summary of the compliance approach to nuclear motions (Decius 1963; Jones and Ryan 1970). The inverse of the nuclear force constant matrix $\mathbf{H}$,

$$\mathbf{H} = \left\{ H_{s,s'} = \frac{\partial^2 W(N, \boldsymbol{Q})}{\partial Q_s\, \partial Q_{s'}} = -\left(\frac{\partial F_{s'}}{\partial Q_s}\right)_{Q_{t \neq s}} \right\}, \tag{14.42}$$

in the purely geometric $\boldsymbol{Q}$-representation, defines the geometric compliance matrix of the “inverse” $\boldsymbol{F}$-representation:

$$\mathbf{G} = \frac{\partial^2 \Theta(N, \boldsymbol{F})}{\partial \boldsymbol{F}\, \partial \boldsymbol{F}} = -\mathbf{H}^{-1} = \left\{ G_{s,s'} = \left(\frac{\partial^2 \Theta(N, \boldsymbol{F})}{\partial F_s\, \partial F_{s'}}\right) = \left(\frac{\partial Q_{s'}}{\partial F_s}\right)_{\{F_{t \neq s}\}} \right\}. \tag{14.43}$$

Here $\Theta(N, \boldsymbol{F})$ [see (14.20)] stands for the Legendre transform of the BO potential-energy surface $W(N, \boldsymbol{Q})$, in which the nuclear coordinates $\boldsymbol{Q}$ are replaced by the corresponding forces $\boldsymbol{F}$ in the list of the parameters of state. Indeed, for the fixed number of electrons N,

$$[d\Theta(N,\boldsymbol{F})]_N = \boldsymbol{Q}\, d\boldsymbol{F}^{\mathrm{T}} \quad \text{and} \quad [\partial^2\Theta(N,F)/\partial F_s \partial F_{s'}]_N = (\partial Q_{s'}/\partial F_s)_{N,\boldsymbol{F}'}. \tag{14.44}$$

The constraint $\boldsymbol{F}' = \{F_{t\neq s}\} = \boldsymbol{0}'$ in these derivatives implies that the remaining part of the nuclear frame is free to relax the atomic positions until the forces associated with these geometrical degrees of freedom exactly vanish, thus marking the minimum of the system energy with respect to $\{Q_{t\neq s}\}$.

The ratio of the matrix elements in sth row of $\mathbf{G}$, $\boldsymbol{G}_s = \{G_{s,s'},\ s' = 1, 2, \ldots\}$ to the diagonal element $G_{s,s}$ then determines kth vector of the nuclear (geometric) interaction constants:

$$(s')_s = G_{s,s'}/G_{s,s} = \left(\frac{\partial Q_{s'}}{\partial F_s}\right)_{F_{t\neq s}} \left(\frac{\partial Q_s}{\partial F_s}\right)^{-1}_{F_{t\neq s}} = \left(\frac{\partial Q_{s'}}{\partial Q_s}\right)_{F_{t\neq s}}, \quad s' = 1,\ 2, \ldots \tag{14.45}$$

These indices describe the minimum-energy responses in the remaining nuclear variables $\{Q_{s'\neq s}\}$, i.e., for the vanishing forces $\boldsymbol{F}'_s = \{F_{t\neq s} = 0\} = \boldsymbol{0}'$, per unit displacement of sth nuclear coordinate. They thus determine the sth geometric MEC. This compliant concept can be used to predict the equilibrium responses of the system geometric structure to a displacement (perturbation) of selected sth nuclear coordinate ΔQ_s from the initial (equilibrium) geometry of the molecule, which account for all mutual couplings between geometric coordinates:

$$\Delta\boldsymbol{Q}(\Delta Q_s)\Big|_{\boldsymbol{F}'_s=\boldsymbol{0}'} = \{(s')_s \Delta Q_s\}. \tag{14.46}$$

In this section, we shall discuss several related concepts within the compliant description of the combined electronic–nuclear treatment of molecular systems. As we have already remarked before, there are two types of geometrical constraints, which can be imposed on the molecule: that of the rigid geometry $\boldsymbol{Q}$ and the condition of the vanishing forces $\boldsymbol{F} = \boldsymbol{0}$, i.e., of the system equilibrium (relaxed) geometry. The electronic degree of freedom N can be treated in a similar way: it can be held "frozen" in the closed molecular system scenario, or be allowed to relax in the open system, in contact with the electronic reservoir.

The fully relaxed description amounts to the compliant description of the molecular electronic and geometric structures, in which one allows the system to relax all its remaining (electronic and nuclear) degrees of freedom in response to the probing displacements in the system number of electrons or positions of its constituent atoms. The (N,$\boldsymbol{F}$)- and (μ,$\boldsymbol{F}$)-representations correspond to the *nuclear* compliant treatment of the molecular geometrical structure, while the (N,$\boldsymbol{Q}$)- and (μ,$\boldsymbol{Q}$)-representations adopt the rigid geometry approach. These Legendre-transformed approaches to the electronic/geometric representations of molecular states thus provide the complete set of quantities, which can be used to monitor (or index) the electronic–geometric couplings in molecular systems, covering both the externally open and closed molecular systems.

The MEC can be also introduced in the combined electron–nuclear treatment of the geometric representations of the molecular structure (Nalewajski 1995b, 1999, 2000a, 2006f; Nalewajski and Korchowiec 1997; Nalewajski et al. 1996a, 2008). Consider, e.g., the generalized interaction constants defined by the combined softness matrix $\mathbf{S}$. The ratios of the matrix elements in $\boldsymbol{S}_{\mu,\boldsymbol{F}} = \{S_{\mu,s'}\}$ to $S_{\mu,\mu}$ define the following interaction constants between the nuclear coordinates and the system average number of electrons:

$$(s')_N = S_{\mu,s'}/S_{\mu,\mu} = (\partial Q_{s'}/\partial\mu)_{\boldsymbol{F}=\boldsymbol{0}}/(\partial N/\partial\mu)_{\boldsymbol{F}=\boldsymbol{0}} = (\partial Q_{s'}/\partial N)_{\boldsymbol{F}=\boldsymbol{0}} \equiv V_{N,s'}. \tag{14.47}$$

They reflect the minimum-energy responses of the system geometrical coordinates per unit displacement in the system number of electrons thus defining the associated MEC:

$$\Delta\boldsymbol{Q}(\Delta N)|_{\boldsymbol{F}=0} = \{(s')_N \Delta N\}. \tag{14.48a}$$

These coefficients are thus equivalent to GFF vector of (14.27),

$$\begin{aligned}\boldsymbol{f} = \{(s')_N\} &= (\partial\boldsymbol{Q}/\partial N)_{\boldsymbol{F}=\boldsymbol{0}} = \boldsymbol{V}_{N,\boldsymbol{F}} \\ &= (\partial\mu/\partial\boldsymbol{F})_N^{\mathrm{T}} = \boldsymbol{V}_{\boldsymbol{F},N}^{\mathrm{T}} = (\partial\boldsymbol{Q}/\partial\mu)_{\boldsymbol{F}=\boldsymbol{0}}/(\partial N/\partial\mu)_{\boldsymbol{F}=\boldsymbol{0}} = \mathbf{S}/S^{rel}.\end{aligned} \tag{14.49}$$

They can be interpreted as an alternative set of NFF indices, which diagnose the normalized effect of changing the oxidation state of the molecular system as a whole on its geometry. These indices define the minimum energy coordinate of (14.48a), grouping responses in nuclear coordinates due to a finite inflow/outflow of electrons, $\Delta N \neq 0$:

$$\Delta\boldsymbol{Q}(\Delta N)|_{\boldsymbol{F}=\boldsymbol{0}} = \Delta N\ \boldsymbol{f}. \tag{14.48b}$$

It should be also realized that NFF can be interpreted as MEC reflecting the rigid geometry response in forces per unit displacement in the system number of electrons:

$$\begin{aligned}\boldsymbol{\varphi} = \{(F_s)_N\} &= (\partial\boldsymbol{F}/\partial N)_{\boldsymbol{Q}} = \boldsymbol{H}_{N,\boldsymbol{Q}} \\ &= (\partial\mu/\partial\boldsymbol{Q})_N^{\mathrm{T}} = \boldsymbol{H}_{\boldsymbol{Q},N}^{\mathrm{T}} = (\partial\boldsymbol{F}/\partial\mu)_{\boldsymbol{Q}}/(\partial N/\partial\mu)_{\boldsymbol{Q}} = \mathbf{s}/S.\end{aligned} \tag{14.50}$$

Here, the geometric softnesses $\mathbf{s}$ represent the *rigid* geometry interactions between forces $\boldsymbol{F}$ and the system chemical potential.

The remaining interaction constants defined in this representation are given by the following ratios of the molecular compliants:

$$(N)_{s,\mu} = \mathsf{S}_{\mu,s}/\mathsf{S}_{s,s} = (\partial N/\partial Q_s)_{\mu,\boldsymbol{F}'_s=\boldsymbol{0}'} \quad \text{and}$$
$$(s')_{s,\mu} = \mathsf{S}_{s,s'}/\mathsf{S}_{s,s} = (\partial Q_{s'}/\partial Q_s)_{\mu,\boldsymbol{F}'_s=\boldsymbol{0}'}. \tag{14.51}$$

In the open molecule, coupled to an external electron reservoir which fixes the system chemical potential, they combine the minimum-energy responses in the system number of electrons and the remaining nuclear coordinates to a unit displacement in Q_s. The associated minimum-energy coordinates,

$$\Delta N(\Delta Q_s)\Big|_{\boldsymbol{F}'_s=\boldsymbol{0}'} = \{(N)_{s,\mu}\Delta Q_s\} \quad \text{and} \quad \Delta\boldsymbol{Q}(\Delta Q_s)\Big|_{\mu,\boldsymbol{F}'_s=\boldsymbol{0}'} = \{(s')_{s,\mu}\Delta Q_s\}, \tag{14.52}$$

add to a variety of descriptors of the electronic and geometric structures of molecular systems. The $\{(N)_{s,\mu}\}$ coupling constants can be used to probe trends in the chemical oxidation/reduction of the open molecule, which follows a given geometrical deformation of the molecule. These probing displacements allow one to identify nuclear changes, which are the most effective in bringing about this electronic transformation of the molecule. The other set $\{(s')_{s,\mu}\}$ tests consequences of a hypothetical perturbation of nuclear positions in the open molecule, thus facilitating a search for the most effective geometric manipulation of the molecular system in question, which is required to bring about the desired overall change in the system nuclear framework.

The *partial* compliant matrix $\mathbf{V}$ of the $(N,\boldsymbol{F})$-representation defines analogous interaction constants for the N-controlled (externally *closed*) molecules:

$$(s')_{\mu} = \mathsf{V}_{N,s'}/\mathsf{V}_{N,N} = (\partial Q_{s'}/\partial N)_{\boldsymbol{F}=\boldsymbol{0}}/(\partial\mu/\partial N)_{\boldsymbol{F}=\boldsymbol{0}} = (\partial Q_{s'}/\partial\mu)_{\boldsymbol{F}=\boldsymbol{0}} = \mathsf{S}_{\mu,s'}, \tag{14.53}$$

where $\{\mathsf{V}_{N,s'}\} \in \boldsymbol{\mathsf{V}}_{N,\boldsymbol{F}}$, and

$$(\mu)_{s,N} = \mathsf{V}_{s,N}/\mathsf{V}_{s,s} = (\partial\mu/\partial F_s)_N/(\partial Q_s/\partial F_s)_N = (\partial\mu/\partial Q_s)_{N,\boldsymbol{F}'_s=\boldsymbol{0}'},$$
$$(s')_{s,N} = \mathsf{V}_{s',s}/\mathsf{V}_{s,s} = (\partial Q_{s'}/\partial F_s)_{N,\boldsymbol{F}'_s=\boldsymbol{0}'}/(\partial Q_s/\partial F_s)_{N,\boldsymbol{F}'_s=\boldsymbol{0}'} = (\partial Q_{s'}/\partial Q_s)_{N,\boldsymbol{F}'_s=\boldsymbol{0}'}, \tag{14.54}$$

with $\{\mathsf{V}_{s,N}\} \in \boldsymbol{\mathsf{V}}_{\boldsymbol{F},N}$. The corresponding minimum-energy coordinates

$$\Delta\boldsymbol{Q}(\Delta\mu)\big|_{N,\boldsymbol{F}=\boldsymbol{0}} = \{(s')_{\mu}\Delta\mu\},$$
$$\Delta\mu(\Delta Q_s)\Big|_{N,\boldsymbol{F}'_s=\boldsymbol{0}'} = \{(\mu)_{s,N}\Delta Q_s\},$$
$$\Delta\boldsymbol{Q}(\Delta Q_s)\Big|_{N,\boldsymbol{F}'_s=\boldsymbol{0}'} = \{(s')_{s,N}\Delta Q_s\}, \tag{14.55}$$

then reflect the equilibrium responses in the system chemical potential and geometrical coordinates due to finite shifts in the system chemical potential or selected geometrical coordinates.

Finally, in the $(\mu,\boldsymbol{Q})$-representation, in which the *partial* compliant matrix **G** is defined, one obtains the following coupling constants:

$$(F_s)_N = G_{\mu,s}/G_{\mu,\mu} = (\partial F_s/\partial \mu)_{\boldsymbol{Q}}/(\partial N/\partial \mu)_{\boldsymbol{Q}} = (\partial F_s/\partial N)_{\boldsymbol{Q}} = \varphi_s; \quad (14.56)$$

$$\begin{aligned}(N)_{F_s,\mu} &= G_{s,\mu}/G_{s,s} = (\partial N/\partial Q_s)_\mu/(\partial F_s/\partial Q_s)_\mu = (\partial N/\partial F_s)_{\mu,\boldsymbol{Q}_s'},\\ (F_{s'})_{F_s,\mu} &= G_{s',\mu}/G_{s,s} = (\partial F_{s'}/\partial Q_s)_{\mu,\boldsymbol{Q}_s'}/(\partial F_s/\partial Q_s)_{\mu,\boldsymbol{Q}_s'} = (\partial F_{s'}/\partial F_s)_{\mu,\boldsymbol{Q}_s'}.\end{aligned} \quad (14.57)$$

These interaction constants determine the associated MEC:

$$\begin{aligned}\Delta \boldsymbol{F}(\Delta N)\big|_{\mu,\boldsymbol{Q}} &= \{(F_s)_N \Delta N\},\\ \Delta \boldsymbol{F}(\Delta F_s)\big|_{\mu,\boldsymbol{Q}_s'} &= \{(F_{s'})_{F_s,\mu}\Delta F_s\},\\ \Delta N(\Delta F_s)\big|_{\mu,\boldsymbol{Q}_s'} &= \{(N)_{F_s,\mu}\Delta F_s\}.\end{aligned} \quad (14.58)$$

14.5 Illustrative Application to Conformational Changes

Recent numerical calculations (Nalewajski et al. 2008) of the joint electronic–nuclear compliants for a selection of polyatomics including H_2O, NO_2, H_2O_2, ClF_3, and NH_2CHO (Fig. 14.1) have generated representative coupling quantities

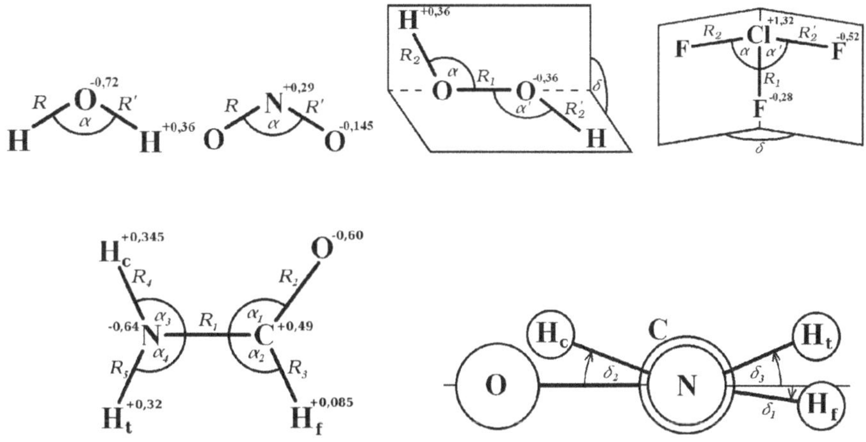

Fig. 14.1 The internal coordinates in five representative molecular systems and the Mulliken net charges of bonded atoms (from HF calculations). The last diagram defines the dihedral angles in formamide, relative to the NCO plane, determining the out-of-plane displacements of the *cis* (H_c), *trans* (H_t) and *formyl* (H_f) hydrogens

(a.u.) and MEC reported in Tables 14.1–14.4. These molecules exhibit a variety of internal geometric degrees of freedom, bond lengths, and angles, which are specified in the Fig. 14.1. In what follows we shall discuss some of these results, generated using the simplest Hartree–Fock (HF) theory in the extended 6-31++G** basis set of Gaussian orbitals, including the *split* valence and

Table 14.1 Comparison of the molecular hardness and softness quantities for the rigid and relaxed geometries. The reported relaxed ($^{rel.}$) quantities are averages of predictions using the $\Delta N = +1$ and $\Delta N = -1$ estimates of NFF; the same convention applies to the molecular compliant data reported in the remaining tables in this section

Molecule	η	$\eta^{rel.}$	S	$S^{rel.}$
H_2O	0.448	0.446	2.231	2.240
NO_2	0.422	0.319	2.369	3.133
H_2O_2	0.468	0.394	2.139	2.541
ClF_3	0.471	0.395	2.125	2.533
NH_2CHO	0.360	0.355	2.779	2.815

Table 14.2 Comparison of selected molecular Fukui function and softness compliants

Compliant	H_2O	NO_2	H_2O_2	ClF_3	NH_2CHO
$(F_R)_N = \varphi_R$ or $\varphi_{R_1} = (F_{R_1})_N$	−0.018	0.112	0.060	0.147	0.024
$\varphi_{R_2} = (F_{R_2})_N$			−0.015	0.063	−0.035
$(F_\alpha)_N = \varphi_\alpha$ or $\varphi_{\alpha_2} = (F_{\alpha_2})_N$	−0.012	−0.171	0.000	0.022	−0.019
$(F_R)_\mu = s_R$ or $s_{R_1} = (F_{R_1})_\mu$	−0.040	0.265	0.128	0.312	0.067
$s_{R_2} = (F_{R_2})_\mu$			−0.031	0.134	−0.099
$(F_\alpha)_\mu = s_\alpha$ or $s_{\alpha_1} = (F_{\alpha_1})_\mu$	−0.028	−0.405	0.000	0.046	−0.005
$(R)_\mu = S_R$ or $S_{R_1} = (R_1)_\mu$	−0.058	0.368	0.251	0.884	0.169
$S_{R_2} = (R_2)_\mu$			−0.039	0.489	−0.134
$(\alpha)_\mu = S_a$ or $S_{\alpha_1} = (\alpha_1)_\mu$	−0.144	−1.407	0.153	0.019	0.044
$(R)_N = f_R$ or $f_{R_1} = (R_1)_N$	−0.026	0.117	0.099	0.349	0.060
$f_{R_2} = (R_2)_N$			−0.015	0.193	−0.048
$(\alpha)_N = f_\alpha$ or $f_{\alpha_2} = (\alpha_2)_N$	−0.064	−0.449	0.060	0.008	−0.076
$(\delta)_N = f_\delta$ or $f_{\delta_1} = (\delta_1)_N$			−2.672	0.717	0.000

Table 14.3 Selected interaction constants for the closed and open H_2O and NO_2

	$(R')_R$	$(\alpha)_R$	$(R)_\alpha$	$(R')_{R,\mu}$	$(\alpha)_{R,\mu}$	$(R)_{\alpha,\mu}$	$(\mu)_{R,N}$	$(\mu)_{\alpha,N}$	$(N)_{R,\mu}$	$(N)_{\alpha,\mu}$
	$(R')_{R,N}$	$(\alpha)_{R,N}$	$(R)_{\alpha,N}$							
H_2O	0.017	−0.168	−0.044	0.018	−0.165	−0.044	0.016	0.011	−0.037	−0.024
NO_2	−0.243	−0.112	−0.051	−0.198	−0.248	−0.094	−0.104	0.182	0.313	−0.455

Table 14.4 Electronic–nuclear coupling constants for H_2O_2 and ClF_3

	$(\mu)_{R_1,N}$	$(\mu)_{R_2,N}$	$(\mu)_{\alpha,N}$	$(\mu)_{\delta,N}$	$(N)_{R_1,\mu}$	$(N)_{R_2,\mu}$	$(N)_{\alpha,\mu}$	$(N)_{\delta,\mu}$
H_2O_2	−0.039	−0.009	−0.014	0.024	0.098	−0.024	0.037	−0.053
ClF_3	−0.133	−0.044	−0.004	−0.148	0.301	0.109	0.010	0.423

polarization functions. In all derivative properties the bond lengths are expressed in a.u., while angles are measured in radians; all these quantities correspond to the ground state equilibrium geometries in the adopted basis set.

This compliant analysis has used the analytical forces and geometrical Hessians, and the *finite* difference estimates of the corresponding N-derivatives. The NFF has been calculated for both the electron-accepting ($\Delta N = +1$) and electron-donating ($\Delta N = -1$) processes, when the system acts as a Lewis acid and base, relative to the attacking N and E agents, respectively. The Mulliken scheme for the neutral system approached by R agent of the unbiased N-derivative given by the arithmetic average of these two estimates has also been examined. These R-estimates are reported in Tables 14.1–14.4. The global hardness, which measures the curvature of the ground state energy surface along the canonical electronic coordinate N, has been similarly estimated by interpolating the energies for the set of hypothetical electronic displacements $\Delta N = (-1, 0, +1)$.

In Table 14.1, we have compared the hardness and softness descriptors for the geometrically rigid and relaxed molecules, respectively. As intuitively expected, relaxing the nuclear positions decreases the electronic hardness (increases softness) of the molecular system. This electronic softening reflects the moderation influence (the LeChâtelier–Braun principle) on the chemical potential response to the primary (electronic) perturbation defining the derivative by the indirectly induced adjustments in the system geometry. A similar "softening" influence due to the system external opening is observed in the electronically relaxed geometric (diagonal) force constants (Nalewajski et al. 2008).

Let us examine these electronic–nuclear coupling effects in more detail. The moderating exchange of electrons between the molecule and its hypothetical electron reservoir determines the effects of the electronic–nuclear coupling in the open molecular systems. Let us assume the initial electronic and geometric equilibrium in such an initially open system: $\mu^0 = \mu_{res.}$ and $\boldsymbol{F}^0 = \boldsymbol{0}$. The LeChâtelier stability criteria of these two (decoupled) facets of the molecular structure require that the conjugate "forces" $\Delta\mu(\Delta N)$ and $\{\Delta F_s(\Delta Q_s)\}$ created by the primary electronic ($\Delta N > 0$) or nuclear $\{\Delta Q_s > 0\}$ displacements, $\Delta\mu(\Delta N) = \eta\, \Delta N$ and $\Delta F_s(\Delta Q_s) = H_{s,s}\, \Delta Q_s$, will subsequently trigger the (*directly* coupled) spontaneous responses of the system, $\delta N(\Delta N)$ and $\delta Q_s(\Delta Q_s)$, which act in the direction to restore the initial equilibrium (see also: Callen 1962). Therefore, these responses must diminish the forces created by the primary displacement, when the hypothetical internal and external barriers effecting these displacements are lifted:

$$\delta\mu[\delta N(\Delta N)] = \eta\, \delta N(\Delta N) = -\Delta\mu(\Delta N) = -\eta\, \Delta N \quad \text{and}$$
$$\delta F_s[\delta Q_s(\Delta Q_s)] = H_{s,s}\delta Q_s(\Delta Q_s) = -\Delta F_s(\Delta Q_s) = -H_{s,s}\Delta Q_s,$$

or $\delta N(\Delta N) = -\Delta N$ and $\delta Q_s(\Delta Q_s) = -\Delta Q_s$. This restoring character is assured by the positive character of the electronic hardness (see Sect. 13.6.3), $\eta > 0$, and of the diagonal nuclear force constants, $H_{s,s} > 0$, since $\Delta\mu(\Delta N) > 0$ implies $\delta N(\Delta N) < 0$, while $\Delta F_s(\Delta Q_s) > 0$ gives rise to $\delta Q_s(\Delta Q_s) < 0$.

However, due to the electron–nuclear coupling in molecules a given displacement in one aspect of the molecular structure creates extra forces in its complementary aspect: $\Delta\mu(\Delta Q_s) = \varphi_s \Delta Q_s$ and $\Delta F_s(\Delta N) = \varphi_s \Delta N$. They trigger the *indirectly* coupled, spontaneous relaxations $\delta N(\Delta Q_s)$ and $\delta Q_s(\Delta N)$, which also act towards diminishing the *directly* coupled forces $\Delta\mu(\Delta N) > 0$ and $\Delta F_s(\Delta Q_s) > 0$ (the LeChâtelier–Braun principle):

$$\delta\mu[\delta N(\delta Q_s)] = \varphi_s \delta Q_s < 0 \quad \text{and} \quad \delta F_s[\delta Q_s(\delta N)] = \varphi_s \delta N < 0.$$

Hence, these *indirectly* induced electronic and/or nuclear relaxations must exhibit the opposite signs with respect to the corresponding NFF indices.

In Table 14.2, we have compared the geometric softnesses $\boldsymbol{s}$ [in $(\mu,\boldsymbol{Q})$-representation] and $\boldsymbol{S}$ [in $(\mu,\boldsymbol{F})$-representation], as well as the alternaltive Fukui function indices: NFF $\boldsymbol{\varphi}$ [in the $(N,\boldsymbol{Q})$-representation] and GFF $\boldsymbol{f}$ [in the $(N,\boldsymbol{F})$-representation]. They measure the electronic–nuclear interaction in the externally open or closed molecules. It follows from this table that the signs of the given NFF index and the corresponding softness component are the same. Indeed, the former represents the scaled version of the latter, with the relevant global hardness (positive) providing the scalling factor, so that these two sets of coupling quantities in fact carry the same physical description of molecular responses. As also explicitly indicated in Table 14.2, the reported quantities represent the relevant compliant constants. The $\boldsymbol{s}$ and $\boldsymbol{\varphi}$ vectors collect the *force* compliants in the open and closed molecular systems, respectively, while $\boldsymbol{S}$ and $\boldsymbol{f}$ data constitute the related *coordinate* compliants.

The selected MEC data for the two triatomic molecules are listed in Table 14.3. In the decoupled treatment the interaction constant $(R')_R = (R')_{R,N}$ reflects the equilibrium linear response in R' per unit displacement in R, $(\alpha)_R = (\alpha)_{R,N}$ measures a similar response in the bond angle created by such "normalized" bond elongation, while $(R)_\alpha = (R)_{\alpha,N}$ stands for the bond length readjustment per unit (1 rad) change in the bond angle. It follows from these purely geometric entries that in the ground state of the water molecule an increase in one bond length generates a small elongation of the other bond and a decrease in the bond angle. The latter coupling effect is also reflected by the negative character of the $(R)_\alpha$ index, which implies a bond shortening following the primary increase in the bond angle. The opposite sign of the coupling constants between two bond lengths is detected in NO_2.

Consider next the effects of the electronic opening on the equilibrium responses of geometric parameters to such geometric displacements. The corresponding electronically relaxed compliants, of the open molecule, are listed in the data columns 4–6 of Table 14.3. It follows from this comparison of the HF results for water molecule that $(s')_{s,\mu} > (s')_{s,N}$, for $Q_s \neq Q_{s'} = \{R, R', \alpha\}$. Therefore, in this approximation an elongation of one bond in response to lengthening of the other bond becomes more emphasized in the open molecule. Indeed, a reference to Table 14.2 indicates that $\varphi_R < 0$ and $\varphi_\alpha < 0$ imply an inflow of electrons $\delta N(\Delta Q_s > 0) > 0$ from the reservoir, for $Q_s = (R, \alpha)$, which starts populating the antibonding MO, thus giving rise to an extra weakening of the other bond R' and hence its elongation.

The final four columns in Table 14.3 measure the effect of the specified geometrical displacement on the electronic state parameters. The $\{(N)_{s,\mu}\}$ indices show that in H_2O the hypothetical bond elongation or increase in the bond angle both create an outflow of electrons from the system to the reservoir, in accordance with the signs of the previously reported NFF indices. The $(\mu)_{s,N}$ indices reflect a direct effect of a hypothetical shift in the coordinate Q_s of the closed system on its chemical potential, when the remaining geometrical degrees of freedom are fully relaxed. As shown in Table 14.3 both these indices for $Q_s = (R, \alpha)$ are positive in water molecule. In other words, longer bonds and larger angle in this system both imply an increase in the system chemical potential. In NO_2 the opposite bond elongation effect is predicted. One also detects changes in the signs of $(N)_{R,\mu}$ and $(R')_{R,\mu}$ indices of the open NO_2, compared with H_2O. It thus follows from these interaction constants that elongating one bond in NO_2 results in an inflow of electrons to this molecule and shortening of the other bond.

Selected electronic compliants for H_2O_2 and ClF_3, due to geometric perturbations are reported in Table 14.4. In the closed molecules an increase in the bond lengths and angles is predicted to lower the system electronic chemical potential, while the opposite effect due to the dihedral angle is detected. In the externally *open* (electronically relaxed) ClF_3 the same perturbations generate an electron inflow to the molecular system from the reservoir, while in H_2O_2 an elongation of $R_2 = R(\text{O–H})$ and opening of the dihedral angle both trigger an electron outflow from the molecule.

14.6 Use of Compliant Constants as Reactivity Indices

The four geometric representations describe alternative scenarios encountered in the theory of chemical reactivity. For example, the closed reactants, geometry rigid or relaxed, in the opening stage of a reaction in the gas phase can be indexed by the derivative properties defined in the $(N,\boldsymbol{Q})$- and $(N,\boldsymbol{F})$-approaches, while the properties of the chemisorbed (externally open) reactants of the heterogeneous catalysis can be characterized using descriptors generated within the $(\mu,\boldsymbol{Q})$- and $(\mu,\boldsymbol{F})$-frameworks.

Both the EP and EF perspectives are covered by the canonical $(N,\boldsymbol{Q})$ representation and its inverse, the generalized softness representation $(\mu,\boldsymbol{F})$, respectively. Therefore, speculative considerations about the electronic or nuclear origins of the primary causes of chemical reactions can be enhanced and quantified using the present development. This information can be applied indirectly – by using the respective sets of CS, or directly – in terms of the geometric MEC. For example, the components of the fully relaxed MEC, defined in the totally inverted $(\mu,\boldsymbol{F})$-representation, provide the information about equilibrium responses in the effective oxidation state and geometry of the chemisorbed reactants per unit displacement $\Delta\mu$

in the electronic chemical potential of the catalyst, or in the selected molecular deformation. This should facilitate an ultimate identification of the crucial electronic/geometrical requirements for the desired reaction pathway, thus aiding a search for the most effective catalyst of the surface reaction of interest.

Clearly, the molecular compliants can be also used directly, as the *one*-reactant criteria of chemical reactivity, to diagnose the preferred sites for attacks by small agents, when the molecule is a part of a reactive system. However, combinations of these molecular descriptors can be also applied in the interreactant *decoupled* approach, e.g., in the D(*B*)—A(*A*) complexes, when the basic(*B*)/acidic(*A*) characters of the two subsystems are known beforehand. Indeed, the $\Delta N_{\mathrm{A}} > 0$ and $\Delta N_{\mathrm{D}} < 0$ displacements of the two reactants are then predetermined by their electronegativity differences, and so are the associated responses in the chemical potentials to these primary perturbations: $\Delta\mu_{\mathrm{A}} > 0$ and $\Delta\mu_{\mathrm{D}} < 0$. These displacements can be subsequently applied to predict the geometrical changes of the two mutually open reactants at the CT stage of the reaction, by using the relevant $(s)_N$ or $(s)_\mu$ compliants, which fully account for the relaxation of the remaining, unconstrained degrees of freedom of the DA complex.

The truly *two*-reactant, intersubsystem-*coupled* approach can be also envisaged, but the relevant compliant and MEC data would eventually require extra calculations on the whole reactive system *A----B*, with the internal coordinates $\boldsymbol{Q}$ now including those specifying the internal geometries of two subsystems as well as coordinates defining their mutual separation/orientation in the reactive supersystem. The *two*-reactant Hessian would then unite the respective blocks of the molecular tensors introduced in Sect. 14.2.

The supersystem relations between perturbations and responses in the canonical geometric representation would then read:

$$\begin{aligned}(\Delta\mu_A, \Delta\mu_B, -\Delta\boldsymbol{F}) \equiv (\Delta\boldsymbol{\mu}, -\Delta\boldsymbol{F}) &= (\Delta N_A, \Delta N_B, \Delta\boldsymbol{Q})\ \mathbf{H}(A\text{---}B)\\ &\equiv (\Delta\boldsymbol{N}, \Delta\boldsymbol{Q})\ \mathbf{H}(A\text{---}B),\end{aligned} \tag{14.59}$$

where the principal electronic–nuclear Hessian of the whole reactive system includes as diagonal blocks the hardnesses of the separate reactants, $\{\eta_X = \eta_{X,X}\}$, $X = A$, B, and the geometric Hessian $\mathbf{H}$ of the whole reactive system. The *off*-diagonal hardnesses $\eta_{A,B} = \eta_{B,A}$ measure the (rigid geometry) response in the chemical potential of one reactant per unit shift in the number of electrons on the other reactant, while the rectangular NFF matrix

$$\boldsymbol{\varphi} = \begin{bmatrix} \boldsymbol{\varphi}_A \\ \boldsymbol{\varphi}_B \end{bmatrix}, \tag{14.60}$$

determines the coupling between the electronic and geometric degrees of freedom in the combined Hessian of this bimolecular reactive system:

$$\mathbf{H}(A\text{---}B) = \begin{bmatrix} \left(\frac{\partial \mu_A}{\partial N_A}\right)_{\boldsymbol{Q}} & \left(\frac{\partial \mu_B}{\partial N_A}\right)_{\boldsymbol{Q}} & -\left(\frac{\partial \boldsymbol{F}}{\partial N_A}\right)_{\boldsymbol{Q}} \\ \left(\frac{\partial \mu_A}{\partial N_B}\right)_{\boldsymbol{Q}} & \left(\frac{\partial \mu_B}{\partial N_B}\right)_{\boldsymbol{Q}} & -\left(\frac{\partial \boldsymbol{F}}{\partial N_B}\right)_{\boldsymbol{Q}} \\ \left(\frac{\partial \mu_A}{\partial \boldsymbol{Q}}\right)_{\boldsymbol{N}} & \left(\frac{\partial \mu_B}{\partial \boldsymbol{Q}}\right)_{\boldsymbol{N}} & -\left(\frac{\partial \boldsymbol{F}}{\partial \boldsymbol{Q}}\right)_{\boldsymbol{N}} \end{bmatrix} = \begin{bmatrix} \eta_A & \eta_{A,B} & -\boldsymbol{\varphi}_A \\ \eta_{B,A} & \eta_B & -\boldsymbol{\varphi}_B \\ -\boldsymbol{\varphi}_A^{\mathrm{T}} & -\boldsymbol{\varphi}_B^{\mathrm{T}} & \mathbf{H} \end{bmatrix} = \begin{bmatrix} \boldsymbol{\eta} & -\boldsymbol{\varphi} \\ -\boldsymbol{\varphi}^{\mathrm{T}} & \mathbf{H} \end{bmatrix}. \tag{14.61}$$

In a practical implementation of this combined treatment of the electronic and nuclear *state* variables one could use as much as possible the intrareactant data generated in calculations on single (separated) reactants or attempt a simple semi-empirical modeling described in the next section.

The fully inverted compliance matrix, $\mathbf{S}(A\text{---}B) = \mathbf{H}(A\text{---}B)^{-1}$, which determines the inverse transformation

$$(-\Delta N_A, -\Delta N_B, \Delta \boldsymbol{Q}) \equiv (-\Delta \boldsymbol{N}, \Delta \boldsymbol{Q}) = (\Delta\mu_A, \Delta\mu_B, \Delta \boldsymbol{F})\mathbf{S}(A\text{---}B) \equiv (\Delta\boldsymbol{\mu}, \Delta\boldsymbol{F})\,\mathbf{S}(A\text{---}B), \tag{14.62}$$

exhibits the following block structure:

$$\mathbf{S}(A\text{---}B) = \begin{bmatrix} -\left(\frac{\partial N_A}{\partial \mu_A}\right)_{\boldsymbol{F}} & -\left(\frac{\partial N_B}{\partial \mu_A}\right)_{\boldsymbol{F}} & \left(\frac{\partial \boldsymbol{Q}}{\partial \mu_A}\right)_{\boldsymbol{F}} \\ -\left(\frac{\partial N_A}{\partial \mu_B}\right)_{\boldsymbol{F}} & -\left(\frac{\partial N_B}{\partial \mu_B}\right)_{\boldsymbol{F}} & \left(\frac{\partial \boldsymbol{Q}}{\partial \mu_B}\right)_{\boldsymbol{F}} \\ -\left(\frac{\partial N_A}{\partial \boldsymbol{F}}\right)_{\boldsymbol{\mu}} & -\left(\frac{\partial N_B}{\partial \boldsymbol{F}}\right)_{\boldsymbol{\mu}} & \left(\frac{\partial \boldsymbol{Q}}{\partial \boldsymbol{F}}\right)_{\boldsymbol{\mu}} \end{bmatrix} = \begin{bmatrix} -S_A^{rel} & -S_{A,B}^{rel} & \boldsymbol{\varsigma}_A \\ -S_{B,A}^{rel} & -S_B^{rel} & \boldsymbol{\varsigma}_B \\ \boldsymbol{\varsigma}_A^{\mathrm{T}} & \boldsymbol{\varsigma}_B^{\mathrm{T}} & \mathbf{G}^{rel} \end{bmatrix} = \begin{bmatrix} -\mathbf{S}^{rel} & \boldsymbol{\varsigma} \\ \boldsymbol{\varsigma}^{\mathrm{T}} & \mathbf{G}^{rel} \end{bmatrix}. \tag{14.63}$$

Here, the geometrically relaxed softness matrix $\mathbf{S}^{rel} = (\partial \boldsymbol{N}/\partial \boldsymbol{\mu})_{\boldsymbol{F}}$ groups the equilibrium responses in the subsystem numbers of electrons following the displacements in the chemical potentials of their (separate) electron reservoirs, the relaxed geometric softness matrix $\boldsymbol{\varsigma}^{\mathrm{T}} = [\boldsymbol{\varsigma}_A^{\mathrm{T}}, \boldsymbol{\varsigma}_B^{\mathrm{T}}]$, where $\boldsymbol{\varsigma}_X = (\partial \boldsymbol{Q}/\partial \mu_X)_{\boldsymbol{F}}$, groups the related adjustments in the geometry of the reactive system, while the electronically relaxed geometric compliant matrix $\mathbf{G}^{rel}$ collects the responses in the internal geometric coordinates to displacements in forces of the externally open reactants coupled to their (separate) electron reservoirs.

Obviously, the two remaining Legendre-transformed representations would similarly generate the associated descriptors of the partially relaxed (electronically or geometrically) reactive systems.

14.7 Modeling Couplings in Collinear Atom–Diatom Collisions

Typical applications of the DFT-based hardness/softness (FF) descriptors as reactivity criteria are limited to rather simple *single*-reactant scenarios, by probing regional preferences of a molecular reactant to an attack by the selected (small) approaching agent. In such applications the charge distribution, ESP and CS of the attacked molecule can be used with some success as guides to the molecular interactive behavior (e.g.: Geerlings et al. 2003; Johnson et al. 2011; Politzer and Murray 2009; Ayers et al. 2009). In a more realistic, *two*-reactant approach, one takes into account response properties of both reactants in the bimolecular reactive system, recognizing differences in their matching/coupling for alternative mutual orientations of the two molecular reactants (e.g., Nalewajski 1997a,b; Nalewajski et al. 1996a; Nalewajski and Korchowiec 1997).

In this section an attempt will be made to use this more adequate treatment in describing the simplest atom exchange reaction in the triatomic (collinear) collision system $\mathcal{R} = A\text{---}B, A = \mathrm{A{-}\!\!-B}, B = \mathrm{C}$ of the concerted *bond-breaking–bond-forming* processes (Nalewajski 2010g):

$$\mathrm{A{-}\!\!-B} + \mathrm{C} \rightarrow [\mathrm{A{-}{-}{-}{-}B{-}{-}{-}{-}C}]^{\dagger} \rightarrow \mathrm{A} + \mathrm{B{-}\!\!-C}. \tag{14.64}$$

As we have shown in the preceding section, the combined approach to the equilibrium states in this simple reactive system as a whole and in its diatomic reactants/products requires an adequate analytical representation of the hardness and NFF derivatives of the molecular energy. In what follows we shall briefly examine possibilities for modeling such descriptors.

This prototype reactive system should allow one to probe several electronic–nuclear coupling processes as functions of the reaction progress variable say $R \equiv R_{\mathrm{BC}}$, e.g., between the reactant *polarization* B $\rightarrow$ A, measured by the intramolecular CT, $q = \Delta N_{\mathrm{A}} = N_{\mathrm{A}} - N_{\mathrm{A}}^{0} = -\Delta N_{\mathrm{B}}$, where N_i^0 stands for the electron population of the isolated atom i = A, B, and the substrate bond length $r \equiv R_{\mathrm{AB}}$, or between the equilibrium amount of the interreactant CT AB$\rightarrow$C, $Q = \Delta N_{\mathrm{C}} = N_{\mathrm{C}} - N_{\mathrm{C}}^{0} = -\Delta N_{\mathrm{AB}}$ and the diatomic bond length r or the interreactant separation R. All these interactions between electronic and geometric aspects of the molecular structure can be approached using both the EF and EP perspectives. In the canonical representation the relations between the system perturbations and responses are determined by the combined electronic/nuclear Hessian of (14.61), while the inverse transformation is effected using the generalized softness matrix defined by the Hessian inverse of (14.63), which combines the associated compliants of the reactive system.

The following development focuses mainly on establishing the analytical, model expressions for the key NFF descriptors of this simple collision complex. In modeling these crucial coupling quantities we shall neglect contributions to chemical potentials of bonded atoms due to a change in their effective external potentials introduced by the presence of the remaining AIM. In such a simplified approach the geometric influence on charge sensitivities originates solely from the internuclear dependence of the coupling AIM hardnesses and the electron flows they effect at the given stage of the reactant approach (Nalewajski 2010g).

14.7.1 Derivative Descriptors of Collinear Reactive System

The relevant *two*-reactant Hessian unites the respective blocks introduced in Sect. 14.2. This principal electronic–nuclear Hessian of the whole reactive system now includes the diagonal blocks of the condensed hardnesses of the two reactants in $\mathcal{R}$,

$$\begin{aligned}\boldsymbol{\eta}^{\mathcal{R}} = \{\eta_{X,Y} &= \partial^2 E(N_A, N_B, \boldsymbol{Q})/\partial N_X \partial N_Y = \partial \mu_Y(N_A, N_B, \boldsymbol{Q})/\partial N_X \\ &= \eta_{Y,X} = \partial \mu_X(N_A, N_B, \boldsymbol{Q})/\partial N_Y\}, \qquad X, Y = (A, B), \end{aligned} \tag{14.65}$$

with $\eta_{X,X} \equiv \eta_X$, and of the purely geometric Hessian of the *second* partials (force constants) of the BO potential with respect to nuclear degrees of freedom:

$$\mathbf{H} = \{\partial^2 W(N_A, N_B, \boldsymbol{Q})/\partial \boldsymbol{Q}\, \partial \boldsymbol{Q}\} = -\partial \boldsymbol{F}(N_A, N_B, \boldsymbol{Q})/\partial \boldsymbol{Q}, \tag{14.66}$$

as well as the *off*-diagonal blocks of the associated NFF descriptors of (14.60).

The canonical relation between the perturbations in the state variables (ΔN_A, ΔN_B, $\Delta \boldsymbol{Q}$) and the coupled responses of their energy ("intensive") conjugates ($\Delta\mu_A$, $\Delta\mu_B$, $-\Delta \boldsymbol{F}$) in the geometric representation (N_A, N_B, $\boldsymbol{Q}$) then reads:

$$\begin{aligned}(\Delta\mu_A, \Delta\mu_B, -\Delta \boldsymbol{F}) \equiv (\Delta\boldsymbol{\mu}, -\Delta \boldsymbol{F}) &= (\Delta N_A, \Delta N_B, \Delta \boldsymbol{Q})\mathbf{H}(A\text{---}B) \\ &\equiv (\Delta \boldsymbol{N}, \Delta \boldsymbol{Q})\mathbf{H}(A\text{---}B).\end{aligned} \tag{14.67}$$

The fully inverted compliance matrix, $\mathbf{S}(A\text{---}B) = \mathbf{H}(A\text{---}B)^{-1}$, similarly determines the transformation of displacements in "intensities" ($\Delta\boldsymbol{\mu}$, $\Delta \boldsymbol{F}$) into the associated responses in the canonical electronic/geometrical state parameters ($-\Delta \boldsymbol{N}$, $\Delta \boldsymbol{Q}$):

$$\begin{aligned}(-\Delta N_A, -\Delta N_B, \Delta \boldsymbol{Q}) \equiv (-\Delta \boldsymbol{N}, \Delta \boldsymbol{Q}) &= (\Delta\mu_A, \Delta\mu_B, \Delta \boldsymbol{F})\mathbf{S}(A\text{---}B) \\ &\equiv (\Delta\boldsymbol{\mu}, \Delta \boldsymbol{F})\mathbf{S}(A\text{---}B).\end{aligned} \tag{14.68}$$

In the *two*-reactant resolution, this generalized softness matrix exhibits the block structure of (14.63):

$$\mathbf{S}(A\text{---}B) = \begin{bmatrix} -S_A^{rel} & -S_{A,B}^{rel} & \boldsymbol{\mathcal{S}}_A \\ -S_{B,A}^{rel} & -S_B^{rel} & \boldsymbol{\mathcal{S}}_B \\ \boldsymbol{\mathcal{S}}_A^{\mathrm{T}} & \boldsymbol{\mathcal{S}}_B^{\mathrm{T}} & \mathbf{G}^{rel} \end{bmatrix} \equiv \begin{bmatrix} -\mathbf{S}^{rel} & \boldsymbol{\mathcal{S}} \\ \boldsymbol{\mathcal{S}}^{\mathrm{T}} & \mathbf{G}^{rel} \end{bmatrix}. \tag{14.69}$$

We recall that the geometrically relaxed softness matrix $\mathbf{S}^{rel}$ groups the equilibrium, fully relaxed responses in the subsystem numbers of electrons, following the displacements in the chemical potentials of their (separate) electron reservoirs,

$$S_X^{rel} = -\partial N_X(\mu_A, \mu_B, \boldsymbol{F})/\partial \mu_X, \qquad S_{X,Y}^{rel} = -\partial N_Y(\mu_A, \mu_B, \boldsymbol{F})/\partial \mu_X, \\ X, Y = (A, B), \tag{14.70}$$

the relaxed geometric softness matrix $\boldsymbol{\mathcal{S}}^{\mathrm{T}} = \left[\boldsymbol{\mathcal{S}}_A^{\mathrm{T}}, \boldsymbol{\mathcal{S}}_B^{\mathrm{T}}\right]$ groups the related adjustments in the geometry of the reactive system,

$$\boldsymbol{\mathcal{S}}_X = \partial \boldsymbol{Q}(\mu_A, \mu_B, \boldsymbol{F})/\partial \mu_X = -[\partial N_X(\mu_A, \mu_B, \boldsymbol{F})/\partial \boldsymbol{F}]^{\mathrm{T}}, \quad X, Y = (A, B), \tag{14.71}$$

and the geometric compliant matrix,

$$\mathbf{G}^{rel} = \partial \boldsymbol{Q}(\mu_A, \mu_B, \boldsymbol{F})/\partial \boldsymbol{F}, \tag{14.72}$$

collects responses in the internal geometric coordinates to displacements in forces, of the externally open reactants coupled to their (separate) electron reservoirs.

There are two independent geometrical degrees of freedom $\boldsymbol{Q} = (r, R)$ in the collinear complex of (14.64). The second variable, treated as the *parameter* measuring the progress of the mutual approach by reactants A = A—B and B = C, can be regarded as an approximation of the MEP for the fully relaxed reactants in the electronic and (*internal*) geometric degrees of freedom. The latter can be realistically modeled by the trajectory conserving the overall phenomenological bond "order" of the two single bonds being *formed* and *broken*, e.g., in the hydrogen exchange (B = H) reaction AH + C = A + HC,

$$n_{\mathrm{A,B}} + n_{\mathrm{B,C}} = 1, \\ \left\{ n_{\mathrm{X,Y}} \to 0 \quad \text{for} \quad R_{\mathrm{X,Y}} \to \infty, \quad n_{\mathrm{X,Y}} = 1 \quad \text{for} \quad R_{\mathrm{X,Y}} = R_{\mathrm{X,Y}}^{e} \right\}, \tag{14.73}$$

where $R_{\mathrm{X,Y}}^{e} = R_{\mathrm{X,Y}}(n_{\mathrm{X,Y}} = 1)$ stands for the equilibrium length of X—Y bond, in the Bond-Energy–Bond-Order (BEBO) method (Johnston and Parr 1963) using Pauling's logarithmic relation between this bond order measure and the internuclear distance (Pauling 1947):

$$R_{\mathrm{X,Y}}(n_{\mathrm{X,Y}}) = R_{\mathrm{X,Y}}(1) - 0.026 \ln n_{\mathrm{X,Y}} \qquad \text{or} \\ R_{\mathrm{X,Y}}(n_{\mathrm{X,Y}}) - R_{\mathrm{X,Y}}(1) \equiv \Delta R_{\mathrm{X,Y}}(n_{\mathrm{X,Y}}) = -0.026 \ln n_{\mathrm{X,Y}}, \tag{14.74}$$

predicting an increase $\Delta R_{\mathrm{X,Y}}$ (in 10^{-9} m) in the bond length for fractional bond orders $n_{\mathrm{X,Y}} < 1$.

Since the collision complex is closed as a whole, there are only two independent electron population variables in the triatomic system: $\boldsymbol{N} = (q, Q)$. For the given value of the reaction coordinate R, they measure the intradiatomic polarization $q(R)$

and the interreactant transfer of electrons $Q(R)$. Thus, for the fixed separation R between reactants, the list of state *variables* in the principal geometrical representation reads: $[q(R), Q(R), r(R)] = [\boldsymbol{N}(R), r(R)]$. This parametric treatment of the reaction coordinate gives rise to the following combined Hessian:

$$\mathbf{H}[\mathrm{AB{-}{-}{-}C}(R)] \equiv \begin{bmatrix} \eta_q(R) & \eta_{q,Q}(R) & -\varphi_q(R) \\ \eta_{Q,q}(R) & \eta_Q(R) & -\varphi_Q(R) \\ -\varphi_q(R) & -\varphi_q(R) & k(R) \end{bmatrix} \equiv \begin{bmatrix} \boldsymbol{\eta}^{eff.}(R) & -\boldsymbol{\varphi}^{eff.}(R)^{\mathrm{T}} \\ -\boldsymbol{\varphi}^{eff.}(R) & k(R) \end{bmatrix}, \tag{14.75}$$

where $k(R) = \partial^2 W[\boldsymbol{N}^{eq.}, r]/\partial r^2|_R \equiv -\partial F_r[\boldsymbol{N}^{eq.}, r]/\partial r|_R$ denotes the force constant of the diatomic reactant AB, calculated for the equilibrium charge distribution $\boldsymbol{N}^{eq.}$. Here, the *effective* (in situ) hardness matrix associated with these collective charge displacements $\boldsymbol{N} = (q, Q)$,

$$\boldsymbol{\eta}^{eff.}(R) = \partial^2 E[\boldsymbol{N}, r(R)]/\partial \boldsymbol{N}\partial \boldsymbol{N}\} \equiv \partial \boldsymbol{\mu}^{eff.}[\boldsymbol{N}, r(R)]/\partial \boldsymbol{N}, \tag{14.76}$$

where $\boldsymbol{\mu}^{eff.} = (\mu_q = \partial E/\partial q,\ \mu_Q = \partial E/\partial Q)$ groups the associated chemical potential descriptors. The effective NFF matrix also contains two components:

$$\begin{aligned} \boldsymbol{\varphi}^{eff.}(R) &= -\partial^2 W[\boldsymbol{N}, r]/\partial r\, \partial \boldsymbol{N}|_R\} \equiv -\partial \boldsymbol{\mu}^{eff.}[\boldsymbol{N}, r]/\partial r|_R = [\partial F_r[\boldsymbol{N}, r(R)]/\partial \boldsymbol{N}]^{\mathrm{T}} \\ &= [\varphi_q = -\partial \mu_q/\partial r = \partial F_r/\partial q, \quad \varphi_Q = -\partial \mu_Q/\partial r = \partial F_r/\partial Q]. \end{aligned} \tag{14.77}$$

One can use either of the above *cross*-derivatives to model the bond length dependence of these two NFF descriptors. However, within reactivity theory the geometric derivative of the chemical potential appears to be more convenient for modeling purposes, since the *finite* difference estimates of the geometrical force with respect to the electronic (flow) variables escape simple analytical representation and appears computationally forbidding.

It should be realized that the internal flow q in AB determines the in situ chemical potential difference $\mu_q = \mu_A - \mu_B = \mu_{CT}(B \to A)$, while the second electron transfer Q (internal in ABC) similarly defines the other in situ chemical potential $\mu_Q = \mu_C - \mu_{AB} = \mu_{CT}(AB \to C)$. These populational "forces" exactly vanish for the equilibrium amounts of CT, when the chemical potentials of the constituent fragments involved in the exchange of electrons are exactly equalized. These (electronic) equilibrium criteria $\mu_q = \mu_Q = 0$ mark the minimum of the electronic energy with respect to electron flows in the molecular systems in question. However, for any point on the MEP of the mutual approach by both reactants, for simplicity assumed at this point as identical with the $R = R_{BC}$ coordinate of the model collision system, one also relaxes the remaining (r) geometrical degrees of freedom of the reactive system, which implies $F_r(R) = 0$.

Therefore, in the present collinear model only the geometrical force F_R along R remains unrelaxed for the specified separation between reactants.

To determine effects of the electron–nuclear coupling, which accompany the mutual approach by reactants in this model reactive system, one requires an adequate analytical representation of the R-dependence of the matrix elements in the effective condensed matrices of (14.75). A more realistic approach should also include a dependence of the A—B force constant upon the current position of the approaching atom C, $k(R)$, which can be generated in the BEBO approximation. It is the main objective of the next section to model these dependencies using the relevant EE equations and the previously proposed analytical representation of the canonical hardness matrix of AIM, $\boldsymbol{\eta}^{\mathrm{AIM}} = \{\eta_{i,j}\}$ (Nalewajski et al. 1988), modeled by the corresponding tensor of the *valence* shell electron repulsion integrals $\{\gamma_{i,j}\}$, taken from the semiempirical SCF MO theories (Pariser 1953; Pariser and Parr 1953; Pople 1953): $\eta_{i,j} \cong \gamma_{i,j}$.

14.7.2 Modeling Electronic and Nuclear Fukui Functions

In the familiar finite difference approximation the energy $E_i(N_i)$ of an open atom i is represented as quadratic function of displacements ΔN_i in its number of electrons N_i, or its effective charge $Q_i = \mathrm{Z}_i - N_i$ (see Sect. 7.3.4), with Z_i standing for the atomic number of the nucleus:

$$\begin{aligned} E_i(N_i) &\cong E_i^0 + (\partial E_i/\partial N_i)|_0 \Delta N_i + {}^1\!/\!_2(\partial^2 E_i/\partial N_i^2)|_0 (\Delta N_i)^2 \\ &= E_i^0 + \mu_i^0 \Delta N_i + {}^1\!/\!_2 \eta_i^0 (\Delta N_i)^2; \end{aligned} \qquad (14.78)$$

here the symbol $|_0$ implies that the derivative is taken for the neutral atom, when $N_i = \mathrm{Z}_i$. This quadratic (Mulliken) interpolation (e.g., Baekelandt et al. 1993) of the energies of the neutral atom and its *singly* charged ions then expresses the two atomic derivatives in the preceding equation in terms of the atom ionization potential I_i and its electron affinity A_i:

$$\mu_i^0 = -{}^1\!/\!_2(A_i + I_i) < 0 \quad \text{and} \quad \eta_i^0 = I_i - A_i \approx \gamma_{i,i} > 0. \qquad (14.79)$$

Thus, the Pariser (1953) formula for the *one*-center electron repulsion integral in semiempirical SCF MO theories (e.g., Pariser and Parr 1953), $\gamma_{i,i} \approx I_i - A_i$, also provides adequate (diagonal) elements of the associated AIM hardnesses: $\eta_{i,i} = \eta_i^0 \approx \gamma_{i,i}$.

The electron population coupling between a given pair (i, j) of AIM is similarly reflected by the corresponding *off*-diagonal hardness,

$$\eta_{i,j} = \partial^2 E[\{N_k\}, \boldsymbol{Q}]/\partial N_i \partial N_j = \partial \mu_i[\{N_k\}, \boldsymbol{Q}]/\partial N_j = \partial \mu_j[\{N_k\}, \boldsymbol{Q}]/\partial N_i = \eta_{j,i}, \qquad (14.80)$$

which was also shown (Nalewajski et al. 1988) to be realistically represented by the (*two*-center) *valence* shell electron repulsion integral of the semiempirical SCF MO theories: $\eta_{i,j} \cong \gamma_{i,j}$. For example, the familiar Ohno (1967) interpolation formula gives the following dependence of this coupling hardness on the separation $R_{i,j} = |\boldsymbol{R}_i - \boldsymbol{R}_j|$ between the two bonded atoms,

$$\eta_{i,j}(R_{i,j}) \cong \gamma_{i,j}(R_{i,j}) = [(a_{i,j})^2 + (R_{i,j})^2]^{-1/2}, \tag{14.81}$$

with the constant $a_{i,j} = ½(\eta_{i,i} + \eta_{j,j})$ then recovering the limiting value $\eta_{i,j}(R_{i,j} \to 0) = \eta_{i,i} = \eta_{j,j}$. The associated geometric (distance) derivative then reads:

$$\begin{aligned} d\eta_{i,j}(R_{i,j})/dR_{i,j} &\cong d\gamma_{i,j}(R_{i,j})/dR_{i,j} = -\{R_{i,j}/[(a_{i,j})^2 + (R_{i,j})^2]\}\gamma_{i,j}(R_{i,j}) \\ &\equiv g_{i,j}(R_{i,j})\gamma_{i,j}(R_{i,j}). \end{aligned} \tag{14.82}$$

With this background in modeling the canonical hardness data of AIM we can now turn to the effective derivative quantities of (14.75)–(14.77). One first observes that the hardness tensor of the constituent AIM in $\mathcal{R}$ depends on the current values of two geometric variables of the collinear complex:

$$\begin{aligned} \boldsymbol{\eta}^{\mathrm{AIM}}(r,R) &= \begin{bmatrix} \eta_{\mathrm{A,A}} & \eta_{\mathrm{A,B}}(r) & \eta_{\mathrm{A,C}}(r+R) \\ \eta_{\mathrm{B,A}}(r) & \eta_{\mathrm{B,B}} & \eta_{\mathrm{B,C}}(R) \\ \eta_{\mathrm{C,A}}(r+R) & \eta_{\mathrm{C,B}}(R) & \eta_{\mathrm{C,C}} \end{bmatrix} \\ &\equiv \begin{bmatrix} \boldsymbol{\eta}_{\mathrm{AB}}(r) & \boldsymbol{\eta}_{\mathrm{AB,C}}(r,R) \\ \boldsymbol{\eta}_{\mathrm{C,AB}}(r,R) & \eta_{\mathrm{C,C}} \end{bmatrix}. \end{aligned} \tag{14.83}$$

Consider first the chemical potential (electronegativity) equalization problem in the diatomic reactant AB. The upper diagonal block $\boldsymbol{\eta}_{\mathrm{AB}}(r)$ of $\boldsymbol{\eta}^{\mathrm{AIM}}(r, R)$ for this molecular fragment and the initial levels of the chemical potentials $\boldsymbol{\mu}_{\mathrm{AB}} = (\mu_{\mathrm{A}}^0, \mu_{\mathrm{B}}^0)$ together determine the resultant chemical potential of the separated AB, equalized due to the interatomic CT, $q(r) = \Delta N_{\mathrm{A}}(r) = -\Delta N_{\mathrm{B}}(r)$,

$$\begin{aligned} \mu_{\mathrm{AB}}(r) = \mu_{\mathrm{A}}(r) &= \mu_{\mathrm{A}}^0 + [\eta_{\mathrm{A,A}} - \eta_{\mathrm{B,A}}(r)]\, q(r) \equiv \mu_{\mathrm{A}}^0 + \eta_{\mathrm{A}}{}^{\mathrm{AB}}(r)q(r) \\ = \mu_{\mathrm{B}}(r) &= \mu_{\mathrm{B}}^0 - [\eta_{\mathrm{B,B}} - \eta_{\mathrm{A,B}}(r)]\, q(r) \equiv \mu_{\mathrm{B}}^0 - \eta_{\mathrm{B}}{}^{\mathrm{AB}}(r)q(r). \end{aligned} \tag{14.84}$$

These relations also determine the optimum amount of this CT component,

$$q(r) = (\mu_{\mathrm{B}}^0 - \mu_{\mathrm{A}}^0)/[\eta_{\mathrm{A}}{}^{\mathrm{AB}}(r) + \eta_{\mathrm{B}}{}^{\mathrm{AB}}(r)] \equiv -\mu_{\mathrm{CT}}{}^0/\eta_{\mathrm{CT}}^{\mathrm{AB}}(r). \tag{14.85}$$

The equalized level of the global/AIM chemical potentials in AB can be thus expressed as the weighted average of the initial chemical potentials of both atoms:

$$\mu_{AB}(r) = \mu_A(r) = \mu_B(r) = \mu_A^0\left[\eta_B{}^{AB}(r)/\eta_{CT}^{AB}(r)\right] + \mu_B^0\left[\eta_A{}^{AB}(r)/\eta_{CT}^{AB}(r)\right]$$
$$\equiv \mu_A^0 f_A{}^{AB}(r) + \mu_B^0 f_B{}^{AB}(r). \tag{14.86}$$

It should be again emphasized that the approximation $\mu_{CT} \approx \mu_{CT}^0$ implies a neglect of shifts in the AIM chemical potentials $\{\mu_i^+\}$, relative to the corresponding separated-atom levels $\{\mu_i^0\}$, generated by the effective external potential due to the presence of the bonding partner(s): $\mu_i^+ \approx \mu_i^0$. This assumption is consistently retained in the present modeling. Inclusion of such contributions would ultimately require the explicit knowledge of the corresponding electronic *Fukui functions* (FF) of the interacting atoms,

$$f_i(\boldsymbol{r}) = \frac{\partial^2 E_i[N_i, v_i]}{\partial N_i \partial v_i(\boldsymbol{r})} = \frac{\partial \mu_i}{\partial v_i(\boldsymbol{r})} = \frac{\partial \rho_i(\boldsymbol{r})}{\partial N_i} \approx \rho_i^F(\boldsymbol{r}), \tag{14.87}$$

which can be approximated by the relevant *Frontier Electron* (FE) densities $\{\rho_i^F(\boldsymbol{r})\}$.

It is also of interest to determine other CS of the diatomic reactant. The inverse of the hardness matrix in atomic resolution generates the associated softness matrix [see (13.111)–(13.114)],

$$\boldsymbol{\sigma}_{AB} = \boldsymbol{\eta}_{AB}^{-1} = \frac{1}{\det \boldsymbol{\eta}_{AB}}\begin{bmatrix} \eta_{B,B} & -\eta_{A,B} \\ -\eta_{B,A} & \eta_{A,A} \end{bmatrix},$$
$$\det \boldsymbol{\eta}_{AB} = \eta_{A,A}\eta_{B,B} - \eta_{A,B}\eta_{B,A}, \tag{14.88}$$

which in turn gives rise to the associated AIM softnesses in AB:

$$\boldsymbol{S}^{AB} = \left[S_A^{AB} = \frac{\partial N_A}{\partial \mu_{AB}},\ S_B^{AB} = \frac{\partial N_B}{\partial \mu_{AB}}\right] = \frac{1}{\det \boldsymbol{\eta}_{AB}}\left[\eta_B^{AB}, \eta_A^{AB}\right], \tag{14.89}$$

and the system global softness $S_{AB} = \partial N_{AB}/\partial \mu_{AB}$, inverse of the global hardness $\eta_{AB} = \partial \mu_{AB}/\partial N_{AB} = 1/S_{AB}$,

$$S_{AB} = S_A^{AB} + S_B^{AB} = \eta_{CT}^{AB}/\det \boldsymbol{\eta}_{AB},$$
$$\eta_{AB} = \det \boldsymbol{\eta}_{AB}/\eta_{CT}^{AB} = (\eta_{A,A}\eta_{B,B} - \eta_{A,B}\eta_{B,A})/(\eta_{A,A} + \eta_{B,B} - \eta_{A,B} - \eta_{B,A}). \tag{14.90}$$

The corresponding FF of the diatomic constituent AIM are then given by the associated softness ratios:

$$f^{AB} = \left[f_A^{AB} = \frac{\partial N_A}{\partial N_{AB}} = \frac{S_A^{AB}}{S_{AB}},\ f_B^{AB} = \frac{\partial N_B}{\partial N_{AB}} = \frac{S_B^{AB}}{S_{AB}}\right] = \frac{1}{\eta_{CT}^{AB}}\left[\eta_B^{AB}, \eta_A^{AB}\right], f_A^{AB} + f_B^{AB} = 1. \tag{14.91}$$

These FF of bonded atoms in AB are seen in (14.86) to provide "weights" in the expression for the equalized chemical potential in terms of the initial chemical potentials of both atoms.

We have thus determined the effective bond length dependence of the global chemical potential in the diatomic reactant AB, generated by the r-dependence of the adopted model *off*-diagonal hardnesses. This function determines the associated NFF index:

$$\begin{aligned}\varphi_r(\mathrm{AB}) &= -\partial\mu_{\mathrm{AB}}(r)/\partial r = [f_{\mathrm{A}}^{\mathrm{AB}}(r) - f_{\mathrm{B}}^{\mathrm{AB}}(r)]\, q(r)[\partial\eta_{\mathrm{A,B}}(r)/\partial r] \\ &= \mu_{\mathrm{CT}}{}^{0}(\eta_{\mathrm{A,A}} - \eta_{\mathrm{B,B}})g_r(r)\, \eta_{\mathrm{A,B}}(r)/[\eta_{\mathrm{CT}}^{\mathrm{AB}}(r)]^2 > 0. \end{aligned} \tag{14.92}$$

The positive character of this derivative results from observing that $\eta_{\mathrm{A,B}}(r) > 0$ and $g_r(r) < 0$, and recognizing the sign combination of remaining factors. For example, assuming the relative acidic character of A and basic of B, i.e., $q > 0$, implies: $\mu_{\mathrm{CT}}^{0} < 0$, $\eta_{\mathrm{A,A}} - \eta_{\mathrm{B,B}} > 0$ and hence $f_{\mathrm{A}}^{\mathrm{AB}} - f_{\mathrm{B}}^{\mathrm{AB}} < 0$. In the last inequality we have additionally used the diatomic charge stability condition: $\eta_{\mathrm{CT}}^{\mathrm{AB}} > 0$.

To summarize, the negative chemical potential (electronegativity) of the diatomic increases with bond elongation, as indeed expected intuitively from the diminished effects of the chemical bonding in such species.

Let us now turn to the global equalization of the chemical potential in the whole triatomic collision complex. For definiteness, we assume the amount of CT (AB) $\rightarrow$ C, $Q = \Delta N_{\mathrm{C}} = -\Delta N_{\mathrm{AB}}$. This outflow of electrons from AB redistributes itself inside this diatomic reactant in accordance with the internal FF indices of (14.91),

$$\Delta N_{\mathrm{A}}(Q) = -f_{\mathrm{A}}{}^{\mathrm{AB}}Q, \quad \Delta N_{\mathrm{B}}(Q) = -f_{\mathrm{B}}{}^{\mathrm{AB}}Q, \tag{14.93}$$

thus establishing the final equilibrium shifts in the electron populations of AIM in $\mathcal{R}$ as a whole:

$$\Delta N_{\mathrm{A}}(q,Q) = q - f_{\mathrm{A}}{}^{\mathrm{AB}}Q, \quad \Delta N_{\mathrm{B}}(q,Q) = -q - f_{\mathrm{B}}{}^{\mathrm{AB}}Q, \quad \Delta N_{\mathrm{C}}(q,Q) = Q. \tag{14.94}$$

Since we have already determined the equalized level of the chemical potential in AB, one can directly apply the previous formulas by treating this diatomic as whole (combined) unit in the reaction complex [(AB)- - -C]. The effective hardness coupling between these two complementary subsystems of $\mathcal{R}$ is then determined by the interfragment hardnesses:

$$\eta_{\mathrm{AB,C}} = \partial\mu_{\mathrm{C}}/\partial N_{\mathrm{AB}} = \eta_{\mathrm{C,AB}} = \partial\mu_{\mathrm{AB}}/\partial N_{\mathrm{C}}.$$

They can be obtained by combining the corresponding matrix elements of $\boldsymbol{\eta}^{\mathrm{AIM}}(r, R)$ (Nalewajski 1989a):

$$\eta_{\mathrm{AB,C}}(r,R) = \frac{\partial \mu_{\mathrm{AB}}}{\partial N_{\mathrm{C}}} = f_{\mathrm{A}}^{\mathrm{AB}} \frac{\partial \mu_{\mathrm{A}}}{\partial N_{\mathrm{C}}} + f_{\mathrm{B}}^{\mathrm{AB}} \frac{\partial \mu_{\mathrm{B}}}{\partial N_{\mathrm{C}}}$$
$$= f_{\mathrm{A}}^{\mathrm{AB}}(r)\, \eta_{\mathrm{A,C}}(r+R) + f_{\mathrm{B}}^{\mathrm{AB}}(r)\, \eta_{\mathrm{B,C}}(R). \qquad (14.95)$$

The effective hardnesses of these two molecular fragments in $\mathcal{R}$,

$$\eta_{\mathrm{AB}}^{\mathrm{ABC}} = \eta_{\mathrm{AB}} - \eta_{\mathrm{C,AB}}, \quad \eta_{\mathrm{C}}^{\mathrm{ABC}} = \eta_{\mathrm{C,C}} - \eta_{\mathrm{AB,C}}, \qquad (14.96)$$

subsequently determine the effective in situ hardness for the interreactant CT, (AB) $\rightarrow$ C,

$$\eta_{\mathrm{CT}}^{\mathrm{ABC}}(r,R) = \eta_{\mathrm{AB}}^{\mathrm{ABC}}(r,R) + \eta_{\mathrm{C}}^{\mathrm{ABC}}(r,R). \qquad (14.97)$$

Finally, the optimum amount of this exchange of electrons,

$$Q(r,R) = -[\mu_{\mathrm{C}}^{0} - \mu_{\mathrm{AB}}(r)]/\eta_{\mathrm{CT}}^{\mathrm{ABC}}(r,R) \equiv -\mu_{\mathrm{CT}}^{\mathrm{ABC}}(r,R)/\eta_{\mathrm{CT}}^{\mathrm{ABC}}(r,R), \qquad (14.98)$$

determines the equalized level of the chemical potential in $\mathcal{R}$ as a whole:

$$\mu_{\mathrm{ABC}}(r,R) = \mu_{\mathrm{AB}}(r,R) = \mu_{\mathrm{AB}}(r) - [\eta_{\mathrm{AB}}(r) - \eta_{\mathrm{C,AB}}(r,R)]Q(r,R)$$
$$\equiv \mu_{\mathrm{AB}}(r) - \eta_{\mathrm{AB}}^{\mathrm{ABC}}(r,R)Q(r,R)$$
$$= \mu_{\mathrm{C}}(R) = \mu_{\mathrm{C}}^{0} + [\eta_{\mathrm{C,C}} - \eta_{\mathrm{AB,C}}(r,R)]Q(r,R)$$
$$\equiv \mu_{\mathrm{C}}^{0} + \eta_{\mathrm{C}}^{\mathrm{ABC}}(r,R)Q(r,R) \equiv \mu_{\mathrm{C}}(r,R)$$
$$= \mu_{\mathrm{A}}(r,R) = \mu_{\mathrm{B}}(r,R). \qquad (14.99)$$

These EE equations can be alternatively formulated in terms of the associated FF for the two complementary fragments involved in the Q flow of electrons:

$$\mu_{\mathrm{ABC}}(r,R) = \mu_{\mathrm{AB}}(r)[\eta_{\mathrm{AB}}^{\mathrm{ABC}}(r,R)/\eta_{\mathrm{CT}}^{\mathrm{ABC}}(r,R)] + \mu_{\mathrm{C}}^{0}[\eta_{\mathrm{C}}^{\mathrm{ABC}}(r,R)/\eta_{\mathrm{CT}}^{\mathrm{ABC}}(r,R)]$$
$$\equiv \mu_{\mathrm{AB}}(r) f_{\mathrm{AB}}^{\mathrm{ABC}}(r,R) + \mu_{\mathrm{C}}^{0} f_{\mathrm{C}}^{\mathrm{ABC}}(r,R). \qquad (14.100)$$

This effective dependence of the triatomic chemical potential upon the two geometric parameters then allows one to derive the relevant NFF in this model collision complex. For example, the triatomic analog of the NFF derivative of (14.92), for the diatomic fragment in the collision complex, now reads:

$$\varphi_r(\mathrm{ABC}) = -\partial \mu_{\mathrm{ABC}}(r,R)/\partial r = -\partial \mu_{\mathrm{AB}}(r,R)/\partial r$$
$$= \varphi_r(\mathrm{AB}) f_{\mathrm{AB}}^{\mathrm{ABC}}(r,R) - \mu_{\mathrm{AB}}(r)\left(\partial f_{\mathrm{AB}}^{\mathrm{ABC}}(r,R)/\partial r\right)$$
$$- \mu_{\mathrm{C}}^{0}\left(\partial f_{\mathrm{C}}^{\mathrm{ABC}}(r,R)/\partial r\right). \qquad (14.101)$$

The previous, diatomic result is seen to be recovered when the whole system is limited to this diatomic reactant alone, i.e., for $f_{AB}^{ABC} = 1$ and $f_{C}^{ABC} = 0$.

14.7.3 Sensitivities for Collective Charge Displacements

We now turn to molecular sensitivities associated with the collective charge exchanges q and Q. Above, we have already identified the relevant in situ chemical potential differences, $\mu_{CT}(B \rightarrow A) \equiv \mu_q = \mu_A - \mu_B$ and $\mu_{CT}(AB \rightarrow C) \equiv \mu_Q = \mu_C - \mu_{AB}$, defining the row vector $\boldsymbol{\mu}^{eff.} = (\mu_q, \mu_Q)$, as the associated partial energy derivatives with respect to these two amounts of the internal CT in the triatomic system in question. They exactly vanish when the two fragments involved in the electron transfer equalize their chemical potentials, i.e., when they reach their mutual equilibrium state.

In order to identify the diagonal elements of the effective hardness matrix $\boldsymbol{\eta}^{eff.}(R)$ (14.75) and (14.76), we refer to the EE equations (14.84) and (14.99) which directly give:

$$\begin{aligned} \eta_{q,q} &= \partial\mu_q/\partial q = \partial\mu_A/\partial q - \partial\mu_B/\partial q = \eta_A^{AB} + \eta_B^{AB} = \eta_{CT}^{AB}, \\ \eta_{Q,Q} &= \partial\mu_Q/\partial Q = \partial\mu_C/\partial Q - \partial\mu_{AB}/\partial Q = \eta_C^{ABC} + \eta_{AB}^{ABC} = \eta_{CT}^{ABC}. \end{aligned} \qquad (14.102)$$

The matrix *off*-diagonal elements, equal by the Maxwell relation,

$$\eta_{q,Q} = \partial\mu_Q/\partial q = \eta_{Q,q} = \partial\mu_q/\partial Q = \partial^2 E/\partial q \partial Q, \qquad (14.103)$$

also follow from the corresponding expressions for displacements in the underlying chemical potentials. Let us examine the first derivative in the preceding equation. One observes that the shift in μ_C due to the internal CT $q = dN_A = -dN_B$ in AB reads:

$$d\mu_C(q) = q(\eta_{A,C} - \eta_{B,C}). \qquad (14.104)$$

The associated displacement in the *equilibrium* chemical potential of AB,

$$d\mu_{AB}(q) = \eta_{AB} dN_{AB}(q) = 0,$$

since the *internal* flows in AB do not modify the fragment overall number of electrons: $dN_{AB}(q) = 0$. Thus, $d\mu_Q(q) = d\mu_C(q)$ and hence

$$\eta_{q,Q} = \eta_{A,C} - \eta_{B,C}. \qquad (14.105)$$

It can be directly verified that the same answer follows from the second derivative in (14.103). Let us examine the displacements in the chemical potentials μ_A

and $\mu_{\rm B}$ defining $\mu_q = \mu_{\rm A} - \mu_{\rm B}$, due to the second CT, $Q = dN_{\rm C} = -dN_{\rm AB}$. A reference to (14.91) and (14.93) gives

$$d\mu_{\rm A}(Q) = dN_{\rm A}(Q)\eta_{\rm A,A} + dN_{\rm B}(Q)\eta_{\rm B,A} + Q\eta_{\rm C,A} = Q\left(\eta_{\rm C,A} - f_{\rm A}^{\rm AB}\eta_{\rm A,A} - f_{\rm B}^{\rm AB}\eta_{\rm B,A}\right),$$
$$d\mu_{\rm B}(Q) = dN_{\rm A}(Q)\eta_{\rm A,B} + dN_{\rm B}(Q)\eta_{\rm B,B} + Q\eta_{\rm C,B} = Q\left(\eta_{\rm C,B} - f_{\rm A}^{\rm AB}\eta_{\rm A,B} - f_{\rm B}^{\rm AB}\eta_{\rm B,B}\right). \tag{14.106}$$

Therefore,

$$\begin{aligned} d\mu_q(Q) &= Q\left[\eta_{\rm C,A} - \eta_{\rm C,B} + f_{\rm A}^{\rm AB}(\eta_{\rm A,B} - \eta_{\rm A,A}) + f_{\rm B}^{\rm AB}(\eta_{\rm B,B} - \eta_{\rm B,A})\right] \\ &= Q\left[\eta_{\rm C,A} - \eta_{\rm C,B} - f_{\rm A}^{\rm AB}\eta_{\rm A}^{\rm AB} + f_{\rm B}^{\rm AB}\eta_{\rm B}^{\rm AB}\right] = Q(\eta_{\rm C,A} - \eta_{\rm C,B}), \end{aligned} \tag{14.107}$$

and hence

$$\eta_{Q,q} = \partial\mu_q/\partial Q = \eta_{\rm C,A} - \eta_{\rm C,B}. \tag{14.108}$$

Since the two (collective) degrees of freedom of the electron distribution in the reactive system ABC are mutually coupled by the nonvanishing hardness $\eta_{Q,q}$, the displacement in one of these variables generates an induced force associated with the other variable. In other words, any of these flows triggers the associated adjustment in the other electron transfer of the reactive system.

Let us now turn to the effective NFF indices of (14.77). It follows from the EE equations for AB that at this level of modeling the r-dependence of the two AIM chemical potentials $\mu_{\rm A}(r) = \mu_{\rm A}^0 + \eta_{\rm A}^{\rm AB}(r)q(r)$ and $\mu_{\rm B}(r) = \mu_{\rm B}^0 - \eta_{\rm B}^{\rm AB}(r)q(r)$, which determine the equilibrium value of $\mu_q(r) = \mu_{\rm A}(r) - \mu_{\rm B}(r)$, originates from both the optimum amount $q(r)$ of the B → A CT and the effective atomic hardnesses of bonded atoms in AB. The resulting chemical potential difference for this *equilibrium* CT $q = q(r)$ in the separated diatomic AB must exactly vanish (the equilibrium criterion),

$$\mu_q^{eq.}(r) = \mu_{\rm CT}^0 + \eta_{\rm CT}^{\rm AB}(r)q(r) = 0. \tag{14.109}$$

Indeed, for the equilibrium distribution of electrons in AB, when $dN_{\rm AB} = 0$, the hypothetical *internal* flow q does not affect the fragment equilibrium chemical potential. For the fixed external potential the latter can be modified only by the net inflow/outflow of electrons: $d\mu_{\rm AB} = \eta_{\rm AB}\, dN_{\rm AB}$. This observation implies the vanishing equilibrium coupling between r and $\mu_q^{eq.}$:

$$\varphi_q^{eq.} = -\partial\mu_q^{eq.}/\partial r = 0. \tag{14.110}$$

However, when the current CT q is treated as an *independent* charge variable, which can assume any admissible nonequilibrium value, as is indeed the case in the defining derivatives of the combined Hessian of (14.75), the differentiation of $\mu_q(r) = \mu_{CT}^0 + \eta_{CT}^{AB}(r)q$ gives:

$$\varphi_q(q) = -\partial\mu_q/\partial r = -q\left(\partial\eta_{CT}^{AB}/\partial r\right) = 2q(\partial\eta_{A,B}/\partial r). \tag{14.111}$$

Since $\partial\eta_{A,B}/\partial r < 0$, this derivative further implies that this quantity, reflecting the coupling between the population force and the internuclear distance, exhibits the opposite sign to that of the current amount of CT q, thus acting in direction to restore the electronic equilibrium.

Consider next the diatomic fragment AB in the whole, triatomic system. The driving force μ_q behind the q flow is then additionally shifted by

$$d\mu_q(Q) = Q\eta_{Q,q}(r,R) = Q[\eta_{C,A}(r+R) - \eta_{C,B}(R)].$$

Therefore, when Q is treated as an arbitrary, r-independent populational variable, the NFF index of (14.111) is then modified by the derivative of the effective coupling hardness $\eta_{Q,q}(r, R)$:

$$\begin{aligned}\varphi_q(q,Q) &= \varphi_q(q) - Q[\partial\eta_{Q,q}(r,R)/\partial r]\\ &= \varphi_q(q) - Q[\partial\eta_{C,A}(R_{A,C})/\partial R_{A,C}].\end{aligned} \tag{14.112}$$

Clearly, the derivative of (14.110), calculated for the equilibrium internal CT $q = q(r, R)$ which equalizes the chemical potentials of A and B in the presence of the third atom C, remains unchanged since $\mu_q^{eq.}(r,R) = \mu_A(r,R) - \mu_B(r,R)$ identically vanishes for all values of the two geometrical variables.

It should be also stressed that the global chemical potential $\mu_{AB}(r)$ of the isolated AB differs from the corresponding level $\mu_{AB}(r,R,Q=0)$ characterizing this diatomic fragment in the presence of C, when these complementary parts are considered to be mutually closed in (A¦B|C); here the vertical *broken* line between A and B again signifies their freedom to exchange electrons, $q \neq 0$, while the vertical *solid* line separating AB and C prevents such hypothetical flows of electrons, $Q = 0$. Indeed, the presence of C modifies the effective external potential of AB. However, since this direct influence is neglected in the present level of approximation, only a finite charge flow $Q = -dN_{AB}$ can modify $\mu_{AB}(r,R,Q)$ relative to $\mu_{AB}(r)$:

$$\mu_{AB}(r,R,Q) = \mu_{AB}(r) - \eta_{AB}^{ABC}(r,R)Q. \tag{14.113}$$

We conclude this section with a short derivation of the remaining effective NFF component:

$$\varphi_Q = -\frac{\partial\mu_Q}{\partial r} = \frac{\partial F_r}{\partial Q}. \tag{14.114}$$

We use the first of the defining derivatives to model this quantity, which couples the electronic and nuclear variables in the whole triatomic system. The Q-dependence of the two chemical potentials defining $\mu_Q = \mu_C - \mu_{AB}$ gives

$$\mu_Q(r,R) = \mu_{CT}^{ABC}(r,R) + Q\eta_{CT}^{ABC}(r,R). \tag{14.115}$$

The interreactant equilibrium condition $\mu_Q^{eq.}(r,R) = 0$ then directly gives the associated optimum amount of CT. Clearly, in such a global equilibrium state of (A¦B¦C), for the equilibrium $Q = Q(r, R)$,

$$\varphi_Q^{eq.} = -\partial\mu_Q^{eq.}/\partial r = 0. \tag{14.116}$$

However, when Q is regarded as an independent state variable, the defining derivative of (14.114) gives:

$$\begin{aligned} -\frac{\partial\mu_Q}{\partial r} &= -\frac{\partial\mu_{CT}^{ABC}}{\partial r} - Q\frac{\partial\eta_{CT}^{ABC}}{\partial r} = \frac{\partial\mu_{AB}(r,R,Q)}{\partial r} - Q\frac{\partial\eta_{CT}^{ABC}}{\partial r} \\ &= \frac{\partial\mu_{AB}(r)}{\partial r} + Q\frac{\partial\eta_{AB}^{ABC}}{\partial r} - Q\left(\frac{\partial\eta_C^{ABC}}{\partial r} + \frac{\partial\eta_{AB}^{ABC}}{\partial r}\right) \\ &= -\varphi_r(AB) - Q\frac{\partial\eta_C^{ABC}}{\partial r} = -\varphi_r(AB) + Q\frac{\partial\eta_{AB,C}}{\partial r}, \end{aligned} \tag{14.117}$$

where:

$$\frac{\partial\eta_{AB,C}}{\partial r} = \frac{1}{\left(\eta_{CT}^{AB}\right)^2}\left(\frac{\partial\eta_{A,B}}{\partial R_{A,B}}\right)(\eta_{A,A} - \eta_{B,B})(\eta_{B,C} - \eta_{A,C}) + \frac{\eta_B^{AB}}{\eta_{CT}^{AB}}\left(\frac{\partial\eta_{A,C}}{\partial R_{A,C}}\right). \tag{14.118}$$

The preceding derivative reflects the r-dependence of the hardness coupling between the two reactants. In an attempt to physically interpret this relation, one first observes that for the electronically stable diatomic $\eta_{CT}^{AB} > 0$ and $\eta_B^{AB} > 0$. Moreover, the geometric derivatives of diatomic hardnesses are both negative, $d\eta_{i,j}(R_{i,j})/dR_{i,j} < 0$, since the *two*-center electron repulsion diminishes with increasing internuclear distance. In the collinear collision complex $\eta_{B,C} - \eta_{A,C} > 0$, while the hardness difference $(\eta_{A,A} - \eta_{B,B})$ is shaped by the relative acidic/basic character of both atoms in the diatomic reactant. For identical atoms, e.g., in the hydrogen exchange reaction H—H + X → H + H—X, the first term in (14.118) identically vanishes, so that the second (negative) term implies that an increase in the bond length r of the diatomic reactant decreases its hardness coupling to the approaching atom. A similar prediction follows when A and B, respectively, denote the *acidic* and *basic* parts of the diatomic reactant, when $\eta_{A,A} - \eta_{B,B} > 0$, since both contributions in the preceding equation are then negative. In both these cases the

NFF of (14.117) becomes negative. Therefore, in this approximation the electronegativity difference $\chi_Q = -\mu_Q$, the driving force of the interreactant charge flow Q, diminishes with an elongation of the diatomic reactant.

14.7.4 Couplings Along the Minimum-Energy Path

In the preceding sections we have treated one interatomic distance, e.g., $R_{AB} = r$ as the internal geometric variable, while the other independent bond length R has been regarded as the *reaction coordinate*, for the mutual approach by the reaction substrates in the reactant (*entrance*) valley of the molecular PES $W(R_{AB} = r,\ R_{BC} = R,\ R_{AC} = r + R)$. In the strong interaction region, between the reactant and product (*exit*) valleys, which includes the transition-state complex $[\text{A----B----C}]^{\ddagger}$, these roles are reversed, with R_{BC} now representing the internal geometric variable of the diatomic product and R_{AB} approximating the reaction coordinate of the product departure.

For atom exchange reactions, in which the single bond is broken/formed in a concerted fashion, (14.73) and (14.74) realistically determine the reaction progress along MEP on the collinear PES in terms of the reaction *bond order* coordinate $n \equiv n_{B,C} \in [0, 1]$, where $n = 0$, i.e., $n_{A,B} = 1 - n = 1$, corresponds to separated reactants and $n = 1$ $(n_{A,B} = 0)$ represents the separated products. The MEP $\mathcal{Q}(n) = [r(n), R(n)] \equiv \boldsymbol{Q}(n)$, parametrically defined by the corresponding Pauling relations, in 10^{-9} m,

$$\begin{aligned} r(n) &= R^e_{A,B} - 0.026\ln(1 - n) \quad \text{and} \\ R(n) &= R^e_{B,C} - 0.026\ln n, \end{aligned} \tag{14.119}$$

or the associated MEP-trajectory equation (14.73),

$$\exp(-38.46\Delta R_{A,B}) + \exp(-38.46\Delta R_{B,C}) = 1, \tag{14.120}$$

determines the corresponding BEBO energy profile $W(n) \equiv W\{\boldsymbol{N}^{eq.}[\boldsymbol{Q}(n)], \boldsymbol{Q}(n)\} \equiv W[\boldsymbol{Q}(n)]$, along this MEP-cut, which is known to adequately represent the reaction energy curve; here $Q(n)$ measures the arc length along the MEP trajectory $\mathcal{Q}(n)$ in the geometrical plane $\boldsymbol{Q} = (r, R)$, with $Q(n = 0) \equiv 0$ identifying the separated reactants and $Q(n = 1) \to \infty$ corresponding to the separated products. Each value of the bond order n then determines the local (normal) coordinate system $[t(n), p(n)]$ consisting of the *tangent* $[t(n)]$ and *normal* $[p(n)]$ directions on the BEBO MEP.

It is of interest in the theory of chemical reactivity to examine the coupling between the electronic and geometrical degrees of freedom along this trajectory. In the present collinear collision complex it is determined by the associated 3×3 Hessian:

$$\mathbf{H}[\text{AB---C}(n)] = \begin{bmatrix} \boldsymbol{\eta}^{eff\cdot}(n) & -\boldsymbol{\varphi}^{eff\cdot}(n)^{\mathrm{T}} \\ -\boldsymbol{\varphi}^{eff\cdot}(n) & k(n) \end{bmatrix}. \quad (14.121)$$

Here,

$$\boldsymbol{\eta}^{eff\cdot}(n) = \partial^2 W/\partial \boldsymbol{N}\partial \boldsymbol{N}|_{\mathscr{Q}(n)} = \partial^2 E/\partial \boldsymbol{N}\partial \boldsymbol{N}|_{\mathscr{Q}(n)}$$

determines the effective populational Hessian of (14.76) for the current point $\mathscr{Q}(n)$ on the BEBO MEP trajectory, while

$$k(n) = \partial^2 W/\partial p^2|_{\mathscr{Q}(n)} = -\partial F_p/\partial p|_{\mathscr{Q}(n)}, \quad (14.122)$$

denotes the force constant of the local *normal* stretching mode *p*, orthogonal to the reaction coordinate *t*. Accordingly, the NFF input $\boldsymbol{\varphi}^{eff\cdot}(n)$ in the MEP Hessian of (14.121) is then defined by the *normal* derivative of $\boldsymbol{\mu}^{eff\cdot}(n)$:

$$\boldsymbol{\varphi}^{eff\cdot}(n) = -\partial \boldsymbol{\mu}^{eff\cdot}/\partial p|_{\mathscr{Q}(n)} = \left[-\partial \mu_q/\partial p|_{\mathscr{Q}(n)}, -\partial \mu_Q/\partial p|_{\mathscr{Q}(n)}\right]. \quad (14.123)$$

Let us recall that for the current point $\mathscr{Q}^0 = [r^0 = r(n^0), R^0 = R(n^0)] = \boldsymbol{Q}(n^0)$ on the MEP of (14.119), the equations defining the (mutually perpendicular) tangent $t(n)$ and normal $p(n)$ directions $\boldsymbol{P}(n) = [t(n), p(n)]$ in the geometrical plane $\boldsymbol{Q} = (r, R)$, respectively, read:

$$\begin{aligned} R'(n^0)(r - r^0) - r'(n^0)(R - R^0) &= 0, \\ r'(n^0)(r - r^0) + R'(n^0)(R - R^0) &= 0, \end{aligned} \quad (14.124)$$

where $R'(n^0) = dR(n)/dn|_{n^0}$ and $r'(n^0) = dr(n)/dn|_{n^0}$. These derivative coefficients subsequently define the local transformation

$$\mathbf{O}(n) = [\boldsymbol{O}_t(n), \boldsymbol{O}_p(n)] \equiv \partial \boldsymbol{P}(n)/\partial \boldsymbol{Q} \quad (14.125)$$

of the original geometrical variables $\boldsymbol{Q} = (r, R)$ to the associated MEP-based directions $\boldsymbol{P}(n) = [t(n), p(n)]$ defined by the corresponding columns $[\boldsymbol{O}_t(n), \boldsymbol{O}_p(n)]$ of $\mathbf{O}(n)$:

$$\begin{aligned} \boldsymbol{P}(n) &= \boldsymbol{Q}\begin{bmatrix} R'(n) & r'(n) \\ -r'(n) & R'(n) \end{bmatrix} \equiv \boldsymbol{Q}\mathbf{O}(n) \\ &= [\boldsymbol{Q}\boldsymbol{O}_t(n), \boldsymbol{Q}\boldsymbol{O}_p(n)] = [t(n), p(n)]. \end{aligned} \quad (14.126)$$

The inverse transformation,

$$\mathbf{O}^{-1}(n) \equiv \partial \boldsymbol{Q}/\partial \boldsymbol{P}(n) = \frac{1}{[R'(n)]^2 + [r'(n)]^2}\begin{bmatrix} R'(n) & -r'(n) \\ r'(n) & R'(n) \end{bmatrix}$$
$$\equiv \begin{bmatrix} \boldsymbol{O}_t^{-1}(n) = \partial \boldsymbol{Q}/\partial t(n) \\ \boldsymbol{O}_p^{-1}(n) = \partial \boldsymbol{Q}/\partial p(n) \end{bmatrix}, \tag{14.127}$$

then determines CS along the MEP from the corresponding quantities defined in the original $\boldsymbol{Q}$-representation. For example, the appropriate *chain* rule manipulation of derivatives gives:

$$\begin{aligned} \boldsymbol{\varphi}^{eff.}(n) &= -\partial \boldsymbol{\mu}^{eff.}/\partial p|_{\boldsymbol{P}(n)} = -(\partial \boldsymbol{Q}/\partial p)|_{\boldsymbol{P}(n)}(\partial \boldsymbol{\mu}^{eff.}/\partial \boldsymbol{Q})|_{\boldsymbol{P}(n)} = \boldsymbol{O}_p^{-1}(n)\boldsymbol{\varphi}^{eff.}[\boldsymbol{P}(n)], \\ F_p(n) &= -\partial W/\partial p|_{\boldsymbol{P}(n)} = -(\partial \boldsymbol{Q}/\partial p)|_{\boldsymbol{P}(n)}(\partial W/\partial \boldsymbol{Q})|_{\boldsymbol{P}(n)} = \boldsymbol{O}_p^{-1}(n)\,\boldsymbol{F}[\boldsymbol{P}(n)], \\ k(n) &= \partial^2 W/\partial p^2|_{\boldsymbol{P}(n)} = (\partial \boldsymbol{Q}/\partial p)|_{\boldsymbol{P}(n)}(\partial^2 W/\partial \boldsymbol{Q}\,\partial \boldsymbol{Q})|_{\boldsymbol{P}(n)}(\partial \boldsymbol{Q}/\partial p)^{\mathrm{T}}|_{\boldsymbol{P}(n)} \\ &= \boldsymbol{O}_p^{-1}(n)\mathbf{H}[\boldsymbol{P}(n)][\boldsymbol{O}_p^{-1}(n)]^{\mathrm{T}}. \end{aligned} \tag{14.128}$$

Here, the square matrix of NFF for $\boldsymbol{P} = \boldsymbol{P}(n)$ reads,

$$\boldsymbol{\varphi}^{eff.}(\boldsymbol{P}) = -\frac{\partial^2 W}{\partial \boldsymbol{Q}\,\partial \boldsymbol{N}}\bigg|_{\boldsymbol{P}} = \left(\frac{\partial \boldsymbol{F}}{\partial \boldsymbol{N}}\bigg|_{\boldsymbol{P}}\right)^{\mathrm{T}} = -\left(\frac{\partial \boldsymbol{\mu}^{eff.}}{\partial \boldsymbol{Q}}\bigg|_{\boldsymbol{P}}\right)$$
$$= \begin{bmatrix} \varphi_{r,q}(\boldsymbol{P}) = -(\partial \mu_q/\partial r)|_{\boldsymbol{P}} & \varphi_{r,Q}(\boldsymbol{P}) = -(\partial \mu_Q/\partial r)|_{\boldsymbol{P}} \\ \varphi_{R,q}(\boldsymbol{P}) = -(\partial \mu_q/\partial R)|_{\boldsymbol{P}} & \varphi_{R,Q}(\boldsymbol{P}) = -(\partial \mu_Q/\partial R)|_{\boldsymbol{P}} \end{bmatrix}, \tag{14.129}$$

and the force constant tensor for the original geometric variables $\boldsymbol{Q} = (r, R)$,

$$\mathbf{H}(\boldsymbol{P}) = \frac{\partial^2 W}{\partial \boldsymbol{Q}\,\partial \boldsymbol{Q}}\bigg|_{\boldsymbol{P}} = -\frac{\partial \boldsymbol{F}}{\partial \boldsymbol{Q}}\bigg|_{\boldsymbol{P}}$$
$$= \begin{bmatrix} k_{r,r}(\boldsymbol{P}) = -(\partial F_r/\partial r)|_{\boldsymbol{P}} & k_{r,R}(\boldsymbol{P}) = -(\partial F_R/\partial r)|_{\boldsymbol{P}} \\ k_{R,r}(\boldsymbol{P}) = -(\partial F_r/\partial R)|_{\boldsymbol{P}} & k_{R,R}(\boldsymbol{P}) = -(\partial F_R/\partial R)|_{\boldsymbol{P}} \end{bmatrix}, \tag{14.130}$$

where the force vector contains two components:

$$\boldsymbol{F}(\boldsymbol{P}) = [-\partial W/\partial r|_{\boldsymbol{P}} = F_r(\boldsymbol{P}), -\partial W/\partial R|_{\boldsymbol{P}} = F_R(\boldsymbol{P})]. \tag{14.131}$$

In the force constant matrix $k_{R,r} = k_{r,R}$ determines the equilibrium coupling between the electronically decoupled geometrical degrees of freedom. The inverse of the force constant matrix, the geometrical *softness* matrix, $\mathbf{H}^{-1} \equiv \mathbf{G}(\boldsymbol{P})$, combines the system purely geometric compliants along MEP.

Finally, instead of separating one geometric variable $t(n)$ as a parametric measure of the reactant/product proximity along the BEBO MEP, one could explicitly include the effects of its coupling to the electron-transfer variables $\boldsymbol{N} = (q, Q)$

and the other geometric variable $p(n)$. This generalized approach calls for the full 4×4 Hessian matrix of the collinear reactive system with respect to both the electronic $\boldsymbol{N}$ and geometric $\boldsymbol{Q} = (r, R)$ or $\boldsymbol{P} = (t, p)$ variables in $\mathcal{R}$,

$$\mathbf{H}^{\mathrm{MEP}}[\mathrm{AB}\text{---}\mathrm{C}(n)] = \begin{bmatrix} \left.\frac{\partial^2 W}{\partial \boldsymbol{N}\, \partial \boldsymbol{N}}\right|_{\boldsymbol{P}(n)} & \left.\frac{\partial^2 W}{\partial \boldsymbol{N}\, \partial \boldsymbol{P}}\right|_{\boldsymbol{P}(n)} \\ \left.\frac{\partial^2 W}{\partial \boldsymbol{P}\, \partial \boldsymbol{N}}\right|_{\boldsymbol{P}(n)} & \left.\frac{\partial^2 W}{\partial \boldsymbol{P}\, \partial \boldsymbol{P}}\right|_{\boldsymbol{P}(n)} \end{bmatrix} = \begin{bmatrix} \boldsymbol{\eta}^{eff.}(n) & -[\boldsymbol{\varphi}^{\mathrm{MEP}}(n)]^{\mathrm{T}} \\ -\boldsymbol{\varphi}^{\mathrm{MEP}}(n) & \mathbf{H}^{\mathrm{MEP}}(n) \end{bmatrix}, \tag{14.132}$$

where:

$$\boldsymbol{\varphi}^{\mathrm{MEP}}(n) = \mathbf{O}^{-1}(n)\boldsymbol{\varphi}^{eff.}(\boldsymbol{P}) \quad \text{and} \quad \mathbf{H}^{\mathrm{MEP}}(n) = \mathbf{O}^{-1}(n)\mathbf{H}(\boldsymbol{P})\mathbf{O}^{-1}(n)^{\mathrm{T}}. \tag{14.133}$$

Derivatives in the first row of $\boldsymbol{\varphi}^{eff.}(\boldsymbol{P})$ in (14.129) constitute vector $\boldsymbol{\varphi}^{eff.}$ already discussed before: $\boldsymbol{\varphi}^{eff.} = (\varphi_{r,q} = \varphi_q, \varphi_{r,Q} = \varphi_Q)$. The second-row derivatives in this NFF matrix can be similarly modeled using the relevant equations for displacements in the fragment chemical potentials, which have been developed already in preceding sections. For example, when Q is regarded as the independent *state* variable, differentiating (14.115) with respect to R gives,

$$\begin{aligned} \varphi_{R,Q} &= -\partial\mu_Q(r,R)/\partial R = -[\partial\mu_{\mathrm{CT}}^{\mathrm{ABC}}(r,R)/\partial R] - Q[\partial\eta_{\mathrm{CT}}^{\mathrm{ABC}}(r,R)/\partial R] \\ &= Q\{[\partial\eta_{\mathrm{AB}}^{\mathrm{ABC}}(r,R)/\partial R] - [\partial\eta_{\mathrm{CT}}^{\mathrm{ABC}}(r,R)/\partial R]\} = Q\left[\partial\eta_{\mathrm{AB,C}}^{\mathrm{ABC}}(r,R)/\partial R\right], \end{aligned} \tag{14.134}$$

where the final derivative follows from the model expression for the interreactant coupling hardness of (14.95):

$$\partial\eta_{\mathrm{AB,C}}^{\mathrm{ABC}}(r,R)/\partial R = f_{\mathrm{A}}^{\mathrm{AB}}[\partial\eta_{\mathrm{A,C}}/\partial R_{\mathrm{A,C}}] + f_{\mathrm{B}}^{\mathrm{AB}}[\partial\eta_{\mathrm{B,C}}/\partial R_{\mathrm{B,C}}] < 0. \tag{14.135}$$

Thus, for a positive amount of the interreactant CT, $Q > 0$, when atom C acts as the *acidic* partner of the *basic* diatomic AB, $\mu_Q < 0$, an increase in the interreactant separation R lowers the flow electronegativity $\chi_Q = -\mu_Q$. The opposite trend is predicted for the acidic AB with respect to C, when $\mu_Q > 0$ and hence $Q < 0$.

The R-derivative of $\mu_q(q, Q)$ in the second row of $\boldsymbol{\varphi}^{eff.}(\boldsymbol{P})$ similarly follows from the fragment chemical potential displacements for the specified charges q and Q. It follows from EE equation (14.84) that the given internal flow q gives rise to

$$\mu_q(q) = \mu_{\mathrm{CT}}^0 + q\eta_{\mathrm{CT}}^{\mathrm{AB}}(r), \tag{14.136}$$

while the subsequent interreactant flow Q modifies this chemical potential difference by the following terms:

$$\mu_q(q,Q) = \mu_q(q) + Q\left[f_{\mathrm{B}}^{\mathrm{AB}}(r)\eta_{\mathrm{B}}^{\mathrm{AB}}(r) - f_{\mathrm{A}}^{\mathrm{AB}}(r)\eta_{\mathrm{A}}^{\mathrm{AB}}(r) + \eta_{\mathrm{C,A}}(r+R) - \eta_{\mathrm{C,B}}(R)\right]. \tag{14.137}$$

Hence, the model expression for the NFF of interest reads:

$$\varphi_{R,q} = -\partial\mu_q/\partial R = Q\left(\frac{\partial\eta_{\mathrm{C,B}}}{\partial R_{\mathrm{C,B}}} - \frac{\partial\eta_{\mathrm{C,A}}}{\partial R_{\mathrm{C,A}}}\right). \tag{14.138}$$

In each of the above approaches the compliant sensitivities of interest are given by the corresponding elements of Hessian's inverse, which defines the generalized softness matrix. Such compliant descriptors give rise to the MEC measuring the equilibrium responses in the remaining *state* variables, per unit shift in the parameter of interest (electronic or nuclear) (see Sect. 14.4).

14.8 Conclusion

All chemical or conformational changes involve both the nuclear displacements and the concomitant electron redistributions. At a given stage of the system displacement, depending on what is considered a "perturbation" and what the equilibrium response to it, the EF or the EP approaches can be adopted. We have presented above the linear-response theoretical framework which covers both these perspectives in reactivity theory. This development has demonstrated that the range of applications of molecular CS is not limited to somewhat oversimplified *single*-reactant problems. They have been shown to be fully capable of covering also the prototype *two*-reactant scenarios.

The coupling between electronic and geometrical structures of molecular systems is embodied in the adiabatic PES. In this development both the molecular compliants reflecting the electronic and/or nuclear adjustments have been determined in the combined treatment of the generalized linear responses of molecular systems, which simultaneously admits the electronic and/or nuclear relaxation of a molecule.

We have also reported illustrative numerical values of alternative derivative quantities describing the molecular responses to both the electronic and nuclear perturbations, within the geometric Legendre-transformed representations defining the EP and EF perspectives on the molecular structure, in which the geometric coordinates $\boldsymbol{Q}$ replace the external potential $v(\boldsymbol{r};\ \boldsymbol{Q})$ in the list of the system state parameters. A brief survey of the derivative descriptors of the externally closed and open molecular systems has been given and the basic relations between displacements of the representation state *parameters* (perturbations) and responses in the conjugate (unconstrained) state *variables* have been summarized for both the rigid and relaxed geometry cases. Specific quantities reflecting the interaction between the geometrical and electronic structures of molecular systems and the

MEC components have been identified and their physical content has been commented upon examined.

The relaxed (compliance) quantities of both the electronic and/or nuclear origin measure the generalized "softnesses" of molecules, which complement the corresponding "hardness" data. Indeed, the electronic softness (electronically relaxed quantity defined for the *rigid* geometry $\boldsymbol{Q}$) and the purely nuclear compliants (geometrically relaxed defined for the closed system, at constant N) are examples of such complementary quantities to more familiar electronic hardness and the nuclear force constant descriptors, respectively. In the *decoupled* treatment of these complementary facets of the molecular structure one neglects the explicit interaction between the electronic (N) and nuclear ($\boldsymbol{Q}$) degrees of freedom, or between their partial energy conjugates, the electronic chemical potential μ, attributed to an external electron reservoir, and the forces $\boldsymbol{F}$ acting on the system nuclei.

In the principal (N,$\boldsymbol{Q}$)-representation this interaction is measured by the NFF. Together with the electronic hardness tensor and geometric Hessian it defines the generalized matrix of the system electronic–nuclear "force constants." By its partial or total inversions all the molecular compliance data have been determined. Such a coupled description of these complementary, electronic and nuclear aspects of the molecular structure provides a more complete treatment of the adiabatic linear responses in molecules, which addresses all alternative scenarios in the theory of chemical reactivity. For example, the MEC reflecting the electronic–nuclear interaction provide a semiquantitative measures of responses in quantities describing one facet of the molecular structure per unit displacements in quantities describing the other, complementary aspect.

The HF results generated for representative polyatomic molecules have used the N-derivatives estimated by finite differences, while the $\boldsymbol{Q}$-derivatives have been calculated analytically by standard methods of quantum chemistry. We have examined the effects of the electronic and/or nuclear relaxations on specific charge sensitivities used in the theory of chemical reactivity, e.g., the hardness, softness, and FF descriptors. New concepts of the geometric Fukui functions and related softnesses, which include the effects of molecular electronic and/or nuclear relaxations, have also been introduced.

In particular, the electronic–geometric interaction has been examined by comparing the corresponding rigid and relaxed hardness/softness and FF data. Among others, these compliants reflect the influence of the nuclear relaxation on the system electronic hardnesses and softnesses, and the effect of the electronic relaxation on the nuclear force constants and vibration compliants. Of particular importance are the components of MEC, which provide the ground state "matching" relations between the hypothetical perturbations of molecular systems and their conjugated equilibrium responses. They should allow one to identify the electronic and/or nuclear perturbations, which are most efficient in facilitating the desired chemical reaction or conformational change.

Finally, a possible use of these coupling constants as reactivity indices has been commented upon in both the *one*- and *two*-reactant approaches. In the interreactant decoupled applications the molecular complaints from calculations on separate reactants can be used directly to semiquantitatively predict the intrareactant effects resulting from the interreactant CT. The building blocks of the combined electronic–geometric Hessian for the *two*-reactant system have been also discussed. However, in such a treatment of reactants, the additional calculations on the reactive system as a whole are required. The corresponding blocks of the generalized compliance matrix have also been identified.

An illustrative example of such a coupled approach to a simple model of reactive collisions in the collinear triatomic system has been presented. The independent (canonical) electronic and geometric state parameters have been identified and explicit analytical expressions for the crucial NFF data, which reflect the coupling between the system electronic and geometric degrees of freedom, have been derived using an adequate semiempirical representation of the atomic hardnesses by the corresponding *valence* shell electron repulsion integrals. This treatment takes into account all physically important interactions between the system electronic and geometrical structures and is capable of generating all relevant *two*-reactant reactivity probes in both the NF and EP perspectives, in the chemical softness and hardness representations, respectively. By additionally using the analytical BEBO trajectory conserving the overall "bond order" of Pauling, which constitutes a realistic approximation of the collinear MEP in such systems, one can also generate the response descriptors along this reaction coordinate.

References

Ayers PW, Yang W, Bartolotti LJ (2009) In: Chattaraj PK (ed) Chemical reactivity theory: a density functional view. CRC, Boca Raton, p 255

Baekelandt BG, Mortier WJ, Schoonheydt RA (1993) In: Sen KD (ed) Hardness, structure and bonding, vol 80. Springer, Berlin, p 187

Baekelandt BG, Janssens GOA, Toufar H, Mortier WJ, Schoonheydt RA, Nalewajski RF (1995) J Phys Chem 99:9784

Callen HB (1962) Thermodynamics: an introduction to the physical theories of equilibrium thermostatics and irreversible thermodynamics. Wiley, New York

Chattaraj PK (ed) (2009) Chemical reactivity theory – a density functional view. CRC, Boca Raton

Cohen MH (1996) In: Nalewajski RF (ed) Topics in current chemistry 183: Density functional theory IV – theory of chemical reactivity, Springer, Berlin, p. 143

Cohen MH, Ganguglia-Pirovano MV, Kurdnovský J (1994) J Chem Phys 101:8988

Cohen MH, Ganguglia-Pirovano MV, Kurdnovský J (1995) J Chem Phys 103:3543

Decius JC (1963) J Chem Phys 38:241

Geerlings P, De Proft F, Langenaeker W (2003) Chem Rev 103:1793

Gutmann V (1978) The donor-acceptor approach to molecular interactions. Plenum, New York

Johnson PA, Bartolotti LJ, Ayers P, Fievez T, Geerlings P (2011) Charge density and chemical reactions: a unified view from conceptual DFT. In: Gatti C, Macchi P (eds) Modern charge-density analysis. Springer, Berlin (in press)

Johnston HS, Parr CJ (1963) J Am Chem Soc 85:2544

Jones LH, Ryan RR (1970) J Chem Phys 52:2003

Nalewajski RF (1985) J Phys Chem 89:2831

Nalewajski RF (1989a) J Phys Chem 93:2658
Nalewajski RF (1995b) In: Gross EKU and Dreizler RM (eds) Proceedings of the NATO ASI on density functional theory. Plenum, New York, p 339
Nalewajski RF (1997a) Int J Quantum Chem 61:181
Nalewajski RF (1997b) In: Mortier WM, Schoonheydt RA (eds) Developments in the theory of chemical reactivity and heterogeneous catalysis. Research Signpost, Trivandrum, p 135
Nalewajski RF (1998a) Polish J Chem 72:1763
Nalewajski RF (1999) Phys Chem Chem Phys 1:1037
Nalewajski RF (2000a) Comput Chem 24:243
Nalewajski RF (2003a) Adv Quant Chem 43:119
Nalewajski RF (2006f) Adv Quant Chem 51:235
Nalewajski RF (2009a) In: Chattaraj PK (ed) Chemical reactivity theory: a density functional view. CRC. Boca Raton, p. 453
Nalewajski RF (2010g) J Math Chem 48:752
Nalewajski RF, Korchowiec J (1997) Charge sensitivity approach to electronic structure and chemical reactivity. World-Scientific, Singapore
Nalewajski RF, Michalak A (1995) Int J Quantum Chem 56:603
Nalewajski RF, Sikora O (2000) J Phys Chem A 104:5638
Nalewajski RF, Korchowiec J, Zhou Z (1988) Int J Quantum Chem Symp 22:349
Nalewajski RF, Korchowiec J, Michalak A (1996a) Top Curr Chem 183:25
Nalewajski RF, Błażewicz D, Mrozek J (2008) J Math Chem 44:325
Ohno K (1967) Adv Quant Chem 3:239
Pariser R (1953) J Chem Phys 21:568
Pariser R, Parr RG (1953) J Chem Phys 21:767
Parr RG, Yang W (1984) J Am Chem Soc 106:4049
Parr RG, Yang W (1989) Density-functional theory of atoms and molecules. Oxford University Press, New York
Pauling L (1947) J Am Chem Soc 69:542
Politzer P, Murray JS (2009) In: Chattaraj PK (ed) Chemical reactivity theory: a density functional view. CRC, Boca Raton, p 243
Pople JA (1953) Trans Faraday Soc 49:1375
Sanderson RT (1951) Science 114:670
Sanderson RT (1976) Chemical bonds and bond energy, 2nd edn. Academic, New York
Swanson BI (1976) J Am Chem Soc 98:3067
Swanson BI, Satija SK (1977) J Am Chem Soc 99:987
Tisza L (1977) Generalized thermodynamics. MIT Press, Cambridge

Chapter 15
Qualitative Approaches to Reactivity Phenomena

Abstract Qualitative explorations of elementary mechanisms of chemical reactions are outlined and simple modeling of reactivity preferences is summarized. The familiar energy profiles along the reaction intrinsic coordinate can be analyzed by examining the associated changes in the *reaction force* and the information content of the electron probability distributions in elementary collisions. Such Shannon and Fisher information "signatures" of the reaction mechanism in both the *position* and *momentum* spaces, respectively, have recently been examined for the representative *disconcerted* (abstraction) and *concerted* (nucleophilic substitution) reactions. The momentum-space entropy/information data have been shown to identify regions where the bond-breaking and bond-forming processes really occur. These features are not revealed by the MEP energy profile and the density of electrons at the Transition-State complex alone.

Selected *single*-reactant approaches to molecular reactivity problem are briefly outlined in the complementary EF (*external-potential*-based) and EP (*density*-based) perspectives. This simplified, qualitative approach, which in the bimolecular reactive system neglects the actual size of the perturbation created by the presence of the attacking agent, is sometimes called the *Conceptual DFT*. It is based upon the basic DFT premise that the electronic density and its responses (derivatives) are sufficient to understand and determine the chemical reactivity trends of molecules, and provides a novel theoretical framework in which one can discuss and ultimately understand the reactivity phenomena. It uses as reactivity indicators the generalized polarizabilities of molecules, reflected by corresponding CS measuring the linear responses to normalized perturbations. Their combinations determine several reactivity indices for specific reactivity problems, e.g., the *electrophilicity/nucleophilicity* indicators of molecular systems and their parts.

The conceptual and interpretative advantages of using in quantum mechanics the separate eigenvalue problems of the *external* and *internal* parts of the matrix representations of physical quantities are stressed. In the reactant resolution of the bimolecular reactive system these two components respectively combine the

R.F. Nalewajski, *Perspectives in Electronic Structure Theory*,
DOI 10.1007/978-3-642-20180-6_15,

diagonal and *off-diagonal* blocks of matrix elements. Their separate diagonalizations imply the corresponding internal and external decouplings, respectively, in terms of the associated collective ("normal") modes of the reactive system. This decoupling scheme has been successfully used in several interpretations of the molecular electronic structure and reactivity. The external eigensolutions are examined in some detail, and their recent applications in both CSA – to extract the most important electron-transfer effects between constituent atoms in model chemisorption systems, and in the MO theory – to precisely identify the interorbital flows of electrons, are mentioned. The associated grouping relations, for combining the partial external/internal eigensolutions, i.e., the complementary subrotations of the basis set vectors into the eigenvectors of the whole matrix, are also derived. The maximum-hardness and HSAB principles of structural chemistry and other applications of the hardness/softness concepts in reactivity phenomena are summarized.

15.1 Introduction

The rearrangements of electron densities accompanying chemical reactions generate the associated changes in the information content of the associated probability distributions. Their entropic/information descriptors provide complementary insights into the critical *bond-forming–bond-breaking* stages of the reaction mechanism (Esquivel et al. 2009a, b; López-Rosa et al. 2010; López-Rosa 2010), which are not revealed by the reaction energy profile along the reaction path on the system PES, which only identifies the key stationary points along MEP: the *minima* attributed to equilibrium structures of reactants/products and *saddle-points* related to *transition states* (TS) (Schlegel 1987). The latter are widely explored and characterized in computational quantum chemistry in terms of the geometric energy gradients and second derivatives defining the Hessian over nuclear positions. These derivative quantities also allow for the MEP-following on the complicated, *many*-dimensional reaction surfaces, from one stationary point to another. The minima and saddle points on the reaction PES have been fully characterized through the first (gradient) and second (Hessian) derivatives over nuclear positions (e.g., Fukui 1981; Ganzález and Schlegel 1990; Fan and Ziegler 1992; Ganzález-García et al. 2006; Ishida et al. 1977).

Understanding TS in *chemical* terms and predicting their influence on the reaction dynamics and kinetics (Eyring 1935; Wigner 1938; see also: Eyring et al. 1958) still remains one of the basic goals of the physical organic chemistry. The entropy/information concepts have also been employed in studying the Heisenberg-like uncertainty products, in context of the quantum-mechanical relations (Białynicki-Birula and Mycielski 1975, 1976; Angulo 1993; López-Rosa et al. 2009; Angulo et al. 2010; López-Rosa 2010), which state limitations to perform measurements on microsystems without disturbing them, and in molecular sciences on the Hammond postulate (Nalewajski 2006g; Nalewajski and Broniatowska 2003b), molecular/atomic complexity, similarity or dissimilarity (e.g., Nalewajski and Świtka 2002;

Nalewajski et al. 2002; Nalewajski and Broniatowska 2003a, b; Nalewajski 2006g, 2009h; López-Rosa 2010), and in studies of the reaction dynamics (Levine 1978).

A variety of density descriptors have been proposed to follow the course of a chemical reaction, to determine and understand TS densities, in order to explain the path preferences and to formulate simple, qualitative models of chemical reactivity. For example, Bader's (1990, 1991) exploration of the Laplacian of the density revealed that the reactants tend to align themselves in the complementary way so as to match the regions of charge-*depletion* in one of them with the charge-*concentration* sites in another. A similar reactivity pattern is revealed by maps of the molecular ESP. This *first*-order criterion calls for the most favorable matching of the electrophilic (electron-deficient) sites of one reactant, identified by basins of the positive (proton-*repulsive*) values of ESP, with the nucleophilic (electron-rich) sites of the other reactant, which exhibit the negative (proton-attractive) values of ESP. It should be further realized that, by the Hellmann–Feynman theorem, the electron density itself measures the *first*-order response of the energy to changes in the external potential, so that together with ESP the responses in the electron density predict regioselectivity trends in chemical reactions.

Several reactivity indicators introduced in preceding sections have also proven useful in understanding/predicting chemical reactivity (e.g., Chandra 1999; Geerlings et al. 2003; Nalewajski 1995a, b, 1997a, b, 2003a, 2006g; Nalewajski and Korchowiec 1997; Nalewajski et al. 1996a; Parr and Yang 1989; Senet 2009). The hardness and softness concepts have been used to justify molecular stability trends of the HSAB principle (Parr and Pearson 1983; Nalewajski 1984), with harder molecules implying less reactive species. The softness index quantifies the system global polarizability and its ability to accommodate the excess charge. Thus, soft, easy to polarize molecules are expected to be more reactive, with the soft acids being capable of acquiring more additional electronic charge in the bond CT processes.

The course of a single- or multi-step conversion of chemical species involves the right sequence of elementary *bond-breaking–bond-forming* steps of the overall chemical change in question, i.e., a description of TS involved, reactive complexes, kinetics, catalysis, stereochemistry, etc. The chemical mechanism thus implies the knowledge of the order in which molecules react, as well as a conceptual description of its elementary steps in both the synchronous (single-step) and nonsynchronous (two-step) chemical transformations.

Contrary to Dewar's (1984) intuitive argument that in multibond processes the synchronicity should be normally prohibited, such Woodward–Hoffman allowed reactions are predicted to be synchronous, with all bond-breaking and bond-forming processes taking place simultaneously. The nonperfectly synchronized features are also observed (e.g., Bernasconi 1992). They are reflected by the structures of TS involved, with nonequal progress being made by the underlying elementary processes of the bond formation/cleavage, charge localization/delocalization, etc. (e.g., Chandra and Sreedhara-Rao 1996; Chandra 1999).

As we have already mentioned above, the IT probes of chemical bonds and reactions, including a detailed description of the reaction mechanism, still represents a challenge in theoretical chemistry. Such investigations of the stereochemical course of chemical reactions (e.g., Schlegel 1987) can be routinely performed in both the position, momentum, and the combined *dual* (phase) spaces. In the MO approaches of both HF and KS theories, the electron density is given by the sum of densities $\{\varphi_i^\sigma(\boldsymbol{r})\}$ due to each occupied SMO,

$$\rho(\boldsymbol{r}) = \sum\nolimits_\sigma \sum\nolimits_i |\varphi_i^\sigma(\boldsymbol{r})|^2 = \sum\nolimits_\sigma \sum\nolimits_i \rho_i^\sigma(\boldsymbol{r}). \tag{15.1}$$

The associated MO in the momentum space $\{\phi_i^\sigma(\boldsymbol{p})\}$ are the Fourier transforms of the corresponding orbitals in the position representation (e.g., Rawlings and Davidson 1958; Gadre and Balanarayan 2009), giving rise to the respective density contributions $\left\{\pi_i^\sigma(\boldsymbol{p}) = |\phi_i^\sigma(\boldsymbol{p})|^2\right\}$,

$$\pi(\boldsymbol{p}) = \sum\nolimits_\sigma \sum\nolimits_i \pi_i^\sigma(\boldsymbol{p}). \tag{15.2}$$

The transformation of electron densities from one space to another is computationally straightforward, since the analytical Fourier transforms of the basic functions used in molecular calculations are known (e.g., Kaijser and Smith 1997).

Another interesting way to analyze chemical processes around the TS invokes the concept of the *reaction force* of Politzer and collaborators (e.g., Toro-Labbé et al. 2009). In a *one*-step chemical reaction the reactants and products are typically separated by the energy barrier in the reaction energy profile along the MEP. The latter can be rigorously established by the classical trajectories of the lowest potential energy paths leading from the TS structure to reactants and products. When this path is formulated, it determines in the mass-weighted Cartesian coordinates the so-called *Intrinsic Reaction Coordinate* (IRC) $\boldsymbol{R}_c$ (Fukui 1981; Ganzález and Schlegel 1990), with the associated IRC cut of the BO potential $W(\boldsymbol{R}_c)$ determining the energy profile of the reaction under consideration.

In the endothermic direction, for the positive *reaction energy*, $\Delta W_r =$ W(products) $-$ W(reactants) > 0, the reaction force,

$$\boldsymbol{F}(\boldsymbol{R}_c) = -\partial W(\boldsymbol{R}_c)/\partial \boldsymbol{R}_c, \tag{15.3}$$

is negative in the entrance (reactant) valley of PES, before TS at $\boldsymbol{R}_\beta$, where it exactly vanishes, and becomes positive beyond this critical point, in the exit (product) valley of the BO potential. In the first, preparatory stage, from the separated reactants to the minimum of the (retarding) reaction force at $\boldsymbol{R}_\alpha$, the reactants undergo structural distortions such as bond-stretchings, angle-bending, rotations, etc. The negative $\boldsymbol{F}(\boldsymbol{R}_c)$ in thus indicative of the resistance of the system to such changes, since overcoming this retarding force requires an extra (activation) energy.

In the second reaction region, from the minimum force at $\boldsymbol{R}_\alpha$ (in entrance valley) to its maximum at $\boldsymbol{R}_\gamma$ (in the exit valley), most of the electronic/geometrical transformations toward the reaction products take place: new bonds begin to form, and these displacements in the system structure gradually overcome the retarding force to produce the positive (driving) force beyond TS, which reaches its maximum at $\boldsymbol{R}_\gamma$, the distorted state of products. The retarding and driving contributions to the reaction force exactly balance at TS, where the energy profile exhibits the maximum.

The system structures corresponding to the minimum and maximum of the reaction force thus provide a natural division of the RC into three parts: the entrance region (reactants $\rightarrow \alpha$) of the structural preparation and the mutual approach of reactants, the true reaction area around TS ($\alpha \rightarrow \gamma$), and the exit region ($\gamma \rightarrow$ products) of the structural relaxation and the mutual departure of products. Notice, that in the (reactants $\rightarrow \alpha$) zone the reaction force (15.3) is conducive for the mutual approach by the two reactants.

This division also allows one to express the transition energy between these two reference points on IRC as the sum of the retarding (positive) energy contribution,

$$\Delta W(\alpha \rightarrow \gamma) = \Delta W(\alpha \rightarrow \beta) + \Delta W(\beta \rightarrow \gamma) = -\int_\alpha^\beta \boldsymbol{F}(\boldsymbol{R}_c) \cdot d\boldsymbol{R}_c - \int_\beta^\gamma \boldsymbol{F}(\boldsymbol{R}_c) \cdot d\boldsymbol{R}_c, \tag{15.4}$$

with the final release of the energy due to the relaxation of the distorted products:

$$\Delta W(\gamma \rightarrow \text{products}) = -\int_\gamma^{\text{products}} \boldsymbol{F}(\boldsymbol{R}_c) \cdot d\boldsymbol{R}_c. \tag{15.5}$$

One can similarly partition the activation energy, $\Delta W_{act.} = \Delta W(\text{reactants} \rightarrow \beta)$ as composed of contributions corresponding to the distortion of reactants and the first zone of the reaction region:

$$\Delta W_{act.} = \Delta W(\text{reactants} \rightarrow \beta) = \Delta W(\text{reactants} \rightarrow \alpha) + \Delta W(\alpha \rightarrow \beta), \tag{15.6}$$

The reaction force concept also enters the virial partitioning of the reaction energy profile relative to the *separated*-reactants level, $\Delta W(\boldsymbol{R}_c) = W(\boldsymbol{R}_c) - W(\text{reactants}) = \Delta T(\boldsymbol{R}_c) + \Delta V(\boldsymbol{R}_c)$, into the associated (electronic) kinetic- and potential-energy components (Nalewajski 1980),

$$\Delta T(\boldsymbol{R}_c) = -\Delta E(\boldsymbol{R}_c) + \boldsymbol{F}(\boldsymbol{R}_c) \cdot \boldsymbol{R}_c \quad \text{and} \quad \Delta V(\boldsymbol{R}_c) = 2\Delta E(\boldsymbol{R}_c) - \boldsymbol{F}(\boldsymbol{R}_c) \cdot \boldsymbol{R}_c. \tag{15.7}$$

This division allows one to examine the influence of the activation energy (barrier height) and of the reaction energy on changes these two energy components undergo in transition states.

15.2 Information Probes of Elementary Reaction Mechanisms

Interesting new results in the IT studies of the elementary reaction mechanism have been recently obtained in the Granada group (Esquivel et al. 2009a, b; López-Rosa et al. 2010; López-Rosa 2010). Both the *global* (Shannon) and *local* (Fisher) information measures have been used in these recent investigations of the course of representative reactions of the radical abstraction of hydrogen (two-step mechanism),

$$H^{\bullet} + H_2 \rightarrow H_2 + H^{\bullet}, \tag{15.8}$$

which requires extra energy to proceed, and the nucleophilic-substitution process of the hydride exchange (S_N2, one-step mechanism):

$$H^- + CH_4 \rightarrow CH_4 + H^-. \tag{15.9}$$

The abstraction process proceeds by homolysis and is kinetically of the *first-*order (S_N1-like). It involves two steps: formation of new radicals, via the homolytic cleavage of the nonpolar, perfectly covalent bond in H_2 at absence of any electrophile or nucleophile, which could initiate the heterolytic pattern, and the subsequent recombination of a new radical with another radical species. The hydride exchange is an example of the kinetically second-order, the first-order in both the *incoming* (nucleophile) and *leaving* (nucleofuge) hydride groups. It proceeds via the familiar Walden-inversion TS in a single, concerted reaction step.

The central quantities of these IT analyses are the Shannon entropies in both the position (r) and momentum (p) spaces,

$$S_r = -\int \rho(\boldsymbol{r}) \ln \rho(\boldsymbol{r})\, d\boldsymbol{r}, \quad S_p = -\int \pi(\boldsymbol{p}) \ln \pi(\boldsymbol{p})\, d\boldsymbol{p}, \tag{15.10}$$

and the related (classical) Fisher information measures:

$$I_r = \int |\nabla\rho(\boldsymbol{r})|^2/\rho(\boldsymbol{r})\, d\boldsymbol{r}, \quad I_p = \int |\nabla\pi(\boldsymbol{p})|^2/\pi(\boldsymbol{p})\, d\boldsymbol{p}. \tag{15.11}$$

The reaction profiles of these entropy/information probes have uncovered the presence of additional features of the two reaction mechanisms by revealing the chemically important regions where the bond-forming and bond-breaking actually occur. These additional features cannot be directly identified from the energy profile alone and from the structure of the TS densities involved. Consistency of predictions resulting from the global (Shannon) (Esquivel et al. 2009a, b) and local (Fisher) (López-Rosa et al. 2010; López-Rosa 2010) information measures has additionally confirmed a more universal and unbiased character of these findings.

Indeed, either of the two complementary Shannon entropies for the model radical abstraction reaction displays a reacher structure than the associated energy profile, which only exhibits one maximum at the TS point on the reaction coordinate. The position entropy S_r displays a local maximum at this TS structure and two minima in its close proximity, whereas the momentum entropy S_p exhibits the global minimum at TS complex and two maxima at points slightly more distant from TS than the corresponding positions of the S_r minima.

It follows from these entropy curves that the approach of the hydrogen molecule by the incoming hydrogen in the proximity of TS first localizes ρ in preparation for the bond rupture, which also implies an associated increase in the kinetic energy (delocalization of π). This preparatory stage is identified by the local minima of S_r (maxima of S_p). Next, when the structure relaxes and the new bond is formed at TS, the position (momentum) densities become more delocalized (localized), which is indeed manifested by the corresponding maximum (minimum) of S_r (S_p). The bond-breaking process requires energy, as indeed witnessed by an earlier maximum of S_p in the entrance valley of PES, which is subsequently dissipated by relaxing the structure at TS. In other words, the reaction complex first gains the energy required for the bond dissociation, and then the position-space density gets localized to facilitate the bond cleavage, which in turn induces the energy/density relaxation toward the TS structure.

Therefore, the entropy representation of the reaction mechanism reveals the whole complexity of this transformation, while the associated MEP profile only localizes the transition state, missing the crucial transitory localization/delocalization and relaxational phenomena involved in this two-step process.

The corresponding S_r (S_p) plot for the hydride-exchange S_N2 process again exhibits the maximum (minimum) at TS, with two additional minima (maxima) in its vicinity, where the bond-breaking is supposed to occur. These additional, pre- and post-TS features are symmetrically placed in the entrance and exit valleys, relative to TS structure, but now they appear at roughly the same values of the *intrinsic reaction coordinate* (IRC) in both the *position*- and *momentum*-representations, in contrast to the two-stage abstraction mechanism, where in the *entrance* valley the p-space maximum of the Shannon entropy has preceded the associated minimum observed in the r-space. This simultaneous r-localization (p-delocalization) may be indicative of the single-step mechanism in which the approach of the nucleophile is perfectly synchronized with the concomitant departure of the nucleofuge, so that the bond-forming and bond-breaking occur in a concerted manner.

It should be observed that both displacements increase the system energy. First, as the nucleophile approaches, this energy is required to overcome the repulsion between reactants and create the *position*-localization (*momentum*-delocalization) facilitating the bond-weakening. As the reaction progresses forward, the energy continues to grow toward the maximum at TS, when the sufficient threshold of the new chemical bond has already been reached to start the structure relaxation inducing the reverse r-delocalization (p-localization) processes leading to the S_r (S_p) maximum (minimum) at TS. This synchronous transformation picture is indeed customarily associated with this particular reaction.

The same sequences of the chemical events are seen in the complementary Fisher-information analysis of the structural features of distributions in both spaces. For the hydrogen abstraction reaction one observes with the progress of IRC toward the TS that, relative to the separated reactants reference, both I_r and I_p at first decrease their values thus marking a lower average gradient content of the associated probability amplitudes (wave functions) in both spaces, i.e., a more regular/uniform distribution (less structure, "order"). The I_p profile is seen to exhibit a faster decay toward the local minimum preceding TS, where I_p reaches the maximum value. These more uniform momentum densities also correspond to the local maxima of the system chemical hardness, with TS marking the local minimum of the latter. The I_r monotonically decreases toward the minimum value at TS, thus missing the additional extrema observed in the S_r plot, which have been previously associated with the bond homolysis. Thus the I_r is not capable of describing the bond-breaking/forming processes, which is clearly uncovered by I_p.

The disconcerted manner of the elementary bond-forming and bond-breaking is directly seen in the corresponding bond-length plots: the breaking of the bond occurs first, and then the system stabilizes by forming the TS structure. Additional insight into the density reconstruction in this homolytic bond rupture comes from examining the corresponding plots of the system dipole moment, reflecting the charge distortion during the reaction progress. The observed behavior of these functions is opposite to I_p. Therefore, in regions of the minimum I_p the dipole moment reaches the maximum value and vice versa. As intuitively expected, the dipole moment identically vanishes at both the TS and separated reactants/products.

Finally, let us turn to the Fisher-information analysis of the hydride-exchange reaction (15.9) involving the heterolytic bond cleavage, with an exchange of charge between reactants. The corresponding I_r and I_p functions of the reaction coordinate now display similar behavior, both exhibiting maxima at TS, where the Shannon entropies have indicated a more delocalized position density and a relatively localized momentum density. One also observes two minima of I_r and I_p in the proximity of TS. These IRC values coincide with the bond-breaking/forming regions, and the change in the curvature of the bond-elongation curves for the incoming and outgoing nucleophile marks the start of an increase in the gradient content of the momentum density, toward the maximum value at TS structure.

15.3 Chemical Reactivity Indices

In Sect. 7.3.7 we have introduced the key electronic CS used as criteria in a qualitative approaches to predict reactivity trends. They can be defined in all alternative perspectives of Chap. 13, including the externally closed or open molecules or their fragments, in both the v-based EF perspective of the BO approximation and the complementary ρ-based approaches of the EP perspective of DFT. These generalized responses of molecular systems are capable of determining all relevant linear responses of molecular and reactive systems one encounters in real reactivity phenomena (e.g., Baekelandt et al. 1993; Gatti and Macchi 2011; Geerlings et al. 2003; Nalewajski 1993a, 1997a, 2002d, 2003a, 2006g, 2009a; Nalewajski and Korchowiec 1997; Nalewajski et al. 1996a). They also generate measures of their mutual interaction in these processes, e.g., the effective electronic–nuclear coupling in reactive systems (Chap. 14), thus accounting for the crucial dependencies between electron distributions and molecular geometry in all admissible reactivity scenarios. These derivative descriptors in the EF and EP representations for both the mutually/externally open and closed reactants have already been systematically surveyed in Chaps. 7, 13, and 14.

Several additional reactivity criteria, which combine derivatives from different representations, have also been designed to correlate specific reactivity phenomena. The most popular among them is the *electrophilicity* index of Parr et al. (1999). It offers a modern (absolute) scale of the classical concepts of electrophilicity/nucleophilicity of Ingold (1933, 1934), who proposed the first global electrophilicity scale based upon the valence-electron theory of Lewis, to describe both the electron-deficient (electrophilic, acidic) and electron-rich (nucleophilic, basic) chemical species, and the related experimental scales of Mayr et al. (Mayr and Patz 1994; Roth and Mayr 1995) providing measures of a relative reactivity of an electrophile/nucleophile.

The absolute scale (Parr et al. 1999; Chattaraj et al. 2006, 2007; Liu 2009) quantifies the negative CT energy, when the system M with chemical potential $\mu_{\rm M} = \mu$ and hardness $\eta_{\rm M} = \eta_{\rm M,M} = \eta$ is brought into contact with the *perfect* (macroscopic) *donor* represented by the infinitely soft reservoir $\mathfrak{r}$ exhibiting the vanishing chemical potential, $\mu_{\mathfrak{r}} = 0$, and zero hardness, $\eta_{\mathfrak{r}} = \eta_{\mathfrak{r},\mathfrak{r}} = 0$. For such a hypothetical scenario, when the hardness coupling between the molecule and its reservoir identically vanishes, $\eta_{\mathfrak{r},\rm M} = \eta_{\rm M,\mathfrak{r}} = 0$, one obtains from (7.191)–(7.193): $\mu_{\rm CT} = \mu$, $\eta_{\rm CT} = \eta$, $N_{\rm CT} = -\mu/\eta$, and hence the electrophilicity index [see (7.195)]:

$$\omega = -\Delta^{(1+2)}E(N_{\rm CT}) = \mu^2/(2\eta) > 0. \tag{15.12}$$

By analogy to the electric power V^2/R in classical electrostatics, where V stands for the electric potential and R denotes the resistance, this electrophile descriptor can be also viewed as a measure of the *electrophilicity power*, when one adopts the electrophilic "potential" $V_E = \mu$ and the associated "resistance" measure $R_E = 2\eta$.

This electrophilicity index thus reflects a property of the molecule being electrophilic. It provides an absolute measure of an overall reactivity index of an *electrophile*, the (electron deficient) reagent attracted to electrons on its partner, that participate in chemical reactions by *accepting* electrons (Lewis' acid A) to form a bond with the *nucleophile*. The latter represents the (electron-rich) chemical species attracted to the unshielded nuclei of the electrophile, thus acting as the Lewis base B in the DA pair. Most electrophiles (acidic reactants) are positively charged, exhibit a positively charged atom, or their valence shells are short of the octet of electrons.

Accordingly, the nucleophiles (basic agents) usually exhibit negative charges, the negatively charged AIM, or possess an accessible free pair of valence electrons.

In a similar way one can define the electrophilicity concept of the electrophile A relative to nucleophilic partner B, when (see Sect. 7.3.4) $\mu_{CT}(A, B) \approx \mu_A^0 - \mu_B^0, \eta_{CT}(A, B) = \eta_A + \eta_B - 2\eta_{A,B} \approx \eta_A^0 + \eta_B^0$, and $N_{CT}(A, B) = -\mu_{CT}(A, B)/\eta_{CT}(A, B)$. It then directly follows from Eq. (7.195):

$$\begin{aligned}\omega(A, B) &= -\Delta^{(1+2)}E[N_{CT}(A, B)] = [\mu_{CT}(A, B)]^2/[2\eta_{CT}(A, B)] \\ &\approx (\mu_A^0 - \mu_B^0)^2/[2(\eta_A^0 + \eta_B^0)].\end{aligned} \tag{15.13}$$

The electrophilicity index shows a strong statistical significance in correlations with both the experimental measures of this quantity and the electroaffinity data (Liu 2009). In chemistry textbooks the electrophilicity comes together with its companion index of the *nucleophilicity*. It has to be defined (e.g., Jaramillo et al. 2006; Liu and Parr 1997) in the partly inverted, compliant representation (μ, v) (see Sect. 7.3.7) in which μ replaces N as an independent state-parameter. For the fixed external potential the second-order expansion of the system *grand*-potential (7.226) then gives:

$$\Delta^{(1+2)}\Omega[\mu, v] = -N\Delta\mu - \frac{1}{2}S(\Delta\mu)^2, \tag{15.14}$$

where $S = 1/\eta$ denotes the system global softness. By minimizing this quadratic function with respect to $\Delta\mu$ one finds

$$\Delta\mu = -N/S \quad \text{and} \quad \Delta^{(1+2)}\Omega = N^2/(2S).$$

Hence the absolute (positive) nucleophilicity index reads:

$$\omega^- = \Delta^{(1+2)}\Omega(\Delta\mu) = N^2/(2S) > 0. \tag{15.15}$$

This *nucleophilicity power* thus corresponds to the nucleophilic (population) "potential" $V_N = -N$ and the corresponding "resistance" $R_N = 2S$. It should be observed that substituting into ω^- the optimum CT relative to the infinitely soft reservoir of the zero chemical potential, $N = N_{CT} = -\mu/\eta$, recovers the electrophilicity index of (15.12):

$$\omega^-(N_{CT}) = (N_{CT})^2/(2S) = \mu^2/(2\eta) = \omega. \tag{15.16}$$

The partner-dependent, relative measure $\omega^-(A, B)$, a companion index of $\omega(A, B)$, has been similarly defined (Jaramillo et al. 2006) in terms of a change in the grand-potential of (15.14) accompanying the internal CT B $\rightarrow$ A,

$$N_{\mathrm{CT}}(\mathrm{A},\mathrm{B}) = -\mu_{\mathrm{CT}}(\mathrm{A},\mathrm{B})/\eta_{\mathrm{CT}}(\mathrm{A},\mathrm{B}) \approx -(\mu_{\mathrm{A}}^{0} - \mu_{\mathrm{B}}^{0})/(\eta_{\mathrm{A}}^{0} + \eta_{\mathrm{B}}^{0}),$$

or equivalently, the associated change in the relative chemical potential of the two reactants:

$$\Delta\mu = \mu_{\mathrm{CT}}(\mathrm{A},\mathrm{B}) = -N_{\mathrm{CT}}(\mathrm{A},\mathrm{B})/S_{\mathrm{CT}}(\mathrm{A},\mathrm{B}),$$

$$\begin{aligned}\omega^{-}(\mathrm{A},\mathrm{B}) &= \Delta^{(1+2)}\Omega[\mu_{\mathrm{CT}}(\mathrm{A},\mathrm{B})] = [N_{\mathrm{CT}}(\mathrm{A},\mathrm{B})]^{2}/[2S_{\mathrm{CT}}(\mathrm{A},\mathrm{B})]\\ &\cong (\mu_{\mathrm{A}}^{0} - \mu_{\mathrm{B}}^{0})^{2}\eta_{\mathrm{CT}}(\mathrm{A},\mathrm{B})/[2(\eta_{\mathrm{A}}^{0} + \eta_{\mathrm{B}}^{0})^{2}]\\ &\approx (\mu_{\mathrm{A}}^{0} - \mu_{\mathrm{B}}^{0})^{2}\eta(\mathrm{A})/[2(\eta_{\mathrm{A}}^{0} + \eta_{\mathrm{B}}^{0})^{2}].\end{aligned} \quad (15.17)$$

Several alternative measures of the nucleophilicity have also been proposed (e.g., Chattaraj and Maiti 2001; Cedillo et al. 2007) together with some local extensions of both electro- and nucleophilicity concepts (Chamorro et al. 2003; Chattaraj et al. 2001, 2003, 2005; Perez et al. 2002; Roy et al. 2005). For example, to describe the electrophilic character of a reactive site at $\boldsymbol{r}$ within a molecule, the local electrophilicity index has been introduced:

$$\omega(\boldsymbol{r}) = \omega f^{+}(\boldsymbol{r}); \quad (15.18)$$

here $f^{+}(\boldsymbol{r})$ is the electronic FF for the nucleophilic attack, which is defined by the difference between the ground-state densities of the anion (ρ_{N+1}) and neutral species $\rho_N(\boldsymbol{r})$ approximated by the appropriate FE density of the latter:

$$f^{+}(\boldsymbol{r}) = \rho_{N+1}(\boldsymbol{r}) - \rho_{N}(\boldsymbol{r}) \approx \rho_{N}^{\mathrm{LUMO}}(\boldsymbol{r}). \quad (15.19)$$

The corresponding local nucleophilicity index would similarly involve the FF for the electrophilic attack, approximated by the other FE density:

$$f^{-}(\boldsymbol{r}) = \rho_{N}(\boldsymbol{r}) - \rho_{N-1}(\boldsymbol{r}) \approx \rho_{N}^{\mathrm{HOMO}}(\boldsymbol{r}), \quad (15.20)$$

$$\omega^{-}(\boldsymbol{r}) = \omega^{-}f^{-}(\boldsymbol{r}). \quad (15.21)$$

There are several alternative combined reactivity descriptors, which have been proposed in the chemical literature, such as electro- and nucleofugality (Ayers et al. 2005a, b) *leaving-group* indicators, *philicity* indices, etc., designed for the key reactivity phenomena. Such "derived" reactivity indicators are particularly important for the *Quantitative Structure–Property Relationships* (QSPR) by capturing in a single index the phenomenon in question.

15.4 Internal and External Eigenvalue Problems

Consider the finite, n-dimensional matrix representation $\mathbf{H} = \langle \boldsymbol{\chi} | \hat{\mathrm{H}} | \boldsymbol{\chi} \rangle$ of the physical quantity H corresponding to the quantum-mechanical (Hermitian) operator $\hat{\mathrm{H}}$, defined by the orthonormal basis vectors $|\boldsymbol{\chi}\rangle = \{|1\rangle, \ldots, |n\rangle\}$ in the molecular Hilbert space. In several computational and/or interpretative applications of the quantum theory one partitions this set into two or several complementary subsets, say, $|\boldsymbol{\chi}\rangle = |\boldsymbol{\chi}^A, \boldsymbol{\chi}^B\rangle = \{|\boldsymbol{\chi}^X\rangle\}$, of dimensions a and b, respectively, $n = a + b$. For example, these subspaces may represent AO originating from atoms $X = (\mathrm{A}, \mathrm{B})$ of a diatomic molecule M = A—B, or they may constitute some arbitrary basis functions of the complementary molecular fragments $X = (A, B)$ of the combined system, e.g., reactants in the bimolecular system $R = A$----B. Such a division of the representation basis set uniquely identifies the associated blocks of $\mathbf{H} = \{\mathbf{H}^{X,Y} = \langle \boldsymbol{\chi}^X | \hat{\mathrm{H}} | \boldsymbol{\chi}^Y \rangle\}$.

This partition, be it in a different CI scenario, is also invoked in the context of Löwdin's (1951, 1962, 1963) *partitioning technique* for solving the eigenvalue equation of the system Hamiltonian, originating from the variation of the Rayleigh–Ritz functional for the expectation value of the system energy expressed in terms of the CI coefficients, when the two subsets of electronic configurations determine the complementary components of the system wave-function. Another subject, in which one encounters such a separation, involves alternative decoupling schemes of the atomically resolved hardness tensor (Nalewajski and Korchowiec 1997; Nalewajski and Michalak 1996, 1998; Nalewajski et al. 1994b, 1996a) of the N-atomic reactive system consisting of reactants A and B, $\boldsymbol{\eta} = \{\boldsymbol{\eta}^{X,Y} = \{\eta_{x,y}, x \in X, y \in Y\}, (X, Y) = A, B\}$, used in a search for the most compact representation of the CT phenomena in the externally open or closed molecular systems.

The diagonalization of $\boldsymbol{\eta}$, determining its principal-axes representation, amounts to the matrix total decoupling and leads to the independent (collective) channels for electron displacements in the system as a whole, called the *Populational Normal Modes* (PNM) (Nalewajski 1988, 1993a, b, 1995a, b, 1997a, b; Nalewajski and Korchowiec 1997; Nalewajski et al. 1996a). The reactant resolution naturally divides the atomic population "modes" into groups corresponding to constituent atoms of each subsystem. Therefore, the molecular hardness matrix can be partitioned into its *internal* (i) and *external* (e) parts, which respectively combine only the intrafragment (diagonal) or the interfragment (*off*-diagonal) blocks of the hardness matrix elements between the system constituent AIM:

$$\boldsymbol{\eta} = \boldsymbol{\eta}^i + \boldsymbol{\eta}^e, \quad \boldsymbol{\eta}^i = \{\boldsymbol{\eta}^{X,X}\delta_{X,Y}\}, \quad \boldsymbol{\eta}^e = \{\boldsymbol{\eta}^{X,Y}(1 - \delta_{X,Y})\}. \tag{15.22}$$

The interfragment decoupled, internal part $\boldsymbol{\eta}^i$ of $\boldsymbol{\eta}$ can be viewed as representing the "unperturbed" (*zeroth*-order) "*promolecular*" system $R^0 = (A|B)$, consisting of the separate molecular fragments (subsets of bonded atoms), before formation of the interfragment chemical bonds in R, $\boldsymbol{\eta}^i \equiv \boldsymbol{\eta}^0$. It can be regarded as the reference for the displacement aspect of the hardness matrix in the *molecular* system "perturbed" by $\boldsymbol{\eta}^e$: $\boldsymbol{\eta} = \boldsymbol{\eta}^i + \boldsymbol{\eta}^e \equiv \boldsymbol{\eta}^0 + \Delta\boldsymbol{\eta}$.

The eigensolutions of $\boldsymbol{\eta}^i$, called *Internal Normal Modes* (INM), then constitute the decoupled channels for the charge redistribution in the separate fragments of the system "promolecule." In the chemical reactivity applications one similarly introduces the *Inter-Subsystem Modes* (ISM), also called the *Inter-Reactant Modes* (IRM), representing the eigensolutions of $\boldsymbol{\eta}^e$ (Nalewajski and Korchowiec 1997; Nalewajski et al. 1996a). These *externally* decoupled channels have been found to generate an attractive framework for the most compact description of the charge flows accompanying the formation of chemical bond(s) (Nalewajski and Korchowiec 1997; Nalewajski and Michalak 1996, 1998; Nalewajski et al. 1994b, 1996a).

To summarize, these two partial eigenvalue problems focus on the complementary aspects of the charge displacements in a transition from the initial (promolecular) to final (molecular) system: the internal modes characterize the independent channels of the charge polarization inside each of the mutually uncoupled (separate, nonbonded) fragments of the promolecule, while the external problem describes the CT processes relative to this initial reference state. In other words, the external (interaction) channels focus solely on the interfragment bonding, i.e., the *displacement* aspect of the molecular electronic structure, while the internal channels deal with the intrafragment bonds, already present in the promolecule, and as such they only characterize the *initial* (polarization) stage of the bond-formation process in the reaction complex as a whole.

One should also mention at this point the partial transformations giving rise to the internal and external decouplings of the whole hardness tensor (Nalewajski and Korchowiec 1997; Nalewajski et al. 1994b, 1996a), rather than of its internal and external parts, which together amount to its total decoupling:

$$\mathbf{C}^\dagger\boldsymbol{\eta}\mathbf{C} = \mathbf{c}(\text{diagonal}) = \{\mathbf{c}^X\delta_{X,Y}\}, \quad \mathbf{C}^\dagger\mathbf{C} = \mathbf{I}. \tag{15.23}$$

Clearly, the *internal decoupling* transformation of $\boldsymbol{\eta}$, which leads to the diagonalization of only the intrafragment blocks of $\boldsymbol{\eta}$, must be identical with that diagonalizing $\boldsymbol{\eta}^i$. By definition, the *external decoupling* of the whole hardness tensor leads to the vanishing *off*-diagonal (coupling) blocks in the transformed tensor. However, since this requirement does not specify the transformation uniquely, the additional *Maximum Overlap Criterion* (MOC) has been used to generate the localized externally decoupled modes in this scheme.

Yet another example of the complementary internal/external perspectives on the charge reorganization in molecules and molecular complexes involves the recently proposed *Natural Orbitals for Chemical Valence* (NOCV) (Mitoraj 2007; Mitoraj and Michalak 2005, 2007; Mitoraj et al. 2006, 2007), which have been successfully applied to interpret the metal–ligand bonds of coordination chemistry. They represent the eigenfunctions of the *chemical-valence* operator (Nalewajski et al. 1997), $\hat{\mathrm{V}} = \hat{\mathrm{P}} - \hat{\mathrm{P}}^0$, defined by the difference between the molecular ($\hat{\mathrm{P}}$) and promolecular

$(\hat{P}^0)$ projections on their respective occupied orbitals, with the occupied AO/MO of the separate fragments now determining the relevant "promolecular" reference. These projection operators respectively identify the occupied subspaces of SO of the system as a whole, and of the AO/MO of the separate fragments of the promolecule. Their AO representations in the LCAO MO theory thus define the associated density (CBO) matrices $\mathbf{P} = \langle \boldsymbol{\chi}|\hat{P}|\boldsymbol{\chi}\rangle$ and $\mathbf{P}^0 = \langle \boldsymbol{\chi}|\hat{P}^0|\boldsymbol{\chi}\rangle$. Hence the overall displacement of **P** due to formation of the interfragment bonds:

$$\Delta\mathbf{P} = \mathbf{P} - \mathbf{P}^0 = \langle \boldsymbol{\chi}|\hat{V}|\boldsymbol{\chi}\rangle.$$

The subsets of basic functions originating from the separate fragments A and B, $|\boldsymbol{\chi}\rangle = |\boldsymbol{\chi}^A, \boldsymbol{\chi}^B\rangle$, then partition these matrices into the corresponding subsystem-resolved blocks:

$$\begin{aligned}
\mathbf{P} &= \{\mathbf{P}^{X,Y}\} = \mathbf{P}^0 + \Delta\mathbf{P},\\
\mathbf{P}^0 &\cong \{\mathbf{P}^{X,X}\delta_{X,Y}\} \equiv \mathbf{P}^i \quad \text{and}\\
\Delta\mathbf{P} &= \{\Delta\mathbf{P}^{X,Y}\} = \{\Delta\mathbf{P}^{X,Y}(1-\delta_{X,Y}) = \mathbf{P}^{X,Y}(1-\delta_{X,Y})\} \equiv \mathbf{P}^e.
\end{aligned}$$

Therefore, the AO representation of the chemical-valence operator is proportional to the external part of the CBO matrix, so that its eigensolutions (NOCV) represent the *bond-order* ISM. They have been shown to generate an attractive framework for describing the interorbital flows of electrons, which accompany the formation of chemical bonds between A and B, capable of the precise separation of the *forward-* and *back*-donations in the coordination bonds (Mitoraj 2007; Mitoraj and Michalak 2005, 2007; Mitoraj et al. 2006, 2007).

Consider again a general n-dimensional representation of the quantum-mechanical (Hermitian) operator $\hat{H}$ attributed to the physical quantity H, $\mathbf{H} = \langle \boldsymbol{\chi}|\hat{H}|\boldsymbol{\chi}\rangle = \{H_{i,j} = \langle i|\hat{H}|j\rangle\}$, e.g., the Hamiltonian operator $\hat{H}(N)$ of an N − electron system corresponding to electronic energy E. To simplify the following analysis, we assume the orthonormal metric of the basis set $|\boldsymbol{\chi}\rangle = \{|i\rangle\}$, $\langle i|j\rangle = \delta_{i,j}$, e.g., that characterizing the orthogonalized AO, the MO configuration functions in the CI expansion of the system wave-function, or the underlying independent atomic modes in CSA (Nalewajski 1988, 1993a, b, 1995a, b, 1997a, b; Nalewajski and Korchowiec 1997; Nalewajski et al. 1996a), in terms of which various hardness decoupling schemes have been formulated.

The overall diagonalization problem $\mathbf{C}^\dagger\mathbf{H}\mathbf{C} = \mathbf{E}(\text{diagonal}) = \{E_\alpha\delta_{\alpha,\beta}\}$, where the columns of the (unitary) transformation matrix

$$\mathbf{C} = \left\{ \boldsymbol{C}_\alpha^{\mathrm{T}} = \langle \boldsymbol{\chi} \mid \alpha\rangle = \begin{bmatrix} \langle \boldsymbol{\chi}^A \mid \alpha\rangle = \boldsymbol{C}_\alpha^{A,\mathrm{T}} \\ \langle \boldsymbol{\chi}^B \mid \alpha\rangle = \boldsymbol{C}_\alpha^{B,\mathrm{T}} \end{bmatrix} \right\}, \quad \mathbf{C}^\dagger\mathbf{C} = \mathbf{I}, \tag{15.24}$$

group the basis set projections for αth eigensolution corresponding to E_α, gives the following eigenvalue equation for this specific solution, written in the block form:

$$\begin{bmatrix} \mathbf{H}^{A,A} & \mathbf{H}^{A,B} \\ \mathbf{H}^{B,A} & \mathbf{H}^{B,B} \end{bmatrix} \begin{bmatrix} \boldsymbol{C}_\alpha^{A,\mathrm{T}} \\ \boldsymbol{C}_\alpha^{B,\mathrm{T}} \end{bmatrix} = E_\alpha \begin{bmatrix} \boldsymbol{C}_\alpha^{A,\mathrm{T}} \\ \boldsymbol{C}_\alpha^{B,\mathrm{T}} \end{bmatrix} \quad \text{or}$$
$$\mathbf{H}^{A,A}\boldsymbol{C}_\alpha^{A,\mathrm{T}} + \mathbf{H}^{A,B}\boldsymbol{C}_\alpha^{B,\mathrm{T}} = E_\alpha \boldsymbol{C}_\alpha^{A,\mathrm{T}}, \quad \mathbf{H}^{B,A}\boldsymbol{C}_\alpha^{A,\mathrm{T}} + \mathbf{H}^{B,B}\boldsymbol{C}_\alpha^{B,\mathrm{T}} = E_\alpha \boldsymbol{C}_\alpha^{B,\mathrm{T}}. \tag{15.25}$$

It forms the basis of the Löwdin *partitioning technique*. By using one of these equations to express one component of the eigenvector in terms of the other, e.g.,

$$\boldsymbol{C}_\alpha^{B,\mathrm{T}} = (E_\alpha \mathbf{I}^B - \mathbf{H}^{B,B})^{-1}\mathbf{H}^{B,A}\boldsymbol{C}_\alpha^{A,\mathrm{T}}, \tag{15.26}$$

and substituting the result into the other (15.25) then gives the effective eigenvalue problem for the chosen independent component:

$$[\mathbf{H}^{A,A} + \mathbf{H}^{A,B}(E_\alpha \mathbf{I}^B - \mathbf{H}^{B,B})^{-1}\mathbf{H}^{B,A}]\boldsymbol{C}_\alpha^{A,\mathrm{T}} = E_\alpha \boldsymbol{C}_\alpha^{A,\mathrm{T}}. \tag{15.27}$$

This technique has been designed to determine the eigensolutions of (15.25) and as such it is widely applied in contemporary quantum chemistry (McWeeny 1989). For the analytical purposes, however, the reverse procedure can be also applied, e.g., of using the known solutions of (15.25) to infer the properties of matrices (Bochevarov and Sherrill 2007).

This partitioning also provides a framework for discussing the solutions of the partial internal and external eigenvalue problems in the subsystem resolution (Nalewajski and Korchowiec 1997; Nalewajski and Michalak 1996, 1998; Nalewajski et al. 1994b, 1996a):

$$\begin{aligned} &\mathbf{O}^\dagger \mathbf{H}^i \mathbf{O} = \mathbf{h} = \{h_m \delta_{m,n}\} = \{\mathbf{h}_X \delta_{X,Y}\}, \quad \mathbf{O}^\dagger \mathbf{O} = \mathbf{I}, \\ &\quad \mathbf{H}^i = \{\mathbf{H}^{X,X}\delta_{X,Y}\}, \quad \mathbf{O} = \{\mathbf{O}^X \delta_{X,Y}\}, \quad \mathbf{O}^X = \{\boldsymbol{O}_m^{X,\mathrm{T}} = \langle \boldsymbol{\chi}^X \mid m \rangle\}, \end{aligned} \tag{15.28}$$

$$\begin{aligned} &\mathbf{U}^\dagger \mathbf{H}^e \mathbf{U} = \varepsilon = \{\varepsilon_k \delta_{k,l}\}, \\ &\quad \mathbf{H}^e = \{\mathbf{H}^{X,Y}(1-\delta_{X,Y})\}, \quad \mathbf{U} = \{\boldsymbol{U}_k^{\mathrm{T}}\}, \quad \mathbf{U}^\dagger \mathbf{U} = \mathbf{I}. \end{aligned} \tag{15.29}$$

The former determines the decoupled eigensolutions of the two diagonal blocks in **H**,

$$\mathbf{H}^{A,A}\boldsymbol{O}_m^{A,\mathrm{T}} = h_m^A \boldsymbol{O}_m^{A,\mathrm{T}} \quad \text{and} \quad \mathbf{H}^{B,B}\boldsymbol{O}_n^{B,\mathrm{T}} = h_n^B \boldsymbol{O}_n^{B,\mathrm{T}}, \tag{15.30}$$

while the solutions of the latter satisfy the coupled equations:

$$\begin{bmatrix} \mathbf{0}^{A,A} & \mathbf{H}^{A,B} \\ \mathbf{H}^{B,A} & \mathbf{0}^{B,B} \end{bmatrix} \begin{bmatrix} \boldsymbol{U}_k^{A,\mathrm{T}} \\ \boldsymbol{U}_k^{B,\mathrm{T}} \end{bmatrix} = \varepsilon_k \begin{bmatrix} \boldsymbol{U}_k^{A,\mathrm{T}} \\ \boldsymbol{U}_k^{B,\mathrm{T}} \end{bmatrix} \quad \text{or}$$
$$\mathbf{H}^{A,B}\boldsymbol{U}_k^{B,\mathrm{T}} = \varepsilon_k \boldsymbol{U}_k^{A,\mathrm{T}} \quad \text{and} \quad \mathbf{H}^{B,A}\boldsymbol{U}_k^{A,\mathrm{T}} = \varepsilon_k \boldsymbol{U}_k^{B,\mathrm{T}}. \tag{15.31}$$

A straightforward elimination of one component in the two preceding equations,

$$U_k^{A,\mathrm{T}} = \varepsilon_k^{-1}\mathbf{H}^{A,B}U_k^{B,\mathrm{T}}, \quad U_k^{B,\mathrm{T}} = \varepsilon_k^{-1}\mathbf{H}^{B,A}U_k^{A,\mathrm{T}}, \tag{15.32}$$

then gives the following effective external eigenvalue problems for each subsystem,

$$\begin{aligned} \mathbf{H}^{A,B}\mathbf{H}^{B,A}U_k^{A,\mathrm{T}} &\equiv \mathbf{F}_A^{eff}U_k^{A,\mathrm{T}} = \varepsilon_k^2 U_k^{A,\mathrm{T}} \quad \text{and} \\ \mathbf{H}^{B,A}\mathbf{H}^{A,B}U_k^{B,\mathrm{T}} &\equiv \mathbf{F}_B^{eff}U_k^{B,\mathrm{T}} = \varepsilon_k^2 U_k^{B,\mathrm{T}}. \end{aligned} \tag{15.33}$$

These vectors determine both components of kth eigenvector in

$$\mathbf{U} = \left\{ U_k^{\mathrm{T}} = \langle \boldsymbol{\chi} \mid k \rangle = \begin{bmatrix} \langle \boldsymbol{\chi}^A \mid k \rangle = U_k^{A,\mathrm{T}} \\ \langle \boldsymbol{\chi}^B \mid k \rangle = U_k^{B,\mathrm{T}} \end{bmatrix} \right\}, \tag{15.34}$$

with the symmetrical effective external operators of molecular fragments defined by the respective diagonal blocks of the squared external part of $\mathbf{H}$, $(\mathbf{H}^{\mathrm{e}})^2$:

$$(\mathbf{F}_X^{eff})^{\mathrm{T}} = \mathbf{F}_X^{eff} = (\mathbf{H}^e)^2_{X,X}, \quad X = A, B.$$

Multiplying from the left (15.31) by $\mathbf{H}^{B,A}$ and $\mathbf{H}^{B,A}$, respectively, also gives

$$\mathbf{F}_B^{eff}U_k^{B,\mathrm{T}} = \varepsilon_k \mathbf{H}^{B,A}U_k^{A,\mathrm{T}}, \qquad \mathbf{F}_A^{eff}U_k^{A,\mathrm{T}} = \varepsilon_k \mathbf{H}^{A,B}U_k^{B,\mathrm{T}}, \tag{15.35}$$

and hence the alternative expressions for one component in terms of the other:

$$U_k^{A,\mathrm{T}} = \varepsilon_k (\mathbf{F}_A^{eff})^{-1}\mathbf{H}^{A,B}U_k^{B,\mathrm{T}}, \quad U_k^{B,\mathrm{T}} = \varepsilon_k (\mathbf{F}_B^{eff})^{-1}\mathbf{H}^{B,A}U_k^{A,\mathrm{T}}. \tag{15.36}$$

It follows from (15.33) that the two subsystem components (U_k^A, U_k^B) of the external mode U_k are the eigenvectors of the effective fragment-operators $F_A^{eff}(a \times a)$ and $F_B^{eff}(b \times b)$, where for definiteness we assume $a \leq b$, which exhibit the same squared external eigenvalue ε_k^2. Together these operators define the block-diagonal (externally decoupled) effective operator for the system as a whole,

$$\mathbf{F}^{eff} = \{\mathbf{F}_{eff}^X \delta_{X,Y}\} = (\mathbf{H}^e)^2,$$

the eigenvalue equation of which reads:

$$\mathbf{F}^{eff}U_k^{\mathrm{T}} = (\mathbf{H}^e)^2 U_k^{\mathrm{T}} = \varepsilon_k^2 U_k^{\mathrm{T}} \quad \text{or} \quad \mathbf{U}^\dagger(\mathbf{H}^e)^2\mathbf{U} = \boldsymbol{\varepsilon}^2 = \{\varepsilon_k^2 \delta_{k,l}\}. \tag{15.37}$$

Therefore, the external modes $\mathbf{U}$ can be determined via the diagonalization of the already intersubsystem decoupled operator $(\mathbf{H}^{\mathrm{e}})^2$, which amounts to the separate subsystem diagonalization (eigenvalue) problems of (15.33). In the *delocalized* modes, for which $\varepsilon_k^2 > 0$, the a eigenvectors $\{U_k^A\}$ of $\mathbf{F}_A^{eff}$ will combine with their b conjugates $\{U_k^B\}$ among the eigenvectors of $\mathbf{F}_B^{eff}$, which exhibit the same squared external eigenvalue ε_k^2, whereas the remaining $b - a \geq 0$ eigenvectors of $\mathbf{F}_B^{eff}$ will combine with the zero component in a, thus giving rise to the *B-localized* (l) channels $U_l = (0_l^A, U_l^B)$, which are inactive in the displacement processes induced by the perturbation $\mathbf{H}^e = \Delta\mathbf{H} \equiv \mathbf{H} - \mathbf{H}^i$. It can be verified using (15.29) that the external eigenvalues of such localized external modes identically vanish:

$$\varepsilon_l = U_l^* \mathbf{H}^e U_l^{\mathrm{T}} = 0.$$

In the hardness-decoupling scenario of the reactive system R = A- - - -B the B-localized solutions can only internally polarize B, playing no part in the CT between the two reactants. Similarly, in the NOCV eigenvalue problem for $\Delta\mathbf{P}$ such natural orbitals do not participate in the interfragment redistribution of electrons, which accompanies the bond-formation process. This general structure of $\mathbf{U}$ is summarized in Fig. 15.1.

Pairs of the mutually orthogonal, $\mathbf{F}^e$-degenerate modes (U_k, U_{-k}), $k = 1, 2, .., A$, for the complementary eigenvalues ($\varepsilon_k > 0$, $-\varepsilon_k < 0$) of $\mathbf{H}^e$, respectively,

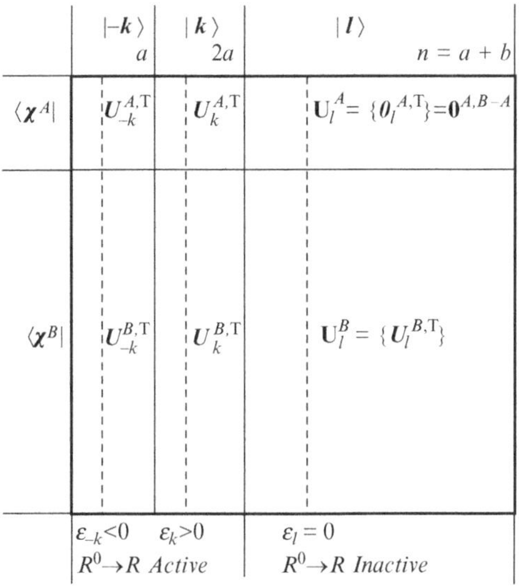

Fig. 15.1 The block structure of the eigenvector matrix $\mathbf{U} = \langle\boldsymbol{\chi}|\boldsymbol{\xi}\rangle$ determining the *external* modes $|\boldsymbol{\xi}\rangle = |\boldsymbol{\chi}\rangle\mathbf{U} = (|-\boldsymbol{k}\rangle, |\boldsymbol{k}\rangle, |\boldsymbol{l}\rangle)$, including the *delocalized* vectors $|-\boldsymbol{k}\rangle$ and $|\boldsymbol{k}\rangle$, determined by the complementary columns of expansion coefficients $(U_{-k}^{\mathrm{T}}, U_k^{\mathrm{T}})$, $k = 1, 2, \ldots, a$, for the negative and positive eigenvalues, respectively, which are active in the $R^0 \to R$ transition, and the *B-localized* vectors $|\boldsymbol{l}\rangle$, which remain inactive in the chemical change induced by the intersubsystem coupling

$$\mathbf{H}^e \boldsymbol{U}_k^{\mathrm{T}} = \varepsilon_k \boldsymbol{U}_k^{\mathrm{T}} \quad \text{and} \quad \mathbf{H}^e \boldsymbol{U}_{-k}^{\mathrm{T}} = -\varepsilon_k \boldsymbol{U}_{-k}^{\mathrm{T}} \equiv \varepsilon_{-k} \boldsymbol{U}_{-k}^{\mathrm{T}}, \tag{15.38}$$

are represented by the symmetric ("bonding") and antisymmetric ("antibonding") combinations of the subsystem components $\boldsymbol{U}_k^A$ and $\boldsymbol{U}_k^B$, which constitute the independent, mutually orthogonal eigensolutions of $\mathbf{F}^{eff} = (\mathbf{H}^e)^2$ corresponding to the same squared external eigenvalue ε_k^2:

$$\boldsymbol{U}_k = (\boldsymbol{U}_k^A, \boldsymbol{U}_k^B), \quad \boldsymbol{U}_{-k} = (\boldsymbol{U}_k^A, -\boldsymbol{U}_k^B) \equiv (\boldsymbol{U}_{-k}^A, \boldsymbol{U}_{-k}^B). \tag{15.39}$$

The external modes (eigenvectors of $\mathbf{H}^e$) active in the transition $R^0 \rightarrow R$ exhibit intermediate degrees of the interfragment delocalization between the *localized* internal modes of both fragments [eigenvectors of $\mathbf{H}^i$ (15.28) and (15.30)] and the fully delocalized molecular modes [eigenvectors of $\mathbf{H} = \mathbf{H}^i + \mathbf{H}^e$ (15.25)]. For example, the IRM in illustrative chemisorption complexes have been found to be more localized, compared to PNM of the whole reactive system (Nalewajski and Korchowiec 1997; Nalewajski and Michalak 1996, 1998; Nalewajski et al. 1996a). Similarly, the NOCV exhibit a relative localized character compared to less polarized canonical MO (Mitoraj 2007; Mitoraj and Michalak 2005, 2007; Mitoraj et al. 2006, 2007).

The interpretative advantage of the external modes of the Hermitian operators, however, lies in their applications to the interaction/coupling between a small subsystem A and its much larger molecular environment B in R, when $b >> a$. Then, only a small number of $2a$ external modes (see Fig. 15.1) is *active* in the promolecule → molecule transition, with a large number of the remaining $n - 2a$ *inactive* (B-localized) modes describing the response of the environment to this chemical displacement. Therefore, the external eigenvectors give a much more compact description of this bond-formation processes between a small and large reactants, compared to the eigenvectors of the whole matrix (Nalewajski and Korchowiec 1997; Nalewajski and Michalak 1996, 1998; Nalewajski et al. 1996a).

15.5 Complementary Decoupling Schemes of Molecular Hardness Tensor

Yet another issue in CSA, which deals with the internal and external decoupling of the symmetric hardness tensor $\boldsymbol{\eta} = \boldsymbol{\eta}^i + \boldsymbol{\eta}^e$ or its separate internal ($\boldsymbol{\eta}^i$) and external ($\boldsymbol{\eta}^e$) contributions, involves the complementary *partial* transformations of the representation basic vectors (atomic modes), the product of which brings about the complete diagonalization (decoupling) of the whole hardness matrix (15.23). For example, the initial (*internal*) decoupling transformation $\mathbf{R} = \{\mathbf{R}^X \delta_{X,Y}\}$ of $\boldsymbol{\eta}^i$ (15.28),

$$\mathbf{R}^{\dagger}\boldsymbol{\eta}^{i}\mathbf{R} = \mathbf{a}(\text{diagonal}), \qquad \mathbf{R}^{\dagger}\mathbf{R} = \mathbf{I}, \tag{15.40}$$

defines the associated complementary "rotation" $\mathbf{C_R}$, such that $\mathbf{RC_R} = \mathbf{C}$ or $\mathbf{C_R} = \mathbf{R}^{\dagger}\mathbf{C}$. Similarly, the initial external transformation $\mathbf{W}$, which decouples $\boldsymbol{\eta}^{e}$ (15.29),

$$\mathbf{W}^{\dagger}\boldsymbol{\eta}^{e}\mathbf{W} = \mathbf{b}(\text{diagonal}), \qquad \mathbf{W}^{\dagger}\mathbf{W} = \mathbf{I}, \tag{15.41}$$

generates the complementary transformation $\mathbf{C_W}$ in $\mathbf{C} = \mathbf{WC_W}$: $\mathbf{C_W} = \mathbf{W}^{\dagger}\mathbf{C}$. All these unitary matrices represent the associated "rotations" of the underlying (atomic) basis vectors (Nalewajski 1991, 1992, 1993a; Nalewajski and Korchowiec 1989a, b, 1991, 1997; Nalewajski and Michalak 1996, 1998; Nalewajski et al. 1988, 1994b, 1996a), giving rise to the collective (delocalized) charge displacement modes of molecular fragments in M = (*A*, *B*) or AIM in the whole molecular system M.

The initial internal decoupling $\mathbf{R}$ brings about the diagonalization of only the intrafragment blocks of $\boldsymbol{\eta}$. This transformation generates the (nondiagonal) intermediate hardness tensor:

$$\boldsymbol{\eta}_{int} = \mathbf{R}^{\dagger}\boldsymbol{\eta}\mathbf{R} = \mathbf{R}^{\dagger}(\boldsymbol{\eta}^{i} + \boldsymbol{\eta}^{e})\mathbf{R} = \mathbf{a} + \mathbf{R}^{\dagger}\boldsymbol{\eta}^{e}\mathbf{R}, \tag{15.42}$$

the eigenvalue problem of which defines the complementary rotation $\mathbf{C_R}$ [see (15.23)]:

$$\mathbf{C}_{\mathbf{R}}^{\dagger}\boldsymbol{\eta}_{int}\mathbf{C_R} = \mathbf{C}_{\mathbf{R}}^{\dagger}\mathbf{a}\mathbf{C_R} + \mathbf{C}^{\dagger}\boldsymbol{\eta}^{e}\mathbf{C} = \mathbf{C}^{\dagger}\boldsymbol{\eta}\mathbf{C} = \mathbf{c}(\text{diagonal}) \equiv \{\mathbf{c}^{X}\delta_{X,Y}\}. \tag{15.43}$$

In the same way the initial external decoupling transformation $\mathbf{W}$ generates the associated intermediate hardness tensor

$$\boldsymbol{\eta}_{ext} = \mathbf{W}^{\dagger}\boldsymbol{\eta}\mathbf{W} = \mathbf{W}^{\dagger}(\boldsymbol{\eta}^{i} + \boldsymbol{\eta}^{e})\mathbf{W} = \mathbf{W}^{\dagger}\boldsymbol{\eta}^{i}\mathbf{W} + \mathbf{b}, \tag{15.44}$$

the eigenvalue problem of which determines its complementary rotation $\mathbf{C_W}$:

$$\mathbf{C}_{\mathbf{W}}^{\dagger}\boldsymbol{\eta}_{ext}\mathbf{C_W} = \mathbf{C}_{\mathbf{W}}^{\dagger}\mathbf{b}\mathbf{C_W} + \mathbf{C}^{\dagger}\boldsymbol{\eta}^{i}\mathbf{C} = \mathbf{C}^{\dagger}\boldsymbol{\eta}\mathbf{C} = \mathbf{c}. \tag{15.45}$$

Expressing the overall eigenvalue problem of (15.23) in terms of these partial internal and external decoupling transformations and their respective complementary rotations gives:

$$\begin{aligned}\mathbf{C}^{\dagger}\boldsymbol{\eta}\mathbf{C} &= \mathbf{C}^{\dagger}(\boldsymbol{\eta}^{i} + \boldsymbol{\eta}^{e})\mathbf{C} = \mathbf{C}_{\mathbf{R}}^{\dagger}(\mathbf{R}^{\dagger}\boldsymbol{\eta}^{i}\mathbf{R})\mathbf{C_R} + \mathbf{C}_{\mathbf{W}}^{\dagger}(\mathbf{W}^{\dagger}\boldsymbol{\eta}^{e}\mathbf{W})\mathbf{C_W}\\ &= \mathbf{C}_{\mathbf{R}}^{\dagger}\mathbf{a}\mathbf{C_R} + \mathbf{C}_{\mathbf{W}}^{\dagger}\mathbf{b}\mathbf{C_W} = \mathbf{c}.\end{aligned} \tag{15.46}$$

This equation partitions the overall diagonalization of the hardness matrix into contributions involving the complementary rotations of the tensor internal and external eigenvalues $\mathbf{a}$ and $\mathbf{b}$: the former are transformed using the external complement $\mathbf{C_R}$ of the primary internal decoupling transformation, while the latter require the internal complement $\mathbf{C_W}$ of the primary external transformation. This grouping relation, for combining the results of the external and internal eigenvalue problems into eigensolutions of the whole matrix, complements (15.27)] of Löwdin's partitioning technique.

We finally observe that the external decoupling can be also formulated with reference to the *total* hardness tensor (Nalewajski and Korchowiec 1997; Nalewajski et al. 1994b, 1996a), by requiring the vanishing *off*-diagonal coupling blocks in the transformed hardness matrix. However, this requirement does not specify the transformation uniquely, so that additional *Maximum Overlap Criterion* (MOC) has been used in the past to fully specify such a scheme. Another example of such a transformation is provided by the product transformation $\tilde{\mathbf{U}} = \mathbf{CR}$, which "removes" the *off*-diagonal charge couplings in the resulting hardness tensor (Nalewajski et al. 1994b, 1996a):

$$\mathbf{R}^{\dagger}(\mathbf{C}^{\dagger}\boldsymbol{\eta}\mathbf{C})\mathbf{R} = \mathbf{R}^{\dagger}\mathbf{cR} = \{\mathbf{R}^{X\dagger}\mathbf{c}^{X}\mathbf{R}^{X}\delta_{X,Y}\}. \qquad (15.47)$$

15.6 Hardness/Softness Principles in Chemistry

There are several general rules governing the behavior/stability of molecular systems, which have been formulated in terms of the system global hardness/softness quantities established in the *Conceptual DFT* (CDFT) approach. The most notable is the *HSAB* principle of Pearson (1973, 1997), that "Hard (H) acids (A) prefer to coordinate hard bases (B) and soft (S) acid prefer to coordinate soft bases for both their thermodynamic and kinetic properties," which has been subsequently justified theoretically within CDFT (Parr and Pearson 1983; Nalewajski 1984). The related *Symbiosis HSAB* rule of Jørgensen (1964), formulated for transition metal complexes, that H(S) ligands in the metal coordination sphere enhance tendency of the central metal ion to coordinated more H(S) ligands, can be also justified using the hardness/softness descriptors of molecules (Nalewajski 1989b, 1990). Other reactivity trends, e.g., *trans* (*cis*) directing influence of ligands (Basolo and Pearson 1958; Chatt et al. 1955; Cotton and Wilkinson 1960) can be also qualitatively predicted within CDFT description (Nalewajski 1993a). Another famous example is the *Maximum Hardness* (MH) principle, that there seems to be a rule of nature that molecules arrange themselves so as to be as hard as possible (Parr and Chattaraj 1991). Theoretical proof of this rule makes use of the FDT of the classical statistical mechanics for the *grand*-ensemble (at constant chemical potential and temperature). It has also been shown that the most stable metallic cluster may be characterized as exhibiting highest value of the chemical hardness (Alonso and Balbás 1993). Of similar physical content is the minimum-electrophilicity principle (Noorizadeh 2007; Chattaraj 2007), that a natural direction of a chemical reaction is toward a state of the minimum electrophilicity. In other words, at equilibrium chemical systems generate species with the lesser electrophilicity, with more stable isomers corresponding to lesser electrophilicity values.

Since the chemical hardness measures the HOMO–LUMO gap in MO energies [see Fig. 7.1 and (7.224)], the structural importance of one is synonymous with importance of the other. One of the notable applications of this idea is to provide a quantitative measure of *aromaticity* (Zhou et al. 1988; Zhou and Parr 1989, 1990; Zhou and Novangul 1990), the elusive characteristics of several cyclical compounds exhibiting highly conjugated π-bonds, giving rise to enhanced electron delocalization in closed circuits in two or three dimensions (e.g., Krygowski and Cyrański 2001; Matito et al. 2009; Zubarev et al. 2009). This property is not directly measurable and is usually characterized indirectly by complementary measurable properties (structural, magnetic, energetic, electronic, etc.) that reflect some manifestations of the system aromatic character. The aromatic compounds are thermodynamically more stable, when compared to their nonaromatic acyclic counterparts. They exhibit higher delocalization energy, a preference for the hydrogen-substitution reactions, compared to addition reactions, which imply a destruction of the conjugated π-electron system. For example, the *activation-hardness* concept, marking the hardness change from reactants to TS of the electrophilic aromatic substitution reaction, has been established. Other things being equal, this process is facilitated by the softer reactants and harder TS (smaller activation-hardness).

The FF index, the relative (unity-normalized) local softness descriptor, $f(\boldsymbol{r}) = s(\boldsymbol{r})/S$ (Parr and Yang 1984; Yang et al. 1984; Ayers and Levy 2000) carries within DFT the same physical information as do the *frontier-electron* (FE) densities of Fukui (1975, 1987). The three different types of this local reactivity criterion apply to different types of reactivity phenomena, with $f^{+}(\boldsymbol{r})$ (15.19) governing the nucleophilic attack, $f^{-}(\boldsymbol{r})$ (15.20) governing the electrophilic attack, and their arithmetic average being used as an adequate indicator of the radical attack preference (e.g., Ayers et al. 2009). Specific algorithms for calculating DFT-based reactivity indices have been developed (Nalewajski and Korchowiec 1997; Nalewajski et al. 1996a; Michalak et al. 1999; Ayers 2001; De Proft and Tozer 2009).

The site selectivity for metals in chemisorption and catalysis, determined by low-energy density fluctuations (Falicov and Somorjai 1985), can be also regarded as reflecting trends in the local softness (Yang and Parr 1985). Fundamental for the science of chemistry also are the variational principles for the chemical reactivity (Ayers and Parr 2000, 2001).

15.7 Conclusion

The reaction force approach of Politzer and collaborators has been summarized and the associated partition of the reaction and activation energies of the IRC energy profile has been introduced. The displacements in the electronic structure of reactants in the chemical reaction can be also characterized using the bond-order and valence concepts (Lendvay 2009). The information descriptors of elementary chemical reactions recently generated in the Granada group for the complementary position and momentum-space descriptions have been shown to extract more details of the reaction mechanism compared to the reaction energy profile alone. They identify the regions where the bond-breaking and bond-forming actually occur. The distinction between the concerted and disconcerted reactions can be properly diagnosed using these entropy/information plots which thus provide sensitive probes into the mechanisms of elementary chemical processes.

In addition to CS introduced in preceding chapters, examples of the combined reactivity indicators, e.g., the electrophilicity and nucleophilicity descriptors of the molecular system as a whole and its constituent parts, have been briefly summarized. Such derived reactivity indices for specific reactivity problems are useful in the QSPR analysis of reactivity features by capturing in a single concept the phenomenon in question. The maximum hardness and related general principles of the molecular electronic structure and their theoretical justifications in the CDFT have been summarized and the use of the hardness/softness (FF) concepts in tackling the aromaticity problem and the *single*-reactant reactivity preferences has been surveyed.

As we have already emphasized in the preceding chapter, the complete set of CS for the current representation under consideration, which recognizes the actual experimental conditions, is needed to describe the *real* reactivity mechanisms in bimolecular systems. Indeed, the system responses to displacements in the hypothetical macroscopic (infinitely soft) reservoir can be hardly treated as reflections of the real molecular responses to the presence of another molecule. In a sense CS, the *normalized* responses (per unit perturbations) can be only regarded as determining the right chemical "vocabulary" from which a sensible "sentences" about true reactivity events can be formulated. As we have also seen in the illustrative example of Sect. 14.6, in order to connect even to the simplest bimolecular reactivity in the collinear collision between an atom and diatomic molecule, the CSA description calls for a quite complicated "language" using many such elementary chemical concepts, which remain strongly coupled along the reaction coordinate.

To summarize, only the complete sensitivity description of real reactivity phenomena, including all interactions between the system electronic and geometric degrees-of-freedom, say at the *second*-order of the perturbational (Taylor) expansion, can safeguard the adequacy of such an approach to molecular behavior. Thus, selecting a single response as indicative of the particular reactivity trend (phenomenon) represents a risky and somewhat biased approximate strategy, which cannot guarantee a preservation of the right energetical hierarchy of the compared effects. As a result the list of successes of such a simplified, qualitative approach is accompanied by a substantial body of its failures.

From the purist point of view such "theories" should be more appropriately classified as simple theoretical *models* of reactivity. These simplified, qualitative approaches, which neglect the actual size of the perturbation created by the presence of the attacking agent in the bimolecular reactive system, are sometimes classified as CDFT, since they are based upon the basic DFT premise that the electronic density and its standard responses (derivatives) are sufficient to understand and determine the chemical reactivities of molecules. They nonetheless allow chemists to unite diverse reactivity behaviors in general structural principles, e.g., the HSAB, maximum-hardness/minimum-electrophilicity and symbiosis rules of chemistry.

Indeed, for a family of related compounds/reactions, i.e., comparable reaction stimuli (perturbations), when some *first/second*-order terms can be regarded as being approximately equal, CS can provide a decent (single-reactant) indicators of reactivity preferences by qualitatively reflecting relative trends in the corresponding key energy contributions. In this section we have summarized selected reactivity indices which have been widely applied by chemists to introduce some "order" and classification into the complicated realm of reactivity phenomena. We have already mentioned before that the chemical hardness and softness ideas have been used as measures of the system overall stability and reactivity, respectively, with the local softness or FF index providing a local reactivity indicator. Such a use of CS descriptors in CDFT in diagnosing the reactivity preferences has been recently surveyed in several monographs (e.g., Sen and Jørgensen 1987, Sen 1993; Nalewajski and Korchowiec 1997; Chattaraj 2009; Gatti and Macchi 2011) and review articles (e.g., Chattaraj and Parr 1993; Nalewajski 1993a, 1994, 1995b, 1997a, b; Nalewajski et al. 1996a; Geerlings et al. 2003; Johnson et al. 2011).

The partial eigenvalue problems of the internal and external parts of the matrix representations of quantum-mechanical operators have been shown to generate an attractive framework for separating the *internal* effects of the reactant polarization from the associated *external* CT phenomena occurring between these complementary subsystems in the bimolecular reactive systems. It also provides the most compact representation of the charge-couplings between AIM, the electron-transfer effects in molecules, and offers an insightful hierarchy of the intra- and interfragment modes in reactive systems.

The qualitative approaches presented in this section enrich the diversity of the available tools for probing and understanding the reaction preferences and the evolution of forces of change, when the reaction progresses along IRC. The set of CS reactivity criteria they generate combine versatile descriptors of various reactivity phenomena, including those reflecting the internal promotion and the coupling between the electronic and geometric structures of molecular reactants.

References

Alonso JA, Balbás LC (1993) In: Sen K (ed) Struct Bond 80: Chemical hardness. Springer, Berlin, p 229
Angulo JC (1993) J Phys A 26:6493
Angulo JC, López Rosa S, Antolín J (2010) Int J Quantum Chem (in press)
Ayers PW (2001) Theor Chem Acc 106:271
Ayers PW, Levy M (2000) Theor Chem Acc 103:353
Ayers PW, Parr RG (2000) J Am Chem Soc 122:2010
Ayers PW, Parr RG (2001) J Am Chem Soc 123:2007

Ayers PW, Anderson JSM, Bartolotti LJ (2005a) Int J Quantum Chem 101:520
Ayers PW, Anderson JSM, Rodriguez JI, Jawed Z (2005b) Phys Chem Chem Phys 7:1918
Ayers PW, Yang W, Bartolotti LJ (2009) In: Chattaraj PK (ed) Chemical reactivity theory: a density functional view. CRC, Boca Raton, p 255
Bader RFW (1990) Atoms in molecules. Oxford University Press, New York (and references therein)
Bader RFW (1991) Chem Rev 91:893
Baekelandt BG, Mortier WJ, Schoonheydt RA (1993) In: Sen KD (ed) Struct Bond 80: Chemical hardness. Springer, Berlin, p 187
Basolo F, Pearson RG (1958) Mechanisms of inorganic reactions. Wiley, New York
Bernasconi CF (1992) Acc Chem Res 25:9
Białynicki-Birula I, Mycielski J (1975) Commun Math Phys 44:129
Białynicki-Birula I, Mycielski J (1976) Ann Phys 100:62
Bochevarov AD, Sherrill CD (2007) J Math Chem 42:59
Cedillo A, Contreras R, Galván M, Aizman A, Andrés J, Safont VS (2007) J Phys Chem A 111:2442
Chamorro E, Chattaraj PK, Fuentalba P (2003) J Phys Chem A 107:7068
Chandra AK (1999) Proc Indian Acad Sci Chem Sci III:589
Chandra AK, Sreedhara-Rao V (1996) Int J Quantum Chem 58:57
Chatt J, Duncanson A, Venanzi LM (1955) J Chem Soc 4456
Chattaraj PK (2007) Indian J Phys 81:871
Chattaraj PK (ed) (2009) Chemical reactivity theory – a density functional view. CRC, Boca Raton
Chattaraj PK, Maiti B (2001) J Phys Chem A 105:169
Chattaraj PK, Parr RG (1993) In: Sen KD (ed) Struct Bond 80: Chemical hardness; structure and bonding, vol 80. Springer, Berlin, p 11
Chattaraj PK, Maiti B, Sarkar U (2001) J Phys Chem A 105:169
Chattaraj PK, Maiti B, Sarkar U (2003) J Phys Chem A 107:4973
Chattaraj PK, Roy DR, Elango M, Subramanian V (2005) J Phys Chem A 109:9590
Chattaraj PK, Sarkar U, Roy DR (2006) Chem Rev 106:2065
Chattaraj PK, Sarkar U, Roy DR (2007) J Chem Educ 84:354
Cotton FA, Wilkinson G (1960) Advanced inorganic chemistry. Interscience, New York
De Proft F, Tozer DJ (2009) In: Chattaraj PK (ed) Chemical reactivity theory: a density functional view. CRC, Boca Raton, p 539
Dewar MJS (1984) J Am Chem Soc 106:209
Esquivel RO, Flores-Gallegos N, Iuga C, Carrera E, Angulo JC, Antolín J (2009a) Phenomenological description of selected elementary chemical reaction mechanisms: an information theoretical study. Phys Lett A (in press)
Esquivel RO, Flores-Gallegos N, Iuga C, Carrera E, Angulo JC, Antolín J (2009b) Theor Chem Acc 124:445
Eyring H (1935) J Chem Phys 3:107
Eyring H, Walter J, Kimball GE (1958) Quantum chemistry. Wiley, New York
Falicov LM, Somorjai GA (1985) Proc Natl Acad Sci USA 82:2207
Fan L, Ziegler T (1992) J Am Chem Soc 114:10890
Fukui K (1975) Theory of orientation and stereoselection. Springer, Berlin
Fukui K (1981) Acc Chem Res 14:363
Fukui K (1987) Science 218:747
Gadre SR, Balanarayan P (2009) In: Chattaraj PK (ed) Chemical reactivity theory: a density functional view. CRC, Boca Raton, p 55
Ganzález C, Schlegel HB (1990) J Phys Chem 94:5523
Ganzález-García N, Pu J, Ganzález-Lofont A, Lluch JM, Truhlar DG (2006) J Chem Theory Comput 2:895
Gatti C, Macchi P (2011) (eds) Modern charge-density analysis. Springer, Berlin (in press)
Geerlings P, De Proft F, Langenaeker W (2003) Chem Rev 103:1793

Ingold CK (1933) J Chem Soc 1120
Ingold CK (1934) Chem Rev 15:225
Ishida K, Morokuma K, Komornicki A (1977) J Chem Phys 66:2153
Jaramillo P, Perez P, Contreras R, Tiznado W, Fuentalba P (2006) J Phys Chem A 110:8181
Johnson PA, Bartolotti LJ, Ayers P, Fievez T, Geerlings P (2011) Charge density and chemical reactions: a unified view from conceptual DFT. In: Gatti C, Macchi P (eds) Modern charge-density analysis. Springer, Berlin (in press)
Jørgensen CK (1964) Inorg Chem 3:1201
Kaijser P, Smith VH Jr (1997) Adv Quant Chem 10:37
Krygowski TM, Cyrański MK (2001) Chem Rev 101:1385
Lendvay G (2009) In: Chattaraj PK (ed) Chemical reactivity theory: a density functional view. CRC, Boca Raton, p 303
Levine RD (1978) Annu Rev Phys Chem 29:59
Liu S (2009) In: Chattaraj PK (ed) Chemical reactivity theory: a density functional view. CRC, Boca Raton, p 179
Liu S, Parr RG (1997) J Chem Phys 106:5578
López-Rosa S (2010) PhD Thesis, University of Granada, Granada, Spain
López-Rosa S, Angulo JC, Antolín J (2009) Physica A 388:2081
López-Rosa S, Esquivel RO, Angulo JC, Antolín J, Dehesa JS, Flores Gallegos N (2010) J Chem Theory Comput 6:145
Löwdin P-O (1951) J Chem Phys 19:1396
Löwdin P-O (1962) J Math Phys 3:969
Löwdin P-O (1963) J Mol Spectrosc 10:12
Matito E, Poater J, Solà M, Schleyer PvR (2009) In: Chattaraj PK (ed) Chemical reactivity theory: a density functional view. CRC, Boca Raton, p. 419
Mayr H, Patz M (1994) Angew Chem Int Ed Engl 33:938
McWeeny R (1989) Methods of molecular quantum mechanics. Academic, London
Michalak A, De Proft F, Geerlings P, Nalewajski RF (1999) J Phys Chem A 103:762
Mitoraj M (2007) Ph.D. Thesis, Jagiellonian University, Krakow, Poland
Mitoraj M, Michalak A (2005) J Mol Model 11:341
Mitoraj M, Michalak A (2007) J Mol Model 13:347
Mitoraj M, Zhu H, Michalak A, Ziegler T (2006) J Org Chem 71:9208
Mitoraj M, Zhu H, Michalak A, Ziegler T (2007) Organometallics 16:1627
Nalewajski RF (1980) Chem Phys 50:127
Nalewajski RF (1984) J Am Chem Soc 106:944
Nalewajski RF (1988) Z Naturforsch 43a:65
Nalewajski RF (1989b) In: Popielawski J (ed) Proceedings of the international symposium on the dynamics of systems with chemical reactions, Świdno, 6–10 June 1988. World Scientific, Singapore, p 325
Nalewajski RF (1990) Acta Phys Polon A 77:817
Nalewajski RF (1991) Int J Quantum Chem 40:265; Errata (1992) 43:443
Nalewajski RF (1992) Int J Quantum Chem 42:243
Nalewajski RF (1993a) Struct Bond 80:115
Nalewajski RF (1993b) J Mol Catal 82:371
Nalewajski RF (1994) Int J Quantum Chem 49:675
Nalewajski RF (1995a) Int J Quantum Chem 56:453
Nalewajski RF (1995b) In: Gross EKU and Dreizler RM (eds) Proceedings of the NATO ASI on density functional theory. Plenum, New York, p 339
Nalewajski RF (1997a) Int J Quantum Chem 61:181
Nalewajski RF (1997b) In: Mortier WM, Schoonheydt RA (eds) Developments in the theory of chemical reactivity and heterogeneous catalysis. Research Signpost, Trivandrum, p 135
Nalewajski RF (2002d) In: Sen KD (ed) Reviews of modern quantum chemistry: a celebration of the contributions of Robert G. Parr, vol 2. World Scientific, Singapore, p 1071

Nalewajski RF (2003a) Adv Quant Chem 43:119
Nalewajski RF (2006g) Information theory of molecular systems. Elsevier, Amsterdam
Nalewajski RF (2009a) In: Chattaraj PK (ed) Chemical reactivity theory: a density functional view. Taylor and Francis, London, p 453
Nalewajski RF (2009h) J Math Chem 45:607
Nalewajski RF, Broniatowska E (2003a) J Phys Chem A 107:6270
Nalewajski RF, Broniatowska E (2003b) Chem Phys Lett 376:33
Nalewajski RF, Korchowiec J (1989a) Acta Phys Polon A76:747
Nalewajski RF, Korchowiec J (1989b) J Mol Catal 54:324
Nalewajski RF, Korchowiec J (1991) J Mol Catal 68:123
Nalewajski RF, Korchowiec J (1997) Charge sensitivity approach to electronic structure and chemical reactivity. World-Scientific, Singapore
Nalewajski RF, Michalak A (1996) J Phys Chem 100:20076
Nalewajski RF, Michalak A (1998) J Phys Chem 102:636
Nalewajski RF, Świtka E (2002) Phys Chem Chem Phys 4:4952
Nalewajski RF, Korchowiec J, Zhou Z (1988) Int J Quantum Chem Symp 22:349
Nalewajski RF, Korchowiec J, Michalak A (1994b) Proc Indian Acad Sci Chem Sci 106:353
Nalewajski RF, Korchowiec J, Michalak A (1996a) Top Curr Chem 183:25
Nalewajski RF, Mrozek J, Michalak A (1997) Int J Quantum Chem 61:589
Nalewajski RF, Świtka E, Michalak A (2002) Int J Quantum Chem 87:198
Noorizadeh S (2007) J Phys Org Chem 20:514
Parr RG, Chattaraj PK (1991) J Am Chem Soc 113:1854
Parr RG, Pearson RG (1983) J Am Chem Soc 105:7512
Parr RG, von Szentpaly LV, Liu S (1999) J Am Chem Soc 121:1922
Parr RG, Yang W (1984) J Am Chem Soc 106:4049
Parr RG, Yang W (1989) Density-functional theory of atoms and molecules. Oxford University Press, New York
Pearson RG (1973) Hard and soft acids and bases. Dowden, Hutchinson, and Ross, Stroudsburg
Pearson RG (1997) Chemical hardness, applications from molecules to solids. Wiley-VCH, Weindheim
Perez P, Toro-Labbe A, Aizman A, Contreras R (2002) J Org Chem 67:4747
Rawlings DG, Davidson ER (1958) J Phys Chem 89:969
Roth M, Mayr H (1995) Angew Chem Int Ed Engl 34:2250
Roy RK, Usha V, Paulovic J, Hirao K (2005) J Phys Chem A 109:4601
Schlegel HB (1987) Adv Chem Phys 67:249
Sen KD (ed) (1993) Struct Bond 80: Chemical hardness. Springer, Berlin
Sen KD, Jørgensen CK (eds) (1997) Struct Bond 66: Electronegativity. Springer, Berlin
Senet P (2009) In: Chattaraj PK (ed) Chemical reactivity theory: a density functional view. CRC, Boca Raton, p 331
Toro-Labbé A, Gutiérrez-Oliva S, Politzer P, Murray JS (2009) In: Chattaraj PK (ed) Chemical reactivity theory: a density functional view. CRC, Boca Raton, p 293 and references therein
Wigner EP (1938) Trans Faraday Soc 34:29
Yang W, Parr RG (1985) Proc Natl Acad Sci USA 82:6723
Yang W, Parr RG, Pucci R (1984) J Chem Phys 81:2862
Zhou Z, Novangul HV (1990) J Phys Org Chem 3:784
Zhou Z, Parr RG (1989) J Am Chem Soc 111:7371
Zhou Z, Parr RG (1990) J Am Chem Soc 112:5720
Zhou Z, Parr RG, Garst JF (1988) Tertrahedron Lett 29:4843
Zubarev DY, Sergeeva AP, Boldyrev AI (2009) In: Chattaraj PK (ed) Chemical reactivity theory: a density functional view. CRC, Boca Raton, p. 439

References

Abramson N (1963) Information theory and coding. McGraw-Hill, New York

Adamo C, Barone V (1998) J Chem Phys 108:664

Ahlrichs R, Lischka H, Staemmler V, Kutzelnigg W (1975a) J Chem Phys 62:1225

Ahlrichs R, Driessler F, Lischka H, Staemmler V, Kutzelnigg W (1975b) J Chem Phys 62:1235

Ahlrichs R, Penco R, Scoles G (1977) Chem Phys Lett 19:119

Albright TA, Burdett JK, Whangbo M-H (1985) Orbital interactions in chemistry. Wiley-Interscience, New York

Alonso JA, Balbás LC (1993) In: Sen K (ed) Hardness: structure and bonding, vol 80. Springer, Berlin, p 229

Andersson Y, Langreth DC, Lundqvist BI (1996) Phys Rev Lett 76:102

Anex B, Shull H (1964) In: Löwdin P-O, Pullman B (eds) Molecular orbitals in chemistry, physics and biology. Academic, New York, p 227

Angulo JC (1993) J Phys A 26:6493

Angulo JC, López Rosa S, Antolín J (2010) Int J Quantum Chem (in press)

Arai T (1957) J Chem Phys 26:435; (1960) Rev Mod Phys 32:370

Aryasetiawan F, Miyake T, Terakura K (2002) Phys Rev Lett 88:166401

Ash RB (1965) Information theory. Interscience, New York

Aslangul C, Constanciel R, Daudel R, Kottis P (1972) Adv Quant Chem 6:94

Atkins PW (1983) Molecular quantum mechanics. Oxford University Press, Oxford

Ayers PW (2000a) Proc Natl Acad Sci USA 97:1959

Ayers PW (2000b) JChem Phys 113:10886

Ayers PW (2001) Theor Chem Acc 106:271

Ayers PW, Cedillo A (2009) In: Chattaraj PK (ed) Chemical reactivity theory: a density functional view. CRC, Boca Raton, p 269

Ayers PW, Levy M (2000) Theor Chem Acc 103:353

Ayers PW, Parr RG (2000) J Am Chem Soc 122:2010

Ayers PW, Parr RG (2001) J Am Chem Soc 123:2007

Ayers PW, Anderson JSM, Bartolotti LJ (2005a) Int J Quantum Chem 101:520

Ayers PW, Anderson JSM, Rodriguez JI, Jawed Z (2005b) Phys Chem Chem Phys 7:1918

Ayers PW, Yang W, Bartolotti LJ (2009) In: Chattaraj PK (ed) Chemical reactivity theory: a density functional view. CRC, Boca Raton, p 255

Bader RF (1960) Mol Phys 3:137

Bader RFW (1990) Atoms in molecules. Oxford University Press, New York (and references therein)

Bader RFW (1991) Chem Rev 91:893

Bader RF, Bandrauk AD (1968) J Chem Phys 49:1666

R.F. Nalewajski, *Perspectives in Electronic Structure Theory*,
DOI 10.1007/978-3-642-20180-6,

Baekelandt BG, Mortier WJ, Schoonheydt RA (1993) In: Sen KD (ed) Hardness, structure and bonding, vol 80. Springer, Berlin, p 187
Baekelandt BG, Janssens GOA, Toufar H, Mortier WJ, Schoonheydt RA, Nalewajski RF (1995) J Phys Chem 99:9784
Balint-Kurti GG, Karplus M (1973) Orbital theories of molecules and solids. Clarendon, Oxford
Barriol J (1971) Elements of quantum mechanics with chemical applications. Barnes and Noble, New York
Bartlett RJ (1981) Annu Rev Phys Chem 32:359
Bartlett RJ (1989) J Phys Chem 93:1697
Bartlett RJ (1995) In: Yarkony DR (ed) Modern electronic structure theory, Part II. World Scientific, Singapore, p 1047
Bartlett RJ (2000) In: Keinan E, Schechter I (eds) Chemistry for the 21 Century. Wiley, Weinheim
Bartlett RJ, Grabowski I, Hirata S, Ivanov S (2005a) J Chem Phys 122:034104
Bartlett RJ, Lotrich VF, Schweigert IV (2005b) J Chem Phys 123:062205
Bartolotti LJ (1981) Phys Rev A 24:1661
Bartolotti LJ (1982) Phys Rev A 26:2243
Bartolotti LJ (1984) J Chem Phys 80:5687
Bartolotti LJ (1987) Phys Rev A 36:4492
Basolo F, Pearson RG (1958) Mechanisms of inorganic reactions. Wiley, New York
Becke AD (1985) Int J Quantum Chem 27:585
Becke AD (1988) Phys Rev A 38:3098
Becke AD (1993) J Chem Phys 98:5648
Becke AD (1996) J Chem Phys 104:1040
Becke AD (1997) J Chem Phys 107:8554
Becke AD, Edgecombe KE (1990) J Chem Phys 92:5397
Becke AD, Johnson ER (2007) J Chem Phys 127:124108
Becke AD, Roussel MR (1989) Phys Rev A 39:3761
Berkowitz M, Parr RG (1988) J Chem Phys 88:2554
Bernasconi CF (1992) Acc Chem Res 25:9
Bernstein RB (1982) Chemical dynamics via molecular beam and laser techniques. Clarendon, Oxford
Berrondo M, Gościnski O (1975) Int J Quantum Chem Symp 9:67
Białynicki-Birula I, Mycielski J (1975) Commun Math Phys 44:129
Białynicki-Birula I, Mycielski J (1976) Ann Phys 100:62
Bingel WA, Kutzelnigg W (1970) Adv Quant Chem 5:201
Bobrowicz FB, Goddard WA III (1977) In: Schaefer III HF (ed) Modern theoretical chemistry, vol 3. Plenum, New York, p 79
Bochevarov AD, Sherrill CD (2007) J Math Chem 42:59
Boese AD, Handy NC (2001) J Chem Phys 114:5487
Boese AD, Doltsinis NL, Handy NC, Sprik M (2000) J Chem Phys 112:1670
Boese AD, Martin JML, Klopper W (2007) J Phys Chem A 111:11122
Bohm D (1980) Causality and chance in modern physics. University of Pennsylvania Press, Philadelphia
Born M (1964) Natural philosophy of cause and chance. Dover, New York
Born M, Oppenheimer JR (1927) Ann Phys 64:457
Boys SF (1968) Proc R Soc Lond A 200:542
Boys SF, Bernardi F (1970) Mol Phys 19:553
Boys SF, Foster J (1960) Rev Mod Phys 32:305
Brandow BH (1967) Rev Mod Phys 39:771
Brillouin L (1933) Actualities Sci et Ind No 71. Hermann et Cie, Paris
Brillouin L (1934) Actualites Sci et Ind No. 159. Hermann et Cie, Paris
Brillouin L (1956) Science and information theory. Academic, New York
Brueckner KA (1955) Phys Rev 97:1353; 100:36

Brueckner KA, Levinson CA (1955) Phys Rev 97:1344
Brueckner KA, Eden RJ, Francis NC (1955) Phys Rev 93:1445
Buckley P, Peat FD (eds) (1979) A question of physics: conversations in physics and biology. University of Toronto Press, Toronto
Buijse MA, Baerends EJ (2002) Mol Phys 100:401
Bukowski R, Podeszwa R, Szalewicz K (2005) Chem Phys Lett 414:111
Bukowski R, Cencek W, Jankowski P, Jeziorska M, Jeziorski B, Kucharski SA, Lotrich VF, Misquitta AJ, Moszyński R, Patkowski K, Podeszwa R, Rybak S, Szalewicz K, Williams HL, Wheatley RJ, Wormer PES, Żuchowski PS (2008) SAPT2008: an *ab initio* program for *many*-body symmetry-adapted perturbation theory calculations of intermolecular interaction energies. University of Delaware and University of Warsaw, Delaware, Warsaw
Callen HB (1962) Thermodynamics: an introduction to the physical theories of equilibrium thermostatics and irreversible thermodynamics. Wiley, New York
Callen HB, Welton TA (1951) Phys Rev 83:34
Capitani JF, Nalewajski RF, Parr RG (1982) J Chem Phys 76:568
Carlson BC, Keller JM (1961) Phys Rev 121:659
Carr R, Parrinello M (1985) Phys Rev Lett 55:2471
Carter EA, Goddard WA III (1988) J Chem Phys 88:3132
Casida ML (1996) In: Seminario JM (ed) Recent developments and applications in modern density functional theory. Elsevier, Amsterdam, p 391
Casida ML, Wesołowski TA (2004) Int J Quantum Chem 96:577
Cedillo A, Chattaraj PK, Parr RG (2000) Int J Quantum Chem 77:403
Cedillo A, Contreras R, Galván M, Aizman A, Andrés J, Safont VS (2007) J Phys Chem A 111:2442
Ceperley DM, Alder BJ (1980) Phys Rev Lett 45:566
Chakarova-Käck SD, Schröder E, Lundqvist BI, Langreth DC (2006) Phys Rev Lett 96:146107
Chamorro E, Chattaraj PK, Fuentalba P (2003) J Phys Chem A 107:7068
Chandra AK (1999) Proc Indian Acad Sci Chem Sci III:589
Chandra AK, Sreedhara-Rao V (1996) Int J Quantum Chem 58:57
Chandra AK, Michalak A, Nguyen MT, Nalewajski RF (1998) J Phys Chem A 102:10182
Chapman NB, Shorter J (eds) (1972) Advances in linear free energy relationships. Plenum, New York
Chatt J, Duncanson A, Venanzi LM (1955) J Chem Soc 4456
Chattaraj PK (2007) Indian J Phys 81:871
Chattaraj PK (ed) (2009) Chemical reactivity theory – a density functional view. CRC, Boca Raton
Chattaraj PK, Maiti B (2001) J Phys Chem A 105:169
Chattaraj PK, Parr RG (1993) In: Sen KD (ed) Chemical hardness; structure and bonding, vol 80. Springer, Berlin, p 11
Chattaraj PK, Maiti B, Sarkar U (2001) J Phys Chem A 105:169
Chattaraj PK, Maiti B, Sarkar U (2003) J Phys Chem A 107:4973
Chattaraj PK, Roy DR, Elango M, Subramanian V (2005) J Phys Chem A 109:9590
Chattaraj PK, Sarkar U, Roy DR (2006) Chem Rev 106:2065
Chattaraj PK, Sarkar U, Roy DR (2007) J Chem Educ 84:354
Christoffersen RE (1989a) Basic principles and techniques of molecular quantum mechanics. Springer, Berlin
Christoffersen RE (1989b) Basic principles and techniques of molecular quantum mechanics. Springer, New York
Christoffersen RE, Shull H (1968) J Chem Phys 48:1790
Ciosłowski J (1988) Phys Rev Lett 60:2141
Ciosłowski J (1991) J Am Chem Soc 113:6756
Ciosłowski J, Mixon ST (1993) J Am Chem Soc 115:1084
Ciosłowski J, Pernal K (2004a) J Chem Phys 120:10364
Ciosłowski J, Pernal K (2004b) Phys Rev B 71:113103

Ciosłowski J, Pernal K (2005) Phys Rev B 71:113103
Ciosłowski J, Pernal K, Buchowiecki M (2003) J Chem Phys 119:6443
Čižek J (1966) J Chem Phys 45:4256
Čižek J (1969) Adv Chem Phys 14:35
Čižek J, Paldus J (1980) Phys Script 21:251
Clementi E, Roetti C (1974) At Data Nucl Data Tables 14:177
Coester F (1958) Nucl Phys 1:421
Coester F, Kümmel H (1960) Nucl Phys 17:477
Cohen MH (1996) Density functional theory IV: theory of chemical reactivity. In: Nalewajski RF (ed) Topics in current chemistry, vol 183. Springer, Berlin, p 143
Cohen MH, Ganguglia-Pirovano MV, Kurdnovský J (1994) J Chem Phys 101:8988
Cohen MH, Ganguglia-Pirovano MV, Kurdnovský J (1995) J Chem Phys 103:3543
Cohen-Tannoudji C, Diu B, Laloë F (1977) Quantum mechanics. Hermann, Wiley, Paris
Coleman AJ (1963) Rev Mod Phys 35:668
Coleman AJ (1981) In: Deb BM (ed) The force concept in chemistry. Van Nostrand, Reinhold, New York, p 418
Coleman AJ, Erdahl RM (1968) (Eds) Reduced density matrices with applications to physical and chemical systems. Queen's Papers on Pure and Applied Mathematics, No. 11. Queens University, Kingston
Colle R, Salvetti O (1975) Theor Chim Acta (Berl) 37:329
Colonna F, Savin A (1999) J Chem Phys 110:2828
Condon EU (1930) Phys Rev 36:1121
Connolly JWD (1977) In: Segal GA (ed) Semiempirical methods of electronic structure calculations, Part A: Techniques. Plenum, New York, p 105
Cooper DL (ed) (2002) Valence bond theory. Elsevier, Amsterdam
Cooper DL, Gerratt J, Raimondi M (1981) Chem Rev 91:929
Cooper DL, Gerratt J, Raimondi M (1987) Adv Chem Phys 69:319
Cooper DL, Gerratt J, Raimondi M (1988) Int Rev Phys Chem 7:59
Cooper DL, Gerratt J, Raimondi M (1990a) In: Klein DJ, Trinastič N (eds) Valence bond theory and chemical structure. Elsevier, Amsterdam, p 287
Cooper DL, Gerratt J, Raimondi M (1990b) In: Gutman I, Cyvin SJ (eds) Advances in the theory of benzenoid hydrocarbons. Top Curr Chem 153:41
Cortona P (1991) Phys Rev B 44:8454
Cotton FA, Wilkinson G (1960) Advanced inorganic chemistry. Interscience, New York
Coulson CA, Fischer I (1949) Philos Mag 40:386
Coulson CA, Longuet-Higgins HC (1947a) Proc R Soc A Lond 191:39
Coulson CA, Longuet-Higgins HC (1947b) Proc R Soc A Lond 192:16
Daudel R (1969) The fundamentals of theoretical chemistry. Pergamon, Oxford
Daudel R (1974) The quantum theory of the chemical bond. D. Reidel, Dordrecht
Davidson ER (1969) J Math Phys 10:725
Davidson ER (1972a) Rev Mod Phys 44:451
Davidson ER (1972b) Adv Quant Chem 6:235
Davidson ER (1974) In: Daudel R, Pullman B (eds) The world of quantum chemistry. Reidel, Dordreecht, p 17
Davidson ER (1976) Reduced density matrices in quantum chemistry. Academic, New York
Davydov AS (1965) Quantum mechanics. Pergamon, Oxford
De Proft F, Tozer DJ (2009) In: Chattaraj PK (ed) Chemical reactivity theory: a density functional view. CRC, Boca Raton, p 539
Deb BM, Ghosh SK (1982) J Chem Phys 77:342
Decius JC (1963) J Chem Phys 38:241
Dewar MJS (1969) Molecular orbital theory of organic chemistry. McGraw Hill, New York
Dewar MJS (1984) J Am Chem Soc 106:209
Dewar MJS, Dougherty RC (1975) The PMO theory of organic chemistry. Plenum, New York

Dion M, Rydberg H, Schröder E, Langreth DC, Lundqvist BI (2004) Phys Rev Lett 92:246401
Dirac PAM (1930) Proc Cambr Philos Soc 26:376
Dirac PAM (1967) The principles of quantum mechanics. Clarendon, Oxford
Dobson JF, Dinte BP (1996) Phys Rev Lett 76:1780
Dobson JF, Wang J (2000) Phys Rev B 62:10038
Dobson JF, Wang J, Dinte BP, McLennan K, Lee HM (2005) Int J Quantum Chem 101:579
Doniach S, Sondheim EH (1982) Green's functions for solid state physicists. Addison-Wesley, Redwood City
Donnelly RA (1979) J Chem Phys 71:2874
Donnelly RA, Parr RG (1978) J Chem Phys 69:4431
Dreizler RM, da Providencia J (1985) (eds) Density functional methods in physics. NATO ASI vol 123. Plenum, New York
Dreizler RM, Gross EKU (1990) Density functional theory: an approach to the quantum many-body problem. Springer, Berlin
Dunning TH Jr (1984) J Phys Chem 88:2469
Dunning TH Jr, Hay PJ (1977) In: Schaefer III HF (ed) Methods of electronic structure, vol 3. Plenum, New York, p 1
Edmiston C, Ruedenberg K (1963) Rev Mod Phys 32:457
Ellison FO (1965) J Chem Phys 43:3654
Ellison FO, Slezak JA (1969) J Chem Phys 50:3942
Ellison FO, Wu AA (1967) J Chem Phys 47:4408
Ellison FO, Wu AA (1968) J Chem Phys 48:1103
Elstner M, Hobza P, Frauenheim T, Suhai S, Kaxiras E (2001) J Chem Phys 114:5149
Engel E (2003) In: Fiolhais FNC, Marques MAL (eds) Lecture notes in physics, vol 620. Springer, Berlin, p 56
Engel E, Dreizler RM (1996) Density functional theory ii: relativistic and time dependent extensions. In: Nalewajski RF (ed) Topics in current chemistry, vol. 181. Springer, Berlin, p 1
Engel E, Müller H, Speicher C, Dreizler RM (1995) Density functional theory, NATO ASI, vol 337. Plenum, New York, p 65
Engel E, Höck A, Dreizler RM (2000) Phys Rev A 61:032502
Epiotis ND (1978) Theory of organic reactions. Springer, Berlin
Epstein ST, Hurley AC, Wyatt RE, Parr RG (1967) J Chem Phys 47:1275
Ernzerhof M (1996) Chem Phys Lett 263:499
Ernzerhof M, Perdew JP, Burke K (1996) Density functional theory I: functionals and effective potentials. In: Nalewajski RF (ed) Topics in current chemistry, vol 180. Springer, Berlin, p 1
Esquivel RO, Rodriquez AL, Sagar RP, Hõ M, Smith VH Jr (1996) Phys Rev A 54:259
Esquivel RO, Flores-Gallegos N, Iuga C, Carrera E, Angulo JC, Antolín J (2009a) Phenomenological description of selected elementary chemical reaction mechanisms: an information theoretical study. Phys Lett A (in press)
Esquivel RO, Flores-Gallegos N, Iuga C, Carrera E, Angulo JC, Antolín J (2009b) Theor Chem Acc 124:445
Eyring H (1935) J Chem Phys 3:107
Eyring H, Walter J, Kimball GE (1958) Quantum chemistry. Wiley, New York
Falicov LM, Somorjai GA (1985) Proc Natl Acad Sci USA 82:2207
Fan L, Ziegler T (1992) J Am Chem Soc 114:10890
Fermi E (1928) Z Phys 48:73
Feynman RP (1939) Phys Rev 56:340
Fiethen A, Heßelmann A, Schütz M (2008) J Am Chem Soc 130:1802
Fisher RA (1922) Philos Trans R Soc A (Lond) 222:309
Fisher RA (1925) Proc Camb Philos Soc 22:700
Fisher RA (1959) Statistical methods and scientific inference, 2nd edn. Oliver and Boyd, London
Flügge S (1974) Practical quantum mechanics. Springer, New York
Fock VA (1930) Z Phys 61:126

Fock VA (1986) Fundamentals of quantum mechanics. Mir, Moscow
Freed KF (1971) Annu Rev Phys Chem 22:313
Frieden BR (2000) Physics from the Fisher information – a unification. Cambridge University Press, Cambridge
Frieden BR, Soffer BH (2010) Weighted fisher informations, their derivation and use. Phys Lett A (in press)
Froese-Fischer C (1977) The Hartree-Fock method for atoms. Wiley, New York
Fuchs M, Gonze X (2002) Phys Rev B 65:235109
Fuentalba P, Guerra D, Savin A (2009) In: Chattaraj PK (ed) Chemical reactivity theory: a density functional view. CRC, Boca Raton, p 281
Fujimoto H, Fukui K (1974) In: Klopman G (ed) Chemical reactivity and reaction paths. Wiley-Interscience, New York, p 23
Fukui K (1975) Theory of orientation and stereoselection. Springer, Berlin
Fukui K (1981) Acc Chem Res 14:363
Fukui K (1987) Science 218:747
Furche F (2001) Phys Rev B 64:195120
Furche F (2008) J Chem Phys 129:114105
Gadre SR (1984) Phys Rev A 30:620
Gadre SR (2002) In: Sen KD (ed) Reviews of modern quantum chemistry: a celebration of the contributions of Robert G. Parr, vol 1. World Scientific, Singapore, p 108
Gadre SR, Balanarayan P (2009) In: Chattaraj PK (ed) Chemical reactivity theory: a density functional view. CRC, Boca Raton, p 55
Gadre SR, Bendale RD (1985) Int J Quantum Chem 28:311
Gadre SR, Sears SB (1979) J Chem Phys 71:4321
Gadre SR, Shirsat RN (2000) Electrostatics of atoms and molecules. Universities Press, Hyderabad
Gadre SR, Bendale RD, Gejii SP (1985a) Chem Phys Lett 117:138
Gadre SR, Sears SB, Chakravorty SJ, Bendale RD (1985b) Phys Rev A 32:2602
Galli G, Parinello M (1992) Phys Rev Lett 69:3547
GAMESS General Atomic and Molecular Electronic Structure System; Schmidt MW, Balridge KK, Boatz JA, Elbert ST, Gordon MS, Jensen JH, Koseki S, Matsunaga N, Nguyen KA, Su S, Windus TL, Dupuis M, Montgomery JA (1993) J Comput Chem 14:1347
Ganzález C, Schlegel HB (1990) J Phys Chem 94:5523
Ganzález-García N, Pu J, Ganzález-Lofont A, Lluch JM, Truhlar DG (2006) J Chem Theory Comput 2:895
Gatti C, Macchi P (2011) (eds) Modern charge-density analysis. Springer, Berlin (in press)
GAUSSIAN 03 (Revision D.01); Frisch MJ, Trucks GW, Schlegel HB, Scuseria GE, Robb MA, Cheeseman JR, Montgomery Jr JA, Vreven T, Kudin KN, Burant JC, Millam JM, Iyengar SS, Tomasi J, Barone V, Mennucci B, Cossi M, Scalmani G, Rega N, Petersson GA, Nakatsuji H, Hada M, Ehara M, Toyota K, Fukuda R, Hasegawa J, Ishida M, Nakajima T, Honda Y, Kitao O, Nakai H, Klene M, Li X, Knox JE, Hratchian HP, Cross JB, Bakken V, Adamo C, Jaramillo J, Gomperts R, Stratmann RE, Yazyev O, Austin AJ, Cammi R, Pomelli C, Ochterski JW, Ayala PY, Morokuma K, Voth GA, Salvador P, Dannenberg JJ, Zakrzewski VG, Dapprich S, Daniels AD, Strain MC, Farkas O, Malick DK, Rabuck AD, Raghavachari K, Foresman JB, Ortiz JV, Cui Q, Baboul AG, Clifford S, Cioslowski J, Stefanov BB, Liu G, Liashenko A, Piskorz P, Komaromi I, Martin RL, Fox DJ, Keith T, Al-Laham MA, Peng CY, Nanayakkara A, Challacombe M, Gill PMW, Johnson B, Chen W, Wong MW, Gonzalez C, Pople JA (2004) Gaussian Inc, Wallingford, CT
Gázquez JL (1993) In: Sen KD (ed) Hardness: structure and bonding, vol 80. Springer, Berlin, p 27
Gázquez JL (2009) In: Chattaraj PK (ed) Chemical reactivity theory: a density functional view. CRC, Boca Raton, p 7
Gázquez JL, Vela A, Galvàn M (1987) In: Sen KD and Jørgensen, C.K (eds) Electronegativity; structure and bonding, vol 66. Springer, Berlin, p 79

Geerlings P, De Proft F, Langenaeker W (eds) (1999) Density functional theory: a bridge between chemistry and physics. VUB University Press, Brussels
Geerlings P, De Proft F, Langenaeker W (2003) Chem Rev 103:1793
Gelfand IM, Fomin SV (1963) Calculus of variations. Englewood Cliffs, Prentice-Hall
Gell-Mann M, Brueckner KA (1957) Phys Rev 106:364
Gerber IC, Ángyán JG (2007) J Chem Phys 126:044103
Gerratt J (1974) A specialist periodical report: theoretical chemistry, vol 1 – Quantum chemistry. The Chemical Society, Burlington House, London, p 60
Ghirinhelli LM, Delle Site L, Mosna RA, Hamilton IP (2010) Information theoretic approach to kinetic-energy functionals: the nearly uniform electron gas. J Math Chem (in press)
Ghosh SK (2009) In: Chattaraj PK (ed) Chemical reactivity theory: a density functional view. CRC, Boca Raton, p 71
Ghosh SK, Berkowitz M (1985) J Chem Phys 83:2976
Ghosh SK, Deb BM (1982) Chem Phys 71:295
Ghosh SK, Deb BM (1983a) Theor Chim Acta (Berl.) 62:209
Ghosh SK, Deb BM (1983b) J Mol Struct 103:163
Ghosh SK, Berkowitz M, Parr RG (1984) Proc Natl Acad Sci USA 81:8028
Gilbert TL (1975) Phys Rev B 12:2111
Gilchrist TL, Storr RC (1972) Organic reactions and orbital symmetry. Cambridge University, Cambridge
Gill GB, Willis MR (1974) Pericyclic reactions. Chapman and Hall, New York
Goddard WA III (1967) Phys Rev 157:81
Goddard WA III, Harding LB (1978) Annu Rev Phys Chem 29:363
Goddard WA III, Ladner RD (1971) J Am Chem Soc 93:6750
Goddard WA III, Dunning TH Jr, Hunt WJ, Hay PJ (1973) Acc Chem Res 6:368
Goedecker S, Umrigar CJ (1998) Phys Rev Lett 81:866
Goldstone J (1957) Proc R Soc (London) A239:267
Goll E, Leininger T, Manby FR, Mitrushchenkov A, Werner H-J, Stoll H (2008) Phys Chem Chem Phys 10:3353
Gombas P (1949) The statistical theory of atoms and its applications. Springer, Vienna
Gombas P (1967) Pseudopotentials. Springer, Vienna
Gopinathan MS, Jug K (1983) Theor Chim Acta (Berl.) 63:497, 511
Gordon RG, Kim YS (1972) J Chem Phys 56:3122
Görling A (1999) Phys Rev Lett 83:5459
Görling A, Levy M (1993) Phys Rev B 47:13101
Görling A, Levy M (1994) Phys Rev A 50:196
Grabo T, Kreibich T, Kurth S, Gross EKU (1999) In: Anisimov VI (ed) Strong Coulomb correlations in electronic structure: beyond the local density approximation. Gordon & Breach, Amsterdam, p 1
Grabowski I (2008) Int J Quantum Chem 108:2076
Grabowski I, Lotrich V (2005) Mol Phys 103:2085
Grabowski I, Hirata S, Ivanov S, Bartlett RJ (2002) J Chem Phys 116:4415
Grabowski I, Lotrich V, Bartlett RJ (2007) J Chem Phys 127:154111
Grimme S (2004) J Comput Chem 25:1463
Grimme S (2006a) JComput Chem 27:1787
Grimme S (2006b) J Chem Phys 124:034108
Gritsenko O, Pernal K, Baerends EJ (2005) J Chem Phys 122:204102
Gross EKU, Dreizler RM (1995) Density functional theory, NATO ASI, vol 337. Plenum, New York
Gross EKU, Kurth S (1993) In: Malli GL (ed) Relativistic and electron correlation effects in molecules and solids, NATO ASI series. Plenum, New York
Gross EKU, Oliveira LN, Kohn W (1988) Phys Rev A 37:2805

Gross EKU, Dobson JF, Petersilka M (1996) Density functional theory II: relativistic and time dependent extensions. In: Nalewajski RF (ed) Topics in current chemistry, vol 181. Springer, Berlin, p 81
Gunnarson O, Lundqvist BI (1976) Phys Rev B 13:4274
Gutmann V (1978) The donor-acceptor approach to molecular interactions. Plenum, New York
Gutowski M, Piela L (1988) Mol Phys 64:337
Gyftopoulos EP, Hatsopoulos GN (1965) Proc Natl Acad Sci USA 60:786
Hagstrom S, Shull H (1963) Rev Mod Phys 35:624
Halevi EA (1992) Orbital symmetry and reaction mechanism – the orbital correspondence in maximum symmetry view. Springer, Berlin
Hammett LP (1935) Chem Rev 17:125
Hammett LP (1937) J Am Chem Soc 59:96
Hammond GS (1955) J Am Chem Soc 77:334
Harl J, Kresse G (2008) Phys Rev B 77:045136
Harrimann JE (1980) Phys Rev A 24:680
Harris FE (1967) Adv Chem Phys 13:205
Harris J, Jones RO (1974) J Phys F 4:1170
Hartley RVL (1928) Bell Syst Tech J 7:535
Hartree DR (1928) Proc Camb Philos Soc 24:89
Heisenberg W (1949) The physical principles of the quantum theory. Dover, New York
Heisenberg W (1958) Physics and philosophy: the revolution in modern science. Dover, New York
Heitler W, London F (1927) Z Physik 44:455–472; for an English translation see: Hettema H (2000) Quantum chemistry classic scientific paper. World Scientific, Singapore
Hellmann H (1935) J Chem Phys 3:61
Hellmann H (1937) Einführung in die quantenchemie. Deuticke, Leipzig
Hepburn J, Scoles G, Penco R (1975) Chem Phys Lett 36:451
Herring C (1940) Phys Rev 57:1169
Heßelmann A, Jansen G (2003) Chem Phys Lett 367:778
Heßelmann A, Schütz M (2006) J Am Chem Soc 128:11730
Heßelmann A, Jansen G, Schütz M (2005) J Chem Phys 122:014103
Hiberty PC (1997) In: Davidson ER (ed) Modern electronic structure theory and applications in organic chemistry. World Scientific, River Edge, NJ, p 289
Hiberty PC, Shaik S (2002a) In: Cooper DL (ed) Valence bond theory. Elsevier, Amsterdam, p 187
Hiberty PC, Shaik S (2002b) Theor Chem Acc 108:225
Hiberty PC, Flament JP, Noizet E (1992) Chem Phys Lett 189:259
Hiberty PC, Humbel S, Archirel P (1994a) J Phys Chem 98:11697
Hiberty PC, Humbel S, Byrman CP, van Lenthe JH (1994b) J Chem Phys 101:5969
Hirata S, Ivanov S, Grabowski I, Bartlett RJ, Burke K, Talman JD (2001) J Chem Phys 115:1635
Hirshfeld FL (1977) Theor Chim Acta (Berl) 44:129
Hirst DH (1985) Potential energy surfaces: molecular structure and reaction dynamics. Taylor and Francis, London
Hõ M, Sagar RB, Schmier H, Weaver DF, Smith VH Jr (1995) Int J Quantum Chem 53:627
Hohenberg P, Kohn W (1964) Phys Rev 136B:864
Hunt WJ, Hay PJ, Goddard WA III (1972) J Chem Phys 57:738
Hurley AC (1976) Electron correlation in small molecules. Academic Press, New York
Hurley AC, Lennard-Jones JE, Pople JA (1953) Proc R Soc Lond A220:446
Huzinaga S, Andzelm J, Kłobukowski M, Radzio-Andzelm E, Sakar Y, Tatewaki H (1984) Gaussian basis sets for molecular calculations. Elsevier, New York
Hylleraas EA (1828) Z Physik 48:469
Hylleraas EA, Undheim B (1930) Z Physik 65:759
Iczkowski RP, Margrave JL (1961) J Am Chem Soc 83:3547
Ingold CK (1933) J Chem Soc 1120
Ingold CK (1934) Chem Rev 15:225

Ishida K, Morokuma K, Komornicki A (1977) J Chem Phys 66:2153
Ivanov S, Hirata S, Bartlett RJ (1999) Phys Rev Lett 83:5455
Jackels CF, Davidson ER (1976) J Chem Phys 64:2908
Janak JF (1978) Phys Rev B 18:7165
Jansen G (2009) Density functional theory approach to intermolecular interactions. Lecture Notes
Jansen G, Heßelmann A (2001) J Phys Chem A 105:11156
Jaramillo P, Perez P, Contreras R, Tiznado W, Fuentalba P (2006) J Phys Chem A 110:8181
Jaynes ET (1957a) Phys Rev 106:620
Jaynes ET (1957b) Phys Rev 108:171
Jaynes ET (1985) In: Smith CR, Grandy WT (eds) Maximum entropy and Bayesian methods in inverse problems. Reidel, Dordrecht
Jeziorski B, Szalewicz K (2002) In: Wilson S (ed) Handbook of molecular physics and quantum chemistry, vol 3, Part 2. Wiley, New York, p 232
Jeziorski B, Moszyński R, Szalewicz K (1994) Chem Rev 94:1887
Johnson CD (1973a) The Hammett equation. Cambridge University, London
Johnson KH (1973b) Adv Quant Chem 7:143
Johnson CD (1975) Chem Rev 75:755
Johnson CS, Pedersen LG (1986) Problems and solutions in quantum chemistry and physics. Dover, New York
Johnson PA, Bartolotti LJ, Ayers P, Fievez T, Geerlings P (2011) Charge density and chemical reactions: a unified view from conceptual DFT. In: Gatti C, Macchi P (eds) Modern charge-density analysis. Springer, Berlin (in press)
Johnston HS, Parr CJ (1963) J Am Chem Soc 85:2544
Jones RAY (1979) Physical and mechanistic organic chemistry. Cambridge University, Cambridge
Jones LH, Ryan RR (1970) J Chem Phys 52:2003
Jørgensen CK (1964) Inorg Chem 3:1201
Jørgensen P, Simons J (1981) Second quantization–based methods in quantum chemistry. Academic, New York
Jug K, Gopinathan MS (1990) In: Maksić ZB (ed) Theoretical models of chemical bonding, vol 2. Springer, Heidelberg, p 77
Jug K, Köster AM (1990) J Am Chem Soc 112:6772
Jurečka P, Hobza P (2003) J Am Chem Soc 125:15608
Jurečka P, Sponer J, Cerny J, Hobza P (2006) Phys Chem Chem Phys 8:1985
Kaijser P, Smith VH Jr (1997) Adv Quant Chem 10:37
Karplus M, Grant DM (1959) 11:409
Kato T (1957) Commun Pure Appl Math 10:151
Kelly HP (1969) Adv Chem Phys 14:129
Khinchin AI (1957) Mathematical foundations of the information theory. Dover, New York
Klopman G (1968) J Am Chem Soc 90:223
Klopman G (ed) (1974a) Chemical reactivity and reaction paths. Wiley-Interscience, New York
Klopman G (1974b) In: Klopman G (ed) Chemical reactivity and reaction paths. Wiley-Interscience, New York, p 55
Kohanoff J, Gidopoulos NI (2003) In: Wilson S (ed) Handbook of molecular physics and quantum chemistry, vol 2: Molecular electronic structure. Wiley, New York, p 532
Kohn W (1986) Phys Rev A 34:737
Kohn W (1993) Chem Phys Lett 208:167
Kohn W (1995) In: Gross EKU, Dreizler RM (eds) Density functional theory. Pleum, New York, p 3
Kohn W, Sham LJ (1965) Phys Rev 140A:1133
Kohn W, Meir Y, Makarov DE (1998) Phys Rev Lett 80:4153
Kołos W, Wolniewicz L (1964) J Chem Phys 41:3663
Kołos W, Wolniewicz L (1965) J Chem Phys 43:2429

Kołos W, Wolniewicz L (1968) J Chem Phys 49:404
Korchowiec J, Uchimaru T (1998) J Phys Chem A 102:10167
Kraka E, Cremer D (1990) In: Maksić ZB (ed) Theoretical models of chemical bonding, Part 3 – The concept of the chemical bond. Springer, Berlin, p 453
Krieger JB, Li Y, Iafrate GJ (1990) Phys Lett A 146:256
Krieger JB, Li Y, Iafrate GJ (1992) Phys Rev A 45:101
Krieger JB, Li Y, Iafrate GJ (1995) In: Gross EKU and Dreizler RM (eds) Density functional theory. Pleum, New York, p 191
Kryachko ES, Ludeña EV (1890) Energy density functional theory of many-electron systems. Kluwer, Dordrecht
Krygowski TM, Cyrański MK (2001) Chem Rev 101:1385
Kucharski SA, Bartlett RJ (1986) Adv Quant Chem 18:281
Kullback S (1959) Information theory and statistics. Wiley, New York
Kullback S, Leibler RA (1951) Ann Math Stat 22:79
Kümmel H (1969) Nucl Phys 22:177
Kümmel S, Kronik L (2008) Rev Mod Phys 80:1
Kümmel S, Perdew JP (2003) Phys Rev B 68:035103
Kuntz PJ (1979) In: Bernstein RB (ed) Atom-molecule collision theory – a guide for the experimentalist. Plenum, New York, p 79
Kutzelnigg W (1963) Theor Chim Acta (Berl.) 1:327
Kutzelnigg W (1977) In: Schaefer III HF (ed) Methods of electronic structure theory, vol 3. Plenum, New York, p 129
Langhoff SR, Davidson ER (1974) Int J Quantum Chem 8:61
Langreth DC, Mehl MJ (1981) Phys Rev Lett 47:446
Langreth DC, Perdew JP (1975) Solid State Commun 17:1425
Langreth DC, Perdew JP (1977) Phys Rev B 15:2884
Lee C, Yang W, Parr RG (1988) Phys Rev B 37:785
Leininger T, Stoll H, Werner H-J, Savin A (1997) Chem Phys Lett 275:151
Lendvay G (2009) In: Chattaraj PK (ed) Chemical reactivity theory: a density functional view. CRC, Boca Raton, p 303
Levine RD (1978) Annu Rev Phys Chem 29:59
Levine IN (1983) Quantum chemistry. Allyn and Bacon, Boston
Levy M (1979) Proc Natl Acad Sci USA 76:6062
Levy M (1982) Phys Rev A 26:1200
Levy M (1987) In: March NH, Deb BM (eds) Single particle density in physics and chemistry. Academic, London, p 45
Levy M, Perdew JP (1985) In: Dreizler RM, da Providencia J (eds) Density functional methods in physics. Plenum, New York, p 11
Levy M, Perdew JP (1993) Phys Rev B 48:11638
Levy M, Perdew JP (1997) Phys Rev B 55:13321
Lewis GN (1916) J Am Chem Soc 38:762
Lieb EH (1982) In: Feshbach M, Shimony A (eds) Physics as natural philosophy; Essays in Honor of Laszlo Tisza on His 75th birthday. MIT, Cambridge, p 111. For a revised version see: Lieb EH (1983) Int J Quantum Chem 24:243; In: Dreizler RM and da Providencia J (eds) Density functional methods in physics. Plenum, New York, p 31
Lin I-C, Röthlisberger U (2008) Phys Chem Chem Phys 10:2730
Lin I-C, Continho-Neto MD, Felsenheimer C, von Lilienfeld OA, Tavernelli IA, Röthlisberger U (2007) Phys Rev B 75:205131
Linderberg J (1977) Int J Quantum Chem 12(suppl 1):267
Liu S (2009) In: Chattaraj PK (ed) Chemical reactivity theory: a density functional view. CRC, Boca Raton, p 179
Liu S, Parr RG (1997) J Chem Phys 106:5578
London F (1928) Z Phys 46:455

Longuet-Higgins HC (1961) Adv Spectrosc 2:429
López-Rosa S (2010) PhD Thesis, University of Granada, Granada, Spain
López-Rosa S, Angulo JC, Antolín J (2009) Physica A 388:2081
López-Rosa S, Esquivel RO, Angulo JC, Antolín J, Dehesa JS, Flores Gallegos N (2010) J Chem Theory Comput 6:145
Lotrich V, Bartlett RJ, Grabowski I (2005) Chem Phys Lett 405:43
Löwdin P-O (1950) J Chem Phys 16:365
Löwdin P-O (1951) J Chem Phys 19:1396
Löwdin P-O (1955a) Phys Rev 97:1474
Löwdin P-O (1955b) Phys Rev 97:1490
Löwdin P-O (1956) Adv Phys 5:111
Löwdin P-O (1959) In: Advances in chemical physics, vol 2. Prigogine, Ith edn. Interscience, New York, p 207
Löwdin P-O (1962) J Math Phys 3:969
Löwdin P-O (1963) J Mol Spectrosc 10:12
Löwdin P-O, Shull H (1956) Phys Rev 101:1730
Lundqvist BI, Andersson Y, Shao H, Chan S, Langreth DC (1995) Int J Quantum Chem 57:247
MacDonald JKL (1933) Phys Rev 43:830
Macke W (1955a) Ann Phys (Leipzig) 17:1
Macke W (1955b) Phys Rev 100:992
Mahanty J, Ninham BW (1976) Dispersion forces. Academic, London
Manne R (1977) Int J Quantum Chem S11:175
March NH (1975) Self-consistent fields in atoms. Pergamon, Oxford
March NH (1982) Phys Rev A 26:1845
Marcus RA (1968) J Phys Chem 72:891–899
Marcus RA (1969) J Am Chem Soc 91:7224
Marini A, Gárcia-González P, Rubio A (2006) Phys Rev Lett 96:136404
Marques MAL, Gross EKU (2004) Annu Rev Phys Chem 55:427
Mathai AM, Rathie PM (1975) Basic concepts in information theory and statistics: axiomatic foundations and applications. Wiley, New York
Matito E, Poater J, Solà M, Schleyer PvR (2009) In: Chattaraj PK (ed) Chemical reactivity theory: a density functional view. Taylor and Francis, London, p 419
Mattuck RD (1976) A guide to Feynman diagrams in the many-body problem, 2nd edn. Mc-Graw-Hill, New York
Mayer I (1983) Chem Phys Lett 97:270
Mayer I (1985) Theor Chim Acta (Berl.) 67:315
Mayr H, Patz M (1994) Angew Chem Int Ed Engl 33:938
McAadon MH, Goddard WA III (1985) J Phys Chem 91:2607
McAadon MH, Goddard WA III (1987) Phys Rev Lett 55:2563
McLachlan AD, Ball MA (1964) Rev Mod Phys 36:844
McNamara JP, Hillier IH (2007) Phys Chem Chem Phys 9:2362
McQuarrie DA (1983) Quantum chemistry. University Science Books and Oxford University Press, London
McWeeny R (1979) Coulson's valence. Oxford University Press, Oxford
McWeeny R (1989) Methods of molecular quantum mechanics. Academic, London
McWeeny R, Ohno KA (1960) Proc R Soc (Lond) A255:367
Mermin ND (1965) Phys Rev A 137:1441
Merzbacher E (1967) Quantum mechanics. Wiley, New York
Messiah A (1961) Quantum mechanics. North Holland, Amsterdam
Meyer W (1977) In: Schaefer III HF (ed) Methods of electronic structure theory, vol 3. Plenum, New York, p 413
Michalak A, De Proft F, Geerlings P, Nalewajski RF (1999) J Phys Chem A 103:762
Misquitta AJ, Szalewicz K (2002) Chem Phys Lett 357:301

Misquitta AJ, Szalewicz K (2005) J Chem Phys 122:214109
Misquitta AJ, Jeziorski R, Szalewicz K (2003) Phys Rev Lett 91:033201
Misquitta AJ, Podeszwa R, Jeziorski B, Szalewicz K (2005) J Chem Phys 123:214103
Mitoraj M (2007) Ph.D. Thesis, Jagiellonian University, Krakow, Poland
Mitoraj M, Michalak A (2005) J Mol Model 11:341
Mitoraj M, Michalak A (2007) J Mol Model 13:347
Mitoraj M, Zhu H, Michalak A, Ziegler T (2006) J Org Chem 71:9208
Mitoraj M, Zhu H, Michalak A, Ziegler T (2007) Organometallics 16:1627
Moffitt W (1951) Proc R Soc (London) A210:245
Møller C, Plessett MS (1934) Phys Rev 46:618
Morgado C, Vincent MA, Hillier IH, Shan X (2008) Phys Chem Chem Phys 9:448
Mortier WM, Schoonheydt RA (eds) (1997) Developments in the theory of chemical reactivity and heterogeneous catalysis. Research Signpost, Trivandrum
Moss BJ, Goddard WA III (1975) J Chem Phys 63:3523
Mrozek J, Nalewajski RF, Michalak A (1998) Polish J Chem 72:1779
Müller AMK (1984) Phys Lett A 105:446
Mulliken RS (1934) J Chem Phys 2:782
Mulliken RS (1935) J Chem Phys 3:573
Mulliken RS (1955) J Chem Phys 23:1833, 1841, 2338, 2343
Mulliken RS (1962) J Chem Phys 36:3428
Murphy DR (1981) Phys Rev A 24:1682
Murphy DR, Parr RG (1979) Chem Phys Lett 60:377
Murray JS, Sen K (eds) (1996) Molecular electrostatic potentials: concepts and applications. Elsevier, Amsterdam
Murrell JN, Bosanac SD (1989) Introduction to the theory of atomic and molecular collisions. Wiley, New York
Murrell JN, Carter S, Farantos SC, Huxley P, Varandas AJC (1984) Molecular potential energy functions. Wiley, New York
Mycielski J, Białynicki-Birula I (1975) Commun Math Phys 44:129
Nagy Á (2003) J Chem Phys 119:9401
Nagy Á, Parr RG (2000) J Mol Struct (Theochem) 501:101
Nagy Á, Parr RG (1994) Proc Indian Acad Sci Chem Sci 106:217
Nagy Á, Parr RG (1996) Int J Quantum Chem 58:323
Nakatsuji H (1973) J Am Chem Soc 95:345
Nakatsuji H (1974a) J Am Chem Soc 96:24
Nakatsuji H (1974b) J Am Chem Soc 96:30
Nalewajski RF (1980) Chem Phys 50:127
Nalewajski RF (1983) J Chem Phys 78:6112
Nalewajski RF (1984) J Am Chem Soc 106:944
Nalewajski RF (1985) J Phys Chem 89:2831
Nalewajski RF (1988) Z Naturforsch 43a:65
Nalewajski RF (1989a) J Phys Chem 93:2658
Nalewajski RF (1989b) In: Popielawski J (ed) Proceedings of the international symposium on the dynamics of systems with chemical reactions, Świdno, 6–10 June 1988. World Scientific, Singapore, p 325
Nalewajski RF (1990) Acta Phys Polon A 77:817
Nalewajski RF (1991) Int J Quantum Chem 40:265; Errata (1992) 43:443
Nalewajski RF (1992) Int J Quantum Chem 42:243
Nalewajski RF (1993a) Struct Bond 80:115
Nalewajski RF (1993b) J Mol Catal 82:371
Nalewajski RF (1994) Int J Quantum Chem 49:675
Nalewajski RF (1995a) Int J Quantum Chem 56:453

Nalewajski RF (1995b) In: Gross EKU and Dreizler RM (eds) Proceedings of the NATO ASI on density functional theory. Plenum, New York, p 339
Nalewajski RF (1996a) (ed) Density functional theory I–IV. Topics in current chemistry, vols 180–183. Springer, Berlin
Nalewajski RF (1996b) (ed) Density functional theory IV: theory of chemical reactivity. Topics in current chemistry, vol 183. Springer, Berlin
Nalewajski RF (1997a) Int J Quantum Chem 61:181
Nalewajski RF (1997b) In: Mortier WM, Schoonheydt RA (eds) Developments in the theory of chemical reactivity and heterogeneous catalysis. Research Signpost, Trivandrum, p 135
Nalewajski RF (1998a) Polish J Chem 72:1763
Nalewajski RF (1998b) Int J Quantum Chem 69:591
Nalewajski RF (1999) Phys Chem Chem Phys 1:1037
Nalewajski RF (2000a) Comput Chem 24:243
Nalewajski RF (2000b) Int J Quantum Chem 78:168
Nalewajski RF (2000c) J Phys Chem A 104:11940
Nalewajski RF (2000d) Top Catal 11–12:469
Nalewajski RF (2001) Adv Quant Chem 38:217
Nalewajski RF (2002a) Phys Chem Chem Phys 4:1710
Nalewajski RF (2002b) Int J Mol Sci 3:237
Nalewajski RF (2002c) Acta Chim Phys Debr 34–35:131
Nalewajski RF (2002d) In: Sen KD (ed) Reviews of modern quantum chemistry: a celebration of the contributions of Robert G. Parr, vol 2. World Scientific, Singapore, p 1071
Nalewajski RF (2002e) Chem Phys Lett 353:143
Nalewajski RF (2002f) In: Barone V, Bencini A, Fantucci P (eds) Recent advances in density functional methods, part III. World Scientific, Singapore, p 257
Nalewajski RF (2003a) Adv Quant Chem 43:119
Nalewajski RF (2003b) J Phys Chem A 107:3792
Nalewajski RF (2003c) Chem Phys Lett 372:28
Nalewajski RF (2003d) Chem Phys Lett 375:196
Nalewajski RF (2003e) Mol Phys 101:2369
Nalewajski RF (2003f) Chem Phys Lett 367:414
Nalewajski RF (2004a) Ann Phys (Leipzig) 13:201
Nalewajski RF (2004b) Chem Phys Lett 386:265
Nalewajski RF (2004c) Mol Phys 102:531
Nalewajski RF (2004d) Mol Phys 102:547
Nalewajski RF (2004e) Struct Chem 15:395
Nalewajski RF (2005a) Theor Chem Acc 114:4
Nalewajski RF (2005b) Mol Phys 103:451
Nalewajski RF (2005c) J Math Chem 38:43
Nalewajski RF (2005d) Chem Phys Lett 410:335
Nalewajski RF (2005e) J Chem Sci 117:455
Nalewajski RF (2006a) Mol Phys 104:365
Nalewajski RF (2006b) Mol Phys 104:493
Nalewajski RF (2006c) Mol Phys 104:1977
Nalewajski RF (2006d) Mol Phys 104:2533
Nalewajski RF (2006e) Mol Phys 104:3339
Nalewajski RF (2006f) Adv Quant Chem 51:235
Nalewajski RF (2006g) Information theory of molecular systems. Elsevier, Amsterdam
Nalewajski RF (2006h) Mol Phys 104:255
Nalewajski RF (2007) J Phys Chem A 111:4855
Nalewajski RF (2008a) J Math Chem 43:265
Nalewajski RF (2008b) J Math Chem 43:780
Nalewajski RF (2008c) J Math Chem 44:414

Nalewajski RF (2008d) J Math Chem 44:802
Nalewajski RF (2008e) Int J Quantum Chem 108:2230
Nalewajski RF (2009a) In: Chattaraj PK (ed) Chemical reactivity theory: a density functional view. Taylor and Francis, London, p 453
Nalewajski RF (2009b) J Math Chem 45:709
Nalewajski RF (2009c) J Math Chem 45:776
Nalewajski RF (2009d) J Math Chem 45:1041
Nalewajski RF (2009e) Int J Quantum Chem 109:425
Nalewajski RF (2009f) Int J Quantum Chem 109:2495
Nalewajski RF (2009g) Adv Quant Chem 56:217
Nalewajski RF (2009h) J Math Chem 45:607
Nalewajski RF (2010a) J Math Chem 47:667
Nalewajski RF (2010b) J Math Chem 47:692
Nalewajski RF (2010c) J Math Chem 47:709
Nalewajski RF (2010d) J Math Chem 47:808
Nalewajski RF (2010e) J Math Chem 47:1068
Nalewajski RF (2010f) Information origins of the chemical bond. Nova Science, New York
Nalewajski RF (2010g) J Math Chem 48:752
Nalewajski RF (2010h) Information perspective on molecular electronic structure. In: Mathematical chemistry. Nova Science, New York (in press)
Nalewajski RF (2010i) Information tools for probing chemical bonds. In: Putz M (ed) Chemical information and computation challenges in 21st: a celebration of 2011 international year of chemistry. Nova Science, New York (in press)
Nalewajski RF (2011a) J Math Chem 49:371
Nalewajski RF (2011b) J Math Chem 49:546
Nalewajski RF (2011c) J Math Chem 49:592
Nalewajski RF (2011d) J Math Chem 49:806
Nalewajski RF, Broniatowska E (2003a) J Phys Chem A 107:6270
Nalewajski RF, Broniatowska E (2003b) Chem Phys Lett 376:33
Nalewajski RF, Broniatowska E (2005) Int J Quantum Chem 101:349
Nalewajski RF, Broniatowska E (2007) Theor Chem Acc 117:7
Nalewajski RF, Gurdek P (2010) On the implicit bond-dependency origins of bridge interactions. J Math Chem (submitted)
Nalewajski RF, Jug K (2002) In: Sen KD (ed) Reviews of modern quantum chemistry: a celebration of the contributions of Robert G. Parr, vol 1. World Scientific, Singapore, p 148
Nalewajski RF, Koniński M (1988) Acta Phys Polon A 74:255
Nalewajski RF, Korchowiec J (1989a) Acta Phys Polon A76:747
Nalewajski RF, Korchowiec J (1989b) J Mol Catal 54:324
Nalewajski RF, Korchowiec J (1991) J Mol Catal 68:123
Nalewajski RF, Korchowiec J (1997) Charge sensitivity approach to electronic structure and chemical reactivity. World-Scientific, Singapore
Nalewajski RF, Loska R (2001) Thoer Chem Acc 105:374
Nalewajski RF, Michalak A (1995) Int J Quantum Chem 56:603
Nalewajski RF, Michalak A (1996) J Phys Chem 100:20076
Nalewajski RF, Michalak A (1998) J Phys Chem 102:636
Nalewajski RF, Mrozek J (1994) Int J Quantum Chem 51:187
Nalewajski RF, Mrozek J (1996) Int J Quantum Chem 57:377
Nalewajski RF, Parr RG (1982) J Chem Phys 77:399; the extremum principle in Eqs. (69) and (70) of this paper is a maximum principle, not a minimum principle
Nalewajski RF, Parr RG (2000) Proc Natl Acad Sci USA 97:8879
Nalewajski RF, Parr RG (2001) J Phys Chem A 105:7391
Nalewajski RF, Sikora O (2000) J Phys Chem A 104:5638
Nalewajski RF, Świtka E (2002) Phys Chem Chem Phys 4:4952

Nalewajski RF, Korchowiec J, Zhou Z (1988) Int J Quantum Chem Symp 22:349
Nalewajski RF, Köster AM, Jug K (1993) Theor Chim Acta (Berl.) 85:463
Nalewajski RF, Formosinho SJ, Varandas AJC, Mrozek J (1994a) Int J Quantum Chem 52:1153
Nalewajski RF, Korchowiec J, Michalak A (1994b) Proc Indian Acad Sci Chem Sci 106:353
Nalewajski RF, Korchowiec J, Michalak A (1996a) Top Curr Chem 183:25
Nalewajski RF, Mrozek J, Mazur G (1996b) Can J Chem 100:1121
Nalewajski RF, Mrozek J, Michalak A (1997) Int J Quantum Chem 61:589
Nalewajski RF, Świtka E, Michalak A (2002) Int J Quantum Chem 87:198
Nalewajski RF, Köster AM, Escalante S (2005) J Phys Chem A 109:10038
Nalewajski RF, Błażewicz D, Mrozek J (2008) J Math Chem 44:325
Nalewajski RF, de Silva P, Mrozek J (2010a) Kinetic-energy/Fisher-information indicators of chemical bonds. In: Wang A, Wesołowski TA (eds) Kinetic energy functional. World Scientific, Singapore (in press)
Nalewajski RF, de Silva P, Mrozek J (2010b) J Mol Struct THEOCHEM 954:57
Nalewajski RF, Szczepanik D, Mrozek J (2010c) Bond differentiation and orbital decoupling in the orbital communication theory of the chemical bond. Adv Quant Chem (in press)
Nesbet RK (1965) Adv Chem Phys 9:321
Nguyen-Dang TT, Ludeña EV, Tall Y (1985) J Mol Struct THEOCHEM 120:247
Noorizadeh S (2007) J Phys Org Chem 20:514
Nyquist H (1928) Phys Rev 32:110
Ohno K (1967) Adv Quant Chem 3:239
Oliveira LN, Gross EKU, Kohn W (1988) Phys Rev A 37:2821
Oliver GL, Perdew JP (1979) Phys Rev A 20:397
Olsson MHM, Hong GY, Warshel A (2004) J Am Chem Soc 96:577
Paldus J, Čižek J (1973) In: Smith D, McRae WB (eds) Energy, structure and reactivity. Wiley, New York, p 198
Paldus J, Čižek J (1975) Adv Quant Chem 9:105
Pariser R (1953) J Chem Phys 21:568
Pariser R, Parr RG (1953) J Chem Phys 21:767
Parks JM, Parr RG (1958) J Chem Phys 28:335
Parks JM, Parr RG (1960) J Chem Phys 32:1657
Parr RG (1963) The quantum theory of molecular electronic structure. Benjamin, New York
Parr RG, Chattaraj PK (1991) J Am Chem Soc 113:1854
Parr RG, Pearson RG (1983) J Am Chem Soc 105:7512
Parr RG, Yang W (1984) J Am Chem Soc 106:4049
Parr RG, Yang W (1989) Density-functional theory of atoms and molecules. Oxford University Press, New York
Parr RG, Donnelly RA, Levy M, Palke WE (1978) J Chem Phys 69:4431
Parr RG, Ayers PW, Nalewajski RF (2005) J Phys Chem A 109:3957
Pauling L (1947) J Am Chem Soc 69:542
Pauling L (1949) Proc R Soc Lond A196:343
Pearson RG (1973) Hard and soft acids and bases. Dowden, Hutchinson, and Ross, Stroudsburg
Pearson RG (1976) Symmetry rules for chemical reactions: orbital topology and elementary processes. Wiley, New York
Pearson RG (1988) Inorg Chem 27:734
Pearson RG (1997) Chemical hardness, applications from molecules to solids. Wiley-VCH, Weindheim
Penrose R (1989) The emperor's new mind: concerning computers, minds, and the laws of physics. Oxford University Press, New York, p 225
Perdew JP (1985) In: Dreizler RM, da Providencia J (eds) Density functional methods in physics. Plenum, New York, p 265
Perdew JP (1991) In: Ziesche P, Eschrig H (eds) Electronic structure of solids '91. Akademie, Berlin, p 11

Perdew JP (1999) In: Geerlings P, De Proft F, Langenaeker W (eds) Density functional theory – a bridge between chemistry and physics. VUB University Press, Brussels, p 87
Perdew JP, Levy M (1983) Phys Rev Lett 51:1884
Perdew JP, Levy M (1984) In: Langreth D, Suhl H (eds) Many-body phenomena at surfaces. Academic, Orlando, p 71
Perdew JP, Norman MR (1982) Phys Rev B 26:5445
Perdew JP, Wang Y (1986) Phys Rev B 33:8800
Perdew JP, Wang Y (1989) Phys Rev B 40:3399
Perdew JP, Wang Y (1991) Phys Rev B 45:13244
Perdew JP, Zunger A (1981) Phys Rev B 23:5048
Perdew JP, Parr RG, Levy M, Balduz JL (1982) Phys Rev Lett 49:1691
Perdew JP, Burke K, Ernzerhoh M (1996a) Phys RevLett 77:3865
Perdew JP, Burke K, Wang Y (1996b) Phys Rev B 54:16533
Perdew JP, Ernzerhof M, Burke K (1996c) J Chem Phys 105:9982
Perdew JP, Burke K, Ernzerhoh M (1997) Phys Rev Lett 78:1396
Perdew JP, Burke K, Wang Y (1998) Phys Rev B 57:14999
Perdew JP, Ruzsinsky A, Tao J, Staroverov VN, Scuseria GE, Csonka GI (2005) J Chem Phys 123:062201
Perez P, Toro-Labbe A, Aizman A, Contreras R (2002) J Org Chem 67:4747
Perez-Jorda JM, Becke AD (1995) Chem Phys Lett 233:134
Pernal K (2005) Phys Rev Lett 94:233002
Pernal K, Baerends EJ (2006) J Chem Phys 124:014102
Pernal K, Ciosłowski J (2004) J Chem Phys 120:5987
Pernal K, Ciosłowski J (2005) Chem Phys Lett 412:71
Petersilka M, Gossmann UJ, Gross EKU (1996) Phys Rev Lett 76:1212
Pfeiffer PE (1978) Concepts of probability theory. Dover, New York
Phillips JC, Kleinman L (1959) Phys Rev 116:287
Phillips JC, Kleinman L (1960) Phys Rev 118:1153
Podeszwa R, Szalewicz K (2007) J Chem Phys 126:194101
Podeszwa R, Bukowski R, Szalewicz K (2006) J Chem Theo Comp 2:400
Poirier R, Kari R, Csizmadia IG (1985) Handbook of Gaussian basis sets. Elsevier, New York
Politzer P, Murray JS (2009) In: Chattaraj PK (ed) Chemical reactivity theory: a density functional view. CRC, Boca Raton, p 243
Politzer P, Truhlar D (eds) (1981) Chemical applications of atomic and molecular electrostatic potentials. Plenum, New York
Pople JA (1953) Trans Faraday Soc 49:1375
Pople JA (1976) In: Schaefer III HF (ed) Modern theoretical chemistry, vol 4. Plenum, New York
Pueckert V (1978) J Phys C11:4945
Rajchel Ł, Żuchowski PS, Szczęśniak MM, Chałasiński G (2009a) A DFT approach to non-covalent interactions *via* monomer polarization and Pauli blockade. *arXiv*, 0908.0798v [physics.chem.-ph] (in press)
Rajchel Ł, Żuchowski PS, Szczęśniak MM, Chałasiński G (2009b) Derivation of the supermolecular interaction energy from the monomer densities in the density functional theory. *arXiv*:0908.0798v [physics.chem.-ph] (in press)
Ramana MV, Rajagopal AK (1983) Adv Chem Phys 54:231
Rapcewicz K, Ashcroft NW (1991) Phys Rev B 44:4032
Rappé AK, Goddard WA III (1982) J Am Chem Soc 104(297):448
Rawlings DG, Davidson ER (1958) J Phys Chem 89:969
Romera E, Dehesa JS (2004) J Chem Phys 120:8906
Roos BO (1972) Chem Phys Lett 15:153
Roos BO, Siegbahn EEM (1977) In: Schaefer III HF (ed) Methods of electronic structure theory. Plenum, New York, p 277
Roth M, Mayr H (1995) Angew Chem Int Ed Engl 34:2250

Roy RK, Usha V, Paulovic J, Hirao K (2005) J Phys Chem A 109:4601
Rumer G (1932) Göttinger Nachr 27:337
Runge E, Gross EKU (1984) Phys Rev Lett 52:997
Salem L (1968a) J Am Chem Soc 90:543
Salem L (1968b) J Am Chem Soc 90:553
Salem L (1969) Chem Br 5:449
Sánchez-Garcia E, Mardyukov A, Tekin A, Crespo-Otero R, Montero LA, Sander W, Jansen G (2008) J Chem Phys 343:168
Sanderson RT (1951) Science 114:670
Sanderson RT (1976) Chemical bonds and bond energy, 2nd edn. Academic, New York
Savin A, Flad H-J (1995) Int J Quantum Chem 56:327
Savin A, Nesper R, Wengert S, Fässler TF (1997) Angew Chem Int Ed Engl 36:1808
Schlegel HB (1987) Adv Chem Phys 67:249
Scuseria GE, Henderson TM, Sorensen DC (2008) J Chem Phys 129:231101
Sears SB (1980) Ph.D Thesis, The University of North Carolina at Chapel Hill, North Carolina
Sears SB, Parr RG, Dinur U (1980) Israel J Chem 19:165
Seidl M, Perdew JP, Kurth S (2000) Phys Rev A 62:012502
Sen KD (ed) (1993) Chemical hardness: structure and bonding, vol 80. Springer, Berlin
Sen KD, Jørgensen CK (eds) (1987) Electronegativity: structure and bonding, vol 66. Springer, Berlin
Sen KD, Antolín J, Angulo JC (2007) Phys Rev A 76:032502
Senet P (1996) J Chem Phys 105:6471
Senet P (1997) J Chem Phys 107:2516
Senet P (2009) In: Chattaraj PK (ed) Chemical reactivity theory: a density functional view. CRC, Boca Raton, p 331
Shaik S (1989) In: Bertran J, Czismadia IG (eds) New theoretical concepts for understanding organic reactions, NATO ASI Series, vol. C267. Kluwer Academic, Dordrecht, p 165
Shaik S, Hiberty PC (1991) In: Maksić ZB (ed) Theoretical models of chemical bonding, vol 4, pp 269
Shaik S, Hiberty PC (1995) Adv Quant Chem 26:100
Shaik S, Hiberty PC (2004) In: Lipkowitz KB, Larter L, Cundari TR (eds) Reviews in computational chemistry, vol 20, p 1
Shaik S, Shurki A, Danovich D, Hiberty P (2001) Chem Rev 101:1501
Shaik S, Danovich D, Wu W, Hiberty PC (2009) Nat Chem 1:443
Shannon CF (1948) Bell Syst Technol J 27(379):623
Shannon CE, Weaver W (1949) The mathematical theory of communication. University of Illinois, Urbana
Sharp RT, Horton GK (1953) Phys Rev 90:317
Shavitt I (1963) Methods Comput Phys 3:1
Shavitt I (1977) In: Schaefer III HF (ed) Methods of electronic structure theory. Plenum, New York, p 189
Shavitt I, Rosenberg BJ, Palalikit S (1976) Int J Quantum Chem Symp 10:33
Shull H (1959) J Chem Phys 30:1405
Shull H (1960) J Am Chem Soc 82:1287
Shull H (1962) J Phys Chem 66:2320
Shull H (1964) J Am Chem Soc 86:1469
Shull H, Löwdin P-O (1958) Phys Rev 110:1466
Shull H, Prosser F (1964) J Chem Phys 40:233
Silvi B, Savin A (1994) Nature 371:683
Simonetta M (1968) In: Rich A, Davidson N (eds) Structural chemistry and molecular biology. Freeman, San Francisco
Sinanoĝlu O (1964) Adv Chem Phys 6:315

Sironi M, Raimondi M, Martinazzo R, Gianturco FA (2002) In: Cooper DL (ed) Valence bond theory. Elsevier, Amsterdam, p 261
Slater JC (1929) Phys Rev 34:1293
Slater JC (1931) Phys Rev 38:1109
Slater JC (1951) Phys Rev 81:385
Slater JC (1960) Quantum theory of atomic structure, vols 1 and 2. McGraw-Hill, New York
Slater JC (1974) Quantum theory of molecules and solids, vol 4: The self-consitent field for molecules and solids. McGraw-Hill, New York
Stoll H, Savin A (1985) In: Dreizler RM, da Providencia J (eds) Density functional methods in physics. Plenum, New York, p 177
Stone AJ (1978) In: Dixon RN, Thomson C (eds) Specialist periodical reports: theoretical chemistry, vol. 3. Bartholomew, Dorking, p 39
Surjan PR (1989) Second quantized approach to quantum chemistry. Springer, Berlin
Swanson BI (1976) J Am Chem Soc 98:3067
Swanson BI, Satija SK (1977) J Am Chem Soc 99:987
Szabo A, Ostlund NS (1982) Modern quantum chemistry: introduction to advanced electronic structure theory. Macmillan, New York
Szalewicz K, Bukowski R, Jeziorski B (2005) In: Dykstra CE, Frenking G, Kim KS, Scuseria GE (eds) Theory and applications of computational chemistry: the first 40 years. A volume of technical and historical perspectives. Elsevier, Amsterdam, p 919
Szasz L (1985) Pseudopotential theory of atoms and molecules. Wiley, New York
Szasz L, Berrios-Pagan I, McGinn G (1975) Z Naturforsch 30a:1516
Talman JD, Shadwick WF (1976) Phys Rev A 14:36
Tekin A, Jansen G (2007) Phys Chem Chem Phys 9:1680
Teller E (1962) Rev Mod Phys 34:627
Theophilou A (1979) J Phys C 12:5419
Thomas LH (1927) Proc Camb Philos Soc 23:542
Tisza L (1977) Generalized thermodynamics. MIT Press, Cambridge
Toro-Labbé A, Gutiérrez-Oliva S, Politzer P, Murray JS (2009) In: Chattaraj PK (ed) Chemical reactivity theory: a density functional view. CRC, Boca Raton, p 293 and references therein
Tully JC (1977) In: Segal GA (ed) Semi-empirical methods of electronic structure calculations, Part A. Plenum, New York, p 173
Tully JC (1980) In: Lawley KP (ed) Potential energy surfaces. Wiley, Chichester, p 63
Tuttle T, Thiel W (2008) Phys Chem Chem Phys 10:2159
Ullrich CA, Grossmann UJ, Gross EKU (1995) Phys Rev Lett 74:872
Valone SM (1980a) J Chem Phys 73:1344
Valone SM (1980b) J Chem Phys 73:4653
van der Waerden BL (ed) (1968) Sources of quantum mechanics. Dover, New York
van Leeuwen R, Gritsenko OV, Baerends EJ (1996) Density functional theory I: functionals and effective potentials. In: Nalewajski RF (ed) Topics in current chemistry, vol 180. Springer, Berlin, p 107
van Lenthe JH, Balint-Kurti GG (1983) J Chem Phys 78:5699
van Mourik T, Gdanitz RJ (2002) J Chem Phys 116:9620
van Voorhis T, Scuseria GE (1998) J Chem Phys 109:400
Volterra V (1959) Theory of functionals. Dover, New York
von Barth U, Hedin L (1972) J Phys C 5:1629
von Lilienfeld OA, Tavernelli I, Röthlisberger U, Sebastianini D (2004) Phys Rev Lett 93:15300490
von Lilienfeld OA, Tavernelli I, Röthlisberger U, Sebastianini D (2005) Phys Rev B 71:195119
von Parr RG, Szentpaly LV, Liu S (1999) J Am Chem Soc 121:1922
von Weizsäcker CF (1935) Z Phys 96:431
Vosko SH, Wilk L, Nusair M (1980) Can J Phys 58:1200
Voter AF, Goddard WA III (1981a) Chem Phys 57:253

Voter AF, Goddard WA III (1981b) J Chem Phys 75:3638
Wang YA, Carter EA (2000) In: Schwartz (ed) Theoretical methods in condensed phase chemistry. Kluwer, Dordrecht, p 117
Wesołowski TA (2004a) J Am Chem Soc 126:11444
Wesołowski TA (2004b) Chimia 58:311
Wesołowski TA, Tran F (2003) J Chem Phys 118:2072
Wesołowski TA, Warshel A (1993) J Phys Chem 97:8050
Wesołowski TA, Weber J (1998) Chem Phys Lett 248:71
Wesołowski TA, Muller RP, Warshel A (1995) J Phys Chem 100:15444
Wiberg KB (1968) Tetrahedron 24:1083
Wigner EP (1934) Phys Rev 46:1002
Wigner EP (1938) Trans Faraday Soc 34:29
Williams HL, Chabalowski CF (2001) J Phys Chem A 105:646
Woodward RB, Hoffmann R (1969) Angew Chem Int Ed Engl 8:781
Woodward RB, Hoffmann R (1970) The conservation of orbital symmetry. Academic, London
Woodward RB, Hoffmann R (1971) The conservation of orbital symmetry. Verlag Chemie, Weinheim
Wu Q, Yang W (2002) J Chem Phys 116:515
Wu X, Vargas MC, Nayak S, Lotrich V, Scoles G (2001) J Chem Phys 115:8748
Wu W, Song L, Cao Z, Zhang Q, Shaik S (2002) J Phys Chem A 106:2721
Xu X, Goddard WA III (2004) Proc Natl Acad Sci USA 101:2673
Yan Z, Perew JP, Kurth S (2000) Phys Rev B 61:16430
Yáñez RJ, Angulo JC, Dehesa SJ (1995) Int J Quantum Chem 56:489
Yang W (1987) Phys Rev Lett 59:1569
Yang W (1988) Phys Rev A 38(5494):5504
Yang W (1992) Phys Rev Lett 66:1438
Yang W, Harriman JE (1986) J Chem Phys 84:3323
Yang W, Parr RG (1985) Proc Natl Acad Sci USA 82:6723
Yang W, Parr RG, Pucci R (1984) J Chem Phys 81:2862
Yasuda K (2001) Phys Rev A 63:032517
Yourgrau W, van der Merve A (eds) (1979) Perspectives in quantum theory. Dover, New York
Zangwill A, Soven P (1980) Phys Rev A21:1561
Zaremba E, Kohn W (1976) Phys Rev B 13:2270
Zhao Y, Truhlar DG (2006) J Chem Phys 125:194101
Zhao Y, Truhlar DG (2008) Theor Chem Acc 120:215
Zhao Y, Schultz NE, Truhlar DG (2005) J Chem Phys 123:161103
Zhao Y, Schultz NE, Truhlar DG (2006) J Chem Theor Comput 2:364
Zhou Z, Novangul HV (1990) J Phys Org Chem 3:784
Zhou Z, Parr RG (1989) J Am Chem Soc 111:7371
Zhou Z, Parr RG (1990) J Am Chem Soc 112:5720
Zhou Z, Parr RG, Garst JF (1988) Tertrahedron Lett 29:4843
Ziesche P (1995) Int J Quantum Chem 56:363
Zimmerli U, Parrinello M, Koumoutsakos P (2004) J Chem Phys 120:2693
Zubarev DY, Sergeeva AP, Boldyrev AI (2009) In: Chattaraj PK (ed) Chemical reactivity theory: a density functional view. Taylor and Francis, London, p 439
Zumbach G, Maschke K (1983) Phys Rev A 28:544; Errata (1984) Phys Rev A 29:1585
Zumbach G, Maschke K (1985) J Chem Phys 82:5604

Index

R.F. Nalewajski, *Perspectives in Electronic Structure Theory*,
DOI 10.1007/978-3-642-20180-6, © Springer-Verlag Berlin Heidelberg 2012

E

P

S

CPSIA information can be obtained at www.ICGtesting.com
Printed in the USA
LVOW01*0804120813

347435LV00001B/11/P

9 783642 201790